Volume 1

Encyclopedia of Energy Engineering *and* Technology

Encyclopedias from Taylor & Francis Group

Agriculture Titles

Dekker Agropedia Collection (Eight Volume Set)
ISBN: 0-8247-2194-2   13-Digit ISBN: 978-0-8247-2194-7

Encyclopedia of Agricultural, Food, and Biological Engineering
Edited by Dennis R. Heldman
ISBN: 0-8247-0938-1   13-Digit ISBN: 978-0-8247-0938-9

Encyclopedia of Animal Science
Edited by Wilson G. Pond and Alan Bell
ISBN: 0-8247-5496-4   13-Digit ISBN: 978-0-8247-5496-9

Encyclopedia of Pest Management
Edited by David Pimentel
ISBN: 0-8247-0632-3   13-Digit ISBN: 978-0-8247-0632-6

Encyclopedia of Pest Management, Volume II
Edited by David Pimentel
ISBN: 1-4200-5361-2   13-Digit ISBN: 978-1-4200-5361-6

Encyclopedia of Plant and Crop Science
Edited by Robert M. Goodman
ISBN: 0-8247-0944-6   13-Digit ISBN: 978-0-8247-0944-0

Encyclopedia of Soil Science, Second Edition (Two Volume Set)
Edited by Rattan Lal
ISBN: 0-8493-3830-1   13-Digit ISBN: 978-0-8493-3830-4

Encyclopedia of Water Science
Edited by B.A. Stewart and Terry Howell
ISBN: 0-8247-0948-9   13-Digit ISBN: 978-0-8247-0948-8

Chemistry Titles

Encyclopedia of Chromatography, Second Edition (Two Volume Set)
Edited by Jack Cazes
ISBN: 0-8247-2785-1   13-Digit ISBN: 978-0-8247-2785-7

Encyclopedia of Surface and Colloid Science, Second Edition (Eight Volume Set)
Edited by P. Somasundaran
ISBN: 0-8493-9615-8   13-Digit ISBN: 978-0-8493-9615-1

Encyclopedia of Supramolecular Chemistry (Two Volume Set)
Edited by Jerry L. Atwood and Jonathan W. Steed
ISBN: 0-8247-5056-X   13-Digit ISBN: 978-0-8247-5056-5

Engineering Titles

Encyclopedia of Chemical Processing (Five Volume Set)
Edited by Sunggyu Lee
ISBN: 0-8247-5563-4   13-Digit ISBN: 978-0-8247-5563-

Encyclopedia of Corrosion Technology, Second Edition
Edited by Philip A. Schweitzer, P.E.
ISBN: 0-8247-4878-6   13-Digit ISBN: 978-0-8247-4878-

Dekker Encyclopedia of Nanoscience and Nanotechnology (Five Volume Set)
Edited by James A. Schwarz, Cristian I. Contescu, and Karol Putyera
ISBN: 0-8247-5055-1   13-Digit ISBN: 978-0-8247-5055-

Encyclopedia of Optical Engineering (Three Volume Set)
Edited by Ronald G. Driggers
ISBN: 0-8247-0940-3   13-Digit ISBN: 978-0-8247-0940-

Business Titles

Encyclopedia of Library and Information Science, Second Edition (Four Volume Set)
Edited by Miriam Drake
ISBN: 0-8247-2075-X   13-Digit ISBN: 978-0-8247-2075-

Encyclopedia of Public Administration and Public Policy (Two Volume Set)
Edited by Jack Rabin
ISBN: 0-8247-4748-8   13-Digit ISBN: 978-0-8247-4748-

Coming Soon

Encyclopedia of Public Administration and Public Policy, Second Edition
Edited by Jack Rabin
ISBN: 1-4200-5275-6   13-Digit ISBN: 978-1-4200-5275-

Encyclopedia of Water Science, Second Edition (Two Volume Set)
Edited by Stanley W. Trimble
ISBN: 0-8493-9627-1   13-Digit ISBN: 978-0-8493-9627-

Encyclopedia of Wireless and Mobile Communications (Three Volume Set)
Edited by Borko Furht
ISBN: 1-4200-4326-9   13-Digit ISBN: 978-1-4200-4326-

These titles are available both in print and online. To order, visit:
www.crcpress.com
Telephone: 1-800-272-7737
Fax: 1-800-374-3401
E-Mail: orders@crcpress.com

Volume 3

Photovoltaic Systems / 1147
Physics of Energy / 1160
Pricing Programs: Time-of-Use and Real Time / 1175
Psychrometrics / 1184
Public Policy for Improving Energy Sector Performance / 1193
Public Utility Regulatory Policies Act (PURPA) of 1978 / 1201
Pumped Storage Hydroelectricity / 1207
Pumps and Fans / 1213
Radiant Barriers / 1227
Reciprocating Engines: Diesel and Gas / 1233
Regulation: Price Cap and Revenue Cap / 1245
Regulation: Rate of Return / 1252
Renewable and Decentralized Energy Options: State Promotion in the U.S. / 1258
Renewable Energy / 1265
Residential Buildings: Heating Loads / 1272
Run-Around Heat Recovery Systems / 1278
Savings and Optimization: Case Studies / 1283
Savings and Optimization: Chemical Process Industry / 1302
Six Sigma Methods: Measurement and Verification / 1310
Solar Heating and Air Conditioning: Case Study / 1317
Solar Thermal Technologies / 1321
Solar Water Heating: Domestic and Industrial Applications / 1331
Solid Waste to Energy by Advanced Thermal Technologies (SWEATT) / 1340
Space Heating / 1357
Steam and Hot Water System Optimization: Case Study / 1366
Steam Turbines / 1380
Sustainability Policies: Sunbelt Cities / 1389
Sustainable Building Simulation / 1396
Sustainable Development / 1406
Thermal Energy Storage / 1412

Volume 3 (Continued)

Thermodynamics / 1422
Tradable Certificates for Energy Savings / 1433
Transportation Systems: Hydrogen-Fueled / 1441
Transportation: Location Efficiency / 1449
Tribal Land and Energy Efficiency / 1456
Underfloor Air Distribution (UFAD) / 1463
Utilities and Energy Suppliers: Bill Analysis / 1471
Utilities and Energy Suppliers: Business Partnership Management / 1484
Utilities and Energy Suppliers: Planning and Portfolio Management / 1491
Utilities and Energy Suppliers: Rate Structures / 1497
Walls and Windows / 1513
Waste Fuels / 1523
Waste Heat Recovery / 1536
Waste Heat Recovery Applications: Absorption Heat Pumps / 1541
Water and Wastewater Plants: Energy Use / 1548
Water and Wastewater Utilities / 1556
Water Source Heat Pump for Modular Classrooms / 1564
Water-Augmented Gas Turbine Power Cycles / 1574
Water-Using Equipment: Commercial and Industrial / 1587
Water-Using Equipment: Domestic / 1596
Wind Power / 1607
Window Energy / 1616
Window Films: Savings from IPMVP Options C and D / 1626
Window Films: Solar-Control and Insulating / 1630
Window Films: Spectrally Selective versus Conventional Applied / 1637
Windows: Shading Devices / 1641
Wireless Applications: Energy Information and Control / 1649
Wireless Applications: Mobile Thermostat Climate Control and Energy Conservation / 1660

Volume 1

Encyclopedia of Energy Engineering *and* Technology

Edited by
Barney L. Capehart
University of Florida
Gainesville, USA

CRC Press
Taylor & Francis Group
Boca Raton London New York

CRC Press is an imprint of the
Taylor & Francis Group, an informa business

CRC Press
Taylor & Francis Group
6000 Broken Sound Parkway NW, Suite 300
Boca Raton, FL 33487-2742

© 2007 by Taylor & Francis Group, LLC, except as noted on the opening page of the entry. All rights reserved.
CRC Press is an imprint of Taylor & Francis Group, an Informa business

No claim to original U.S. Government works
Printed in the United States of America on acid-free paper
10 9 8 7 6 5 4 3 2 1

International Standard Book Number-13: 978-0-8493-3653-9 (Hardcover)

This book contains information obtained from authentic and highly regarded sources. Reprinted material is quoted with permission, and sources are indicated. A wide variety of references are listed. Reasonable efforts have been made to publish reliable data and information, but the author and the publisher cannot assume responsibility for the validity of all materials or for the consequences of their use.

No part of this book may be reprinted, reproduced, transmitted, or utilized in any form by any electronic, mechanical, or other means, now known or hereafter invented, including photocopying, microfilming, and recording, or in any information storage or retrieval system, without written permission from the publishers.

For permission to photocopy or use material electronically from this work, please access www.copyright.com (http://www.copyright.com/) or contact the Copyright Clearance Center, Inc. (CCC) 222 Rosewood Drive, Danvers, MA 01923, 978-750-8400. CCC is a not-for-profit organization that provides licenses and registration for a variety of users. For organizations that have been granted a photocopy license by the CCC, a separate system of payment has been arranged.

Trademark Notice: Product or corporate names may be trademarks or registered trademarks, and are used only for identification and explanation without intent to infringe.

Library of Congress Cataloging-in-Publication Data

Encyclopedia of energy engineering and technology / edited by Barney L. Capehart.
    p. cm.
    Includes bibliographical references and index.
    ISBN-13: 978-0-8493-3653-9 (alk. paper : set)
    ISBN-10: 0-8493-3653-8 (alk. paper : set)
    ISBN-13: 978-0-8493-5039-9 (alk. paper : v. 1)
    ISBN-10: 0-8493-5039-5 (alk. paper : v. 1)
    [etc.]
    1. Power resources--Encyclopedias. 2. Power (Mechanics)--Encyclopedias. I. Capehart, B. L. (Barney L.)

TJ163.2.E385 2007
621.04203--dc22
    2007007878

Visit the Taylor & Francis Web site at
http://www.taylorandfrancis.com

and the CRC Press Web site at
http://www.crcpress.com

*In memory of my mother and father
and
for my grandchildren Hannah and Easton.
May their energy future be efficient and sustainable.*

Editorial Advisory Board

Barney L. Capehart, Editor

*Department of Industrial and Systems Engineering,
University of Florida, Gainesville, Florida, U.S.A.*

Mr. Larry B. Barrett
President, Barrett Consulting Associates, Colorado Springs, Colorado, U.S.A.

Dr. Sanford V. Berg
Former Director, Public Utility Research Center, University of Florida, Gainesville, Florida, U.S.A.

Dr. David L. Block, Director Emeritus
Florida Solar Energy Center, Cocoa, Florida, U.S.A.

Dr. Marilyn A. Brown
Director, Energy Efficiency, Reliability, and Security Program, Oak Ridge National Laboratory, Oak Ridge, Tennessee, U.S.A.

Dr. R. Neal Elliott, P.E.
Director of Industrial Energy Programs, American Council for an Energy Efficient Economy, Washington, D.C., U.S.A.

Mr. Clark W. Gellings
Vice President of Innovation, Electric Power Research Institute (EPRI), Palo Alto, California, U.S.A.

Dr. David Goldstein
Head of Energy Efficiency Section, Natural Resources Defense Council, San Francisco, California, U.S.A.

Dr. Alex E. S. Green
Professor Emeritus, College of Engineering, University of Florida, Gainesville, Florida, U.S.A.

Dr. David L. Greene
Corporate Fellow, Oak Ridge National Laboratory, Knoxville, Tennessee, U.S.A.

Dr. Jay Hakes
Director, Jimmy Carter Presidential Library, and Former Director, US Energy Information Agency, Atlanta, Georgia, U.S.A.

Dr. Richard F. Hirsh
Professor of History of Technology, Virginia Polytechnic University, Blacksburg, Virginia, U.S.A.

Mr. Ronald E. Jarnagin
Research Scientist, Pacific Northwest National Laboratory, Richland, Washington, U.S.A.

Dr. Xianguo Li, P.Eng.
Editor, International Journal of Green Energy, and Professor, Department of Mechanical Engineering, University of Waterloo, Waterloo, Ontario, Canada

Mr. Steve Nadel
Executive Director, American Council for an Energy Efficient Environment (ACEEE), Washington, D.C., U.S.A.

Mr. Dan Parker
Principal Research Scientist, Florida Solar Energy Center, Cocoa, Florida, U.S.A.

Mr. Steven A. Parker
Senior Engineer and Director of New Technology Applications, Pacific Northwest National Laboratory, Richland, Washington, U.S.A.

Mr. F. William Payne
Editor Emeritus, Strategic Planning for Energy and the Environment, Marble Hill, Georgia, U.S.A.

Mr. Neil Petchers
President, NORESCO, Westborough, Massachusetts, U.S.A.

Dr. Stephen A. Roosa, CEM, CIAQP, CMVP, CBEP, CDSM, MBA
Performance Engineer, Energy Systems Group, Inc., Louisville, Kentucky, U.S.A.

Dr. Wayne H. Smith
Interim Dean for Research and Interim Director, Florida Agricultural Experiment Station, University of Florida, Gainesville, Florida, U.S.A.

Prof. James S. Tulenko
Laboratory for Development of Advanced Nuclear Fuels and Materials, University of Florida, Gainesville, Florida, U.S.A.

Dr. Dan Turner
Director, Energy Systems Laboratory, Texas A&M University, College Station, Texas, U.S.A.

Dr. Wayne C. Turner
Editor, Energy Management Handbook, and Regents Professor, Industrial Engineering and Management, Oklahoma State University, Stillwater, Oklahoma, U.S.A.

Dr. Jerry Ventre
Former Research Scientist, Florida Solar Energy Center, Oviedo, Florida, U.S.A.

Dr. Richard Wakefield
Vice President, Transmission & Regulatory Services, KEMA, Inc., Fairfax, Virginia, U.S.A.

Contributors

Bill Allemon / *North American Energy Efficiency, Ford Land, Dearborn, Michigan, U.S.A.*

Paul J. Allen / *Reedy Creek Energy Services, Walt Disney World Co., Lake Buena Vista, Florida, U.S.A.*

Fatouh Al-Ragom / *Building and Energy Technologies Department, Kuwait Institute for Scientific Research, Safat, Kuwait*

Kalyan Annamalai / *Department of Mechanical Engineering, College Station, Texas, U.S.A.*

John Archibald / *American Solar Inc., Annandale, Virginia, U.S.A.*

Senthil Arumugam / *Enerquip, Inc., Medford, Wisconsin, U.S.A.*

M. Asif / *School of the Built and Natural Environment, Glasgow Caledonian University, Glasgow, Scotland, U.K.*

Essam Omar Assem / *Arab Fund for Economic and Social Development, Arab Organizations Headquarters Building, Shuwaikh, Kuwait*

U. Atikol / *Energy Research Center, Eastern Mediterranean University, Magusa, Northern Cyprus*

Rick Avery / *Bay Controls, LLC, Maumee, Ohio, U.S.A.*

Lu Aye / *International Technologies Centre (IDTC), Department of Civil and Environmental Engineering, The University of Melbourne, Victoria, Australia*

M. Babcock / *EMO Energy Solutions, LLC, Vienna, Virginia, U.S.A.*

Christopher G. J. Baker / *Chemical Engineering Department, Kuwait University, Safat, Kuwait*

K. Baker / *EMO Energy Solutions, LLC, Vienna, Virginia, U.S.A.*

Peter Barhydt / *Arctic Combustion Ltd., Mississauga, Ontario, Canada*

Larry B. Barrett / *Barrett Consulting Associates, Colorado Springs, Colorado, U.S.A.*

Rosemarie Bartlett / *Pacific Northwest National Laboratory, Richland, Washington, U.S.A.*

Fred Bauman / *Center for the Built Environment, University of California, Berkeley, California, U.S.A.*

David Beal / *Florida Solar Energy Center, Cocoa, Florida, U.S.A.*

Sanford V. Berg / *Director of Water Studies, Public Utility Research Center, University of Florida, Gainesville, Florida, U.S.A.*

Paolo Bertoldi / *European Commission, Directorate General JRC, Ispra (VA), Italy*

Asfaw Beyene / *Department of Mechanical Engineering, San Diego State University, San Diego, California, U.S.A.*

Ujjwal Bhattacharjee / *Energy Engineering Program, University of Massachusetts—Lowell, Lowell, Massachusetts, U.S.A.*

David Bisbee / *Customer Advanced Technologies Program, Sacramento Municipal Utility District (SMUD), Sacramento, California, U.S.A.*

John O. Blackburn / *Department of Economics, Duke University, Durham, North Carolina, U.S.A.*

Richard F. Bonskowski / *Office of Coal, Nuclear, Electric and Alternate Fuels, Energy Information Administration, U.S. Department of Energy, Washington, D.C., U.S.A.*

Robert G. Botelho / *Energy Management International, Seekonk, Massachussetts, U.S.A.*

Mark T. Brown / *Department of Environmental Engineering Sciences, University of Florida, Gainesville, Florida, U.S.A.*

Michael L. Brown / *Energy and Environmental Management Center, Georgia Institute of Technology, Savannah, Georgia, U.S.A.*

Alexander L. Burd / *Advanced Research Technology, Suffield, Connecticut, U.S.A.*

Galina S. Burd / *Advanced Research Technology, Suffield, Connecticut, U.S.A.*

James Call / *James Call Engineering, PLLC, Larchmont, New York, U.S.A.*

Norm Campbell / *Energy Systems Group, Newburgh, Indiana, U.S.A.*

Barney L. Capehart / *Department of Industrial and Systems Engineering, University of Florida College of Engineering, Gainesville, Florida, U.S.A.*

Lynne C. Capehart / *Consultant and Technical Writing Specialist, Gainesville, Florida, U.S.A.*

Cristián Cárdenas-Lailhacar / *Department of Industrial and Systems Engineering, University of Florida, Gainesville, Florida, U.S.A.*

Jack Casazza / *American Education Institute, Springfield, Virginia, U.S.A.*

Yunus A. Cengel / *Department of Mechanical Engineering, University of Nevada, Reno, Nevada, U.S.A.*

Guangnan Chen / *Faculty of Engineering and Surveying, University of Southern Queensland, Toowoomba, Queensland, Australia*

David E. Claridge / *Department of Mechanical Engineering, and Associate Director, Energy Systems Laboratory, Texas A & M University, College Station, Texas, U.S.A.*

James A. Clark / *Clark Energy, Inc., Broomall, Pennsylvania, U.S.A.*

Gregory Cmar / *Interval Data Systems, Inc., Watertown, Massachusetts, U.S.A.*

Kevin C. Coates / *Transportation & Energy Policy, Coates Consult, Bethesda, Maryland, U.S.A.*

Bruce K. Colburn / *EPS Capital Corp., Villanova, Pennsylvania, U.S.A.*

Stephany L. Cull / *RetroCom Energy Strategies, Inc., Elk Grove, California, U.S.A.*

Rick Cumbo / *Operations and Maintenance Manager, Armstrong Service, Inc., Orlando, Florida, U.S.A.*

Leonard A. Damiano / *EBTRON, Inc., Loris, South Carolina, U.S.A.*

James W. Dean / *Office of Strategic Analysis and Governmental Affairs, Florida Public Service Commission, Tallahassee, Florida, U.S.A.*

Steve DeBusk / *Energy Solutions Manager, CPFilms Inc., Martinsville, Virginia, U.S.A.*

Harvey E. Diamond / *Energy Management International, Conroe, Texas, U.S.A.*

Craig DiLouie / *Lighting Controls Association, National Electrical Manufacturers Association, Rosslyn, Virginia, U.S.A.*

Ibrahim Dincer / *Faculty of Engineering and Applied Science, University of Ontario Institute of Technology (UOIT), Oshawa, Ontario, Canada*

Michael R. Dorrington / *Cameron's Compression Systems Division, Houston, Texas, U.S.A.*

Steve Doty / *Colorado Springs Utilities, Colorado Springs, Colorado, U.S.A.*

David S. Dougan / *EBTRON, Inc., Loris, South Carolina, U.S.A.*

John Duffy / *Department of Mechanical Engineering, University of Massachussetts—Lowell, Lowell, Massachussetts, U.S.A.*

Philip Fairey / *Florida Solar Energy Center, Cocoa, Florida, U.S.A.*

Carla Fair-Wright / *Maintenance Technology Services, Cameron's Compression Systems Division, Houston, Texas, U.S.A.*

Ahmad Faruqui / *The Brattle Group, San Francisco, California, U.S.A.*

J. Michael Fernandes / *Mobile Platforms Group, Intel Corporation, Santa Clara, California, U.S.A.*

Steven Ferrey / *Suffolk University Law School, Boston, Massachusetts, U.S.A.*

Andrew R. Forrest / *Government Communications Systems Division, Harris Corporation, Melbourne, Florida, U.S.A.*

R. Scot Foss / *IR Air Solutions, Davidson, North Carolina, U.S.A.*

Jackie L. Francke / *Geotechnika, Inc., Longmont, Colorado, U.S.A.*

Fred Freme / *Office of Coal, Nuclear, Electric and Alternate Fuels, Energy Information Administration, U.S. Department of Energy, Washington, D.C., U.S.A.*

Dwight K. French / *Energy Consumption Division, Energy Information Administration, U.S. Department of Energy, Washington, D.C., U.S.A.*

Kevin Fuller / *Interval Data Systems, Inc., Watertown, Massachusetts, U.S.A.*

Kamiel S. Gabriel / *University of Ontario Institute of Technology, Oshawa, Ontario, Canada*

Clark W. Gellings / *Electric Power Research Institute (EPRI), Palo Alto, California, U.S.A.*

Geoffrey J. Gilg / *Pepco Energy Services, Inc., Arlington, Virginia, U.S.A.*

Bill Gnerre / *Interval Data Systems, Inc., Watertown, Massachusetts, U.S.A.*

Fredric S. Goldner / *Energy Management and Research Associates, East Meadow, New York, U.S.A.*

Dan Golomb / *Department of Environmental, Earth and Atmospheric Sciences, University of Massachusetts—Lowell, Lowell, Massachusetts, U.S.A.*

John Van Gorp / *Power Monitoring and Control, Schneider Electric, Saanichton, British Columbia, Canada*

Alex E. S. Green / *Green Liquids and Gas Technologies, Gainesville, Florida, U.S.A.*

David C. Green / *Green Management Services, Inc., Fort Myers, Florida, U.S.A.*

L. J. Grobler / *School for Mechanical Engineering, North-West University, Potchefstroom, South Africa*

H. M. Güven / *Energy Research Center, Eastern Mediterranean University, Magusa, Northern Cyprus*

Mark A. Halverson / *Pacific Northwest National Laboratory, West Halifax, Vermont, U.S.A.*

Shirley J. Hansen / *Hansen Associates, Inc., Gig Harbor, Washington, U.S.A.*

Susanna S. Hanson / *Global Applied Systems, Trane Commercial Systems, La Crosse, Wisconsin, U.S.A.*

Glenn C. Haynes / *RLW Analytics, Middletown, Connecticut, U.S.A.*

Warren M. Heffington / *Industrial Assessment Center, Department of Mechanical Engineering, Texas A&M University, College Station, Texas, U.S.A.*

Arif Hepbasli / *Department of Mechanical Engineering, Faculty of Engineering, Ege University, Izmir, Turkey*

Keith E. Herold / *Department of Mechanical Engineering, University of Maryland, College Park, Maryland, U.S.A.*

James G. Hewlett / *Energy Information Administration, U.S. Department of Energy, Washington, D.C., U.S.A.*

Richard F. Hirsh / *Department of History, Virginia Tech, Blacksburg, Virginia, U.S.A.*

Kevin Hoag / *Engine Research Center, University of Wisconsin-Madison, Madison, Wisconsin, U.S.A.*

Nathan E. Hultman / *Science, Technology, and International Affairs, Georgetown University, Washington, D.C., U.S.A.*

W. David Hunt / *Pacific Northwest National Laboratory, Washington, D.C., U.S.A.*

H. A. Ingley III / *Department of Mechanical Engineering, University of Florida, Gainesville, Florida, U.S.A.*

Nevena H. Iordanova / *Senior Utility Systems Engineer, Armstrong Service, Inc., Orlando, Florida, U.S.A.*

Charles P. Ivey / *Aiken Global Environmental/FT Benning DPW, Fort Benning, Georgia, U.S.A.*

Mark A. Jamison / *Public Utility Research Center, University of Florida, Gainesville, Florida, U.S.A.*

Somchai Jiajitsawat / *Department of Mechanical Engineering, University of Massachussetts—Lowell, Lowell, Massachusetts, U.S.A.*

Gholamreza Karimi / *Department of Mechanical Engineering, University of Waterloo, Waterloo, Ontario, Canada*

Janey Kaster / *Yamas Controls, Inc., South San Francisco, California, U.S.A.*

Sila Kiliccote / *Commercial Building Systems Group, Lawrence Berkeley National Laboratory, Berkeley, California, U.S.A.*

Birol I. Kilkis / *Green Energy Systems, International LLC, Vienna, Virginia, U.S.A.*

Ronald L. Klaus / *VAST Power Systems, Elkhart, Indiana, U.S.A.*

Milivoje M. Kostic / *Department of Mechanical Engineering, Northern Illinois University, DeKalb, Illinois, U.S.A.*

Athula Kulatunga / *Department of Electrical and Computer Engineering Technology, Purdue University, West Lafayette, Indiana, U.S.A.*

James W. Leach / *Industrial Assessment Center, Mechanical and Aerospace Engineering, North Carolina State University, Raleigh, North Carolina, U.S.A.*

Larry Leetzow / *World Institute of Lighting and Development Corp./Magnaray International Division, Sarasota, Florida, U.S.A.*

Jim Lewis / *Obvius LLC, Hillsboro, Oregon, U.S.A.*

Xianguo Li / *Department of Mechanical Engineering, University of Waterloo, Waterloo, Ontario, Canada*

Todd A. Litman / *Victoria Transport Policy Institute, Victoria, British Columbia, Canada*

Mingsheng Liu / *Architectural Engineering Program, Director, Energy Systems Laboratory, Peter Kiewit Institute, University of Nebraska—Lincoln, Omaha, Nebraska, U.S.A.*

Robert Bruce Lung / *Resource Dynamics Corporation, Vienna, Virginia, U.S.A.*

Alfred J. Lutz / *AJL Resources, LLC, Philadelphia, Pennsylvania, U.S.A.*

David MacPhaul / *CH2M HILL, Gainesville, Florida, U.S.A.*

Richard J. Marceau / *University of Ontario Institute of Technology, Oshawa, Ontario, Canada*

Rudolf Marloth / *Department of Mechanical Engineering, Loyola Marymount University, Los Angeles, California, U.S.A.*

Sandra B. McCardell / *Current-C Energy Systems, Inc., Mills, Wyoming, U.S.A.*

William Ross McCluney / *Florida Solar Energy Center, University of Central Florida, Cocoa, Florida, U.S.A.*

Donald T. McGillis / *Pointe Claire, Québec, Canada*

Aimee McKane / *Lawrence Berkeley National Laboratory, Washington, D.C., U.S.A.*

D. Paul Mehta / *Department of Mechanical Engineering, Bradley University, Peoria, Illinois, U.S.A.*

Zohrab Melikyan / *HVAC Department, Armenia State University of Architecture and Construction, Yerevan, Armenia*

Ingrid Melody / *Florida Solar Energy Center, Cocoa, Florida, U.S.A.*

Mark Menezes / *Emerson Process Management (Rosemount), Mississauga, Ontario, Canada*

Adnan Midilli / *Department of Mechanical Engineering, Faculty of Engineering, Nigde University, Nigde, Turkey*

Ruth Mossad / *Faculty of Engineering and Surveying, University of Southern Queensland, Toowoomba, Queensland, Australia*

Naoya Motegi / *Commercial Building Systems Group, Lawrence Berkeley National Laboratory, Berkeley, California, U.S.A.*

Alex V. Moultanovsky / *Automotive Climate Control, Inc., Elkhart, Indiana, U.S.A.*

Martin A. Mozzo / *M and A Associates, Inc., Robbinsville, New Jersey, U.S.A.*

T. Muneer / *School of Engineering, Napier University, Edinburgh, Scotland, U.K.*

Greg F. Naterer / *University of Ontario Institute of Technology, Oshawa, Ontario, Canada*

Robert R. Nordhaus / *Van Ness Feldman, PC, Washington, D.C., U.S.A.*

Michael M. Ohadi / *The Petroleum Institute, Abu Dhabi, United Arab Emirates*

K. E. Ohrn / *Cypress Digital Ltd., Vancouver, British Columbia, Canada*

Svetlana J. Olbina / *Rinker School of Building Construction, University of Florida, Gainesville, Florida, U.S.A.*

Eric Oliver / *EMO Energy Solutions, LLC, Vienna, Virginia, U.S.A.*

Mitchell Olszewski / *Oak Ridge National Laboratory, Oak Ridge, Tennessee, U.S.A.*

Bohdan W. Oppenheim / *U.S. Department of Energy Industrial Assessment Center, Loyola Marymount University, Los Angeles, U.S.A.*

Leyla Ozgener / *Department of Mechanical Engineering, Celal Bayar University, Manisa, Turkey*

Graham Parker / *Battelle Pacific Northwest National Laboratory, U.S. Department of Energy, Richland, Washington, U.S.A.*

Steven A. Parker / *Pacific Northwest National Laboratory, Richland, Washington, U.S.A.*

Ken E. Patterson II / *Advanced Energy Innovations—Bringing Innovation to Industry, Menifee, California, U.S.A.*

Klaus-Dieter E. Pawlik / *Accenture, St Petersburg, Florida, U.S.A.*

David E. Perkins / *Active Power, Inc., Austin, Texas, U.S.A.*

Jeffery P. Perl / *Chicago Chem Consultants Corp., Chicago, Illinois, U.S.A.*

Neil Petchers / *NORESCO Energy Consulting, Shelton, Connecticut, U.S.A.*

Eric Peterschmidt / *Honeywell Building Solutions, Honeywell, Golden Valley, Minnesota, U.S.A.*

Robert W. Peters / *Department of Civil, Construction & Environmental Engineering, University of Alabama at Birmingham, Birmingham, Alabama, U.S.A.*

Mark A. Peterson / Sustainable Success LLC, Clementon, New Jersey, U.S.A.

Mary Ann Piette / Commercial Building Systems Group, Lawrence Berkeley National Laboratory, Berkeley, California, U.S.A.

Wendell A. Porter / University of Florida, Gainesville, Florida, U.S.A.

Soyuz Priyadarsan / Texas A&M University, College Station, Texas, U.S.A.

Sam Prudhomme / Bay Controls, LLC, Maumee, Ohio, U.S.A.

Jianwei Qi / Department of Mechanical Engineering, University of Maryland, College Park, Maryland, U.S.A.

Ashok D. Rao / Advanced Power and Energy Program, University of California, Irvine, California, U.S.A.

Ab Ream / U.S. Department of Energy, Washington, D.C., U.S.A.

Rolf D. Reitz / Engine Research Center, University of Wisconsin-Madison, Madison, Wisconsin, U.S.A.

Rich Remke / Commercial Systems and Services, Carrier Corporation, Syracuse, New York, U.S.A.

Silvia Rezessy / Energy Efficiency Advisory, REEEP International Secretariat, Vienna International Centre, Austria, Environmental Sciences and Policy Department, Central European University, Nador, Hungary

James P. Riordan / Energy Department, DMJM + Harris / AECOM, New York, U.S.A.

Vernon P. Roan / University of Florida, Gainesville, Florida, U.S.A.

Stephen A. Roosa / Energy Systems Group, Inc., Louisville, Kentucky, U.S.A.

Michael Ropp / South Dakota State University, Brookings, South Dakota, U.S.A.

Marc A. Rosen / Faculty of Engineering and Applied Science, University of Ontario Institute of Technology, Oshawa, Ontario, Canada

Christopher Russell / Energy Pathfinder Management Consulting, LLC, Baltimore, Maryland, U.S.A.

Hemmat Safwat / Project Development Power and Desalination Plants, Consolidated Contractors International Company, Athens, Greece

Abdou-R. Sana / Montreal, Québec, Canada

Diane Schaub / Industrial and Systems Engineering, University of Florida, Gainesville, Florida, U.S.A.

Diana L. Shankle / Battelle Pacific Northwest National Laboratory, U.S. Department of Energy, Richland, Washington, U.S.A.

S. A. Sherif / Department of Mechanical and Aerospace Engineering, Wayne K. and Lyla L. Masur HVAC Laboratory, University of Florida, Gainesville, Florida, U.S.A.

Brian Silvetti / CALMAC Manufacturing Corporation, Fair Lawn, New Jersey, U.S.A.

Carey J. Simonson / Department of Mechanical Engineering, University of Saskatchewan, Saskatoon, Saskatchewan, Canada

Amy Solana / Pacific Northwest National Laboratory, Portland, Oregon, U.S.A.

Paul M. Sotkiewicz / Public Utility Research Center, University of Florida, Warrington College of Business, Gainesville, Florida, U.S.A.

Amanda Staudt / Board on Atmospheric Sciences and Climate, The National Academies, Washington, D.C., U.S.A.

Nick Stecky / NJS Associates, Denville, New Jersey, U.S.A.

Kate McMordie Stoughton / Battelle Pacific Northwest National Laboratory, U.S. Department of Energy, Richland, Washington, U.S.A.

Therese Stovall / Oak Ridge National Laboratory, Oak Ridge, Tennessee, U.S.A.

Gregory P. Sullivan / Battelle Pacific Northwest National Laboratory, U.S. Department of Energy, Richland, Washington, U.S.A.

John M. Sweeten / Texas Agricultural Experiment Station, Amarillo, Texas, U.S.A.

Thomas F. Taranto / ConservAIR Technologies, LLP, Baldwinsville, New York, U.S.A.

Michael Taylor / Honeywell Building Solutions, Honeywell, St. Louis Park, Minnesota, U.S.A.

Albert Thumann / Association of Energy Engineers, Atlanta, Georgia, U.S.A.

Robert J. Tidona / RMT, Incorporated, Plymouth Meeting, Pennsylvania, U.S.A.

Jill S. Tietjen / Technically Speaking, Inc., Greenwood Village, Colorado, U.S.A.

Lorrie B. Tietze / Interface Consulting, LLC, Castle Rock, Colorado, U.S.A.

Greg Tinkler / RLB Consulting Engineers, Houston, Texas, U.S.A.

Alberto Traverso / *Dipartimento di Macchine Sistemi Energetici e Trasporti, Thermochemical Power Group, Università di Genova, Genova, Italy*

Douglas E. Tripp / *Canadian Institute for Energy Training, Rockwood, Ontario, Canada*

James S. Tulenko / *Laboratory for Development of Advanced Nuclear Fuels and Materials, University of Florida, Gainesville, Florida, U.S.A.*

W. D. Turner / *Director, Energy Systems Laboratory, Texas A & M University, College Station, Texas, U.S.A.*

Wayne C. Turner / *Industrial Engineering and Management, Oklahoma State University, Stillwater, Oklahoma, U.S.A.*

S. Kay Tuttle / *Process Heating, Boilers and Foodservices, Duke Energy, Charlotte, North Carolina, U.S.A.*

Robert E. Uhrig / *University of Tennessee, Knoxville, Tennessee, U.S.A.*

Sergio Ulgiati / *Department of Sciences for the Environment, Parthenope, University of Naples, Napoli, Italy*

Francisco L. Valentine / *Premier Energy Services, LLC, Columbia, Maryland, U.S.A.*

Ing. Jesús Mario Vignolo / *Instituto Ingeniería Eléctrica, Universidad de la República, Montevideo, Uruguay*

Paul L. Villeneuve / *University of Maine, Orono, Maine, U.S.*

T. G. Vorster / *School for Mechanical Engineering, North-West University, Potchefstroom, South Africa*

James P. Waltz / *Energy Resource Associates, Inc., Livermore, California, U.S.A.*

Devra Bachrach Wang / *Natural Resources Defense Council, San Francisco, California, U.S.A.*

John W. Wang / *Department of Mechanical Engineering, University of Massachussetts—Lowell, Lowell, Massachussetts, U.S.A.*

David S. Watson / *Commercial Building Systems Group, Lawrence Berkeley National Laboratory, Berkeley, California, U.S.A.*

William D. Watson / *Office of Coal, Nuclear, Electric and Alternate Fuels, Energy Information Administration, U.S. Department of Energy, Washington, D.C., U.S.A.*

Marty Watts / *V-Kool, Inc., Houston, Texas, U.S.A.*

Tom Webster / *Center for the Built Environment, University of California, Berkeley, California, U.S.A.*

Michael K. West / *Building Systems Scientists, Advantek Consulting, Inc., Melbourne, Florida, U.S.A.*

Robert E. Wilson / *ConservAIR Technologies Co., LLP, Kenosha, Wisconsin, U.S.A.*

Michael Burke Wood / *Ministry of Energy, Al Dasmah, Kuwait*

Kurt E. Yeager / *Galvin Electricity Initiative, Palo Alto, California, U.S.A.*

Contents

Topical Table of Contents xxi
Foreword xxix
Preface xxxi
Common Energy Abbreviations xxxiii
Thermal Metric and Other Conversion Factors xxxvii
About the Editor xxxix

Volume 1

Accounting: Facility Energy Use / *Douglas E. Tripp*	1
Air Emissions Reductions from Energy Efficiency / *Bruce K. Colburn*	9
Air Quality: Indoor Environment and Energy Efficiency / *Shirley J. Hansen*	18
Aircraft Energy Use / *K. E. Ohrn*	24
Alternative Energy Technologies: Price Effects / *Michael M. Ohadi and Jianwei Qi*	31
ANSI/ASHRAE Standard 62.1-2004 / *Leonard A. Damiano and David S. Dougan*	50
Auditing: Facility Energy Use / *Warren M. Heffington*	63
Auditing: Improved Accuracy / *Barney L. Capehart and Lynne C. Capehart*	69
Auditing: User-Friendly Reports / *Lynne C. Capehart*	76
Benefit Cost Analysis / *Fatouh Al-Ragom*	81
Biomass / *Alberto Traverso*	86
Boilers and Boiler Control Systems / *Eric Peterschmidt and Michael Taylor*	93
Building Automation Systems (BAS): Direct Digital Control / *Paul J. Allen and Rich Remke*	104
Building Geometry: Energy Use Effect / *Geoffrey J. Gilg and Francisco L. Valentine*	111
Building System Simulation / *Essam Omar Assem*	116
Carbon Sequestration / *Nathan E. Hultman*	125
Career Advancement and Assessment in Energy Engineering / *Albert Thumann*	131
Climate Policy: International / *Nathan E. Hultman*	137
Coal Production in the U.S. / *Richard F. Bonskowski, Fred Freme, and William D. Watson*	143
Coal Supply in the U.S. / *Jill S. Tietjen*	156
Coal-to-Liquid Fuels / *Graham Parker*	163
Cold Air Retrofit: Case Study / *James P. Waltz*	172
Combined Heat and Power (CHP): Integration with Industrial Processes / *James A. Clark*	175
Commissioning: Existing Buildings / *David E. Claridge, Mingsheng Liu, and W. D. Turner*	179
Commissioning: New Buildings / *Janey Kaster*	188
Commissioning: Retrocommissioning / *Stephany L. Cull*	200
Compressed Air Control Systems / *Bill Allemon, Rick Avery, and Sam Prudhomme*	207
Compressed Air Energy Storage (CAES) / *David E. Perkins*	214
Compressed Air Leak Detection and Repair / *Robert E. Wilson*	219

Compressed Air Storage and Distribution / *Thomas F. Taranto*	226
Compressed Air Systems / *Diane Schaub*	236
Compressed Air Systems: Optimization / *R. Scot Foss*	241
Cooling Towers / *Ruth Mossad*	246
Data Collection: Preparing Energy Managers and Technicians / *Athula Kulatunga*	255
Daylighting / *William Ross McCluney*	264
Demand Response: Commercial Building Strategies / *David S. Watson, Sila Kiliccote, Naoya Motegi, and Mary Ann Piette*	270
Demand Response: Load Response Resources and Programs / *Larry B. Barrett*	279
Demand-Side Management Programs / *Clark W. Gellings*	286
Desiccant Dehumidification: Case Study / *Michael K. West and Glenn C. Haynes*	292
Distributed Generation / *Paul M. Sotkiewicz and Ing. Jesús Mario Vignolo*	296
Distributed Generation: Combined Heat and Power / *Barney L. Capehart, D. Paul Mehta, and Wayne C. Turner*	303
District Cooling Systems / *Susanna S. Hanson*	309
District Energy Systems / *Ibrahim Dincer and Arif Hepbasli*	316
Drying Operations: Agricultural and Forestry Products / *Guangnan Chen*	332
Drying Operations: Industrial / *Christopher G. J. Baker*	338
Electric Motors / *H. A. Ingley III*	349
Electric Power Transmission Systems / *Jack Casazza*	356
Electric Power Transmission Systems: Asymmetric Operation / *Richard J. Marceau, Abdou-R. Sana, and Donald T. McGillis*	364
Electric Supply System: Generation / *Jill S. Tietjen*	374
Electricity Deregulation for Customers / *Norm Campbell*	379
Electricity Enterprise: U.S., Past and Present / *Kurt E. Yeager*	387
Electricity Enterprise: U.S., Prospects / *Kurt E. Yeager*	399
Electronic Control Systems: Basic / *Eric Peterschmidt and Michael Taylor*	407
Emergy Accounting / *Mark T. Brown and Sergio Ulgiati*	420
Emissions Trading / *Paul M. Sotkiewicz*	430
Energy Codes and Standards: Facilities / *Rosemarie Bartlett, Mark A. Halverson, and Diana L. Shankle*	438
Energy Conservation / *Ibrahim Dincer and Adnan Midilli*	444
Energy Conservation: Industrial Processes / *Harvey E. Diamond*	458
Energy Conservation: Lean Manufacturing / *Bohdan W. Oppenheim*	467
Energy Conversion: Principles for Coal, Animal Waste, and Biomass Fuels / *Kalyan Annamalai, Soyuz Priyadarsan, Senthil Arumugam, and John M. Sweeten*	476
Energy Efficiency: Developing Countries / *U. Atikol and H. M. Güven*	498
Energy Efficiency: Information Sources for New and Emerging Technologies / *Steven A. Parker*	507
Energy Efficiency: Low Cost Improvements / *James Call*	517
Energy Efficiency: Strategic Facility Guidelines / *Steve Doty*	524
Energy Information Systems / *Paul J. Allen and David C. Green*	535
Energy Management: Organizational Aptitude Self-Test / *Christopher Russell*	543
Energy Master Planning / *Fredric S. Goldner*	549
Energy Project Management / *Lorrie B. Tietze and Sandra B. McCardell*	556
Energy Service Companies: Europe / *Silvia Rezessy and Paolo Bertoldi*	566

Volume 2

Energy Star® Portfolio Manager and Building Labeling Program / *Bill Allemon*	573
Energy Use: U.S. Overview / *Dwight K. French*	580
Energy Use: U.S. Transportation / *Kevin C. Coates*	588
Energy: Global and Historical Background / *Milivoje M. Kostic*	601
Enterprise Energy Management Systems / *Gregory Cmar, Bill Gnerre, and Kevin Fuller*	616
Environmental Policy / *Sanford V. Berg*	625
Evaporative Cooling / *David Bisbee*	633
Exergy: Analysis / *Marc A. Rosen*	645
Exergy: Environmental Impact Assessment Applications / *Marc A. Rosen*	655
Facility Air Leakage / *Wendell A. Porter*	662
Facility Energy Efficiency and Controls: Automobile Technology Applications / *Barney L. Capehart and Lynne C. Capehart*	671
Facility Energy Use: Analysis / *Klaus-Dieter E. Pawlik, Lynne C. Capehart, and Barney L. Capehart*	680
Facility Energy Use: Benchmarking / *John Van Gorp*	690
Facility Power Distribution Systems / *Paul L. Villeneuve*	699
Federal Energy Management Program (FEMP): Operations and Maintenance Best Practices Guide (O&M BPG) / *Gregory P. Sullivan, W. David Hunt, and Ab Ream*	707
Fossil Fuel Combustion: Air Pollution and Global Warming / *Dan Golomb*	715
Fuel Cells: Intermediate and High Temperature / *Xianguo Li and Gholamreza Karimi*	726
Fuel Cells: Low Temperature / *Xianguo Li*	733
Geothermal Energy Resources / *Ibrahim Dincer, Arif Hepbasli, and Leyla Ozgener*	744
Geothermal Heat Pump Systems / *Greg Tinkler*	753
Global Climate Change / *Amanda Staudt and Nathan E. Hultman*	760
Green Energy / *Ibrahim Dincer and Adnan Midilli*	771
Greenhouse Gas Emissions: Gasoline, Hybrid-Electric, and Hydrogen-Fueled Vehicles / *Robert E. Uhrig*	787
Heat and Energy Wheels / *Carey J. Simonson*	794
Heat Exchangers and Heat Pipes / *Greg F. Naterer*	801
Heat Pipe Application / *Somchai Jiajitsawat, John Duffy, and John W. Wang*	807
Heat Pumps / *Lu Aye*	814
Heat Transfer / *Yunus A. Cengel*	822
High Intensity Discharge (HID) Electronic Lighting / *Ken E. Patterson II*	830
HVAC Systems: Humid Climates / *David MacPhaul*	839
Hybrid-Electric Vehicles: Plug-In Configuration / *Robert E. Uhrig and Vernon P. Roan*	847
Independent Power Producers / *Hemmat Safwat*	854
Industrial Classification and Energy Efficiency / *Asfaw Beyene*	862
Industrial Energy Management: Global Trends / *Cristián Cárdenas-Lailhacar*	872
Industrial Motor System Optimization Projects in the U.S. / *Robert Bruce Lung, Aimee McKane, and Mitchell Olszewski*	883
Insulation: Facilities / *Wendell A. Porter*	890
Integrated Gasification Combined Cycle (IGCC): Coal- and Biomass-Based / *Ashok D. Rao*	906
IntelliGrid^{SM} / *Clark W. Gellings and Kurt E. Yeager*	914
International Performance Measurement and Verification Protocol (IPMVP) / *James P. Waltz*	919
Investment Analysis Techniques / *James W. Dean*	930

LEED-CI and LEED-CS: Leadership in Energy and Environmental Design for Commercial Interiors and Core and Shell / *Nick Stecky* 937

LEED-NC: Leadership in Energy and Environmental Design for New Construction / *Stephen A. Roosa* 947

Life Cycle Costing: Electric Power Projects / *Ujjwal Bhattacharjee* 953

Life Cycle Costing: Energy Projects / *Sandra B. McCardell* 967

Lighting Controls / *Craig DiLouie* 977

Lighting Design and Retrofits / *Larry Leetzow* 993

Liquefied Natural Gas (LNG) / *Charles P. Ivey* 1001

Living Standards and Culture: Energy Impact / *Marc A. Rosen* 1009

Maglev (Magnetic Levitation) / *Kevin C. Coates* 1018

Management Systems for Energy / *Michael L. Brown* 1027

Manufacturing Industry: Activity-Based Costing / *J. Michael Fernandes, Barney L. Capehart, and Lynne C. Capehart* 1042

Measurement and Verification / *Stephen A. Roosa* 1050

Measurements in Energy Management: Best Practices and Software Tools / *Peter Barhydt and Mark Menezes* 1056

Mobile HVAC Systems: Fundamentals, Design, and Innovations / *Alex V. Moultanovsky* 1061

Mobile HVAC Systems: Physics and Configuration / *Alex V. Moultanovsky* 1070

National Energy Act of 1978 / *Robert R. Nordhaus* 1082

Natural Energy versus Additional Energy / *Marc A. Rosen* 1088

Net Metering / *Steven Ferrey* 1096

Nuclear Energy: Economics / *James G. Hewlett* 1101

Nuclear Energy: Fuel Cycles / *James S. Tulenko* 1111

Nuclear Energy: Power Plants / *Michael Burke Wood* 1115

Nuclear Energy: Technology / *Michael Burke Wood* 1125

Performance Contracting / *Shirley J. Hansen* 1134

Performance Indicators: Industrial Energy / *Harvey E. Diamond and Robert G. Botelho* 1140

Volume 3

Photovoltaic Systems / *Michael Ropp* 1147

Physics of Energy / *Milivoje M. Kostic* 1160

Pricing Programs: Time-of-Use and Real Time / *Ahmad Faruqui* 1175

Psychrometrics / *S. A. Sherif* 1184

Public Policy for Improving Energy Sector Performance / *Sanford V. Berg* 1193

Public Utility Regulatory Policies Act (PURPA) of 1978 / *Richard F. Hirsh* 1201

Pumped Storage Hydroelectricity / *Jill S. Tietjen* 1207

Pumps and Fans / *L. J. Grobler and T. G. Vorster* 1213

Radiant Barriers / *Ingrid Melody, Philip Fairey, and David Beal* 1227

Reciprocating Engines: Diesel and Gas / *Rolf D. Reitz and Kevin Hoag* 1233

Regulation: Price Cap and Revenue Cap / *Mark A. Jamison* 1245

Regulation: Rate of Return / *Mark A. Jamison* 1252

Renewable and Decentralized Energy Options: State Promotion in the U.S. / *Steven Ferrey* 1258

Renewable Energy / *John O. Blackburn* 1265

Residential Buildings: Heating Loads / *Zohrab Melikyan* 1272

Run-Around Heat Recovery Systems / *Kamiel S. Gabriel* 1278

Savings and Optimization: Case Studies / *Robert W. Peters and Jeffery P. Perl* 1283

Savings and Optimization: Chemical Process Industry / *Jeffery P. Perl and Robert W. Peters*	1302
Six Sigma Methods: Measurement and Verification / *Carla Fair-Wright and Michael R. Dorrington*	1310
Solar Heating and Air Conditioning: Case Study / *Eric Oliver, K. Baker, M. Babcock, and John Archibald*	1317
Solar Thermal Technologies / *M. Asif and T. Muneer*	1321
Solar Water Heating: Domestic and Industrial Applications / *M. Asif and T. Muneer*	1331
Solid Waste to Energy by Advanced Thermal Technologies (SWEATT) / *Alex E. S. Green*	1340
Space Heating / *James P. Riordan*	1357
Steam and Hot Water System Optimization: Case Study / *Nevena Iordanova and Rick Cumbo*	1366
Steam Turbines / *Michael Burke Wood*	1380
Sustainability Policies: Sunbelt Cities / *Stephen A. Roosa*	1389
Sustainable Building Simulation / *Birol I. Kilkis*	1396
Sustainable Development / *Mark A. Peterson*	1406
Thermal Energy Storage / *Brian Silvetti*	1412
Thermodynamics / *Ronald L. Klaus*	1422
Tradable Certificates for Energy Savings / *Silvia Rezessy and Paolo Bertoldi*	1433
Transportation Systems: Hydrogen-Fueled / *Robert E. Uhrig*	1441
Transportation: Location Efficiency / *Todd A. Litman*	1449
Tribal Land and Energy Efficiency / *Jackie L. Francke and Sandra B. McCardell*	1456
Underfloor Air Distribution (UFAD) / *Tom Webster and Fred Bauman*	1463
Utilities and Energy Suppliers: Bill Analysis / *Neil Petchers*	1471
Utilities and Energy Suppliers: Business Partnership Management / *S. Kay Tuttle*	1484
Utilities and Energy Suppliers: Planning and Portfolio Management / *Devra Bachrach Wang*	1491
Utilities and Energy Suppliers: Rate Structures / *Neil Petchers*	1497
Walls and Windows / *Therese Stovall*	1513
Waste Fuels / *Robert J. Tidona*	1523
Waste Heat Recovery / *Martin A. Mozzo*	1536
Waste Heat Recovery Applications: Absorption Heat Pumps / *Keith E. Herold*	1541
Water and Wastewater Plants: Energy Use / *Alfred J. Lutz*	1548
Water and Wastewater Utilities / *Rudolf Marloth*	1556
Water Source Heat Pump for Modular Classrooms / *Andrew R. Forrest and James W. Leach*	1564
Water-Augmented Gas Turbine Power Cycles / *Ronald L. Klaus*	1574
Water-Using Equipment: Commercial and Industrial / *Kate McMordie Stoughton and Amy Solana*	1587
Water-Using Equipment: Domestic / *Amy Solana and Kate McMordie Stoughton*	1596
Wind Power / *K. E. Ohrn*	1607
Window Energy / *William Ross McCluney*	1616
Window Films: Savings from IPMVP Options C and D / *Steve DeBusk*	1626
Window Films: Solar-Control and Insulating / *Steve DeBusk*	1630
Window Films: Spectrally Selective versus Conventional Applied / *Marty Watts*	1637
Windows: Shading Devices / *Svetlana J. Olbina*	1641
Wireless Applications: Energy Information and Control / *Jim Lewis*	1649
Wireless Applications: Mobile Thermostat Climate Control and Energy Conservation / *Alexander L. Burd and Galina S. Burd*	1660

Topical Table of Contents

I. Energy, Energy Sources, and Energy Use

Energy: Historical and Technical Background

Energy Conservation / *Ibrahim Dincer and Adnan Midilli*	444
Energy Efficiency: Developing Countries / *U. Atikol and H. M. Güven*	498
Energy Use: U.S. Overview / *Dwight K. French*	580
Energy Use: U.S. Transportation / *Kevin C. Coates*	588
Energy: Global and Historical Background / *Milivoje M. Kostic*	601
Living Standards and Culture: Energy Impact / *Marc A. Rosen*	1009
National Energy Act of 1978 / *Robert R. Nordhaus*	1082
Natural Energy versus Additional Energy / *Marc A. Rosen*	1088
Physics of Energy / *Milivoje M. Kostic*	1160
Public Policy for Improving Energy Sector Performance / *Sanford V. Berg*	1193
Public Utility Regulatory Policies Act (PURPA) of 1978 / *Richard Hirsh*	1201

Fossil Fuels

Coal Production in the U.S. / *Richard F. Bonskowski, Fred Freme, and William D. Watson*	143

Nuclear Energy

Nuclear Energy: Fuel Cycles / *James S. Tulenko*	1111
Nuclear Energy: Power Plants / *Michael Burke Wood*	1115
Nuclear Energy: Technology / *Michael Burke Wood*	1125

Renewable Energy

Alternative Energy Technologies: Price Effects / *Michael M. Ohadi and Jianwei Qi*	31
Biomass / *Alberto Traverso*	86
Geothermal Energy Resources / *Ibrahim Dincer, Arif Hepbasli, and Leyla Ozgener*	744
Green Energy / *Ibrahim Dincer and Adnan Midilli*	771
Photovoltaic Systems / *Michael Ropp*	1147
Renewable and Decentralized Energy Options: State Promotion in the U.S. / *Steven Ferrey*	1258
Renewable Energy / *John O. Blackburn*	1265
Solar Thermal Technologies / *M. Asif and T. Muneer*	1321
Solid Waste to Energy by Advanced Thermal Technologies (SWEATT) / *Alex E. S. Green*	1340
Wind Power / *K. E. Ohrn*	1607

Energy Storage and Derived Energy

Derived Energy

Coal-to-Liquid Fuels / *Graham Parker*	163
Transportation Systems: Hydrogen-Fueled / *Robert E. Uhrig*	1441

I. Energy, Energy Sources, and Energy Use (cont'd.)

Derived Energy (cont'd.)

Energy Storage

Compressed Air Energy Storage (CAES) / *David E. Perkins*	214
Pumped Storage Hydroelectricity / *Jill S. Tietjen*	1207

Fuel Cells

Fuel Cells: Intermediate and High Temperature / *Xianguo Li and Gholamreza Karimi*	733
Fuel Cells: Low Temperature / *Xianguo Li*	744

II. Principles of Energy Analysis and Systems/Economic Analysis of Energy Systems

Energy Systems Analysis

Accounting: Facility Energy Use / *Douglas E. Tripp*	1
Auditing: Facility Energy Use / *Warren M. Heffington*	63
Building System Simulation / *Essam Omar Assem*	116
Emergy Accounting / *Mark T. Brown and Sergio Ulgiati*	420
Exergy: Analysis / *Marc A. Rosen*	645
Facility Energy Use: Analysis / *Klaus-Dieter E. Pawlik, Lynne C. Capehart, and Barney L. Capehart*	680

Financial and Economic Analysis Principles

Benefit Cost Analysis / *Fatouh Al-Ragom*	81
Investment Analysis Techniques / *James W. Dean*	930
Life Cycle Costing: Electric Power Projects / *Ujjwal Bhattacharjee*	953
Life Cycle Costing: Energy Projects / *Sandra B. McCardell*	967

Principles of Electric Energy Systems

Electric Power Transmission Systems: Asymmetric Operation / *Richard J. Marceau, Abdou-R. Sana, and Donald T. McGillis*	364
Facility Power Distribution Systems / *Paul L. Villeneuve*	699

Principles of Thermal Energy Systems

Energy Conversion: Principles for Coal, Animal Waste, and Biomass Fuels / *Kalyan Annamalai, Soyuz Priyadarsan, Senthil Arumugam, and John M. Sweeten*	476
Heat Transfer / *Yunus A. Cengel*	822
Psychrometrics / *S. A. Sherif*	1184
Thermodynamics / *Ronald L. Klaus*	1422

III. Utilities, Suppliers of Energy, and Utility Regulation

Electric Supply System

Electric Power Transmission Systems / *Jack Casazza*	356
Electric Supply System: Generation / *Jill S. Tietjen*	374
Independent Power Producers / *Hemmat Safwat*	854
Integrated Gasification Combined Cycle (IGCC): Coal- and Biomass-Based / *Ashok D. Rao*	906
IntelliGrid^SM / *Clark W. Gellings and Kurt E. Yeager*	914
Nuclear Energy: Economics / *James G. Hewlett*	1101

Fuel Supply System

Coal Supply in the U.S. / *Jill S. Tietjen*	156
Liquefied Natural Gas (LNG) / *Charles P. Ivey*	1001

Utilities: Overview

District Energy Systems / *Ibrahim Dincer and Arif Hepbasli*	316
Electricity Enterprise: U.S., Past and Present / *Kurt E. Yeager*	387
Electricity Enterprise: U.S., Prospects / *Kurt E. Yeager*	399
Utilities and Energy Suppliers: Bill Analysis / *Neil Petchers*	1471
Utilities and Energy Suppliers: Rate Structures / *Neil Petchers*	1497
Water and Wastewater Utilities / *Rudolf Marloth*	1556

Utility Regulatory Issues

Demand Response: Load Response Resources and Programs / *Larry B. Barrett*	279
Demand-Side Management Programs / *Clark W. Gellings*	286
Electricity Deregulation for Customers / *Norm Campbell*	379
Net Metering / *Steven Ferrey*	1096
Pricing Programs: Time-of-Use and Real Time / *Ahmad Faruqui*	1175
Regulation: Price Cap and Revenue Cap / *Mark A. Jamison*	1245
Regulation: Rate of Return / *Mark A. Jamison*	1252
Utilities and Energy Suppliers: Planning and Portfolio Management / *Devra Bachrach Wang*	1491

IV. Facilities and Users of Energy

Building Envelope

Facility Air Leakage / *Wendell A. Porter*	662
Insulation: Facilities / *Wendell A. Porter*	890
Radiant Barriers / *Ingrid Melody, Philip Fairey, and David Beal*	1227
Walls and Windows / *Therese Stovall*	1513
Window Energy / *William Ross McCluney*	1616
Window Films: Savings from IPMVP Options C and D / *Steve DeBusk*	1626

IV. Facilities and Users of Energy (*cont'd.*)

Building Envelope (cont'd.)

Window Films: Solar-Control and Insulating / *Steve DeBusk*	1630
Window Films: Spectrally Selective versus Conventional Applied / *Marty Watts*	1637
Windows: Shading Devices / *Svetlana J. Olbina*	1641

Compressed Air Systems

Compressed Air Control Systems / *Bill Allemon, Rick Avery, and Sam Prudhomme*	207
Compressed Air Leak Detection and Repair / *Robert E. Wilson*	219
Compressed Air Storage and Distribution / *Thomas F. Taranto*	226
Compressed Air Systems / *Diane Schaub*	236
Compressed Air Systems: Optimization / *R. Scot Foss*	241

Facility Controls and Information Systems

Building Automation Systems (BAS): Direct Digital Control / *Paul J. Allen and Rich Remke*	104
Electronic Control Systems: Basic / *Eric Peterschmidt and Michael Taylor*	407
Energy Information Systems / *Paul J. Allen and David C. Green*	535
Enterprise Energy Management Systems / *Gregory Cmar*	616

Heat Recovery

Heat and Energy Wheels / *Carey J. Simonson*	794
Heat Exchangers and Heat Pipes / *Greg F. Naterer*	801
Heat Pipe Application / *Somchai Jiajitsawat, John Duffy, and John W. Wang*	807
Run-Around Heat Recovery Systems / *Kamiel S. Gabriel*	1278
Waste Heat Recovery / *Martin A. Mozzo Jr.*	1536
Waste Heat Recovery Applications: Absorption Heat Pumps / *Keith E. Herold*	1548

HVAC Systems

Cold Air Retrofit: Case Study / *James P. Waltz*	172
Cooling Towers / *Ruth Mossad*	246
Desiccant Dehumidification: Case Study / *Michael K. West and Glenn C. Haynes*	292
District Cooling Systems / *Susanna S. Hanson*	309
Evaporative Cooling / *David Bisbee*	633
Geothermal Heat Pump Systems / *Greg Tinkler*	753
Heat Pumps / *Lu Aye*	814
HVAC Systems: Humid Climates / *David MacPhaul*	839
Pumps and Fans / *L. J. Grobler and T. G. Vorster*	1213
Thermal Energy Storage / *Brian Silvetti*	1412
Underfloor Air Distribution (UFAD) / *Tom Webster and Fred Bauman*	1463

Industrial Facilities

Combined Heat and Power (CHP): Integration with Industrial Processes / *James A. Clark*	175
Drying Operations: Agricultural and Forestry Products / *Guangnan Chen*	332
Drying Operations: Industrial / *Christopher G.J. Baker*	338
Energy Conservation: Industrial Processes / *Harvey E. Diamond*	458
Energy Conservation: Lean Manufacturing / *Bohdan W. Oppenheim*	467
Industrial Classification and Energy Efficiency / *Asfaw Beyene*	862
Industrial Energy Management: Global Trends / *Cristián Cárdenas-Lailhacar*	872
Industrial Motor System Optimization Projects in the U.S. / *Robert Bruce Lung, Aimee McKane, and Mitchell Olszewski*	883
Performance Indicators: Industrial Energy / *Harvey E. Diamond and Robert G. Botelho*	1140
Savings and Optimization: Case Studies / *Robert W. Peters and Jeffery P. Perl*	1283
Savings and Optimization: Chemical Process Industry / *Jeffery P. Perl and Robert W. Peters*	1302
Waste Fuels / *Robert J. Tidona*	1523

Lighting and Motors

Daylighting / *William Ross McCluney*	264
Electric Motors / *H. A. Ingley III*	349
High Intensity Discharge (HID) Electronic Lighting / *Ken E. Patterson II*	830
Lighting Controls / *Craig DiLouie*	977
Lighting Design and Retrofits / *Larry Leetzow*	993

On Site Electric Generation

Distributed Generation / *Paul M. Sotkiewicz and Ing. Jesús Mario Vignolo*	296
Distributed Generation: Combined Heat and Power / *Barney L. Capehart and D. Paul Mehta, and Wayne C. Turner*	303
Reciprocating Engines: Diesel and Gas / *Rolf D. Reitz and Kevin Hoag*	1233
Water-Augmented Gas Turbine Power Cycles / *Ronald L. Klaus*	1574

Space Heating, Water Heating, and Water Use

Residential Buildings: Heating Loads / *Zohrab Melikyan*	1272
Solar Heating and Air Conditioning: Case Study / *Eric Oliver, K. Baker, M. Babcock, and John Archibald*	1317
Solar Water Heating: Domestic and Industrial Applications / *M. Asif and T. Muneer*	1331
Space Heating / *James P. Riordan*	1357
Water Source Heat Pump for Modular Classrooms / *Andrew R. Forrest and James W. Leach*	1564
Water-Using Equipment: Commercial and Industrial / *Kate McMordie Stoughton and Amy Solana*	1587
Water-Using Equipment: Domestic / *Amy Solana and Kate McMordie Stoughton*	1596

Steam Boilers and Steam Systems

Boilers and Boiler Control Systems / *Eric Peterschmidt and Michael Taylor*	93
Steam and Hot Water System Optimization: Case Study / *Nevena Iordanova and Rick Cumbo*	1366
Steam Turbines / *Michael Burke Wood*	1380

V. Energy Management

Commissioning

Commissioning: Existing Buildings / *David E. Claridge, Mingsheng Liu, and W. D. Turner* . . .	179
Commissioning: New Buildings / *Janey Kaster*	188
Commissioning: Retrocommissioning / *Stephany L. Cull*	200

Energy Auditing and Benchmarking

Auditing: Improved Accuracy / *Barney L. Capehart and Lynne C. Capehart*	69
Auditing: User-Friendly Reports / *Lynne C. Capehart*	76
Building Geometry: Energy Use Effect / *Geoffrey J. Gilg and Francisco L. Valentine*	111
Facility Energy Use: Benchmarking / *John Van Gorp*	690
Federal Energy Management Program (FEMP): Operations and Maintenance Best Practices Guide (O&M BPG) / *Gregory P. Sullivan, W. David Hunt, and Ab Ream* .	707
Manufacturing Industry: Activity-Based Costing / *J. Michael Fernandes, Barney L. Capehart, and Lynne C. Capehart*	1042

Energy Codes and Standards

Air Quality: Indoor Environment and Energy Efficiency / *Shirley J. Hansen*	18
ANSI/ASHRAE Standard 62.1 - 2004 / *Leonard A. Damiano and David Dougan*	50
Energy Codes and Standards: Facilities / *Rosemarie Bartlett, Mark A. Halverson, and Diana L. Shankle*	438

Energy Management Programs for Facilities

Career Advancement and Assessment in Energy Engineering / *Albert Thumann*	131
Data Collection: Preparing Energy Managers and Technicians / *Athula Kulatunga*	255
Energy Efficiency: Information Sources for New and Emerging Technologies / *Steven A. Parker*	507
Energy Management: Organizational Aptitude Self-Test / *Christopher Russell*	543
Energy Master Planning / *Fredric S. Goldner*	549
Management Systems for Energy / *Michael L. Brown*	1027

Energy Savings Projects

Demand Response: Commercial Building Strategies / *David S. Watson, Sila Kiliccote, Naoya Motegi, and Mary Ann Piette*	270
Energy Efficiency: Low Cost Improvements / *James Call*	517
Energy Efficiency: Strategic Facility Guidelines / *Steve Doty*	524
Energy Project Management / *Lorrie B. Tietze and Sandra B. McCardell*	556
Facility Energy Efficiency and Controls: Automobile Technology Applications / *Barney L. Capehart and Lynne C. Capehart*	671
Tribal Land and Energy Efficiency / *Jackie L. Francke and Sandra B. McCardell*	1456
Utilities and Energy Suppliers: Business Partnership Management / *S. Kay Tuttle*	1484
Wireless Applications: Mobile Thermostat Climate Control and Energy Conservation / *Alexander L. Burd and Galina S. Burd*	1660

Measurement and Verification of Energy Savings

International Performance Measurement and Verification Protocol (IPMVP) / *James P. Waltz*	919
Measurement and Verification / *Stephen A. Roosa*	1050
Measurements in Energy Management: Best Practices and Software Tools / *Peter Barhydt and Mark Menezes*	1056
Six Sigma Methods: Measurement and Verification / *Carla Fair-Wright and Michael R. Dorrington*	1310
Wireless Applications: Energy Information and Control / *Jim Lewis*	1649

Performance Contracting

Energy Service Companies: Europe / *Silvia Rezessy and Paolo Bertoldi*	566
Performance Contracting / *Shirley J. Hansen*	1134

VI. Current Energy and Environmental Issues

Environmental Regulation: Overview

Air Emissions Reductions from Energy Efficiency / *Bruce K. Colburn*	9
Carbon Sequestration / *Nathan E. Hultman*	125
Climate Policy: International / *Nathan E. Hultman*	137
Emissions Trading / *Paul M. Sotkiewicz*	430
Environmental Policy / *Sanford V. Berg*	625
Exergy: Environmental Impact Assessment Applications / *Marc A. Rosen*	655
Fossil Fuel Combustion: Air Pollution and Global Warming / *Dan Golomb*	715
Global Climate Change / *Amanda Staudt and Nathan E. Hultman*	760
Tradable Certificates for Energy Savings / *Silvia Rezessy and Paolo Bertoldi*	1433

Sustainable Buildings

Energy Star® Portfolio Manager and Building Labeling Program / *Bill Allemon*	573
LEED-CI and LEED-CS: Leadership in Energy and Environmental Design for Commercial Interiors and Core and Shell / *Nick Stecky*	937
LEED-NC: Leadership in Energy and Environmental Design for New Construction / *Stephen A. Roosa*	947
Sustainability Policies: Sunbelt Cities / *Stephen A. Roosa*	1389
Sustainable Building Simulation / *Birol I. Kilkis*	1396
Sustainable Development / *Mark A. Peterson*	1406

VII. Transportation and Other Energy Uses

Other Energy Uses

Water and Wastewater Plants: Energy Use / *Alfred J. Lutz*	1548

VII. Transportation and Other Energy Uses (*cont'd.*)

Transportation

Aircraft Energy Use / *K. E. Ohrn*	24
Greenhouse Gas Emissions: Gasoline, Hybrid-Electric, and Hydrogen-Fueled Vehicles / *Robert E. Uhrig*	787
Hybrid-Electric Vehicles: Plug-In Configuration / *Robert E. Uhrig and Vernon P. Roan*	847
Maglev (Magnetic Levitation) / *Kevin C. Coates*	1018
Mobile HVAC Systems: Fundamentals, Design, and Innovations / *Alex V. Moultanovsky*	1061
Mobile HVAC Systems: Physics and Configuration / *Alex V. Moultanovsky*	1070
Transportation: Location Efficiency / *Todd A. Litman*	1449

Foreword

The Association of Energy Engineers (AEE) is proud to be a sponsor of the *Encyclopedia of Energy Engineering and Technology, Three-Volume Set*, edited by Dr. Barney L. Capehart. In 2007 AEE is celebrating its 30th anniversary and it is a fitting tribute that the *Encyclopedia of Energy Engineering and Technology* is published at this time. AEE defined the energy engineering profession and this comprehensive work details the core elements for success in this field.

Dr. Capehart has performed a monumental task of facilitating over 300 researchers and practitioners who have contributed to this three-volume set. These distinguished authorities share a wealth of knowledge on approximately 190 topics. Dr. Capehart, through his training and publications, has significantly impacted the energy engineering profession and has helped make it what it is today.

Global climate change concerns and unstable energy prices have raised the importance of energy engineering. This encyclopedia will be a valuable tool in assisting energy engineers to reach their potential. The Association of Energy Engineers (AEE) and our network of 8,000 members in 77 countries would like to thank Dr. Barney Capehart and the numerous volunteers who have made this work possible.

Albert Thumann, P.E., CEM
Executive Director
The Association of Energy Engineers (AEE)

The Association of Energy Engineers

> The Association of Energy Engineers' mission is to promote the scientific and educational interests of those engaged in the energy industry and to foster action for sustainable development.

The Association of Energy Engineers (AEE) is proud to sponsor the Encyclopedia of Energy Engineering and Technology. The Encyclopedia of Engineering and Technology is an important contribution to the field of Energy Engineering and will be invaluable to the industry.

AEE provides a gateway for information on the dynamic field of energy efficiency, renewable energy and global warming. Celebrating its 30th year, AEE is in the forefront of energy engineering technology transfer. With a full array of information and outreach programs from technical seminars, conferences, books and journals to critical buyer-seller networking trade shows.

AEE presents three important industry events every year. These events take place across the U.S. to allow energy professionals from all regions to attend. The following events bring attendees from across the U.S. and around the world together to network and to discuss the most up-to-date technologies and innovations affecting the industry:

Globalcon
www.GLOBALCONevent.com

West Coast Energy Management Congress (EMC)
www.energyevent.com

World Energy Engineering Congress (WEEC)
www.energycongress.com

The Association also offers a variety of information resource tools. As a growing membership organization, the overall strength of AEE is highlighted by a strong membership base of over 8,000 professionals and recognized certification programs, including Certified Energy Managers (CEMs), Lighting Professionals, Indoor Air Quality Professionals, and Cogeneration Professionals. The Association's network of 67 local and regional chapters has powerful grassroots agendas and members. Further, AEE's roster of corporate members is a veritable "Who's Who" of the commercial, industrial, institutional, governmental, energy services, and utility sectors.

The Association of Energy Engineers is pleased that an Encyclopedia of this magnitude has been created for use by industry professionals and advocates.

For more details:
www.aeecenter.org

ASSOCIATION OF ENERGY ENGINEERS

Phone: 770-447-5083
Fax: 770-446-3969

4025 Pleasantdale Rd, Suite 420
Atlanta, GA 30340

www.aeecenter.org

Preface

Energy engineers and technologists have made efficient and cost effective devices for many years which provide the energy services society wants and expects. From air conditioners to waste fuels, energy engineers and technologists continue to make our lives comfortable and affordable using limited resources in efficient and renewable ways.

Over 300 researchers and practitioners, through 190 entries, provide ready access to the basic principles and applications of energy engineering, as well as advanced applications in the technologies of energy production and use. The global supply of energy is increasingly being stressed to provide for an expanding world population. Energy efficiency, energy conservation through energy management, and use of renewable energy sources are three of the major strategies that in the future will help provide the energy and energy services for the world's population and the world's economy.

This unique reference contains state-of-the-art progress on the most important topics in this field of energy engineering and technology. All entries in the encyclopedia have been written by experts in their specialties, and have been reviewed by subject matter authorities. This distinguished group of experts share a wealth of knowledge on topics such as:

- Energy, energy supplies and energy use
- Renewable and alternative energy sources
- Technical, economic and financial analysis of energy systems
- Energy uses in buildings and industry
- Energy efficiency and energy conservation opportunities and projects
- Commissioning, benchmarking, performance contracting, and measurement and verification
- Environmental regulation and public policy for energy supply and use
- Global climate change and carbon control
- Sustainable buildings and green development
- Hybrid electric and hydrogen fueled vehicles and maglev transportation

The *Encyclopedia of Energy Engineering and Technology, Three-Volume Set,* is a key reference work for professionals in academia, business, industry and government, as well as students at all levels. It should be regularly consulted for basic and advanced information to guide students, scholars, practitioners, the public, and policy makers. Contributions address a wide spectrum of theoretical and applied topics, concepts, methodologies, strategies, and possible solutions.

The On-Line Edition is a dynamic resource that will grow as time and knowledge progress. Suggestions for additional content are welcomed by the editor, and new authors should contact me at the e-mail address listed below.

As editor, I would like to thank the people who worked hard to initiate this encyclopedia, and make the project a success. Thanks go to Russell Dekker at Marcel Dekker, and Al Thumann, the Executive Director of the Association of Energy Engineers (AEE) for getting the project going. I appreciate their confidence in my ability to accomplish this immense project. Part of the purpose of this project is to help provide professional and educational support for new people coming into our area of energy engineering. A profession can only succeed and grow if new people have a resource to learn about a new area, and find this area interesting and exciting for their careers.

Directors Oona Schmidt and Claire Miller were both excellent in helping to organize and specify the work that had to be done to get the encyclopedia started and on track for completion. Editorial Assistants Andrea Cunningham, Lousia Lam, and Marisa Hoheb provided daily help to me, and kept all of the records and contacts with the authors. Their help was invaluable.

Preparation of this modern compendium on energy engineering and technology has only been possible through the commitment and hard work of hundreds of energy engineers from around the globe. I want to thank all of the authors for their outstanding efforts to identify major topics of interest for this project, and to write interesting and educational articles based on their areas of expertise. Many of the authors also served a dual function of both writing their own articles, and reviewing the submissions of other authors. Another important group of people were those on the Editorial Board who helped submit topics, organizational ideas, and lists of potential authors for the encyclopedia. This Board was a great help in getting the actual writing of articles started, as well as many of the Editorial Board members also contributed articles themselves.

I would also like to thank my wife Lynne for her continuing support of all of my projects over the years. And finally, I would like to dedicate this encyclopedia to my grandchildren Hannah and Easton. They are a part of the future that I hope will be using the efficient and sustainable energy resources presented in this encyclopedia.

Barney L. Capehart, Editor
Capehart@ise.ufl.edu

Common Energy Abbreviations

The following abbreviations were provided by the Energy Information Administration's National Energy Information Center. The Energy Information Administration is a statistical agency of the U.S. Department of Energy.

AC	alternating current
AFUDC	allowance for funds used during construction
AFV	alternative-fuel vehicle
AGA	American Gas Association
ANSI	American National Standards Institute
API	American Petroleum Institute
ASTM	American Society for Testing and Materials
bbl	barrel(s)
bbl/d	barrel(s) per day
bcf	billion cubic feet
BLS	Bureau of Labor Statistics within the U.S. Department of Labor
BOE	barrels of oil equivalent (used internationally)
Btu	British thermal unit(s)
BWR	boiling-water reactor
C/gal	cents per gallon
CAFE	corporate average fuel economy
CARB	California Air Resources Board
CDD	cooling degree-days
CERCLA	Comprehensive Environmental Response, Compensation, and Liability Act
CF	cubic foot
CFC	chlorofluorocarbon
CFS	cubic feet per second
CH_4	Methane
CHP	combined heat and power
CNG	compressed natural gas
cnt	cent
CO	carbon monoxide
CO_2	carbon dioxide
CPI	consumer price index
CWIP	construction work in progress
DC	direct current
DOE	Department of Energy
DRB	demonstrated reserve base
DSM	demand-side management
E85	A fuel containing a mixture of 85 percent ethanol and 15 percent gasoline
E95	A fuel containing a mixture of 95 percent ethanol and 5 percent gasoline
EAR	estimated additional resources
EIA	Energy Information Administration
EIS	Environmental Impact Statement
EOR	enhanced oil recovery
EPA	Environmental Protection Agency
EPACT	Energy Policy Act
EU	European Union
EWG	exempt wholesale generator
FASB	Financial Accounting Standards Board
FBR	fast breeder reactor

FERC	Federal Energy Regulatory Commission
FGD	flue-gas desulfurization
F.O.B	free on board
FPC	Federal Power Commission
FRS	Financial Reporting System
gal	gallon
GDP	gross domestic product
GNP	gross national product
GVW	gross vehicle weight
GW	gigawatt
GWe	gigawatt-electric
GWh	gigawatthour
GWP	global warming potential
HCFC	hydrochlorofluorocarbon
HDD	heating degree-days
HFC	hydrofluorocarbon
HID	high-intensity discharge
HTGR	high temperature gas-cooled reactor
HVAC	heating, ventilation, and air-conditioning
IEA	International Energy Agency
IOU	investor-owned utility
IPP	independent power producer
ISO	independent system operator
kVa	kilovolt-Ampere
kW	kilowatt
kWe	kilowatt-electric
kWh	kilowatthour
lb	pound
LDC	local distribution company
LEVP	Low Emissions Vehicle Program
LHV	lower heating value
LIHEAP	Low-Income Home Energy Assistance Program
LNG	liquefied natural gas
LPG	liquefied petroleum gases
LRG	liquefied refinery gases
LWR	light water reactor
M	thousand
Mcf	one thousand cubic feet
MECS	Manufacturing Energy Consumption Survey
MM	million (10^6)
MMbbl/d	one million (10^6) barrels of oil per day
MMBtu	one million (10^6) British thermal units
MMcf	one million (10^6) cubic feet
MMgal/d	one million (10^6) gallons per day
MMst	one million (10^6) short tons
MPG	miles per gallon
MSA	metropolitan statistical area
MSHA	Mine Safety and Health Administration
MSW	municipal solid waste
MTBE	methyl tertiary butyl ether
MW	megawatt
MWe	megawatt electric
MWh	megawatthour
N_2O	Nitrous oxide
NAAQS	National Ambient Air Quality Standards
NAICS	North American Industry Classification System

NARUC	National Association of Regulatory Utility Commissioners
NERC	North American Electric Reliability Council
NGL	natural gas liquids
NGPA	Natural Gas Policy Act of 1978
NGPL	natural gas plant liquids
NGV	natural gas vehicle
NOAA	National Oceanic and Atmospheric Administration
NOPR	Notice of Proposed Rulemaking
NO$_x$	nitrogen oxides
NRECA	National Rural Electric Cooperative Association
NUG	nonutility generator
NURE	National Uranium Resource Evaluation
NYMEX	New York Mercantile Exchange
O$_3$	Ozone
O&M	operation and maintenance
OCS	Outer Continental Shelf
OECD	Organization for Economic Cooperation and Development
OEM	original equipment manufacturers
OPEC	Organization of Petroleum Exporting Countries
OPRG	oxygenated fuels program reformulated gasoline
OTEC	ocean thermal energy conversion
PADD	Petroleum Administration for Defense Districts
PBR	pebble-bed reactor
PBR	performance-based rates
PCB	polychlorinated biphenyl
PFCs	perfluorocarbons
PGA	purchased gas adjustment
PPI	Producer Price Index
PUD	public utility district
PUHCA	Public Utility Holding Company Act of 1935
PURPA	Public Utility Regulatory Policies Act of 1978
PV	photovoltaic
PVC	photovoltaic cell; polyvinyl chloride
PWR	pressurized-water reactor
QF	qualifying facility
QUAD	quadrillion Btu: 10^{15} Btu
RAC	refiners' acquisition cost
RAR	reasonable assured resources
RBOB	Reformulated Gasoline Blendstock for Oxygenate Blending
RDF	refuse-derived fuel
REA	Rural Electrification Administration
RECS	Residential Energy Consumption Survey
RFG	reformulated gasoline
RSE	relative standard error
RVP	Reid vapor pressure
SEER	seasonal energy efficiency ratio
SF$_6$	sulfur hexafluoride
SI	International System of Units (Système international d'unités)
SIC	Standard Industrial Classification
SNG	synthetic natural gas
SO$_2$	sulfur dioxide
SPP	small power producer
SPR	Strategic Petroleum Reserve
SR	speculative resources
T	trillion 10^{12}
TVA	Tennessee Valley Authority

TW	Terawatt
U₃O₈	Uranium oxide
UF₆	uranium hexaflouride
ULCC	ultra large crude carrier
UMTRA	Uranium Mill Tailings Radiation Control Act of 1978
USACE	U.S. Army Corps of Engineers (sometimes shortened to USCE in EIA tables)
USBR	United States Bureau of Reclamation
V	Volt
VAWT	vertical-axis wind turbine
VIN	vehicle identification number
VLCC	very large crude carrier
VMT	vehicle miles traveled
VOC	volatile organic compound
W	watt
WACOG	weighted average cost of gas
Wh	watt hour
WTI	West Texas Intermediate

Thermal Metric and Other Conversion Factors

The following tables appeared in the January 2007 issue of the Energy Information Administration's *Monthly Energy Review*. The Energy Information Administration is a statistical agency of the U.S. Department of Energy.

Table 1 Metric Conversion Factors

These metric conversion factors can be used to calculate the metric-unit equivalents of values expressed in U.S. Customary units. For example, 500 short tons are the equivalent of 453.6 metric tons (500 short tons × 0.9071847 metric tons/short ton = 453.6 metric tons).

Type of Unit	U.S. Unit		Equivalent in Metric Units	
Mass	1 short ton (2,000 lb)	=	0.907 184 7	metric tons (t)
	1 long ton	=	1.016 047	metric tons (t)
	1 pound (lb)	=	0.453 592 37[a]	kilograms (kg)
	1 pound uranium oxide (lb U^{3}O^8)	=	0.384 647[b]	kilograms uranium (kgU)
	1 ounce, avoirdupois (avdp oz)	=	28.349 52	grams (g)
Volume	1 barrel of oil (bbl)	=	0.158 987 3	cubic meters (m^3)
	1 cubic yard (yd^3)	=	0.764 555	cubic meters (m^3)
	1 cubic foot (ft^3)	=	0.028 316 85	cubic meters (m^3)
	1 U.S. gallon (gal)	=	3.785 412	liters (L)
	1 ounce, fluid (fl oz)	=	29.573 53	milliliters (mL)
	1 cubic inch (in^3)	=	16.387 06	milliliters (mL)
Length	1 mile (mi)	=	1.609 344[a]	kilometers (km)
	1 yard (yd)		0.914 4[a]	meters (m)
	1 foot (ft)	=	0.304 8[a]	meters (m)
	1 inch (in)	=	2.54[a]	centimeters (cm)
Area	1 acre	=	0.404 69	hectares (ha)
	1 square mile (mi^2)	—	2.589 988	square kilometers (km^2)
	1 square yard (yd^2)	=	0.836 127 4	square meters (m^2)
	1 square foot (ft^2)	=	0.092 903 04[a]	square meters (m^2)
	1 square inch (in^2)	=	6.451 6[a]	square centimeters (cm^2)
Energy	1 British thermal unit (Btu)[c]	=	1,055.055 852 62[a]	joules (J)
	1 calorie (cal)	=	4.186 8[a]	joules (J)
	1 kilowatthour (kWh)	=	3.6[a]	megajoules (MJ)
Temperature[d]	32 degrees Fahrenheit (°F)	=	0[a]	degrees Celsius (°C)
	212 degrees Fahrenheit (°F)	=	100[a]	degrees Celsius (°C)

[a]Exact conversion.
[b]Calculated by the Energy Information Administration.
[c]The Btu used in this table is the International Table Btu adopted by the Fifth International Conference on Properties of Steam, London, 1956.
[d]To convert degrees Fahrenheit (°F) to degrees Celsius (°C) exactly, subtract 32, then multiply by 5/9.

Notes: • Spaces have been inserted after every third digit to the right of the decimal for ease of reading. • Most metric units belong to the International System of Units (SI), and the liter, hectare, and metric ton are accepted for use with the SI units. For more information about the SI units, see http://physics.nist.gov/cuu/Units/index.html.
Web Page: http://www.eia.doe.gov/emeu/mer/append_b.html.
Sources: • General Services Administration, Federal Standard 376B, *Preferred Metric Units for General Use by the Federal Government* (Washington, D.C., January 1993), pp. 9-11, 13, and 16. • U.S. Department of Commerce, National Institute of Standards and Technology, Special Publications 330, 811, and 814. • American National Standards Institute/ Institute of Electrical and Electronic Engineers, ANSI/IEEE Std 268-1992, pp. 28 and 29.

Table 2 Metric Prefixes
The names of multiples and subdivisions of any unit may be derived by combining the name of the unit with prefixes as below.

Unit Multiple	Prefix	Symbol	Unit Subdivision	Prefix	Symbol
10^1	deka	da	10^{-1}	deci	D
10^2	hecto	h	10^{-2}	centi	c
10^3	kilo	k	10^{-3}	milli	m
10^6	mega	M	10^{-6}	micro	μ
10^9	giga	G	10^{-9}	nano	n
10^{12}	tera	T	10^{-12}	pico	p
10^{15}	peta	P	10^{-15}	femto	f
10^{18}	exa	E	10^{-18}	atto	a
10^{21}	zetta	Z	10^{-21}	zepto	z
10^{24}	yotta	Y	10^{-24}	yocto	y

Web Page: http://www.eia.doe.gov/emeu/mer/append_b.html.
Source: U.S. Department of Commerce, National Institute of Standards and Technology, *The International System of Units (SI)*, NIST Special Publication 330, 1991 Edition (Washington, D.C., August 1991), p. 10.

Table 3 Other Physical Conversion Factors
The factors below can be used to calculate equivalents in various physical units commonly used in energy analyses. For example, 10 barrels are the equivalent of 420 U.S. gallons (10 barrels x 42 gallons/barrel = 420 gallons).

Energy Source	Original Unit		Equivalent in Final Units	
Petroleum	1 barrel (bbl)	=	42[a]	U.S. gallons (gal)
Coal	1 short ton	=	2,000[a]	pounds (lb)
	1 long ton	=	2,240[a]	pounds (lb)
	1 metric ton (t)	=	1,000[a]	kilograms (kg)
Wood	1 cord (cd)	=	1.25[b]	shorts tons
	1 cord (cd)	=	128[a]	cubic feet (ft^3)

[a]Exact conversion.
[b]Calculated by the Energy Information Administration.
Web Page: http://www.eia.doe.gov/emeu/mer/append_b.html.
Source: U.S. Department of Commerce, National Institute of Standards and Technology, *Specifications, Tolerances, and Other Technical Requirements for Weighing and Measuring Devices*, NIST Handbook 44, 1994 Edition (Washington, D.C., October 1993), pp. B-10, C-17 and C-21.

About the Editor

Dr. Barney L. Capehart is a Professor Emeritus of Industrial and Systems Engineering at the University of Florida, Gainesville, Florida. He has BS and Master's degrees in Electrical Engineering and a Ph.D. in Systems Engineering. He taught at the University of Florida from 1968 until 2000, where he taught a wide variety of courses in energy systems analysis, energy efficiency and computer simulation. For the last 25–30 years, energy systems analysis has been his main area of research and publication. He is the co-author or chapter contributor on six books on energy topics, the author of over 50 energy research articles in scholarly journals, and has been involved in 11–12 funded energy research projects totaling over $2,000,000.

He has performed energy efficiency and utility research projects for the Florida Governor's Energy Office, the Florida Public Service Commission, and several utilities in the state of Florida. In addition, he has served as an Expert Witness in the development of the Florida Energy Efficiency and Conservation Act electric growth goals, the Florida Rules for Payments to Cogenerators, and the passage of the Florida Appliance Efficiency Standards Act. He is one of the state's leading experts on electric utility Demand Side Management programs for reducing customer costs and for increasing the efficiency of customer end-use.

He also has broad experience in the commercial/industrial sector, with his management of the University of Florida Industrial Assessment Center from 1990 until 1999. This center performs audits for small and medium sized manufacturing plants, and is funded by the U.S. Department of Energy. He has personally conducted over 100 audits of industrial facilities, and has helped students conduct audits of hundreds of office buildings, small businesses, government facilities, and apartment complexes. He regularly taught a University of Florida course on Energy Management to about fifty engineering students each year. He currently teaches Energy Management seminars around the country and around the world for the Association of Energy Engineers (AEE). In addition, he has been a prolific author of newspaper articles and has appeared on numerous radio and TV talk shows to discuss his work on energy efficiency and reducing people's electric bills. He is also a popular speaker at civic, professional, educational, and government meetings to talk about the benefits of energy efficiency.

Dr. Capehart is a Fellow of the Institute of Electrical and Electronic Engineers, the American Association for the Advancement of Science, and the Institute of Industrial Engineers. He is also a Member of the Hall of Fame of the AEE. He is listed in Who's Who in the World and Who's Who in America. In 1988 he was awarded the Palladium Medal by the American Association of Engineering Societies for his work on energy systems analysis and appliance efficiency standards. He has conducted many energy research projects, and has published over fifty journal articles. He was the Editor of the *International Journal of Energy Systems* from 1985–1988.

He is the editor of *Information Technology for Energy Managers—Understanding Web Based Energy Information and Control Systems*, Fairmont Press, 2004; senior co-editor of *Web Based Energy Information and Control Systems*, Fairmont Press, 2005; senior co-editor of *Web Based Enterprise Energy Management and BAS Systems*, Fairmont Press, 2007; senior co-author of the fifth edition of the *Guide to Energy Management*, Fairmont Press, 2006; and author of the chapter on *Energy Management* in the *Handbook of Industrial Engineering, Second Edition*, by Gavriel Salvendy. He also wrote the chapter on *Energy Auditing* for the *Energy Management Handbook, Sixth Edition* by Wayne C. Turner.

Encyclopedia of
Energy Engineering and Technology
First Edition

Volume 1
Pages 1 through 572
Acc–Energy Serv

Accounting: Facility Energy Use

Douglas E. Tripp
Canadian Institute for Energy Training, Rockwood, Ontario, Canada

Abstract
Energy Accounting is the management technique that quantitatively monitors energy consumption, relates consumption to key independent variables such as production and weather, and assesses energy performance or efficiency over time and against relevant benchmarks. The successful practice of energy accounting is predicated on the identification of the right kinds of data to be collected, the use of appropriate statistical methods to correlate consumption to the independent variables, and the reporting of the right information to the right people in the organization. Energy Monitoring and Targeting (M&T) is a technique for energy performance analysis that overcomes possible deficiencies in the traditional performance indices or energy intensity.

INTRODUCTION

Energy Accounting is an essential component of effective energy management, just as financial accounting is essential to organizational management. In order to gain the full benefit of energy management, organizations need to be able to monitor their energy consumption, relate consumption to the independent variables that drive it, compare the energy performance of their plants and buildings to themselves over time and to other similar facilities, and assess the impact of energy saving measures.

This description of energy accounting is intended to (1) provide insight into the basic principles and methods of energy accounting and (2) expand the conventional understanding of energy accounting to include the analysis technique commonly known as Monitoring and Targeting.

Monitoring, recording, and reporting on gross energy consumption are all straightforward. Complications arise, however, when performance indices enter the picture. Indices in the industrial environment referred to collectively as energy intensity relate energy consumption to measures of production, for example GJ/tonne, MBtu/ton, and so on. Performance indices that are relevant to building operations typically relate energy consumption to conditioned floor area. For reasons developed in this article, both indices can be misleading; a more detailed energy performance model yields far greater insight into energy performance and overcomes the potential difficulties that accompany the use of energy intensity values.

This entry offers some working language for energy accounting, identifies the kinds of data that should be collected and analyzed, describes a number of approaches to energy performance assessment, and develops the basic principles and techniques of Monitoring and Targeting as the recommended approach to energy accounting.

OVERVIEW OF ENERGY ACCOUNTING

Defining Energy Accounting

The adage, "if you don't measure it, you can't manage it" clearly applies to energy use. Just as financial accounting is necessary for effective management of an organization, energy accounting is a key element of energy management. Simply put, energy accounting is a system to measure, record, analyze, and report energy consumption and cost on a regular basis.[1]

Energy accounting systems typically consist of three parts: (1) a system to routinely monitor energy consumption and the variables that influence consumption, (2) an energy use record and reporting system, and (3) a performance measure.[2]

The effectiveness of the energy accounting system depends on the rigor with which consumption patterns are analyzed and correlated to independent variables such as production and weather. In many organizations, the energy accounting process is integrated with a statistical analysis methodology often referred to as Monitoring and Targeting. This is addressed in more detail in "Energy Monitoring, Targeting, and Reporting".

Depending on the goals of the organization, the accounting system may achieve some or all of the following objectives:

1. Track, record, and attribute energy consumption and costs.
2. Verify energy billings and troubleshoot errors.
3. Provide a basis for prioritizing energy capital investments.

Keywords: Accounting; Energy performance indicators; Energy intensity; Specific energy consumption; Degree days; Monitoring and targeting.

4. Provide a basis for energy budgets as part of the overall budgeting process.
5. Identify unaccounted for energy waste.
6. Identify opportunities for performance improvement and evaluate the impact of performance improvement measures.
7. Optimize energy purchase practices.[3]

Energy Accounting and Energy Management

While many definitions of energy management are used, one that captures the essence of this organizational activity is:

> The judicious and effective use of energy to maximize profits (minimize costs) and enhance competitive positions.[4]

Energy management is consistent with other dimensions of continuous improvement. Key functions that comprise energy management include:

- Purchase or supply energy at the lowest possible cost.
- Ensure that energy is used at the highest possible efficiency.
- Utilize the most appropriate technology—from a business case perspective—to meet organizational needs.

Energy management is, or should be, an integral part of overall organizational management, and is practised by most of today's leading organizations in all sectors. Investments in energy management are generally sound, offering attractive returns to plant and building owners.[5]

Energy accounting is one of the tools employed, along with a variety of others, including policy and planning, training, communicating, investment appraisal, and operations and maintenance.

Methods of Energy Accounting

The methods employed in energy accounting depend on the nature of the organization's facilities, e.g., industrial plant or commercial building, and its objectives for the accounting program. In all cases, however, comparison of energy performance over time, and perhaps from site to site or building to building, is a likely element of the analysis.

A critical challenge when comparison of energy consumption is being carried out is the need to adjust consumption data:

- For changes in weather—heating or cooling degree days—over time or from place to place.
- For varying levels of activity, e.g., production or occupancy.
- For changes in space utilization in the facility, e.g., changes in conditioned floor space in a building.

Ultimately the methods described in "Energy Monitoring, Targeting, and Reporting" provide a rigorous statistical basis for these adjustments, but in many cases, other approaches are employed for performance comparisons. They include:

- *Present-to-past comparison*, in which energy consumption for a given period, i.e., a specific month, quarter or year, is compared to the same period in a previous or base year. Since there is no attempt to adjust for changes in weather from year to year, this comparison is rough at best, especially when applied to heating and cooling loads.
- *Multiple year monthly average*, in which the base for comparison is the average consumption of several years for the period in question; again, no adjustment is made for changes in weather, but the assumption is that variation in weather is eliminated as a factor by averaging several years (a flawed assumption if climate change patterns result in a trend in weather rather than random variation).
- *Heating/cooling degree day (HDD and CDD) adjustment*, in which average temperature and degree-day data are used to adjust heating and cooling energy consumption to a common base for year to year or period to period comparison.
- *Correction for changing conditioned floor area*, in which it is assumed that energy consumption will increase or decrease proportionately to increased or decreased conditioned area.[6]
- *Adjustment for changing production*, in which energy intensity (energy consumed per unit of production) is used to scale consumption up or down for comparison. While it is necessary to adjust for production, this approach is not recommended for reasons that are addressed in "Problems with Energy Performance Indicators".

Energy Accounting Tools

A commonly held view is that significant expenditures in metering, data collection systems, and software applications must be made before energy accounting can be done. However, many organizations have discovered that manual collection of data from existing meters and records, and manual analysis using methods such as those described in "Energy monitoring, Targeting, and Reporting" can yield useful results.

Manual energy accounting can be greatly facilitated with the use of commercial spreadsheet programs to automate the numerous calculations that may be required, such as energy intensity and energy consumption per square foot per HDD. As well, other embedded functions such as graphing, regression, averaging, and others, can be helpful for analysis and presentation of results.

Energy Accounting Software

As the energy accounting system becomes more sophisticated and complex, or in the case of larger or multi-site organizations, commercial energy accounting software packages are available. These packages make it easier to input data, carry out analysis, and generate reports. They also typically incorporate weather and floor area corrections, and may enable the direct download of energy consumption, demand, and cost data from service meters or utility-based web sites.[7]

A number of accounting software packages are available commercially. The following list provides some examples (no effort has been taken to ensure that this list is complete or to verify the functionality of the packages):

- Envision™, a stand-alone hardware and software package for tracking energy use; Energard Corporation, Redmond, WA, www.energard.com
- Faser™, software for tracking, analyzing, and reporting utility bill data; OmniComp Inc., Houston, TX, www.omni-comp.com
- Metrix™, an energy accounting system that focuses on energy projects and the savings related to the projects; SRC Systems Inc., Berkeley, CA, www.src-systems.com
- Utility Manager™, software that targets the commercial and public sector markets, including school districts and local governments; Illinova Energy Partners, Oak Brook, IL, www.illinova.com/iep.nsf/web/IEPSubsidiaryHome[8]
- Meter Manager™, a system that combines a utilities supervision function with sub-metering and aggregated metering; Carma Industries, Peterborough, ON, www.carmaindustries.com
- Global Mvo Asset Manager™, a multi-purpose software package combining metering and sub-metering technologies with on-site and remote measuring, verification, and operational capabilities; Global Facman Enterprises Inc., 12180 Chemin du Golf, Montreal, QC, www.globalmvo.com[9]

STRUCTURAL CONSIDERATIONS—ENERGY ACCOUNT CENTERS

Organizations typically identify cost centers for financial management. For similar reasons, including the assignment of accountability to line managers, energy account centers (EAC) are helpful in organizing for energy accounting.

Energy account centers work by identifying geographically definable areas of management accountability, installing meters on energy utilities, and energy-based utilities (e.g., steam from a central plant, compressed air, etc.) at the point of entry to the EAC department or operating unit, and providing consumption reports as a component of the management information system.[10]

There are constraints and guidelines that apply to the selection of energy account centers that do not apply necessarily to financial cost centers. If possible, selection should be based on the following criteria:

- If sub-metering is required, the potential cost savings from energy reduction justify the cost of installing new metering.
- Energy consumption can be measured.
- Ownership of the energy account center can be established. The center might be a production department, a single meter, an aggregate of several meters, or other possibilities.
- An activity variable is identifiable. In the industrial sector, the variable may be production associated with a production unit that is established as an energy account center, whereas in certain building sectors, the variable may be occupancy rates (more about this in "Energy Monitoring, Targeting, and Reporting").
- There is a linkage between the account center and the organizational structure. Accountability can be assigned to an appropriate manager, and reporting can be integrated fully in the management information system.[11]

KEY ACTIVITIES IN ENERGY ACCOUNTING

Tabulation of Data

Energy consumption data is available from accounting records. Utility and fuel supplier invoices contain valuable information about consumption that can be tabulated. For various reasons, fuel and electricity consumption data must be treated separately.

For electricity, the following data should be collected and tabulated: billing month, number of days in the billing period, demand, and energy. From these data, a number of derived factors should be determined and tabulated, including daily energy, energy cost as a percentage of the total, demand cost as a percentage of the total, total cost, blended or average cost per kWh, and load factor as defined by Eq. 1:

Load factor (%)

$$= \frac{\text{kWh used in period}}{\text{Peak kW} \times 24\,\text{h per day} \times \text{\#days in period}} \times 100 \quad (1)$$

For fuels, data should be recorded in physically measurable units (cubic feet, gallons, etc.) rather than dollars that can fluctuate over time (e.g., via utility rate changes, product price changes). Where two different energy sources feed thermal energy data into the same system, it may be necessary to convert them to a common unit. In a spreadsheet program, units can be converted as needed after the quantities are entered in their original units.

In addition to energy use data, data on the factors that influence energy usage are collected and tabulated, including production quantities, outside air temperature, time the facility is occupied, and so on.

Calculation of Energy Performance Indicators

A key component of energy accounting is the determination of energy performance indicators that enable comparison of energy efficiency over time, from site to site, and against appropriate benchmarks. A number of indicators are commonly used to relate consumption to measures of activity, weather factors, and facility size, for example Btu/unit of production, Btu/degree day, Btu/ft^2, or combinations of these. Others that may be useful include Btu/sales dollar, energy spend/sales or profit or value added, Btu/direct labor cost.[12]

Several of these indicators find common use in specific sectors, as indicated in the following paragraphs.

Energy Utilization Index[13]

The energy utilization index (EUI) is used in the building sector, usually based on annual energy consumption related to conditioned floor space. The basic factor in Btu/ft^2, kWh/m^2, or some other appropriate unit is normalized by adjusting for operating hours, weather, etc. for energy type, or for energy use (i.e., heat, lighting, air conditioning, etc.).

Normalized Performance Indicator[14]

The basic performance factor or EUI in Btu/ft^2, kWh/m^2, or some other appropriate unit is normalized by adjusting for operating hours, weather, etc. for energy type, or for energy use (i.e., heat, lighting, air conditioning, etc.). The normalized performance indicator (NPI) is used for comparison of buildings of similar type.

Specific Energy Consumption[15]

This indicator is an energy intensity used in industry to relate consumption to production, expressed, for example, as MBtu/ton, GJ/unit, or other appropriate units. While commonly used, there is a real danger that these indicators can be misleading. Variation in the specific energy consumption (SEC) may be due to economies of scale, production problems, weather, or other factors that are not related to energy management. As well, it is important to consider the fixed and variable components of energy consumption, as discussed in "Problems with Energy Performance Indicators".

Energy Balance

Just as financial accounting involves the reconciliation of revenues and expenses, so energy accounting can (and should) reconcile energy inputs and outputs. Secure in the First Law of Thermodynamics, the principle that all energy can be accounted for, since it cannot be created or destroyed, enables the energy manager to balance inputs and outputs.

Inputs are relatively easily calculated on the basis of purchased energy, although energy exchanges between the facility and the environment may also need to be included, such as air infiltration/exfiltration in buildings, solar gain, and so on. Methods exist for calculating these.

Outputs or end-uses require an energy load inventory— for both electrical and thermal loads. While time-consuming, the preparation of a load inventory involves:

- The counting and tabulation of electrical devices in the facility, including their nameplate or measured demand and energy consumption and times of operation.
- The measurement or calculation of all thermal loads, such as burners in boilers and furnaces, steam or hot water flow, ventilation air flow, fluids to drain, heat loss and heat gain through the facility envelope, and so on.
- The calculation of total consumption based on the electrical and thermal inventories.
- Finding either the electrical load or thermal load for systems with varying loads is not always easy. Actual measurements or simulations may be needed to find loads of motors, HVAC, and boiler systems.

Total energy consumed should balance with total energy purchased within reasonable limits of error. If that is not the case, an unaccounted for loss may be awaiting discovery.

ENERGY MONITORING, TARGETING, AND REPORTING

An important element of Energy Accounting is the determination of the functional relationships between consumption and the independent variables that drive consumption. While often viewed as an issue separate from energy accounting, Energy Monitoring, Targeting and Reporting (MT&R) is an analysis technique that yields these functional relationships. As noted below, it is also a technique that provides a sound basis for energy budgeting, which is clearly part of the accounting process.

Working Definitions

By definition, MT&R is the activity that uses information on energy consumption as a basis for control and managing consumption downward. The three component activities are distinct yet inter-related:

- *Monitoring* is the regular collection of information on energy use. Its purpose is to establish a basis of management control, to determine when and why energy consumption is deviating from an established pattern, and to serve as a basis for taking management action where necessary. Monitoring is essentially aimed at preserving an established pattern.
- *Targeting* is the identification of levels of energy consumption, which are desirable as a management objective to work toward.
- *Reporting* involves putting the management information generated from the Monitoring process in a form that enables ongoing control of energy use, the achievement of reduction targets, and the verification of savings.

Monitoring and Targeting have elements in common, and they share much of the same information. As a rule, however, Monitoring comes before Targeting, because without Monitoring, you cannot know precisely where you are starting from or decide if a target has been achieved. The Reporting phase not only supports management control, but also provides for accountability in the relationship between performance and targets.

Energy Monitoring, Targeting and Reporting is consistent with other continuous improvement techniques applied in organizations, and should be viewed as an ongoing, cyclical process.

Energy Monitoring

There are two essential steps in energy Monitoring; they are:

1. The determination of a functional relationship between consumption and the independent variables that drive consumption (or what can be termed an energy performance model), typically production in the manufacturing environment, weather and occupancy in the buildings sector, or combinations of these and other variables.
2. The comparison of actual consumption to that predicted by the energy performance model.

The Energy Performance Model

Various methods of developing an energy performance model are used. Often linear regression produces a useful model relating consumption to production in manufacturing, or consumption to degree-days in buildings. In other instances, multi-variant regression on both production and degree-days or other combinations of variables is required to generate a useful model.

Energy used in production processes typically heats, cools, changes the state of, or moves material. While it is impossible to generalize, as industrial processes are both complex and widely varied, a theoretical assessment of specific processes gives reason to expect that energy plotted against production will produce a straight line of the general form:

$$y = mx + c \qquad (2)$$

where c, the intercept (and zero production energy consumption), and m, the slope, are empirical coefficients, characteristic of the system being analyzed.

In the case of heating and cooling loads in buildings, a theoretical relationship between energy and degree-days typically takes the form of Eq. 3:

$$H = (UA + C_P NV) \times \text{degree-days} + c \qquad (3)$$

where:

- H is the heat added to or removed from the building per unit of time.
- U is the heat transfer coefficient of the building envelope, taking into account its components such as glazing, interior wall finish, insulation, exterior wall, etc.
- A is the external area of the building envelope.
- C_p is the specific heat of air.
- N is the number of air changes per unit of time.
- V is the volume of the building being ventilated.

U, A, C_p, N, and V are all characteristic constants of the building. Eq. 3 is the equation of a straight line when H is plotted against degree-days, having a slope $= (UA + C_p NV)$ and an intercept on the y-axis $= c$. This constant c is the 'no load' energy consumed, no matter the weather conditions, by such things as office equipment, the losses from the boiler, lighting, and people.

Fig. 1 illustrates a performance model obtained from a consumption–production regression. In this case, the energy consumed in MMBtu is equal to 2.0078 times the production in pounds plus a constant 64,966 MMBtu (the intercept of the regression line) for a baseline period that is considered to represent consistent performance. It is important to recognize that total consumption typically consists of at least these two components:

- A variable, production-dependent load.
- A constant, production-independent load.

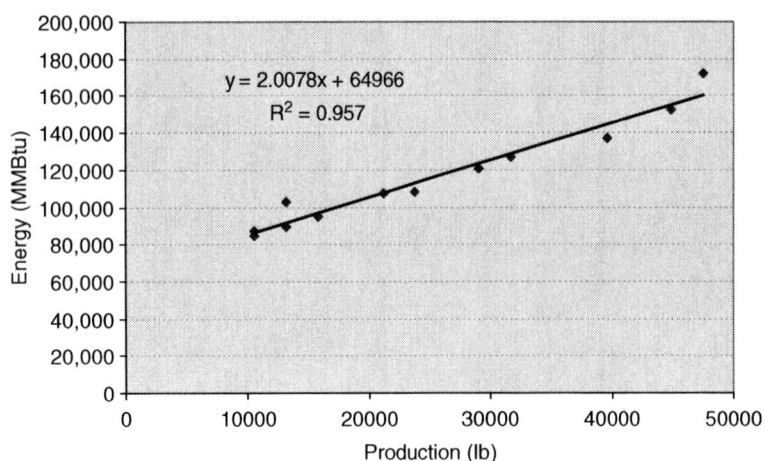

Fig. 1 Regression analysis of baseline for food processing plant energy consumption.

Similar models can be produced for buildings, in which case degree-days rather than production may be the independent variable.

In addition to providing a basis for the reduction of energy waste, the energy performance model also provides the means of determining the energy budget for a projected level of industrial activity or projected weather conditions in a future period.

Cumulative Sum

The comparison of actual and theoretical or predicted energy consumption uses is called cumulative sum of differences (CUSUM) analysis. In CUSUM analysis, a cumulative sum of the differences between the theoretical energy consumption calculated from the energy performance model and the actual consumption is calculated ($\sum$[theoretical − actual]). A time series plot of CUSUM values is illustrated in Fig. 2.

The CUSUM graph yields the following kinds of information:

- Changes in slope represent changes in energy performance.
- A downward (negative) slope represents consumption less than that predicted by the energy performance model, and vice versa.
- Cumulative energy savings or losses (in comparison to the energy performance model or baseline) at any point are equal to the ordinate of the point in question on the CUSUM curve.

Targeting

Based on the information derived from energy Monitoring, it is possible to set reduction targets in the form of energy performance models that:

- Represent best historical performance.
- Incorporate specified reductions to the fixed and variable components of total load.
- Eliminate periods of poor performance to establish a basis for future performance.
- Are defined by other similar criteria.

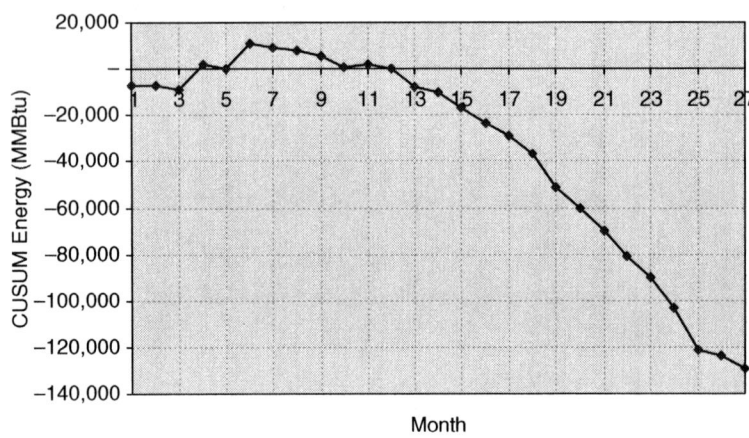

Fig. 2 Cumulative sum (CUSUM) graph for food processing plant.

Problems with Energy Performance Indicators

Especially in the industrial sector, energy performance is often expressed in terms of an energy intensity indicator, as discussed in "Energy Monitoing". The energy performance model resulting from Monitoring analysis makes it evident that there are serious limitations to energy intensity indicators in providing an accurate measure of performance, as Fig. 3 illustrates.

As illustrated, it is possible to look at a performance point (1) in terms of its energy intensity, represented by the solid line, or as a point on the true energy performance model line represented as a dashed line. Points above the energy intensity line by definition have higher energy intensity values, and vice versa. Similarly, points above the energy performance model line represent worse energy efficiency, while points below represent improved energy efficiency.

If production were changed such that the performance point moved from (1) to (2), it would appear that energy intensity has decreased, that is, improved; however, quite the opposite is true when the real performance model is considered. Conversely, a decrease in production that changed the performance point from (1) to (3) would appear to worsen performance when indicated by the energy intensity, whereas, again, quite the opposite is true.[16]

It follows that the only reliable indicator of performance is that derived from the energy performance model; the simple, and widely used, energy intensity value must be viewed with real caution.

Reporting

Reporting within a Monitoring and Targeting system has a number of functions:

- To create motivation for energy saving actions.
- To report regularly on performance.
- To monitor overall utility costs.
- To monitor cost savings.

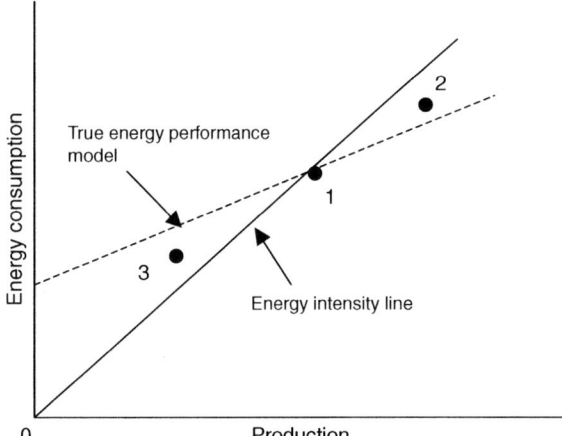

Fig. 3 The problem with energy intensity.

Within most organizations, the need for the type of information generated by a Monitoring and Targeting system varies with level and responsibility. Typically, as the need moves from the operational level in the plant to the senior management level, the requirement for detail diminishes, as does the frequency of Reporting. Operations staff need energy control information to stimulate specific energy savings actions. Senior managers need summary information with which to guide the organization's energy management effort. One report for all will not result in actions being undertaken and decisions being made.[17]

CONCLUSION

Successful organizations include energy management in their management information systems. At the very least, they track consumption, identify and respond to trends, base energy purchase strategies on detailed knowledge of their consumption patterns, and determine performance indices for comparison to their own facilities over time and benchmarks for other similar facilities.

Since the commonly used energy performance indices tend to be energy intensities (i.e., consumption per unit of production, consumption per unit of floor space, consumption cost as a fraction of total product cost, and others) their usefulness in driving performance improvement may be limited.

Energy Monitoring and Targeting is an approach to performance assessment that provides a basis for managing energy consumption downwards. It is based on the determination of an energy performance model that takes into account the fixed and variable components of total consumption. Energy Monitoring and Targeting can, and should, be the basis for effective energy accounting.

REFERENCES

1. California Energy Commission. *Energy Accounting: A Key Tool in Managing Energy Costs*. January 2000; P400-00-001B, 1.
2. Capehart, B.L.; Turner, W.C.; Kennedy, W.J. *Guide to Energy Management*, 4th Ed.; The Fairmont Press, Inc.: Lilburn, GA, 2003; 24.
3. California Energy Commission. *Energy Accounting: A Key Tool in Managing Energy Costs*, January 2000, P400-00-001B; 2–3.
4. Capehart, B.L.; Turner, W.C.; Kennedy, W.J. *Guide to Energy Management*, 4th Ed.; The Fairmont Press, Inc.: Lilburn GA, 2003; 1.
5. Natural Resources Canada (formerly Energy, Mines and Resources Canada). *Energy Accounting*; Canadian

Government Publishing Centre, Ottawa, Canada, 1989; ISBN 0-662-14152-0, 1.
6. California Energy Commission, *Energy Accounting: A Key Tool in Managing Energy Costs*, January 2000, P400-00-001B; 13–14.
7. California Energy Commission, *Energy Accounting: A Key Tool in Managing Energy Costs*, January 2000, P400-00-001B; 15–17.
8. California Energy Commission, *Energy Accounting: A Key Tool in Managing Energy Costs*, January 2000, P400-00-001B; A-1–A-3.
9. Enbridge Gas Distribution. *Energy Accounting Guide for Energy Savings*, adapted with permission from California Energy Commission. *Energy Accounting: A Key Tool in Managing Energy Costs,* January 2000, Scarborough ON, Canada, www.enbridge.com; 23.
10. ETSU; Cheriton Technology Management Ltd *Good Practice Guide 112: Monitoring and Targeting in Large Companies*; Department of the Environment, Transport and the Regions: Oxfordshire, UK, 1998; 2.
11. ETSU, The John Pooley Consultancy *Good Practice Guide 91: Monitoring and Targeting in Large Manufacturing Companies*; Department of the Environment, Transport and the Regions: Oxfordshire, UK, 1994; 8.
12. Capehart, B.L.; Turner, W.C.; Kennedy, W.J. *Guide to Energy Management*, 4th Ed.; The Fairmont Press, Inc.: Lilburn GA, 2003; 30.
13. Capehart, B.L.; Turner, W.C.; Kennedy, W.J. *Guide to Energy Management*, 4th Ed.; The Fairmont Press, Inc.: Lilburn GA, 2003; 26.
14. ETSU, The John Pooley Consultancy, *Good Practice Guide 91: Monitoring and Targeting in Large Manufacturing Companies*, Department of the Environment, Transport and the Regions: Oxfordshire UK; 1994; 14.
15. ETSU, The John Pooley Consultancy, *Good Practice Guide 91: Monitoring and Targeting in Large Manufacturing Companies*, Department of the Environment, Transport and the Regions: Oxfordshire UK; 1994; 14–15.
16. Stephen Dixon. *Private Discussion*, sdixon@knowenergy.com.
17. ETSU, The John Pooley Consultancy, *Good Practice Guide 91: Monitoring and Targeting in Large Manufacturing Companies*, Department of the Environment, Transport and the Regions: Oxfordshire UK; 1994; 20–22.

Air Emissions Reductions from Energy Efficiency

Bruce K. Colburn
EPS Capital Corp., Villanova, Pennsylvania, U.S.A.

Abstract

Energy efficiency has become a popular buzz phrase in the 21st century, but in addition to the pure economic cash savings that can be affected from implementing such work there is also the potential for a cleaner environment. Since becoming operational in February 2005, the Kyoto Treaty calls for a reduction in greenhouse gases (GHG) from all developed countries according to a rather strict time schedule. There is much debate as to whether those targets can be met through existing technologies. One of the obvious solutions to part of this problem is to dramatically increase energy efficiency programs because they permanently reduce the use of electricity and fuels. With the advent of the Kyoto Treaty, a trading mechanism for buying and selling CO_2 credits also can provide some organizations an additional financial incentive. In addition, there are various other undesirable emissions such as SO_x, unburned hydrocarbons, mercury, dioxins, and other undesirable "products of combustion" which are also reduced as a result of energy efficiency.

INTRODUCTION

Energy efficiency refers to the avoidance of waste in the utilization of energy, regardless of its source. Some electricity is not generated from fossil fuels but from hydro, wind, solar, biomass, and geothermal energy. All are considered a precious commodity, so whether fossil fuels or renewable sources are involved, the concept of being as efficient as possible in the use of energy in a facility or industrial process is a reasonable and business-like goal. In the process of evaluating energy efficiency options, a basic guideline is if electricity and thermal fuel usage is reduced, then, in general, a positive environmental condition will simultaneously occur. The simple mechanism of reducing electricity and fuel also reduces the resulting air pollution from most of these sources.[1–4] There is one exception: electricity from a renewable resource. But in practice, renewables currently supply only a small fraction of energy in most countries (Sweden is a notable exception). Regardless, a reduction in the use of energy will reduce the fossil fuels used somewhere in the system. This reduction in fossil fuels reduces the greenhouse gas (GHG) emissions and other products of combustion such as sulphur, mercury, and dioxins, amongst others.

Under the Kyoto Treaty, there are multiple means established for reducing air emissions in the six regulated gases: CO_2, CH_4, N_2O, hydro oxide, hydro fluorocarbons (HFC) and per fluorocarbons (PFC), and sulphur hexaflouride—SF_6. All have been configured in terms of CO_2 "equivalents" for purposes of calculating the avoided atmospheric environmental effects. The overall objective is substantial, permanent, worldwide reductions in total air emissions of CO_2 and other GHG equivalents. To do so, a series of techniques involving not only reductions within a country but the trading of "credits" from implementation by others in the same or other countries has been set up. Examples of these are Emission Trading (ET), Joint Implementation (JI), and Clean Development Mechanism (CDM), and are all focused on the major industrial countries (referred to as "Annex I" countries, which is similar to "Annex B" in the Kyoto Protocol) to implement reductions elsewhere by claiming credit for the business or government which made that reduction possible (hence the "trading" acronym). Similar in approach to the Kyoto Treaty, the United States set up the 1990 Clean Air Act Amendment set up a "cap and trade" system, similar to GHG, for SO_2 emissions.

The ratification of the Kyoto Treaty required that most industrialized countries (the United States not included) reduce their total GHG emissions. The opportunity arose for some organizations to sell these GHG reductions in units of metric tonnes of CO_2 to other organizations that, at the same time, feel they need them to "offset" what is otherwise a reduction requirement. In this way, a business can either invest in energy efficiency by purchasing equipment solely designed to sequester carbon and thus create reductions in net CO_2, or buy CO_2 "offsets" which will be recognized internationally and allow credits against their target reduction requirement. In all cases, capital expenditures of some form are required, but this range of options allows businesses to select the optimum mix. It is noted that although some persons may disagree that global warming exists, the data available suggests otherwise. Fig. 1 illustrates the dramatic rise in

---

Keywords: Energy efficiency; Emissions; Air emission; Greenhouse gases; GHG; Kyoto Treaty; Emissions trading; CO2 reduction; Clean air; Emission reduction.

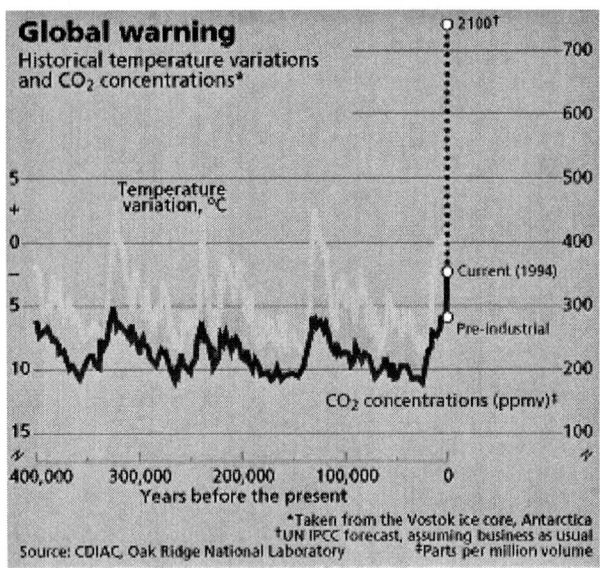

Fig. 1 Global temp and CO_2 vs time.

temperature as the CO_2 concentration increases while Fig. 2 shows the noticeable rise in both CO_2 and CO concentrations in the atmosphere.

Overall, energy cost reduction is the best solution for reducing GHG air emissions because it provides the only mechanism that allows a user to invest capital and reap a direct economic return on that investment while simultaneously receiving CO_2 reductions due to reductions in consumption of electricity and fossil fuels. All other approaches seem to rely on investing in capital, which is a burden cost—it does reduce the total GHG, but with little or no economic return (for example, a baghouse collecting particulate may have some trivial reuse value but it will be nowhere near the amortization investment and annual operating cost). Therefore, one of the best solutions is to invest in energy efficiency and, as a direct result, simultaneously improve the environment through a reduced need for energy.

Emissions measurements in metric tons ("tonnes") of CO_2 equivalents are the internationally accepted norm for

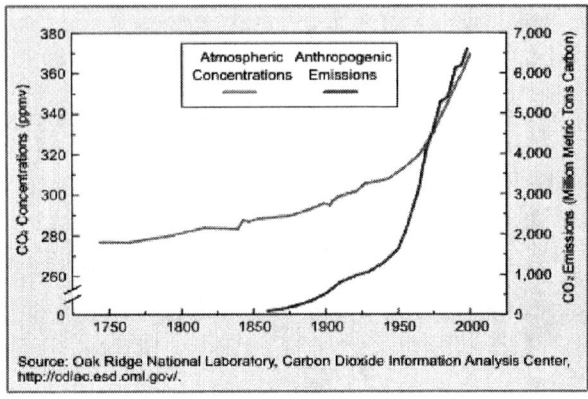

Fig. 2 CO_2 concentration vs time.

reducing the "global warming" problem. Various calculations have been made for GHG reductions, but most depend on such factors as the mix of electric generation from fossil fuels, which varies greatly around the world. For example, until the 21st century in Quebec, Canada, and Brazil, most electricity came from hydroelectric plants, which do not produce GHGs. There is a debate in the science community whether GHGs emitted from decomposing growing plants—which are destroyed when land is dammed up—are, in fact, a GHG "penalty" which should be charged to hydroelectric generation. Therefore, the argument follows that if one saves electricity consumption, one does not "help" the environment by reducing GHG emissions. Thus far, this argument is not widely accepted in the environmental community. However, in practice, all geographic areas have (inefficient, old) fossil fuel plants as "topping" systems at a minimum, to cover the electricity that could not be produced from base-loaded hydro or nuclear energy. In such cases, this energy efficiency would translate to a reduction in operation for these polluting fossil plants as the "last run" unit.

In the United States, decisions on energy efficiency tend to be made solely on the basis of pure economics. In some countries, culture, good sense, and government edict (not necessarily related to cost/benefit) have been the deciding criteria for the operational mix of generating plants. For example, the 'push' in Sweden in the 1930s for very well-insulated buildings has finally paid off financially. For decades, however, it meant only that people were more comfortable in the cold climate as compared to others in surrounding areas, regardless of the amount of utility bills paid. Despite a similarly cold climate, Norway does not possess the same well-insulated buildings as Sweden, uses about twice the energy per capita as its neighbour, and is only now pushing energy efficiency. This difference alone can be seen as cultural issues affecting energy efficiency in a given country and the use of fossil fuels to run the local economy. Japan imports about 97% of its energy, so it focuses strongly on energy efficiency in every aspect of life because it is economically prudent both for business and government to do so. On the opposite spectrum, the United States wastes the most energy and has historically done so for over 100 years, primarily due to historically low energy prices. Because energy was cheap, methods for energy reduction did not find much of a foothold in the United States until the 1970s. These examples illustrate that there are a variety of reasons for being energy-wasteful or energy-efficient, and cost effectiveness has not always entered that equation.

THE HYDROGEN ECONOMY ALTERNATIVE

The previously mentioned issues of emissions and the GHG problem of excessive CO_2 generation are greatly reduced if the input fossil fuel is pure hydrogen, H_2. Hydrogen is clean combustion (effectively no CO_2 formation) compared to natural gas (let alone other fossil fuels) and the result of combustion is almost pure water vapor. This is because when pure hydrogen is burned in the atmosphere, there is not an accompanying carbon molecule to recombine. Issues of CO_2 tend not to be present to the same extent (although there is atmospheric recombination). Therefore, scientists have noted that if automobiles ran on hydrogen, much of the current air pollution generated by transportation would be reduced, as would the GHG emissions of CO_2. However, very little hydrogen exists naturally in nature. Most of it is produced through industrial process means such as stripping hydrogen off of methane (CH_4) or using electrolysis to separate water (H_2O) into pure oxygen and pure hydrogen (which requires substantial energy). As the entire oceans are full of hydrogen (water is H_2O), it would be great if cost-effective methods of separation were available for this "infinite" potential energy source other than pure electrolysis, which is simple and very energy-inefficient.

The greatest benefit of hydrogen use could be in transportation, namely cars and trucks. However, there are serious infrastructure problems at the present time that currently limit the practical use of such technology. In conjunction with the cost of the fuel and lack of refueling systems, few vehicles would be capable of combusting hydrogen. Also, there is very little refining capability in producing large amounts of hydrogen and there is a lack of means to safely transport it around the country. One great benefit of hydrogen is its energy storage capacity in a small space, just as with natural gas or oil. In contrast, electricity is normally stored via chemical batteries or esoteric solutions as pumped hydro storage or compressed air caverns, which are large, very localized, expensive, and relatively inefficient. It is interesting to note, however, that small packages of compressed air storage for electricity is now available (2006) in modules that can fit in a closet and produce UPS power for personal computers and other critical power functions.

Above all else, it should be understood that hydrogen is not available in its natural state in meaningful amounts, and therefore, it virtually always requires energy to "produce" it. In practice, hydrogen is merely a method for "storing" energy—possibly convenient and with high energy density—but an entire infrastructure system of storage, distribution, and fueling would certainly have to be developed. Currently there is no structure, to say nothing of the production plant infrastructure worldwide which is very insignificant at this time. So when it is said that hydrogen combustion produces no emissions, it should also be understood that generating hydrogen causes GHG emissions, unless renewable energy sources are involved. So there may be no net effective GHG emissions reductions using hydrogen as a fuel source, despite the rhetoric of some. This total energy accounting is currently controversial because of arguments that only renewables such as solar and wind can be used to generate hydrogen. However, such arguments forget that without generating the hydrogen, those same renewables could have been used to displace existing fossil fuel production in the first place.

EMISSIONS IMPACTS

In most countries, there are now environmental agencies that oversee the regulation of pollution within its borders. In the United States, the Environmental Protection Agency has that charge and allows individual states to administer their own more severe rules (such as California). In most cases, equipment that directly combusts fossil fuels must secure air permits and have "permission to pollute," which usually involves having limits on the peak rate of generation of certain pollutants as well as total annual pollutants. In cases where the pollution would otherwise be too severe, scrubber systems as "control" devices are mandated to remove sulphur, mercury, or other undesirable products which are deemed unhealthy to the public. However, no country had ever regulated the amount of CO_2 emitted until the Kyoto Treaty. Systems have been developed which can either reduce or capture NO_x (nitrogen oxides, which cause smog), but none have been tested in commercial use yet for pure capture of NO_x or CO_2. With the modern understanding that GHGs are promoting global warming, steps are now underway in most countries to change that paradigm and drastically reduce the CO_2 emissions into the atmosphere through a combination of public awareness and new laws governing air emissions.

In Table 1 (in which all tonnes are metric), a typical set of emissions factors are shown, representing the weight of pollutants per MMBtu in a given category emitted from combustion of three common fossil fuels used today. These fuels are natural gas, #2 Fuel Oil (a form of light diesel), and coal. These emissions figures by themselves seem fairly tame at per million Btu [Higher Heating Value (HHV)], but when applied to a typical large industrial boiler they can lead to a lot of magnitude pollution per year. The second half of the table shows the impact of fossil fuel combustion[5] on some of the critical GHGs as a function of the fossil fuel source. It can be seen that merely performing a fuel switch from coal to natural gas can have a dramatic impact on the CO_2 generated for the same energy output result. This is one of the reasons that natural gas has become so popular as a fuel source.

In Britain and elsewhere in Europe, a new rigid agenda is forthcoming to large industries called the "carbon crunch," namely butting up against the allowable limits of CO_2 emissions for a given nation regardless of its population, economic growth, wealth, or any other parameters. This will soon mean that some corporations

Table 1 Air emissions from burning fossil fuels

Pollutant fuel	CO_2	CO	NO_x	Lead	PM10	Voc	SO_x
	Units of pounds/MMBTU fuel combusted						
Nat gas	121.22	0.082	0.03	0.0000005	0.0076	0.0055	0.0006
#2 Oil	178.4	0.04	0.169	0.0000005	0.016	0.008896	0.2603
Coal	240.5		0.6				1.75

Using 2,000,000 MMBTU natural gas, or #2 fuel oil, we obtain the following:

	Metric tonnes pollutants for 2,000,000 MMBTU combusted						
Nat gas	110,000	74	27	0.0	6.9	5.0	0.5
#2 Oil	161,887	36	153	0.0	14.5	8.1	236.2
Coal	218,240		544				1588.0

will find that if they want to expand production, they will either have to implement major energy efficiency upgrade programs on a scale not seen previously or they will have to pay dearly for CO_2 reductions implemented by them and pay someone for credits through a free market where the highest bidder wins. Recent data suggests that in Britain, a tonnes of CO_2 credit may go for about 20 €, or about $25 USD, which is about five times the estimated value a few years ago—and the carbon crunch has yet to really begin! Not only are CO_2 emissions being capped by country, but they are being lowered—in the next 10 years, the total CO_2 emissions for most Kyoto signators will have to be 10% below their 1990 levels of CO_2 emissions. A notable exception is Russia, which, due to the collapse of the Soviet Empire and the retraction of industry, already has had a reduction in fossil fuel due to small scale "depression" from business downturn while the country tries to adjust to a free market economy.

Another factor in environmental emissions is the effect on marginal GHG emissions from generation due to the implementation of energy efficiency measures. The amount of CO_2 saved varies by region, season, and time of day (TOD) simply due to the mix of current electric generation equipment in place. Some states are attempting to receive credit for the energy efficiency impacts in their State Implementation Plans (SIPs) submitted to the EPA concerning electric power generation; this involves developing local-specific factors for the kWh savings and when they occur (day/night/weekend). At an earlier point, the Alberta Interconnected Systems (AIS) in Canada had their marginal generation emission rate at 0.211 T/MWh, but their average was 0.279 T/MWh. This shows the strange effect that the electrical energy efficiency impact has on their more GHG-friendly plants (it is likely that the base-loaded coal plants were unaffected). Some energy efficiency measures will have very different emissions profiles than others even within the same plant or building simply due to the time impacts of those savings. Examples are lighting, which saves more at night than in the day, and a chiller savings measure, which is likely to save more in the daytime than in the nighttime. In fact, normal energy efficiency savings off-peak have a lower economic value to a project than savings during the daytime hours. However, the emissions reductions impact at night, per MWh, may be larger than by day, because many electric utilities are base-loaded with coal and may actually reduce coal consumption at night. Those reductions by day may reduce only gas-fired equipment consumption. This further demonstrates part of the problem with GHG emissions: at present, there is a reverse effect in that the least cash cost savings can occur in such a way that it discourages saving measures that would otherwise provide the maximum GHG emissions reductions.

ENERGY EFFICIENCY AND ENVIRONMENTAL STEWARDSHIP COMPLEMENT EACH OTHER

No matter how you look at it, permanently reducing the volume of fuels and kWh used reduces the total raw fuel inputs as reducing fossil fuel combustion ultimately reduces air pollution. Differing mixes of fuels occur in different regions of a country for electric power generation, but every country uses electric power that has a meaningful fraction generated from fossil fuels—typically natural gas, oil, and coal. These three fossil fuels cause major air pollution, regardless of the country's current government's environmental regulations on clean combustion or scrubber technology.

The environmental aspects associated with energy efficiency can certainly assist in the reduction of pollution, but it will not be eliminated solely by aggressive energy efficiency.

Virtually every responsible scientist now concedes that the global warming problem is real, and the arguments now tend to be focused on how much human intervention can impact the reduction in global warming and at what rate. In July 2005, the Wisconsin Public Service Commission of the United States (the state regulating body for electric utilities) redefined and allowed energy efficiency investments by the electric utilities to equal footing with investments in generation, transmission, and distribution. This is apparently the first utility-regulating body to recognize legally that reducing kWh or kW has just as much economic value as investments in generating, transmission, and distribution equipment. This provides some rational economic basis for not only allowing but encouraging real energy efficiency by promoting less energy use during a time in which energy use is otherwise growing. Such encouragement, at the very least, could flatten out electric usage over time and allow the utility to reap a return on their investment at their normal return rates. Instead of rewarding a utility for investments in growth, it is essentially rewarding a utility for slowing or stopping growth in electric energy consumption. Several states require utilities to produce integrated resource plans; for example, PacifCorp has agreed to procure all available cost-effective energy efficiency resources before, and in addition to, conventional generation assets. California has a similar provision so as to promote energy efficiency in the electrical generation area. Many other states have alternate programs, but many are geared at attempting to promote some form of energy-efficient and alternate energy solutions within the bounds of cost effectiveness and still be reasonably acceptable by the public at large. This demonstrates that here in the United States, above all else, the states recognize and are responding to public desires, whereas currently the federal government still seems to be considering whether GHG emissions cause global warming. A coalition of north eastern states has banded together to create legislation referred to as "mini-Kyoto," which recognizes and accepts that ignoring the problem of global warming is no longer appropriate. Because the United States federal government has not and will not be leading the charge, those selected states will create their own more rigid rules even though it might initially cause some negative economic impact in those isolated states. The feedback, however, is that over time they will gain advantage by being a leader in the world, regardless of federal guidelines, and their individual states will actually benefit from the improved situation *vis a vis* GHG control.

EU countries have adopted the Kyoto Treaty as well as developed programs and incentives to promote both energy efficiency and emissions limits—it is clear the two are strongly related. Because all EU countries signed the Kyoto Treaty, there is also unanimity of purpose and focus in the EU. Because the United States national government has not ratified the Kyoto Treaty, numerous states are now taking the lead and imposing Kyoto-like rules and regulations out of a common sense of need and purpose.

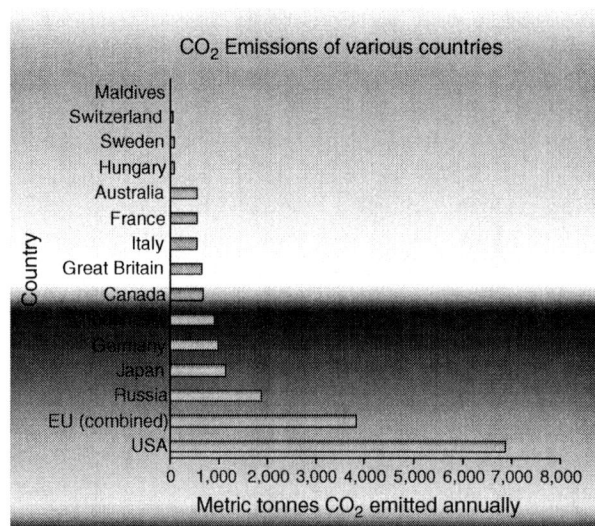

Fig. 3 Annual metric tonnes CO_2 by country-1998.

Fig. 3 shows the huge disparity between total CO_2 emissions by the United States and other countries. This is further clarified by Fig. 4, showing the per capita emissions of CO_2 per country (per country emissions data from UN Millennium Indicators, 2005). In both cases, the United States is the biggest offender of any country by far, and when compared to Europe, is downright profligate (the EU as a whole is less than half the CO_2 per person than the United States or Canada and can be easily traced to cheap energy in North America). The development of the Chicago Climate Exchange is another example of a voluntary program in the United States that has begun without government initiative. Although trading is slow, it is increasing. More companies, cities, and states in the United States are joining to demonstrate their commitment to reducing global warming, even if it costs them and their citizens some additional money in the short-term. The nightmare scenario, however, is for fifty U.S. states to adopt fifty sets of laws that all differ substantially. Remember that California alone has the world's sixth largest economy, so some of these state programs may have tremendous influence in Congress by forcing a

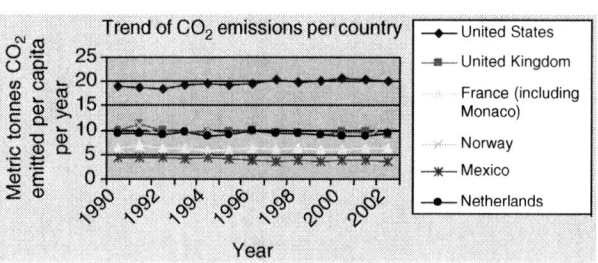

Fig. 4 Per capita CO_2 emissions by country-trends of CO_2 emissions per country.

rational national plan despite of the United State's refusal to ratify the Kyoto Treaty.

In the United States, a program with RECs (Renewable Energy Credits) is one way to reap additional financial benefit out of a renewable electrical energy production project—the RECs can be sold as certificates proving that the recipient has obtained renewable (and hence GHG-free) generation. The state of Pennsylvania has gone even further and identified energy efficiency results in terms of an "Alternate Energy Credit" (AEC) source, which can be sold (similar to the idea that elimination of a Watt-hour is a "negawatthour"). These revenues can thus be initially used to help support the energy efficiency investment and reduce environmental pollution and GHGs. These RECs and AECs are all based on 1 MWh electricity "generation" units. Depending on the location of the source of the electrical power generated, 1 MWh could represent 0.3–0.5 T of CO_2 avoided. Prices for these RECs vary by state or region, depending on the base cost of distributed electricity in a given area. The new AECs are predicted to be as low as \$25/MWh in Pennsylvania due to lower-cost power, while prices in Massachusetts have been about \$55/MWh. This REC credit is in addition to the actual raw KWH sale itself.

CARBON SEQUESTRATION

One area of renewables which has created some controversy is that of biomass—namely the use of grown crops as a fuel source. Such crops absorb CO_2, much as humans breathe oxygen for survival. The concept for emissions credit is that if one combusts biomass, then the CO_2 returned to the atmosphere will be absorbed by other crops, which will then be harvested and consumed as fuel. Similarly, forests and other greenery are considered "sinks" for CO_2, and the focused process of creating such sinks is referred to as "carbon sequestration" (literally grabbing CO_2 from the atmosphere). The concept is that by using renewable crops in boilers, one can generate electricity (or thermal heat for steam for processes) and the CO_2 emitted is actually the "gas" needed by the next round of crops for their breathing source. This creates a mutually circulating fluid wherein the generated CO_2 actually helps support the absorption of CO_2, which helps both industry and the environment. As part of any such technology, appropriate equipment must be used to scrub the exhaust for undesirable particulates and sulphur compounds and to control nitrogen compound discharges. The process, demonstrated in Fig. 5, is not perfect, but from a pure CO_2 viewpoint it is considered "neutral".

Carbon sequestration has led to some major research, including technologies and techniques for capturing the carbon before it enters the atmosphere after fossil fuels or other fuels are combusted and methods to sequester the carbon captured in the ocean.

The practical technologies, some of which exist in a fairly efficient (but not cost-effective) way, include solvent absorption/scrubbing; physical adsorption; gas absorption membrane systems (which themselves require a lot of energy to operate effectively); cryogenic fractionation (supercooling and distilling liquid CO_2 for separation and then some form of disposal as a concentrate); and chemical looping (in which flue exhaust gases are contacted with special metal oxides, which release oxygen for combustion but capture the carbon). By early 2006, a consortium of BP and Scottish & Southern Energy intend to have on-line a new power plant that strips the hydrogen off natural gas

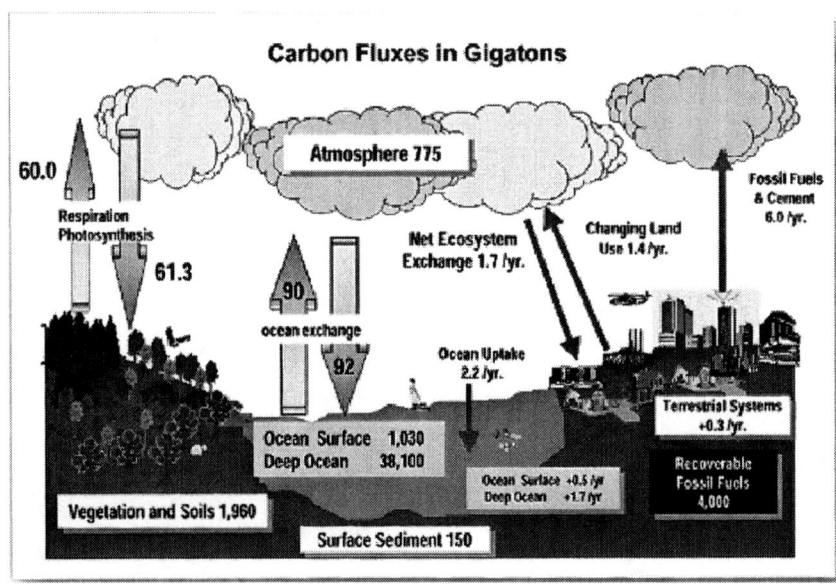

Fig. 5 Carbon sequestration options. (Courtesy US National Energy Technology Laboratory).

and captures the resulting CO_2 at the source before it can be emitted to the atmosphere. This CO_2 would be sequestered as a solid material and then injected into the ocean bottom, and the resulting H_2 would be using for fuel to generate electricity without generation of CO_2 in the exhaust. This would be the first industrial scale demonstration of sequestration. The most common method available presently is to plant more forest area and let the carbon dioxide released from normal combustion be absorbed over time into the growing tree plants, which then produce oxygen from photosynthesis as the trees grow. This "offset mechanism" approach still has the negative effect in that, initially, the CO_2 is discharged first into the atmosphere and only over time is a portion of it absorbed. However, this approach is inefficient because it also requires large plots of land—unless the forested land can grow harvestable cash crops that can be reused on a continuing basis. This is the new focus of some biomass power generation products.

EXAMPLE OF ENERGY EFFICIENCY YIELDING ENVIRONMENTAL BENEFITS AND COST EFFICIENCY

Consider using sludge in an anaerobic digester at a municipal wastewater treatment plant (WWTP) in India simultaneously to cogenerate electricity and heat, use the heat to produce more biogas, more quickly destroy sludge solids (thereby reducing the sludge dewatering process and ultimately reducing the amount of waste solids to be carried away by trucks), and reduce the environmental pollution (which presently flares 100% of the biogas and produces the most emissions). The baseline production of biogas through anaerobic digesters was originally low only because no digester gas was used for heating the sludge via a hot water boiler, due to the mild weather conditions in-country and the original desire for low first-cost construction of such a plant. Controlling operating costs was not a priority originally, but instead the need was for rapidly installing and operating WWTPs where there had originally been none. However, this means that the unheated sludge circulating in the anaerobic digesters does not decompose very quickly, limiting the destruction of solids in the sludge and the amount of gas generated (which then flares). The unheated sludge's average temperature was 24°C, whereas it typically heats to 36°C–40°C for optimum decomposition (36°C being considered optimum due to tradeoffs in energy required to heat the material and biological activity). The plant electrical load was approximately 850 KW (peaks about 950 KW), all from the electric grid, with TOD utility rates.

An ESCO project utilizing cogeneration, peak shaving, and ultrasound to improve the economics was planned for this (unheated) anaerobic digestion process. The energy measure consisted of 750 KW (BF?) cogeneration, a 250 KW peak shaving generator (the actual operation of the peak shaver depends on the actual peak demand during the on-peak electric period), heat recovery for heating the digesters, and an ultrasound system for further breaking down the fibrous materials in the sludge, thus generating more biogas due to additional surface area exposed in the sludge. The end result of such an approach was substantially greater biogas generation on-site and a further dewatering of the sludge and destruction of volatiles. Part of the reason for such a project was the desire of the government to promote energy efficiency (thus lowering operating costs) and demonstrating CO_2 offset benefits (because the flared gas would be used in offsetting other electrical generation). In addition, the resulting sludge was originally pressed and dewatering was accomplished with a fuel oil-fired dryer on-site to further remove water before the final sludge was hauled by truck to a landfill (and with the gas flared in sight of the fuel oil-fired dryer). This trucking also used fossil fuel, and the landfill consumed fuel oil for equipment to distribute the material around. The project reduced the amount of dewatering required and thus decreased the remaining sludge weight slightly by producing the additional biogas through further decomposition. This resulted in even further (secondary) CO_2 reductions through less trucking and landfill activity.

Based on the engineering analysis performed, the total installation cost was determined to be about $2,100,000. Because the existing biogas was naturally generated without heating, the resulting products from the plant was dryer sludge that would have further decomposed at a landfill, letting off additional unconstrained CH_4 (with an ozone damage about 21 times that of the CO_2 generated from combustion). So, not only was there the CO_2 generated from the flare of gas, but also the unburned hydrocarbons given off from decomposition. Being able to generate more gas on-site and using it on-site to offset grid electricity generated by a cogeneration plant does not produce more net CO_2 at the plant site. This is because, in this case, the CH_4 (and associated unburned methane from material decomposition) not generated from combustion under the project would be naturally generated over time through decomposition, but with no accompanying environmental or financial benefits. For illustrative purposes, only the equivalents for NO_x and SO_x are ignored herein. Table 2 shows the results of an energy and environmental efficiency project which also generates meaningful GHG reductions as a direct part of the project, and additional economic enhancements possible due to this environmental benefit on what otherwise might be considered only an energy efficiency project. These economic enhancements include CO_2 credits that could be sold to the EU or elsewhere to buy down the cost of the project.

This information shows how the upfront purchase of future CO_2 credits[6] for avoided CO_2 can be beneficial in

Table 2 Benefits of CO_2 credits for environmental projects

Wastewater treatment plant (WWTP) biogas enhancement and recovery in India (MCF = 1000 cubic ft at standard conditions)		
Life of project, years	15	
Original bio gas generated/year, MCF/year	137,664	
Bio gas generated/year after Project MCF/year	137,664	
Percent methane in biogas	65.0%	
BTU content biogas, BTU/ft^3 HHV	650	
KWH/year generated on site before retrofit	0	
Cogen heat rate (assumes all auxiliaries within), BTU/KWH HHV	15,000	
KWH/year generated on Site after retrofit	5,447,894	
Current grid elec producer cogen heat rate-nat gas, BTU/KWH HHV	8500	
Amount of methane gas for engines, MCF CH_4	89,482	
Amount of inert CO_2 gas, MCF CO_2	48,182	
Hours per year operation	8000	
Amount CO_2 generated by grid cogen plant, tonnes/year	2437	
Amount CO_2 avoided by trucks to landfill, tonnes/yr	128	
Amount CO_2 avoided by equipment at landfill, tonnes/yr	5	
Lifetime CO_2 avoided by Project	38,552	
Implementation cost, USD	$2,106,978	Already has digesters
Implementation cost. Rs	105,348,879	
Savings benefits/year, Rs	21,587,885	
Raw payback	4.9	Years-not attractive
Average cost/KWH, Rs./KWH	3.96	(U.S. 8 cents/KWH)
Necessary sale price to be financiable, Rs	86,351,540	Management Decision
Minimum buydown required for financing-U.S. company grant	18,997,339	Rs. $379,946.78
Cost/tonnes CO_2 avoided, Rs	493	Rs.
Cost/tonne CO_2 avoided, USD/tonne	$9.86	Minimum acceptable
Payback with 9.86/tonne Buyback of CO_2 credits	4.0	
Buydown at $30/tonne	$1,156,559	E.U. Price in 8/05
Net price after applying carbon credits to sale price	$950,419	
Net payback with E.U. free market CO_2 prices 8/05	2.2	Years Very Attractive Investment

assisting the economics of a project. The difficulty in such a case as this is that the avoidance of kWh purchases becomes part of the "savings," but those savings come from elsewhere, even though, with strict measurement and verification protocols, the true benefits can be documented and certified. It also shows how a net sale price of $10/T of CO_2 could dramatically affect the payback of an industrial project in India, and yet that price is less than half of the currently expected price in the EU as of fall 2005. This demonstrates that utilizing the digester gas in a responsible manner cannot only reduce operating costs but also reduce what otherwise is pumped into the atmosphere as additional tonnes of CO_2. It further shows how an EU business investing in a country like India—where the GHG credit limits do not currently apply yet credit is received for these certified GHG credits in the EU—can be beneficial, especially when they could invest about 40% of the unit price and receive equal (per metric tonne) GHG reduction credits.

To put this in perspective with the 493 Rs/T of CO_2 credit and using the annual CO_2 savings and the economic impact for the stated kWh avoided/year, we calculate a net added GHG impact savings of 0.2326 Rs/kWh, which compared to 3.962611 Rs/kWh average price is only about 6% improvement in the annual payback calculation. However, by being able to sell upfront many future years of GHG benefits and then using that money to reduce the net capital cost, there is a dramatic impact on the economic viability of the ECM, going from about 4.9 years simple payout to 4.0 years, pulling the project into the minimum range for selection. So, considering that the GHG credit only created about a 6% increase in annual equivalent, the economic benefit is multiplied many times over, if one can plan for 15 years into the future of reliable operation. It is the long-term years of multiplier that create the dramatic impact as well as the fact that there are currently organizations willing to pay upfront for those future GHG credits on the basis that there is a valid M&V methodology in place for verifying the true future efficiency and operational status of the resulting ECM. If one considers a gas-fired plant today, with U.S. prices of about $15/MMBtu (HHV), the fuel value alone of a generated kWh could be as high as 0.15 USD, but the GHG value of avoiding that kWh at the U.K. price of about $30/T CO_2 is about 1.63 US cents/kWh, or only about 11% of the fuel price. That benefit is converted to a 15-year upfront stream of money and could dramatically cut the cost of the project by 30%–40% due to the upfront nature of capitalizing the future cash flow stream.

CONCLUSIONS

The quest for reductions in environmental pollution—whether it be different chemical compositions such as SO_x, CO_2, NO_x, etc. or whether it be simply heat dumped to the atmosphere (such as a cooling tower does)—shares a common thread in energy efficiency.[7–9] The simple reduction in energy consumption for doing the same job reduces the requirement for energy to be produced. Because the majority of electrical and thermal energy comes from fossil fuel combustion, it simply reduces the need for these fossil fuels in order to achieve the same result. If such efforts were to be accomplished on a vast international scale, it might be possible for the existing fossil fuel pricing scenarios to be meaningfully altered. Also, the reduction in GHGs only helps reduce the global warming situation while at the same time reducing the other air pollutants such as NO_x and SO_x. The beauty of the energy efficiency approach to environmental air pollution and GHG reductions is that the cash cost reductions, which can be achieved by selling the environmental credits in the appropriate market, can help pay for the simultaneous environmental improvements in many cases. The ongoing work in carbon sequestration could find itself clashing somewhat with energy efficiency. If simple inexpensive sequestration were possible, the energy efficiency aspect of environmental pollution might not receive the same level of focus as it does currently due to pollution laws that mandate businesses to invest first in carbon sequestration, even to the exclusion of energy efficiency measures. However, this is unlikely to happen. Fortunately, there is mounting evidence that energy efficiency is growing in importance as a key element in the fight to reduce GHGs because in general it relies on proven technologies with long-term value.

ACKNOWLEDGMENTS

The author would like to express his appreciation to Mr. Kevin Warren, P.E. of Warren Energy Engineering, LLC in Lincoln University, PA for his detailed and insightful comments and for the critique of the above essay. His input, without exception, was incorporated into the manuscript and the author believes it has added clarity and completeness to the topics (however limited in this document due to space constraints).

REFERENCES

1. Haberl, J.; Culp, C.; Yazdani, B.; et al. *Energy Efficiency/ Renewable Energy Impact In the Texas Emissions Reduction Plan (TERP)*, Volume I-Summary Report, Energy Systems Laboratory; Texas A&M University: College Station, TX.
2. *Sector-Based Pollution Prevention: Toxic Reductions Through Energy Efficiency and Conservation Among Industrial Boilers*, July 2002, The Delta Institute, Chicago, IL. (http://www.delta-institute.org).
3. Colburn, B.; Walawalkar, R. *Global Emission Trading: Opportunities and Challenges for ESCOs*, Proceedings of the 2004 WEEC, Austin, TX, September 2004.
4. York, D. *Energy Efficiency and Emissions Trading: Experience from the Clean Air Act Amendments of 1990 for Using Energy Efficiency to Meet Air Pollution Regulations*, Report U034, American Council for an Energy-Efficient Economy: Washington, DC.
5. Mullet, M.R.; Schultz, S.C.; et al. *Wise Rules for Industrial Efficiency: A Tool Kit for Estimating Energy Savings and Greenhouse Gas Emissions Reductions*; U.S. Environmental Protection Agency Report EPA 231-R-98-014, July 1998.
6. New magazine called "Carbon Finance," available through http://www.carbon-financeonline.com.
7. Martin, N.; Anglani, N.; Einstein, D.; Khrushch, M.; Worrell, E.; Price, L.K. *Opportunities to Improve Energy Efficiency and Reduce Greenhouse Gas Emissions in the U.S. Pulp and Paper Industry*; Lawrence Berkeley National Laboratory Report LBNL-46141, July 2000.
8. *Clean Air Through Energy Efficiency*, 2004 Texas SB5 Report From the State Energy Conservation Office (SECO), State of Texas.
9. Pendergast, D. *Science and Technology Development to Integrate Energy Production and Greenhouse Gas Management*, Presented to Americas Nuclear Energy Symposium, October 2004, also http://www.computare.org/Support%20 documents/Publications/Energy%20and%20CO_2%20Management.htm.

Air Quality: Indoor Environment and Energy Efficiency

Shirley J. Hansen
Hansen Associates, Inc., Gig Harbor, Washington, U.S.A.

Abstract

In the days following the oil embargo of 1973, it became common practice to cover outside air intakes. This was just one of many actions taken by the uninformed in the hope of reducing energy consumption. Many of these measures, unfortunately, had a negative impact on the quality of the indoor air. Out of such ignorance came an assumption that energy efficiency (EE) and indoor air quality (IAQ) could not both be served in the same facility.

Over the years, the owner's dilemma regarding IAQ and EE has persisted. Many professional facility managers and real estate managers perceive only two options. There is the constant demand to run facilities as cost effectively as possible, which means that EE should be given a high priority. Unfortunately, many believe that this will result in poor IAQ, which can hurt productivity and/or lose tenants. They fear that a focus on IAQ will drive up their energy costs.

Today, we know that the IAQ risks associated with EE are more perceived than real. Yet fears remain that EE measures may have a negative impact on IAQ. These fears have increased the perception of IAQ risks, created EE sales resistance, and changed the financial dynamics of many projects. Recognizing that these fears exist and need to be treated is a critical first step in serving EE needs. This article addresses those fears and the real relationship between EE and IAQ.

INTRODUCTION

To examine the concerns related to energy efficiency (EE) and indoor air quality (IAQ), and to establish ways to achieve both in a given facility, it is important to:

1. Identify the sources that have linked EE and IAQ and determine whether any causal relationship between the two exists
2. Assess the advantages and disadvantages of ventilation as an IAQ mitigator
3. Consider ways that EE and IAQ might be compatible in a given facility

For years, the second or third paragraph of nearly every IAQ article has mentioned the energy crisis of the 1970s, the resulting tight buildings, and the growing IAQ problems. Readers have been left with the impression that as energy prices soared in the 1970s, owners and facility managers tightened buildings to save money and left occupants sealed in these tight boxes with pollution all about. These fears seem to be substantiated by a report by the National Institute of Safety and Health (NIOSH).

In its early report of investigations to date, NIOSH stated that 52% of the IAQ problems found were due to "inadequate ventilation." Somehow, that got translated to "inadequate outside air." A more careful look at that NIOSH's 52% figure reveals that such a translation misrepresented the findings. The "inadequate ventilation" problems encountered by NIOSH included

- Ventilation effectiveness (inadequate distribution)
- Poor HVAC maintenance
- Temperature and humidity complaints
- Filtration concerns
- Inappropriate energy conservation measures

Inadequate outside air was only one of a long list of problems

National Institute of Safety and Health also pointed out that the 52% figure was based on soft data. To the extent, however, that they represented primary problems in the investigated buildings, the NIOSH findings imparted another critical piece of information that typically is overlooked: of the problems NIOSH found, 48% were not solved by ventilation. National Institute of Safety and Health determined that nearly half of the problems it had investigated were not related to ventilation. If the NIOSH data and problems identified by other investigation teams are considered collectively, it seems safe to surmise that a great many of our indoor air problems cannot be satisfied solely by increasing outdoor air intake.

Somehow, indoor environment thought processes have been permeated by the idea that a tight building is not good and that it uses only recirculated air. Too often, ventilation has been perceived as being the preferred answer—which, of course, has increased energy consumption.

For nearly two decades, ventilation advocates have almost convinced facility managers and consultants that

---

Keywords: Indoor air quality; Tight-building syndrome; Sick-building syndrome; ASHRAE 62; Contaminants; IAQ mitigation.

opening the windows is the only measure needed for the air to get better "naturally." In the interest of both IAQ and EE, a careful look at a broader range of options is needed.

THE "FRESH-AIR" OPTION

If the air outside contains more contaminants than the inside air does, an outside-air solution may not be the answer. Fresh, natural air sounds wholesome, and it seems to be an attractive option. However, that natural air can be heavily polluted. When stepping outside the United Airlines terminal at O'Hare International Airport, for example, even a casual observer can tell that the air outside is much worse than the air inside. There is no "fresh air" for the O'Hare facility people to bring into the terminal. Natural ventilation could be a disaster. Opening the windows is not a viable option.

Hay-fever sufferers also tell us that opening the windows and letting in natural fresh air won't work. Between sniffles, they argue strongly against it.

From another perspective, we should analyze what happens inside when we open the window. What seemed like a good idea can cause a stack effect, in which warm air rises and pressure increases near the ceiling or roof. If we are concerned about a classroom, we could create negative pressure in the basement. Should that school have radon problems, cross-ventilation could cause even more radon to be drawn into the classrooms.

An alternative may be to induce outside air mechanically; that air then can be filtered and diffused through the facility. This method may be helpful, but it is not without problems.

VENTILATION CONSIDERATIONS

Ventilation is not always the answer. If we are to clear the air about the relationship between IAQ and EE, we need to make that statement even stronger. Ventilation is seldom the best answer. Certainly, it is an expensive answer.

The ASHRAE 62 standard is titled "Ventilation for Acceptable Indoor Air Quality." To the uninitiated, that sounds as though ventilation will deliver "acceptable" air. It may not. At the very least, the title implies that an organization as prestigious as the American Society of Heating, Refrigerating, and Air-Conditioning Engineers, Inc. has given its blessing to ventilation as the mitigating strategy.

As the various versions of ASHRAE 62 have been formulated over the years, the idea that most of our IAQ problems can be cured by ventilation has prevailed. ASHRAE 62 has, of course, brought relief to many, many people, who otherwise would have suffered from sick-building syndrome. The ASHRAE 62 standard met a key need during the years when it was very difficult to determine what some of the pollutants were, what their levels of concentration were (or should be), and what their sources were. Investigation and measurement protocols have come a long way since the first ASHRAE 62 standard was written, but we are not there yet. Increased ventilation can continue to give relief to occupants when we aren't quite sure what else to do.

Ventilation, however, is not the preferred treatment for IAQ problems—and it never has been. The U.S. Environmental Protection Agency has been telling us for years that the best mitigating strategy is control at the source.

In the 1970s and 1980s, outside air was reduced so that we wouldn't have to pay the higher energy costs of conditioning air and moving it around. With less outside air, we suddenly became aware of the contaminants that had been there all along. Less outside air meant greater concentrations. Because reduced ventilation was a fairly standard remedy in the 1970s, it is not surprising that the knee-jerk response to the air-quality dilemmas has been to increase ventilation.

Drawing more air into the building and blowing it around, however, has not necessarily solved IAQ problems. Sometimes, in fact, it has made things worse.

LOSING GROUND

Ventilation has created some IAQ problems where they did not previously exist. Two cases in point will help document the problems that the more-ventilation "remedy" fosters.

Relative Humidity

Historically, when construction costs exceeded the budget, one of the first ways to cut costs was to remove the humidifier/dehumidifier equipment from the specs. Today, without those humidifiers or dehumidifiers, it is very hard to correct the negative impact of increased ventilation on relative humidity. To reduce potential indoor pollutants where IAQ problems may not exist, increased ventilation has invited in all the IAQ problems associated with air that is too dry or too humid. With more than 50 years of data on respiratory irritation—even illness—due to dry air, creating drier air in colder climates suggests that we may be exacerbating the problem.

With all that is known about microbiological problems and their relationship to humid air, creating more-humid air in subtropical climates through increased ventilation is a questionable "remedy."

The Dilution Delusion

Increased ventilation thinking has prompted heavy reliance on dilution as the answer. Visualize, for a moment, all those airborne contaminants as a bright neon-orange liquid flowing out of a pipe in an occupied area. Would hosing it down each morning be considered to be a satisfactory solution? We have gained false confidence in dilution because the air pollutants cannot be seen—that does not mean they are less of a problem or that dilution is necessarily the solution.

The problem may not have been eliminated by reducing levels of concentration. There is still a lot that we do not know about chronic low-level exposure to some contaminants. A very real possibility exists that in a couple of decades, science may reveal that solution by dilution was nothing but delusion—a very serious delusion.

DETERMINING THE VALUE OF INCREASED OUTSIDE AIR

Using increased outdoor air as an IAQ mitigating strategy tends to make several other assumptions.

First, it assumes that increased outside air is going to reach the occupants in the building. As recently as the mid-1980s, a study of office ventilation effectiveness by Milt Mechler found that 50% of offices in the United States had ventilation designs that "short-circuited" the air flow. When considering possible treatment for IAQ problems, owners, EE consultants, and energy service companies should look at the facilities' air distribution system. Where are the diffusers? Increasing the outside air may cause a nice breeze across the ceiling, but it may do little for the occupants.

Second, the outdoor-air focus may prompt increased outside air when recirculated cleaned air may be better. Filtration and air cleaning were virtually ignored in ASHRAE's 62-89 standards and have not been treated sufficiently in subsequent work. Bringing in more outside air, which may be better than inside air, can still cost millions and millions of dollars. The fresh-air focus has too often overruled economics when specified filtration of recirculated air could provide the needed IAQ.

When unnecessary fossil fuels are burned to condition and circulate additional outside air, concern is raised about the impact on the quality of the outside air. A study conducted by the author that was reported at the Indoor Air 1991 conference in Helsinki determined that compliance with ASHRAE 62-1989 increased U.S. public schools' energy costs by approximately 20%. This measure not only expended a lot of precious tax revenue, but also offered an indication of the tons of additional pollutants that were put into the air each year.

POLLUTION SOURCES

Historically, the amount of outside air needed in a facility has been gauged by the CO_2 concentration in the air. This has been done because CO_2 is easier to measure than many contaminants are, so it serves as a good surrogate. Because people give off CO_2, it logically followed that the air changes per hour should be based on the number of people in an area. People-pollution thinking partially has its roots in the Dark Ages, when baths were not common and associated body odor was a major concern. Through the years, smoking problems have also led to using the number of people in an area as an air intake barometer. In fact, earlier ASHRAE ventilation standards were often referred to as odor standards.

Total reliance on people pollution has led us away from all the other pollution sources. We have subsequently had the Renaissance, the Industrial Age, the Technological Age, and the Information Age, each contributing new pollution concerns.

New volatile organic compounds (VOCs) are added to the list each year. As we "progress," people pollutants become less of a factor and building materials, furnishings, and "new and improved" equipment take on greater importance. Recent European studies have shown that the building pollutant load is much larger than we expected. When we measure our air intake per occupant, the pollutants created and dispersed to the outdoors by other sources are often overlooked.

Ventilation per occupant does not meet IAQ needs if pollutant sources other than people dominate an area. Laser printers and copiers, as they operate, give off just as many pollutants whether there are two people or 20 people in an office. Bioaerosols released from previously flooded carpet may pose as great a threat if there are 30 people in a room as they will with 300 people in the room. In fact, increased air circulation may draw air up from floor level and increase contaminants at nose level.

The problem has been aggravated by NIOSH's describing energy-efficient buildings as tight buildings. "Sick-building syndrome" and "tight-building syndrome" became synonymous. The idea became so pervasive that it prompted some very energy-inefficient operations. Too often, operable windows have been removed from building designs. Citing such concerns, one Midwest architect designed a ten-story municipal building with all the windows sealed shut. Recognizing the problem, the energy/environmental manager for the city went through and manually changed all those windows to be operable.

Blaming tight buildings gave us charts like the one in Fig. 1, where we were encouraged to compare those minuscule energy savings with the huge personnel losses. The implications were clear: we were trying to save pennies in EE while losing many dollars to lost productivity due to poorer working conditions.

The conclusion seemed to be obvious: there is a direct correlation between EE and IAQ problems.

To prove this hypothesis, however, it is necessary to show that EE buildings have poorer air quality and lower productivity. Or, to state it another way, there is a direct correlation between a tight building and occupant health.

Air Quality: Indoor Environment and Energy Efficiency

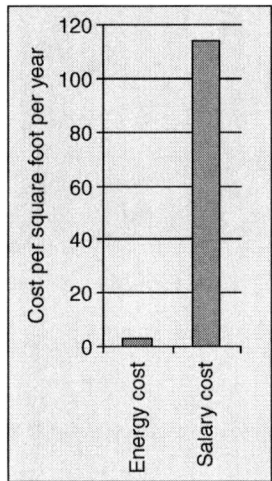

Fig. 1 Cost of energy and salaries in a typical office building.

With a little regression analysis, we ought to be able to build a straight-line relationship: the more energy efficient a building becomes the greater absenteeism and lost productivity become.

As the virtues of tight buildings are weighed, it is easy to forget that those creaky, decrepit old leaky buildings were full of unconditioned, unfiltered, uncontrolled breezes. Drafty buildings were just as apt to cause discomfort as fresh air was.

More than a decade ago, Joseph J. Romm's excellent article "Lean and Clean Management"[4] cited several instances in which EE improved productivity. One example he offered was West Bend Mutual Insurance Company's 40% reduction in energy consumption while it documented a 16% increase in productivity.

VIRTUES OF VENTILATION

Ventilation definitely has its place in an IAQ program. Ventilation can be a good mitigating strategy when the contaminant or its source cannot be determined. Ventilation can also serve as an intermediate step until action can be taken. Further, ventilation may be the best option when source mitigation strategies are simply too costly. Specific applications of ventilation (e.g., localized source control or subslab ventilation to control radon) are valuable control measures. In such instances, more energy may need to be consumed to satisfy IAQ needs.

DISTINGUISHING BETWEEN ENERGY EFFICIENCY AND CONSERVATION

In considering IAQ needs, the distinction between energy conservation and EE becomes critical. By definition, conservation means using less. Further, conservation is still associated with the Emergency Building Temperature Restriction regulations of the 1970s, which led us to equate conservation with deprivation. On the other hand, EE means using the required amount of energy for a healthy, productive workplace or for a process as efficiently as possible.

If we are true to such a definition, it is always possible to have both EE and IAQ.

THE REAL IAQ/EE RELATIONSHIP

There is a surprising relationship between IAQ and EE. First, survey after survey tells us that when utility bills started climbing in the 1970s, the first place where many owners and facility managers found the money to pay those bills was the maintenance budget. This was especially true of institutions on rigid budgets, such as public schools and hospitals. As the utility bills have gone up through the years, those institutions have progressively cut deeper and deeper into maintenance until their deferred maintenance bills have become staggering.

The second relationship between IAQ and EE can also be traced back to energy prices and maintenance. As energy prices climbed, owners bought more sophisticated energy-efficient equipment. Unfortunately, the training of operations and maintenance (O&M) personnel to operate and maintain that equipment did not keep up. Sometimes, the training wasn't offered when the equipment was installed. More often, there was turnover in the O&M personnel, and the new staff did not receive the necessary training.

Keeping these relationships in mind, it's sad to learn that for a long time, we have known that a majority of the IAQ problems found are due to inadequate operations and maintenance. Table 1 offers a review of IAQ problems found by NIOSH, Honeywell's IAQ Diagnostics Group, and the Healthy Buildings Institute (HBI) in the early 1990s. The labels are different, but the commonality of O&M-related problems is very apparent.

THE MUTUAL GOAL OF IAQ AND EE

A careful look at our true goal is needed. Every facility management professional and design professional professes that it is his or her desire to provide owners a facility that has an attractive, healthy, safe, productive environment as cost-effectively as possible. If indeed that is the goal, IAQ and EE are very compatible. They go hand in hand.

Assessing some guidelines of the "1980s" will help bring these two aspects in line. First, let's look at what we call the 80-10-10 rule. Eighty percent of IAQ problems can usually be spotted with an educated eye and a walk-through of a facility. This walk-through may include some very basic measurements (temperature, humidity, CO_2, etc.), but it is not a sophisticated, in-depth investigation. The other 20% of problem facilities require more specialized testing—often exhaustive, expensive

Table 1 Sources of indoor air quality (IAQ) problems

Org.	NIOSH	HONEYWELL	HBI
Bldgs.	529	50	223
Yr.	1987	1989	1989
	Inadequate ventilation (52%)[a]	Operations and maintenance (75%)	Poor ventilation
			No fresh air (35%)
		Energy mgmt.	Inadequate fresh air (64%)
	Inside	Maintenance	Distribution (46%)
	contamination (17%)	Changed loads	
	Outside	Design	Poor filtration
	contamination (11%)	Ventilation/distribution (75%)	Low filter efficiency (57%)
	Microbiological	Filtration (65%)	Poor design (44%)
	contamination (5%)	Accessibility/drainage (60%)	Poor installation (13%)
	Building fabric	Contaminants (60%)	Contaminated systems
	contamination (3%)	Chemical	Excessively dirty
		Thermal	duct work (38%)
		Biological	Condensate trays (63%)
			Humidifiers (16%)

[a]Percentages exceed 100% due to the multifactorial nature of IAQ problems.
Source: From The Fairmont Press (see Ref. 1).

testing, which typically finds only one-half of the remaining problems. To summarize, 80% of IAQ problems are detected through a relatively simple walk-through; 10% are resolved through sophisticated, expensive testing; and nearly 10% remain unresolved.

When considered from the EE perspective, a U.S. Department of Energy study conducted by The Synetics Group (TSG)[5] reported that up to 80% of the savings in an EE program comes from the energy-efficient practices of the O&M staff. What bitter irony! To save money to pay the utility bill, owners cut operations and maintenance. Then they end up with maintenance-related IAQ problems and higher energy bills. So the vicious cycle starts all over again, with more cuts in the maintenance budget.

Fortunately, a positive side to such a vicious cycle can help reverse the situation.

If an IAQ walk-through investigation is paired with a walk-through energy audit, (For more information on audits and auditing, see Hansen and Brown.[2]) it is quite conceivable that the identified future energy savings can pay for the needed IAQ work. One walk-through can identify the IAQ problems and determine ways to finance the mitigation. This approach proves once more how compatible IAQ and EE can be.

EE VS IAQ

In pondering this relationship, it is well to consider what will happen as energy prices continue their upward trend (and they will), if for no other reason than that we need to start calculating the real cost of energy. Whether the increases are due to unrest in the Middle East or increasing demands from China, or whether we start doing a better job of figuring the costs of externalities, prices will trend upward. We are dealing with a finite source and increasing demands. It is a serious miscalculation to assume that fossil fuel prices will have a downward trend.

As the cost of energy goes up, IAQ and EE are apt to be at loggerheads. This does not have to be the case. If IAQ leaders persist in attributing IAQ problems to energy-efficient buildings, as well as relying on more and more outside air for the answer, we lose. For cost-conscious owners, climbing energy costs typically outweigh most IAQ concerns. Only expensive—and often unnecessary—regulations setting outside air requirements can compete against escalating energy prices. The regulated solution may not solve the IAQ problem, but it will definitely increase energy costs—and increase pollution emissions. Sadly, it could all be paid for with money that typically is wasted on energy inefficient operations.[3]

Environmental concerns, higher energy prices, national security issues, and the unnecessary waste of our limited energy resources make increased ventilation a costly answer at the very least. Sustainable development means that we all put our heads together and work for a quality indoor environment, EE, and a quality outdoor environment.

Research has shown that the use of outside air is not always our best option, and that there is nothing wrong with a tight building—provided that a tight building is well designed and well maintained. Tight buildings more readily reveal professional errors. Tight buildings are less forgiving of poor maintenance. A well-designed, well-maintained tight building, however, can provide EE and quality indoor air.

If our ultimate goal is to produce a comfortable, productive indoor environment as cost effectively as possible, EE and IAQ are on the same side. Good managers and effective EE consultants need to have command of both if they are to do their jobs effectively.

REFERENCES

1. Burroughs, H.E.; Hansen, S.J. *Managing Indoor Air Quality*, 3d Ed.; The Fairmont Press: Lilburn, GA, 2005.
2. Hansen, S.J.; Brown, J.W. *Investment Grade Energy Audits: Making Smart Energy Choices*; The Fairmont Press: Lilburn, GA, 2004.
3. Hansen, S.J. *Manual for Intelligent Energy Services*; The Fairmont Press: Lilburn, GA, 2002.
4. Romm, J. Lean and clean management. Energy Environ. Manage. **1995**.
5. The Synetics Group. Evaluation of ICP Grants Program, 1983.

Aircraft Energy Use

K. E. Ohrn
Cypress Digital Ltd., Vancouver, British Columbia, Canada

Abstract

The aviation industry consumes a relatively small amount of the world's fossil fuels. It has a solid record of reducing its consumption and is driven to do so by the large economic impact of fuel costs on the industry. Fuel consumption has been reduced by constant change and improvement in engine and airframe technology, materials, operations planning, aircraft operation, and maintenance practices.

There are incremental gains to be realized in aircraft technology and Air Traffic control technology and procedures. But the predicted rate of industry growth through 2020 will exceed these gains, causing an increase in the industry's overall consumption of fossil fuels. And there do not appear to be any new fuels on the planning horizon.

Several promising areas for breakthrough aircraft technologies have been identified, but all of them are very challenging. Similarly, major gains in Air Traffic Control efficiencies will not be easy to implement.

INTRODUCTION

This article is a broad overview of fuel consumption in the aviation industry. It covers a range of topics, each of which could be expanded considerably. It is intended as an introductory reference for engineers, students, policy-makers, and the general public.

Commercial aviation burns a relatively small 2%–3% of the world's fossil fuels. Military and general aviation accounts for a small and declining proportion of that amount. Since fuel makes up about 12%–17% of an airline's operating costs, the industry has clear economic incentives to reduce consumption. Strong competition in the airline business and its supplier industries has made such progress rapid and effective.

Aviation fuels are comprised mostly of kerosene, which is produced through the distillation of crude oil. Jet turbine fuels account for around 6% of refinery production worldwide. The biggest contribution to the reduction of fuel consumption has been the development of aircraft propulsion from piston-driven propellers to turbofans that use very exotic materials. Air frames have evolved from wood and canvas to aluminium, titanium, and carbon–fiber composites, with significant reduction in weight and an increase in strength.

Once aircraft are in service, operators maintain and operate them effectively in a variety of ways. These include drag reduction programs, flight planning systems, pilot techniques, advanced on-board flight control systems, maintenance, and trend analysis programs.

In the future, incremental gains in fuel efficiency will continue as weight reductions and engine efficiency gains continue, along with the utilization of more sophisticated control systems and manufacturing processes.

Air Traffic Control systems and procedures can be improved, leading to fuel efficiencies through reduced trip distances, and less holding or local maneuvering.

The industry, as a whole, illustrates how far fuel conservation can be taken.

TRENDS IN CONSUMPTION

Total

In 2003,[1] the Transportation sector accounted for 27% of the world's energy consumption, using most of its share as common gasoline. Aircraft account for between 2 and 3% of the fossil fuel burned world-wide, and 6% of petroleum consumption.[2]

The vast majority of aircraft fuel is consumed by the world's 18,000 or so commercial jet aircraft. Fuel consumption by military aircraft is estimated[3] to have dropped from 36% in 1976 to 18% in 1992, and is projected to drop to 7% in 2015 and 3% in 2050.

Most industry associations and observers[4,5] predict a continued growth rate in flight and passenger volumes, averaging at 5% per year for the industry.

Consumption per ATK

The industry has been able to lower its average fuel consumption dramatically in the last 40 years, even as the level of flight activity has soared (Fig. 1).[1,6]

The Available Ton-Kilometer (ATK) is a measure of production capacity for both scheduled and unscheduled passenger and freight aviation. An ATK is defined as the

Keywords: Aircraft; Aviation; Turbine; Fuel; Kerosene.

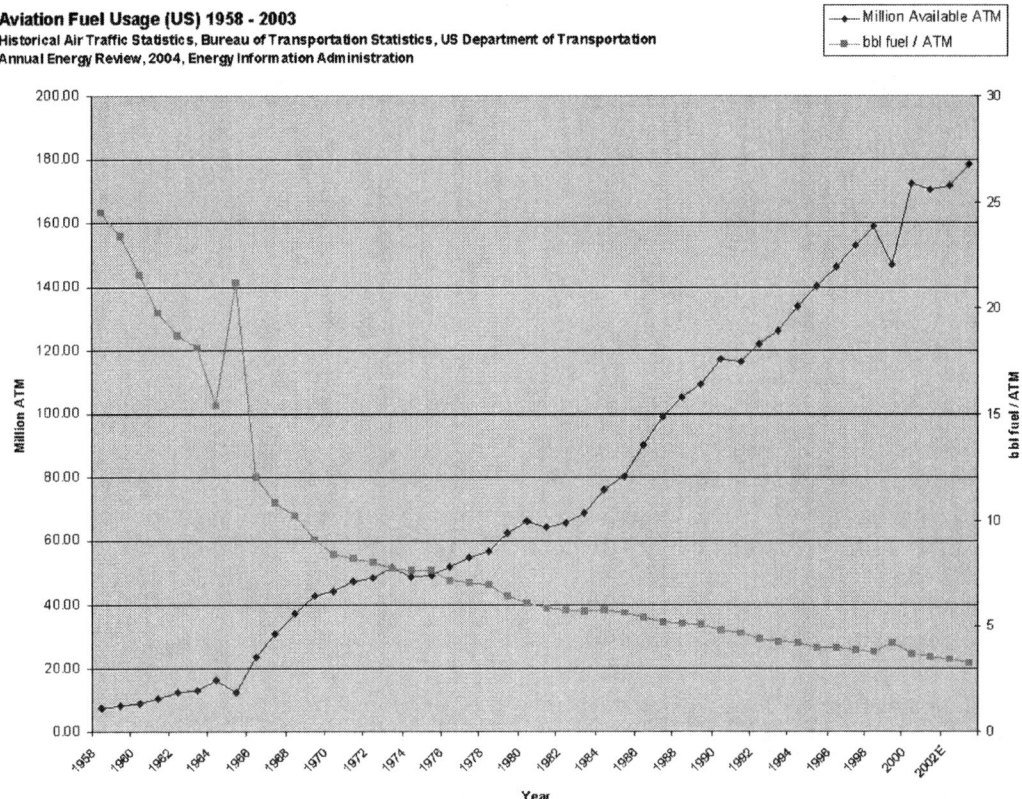

Fig. 1 Aviation fuel consumption per available ton-kilometer (ATK).

number of tons of available aircraft capacity multiplied by the number of kilometers these tons are flown. This measure isolates fuel efficiency discussions of technology and infrastructure from more complex discussions concerning fuel usage per passenger-kilometer, which is more of a market-driven measure. Airlines have several nontechnological paths to pursue in obtaining the most revenue for their fuel dollar. These include keeping aircraft as full as possible, matching aircraft type and schedule to routes and demand, and, thus, spreading fuel costs to cover more passengers and generate greater revenue.

ECONOMIC IMPACT OF ENERGY COSTS ON COMMERCIAL AVIATION

Impetus Towards Conservation

Like many other industries, the energy in aircraft fuel is crucial to airline operations. But fuel is a large percentage (12%–17%) of airline operating costs, usually second only to wages. The percentage varies with the type of carrier and its route structure. Fuel cost varies by as much as 30% between different airports due to transportation costs from refinery to airport, local supplier cost structures, volume discounts, and government tariffs or price support policies.[4]

Given the highly competitive nature of the business, and its high-cost, low-margin characteristics, there is strong reason to pursue fuel efficiency. And the industry has been diligent and successful in its conservation efforts.

In addition to reducing consumption, airlines pursue several strategies to reduce fuel cost. These include fuel tankering, or carrying excess fuel from a low-cost airport to a high-cost one; local supplier negotiations; hedging; and so forth. These strategies extend beyond the scope of this article, but simply reducing consumption still constitutes the best long-term strategy for dealing with fuel costs.

PRODUCTION OF AVIATION FUEL

Crude oil delivered to a refinery is converted into upward of 2000 products,[7] but the most profitable and high-volume products are gasoline, jet fuels, and diesel fuel. Naphtha jet fuels have been used in the military, but phased out in favor of kerosene-based fuels. The major jet fuel types are Jet A, Jet A-1, JP-5, and JP-8.

In terms of overall refinery production, jet fuel accounts for around 6% of output by volume. Only a fairly small number of refineries produce jet fuel.

Sans breakthroughs, kerosene-based fuels seem to be inescapable for the industry.[8]

Jet Fuels

Jet fuel is an output derived from atmospheric distillation, catalytic cracking, and, in some cases, hydro treating sections of the refinery. This depends on the composition of the input crude oil. The final product is a blend of distilled kerosene, which is often upgraded to remove impurities, and heavier hydro and catalytically cracked distillates.[7,9]

The fuel grade (or type) is controlled through strict specifications by American Society for Testing and Materials (ASTM) International, the military, and others.

- *Jet A*—widely available in the U.S.; freeze point −40C; somewhat cheaper and easier to make than other types, helping to ensure wide availability
- *Jet A-1*—widely available outside the U.S.; freeze point −47C. This and Jet A account for the majority of jet fuel usage.[9]
- *Jet B*—blend of gasoline and kerosene (so-called "wide-cut" fuel, which has a range of hydrocarbon compounds in the gasoline and kerosene boiling ranges) as an alternative to Jet A-1 for higher performance in very cold climates
- *JP-4*—military equivalent of Jet B; with additional corrosion inhibition and anti-icing additives (NATO code F-40)
- *JP-5*—military specification (NATO Code F-44)
- *JP-8*—military equivalent to Jet A-1; with additional corrosion inhibition and anti-icing additives (NATO code F-34)

Aviation Gasoline

Gasoline is used in aviation piston engines, and accounts for about 1.1% of the volume of jet fuel, and about 0.2% of the volume of motor fuel[1] in the U.S. ASTM Specification D910 recognizes two octane ratings, 80 and 100, and a low-lead version called 100LL. The product is usually a blend of distillates and alkylates. Normally, the refiner adds tetraethyl lead as required to meet the grade specifications, as well as identifying dyes, which improves safety by allowing different grades of aviation gasoline to be identified by color. The refiner may also add an icing inhibitor, antioxidants, and a few other approved additives, depending on local requirements.

FACTORS IN CONSUMPTION REDUCTION AND CONTROL

Engines (Design History)

According to data produced by Rolls–Royce[10] and International Air Transport Association (IATA),[11] the industry has greatly reduced fuel consumption in the last 50 years. Over 60% of that reduction is due to vast changes in engine technology.

The piston engine which made the 1903 flight of the Wright Brother's first aircraft possible was a water-cooled, four-cylinder, inline design. It weighed around 179 lbs and produced 12 hp.

In 1917, driven by World War I military requirements, the Liberty engine produced 400 hp from an air-cooled V-12 design, and weighed about 790 lbs.

This progress continued through World War II until around 1950, when the Wright R-3350 typified the end-stage of aircraft piston engine development. It produced 3700 hp from 18 cylinders arranged in two radial rows and weighed about 3670 lbs. But the end was in sight due to complexity and limits to the overall power available from a cost-effective piston aircraft engine.

The industry, both military and commercial, had by then turned its attention to the turbine (or "jet") engine. By 1950, Pratt and Whitney had demonstrated the J-57, which produced high thrust with reasonably low fuel consumption. It was used in early jet transports, such as the Boeing 707 and Douglas DC-8. Although very noisy, the engines were acceptably economical to operate and provided the industry much faster aircraft that offered improved travel experiences.

The industry wanted larger engines that generated more power so that much larger aircraft could be built. Engine manufacturers Pratt and Whitney, Rolls–Royce, and General Electric came up with turbofan engines. These engines have cores like traditional turbine engines, but much of their thrust is derived from a large fan located at the front of the engine. This engine can be thought of as an axially-driven, multi-bladed, ducted propeller. This "propeller" provides up to 85% of the engine's thrust. This engine type is characterized by its "bypass ratio," which describes the amount of air coming into the engine that does not go through the central turbine. Typical bypass ratios are from five to one to nine to one. Additionally, these engine designs are much quieter than their predecessors and have much better fuel economy.

A leading engine of this class in 2005 was the General Electric GE90-115B, which weighs 18,300 lbs and produces 120,000 lbs of thrust.

The performance improvements in turbine engine technology are illustrated below. Specific Fuel Consumption (SFC) is a measure of the fuel flow (lbs per hour) of the thrust produced by the engine. The chart shows a sample of representative commercial turbine engines (Fig. 2).

Air Frames (Weight, Composite Materials)

An aircraft in level cruise flight has four forces in balance: lift equals weight and thrust equals drag. Engines provide thrust and consume the aircraft's fuel. For commercial

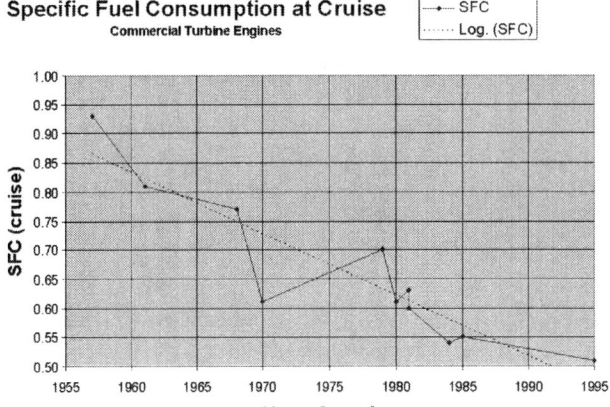

Fig. 2 Specific fuel consumption (SFC) trend chart.

aircraft, drag comes from several sources, but the largest are: induced drag, a by-product of lift, and parasitic drag, which is caused by the air friction and turbulence over the exterior surfaces of the aircraft, as the aircraft moves air out of its way, and by antennae, landing gear, and so on.

Induced drag has a strong relationship to weight: less weight means less lift is required. Induced drag also depends on the design of the wing and its airfoil (wing cross-section) and the angle of attack of the wing. Generally, this drag increases with the square of lift, and the square of aircraft weight.

Any reduction in aircraft weight will directly result in reduced fuel consumption. Reduction in an engine's fuel consumption means that less fuel is required for a given aircraft and payload, hence there is a cumulative effect on its overall efficiency.

Aircraft manufacturers have continuously used a variety of means to reduce aircraft weight. The materials used for aircraft structures have changed from wood and canvas to plywood through various aluminium alloys and, since the 1970s, have included carbon–fiber composites used for simple panels and complex and critical components, such as engine fan blades.

In the Boeing 787 aircraft scheduled for first flight in mid-2007, carbon–fiber resin composites will make up approximately 50% of its structural weight, compared to 12% for the Boeing 777. The planned composite components extend far beyond the usual and into the wings and entire fuselage sections. Aluminium, titanium, and steel constitute the remainder of the structural weight. Airbus Industrie is more conservative, with about 25% composites in the airframe of the A380, which is scheduled to enter service in late 2007.

In the U.S., National Aeronautics and Space Administration (NASA's) Advanced High Temperature Engine Materials Technology Program and the National Integrated High Performance Turbine Engine Technology (IHPTET) Program have investigated and promoted the use of polymer-matrix composites, metal-matrix/intermetallic-matrix, and ceramic-matrix composites for high-temperature parts of aircraft engines. These materials could allow the construction of higher-temperature engines with greater combustion efficiency, all at significantly lower weights. An example is the F136 military engine, which uses a titanium matrix composite in its compressor rotors.

Airlines also work to manage and reduce weight throughout the aircraft and its operation. Excess weight can build up from moisture, dirt, and rubbish in the aircraft, unnecessary supplies, and excess passenger equipment. Boeing[12] estimates that an aircraft will increase in weight by about 0.1%–0.2% per year, leveling off at about 1% in five to ten years. A 1% reduction in weight results in a 0.75 to one percent reduction in trip fuel, depending on the engine type.

Drag Reduction Programs

Drag increases required thrust, so aerodynamic cleanliness is an ongoing challenge.[13,12] Dirt and oil, skin roughness, dents, misaligned fairings, incorrect control rigging, deteriorating seals, mismatched surfaces, and joint gaps (e.g., doors and access panels) all contribute to drag and increased fuel consumption. The most sensitive areas of the aircraft are those where local flow velocities are high and boundary layers are thin: the nose area, the wing leading edges and upper surface, the elevator and rudder leading edges, engine nacelles, and support pylons. If not maintained, a modern transport aircraft can expect a two percent increase in drag within a few years as a result of these factors.

Aircraft Pre-Flight Planning

Before every flight, pilots and operations staff make decisions that affect the overall fuel consumption of each aircraft. Based on knowledge of the aircraft, schedule, payload, and weather, they prepare a load plan and a flight plan. The variables in these plans, and decisions made around them, have a major effect on fuel consumption for the flight.

Center of Gravity (C of G)

Operations dispatchers plan the fuel and cargo load to place the center of gravity within the correct range for safe operation. However, if possible, placing the C of G in the aft portion of this range will result in reduced fuel consumption. This is because when C of G is aft, there is less elevator control surface negative lift required to maintain the correct cruise attitude. This means less lift is required from the wings, resulting in less induced drag. Less negative lift from the tail plane also means less induced drag from this area.

Fuel Quantity

Extra fuel, while comforting to passengers and crew, requires extra fuel burn due to the weight of this extra fuel. A better strategy is to accurately plan the flight to carry the correct amount of fuel and reserves. Elements of this strategy are to:

- Determine the accurate payload and use aircraft weight by tail number if possible.
- Plan the fuel load as required for safety and regulatory requirements, with optimum choice of an alternate airport, careful consideration of the rules that apply to the flight, depending on its origin and destination, and minimal "discretionary" fuel requests.
- If possible, use the re-dispatch technique to minimize contingency fuel requirements.
- Provide accurate, optimized flight planning using the latest origin, destination, and en-route weather information and planning techniques. This involves: choosing a great circle route to reduce distance traveled, if possible; flying pressure patterns and maximizing the use of prevailing wind to reduce enroute flying time; selecting cruise speeds that are, again, an optimum compromise between fuel consumption and schedule performance considerations; and using step climb techniques as required to move to newer altitudes as aircraft weight decreases during the flight.

Pilot Operations Techniques

Pilots can make incremental reductions to fuel consumption through a variety of techniques. They can delay starting the engine until the last minute, after Air Traffic Control (ATC) has issued departure clearances, so that such delays occur at the gate with the engines off. Pilots can minimize the use of the on-board Auxiliary Power Unit, a small turbine that supplies electrical power and compressed air at a higher fuel cost compared to ground power units. Where permitted, the aircraft can also taxi on one engine. Ground operations thrust and braking can then be minimized.

Moreover, pilots can utilize the appropriate flap settings, and retract them as soon as possible to reduce drag. They can follow minimum cost climb profiles whenever possible, but may be thwarted by noise restrictions and ATC congestion problems.

In flight, good control surface trim techniques can save as much as 0.5% in fuel burn by minimizing drag.[12] The appropriate management of air conditioning packs can reduce fuel burn by 0.5%–1.0%. Pilots can use cargo heat and anti-icing judiciously.

There is an optimum point to begin descent into the destination airport. If the plane descends too early, fuel is wasted due to higher consumption while cruising at lower altitudes; if the plane descends too late, the descent speed is too high and energy is wasted. Pilots can delay lowering flaps and landing gear until the last minute: fuel consumption in this high-drag configuration is up to 150% of that in a "cleaner" configuration.

In all of these techniques, safety is the overriding concern. Pilots will always choose a conservative and safe option over a more economical one.

Flight Controls (Autopilot, FMC, W&B)

Modern transport aircraft have significant on-board flight control and management systems that can be used to reduce fuel consumption.

Some Airbus aircraft, for example, have a Fuel Control and Management Computer (FCMC) that can determine the C of G of the aircraft and continuously adjust it toward an optimum position for different flight regimes by pumping fuel to and from an aft-located "trim tank."

Airbus also has a Flight Management Computer (FMC) that can plan step climbs. It also can show the pilots their current optimum altitudes and cruise speeds, in addition to the current actuals, taking into account upper wind forecasts for the flight's planned route. The FMC calculates the optimum top of descent point. When in "managed mode," the FMC uses a "cost index" to account for the carrier's preferences between fuel costs, other direct operating costs, and time savings when calculating cruise altitude and speed.

Control of Engine Maintenance

Boeing[12] recommends several procedures to maintain economical engine operation. These are on-wing washing, which reduces dirt buildup, and bleed air rigging, which compensates for leaks due to system wear. Bleed air is taken from the engine's core and used for a variety of purposes where heated compressed air is needed, such as cabin pressurization and wing de-icing.

Regulatory agencies have mandated significant amounts of on-board data gathering for safety and accident investigation purposes. The industry has found ways to lever this data to provide information about engine health and performance. There are two methods used:

1. Post flight: flight data, gathered manually or electronically, can be loaded into various computer programs after the flight's completion, often on a sampling basis.
2. In-flight: using online data link networks, airlines can downlink in-flight data from the aircraft, among many other types of routine operational reports. ARINC Incorporated of Annapolis, Maryland (GlobalLink) and SITA of Geneva, Switzerland (Aircom) provide Aircraft Communications Addressing and Reporting System (ACARS) services through a world-wide network of satellites,

Very High Frequency (VHF), and High Frequency (HF) ground stations used by airlines and business aircraft operators. Satellite services use four Inmarsat-3 satellites and constitute a global resource for appropriately-equipped aircraft, with the exception of polar regions.

The data are analyzed to determine overall fuel consumption and provide feedback on the success of flight planning and deterioration, if any, of the fuel efficiency of each engine.

This data usually provides the basis for engine trend monitoring, where parameters of interest are compared over time. The onboard computers can also capture short-term "limit exceedance" events, which are gathered on an exception basis.

Airlines, small and large, use in-house software, or software and services provided by many different companies, such as General Electric Aircraft Engines, to perform trend analysis on their engines. This software will predict and characterize trends based on the data provided, including analysis of the combustion efficiency and internal thermodynamics of the hot core sections of the engine. For example, a drop-off in fuel efficiency is probably a sign of wear problems. When certain thresholds of fuel flow, temperature, and so forth are met, the software provides alerts to maintenance staff. In rare cases, an engine may be scheduled for early removal and overhaul. For safety and economic reasons, this is in the operator's best interests. Economic factors include both fuel efficiency and reduced maintenance costs derived as a result of early problem rectification.

THE FUTURE OF GLOBAL AIRCRAFT FUEL CONSUMPTION

Fuel efficiency gains are forecast[3] to be about 2% per year for the foreseeable future. This includes gains from engines, airframes, and operational procedures.

Given a projected airline industry growth rate of about 5% per year, overall industry fuel consumption will continue to rise. If the industry continues to depend upon fossil fuels, it will become more and more expensive and may finally reach a downturn in growth as flights cease to be affordable for tourism and related discretionary travel.

While incremental gains in existing technology are still available, major future gains will depend on breakthrough thinking in airframe design or related technologies.

Incremental Gains

Aircraft designers will continue to reduce aircraft weight through new metals and composites and incremental reduction in the weight of on-board equipment. Active pitch stability features built into fly-by-wire, computer-assisted flight controls (autopilots) could provide a one to three percent reduction in overall fuel efficiency.[8] The continued incorporation of wing-tip devices ("winglets") will reduce induced drag, as will better manufacturing processes, which will smooth exterior surfaces.

Breakthrough Gains

Active systems used to increase laminar flow over the fuselage and wings are very attractive ways to decrease drag, but are fraught with technical challenges.

Fundamentally new designs, such as a blended-wing body, face different challenges, mostly in the realm of passenger acceptance. Similarly, shape-changing wings (morphing-capable) would allow an aircraft to use the most efficient wing size and shape for various flight stages. Coupled with support computers, this could also allow ailerons, rudders, and elevators to be eliminated.

There are potential breakthroughs in materials. Nanotechnology promises to provide materials that are much different and potentially feature orders of magnitude increase in strength-to-weight ratios.

ROLE OF GOVERNMENT

While aircraft technology has been fertile ground for fuel conservation efforts, there are similar efforts underway in other areas. Air Traffic Control is a service that is either regulated by or provided by governments. As such, governments have a large role to play in reducing aircraft fuel consumption.

Clearly, overall trip fuel consumption depends on the length of the flight. If distance traveled and flying time due to holding or local maneuvering can be reduced, optimized, and streamlined, fuel consumption will be reduced. Industry estimates categorize this savings in the six to 12% range over a twenty-year period.[3]

The industry has begun to use Global Positioning Satellites (GPS) to provide optimum point-to-point navigation capabilities. This is in contrast to classic airway navigation, which rarely offers direct or great circle routing, but rather a series of "legs" between fixed-position, ground-based radio navigation stations. When supplemented by a Wide Area Augmentation System (WAAS) and a Local-Area Augmentation System (LAAS), GPS-equipped aircraft can operate in instrument flight conditions for enroute navigation right down to so-called "nonprecision" approaches to the runway. This can reduce distance traveled, fundamentally reducing fuel consumption.

The industry has also continued to move to more advanced ATC systems and procedures. These are intended to streamline airport departure, enroute, and arrival procedures and timing to reduce waste and fuel consumption.

In 1995, the industry began trials of Future Air Navigation System 1 (FANS1) equipment and procedures. This equipment delivers routine ATC information to and from the cockpit via data link, and reduces the use of voice communications, which is a critical bottleneck for air traffic controllers. The industry is moving toward improving arrival and departure sequencing and enroute spacing and increasing flexibility for airline-preferred routing.

Eventually, the industry would like to see a single integrated global air traffic management system to safely optimize the use of scarce airspace (particularly near busy airports).

CONCLUSION

The aviation industry uses a small percentage of the world's energy, but cannot survive in any form without it. The cost of energy in the form of fuel comprises a large percentage of industry operating costs.

In response, the industry has developed considerable expertise and sophistication in monitoring, controlling, and reducing its energy consumption on a per-unit basis. As such, this response serves as an example of how far one can go in pursuit of conservation.

REFERENCES

1. Annual Energy Review 2003. Energy Information Administration, U.S. Department of Energy, DOE/EIA-0384(2003), Washington, DC, U.S., September 2004.
2. Key World Energy Statistics—2004 Edition. International Energy Agency, OECD/IEA—2, rue André-Pascal, 75775 Paris Cedex 16, France or 9, rue de la Fédération, 75739 Paris Cedex 15, France, 2004.
3. Penner, J.E.; Lister, D.H.; Griggs, D.J.; Dokken, D.J.; McFarland, M. Aviation and the Global Atmosphere, Summary for Policymakers, Intergovernmental Panel on Climate Change (IPCC), 1999.
4. Doganis, R. *Flying Off Course, The Economics of International Airlines*, 3rd Ed.; Routledge, Taylor and Francis Group, Oxford, UK, 2002.
5. Technology Update, Aerospace Engineering, March 2003. http://www.sae.org/aeromag/techupdate/03-2003/2-23-2-6.pdf (accessed Sept 20, 2005).
6. Historical Air Traffic Statistics, Annual, Bureau of Transportation Statistics, U.S. Department of Transportation, 400 7th Street, SW, Room 3430, Washington, DC 20590. http://www.bts.gov/programs/airline_information/indicators/airtraffic/annual/1954–1980.html (accessed Sept 20, 2005).
7. Gary, J.H.; Handwerk, G.E. *Petroleum Refining, Technology and Economics*, 3rd Ed.; Marcel Dekker, Inc., 1994.
8. Penner, J.E.; Lister, D.H.; Griggs, D.J.; Dokken, D.J.; McFarland, M. Aviation and the Global Atmosphere, Intergovernmental Panel on Climate Change (IPCC), 1999.
9. Technical Review Aviation Fuels, Chevron Products Company, 2000. http://www.chevron.com/products/prodserv/fuels/bulletin/aviationfuel/pdfs/aviation_fuels.pdf (accessed Sept 20, 2005).
10. Upham, P.; Maughan, J.; Raper, D.; Thomas, C. Towards Sustainable Aviation, Earthscan Publications Ltd, 2003.
11. World Airline Transport Statistics, 2003, IATA.
12. Fuel Conservation, Flight Operations Engineering, Boeing Commercial Airplanes, November 2004. http://www.iata.org/NR/ContentConnector/CS2000/Siteinterface/sites/whatwedo/file/Boeing_Fuel_Cons_Nov04.pdf (accessed Sept 20, 2005).
13. Getting Hand-on Experience with Aerodynamic Deterioration, Airbus Industrie, FAST/Number 21. http://www.content.airbusworld.com/SITES/Customer_services/html/acrobat/fast_21_p15_24.pdf (accessed Sept 20, 2005).

Alternative Energy Technologies: Price Effects

Michael M. Ohadi[*]
The Petroleum Institute, Abu Dhabi, United Arab Emirates

Jianwei Qi
Department of Mechanical Engineering, University of Maryland, College Park, Maryland, U.S.A.

Abstract

The world is now facing the reality that fossil fuels are a finite resource that will be exhausted someday, that global consumption is outpacing the discovery and exploitation of new reserves, and that the global environment is worsening due to increasing greenhouse gas (GHG) emissions caused by traditional fossil fuels. As a result, there is a renewed push for alternative energy technologies, as well as technologies that can enhance recovery, transportation, and energy utilization and conversion efficiencies. In this entry, a review of current alternative energy technologies and their relevance in various energy sectors will be offered, including their most recent progress and the remaining challenges to overcome. Technology barriers and research/development opportunities for further growth in each category are outlined, and future projected growth is discussed in brief.

ALTERNATIVE ENERGY TECHNOLOGIES

World energy consumption is expected to grow continuously over the next two decades. Much of the growth in new energy demand is expected to come from countries of the developing world, such as Asia. At present, developing countries, comprising more than 75% of the world's population, account for only about one-third of the world's electricity consumption, but this is expected to increase rapidly. Fossil fuels (such as oil, natural gas, and coal) have been the world's primary energy source for several decades due to their competitive low prices. However, with the high world oil prices brought on by the oil price shocks after the OPEC oil embargo of 1973–1974 and the Iranian Revolution of 1979, the use of oil for electricity generation has been slowing since the mid-1970s, and alternative energy sources, such as nuclear power, increased rapidly from the 1970s through the mid-1980s. In addition, given the recent increase in prices of fossil fuels and world compliance with carbon emission reduction policies such as the Kyoto Protocol, nonfossil fuels (including nuclear power and renewable energy sources such as hydroelectricity, geothermal, biomass, solar, and wind power) could become more attractive.

Renewable energy resources have served humans for hundreds of years in the form of water wheels, windmills, and biomass fuels during the industrial revolution. Modern efforts to harness these resources increased sharply after the oil crisis in 1970s, which provided incentive to bring these renewable sources to market to produce electricity for all economic sectors, fuels for transportation, and heat for buildings and industrial processes. Theoretically, renewable energy sources can meet the world's energy demand many times over. After two decades of dramatic technical progress, renewable energy technologies now have the potential to become major contributors to the global energy supply. Some of the technologies are already well established while others require further efforts in research, development, and deployment to become economically competitive in the traditionally fossil fuel-dominated market. However, renewable energy technologies can now be considered major components of local and regional energy systems, as they have become both economically viable and environmentally preferable alternatives to fossil fuels. For each energy demand sector, such as electric power, industrial process and building, and transportation fuels, there are several renewable energy technologies being developed. If any one technology fails to meet the technological and economic goals of its demand sector, at least one other technology will be available for that sector.

Biomass Energy

Biomass refers to green plants or almost any organic product derived from plants. It is actually a form of solar energy that is collected and stored as chemical energy by green plants and then converted to more convenient energy forms (i.e., electric energy and thermal energy) or energy carrier fuels in solid, liquid, and gaseous states. Biomass is the only renewable energy resource that can be converted to liquid fuels like oil. Biomass is used in four

---

Keywords: Global warming; Greenhouse gases; Renewable energy; Biomass; Solar energy; Wind energy; Hydrogen; Fuel cells.

[*] On leave from the University of Maryland, College park, where he serves as a Professor of Mechanical Engineering.

main ways: direct combustion, electric power generation, conversion to gas for use as fuel or chemical feedstock, and conversion to liquid fuels. There are abundant biomass resources, including trees and grasses, starch and oil seeds, sawdust, wood waste, agricultural residues, food-processing waste, paper, and municipal solid waste (MSW). Biomass energy commonly refers to both traditional biomass and modern biomass.

Traditional Biomass: traditional biomass is chemical energy directly converted through combustion into thermal energy for heating and cooking. It has been used since humans first discovered fire; it was the first application of renewable energy history. It has taken such forms as fuel wood, animal waste, and crop residue burned in stoves. An estimated 2.4 billion people in developing countries use biomass as their primary fuel for cooking and heating. Traditional biomass provides about 7%–14% of the global primary energy supply and averages 30%–45% of the energy used in developing countries, though some developing countries approach 90%.[1] Today, new biomass stoves and heaters have improved efficiency. About three quads of energy are being provided in the United States today by wood—roughly half the contribution of nuclear power. Municipal solid waste combustion also provides a small amount of process heat. At present, the availability of low-priced wood is the key constraint for its market growth.

Modern Biomass: modern biomass is converted into electricity, transport fuels, or chemicals (e.g., ethanol, methane, and biodiesel) using related process facilities. For example, China and India convert animal and plant wastes into methane for lighting, heating, cooking, and electricity generation, using bacteria to decompose biomass into biogas digesters. Modern biomass accounts for 20% of Brazil's primary energy supply, made possible by significant increases in the past 20 years in the use of ethanol fuels for vehicles and sugarcane waste for power generation. Global annual ethanol production from biomass is estimated at 18 billion liters, 80% of which is in Brazil.[2] Gasification of biomass for methane production may provide a competitive source for the nation's natural gas market. Meanwhile, the conversion of a large portion of MSW and sewage sludge to methane via anaerobic digestion may provide an attractive alternative means of disposing of such wastes.

Biomass Fuels

There are two kinds of biomass-derived liquid fuels for vehicles: ethanol fuel and biodiesel.

Ethanol: ethanol is an alcohol fuel traditionally fermented from corn kernels (corn alcohol). In 2002, 2.13 billion gallons of ethanol were produced in the United States (up 20% from 2001), which was still a small amount compared to U.S. oil imports (less than 2.6%). Ethanol can power specially designed vehicles that run on pure ethanol, or it can be mixed with gasoline or diesel fuel as an additive for use in ordinary vehicles to boost combustion and reduce vehicle emissions. According to the U.S. Environmental Protection Agency, motor vehicle emissions of carbon monoxide can be reduced by 25%–30% with the use of ethanol blended with gasoline. Ethanol–gasoline blends can also reduce ozone levels that contribute to urban smog. In addition, the combustion of ethanol produces 90% less carbon dioxide than gasoline. A blend of 10% ethanol and 90% gasoline has been widely used throughout the nation for many years. Higher level blends of 85 and 95% ethanol are being tested in government fleet vehicles, flexible-fuel passenger vehicles, and urban transit buses. There are already nearly 50,000 such vehicles already in operation, and their use is expected to grow as federal, state, municipal, and private fleet operators seek to comply with the alternative fuel requirements of the Energy Policy Act of 1992 and the Clean Air Act Amendments of 1990.

However, market issues related to ethanol production efficiency, cost competition with gasoline, the commercial viability and costs of specially designed ethanol-only vehicles, the fuel distribution infrastructure, and ratios of ethanol to gasoline in gasohol blending a challenge the attractiveness of ethanol as an alternative fuel. In addition, corn requires high amounts of energy (in the forms of fertilizer, farm equipment fuel, and coal–fire electricity) to grow, harvest, and process. Some research has shown that ethanol consumes more energy than it produces when traditional methods of production are utilized. As a result, renewable energy research has turned its focus toward a new biotech method of producing ethanol—termed bioethanol—from cellulosic biomass, such as agriculture waste products and MSW. Feedstocks include corn husks, rice straws, rice hulls, wood chips, sugarcane, forest thinnings (which also prevent wildfires), waste newspaper, and grasses and trees cultivated as energy crops. Bioethanol requires less energy to produce than ethanol and uses materials that are currently burned or buried. The biological production of ethanol involves hydrolysis of fibrous biomass, using enzymes or acid catalysts, to form soluble sugars, followed by microbial conversion of sugars to ethanol. As a result of technical advances, such as the genetic engineering of specialized enzymes and microbes, the cost of bioethanol production in the lab was decreased from $3.60/gallon in 1980 to about $1.20 in the 1990s.[3] Ultimately, the goal is for bioethanol to become competitive with gasoline in price. Research focuses on producing low-cost enzymes to break down cellulose, improving microorganism performance, producing suitable energy crops, and demonstrating ethanol production from a variety of biomass feedstocks.

Unfortunately, the transition of bioethanol from the laboratory to the highway has been slow. Though biomass is a renewable resource, ethanol is limited by available land. According to a recent Department of Energy (DOE)

Table 1 Four sources required for production of five quads of bioethanol

Source	Portion of ethanol (%)	Note
Cropland	44	Using 10% of current total U.S. cropland, including conservation reserve program acreage
Grassland	19	Using 10% of current grassland
Agricultural waste	25	Using 100% agriculture waste
Waste wood	12	Using 100% waste wood

report, five quads of bioethanol would be needed to provide 45% of the fuel used in gasoline vehicles,[4] which would use current cropland and grassland to produce 63% of bioethanol—as shown in Table 1.

The EIA reports that of the 10.38 MBPD of motor fuel consumed by motor vehicles in the United States in 1999, 0.28 MBPD (2.7%) was comprised of alternative or replacement fuels. More than 90% of this consisted of methyl-tertiary butyl ester (MTBE) (0.2 MBPD) and ethanol (0.06 MBPD) blended with gasoline. Alternative fuels, such as compressed natural gas, methanol, and LPG comprise only 0.02 MBPD.[5] However, MTBE, a petroleum-based oxygenate additive for reformulated gasoline, is toxic and can threaten the safety of community water supplies. When compared with MTBE, ethanol provides extra oxygen, increasing combustion temperatures and efficiency, and lower emissions. Ethanol is expected to replace MTBE as the oxygenate for reformulated gasoline in the near future. As a result, ethanol use increased from 133 trillion Btu in 2001 to 156 trillion Btu in 2002, and surged to 220 trillion Btu in 2003 (see Fig. 1). Production is projected to increase to 278,000 barrels per day in 2025, with about 27% of the growth from conversion of cellulosic biomass (such as wood and agricultural residues) due to rapid improvement in the technology.[6]

Today, more than 60% of Brazil's sugarcane production goes to produce ethanol. Technological advances have continued to improve the economic competitiveness of ethanol and gasohol relative to conventional gasoline, although the price of oil and competitive forces in global automotive technology greatly affects ethanol's prospects. In 2000, over 40% of automobile fuel consumption and 20% of total motor vehicle fuel consumption in Brazil was ethanol, displacing the equivalent of 220,000 barrels of oil per day. Moreover, ethanol is not the only fuel that can be produced from biomass. About 1.2 billion gallons of methanol, currently made from natural gas, are sold in the United States annually, with about 38% of this used in the transportation sector. (The rest is used to make solvents and chemicals). Methanol can also be produced from biomass through thermochemical gasification.

Biodiesel: biodiesel is defined as the mono-alkyl esters of fatty acids processed from any vegetable oil or animal fat. Biodiesel is an alternative fuel for diesel engines that is receiving great attention around the world because it is renewable, and 100% biodiesel eliminates sulfur emissions and reduces particulate matter and other pollutants by 50%. It does increase emissions of one smog-producing pollutant, nitrogen oxide, NO_x, but this can be solved by adjusting engine timing. Biodiesel generally has a lower heating value that is 12% less than No. 2 diesel fuel on a weight basis (16,000 Btu/lb compared with 18,300 Btu/lb). Because biodiesel has a higher density, the lower heating value is only 8% less on a volume basis (118,170 Btu/gallon for biodiesel compared with 129,050 Btu/gallon for No. 2 diesel fuel). Biodiesel can be used in its pure form or in blends with diesel fuel in diesel engines with no modification. In 2000, heavy trucks used 30% as much fuel as light vehicles. Diesel fuel, currently produced from petroleum, is also being produced in limited quantities from soybeans, but research has shown that diesel fuel can also be produced from less costly and more abundant sources, such as the natural oils occurring in algae and pyrolysis of biomass, other vegetable oils (such as corn oil, canola oil, cottonseed oil, mustard oil, and palm oil), animal fats, and waste oils. Total annual production of U.S. fats and oils is about 35.3 billion pounds per year (less than the equivalent of 4.64 billion gallons of biodiesel).[8] The dominant factor in biodiesel production is the feedstock cost, with capital cost contributing only about 7% of the product cost. Sales of biodiesel fuels have exploded thirtyfold since 1999 to 15 million gallons and over 33 million gallons in 2000, per

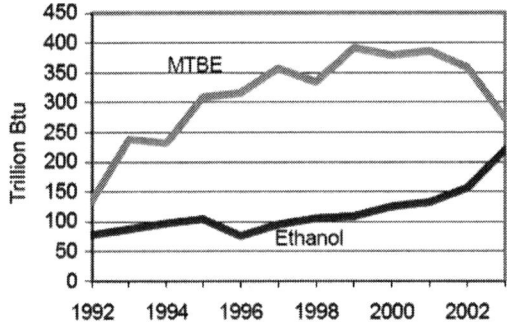

Fig. 1 Ethanol and MTBE consumption in the transportation sector (1992–2003).
Source: From Energy Information Administration (see Ref. 7).

the EIA. The United States has the capacity to produce much more biodiesel, and its output is projected to grow by 1.8% per year.[9]

United States farms and fields have yielded some homegrown energy choices, like biodiesel and ethanol, but it is still hard for them to compete with fossil fuel, with its relatively low prices and robust infrastructure, in the market. Currently, a subsidy is offered by the Department of Agriculture's Commodity Credit Corporation for the promotion and production of biodiesel. However, biodiesel is as much as twice the cost of petroleum. It is obvious that biodiesel could not completely replace petroleum-based diesel fuel in the near future. Even with the unrealistic scenario that all of the vegetable oil and animal fat currently produced were used to produce biodiesel, only about 15% of the current demand for on-highway diesel fuel could be replaced.

Solar Energy

The sun's energy is the primary source for most energy forms found on the earth. Nature's energy resources are confined to two categories: earth-stored fossil residues and nuclear isotopes that are limited by the finite amounts that exist on the earth, and the radiation flux of solar energy that is clean, abundant, and renewable. Although solar energy holds tremendous potential to benefit the world by diversifying its energy supply, reducing dependence on fossil fuels, improving the quality of the global environment, and stimulating the economy by creating jobs in the manufacture and installation of solar energy systems, solar energy's economic utility is limited by the finite rate at which the sun's energy can be captured, concentrated, stored, and/or converted for use in the highest value energy forms, and by the land areas that societies can dedicate to harness it. The amount of solar energy received across U.S. latitudes is approximately 22 quads per year per 4000 km^2 (about a million acres) on average.[10] Thus, about 40–80 thousand km^2 of land—roughly two to four times the size of Massachusetts—could supply about 20 quads, or 20%–25%, of today's total U.S. energy requirements (currently, PV solar cells convert 10%–20% of incident radiation directly to electricity).

Solar Energy Prospects

According to the outline of the U.S. DOE energy technologies program,[11] solar energy will increase the world's energy supply and enhance the reliability of the global energy infrastructure, thus creating a more stable environment for economic growth. The distributed, modular characteristics of solar energy offer tremendous flexibility for both grid-connected and off-grid electricity applications. Distributed energy technologies are expected to supply an increasing share of the electricity market to improve power quality and reliability problems, such as

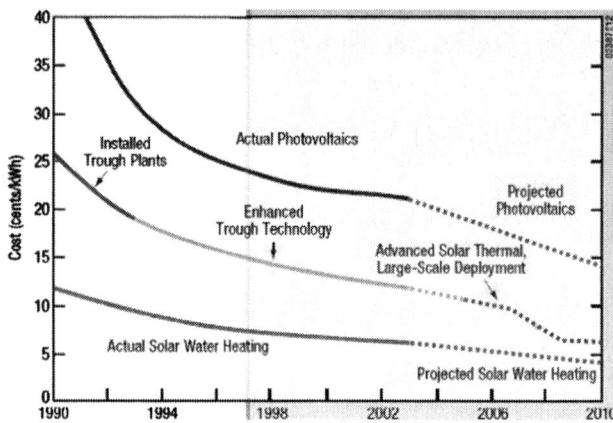

Fig. 2 The cost roadmap of solar energy technology. Source: From U.S. Department of Energy (see Ref. 11).

power outages and disturbances. With improved technology supported by the U.S. DOE, the cost of solar energy has dropped substantially in the past decade and continues to decline. The projected costs (shown as dashed lines in Fig. 2) are based on continuing the proposed budget support for the DOE Solar Program.

The long-term cost goals are even more ambitious. For example, the goal for photovoltaics, which will become an economically competitive alternative to traditional fossil fuel energy, is $0.06 per kilowatt-hour (kWh) in 2020.

Wind Energy

Wind is a form of solar energy. Winds are caused by the uneven heating of the atmosphere by the sun, the irregularities of the earth's surface, and rotation of the earth. Wind flow patterns are modified by the earth's terrain, bodies of water, and vegetation. Wind energy or wind power describes the process by which the wind is used to generate mechanical power or electricity. This mechanical power can be used for specific tasks (such as grinding grain or pumping water) or fed to a generator that can convert this mechanical power into electricity. Since early recorded history, people have been harnessing the energy of the wind. Wind propelled boats along the Nile River as early as 5000 B.C.; by 200 B.C., simple windmills in China were pumping water.[2] Commonly called wind turbines, machines that convert the kinetic energy in the wind into electricity through a generator appeared in Denmark as early as 1890. In the 1940s, the largest wind turbine of the time began operating on a Vermont hilltop known as Grandpa's Knob, which rated at 1.25 MW in winds of about 30 mph and fed electric power to the local utility network for several months during World War II.[12]

The popularity of using wind energy has always fluctuated with the price of fossil fuels. When fuel prices fell after World War II, interest in wind turbines declined. But when the price of oil skyrocketed in the 1970s, so did worldwide interest in wind turbine generators. The rapid progress in wind turbine technology has refined old ideas and introduced new ways of converting wind energy into useful power. Many of these approaches have been demonstrated in "wind farms" or "wind power plants" (groups of turbines), which feed electricity into the utility grid in the United States and Europe. Since the 1970s, wind energy has expanded its role in electricity generation. The worldwide installed capacity of grid-connected wind power has now exceeded 40 GW, corresponding to an investment of approximately $40 billion.[12] A demand for clean, diverse sources of electricity and state and federal incentives to stimulate the market have contributed to wind energy's growth in the United States. Wind energy installations in the United States increased during the past decade from about 1800 MW in 1990 to more than 6000 MW at the end of 2003, enough to power almost three million average homes. The average U.S. wind energy growth rate for the past five years was 24%. This growth can be attributed to a greatly reduced cost of production, from 80 cents (current dollars) per kWh in 1980 to cents per kWh in 2002. The global wind energy installed capacity has increased exponentially over a 25-year period, and in the process, the cost of energy (COE) from wind power plants has been reduced by an order of magnitude, becoming very close in cost to power from modern combined-cycle power plants in some locations. According to the American Wind Energy Association, as much as 13,500 additional megawatts of wind capacity may be installed worldwide in the next decade.[23] Wind energy is the world's fastest growing energy source and will power industry, businesses, and homes with clean, renewable electricity for many years to come.

Wind Energy Technology Development

The considerable potential of wind energy was not tapped before the 1980s because the wind turbine technology was not competitive with most central fossil fuel-fired generating stations. But over the past two decades, the rapid progress in wind turbine technologies has led to more cost-effective wind turbines that are more efficient in producing electricity. The progress was mainly motivated by the oil embargoes and fuel price escalations of the 1970s and more recently by environmental concerns. The goal is to develop cost-effective, low wind-speed turbines for Class sites (13-mph average annual wind speed) that can produce electricity onshore for $0.03/kWh and offshore for $0.05/kWh by the end of 2012. This will open up 20 times more U.S. land for wind energy development, and because many of these sites tend to be closer to urban load centers, the problem of transmission line expansion will be greatly simplified. But the current turbine designs are not well suited to low wind-speed sites and have only limited potential to achieve lower energy costs. If such technology can be successfully developed, the wind resources across the Great Plain states could potentially generate more electricity than is currently consumed by the entire nation.[13] Although wind power plants have relatively little impact on the environment compared to other conventional power plants, there is some concern over the noise produced by the rotor blades, aesthetic (visual) impacts, and the danger to birds, which have been killed by flying into the rotors. Most of these problems have been resolved or greatly reduced through technological advances or by properly siting wind plants.

Future technology improvements for low-speed wind technology must address three principal areas:

- Turbine rotor diameters must be larger to harvest the lower-energy winds from a larger inflow area without increasing the cost of the rotor.
- Towers must be taller to take advantage of the increased wind speed at greater heights.
- Generation equipment and power electronics must be more efficient to accommodate sustained light wind operation at lower power levels without increasing electrical system costs.

Hydrogen and Fuel Cells

Hydrogen is the simplest element; an atom consists of only one proton and one electron. It is also the most plentiful element in the universe. Although in many ways hydrogen is an attractive replacement for fossil fuels, it does not occur in nature as the fuel H_2. Rather, it occurs in chemical compounds, such as water or hydrocarbons that must be chemically transformed to yield H_2. Hydrogen is found in water and hydrocarbon organic compounds that make up many of our fuels, such as gasoline, coal, natural gas, methanol, and propane. Although President Bush has called hydrogen a "pollution free" technology, extracting hydrogen from its most common source, water, requires electricity that may come from fossil fuels, such as coal or nuclear energy. Hydrogen, like electricity, is a carrier of energy, and like electricity, it must be produced from a natural resource. Nevertheless, it promises substantial contributions to global energy supplies and minimal environmental impact in the long term.

Hydrogen

Hydrogen is the simplest chemical fuel (essentially a hydrocarbon without the carbon) that makes a highly efficient, clean-burning energy carrier and a secondary form of energy that has to be produced like electricity. When hydrogen is used to power a special battery called a fuel cell, its only waste product is water. Hydrogen-powered fuel cells and engines could become as common as the gasoline and diesel engines of the late 20th century and could power cars, trucks, buses, and other vehicles, as well as homes, offices, and factories. Hydrogen has the potential to fuel transportation vehicles with zero emissions, provide process heat for industrial processes, supply domestic heat through cogeneration, help produce electricity for centralized or distributed power systems, and provide a storage medium for electricity from renewable energy sources. Some envision an entire economy based on hydrogen in the future.[14] At present, most of the world's hydrogen is produced from natural gas by a process called steam reforming. However, producing hydrogen from fossil fuels would not be an advancement because steam reforming does not reduce the use of fossil fuels, but rather shifts them from end use to an earlier production step; in other words, steam reforming would still release carbon to the environment in the form of CO_2. Thus, to achieve the benefits of the hydrogen economy, the hydrogen must be produced more cost effectively from nonfossil resources, such as water, using a renewable energy source like wind or solar. Although the potential benefits of a hydrogen economy are significant, many barriers to commercialization—technical challenges, and otherwise—must be overcome before hydrogen can offer a competitive alternative for consumers.

Commercial Barriers. The commercial barriers to the widespread use of hydrogen are the high cost of hydrogen production, low availability of hydrogen production systems, the challenge of providing safe production and delivery systems (i.e., economical storage and transportation technologies), and public acceptance.

- *Hydrogen Storage.* Hydrogen has a low energy density in terms of volume, making it difficult to store amounts adequate for most applications in a reasonably-sized space. This is a particular problem for hydrogen-powered fuel cell vehicles, which must store hydrogen in compact tanks. Hydrogen is currently stored in tanks as a compressed gas or cryogenic liquid. The tanks can be transported by truck or the compressed gas can be sent across distances of less than 50 miles by pipeline. Other options are to store hydrogen in a cryogenic liquid state or solid state. Technologies that store hydrogen in a solid state are inherently safer and have the potential to be more efficient than gas or liquid storage. These are particularly important for vehicles with on-board storage of hydrogen. High-pressure storage tanks are currently being developed, and research is being conducted into the use of solid-state storage technologies, such as metal hydrides, which involve chemically reacting the hydrogen with a metal; carbon nanotubes, which take advantage of the gas-on-solids adsorption of hydrogen and retain high concentrations of hydrogen; and glass microspheres, which rely on changes in glass permeability with temperature to fill the microspheres with hydrogen and trap it there. However, the statistical cost, durability, fast-fill, discharge performance, and structural integrity data of hydrogen storage systems must be improved before proceeding with commercialization.
- *Safety, Codes, and Standards.* Hydrogen, like gasoline or any other fuel, has safety risks and must be handled with due caution. Unlike the handling of gasoline, handling hydrogen will be new to most consumers. Therefore, developers must optimize new fuel storage and delivery systems for safe everyday use, and consumers must become familiar with hydrogen's properties and risks. Codes and standards are needed to ensure safety as well as to commercialize hydrogen as a fuel.
- *Public Acceptance.* Finally, public acceptance of hydrogen depends not only on its practical and commercial appeal, but also on its record of safety in widespread use. Because a hydrogen economy would be a revolutionary change from the world we know today, educating the general public, training personnel in the handling and maintenance of hydrogen system components, adopting codes and standards, and developing certified procedures and training manuals for fuel cells and safety standards will foster hydrogen's acceptance as a fuel.

Technology Roadmap (Present-2030). Technical challenges for hydrogen commercialization include cost-effective, energy-efficient production technologies and safe, economical storage and transportation technologies. The U.S. DOE has provided a national version of America's transition to a hydrogen economy to 2030 or beyond.[15] This technology roadmap consists of three steps. The first step toward a clean energy future will focus on technology development and initial market penetration to build on well-known commercial processes for producing, storing, transporting, and using hydrogen. In the mid-term, as hydrogen use increases and hydrogen markets grow, the expansion of the market and infrastructure investment will make the cost of hydrogen and fuel cell economically competitive with traditional fossil fuels. For the long term, when the market and infrastructure are more fully developed, wider uses of more cost-effective advanced technologies will be an important step toward a hydrogen economy.[15]

Today, large centralized steam methane reformers are used to produce hydrogen for chemical industries. This will continue to be the likely choice for meeting increased hydrogen demand in the near term. Electrolyzers are also used to produce the high-purity hydrogen needed for electronics manufacturing and other specialty uses. Compressed hydrogen tanks are available today, although the low energy density of hydrogen means large tanks are needed. As a liquid, hydrogen's energy density is substantially improved, but boil-off losses are a concern. Today, hydrogen is transported by pipeline or over the road in cylinders, tube trailers, and cryogenic tankers.

A small amount is shipped by rail car or barge. Hydrogen has also long been used in the space program as a propellant for the space shuttle and in the on-board fuel cells that provide the shuttle's electric power. New combustion equipment is being designed specifically for hydrogen in turbines and engines, and vehicles with hydrogen internal combustion engines have been demonstrated. Also being tested is the combustion of hydrogen–natural gas blends to improve the yield of natural gas reforming in an effort to lower cost and raise efficiency. Fuel cells are in various stages of development for transportation, stationary, and portable applications. Incremental advances of current technologies provide a low-risk commercial entry into the hydrogen economy.

Fuel Cells

The widespread use of hydrogen as an energy source in the world could help address concerns about energy security, global climate change, and air quality. Fuel cells are an important enabling technology for a future hydrogen economy and have the potential to revolutionize power generation, offering cleaner, more efficient alternatives to the combustion of gasoline and other fossil fuels. Fuel cells promise to be a safe and effective way to use hydrogen for both vehicles and electricity generation. Although these applications would ideally run off pure hydrogen, in the near term they are likely to be fueled with natural gas, methanol, or even gasoline. If the fuel cell is to become the modern steam engine, basic research must provide breakthroughs in understanding, materials, and design to make a hydrogen-based energy system a vibrant and competitive force. Fuel cell technology is not a new invention. Actually, fuel cell development predated the internal combustion engine, but lacked a commercial venue until NASA decided to incorporate fuel cells in spacecrafts during the 1960s. Phosphoric acid fuel cells are already commercially available and can generate electricity in 200-kW capacities selling for $3/W, using natural gas as the source of hydrogen; molten carbonate has also been demonstrated at large (2-MW) capacities.

A fuel cell works like a battery but does not run down or need recharging. Fuel cells convert hydrogen—hydrogen gas or hydrogen reformed within the fuel cell from natural gas, alcohol fuels, or some other source—directly into electrical energy with no combustion. They will produce electricity and heat as long as fuel (hydrogen) is supplied. A fuel cell consists of two electrodes, a negative electrode (or anode) and a positive electrode (or cathode), sandwiched around an electrolyte. Hydrogen is fed to the anode, and oxygen is fed to the cathode. Activated by a catalyst, hydrogen atoms separate into protons and electrons, which take different paths to the cathode. The electrons go through an external circuit, creating a flow of electricity. The protons migrate through the electrolyte to the cathode, where they reunite with oxygen and the electrons to produce water and heat.

Fuel cell technologies are significantly more energy efficient than combustion-based power generation technologies. A conventional combustion-based power plant typically generates electricity at efficiencies of 33%–35%, while fuel cell plants can generate electricity at efficiencies of up to 60%. When fuel cells are used to generate electricity and heat (cogeneration) they can reach efficiencies of up to 85%. Internal-combustion engines in today's automobiles convert less than 30% of the energy in gasoline into power that moves the vehicle. Vehicles using electric motors powered by hydrogen fuel cells will be much more energy efficient, utilizing 40%–60% of the fuel's energy.[28]

Technology Challenges. Although NASA has used hydrogen fuel cells for space missions since the 1960s, terrestrial applications are still in their infancy. The lack of an economical process for hydrogen production and suitable storage methods are two of the greatest obstacles to commercialization, especially in the transportation sector. Research goals include developing technologies to produce hydrogen from sunlight and water and biomass; developing low-cost and low-weight hydrogen storage technologies for both stationary and vehicle-based applications, such as carbon nanotubes and metal hydrides; and developing codes and standards to enable the widespread use of hydrogen technologies. Technological development is addressing the following key challenges in the commercialization of fuel cell and hydrogen infrastructure technologies[28]:

- *Cost.* Cost is the greatest challenge to fuel cell development and adaptation, and it is a factor in almost all other fuel cell challenges, as well. Materials and manufacturing costs are high for fuel cell components (i.e., catalysts, membranes, bipolar plates, and gas diffusion layers). Statistical data for fuel cell vehicles that are operated under controlled, real-world conditions are very limited and often proprietary. For example, some fuel cell designs require expensive, precious-metal catalysts, and others require costly materials that are resistant to extremely high temperatures. Currently, the costs for automotive internal combustion engine power plants are about $25–$35/kW. The targeted cost for a fuel cell system to be competitive in transportation applications is around $30/kW, and the acceptable price point for stationary power systems is considerably higher (i.e., $400–$750/kW for widespread commercialization and as much as $1000/kW for initial applications).
- *Durability and Reliability.* All fuel cells are prone, in varying degrees, to catalyst poisoning, which decreases fuel cell performance and longevity. The durability and reliability of fuel cell systems operating over automotive drive cycles has not been established. Vehicle

fuel cell power systems will be required to achieve the same level of durability and reliability of current engines over the full range of vehicle operating conditions at 40°C–80°C (i.e., 5000 h lifespan or about 150,000 miles equivalent). Stationary fuel cells must achieve greater than 40,000 h of reliable operation at −35 to 40°C to meet market requirements. Fuel cell component degradation and failure mechanisms are not well understood. The cycle life of hydride storage systems also needs to be evaluated in real-world circumstances.

- *System Size.* The volume and weight of current fuel cell systems are too high to meet the packaging requirements for transportation or stationary applications. System volume minimization and weight reduction will focus not only on the fuel cell stack, but also on the ancillary components and major subsystems making up the balance of power system (e.g., fuel processor, compressor/expander, and sensors).
- *Air, Water, and Thermal Management.* Fuel cell performance and efficiency must meet or exceed that of competing technologies in order to be commercially accepted. Today's compressor technologies are not suitable for fuel cell applications that need low power consumption and less packaging volume. Vehicle fuel cell systems must start rapidly from any ambient condition with minimal fuel consumption. Cost-effective thermal and water management technologies are needed, including heat recovery and water utilization, cooling, and humidification. The low operating temperature of proton exchange membrane (PEM) fuel cells results in a relatively small difference between the operating and ambient temperatures, which need advanced technologies to allow high operating temperatures and to improve combined heat and power system performance.

Nuclear Energy

Nuclear technology uses the energy released by splitting the atoms of certain elements. It was first developed in the 1940s, and during the Second World War, research initially focused on producing bombs by splitting the atoms of either uranium or plutonium. Only in the 1950s did attention turn to peaceful applications of nuclear fission, notably power generation. Today, the world produces as much electricity from nuclear energy as it did from all sources combined in 1960. Civil nuclear power, with some 440 commercial nuclear power reactors in 30 countries, can now exceed 12,000 reactor years of experience, and nuclear power supplies 16% of global needs with a total installed capacity of over 360,000 MWe.[18] This is more than three times the total generating capacity of France or Germany from all sources. The economics of nuclear power may be more favorable in countries where other energy fuels (mostly imported) are relatively expensive. In 2002, nineteen countries depended on nuclear power for at least 20% of their electricity generation (Fig. 3), while three quarters of both France's and Lithuania's power are derived from nuclear energy.

However, accidents at Three Mile Island in the United States in 1979 and at Chernobyl in the Soviet Union in 1986 pushed public opinion and national energy policies away from nuclear power as a source of electricity. But, after nearly two decades of the antinuclear tide, nuclear energy today is at a turning point. Nuclear power plant

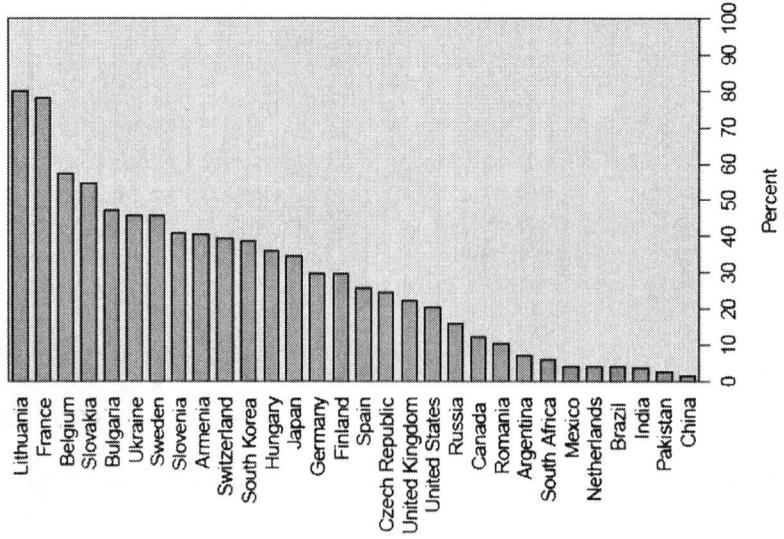

Fig. 3 Nuclear shares of national electricity generation, 2002.
Source: From IAEA, International Atomic Energy Agency (see Ref. 16).

performance has shown a steady improvement over the past 10–15 years: one quarter of the world's reactors have load factors of more than 90% and almost two-thirds do better than 75%, compared to about a quarter of them in 1990.

Nuclear power is the only mature, noncarbon electricity generation technology that can significantly contribute to the long-term, globally sustainable energy mix. Besides providing electricity, nuclear energy contributes to a number of policy goals, including achieving energy independence, keeping the air clean, and reducing carbon emissions. European countries have begun construction of a nuclear reactor, and six more are likely to be constructed in the next decade. The U.S. Nuclear Power 2010 program aims to build new nuclear power plants in the United States by the end of the decade, and expects that the advanced reactor designs will produce electricity in the range of $1000–$1200 per kW of electricity. The highest growth in nuclear generation is expected in the developing world, where consumption of electricity from nuclear power is projected to increase by 4.1% per year between 2001 and 2025. Developing Asia, in particular, is expected to see the largest increase in nuclear generating capacity, accounting for 95% of the total increase in nuclear power capacity for the developing world. Of the 44 gigawatts of additional installed nuclear generating capacity projected for developing Asia, 19 gigawatts are projected for China, 15 gigawatts are projected for South Korea, and 6 gigawatts are projected for India.[18]

Nuclear energy is, in many places, competitive with fossil fuel for electricity generation, despite relatively high capital costs and the need to internalize all waste disposal and decommissioning costs. If the social, health, and environmental costs of fossil fuels are also taken into account, nuclear energy is superior. A 2004 report from the University of Chicago, funded by the U.S. DOE, compares the power costs of future nuclear, coal, and gas-fired power generation in the United States. Various nuclear options are covered, and for ABWR or AP1000, they range from 4.3 to 5.0 c/kWh on the basis of overnight capital costs of $1200–$1500/kW, a 60-year plant life, a 5-year construction period, and 90% capacity. Coal yields 3.5–4.1 c/kWh and gas (CCGT) 3.5–4.5 c/kWh, depending greatly on fuel price.[17] When considering a minimal carbon control cost impact of 1.5 c/kWh for coal and 1.0 c/kWh for gas superimposed on the above figures, nuclear is even more competitive.

Overview of Nuclear Power Technology

The principles for using nuclear power to produce electricity are the same for most types of reactors. The energy released from continuous fission of the fuel atoms is harnessed as heat in either gas or water and it is used to produce steam. The steam is used to drive the turbines, which produce electricity (as in most fossil fuel plants). In most naval reactors, steam drives a turbine directly for propulsion.

There are several components common to most types of reactors:

- *Fuel.* Pellets of uranium oxide (UO_2) are usually arranged in tubes to form fuel rods. The rods are arranged into fuel assemblies in the reactor core.
- *Moderator.* This slows down the neutrons released from fission so that they cause more fission. The moderator is usually water, but may be heavy water or graphite.
- *Control Rods.* These are made from neutron-absorbing material, such as cadmium, hafnium, or boron, and are inserted into or withdrawn from the core to control the rate of reaction, or to halt it.
- *Coolant.* A coolant, such as a liquid or gas, circulates through the core so as to transfer heat from it. In light water reactors, the moderator also functions as coolant.
- *Pressure Vessel/Pressure Tubes.* This is a robust steel vessel containing the reactor core and moderator/coolant, but it may also be a series of tubes holding the fuel and conveying the coolant through the moderator.
- *Steam Generator.* This is the part of the cooling system where the heat from the reactor is used to make steam for the turbine.
- *Containment.* The packaging structure around the reactor core is designed to protect it from outside intrusion and to protect those outside from the effects of radiation in case of any malfunction inside. It is typically a meter-thick concrete and steel structure.

Types of Nuclear Reactors

Most nuclear electricity is generated using just two kinds of reactors that were developed in the 1950s and have been improved since. In the United States, Westinghouse designed the first fully commercial pressurized water reactor (PWR) of 250 MWe, Yankee Rowe, which started up in 1960 and operated until 1992. Meanwhile, Argonne National Laboratory developed the first boiling water reactor (BWR). The first of this type, the Dresden-1 of 250 MWe, was designed by General Electric and started earlier in 1960. A prototype BWR, Vallecitos, ran from 1957 to 1963. By the end of the 1960s, orders were being placed for PWR and BWR reactor units of more than 1000 MWe.[19]

Pressurized Water Reactors utilize pressurized water as a moderator and coolant. The fuel, ceramic uranium dioxide, is typically encased in long zirconium alloy tubes. The uranium-235 is enriched from its original 0.7% abundance to 3.5%–5.0%.

Boil Water Reactors are similar to PWRs, except that the coolant water is allowed to boil, and steam passes from

Table 2 Overview of commercial nuclear power reactors

Reactor type	Main countries	Number	GWe	Fuel	Coolant	Moderator
Pressurized water reactor (PWR)	United states, France, Japan, Russia	268	249	Enriched UO_2	Water	Water
Boiling water reactor (BWR)	United states, Japan, Sweden	94	85	Enriched UO_2	Water	Water
Gas-cooled reactor (magnox &AGR)	UK	23	12	Natural U (metal), enriched UO_2	CO_2	Graphite
Pressurized heavy water reactor "CANDU" (PHWR)	Canada	40	22	Natural UO_2	Heavy water	Heavy water
Light water graphite reactor (RBMK)	Russia	12	12	Enriched UO_2	Water	Graphite
Fast neutron reactor (FBR)	Japan, France, Russia	4	1	PuO_2 and UO_2	Liquid sodium	None
	TOTAL	441	381			

Source: From Nuclear Engineering (see Ref. 19).

the top of the reactor directly to the turbine. Currently, more than 90 of these are operating throughout the world.

Advanced Gas-Cooled Reactors (AGRs) are the second generation of British gas-cooled reactors, using graphite moderators and carbon dioxide as coolant. The fuel is uranium oxide pellets, enriched to 2.5%–3.5%, in stainless steel tubes. The carbon dioxide circulates through the core, reaching 650°C, and then past steam generator tubes outside the core, but still inside the concrete and steel pressure vessel. Control rods penetrate the moderator, and a secondary shutdown system involves injecting nitrogen into the coolant.

Pressurized Heavy Water Reactors (PHWRs) are a Canadian reactor development headed down a quite different track, using natural uranium fuel and heavy water as a moderator and coolant. The first unit started up in 1962.

Fast Neutron Reactors (FBRs) (there is only one in commercial service) do not have a moderator and utilize fast neutrons, generating power from plutonium, while simultaneously making more of it from the U-238 isotope in or around the fuel. While these reactors get more than 60 times as much energy from the original uranium compared to normal reactors, they are expensive to build and must contend with resource scarcity before coming into their own.

Several different types of reactors in current commercial nuclear power plants are summarized in Table 2.

Several generations of reactors can be commonly distinguished. Generation I reactors were developed in the 1950s–1960s, and relatively few are still running today. They mostly used natural uranium fuel and graphite moderators. Generation II reactors developed in the 1970s–1980s are typified by the present U.S. fleet and most are in operation elsewhere. They typically use enriched uranium fuel and are mostly cooled and moderated by water. Pressurized water reactors and BWRs are known as light-water reactors. Around the world, with few exceptions, other countries have chosen light-water designs for their nuclear power programs, so that, today, 65% of the world capacity is PWR and 23% BWR.

Generation III reactors are advanced reactor designs developed in the 1990s. With enhanced safety, these reactors are more economical to build, operate, and maintain than the previous generation. The first Generation III system, a General Electric-designed advanced BWR, started operating in 1996 at the Kashiwazaki-Kariwa Nuclear Power Station in Japan. More than a dozen Generation III advanced reactor designs are in various stages of development. Some have evolved from the PWR, BWR, and CANDU designs above; some are more radical departures. The best-known radical new design is the Pebble Bed Modular Reactor (PBMR), which uses high-temperature helium to cool the reactor and drive the turbine directly. One of the limitations of current light-water reactor technology is that the thermal efficiency that can be achieved is limited to the achieved maximum temperature of 350°C. The PBMR is designed to achieve at least 900°C, which will give a thermal efficiency of up to 44%. This translates into roughly one-third more output than a conventional PWR. The first PBMR is currently planned for commercial operation in South Africa by around 2010.

Future Nuclear Power Technology

Generation I–III reactors recycle plutonium (and possibly uranium), while the future of nuclear reactors, known as Generation IV systems, have revolutionary reactor and fuel cycle systems.[29] Most will have closed fuel cycles that burn the long-lived actinides that form part of the spent fuel so that fission products are the only high-level waste. Many will also be fast neutron reactors. Six new designs, including three PWRs, three BWRs, and two high temperature gas-cooled reactors (HTRs), were identified for further study by the Generation IV International Forum (GIF), which was initiated in 2000 by a group of nine countries: Argentina, Brazil, Canada, France, Japan, South Africa, South Korea, the U.K., and the U.S. Switzerland became a member of the forum in February 2002, and the European Atomic Energy Community joined in July 2003. GIF is an international initiative whose goal is to develop nuclear energy systems that can supply future global needs for electricity, hydrogen, and other products. The aim is to deploy these systems no later than 2030 to provide competitively priced and reliable energy products, while satisfactorily addressing nuclear safety, waste, proliferation, and physical protection concerns. All six revolutionary nuclear reactor technology concepts identified for development by GIF operate at higher temperatures than the Generation II and III reactors currently in operation. The new systems range from a supercritical water-cooled reactor (SCWR), which operates at 510°C–550°C, to a helium-cooled very high-temperature gas reactor (VHTR), which has an operating temperature of 1000°C. Three of the six Generation IV concepts are fast reactor systems that are cooled either by helium gas, lead, or sodium. All use depleted uranium as a fuel. The main technological challenges are addressed as follows:

- Generation IV technology must address the high-level waste from fission reactions. This waste includes heavy nuclides–actinides such as neptunium, americium, and curium—that remain highly radioactive for tens of thousands of years. The helium-, lead-, and sodium-cooled fast reactors are designed to have closed fuel cycles in which the actinides are separated from the spent fuel and returned to the fission reactors. Supercritical water-cooled reactors are the only one of the six Generation IV technologies that are cooled by water. Supercritical water-cooled reactors are designed to be a thermal reactor in the intermediate term, using enriched UO_2 as a fuel with a once-through fuel cycle. However, the ultimate goal is to build them as a fast neutron reactor with full actinide recycling. Very high-temperature gas reactors have an open fuel cycle. They will employ enriched UO_2 as a fuel, possibly in the form of pebbles coated with a graphite moderator like those required for PBMR.
- In the longer term, uranium resource availability could also become a limiting factor. Thus, one challenge to long-term, widespread deployment of Generation IV nuclear energy systems is that their fuel cycles must minimize the production of long-lived radioactive wastes while conserving uranium resources.
- Very high-temperature gas reactors, helium- and lead-cooled fast reactors and the molten salt reactor are all designed to generate electricity and also to operate at sufficiently high temperatures to produce hydrogen by thermochemical water cracking. Thermochemical hydrogen production can be achieved at temperatures of less than 900°C, using processes such as the sulfur–iodine cycle in which sulfur dioxide and iodine are added to water, resulting in an exothermic reaction that creates sulfuric acid and hydrogen iodide. At 450°C, the HI decomposes to iodine (which is recycled) and hydrogen. Sulfuric acid decomposes at 850°C, forming sulfur dioxide (which is recycled), water, and oxygen. The only feeds to the process are water and high-temperature heat, typically 900°C, and the only products are hydrogen, oxygen, and low-grade heat. Because of its potential to produce cheap and green hydrogen, VHTR technology has been given high priority by the U.S. DOE in its Next Generation Nuclear Plant (NGNP), which aims to make both electricity and hydrogen at very high levels of efficiency and near zero emissions.

Nuclear Energy Challenges

The main challenges nuclear power development faces in developed countries are economic and political, while the main issues in developing countries are a lack of adequate infrastructure, particularly in the back-end of the fuel cycle, a lack of expertise in nuclear technology and its safety culture, as well as financial issues. Developing countries are more likely to profit from the enhanced and passive safety features of the new generation of reactors, which have a stronger focus on the effective use of intrinsic characteristics, simplified plant design, and easy construction, operation, and maintenance. Public concerns about nuclear safety, national security, nuclear waste, and nonproliferation are also hindering the development of nuclear power. If nuclear power is to contribute in significant ways to meeting future energy demands, these issues, real or perceived, must be addressed.

Nuclear Safety. The International Nuclear Safety Advisory Group (INSAG) has suggested requiring that future nuclear plants be safer by a factor of 10 than the targets set for existing reactors (i.e., targets of 10^{-5}/year for core damage and 10^{-6}/year for large radioactive releases for future plants).[20]

There are four primary goals for safety issues in nuclear energy development:

- The first is reactivity control, which is the process of stopping fission reactions. The lack of reactivity control caused the Chernobyl reactor accident in 1986.
- The second safety goal is to reliably remove decay heat, which is the heat generated by radioactive decay of the fission products that continue to be produced even after fission reactions stop. Decay heat, if not removed, can result in overheating of and damage to the fuel, such as occurred in the Three Mile Island accident in Pennsylvania in March 1979.
- The third goal is to provide multiple barriers to contain the radioactive material. The barriers include the fuel cladding, the reactor vessel, and the containment building.
- The fourth goal is plant safety sufficient to eliminate the need for detailed evacuation plans, emergency equipment, and periodic emergency evacuation drills.

Although statistics comparing the safety of nuclear energy with alternative means of generating electricity show nuclear to be the safest, vast efforts to enhance the safety of advanced Generation III+ reactor designs and the revolutionary Generation IV technologies continue. These designs incorporate what are known as passive safety systems. Evolutionary designs explore many avenues to increased safety, including using modern control technology, simplifying safety systems, making use of passive designs, and extending the required response times for safety systems actuation and operator action. One example of an advanced reactor with passive safety systems is the economic simplified boiling-water reactor (ESBWR), which was developed by General Electric from its advanced boiling-water reactor design and is at the preapplication stage for NRC design certification. The Westinghouse AP1000 light-water reactor features advanced passive safety systems with fewer components than conventional PWRs. For example, compared with a conventional 1000-MW PWR, AP1000 has 50% fewer valves, 35% fewer pumps, 80% less pipe, and 85% less cable. The reactor has a modular design that will reduce construction time to as little as three years from the time the concrete is first poured to the time that fuel is loaded into the core.

Table 3 summarizes some key technical challenges and potential technical responses for a potential nuclear power increase in capacity while at the same time improving economic attractiveness.

However, in spite of the evolutionary improvements in safety, support for nuclear power has not increased in Western Europe and North America. One explanation is that, given there have been significant accomplishments in the area of safety, other issues (such as spent fuel and nuclear waste management) have replaced improvement of existing safety features as the greatest challenges to the future development of nuclear power.

Spent fuel and nuclear waste management. Nuclear energy produces both operational and decommissioning wastes, which must be contained and managed. Spent fuel can be safely stored for long periods in water-filled pools or dry facilities, some of which have been in operation for 30 years. Although the volumes of nuclear waste are small compared to waste from other forms of electricity generation, spent fuel and radioactive wastes from nuclear power plants and spent fuel reprocessing—and ultimately from plant decommissioning—still need to be managed safely. For now, most high-level waste from commercial nuclear power is either stored on-site or transported to interim storage sites. In fact, nuclear power is the only energy-producing industry that takes full responsibility for all its waste and figures these costs into its product, which is a key factor in sustainability. Desirable features of innovative nuclear fuel cycles are economic competitiveness; reduction of nuclear waste and the hazards associated

Table 3 Nuclear power development—key challenges and potential technical responses

Challenges	Availability of a technological response
Safety area eliminate severe core damage	Reactor and nuclear power plant designs based on maximum use of passive safety features
Waste management minimize the volume and toxicity of nuclear waste	Multirecycling of most dangerous long-lived radioactive elements in reactors with fast neutron spectrum
Nonproliferation furtherance of nonproliferation aims, namely that nuclear materials cannot be easily acquired or readily converted for nonpeaceful purposes	Recycling of fissile materials, together with radioactive minor actinides; integral nuclear fuel cycle without the stockpiling of Pu
Resource base increase the resource base more than tenfold	Multirecycling of plutonium in fast reactors

with its long-term storage; furtherance of nonproliferation aims, namely that nuclear materials cannot be easily acquired or readily converted for nonpeaceful purposes; and improved efficiency in resource use. Table 4 gives examples of recent work in innovative nuclear fuel cycles. Although no large-scale programs on innovative nuclear fuel cycles are being implemented at present, some countries are investigating the necessary steps for change in the current situation.

Experience with both storage and transport over half a century clearly shows that there is no technical problem in managing civil nuclear waste sans environmental impact. Shortage of capacity for spent fuel storage is today's eminent issue in several countries where long-term waste disposal policy remains unsettled. The greatest concern over the storage of high-level nuclear waste is that over such a long period of time, the containers in which waste is stored could eventually leak. However, the scientific and technical communities generally feel confident that technical solutions for spent fuel and nuclear waste conditioning and disposal already exist. The question has become politicized and focused on final disposal because high-level nuclear waste must be stored for thousands of years, but there is general consensus that geologic disposal of spent fuel or high-level radioactive waste from reprocessing can be carried out safely in stable, deep geologic formations. However, site selection remains a major political issue in most countries developing such facilities, and no such commercial facility has yet been authorized. Although most nations have identified potential underground storage sites and conducted geological and geophysical tests as to their suitability, no underground storage site has progressed beyond the planning stage. In the United States, which is perhaps the most advanced in the planning stage, President Bush authorized the construction of a nuclear waste repository at Yucca Mountain in Nevada in 2002.

Regardless of whether particular wastes remain a problem for centuries, millennia, or forever, there is a clear need to address the question of their safe disposal. Therefore, technological breakthroughs over a range of reactors and a range of reactor characteristics are now needed to cope with emerging issues, such as nonproliferation, environmental mitigation, economics, and enhanced safety and security needs.

ALTERNATIVE ENERGY PROSPECTS

It has been less than 100 years since fossil fuels became the predominant sources of energy in the world with the discovery of oil, the development of natural gas fields, and the widespread distribution of electricity from coal-powered central power plants. From the dawn of human civilization until about 100 years ago, human and animal muscle and wood, with lesser amounts of solar, wind,

Table 4 Innovative technologies related to nuclear fuel cycle

Attribute	Process and system	Relevant countries	Features
Fuel composition and process	Pyro-process	Japan, Russia, United States	Nuclear waste volume is smaller and process facility is simpler than those of wet process (expected economic and environmental advantages)
	Vibro-packed fuel	Russia, Switzerland	Fuel particle is directly produced from acid solution from reprocessing (economic merit is expected to compare to powder technology)
	DUPIC system	Canada, Rep. of Korea	Plutonium is not separated from PWR spent fuel (proliferation resistance is expected)
	Thorium fuel (Th–U, Th–Pu)	India, United States	Th resource is abundant. Fuel with Th–^{233}U composition generates less MA than U–Pu fuel
	Inert-matrix fuel	France, Japan, Switzerland	Due to chemically stable oxide, spent fuel is regarded as waste form (environmental mitigation)
Partitioning and transmutation (P–T) system	Accelerator driven system	France, Japan, United States	High neutron energy produced in accelerator destroys MA, LLFP. Sub-critical core enhances safety
	P–T system with fast reactor (FR)	Japan, Russia	Existing FR technology is applied for destruction of MA, LLFP
Fast reactor and fuel cycle system	Pb (+Bi) FR	Russia	Enhanced resource utilization, proliferation resistance, safety, and waste features

Source: From The National Academies Press (see Ref. 21).

hydro, and geothermal power, were the predominant sources of energy used by mankind. Can the renewable resources that sustained early civilization be harvested more cost effectively to meet a significant portion of the much higher demands of today's society?

Many regions of the world are rich in renewable resources. Winds in the United States contain energy equivalent to 40 times the amount of energy the nation uses. The total sunlight falling on the country is equivalent to 500 times America's energy demand. And accessible geothermal energy adds up to 15,000 times the national demand.[27] There are, of course, limits to how much of this potential can be used because of competing land uses, competing costs from other energy sources, and limits to the transmission systems needed to bring energy to end users. Moreover, the market penetration potential of renewable energy technologies will face a situation confronting any new technology and institutional constraints that attempt to resist an entrenched technology. But renewable energy technologies have made great progress in cost reduction, increased efficiency, and increased reliability in the past 30 years, as well as increasing contributions to the world's energy supply. In 2000, the global electricity capacity generated by renewable energy sources accounted for about 30% (102 GW) of the total electric power capacity (see Table 5).[2] Most of this energy came from hydroelectric power. However, other nonhydro renewable energy resources such as biomass, alcohol fuel, wind, geothermal, and solar energies have increased due to technological innovations and cost reductions and are becoming viable, commercially competitive, and environmentally preferable alternatives to traditional fossil fuels. While the global resources of fossil fuels gradually decline and environmental concerns associated with fossil fuels increase, the benefits of renewable energy are undeniably attractive.

Renewable energy sources have historically had a difficult time breaking into the existing markets dominated by traditional, large-scale, fossil fuel-based systems. This is partly because renewable and other new energy technologies are only now being mass-produced and have previously had high capital costs relative to more conventional sources because fossil fuel-based power systems have benefited from a range of subsidies, cheap raw materials, mass volume production over a span of many years, and a mature infrastructure. Moreover, the alternative energy sector has lacked appeal to investors in the United States because of heavy regulation, low growth, long-time return, and lack of support for innovative new companies in established energy markets.

The push to develop renewable and other clean energy technologies is no longer being driven solely by environmental concerns. Rather, these technologies are now also becoming economically competitive (Fig. 4). With continued research and deployment, many renewable-energy technologies are expected to continue their steep reductions in cost for such programs as wind, solar thermal, and photovoltaic energy. Several could become competitive over the next decade or two, either directly or in distributed utilities. Wind turbines, in particular, could even become broadly competitive with gas-fired combined-cycle systems within the next ten years in places where there are winds of medium or high quality. One disadvantage of renewable energy systems has been the intermittent nature of some sources, such as wind and solar energy. But this problem is not insurmountable—one solution is to develop diversified systems that maximize the contribution of renewable energy sources to meet daily needs, but also to use clean natural gas and/or biomass-based power generation to provide base-load power for the peak times in energy use, such as evening air conditioning or heating demand. Because of the changing U.S. electricity marketplace, remote or distributed markets for renewable electricity, as discussed above, appear to be more promising today than centralized electricity markets. Renewable energy sources are pinning their hopes on breakthroughs in the development of small, stationary and portable fuel cells and on the fast growing

Table 5 Renewable grid-based electricity generation capacity installed as of 2000

Technology	All countries (MW)	Developing countries (MW)
Total world electric power capacity	3400,000	150,000
Large hydropower	680,000	260,000
Small hydropower	43,000	25,000
Biomass power	32,000	17,000
Wind power	18,000	1,700
Geothermal power	8,500	3,900
Solar thermal power	350	0
Solar photovoltaic power (grid)	250	0
Total renewable power capacity	102,000	48,000

Source: From Renewable Energy Publications (see Ref. 2).

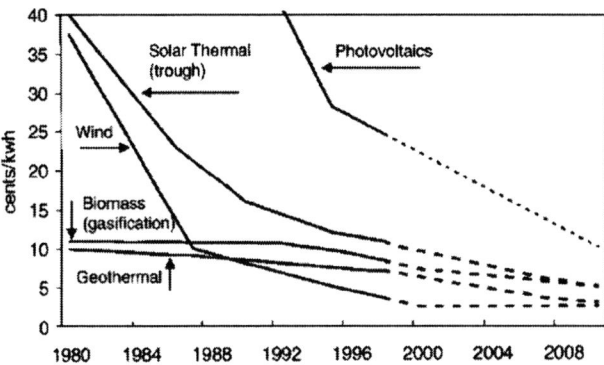

Fig. 4 The renewable electricity cost.
Source: From National Renewable Energy Laboratory (see Ref. 22).

market for them as minipower plants for use in factories, offices, retail stores, homes, and automobiles.

However, despite the significant benefits of alternative energy sources, according to the recent AEO2004 forecast, which assumes the world's oil supply peak will not occur before 2025, petroleum products are predicted to dominate energy use in the transportation sector. Energy demand for transportation is projected to grow from 26.8 quadrillion Btu in 2002 to 41.2 quadrillion Btu in 2025 (Fig. 5).

According to the forecast, motor gasoline use will increase by 1.8% per year from 2002 to 2025, when it will make up to 60% of transportation energy use. Alternative fuels are projected to displace only 136,800 barrels of oil equivalent per day in 2010 and 166,500 barrels per day in 2025 (2.1% of light-duty vehicle fuel consumption) in response to current environmental and energy legislation intended to reduce oil use. Gasoline's share of the demand is expected to be sustained by low gasoline prices and slower fuel efficiency gains for conventional light-duty vehicles (cars, vans, pickup trucks, and sport utility vehicles) than were achieved during the 1980s.

Therefore, even though renewable energy technologies have advanced dramatically during the past 30 years, fossil fuels seem slated to dominate the energy supply, and oil will remain the world's foremost source of primary energy consumption throughout the 2001–2025 period (at 39%) despite expectations that countries in many parts of the world will be switching from oil to natural gas and other fuels for their electricity generation. Robust growth in transportation energy use—overwhelmingly fueled by petroleum products—is expected to continue over the 24-year forecast period as shown in Fig. 6.[19] For this reason, oil is projected to retain its predominance in the global energy mix, notwithstanding increases in the penetration of new technologies, such as hydrogen-fueled vehicles.

The same trend of fossil fuel dominance over a slowly growing market in renewable energy can be found in the United States in the past five years. Renewable energy consumption in 2003 grew only 3% to 6.1 quadrillion Btu.[26] Overall, renewable energy contributed only 6% of the nation's total energy supply. Current levels of renewable energy use represent only a tiny fraction of what could be developed, but the U.S. energy infrastructure is mainly based on the consumption of fossil fuels. As shown in Fig. 7, solar and wind only accounted for 3% of total renewable energy use in 2003—less than 0.2% of the U.S. energy supply. Without government policies or programs—such as environmental laws aimed at limiting or reducing pollutants from the combustion of fossil fuel consumption and encouraging the use of nonfossil fuels—consumption of fossil fuels like oil, natural gas, and coal are expected to supply most of the primary energy needed to meet the projected demand for end-use consumption.

In 2001, the U.S. DOE produced a 50-year perspective for future U.S. Highway Energy Use.[4] The DOE assumed the demand for world oil products would grow at 2% per year. After 2020, when conventional oil production is assumed to peak—once 50% of ultimate resources have been produced and begin a continual decline—the gap

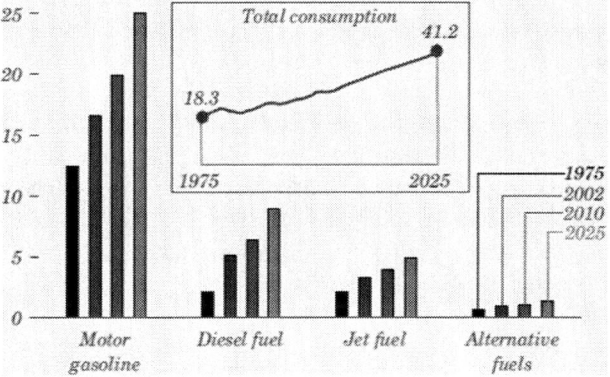

Fig. 5 Transportation energy consumption by fuel (quads).
Source: From U.S. Department of Energy (see Ref. 6).

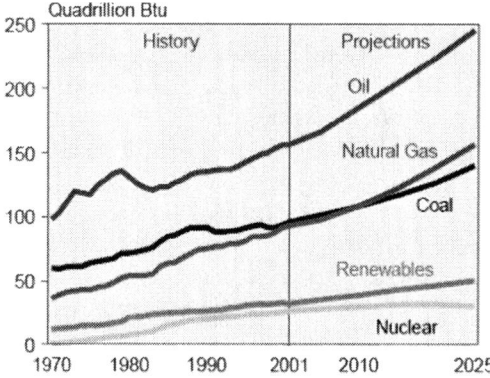

Fig. 6 World marketed energy consumption by energy source, 1970–2025.
Source: From Washington Times Corporation (see Ref. 25).

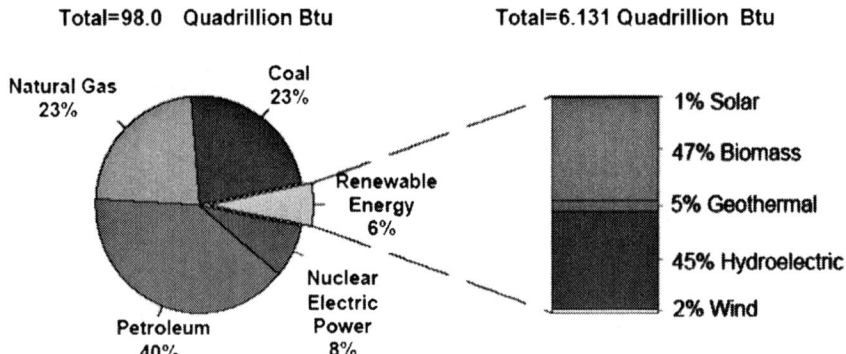

Fig. 7 Renewable energy consumption in the U.S.'s energy supply, 2003.
Source: From U.S. Department of Energy (see Ref. 26).

between continuing demand growth and declining production could be around the equivalent of 50 billion barrels of oil (145 MBPD) by 2050, or almost twice the current conventional oil production. Although there is considerable uncertainty, the report suggested that the United States should start transportation's energy transition immediately because the time needed to fully implement a new vehicle technology in all vehicles on the road will be 30 years or more, and fully implementing a new fuel will take even longer.

The market penetration of alternative renewable fuels, such as biofuel/biodiesel or hydrogen (see Table 6), are dependent on their cost-effective production, storage, delivery, safety, and customer acceptance, as well as related vehicle technology development, such as HEVs, battery-electric vehicles, and fuel cells.

Obviously, the transition to energy diversity with significant proportions of renewable energy and nuclear energy will change the world's use of energy, its economy, and the environment. A recent study on the costs and benefits of gradually increasing the share of the nation's electricity from wind, biomass, geothermal and solar energy, as proposed in renewable portfolio standards (RPS), range from a 4% reduction in carbon dioxide emissions in 2010 to a 20% reduction in 2020, which would freeze electricity-sector carbon dioxide emissions at year 2000 levels through 2020 at a modest cost of $18 per ton reduced.[27] In particular, the plains, western, and mid-Atlantic states are projected to generate more than 20% of their electricity from a diverse mix of renewable energy technologies. By contrast, carbon dioxide emissions are projected to grow 24% over the same period under a business-as-usual scenario. Renewable energy and energy efficiency are not only affordable, but their expanded use will also open new areas of innovation and business, creating opportunities and a fair marketplace for the transition to a clean energy economy, all of which will require leadership and vision from both government and industry. However, in the absence of government policies for both technology push and market pull, those renewable energy technologies will be not widely commercialized, and it will be difficult to extend the use of renewable sources on a large scale because most renewable energy sources are not expected to compete economically with fossil fuels in the mid-term if there are no significant changes in the cheaply-priced fossil fuel market.

CONCLUSIONS

Over the past 30 years, renewable energy technologies with unique environmental and social benefits have made significant progress as alternative energy options to fossil fuels. Of the many alternative sustainable energy technologies that have been developed, some are already making large inroads into the marketplace, such as nuclear power, biomass, hydropower, geothermal, wind, and solar power. Other technologies, perhaps those most beneficial to a sustainable future, require more efforts in research, development, and deployment before they can become economically viable and technically feasible. The recent rise in demand for energy, which is mostly due to increased demand in China and India as part of the expansion of their modern economy, coupled with the geopolitical aspects of the energy sector, has caused record high oil and gas prices across the globe. To confront the limited availability of what has been thus far viewed as inexpensive fossil fuels and growing energy demand from developing countries, the world is now necessarily on the brink of an energy transition from fossil fuels to clean renewable energy sources. However, the global energy infrastructure is mainly based on the consumption of fossil fuels, with very little being done to reduce dependence on these energy resources until now. Although after three decades of development, renewable energy has been proven technically and economically feasible, the timing of the transition will depend on government polices and the will of the customer. For example, current nuclear energy is technically and economically feasible to meet market demand, but certain social and safety issues remain that need to be addressed through a collaborative approach of

Table 6 Comparisons of gasoline and alternative transport fuels

	Gasoline	Biodiesel (B20)	Compressed natural gas (CNG)	Ethanol (E85)	Hydrogen
Chemical structure	C_4–C_{12}	Methyl esters of C_{16}–C_{18} fatty acids	CH_4	CH_3CH_2OH	H_2
Main fuel source	Crude oil	Soy bean oil, waste cooking oil, animal fats, and rapeseed oil	Underground reserves	Corn, grains, or agricultural waste	Natural gas, methanol, and other energy sources
Energy content per gallon	109,000–125,000 Btu	117,000–120,000 Btu (compared to diesel #2)	33,000–38,000 Btu @ 3,000 psi; 38,000–44,000 @ 3,600 psi	~80,000 Btu	Gas: ~6,500 Btu @ 3 kpsi; ~16,000 Btu @ 10 kpsi Liquid: ~30,500 Btu
Price per gallon/ gasoline gallon equivalent (GGE)	$1.99	$2.06	$1.40	$2.28	$5–$11 (2003) $1.5–$2.25 (2015)
Environmental impacts of burning fuel	Produces harmful emissions; however, gasoline and gasoline vehicles are rapidly improving and emissions are being reduced	Reduces particulate matter and global warming gas emissions compared to conventional diesel; however, NO_x emissions may be increased	CNG vehicles can demonstrate a reduction in ozone-forming emissions (CO and NO_x) compared to some conventional fuels; however, HC emissions may be increased	E-85 vehicles can demonstrate a 25% reduction in ozone-forming emissions (CO and NO_x) compared to reformulated gasoline	Zero regulated emissions for fuel cell-powered vehicles, and only NO_x emissions possible for internal combustion engines operating on hydrogen
Energy security impacts	Manufactured using imported oil, which is not an energy secure option	Biodiesel is domestically produced and has a fossil energy ratio of 3.3 to 1, which means that its fossil energy inputs are similar to those of petroleum	CNG is domestically produced. The United States has vast natural gas reserves	Ethanol is produced domestically and it is renewable	Hydrogen can help reduce U.S. dependence on foreign oil by being produced by renewable resources
Fuel availability	Available at all fueling stations	Available in bulk from an increasing number of suppliers. There are 22 states that have some biodiesel stations available to the public	More than 1,100 CNG stations can be found across the country. California has the highest concentration of CNG stations. Home fueling will be available in 2003	Most of the E-85 fueling stations are located in the Midwest, but in all, approximately 150 stations are available in 23 states	There are only a small number of hydrogen stations across the country. Most are available for private use only

(Continued)

Table 6 Comparisons of gasoline and alternative transport fuels *(Continued)*

	Gasoline	Biodiesel (B20)	Compressed natural gas (CNG)	Ethanol (E85)	Hydrogen
Maintenance issues		Hoses and seals may be affected with higher-percent blends, lubricity is improved over that of conventional diesel fuel	High-pressure tanks require periodic inspection and certification	Special lubricants may be required. Practices are very similar, if not identical, to those for conventionally fueled operations	When hydrogen is used in fuel cell applications, maintenance should be very minimal
Safety issues (Without exception, all alternative fuel vehicles must meet today's OEM safety standards)	Gasoline is a relatively safe fuel because people have learned to use it safely. Gasoline is not biodegradable though, so a spill could pollute soil and water	Less toxic and more biodegradable than conventional fuel, can be transported, delivered, and stored using the same equipment as for diesel fuel	Pressurized tanks have been designed to withstand severe impact, high external temperatures, and automotive environmental exposure	Ethanol can form an explosive vapor in fuel tanks. In accidents; however, ethanol is less dangerous than gasoline because its low evaporation speed keeps alcohol concentration in the air low and nonexplosive	Hydrogen has an excellent industrial safety record; codes and standards for consumer vehicle use are under development

Source: From U.S. Department of energy (see Ref. 24).

the world community. The recently rising prices of conventional sources of energy give the strongest economic argument in favor of the expanded market of renewable energy sooner rather than later. The evolution of any new energy technology will take a long time (more than 20 years) to be well accepted and established in the market. Energy substitution will begin in earnest when the costs of energy production by alternative methods are lower than the prevailing prices of conventional sources and when consumers are convinced there will be no reversal in price and supply trends. Although greater energy utilization efficiencies have proven to be effective conservation options in reducing energy consumption over past decades, they cannot and should not be viewed as an ultimate remedy or solution. In the long run, renewable energy and advanced nuclear energy seem to be the best options for meeting the increasing clean energy demand.

REFERENCES

1. Clemmer, S.L.; Donovan, D.; Nogee, A. *Clean Energy Blueprint: A Smarter National Energy Policy for Today and the Future*; Union of Concerned Scientists and Tellus Institute: Cambridge, MA, June 2001.
2. Martinot, E.; Chaurey, A.; Lew, D.; Moreira, J.R.; Wamukonya, N. Renewable energy markets in developing countries. Ann. Rev. Energy Env. **2002**, *27*, 309–348.
3. Zhang, M.; Eddy, C.; Deanda, K.; Finkelstein, M.; Picataggio, S. Metabolic engineering of a pentose metabolism pathway in ethanologenic zymomonas mobilis. Science **1995**, *267*, 240–243.
4. Birky, A. et al. Future U.S. highway energy use: a fifty year perspective (draft 5/3/01). *Office of Transportation Technologies, Energy Efficiency and Renewable Energy*; U.S. Department of Energy: Washington, DC.
5. Energy Information Administration. Annual Energy Review 1999, DOE/EIA-0384 (99), Washington, DC, 1999.
6. AEO2004 The Annual Energy Outlook 2004 (AEO2004). *Energy Information Administration, Office of Integrated Analysis and Forecasting*. U.S. Department of Energy: Washington, DC.
7. Energy Information Administration. *Alternatives to Traditional Transportation Fuels, 2003. Estimated Data*; Washington, DC, Febraury 2004.
8. Pearl, G.G. Animal fat potential for bioenergy use. *Bioenergy 2002*, The 10th Biennial Bioenergy Conference, Boise, ID, September 22–26, 2002.
9. EIA, Assumptions to the Annual Energy Outlook 2004, Report No: DOE/EIA-0554, February 2004.
10. Weiss, S.B. Can. J. Forest Res. **2000**, *30*, 1953.

11. DOE. Solar Energy Technologies Program—Multi-Year Technical Plan 2003–2007 and Beyond, Report No: DOE/GO-102004-1775, 2004.
12. Web site for the European Wind Energy Association, http://www.ewea.org/ (accessed May 19, 2004).
13. Schwartz, M.N.; Elliott, D.L. Arial Wind Resource Assessment of the United States. In *Alternative Fuels and the Environment*; Lewis Publishers: Boca Raton, FL, 1994; [Chapter 17].
14. DOE. Strategic Plan for DOE Hydrogen Program, DOE/GO-10 098-532, U.S. Dept. of Energy, 1998.
15. DOE. A National Version of America's Transition to A Hydrogen Economy–To 2003 and Beyond, National Hydrogen Version Meeting, Washington, DC, November 15–16, 2001.
16. IAEA, International Atomic Energy Agency, Reference Data Series 2, *Power Reactor Information System*, Web site www.iaea.org/programmes/a2/.
17. University of Chicago, *The Economic Future of Nuclear Power*, August 2004.
18. IEO. International energy outlook 2004. *Energy Information Administration, Office of Integrated Analysis and Forecasting*, U.S. Department of Energy: Washington, DC, April, 2004.
19. Nuclear Engineering International Handbook 2005.
20. Basic Safety Principles for Nuclear Power Plants (INSAG-12). 75-INSAG-3 Rev 1, IAEA, 1999.
21. Fukuda, K.; Bonne, A.; Mourogov, V.M. Global view on Nuclear Fuel Cycle—Challenges for the 21st century. Proceedings of the International Conference on Future Nuclear Systems Global'99, Jackson Hole, Wyoming, U.S.A.
22. Bull, S.R. Renewable energy today and tomorrow. Proc. IEEE **2001**, *89* (8), 1216–1226.
23. American Wind Energy Association, Global Wind Energy Market Report, Washington, DC, March 12, 2003.
24. DOE, The 12th Clean Cities Alternative Fuel Price Report, June 29, 2004 (http://www.eere.energy.gov/afdc/resources/pricereport/).
25. Tanenbaum, B.S. Fossil Fuels and Energy Independence, From The World &I, Washington Times Corporation, 2002; pp. 148–155.
26. Energy Information Administration (EIA), Renewable Energy Trends With Preliminary Data For 2003, 2004.
27. Clemmer, S.; Nogee, A.; Brower, M.C. A Powerful Opportunity: Making Renewable Electricity the Standard, Union of Concerned Scientists, November 1998.
28. DOE Hydrogen Program. The Hydrogen, Fuel Cells & Infrastructure Technologies Program Multi-Year Research, Development and Demonstration Plan, February 2005 (http://www.hydrogen.energy.gov/).
29. Generation IV International Forum (GIF) Online Reports, A Technology Roadmap for Generation IV Nuclear Energy Systems, December 2002 (http:gen-iv.ne.doe.gov/GENIVDOCUMENTs.asp).

ANSI/ASHRAE Standard 62.1-2004

Leonard A. Damiano

David S. Dougan
EBTRON, Inc., Loris, South Carolina, U.S.A.

Abstract

American National Standards Institute (ANSI)/American Society of Heating and Refrigeration and Air-Conditioning Engineers, Inc. (ASHRAE) Standard 62.1-2004 is a short but often misunderstood document outlining ventilation requirements intended to provide acceptable indoor air quality (IAQ) for new buildings or those with major renovations. Because of the rate-based nature of both procedures allowed for compliance, this analysis focuses on the practical needs of reliable intake rate control and the risks of indirect controls. Design recommendations offered are intended to increase the potential for both predictable compliance and the flexibility to accommodate future changes while providing the greatest control reliability with the most energy-efficient methods.

INTRODUCTION

This is a summary of American Society of Heating and Refrigeration and Air-Conditioning Engineers, Inc. (ASHRAE) Standard 62.1-2004, ventilation for acceptable indoor air quality (IAQ) in Commercial, Institutional, Industrial, and High rise residential buildings,[1] as it impacts and is influenced by ventilation control requirements, methods, and equipment. Operational implementation of these requirements can have a sizeable influence on energy usage when applied improperly or incompletely. Operational precision and design reliability are essential for energy minimization when compliance with 62.1 and energy codes are simultaneous goals.

This is not a condensed version and this does not cover all requirements of the Standard. Designers are strongly encouraged to read the entire 19-page document (a total of 44 pages with appendices). The Standard cannot be understood or properly applied without considering the relationships and interdependencies of requirements of the document.

This American National Standards Institute (ANSI)-approved standard has been developed by a Standing Standards Project Committee (SSPC) of the ASHRAE under a 'continuous maintenance' protocol. At any point in time, the 'official' Standard is comprised of both the most recently published parent document and all current addenda. The latest parent document was republished earlier this year to combine 17 addenda that had been approved subsequent to the original release of the 62-2001 'parent' document in January 2002. The result is a final version that is substantially different from the basic ventilation standard we have used since 1989.[1]

In 2003, the scope of the Standard officially changed and a separate ASHRAE committee was formed to address the specific needs of low-rise residential buildings. The existing Standard became known as 62.1 and the new residential standard became 62.2.

The promised *62.1 User's Manual* was recently published in December 2005. Work continues on a *Guideline 19P*, which is intended to provide design guidance for methods that exceed the minimum requirements of the Standard. Both of these supplemental documents should assist the designer and the facility operator in their understanding of and compliance with the Standard.

ANALYSIS AND RECOMMENDATIONS

Our discussion of Standard 62.1 will try to mimic the structure of the Standard, provide recommendations for compliance, and highlight methods and assumptions to avoid. Our objectives have determined the content.

The Standard's "Purpose" and "Scope" are covered in Sections 1 and 2. To comply with the Standard, designers of mechanical ventilation systems are tasked to provide specific minimum rates of acceptable outdoor air to the breathing level of the occupied structures. In doing so, an acceptable indoor environment may be achieved providing improved occupant productivity and health. The procedures allowed for compliance with our national standard on ventilation are prescriptive or performance-based. Their selection and application should be evaluated for IAQ risk by the design practitioner.

Keywords: ASHRAE; Standard 62.1; Indoor air quality; IAQ; Ventilation; Airflow control; Building pressurization; Moisture management.

Definitions

Section 3 addresses the definition of terms used within the Standard. Noteworthy is the Standard's definition of "acceptable IAQ," which is defined as:

> ...air in which there are no known contaminants at harmful concentrations as determined by cognizant authorities and with which a substantial majority (80% or more) of the people exposed do not express dissatisfaction.[1]

This means that 62.1, like all ASHRAE Standards, assumes that one out of five occupants (20%) might not be satisfied with the results of compliance and might express dissatisfaction with the IAQ, even if the Standard is followed perfectly. Many sources have concluded that the majority of Heating Ventilation Air Conditioning (HVAC) systems designed in the United States do not meet the minimum ventilation rates prescribed during operation. In which case, the actual occupant dissatisfaction level is exponentially greater in practice.[2] It is not uncommon for rates to fall below levels that result in occupant dissatisfaction significantly greater than 50%. Many systems cannot meet the minimum airflow requirements at the occupied space during operation because of design choices and equipment limitations or due to the dynamic nature of mechanical ventilation systems and the constant external forces acting on the building envelope.

The impacts from these continuously changing external conditions are not limited to variable air volume (VAV) systems.[2] Outdoor airflow rates will also vary for systems that provide a constant volume of supply air (CAV) to the conditioned space, as a result of:

1. Changes in wind or stack conditions on the intake system,[3]
2. Changes in filter loading, or
3. Changes in airflow requirements during an economizer cycle.

The lack of specific guidelines to overcome the effect of changing system dynamics on ventilation rates and air distribution for today's HVAC systems are partially to blame for many design deficiencies observed.

Unlike thermal comfort, the effect of IAQ is difficult to measure (The work item that caused the most controversy was an attempt to standardise design criteria for the indoor environment). The criteria developed in the process have been published as a CEN Technical Report CR 1752. It specifies the levels of temperature, air velocity, noise, and ventilation for occupied spaces. Values are given for three categories of environmental quality: A—a high level of expectation, B—a medium level and C—a moderate level.

Supporting information is given on the derivations of the specified values of the parameters as well as to enable alternatives, such as different clothing levels, to be accommodated in the design assumptions. The most debatable section is on IAQ. Here, prominence is given to the evaluation of the required ventilation rate for comfort based on perceived air quality, the method developed by Professor Fanger and his colleagues in Denmark. While some data is presented, it is acknowledged that more research is needed to provide reliable information on pollution loads from materials and on the additive effects of emissions from multiple sources (Source: http://www.aivc.org/frameset/frameset.html?../Air/20_3/jackman.html~mainFrame accessed on June 2005). Many believe that the outdoor air levels specified by ASHRAE are too low and should actually be increased, as indicated by published research and reflected in European standards from CEN Technical Committee 156 and their publication CR1752.[4]

Outdoor Air Quality

Section 4 of the Standard describes a three-step process to evaluate outdoor air for acceptability. One of those steps requires examination of both the regional and the local air quality by the building owner. The section also specifies the documentation required to support the conclusions of this preliminary review.

If the outdoor air quality is found to be unsuitable per Section 4, then treatment may be required as indicated in Section 6.2.1. Outdoor air treatment involves the removal of the particulates and gases encountered that are in excess of the minimum standards cited by cognizant authorities in Section 4.1.

Systems and Equipment

Section 5 specifies the minimum systems and equipment required under Standard 62.1. Section 5.4 states:

> Mechanical ventilation systems shall include controls, manual or automatic, that enable the fan system to operate whenever the spaces served are occupied. The system shall be designed to *maintain the minimum outdoor airflow* as required by Section 6 *under any load condition*. Note: VAV systems with fixed outdoor air damper positions must comply with this requirement at minimum supply airflow.[1]

The Standard recognizes that changes in mixed air plenum pressure, up to 0.5 in. WG [125 Pa] variation on VAV systems, can significantly influence outdoor air intake flow rates. However, it neglects the significant influence of external pressure variations on all systems that result from changes in wind and stack pressures, which often exceeds 0.5 in. WG [125 Pa]. Therefore, providing the minimum outdoor airflow defined in Section 6 'effectively' requires a dynamic control alternative for compliance—possibly the use of permanent devices capable of maintaining outdoor airflow rates.

Not mentioning airflow measurement is analogous to ignoring the requirement for temperature measuring devices to maintain continuous temperature control. Because many systems, especially VAV, have thermal load requirements that differ from their ventilation needs, the requirements of this section can be more sustainable if the multispace Eqs. 6-1 to 6-8 are calculated for the design supply flows to individual zones using the minimum outdoor air requirements to each zone. In order to achieve the industry "standard of care" in professional HVAC design, the mechanical engineer is required to determine which zones may become 'critical' and that the worst 'critical zone' is at its minimum supply airflow. Even with the average reduction potentials due to the new ventilation rate (Table 6-1), this could still impose a severe energy penalty to many VAV system designs.

Because of the requirements set forth in the Standard for compliance "under any load condition," it is necessary to maintain a constant rate of outdoor airflow in dynamic systems. Logically, Section 5 should require continuous airflow measurement at the intake of all air-handling units with automatic controls that function to provide a building or space with a constant rate of outdoor air, regardless of the system size or type. Doing so would alleviate several practical issues, clarify application and compliance questions in Section 6.2.7, Dynamic Reset; Section 7.2.2, Air Balancing; and Section 8.4.1.8, Outdoor Airflow Verification. A continuous measurement requirement was explicitly stated in the draft Standard 62-89R before Standard 62 became politicized, which was much more complicated and vague to the point of confusion.

American Society of Heating and Refrigeration and Air-Conditioning Engineer's new *62.1 User's Manual* provides more insight into the appropriate, if not encouraged, use of permanent instrumentation for continuous measurement, by explaining further:

> For VAV systems that can operate under a wide range of operating conditions, the system must be designed to provide the minimum outdoor air rates under all reasonably anticipated operating conditions (see Ref. 5, pp. 6–33).

> To comply [with Std. 62.1], most VAV systems will need to be designed with outdoor airflow sensors and modulating dampers or injection fans (see Ref. 5, pp. 5–10). In most cases, an active control system must be provided at the air intake and sometimes at the zone level to ensure minimum rates are maintained... Note that a fixed-speed, outdoor air fan without control devices will not maintain rates within the required accuracy (see Ref. 5, pp. 5–11).

We are encouraged to use direct measurement feedback for continuous control on all VAV designs, even those using a powered outdoor air system (i.e., injection fan, heat recovery ventilators (HRV)/energy recovery ventilators (ERV), smaller dedicated outdoor air systems (DOAS), etc.). Although not contained in the society's 'minimum' standard, ASHRAE is highlighting the potential source of problems and the more obvious means to avoid them.

We believe and recommend that Section 5 of the Standard should encourage the use of airflow measuring devices in the supply air to critical zones of VAV systems, allowing not only for improved operating savings for continuous verification of compliance and as a diagnostic tool but also to reset intake rates based on input from the continuous calculation of the multispace equations defined in Section 6.2. Although this may sound impractical to some designers, the technology is available and surprisingly cost effective, especially when considering the potential benefits in occupant productivity and health, reduced potential liability, and the energy savings available.

Pressurization and Mold

Addendum 62x (62-2001) was approved in 2004 for inclusion with the Standard, but problems identified immediately after publication generated addendum 62.1a (Table 1), which was just published in a supplement to the 2004 document during this writing (May 2006).

The proposed addendum only addresses positive pressure during periods of dehumidification.

Moisture is a prerequisite for mold and fungal growth and the condition should be avoided. Whenever the temperature of a building envelope is lower than the dew point of air migrating across it, there will be condensation and the potential for mold. Costs for design improvements or preventive actions prior to mold conditions can range from an additional \$1.50 to \$15/ft^2 [\$16–\$161/m^2]. Costs for mold remediation and repair can range from \$30 to \$65/ft^2 [\$323–\$700/m^2] [6]. That effectively equals a penalty for inaction ranging from 4 to 20 times that of the cost for prevention.

The *62.1 User's Manual* also comments on pressure effects and 'pressurization flow.'

> Positive pressure in hot, humid climates can also reduce interstitial moisture and resultant fungal microbial growth.

Table 1 Section 5.10

5.10 **Dehumidification Systems.** Mechanical air-conditioning systems with dehumidification capability shall be designed to comply with the following:

5.10.1 **Relative Humidity.** Occupied space relative humidity shall be designed to be limited to 65% or less at either of the two following design conditions:1) at the peak outdoor dew point design conditions and at the peak indoor design latent load, or when system performance is analyzed with outdoor air at the dehumidification design condition (that is, design dew point and mean coincident dry-bulb temperature) and with the space interior loads (both sensible and latent) at cooling design values and space solar loads at zero.
    Note: System configuration and/or climatic conditions may adequately limit space relative humidity at these conditions without additional humidity-control devices. The specified conditions challenge the system dehumidification performance with high outdoor latent load and low space sensible heat ratio.
    Exception: Spaces where process or occupancy requirements dictate higher humidity conditions, such as kitchens, hot tub rooms that contain heated standing water, refrigerated or frozen storage rooms and ice rinks, and/or spaces designed and constructed to manage moisture, such as shower rooms, pools and spas.

5.10.2 **Exfiltration.** For a building, the design minimum outdoor air intake shall be greater than the design maximum exhaust airflow when the mechanical airconditioning systems are dehumidifying.
    Exception: Where excess exhaust is required by process considerations and approved by the authority having jurisdiction, such as in certain industrial facilities.
    Note: Although individual zones within the building may be neutral or negative with respect to outdoors or to other zones, net positive mechanical intake airflow for the building as a whole reduces infiltration of untreated outdoor air. [1]

Stack and wind effects can cause large regions, such as entire levels or façades, to be negatively pressurized.[6]

A building that is excessively pressurized may cause damage to the structural integrity of the building envelope.[6]

Noting the proliferation of mold in buildings, the ASHRAE board issued *Minimizing Indoor Mold Through Management of Moisture in Building Systems* in June 2005, stating that sound moisture management should take precedence over energy cost savings.[7] This Position Paper outlines recommendations for the management of moisture in buildings by describing issues related to the topic and highlighting resources available through the society. This policy statement will filter through the society's organization and eventually impact technical programs, research, and standards.

Some of their recommendations for proper moisture management include:

- Building and system design, operation, and maintenance provide for drying of surfaces and materials prone to moisture accumulation under normal operating conditions.
- Mechanical system design should properly address ventilation air.
- The sequence of operation for the HVAC system should contain appropriate provisions to manage humidity, control pressurization, and monitor critical conditions.[7]

The flaw of ASHRAE's position is in ignoring the potential for high humidity alone to provide sufficient moisture content for mold growth. Studies have shown that in temperatures between 30 and 86°F [−1.1°C–30°C] with a minimum relative humidity of 70% RH (noncondensing infiltration), mold growth has appeared on plasterboard, brick, and concrete within 3 days. At 65.3°F [18.5°C] (with adequate RH and "inadequate" substrate), mold grows on building materials after 6 h. It was shown to take only 1 h to grow with "adequate" substrate.[2,8]

Part of the solution to preventing infiltration of unfiltered and unconditioned humid air appears to be simple. In 1996, the Florida Solar Energy Center first published a case study which identified that an extremely small negative pressure differential created conditions that lead to mold problems in a small commercial building. This was later supported by a 2002 ASHRAE journal article whose recommendations indicated that differential pressures as low as +0.004–+0.008 in. WG [1–2 Pa] will prevent moisture infiltration problems.[9] This counterflow overcomes most of the natural pressures that power moisture migration, namely vapor, temperature (stack), and wind pressures. Those periods when pressurization flow is insufficient to counter infiltration are generally limited in duration. Thereafter, the flow of air to the direction of higher dew point temperature can remove any residual moisture in the wall cavity.

Control precision and control stability should be key objectives when energy usage is to be minimized and dynamic control of space pressurization is used.

'Building pressure' is accomplished by creating a pressurization flow. Anything that changes the pressurization flow will result in fluctuations in building pressure. The pressurization flow is generally influenced by the HVAC system by controlling the volumetric differential of either the intake/relief air or the supply/return air. Heating ventilation air conditioning system control strategies that ignore these relationships or poorly implement them are widely known to have inherent pressurization problems.

Regardless of the system design, an effective method for maintaining pressurization flow is to monitor and control these airflow differentials. Typically, pressurization airflow (Q_P) is maintained at a fixed differential by an independent control loop; independent of the supply airflow rate required for temperature control, which uses a separate control loop. The airflow relationship is as follows:

$$Q_P = Q_{SA} - (Q_{RA} + Q_{EX(local)}), \text{ or}$$
$$Q_P = Q_{OA} - (Q_{EX(ahu)} + Q_{EX(local)})$$

where Q_P, pressurization airflow; Q_{SA}, supply airflow; Q_{RA}, return airflow; $Q_{EX(ahu)}$, exhaust/relief at AHU; $Q_{EX(local)}$, sum of local exhausts for zones served by AHU.[2]

The Standard must eventually address wind and stack effect and provide design guidelines that reflect conditions that influence buildings in their normal, native environment. In addition, increased humidity combined with wind- and stack-driven infiltration during periods when the ventilation system is not operating may be a significant factor influencing mold and fungal growth, e.g., offices and schools during closures. Designers and building operators should consider a limited night setback mode with provisions for humidity and pressurization flow controls. Such provisions would also tend to compensate for the building-generated contaminants by supplying a base ventilation rate, sufficient for minimal pressurization flow.

There is potential for condensation to occur under a positive pressure environment during periods of humidification in cold climates because the dew point of the air within a building could potentially be greater than the temperature of the building envelope. Maintaining a building at 'net neutral' pressure would be more appropriate under these conditions. 'Net neutral' control requires even more precise instrumentation and measurement directionality.

The widespread use of ERV in some geographic areas has decreased the amount of outdoor air available to pressurize a building and decreased the margin of error in control to maintain it. Although outdoor airflow rates into many buildings have increased with the use of ERVs, there is a strong potential for an increase in building pressurization problems which could lead to increased mold and fungal growth. Designers should exercise caution when implementing strategies that rely on ERV units for outdoor air and result in building design pressures that are close to net neutral. Wind, stack, and filter loading can easily result in depressurized buildings and increased condensation within the hidden cavities of the building envelope.[2]

Procedures

Section 6, Procedures, is the heart of the Standard. For compliance, designers must claim using either the ventilation rate procedure (VRP) or the indoor air quality procedure (IAQP) to determine the minimum dilution ventilation rate required for their design. Designers and operators cannot selectively ignore the parts they do not like. The entire procedure must apply. Parts from each procedure cannot be combined to achieve ventilation rates lower than those determined by the VRP alone. Great care should be given to the selection between these procedures.

The VRP, as defined in Section 6.1.1, "is a prescriptive procedure in which outdoor air intake rates are determined based on space type/application, occupancy level and floor area."[1] The key phrase is "prescriptive procedure in which outdoor air intake rates are determined." Very simply, this implies the need for some form of airflow measurement.

The alternative, IAQP in Section 6.1.2, "is a design procedure in which outdoor air intake rates and other system design parameters are based on an analysis of contaminant sources, contaminant concentration targets and *perceived* acceptability targets."[1] Any analysis of this procedure quickly reveals the clear discussion of airflow rate requirements based on varying contaminant levels.

Ventilation Rate Procedure

The VRP detailed in Section 6.2 is rate based. There is no question about this. In fact, the entire Standard is rate based, including the IAQP, which only provides the means to calculate allowable reductions in the design ventilation rate from those in Table 6-1 (see also Appendix D). The explicit statement to this effect was removed last year in an effort to render the language in the Standard more 'code enforceable.' Designers claiming compliance with the VRP must be able to document and substantiate that minimum intake rates are maintained during operation and "under all load conditions." They are not just 'capable of' maintaining—they are 'maintained' at no less than the higher of either code-required levels or those indicated by Table 6-1, and calculations in Section 6.2 of the Standard.

Once the outdoor air is determined to be acceptable or has been treated for use indoors, we can begin to determine how much is needed under our specific design situation. We can simplify the relationships and view the VRP with the aid of this partial flow chart from the new *62.1 User's Manual* (Fig. 1).

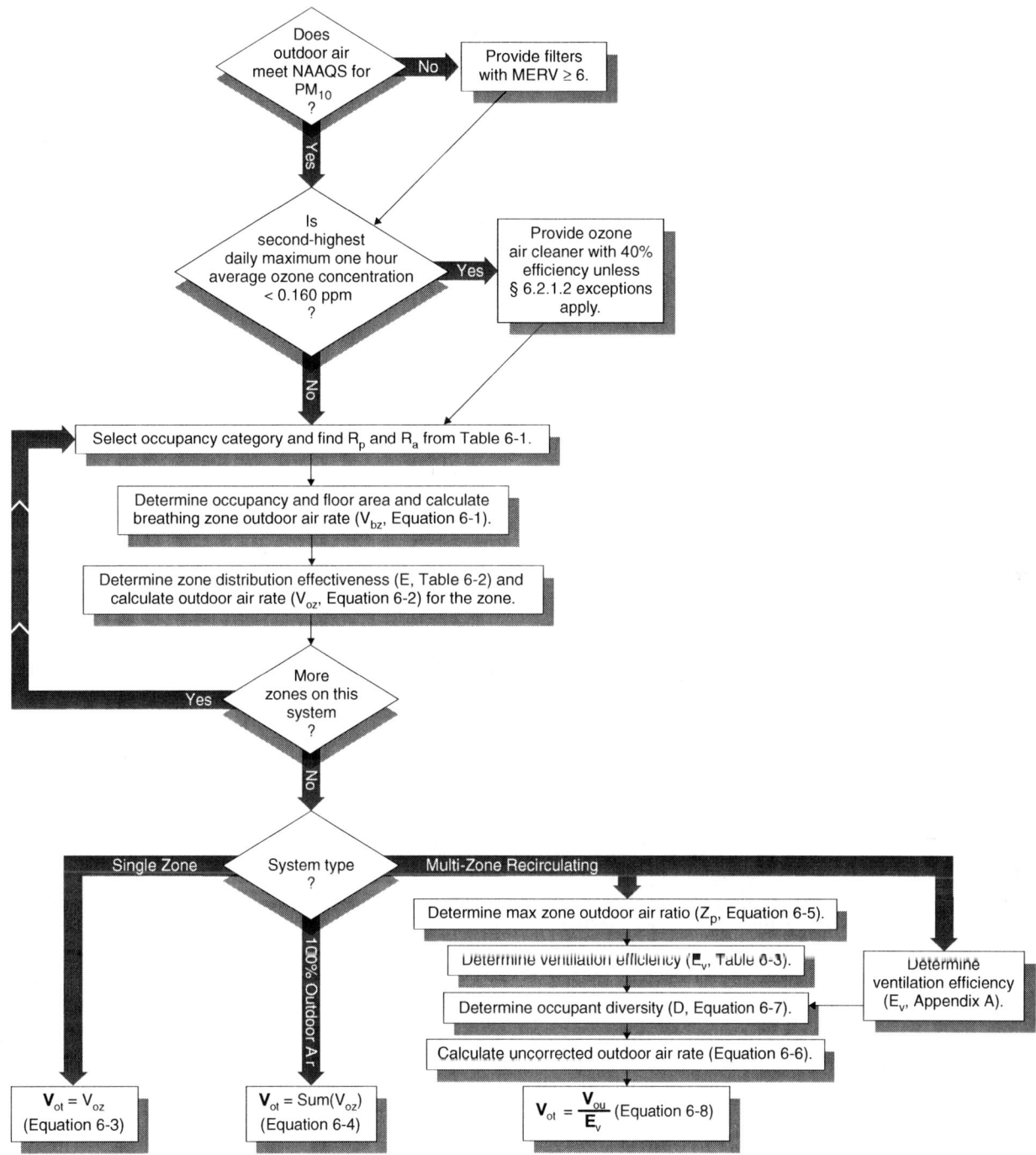

Fig. 1 Ventilation rate procedure (VRP) flow chart.
Source: From ASHRAE Standard 62.1 User's Manual (see Ref. 5).

First, we must calculate outdoor airflow requirements for the zone (V_{oz}), as detailed in Sections 6.2.2.1–6.2.2.3, which can be summarized with their corresponding equations and reference numbers below:

Calculate breathing − zone outdoor airflow:

$$V_{bz} = R_p P_z + R_a A_z \qquad (1)$$

Determine zone air distribution effectiveness:

$$E_z = \text{Table 6-2}$$

Calculate zone outdoor airflow at diffusers:

$$V_{oz} = V_{bz}/E_z \qquad (2)$$

Then, determine the outdoor airflow requirements for the system (V_{ot}) and calculate minimum outdoor air intake flow. We are given three general system types to choose from:

Single – zone systems: $\quad V_{ot} = V_{oz}$ (3)
100% OA systems: $\quad V_{ot} = \sum_{all\ zones} V_{oz}$ (4)

and

Multiple – zone recirculating systems

Outside Air Intake: $\quad V_{ot} = V_{ou}/E_v$ (8)

In Multizone recirculation systems (V_{ou} and E_v), the variables needed to solve for V_{ot} are determined in Sections 6.2.5.1–6.2.5.4 and summarized below, but will be examined in more detail later.

Calculate the Zone primary outdoor air fraction:

$$Z_p = V_{oz}/V_{pz} \quad (5)$$

Determine the Uncorrected outdoor air intake:

$$V_{ou} = D\sum_{all\ zones} R_p P_z + \sum_{all\ zones} R_a A_z \quad (6)$$

and

Accounting for Occupant Diversity:

$$D = P_s / \sum_{all\ zones} P_z \quad (7)$$

One subtle change in the updated Standard includes the characterization, usage, and definition of "breathing zone." Plus, separate components are included to address both occupant and building-generated contaminants. Rates are no longer determined solely on occupancy "per person."

Total intake rates at the air handler can be directly determined with handheld instruments used in accordance with prescribed standards or by using an appropriate and permanently installed airflow measuring device. Total intake rates may be indirectly estimated by several other means (i.e., supply/return differential calculation, temperature balance, mass balance, steady-state CO_2 concentration, etc.). However, the uncertainty of indirect techniques introduces a significant level of risk (see Ref. 5, pp. 5–12).[2,10] The designer, facility owner, and occupants should carefully consider the method employed prior to implementation of any CO_2-based demand controlled ventilation (DCV) scheme as the sole method of intake rate determination.

The new VRP in Section 6.2 recognizes the magnitude of building-generated pollutants and subsequently added the "building component" in the zone ventilation equation. Table 6-1 and the accompanying notes specify outdoor air requirements for specific applications. Eq. 1 (Table 2) is now based on the combination of ventilation rates per person (as CFM/p) PLUS ventilation rates per floor area (as CFM/ft^2). Therefore, systems that meet these requirements:

1. The minimum requirements of Table 6-1 combined with
2. The calculated volume of outdoor air required by Section 6.2 and
3. The outdoor air quality requirements set forth in Section 4
4. While "under any load condition,"

can claim that their ventilation system complies with the Standard through the VRP.

Under ideal and very specific conditions, CO_2 levels can only reflect the rate that outdoor air enters the building on a per person basis, through any and all openings. Therefore, CO_2-based DCV with single 'ppm' set point control cannot be implemented under the new requirements of Standard 62.1 unless it is applied with excessive conservatism and the accompanying increase in energy usage. Otherwise, it will invariably underventilate spaces, overventilate spaces, or require that the Standard be rewritten or interpreted in such a way to allow the potentially large airflow errors that will result from using CO_2 sensor input (Table 3).

Calculated using the concentration balance formula in ASHRAE 62.1 Appendix C, at Various Population Densities in an Office Space[2]

Table 2 Steady-state CO_2 differentials

# People	Required total OA (CFM)	CFM/person	CO_2 rise [C_i–C_o] (ppm)	Comments
7 (+17%)	95	13.5	807	Underventilated
6 (base)	90	15	700	Using 700 ppm set point
5 (−17%)	85	17	644	Overventilated
3 (−50%)	75	26	438	

Area, 1000 ft^2; Total OA CFM required, 0.06 CFM/ft^2 + 5 CFM/person.
Source: From American Society of Heating, Refrigeration and Air-Conditioning Engineers, Inc. (see Ref. 1).

Table 3 Section 6.2.2.1

> **6.2.2.1 Breathing Zone Outdoor Airflow.** The design outdoor airflow required in the breathing zone of the occupiable space or spaces in a zone, i.e., the breathing zone outdoor airflow (V_{bz}), shall be determined in accordance with Equation 6-1.
>
> $$V_{bz} = R_p P_z + R_a A_z \qquad (6\text{-}1)$$
>
> where:
> A_z = zone floor area: the net occupiable floor area of the zone m^2, (ft^2).
> P_z = zone population: the largest number of people expected to occupy the zone during typical usage. If the number of people expected to occupy the zone fluctuates, P_z may be estimated based on averaging approaches described in Section 6.2.6.2.
>   **Note**: If P_z cannot be accurately predicted during design, it shall be an estimated value based on the zone floor area and the default occupant density listed in Table 6-1.
> R_p = outdoor airflow rate required per person as determined from Table 6-1.
>   **Note**: These values are based on adapted occupants.
> R_a = outdoor airflow rate required per unit area as determined from Table 6-1.
>
> **Note**: Equation 6-1 is the means of accounting for people-related sources and area-related sources for determining the outdoor air required at the breathing zone. The use of Equation 6-1 in the context of this standard does not necessarily imply that simple addition of sources can be applied to any other aspect of indoor air quality.[1]

The new *62.1 User's Manual* is really very well done. It will improve the readers' understanding of the Standard and provide needed design guidance with its examples. However, it does make a number of assertions that will leave you scratching your head. Here is one excerpt from the appendix on CO_2:

> This appendix describes how CO_2 concentration may be used to control *the occupant component* of the ventilation rate (see Ref. 6, pp. A–1).

This is a real problem for practitioners. The document intentionally leaves the details of how to address the 'building component' of the total rate required to the imagination of the designer wanting to use CO_2. It should advise the reader of at least the most correct method to use. If it cannot, the appendix should be deleted or it should not suggest that it offers guidance to "users."

It also appears that the fundamental basis for the Standard, maintaining minimum intake rates, has been transformed by the Appendix to the *User's Manual* to become the maintenance of "an acceptable bioeffluent concentration."[6] ASHRAE has published interpretations to the Standard, insisting that sensing for a single contaminant ignores potentially high levels of all other contaminants. To do so is to ignore all other design, environmental, or operational factors that impact the actual variable to be controlled—ventilation rates. Sensing an "indicator" of proper ventilation and using it for direct control raises liability issues and ignores what is really occurring at the outdoor air intake.

Indirect measurements for control typically carry such a large degree of uncertainty that one can never be secure that the controlled variable (ventilation rates) will not drop below or substantially exceed the mandated minimums under operating conditions.

There is too much risk. Compliance with the intelligent motor control (IMC) using CO_2 is questionable. Compliance with ASHRAE 62.1 is questionable. Why use it when there are other more reliable methods available to accomplish the same function or similar results?

Multispace 'Equations' Become a Design 'Procedure'

Once the breathing zone outdoor air requirement is determined, the Standard requires an adjustment based on the distribution system's efficiency and effectiveness. This makes complete sense because the air must reach the breathing zone to be effective.

Multizone recirculation systems are not as efficient as 100% OA systems and are therefore required to be factored by their approximate and relative inefficiency. We are given two methods to determine this factor:

1. Table 6-3, "default E_v" method
2. Appendix A, "calculated E_v" method

These methods produce significantly different results. The more precise method is contained in Appendix A, and as might be expected, it is more involved. The Table's conciseness requires it to be more conservative and therefore not as efficient in many situations.

For each multiple zone recirculating system (VAV or CAV), the primary outdoor airflow fraction must be calculated for all zones that may become 'critical' (only one zone can be critical on CAV systems). The "critical zone" is defined as the zone that has the highest percentage of outdoor air required in the primary air stream. When analyzing a VAV system dynamically, treat it as a CAV system.

As an example, if the supply air distribution system is located close to the return air, a short circuit is generally created. The Standard requires designers to use a zone air distribution effectiveness (E_v) of 0.5, which essentially doubles the amount of outdoor air required. In contrast to this example, a system with a ceiling supply and a ceiling return has a zone distribution effectiveness of 1.0 during cooling and 0.8 during heating. Therefore, the outdoor air setpoint must be reset seasonally or the more conservative factor used.

Systems that provide a variable supply of air volume to the conditioned space are influenced by everything previously discussed. In addition, outdoor airflow rates will vary as a result of changes in mixed air plenum pressure. If the design did not assume the worst-case scenario when the outdoor airflow rate for the air handler was determined, outdoor airflow rates on VAV systems may need to be reset based on calculations of the multispace equations (Eqs. 6-5 to 6-8, defined in Table 4), in order to avoid potentially excessive over ventilation and the associated energy penalty.

Advanced VAV control strategies can satisfy the requirements of Sections 6.2.5.1–6.2.5.4 dynamically and therefore more efficiently than static strategies. This can be accomplished by automatically determining the critical zone fraction to continuously calculate the corrected fraction of outdoor air. The calculation requires that the total supply airflow rate be continuously measured and that the airflow rate of the critical zones is measured with permanent airflow measuring devices capable of accurate measurement.

Airflow sensors provided with VAV boxes should not be used for these calculations. Although the original equipment manufacturer (OEM) devices may be adequate in modulating a terminal box for thermal comfort, the combination of typically poor inlet conditions, low quality airflow pickups, and low cost pressure sensors in the direct digital control (DDC) controller will not result in the measurement accuracy necessary for proper calculation of Eqs. 6-1 and 6-5 to 6-8. Conservative mathematical modeling has demonstrated that typical VAV box measurement performance

Table 4 Section 6.2.5

6.2.5 **Multiple-Zone Recirculating Systems.** When one air handler supplies a mixture of outdoor air and recirculated return air to more than one zone, the outdoor air intake flow (V_{ot}) shall be determined in accordance with section 6.2.5.1 through 6.2.5.4 [Equations 6-5 through 6-8].

6.2.5.1 **Primary Outdoor Air Fraction.** When Table 6-3 is used to determine system ventilation efficiency, the zone primary outdoor air fraction (Z_p) shall be determined in accordance with Equation 6-5.

$$Z_p = V_{oz} / V_{pz} \qquad (6\text{-}5)$$

where V_{pz} is the zone primary airflow, i.e., the primary airflow to the zone from the air handler including outdoor air and recirculated return air.
Note: For VAV systems, V_{pz} is the minimum expected primary airflow for design purposes.

6.2.5.2 **System Ventilation Efficiency.** The system ventilation efficiency (E_v) shall be determined using Table 6-3 or Appendix A.

6.2.5.3 **Uncorrected Outdoor Air Intake.** The design uncorrected outdoor air intake (V_{ou}) shall be determined in accordance with Equation 6-6.

$$V_{ou} = D \text{ all zones } R_p P_z + \text{ all zones } R_a A_z \qquad (6\text{-}6)$$

The occupant diversity, D, may be used to account for variations in occupancy within the zones served by the system. The occupancy diversity is defined as:

$$D = P_s / \text{all zones } P_z \qquad (6\text{-}7)$$

where the system population (P_s) is the total population in the area served by the system. Alternative methods may be used to account for population diversity when calculating V_{ou}, provided that the resulting value is no less than that determined by Equation 6-6.
Note: The uncorrected outdoor air intake (V_{ou}) is adjusted for diversity but uncorrected for ventilation efficiency.

6.2.5.4 **Outdoor Air Intake.** The design outdoor air intake flow (V_{ot}) shall be determined in accordance with Equation 6-8.

$$V_{ot} = V_{ou} / E_v \qquad (6\text{-}8)\text{"} \; [1]$$

can be statistically exceeded by boxes without a measurement device.[2] Accurate airflow measuring devices having a total installed accuracy better than 5% of reading at maximum system turndown should be installed in the supply ducts for critical zones.

Increased precision allows terminal box selection for optimum energy performance as well as improved sound performance. Accordingly, a recently proposed ASHRAE TC1.4 Research Work Statement claims that using VAV box sensors and controllers allowing "a 20% minimum airflow setpoint can save $0.30/ft^2-yr. [in energy costs]. Multiplied across the millions [billions?] of square feet of commercial space served by VAV boxes, the potential economic, health and productivity benefits are significant."

Consider, for example, 1.5 million ft^2 of occupied area in a single 50+ story Manhattan high-rise office building. The savings potential equates to $450,000 per year in energy for one building. The 1999 Commercial Buildings Energy Consumption Survey (CBECS) showed 12 of 67.3 billion total ft^2 of commercial space as offices (2003 Report by Iowa Energy Center, 1999 CBECS, http://www.eia.doe.gov/emeu/efficiency/cbecstrends/pdf/cbecs_trends_5a.pdf). If we assume that 28% ((using 70% of 40% VAV offices) http://www.buildingcontrols.org/download/untracked/NBCIP_PNWD_3247%20F.pdf accessed July 2005 of VAV office buildings are capable of being upgraded, the United States would be looking at a conservative (50% of $0.30/ft^2) potential annual energy savings of $250 million per year. With total commercial space projected by department of energy (DOE) to increase 46% to 105 billion ft^2 by 2025 and all future VAV office designs capable of saving $0.30/ft^2, yes, the potential is understatedly "significant."

The results from these multispace equations can provide wide variations in outdoor airflow requirements in some systems. Increasing the critical zone supply flow while using reheat can reduce total outdoor airflow rates and overall energy usage. This method has been simulated using the multispace equation from Standard 62-2001 at Penn State University, with published results showing greater energy efficiency than the same system supplying the maximum, worst-case V_{ot} continuously. The basic variables, relationships, and end results should be the same using the VRP of Standard 62.1.

Then, the VRP continues and provides us with additional options to help make the design more specific to the designer's needs and to the demands of the situation. You may...

- Design using the short-term "averaged" population rather than the peak—Table 5 (9)
- Operate (and dynamically reset requirements) using "current" population data—Table 6, "DCV"

The peak population value may be used as the design value for P_z. Alternatively, time-averaged population determined as described in Section 6.2.6.2 may be used to determine P_z.

Outdoor airflow rates can also be reduced if the critical zones have variable occupancy or other unpredictably variable (dynamic) conditions. Changes in occupancy (or ventilation 'demand') can be detected in many ways, as indicated in the 'note' below. Therefore, DCV systems functioning to dynamically adjust the outdoor air intake setpoint should not be limited to CO_2 measurement input alone.

Section 6.2.7 (Table 6) on Dynamic Reset addresses conditions when the ventilation control system...

Table 5 Section 6.2.6

> **6.2.6 Design for Varying Operating Conditions.**
> **6.2.6.1 Variable Load Conditions.** Ventilation systems shall be designed to be capable of providing the required ventilation rates in the breathing zone whenever the zones served by the system are occupied, including all full- and part-load conditions.[1]

Table 6 Section 6.2.7

> "...may be designed to reset the design *outdoor air intake flow* (V_{ot}) and/or space or zone airflow as operating conditions change. These conditions include but are not limited to:
> 1. <u>Variations in occupancy</u> or ventilation airflow in one or more individual zones for which ventilation airflow requirements will be reset.
>   **Note**: Examples of measures for estimating such variations include: occupancy scheduled by time-of-day, a direct count of occupants, or an estimate of occupancy or ventilation rate per person using occupancy sensors such as those based on indoor CO_2 concentrations.
> 2. <u>Variations in the efficiency</u> with which outdoor air is distributed to the occupants under different ventilation system airflows and temperatures.
> 3. A <u>higher fraction of outdoor air in the air supply</u> due to intake of additional outdoor air for free cooling or exhaust air makeup.[1]

There is no indication in the Standard of how to implement Dynamic Reset with CO_2, which was left to be addressed by the *User's Manual*. The *User's Manual* included an appendix intended to address the continuing questions by users and designers: "How can we apply CO_2 measurements to DCV and comply with the VRP of Standard 62.1?" To this end, the appendix hedged sufficiently to avoid answering this question directly, underscoring the potential problems in applying DCV and simultaneously employing outdoor airflow measurement.

This section (Section 6.2.7), independent of the *User's Manual*, would lead one to believe that CO_2 is a method of "counting" and not an input to be used for direct ventilation control. Any counting method can be used to reset a flow rate established and controlled by some other measurement means. Intake rates should not be indirectly determined by the "counting" method.

Indoor Air Quality Procedure

Section 6.3 Indoor Air Quality Procedure begins...

> The Indoor Air Quality Procedure is a performance-based design approach in which the building and its ventilation system are designed to maintain the concentrations of specific contaminants at or below certain limits identified during the building design and to achieve the design target level of *perceived* IAQ *acceptability* by building occupants and/or visitors....[1]

The concept of providing "performance-based" solutions is desirable in principle. However, there are numerous risks associated with both the quantitative and subjective evaluations provided within the IAQ procedure that every designer should understand.

Because there are numerous contaminants that either will not be detected or for which "definite limits have not been set," this portion of the procedure has significant risks associated with it. It is unlikely that all contaminants of concern will be evaluated or reduced to acceptable levels. It is also not practical to measure all potential contaminants, and in some cases, such as with fungus or mold, measurement may not be possible.

Table 7 combined with Table 8 (b) and (c) emphasize the risk associated with the Indoor Air Quality procedure. The uncertainty of using "subjective occupant evaluations" together with the admittedly limited listings in Appendix B-SUMMARY OF SELECTED AIR QUALITY GUIDELINES, may be too great for many designers to 'claim' this procedure for compliance.

However, the concept of controlling the source of the contaminants makes perfect sense and is, in our opinion, more properly utilized under design approach Table 8 (d).

Because airflow rates are typically reduced in the IAQP, the measurement and control of intake rates are even more critical, especially on systems where the thermal load change is independent of the occupants and their activities. In addition, caution should be exercised when reducing outdoor airflow rates because it is also required to maintain proper building pressure, helps to minimize energy use, improves comfort control, and prevents mold growth within wall cavities.

Design Documentation Procedures

Section 6.4, Design Documentation Procedures, states:

> Design criteria and assumptions shall be documented and should be made available for operation of the system

Table 7 Section 6.3.1.3

6.3.1.3 **Perceived Indoor Air Quality.** The criteria to achieve the design level of acceptability shall be specified in terms of the percentage of building occupants and/or visitors expressing satisfaction with perceived indoor air quality."

Table 8 Section 6.3.1.4

6.3.1.4 **Design Approaches.** Select one or a combination of the following design approaches to determine minimum space and system outdoor airflow rates and all other design parameters deemed relevant (e.g., air cleaning efficiencies and supply airflow rates).
 (a) Mass balance analysis. The steady-state equations in Appendix D, which describe the impact of air cleaning on outdoor air and recirculation rates, may be used as part of a mass balance analysis for ventilation systems serving a single space.
 (b) Design approaches that have proved successful in similar buildings...
 (c) Approaches validated by contaminant monitoring and subjective occupant evaluations in the completed building. An acceptable approach to subjective evaluation is presented in Appendix B, which may be used to validate the acceptability of perceived air quality in the completed building.[1]
 (d) Application of one of the preceding design approaches (a,b, or c) to specific contaminants and the Ventilation Rate Procedure would be used to determine the design ventilation rate of the space and the IAQ Procedure would be used to address the control of the specific contaminants through air cleaning or some other means.[1]

within a reasonable time after installation. See Sections 4.3, 5.2.3, 5.17.4 and 6.3.2 regarding *assumptions that should be detailed* in the documentation.[1]

Within Section 5.2.3, Ventilation Air Distribution requires us to:

...specify minimum requirements for air balance testing or reference applicable national standards for measurement and balancing airflow.[1]

Providing permanently installed instruments and controls that result in and verify compliance with ASHRAE Standard 62.1 is perhaps one of the best reasons to provide such devices as part of any HVAC system design. Continuous data inputs may also be used to aid start-up, test and balance, commissioning, and measurement and verification (M&V) for energy usage and ongoing diagnostics. More precise and more reliable control could be viewed as a bonus.

Construction and System Start-Up

Section 7 addresses the construction and start-up phases of the project and has been included because a significant number of documented IAQ cases were a result of activities which took place during these phases of the project. The construction phase, addressed in Section 7.1 of the Standard, applies to "ventilation systems and the spaces they serve in new buildings and additions to or alterations in existing buildings."[1] The Standard addresses both the protection of materials and protection of occupied areas.

Mechanical barriers are specified to protect occupied areas from construction generated contaminants. In addition, the HVAC system must be able to maintain occupied spaces at positive pressures with respect to the construction areas. In many cases, the HVAC system does not have the adequate capacity and controls to provide a barrier to the migration of contaminants using positive pressurization flow. Designers must consider the condition of the existing ventilation system and its ability to maintain a pressurized environment for spaces expected to continue occupancy, prior to initiating physical construction activities at the site.

The start-up phase, covered in Section 7.2, provides guidelines for air balancing, testing of drain pans, ventilation system start-up, testing of damper controls, and documentation requirements.

Section 7.2.2, Air Balancing, requires that systems be balanced, "at least to the extent necessary to verify conformance with the total outdoor air flow and space supply air flow requirements of this standard."[1] Unfortunately, the airflow rates of the system will vary after this activity has occurred, in most systems, for reasons discussed in the analysis of Section 5, Systems and Equipment. When applied in accordance with the manufacturer's recommendations, some airflow measuring devices only require the verification of operation by test and balance professionals. This TAB "snapshot" of airflow rates is analogous to providing a one-time setup for temperature control, which obviously would not be very effective. Providing permanently mounted airflow measuring stations would also support compliance with and reduce the time required to supply the documentary requirements for ventilation set forth in Section 7.2.6 (c).

Operations and Maintenance

All systems constructed or renovated after the date the Standard's 2001 version parent document was originally adopted are required to be operated and maintained in accordance with the provisions set forth in these newer sections of the Standard. It is important to recognize that if the building is altered or if its use is changed, the ventilation system must be reevaluated. Buildings that are likely to be changed or altered during their life spans should consider including a robust HVAC system design that takes into account changes in airflow rate requirements imposed by this Standard. Of course, provisions for permanently mounted airflow measurement devices and controls would significantly reduce both the cost and time associated with such changes as long as the HVAC load capacity could accommodate future requirements.

Section 8.4.1.7 addresses sensors. "Sensors whose primary function is dynamic minimum outdoor air control, such as flow stations..."[1] is discussed in this section even though they were not mentioned under Section 5, Systems and Equipment. Section 8.4.1.7 requires that sensors have their accuracy verified "once every six months or periodically in accordance with the Operations and Maintenance Manual."[1] The Operations and Maintenance Manual for some airflow measuring devices does not recommend periodic recalibration. Permanently calibrated airflow instrumentation has a significant advantage over other airflow measuring technologies and CO_2 sensors, whose transmitters are subject to frequent adjustments, zeroing, or regular calibrations to correct for analog electronic circuitry and sensor drift.

However, Section 8.4.1.8 (Outdoor Air Flow Verification) only requires the verification of airflow rates "once every five years."[1] Because external and system factors change continuously, clearly influencing outdoor airflow rates, this requirement does little to assure that proper ventilation rates are maintained under normal operation at different times of the day and the year. It effectively places the burden of verification of new building/system performance on the building operator, who is often not in a position to make such a determination.

This apparent contradiction with Section 8.4.1.7 will likely be examined by the ASHRAE SSPC62.1 committee in the near future. Permanent outdoor airflow measuring stations would provide continuous verification and the necessary control inputs to maintain ventilation requirements, automatically minimizing intake rates for energy usage and preventing other control inputs from causing a maximum intake limit from being exceeded.

CONCLUSIONS

American Society of Heating and Refrigeration and Air-Conditioning Engineers Standard 62.1 prescribes ventilation rates for acceptable IAQ. It should be clear to the building operator and the design professional that the dynamic nature of mechanical ventilation requires dynamic control to insure the continuous maintenance of specific predetermined conditions. As a rate-based standard, continuous airflow measurement should logically be a central component of any effective control strategy to assure acceptable IAQ and minimize the costs of energy to provide it.

REFERENCES

1. ASHRAE Standard 62.1-2004. *Ventilation for Acceptable Indoor Air Quality in Commercial, Institutional, Industrial and High Rise Residential Buildings*; American Society of Heating, Refrigeration and Air-Conditioning Engineers, Inc.: Atlanta, GA, 2005.
2. Dougan, D.S. IAQ by design and airflow measurement for acceptable indoor air quality. *Proceedings from IAQ Seminars*; EBTRON, Inc.: Loris, SC, 1999–2005.
3. Solberg, D.P.W.; Dougan, D.S.; Damiano, L.A. Measurement for the control of fresh air intake. ASHRAE J. **1990**, *32* (1), 46–51.
4. CEN (European Committee for Standardisation) Design criteria and the indoor environment. In *The Technical Report CR1752*; TC156 Ventilation for Buildings: Brussels, 1998–2004.
5. ASHRAE. 62.1 *User's Manual*; American Society of Heating, Refrigeration and Air-Conditioning Engineers, Inc.: Atlanta, GA, 2005; Dec 2005.
6. Bailey, R. Personal Correspondence. Chair ASHRAE TC1.12 Moisture Management in Buildings, Bailey Engineering: Jupiter, FL, April 13, 2005.
7. ASHRAE Board of Directors. Minimizing Indoor Mold Through Management of Moisture in Building Systems. Position Paper, American Society of Heating, Refrigeration and Air-Conditioning Engineers, Inc.: Atlanta, GA, June 2005.
8. Krus, M.; Sedlbauer, K.; Zillig, W.; Křnzel, H.M. *A New Model for Mould Prediction and Its Application on a Test Roof*; Fraunhofer Institute for Building Physics, Branch Institute Holzkirchen: Munchen, 1999; (www.hoki.ibp.fhg.de/ibp/publikationen/konferenzbeitraege/pub1_39.pdf accessed March 2005).
9. Brennan, T.; Cummings, J.; Lstiburek, J. Unplanned airflows and moisture problems. ASHRAE J. **2002**, *44* (11), 48.
10. Krarti, M.; Brandemuehi, M.J.; Schroeder, C.; Jeannet, E. *Techniques for measuring and controlling outside air intake rates in variable air volume systems*, Final Report on ASHRAE RP-980; University of Colorado: Boulder, CO, 1999.

Auditing: Facility Energy Use

Warren M. Heffington
Industrial Assessment Center, Department of Mechanical Engineering, Texas A&M University, College Station, Texas, U.S.A.

Abstract

This entry discusses the energy audit and assessment processes, including the analysis of utility data, walk-through assessments, detailed assessments, and the reporting of results. This article will also provide a brief history of energy auditing, as well as a look toward the future.

INTRODUCTION

Energy assessments are associated with energy conservation programs and energy efficiency improvements. However, cost containment (usually associated with utility cost savings) is always the most sought after result. Changes that reduce operating expenses sometimes save little energy (e.g., correcting a power factor) or may increase energy use (e.g., replacing purchased fuel with opportunity fuel). Often known as energy audits, an older term that is being replaced by "energy assessments," these analyses of facility or system performance are aimed at improving performance and containing operating costs. Facilities that may be audited for energy use can be commercial or institutional buildings, industrial plants, and residences; examples of the systems that may be audited include air conditioning, pumping, and industrial production lines. The end of this article contains important references on the subject of energy assessment.[1,2]

Energy assessments today owe much to the National Energy Conservation Policy Act of 1978. This act resulted in the Institutional Conservation Program, a nationwide program of the U.S. Department of Energy (DOE) which called for matching grant funding to implement energy efficiency retrofits in secondary schools and hospitals. Directed by state energy offices, energy assessments predicted savings from these retrofits. A corresponding national DOE program is the Industrial Assessment Center (IAC) program,[3] which provides assessments of manufacturing plants and also educates university students about energy conservation. Sometimes pollution prevention, waste reduction, and productivity improvement projects are included with the energy assessments because the techniques for analyzing such projects are similar. Other DOE-sponsored assessment programs include the Federal Energy Management Program[4] for federal facilities and plant-wide assessments for private industries have also led to informative case studies and energy-efficient technology transfer.

The following section discusses the process of conducting an energy assessment, including pre-assessment activities, the actual assessment, and reporting of findings.

ASSESSMENTS

The energy assessment process may be conducted by a single user of a relatively unsophisticated software package that focuses on a few simple energy issues or a team of professional engineers who analyze problems using sophisticated software and complex calculations. The individual energy assessor might conduct assessments of residences as a city or utility service, and provide simple printouts of recommended changes for homeowners with some savings data and perhaps cost estimates. The more expensive and detailed assessments by teams of engineers may be reserved for large buildings and industrial complexes. In those cases, the feedback is often in the form of a formal technical report containing conceptual designs that allow capital decisions to be evaluated and made about proceeding with further analysis and planning.

The main objective of all assessments is to identify, quantify, and report on cost containment projects that can be implemented in a facility or system. These projects go by several names: energy conservation opportunity (ECO), energy conservation measure (ECM), energy cost reduction opportunity (ECRO), assessment recommendation (AR), and others. These projects are the heart of any energy analysis.

Assessments are characterized by the level of effort required and by the capital necessary to accomplish the recommended measures. The simplest type is a walk-through or scoping assessment and may result in no more than a list of possible projects for consideration. More complex, detailed assessments result in a formal technical report showing calculations of savings and

Keywords: Energy audits; Energy assessments; Energy conservation; Demand-side management; Energy audit reporting; Energy assessment reporting.

implementation costs. Walkthroughs and detailed assessments may be subdivided into projects requiring little capital and capital intensive projects. These low-capital projects are sometimes tasks that employees such as maintenance personnel should be accomplishing as part of their regular duties, sometimes known as maintenance and operation (M&O) opportunities,. ASHRAE designates the assessments and projects by levels: Level I—Walkthrough Assessment, Level II—Energy Survey and Analysis, and Level III—Detailed Analysis of Capital Intensive Modifications.[5] The most common measure of financial merit is simple payback although life cycle costing may be used and is required by some federal programs under the Federal Energy Management Program.[6]

ASSESSMENT PROCEDURE

Pre-Assessment Activities

Pre-assessment visit activities allow the assessment team to learn about the facility prior to an actual visit and should include obtaining the facility's energy consumption history. Often important clues about cost-containment projects come from historical information available from the facility or system operators' records for all significant energy streams. All assessments should be preceded by a review of at least 12 consecutive months of bills. In the case of energy, both energy in common units such as kilowatt hour (kWh), thousand cubic feet (MCF), or million British thermal unit (Btu) and cost in dollars should be graphed so that fluctuations in usages and costs are visible. The American National Standards Institute provides examples of these graphs.[7] Electrical demand in kilowatt (kW) or kilovolt ampere (kVA) should be plotted also, as should load factors and energy use, and energy cost indices in the case of buildings. Cost savings measures that can be identified from this review may result from inappropriate tariffs and tax charges, for example. This information gives an indication of possible projects to pursue during future phases of the assessment.

If pollution and waste are to be considered, such things as annual waste summaries and pollution information, as required by government regulators under the Clean Air Act or other legislation, should be reviewed. Water bills contain information about sewer charges as well as water consumption. Obtaining the tariff schedules for the various charges are important and all monthly charges should be recomputed to assure there are no mistakes. Tariffs in general are not well understood by users, who may not have the time or personnel to review the charges for errors. There are two important results of recalculation. First, the reviewer can reassure the client that the charges are correct (this can be comforting to most clients who have no good idea about the correctness of charges) and the reviewer can identify the rare error.

Second, facility energy bills and associated tariffs provide information so that the assessor can develop the avoided cost of energy (and demand) applicable to each billing account. The avoided cost usually is not evident directly in the tariff and typically the units for avoided costs are the same as the units of pricing. For example, the avoided cost of electrical energy measured in $/kWh may contain the block cost, fuel adjustment, and perhaps other charges. Sometimes the avoided cost is developed by dividing cost for a period by the consumption in the same period. This provides useful results for certain analyses, but such things as demand and fixed periodic charges may cause this approach to yield significant errors when calculating cost savings.

Simple facility layout maps should be provided to show building floorplans, define building areas, and illustrate major equipment locations. Additional layouts of convenient size (e.g., $8\frac{1}{2}'' \times 11''$) are useful for notations by assessment team members during both the walkthrough and the detailed assessment visit.

Walkthrough Assessments

Walkthrough assessments usually have three major outcomes: the identification of potential energy conservation projects; the identification of the effort, including skills, personnel and equipment required for a detailed assessment; and an estimation of the cost of a detailed assessment. The need for data recording may be identified at this time and arrangements to install sensors and loggers can begin. Self-contained, easily-installed loggers may be installed at this time (the walkthrough visit may be a useful time to place logging equipment and other requests for data if a detailed visit will follow at a reasonable time for retrieving the loggers and data).

The walkthrough visit also provides an opportunity to obtain plant information that will be valuable in conducting the detailed assessment. Such data includes building and industrial process line locations and functions, and the locations of major equipment, mechanical rooms, or electrical control centers, meters, and transformers. Major energy users and problems often can be identified during a walkthrough visit. A manager or senior employee familiar with processes, systems, and energy consumption should accompany the energy expert during the walkthrough assessment.

On occasion, the walkthrough assessment may reveal that further detailed assessment work would not be particularly fruitful, and thus it may end the assessment process. If a detailed assessment is to follow, the result of the walkthrough should be a letter-type report with a general identification of potential projects and an effort or cost estimate for the detailed assessment. Generally, there will be no calculation of savings in this initial report. The walkthrough (in addition to its preceding discussions) provides an opportunity to obtain information about any previous energy assessments. Plant personnel should also be asked about any in-house studies to save energy, reduce waste, and increase productivity that may have been undertaken.

Detailed Assessments

The goal of detailed assessments is to gather accurate technical data that will allow the assessment team to prepare a formal, technical report describing projects that can be implemented in the facility to contain costs. The effectiveness of cost saving projects generally depends on calculation of reductions in energy use, demand, and, if waste and productivity issues are also considered, then the cost effects of reducing waste or pollution and increasing productivity should also be calculated. Such calculations require carefully obtained, accurate data. The detailed assessment often involves a team of engineers and technicians, accompanied, when necessary, by facility personnel, compared to the walkthrough assessment which is usually performed by fewer persons.

Even for one-day visits such as those exemplified by DOE's IAC program, the visit can be divided into a tour of the plant equivalent to a walkthrough, and a period of detailed data gathering, equivalent to a detailed assessment, as personnel revisit key areas of the plant to gather data and observe processes.

At this time, data necessary to compose a good description of the facility should be collected if the preliminary activities and walkthrough have not satisfied this need. A good assessment report contains a facility description, complete with facility layout showing the location of major areas and equipment. The projects described in the report should refer to the applicable facility description section to maintain continuity and perspective.

The major data obtained, however, pertains to the cost containment projects to be described in the detailed assessment report. Energy projects fall into four major areas, the first three technical and the final one administrative in nature:

- Turning off equipment that is used or idling unnecessarily
- Replacing equipment with more efficient varieties
- Modifying equipment to operate more efficiently
- Administrative projects dealing with energy procurement

Often data for the administrative projects is obtained from contracts, billing histories, and interviews with managers and energy suppliers. Data for the administrative projects are relatively easier to obtain than that for the technical projects, and may in fact be obtained as part of preassessment or walkthrough activities.

Technical data for equipment turn-off, replacement, or modification generally is more difficult to obtain, and mostly will be obtained during the detailed assessment phase. It involves gathering equipment size data such as power or other capacity ratings, efficiencies, load factors (which may be a function of time), and operating hours. Size data is relatively easy to obtain from nameplate or other manufacturer specification information. Estimates of equipment sizes and efficiencies often are unreliable. Every major piece of equipment and system should be considered for possible savings, and equipment should be inspected for condition and to determine if it is operating properly. Efficiencies, while often not directly obtainable from data available in the facility, may be obtained from the manufacturer. Efficiency data as a function of load is desirable, because efficiency generally varies with load.

The most difficult and unreliable data to collect often are load factors and operating hours. Equipment loading is often estimated because measurements are expensive and difficult to make. Installing measurement equipment may require shutdown of the equipment being monitored, which may present production or operating problems. If the load varies with time, then measurements should be made for a long enough period to cover all possible loadings. If the load variation is predictable, covering one period is sufficient. If the variation is unpredictable, then the measurement only gives data useful to show what a turn-off, equipment change, or modification would have saved for that period. For equipment with a constant load, a one-time measurement will be adequate, but even a one-time measurement may be difficult to obtain. Because of the difficulty measuring and unpredictability of loading, load factors are often estimated. Savings are proportional to these load factors, and thus erroneous estimates of load factors can lead to large errors in savings estimates. Whenever possible, load factors should be based on measured data.

Operating hours can also be unreliable because they are often obtained from interviews with operators or supervisors, who may not understand exactly what is being asked, or who simply may not have accurate information. Time of operation errors are difficult to eliminate. Gross errors such as those that may be made about equipment turn-off after normal operating hours can sometimes be eliminated by an after-hours visit by the assessment team. Measurements of annual equipment operating hours are useful because project savings usually are measured on an annual basis. Only rarely is this possible (for example, a project subject to measurement and verification for performance contracting purposes might yield a year of measured data, but such projects are relatively rare). Relatively inexpensive measurements can be made of operating hours for short periods by self-contained loggers that measure equipment on/off times. The operating schedule for each building or production area is needed, including information about breaks in operation such as lunch and information about holidays.

Technical and physical data useful in identifying projects, calculating savings, designing conceptual projects or management procedures to capture the savings, estimating implementation costs, and composing the facility description may be obtained in any phase of the assessment process. Some information will have been obtained prior to the walkthrough assessment visit, but during the walkthrough, additional information can be obtained about operation, function, production, and building information. Function or production information for the various areas of the facility is needed. For all types of facilities, general construction information about the buildings should be gathered, including wall and roof types; wall and roof heights and lengths; and lighting and HVAC system types, numbers of units, controls, and operating hours. An inventory of equipment should also be assembled.

ASSESSMENT REPORTING

All energy assessment reports have a similar format. There should be an executive summary, a recommendation section, a facility description, and an energy consumption analysis, though not necessarily in that order. The recommendations are the heart of the report. Although they often appear after other major sections in older reports, modern usage has them appearing earlier, after the executive summary. The executive summary generally appears first, but is written last to summarize the other sections. Other parts such as disclaimers, acknowledgements, and appendices may be used as necessary.

Recommended Projects

Assessment recommendations are the most important part of the report and contain the technical analysis of the facility's energy usage. These recommendations should be clear, with the source of data documented and the analysis technique described with sample calculations. The details of manual calculations should be given and subject to independent verification. Analyses whose most important calculation is multiplying some portion of the utility bill by a "rule of thumb" percentage to establish savings for a project are not suitable.

Just as for energy analysis reports, individual assessment recommendations usually follow a fairly uniform outline. There is a short description followed by a summary of energy and cost savings, and of the implementation cost and financial measure of merit (often simple payback). Akin to the facility description section of an assessment report, there is an observation section which describes the existing situation, problems, and recommended changes, including designs or management techniques proposed to capture the savings. Then there is a calculation section which shows the results in energy, demand, and cost savings, and an implementation cost section, which develops the implementation cost and the financial measure of merit, either payback or some more sophisticated method.

If computer analysis is used to calculate savings, the program should be a recognized program applied by a trained user. Inputs and outputs should be carefully considered for accuracy and feasibility. If locally developed spreadsheets are used, then a sample calculation should be given in the body of the report for each important result. Avoid "black-box" analyses where little or nothing is known about the algorithm being applied in user-developed computer programs or spreadsheets that are not widely recognized.

For some assessment recommendations (e.g., a simple turn-off recommendation for equipment operating unnecessarily after hours), implementation can follow without further study. However, for more complex projects (e.g., replacing an air-blown material mover with a mechanical conveyor), additional design and engineering may be required before implementation. The design level of energy assessment reports is conceptual, not detailed, in nature. The cost analysis in the implementation section for such conceptual designs should give management sufficient information to decide whether to go forward with the project. Also, if additional design and engineering are required, then an investment in further study will be needed before a final decision.

When the result of an assessment recommendation is dependent upon the accomplishment of a separate recommendation, then the recommendation should be calculated in both the independent and dependent mode. For example, if an assessment recommendation reduces natural gas consumption in a facility, saving energy and money, and if in the same analysis a recommendation is made to change suppliers to achieve a less expensive cost of natural gas energy, then the project to change suppliers depends on the accomplishment of the project to reduce gas consumption; if it is not accomplished, the savings due to the supplier change will be greater. The effect of the project to change suppliers should be calculated both as though the consumption reduction would not occur (the independent case) and as if the consumption reduction would occur. In the latter, dependent case, the effect of changing suppliers is affected by the reduction in consumption. In this case, savings from the project to reduce consumption also depends on making the supplier change, and that project can be calculated in both the dependent and independent modes. To avoid a complicated mix of projects and calculations, in some cases where a single large project is composed of several smaller, related projects, a hierarchy of projects and dependencies can be established to guide the dependency calculations. The $100 million LoanSTAR program that placed energy-efficient retrofits in public buildings in the state of Texas pioneered this approach.[8]

Each project should refer to the plant description for the location of systems and for information about equipment; similarly, it should refer to the energy consumption section when that area is important to the recommendation.

Assessment recommendations are an acceptable place to show the environmental effect of each recommendation by calculating the carbon equivalent, or NO_x, reduction due to reducing energy use. Emission factors are available for various areas of the country that can be used for these calculations.

Implementation information, including warnings about possible personnel and equipment safety issues should be given in each assessment recommendation. Safety issues should receive prominent, obvious display.

Plant Description

Plant description information is obtained during all phases of the assessment process. This part of the report may be divided into a facility description and a process description, and it provides context for the rest of the report. The facility description describes the buildings, their construction, major production areas, and equipment locations. Transformers and meter locations for energy streams should be shown. A building layout should be included to show major production areas and equipment locations, particularly those important to the energy analysis being described in the report.

The process description should describe each major process that goes on in the plant, starting with the procurement of raw materials giving sources and delivery methods—and proceeding until products are packaged, perhaps warehoused, and shipped. Emphasis should be placed on points in the process where major amounts of energy are consumed and on the machines that are involved in that energy usage. It is important to include a process flow diagram for each process.

At some point, either in the facility or process description, the major energy consuming equipment in the plant should be listed. If waste is to be considered, then points in the process where waste is generated are especially significant, and in that case a table of waste streams, handling equipment, and storage locations should be included.

Energy Consumption Data

Energy analysis data can be obtained for composing this important section of the report before the plant is visited. For an assessment that is broad in scope, the manufacturer should be asked to provide copies of a minimum of 12 consecutive months of energy bills for analysis. Sometimes even more data is desirable for a detailed audit. Shorter periods will not cover a complete annual weather cycle. Industrial analyses, unlike analyses of commercial and institutional buildings, usually are independent of weather. However, some industrial projects are weather dependent. Particularly in the case of industrial buildings which are fully climate controlled, weather may be an important factor, and thus bills representing the full 12 months of usage are an extremely helpful amount to seek. They do not need to be for an annual or particular fiscal period from the standpoint of technical analysis, but they should be consecutive and as recent as possible. Copies of actual bills should be sought because summaries likely will not give information such as the applicable tariff, any taxes, late fees, and demand and power factor data. For a broad-based assessment, bills for all important energy streams should be obtained. For special emphasis on equipment or areas of production, only bills for relevant energy streams should be considered.

From the bills, relevant tables and graphs of energy consumption, demand, power factor, and costs should be prepared for each account or meter. Careful review of this data often reveals important data about such things as demand control possibilities, equipment usage, and production changes. It provides a starting point for in-plant discussions with management about cost reduction possibilities. For plants that do not have sufficient personnel to review and consider tariffs and bills carefully, such problems as incorrect billing for state sales tax, late fees, inappropriate tariffs, and the occasional error in billings may be revealed to the energy auditor before the first plant visit.

At this stage, the avoided cost of energy (and demand, if applicable) for each energy stream can be determined. Avoided cost, sometimes incorrectly called marginal cost, is the amount that the plant will save (avoid paying) if energy use or demand is reduced by one unit. Electricity probably will have a demand component to its billing as well as an energy cost. Demand charges for natural gas [e.g., in $/(MCF/day)/billing period], for steam [e.g., in $/(1000 lb/day)/billing period], and possibly for other energy streams are infrequent. Common energy costs that are often used for rapid analyses and comparisons, such as the cost per unit of electrical energy obtained by dividing total electric cost in a period by the total electric energy consumed in kWh (a common method that blends energy, demand, and all other charges such as the customer or meter charge into the result) should be developed in this section.

This data can also be used to analyze electrical load factors of the types used by utilities (electric load factor is the consumption of energy in a billing period divided by the peak demand and the total number of hours in the billing period). The resulting values can be compared to nominal values. In addition, a load factor called the production load factor can be developed if the operating hours of the main production area of the plant are available (production load factor may be called operating load factor and is the consumption of energy in a billing period divided by the peak demand and the number of operating hours in the billing period). In this case, the computed load factor can be compared to unity, which would represent the best possible use of the plant's equipment during operating hours. Commonly, no plant or system actually achieves unity, so a production load factor of 75 or 85% may be considered good. As a diagnostic tool, if the production load factor exceeds unity, and nothing in the plant should be operating outside the main production hours, then unnecessary energy consumption after hours is indicated.

Executive Summary

This section of the assessment summarizes the report, and particularly the assessment recommendations. A table summarizing each assessment recommendation in the order that they occur in the report is very useful. Most often, these appear in descending order of estimated annual savings. The executive summary should give overall information about the total energy consumption and cost, as well as the common unit costs of energy that often are used for comparisons. Summary information about wastes and pollution prevention can be shown here as well.

The executive summary is the place to give and emphasize information that will enhance implementation of the recommended projects, and to provide warnings about safety and other implementation issues that may be repeated in the individual assessment recommendations.

CONCLUSION

The energy assessment process, including reporting, has been covered. For large facilities desiring to cut costs, the process that has been described will likely continue as a part of the cost-cutting process for a long time. Self-assessment procedures are available that can be used by smaller organizations that perhaps cannot afford to pay for an independent assessment.

These procedures include manuals to guide calculations[9] and a simple spreadsheet approach designed for use by facility operators untrained in energy assessments.[10] Benchmarking for commercial and institutional buildings is ahead of the industrial sector where important advances are still being made. There has been great diversity (in quality and meaning) of the data available to set standards of comparison for industry. The summer conference of the American Council for an Energy Efficient Economy in 2005 discussed industrial benchmarking[11] and more advances can be expected in this area.

REFERENCES

1. Capehart, B.L.; Spiller, M.B.; Frazier, S. Energy auditing. In *Energy Management Handbook*; Turner, W.C., Ed., Fairmont Press: Lilburn, GA, 2005; 23–39.
2. Energy Use and Management *2003 ASHRAE Handbook, Heating, Ventilating and Air-Conditioning Applications*; American Society of Heating, Refrigeration, and Air-Conditioning Engineers (ASHRAE): Atlanta, Georgia, 2003; 35.1–35.19.
3. http://www.oit.doe.gov/iac/ and http://iac.rutgers.edu/ (accessed August 2005).
4. http://www.eere.energy.gov/femp/ (accessed August 2005).
5. Energy Use and Management *2003 ASHRAE Handbook, Heating, Ventilating and Air-Conditioning Applications*; American Society of Heating, Refrigeration, and Air-Conditioning Engineers (ASHRAE): Atlanta, GA, 2003; 35.14.
6. Fuller, S.K.; Peterson, S.R. Life-cycle costing manual for the federal energy management program In *National Institute of Standard Technology Handbook*, US Government Printing Office: Washington, DC, 1996; Vol.135; http://www.bfrl.nist.gov/oae/publications/handbooks/135/pdf (accessed August 2005).
7. American National Standards Institute *A Management System for Energy*; ANSI/MSE 2000 Standard: New York, 2000.
8. *Texas LoanSTAR Program: Technical Guidelines and Format*, Texas State Energy Conservation Office: Austin, Texas, 1996, 40.
9. Muller, M.R.; Papadaratsakis, K. *Self-Assessment Workbook for Small Manufacturers*; Center for Advanced Energy Systems, Rutgers, the State University of New Jersey: Piscataway, NJ, 2003.
10. Ferland, K.; Kampschroer, N.; Heffington, W. M.; Huitink, D.; Kelley, R. *Texas Size Savings! A Step-by-Step Energy Assessment Guide and Calculator for Small Manufacturers*; Center for Energy and Environmental Resources, University of Texas at Austin, Austin, Texas, 2005.
11. 2005 ACEEE Summer Study on Energy Efficiency in Industry *Cutting the High Cost of Energy*; American Council for an Energy Efficient Economy: New York, 2005; (aceee_publications@aceee.org).

Auditing: Improved Accuracy

Barney L. Capehart
Department of Industrial and Systems Engineering, University of Florida College of Engineering, Gainesville, Florida, U.S.A.

Lynne C. Capehart
Consultant and Technical Writing Specialist, Gainesville, Florida, U.S.A.

Abstract

A frequent criticism of the quality and accuracy of energy audits is that they overestimate the savings potential for many customers. This entry discusses several problem areas that can potentially result in over-optimistic savings projections and suggests ways to increase quality and produce more accurate energy audits. Performing an energy and demand balance is the initial step a careful energy analyst should take when starting to evaluate the energy use at a facility. These balances allow one to determine what the largest energy users are in a facility, to find out whether all energy uses have been identified, and to check savings calculations by determining whether more savings have been identified than are actually achievable. Use of the average cost of electricity to calculate energy savings can give a false picture of the actual savings and may result in over-optimistic savings predictions. This entry discusses how to calculate the correct values from the electricity bills and when to use these values. Finally, the entry discusses several common energy-saving measures that are frequently recommended by energy auditors. Some of these may not actually save as much energy or demand as expected except in limited circumstances. Others have good energy-saving potential but they must be implemented carefully to avoid increasing energy use rather than decreasing it.

INTRODUCTION

Critics of energy audit recommendations often say that auditors overestimate the savings potential available to the customer. The possibility of overestimation concerns utilities who do not want to pay incentives for demand-side management programs if the facilities will not realize the expected results in energy or demand savings. Overestimates also make clients unhappy when their energy bills do not decrease as much as promised. The problem multiplies when a shared savings program is undertaken by the facility and an energy service company. Here, the difference between the audit projections and the actual metered and measured savings may be so significantly different that either there are no savings for the facility or the energy service company makes no profit.

More problems are likely to concern the accuracy of the energy audits for industrial manufacturing facilities and large buildings than for smaller commercial facilities because the equipment and operation of larger facilities is more complex. Based on our auditing experience over the last fifteen years, we have identified a number of areas where problems are likely to occur, and a number of these are presented and discussed. In addition, we have developed a few methods and approaches to dealing with these potential problems and we have found a few ways to initiate our energy audit analyses that lead us to more accurate results. One of these approaches is to collect data on the energy-using equipment in a facility and then to perform both an energy and a demand balance to help insure that we have reasonable estimates of energy uses—and therefore, energy savings—of this equipment.

CALCULATING ENERGY AND DEMAND BALANCES

The energy and demand balances for a facility are an accounting of the energy flows and the power used in the facility. These balances allow the energy analyst to track the energy and power inputs and outputs (uses) and to see if they match. A careful energy analyst should perform an energy and demand balance on a facility before developing and analyzing any energy management recommendations. This way, the analyst can determine what the largest energy users are in a facility, find out whether all or almost all energy uses have been identified, and see whether more savings have been identified than are actually achievable. Making energy use recommendations without utilizing the energy and demand balances is similar to making budget cutting recommendations without knowing exactly where the money is currently being spent.

When we perform an energy survey (audit), we inventory all of the major energy-using equipment in the

Keywords: Audits; Energy audits; Energy analyses; Energy savings; Energy balances; Demand balances; Motor load factors.

facility. Then we list the equipment and estimate its energy consumption and demand using data gathered at the facility such as nameplate ratings of the equipment and operating hours. We develop our energy balance by major equipment categories such as lighting, motors, heating, ventilating and air conditioning (HVAC), air compressors, etc. We also have a category called miscellaneous to account for loads that we did not individually survey, such as copiers, electric typewriters, computers, and other plug loads. We typically allocate 10% of the actual energy use and demand to the miscellaneous category in the demand and energy balances. (For an office building instead of a manufacturing facility, this miscellaneous load might be 15%–20% because of extensive computer and other electronic plug loads.) Then we calculate the energy and demand for each of the other categories.

Lighting

The first major category we analyze is lighting because this is usually the category in which we have the most confidence in knowing the actual demand and hours of use. Thus, we believe that our energy and demand estimates for the lighting system are the most accurate, and can then be subtracted from the total actual use to let us continue to build up the energy and demand balance for the facility. We record the types of lamps and the number of lamps used in each area of the facility and ask the maintenance person to show us the replacement lamps and ballasts used. With this lamp and ballast wattage data together with a good estimate of the hours that the lights are on in the various areas, we can construct what we believe to be a fairly accurate description of the energy and demand for the lighting system. Operational hours of lighting are now easily obtained through the use of inexpensive minidata loggers. These minidata loggers are typically under $100 and can record detailed on-off data for a month or more at five minute intervals.

Air Conditioning

There is generally no other "easy" or "accurate" category to work on, so we proceed to either air conditioning or motors. In most facilities there will be some air conditioning, even if it is just for the offices that are usually part of the industrial or manufacturing facility. Many facilities—particularly here in the hot and humid southeast—are fully air conditioned. Electronics, printing, medical plastics and devices, and many assembly plants are common facilities that we see that are fully air conditioned. Boats, metal products, wood products, and plastic pipe manufacturing facilities are most often not air conditioned. Air conditioning system name plate data is usually available and readable on many units and efficiency ratings can be found from published Air-Conditioning and Refrigeration Institute (ARI) data[1] or from the manufacturers of the equipment. The biggest problem with air conditioning is to get runtime data that will allow us to determine the number of full-load equivalent operating hours for the air conditioning compressors or chillers. From our experience in north and north-central Florida, we use about 2200–2400 h per year of compressor runtime for facilities that have air conditioning which responds to outdoor temperature. Process cooling requirements are much different and typically have much larger numbers of full-load equivalent operating hours. With the equipment size, the efficiency data, and the full-load equivalent operating hours, we can construct a description of the energy and demand for the air conditioning system. Again, the minidata loggers can make a huge contribution to the accuracy of our audits by measuring the compressor and fan runtimes and by estimating the load factor on the compressor by measuring the line current into the compressor motor. As long as the load on the motor is 50% or above, this method gives a reasonably accurate result.

Motors

Turning next to motors, we begin looking at one of the most difficult categories to deal with in the absence of fully metered and measured load factors on each motor in the facility. In a one day plant visit, it is usually impossible to get actual data on the load factors for more than a few motors. Even then, that data is only good for the one day that it was taken. Very few energy auditing organizations can afford the time and effort to make long-term measurements of the load factor on each motor in a large facility. Thus, estimating motor load factors becomes a critical part of the energy and demand balance and also a critical part of the accuracy of the actual energy audit analysis. Using minidata loggers to estimate the motor load factor by measuring the line current is reasonably accurate as long as the motor is 50% loaded at least. Motor name plate data shows the horsepower rating, the manufacturer, and sometimes the efficiency. If not, the efficiency can usually be obtained from the manufacturer or from standard references such as the Energy-Efficient Motor Systems Handbook[2] or from software databases such as MotorMaster, produced by the Washington State Energy Office.[3] We inventory all motors over 1 hp and sometimes try to look at the smaller ones if we have time.

Motor runtime is another parameter that is very difficult to get, but the minidata loggers have solved this problem to a great extent. However, it is still not likely that a minidata logger will be placed on each motor over 1 hp in a large facility. When the motor is used in an application where it is constantly on, it is an easy case. Ventilating fans, circulating pumps, and some process drive motors are often in this class because they run for a known, constant period of time each year. In other cases, facility operating personnel must help provide estimates of motor runtimes.

With data on the horsepower, efficiency, load factor, and runtimes of motors we can construct a detailed table of motor energy and demands to use in our balances. Motor load factors will be discussed further in a later section.

Air Compressors

Air compressors are a special case of motor use with most of the same problems. Some help is available in this category because some air compressors have instruments showing the load factor and some have runtime indicators for hours of use. Otherwise, use of two minidata loggers on each air compressor is required. Most industrial and manufacturing facilities will have several air compressors, and this may lead to some questions as to which air compressors are actually used and for how many hours they are used. If the air compressors at a facility are priority scheduled, it may turn out that one or more of the compressors are operated continuously, and one or two smaller compressors are cycled or unloaded to modulate the need for compressed air. In this case, the load factors on the larger compressors may be unity. Using this data on the horsepower, efficiency, load factor, and runtimes of the compressors, we develop a detailed table of compressor energy use and demand for our energy and demand balances.

Other Process Equipment

Specialized process equipment must be analyzed on an individual basis because it will vary tremendously depending on the type of industry or manufacturing facility involved. Much of this equipment will utilize electric motors and will be covered in the motor category. Other electrically-powered equipment such as drying ovens, cooking ovens, welders, and laser and plasma cutters are nonmotor electric uses and must be treated separately. Equipment name plate ratings and hours of use are necessary to compute the energy and demand for these items. Process chillers are in another special class that is somewhat different from comfort air conditioning equipment because the operating hours and loads are driven by the process requirements and not the weather patterns and temperatures. Minidata loggers will be of significant help with many of these types of equipment.

Checking the Results

Once the complete energy and demand balances are constructed for the facility, we check to see if the cumulative energy/demand for these categories plus the miscellaneous category is substantially larger or smaller than the actual energy usage and demand over the year. If it is, and we are sure we have identified all of the major energy uses, we know that we have made a mistake somewhere in our assumptions. As mentioned above, one area that we have typically had difficulty with is the energy use of motors. Measuring the actual load factors is difficult on a one-day walkthrough audit visit, so we use our energy balance to help us estimate the likely load factors for the motors. We do this by adjusting the load factor estimates on a number of the motors to arrive at a satisfactory level of the energy and demand from the electric motors. Unless we do this, we are likely to overestimate the energy used by the motors, thus over-estimating the energy savings from replacing standard motors with high-efficiency motors.

As an example, we performed an energy audit for one large manufacturing facility with a lot of motors. We first assumed that the load factors for the motors were approximately 80%, based on what the facility personnel told us. Using this load factor gave us a total energy use for the motors of over 16 million kWh/yr and a demand of over 2800 kW. Because the annual energy use for the entire facility was just over 11 million kWh/yr and the demand never exceeded 2250 kW, this load factor was clearly wrong. We adjusted the average motor load factor to 40% for most of the motors, which reduced our energy use to 9 million kWh and the demand to just under 1600 kW. These values are much more reasonable with motors making up a large part of the electrical load of this facility.

After we are satisfied with the energy/demand balances, we use a graphics program to draw a pie chart showing the distribution of energy/demand between the various categories. This allows us to visually represent which categories are responsible for the majority of the energy use. It also allows us to focus our energy-savings analyses on the areas of largest energy use.

PROBLEMS WITH ENERGY ANALYSIS CALCULATIONS

Over the course of performing over 200 large facility energy audits, we have identified a number of problem areas. One lies with the method of calculating energy cost savings (CS)—whether to use the average cost of electricity or break the cost down into energy and demand cost components. Other problems include instances where the energy and demand savings associated with specific energy efficiency measures may not be fully realized or where more research should go into determining the actual savings potential.

On-peak and Off-peak Uses: Overestimating Savings by Using the Average Cost of Electricity

One criticism of energy auditors is that they sometimes overestimate the dollar savings available from various energy efficiency measures. One way overestimation can result is when the analyst uses only the average cost of electricity to compute the savings. Because the average cost of electricity includes a demand component, using this average cost to compute the savings for companies who operate on more than one shift can overstate the dollar savings. This is because the energy cost during the off-peak hours does not include a demand charge. A fairly obvious example of this type of problem occurs when the average cost of electricity is used to calculate savings from installing high-efficiency security lighting. In this instance, there is no on-peak electricity use, but the savings will be calculated as if all the electricity was used on-peak.

The same problem arises when an energy efficiency measure does not result in an expected—or implicitly expected—demand reduction. Using a cost of electricity that includes demand in this instance will again overstate the dollar savings. Examples of energy efficiency measures that fall into this category are: occupancy sensors, photosensors, and adjustable speed drives (ASD). Although all of these measures can reduce the total amount of energy used by the equipment, there is no guarantee that the energy use will only occur during off-peak hours. While an occupancy sensor will save lighting kWh, it will not save any kW if the lights come on during the peak load period. Similarly, an ASD can save energy use for a motor, but if the motor needs its full load capability—as an air-conditioning fan motor or chilled water pump motor might—during the peak load period, the demand savings may not be there. The reduced use of the device or piece of equipment on peak load times may introduce a diversity factor that produces some demand savings. However, in most instances, even these savings will be overestimated by using the average cost of electricity.

On the other hand, some measures can be expected to provide their full demand savings at the time of the facility's peak load. Replacing 40 W T12 fluorescent lamps with 32 W super T8 lamps will provide a verifiable demand savings because the wattage reduction will be constant at all times and will specifically show up during the period of peak demand. Shifting loads to off-peak times should also produce verifiable demand savings. For example, putting a timer or energy management system control on a constant load electric drying oven to insure that it does not come on until the off-peak time will result in the full demand savings. Using high-efficiency motors also seems like it would produce verifiable savings because of its reduced kW load, but in some instances, there are other factors that tend to negate these benefits. This topic is discussed later on in this entry.

To help solve the problem of overestimating savings from using the average cost of electricity, we divide our energy savings calculations into demand savings and energy savings. In most instances, the energy savings for a particular piece of equipment is calculated by first determining the demand savings for that equipment and then multiplying by the total operating hours of the equipment. To calculate the annual CS, we use the following formula:

$$CS = [\text{Demand Savings} \times \text{Average Monthly Demand Rate} \times 12 \text{ mo/yr}] + [\text{Energy Savings Average Cost of Electricity without Demand}]$$

If a recommended measure has no demand savings, then the energy CS is simply the energy savings times the average cost of electricity without demand (or off-peak cost of electricity). This procedure forces us to think carefully about which equipment is used on-peak and which is used off-peak.

To demonstrate the difference in savings estimates, consider replacing a standard 30 hp motor with a high-efficiency motor. The efficiency of a standard 30 hp motor is 0.901 and a high-efficiency motor is 0.931. Assume the motor has a load factor of 40% and operates 8760 h/yr (three shifts). Assume also that the average cost of electricity is \$0.068/kWh, the average demand cost is \$3.79/kW/mo, and the average cost of electricity without demand is \$0.053/kWh. The equation for calculating the demand of a motor is:

$$D = HP \times LF \times 0.746 \times 1/\text{Eff}$$

The savings on demand (or demand reduction) from installing a high-efficiency motor is:

$$\begin{aligned} DR &= HP \times LF \times 0.746 \times (1/\text{Eff}_S - 1/\text{Eff}_H) \\ &= 30\,\text{hp} \times 0.40 \times 0.746\,\text{kW/hp} \times (1/0.901 - 1/0.931) \\ &= 0.32\,\text{kW} \end{aligned}$$

The annual energy savings are:

$$ES = DR \times H = 0.32\,\text{kW} \times 8760\,\text{h/yr} = 2803.2\,\text{kWh/yr}$$

Using the average cost of electricity above, the cost savings (CS_1) is calculated as

$$\begin{aligned} CS_1 &= ES \times (\text{Average cost of electricity}) \\ &= 2803.2\,\text{kWh/yr} \times \$0.068/\text{kWh} = \$190.62/\text{yr} \end{aligned}$$

Using the recommended formula above,

$$\begin{aligned} CS &= [\text{Demand Savings} \times \text{Average Monthly Demand Rate} \times 12\,\text{mo/yr}] + [\text{Energy Savings} \times \text{Average Cost of Electricity without Demand}] \\ &= (0.32\,\text{kW} \times \$3.79/\text{mo} \times 12\,\text{mo/yr}) + (2803.2\,\text{kWh/yr} \times \$0.053/\text{kWh}) \\ &= (\$14.55 + \$148.57)/\text{yr} \\ &= \$163.12/\text{yr} \end{aligned}$$

In this example, using the average cost to calculate the energy CS overestimates the CS by \$27.50 per year, or 17%. Although the actual amount is small for one motor, if this error is repeated for all the motors for the entire facility as well as all other measures that only reduce the demand component during the off-peak hours, then the cumulative error in CS predictions can be substantial.

Motor Load Factors

Many of us in the energy auditing business started off assuming that motors ran at full load or near full load and based our energy consumption analysis and energy-saving analysis on that premise. Most books and publications that give a formula for finding the electrical load of a motor do not even include a term for the motor load factor. However, because experience showed us that few motors actually run at full load or near full load, we were left in a quandary about what load factor to actually use in our calculations because we rarely had good measurements on the actual motor load factor. A recent paper by R. Hoshide shed some light on the distribution of motor load factors provided from his experience.[4] In this paper, Hoshide noted that only about one-fourth of all three-phase motors run with a load factor greater than 60%, with 50% of all motors running at load factors between 30 and 60% and one-fourth running with load factors less than 30%. Thus, those of us who had been assuming that a typical motor load factor was around 70 or 80% had been greatly overestimating the savings from high-efficiency motors, ASD, high-efficiency belts, and other motor-related improvements.

The energy and demand balances discussed earlier also confirm that overall motor loads in most facilities cannot be anywhere near 70%–80%. Our experience in manufacturing facilities has been that motor load factors are more correctly identified as being in the 30%–40% range. With these load factors, we get very different savings estimates and economic results than when we assume that a motor is operating at a 70% or greater load factor as shown in our example earlier.

One place where the motor load factor is critical but often overlooked is in the savings calculations for ASD. Many motor and ASD manufacturers provide easy-to-use software that will determine savings with an ASD if you supply the load profile data. Usually, a sample profile is included that shows calculations for a motor operating at full load for some period of time and at a fairly high overall load factor—i.e., around 70%. If the motor only has a load factor of 50% or less to begin with, the savings estimates from a quick use of one of these programs may be greatly exaggerated. If you use the actual motor use profile with the load factor of 50%, you may find that the ASD will still save some energy and money, but often not as much as it looks like when the motor is assumed to run at the higher load factor. For example, a 20 hp motor may have been selected for use on a 15 hp load to insure that there is a "safety factor." Thus, the maximum load factor for the motor would be 75%. A typical fan or pump in an air conditioning system that is responding to outside weather conditions may only operate at its maximum load about 10% of the time. Because that maximum load here is only 15 hp, the average load factor for the motor might be more like 40%, and will not be even close to 75%.

High-Efficiency Motors

Another interesting problem area is associated with the use of high-efficiency motors. In Hoshide's paper mentioned earlier, he notes that in general, high-efficiency motors run at a faster full load speeds than standard efficiency motors.[4] This means that when a standard motor is replaced by a high-efficiency motor, the new motor will run somewhat faster than the old motor in almost every instance. This is a problem for motors that drive centrifugal fans and pumps because the higher operating speed means greater power use by the motor. Hoshide provides an example where he shows that a high-efficiency motor that should be saving about 5% energy and demand actually uses the same energy and demand as the old motor. This occurs because the increase in speed of the high-efficiency motor offsets the power savings by almost exactly the same 5% due to the cube law for centrifugal fans and pumps.

Few energy auditors ever monitor fans or pumps after replacing a standard motor with a high-efficiency motor; therefore, they have not realized that this effect has cancelled the expected energy and demand savings. Because Hoshide noted this feature of high-efficiency motors, we have been careful to make sure that our recommendations for replacing motors with centrifugal loads carry the notice that it will probably be necessary to adjust the drive pulleys or the drive system so that the load is operated at the same speed in order to achieve the expected savings.

Motor Belts and Drives

We have developed some significant questions about the use of cogged and synchronous belts and the associated estimates of energy savings. It seems fairly well accepted that cogged and synchronous belts do transmit more power dfrom a motor to a load than if standard smooth V-belts are used. In some instances, this should certainly result in some energy savings. A constant torque application like a conveyor drive may indeed save energy with a more efficient drive belt because the motor will be able to supply that torque with less effort. Consider also a feedback-controlled application such as a thermostatically-controlled ventilating fan or a level-controlled pump. In this case, the greater energy transmitted to the fan or pump should result in the task being accomplished faster than if less drive power was supplied and some energy savings should exist. However, if a fan or a pump operates in a nonfeedback application—as is very common for many motors—then there will not be any energy savings.

For example, a large ventilating fan that operates at full load continuously without any temperature or other feedback may not use less energy with an efficient drive belt because the fan may run faster as a result of the drive belt having less slip. Similarly, a pump which operates continuously to circulate water may not use less energy with an efficient drive belt. This is an area that needs some monitoring and metering studies to check the actual results.

Whether or not efficient drive belts result in any demand savings is another question. Because in many cases the motor is assumed to be supplying the same shaft horsepower with or without high-efficiency drive belts, a demand savings does not seem likely in these cases. Possibly using an efficient belt on a motor with a constant torque application that is controlled by an ASD might result in some demand savings. However, for the most common applications, the motor is still supplying the same load, and thus would have the same power demand. For feedback-controlled applications, there might be a diversity factor involved so that the reduced operation times could result in some demand savings, but not the full value otherwise expected. Thus, using average cost electricity to quantify the savings expected from high-efficiency drive belts could well overestimate the value of the savings. Verification of the cases where demand savings are to be expected is another area where more study and data are needed.

Adjustable Speed Drives (ASDs)

We would like to close this discussion with a return to ASDs because these are devices that offer a great potential for savings, but have far greater complexities than are often understood or appreciated. Fans and pumps form the largest class of applications where great energy savings are possible from the use of ASDs. This is a result again of the cube law for centrifugal fans and pumps—where the power required to drive a fan or pump is specified by the cube of the ratio of the flow rates involved. According to the cube law, a reduction in flow to one-half the original value could now be supplied by a motor using only one-eighth of the original horsepower. Thus, whenever an air flow or liquid flow can be reduced, such as in a variable air volume system or with a chilled water pump, there are dramatic savings possible with an ASD. In practice, there are two major problems with determining and achieving the expected savings.

The first problem is the one briefly mentioned earlier—determining the actual profile of the load involved. Simply using the standard profile in a piece of vendor's software is not likely to produce very realistic results. There are so many different conditions involved in fan and pump applications that taking actual measurements is the only way to get a very good idea of the savings that will occur with an ASD. Recent papers have discussed the problems with estimating the loads on fans and pumps and have shown how the cube law itself does not always give a reasonable value.[5–7] The Industrial Energy Center at Virginia Polytechnic Institute and Virginia Power Company have developed an approach where they classify potential ASD applications into eight different groups and then estimate the potential savings from analysis of each system and from measurements of that system's operation.[8] Using both an analytical approach and a few measurements allows them to get a reasonable estimate of the motor load profile and thus a reasonable estimate of the energy and demand savings possible.

The second problem is achieving the savings predicted for a particular fan or pump application. It is not enough just to identify the savings potential and then install an ASD on the fan or pump motor. In most applications, there is some kind of throttling or bypass action that results in almost the full horsepower still being required to drive the fan or pump most of the time. In these applications, the ASD will not save much unless the system is altered to remove the throttling or bypass device and a feedback sensor is installed to tell the ASD what fraction of its speed to deliver. This means that in many air flow systems, the dampers or vanes must be removed so that the quantity of air can be controlled by the ASD changing the speed of the fan motor. In addition, some kind of feedback sensor must be installed to measure the temperature or pressure in the system to send a signal to the ASD or a programmable logic controller (PLC) to alter the speed of the motor to meet the desired condition. The additional cost of the needed alterations to the system and the cost of the control system greatly change the economics of an ASD application compared to the case where only the purchase cost and installation cost of the actual ASD unit is considered.

For example, a dust collector system might originally be operating with a large 150 hp fan motor running continuously to pick up the dust from eight saws. However, because production follows existing orders for the product, sometimes only two, three, or four saws are in operation at a particular time. Thus, the load on the dust collector is much lower at these times than if all eight saws were in use. An ASD is a common recommendation in this case, but estimating the savings is not easy to begin with and once the costs of altering the collection duct system and the cost of adding a sophisticated control system to the ASD is considered, the bottom line result is much different than the cost of the basic ASD with installation. Manual or automatic dampers must be added to each duct at a saw so that it can be shut off when the saw is not running. In addition, a PLC for the ASD must be added to the new system together with sensors added to each damper so that the PLC will know how many saws are in operation and therefore what speed to tell the ASD for the fan to run to meet the dust collection load of that number of saws. Without these system changes and control additions, the

ASD itself will not save any great amount of energy or money. Adding them in might double the cost of the basic ASD and double the payback time that may have originally been envisioned.

Similarly, for a water or other liquid flow application, the system piping or valving must be altered to remove any throttling or bypass valves and a feedback sensor must be installed to allow the ASD to know what speed to operate the pump motor. If several sensors are involved in the application, a PLC may also be needed to control the ASD. For example, putting an ASD on a chilled water pump for a facility is much more involved and much more costly than simply cutting the electric supply lines to the pump motor and inserting an ASD for the motor. Without the system alterations and without the feedback control system, the ASD cannot provide the savings expected.

CONCLUSION

Energy auditing is not an exact science, but a number of opportunities are available for improving the accuracy and the quality of the recommendations. Techniques that may be appropriate for small-scale energy audits can introduce significant errors into the analyses for large complex facilities. We began by discussing how to perform an energy and demand balance for a company. This balance is an important step in doing an energy-use analysis because it provides a check on the accuracy of some of the assumptions necessary to calculate savings potential. We also addressed several problem areas that can result in over-optimistic savings projections and suggested ways to prevent mistakes. Finally, several areas where additional research, analysis, and data collection are needed were identified. Once this additional information is obtained, we can produce better and more accurate energy audit results.[9]

REFERENCES

1. Air-Conditioning and Refrigeration Institute, *ARI Unitary Directory*, published every six months. Also available through www.ARI.org.
2. Nadel, S.; Shepard, M.; Greenberg, S.; Katz, G.; de Almeida, A. *Energy-Efficient Motor Systems: A Handbook on Technology, Program and Policy Opportunities*, 2nd Ed.; American Council for an Energy-Efficient Economy: Washington, DC, 2002.
3. Washington State Energy Office, MotorMaster Electric Motor Selection Software and Database, updated annually. Available from the U.S. Department of Energy, at www.eere.energy.gov/industry 2006.
4. Hoshide, R.K. Electric Motor Do's and Don'ts. Energy Eng. **1994**, *91* (1).
5. Stebbins, W.L. Are you certain you understand the economics for applying ASD systems to centrifugal loads? Energy Eng. **1994**, *91* (1).
6. Vaillencourt, R.R. Simple solutions to VSD pumping measures. Energy Eng. **1994**, *91* (1).
7. Kempers, G. DSM pitfalls for centrifugal pumps and fans. Energy Eng. **1995**, *92* (2).
8. Personal communication with Mr. Mark Webb, Senior Engineer, Virginia Power Company: Roanoke, VA, January 1995.
9. Capehart, B.; Capehart, L. Improving industrial energy audit analysis. In *Proceedings of the 1995 ACEEE Summer Study in Industry*, ACEEE: Washington, DC, August 2005.

BIBLIOGRAPHY

1. Capehart, B.L.; Turner, W.C.; Kennedy, W.G. *Guide to Energy Management*, 5th Ed.; Fairmont Press: Lilburn, GA, 2005.
2. Turner, W.C., Ed. *Energy Management Handbook* 5th Ed., Fairmont Press: Lilburn, GA, 2006.
3. Thumann, A.; Younger, W.J. *Handbook of Energy Audits*, 6th Ed.; Fairmont Press: Lilburn, GA, 2003.
4. Thumann, A. *Plant Engineers and Managers Guide to Energy Conservation*, 7th Ed.; Fairmont Press: Lilburn, GA, 2006.

Auditing: User-Friendly Reports

Lynne C. Capehart
Consultant and Technical Writing Specialist, Gainesville, Florida, U.S.A.

Abstract

Energy audits do not save money and energy for companies unless the recommendations are implemented. Audit reports should be designed to encourage implementation, but often they impede it instead. In this article, the author discusses her experience with writing industrial energy audit reports and suggests some ways to make the reports more user-friendly. The goal in writing an audit report should not be the report itself; rather, it should be to achieve implementation of the report recommendations and thus achieve increased energy efficiency and energy cost savings for the customer.

INTRODUCTION

This article addresses two questions: "Why should an energy audit report be user-friendly?" and "How do you make an audit report user-friendly?" The author answers these questions in the context of sharing experience gained by writing audit reports for industrial clients of the University of Florida Industrial Assessment Center (UF IAC).

At the UF IAC, we had two goals in writing an audit report. Our first goal was to provide our clients with the facts necessary to make informed decisions about our report recommendations. Our second goal, which was as important as the first, was to interest our clients in implementing as many of our recommendations as possible. We found that "user-friendly" audit reports helped us achieve both goals.

WHAT IS A USER-FRIENDLY AUDIT REPORT?

The definition of "user-friendly" is something that is easy to learn or to use. People generally think of the term "user-friendly" related to something like a computer program. A program that is user-friendly is one that you can use with minimal difficulty. We have applied the same term to audit reports to mean a report that communicates its information to the user (reader) with a minimum amount of effort on the reader's part. We operate on the belief that a reader who is busy will not want to spend valuable time struggling to understand what the report is trying to say. If the report is not clear and easy to follow, the reader will probably set it down to read later, and "later" may never come!

Keywords: Audit reports; Energy auditing; Audit recommendations; Report writing; Technical writing.

HOW DO YOU WRITE A USER-FRIENDLY AUDIT REPORT?

From our experience, we have identified a number of key points for successfully writing a user-friendly audit report. These points are summarized below.

Know your audience

The first thing to keep in mind when you start to write anything is to know who your audience is and to tailor your writing to that audience. When writing an industrial audit report, your readers can range from the company president to the head of maintenance. If recommendations affect a number of groups in the company, each group leader may be given a copy of the report. Thus, you may have persons of varying backgrounds and degrees of education reading the report. Not all of them will necessarily have a technical background. The primary decision maker may not be an engineer; the person who implements the recommendations may not have a college degree.

We dealt with this problem by writing a report with three basic sections. Section One was an executive summary that briefly described our recommendations and tabulated our results such as the energy and dollar savings and the simple payback times. Section Two was a brief description of a recommended energy management program for our client. Section Three was a detailed section that we called our technical supplement. This section of our report included the calculations that supported our recommendations and any specific information relating to implementation. (These sections are described more fully later in this article.)

Use a simple, direct writing style

Technical writers often feel compelled to write in a third-person, passive, verbose style. Because energy audit reports are technical in nature, they often reflect this

Auditing: User-Friendly Reports

writing style. Instead, you should write your audit report in clear, understandable language. As noted above, your reader may not have a technical background. Even a reader who does will not be offended if the report is easy to read and understand. Some specific suggestions are:

Simplify your writing by using active voice. Technical writers use passive voice, saying "It is recommended..." or "It has been shown..." rather than "We recommend..." or "We have shown..." Passive voice allows the writer to avoid taking direct responsibility for the recommendations. Be clear and straightforward in your writing by using active voice wherever possible.

Address the report to the reader. Write as if you were speaking directly to the reader. Use the words "you" and "your." Say "your company...," "your electric bill...," etc. Make the report plain and simple. The following examples show how to do this.

- Not: Installation of high-efficiency fluorescent lamps in place of the present lamps is recommended.
- But: Install high-efficiency fluorescent lamps in place of your present lamps.
- Or: We recommend that you install high-efficiency fluorescent lamps in place of your present lamps.
- Not: Twelve air leaks were found in the compressor system during the audit of this facility.
- But: We found twelve air leaks in the compressor system when we audited your facility.
- Or: You have twelve air leaks in your compressor system.

Avoid technical jargon that your reader may not understand. Do not use acronyms such as ECO, EMO, or EMR without explaining them. (Energy Conservation Opportunity, Energy Management Opportunity, Energy Management Recommendation).

Present Information Visually

Often the concepts you need to convey in an audit report are not easy to explain in a limited number of words. To solve this problem, we often used drawings to show what we meant. For example, we had a diagram that showed how to place the lamps in fluorescent lighting fixtures when you are eliminating two of the lamps in a four-lamp fixture and adding reflectors. We also had a diagram showing how a heat pipe works.

We also presented our client's energy use data visually with graphs showing the annual energy and demand usage by month. These graphs gave a picture of use patterns. Any discrepancies in use showed up clearly.

Make Calculation Sections Helpful

The methodology and calculations used to develop specific energy management opportunity recommendations can be very helpful in an audit report. When you include the methodology and calculations, the technical personnel have the opportunity to check the accuracy of your assumptions and your work. Because not every reader wants to wade through pages describing the methodology and showing the calculations, we provided this information in a technical supplement to our audit report. Because this section was clearly labeled as the technical supplement, nontechnical readers could see instantly that this section might be difficult for them to understand and that they could ignore it.

Use Commonly Understood Units

In your report, be sure to use units that your client will understand. Discussing energy savings in terms of BTUs is not meaningful to the average reader. Kilowatt-hours for electricity or therms for natural gas are better units because most energy bills use these units.

Make Your Recommendations Clear

Some writers assume that their readers will understand their recommendation even if it is not explicitly stated. Although the implication may often be clear, better practice is to clearly state your recommendation so that your reader knows exactly what to do.

- Not: Install occupancy sensors in the conference room and restrooms.
- But: You should purchase five occupancy sensors. Install one in the conference room and one in each of the four restrooms.

Explain Your Assumptions

A major problem with many reports is the author's failure to explain the assumptions underlying the calculations. For example, when you use operating hours in a calculation, show how you got the number. "Your facility operates from 7:30 A.M. to 8:00 P.M., 5 days a week, 51 weeks per year. Therefore, we will use 3188 annual operating hours in our calculations."

When you show your basic assumptions and calculations, the reader can make adjustments if those facts change. In the example above, if the facility decided to operate 24 h/day, the reader would know where and how to make changes in operating hours because we had clearly labeled that calculation.

Use a section of your report to list your standard assumptions and calculations. That way, you do not have to repeat the explanations for each of your recommendations. Some of the standard assumptions/calculations that can be included in this section are operating hours, the average cost of electricity, the demand rate, the off-peak cost of electricity, and the calculation of the fraction of air conditioning load attributable to lighting.

Be Accurate and Consistent

The integrity of a report is grounded in its accuracy. This does not just pertain to the correctness of calculations. Clearly, inaccurate calculations will destroy a report's credibility, but other problems can also undermine the value of your report.

Be consistent throughout the report. Use the same terminology so your reader is not confused. Make sure that you use the same values. Do not use two different load factors for the same piece of equipment in different recommendations. For example, you might calculate the loss of energy due to leaks from an air compressor in one recommendation and the energy savings due to replacing the air compressor motor with a high efficiency motor in another recommendation. If you use different load factors or different motor efficiencies in each recommendation, your results will not be consistent or accurate.

Proofread your report carefully. Typographical and spelling errors devalue an otherwise good product. With computer spell checkers, there is very little excuse for misspelled words. Your nontechnical readers are likely to notice this type of error, and they will wonder if your technical calculations are similarly flawed. Textual errors can also sometimes change the meaning of a sentence—if you say "Do not..." instead of "Do...," you have made a major mistake.

REPORT SECTIONS

We found that the following report format met our clients' needs and fit our definition of a user-friendly report.

Executive Summary

The audit report starts with an executive summary, which lists the recommended energy conservation measures and shows the implementation cost and dollar savings amount. This section is intended for the readers who only want to see the bottom line. Although the executive summary can be as simple as a short table, you may add brief text to explain the recommendations and include any special information needed to implement the recommendations. We copied the executive summary on colored paper so that it stood out from the rest of the report.

Energy Management Plan

Following the executive summary, we provided some information to the decision makers on how to set up an energy management program in their facility. We viewed this section as one that encouraged implementation of our report, so we tried to make it as helpful as possible.

Energy Action Plan. In this subsection, we described the steps that a company should consider in order to start implementing our recommendations.

Energy Financing Options. We also included a short discussion of the ways that a company can pay for the recommendations. We covered the traditional use of company capital, loans for small businesses, utility incentive programs, and the shared savings approach of the energy service companies.

Maintenance Recommendations. We did not usually make formal maintenance recommendations in the technical supplement because the savings are not often easy to quantify. However, in this section of the report, we provided energy-savings maintenance checklists for lighting, heating/ventilation/air-conditioning, and boilers.

The Technical Supplement

The technical supplement is the part of the report that contains the specific information about the facility and the audit recommendations. Our technical supplement had two main sections: one included our assumptions and general calculations and the other described the recommendations in detail, including the calculations and methodology. We sometimes included a third section that described measures we had analyzed and determined were not cost-effective or that had payback times beyond the client's current planning horizon.

Standard Calculations and Assumptions

This section was briefly described above when we discussed the importance of explaining assumptions. Here we provided the reader with the basis for understanding many of our calculations and assumptions. We included a short description of the facility: square footage (both air conditioned and unconditioned areas), materials of construction, type and level of insulation, etc. If we were dividing the facility into subareas, we described those areas and assigned each an area number that was then used throughout the recommendation section.

Standard values calculated in this section included operating hours, the average cost of electricity, the demand rate, the off-peak cost of electricity, and the calculation of the fraction of air conditioning load attributable to lighting. When we calculated a value in this section, we labeled the variable with an identifier that remained the same throughout the rest of the report. For example, operating hours was OH wherever it was used; demand rate was DR.

Audit Recommendations

This section contained a discussion of each of the energy management opportunities we had determined were cost-effective. Each energy management recommendation (or EMR) that was capsulized in the executive summary was described in depth here.

Again, we tried to make the EMRs user-friendly. To do this, we put the narrative discussion at the beginning of the recommendation and left the technical calculations for the very end. This way, we allowed the readers to decide for themselves whether they wanted to wade through the specific calculations.

Each EMR started with a table that summarized the energy, demand, and cost savings, implementation cost, and simple payback period. Then we wrote a short narrative section that provided some brief background information about the recommended measure and explained how it should be implemented at this facility. If we were recommending installation of more than one item (lights, motors, air conditioning units, etc.), we often used a table to break down the savings by unit or by area.

The final section of each EMR was the calculation section. Here we explained the methodology that we used to arrive at our savings estimates. We provided the equations and showed how the calculations were performed so that our clients could see what we had done. If they wanted to change our assumptions, they could. If some of the data we had used was incorrect, they could replace it with the correct data and recalculate the results. However, by placing the calculations away from the rest of the discussion rather than intermingling the two, we didn't scare off the readers who needed to look at the other information.

Appendix

We used an appendix for lengthy data tables. For example, we had a motor efficiencies table that we used in several of our EMRs. Instead of repeating it in each EMR, we put it in the appendix. We also included a table showing the facility's monthly energy-use history and a table listing the major energy-using equipment. Similar to the calculation section of the EMRs, the appendix allowed us to provide backup information without cluttering up the main body of the report.

SHORT FORM AUDIT REPORT

Many energy auditors use a short form audit report. A short report is essential when the cost of the audit is a factor. Writing a long report can be time-consuming and it increases the cost of an audit.

The short form report is useful when an on-the-spot audit report is required because the auditor can use a laptop computer to generate it. It is also an excellent format for preliminary audit reports when the company will have to do further analysis before implementing most of the recommendations.

However, some short form audit reports have drawbacks. When a report is ultra-short and only provides the basic numbers, the reader will not have a memory crutch if he returns to the report sometime after the auditor has left. Because some clients do not implement the recommendations immediately but wait until they gather the necessary capital, an ultra-short form report may lose its value. Therefore, some explanatory text is a critical component of a user-friendly short form report. The executive summary described above could serve as a model short form audit report.

FEEDBACK

Customer feedback is as appropriate in energy auditing as in any other endeavor. An easy way to get feedback is to give the customer a questionnaire to evaluate the audit service and the report. In our feedback form, we listed each section of the report and asked the client to rate each section on a scale of 1–10, with 1 being poor and 10 being excellent. We asked for a rating based on whether the section was easy to read and we asked for a second rating of the likelihood that our recommendations would be implemented. (We also asked for any additional comments, but seldom got those.)

The questionnaire must be easy to fill out. If it takes too much time to read and fill out, the clients won't take the time to return it. We used to send the questionnaire along with the report, but those were seldom returned. We decided to wait for a month and then send the questionnaire as a follow-up to the audit. We had a much greater return rate when we used this method.

CONCLUSION

Many audit reports are not user-friendly. Most often they are either lengthy documents full of explanations, justifications, and calculations or they are very short with little backup information. If a report is so long that it intimidates your readers by its very size, they may set it aside to read when they have more time. If it is so short that needed information is lacking, the readers may not believe the results.

Writing a user-friendly audit report is an important step in promoting implementation of audit recommendations. If you adopt some of these report-writing suggestions, you should be able to produce your own successful user-friendly energy audit report.

ACKNOWLEDGMENT

An earlier version of this article appeared in Strategic Planning for Energy and the Environment, and it is used with permission from Fairmont Press.

BIBLIOGRAPHY

1. How to Write in Plain English, http://www.plainenglish.co.uk/plainenglishguide.html.
2. A Short Course on Writing Technical Reports, http://www.technical-writing-course.com/type-of-technical-report.html.
3. Technical Writing and Writing a Technical Report, www.engr.arizona.edu/robotics/Robotics/Technical%20Report%20Guidelines.doc.
4. Short Reports: How to Write Routine Technical Documents, http://jerz.setonhill.edu/writing/technical/reports/index.html.
5. Strunk, W. Jr; White, E.B.; Angell, R. *The Elements of Style*, 4th Ed.; 2000.
6. Philip, R. *Science and Technical Writing: A Manual of Style*, 2nd Ed.; Routledge Study Guides S.: London, 2000.

Benefit Cost Analysis

Fatouh Al-Ragom
Building and Energy Technologies Department, Kuwait Institute for Scientific Research, Safat, Kuwait

Abstract

Benefit cost analysis is one of several methods utilized to evaluate the feasibility of capital investment. The benefit cost analysis calculates the present worth of all benefits, then calculates the present worth of all costs and takes the ratio of the two sums. This ratio is either known as a benefit/cost ratio, a savings/interest ratio, or a profitability index. The benefit cost analysis is an economic decision-making criterion that can measure the economic consequences of a decision over a specified period of time. It is used to evaluate whether an alternative option is economically viable when compared to a base case which is usually the "do nothing" option or it can be used to rank several options that are competing for a limited budget.

INTRODUCTION

The implementation of energy efficiency projects is linked to the allocation of funds. Economic decision-making tools or methods are often required to justify a project's implementation and assess its economical feasibility. Several methods are available, such as life-cycle cost, simple or discounted payback, benefit cost analysis, and internal and adjusted rate-of-return.[1]

The selection of the method depends on many factors and, in fact, more than one method can be technically appropriate for economical decisions. These methods are often utilized to compare several options or to compare an alternative to the current situation usually referred to as the "do nothing" option. The American Society for Testing and Materials (ASTM) provides a guide that details the selection of economic methods for evaluating investments.[2]

This entry describes the economic assessment method known as benefit cost analysis (BCA). This method is simply considered as an attempt to identify and express, in dollar terms, all of the effects of proposed policies or projects and then to relate those effects with endured costs in a simple, dimensionless parameter.

BACKGROUND

Benefit cost analysis is one of the economic assessment methods that can aid policymakers in deciding whether the advantages of a particular course of action are likely to outweigh its drawbacks.

Benefit cost analysis was conceived over 150 years ago by the French engineer Jules Dupuit. The BCA method saw its first widespread use when the United States government wanted to evaluate the impact of water projects in the late 1930s. Since then, it has been used to analyze policies affecting several sectors including public health, transportation, criminal justice, education, defense, and the environment.[7]

The BCA method involves evaluating benefits of a project against its costs in a ratio format that is known as benefit-to-cost-ratio (BCR). Another variation of BCR is known as savings-to-investment-ratio (SIR). The difference between the two is that BCR is used when the focus of the analysis is on benefits (that is, advantages measured in dollars) relative to project costs while SIR is used when the focus of the assessment is on project savings (that is, cost reductions).[1] The BCR is referred to as the profitability index (PI) in some financial publications.[3] The estimation of a project's PI is usually carried out based on the organization's current net value of cash flows.

SIGNIFICANCE OF BCR

The BCR provides a standard for measuring economic performance in a single number that indicates whether a proposed project or system is preferred over a mutually exclusive alternative that serves as the baseline for the economic analysis. In addition, the BCR indicates the discounted dollar benefits (or savings) per dollar of discounted costs. Moreover, it can be used to determine if a given project or system is economically feasible relative to the alternative of not implementing it. Also, when computed on increments of benefits (or savings) and costs, the BCR can be used to determine if one design or size of a system is more economic than another. On the other hand, when funding is limited, the BCR can be used

Keywords: Economical assessment; Decision-making; Resource-allocation decisions; Benefit-to-cost-ratio; Savings-to-investment-ratio; Profitability index.

as an aid to select the most economically efficient set of projects from among several available options that are competing for limited funding. Selecting an efficient set of projects will maximize aggregate net benefits or net savings obtainable for the budget.[1] The BCR and the PI examine cash flows, not accounting profits, and recognize the time-value of money. While these ratios can be accurate predictors of economic efficiency, their accuracy depends on the accuracy of cash flow predictions.

PROCEDURE

Before conducting a BCA, multiple implementation alternatives should be identified in a way that allows for a fair comparison. The constraints or requirements of a successful end solution should be also clearly identified. Additionally, costs and benefits should be put into standard units (usually dollars) so that they can be compared directly. The dollar amounts used in the BCA should all be discounted, that is, expressed in time-equivalent dollars, either in present value or uniform annual value terms.

The BCR is a numerical ratio that indicates the expected economic performance of a project by the size of the ratio. A ratio less than 1.0 indicates a project that is uneconomic, a ratio of 1.0 indicates a project whose benefits or savings are just equal to its costs, and a ratio greater than 1.0 indicates a project that is economic.

The recommended steps for carrying out an economic evaluation using the BCA method are summarized as follows:

- *Identification of objectives, constraints, and alternatives.* The decision-maker's objectives should be clearly specified. This is crucial to defining the problem. Moreover, constraints that limit potential alternatives for accomplishing the objectives should be identified such as economic and environmental limitations. Finally, alternatives that are technically and otherwise feasible in view of the constraints should be identified for consideration.
- *Data collection.* Actual or expected cash flows are needed for all phases of the project including revenues or other benefits; acquisition costs, including costs of planning, design, construction, purchase, installation, site preparation; utility costs, including costs of energy and operating and maintenance costs; repair and replacement costs; salvage values; disposal costs; and insurance costs. Moreover, information is also needed regarding the study period, discount rate, any applicable tax rates and rules, and the terms of financing (if applicable).
- *Expression of cash flow.* A decision should be made about whether to express the discounted cash flow of costs and benefits within each year in present-value dollars or in annual-value dollars. This should also include deciding whether to work in constant dollars using a real discount rate or in current dollars using a nominal discount rate. When using constant dollars, inflation is not included in the estimates of costs and benefits.[5]
- *Compute the BCR or SIR.* In concept, the BCA ratios are simple: benefits (or savings) divided by costs, where all dollar amounts are discounted to current or annual values. The BCR will be computed using the following formula:

$$\text{BCR} = \frac{\sum_{t=0}^{n} B_t/(1+i)^t}{\sum_{t=0}^{n} C_t/(1+i)^t} \quad (1)$$

Where B_t, benefits in period t; C_t, costs in period t; n, life of the investment; i, discount rate

Note that $1/(1+i)^t$ is the discount factor used to calculate the present worth of a single future payment. Thus, to account for all future values, each value is calculated separately and then all values are added together. When evaluating an energy saving opportunity, its benefit at the beginning (at $t=0$) will be zero, while for the following time periods the benefits (savings) will be evaluated separately. Usually, the costs will only be considered at $t=0$ representing the initial cost. If the analysis will be carried out utilizing the uniform series of present worth as in the case of constant annual energy savings, then the following equation can be utilized to estimate the BCR:

$$\text{BCR} = \frac{B((1+i)^n - 1)/i(1+i)^n}{C}$$

$$= \frac{B \times (P/A, i\%, n)}{C} \quad (2)$$

Where B, annual benefits; C, initial cost; $(P/A, i\%, n)$, discount factor for a uniform series of cash flows

It should be noted that the BCR and PI represent the same parameter; and thus the PI can be calculated using Eq. 1. Alternatively Eq. 3 can be also used.

$$\text{PI} = \frac{\text{PV (future cash flows)}}{\text{initial investment}}$$

$$= 1 + \frac{\text{NPV}}{\text{initial investement}} \quad (3)$$

Where PV, Present value; NPV, Net present value of cash = benefits (inflows) − expenses (outflows)

Example 1

A building owner has $12,000 available to utilize for improvements. The owner is considering replacing the old

Table 1 Cash flow details

n (Year)	(1): Annual savings	(2): $1/(1+i)^n$	PV = (1)×(2)
0	0	1	0
1	1765	0.952	1681
2	1765	0.907	1600.9
3	1765	0.864	1524.7
4	1765	0.823	1452.1
5	1765	0.784	1382.9
6	1765	0.746	1317.1
7	1765	0.711	1254.4
8	1765	0.677	1194.6
9	1765	0.645	1137.7
10	1765	0.614	1083.6
			13,629

boiler that has an efficiency of 65% with a new one that has an efficiency of 85%. Assuming that the lifespan of the boiler is 10 years and the discount rate is 5%, evaluate if this investment is cost-effective or not. The boiler uses oil and consumes 6000 gallons/year at a cost of $1.25 per gallon.

This example compares the option of using a new boiler to the "do nothing" option of using the old boiler. Accordingly, due to the higher efficiency of the new boiler, annual operational costs will be lower, assuming that the salvage values of the old boiler and the new one are insignificant.

The annual savings due to the new boiler's efficiency improvement are calculated using the following equation[4]:

Fuel Savings

$$= \frac{\eta_{New} - \eta_{Old}}{\eta_{New}} \times \text{Old Fuel Consumption} \times \text{Fuel Cost} \quad (4)$$

$$\text{Fuel Savings} = \frac{0.85 - 0.65}{0.85} \times 6000 \text{ gallons/year} \times 1.25/\text{gallons} = 1765$$

The cash flows are detailed in Table 1, and are calculated year by year.

An alternative option to estimating the present worth of future savings would be to use a uniform series present worth factor. This is done for ease of calculation. Instead of discounting the amounts in each year and adding them together, the cash flows can be grouped into categories with the same pattern of occurrence and discounted using discount factors. In this case, since there is a uniform savings of $1,765 per year, the present value of the future annual cash flows is calculated using the following equation[8]:

$$PV = A \frac{(1+i)^n - 1}{i(1+i)^n} = A \times (P/A, i\%, n) \quad (5)$$

$$PV = 1765 \times (P/A, 5\%, 10) = 1765 \times 7.722 = 13,629$$

Where PV, present value of future annual cash flows; A, annual savings; (P/A, i%, n), discount factor for uniform series of cash flows. The values of this factor are listed in financial analysis books. In this example, the value of (P/A, 5%, 10) = 7.722.[6]

The BCR using Eq. 2 will be calculated as:

$$BCR = \frac{\$13,629}{\$12,000} = 1.136$$

Since the BCR is more than one, this means that the benefits outweigh the costs; hence, the boiler replacement is financially justified.

- In practice, it is important to formulate the ratio so as to satisfy the investor's objective. This requires attention to the placement of costs in the numerator and denominator. To maximize net benefits from a designated expenditure, it is necessary to place only that portion of costs in the denominator on which the investor wishes to maximize returns. Then, the BCR will be computed using the following formula:

$$BCR = \frac{\sum_{t=0}^{n}(B_t - C_t)/(1+i)^t}{\sum_{t=0}^{n} IC_t/(1+i)^t} \quad (6)$$

Where B_t, benefits in period t; that is, the advantages in revenue or performance (measured in dollars) of the building or system as compared to a mutually exclusive alternative; C_t, costs in period t, excluding investment costs that are to be placed in the denominator for the building or system, minus counterpart costs in period t for a mutually exclusive alternative; IC_t, investment costs in period t on which the investor wishes to maximize the return, minus similar investment costs in period t for a mutually exclusive alternative; i, discount rate.

- Special attention should be paid when placing cost and benefit items into the ratio especially in the case when several projects are compared and a ranking of their cost-effectiveness needs to be established. Changing the placement of a cost item from the denominator (where it increases costs) to the numerator (where it

Table 2 Investment details among alternative projects

Project code	(1) Project description	(2) Project cost PV ($)	(3) Energy savings PV ($)	(4) BCR	(5) Ranking	(6) Net benefits	(7) PI
A	Wall R-10	6,050	12,000	1.983	2	5950	1.983
B	Wall R-15	6,900	13,800	2.000	1	6900	2.000
C	Roof R-15	7,200	5,000	0.694	5	−2200	0.694
D	Roof R-20	7,500	5,700	0.760	4	−1800	0.760
E	Double glass	6,300	4,100	0.651	6	−2200	0.651
F	Reflective glass	7,520	10,000	1.330	3	2480	1.330

Note: (4)=(3)/(2), (6)=(3)−(2) and (7)=1+(6)/(2).

decreases benefits or savings) will affect the relative rankings of competing independent projects, and thereby influence investment decisions.[1]

- Biasing effects, detrimental to economic efficiency, can result from certain formulations of the BCR and SIR ratios. As an example, when comparing competing projects that differ significantly in their maintenance costs, placing maintenance costs in the denominator with investment costs tends to bias the final selection away from projects with relatively high maintenance costs, even though they may offer higher net benefits (profits) than competing projects. Similar biasing effects can occur in the placement of other non-investment costs such as energy or labor costs. This highlights the fact that adding a given amount to the denominator of a ratio reduces the quotient more than does subtracting an identical amount from the numerator. Hence, to eliminate this bias when the objective is to maximize the return on the investment budget, all non-investment costs should be placed in the numerator.[1]

Table 3 Projects ranked by initial cost

Project number	Initial cost	NPV	PI
1	430,000	250,500	1.583
2	380,000	215,000	1.566
3	310,000	40,000	1.129
4	260,000	60,000	1.231
5	230,000	310,000	2.348
6	185,000	105,000	1.568
7	135,000	125,000	1.926
8	110,000	76,000	1.691
9	90,000	122,000	2.356
10	85,000	96,000	2.129

PROJECT SELECTION UNDER CAPITAL RATIONING

When a firm needs to decide whether to invest in certain projects, the net present value (NPV) method is commonly used. In this method, all of the present values of both cash inflows and outflows are accounted for, and the difference in the values represents the NPV. A higher NPV reflects better rewards and higher benefits. If a firm wants to select the best of new projects within a limited budget, then it should choose the set of projects that provides the largest total NPV. Here the BCA can be a useful tool for identifying the best projects to choose under the capital rationing method.[3] This is because the BCA normalizes the evaluation as it measures the NPV per dollar invested regardless of the project size. The following two examples illustrate the selection procedure for projects under capital rationing.

Example 2

In Table 2, several energy conservation measures are assessed in terms of their economic viability. It is required to select the optimum alternative, but also to maintain the budget constraint that states that only $14,000 is available as a capital investment. Assume that the lifespan of the energy conservation measures is 15 years and the discount rate is 5%.

The BCR values are calculated utilizing the present value of money and listed in Table 2. The highest BCR value was that of the R-15 wall insulation (Alternative B), followed by R-10 wall insulation (Alternative A), and then reflective glass (Alternative F). All other measures had BCR values below 1, which means that the returns on investment in these energy conservation measures will not pay for themselves after 15 years. Due to the budget constraint, Alternatives A and F with a total cost of $13,570 represent the optimum alternatives. If the available budget was $15,000 instead of $14,000 then

Table 4 Profitability index for project selection

Project number	Initial cost	NPV	PI	Cumulative Cost	Cumulative NPV	Cost	NPV
9	90,000	122,000	2.36	90,000	122,000	90,000	122,000
5	230,000	310,000	2.35	320,000	432,000	320,000	432,000
10	85,000	96,000	2.13	405,000	528,000	405,000	528,000
7	135,000	125,000	1.93	540,000	653,000	540,000	653,000
8	110,000	76,000	1.69	650,000	729,000	650,000	729,000
1	430,000	250,500	1.58	1,080,000	979,500		
6	185,000	105,000	1.57	1,265,000	1,084,500	835,000	834,000
2	380,000	215,000	1.57	1,645,000	1,299,500	1,215,000	1,049,000
4	260,000	60,000	1.23	1,905,000	1,359,500	1,475,000	1,109,000
3	310,000	40,000	1.13	2,215,000	1,399,500	1,785,000	1,149,000

Alternatives B and F, with a total cost of $14,420, can be selected.

Example 3

A company has a maximum of $1.5 million to invest in retrofitting old buildings to reduce their energy consumption. After evaluating numerous projects, only those that are listed in Table 3 were selected to analyze. Unfortunately, to undertake all these projects a total fund of $2.215 million is required. Accordingly, a decision has to be made to stay within the budget constraint.

The projects in Table 4 are ranked in descending order by their PI. A higher PI is more desirable than the lower one as it creates more NPV per dollar of the initial investment.

Given this scenario (and keeping in mind the budget constraint of $1.5 million), projects 9, 5, 10, 7, 8, 1, and 6 will be selected. The total accumulative cost of these projects amounts to $1.265 million with a cumulative NPV of $1.0845 million. On the other hand, when skipping project 1 and going down the list to include projects 2 and 4, the cumulative initial cost amounts to $1.475 with an NPV of $1.109. The NPV of the second group of projects was higher than that of the first group. This means that the second group will provide a more economic solution.

CONCLUSION

Several points should be considered when conducting BCA.

- The outcome of any analysis will vary depending on the data estimates and assumptions. Thus, it is important to carefully select the assumed values for critical parameters such as the discount rate values to arrive at a realistic solution.
- If the outcome appears particularly sensitive to the value assigned to a given parameter, and the estimate is of poor or unknown quality, the analyst may wish to improve the quality of the data. Sensitivity analysis, a useful technique for identifying critical parameters, should be carried out accordingly.
- Alternatives must be compared over the same study period; otherwise, asset values including costs and savings (benefits) may be repeated.
- NPV and PI will always yield the same decision, though they will not necessarily rank projects in the same order.

REFERENCES

1. American Society for Testing and Materials (ASTM), E964. *Standard Practice for Measuring Benefit-to-Cost and Savings-to-Investment Ratios for Buildings and Building Systems*; American Society for Testing and Materials: West Conshohocken, PA, 2002.
2. American Society for Testing and Materials (ASTM), E1185. *Standard Guide for Selecting Economic Methods for Evaluating Investments in Buildings and Building Systems*; American Society for Testing and Materials: West Conshohocken, PA, 2002.
3. Emery, D.R.; Finnerty, J.D. *Corporate Financial Management*; Prentice-Hall, Inc.: Upper Saddle River, NJ, 1997.
4. Krarti, M. *Energy Audit of Building Systems, an Engineering Approach*; CRC Press: Boca Raton, FL, 2000.
5. McKenzie, A. *Benefit-Cost Analysis for Transportation Projects. Office of Investment Management*; Minnesota Department of Transportation: Washington, DC, 2005.
6. NetMBA.com. Internet Center for Management and Business Administration, Inc. *Finance: Annuities*, http://www.netmba.com/finance/time-value/annuity/ (accessed March 2006).
7. Portney, P.R. *The Concise Encyclopedia of Economics. Benefit-Cost Analysis*. http://www.econlib.org/library/Enc/BenefitCostAnalysis.html (accessed March 2006).
8. Thumann, A.; Wainwright, F. *Financing Energy Projects Deskbook*; The Fairmont Press: Atlanta, GA, 1997.

Biomass

Alberto Traverso
Dipartimento di Macchine Sistemi Energetici e Trasporti, Thermochemical Power Group, Università di Genova, Genova, Italy

Abstract
This entry deals with biomass as an energy source. Different types of biomass are described from the energy perspective, focusing on those more interesting for energy application. The main energy conversion technologies available are outlined, as well as the properties of their main products. Finally, an overview over the benefits that come from biomass exploitation for energy purposes is provided.

INTRODUCTION

This work is organized into seven main sections. The first paragraph provides the reader with a general overview on biomass, including the definition, environmental benefits, energetic properties, and a short list of biomass types that can be used as energy sources. The second paragraph illustrates the mechanical processes needed to produce standardized solid biomass fuels. The third paragraph describes one of the major technologies for converting biomass into energy, which is combustion. The fourth paragraph analyzes pyrolysis and gasification as promising techniques for efficient exploitation of biomass still in a demonstration phase. The fifth paragraph is concerned with biochemical processes for producing biogas and biofuels for transportation. The sixth paragraph outlines the major benefits from biomass exploitation for energy purposes. The seventh paragraph is constituted by the concluding remarks.

GENERALITIES ABOUT BIOMASS

In general, biomass is the substance produced or by-produced by biological processes.

Commonly, biomass refers to the organic matter derived from plants and generated through photosynthesis. Biomass not only provides food but also construction materials, fibers, medicines, and energy. In particular, biomass can be regarded as solar energy stored in the chemical bonds of the organic material. Carbon dioxide (CO_2) from the atmosphere and water absorbed by the plants roots are combined in the photosynthetic process to produce carbohydrates (or sugars) that form the biomass. The solar energy that drives photosynthesis is stored in the chemical bonds of the biomass structural components. During biomass combustion, oxygen from the atmosphere combines with the carbon and hydrogen in biomass to produce CO_2 and water. The process is therefore cyclic because the carbon dioxide is then available to produce new biomass. This is also the reason why bioenergy is potentially considered carbon-neutral, although some CO_2 emissions occur due to the use of fossil fuels during the production and transport of biofuels.

Biomass resources can be classified according to the supply sector, as shown in Table 1.

The chemical composition of plant biomass varies among species. Yet, in general terms, plants are made of approximately 25% lignin and 75% carbohydrates or sugars. The carbohydrate fraction consists of many sugar molecules linked together in long chains or polymers. Two categories are distinguished: cellulose and hemi-cellulose. The lignin fraction consists of non-sugar-type molecules that act as a glue, holding together the cellulose fibers.

The Energy Content of Biomass

Bioenergy is energy of biological and renewable origin, normally derived from purpose-grown energy crops or by-products of agriculture. Examples of bioenergy resources are wood, straw, bagasse, and organic waste. The term bioenergy encompasses the overall technical means through which biomass is produced, converted, and used. Fig. 1 summarizes the variety of processes for energy production from biomass.

The calorific value of a fuel is usually expressed as the higher heating value (HHV) or the lower heating value (LHV). The difference is caused by the heat of evaporation of the water formed from the combustion of hydrogen in the material and the original moisture. Note that the difference between the two heating values depends on the chemical composition of the fuel.

Keywords: Biomass; Renewable energy; Bioenergy; Biomass conversion; Biofuel.

Table 1 Types of biomass for energy use

Supply sector	Type	Example
Forestry	Dedicated forestry	Short rotation plantations (e.g., willow, poplar, eucalyptus)
	Forestry by-products	Wood blocks, wood chips from thinnings
Agriculture	Dry lignocellulosic energy crops	Herbaceous crops (e.g., miscanthus, reed canary-grass, giant reed)
	Oil, sugar, and starch energy crops	Oil seeds for methylesters (e.g., rape seed, sunflower)
		Sugar crops for ethanol (e.g., sugar cane, sweet sorghum)
		Starch crops for ethanol (e.g., maize, wheat)
	Agricultural residues	Straw, prunings from vineyards, and fruit trees
	Livestock waste	Wet and dry manure
Industry	Industrial residues	Industrial waste wood, sawdust from saw mills
		Fibrous vegetable waste from paper industries
Waste	Dry lignocellulosic	Residues from parks and gardens (e.g., prunings, grass)
	Contaminated waste	Demolition wood
		Organic fraction of municipal solid waste
		Biodegradable landfilled waste, landfill gas
		Sewage sludge

Source: From European Biomass Industry Association and DOE Biomass Research and Development Initiative (see Refs. 1, 2).

The most important property of biomass feedstocks with regard to combustion—and to the other thermochemical processes—is the moisture content, which influences the energy content of the fuel. Wood, just after falling, has a typical 55% water content and LHV of approximately 7.1 MJ/kg; logwood after 2–3 years of air-drying may present 20% water content and LHV of 14.4 MJ/kg; pellets show a quite constant humidity content of about 8% with LHV equal to 17 MJ/kg.

MECHANICAL PROCESSES FOR ENERGY DENSIFICATION

Some practical problems are associated with the use of biomass material (sawdust, wood chips, or agricultural residues) as fuel. Those problems are mainly related to the high bulk volume, which results in high transportation costs and requires large storage capacities, and to the high moisture content, which can result in biological degradation as well as in freezing and blocking the in-plant transportation systems. In addition, variations in moisture content make difficult the optimal plant operation and process control. All of those problems may be overcome by standardization and densification. The former consists of processing the original biomass in order to obtain fuels with standard size and heating properties, while the latter consists in compressing the material, which needs to be available in the sawdust size, to give it more uniform properties.

Table 2 reports the main features of pellets, briquettes, and chips.

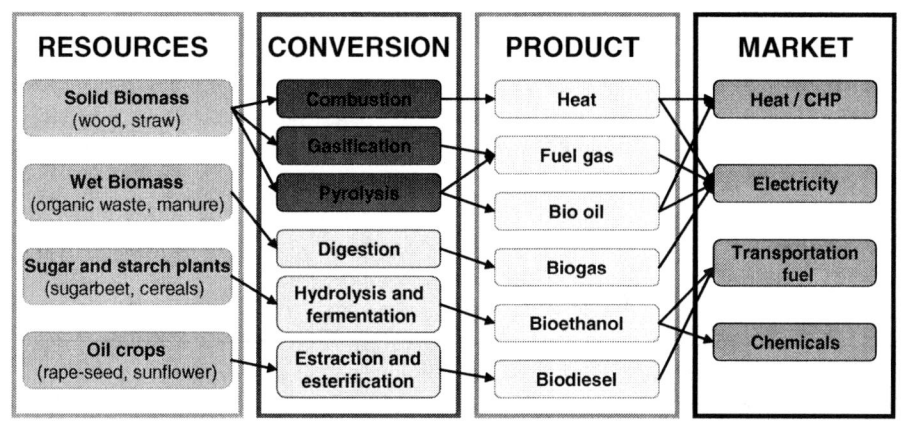

Fig. 1 Processes to convert biomass into useful energy, i.e., bioenergy. Source: From European Biomass Industry Association, Elsevier Applied Science Publishers Ltd, and Risoe National Laboratory (see Refs. 1, 3, and 4).

Table 2 Comparison of different solid wood fuels

	Pellets	Briquettes	Chips
Appearance			
Raw material	Dry and ground wood or agricultural residues	Dry and ground wood or agricultural residues. Raw material can be more coarse than for pelleting due to the larger dimensions of final product	Dry wood logs
Shape	Cylindrical (generally Ø 6–12 mm, with a length 4–5 times the Ø)	Cylindrical (generally Ø 80–90 mm) or parallelepiped (150×70×60 mm)	Irregularly parallelepiped (70×30×3 mm)

BIOMASS COMBUSTION

The burning of wood and other solid biomass is the oldest energy technology used by man. Combustion is a well-established commercial technology with applications in most industrialized and developing countries, and development is concentrated on resolving environmental problems, improving the overall performance with multi-fuel operation, and increasing the efficiency of the power and heat cycles (CHP).

The devices used for direct combustion of solid biomass fuels range from small domestic stoves (1–10 kW) to the large boilers used in power and CHP plants (>5 MW). Intermediate devices cover small boilers (10–50 kW) used for heating in single-family houses, medium-sized boilers (50–150 kW) used for heating in multifamily houses or buildings, and large boilers (150 kW to over 1 MW) used for district heating. Cofiring in fossil-fired power stations enables the advantages of large size plants (>100 MWe) that are not applicable for dedicated biomass combustion due to limited local biomass availability.

To achieve complete burnout and high efficiencies in small-scale combustion, downdraft boilers with inverse flow have been introduced, which apply the two-stage combustion principle. An operation at very low load should be avoided as it can lead to high emissions. Hence, it is recommended to couple log wood boilers to a heat storage tank. Because wood pellets are well suited for automatic heating at small heat outputs (as needed for nowadays buildings), pellet furnaces are an interesting application with growing propagation. Thanks to the well-defined fuel at low water content, pellet furnaces can easily achieve high combustion quality. They are applied both as stoves and as boilers, and they are encountering increasing acceptance especially in urban areas because modern pellet stoves are nowadays efficient home heating appliances. While a conventional fireplace is less than 10% efficient at delivering heat to a house, an average modern pellet stove achieves 80%–90% efficiency. Technology development has led to the application of strongly improved heating systems that are automated and have catalytic gas cleaning equipment. Such systems significantly reduce the emissions from fireplaces and older systems while at the same time improving the efficiency.

Understoker furnaces are mostly used for wood chips and similar fuel with relatively low ash content, while grate furnaces can also be applied for high ash and water content. Special types of furnaces have been developed for straw that has very low density and is usually stored in bales. Other than conventional grate furnaces operated with whole bales, cigar burners and other specific furnaces are in operation. Stationary or bubbling fluidized bed (SFB) as well as circulating fluidized bed (CFB) boilers are applied for large-scale applications and are often used for waste wood or mixtures of wood and industrial wastes, e.g., from the pulp and paper industry.

Co-Combustion

Bioenergy production may be hampered by limitations in the supply or by fuel quality. In those cases, the cofiring of several types of biomass or cofiring biomass with coal ensures flexibility in operation, both technically and economically. Several concepts have been developed:

- *Co-combustion or direct cofiring.* The biomass is directly fed to the boiler furnace, if needed, after physical preprocessing of the biomass such as drying, grinding, or metal removal is applied. This typically takes place in bubbling or CFB combustors. Such technologies can be applied to a wide range of fuels, even for very wet fuels like bark or sludge. Multifuel

fluidized bed boilers achieve efficiencies over 90% while flue gas emissions are lower than for conventional grate combustion due to lower combustion temperatures.
- *Indirect cofiring.* Biomass is first gasified and the fuel gas is cofired in the main boiler. Sometimes the gas has to be cooled and cleaned, which is more challenging and implies higher operation costs.
- *Parallel combustion.* The biomass is burnt in a separate boiler for steam generation. The steam is used in a power plant together with the main fuel.

Problems in Biomass Combustion

Biomass has a number of characteristics that makes it more difficult to handle and combust than fossil fuels. The low energy density is the main problem in handling and transport of the biomass, while the difficulties in using biomass as fuel relates to its content of inorganic constituents. Some types of biomass used contain significant amounts of chlorine, sulfur, and potassium. The salts—KCl and K_2SO_4—are quite volatile, and the release of these components may lead to heavy deposition on heat transfer surfaces, resulting in reduced heat transfer and enhanced corrosion rates. Severe deposits may interfere with operation and cause unscheduled shut downs.

In order to minimize these problems, various fuel pretreatment processes have been considered, including washing the biomass with hot water or using a combination of pyrolysis and char treatment.

THERMOCHEMICAL CONVERSION OF BIOMASS

Pyrolysis and gasification are the two most typical thermochemical processes that do not produce useful energy directly because they convert the original bioenergy feedstock into more convenient energy carriers such as producer gas, oil, methanol, and char.[3]

Pyrolysis

Pyrolysis is a process for thermal conversion of solid fuels, like biomass or wastes, in the complete absence of oxidizing agent (air/oxygen) or with such limited supply that gasification does not occur to any appreciable extent. Commercial applications are either focused on the production of charcoal or the production of a liquid product—the bio-oil and pyro-gas. Charcoal is a very ancient product, even if traditional processes (partial combustion of wood covered by a layer of earth) are very inefficient and polluting. Modern processes such as rotary kiln carbonization are presently used in industry. Bio-oil production (or wood liquefaction) is potentially very interesting as a substitute for fuel oil and as a feedstock for production of synthetic gasoline or diesel fuel. Pyro-gas has higher energy density than gasification gas (syngas) because it has been created without oxygen (and nitrogen, if air is employed), hence it does not contain the gaseous products of partial combustion.

The pyrolysis process takes place at temperatures in the range 400°C–800°C and during this process most of the cellulose and hemicellulose and part of the lignin will disintegrate to form smaller and lighter molecules, which are gases at the pyrolysis temperature. As these gases cool, some of the vapors condense to form a liquid, which is the bio-oil and the tars. The remaining part of the biomass, mainly parts of the lignin, is left as a solid, i.e., the charcoal. It is possible to influence the product mix through a control of heating rate, residence time, pressure, and maximum reaction temperature so that either gases, condensable vapors, or the solid charcoal is promoted.

Gasification

Gasification technology has been developing since the 18th century, and it is still in a development phase.[5,6] Gasification is a conversion process that involves partial oxidation at elevated temperature. It is intermediate between combustion and pyrolysis. In fact, oxygen (or air) is present but it is not enough for complete combustion. This process can start from carbonaceous feedstock such as biomass or coal and convert them into a gaseous energy carrier. The overall gasification process may be split into two main stages: the first is pyrolysis stage, i.e., where oxygen is not present but temperature is high, and here typical pyrolysis reactions take place; the second stage is the partial combustion, where oxygen is present and it reacts with the pyrolyzed biomass to release heat necessary for the process. In the latter stage, the actual gasification reactions take place, which consist of almost complete charcoal conversion into lighter gaseous products (e.g., carbon monoxide and hydrogen) through the chemical oxidizing action of oxygen, steam, and carbon dioxide. Such gases are injected into the reactor near the partial combustion zone (normally, steam and carbon dioxide are mutually exclusive). Gasification reactions require temperature in excess of 800°C to minimize tar and maximize gas production. The gasification output gas, called "producer gas," is composed by hydrogen (18%–20%), carbon monoxide (18%–20%), carbon dioxide (8%–10%), methane (2%–3%), trace amounts of higher hydrocarbons like ethane and ethene, water, nitrogen (if air is used as oxidant agent), and various contaminants such as small char particles, ash, tars, and oils. The incondensable part of producer gas is called "syngas" and it represents the useful product of gasification. If air is used, syngas has a high heating value in the order of 4–7 MJ/m^3, which is exploitable for boiler, engine, and turbine operation, but due to its low energy density, it is not suitable for pipeline transportation. If pure oxygen is used, the syngas high heating value almost doubles (approximately 10–18 MJ/m^3 high heating value), hence such a syngas is suitable for limited pipeline distribution as well as for conversion to liquid fuels (e.g., methanol and gasoline). However, the most common technology is the air gasification because it avoids the costs and the hazards of oxygen production and usage. With air gasification, the syngas efficiency—describing the energy content of the cold gas stream in relation to that of the input biomass stream—is in the order of 55%–85%, typically 70%.

Comparison of Thermal Conversion Methods of Biomass

Table 3 reports a general overview on specific features of the conversion technologies analyzed here, showing the related advantages and drawbacks.

BIOCHEMICAL CONVERSION OF BIOMASS

Biochemical conversion of biomass refers to processes that decompose the original biomass into useful products. Commonly, the energy product is either in the liquid or in gaseous forms, hence it is called "biofuel" or "biogas," respectively. Biofuels are very promising for transportation sector, while biogas is used for electricity and heat production. Normally, biofuels are obtained from dedicated crops (e.g., biodiesel from seed oil), while biogas production results from concerns over environmental issues such as the elimination of pollution, the treatment of waste, and the control of landfill greenhouse gas emissions.

Biogas from Anaerobic Digestion

The most common way to produce biogas is anaerobic digestion of biomass. Anaerobic digestion is the bacterial breakdown of organic materials in the absence of oxygen. This biochemical process produces a gas called biogas, principally composed of methane (30%–60% in volume) and carbon dioxide. Such a biogas can be converted to energy in the following ways:

- Biogas converted by conventional boilers for heating purposes at the production plant (house heating, district heating, industrial purposes).
- Biogas for combined heat and power generation.
- Biogas and natural gas combinations and integration in the natural gas grid.
- Biogas upgraded and used as vehicle fuel in the transportation sector.
- Biogas utilization for hydrogen production and fuel cells.

An important production of biogas comes from landfills. Anaerobic digestion in landfills is brought about by the microbial decomposition of the organic matter in refuse. Landfill gas is on average 55% methane and 45% carbon dioxide. With waste generation increasing at a faster rate than economic growth, it makes sense to recover the energy from that stream through thermal or fermentation processes.

Biofuels for Transport

A wide range of chemical processes may be employed to produce liquid fuels from biomass. Such fuels can find a very high level of acceptance by the market thanks to their relatively easy adaptation to existing technologies (i.e., gasoline and diesel engines). The main potential biofuels are outlined below.

- Biodiesel is a methyl-ester produced from vegetable or animal oil to be used as an alternative to conventional petroleum-derived diesel fuel. Compared to pure vegetable or animal oil, which can be used in adapted diesel engines as well, biodiesel presents lower viscosity and slightly higher HHV.
- Pure vegetable oil is produced from oil plants through pressing, extraction, or comparable procedures, crude or refined but chemically unmodified. Usually, it is

Table 3 Qualitative comparison of technologies for energy conversion of biomass

Process	Technology	Economics	Environment	Market potential	Present deployment
Combustion–heat	+++	€	+++	+++	+++
Combustion–electricity	++(+)	€€	++(+)	+++	++
Gasification	+(+)	€€€	+(++)	+++	(+)
Pyrolysis	(+)	€€€€	(+++)	++(+)	(+)

+, low; +++, high; €, cheap; €€€€, expensive.
Source: From European Biomass Industry Association and Risoe National Laboratory (see Refs. 1, 4).

compatible with existing diesel engines only if blended with conventional diesel fuel at rates not higher than 5%–10% in volume. Higher rates may lead to emission and engine durability problems.

- Bioethanol is ethanol produced from biomass or the biodegradable fraction of waste. Bioethanol can be produced from any biological feedstock that contains appreciable amounts of sugar or other matter that can be converted into sugar, such as starch or cellulose. Also, ligno-cellulosic materials (wood and straw) can be used, but their processing into bioethanol is more expensive. Application is possible to modified spark ignition engines.
- Bio-ETBE (ethyl-tertio-butyl-ether) is ETBE produced on the basis of bioethanol. Bio-ETBE may be effectively used to enhance the octane number of gasoline (blends with petrol gasoline).
- Biomethanol is methanol produced from biomass. Methanol can be produced from gasification syngas (a mixture of carbon monoxide and hydrogen) or wood dry distillation (old method with low methanol yields). Most all syngas for conventional methanol production is produced by the steam reforming of natural gas into syngas. In the case of biomethanol, a biomass is gasified first to produce a syngas from which the biomethanol is produced. Application is possible to spark ignition engines and fuel cells. Compared to ethanol, methanol presents more serious handling issues because it is corrosive and poisonous for human beings.
- Bio-MTBE (methyl-tertio-butyl-ether) is a fuel produced on the basis of biomethanol. It is suitable for blends with petrol gasoline.
- Biodimethylether (DME) is dimethylether produced from biomass. Bio-DME can be formed from syngas by means of oxygenate synthesis. It has emerged only recently as an automotive fuel option. Storage capabilities are similar to those of LPG. Application is possible to spark ignition engines.

BENEFITS FROM BIOMASS ENERGY

There is quite a wide consensus that, over the coming decades, modern biofuels will provide a substantial source of alternative energy. Nowadays, biomass already provides approximately 11%–14% of the world's primary energy consumption (data varies according to sources).

There are significant differences between industrialized and developing countries. In particular, in many developing countries, bioenergy is the main energy source—even if used in very low-efficiency applications (e.g., cooking stoves have an efficiency of about 5%–15%). Furthermore, inefficient biomass utilization is often associated with the increasing scarcity of hand-gathered wood, nutrient depletion, and the problems of deforestation and desertification.

One of the key drivers to bioenergy deployment is its positive environmental benefit, in particular regarding the global balance of green house gas (GHG) emissions. This is not a trivial matter, because biomass production and use are not entirely GHG neutral. In general terms, the GHG emission reduction as a result of employing biomass for energy reads as reported in Table 4.

Bioenergy is a decentralized energy option whose implementation presents positive impacts on rural development by creating business and employment opportunities. Jobs are created all along the bioenergy chain, from biomass production or procurement to its transport, conversion, distribution, and marketing.

Bioenergy is a key factor for the transition to a more sustainable development.

CONCLUSIONS

Biomass refers to a very wide range of substances produced by biological processes. In the energy field, special focus has been and will be placed on vegetable biomass such as wood and agricultural by-products because of the energy potential as well as economic and environmental benefits. Size and humidity standardization of biomass is a necessary step to make it suitable for effective domestic and industrial exploitation. Chips, briquettes, and pellets are modern examples of standard solid fuels.

Biomass can be converted to energy in three may pathways: combustion, thermochemical processing, and biochemical processing. The combustion of solid biomass for the production of heat or electricity is the most viable technology, while pyrolysis and gasification still face economic and reliability issues. Among biochemical processes, anaerobic digestion is often used to reduce the environmental impact of hazardous waste and landfills. Biochemical processes are also

Table 4 Benefits in reduction of green houses gas emissions

+	Avoided mining of fossil resources
−	Emission from biomass production
+	Avoided fossil fuel transport (from producer to user)
−	Emission from biomass fuel transport (from producer to user)
+	Avoided fossil fuel utilization

+, positive; −, neutral.
Source: From Risoe National Laboratory (see Ref. 4).

concerned with the conversion of biomass into useful fuels for transportation, such as biodiesel, bioethanol, biomethanol, and others. All of them can effectively contribute to the transition to a more sustainable transportation system at zero GHG emissions.

Biomass represents a viable option for green energy resources of the 21st century.

ACKNOWLEDGMENTS

The author wishes to thank Professor A. F. Massardo of the University of Genoa for the invaluable help during the last years of research activity.

REFERENCES

1. European Biomass Industry Association, http://www.eubia.org (accessed on November 21, 2006).
2. DOE Biomass Research and Development Initiative, http://www.brdisolutions.com (accessed on November 21, 2006).
3. Overend, R.P.; Milne, T.A.; Mudge, L.K. *Fundamentals of Thermochemical Biomass Conversion*; Elsevier Applied Science Publishers Ltd: NewYork, 1985.
4. Risoe National Laboratory, Denmark, http://www.risoe.dk (accessed on November 21, 2006).
5. Bridgewater, A.V. The technical and economic feasibility of biomass gasification for power generation. Fuel J. **1995**, *74* (6–8), 557–564.
6. Franco, A.; Giannini, N. Perspectives for the use of biomass as fuel in combined cycle power plants. Int. J. Thermal Sci. **2005**, *44* (2), 163–177.

Boilers and Boiler Control Systems

Eric Peterschmidt
Honeywell Building Solutions, Honeywell, Golden Valley, Minnesota, U.S.A.

Michael Taylor
Honeywell Building Solutions, Honeywell, St. Louis Park, Minnesota, U.S.A.

Abstract

Many commercial and industrial facilities use boilers to produce steam or hot water for space heating or for process heating. Boilers are typically major users of energy, and any person involved in energy management needs to know how a boiler works and how the performance of a boiler can be maintained or improved. This article outlines the types of boilers used to heat facilities, while providing an overview of basic boiler controls and the parameters that affect energy efficiency.

INTRODUCTION

A boiler is a closed vessel intended to heat water and produce hot water or steam through the combustion of a fuel or through the action of electrodes or electric resistance elements. Many commercial and industrial facilities use boilers to produce steam or hot water for space heating or for process heating. Boilers are typically major users of energy, and any person involved in a facility's energy management needs to know how a boiler works and how the performance of a boiler can be maintained or improved. In particular, it is important to know what parameters of a boiler system are the most important. For fossil fuel-fired boilers, the combustion efficiency is the major parameter of interest; this is most often controlled by providing the optimum amount of combustion air that is mixed with the fuel. Thus, understanding boiler control systems is extremely important. Steam and hot water boilers are available in standard sizes from very small boilers for apartments and residences to very large boilers for commercial and industrial uses.

BOILER TYPES

Boilers are classified by water temperature or steam pressure. They are further classified by the type of metal used in construction (cast iron, steel, or copper), by the type of fuel or heat element (oil, gas, or electricity), or by the relationship of fire or water to the tubes (i.e., firetube or watertube).

- Low-pressure boilers are those designed to produce steam up to 15 psig or hot water up to 250°F with pressures up to 160 psig.
- Medium- and high-pressure boilers produce steam above 15 psig or hot water above 160 psig or 250°F or both.

Boilers are typically constructed of cast iron or welded steel. Cast iron boilers (Fig. 1) are made of individually cast sections and are joined together using screws or nuts and tie rods or threaded rivets. The number of sections can be varied to provide a range of capacities.

Steel boilers come in a wide variety of configurations. They are factory-assembled and welded and shipped as a unit. Fig. 2 illustrates a firetube boiler. The fire and flue gases are substantially surrounded by water. The products of combustion pass through tubes to the back then to the front and once more to the back before finally exiting at the front. This makes it a four-pass boiler. Firetube boilers are manufactured in many other configurations such as:

- External firebox—The firebox is not surrounded by water.
- Dry back—Firetubes are directly available from cleanout doors at the back of boiler.
- Scotch–Marine—Employs low water volume and has a fast response.

Watertube boilers are steel body boilers used for high capacity requirements of more than 2 million Btu per hour (Btu/h). Watertube boilers use a water-cooled firebox which prolongs the life of furnace walls and refractories.

Keywords: Boiler; Boiler controls; Automatic temperature controls; Water/steam; Combustion; Heating; Safeguard control; Energy efficiency.

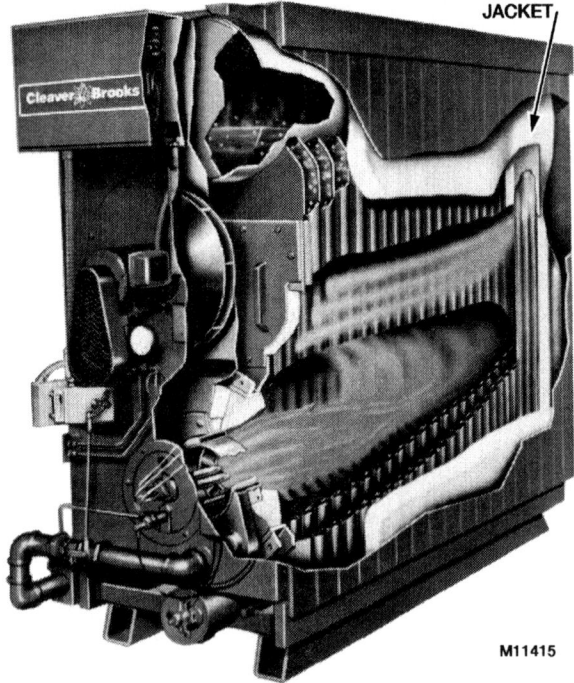

Fig. 1 Typical cast iron boiler (watertube).

Modular boilers are small, hot water boilers rated from 200,000 to 900,000 Btu/h input. These boilers are available with gross efficiencies of 85% or higher. Fig. 3 shows the features of a typical modular boiler. These boilers are often used in tandem to provide hot water for space heating and/or domestic hot water. For example, if the designed heating load were 2 million Btu/h, four 600,000 Btu/h (input) modular boilers might be used. If the load were 25% or less on a particular day, only one boiler would fire and cycle on and off to supply the load. The other three boilers would remain off with no water flow. This reduces the flue and jacket (covering of the boiler) heat losses.

Some modular boilers have a very small storage capacity and very rapid heat transfer so water flow must be proven before the burner is started.

Electric boilers heat water or produce steam by converting electrical energy to heat using either resistance elements or electrodes. Electric boilers are considered to be 100% efficient since all power that is consumed directly produces hot water or steam. Heat losses through the jacket and insulation are negligible and there is no flue.[1]

Electrode boilers (as seen in Fig. 4) have electrodes immersed in the water. Electrical current passes through the water between electrodes, and this current and the resistance of the water results in generated heat. Electrode boilers are available in sizes up to 11,000 kW. Resistance boilers have the resistance (heating) elements immersed in, but electrically insulated from, the water and are manufactured in sizes up to 3,000 kW. Electric elements and electrodes are usually grouped to provide four or more stages of heating. A step controller responds to steam pressure or hot water temperature, activating each stage of heating as required to heat the building.

BOILER RATINGS AND EFFICIENCY

Boilers can be rated in several ways. Fig. 5 shows the commonly used ratings and terms. The terms Btu/h (Btu per hour) and MBtu/h or MB/H (1000 Btu/h) indicate the

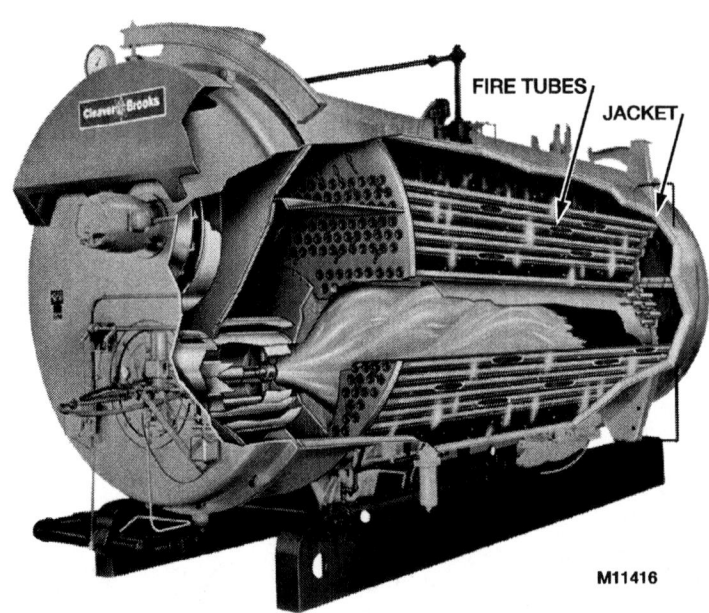

Fig. 2 Typical firetube boiler.

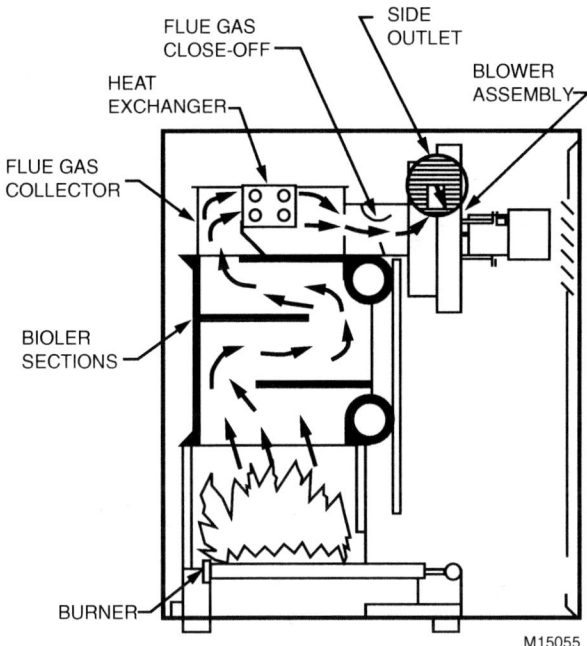

Fig. 3 High efficiency modular boiler.

boiler's input rate. Input ratings are usually shown on the boiler's (or burner's) nameplate. The terms bhp (boiler horse power), EDR (equivalent direct radiation), and pounds per hour (of steam) indicate the boiler's output rate.

Gross efficiency of the boiler is the output (steam or water heat content and volume) divided by the fuel input (measured by a fuel meter at steady-state firing conditions). The combustion efficiency, as indicated by flue gas conditions, does not take into account jacket, piping, and other losses, so it is always higher than the gross efficiency.

A testing procedure issued by the U.S. Department of Energy in 1978 measures both on-cycle and off-cycle losses based on a laboratory procedure involving cyclic conditions. The result is called the AFUE (Annual Fuel Utilization Efficiency) rating, or seasonal efficiency, which is lower than gross efficiency.

COMBUSTION IN BOILERS

When gas, oil, or another fuel is burned, several factors must be considered if the burning process is to be safe, efficient, and not impact the environment. The burning process must adhere to the following guidelines:

1. Provide enough air so that combustion is complete, and undesirable amounts of carbon monoxide or other pollutants are not generated.
2. Avoid excess air in the fuel-air mixture which would result in low efficiency.
3. Completely mix the air with fuel before introducing the mixture into the firebox.
4. Provide safety controls so that fuel is not introduced without the presence of an ignition flame or spark and so that flame is not introduced in the presence of unburned fuel.
5. Avoid water temperatures below the dewpoint of the flue gas to prevent condensation on the fireside of the boiler.

Combustion can be monitored by flue gas analysis. For large boilers, over 1,000,000 Btu/h, the analysis is typically continuous. For small boilers, flue gas is analyzed periodically using portable instruments. Flue gas composition analysis routinely measures the percentage of CO_2 (carbon dioxide) or O_2 (oxygen), but generally not both. Ideal CO_2 concentration is in the 10%–12% range. The percentage of oxygen remaining is the most reliable indication of complete combustion. The ideal O_2 concentration in the flue gas is in the 3%–5% range. Lower concentrations are impractical and often unsafe. Higher O_2 concentrations mean that an excessive quantity of air is being admitted to the combustion chamber and must be heated by the fuel. This excess air passes through the boiler too quickly for the heat to be efficiently transferred to the water or steam, and thereby reduces the combustion efficiency. CO_2 measuring instruments are simpler and cost less than O_2 measuring instruments.

The CO_2 or O_2 concentration, plus the stack temperature, provides a burner combustion efficiency in percent—either directly or by means of charting. This combustion efficiency indicates only the amount of heat extracted from

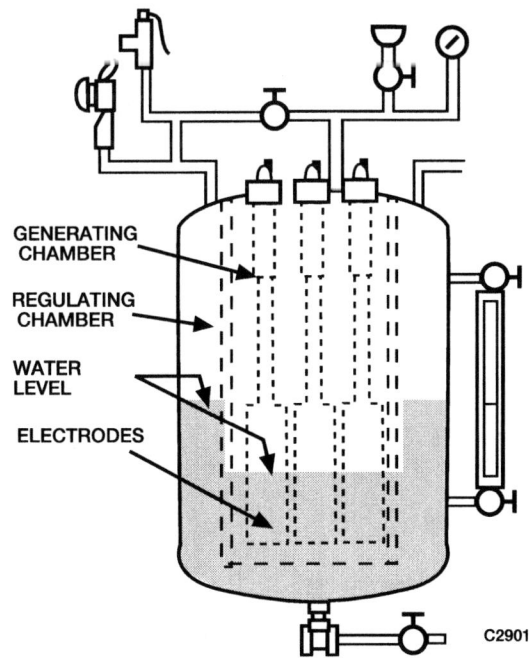

Fig. 4 Electrode steam boiler.

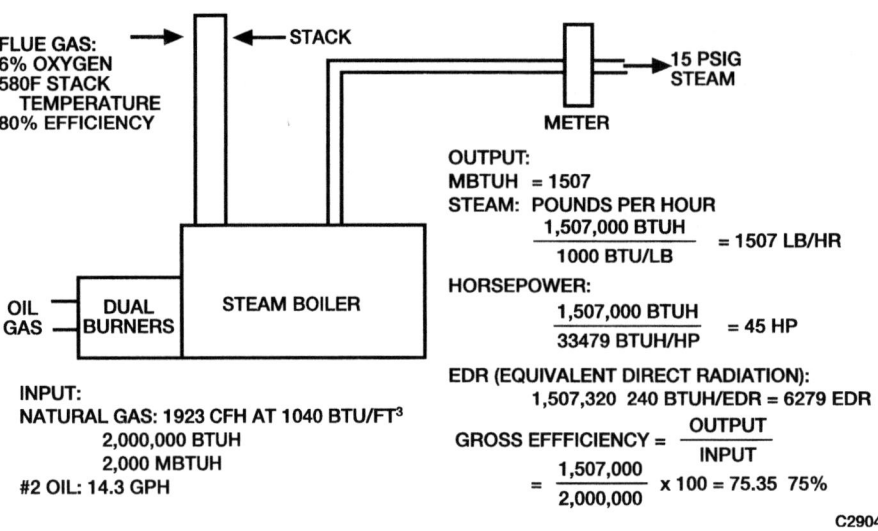

Fig. 5 Boiler ratings and efficiency.

the fuel. It does not account for excess heating of combustion air, or losses from leaks or the boiler jacket, among other factors.

For oil-fired boilers, the oil burners are usually of the atomizing variety, that is, they provide a fine spray of oil. Several types of these oil burners exist:

- Gun type burners spray oil into a swirling air supply.
- Horizontal, rotary burners use a spinning cup to whirl oil and air into the furnace.
- Steam- or air-atomizing burners use high pressured air or 25 psig steam to break up the oil into fine droplets.

For modulating or high/low flame control applications, the rotary or steam/air-atomizing burners are most common.

For natural gas-fired boilers, the two typical types of gas burners are the atmospheric injection burner and the power type burner. The atmospheric injection burner uses a jet of gas to aspirate combustion air and is commonly used in home gas furnaces and boilers. The raw-gas ring burner (refer to Fig. 6) is an atmospheric injection burner. Power burners (refer to Fig. 7) use a forced-draft fan to thoroughly mix air and gas as they enter the furnace. Common power burner applications are in the commercial and industrial sectors.

BASIC BOILER CONTROLS

Boilers have to provide steam or hot water whenever heat is needed. A conventional BMCS (boiler management control system) is often set to provide a continuous hot water or steam supply between October and May at anytime the OA (outside air) temperature drops to 60°F for more than 30 min and an AHU (air handling unit) is calling for heat. The BCMS should include a software on/off/auto function. Unlike chillers, boilers can be left enabled at no-load conditions, during which time the water temperature will be held at the designed temperature. Frequent warm-up and shut-down of boilers causes stress buildup. Boiler manufacturers' recommendations provide specific guidelines in this area of operation.

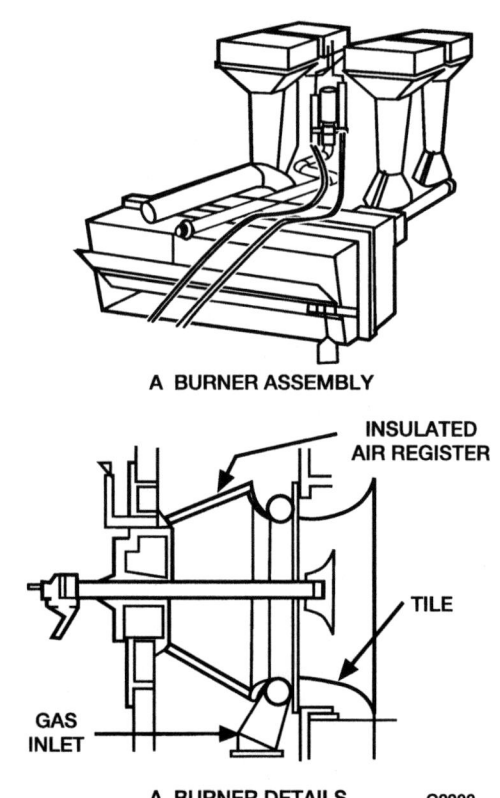

Fig. 6 Raw gas ring burner.

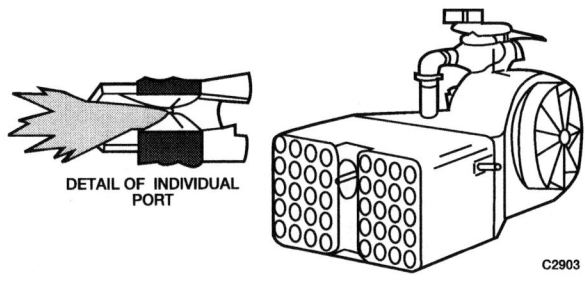

Fig. 7 Multiport forced-draft gas burner.

Unless a low-limit for water temperature is used, hot water boiler burners are not controlled to provide water temperatures based on outdoor temperatures, because the reset schedules require water temperatures to be supplied below the dewpoint temperature of the flue gas. Some boilers require incoming water temperatures to be above 140°F before going to high-fire. In this case, if a building is using a hot water system and the boiler is locked into low-fire because the incoming water is too cold, the system may never recover.

The following are three ways to control the output of a commercial boiler:

1. On/off (cycling) control
2. High-fire/low-fire control
3. Modulating control

On/off (cycling) control is most common for small boilers up to 1,000,000 Btu/h capacity. The oil or gas burner cycles on and off to maintain steam pressure or water temperature. Cycling control causes losses in efficiency because of the cooling (which is necessary for safety) of the fireside surfaces by the natural draft from the stack during the off, pre-purge and post-purge cycles.

High-fire/low-fire burners provide fewer off-cycle losses since the burner shuts off only when loads are below the low-fire rate of fuel input.

Modulating control is used on most large boilers because it adjusts the output to match the load whenever the load is greater than the low-fire limit, which is usually not less than 15% of the full load capacity. Steam pressure or hot water temperature is measured to determine the volume of gas or oil admitted to the burner.

Boiler firing and safety controls are boiler-manufacturer furnished and code approved. A BMCS usually enables a boiler to fire, provides a setpoint, controls pumps and blending valves, and monitors operation and alarms.

Combustion control regulates the air supplied to a burner to maintain a high gross efficiency in the combustion process. More sophisticated systems use an oxygen sensor in the stack to control the amount of combustion air supplied. Smoke density detection devices can be used in the stack to limit the reduction of air so stack gases stay within smoke density limits. A continuous reading and/or recording of flue gas conditions—O_2 concentration percentage, stack temperature—is usually included in the control package of large boilers.

A simple combustion control system contains a linkage that readjusts the air supply from the same modulating motor that adjusts the fuel supply (refer to Fig. 8). There may be a provision to stop the flow of air through the fluebox during the off-cycles.

Flame Safeguard Control

Flame safeguard controls are required on all burners. Flame controls for large burners can be very complicated while controls for small burners such as a residential furnace are relatively simple. The controls must provide foolproof operation—i.e., they must make it difficult or impossible to override any of the safety features of the system. The controls also should be continuously self checked. For commercial and industrial burners, the flame

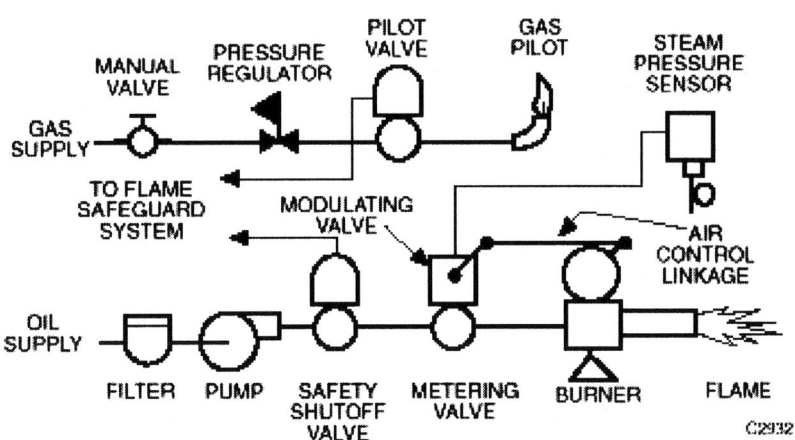

Fig. 8 Combustion control for rotary oil burner.

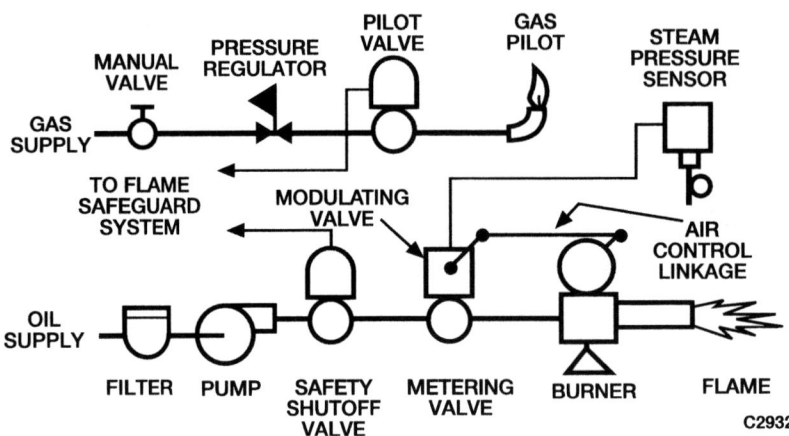

Fig. 9 Simple flame safeguard for a gas furnace.

safeguard control generally goes through a series of operations similar to the following.

- Purge the firebox of unburned fuel vapor (prepurge).
- Light the pilot.
- Verify that the pilot is lit.
- Open the main fuel valve.
- Verify that the flame is present as soon as fuel is introduced.
- Cut off the fuel supply promptly if the flame fails.
- Purge the firebox of any unburned fuel after each on-cycle (post-purge).

The key to any flame safeguard system is a reliable and fast means of detecting the presence or absence of a flame. Methods of detection include:

- Response of a bimetal sensor to heat (slow response).
- Response of a thermocouple to heat (slow response).
- Flame conductivity (fast, but not reliable response)
- Flame rectification (fast, reliable response).
- Ultraviolet flame detection (fast, reliable response).
- Lead sulfide (photo) cells (fast, reliable response if a flame frequency check is included).

Some sensors can potentially malfunction because of short circuits, hot refractories, or external light sources. Other sensors, like flame rectification and ultraviolet detection, respond to flame only. Flame safeguard systems must be approved by Underwriter's Laboratory (UL) or Factory Mutual for specific applications. Fig. 9 shows a flame safeguard system commonly applied to small gas boilers or furnaces. The flame of the gas pilot impinges on a thermocouple which supplies an electric current to keep the pilotstat gas valve open. If the pilot goes out or the thermocouple fails, the pilotstat valve closes or remains closed preventing gas flow to the main burner and pilot. The pilotstat must be manually reset.

Fig. 10 shows how flame safeguard controls are integrated with combustion controls on a small, oil-fired steam boiler. The ultraviolet (UV) flame detector is located where it can see the flame and will shutdown the burner when no flame is present.

In addition to the combustion, safety, and flame safeguard controls shown in Fig. 10, larger burners often provide additional measuring instrumentation such as:

- Percentage of O_2 or CO_2 in flue gas (to monitor combustion efficiency)
- Flue gas temperature
- Furnace draft (in inches of water) column
- Steam flow with totalizer or hot water Btu with totalizer
- Oil and/or gas flow with totalizer
- Stack smoke density

CONTROL OF MULTIPLE BOILER SYSTEMS

Basic boiler connections for a three-zone hot water system are shown in Fig. 11. In this system, two boilers are connected in parallel. Hot water from the top of the boilers moves to the air separator which removes any entrapped air from the water. The expansion tank connected to the separator maintains the pressure in the system. The tank is about half full of water under normal operating conditions. Air pressure in the tank keeps the system pressurized and allows the water to expand and contract as the system water temperature varies. Water from the boiler moves through the separator to the three zone pumps, each of which is controlled by its own thermostat. In some systems, each zone may have a central pump and a valve. Return water from each zone goes back to the boiler in the return line. Several variations are possible within this type system, but the process is the same. There is no minimum boiler water flow limit in this example.

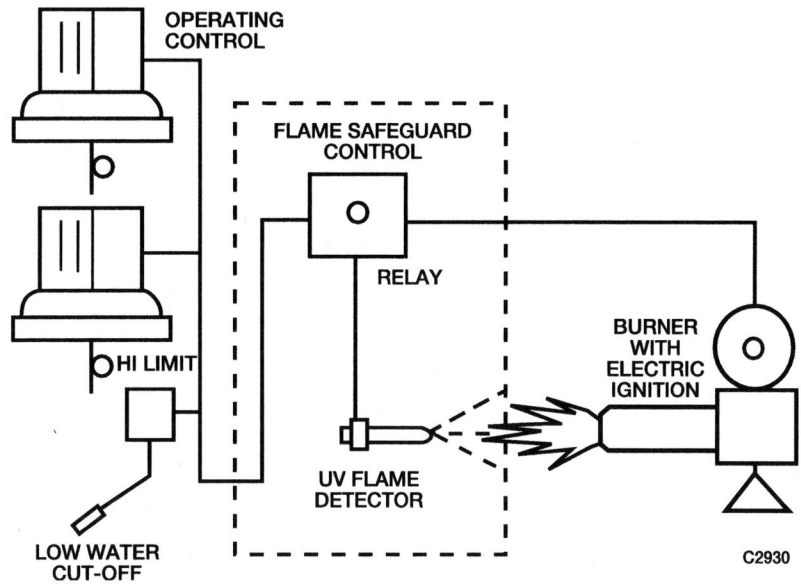

Fig. 10 Combustion controls with flame safeguard circuit.

The Dual Boiler Plant Control example in Fig. 12 is a dual boiler plant with high-fire/low-fire controlled boilers. A minimum incoming water temperature of 145°F is required prior to high-fire, water flow must be maintained when the boiler is enabled, and a secondary hot water reset schedule of 110°F water at 55°F OA temperature and 180°F water at 5°F OA temperature. These concepts adapt well for single- or multiple-boiler systems.

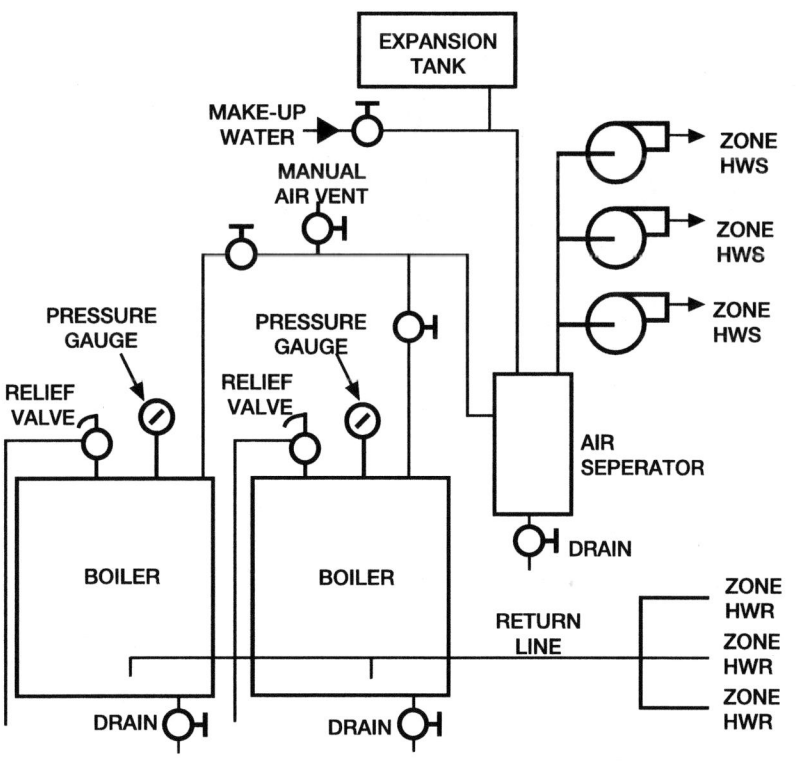

Note: The primary/secondary decoupler is sized for the full secondary flow, and like the chiller plant decoupler, should be a minimum of 6 pipe diameters in length. Unlike the chiller decoupler, normal flow may occur in either direction.

Fig. 11 Typical piping for multiple-zone heating system.

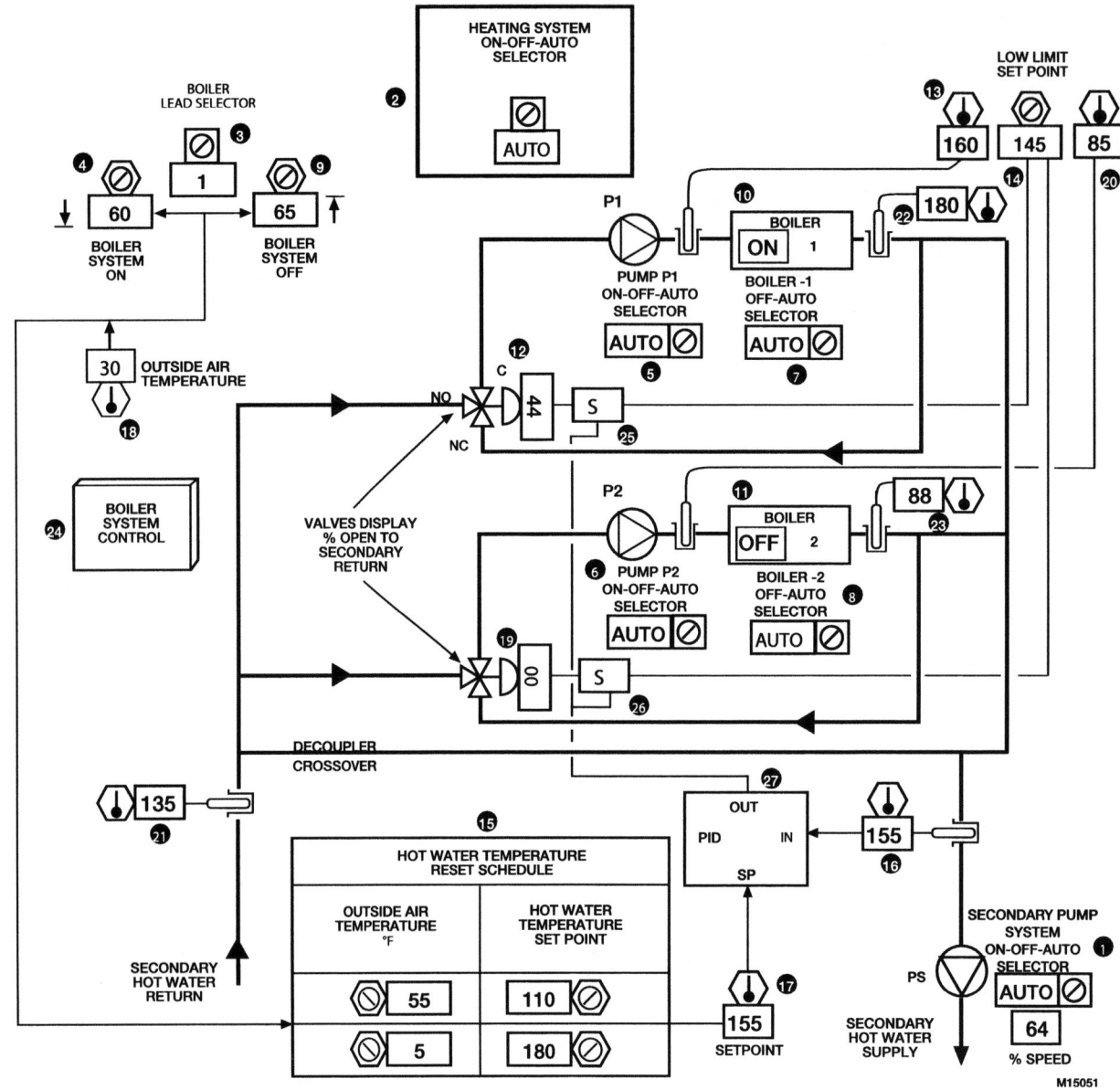

Fig. 12 Dual boiler plant control graphic.

Functional Description

Item no.	Function
1	On/off/auto function for secondary pumping system
2	On/off/auto function for heating system
3	Selects the lead boiler
4	Heating system start point (OA temperature)
5, 6	On/off/auto function for primary pumps
7, 8	Off/auto function for boilers
9	Heating system stop point (OA temperature)
10, 11	Operator information
12–14	Valve modulates to prevent incoming water from dropping below the low-limit setpoint (145°F)

Item no.	Function
15–18	Secondary water setpoint reset from OA
19, 20	Valve modulates to prevent incoming water from dropping below the low-limit setpoint (145°F)
21–23	Operator information
24	Icon, selects the Boiler System Control dynamic display (as seen in Fig. 13)
25, 26	Software signal selection functions, allows valve to control secondary HW temperature, subject to boiler low-limits
27	OA reset valve control PID

Features

1. Full flow through operating boilers
2. Minimum temperature limit on the boiler's incoming water
3. Variable-flow secondary system with full boiler flow
4. Automatic boiler staging
5. User-friendly monitoring and adjustment

Conditions for Successful Operation

1. Control network, software, and programming to advise the heating plant controller of secondary fan and water flow demands.
2. Interlock and control wiring coordinated with the boiler manufacturer.
3. Control in accord with boiler manufacturer's recommendations.
4. Proper setpoint and parameter-project-specific settings.

Specification

The heating plant shall operate under automatic control anytime the secondary pump's on/off/auto function is not "OFF," subject to a heating system's on/off/auto software function. The lead boiler, as determined by a software-driven lead-boiler-selection function, shall be enabled anytime the date is between October 1 and May 1, the OA temperature drops below 60°F for greater than 30 min, and an AHU is calling for heat. Each boiler's primary pump shall have a software on/off/auto function, and each boiler shall have a software auto/off function. The heating

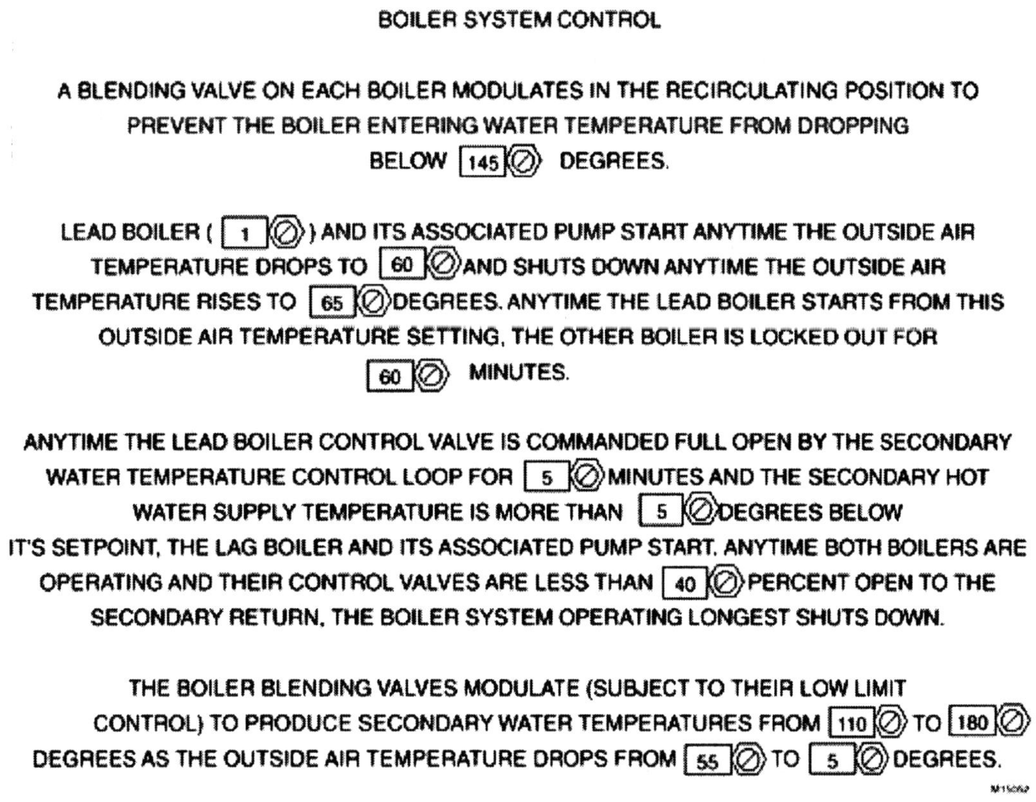

BOILER SYSTEM CONTROL

A BLENDING VALVE ON EACH BOILER MODULATES IN THE RECIRCULATING POSITION TO PREVENT THE BOILER ENTERING WATER TEMPERATURE FROM DROPPING BELOW 145 DEGREES.

LEAD BOILER (1) AND ITS ASSOCIATED PUMP START ANYTIME THE OUTSIDE AIR TEMPERATURE DROPS TO 60 AND SHUTS DOWN ANYTIME THE OUTSIDE AIR TEMPERATURE RISES TO 65 DEGREES. ANYTIME THE LEAD BOILER STARTS FROM THIS OUTSIDE AIR TEMPERATURE SETTING, THE OTHER BOILER IS LOCKED OUT FOR 60 MINUTES.

ANYTIME THE LEAD BOILER CONTROL VALVE IS COMMANDED FULL OPEN BY THE SECONDARY WATER TEMPERATURE CONTROL LOOP FOR 5 MINUTES AND THE SECONDARY HOT WATER SUPPLY TEMPERATURE IS MORE THAN 5 DEGREES BELOW IT'S SETPOINT, THE LAG BOILER AND ITS ASSOCIATED PUMP START. ANYTIME BOTH BOILERS ARE OPERATING AND THEIR CONTROL VALVES ARE LESS THAN 40 PERCENT OPEN TO THE SECONDARY RETURN, THE BOILER SYSTEM OPERATING LONGEST SHUTS DOWN.

THE BOILER BLENDING VALVES MODULATE (SUBJECT TO THEIR LOW LIMIT CONTROL) TO PRODUCE SECONDARY WATER TEMPERATURES FROM 110 TO 180 DEGREES AS THE OUTSIDE AIR TEMPERATURE DROPS FROM 55 TO 5 DEGREES.

Fig. 13 Boiler system control dynamic display.

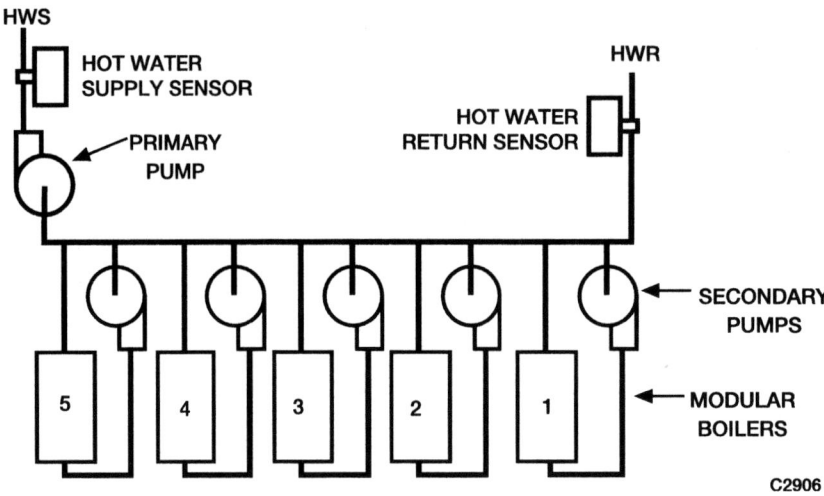

Fig. 14 Typical primary–secondary piping for modular boilers.

plant shall be disabled anytime the OA temperature rises to 65°F for greater than 1 min and after May 1.

Anytime the boiler plant is enabled, the lead boiler's primary pump shall start and, as flow is proven, the boiler shall fire under its factory controls to maintain 180°F. If the lead boiler's status does not change to "on," or if flow is not proven within 5 min, the lag boiler shall be enabled.

During boiler operation, a three-way blending valve shall position to place the boiler flow in a recirculating mode until the water entering the boiler exceeds a low-limit value of 145°F, at which time the blending valve shall modulate to maintain the secondary water temperature between 110 and 180°F as the OA temperature varies from 55 to 5°F.

The lag boiler shall be locked out from operation for 60 min after the lead boiler starts. Thereafter, anytime one boiler control valve is commanded full open by the secondary temperature control loop for greater than 5 min and the secondary water temperature is a temperature less than 5°F below the secondary water temperature setpoint, the "off" (lag) boiler pump shall start. And, upon proving flow, the "off" boiler shall be enabled to fire under its factory controls to maintain 180°F. The just-started boiler's blending valve shall be controlled by an incoming 145°F water temperature low-limit sensor and setpoint similar to the lead boiler's, and subsequently, in unison with the other boiler's blending valve to maintain the reset, secondary hot-water temperature.

Anytime both boilers are operating and their control valves are less than 40% open to the secondary return line, the boiler and pump that has run longest shall shut down.

Modular Boilers

Modular boilers provide heat over a large range of loads and avoid standby and other losses associated with operating large boilers at small loads. Fig. 14 shows a primary-secondary piping arrangement where each modular boiler has its own pump. The boiler pump is on when the boiler is on.

Boilers that are off have no flow and are allowed to cool. Each boiler that is on operates at or near full capacity. Avoiding intermittent operation prevents losses up the stack or to the surrounding area when the boiler is off.

Normal control of modular boilers cycles one of the on-line boilers to maintain water temperature in the supply main to meet load requirements. The supply main's control sensor cycles the boilers in sequence. If the load increases beyond the capacity of the boilers that are on-line, an additional boiler is started. The lead (cycling) boiler can be rotated on a daily or weekly basis to equalize wear among all boilers or when using digital controls, the program can start the boiler that has been off the longest.

CONCLUSION

In many facilities, boilers represent the most significant pieces of energy consuming equipment. Understanding how a boiler functions, and how it can best be controlled may lead to large energy savings for residential, commercial, and industrial complexes.

REFERENCES

1. *Engineering Manual of Automatic Controls*, Honeywell International: Golden Valley, MN, 2005; available online at www.buildingcontrolworkbench.com.

BIBLIOGRAPHY

1. Boilers and Fired Systems. Chapter 5 in *Energy Management Handbook*, 5th Ed., Turner, W.C., Ed., Fairmont Press: Lilburn, GA, 2006.
2. Petchers, N. Boilers. *Combined Heating, Cooling, and Power Handbook*; Fairmont Press: Lilburn, GA, 2003; (Chapter 7).
3. Boilers, Chapter 27, *ASHRAE Handbook of HVAC Sytems and Equipment*; American Society of Heating, Air Conditioning, and Refrigerating Engineers: Atlanta, GA, 2004.
4. Heselton Ken, *Boiler Operators Handbook*, Fairmont Press: Lilburn, GA.
5. Harry Taplin. *Boiler Plant and Distribution System Optimization Manual*; Fairmont Press: Lilburn, GA.

Building Automation Systems (BAS): Direct Digital Control

Paul J. Allen
Reedy Creek Energy Services, Walt Disney World Co., Lake Buena Vista, Florida, U.S.A.

Rich Remke
Commercial Systems and Services, Carrier Corporation, Syracuse, New York, U.S.A.

Abstract

This chapter is designed to help energy managers understand some of the fundamental concepts of Building Automation Systems (BAS). A BAS is used to control energy consuming equipment—primarily for heating, ventilating and air conditioning (HVAC) equipment and lighting controls. We thoroughly examine each component of a BAS in today's BAS technology and what a BAS might look like in the future. The BAS of tomorrow will rely heavily on the Web, TCP/IP, high-speed data networks, and enterprise level connectivity.

INTRODUCTION

The combination of low-cost, high-performance microcomputers together with the emergence of high-capacity communication lines, networks, and the Internet has produced explosive growth in the use of Web-based technology for direct digital control (DDC) building automation systems (BAS).[1] Many of these current BAS systems use a proprietary information structure and communications protocol that greatly limits the plug-and-play application and addition of interchangeable components in the system. Control solutions such as BACnet and LonWorks have helped this situation somewhat, but they have also introduced their own levels of difficulties. The BAS of the future will integrate state-of-the-art information technology (IT) standards used widely on the Internet today. These new IT-based systems are rapidly overtaking the older BAS systems. All of the established BAS companies are quickly developing ways to interface their systems using IT standards to allow the use of Web browsers such as Internet Explorer and Netscape Navigator.

This article will examine all facets of a BAS, from field controllers to the front-end interface. The emphasis is on understanding the basic BAS components and protocols first, and then examining what a BAS might look like in the future based on the influence of IT standards. Finally, this article will discuss upgrade options for legacy BAS systems and BAS design strategies.

Even though we will be referring exclusively to the term BAS in this chapter, the building automation controls industry also uses the following terms interchangeably: direct digital control (DDC), energy management system (EMS), building automation and control system (BACS), and building management system (BMS).

THE BASICS OF TODAY'S BAS

At a minimum, a BAS is used to control functions of a heating, ventilating, and air conditioning (HVAC) system, including temperature and ventilation, as well as equipment scheduling. Additional basic features include the monitoring of utility demand, energy use, building conditions, climatic data, and equipment status. Even basic BAS are generally expected to perform control functions that include demand limiting and duty cycling of equipment. Building automation systems report outputs can show the facility utility load profiles, the trends and operation logs of equipment, and the generation of maintenance schedules.

More elaborate BAS can integrate additional building systems—such as video surveillance, access control, lighting control, and interfacing—with the fire and security systems. However, in large organizations and on campuses today, it is still more common to see dedicated systems for these additional building systems due to divisions in management functional responsibility, code issues, and the features and performance of dedicated systems.

Today's BAS are expected to receive and process more sophisticated data on equipment operation and status, such as data from vibration sensors on motors, ultrasonic sensors on steam traps, infrared sensors in equipment rooms, and differential pressure sensors for filters. Top-of-the-line BAS today also have additional capabilities, such as chiller/boiler plant optimization, time schedule and setpoint management, alarm management, and tenant billing to name a few. Most BAS manufacturers today have started to offer some form of Web-based access to their existing control systems and are actively developing Web-based capability for their future products.

Keywords: Building automation systems, BAS; Direct digital control, DDC; Internet; LonWorks; BACnet.

CONTROLLER-LEVEL HARDWARE AND SOFTWARE

Controller Hardware

Building automation systems controllers are used to provide the inputs, outputs, and global functions required to control the mechanical and electrical equipment. Most BAS manufacturers provide a variety of controllers tailored to suit the specific need. Shown below is a list of the most common BAS controllers.

Communications interface: Provides the communication interface between the operator workstation and the lower-tier controller network. On a polling controller network, a communications interface is used to transfer data between the controllers.

Primary controller: Provides global functions for the BAS control network that can include real-time clocks, trend data storage, alarms, and other higher-level programming support. Some BAS manufacturers combine all these functions into one primary controller, while other manufacturers have separate controllers that are dedicated to each global function.

Secondary controller: Contains the control logic and programs for the control application. Secondary controllers usually include some on-board input/output (I/O) and may interface to expansion modules for additional I/O. Inputs include temperatures, relative humidity, pressures, and fan and pump status. Outputs include on/off and valve or damper control. Also included in this group are application-specific controllers that have limited capability and are designed for a specific task. Examples include controllers for variable air volume terminal unit (VAV) boxes, fan coil units, and multistage cooling and heating direct-expansion (DX) air conditioning systems.

For further reference, the Iowa Energy Center has an excellent Web site (www.ddc-online.org) that shows a complete overview of the designs, installations, operation, and maintenance of most BAS on the market today.

Controller Programming

Building automation systems controllers typically contain software that can control output devices to maintain temperature, relative humidity, pressure, and flow at a desired setpoint. The software programming can also adjust equipment on/off times based on a time-of-day and day-of-week schedule to operate only when needed.

The software used to program the controllers varies by BAS manufacturer and basically falls into three categories:

1. Fill-in-the-blank programming using standard algorithms
2. Line-by-line custom programming
3. Graphical custom programming.

Fill-in-the-blank: This type of programming uses precoded software algorithms that operate in a consistent, standard way. The user fills in the algorithm configuration parameters by entering the appropriate numbers in a table. Typically, smaller control devices use this type of programming, like those that control a fan coil or VAV box controller. These devices all work the same way and have the same inputs and outputs.

A few manufacturers have used fill-in-the-blank programming for devices that are more complex with which a variety of configurations can exist, such as air handlers. Standard algorithms are consistent for each individual component. As an example, the chilled-water valve for an air-handling unit is programmed using the same standard algorithm with only the configuration parameters adjusted to customize it for the particular type of valve output and sensor inputs. Programming all of the air-handler devices using the appropriate standard algorithm makes the air-handling unit work as a system.

The advantage of fill-in-the-blank standard algorithms is that they are easy to program and are standard. The downside is that if the standard algorithm does not function as desired, or if a standard algorithm is not available, the system requires development of a custom program.

Line-by-line custom programming: Control programs are developed from scratch and are customized to the specific application using the BAS manufacturer's controls programming language. In most cases, programs can be reused for similar systems with modifications as needed to fit the particular application.

The advantage of line-by-line custom programs is that technicians can customize the programs to fit any controls application. The disadvantage is that each program is unique, and troubleshooting control problems can be tedious, because each program must be interrogated line by line.

Graphical custom programming: Building automation systems manufacturers developed this method to show the control unit programs in a flowchart style, thus making the programming tasks more consistent and easier to follow and troubleshoot.

Below are some additional issues to consider regarding control unit programming:

- Can technicians program the control units remotely (either network or modem dial-in), or must they connect directly to the control unit network at the site?
- Does the BAS manufacturer provide the programming tools needed to program the control units?
- Is training available to learn how to program the control units? How difficult is it to learn?
- How difficult is it to troubleshoot control programs for proper operation?

Controller Communications Network

The BAS controller network varies depending on the manufacturer. Several of the most common BAS controller networks used today include RS-485, Ethernet, attached resource computer network (ARCNET), and LonWorks.

RS-485. This network type was developed in 1983 by the Electronic Industries Association (EIA) and the Telecommunications Industry Association (TIA). The EIA once labeled all its standards with the prefix "RS" (recommended standard). An RS-485 network is a half-duplex, multidrop network, which means that multiple transmitters and receivers can exist on the network.

Ethernet: The Xerox Palo Alto Research Center (PARC) developed the first experimental Ethernet system in the early 1970s. Today, Ethernet is the most widely used local area network (LAN) technology. The original and most popular version of Ethernet supports a data transmission rate of 10 Mb/s. Newer versions of Ethernet called "Fast Ethernet" and "Gigabit Ethernet" support data rates of 100 Mb/s and 1 Gb/s (1000 Mb/s).

ARCNET: A company called Datapoint originally developed this as an office automation network in the late 1970s. The industry referred to this system as ARC (Attached Resource Computer) and the network that connected these resources as ARCNET. Datapoint envisioned a network with distributed computing power operating as one larger computer.

LonWorks: This network type was developed by Echelon Corporation in the 1990s. A typical node in a LonWorks control network performs a simple task. Devices such as proximity sensors, switches, motion detectors, relays, motor drives, and instruments may all be nodes on the network. Complex control algorithms, such as running a manufacturing line or automating a building, are performed through the LonWorks network.

Controller Communications Protocol

A communications protocol is a set of rules or standards governing the exchange of data between BAS controllers over a digital communications network. This section describes the most common protocols used in a BAS.

BACnet: A data communication protocol for building automation and control networks is a standard communication protocol developed by the American Society of Heating, Refrigerating, and Air-Conditioning Engineers (ASHRAE) specifically for the building controls industry. It defines how applications package information for communication between different BAS. The American National Standards Institute (ANSI) has adopted it as a standard (ASHRAE/ANSI 135-2001).

LonTalk: An interoperable protocol developed by Echelon Corporation and named as a standard by the Electronics Industries Alliance (ANSI/EIA-709.1-A-1999). Echelon packages LonTalk on its "neuron chip," which is embedded in control devices used in a LonWorks network.

Proprietary RS-485: The protocol implemented on the RS-485 network is usually proprietary and varies from vendor to vendor. The Carrier Comfort Network (CCN) is an example of a proprietary RS-485 communications protocol.

Modbus: In 1978, Modicon developed the Modbus protocol for industrial control systems. Modbus variations include Modbus ASCII, Modbus RTU, Intel® Modbus RTU, Modbus Plus, and Modbus/IP. Modbus protocol is the single most-supported protocol in the industrial controls environment.

TCP/IP: Transmission Control Protocol/Internet Protocol (TCP/IP) is a family of industry-standard communications protocols that allow different networks to communicate. It is the most complete and accepted enterprise networking protocol available today, and it is the communications protocol of the Internet. An important feature of TCP/IP is that it allows dissimilar computer hardware and operating systems to communicate directly.

ENTERPRISE-LEVEL HARDWARE AND SOFTWARE

Client Hardware and Software

Normally, a personal computer (PC) workstation provides operator interface into the BAS. The PC workstation may or may not connect to a LAN. If a server is part of the BAS, the PC workstation would need LAN access to the server data files and graphics. Some smaller BAS use standalone PCs that have all the BAS software and configuration data loaded on each PC. Keeping the configuration data and graphics in sync on each PC becomes problematic with this design.

A graphical user interface (GUI) is one of the client-side software applications that provide a window into the BAS. The GUI usually includes facility floor plans that link to detailed schematic representations and real-time control points of the building systems monitored by the BAS. The GUI allows technicians to change control parameters such as setpoints and time schedules, or to override equipment operations temporarily. Other client-side software applications include the following:

- Alarm monitoring
- Password administration
- System setup configuration
- Report generation
- Control-unit programming and configuration.

Server Hardware and Software

Servers provide scalability, centralized global functions, data warehousing, multiuser access, and protocol translations for a midsize to large BAS. Servers have become more prominent in the BAS architecture as the need has grown to integrate multivendor systems, publish and analyze data over an intranet or extranet, and provide multiuser access to the BAS. While having a central server on a distributed BAS may seem contradictory, in reality, a server does not take away from the standalone nature of a distributed control system. Servers enhance a distributed control system by providing functions that applications cannot perform at the controller level. In fact, a BAS may have several servers distributing tasks such as Web publishing, database storage, and control system communication.

Servers provide the ability to control a BAS globally. Facilitywide time scheduling, load shedding, and setpoint resets are examples of global functions a BAS server can perform. Because these types of functions are overrides to the standard BAS controller-level programs, having them reside in the server requires that steps be taken to ensure continued control system operation should the server go down for any length of time. The distributed BAS should have the ability to "time out" of a server override if communications with the server is lost. When the server comes back online, the BAS should have rules that govern whether the override should still be in effect, start over, or cancel. Servers also can perform computational tasks, offloading this work from the BAS control units.

BAS DESIGN ISSUES

Aside from the impact that IT will have on future EMS, there are some fundamental characteristics that owners have always desired and will continue to desire from a BAS:

- Having a single-seat user interface
- Compatible with existing BAS
- Easy to use
- Easily expandable
- Competitive and low cost
- Owner maintainable.

There have been several changes made by the BAS industry to help satisfy some of these desires. The creation of "open" protocols such as LonWorks and BACnet has made field panel interoperability plausible. The development of overlay systems that communicate with multiple BAS vendor systems has made a single-seat operation possible. However, each has introduced its own levels of difficulties and additional cost.

Users that master their BAS are more likely to be successful than users that delegate the responsibility to someone else. The BAS vendor should be a partner with the owner in the process, not the master.

New-Facility BAS Design

There are two strategies available for the design and specification of BAS for new facilities:

1. Specifying a multivendor interoperable BAS
2. Standardizing on one BAS manufacturer's system.

Specifying a multivendor interoperable BAS is probably the most popular choice among the facility design community. The engineer's controls design is more schematic and the specifications are more performance based when this approach is used. In other words, the engineer delegates the responsibility of the detailed BAS design to the temperature controls contractor because the engineer does not actually know which BAS vendor will be selected. Even if only one BAS vendor was selected, it is very rare that the engineer would be intimately knowledgeable with this system anyway. Therefore, the resulting BAS design is by nature somewhat vague and entirely performance based. The key to making this approach successful is in the details of the performance specification, which is not a trivial task. Competition results from multiple BAS vendors bidding on the entire BAS controls installation.

The second approach is based on standardizing on one BAS manufacturers system. To create competition and keep installation cost low, the engineer must create the BAS design as part of his or her design documents and prescriptively specify all components of the BAS. This allows multiple temperature control contractors to bid on the BAS installation (wire, conduit, and sensor actuators)— everything outside of the BAS field panel. Everything inside the BAS field panel is owner furnished. Contractors familiar with the owners' BAS, or the owners' own technicians, perform the controller wire termination, programming, and startup. This approach is successful when all parties work together. The design engineer must produce a good BAS design. The temperature controls contractor must install the field wire, conduit, sensors, and actuators properly. Finally, the BAS contractor must terminate and program the BAS panel correctly. A successful project is a system that integrates seamlessly with the owners' existing BAS.

Upgrading an Existing BAS

Most users already own and operate a legacy BAS that they might desire to upgrade from a standalone BAS to a network-based system.[2] The benefits of a network-based BAS appear as better standard operational practices and procedures, opportunities to share cost-savings programs and strategies, and wider access to building control processes. The key to justifying the costs associated with networking a BAS is that it can be done at a reasonable cost and is relatively simple to implement and operate.

There are three main strategies available when upgrading a BAS from a standalone system to a network-based system:

1. Remove the existing BAS and replace it with a new network-based BAS.
2. Update the existing BAS with the same manufacturer's latest network-based system.
3. Install a BAS interface product that networks with an existing BAS.

The first upgrade strategy is simply to replace the existing BAS with a newer network-based BAS that has been established as a standard within your company. The cost for this option is solely dependent on the size of the BAS that will be replaced. However, this approach might be justified if the existing BAS requires high annual maintenance costs or has become functionally obsolete.

The second upgrade strategy available is to contact the original BAS manufacturer and request a proposal for its upgrade options. Most BAS manufacturers have developed some form of Ethernet network connectivity. Typically, some additional hardware and software is required to make the system work on an Ethernet network. The cost for this might be very reasonable, or it could be very expensive. It all depends on how much change is required and on the associated hardware, software, and labor cost to make it all work.

The third upgrade strategy involves the installation of a new network-based system that is specifically designed to interface with different BAS systems. These systems typically have dedicated hardware that connects to the BAS network and software drivers that communicate with the existing BAS controllers. The new BAS interface controllers also have an Ethernet connection so they can communicate on the corporate LAN. Users view the BAS real-time data by using Web browser software on their PC. The advantage of this strategy is that a multitude of different BAS systems can be interfaced. The disadvantage is that the existing BAS software must still be used to edit or add new control programs in the existing BAS field controllers.

FUTURE TRENDS IN BAS

The future of DDC in BAS can be found on the Web. Most BAS manufacturers see the need to move their products to the Internet tremendous economies of scale and synergies can be found there. Manufacturers no longer have to create the transport mechanisms for data to flow within a building or campus they just need to make sure their equipment can utilize the network data paths already installed or designed for a facility. Likewise, with the software to display data to users, manufacturers that take advantage of presentation-layer standards such as Hypertext markup language (HTML) and Java can provide the end user a rich, graphical, and intuitive interface to their BAS using a standard Web browser.

So where do we go from here?

Faster, Better, Cheaper

Standards help contain costs by not reinventing the wheel every time a new product is developed or brought to market. While there is a risk of stagnation or at least uninspired creativity using standards, Internet standards have yet to fall into this category, due to the large consumer demand for rich content on the Internet. A BAS, even at its most extensive implementation, will use only a tiny subset of the tools available for creating content on the Internet.

When a BAS manufacturer does not have to concentrate on the transport mechanism of data or the presentation of that data, new products can be created at a lower cost and more quickly. When the user interface is a Web browser, building owners can foster competition among manufacturers because each BAS system is inherently compatible with any competitors at the presentation level. All that separates one BAS from another in a Web browser is a hyperlink.

Another area where costs will continue to fall in using Internet standards is the hardware required to transport data within a building or a campus. Off-the-shelf products such as routers, switches, hubs, and server computers make the BAS just another node of the IT infrastructure. Standard IT tools can be used to diagnose the BAS network, generate reports of BAS bandwidth on the intranet, and back up the BAS database.

Owners will reap the benefits of Internet standards through a richer user interface, more competition among BAS providers, and the ability to use their IT infrastructure to leverage the cost of transporting data within a facility.

The Enterprise

Extensible markup language (XML): XML is an Internet standard that organizes data into a predefined format for the main purpose of sharing between or within computer systems. What makes XML unique is that data tags within the XML document can be custom or created on the fly and, unlike HTML tags, are not formatted for presenting the data graphically. This makes XML a great choice for machine-to-machine (M2M) communication.

Why is M2M so important? Because the next wave of BAS products will include "hooks" into other Internet-based systems. Building automation systems have done a great job of integrating building-related components together. BACnet, LonWorks, and Modbus provide the capability of connecting disparate building components

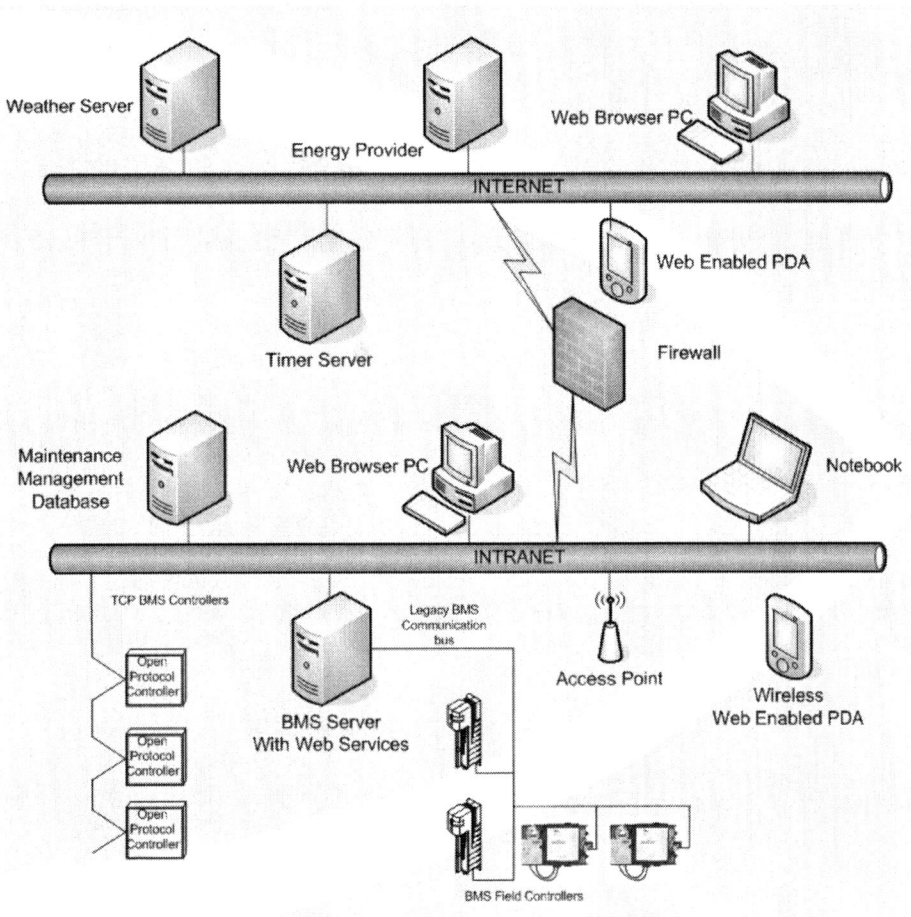

Fig. 1 Future building automation systems (BAS) network schematic.

made by different manufacturers, so that a lighting control panel can receive a photocell input from a rooftop building controller or a variable-frequency drive can communicate an alarm on the BAS when a failure occurs.

The future will require a BAS to connect to enterprise-level systems, not just building-level systems. This is where M2M and Web services come into play. Web services can be thought of as plug-ins allowing a BAS to communicate with a Web-based system or server. An example of this would be time synchronization. The Internet has many time servers that can provide the exact local time as well as greenwich mean time (GMT). A BAS can have a Web service that would plug into the BAS, synchronizing all of the time clocks within a facility with the atomic clock in Boulder, Colorado. Another example would be obtaining the outside air temperature from the local weather service. Instead of the BAS just measuring the outside air temperature at a local controller, a Web service could provide the outside air temperature, humidity, barometric pressure, and any other weather-related data. Now the BAS can make more intelligent decisions on using outdoor air for comfort cooling, determining wet-bulb setpoints for cooling towers, or even announcing that a storm is imminent.

More enticing than connecting to weather and time servers is the promise of connecting to a facility's enterprise data. The BAS of the future must become an integral part of the decision-making for allocating personnel, budgeting maintenance and upgrades, purchasing energy, and billing those that use the energy. Most larger facilities have departments that provide these types of services, yet the BAS has always stood alone, providing input through exported reports, system alarms, or human analysis. Enterprise-level integration would create Web services to connect directly to these systems, providing the data necessary for making informed decisions about capital investments, energy, or personnel. See Fig. 1 for what a BAS might look like in the future.

The good news is that XML and Web services have gained market acceptance to become the standards for enterprise-level connectivity. The bad news is that this is still in its infancy for most BAS vendors. It is a very costly effort to create an enterprise-level Web service today. Even though Web services are supported by Microsoft

Corporation, Apple Computer, Sun Microsystems, and others, they can still be custom solutions tailored to a specific accounting, maintenance management, or energy procurement system. For Web services to become mainstream in the BAS world, common services will need to be created that can be used by all BAS vendors. In addition, for Web services to be implemented properly in facilities, the skill set for BAS programmers and installers will need to include XML and a basic understanding of IP. If facility managers and technicians are to be able to make changes, adjustments, and enhancements to their enterprise system, they too will require this skill set.

The future will also need to better define the decision logic and troubleshooting tools when implementing Web services. When the BAS sends duplicate alerts to a maintenance management system, where does the logic reside to send only one technician to the trouble call? This is currently undefined. Standard tools for testing scenarios online and offline need to be developed. Even though Web services typically rely on XML, which is a self-documenting standard, XML can be very verbose. Tools need to be created to help technicians discover and correct errors quickly. When a facility decides to change its accounting system to a newer version or a different vendor, will the BAS be able to adapt? Conversion and upgrade tools need to also be considered when defining BAS Web services.

Even without all the tools identified, enterprise-level connectivity is moving ahead rapidly. The benefits of integrating BAS data within a facility's other systems can outweigh the immediate need for a complete set of tools. Web services through XML place the BAS directly into the facility data infrastructure. That's a good place to be for an energy manager wanting to maximize the investment in a facility's BAS.

CONCLUSION

The BAS of old relied heavily on a collection of separate systems that operated independently, often with proprietary communication protocols that made expansion, modification, updates, and integration with other building or plant information and control systems very cumbersome, if not impossible. Today the BAS is expected not only to handle all of the energy- and equipment-related tasks, but also to provide operating information and control interfaces to other facility systems, including the total facility or enterprise management system.

Measuring, monitoring, and maximizing energy savings are fundamental tasks of all BAS, and are the primary justification for many BAS installations. Improving facility operations in all areas, through enterprise information and control functions, is fast becoming an equally important function of the overall BAS or facility management system. The Web provides the means to share information more easily, quickly, and cheaply than ever before. There is no doubt that the Web is having a huge impact on the BAS industry. The BAS of tomorrow will rely heavily on the Web, TCP/IP, high-speed data networks, and enterprise-level connectivity. If you have not done so already, it is a good time for energy managers to get to know their IT counterparts at their facilities, along with those in the accounting and maintenance departments. The future BAS will be here sooner than you think. Get ready—and fasten your seat belts!

REFERENCES

1. Capehart, B.; Allen, P.; Remke, R.; et al. *IT Basics for Energy Managers—The Evolution of Building Automation Systems Toward the Web*; Information Technology for Energy Managers, The Fairmont Press, Inc.: Lilburn, GA, 2004.
2. Allen, P.; Remke, R.; Tom, S. *Upgrade Options for Networking Energy Management Systems*; Information Technology for Energy Managers, The Fairmont Press, Inc.: Lilburn, GA, 2005; Vol. 2.

BIBLIOGRAPHY

1. RS-485, www.engineerbob.com/articles/rs485.pdf
2. Ethernet, www.techfest.com/networking/lan/ethernet.htm
3. ARCNET, www.arcnet.com
4. LonWorks and LonTalk, www.echelon.com/products/Core/default.htm
5. BACnet, www.bacnet.org
6. Modbus, www.modbus.org/default.htm
7. XML, www.xml.com
8. Iowa Energy Office, www.ddc-online.org

Building Geometry: Energy Use Effect

Geoffrey J. Gilg
Pepco Energy Services, Inc., Arlington, Virginia, U.S.A.

Francisco L. Valentine
Premier Energy Services, LLC, Columbia, Maryland, U.S.A.

Abstract

Energy service companies (ESCOs) use the energy use index (EUI) as a tool to evaluate a building's potential for reduction in energy use. Select Energy Services, Inc. (SESI) has found that consideration of building geometry is useful in evaluating a building's potential for energy use reduction. Building load and energy-use simulations using Trace® and PowerDOE®, respectively, were conducted to gain insight into how building geometry impacts heating, ventilation, and air-conditioning (HVAC) sizing and energy use. The ratio of gross wall area to gross floor area, A_{wall}/A_{floor}, has been found to be a useful factor to consider when making EUI comparisons. Simulations suggest that buildings with higher A_{wall}/A_{floor} ratios require higher central plant capacities and use more energy per unit area to satisfy the heating and cooling loads. Taking a building's geometry (A_{wall}/A_{floor}) into account while estimating savings potential may produce more accurate results.

INTRODUCTION

Select Energy Services, Inc. (SESI) has conducted a multitude of building evaluations in the course of its performance contracting and design work. Select Energy Services, Inc. has many energy engineers with real-world heating, ventilation, and air-conditioning (HVAC) design experience, which often provides insight into peculiarities. One such peculiarity is "Why do two buildings of the same usage and square footage exhibit energy use indexes (EUIs) significantly different from one another?" In an attempt to answer this question, SESI conducted a series of simulations which focused on building geometry and its contribution to heating and cooling loads and annual energy use.

The tools used in this analysis are Trace® and PowerDOE®. Trace® is a software package published by C.D.S. Software that is used to determine equipment loads. PowerDOE® is published by the Electric Power Research Institute, Inc. as a "front-end" for the U.S. Department of Energy's DOE-2 building energy simulator. PowerDOE® is used to simulate annual building energy use. Both software packages allow easy and economical means to evaluate a building's HVAC capacity requirements and resulting energy use.

ENERGY USE INDEX

Even before setting foot on site, an energy service company (ESCO) can get a preliminary estimate of the potential for energy cost reduction. This can be done by analyzing the fuel and electric rates, looking for credits or rate restructuring, and evaluating EUIs. Energy use index is defined as the ratio of total annual energy used, in kBtus, divided by the square footage of the building.

$$\text{EUI} = \frac{\text{kBtu}}{\text{ft}^2} \qquad (1)$$

The EUI is used as a barometer for estimating the potential for energy savings. However, it must be applied with discretion or an ESCO could pass on a great opportunity or overestimate the potential for energy cost savings.

How the Energy Use Index is Used

Once utility billing data, equipment data, and building square footage have been provided, an EUI evaluation can be conducted. The calculated EUI of the building is compared to an "ideal" EUI. The difference between the building EUI and the ideal EUI is the potential for energy savings (see Fig. 1). However, it is often cost prohibitive to attain the entire EUI differential, so ESCOs often prescribe a maximum, economically attainable, EUI improvement. Fifty percent is often used, but this depends on many factors.

From the above methodology, it is easy to see how comparing EUIs, based solely on square footage, can sometimes result in an inaccurate evaluation of the energy savings potential.

☆ This entry originally appeared as "The Effect of Building Geometry on Energy Use" in *Energy Engineering*, Vol. 100, No. 3, 2004. Reprinted with permission from AEE/Fairmont Press.

Keywords: Building geometry; Energy use index; Performance contracting; Energy simulation; Load calculations; Energy modeling; HVAC; Energy service company.

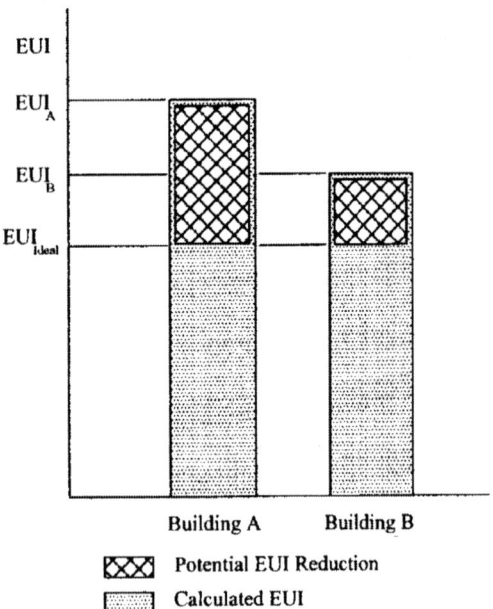

Fig. 1 Energy use comparison.

BUILDING GEOMETRY

Building geometry is an important factor to consider from a design standpoint. It influences heat loss, heat gains, infiltration, and solar gains which influence the heating and cooling load. Typically, the more wall (including windows) area available, the higher the heating and cooling loads. The first 10–15 ft from the exterior wall is considered the perimeter zone. The perimeter zone heating/cooling load is constantly changing because it is under the influence of the weather via the building envelope (walls, windows, and roof) as well as internal loads (occupants, lights, equipment, etc.). Inside this area is the interior zone, which experiences much less heating/cooling load variation (only internal loads). Therefore, if evaluating two buildings of equal floor area, use, occupancy, etc. the building with the larger interior zone will typically require less heating and cooling capacity and use less energy annually.

Building geometry is also an important factor to consider from an energy-use standpoint. If two buildings

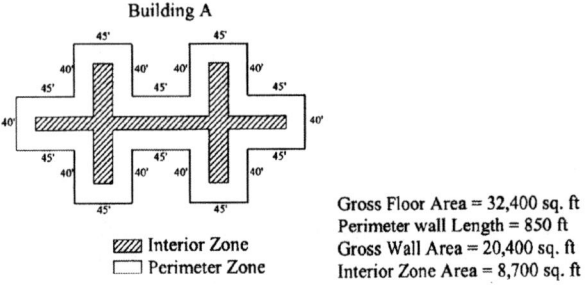

Fig. 2 Building A floor plan.

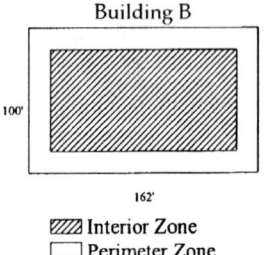

Fig. 3 Building B floor plan.

of same square footage, use, schedule, controls, occupancy, and construction exhibit significantly different EUIs, differences in building geometries may explain why. Because of its greater exposure to environmental conditions, the building with more wall area will likely have the higher EUI. Therefore, it is not uncommon to find that buildings with multiple floors or eccentric shapes use more energy than single-floor, rectangular buildings of the same square footage. For example, Figs. 2 and 3 illustrate two-story buildings, each with 32,400 ft^2 of floor space. Building A has a layout found on many military installations and educational campuses, and its protruding wings resemble radiator fins, both in cross-section and in thermal effect.

Building B is another building of the same square footage, but in the shape of a rectangle, a more compact shape with less exposed surface area.

Given the same building load parameters, Trace® calculations indicate that Building A requires 15% additional cooling capacity and 25% more heating capacity than Building B. PowerDOE® energy use simulations indicate that Building A will use 15% more energy annually than Building B. This example shows that building geometry is indeed an important factor to consider when estimating the potential for energy use reduction based on EUI comparisons. With the realization that building geometry affects energy use, how can it be accounted for, during an EUI comparison?

BUILDING MODELS

To evaluate the effect of geometry on energy use, several hypothetical models were defined, constructed, and analyzed using Trace® and PowerDOE®. The model definitions include such parameters as use (school, office, warehouse), geographical location, schedule, overall heat transfer coefficients (wall, roof, windows), and others. This methodology provides the ability to evaluate the effect of geometry in buildings with distinctly different uses and geometries. Models of the buildings were constructed in Trace® and PowerDOE® using the same building parameters. These programs allow parameters such as orientation, location, and geometry to be changed with a keystroke. A schedule of model assumptions is provided in Appendix A.

RESULTS

After conducting the building load and energy use simulations, a factor has emerged that explains why a building uses more or less energy per square foot than another. This factor takes into account differences in building geometry when evaluating energy use reduction potential. This geometric ratio (GR) is defined as the ratio of gross perimeter wall area (A_{wall}) to gross floor area (A_{floor}).

$$GR = \frac{\sum A_{Wall}}{\sum A_{Floor}} = \frac{A_{Wall}}{A_{Floor}} \qquad (2)$$

Comparisons of Trace® load calculations indicate that buildings with higher wall to floor area ratios require larger heating and cooling plants. The effect of building geometry has been found to be more pronounced as outdoor air requirements decrease. For example, Trace® load calculations (heating only) for a warehouse indicate that with each percentage point increase in the GR, the peak heating requirement increases approximately 1%. Geometry, in this example, has such a significant effect because envelope load is a larger percentage of the total heating plant requirement. Annual energy use predicted by PowerDOE®, on the other hand, is relatively flat for warehouse structures while showing significant geometric effects for school and office-type occupancies. As the outdoor air requirements increase, the contribution of envelope loads to the total heating and cooling load decreases.

Simulations have also shown that it is important to consider the percentage window area per unit wall area. The amount of window in a wall has a significant impact in the overall heat loss/gain of that exposure. Buildings with higher window to wall area ratios typically require larger heating and cooling capacities and use more energy annually per unit floor area than buildings with lower window to wall area ratios.

Trace® and PowerDOE® simulations also indicate that the orientation of a building is also an important factor to consider. If a building has a high (2 or higher) aspect (length/width) ratio and is oriented so that the long sides of the buildings are facing east–west, this building typically requires larger heating and cooling plant capacities and will consume more energy annually. Had this same building been oriented such that a long side was facing north–south, the heating and cooling plant capacity and energy use could have been reduced. Orientation of the long side of the building in a north–south direction also could have permitted more effective use of natural light to reduce lighting energy requirements which can further reduce the cooling load. Therefore, when considering the potential for energy reduction in this particular building, it is advisable to take the building's orientation into account.

Differences in building energy use can also be explained by considering geometry as it relates to the original intended use of the building. On many military installations, especially those associated with airfields, there are numerous single level, high ceiling, marginally insulated buildings whose original intended use and design are not consistent with their present utilization. For example, storage buildings and aircraft hangars are often converted to office space (without upgrading the walls, windows, or roof insulation). As such, they have undergone numerous HVAC retrofits through the years as the hangar/storage space is further converted to office space. Buildings of this type typically use more energy than buildings whose original intended use was that of an office.

The age of a building, in conjunction with geometry, also helps explain differences in EUI. Older buildings have experienced much more wear and tear and typically have higher infiltration rates. Additionally, older buildings are typically constructed with lower R-value materials than contemporary construction. Heating, ventilation, and air-conditioning equipment is typically at, near, or far past its useful life and requires frequent maintenance. In addition, older buildings typically are not insulated as well. Due to space restrictions, many older buildings have rooms or wings that have been added to the original building. This addition can significantly increase the perimeter wall area with only a small increase in square footage. Finally, older buildings do not benefit from recent quality control and construction standards. As a result, it is not uncommon for older buildings to exhibit higher EUIs than similar buildings of recent construction. PowerDOE® modeling has shown that building geometry, as indicated by the GR, has more influence on energy use in older buildings.

How the Geometric Ratio can be Used

When evaluating the potential for energy use reduction, it may be to the ESCO's advantage to take into account the GR of the buildings under consideration. The GR represents the influence of building geometry on energy use and can be used to gauge the effectiveness of certain energy conservation measures (ECMs). For example, ECMs associated with walls and windows such as window film, window replacements, and wall insulation upgrades may have a larger EUI impact in buildings with higher GRs. This is because wall and window conduction and solar gains are a higher percentage of the total HVAC load. Alternately, ECMs such as air-side economizers, lighting retrofits, and roof insulation upgrades may have a larger EUI impact in building with lower GRs. This is because the percentage of contribution of outdoor air, lighting, and roof conduction to the total building load is typically higher for buildings with lower GRs.

CONCLUSIONS

At the conclusion of the calculations and simulations using Trace® and PowerDOE®, SESI believes that it has, at least in small part, contributed to a better understanding of building geometry and its impact on heating and cooling requirements and energy use. This understanding can be used to better estimate the effectiveness of certain ECMs, which can help to avoid underestimation as well as overestimation of potential EUI improvement. This exercise has shown that there is value in considering building geometry while estimating the potential for energy use reduction.

APPENDIX A. MODEL ASSUMPTIONS

Common assumptions

Location	Washington, DC
System type	VAV w/hot water reheat coils (VAV=Variable Air Volume)
Heating thermostat setpoint (°F)	68
Cooling thermostat setpoint (°F)	78
Floor to floor (ft)	12
Plenum height (ft)	3
Floor type	Slab on grade (6 concrete and 12 soil)
Roof type	Flat—built up
Infiltration rate (ft³/ft² wall)	0.15 ft³/ft² wall area
Sensible load (Btu/h per person)	250
Latent load (Btu/h per person)	200
U_{wall} (Btu/(h ft²))	0.1 Btu/(h ft²)
U_{window} (Btu/(h ft²))	0.5 Btu/(h ft²)
U_{roof} (Btu/(h ft²))	0.05 Btu/(h ft²)
Schedule	100% (weekdays, 8 A.M.–5 P.M.)
	20% (weekdays, 5 P.M.–8 A.M.)
	20% (weekends and holidays)
Ignore	Gymnasium, library, locker rooms, rest rooms, cafeteria, lounges, corridor spaces

Schools

Parameter	Model #1	Model #2	Model #3
Length × width (ft)	200 × 80	100 × 80	73 × 73
Number of floors	1	2	3
Orientation (long side)	15 South of due east	15 South of due east	15 South of due east
System type	VAV w/hot water reheat	VAV w/hot water reheat	VAV w/hot water reheat
Window percentage (of wall area)	20	20	20
Occupancy (people/1000 ft²)	50/1000 ft²	50/1000 ft²	50/1000 ft²
Ventilation (ft³/min/person)	15	15	15
Lighting load (W/ft²)	1.5	1.5	1.5
Equipment load (W/ft²)	0.25	0.25	0.25

Offices

Parameter	Model #1	Model #2	Model #3
Length × width (ft)	200 × 80	100 × 80	73 × 73
Number of floors	1	2	3
Orientation (long side)	15 South of due east	15 South of due east	15 South of due east
System type	VAV w/hot water reheat	VAV w/hot water reheat	VAV w/hot water reheat
Window percentage (of wall area)	20	20	20
Occupancy (people/1000 ft²)	7	7	7
Ventilation (ft³/min/person)	20	20	20
Lighting load (W/ft²)	1.5	1.5	1.5
Equipment load (W/ft²)	0.5	0.5	0.5

Warehouse

Parameter	Model #1	Model #2	Model #3
Length × width (ft)	200 × 80	100 × 80	73 × 73
Number of floors	1	2	3
Orientation (long side)	15 South of due east	15 South of due east	15 South of due east

System type	Gas unit heaters	Gas unit heaters	Gas unit heaters
Window percentage (of wall area)	20	20	20
Occupancy (people/1000 ft^2)	5	5	5
Ventilation (ft^3/min/ft^2)	0.05	0.05	0.05
Lighting load (W/ft^2)	1.5 W	1.5	1.5
Equipment load (W/ft^2)	0.25 W	0.25	0.25

Examples

Parameter	Model #1- Building A	Model #2- Building B	Model #2
Length × width (ft)	See Fig. 2	See Fig. 3	100 × 80
Number of floors	3	3	2
Orientation (long side)	15 South of due east	15 South of due east	75
System type	VAV w/hot water reheat	VAV w/hot water reheat	VAV w/hot water reheat
Window percentage (of wall area)	20	20	20
Occupancy (people/1000 ft^2)	50	50	50
Ventilation (ft^3/min/person)	20	20	20
Lighting load (W/ft^2)	1.5	1.5	1.5
Equipment load (W/ft^2)	0.5	0.5	0.5

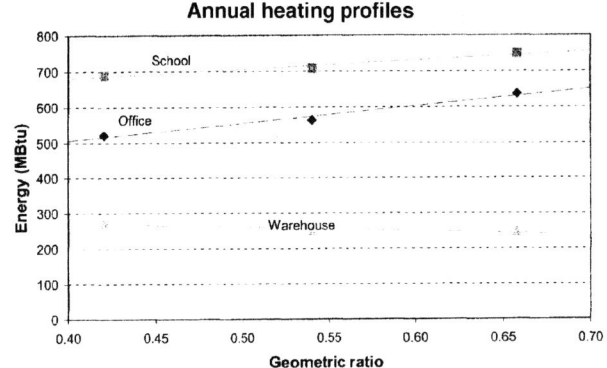

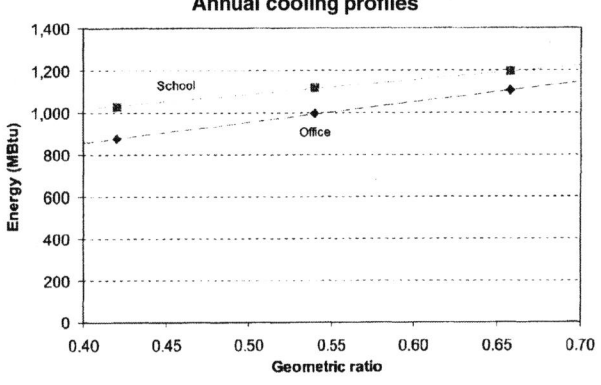

BIBLIOGRAPHY

1. ASHRAE-Fundamentals, American Society of Heating, Refrigeration, and Air-Conditioning Engineers, Inc., 2001.
2. Trane Air Conditioning Manual, The Trane Company, 1996.

Building System Simulation

Essam Omar Assem[1]
Arab Fund for Economic and Social Development, Arab Organizations Headquarters Building, Shuwaikh, Kuwait

Abstract
This entry addresses the topic of building energy system simulation and touches on its importance, benefits, and relevance to the research, architecture, and engineering communities. The entry first gives an illustration of the different energy flow paths that occur in buildings and then describes the main features of the approaches that have evolved over the past four decades in an attempt to reach a solution to those energy flow paths. In addition, an overview of the simulation process in general terms is given, with certain simulation output parameters highlighted. The entry closes with real-life examples to give the reader a sense of some of the important applications of building energy system simulation programs.

INTRODUCTION

In the context of energy systems, the term building system simulation (BSS) refers to the application of computer programs designed specifically to simulate the energy systems that exist within buildings so that a realistic prediction of the thermal performance of a building can be achieved. The most common thermal performance parameters that are referred to by heating, ventilation, and air conditioning (HVAC) engineers are the peak cooling/heating load and energy consumption. The former can be used to determine the size of the HVAC system, whereas the latter can be used to determine the system's energy efficiency. The main advantage of using BSS programs is that the thermal performance of the energy systems in a building can be determined and possibly optimized during the design stage.

For any building, the main physical issues that need to be addressed by the BSS are shown in Fig. 1 in terms of energy flow paths, which can be summarized as follows:

- Surface convection represents the heat flow from an opaque surface to the adjacent air. This process occurs on both the internal and external surfaces of a building wall. The external convection process is influenced by the surface finish, wind speed, and direction, whereas the internal convection process is influenced by the forced air flow injected by the mechanical system serving the building zone and, in the absence of a mechanical system, is influenced by the buoyancy-driven natural convection.

- Inter-surface longwave radiation represents the amount of radiation, which is a function of surface temperatures, emitted from one surface to other visible surfaces, and is influenced by surface emissivity and geometry.

- The intensity of shortwave radiation on a surface is a function of time, surface geometry, and position of the sun relative to the geographical location of the building. The magnitude of shortwave radiation penetrating an opaque surface is influenced by the surface absorptivity and, in the case of transparent surfaces (i.e., glass), is influenced by surface absorptivity, transmissivity, and reflectivity.

- Shading caused by external obstructions to the sun-ray path affects the amount of solar radiation intensity on the external surface of an opaque element and the amount of shortwave radiation penetrating a transparent element, such as a window with overhang. In addition, the amount of solar radiation penetrating a transparent surface and the internal surfaces receiving this radiation are a function of the angle of incidence of the solar radiation and the geometry of the transparent surface, as well as the geometry of the internal surfaces.

- The process of air flow within buildings is affected by the building resistance to the unidirectional air leakage between outside to inside, by the amount of air being circulated among building zones, and also by air circulation within each zone. The amount of air leakage between the outside and the inside of the building is influenced by the external air boundary conditions (such as wind speed, pressure, and air temperature), and by the internal air temperature and pressure.

- There is also the process of casual heat gain from lighting fixtures, equipment, and people. This heat gain will constitute both radiative and convective components.

Keywords: Building energy simulation; Peak load; Annual energy consumption; Thermal comfort; Day lighting; HVAC Simulation.

[1] This entry does not represent the nature of work conducted at the Arab Fund; it was prepared by the author based on his academic and previous work experience.

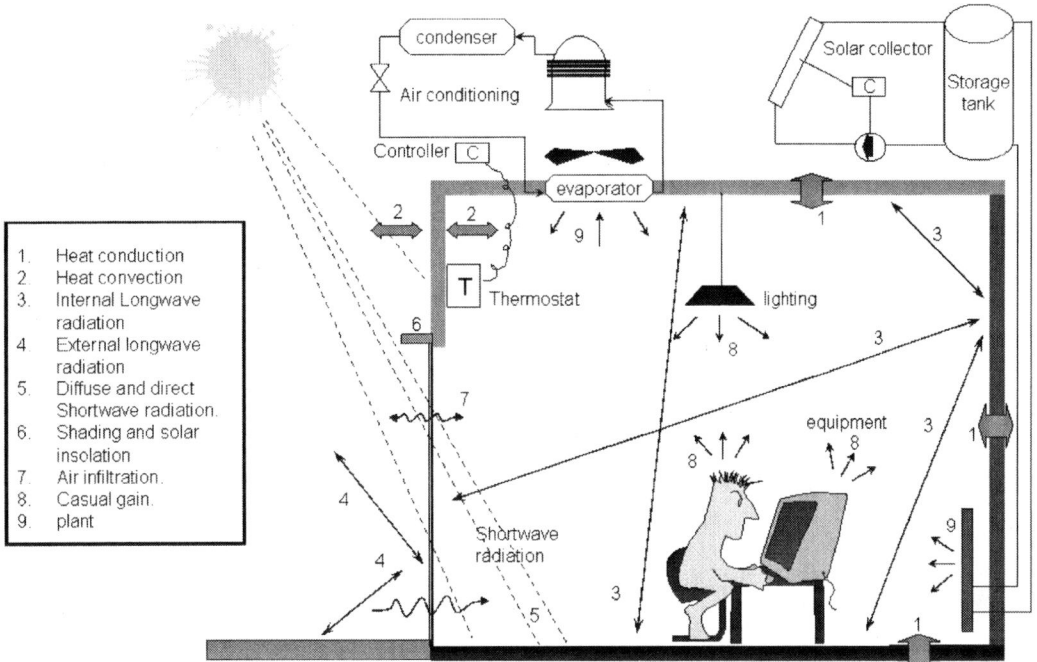

Fig. 1 Energy flow paths in buildings.

- The process of plant interaction with building zones affects their thermal comfort and can be caused by the action of a thermostat sensing and monitoring the zone's air condition. The thermostat setting can be influenced by the building occupants or by the action of an automatic control mechanism.
- Control processes are caused by the response of a plant system to dynamically changing internal conditions. The plant response is influenced by the stream of signals constantly being fed back to the various controllers that actuate the appropriate control variable according to some control action.
- Moisture transfer processes are caused not only by internal generation processes and air migration from outside, but also by the condition of the air being injected into the zone by an air conditioning system.
- In buildings, it is also common to find movable features that can be used to control the effect of certain parameters on the indoor air, such as solar radiation and ambient air. Such features are influenced by social factors and by occupant comfort level.

HISTORICAL DEVELOPMENT

From the early 1960s until the late 1980s, many BSS programs have been developed to tackle the energy flow paths described earlier. These ranged from manual methods, in which the complexity of the real system is reduced to lessen the computational overhead and the input demands on the user, to highly sophisticated approaches in which mathematical models are constructed to represent the interactions observed in reality as closely as possible.

The evolution of design tools,[3] from the traditional to the present-day simulation approach, is summarized in Table 1. First-generation models focused on a simple "handbook" approach, which is piecemeal in that no coupling between the various discrete calculations was made. A steady-state U-value calculation may be conducted to evaluate envelope heat loss, for example; then a lookup table may be used to determine an allowance for zone solar gain; and finally, a degree-day relationship may be used to predict long-term energy requirements. In such calculations, many assumptions are encompassed to simplify their application. In addition, the calculations are based on steady-state, unidirectional wall heat flow, which rarely occurs in the real world.

Second-generation models took into account the temporal aspect of energy flow, especially in the case of elements with large time constants, such as in multilayered constructions. Analytical methods such as the response factor,[10] time, or frequency domain were applied based on the assumption that the building system is linear and possesses heat transfer properties that are time invariant. Modeling integrity remained low because of the decoupling and simplifying assumptions applied to the flow paths. Although the problem encountered with

Table 1 Energy models evolution

Evolution stage	Main features	
1st Generation (traditional)	Handbook orientated	Indicative
	Simplified	Application limited
	Piecemeal	Difficult to use
2nd Generation	Dynamics important	
	Less simplified	
	Still piecemeal	
3rd Generation (current)	Field problem approach	
	Move to numerical methods	
	Integrated view of energy heat and mass transfer considered	
	Better user interface	
	Partial computer aided building design (CABD) integration	
4th Generation (next)	CABD integration	
	Advanced numerical methods	Predictive
	Intelligent knowledge based	Generalized
	Advanced software engineering	Easy to use

first-generation models related to quantifying transient heat conduction through multilayered constructions was overcome by second-generation models, a number of new problems emerged. These problems include the need for fast, powerful computers; an extensive input data set to define building geometry; and more appropriate user interfaces.

In recent years, numerical methods in third-generation models have played an increasingly important role in the analysis of heat transfer problems. Numerical methods can be used to solve complex, time-varying problems of high and low order. All fundamental properties are assumed to be time dependent, and coupling among the temporal processes is accounted for. The method allows for high-integrity modeling. Differential equation sets, representing the dynamic energy and mass balances within combined building and plant networks, are solved simultaneously and repeatedly at each (perhaps variable) time step as a simulation is conducted under the influence of control action. Modeling integrity has been increased so that the system is more representative of reality and is easy to use because of the improved graphical I/O (input/output) user interface. The highly transient nature and dynamic response of HVAC equipment are now fully taken into account.

TYPICAL APPLICATIONS

Building system simulation programs can be applied in many situations; they are used by researchers, engineers, and architects to solve real-life problems. In addition to the calculation of the peak cooling/heating loads and annual energy consumption, the following examples give a sense of other situations in which BSS can be a useful tool:

- Predicting daylight levels within building spaces to minimize the energy consumed by artificial lighting and to improve visual comfort and productivity.
- Predicting the pattern of air distribution within spaces to optimize the air conditioning duct design.
- Predicting the effect of using innovative building envelope designs (such as vacuum-insulated wall panels and dual-skin facades) on the overall performance of buildings and on thermal comfort.
- Given a number of known energy conservation measures, BSS can be used to identify the most cost-effective option so that the capital investments are justified.
- For existing buildings, computer models can be developed and calibrated to determine the energy savings corresponding to alternative energy conservation measures or to alternative building and HVAC systems operational strategies.
- The impact of renewable technologies, such as solar collectors and photovoltaic, on energy consumption can be investigated.
- Application of passive heating/cooling concepts can be predicted easily by using simulation programs, and alterations to the building design can be made accordingly.
- The dynamic response of buildings when subjected to temporal boundary conditions can be investigated.

- Building system simulation are used extensively to develop codes and standards for energy conservation in different classes of buildings.

THE SIMULATION PROCESS

In any BSS program, the whole process can be represented by three main stages: model definition, simulation, and result analysis and interpretation, as indicated in Fig. 2. Depending on the level of detail, the building first has to be described in a manner that is acceptable to the program. The early versions of the DOE-2,[5] for example, required the use of a special description language to define the input parameters associated with the building under consideration. This required the user to learn the language in addition to learning the use of the simulation and output data analysis. To simplify the model definition process, graphical user interfaces (GUI) were later developed so that the errors at the input stage can be minimized. In any way, the input data describing the building and its systems then are fed to the solver (simulator) to perform the simulation according to a predefined schedule in which the simulation period and output parameters are specified. When the output is obtained from a simulation, it can be checked, and if necessary, the input stage may be repeated to rectify any errors made and to repeat the simulation. Even with the sophisticated GUIs available with many BSS programs, a verification procedure must be followed to ensure that the results of the simulation are reasonable and acceptable.

The most common input parameters required to simulate a building system are

- *The weather data.* Usually, this represents hourly data for many parameters, such as the dry bulb temperature; relative humidity; wind speed and direction; and diffuse, direct, and global solar radiation. A typical metrological year can be established to represent the weather pattern for a particular location, using at least 10 years' worth of measured weather data.
- *The building location.* This can be established from the geographical location of the building site in terms of longitude and latitude.
- *The building geometry.* This is determined by defining each external surface's geometry and orientation (walls and roofs).
- *Building materials.* The construction materials used in walls and roofs are defined in terms of layers, including the thermophysical properties such as the density, specific heat capacity, and thermal conductivity for each layer.
- *Casual loads.* Those are internal loads from lighting, appliances, and people, which are specified with a predefined schedule.

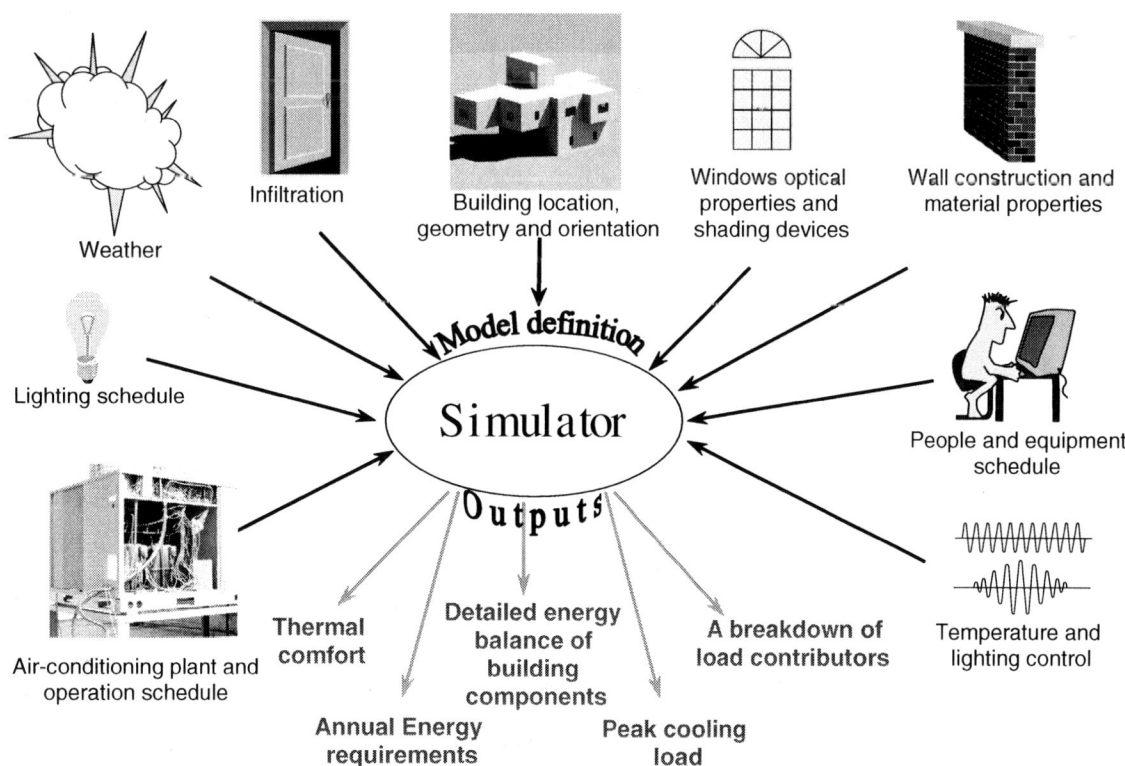

Fig. 2 The simulation process in terms of inputs and output.

- *Shading devices.* They could be used on an opaque surface (walls or roof) and transparent surfaces (windows and glass doors).
- *Transparent surfaces.* Their size, geometry, and optical properties (visible transmittance, solar transmittance, absorptance, and U-value) for both exterior and interior surfaces.
- *Plant.* This is the system used to maintain the internal environment of a building at a certain temperature and humidity, such as the HVAC system.
- *Infiltration.* This relates to the uncontrolled bidirectional flow between the inside of the building and the outside.
- *Control.* The control strategy adopted by the HVAC system can be defined by specifying the properties of the controller and the range of conditions to perform the control. In addition, lighting controls can be specified for situations in which daylight utilization may be required.

The input parameters mentioned above are adequate in most cases to perform the simulation. With some state-of-the-art BSS programs such as the ESP-r,[6] however, it is possible to define additional input data to allow addressing fairly complex issues if deemed necessary. It may be required, for example, to perform three-dimensional conduction analysis of a particular section within a building's construction to estimate the impact of thermal bridging on the heat flow through the building's envelope, which will require the definition of the domain in three-dimensional form.[2] In addition, a more accurate assessment of the infiltration levels and air movement between the building spaces may be necessary to arrive at a more realistic assessment of the thermal performance of the building, in which case an air flow network will need to be specified,[8] or if this was not adequate, a computational fluid dynamics calculation[12] can be incorporated to reach an even higher accuracy level. Furthermore, the effect of environmental systems dynamics can be investigated in detail using BSS programs such as EnergyPlus,[4] ESP-r, HVACSIM+,[9] and TRNSYS.[13] Such BSS programs were designed to study the transient effect of both the HVAC systems and the building, and with some programs, it is even possible to use smaller simulation time steps for the HVAC simulation because the HVAC systems exhibit time constants that are much smaller than the building structure.

When the input parameters are defined, the simulator performs the calculations, based on the adopted mathematical approach, to generate the output required by the user. It is possible to perform the analysis at the whole-system level (i.e., the building and plant systems) or at the system level (the building or plant), or down to the component level (i.e., a cooling coil, a fan, a window, a wall, etc.). Some of the outputs that can be obtained are shown in Fig. 2. These outputs include:

- *Thermal comfort.* This is at the whole-system level and is a very important parameter for many buildings. The simulation output can indicate whether the space in question has an acceptable thermal comfort condition for the occupants.
- *Annual energy requirements.* This is also at the whole-system level, which gives an indication of the total electrical energy consumed by the HVAC system (another source of energy, such as gas or oil, can also be included) and by the lighting and equipment or appliances used in the building.
- *Detailed energy balance of building components.* This is at the component level, for which the output data can be studied to investigate the thermal performance of individual building components such as walls, roof, and windows.
- *Peak load requirements.* This information is important because it is used to size the HVAC systems for a building. It is common to perform the calculation on the building system first to determine the peak cooling/heating load required and then to utilize the output to perform a whole-plant and building analysis to determine the peak electrical demand for the whole plant and building system.
- *Breakdown of load contributors.* This information is useful to identify the major load users so that countermeasures of more significant impact on energy use may be introduced.

To realize the benefits of BSS programs in real-life situations, two examples are discussed in the following sections. The examples focus mainly on how the problem was addressed and also on how to utilize the simulation output so that the set objectives can be achieved. The detailed input data fed to the BSS program to conduct the simulations in the examples were not addressed, because the data are fairly detailed and will not serve the purpose.

EXAMPLE 1

The dayroom area of a hospital in Scotland suffers from high temperature during summer. The dayroom area is joined to the dining-room area and consists of windows with large glazing area. The occupants of the dayroom are elderly patients, who are not always capable of judging thermal stress and so would be unlikely to leave the dayroom when overheating occurs or to take corrective action by, for example, opening a window. The objective of this study was to advise on possible modifications to the hospital ward to better control its radiation and temperature environment.[7]

A base case test was conducted for a typical Glasgow summer weather pattern for the period of Wednesday, July 7 to Thursday, July 8. In this test, it was assumed that all the

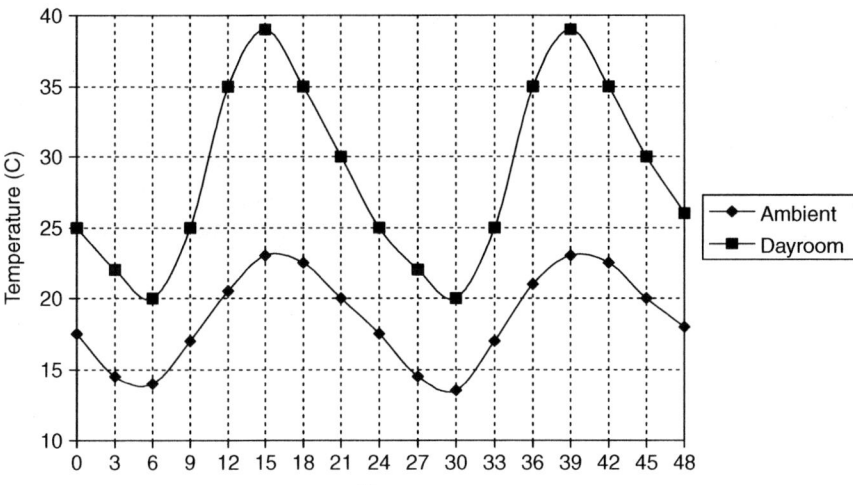

Fig. 3 Base case with no blinds and one air change per hour.

blinds were open in the dayroom and that a fixed infiltration rate of one air change per hour prevailed. Shading analysis was performed to take into account any obstructions and overhangs. The resulting temperatures are shown in Fig. 3. As indicated, the thermal stresses in the dayroom are significant. In reality, it is unlikely that fixed infiltration will occur, because this will depend on many factors, such as the magnitude of open windows, cracks, vents, and so on, as well as changes in pressures and temperatures across the building envelope. For this reason, an air flow network was set up in which the dayroom was represented by nodes on two vertical levels to account for temperature stratification; other air nodes were added to represent wind-induced pressures on the various facades. Then the air flow network nodes were connected by flow components to represent cracks, windows, and doors.

Initially, it was assumed that windows in the dayroom would remain open at all times. The effect of opening and closing of windows in the dining room, however, was considered subsequently by the addition of components to represent flow control. The control action taken was to open a window when the adjacent indoor air node temperature exceeded 20°C. This action was applied to the dining room. An extract fan component was also defined and was set to be thermostatically controlled so that air was extracted from the dayroom's upper point (hottest level) if the upper-level temperature exceeded 25°C, and the hot air was exhausted through the west wall.

A plant was also set up to represent the heating system, which is expected to be operating when room temperatures are below the desired heating set point temperature of 21°C. The drop in temperature below the heating set point occurs because it was assumed that dayroom windows are open all the time and, thus, ambient air free cooling takes place. The plant consisted of oil-filled electric radiators located at the appropriate zones. Each zone has a radiator with a heating capacity sufficient to bring the zone temperature to the heating set point. The dayroom was fitted with a 3000 W electric radiator, whereas the dining room was fitted with a 5000 W unit. Note that these values were arrived at by conducting an initial simulation without the plant. The heating output was split into 40% convective and 60% radiative.

It should be noted that the building, plant, and fluid flow arrangement was arrived at after a number of runs using different control schemes. The results for the predicted room temperature are shown in Fig. 4.

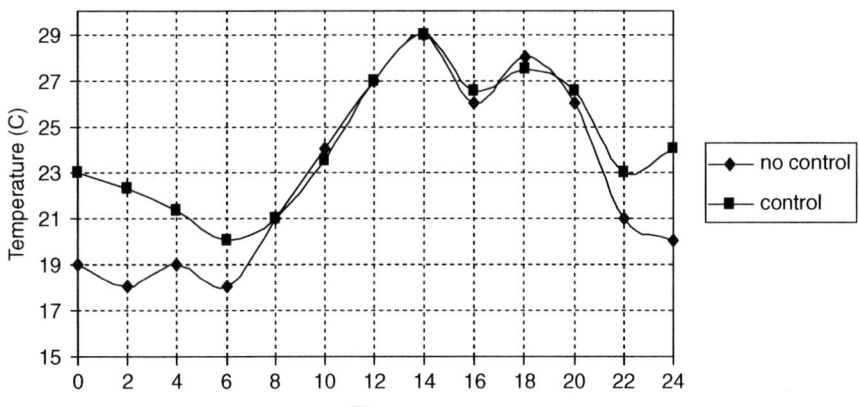

Fig. 4 Predicted dayroom temperature with plant and mass flow network with and without control on dayroom windows.

9:00 h
11:00 h
15:00h
16:00h

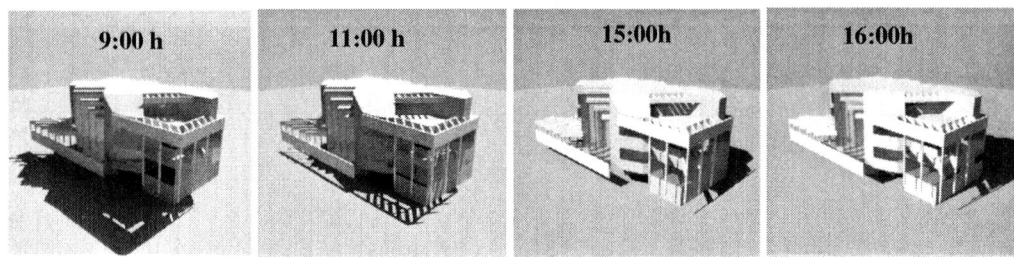

Shading analysis for the assembled villa.

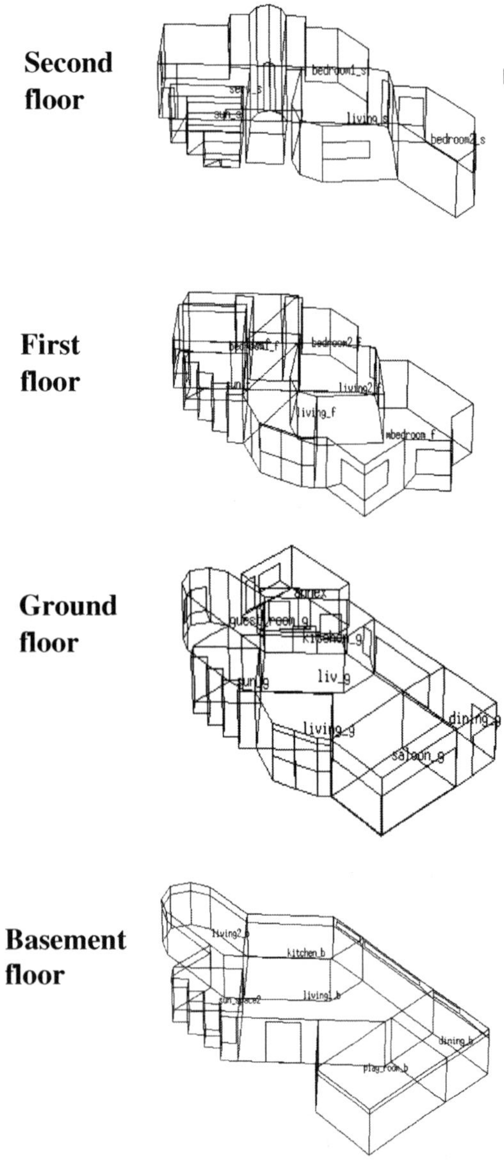

Second floor
First floor
Ground floor
Basement floor

Wire-frame representation of each floor in the villa.

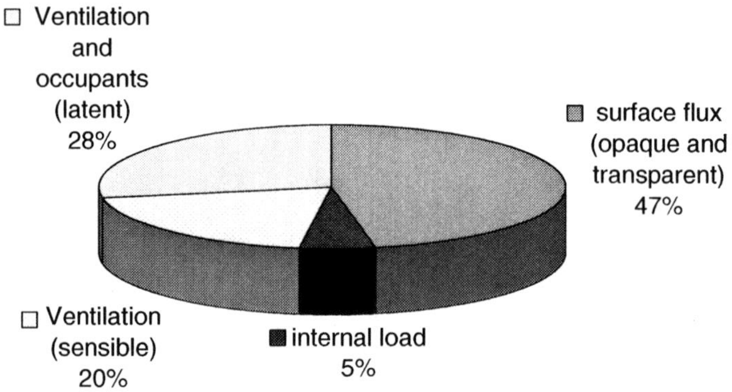

Ventilation and occupants (latent) 28%
surface flux (opaque and transparent) 47%
Ventilation (sensible) 20%
internal load 5%

Peak load components.

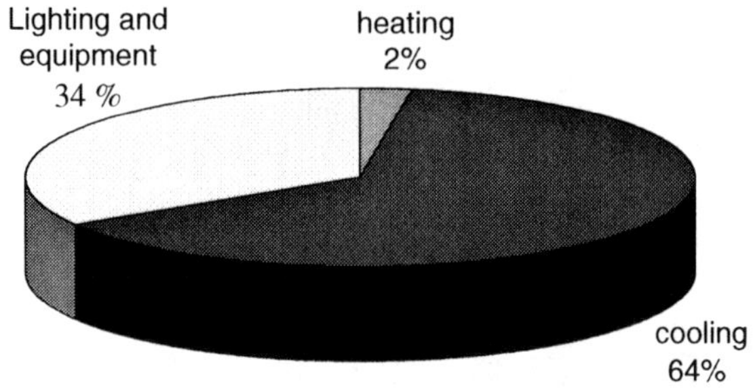

Lighting and equipment 34 %
heating 2%
cooling 64%

Components distribution of annual energy consumption.

Fig. 5 The different stages required, with detailed building energy analysis programs.
Source: From Ref.1.

Fig. 4 shows that for the case with no control, the dayroom temperature is lower than the dining-room temperature when there is no solar radiation. The dayroom temperature increases during the day to a level slightly higher than the dining-room temperature. Heating takes place when the dayroom and dining-room temperatures fall below 21°C. A drop in the dayroom's upper-level temperature occurs in midmorning because that is when the extraction fan is activated by the thermostatic controller. With this building, plant, and flows arrangement, it was possible to decrease the dayroom temperature with respect to the base case. It also was possible to reduce the amount of heating required by the heating system if control also was imposed on the dayroom windows so that they remain closed if the dayroom temperature fell below a certain level. To achieve this, an additional flow component was considered in the simulation to represent flow control so that the window is closed if the dayroom temperature falls below 20°C. The predicted temperatures are also shown in Fig. 4 by the curve with control. The total heating required during the simulation period and in the case where the dayroom windows were left open was 102.2 kWhrs, compared with 39.5 kWhrs for the window-control case.

EXAMPLE 2

A private residential villa located in Kuwait City consists of large glazed areas and complex shading structures. For such villas, solar insulation is a concern because the weather in Kuwait City is typically hot, with a long summer extending over seven months when air conditioning is required. The number of cloudy days during the summer is almost negligible. Therefore, the electricity provider requested the investigation of the thermal performance of the villa when subjected to a number of energy conservation measures.[1] The plot has an area of 800 m^2, with the villa's constructional floor area being 1127 m^2, and is characterized by a sunspace. The external glazing area is 250 m^2 with a glazing to total external wall area ratio of 22%.

Fig. 5 shows that the villa has a complex geometrical shape with a complex shading structure. For this reason, a detailed shading analysis was carried out to account for the effect of the shading structures on solar insolation. In addition, annual simulations were conducted for the cases considered, which include:

    Base case: Use double tinted glass and no insulation on columns and beams.
    Case 1: Add thermal insulation to exposed floors.
    Case 2: Reduce window area by 20%.
    Case 3: Use double reflective glass.
    Case 4: Reduce air change by 50%.
    Case 5: Columns and beams fully insulated.
    Case 6: Use more energy efficient lighting.
    Case 7: Combine case-1, case-3 and case-5.
    Case 8: Combine case-4, and case-7.
    Case 9: Combine case-2, and case-8.
    Case 10: Combine case-2, and case-7.

For each case, the peak cooling demand and annual energy consumption were predicted by the BSS program. The peak cooling load was normalized over the living floor area so that it could be considered in updating the existing limit stipulated in the Kuwaiti code of practice for energy conservation in buildings.[1] It was possible from the simulation results to determine which conservation measure(s) to consider and what alterations were required so that the villa's thermal performance could be improved significantly over the base case design.

In addition, by including the cost factor, a cost/benefit analysis was performed, using the actual cost of implementing the energy conservation measure and using the simulation output to reflect the savings in terms of the initial cost of the air conditioning equipment and the cost of energy over the life cycle of the building. This gave a clear picture of the options that could be considered based on certain criteria (energy, cost, or both). A similar analysis was conducted on a number of other private villas, and the results of the normalized peak cooling load were used to update the limits for such buildings in the code of practice for energy conservation. This limit is of paramount importance because it is referred to by HVAC engineers to determine the maximum capacity of the air conditioning system; the limit can also be verified by the concerned authority for compliance with the code.

CONCLUSION

Building system simulation programs have evolved into more sophisticated tools over the past few decades. The developments made focused not only on the user interface level, but also on the mathematical approach adopted, leading to more accurate and realistic predictions of the thermal response of buildings. Building system simulation have wide applications for engineers, architects, and researchers. They are valuable tools that can be used to investigate the impact of numerous energy measures on the overall performance of buildings. State-of-the-art BSS tools allow for integrated building design, which consequently leads to designs that are optimized for thermal and visual comfort and for energy efficiency.

REFERENCES

1. Assem, E. Energy efficiency for fuelling the world. Proceedings of Energy Conservation in Buildings Workshop, Kuwait, December 13–17, Kwait Times: Kuwait, 2003, 305–328.

2. Ben-Nakhi, A. Development of an integrated dynamic thermal bridging assessment environment. Energy Buildings **2003**, *35* (4), 375–382.
3. Clarke, J.A. The energy kernel system. Energy Buildings **1988**, *10* (3), 259–266.
4. Crawley, D.B., et al. Energy plus: an update. Proceedings of SimBuild 2004 in Boulder, Colorado. International Building Performance Simulation Association: Boulder, Colorado, United States, August 4–6, 2004.
5. DOE-2 *DOE-2 Engineers Manual Version 2.1A*; National Technical Information Service, U.S. Department of Commerce: Springfield, Virginia, 1981.
6. ESP-r. The ESP-r System for Building Energy Simulation User Guide. Version 10 Series, ESRU Manual U02/1, University of Strathclyde: Glasgow, Scotland, October 2002.
7. Hand, J.; Assem, E. *Analysis of invernairn ward dayroom of erskine hospital*. ESRU Project Report, University of Strathclyde: Glasgow, Scotland, 1990.
8. Hensen, J.L.M., *On the Thermal Interaction of Building Structure and Heating and Ventilating Systems*. PhD thesis, Technische Universiteit: Eindhoven, Netherlands, 1991.
9. Clark, D.R., HVACSIM+ Program. Reference Manual. NBSIR 84-2996, National Institute of Standards and Testing, Gaithersberg, Maryland, 1985.
10. Kusuda, T. Thermal response factors for multilayer structures of various heat conduction systems. ASHRAE Trans. **1969**, *75* (1), 246–270.
11. Ministry of Electricity and Water. Code of practice for energy conservation in kuwaiti buildings. Report No. MEW R-6, State of Kuwait, 1983.
12. Negrao COR. *Conflation of Computational Fluid Dynamics and Building Thermal Simulation*. PhD thesis, University of Strathclyde: Glasgow, Scotland, 1995.
13. TRNSYS A transient simulation program. *Solar Energy Laboratory*, University of Wisconsin: Madison, Wisconsin, 1983.

Carbon Sequestration

Nathan E. Hultman
Science, Technology, and International Affairs, Georgetown University, Washington, D.C., U.S.A.

Abstract
Carbon dioxide (CO_2), a byproduct of hydrocarbon combustion and a natural emission from biomass burning, respiration, or decay, is a major greenhouse gas and contributor to anthropogenic climate change. Carbon sequestration describes the processes by which carbon can be either removed from the atmosphere (as CO_2) and stored, or separated from fuels or flue gases and stored. Carbon sequestration can thus be either technological (usually called carbon capture and storage) or biological (biological carbon sequestration). The viability of carbon sequestration depends on the cost of the process and the policy context that determines the value of sequestered carbon.

INTRODUCTION

The increasing likelihood of human-caused changes in climate could lead to undesirable impacts on ecosystems, economies, and human health and well-being. These potential impacts have prompted extensive assessment of options to reduce the magnitude and rate of future climate changes. Since climate changes are derived ultimately from increases in the concentrations of greenhouse gases (GHGs) in the atmosphere, such options must target either (a) reductions in the rate of inflow of GHGs to the atmosphere or (b) the removal of GHGs from the atmosphere once they have been emitted. Carbon sequestration refers to techniques from both categories that result in the storage of carbon that would otherwise be in the atmosphere as CO_2.

CO_2 is often targeted among the other GHGs because it constitutes the vast majority of GHG emissions by mass and accounts for three-fifths of the total anthropogenic contribution to climate change. Human emissions of CO_2 come primarily from fossil fuel combustion and cement production (80%), and land-use change (20%) that results in the loss of carbon from biomass or soil.

The rate of inflow of GHGs to the atmosphere can be reduced by a number of complementary options. For CO_2, mitigation options aim to displace carbon emissions by preventing the oxidation of biological or fossil carbon. These options include switching to lower-carbon fossil fuels, renewable energy, or nuclear power; using energy more efficiently; and reducing the rate of deforestation and land-use change. On the other hand, sequestration options that reduce emissions involve the capture and storage of carbon before it is released into the atmosphere.

CO_2 can also be removed directly from the atmosphere. While the idea of a large-scale, economically competitive method of technologically "scrubbing" CO_2 from the atmosphere is enticing, such technology currently does not exist. Policy has therefore focused on the biological process of carbon absorption through photosynthesis, either through expanding forested lands or, perhaps, enhancing photosynthesis in the oceans. This entry describes both the technological and biological approaches to carbon sequestration.

TECHNOLOGICAL SEQUESTRATION: CARBON CAPTURE AND STORAGE

The technological process of sequestering CO_2 requires two steps: first, the CO_2 must be separated from the industrial process that would otherwise emit it into the atmosphere; and second, the CO_2 must be stored in a reservoir that will contain it for a reasonable length of time. This process is therefore often referred to as carbon capture and storage (CCS) to distinguish it from the biological carbon sequestration that is described later.

Sources of Carbon

The best sites for CCS are defined by the efficiency of the capture technique, the cost of transport and sequestration, and the quantity of carbon available. The large capital requirements for CCS also dictate that large, fixed industrial sites provide the best opportunities. Therefore, although fossil-fueled transportation represents about 20% of current global CO_2 emissions, this sector presents no direct options for CCS at this time. The industrial sector, on the other hand, produces approximately 60% of current CO_2 emissions; most of these emissions come from large point sources which are ideal for CCS, such as power

Keywords: Carbon sequestration; Capture and storage; Carbon sinks; Geologic storage; Climate policy; Emissions trading; Carbon dioxide (CO_2).

stations, oil refineries, petrochemical and gas reprocessing plants, and steel and cement works.[1]

Separation and Capture

Carbon capture requires an industrial source of CO_2; different industrial processes create streams with different CO_2 concentrations. The technologies applied to capture the CO_2 will therefore vary according to the specific capture process.[2–4] Capture techniques can target one of three sources:

- Post-combustion flue gases.
- Pre-combustion capture from gasification from power generation.
- Streams of highly pure CO_2 from various industrial processes.

Post-Combustion Capture

Conventional combustion of fossil fuels in air produces CO_2 streams with concentrations ranging from about 4 to 14% by volume. The low concentration of CO_2 in flue gas means that compressing and storing it would be uneconomical; therefore, the CO_2 needs to be concentrated before storage. Currently, the favored process for this task is chemical absorption, also known as chemical solvent scrubbing. Cooled and filtered flue gas is fed into an absorption vessel with a chemical solvent that absorbs the CO_2. The most common solvent for this process is monothanolamine (MEA). The CO_2-rich solvent is then passed to another reaction vessel called a stripper column. It is then heated with steam to reverse the process, thus regenerating the solvent and releasing a stream of CO_2 with a purity greater than 90%.

Scrubbing with MEA and other amine solvents imposes large costs in energy consumption in the regeneration process; it requires large amounts of solvents since they degrade rapidly; and it imposes high equipment costs since the solvents are corrosive in the presence of O_2. Thus, until solvents are improved in these areas, flue gas separation by this method will remain relatively costly: just the steam and electric load from a coal power plant can increase coal consumption by 40% per net kWh_e. Estimates of the financial and efficiency costs from current technology vary. Plant efficiency is estimated to drop from over 40% to a range between 24 and 37%.[2,5,6] For the least efficient systems, carbon would cost up to $70/t CO_2 and result in an 80% increase in the cost of electricity.[5] Other studies estimate an increase in the cost of electricity of 25%–75% for natural gas combined cycle and Integrated Gasification Combined Cycle (IGCC), and of 60%–115% for pulverized coal.[4] A small number of facilities currently practice flue gas separation with chemical absorption, using the captured CO_2 for urea production, foam blowing, carbonated beverages, and dry ice production. In addition, several developments may improve the efficiency of chemical absorption.

Several other processes have been proposed for flue-gas separation. Adsorption techniques use solids with high surface areas, such as activated carbon and zeolites, to capture CO_2. When the materials become saturated, they can be regenerated (releasing CO_2) by lowering pressure, raising temperature, or applying a low-voltage electric current. A membrane can be used to concentrate CO_2, but since a single pass through a membrane cannot achieve a great change in concentration, this process requires multiple passes or multiple membranes. An alternative use for membranes is to use them to increase the efficiency of the chemical absorption. In this case, a membrane separating the flue gas from the absorption solvent allows a greater surface area for the reaction, thus reducing the size and energy requirements of the absorption and stripper columns. *Cryogenic* techniques separate CO_2 from other gases by condensing or freezing it. This process requires significant energy inputs and the removal of water vapor before freezing.

One of the main limitations to flue-gas separation is the low pressure and concentration of CO_2 in the exhaust. An entirely different approach to post-combustion capture is to dramatically increase the concentration of CO_2 in the stream by burning the fuel in highly enriched oxygen rather than air. This process, called oxyfuel combustion, produces streams of CO_2 with a purity greater than 90%. The resulting flue gas will also contain some H_2O that can be condensed and removed, and the remaining high-purity CO_2 can be compressed for storage. Though significantly simpler on the exhaust side, this approach requires a high concentration of oxygen for the intake air. While this process alone may consume 15% of a plant's electric output, the separated N_2, Ar, and other trace gases also can be sold to offset some of the cost. Oxyfuel systems can be retrofitted onto existing boilers and furnaces.

Pre-Combustion Capture

Another approach involves removing the carbon from fossil fuels before combustion. First, the fuel is decomposed in the absence of oxygen to form a hydrogen-rich fuel called synthesis gas. Currently, this process of gasification is already in use in ammonia production and several commercial power plants fed by coal and petroleum byproducts; these plants can use lower-purity fuels and the energy costs of generating synthesis gas are offset by the higher combustion efficiencies of gas turbines; such plants are called IGCC plants. Natural gas can be transformed directly by reacting it with steam, producing H_2 and CO_2. While the principle of gasification is the same for all carbonaceous fuels, oil and coal require intermediate steps to purify the synthesis fuel and convert the byproduct CO into CO_2.

Gasification results in synthesis gas that contains 35%–60% CO_2 (by volume) at high pressure (over 20 bar). While current installations feed this resulting mixture into the gas turbines, the CO_2 can also be separated from the gas before combustion. The higher pressure and concentration give a CO_2 partial pressure of up to 50 times greater than in the post-combustion capture of flue gases, which enables another type of separation technique of physical solvent scrubbing. This technique is well known from ammonia production and involves the binding of CO_2 to solvents that release CO_2 in the stripper under lower pressure. Solvents in this category include cold methanol, polyethelene glycol, propylene carbonate, and sulpholane. The resulting separated CO_2 is, however, near atmospheric pressure and requires compression before storage (some CO_2 can be recovered at elevated pressures, which reduces the compression requirement). With current technologies, the total cost of capture for IGCC is estimated to be greater than $25 per ton of CO_2; plant efficiency is reduced from 43 to 37%, which raises the cost of electricity by over 25%.[5]

Pre-combustion capture techniques are noteworthy not only for their ability to remove CO_2 from fossil fuels for combustion in turbines, but also because the resulting synthesis gas is primarily H_2. They therefore could be an important element of a hydrogen-mediated energy system that favors the higher efficiency reactions of fuel cells over traditional combustion.[7]

Industrial CO_2 Capture

Many industrial processes release streams of CO_2 that are currently vented into the atmosphere. These streams, currently viewed as simple waste in an economically viable process, could therefore provide capture opportunities. Depending on the purity of the waste stream, these could be among the most economical options for CCS. In particular, natural gas processing, ethanol and hydrogen production, and cement manufacturing produce highly concentrated streams of CO_2. Not surprisingly, the first large-scale carbon sequestration program was run from a previously vented stream of CO_2 from the Sleipner gas-processing platform off the Norwegian coast.

Storage of Captured CO_2

Relatively small amounts of captured CO_2 might be re-used in other industrial processes such as beverage carbonation, mineral carbonates, or commodity materials such as ethanol or paraffins. Yet most captured CO_2 will not be re-used and must be stored in a reservoir. The two main routes for storing captured CO_2 are to inject it into geologic formations or into the ocean. However, all reservoirs have some rate of leakage and this rate is often not well known in advance. While the expected length of storage time is important (with targets usually in the 100–1000 year range), we must therefore also be reasonably confident that the reservoir will not leak more quickly than expected, and have appropriate measures to monitor the reservoir over time. Moreover, transporting CO_2 between the point of capture and the point of storage adds to the overall cost of CCS, so the selection of a storage site must account for this distance as well.

Geologic Sequestration

Geologic reservoirs—in the form of depleted oil and gas reservoirs, unmineable coal seams, and saline formations—comprise one of the primary sinks for captured CO_2. Estimates of total storage capacity in geologic reservoirs could be up to 500% of total emissions to 2050 (Fig. 1).

Captured CO_2 can be injected into depleted oil and gas reservoirs, or can be used as a means to enhance oil recovery from reservoirs nearing depletion. Because they held their deposits for millions of years before extraction, these reservoirs are expected to provide reliable storage for CO_2. Storage in depleted reservoirs has been practiced for years for a mixture of petroleum mining waste gases called "acid gas."

A petroleum reservoir is never emptied of all its oil; rather, extracting additional oil just becomes too costly to justify at market rates. An economically attractive possibility is therefore using captured CO_2 to simultaneously increase the yield from a reservoir as it is pumped into the reservoir for storage. This process is called enhanced oil recovery. Standard oil recovery yields only about 30%–40% of the original petroleum stock. Drilling companies have years of experience with using compressed CO_2, a hydrocarbon solvent, to obtain an additional 10%–15% of the petroleum stock. Thus, captured CO_2 can be used to provide a direct economic

Reservoir Type	Storage Capacity	
	billion tonnes CO_2	% of E
Coal basins	170	8%
Depleted oil reservoirs	120	6%
Gas basins	700	34%
Saline formations		
Terrestrial	5,600	276%
Off-shore	3,900	192%
Total	10,490	517%

Fig. 1 CO_2 Reservoirs. Carbon dioxide storage capacity estimates. E is defined as the total global CO_2 emissions from the years 2000–2050 in IPCC's business-as-usual scenario IS92A. Capacity estimates such as these are rough guidelines only and actual utilization will depend on carbon economics.
Source: From Ref. 8.

benefit along with its placement in a reservoir. This benefit can be used to offset capture costs.

Coal deposits that are not economically viable because of their geologic characteristics provide another storage option. CO_2 pumped into these unmineable coal seams will adsorb onto the coal surface. Moreover, since the coal surface prefers to adsorb CO_2 to methane, injecting CO_2 into coal seams will liberate any coal bed methane (CBM) that can then be extracted and sold. This enhanced methane recovery is currently used in U.S. methane production, accounting for about 8% in 2002. Such recovery can be used to offset capture costs. One potential problem with this method is that the coal, as it adsorbs CO_2, tends to swell slightly. This swelling closes pore spaces and thus decreases rock permeability, which restricts both the reservoir for incoming CO_2 and the ability to extract additional CBM.

Saline formations are layers of porous sedimentary rock (e.g., sandstone) saturated with saltwater, and exist both under land and under the ocean. These layers offer potentially large storage capacity representing several hundred years' worth of CO_2 storage. However, experience with such formations is much more limited and thus the uncertainty about their long-term viability remains high. Moreover, unlike EOR or CBM recovery with CO_2, injecting CO_2 into saline formations produces no other commodity or benefit that can offset the cost. On the other hand, their high capacity and relative ubiquity makes them attractive options in some cases. Statoil's Sleipner project, for example, uses a saline aquifer for storage.

Research and experimentation with saline formations is still in early stages. To achieve the largest storage capacities, CO_2 must be injected below 800 m depth, where it will remain in a liquid or supercritical dense phase (supercritical point at 31°C, 71 bar). At these conditions, CO_2 will be buoyant (a density of approximately 600–800 kg/m^3) and will tend to move upward. The saline formations must therefore either be capped by a less porous layer or geologic trap to prevent leakage of the CO_2 and eventual decompression.[9] Over time, the injected CO_2 will dissolve into the brine and this mixture will tend to sink within the aquifer. Also, some saline formations exist in rock that contains Ca-, Mg-, and Fe-containing silicates that can form solid carbonates with the injected CO_2. The resulting storage as rock is highly reliable, though it may also hinder further injection by closing pore spaces. Legal questions may arise when saline formations, which are often geographically extensive, cross national boundaries or onto marine commons.

Ocean Direct Injection

As an alternative to geologic storage, captured CO_2 could be injected directly into the ocean at either intermediate or deep levels. The oceans have a very large potential for storing CO_2, equivalent to that of saline aquifers ($\sim 10^3$ Gt). While the ocean's surface is close to equilibrium with atmospheric carbon dioxide concentrations, the deep ocean is not because the turnover time of the oceans is much slower (~ 5000 years) than the observed increases in atmospheric CO_2. Since the ocean will eventually absorb much of the atmospheric perturbation, injecting captured CO_2 into the oceans can therefore be seen as simply bypassing the atmospheric step and avoiding the associated climate consequences. Yet little is known about the process or effects—either ecological or geophysical—of introducing large quantities of CO_2 into oceanic water.

At intermediate depths (between 500 and 3000 m), CO_2 exists as a slightly buoyant liquid. At these depths, a stream of CO_2 could be injected via a pipe affixed either to ship or shore. The CO_2 would form a droplet plume, and these droplets would slowly dissolve into the seawater, disappearing completely before reaching the surface. Depressed pH values are expected to exist for tens of km downcurrent of the injection site, though changing the rate of injection can moderate the degree of perturbation. In addition, pulverized limestone could be added to the injected CO_2 to buffer the acidity.

Below 3000 m, CO_2 becomes denser than seawater and would descend to the seafloor and pool there. Unlike intermediate injection, therefore, this method does not lead to immediate CO_2 dissolution in oceanic water; rather, the CO_2 is expected to dissolve into the ocean at a rate of about 0.1 m/y. Deep injection thus minimizes the rate of leakage to the surface, but could still have severe impacts on bottom-dwelling sea life.

The primary obstacles to oceanic sequestration are not technical but relate rather to this question of environmental impacts.[10] Oceanic carbon storage might affect marine ecosystems through the direct effects of a lower environmental pH; dissolution of carbonates on fauna with calcareous structures and microflora in calcareous sediments; impurities such as sulfur oxides, nitrogen oxides, and metals in the captured CO_2; smothering effects (deep injection only); and changes in speciation of metals and ammonia due to changes in pH. Few of these possibilities have been studied in sufficient detail to allow an informed risk assessment. In addition, the legality of dumping large quantities of CO_2 into the open ocean remains murky.

Overall Costs of CCS

The costs of CCS can be measured either as a cost per tonne of CO_2, or, for power generation, a change in the cost of electricity (Fig. 2). The total cost depends on the cost of capture, transport, and storage. Capture cost is mainly a function of parasitic energy losses and the capital cost of equipment. Transport cost depends on distance and terrain. Storage costs vary depending on the reservoir but are currently a few dollars per tonne of CO_2. The variety of

Fossil plant type	Cost of CCS
	¢ per kWh
Natural gas combined cycle	1–2
Pulverized coal	2–3
Coal IGCC	2–4

Fig. 2 Additional costs to power generation from CCS. Approximate capture and storage costs for different approaches to power plant sequestration.
Source: From Refs. 4,5,11,12.

approaches to CCS and the early stages of development make precise estimates of cost difficult, but current technology spans about $25–$85/t CO_2.

BIOLOGICAL SEQUESTRATION: ENHANCING NATURAL CARBON SINKS

The previous sections have described processes by which CO_2 could be technologically captured and then stored. Photosynthesis provides an alternate route to capture and store carbon. Enhancing this biological process is therefore an alternative method of achieving lower atmospheric CO_2 concentrations by absorbing it directly from the air.

Terrestrial Carbon Sinks

Carbon sequestration in terrestrial ecosystems involves enhancing the natural sinks for carbon fixed in photosynthesis. This occurs by expanding the extent of ecosystems with a higher steady-state density of carbon per unit of land area. For example, because mature forest ecosystems contain more carbon per hectare than grasslands, expanding forested areas will result in higher terrestrial carbon storage. Another approach is to encourage the additional storage of carbon in agricultural soils. The essential element in any successful sink enhancement program is to ensure that the fixed carbon remains in pools with long lives.

Afforestation involves planting trees on unforested or deforested land.[13,14] The most likely regions for forest carbon sequestration are Central and South America and Southeast Asia because of relatively high forest growth rates, available land, and inexpensive labor. However, the translation of forestry activities into a policy framework is complex. Monitoring the carbon changes in a forest is difficult over large areas, as it requires not only a survey of the canopy and understory, but also an estimate of the below-ground biomass and soil carbon. Some groups have voiced concern over the potential for disruption of social structures in targeted regions.

Soil carbon sequestration involves increasing soil carbon stocks through changes in agriculture, forestry, and other land use practices. These practices include mulch farming, conservation tillage, agroforestry and diverse cropping, cover crops, and nutrient management that integrates manure, compost, and improved grazing. Such practices, which offer the lowest-cost carbon sequestration, can have other positive effects such as soil and water conservation, improved soil structure, and enhanced soil fauna diversity. Rates of soil carbon sequestration depend on the soil type and local climate, and can be up to 1000 kg of carbon per hectare per year. Management practices can enhance sequestration for 20–50 years, and sequestration rates taper off toward maturity as the soil carbon pool becomes saturated. Widespread application of recommended management practices could offset 0.4 to 1.2 GtC/y, or 5%–15% of current global emissions.[15]

If sinks projects are to receive carbon credits under emissions trading schemes like that in the Kyoto Protocol, they must demonstrate that the project sequestered more carbon than a hypothetical baseline or business-as-usual case. They must also ensure that the carbon will remain in place for a reasonable length of time, and guard against simply displacing the baseline activity to a new location.

Ocean Fertilization

Vast regions of the open ocean have very little photosynthetic activity, though sunlight and major nutrients are abundant. In these regions, phytoplankton are often deprived of trace nutrients such as iron. Seeding the ocean surface with iron, therefore, might produce large phytoplankton blooms that absorb CO_2. As the plankton die, they will slowly sink to the bottom of the ocean, acting to transport the fixed carbon to a permanent burial in the seafloor. While some experimental evidence indicates this process may work on a limited scale, little is known about the ecosystem effects and potential size of the reservoir.[16]

PROSPECTS FOR CARBON SEQUESTRATION

Carbon sequestration techniques—both technological and biological—are elements of a portfolio of options for addressing climate change. Current approaches hold some promise for tapping into the geologic, biologic, and oceanic potential for storing carbon. The costs of some approaches, especially the improved management of agricultural and forest lands, are moderate (Fig. 3).

Sequestration Technique	Cost
	$ per T CO_2
Carbon capture & storage	26–84
Tree planting & agroforestry	10–210
Soil carbon sequestration	6–24

Fig. 3 Costs of carbon sequestration. Estimates for sequestration costs vary widely. Future costs will depend on rates of technological change.
Source: From Refs. 4,5,7,11–14.

Yet these opportunities are not infinite and additional options will be necessary to address rising global emissions. Thus, the higher costs of current technological approaches are likely to drop with increasing deployment and changing market rates for carbon.

Possible developments include advanced CO_2 capture techniques focusing on membranes, ionic (organic salt) liquids, and microporous metal organic frameworks. Several alternative, but still experimental, sequestration approaches have also been suggested. Mineralization could convert CO_2 to stable minerals. This approach seeks, therefore, to hasten what in nature is a slow but exothermic weathering process that operates on common minerals like olivine, forsterite, or serpentines (e.g., through selected sonic frequencies). It is possible that CO_2 could be injected in sub-seafloor carbonates. Chemical looping describes a method for combusting fuels with oxygen delivered by a redox agent instead of by air or purified oxygen; it promises high efficiencies of energy conversion and a highly enriched CO_2 exhaust stream. Research also continues on microbial CO_2 conversion in which strains of microbes might be created to metabolize CO_2 to produce saleable commodities (succinic, malic, and fumeric acids). In addition, the nascent science of monitoring and verifying the storage of CO_2 will be an important element toward improving technical performance and public acceptance of sequestration techniques.

ACKNOWLEDGMENTS

The author thanks the anonymous reviewer for helpful comments on an earlier draft.

REFERENCES

1. Gale, J., Overview of CO_2 Emission Sources, Potential, Transport, and Geographical Distribution of Storage Possibilities. In *Proceedings of IPCC workshop on carbon dioxide capture and storage, Regina, Canada*, Nov 18–21, 2002; 15–29. http://arch.rivm.nl/env/int/ipcc/pages_media/ccs2002.html (accessed April 2005). A revised and updated version of the papers presented at the Regina workshop is now available. See, Metz, B.; Davidson, O.; de Coninck, H.; Loos, M.; Meyer, L., Special Report on Carbon Dioxide Capture and Storage. Intergovernmental Panel on Climate Change: Geneva, http://www.ipcc.ch. (accessed May 2006), 2006.
2. Thambimuthu, K.; Davison, J.; Gupta, M., CO_2 Capture and Reuse, In *Proceedings of IPCC workshop on carbon dioxide capture and storage*, Regina, Canada, Nov 18–21, 2002; 31–52, http://arch.rivm.nl/env/int/ipcc/pages_media/ccs2002.html (accessed April 2005).
3. Gottlicher, G. *The Energetics of Carbon Dioxide Capture in Power Plants*; U.S. Department of Energy Office of Fossil Energy: Washington, DC, 2004; 193.
4. Herzog, H.; Golomb, D. Carbon capture and storage from fossil fuel use. In *Encyclopedia of Energy*; Cleveland, C.J., Ed.; Elsevier Science: New York, 2004; 277–287.
5. National Energy Technology Laboratory *Carbon Sequestration: Technology Roadmap and Program Plan*; U.S. Department of Energy: Washington, DC, 2004.
6. Gibbins, J.R.; Crane, R.I.; Lambropoulos, D.; Booth, C.; Roberts, C.A.; Lord, M., Maximizing the Effectiveness of Post Combustion CO_2 Capture Systems. In *Proceedings of the 7th International Conference on Greenhouse Gas Abatement*, Vancouver, Canada, Sept 5–9, 2004; Document E2-2. http://www.ghgt7.ca (accessed June 2004).
7. Williams, R.H., Decarbonized Fossil Energy Carriers and Their Energy Technology Competitors. In *Proceedings of IPCC workshop on carbon dioxide capture and storage*, Regina, Canada, November, 2002; 119–135. http://arch.rivm.nl/env/int/ipcc/pages_media/ccs2002.html (accessed April 2005).
8. Dooley, J.J.; Friedman, S.J. *A Regionally Disaggregated Global Accounting of CO_2 Storage Capacity: Data and Assumptions*; Battelle National Laboratory: Washington DC, 2004; 15.
9. Kårsted, O., Geological Storage, Including Costs and Risks, in Saline Aquifers. In *Proceedings of IPCC workshop on carbon dioxide capture and storage*, Regina, Canada, Nov 18–21, 2002; 53–60. http://arch.rivm.nl/env/int/ipcc/pages_media/ccs2002.html (accessed April 2005).
10. Johnston, P.; Santillo, D., Carbon Capture and Sequestration: Potential Environmental Impacts, In *Proceedings of IPCC workshop on carbon dioxide capture and storage*, Regina, Canada, Nov 18–21, 2002; 95–110. http://arch.rivm.nl/env/int/ipcc/pages_media/ccs2002.html (accessed April 2005).
11. Dooley, J.J.; Edmonds, J.A.; Dahowski, R.T.; Wise, M.A., Modeling Carbon Capture and Storage Technologies in Energy and Economic Models. In *Proceedings of IPCC workshop on carbon dioxide capture and storage*, Regina, Canada, Nov 18–21, 2002; 161–172. http://arch.rivm.nl/env/int/ipcc/pages_media/ccs2002.html (accessed April 2005).
12. Freund, P.; Davison, J., General Overview of Costs. In *Proceedings of IPCC workshop on carbon dioxide capture and storage*, Regina, Canada, Nov 18–21, 2002; 79–93. http://arch.rivm.nl/env/int/ipcc/pages_media/ccs2002.html (accessed April 2005).
13. Van Kooten, G.C.; Eagle, A.J.; Manley, J.; Smolak, T. How costly are carbon offsets? A meta-analysis of carbon forest sinks. Environ. Sci. Policy **2004**, *7*, 239–251.
14. Richards, K.R.; Stokes, C. A review of forest carbon sequestration cost studies: a dozen years of research. Climatic Change **2004**, *68*, 1–48.
15. Lal, R. Soil carbon sequestration impacts on global climate change and food security. Science **2004**, *304*, 1623–1627.
16. Buesseler, K.O.; Andrews, J.E.; Pike, S.M.; Charette, M.A. The effects of iron fertilization on carbon sequestration in the southern ocean. Science **2004**, *304*, 414–417.

Career Advancement and Assessment in Energy Engineering

Albert Thumann
Association of Energy Engineers, Atlanta, Georgia, U.S.A.

Abstract
The Association of Energy Engineers (AEE) has helped define the profession of energy engineering through continuing education programs and journals. Association of Energy Engineers provides many networking opportunities through local chapters. This entry details continuing education programs available through AEE, the salary structure of energy professionals, and achieving excellence through certifications.

INTRODUCTION

The profession of energy engineering has gained new importance for several important factors. First, global warming is now considered a reality, and there is a need to reduce greenhouse gases by applying energy-efficient technologies. In addition, surging energy prices have caused companies to evaluate how they use energy and to reduce operating costs. A third factor is the reliability of the electric power grid. In the summer of 2006, blackouts occurred in Queens, New York and other parts of the country. Energy engineering professionals play a key role in reducing the need for power generation and distribution lines. The Association of Energy Engineers (AEE) is dedicated to improving the practice of energy engineering through certification programs, continuing education programs, and networking opportunities through 44 chapters throughout the world.

The profession of energy engineering is relatively new. The oil embargo of the early 1970s created a demand for engineers who can apply the latest energy efficiency technologies to reduce demand in buildings and industrial processes. Most colleges do not offer degrees in energy engineering. An individual pursuing this profession usually attends courses presented by universities and associations, and receives on-the-job training. Some universities—such as Texas A&M, Georgia Tech, and the University of Wisconsin—offer graduate courses in energy engineering and management.

THE ASSOCIATION OF ENERGY ENGINEERS

The AEE was founded in 1977 as a 501 c(6) not-for-profit professional society. The purpose of the AEE is to promote the scientific and educational aspects of those engaged in the energy industry. In the 1970s, the profession of energy engineering and energy management was new. The AEE defined the important functions energy engineers and managers perform and played a key role in the professions' development.

One of the AEE's first tasks was to create an authoritative journal that would guide energy engineers in applying new energy-efficient technologies and applications. The *Energy Engineering* journal was born out of this need and currently is edited by noted authority Dr. Wayne Turner.

The AEE recognized that energy engineers need both technical and management skills. Energy engineers need a broad understanding of fuels procurement, commodity, and risk management, as well as organizational and motivational skills. The *Strategic Planning for Energy and the Environment* journal, also edited by Dr. Wayne Turner, was developed by the AEE to meet this need.

To help energy engineers meet the challenges of power reliability and the development of new energy supplies, the AEE launched the *Cogeneration and Distributed Generation* journal, currently edited by Dr. Steven Parker.

Today, the AEE's network includes 8500 members in 77 countries, with chapters in 69 cities.

The AEE presents numerous training and certification programs to help energy engineers reach their potential.

For complete details on programs offered by the AEE, the reader is referred to www.aeecenter.org.

SALARY STRUCTURE OF ENERGY ENGINEERS

The AEE conducted a salary survey of its 8500 members in 2005. The results of the salary survey follow.

AEE INCOME AND SALARY SURVEY

Only those who were employed full time as of January 1, 2005 were surveyed.

Keywords: Profession; Energy engineering; Association of Energy Engineers (AEE); Salary structure of energy engineers; Continuing education programs; Continuing Education Units (CEUs); Scholarship; Professional certification programs.

1. Please input your base salary (to the nearest $10,000) as of January 1, 2004, to January 1, 2005 (exclude bonus, overtime, fees, and income from secondary employment).

Base Salary	Total	Percentage
$20,000 to $30,000	3	0.65%
$30,000 to $40,000	5	1.08%
$40,000 to $50,000	17	3.66%
$50,000 to $60,000	47	10.13%
$60,000 to $70,000	82	17.67%
$70,000 to $80,000	99	21.34%
$80,000 to $90,000	78	16.81%
$90,000 to $100,000	56	12.07%
$100,000 to $110,000	36	7.76%
$110,000 to $120,000	13	2.80%
$120,000 to $130,000	9	1.94%
$130,000 to $140,000	5	1.08%
$140,000 to $150,000	5	1.08%
Over $150,000	9	1.94%

Average Annual Salary: $85,625.00.

2. Please input your additional income (to the nearest $1000) from primary job, such as bonus, overtime, and fees as of January 1, 2004 to January 1, 2005.

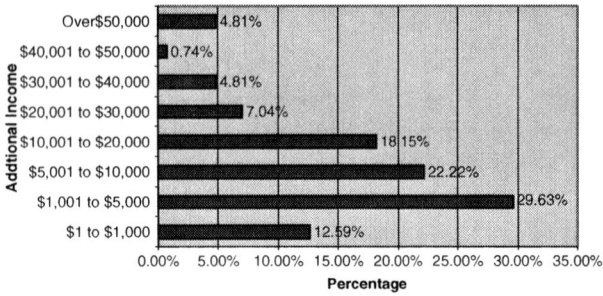

Average Annual Bonus Amount: $14,274.07.

3. Are you a graduate from a 4-year accredited college?

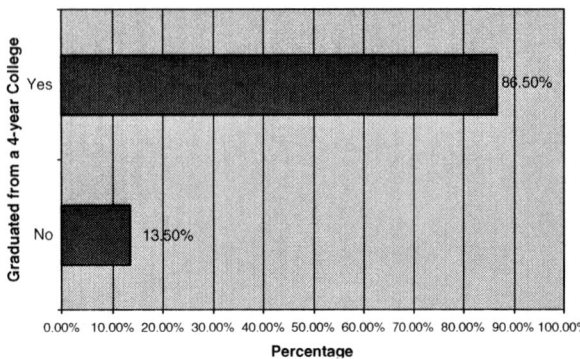

4. Do you have a post-graduate degree from an accredited college?

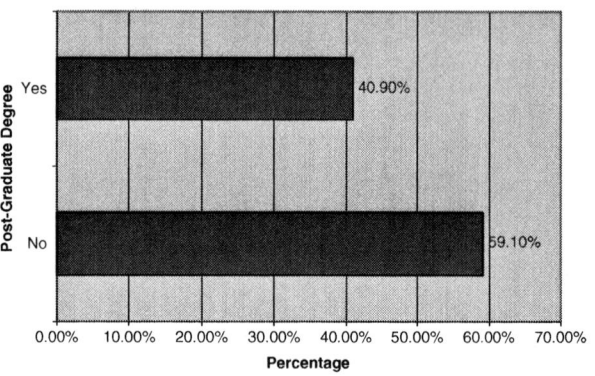

5. Are you a registered Professional Engineer or Architect?

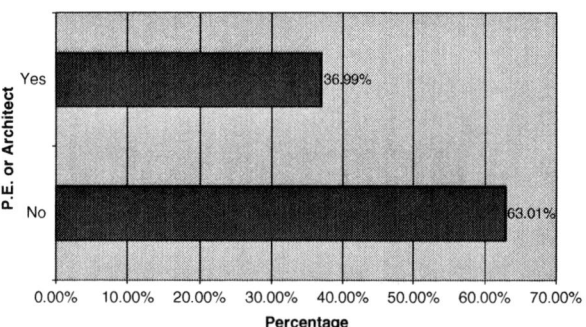

6. Do you hold a valid certification from AEE, such as a Certified Energy Manager (CEM), Energy Manager In Training (EMIT), Certified Lighting Efficiency Professional (CLEP), Certified Power Quality (CPQ), Certified Building Commissioning Professional (CBCP), Distributed Generation Certified Professional (DGCP), Certified Measurement & Verification Professional (CMVP), Certified Demand Side Manager (CDSM), Certified Cogeneration Professional (CCP), Certified Energy Procurement (CEP), Certified Indoor Air Quality Professional (CIAQP), Certified Indoor Air Quality Technician (CIAQT), Certified Testing, Adjusting and Balancing (CTAB) or Certified GeoExchange Designer (CGD)?

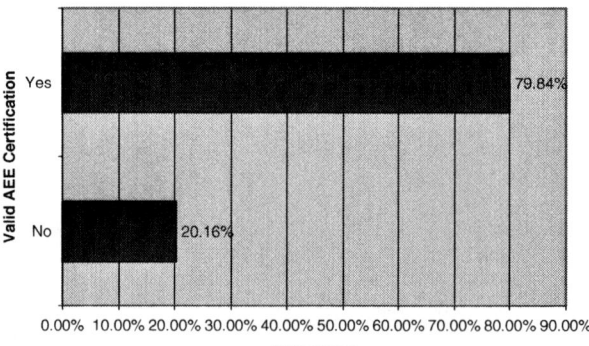

7. Because you have assumed energy management responsibilities at your company, are you:
   A Receiving significantly higher compensation than before?
   B Receiving higher visibility?
   C In a better position for advancement?

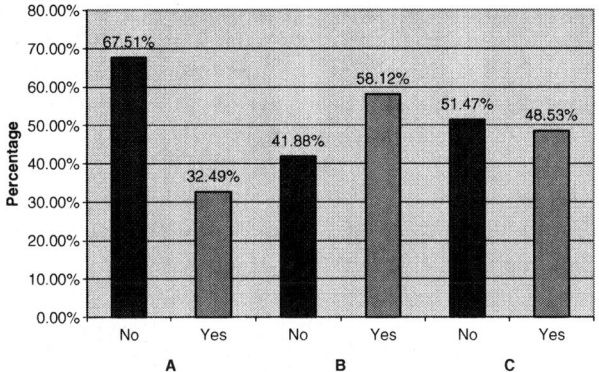

8. Is your company currently:

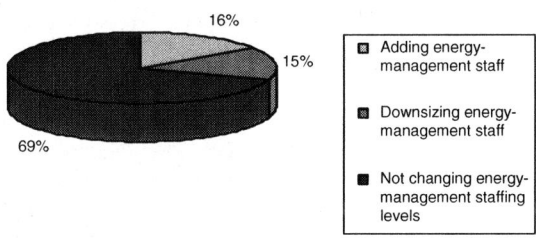

9. How many years of experience do you have?

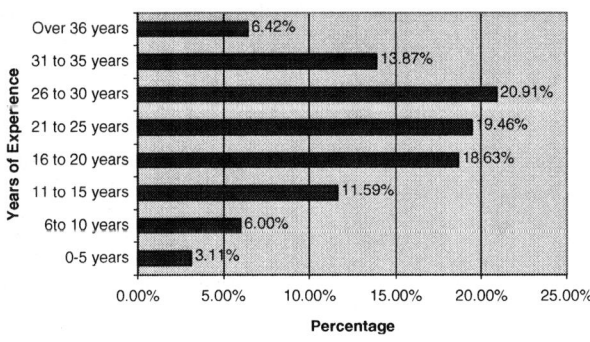

10. How many of those years have you been involved in energy management?

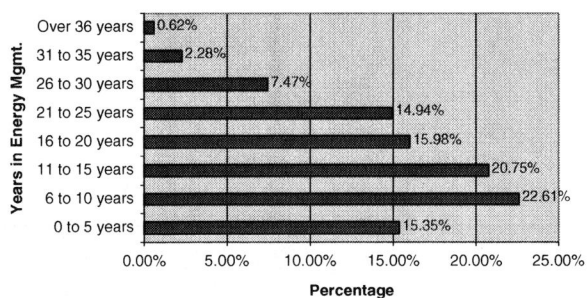

11. How many years have you been a member of the Association of Energy Engineers?

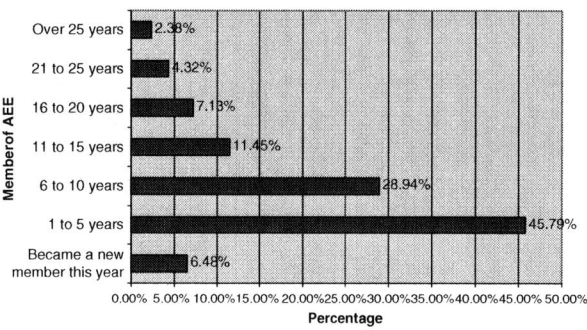

12. Please identify the location where you are employed.

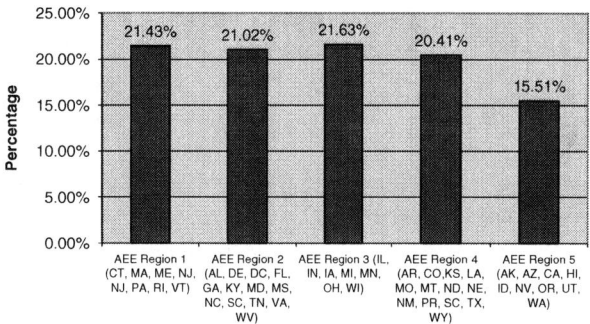

PROFESSIONAL CERTIFICATION PROGRAMS

Most professionals practicing in the energy engineering industry do not have a degree in this field. Specialty certification programs developed by the AEE and other professional organizations offer the following benefits:

- Certification is part of total quality management. When an individual becomes certified in a designated field, his or her professional achievement is recognized in the eyes of colleagues, government agencies, present and prospective employers, and clients.
- Certification establishes a standard of professional competence that is recognized throughout the industry.
- Certification fosters development of the individual's capabilities through encouragement of long-term career goals.
- Certification promotes quality through continuing education to ensure a high level of competence within constantly changing fields.

CERTIFICATION PROGRAMS OFFERED BY THE AEE

The Certified Energy Manager (CEM) program is broad based and emphasizes the technical aspects of what energy managers in government and the private sector need to know. The Certified Energy Manager represents a "who's who" of energy management. Since 1982, more than 6400 individuals have gained the status of CEM.

The Certified Business Energy Professional (BEP) program awards special recognition to those business/marketing and energy professionals who have demonstrated a high level of competence and ethical fitness for business/marketing and energy management-related disciplines, as well as laws governing and affecting energy professionals.

The Certified Lighting Efficiency Professional (CLEP) program was developed to identify specialists in lighting efficiency. This program is also recognized as meeting the certification requirements of the U.S. Environmental Protection Agency's Green Lights Lighting Management Company Ally program.

The Certified Indoor Air Quality Professional (CIAQP) program is designed to meet the growing needs of businesses to identify qualified indoor-air-quality practitioners who are equipped to solve problems created by "sick buildings," as well as facility managers who are responsible for operating healthy buildings while maintaining comfort and reducing energy costs.

The Certified Power Quality Professional (CPQ) program demonstrates the interactions between the power source and sensitive loads in the field of power quality and reliability.

The Certified Geoexchange Designer (CGD) program is designed to recognize professionals who have demonstrated high levels of experience, competence, proficiency, and ethical fitness in applying the principles and practices of geothermal heat pump design and related disciplines. The CGD certification is granted by the AEE and sponsored by the Geothermal Heat Pump Consortium (GHPC). Associated training programs are presented by the International Ground Source Heat Pump Association (IGSHPA).

The Certified Energy Procurement Professional (CEP) program covers the acquisition of both electricity and natural gas from the purchasing/procurement and selling/marketing perspectives.

The Distributed Generation Certified Professional (DGCP) program identifies individuals who have demonstrated high levels of experience, competence, proficiency, and ethical fitness, bringing to their professional activities the full scope of knowledge essential to the effective development and management of today's distributed generation projects.

The Certified Measurement and Verification Professional (CMVP) program was established by the AEE in cooperation with the International Performance Measurement and Verification Protocol (IPMVP), with the dual purpose of recognizing the most qualified professionals in this growing area of the energy industry, as well as raising overall professional standards within the measurement and verification field.

The Certified Building Commissioning Professional (CBCP) program was developed with the dual purpose of recognizing the most highly qualified professionals in this rapidly expanding area within the industry and raising overall professional standards in the building-commissioning field.

The Certified Hotel Environmental Manager (CHEM) program is designed for the multiple purposes of raising the professional standards of those engaged in hotel environmental management, identifying individuals with acceptable knowledge of the principles and practices of hotel environmental management, and awarding those individuals special recognition for demonstrating a high level of competence and ethical fitness in hotel environmental management. Sponsored by the U.S. Agency for International Development (USAID), the CHEM certification is granted by the AEE. The associated training program is presented by PA Consulting Group.

The Emissions Trading Certified Professional (ETC) program is designed to award special recognition to professionals who have demonstrated a high level of knowledge, experience, competence, and ethical fitness covering the full spectrum of activities related to the trading of emissions allowances and evaluation of emissions credits. By obtaining the ETC credential, the individual establishes his or her status as a qualified expert in this growing area of specialized expertise.

THE AEE CERTIFICATION PROCESS

Each program requires the individual to complete an application, which includes requests for the following information:

- Demonstrated experience in the field. Each program specifies the minimum experience levels required. Employee or client verification is also required.
- Documented education or professional registration.
- Professional references to verify experience and qualifications.

In addition, the individual must complete a four-hour open-book examination. The exam consists of multiple-choice and true/false questions. The actual test questions are framed to ascertain both specific knowledge and practical experience. Sample questions and study guides are available from the AEE. The completed application and test score weigh equally in determining whether the individual meets certification requirements;

both are reviewed by the governing board of the specific program.

CONTINUING EDUCATION PROGRAMS OFFERED BY AREA

The AEE offers a wide range of training options. Each training option offers Continuing Education Units (CEUs), which are important for documenting courses successfully completed (one CEU equals ten professional development hours [pdh]). In addition, a certificate of participation is awarded.

In 2005, 27 states required CEUs as a prerequisite for renewal of professional engineering licenses.

Programs offered by the AEE include:

- *Live seminars.* The AEE presents a wide range of courses in cities across the nation. Several seminars are designed to prepare students for the professional certification examinations. Live programs offer an optimum learning environment with ample time to interact with the instructor, as well as colleagues in attendance.
- *In-house seminars.* Most of the live seminars, including professional certification training programs, can also be presented at company facilities around the world.
- *Online real-time training (synchronous).* Students can participate in a real-time seminar at the office or at home with access to the Internet and a telephone. Students communicate with the instructor through a scheduled conference call and view the instructor's Microsoft PowerPoint presentation via the Internet.
- *Self-paced online training (asynchronous).* Each student receives a workbook containing the training materials and examination questions and completes the training at his or her own pace. The students who pass the online examination receive a certificate of course completion and are awarded CEUs. Students can interact with the instructor and fellow students during regularly scheduled chat sessions.
- *24/7 online training (asynchronous).* The course material is accessed online 24 hr a day for up to 30 days. The students who pass the online examination are able to print a certificate of course completion and are awarded CEUs.
- *Conferences and expositions.* The AEE offers three conferences and expositions each year. The purpose is to present the latest technologies and applications by leading experts in the field. The flagship event presented by the AEE is the World Energy Engineering Congress (WEEC).

STUDENT PROGRAMS FOR THE ENERGY ENGINEERING PROFESSION THROUGH THE AEE

The future of the energy engineering profession is in developing new talent through student programs. The AEE encourages student participation as follows:

- *Student membership.* Student members receive all publications and can participate at a reduced rate of $15 annually. Dues are subsidized by the AEE.
- *Student chapters.* Networking with fellow classmates and seasoned professions is accomplished through student chapters across the nation.
- *Scholarships.* To help students further their educations in the field of energy engineering, scholarships are offered. The Foundation of the AEE is a 501 (c)3 not-for-profit corporation. Since its inception, the AEE has awarded $480,000 in scholarships.

VISIONS FOR THE FUTURE

The energy engineering profession continues to grow. The need for energy efficiency has never been greater, due to the following circumstances.

ENERGY SECURITY

- Since 1985, imports of refined petroleum products have increased by 34%. Today, the total import of oil is 56.8%. The volatility in the Middle East and other foreign sources of supply has led to disruptions in the availability of oil and higher prices. By mid-2005, oil prices spiraled to more than $60 per barrel.
- Since 1970, U.S. production of crude oil has declined from 9.6 to 5.8 million barrels per day. While, consumption has increased from 14.7 to 20 million barrels per day. The Organization of Petroleum Exporting Countries (OPEC) has tightened supplies in the past, causing gasoline and oil prices to spiral. The "war on terrorism" will cause further instability in the Middle East and prioritize this nation's need to be energy secure.
- The power blackout of 2003 affecting New York, Detroit, and other major cities indicates that the present transmission grid is congested, outdated, and in dire need of overhaul. According to the U.S. Energy Information Administration, electricity use will increase 22% by 2010, placing further demands on an obsolete transmission grid.
- The long-term price of natural gas in the United States has more than doubled over the past six years. U.S.

natural gas production history shows that new wells are being depleted more quickly all the time; the current decline rate is 28% per year. Although this is partially due to growing demand, it is also due to the fact that the large fields of natural gas are all aging and in terminal decline. Newer natural gas fields tend to be smaller and are produced (and depleted) quickly in the effort to maintain overall production levels. Production from wells drilled in 2003 has been declining at a rate of 23% per year.

ENERGY-EFFICIENT ECONOMY

Energy efficiency improvements have had a major impact on companies' profitability, decreasing energy usage, and reducing greenhouse gasses. Over the past 25 years, per capita use of energy has declined 0.8%. A continuing drive for energy efficiency can help keep prices down and buy the nation time to address critical supply problems.

Energy efficiency programs have saved consumers more than $25 billion a year while improving the quality of life. Energy efficiency saved 70 quadrillion Btu from 1972 to 1999.

Energy efficiency technologies can also give the nation time to rebuild and modernize the electric transmission infrastructure.

UPGRADING TRANSMISSION AND DISTRIBUTION SYSTEMS

The U.S. electric generation and transmission system is 70 years old and is based on technologies from the 1950s. The electric transmission grid consists of approximately 160,000 mi of high-voltage transmission lines and is in dire need of replacement and expansion. The power blackout of 2003 and the conclusions reached by the North American Electric Reliability Council (NERC) indicate that the nation is fast approaching a crisis stage with respect to the reliability of the transmission grids.

The transmission grid has become a "superhighway" for electric utilities to buy and sell power. Congestion from the increased flow of electricity over great distances is now a reality. The Federal Energy Regulatory Commission (FERC) has identified causes of congestion, including:

- insufficient transmission lines to match electricity generated
- inadequate transmission capacity to meet demand

According to the Edison Electric Institute:
"Between 1979 and 1989, transmission capacity grew at a slightly faster rate than the demand for electricity during peak periods. But in the subsequent years, infrastructure needs did not keep up with that demand. To handle the requirements that the transmission system expects over the next 10 years, about 27,000 GW–mi are required; however, only 6000 GW–mi are planned."

CONCLUSION

The AEE has played a vital role in developing the energy engineering profession. Through training and certification programs, professionals have gained new tools to improve the efficiency of buildings and industry.

Energy engineering is a growing profession. The AEE has played a key role in the development of the energy engineering profession and has helped professionals reach their full potential. Through the Foundation of AEE, $480,000 in scholarships have been awarded to help students in the field of energy engineering. The future of energy engineering is exceedingly bright as energy engineers seek out solutions to reduce greenhouse gases.

Climate Policy: International

Nathan E. Hultman
Science, Technology, and International Affairs, Georgetown University, Washington, D.C., U.S.A.

Abstract

Climate change is a long-term problem. Policy to address climate change faces the challenge of motivating collective action on global public goods in a world with no single international authority. International agreements can nevertheless aim to (1) reduce greenhouse gas emissions, possibly through an international emissions trading system (ETS) like that outlined in the Kyoto Protocol; (2) develop new low-emissions technology by providing incentives for cooperation on technology research and implementation; (3) provide adaptation assistance to countries and populations least able to cope with expected changes in climate.

INTRODUCTION

Anthropogenic climate change presents one of society's most vexing policy challenges. Because the problem stems largely from the fossil-fueled global economy, the costs of reducing emissions would accrue immediately and are easily quantified. On the other hand, the costs of potential damages are difficult to estimate and will likely be long-term. In addition, governing the global atmospheric commons requires a large number of actors to agree on and comply with a mechanism of self-restraint; otherwise, even the relatively virtuous would tire of the rest of the world's free-riding. Moreover, understanding the complex risks and uncertainties of climate change is a challenge even for specialists. Communicating this information accurately and effectively to a marginally interested public is harder still.

Large, long-term changes would be required to reduce the risks of climate change. While the absolute amount of greenhouse gases (GHGs) in the atmosphere affects the degree of climate change, the only variable that society can easily control is the rate of GHG emissions. To simply stabilize the absolute amount of atmospheric GHGs, the emission rate must drop to about one-half to one-third of its present levels. Deciding on what concentrations constitute moderately safe levels is challenging, and requires deriving impact estimates (such as temperature change) from possible GHG stabilization levels,[1] as well as from the projected rate of emissions reduction.

An effective international regime to govern climate change policies must balance climate protection goals with the limited enforcement ability inherent in international law.[2–4] Because the atmosphere is a common resource, protecting the climate is in most people's best interest, but no country will want to burden itself unreasonably burden itself with the excess costs of a climate friendly policy unless it believes that other countries are making equivalent sacrifices. An effective climate regime must therefore minimize free-riding. It must also ensure that the participants are complying with their obligations, which requires systems of monitoring and enforcement. Finally, because of our evolving understanding of the science of climate change and its relationship to human societies, a sound policy must remain flexible enough to incorporate new information as it becomes available, with procedures for regular scientific re-evaluation and regular review of the adequacy of the policy.

CLIMATE POLICY OPTIONS

Addressing climate change requires coordination of domestic and international systems to reduce GHG emissions and assist countries in adapting to climate change. International treaties can set guidelines for action[5,6] that are then implemented via domestic legislation.

Reducing Emissions

Several policy mechanisms can address the free-ridership and overexploitation associated with public goods like the climate. Governments can regulate the common resource directly or stipulate specific technological approaches. This command-and-control approach is relatively simple in that the rules can be set by panels of experts, and it has the appearance of fairness, because everybody must attain the same goals. However, in a diverse economy with differing costs of pollution abatement to firms, it can lead to large imbalances in the cost of compliance.

Keywords: Kyoto protocol; Climate change; Emissions trading; Clean development mechanism; Greenhouse gas; Framework convention on climate change; Global warming.

Alternatively, a governmentally set price on the externality could allow producers more flexibility. This price can be set directly as a tax (for example, $10 per ton of carbon dioxide (CO_2)), or indirectly by setting a total emissions limit and allowing entities to trade the rights to emit. These methods—taxes, emissions trading, or a hybrid of the two—can greatly reduce the total cost of compliance with the environmental target. Emissions trading systems come in two forms: in a baseline-and-credit (or permit) system, in which individual projects that result in a reduction of emissions below a pre-agreed baseline are granted credits that can be sold to firms that are not able to meet their reduction obligations. Alternatively, a cap-and-trade system sets an overall cap on emissions and then distributes, free or at auction, the entire amount of emissions allowances out to the producers.

Governments may also implement other policies to address market failures, for example by establishing minimum standards of efficiency or performance, supporting research and development of less-polluting technologies, or even guaranteeing a market for new technologies.

Equity Questions

One contentious question in developing a GHG trading system is how to allocate emissions quotas. Until now, all countries have had free access to the atmosphere; setting limits will inevitably lead to argument about who deserves a bigger slice. Three options for allocating these rights illustrate the policy challenge. The first method, usually called grandfathering, allocates permits according to what various countries or industries have emitted in the recent past. This method causes the minimum economic disruption, but it may also reward inefficient resource use and ignore the benefits that have already accrued to polluters. Alternatively, if one views the atmosphere as a universal resource or a life-support commons, then the quota may be allocated on a per-capita basis so that each person is assigned the same right to using the resource. If, on the other hand, one views the atmosphere as an economic input, the quota might be allocated to each country in proportion to its gross domestic product (GDP). The United States, for example, currently produces 24% of the world's GHG emissions,[7] has 5% of the world's population, and accounts for 21% of the world's GDP.

These simple formulae will likely not be used directly, but they do provide bases for negotiating commitments in the international community. One particularly large divide between developing and developed country positions is how to account for the cumulative GHGs emitted since the beginning of the Industrial Revolution by developed countries.[8] This atmospheric debt represents about 80% of the total anthropogenic GHG contribution, and less-developed countries' contribution will likely not equal that of developed countries until around 2100. These countries argue that richer countries should therefore move first to forestall further emissions. Developed countries often view the situation differently, pointing out that less-developed countries as a group, including China and India, will by 2010 emit GHGs at an annual rate equal to that of the developed world. Indeed China itself will soon be the world's largest GHG emitter, surpassing even the United States.

Whether generated through capture, biological sequestration, or mitigation, trading emissions requires measurement.[9] The most reliable statistics on GHG emissions relate to the burning of fossil fuels. Most countries, especially the industrialized countries that emit the most, keep detailed records of fossil fuel stocks and flows. Therefore, national emissions from fossil fuels are relatively precise and have a high amount of certainty.

Adaptation

Finally, given that some climate change is at this point inevitable, climate policy must encompass not only policies to reduce emissions of GHGs but also policies to enhance the resilience of countries to the expected changes. Often called adaptation measures, such activities include support for diversifying economies away from vulnerable crops, enhancing physical infrastructure, and bolstering institutional capabilities. Unlike policies focusing on GHGs, moreover, the benefits of adaptation accrue relatively quickly as they can immediately reduce suffering from hurricanes or floods regardless of the cause of these events.[10,11] Discussions about adaptation are often linked to questions about liability for climate damages.

EARLY INTERNATIONAL RESPONSE

Although Svante Arrhenius had postulated the existence of the greenhouse effect in 1896,[12] and significant scientific inquiry re-emerged in the 1950s, public concern about anthropogenic climate change was not significant until the late 1980s. Along with other simultaneously emerging global environmental problems like stratospheric ozone depletion and biodiversity loss, climate change moved quickly into the international arena.[13]

The international community's first concrete response was to refer the scientific questions to the World Meteorological Organization (WMO) and the United Nations Environment Programme (UNEP). In 1988, the WMO and UNEP established the Intergovernmental Panel on Climate Change (IPCC), which has since become the major international expert advisory body on climate change.[14] The IPCC divides thousands of experts into three working groups on climate science, impacts of climate change, and human dimensions. It produces

Region	Greenhouse Gas Emissions billion tons CO_2e per year		
	1990	2000	KP Target
World	21.81	23.63	
Developing Countries	6.91	9.64	
Annex I	14.90	13.99	13.46
European Union	3.33	3.28	2.76
United States	4.98	5.76	4.55
Non-EU, Non-US OECD	1.84	2.20	2.05
Russia and Eastern Europe	4.75	2.74	4.19

Fig. 1 Historical emissions and Kyoto Protocol target emissions. Historical emissions from the United States Department of Energy (see Ref. 19) do not include land-use change emissions; Kyoto targets are based on net emissions reported to the UNFCCC and include land-use change emissions. Country-specific targets are available in the Kyoto Protocol text and from the UNFCCC Secretariat From UNFCC Secretariat Internet Resources (see Ref. 15) Developing Countries are defined as countries not included in Annex I.
Source: From Ref. 15.

comprehensive Assessment Reports every 5–6 years that describe the current state of expert understanding on climate, as well as smaller, targeted reports when they are requested by the international community.

THE U.N. FRAMEWORK CONVENTION ON CLIMATE CHANGE

The first international treaty to address climate change was the United Nations Framework Convention on Climate Change (UNFCCC), which entered into force in 1994 and has been ratified by 186 countries, including the United States.[15] Having emerged from the 1992 U.N. Conference on Environment and Development (the "Earth Summit"), the UNFCCC sets broad objectives and areas of cooperation for signatories. As the objective, it states that Parties to the Convention should cooperate to "prevent dangerous anthropogenic interference with the climate system." Here, dangerous is not defined explicitly but is required to include ecosystems, food supply, and sustainable economic development.

The UNFCCC identifies several important principles for guiding future treaty agreements. First, it endorses international equity. Second, it states that all signatories share a "common but differentiated responsibility" to address climate change. All countries must therefore participate, but they are allowed to do so in a way that depends on their domestic situation and historic GHG contributions. Third, the UNFCCC instructs the Parties to apply precaution in cases risking "serious or irreversible damage."

The UNFCCC also defined some emissions reduction goals for richer countries. Specifically, it grouped most developed countries into Annex I Parties (Annex I is a designation in the UNFCCC and reproduced in the Kyoto Protocol as Annex B. Annex I countries are: Australia, Austria, Belgium, Bulgaria, Croatia, Czech Republic, Denmark, Estonia, Finland, France, Germany, Hungary, Iceland, Ireland, Italy, Japan, Latvia, Liechtenstein, Lithuania, Luxembourg, Monaco, Netherlands, New Zealand, Norway, Poland, Portugal, Russian Federation, Slovakia, Slovenia, Spain, Sweden, Switzerland, U.K., Ukraine, U.S.A.) and urged them to stabilize their total emissions of GHGs at 1990 levels by the year 2000. These targets were non-binding, and in retrospect, most countries did not meet these initial goals. Finally, and most importantly, the Framework Convention established a system of national emissions reporting and regular meetings of Parties with the goal of creating subsequent, more significant commitments.[16,17]

Interim Negotiations

Subsequent debates focused, therefore, on negotiating a new treaty (called a Protocol) that could enhance international action. Yet, contentious debate arose over whether developing countries would be required to agree to any reductions in return for caps on Annex I emissions.

Several negotiating blocs were solidified during this period and remain active today. The broadest split between developed and developing countries was already evident in the UNFCCC. Within developing countries, the strongest advocates for action emerged in the Alliance of Small Island States (AOSIS), an association of low-lying coastal countries around the world that are extremely vulnerable to inundation due to sea-level rise. On the other hand, the Organization of Petroleum Exporting Countries (OPEC) has been reluctant to endorse any regulation of their primary export, fossil fuel, which when burned creates the GHG CO_2.

Developed countries also have several blocs: The European Union (EU), which functions as a single legal party to the convention, has tended to favor strong action on climate change, whereas the United States, Japan, Canada, New Zealand, and Australia have been more circumspect. Russia was never an enthusiastic advocate of action on climate, but the collapse of its economy in the 1990s means that its emissions decreased considerably, allowing it some flexibility in negotiating targets. A 1995 agreement (the Berlin Mandate) adopted by the Parties to the UNFCCC stated that developing countries should be exempt from any binding commitments, including caps on their emissions, in the first commitment period. The U.S. Senate disagreed and, in 1997, declared they would not ratify any Protocol to the UNFCCC that did not call for concrete targets from developing countries.

THE KYOTO PROTOCOL

After two years of preliminary negotiations, delegates to the UNFCCC met in Kyoto, Japan in 1997 to complete a more significant treaty calling for binding targets and timetables, eventually agreeing on the Kyoto Protocol to the UNFCCC. Maintaining the principle of the Berlin Mandate, delegates rejected language that required participation by developing countries, thus damping U.S. enthusiasm. Nevertheless, the Kyoto Protocol entered into force in 2005, having been ratified by EU countries, Canada, Japan, Russia, and most developing countries. The United States and Australia are currently not Parties to the Protocol. The Kyoto Protocol builds on the UNFCCC with specific and legally binding provisions.

Targets

First, it set legally binding emissions targets for richer countries. These targets oblige Annex I Parties as a group to reduce their emissions to a level 5.2% below 1990 levels by the target period (Fig. 1). This overall average reflects reductions of 8% for the EU, 7% for the United States, and 6% for Canada; as well as increases of 8% for Australia and 10% for Iceland. Russia, whose emissions had dropped significantly between 1990 and 2000 because of economic contraction, was nevertheless awarded a 0% change, effectively providing an effort-free bonus (often called hot air). The target period is defined as 2008–2012, and countries are allowed to take an average of their emissions over this period for demonstrating compliance.

In addition, an individual country's emissions are defined as a weighted sum of emissions of seven major GHGs: CO_2, methane (CH_4), nitrous oxide (N_2O), hydrofluorocarbons (HFCs), perfluorocarbons (PFCs), chlorofluorocarbons (CFCs), and sulfur hexafluoride (SF_6). These gases are weighted according to a quantity called global warming potential (GWP) that accounts for different heat-trapping properties and atmospheric lifetimes of the seven gases. For a 100-year time horizon, example GWPs are: 1.00 for CO_2 (by definition), 21 for CH_4, 296 for N_2O, 100–1000 for a wide variety of halocarbons, and 22,200 for SF_6.[18] The resulting sum is reported in terms of "carbon-dioxide equivalent" or CO_2e. The Protocol allows countries to calculate a net emissions level, which means they can subtract any GHGs sequestered because of, for example, expansion of forested areas.

Implementation

The Protocol encourages countries to achieve their target primarily through domestic activities, usually called policies and measures. These include improved energy efficiency, increased use of renewable energy, and switching to lower-carbon forms of fossil fuels such as natural gas. In addition, the Protocol allows countries to offset emissions if certain domestic activities serve to absorb and sequester CO_2 from the atmosphere, thus reducing their net contribution to climate change. Allowable carbon sinks projects currently include afforestation, reforestation, forest management, cropland management, grazing land management, and revegetation. Conversely, deforestation is a process that must be counted as a cause of emissions.

The Kyoto Protocol also allows countries to obtain credits from other countries. In particular, it established three market-based mechanisms to provide states with flexibility in meeting their binding emissions reduction targets: emissions trading (ET), joint implementation (JI), and the clean development mechanism (CDM). Despite the different names, these three mechanisms are actually all forms of emissions trading—they create ways to reduce the overall cost of reaching the targets outlined above by allowing lower-cost reductions to be bought and sold on the market. All traded units are denoted in tons of CO_2e.

Emissions Trading (sometimes called Allowance Trading) is a cap-and-trade system under which the Annex I parties are assigned a maximum level of emissions (see Fig. 1), known as their assigned amount. They may trade these rights to emit through a UNFCCC registry. Only developed country Parties may participate in ET. Units of ET are termed *assigned amount units* (AAUs).

Joint Implementation is a baseline-credit system that allows trading of credits arising from projects coordinated between fully developed countries and countries in eastern Europe with economies in transition. This is a so-called project-based system, under which reductions below an independently certified baseline can be sold into the market. Joint implementation units are termed emissions reduction units (ERUs).

The CDM is another baseline-credit system that allows trading of credits arising from projects in developing countries. Another project-based system, the CDM will allow only projects that contribute both to sustainable development and to climate protection. Furthermore, they must provide benefits that would not have occurred in the absence of the CDM (so-called additionality). Post-Kyoto negotiations determined that acceptable projects include those that employ renewable energy, fossil-fuel repowering, small-scale hydroelectric power, and sinks; some projects (e.g., those under 15 MW_e) are also deemed to be "small-scale" and enjoy a streamlined approval process. Projects are subject to a process of public participation. Final acceptance of project proposals rests with the CDM Executive Board, which also approves methodologies and designates operational entities—NGOs, auditors, and other private developers—that implement and verify projects. A levy on each project will fund activities that help poor countries adapt to a changing climate. Clean development mechanism units are called certified emissions reductions (CERs).

Kyoto rules allow AAUs, ERUs, and CERs to be fungible or substitutable for each other. A final unit specific to sinks, called a removal unit (RMU), will be partially separate from this pool since it cannot be banked, or held from one commitment period to the next.

Other Provisions

The allowance assignments and flexibility mechanisms are the most significant elements of the Kyoto Protocol and its associated rules. Other noteworthy commitments include minimizing impacts on developing countries—primarily through funding and technology transfer—and establishing expert teams to develop monitoring, accounting, reporting, and review procedures.

The UNFCCC, Kyoto Protocol, and associated agreements establish three multilateral funds to assist poorer countries. The Adaptation Fund, financed through a levy on CDM transactions, is designed to help countries bolster their institutional and infrastructural capacity to manage changes in climate and damages from weather events. The Climate Change Fund focuses on technology transfer and economic diversification in the energy, transportation, industry, agricultural, forestry, and waste management sectors. Finally, the Least Developed Countries Fund exists to provide additional support in adaptation activities for the very poorest countries. The latter two funds are financed by voluntary contributions.

EUROPEAN UNION EMISSIONS TRADING SYSTEM

In 2005, the EU implemented what is to date the largest operational emissions trading system (ETS) for GHGs.[20,21] The EU–ETS is a cap-and-trade system for all 25 EU member countries, and it covers approximately 12,000 installations in six sectors (electric power; oil refining; coke ovens, metal ore, and steel; cement; glass and ceramics; paper and pulp). The plan regulates only CO_2 emissions until 2007, but thereafter other GHGs will be included. About one-half of the EU's CO_2 emissions will be regulated under the EU–ETS during this first 2005–2007 phase. The initial allocation of credits is based on individual countries' plans; countries are allowed to auction up to 5% of allowances in the first phase and 10% thereafter.

Notably, the EU–ETS contains more rigorous enforcement provisions than Kyoto. Compared to international law, the EU has far greater leverage to enforce legal provisions, and the EU–ETS imposes a steep fine (€40/tCO_2) for non-compliance. The EU–ETS replaces some national-level policies to control GHGs, notably the United Kingdom's pioneering ETS and other voluntary programs.[22] Accordingly, some facilities that had previously taken action to comply with pre-existing national laws are allowed limited exceptions to the EU–ETS.

Through a linking directive,[23] credits generated through Kyoto Protocol CDM or JI projects may be used to fulfill obligations under the EU–ETS, thereby providing an important market for these offsets. However, because of European concerns about the possible negative consequences of biological carbon sequestration (sinks) projects, CDM or JI credits generated within these categories are ineligible for the EU–ETS. The EU–ETS will thus form, by far, the largest trading program in the world and will likely set the standards for subsequent programs in other countries.

VOLUNTARY AND REGIONAL PROGRAMS

Trading programs outside national government legislation have also emerged. British Petroleum was the earliest major corporate adopter of an internal GHG trading system and, subsequently, has had a large consultative role in drafting both the Kyoto and U.K. emissions trading rules. Although the United States has declared its intention to ignore Kyoto,[24] many large American corporations have also adopted internal targets. The Chicago Climate Exchange has organized voluntary commitments from companies in the hope of establishing a position as the dominant American exchange. Yet voluntary programs, whether they derive from corporate or governmental initiatives, are ultimately constrained: often the most egregious polluters choose simply not to volunteer, and those firms that do participate may still not reduce emissions to socially desirable levels.

Many U.S. states, such as California and Oregon, have also passed or are considering legislation that would curb emissions directly or indirectly. While independent state initiatives are valuable in providing domestic innovation for the United States,[25] they are unlikely to add up to a significant reduction in global emissions. In addition, many of the companies most vulnerable to GHG regulation are asking for some guidance on what they can expect from regulators, and a state-by-state patchwork can never replace federal legislation for regulatory certainty.

CONCLUSION

International climate policy faces a period of uncertainty, innovation, and evolution over the next decade. The Kyoto Protocol remains the primary international agreement for addressing climate change globally, despite its imperfections and the continued absence of the United States and Australia. Yet, since Kyoto expires in 2012, attention has turned to negotiating a subsequent agreement that could re-engage the United States, involve China, India, and other developing countries more directly, and address concerns about compliance and enforcement.[26–28] Given the difficulties in forging an immediate, broad international consensus, the EU–ETS will likely foster the most institutional innovation and GHG market development in the near term.

The most likely interim solution, therefore, will consist of multiple, overlapping regimes that link domestic-level emissions reductions into one or more international markets.[29] In this model, the United States could, for example, institute a unilateral domestic program that addresses emissions and then allow Kyoto credits to be admissible for compliance as the EU has done. Additional agreements governing, for example, technology standards or adaptation policy may also emerge. From the perspective of energy, this evolving climate change policy will impose a carbon constraint on energy use,[30,31] most likely through a non-zero cost for GHG emissions.

ACKNOWLEDGMENTS

The author gratefully thanks the two anonymous reviewers for their thorough, detailed, and constructive comments on a previous draft.

REFERENCES

1. Mastrandrea, M.D.; Schneider, S.H. Probabilistic integrated assessment of "dangerous" climate change. Science **2004**, *304* (5670), 571–575.
2. Barrett, S. *Environment and Statecraft: The Strategy of Environmental Treaty-Making*; Oxford University Press: New York, 2003; 360.
3. Brown Weiss, E.; Jacobson, H.K. Getting countries to comply with international agreements. Environment **1999**, *41* (6), 16–20, see also 37–45.
4. Bodansky, D. International law and the design of a climate change regime. In *International Relations and Global Climate Change*; Luterbacher, U., Sprinz, D.F., Eds., MIT Press: Cambridge, Mass, 2001; 201–220.
5. Aldy, J.E.; Barrett, S.; Stavins, R.N. Thirteen plus one: a comparison of global climate policy architectures. Climate Policy **2003**, *3* (4), 373–397.
6. Sandalow, D.B.; Bowles, I.A. Fundamentals of treaty-making on climate change. Science **2001**, *292* (5523), 1839–1840.
7. United States Energy Information Administration. *Emissions of Greenhouse Gases in the United States 2003*, United States Department of Energy: Washington, DC, 2004; 108.
8. Smith, K.R. Allocating responsibility for global warming: the natural debt index. Ambio **1991**, *20* (2), 95–96.
9. Vine, E.; Kats, G.; Sathaye, J.; Joshi, H. International greenhouse gas trading programs: a discussion of measurement and accounting issues. Energy Policy **2003**, *31* (3), 211–224.
10. Pielke, R.A. Rethinking the role of adaptation in climate policy. Global Environ. Change **1998**, *8* (2), 159–170.
11. Wilbanks, T.J.; Kane, S.M.; Leiby, P.N.; Perlack, R.D. Possible responses to global climate change: integrating mitigation and adaptation. Environment **2003**, *45* (5), 28.
12. Rodhe, H.; Charlson, R.; Crawford, E. Svante Arrhenius and the greenhouse effect. Ambio **1997**, *26* (1), 2–5.
13. Bodansky, D. The history of the global climate change regime. In *International Relations and Global Climate Change*; Luterbacher, U., Sprinz, D.F., Eds., MIT Press: Cambridge, Mass, 2001; 23–40.
14. Intergovernmental Panel on Climate Change Secretariat. Sixteen Years of Scientific Assessment in Support of the Climate Convention: Geneva, 2004; 14.
15. United Nations Framework Convention on Climate Change Secretariat. UNFCCC Secretariat Internet Resources: Bonn, Germany, 2005.
16. Molitor, M.R. The United Nations climate change agreements. In *The Global Environment: Institutions, Law, and Policy*; Vig, N.J., Axelrod, R.S., Eds., CQ Press: Washington, DC, 1999; 135–210.
17. Schneider, S.H. Kyoto protocol: the unfinished agenda–An editorial essay. Climatic Change **1998**, *39* (1), 1–21.
18. Intergovernmental panel on climate change In *Climate Change 2001: The Scientific Basis*; Houghton, J.T., Ding, Y., Griggs, D.J., Noguer, M., van der Linden, P.J., Dai, X., Maskell, K., Johnson, C.A., Eds., Cambridge University Press: Cambridge, UK, 2001; 881.
19. United States Energy Information Administration *World Carbon Dioxide Emissions from the Consumption and Flaring of Fossil Fuels, 1980–2000*; United States Department of Energy: Washington, DC, 2002.
20. European Parliament. Directive 2003/87/EC Establishing a Scheme for GHG Emission Allowance Trading within the EU: Brussels, 2003.
21. Pew Center on Global Climate Change *The European Union Emissions Trading Scheme (EU–ETS): Insights and Opportunities*; The Pew Center on Global Climate Change: Washington, DC, 2005; 20.
22. Lecocq, F. *State and Trends of the Carbon Market*; Development Economics Research Group, World Bank: Washington, DC, 2004; 31.
23. European Parliament. The Kyoto Protocol's Project Mechanisms: Amendment of Directive 2003/87/EC: Brussels, 2004.
24. Victor, D.G. *Climate Change: Debating America's Policy Options*; Council on Foreign Relations: New York, 2004; 165.
25. Rabe, B.G. *Statehouse and Greenhouse: The Emerging Politics of American Climate Change Policy*; Brookings Institution Press: Washington, DC, 2004; 212.
26. Stavins, R.N. Forging a more effective global climate treaty. Environment **2004**, 23–30.
27. Beyond Kyoto, B.J. Foreign Affairs **2004**, *83* (4), 20–32.
28. Bodansky, D.; Chou, S.; Jorge-Tresolini, C. *International Climate Efforts Beyond 2012: A Survey of Approaches*; Pew Center on Global Climate Change: Washington, DC, 2004; 62.
29. Stewart, R.B.; Wiener, J.B. Practical climate change policy. Issues Sci. Technol. **2003**, *20* (2), 71–78.
30. Pacala, S.; Socolow, R. Stabilization wedges: solving the climate problem for the next 50 years with current technologies. Science **2004**, *305* (5686), 968.
31. Wirth, T.E.; Gray, C.B.; Podesta, J.D. The future of energy policy. Foreign Affairs **2003**, 132–155.

Coal Production in the U.S.

Richard F. Bonskowski
Fred Freme
William D. Watson
Office of Coal, Nuclear, Electric and Alternate Fuels, Energy Information Administration, U.S. Department of Energy, Washington, D.C., U.S.A.

Abstract

U.S. coal production is historically important. Coal is now used principally as a fuel for electricity generation and other industrial processes. Coal mining practices and technologies have become more productive due to geographic changes in mining areas and increased mine sizes that produce coal at competitive prices. Coal prices respond to occasional surges in demand and to regulatory requirements but, over the long term, they have trended lower in real dollars.

INTRODUCTION

Coal has been an energy source for hundreds of years in the United States. It has helped provide many basic needs, from energy for domestic heating and cooking; to transportation for people, products, and raw materials; to energy for industrial applications and electricity generation. America's economic progress historically is linked to the use of coal from its abundant coal resources.

MILESTONES IN U.S. COAL PRODUCTION

Coal production in the United States grew steadily from the colonial period, fed the Industrial Revolution, and supplied industrial and transportation fuel during the two World Wars (Fig. 1). In 1950 the five major sectors were industrial, residential and commercial, metallurgical, electric power, and transportation with each sector accounting for 5%–25% of total consumption. From the end of World War II to 1960, coal use for rail and water transportation and for space heating declined. Coal demand grew, however, with the postwar growth in American industry and increased electricity generation starting in the early 1960s.

In 1950, U.S. coal production was 560 million short tons (mmst). In 2003, U.S. coal production was 1.07 billion short tons, an average annual increase in coal production of 1.2% per year (Table 1). Of the coal ranks in Table 1, bituminous is relatively high-Btu coal mined mostly in the East and Midwest, subbituminous is medium-Btu coal mined only in the western states, lignite is low-Btu coal principally mined in the Gulf Coast and North Dakota, and anthracite is relatively high-Btu coal mined in small quantities in Pennsylvania. With the growing importance of lower-Btu coals in the production mix over time, the energy content of coal production has not grown as rapidly as its tonnage. To depict general trends, yet allow space for selected details, coal statistics in Table 1 are shown for each year from 1993 to 2003, for every 5 years from 1953 to 1993, and for 1950. A large proportion of U.S. production is consumed domestically, so yearly coal consumption levels track coal production.

In 1973, the Arab oil embargo renewed interest in the vast U.S. coal reserves, as the nation strived to achieve energy independence. The number of coal mines and new mining capacity burgeoned. Between 1973 and 1976, coal production increased by 14.4%, or 86.3 mmst.[1] In 1978, the Power Plant and Industrial Fuel Use Act mandated conversion of most existing oil-burning power plants to coal or natural gas. New research on coal liquefaction and gasification technologies was aimed at replacing imported petroleum and supplementing domestic gas supplies. Those high-cost projects were put on hold when crude oil prices fell several years later, making synthesized coal liquids and gases uneconomic.

The shift of coal production from traditional eastern coalfields to the western United States is the most important development affecting coal markets in the last 30 years. Thick beds of low-sulfur coal with low mining cost are extensive in the Northern Great Plains states of Wyoming, Montana, and North Dakota. Starting in the 1970s, increasingly stringent restrictions on atmospheric emissions of sulfur dioxide at power plants made this coal often the most cost-effective choice for meeting sulfur dioxide limits without the installation of expensive equipment retrofits. In a matter of a few decades, a localized western resource grew to more than half of all

Keywords: Coal production; Coal preparation; Coal price; Coal mining productivity; Technology; Surface; Underground; Employment.

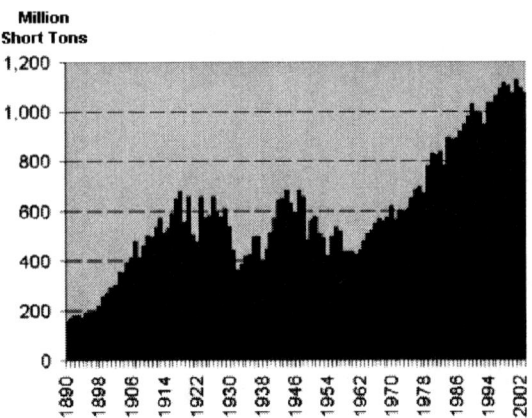

Fig. 1 United States coal production, 1890–2003.
Sources: From Energy Information Administration and Bureau of mines and U.S. geological survey (see Refs. 7 and 22).

U.S. production, from just over 60 mmst in 1973 to 549 mmst in 2003. This growth was accomplished through the deployment of long-distance coal haulage in unit trains (of more than 100 railcars moving coal only, to a single destination) and the exploitation of scale economies in the form of immense western surface coal mines. Average U.S. mine size in 2003 at 0.814 mmst per year far exceeded the average mine size in 1973 of 0.126 mmst per year. The largest U.S. mine, the North Antelope Rochelle Complex in Wyoming, alone produced over 80 mmst in 2003.

In the United States today, coal demand is driven by the electric power sector, which accounts for 90% of consumption (compared to the 19% it represented in 1950). As demand for electricity grew, demand for coal to generate it rose and resulted in increasing coal production. There were years in which coal production declined from the prior year, but, excluding years affected by a major unionized coal strike, annual increases in coal production between 1950 and 2003 outnumber decreases by almost two to one.

HOW U.S. COAL IS PRODUCED—TYPES OF MINING[2]

Growth of U.S. coal production involved expansions and adaptations in both established and evolving mining technologies. The important types of coal mining and technologies are as follows:

- Underground Mining—Extraction of coal from enclosing rock strata by tunneling below the ground surface. Also known as "deep mining," there are three types, based on mode of access. Drift mines, the easiest type to open, tunnel directly into the outcrop of horizontal or slightly sloping coal seams. Shaft mines reach the coal via a vertical shaft. In slope mines, the shaft descends on a gradient to reach the coal (suitable in hilly terrain). The following principal technologies are used in underground mining:
  — Conventional Mining—The traditional method, which employs the "room and pillar" mine layout, leaving massive pillars of undisturbed coal in place to support the overlying rock. It includes undercutting the exposed coal (the face), drilling and blasting the coal with explosive or high-pressure air, loading the broken coal into shuttle cars, and installing supplementary roof supports as needed.
  — Continuous Mining—Uses a mobile machine with forward, toothed cylinders that rotate and gouge coal from the face, where it falls onto a pan, is pulled onto loading belts and is fed to shuttle cars or movable conveyors.
  — Longwall Mining—An automated form of underground coal mining characterized by high recovery and extraction rates, feasible only in relatively flat-lying, thick, uniform coalbeds. A reserve block to be mined, the "panel," averages 1000 ft wide and 10,000 or more ft long, and is prepared by continuous mining of coal to create access tunnels on all four sides. When the longwall machinery is in place, the entire average 1000-ft width, the working face of coal is progressively sheared away, ceiling to floor, in a series of advancing passes. Dislodged coal is continuously removed via a floor-level conveyor system. Mining advances beneath automated movable roof supports within the 10,000-ft coal panel. The roof is allowed to collapse evenly in the mined out areas behind the supports.
- Surface Mining—Excavation of coal, in the most basic case, from outcroppings; or more generally, by removal of the overlying rock and soil (overburden) to expose one or more seams of coal. The following principal techniques are used in surface mining:
  — Strip Mining—An early synonym for surface mining, still widely used; to some people, it may connote irresponsible methods, without land restoration.
  — Contour Mining—A surface method used in sloping terrain, in which one or more coal beds are mined at outcrop by removing overburden to expose the coal beds.
  — Area Mining—A surface method used in flat terrain to expose coal for recovery by excavating long, successive "box cuts" or pits. Overburden excavated from the cut being mined is deposited in the previous, mined-out cut.
  — Auger and Highwall Mining—Mining usually performed within a contour or area mine, in coal in place beneath the "final highwall" (the standing

Table 1 Historical coal production by type of mining and by coal rank, selected years (production in millions of short tons)

Year	Type of mining		U.S. coal production	Bituminous coal production[b]	Subbituminous coal production	Lignite production	Anthracite production
	Underground	Surface[a]					
2003	352.8	719.0	1071.8	541.5	442.6	86.4	1.3
2002	357.4	736.9	1094.3	572.1	438.4	82.5	1.4
2001	380.6	747.1	1127.7	611.3	434.4	80.0	1.9
2000	373.7	700.0	1073.6	574.3	409.2	85.6	4.6
1999	391.8	708.6	1100.4	601.7	406.7	87.2	4.8
1998	417.7	699.8	1117.5	640.6	385.9	85.8	5.3
1997	420.7	669.3	1089.9	653.8	345.1	86.3	4.7
1996	409.8	654.0	1063.9	630.7	340.3	88.1	4.8
1995	396.2	636.7	1033.0	613.8	328.0	86.5	4.7
1994	399.1	634.4	1033.5	640.3	300.5	88.1	4.6
1993	351.1	594.4	945.4	576.7	274.9	89.5	4.3
1988	382.2	568.1	950.3	638.1	223.5	85.1	3.6
1983	300.4	481.7	782.1	568.6	151.0	58.3	4.1
1978	242.8	427.4	670.2	534.0	96.8	34.4	5.0
1973	300.1	298.5	598.6	543.5	33.9	14.3	6.8
1968	346.6	210.1	556.7	545.2	—[b]	—[b]	11.5
1963	309.0	168.2	477.2	458.9	—[b]	—[b]	18.3
1958	297.6	134.0	431.6	410.4	—[b]	—[b]	21.2
1953	367.4	120.8	488.2	457.3	—[b]	—[b]	30.9
1950	421.0	139.4	560.4	516.3	—[b]	—[b]	44.1

[a] Beginning in 2001, includes a small amount of refuse coal recovery.
[b] Subbituminous coal and lignite production were treated as bituminous coal prior to 1973 and cannot be reported separately.
Source: From Energy Information Administration (see Ref. 1).

exposed rock at the location where overburden becomes too thick for economical excavation). Auger mining uses a large-diameter drill to excavate a succession of holes within the plane of the coal bed, recovering the drilled coal. Highwall mining uses remote-controlled cutting machines, known as highwall miners—or underground mining machines, known as thin-seam miners—to mine out successive broad channels of coal from the seam left in place at the highwall.

— Mountaintop Removal (MTR) Mining—An adaptation of area mining to mountainous terrain. Often on massive scales, MTR removes all successive upper layers of rock and broad perimeters of lower rock layers. It recovers about 85%[3] of all upper coal beds contained within the rock layers and large portions of the lower beds. Mountaintop Removal operations may affect the top 250–600 ft of Appalachian peaks and ridges; they have recovered coal from as many as 18 coalbeds.[4] Mountaintop Removal mining creates huge quantities of excavated overburden that are disposed of as fill in upper portions of adjacent valleys. The fill operations are environmentally controversial, but the creation of relatively flat, developable land can be economically beneficial in steep mountainous areas.

COAL MINING TECHNOLOGY TRENDS

In the period since 1973, four distinct trends have dominated U.S. coal mining technology. The overall growth in surface coal mining at the expense of underground coal mining is the first. In 1973, underground and surface mines each accounted for 50% of total coal production. In the next 30 years, the production share from underground mines declined by a third:

Year	Underground percentage (%)	Surface percentage (%)
1973	50	50
1983	38	62
1993	37	63
2003	33	67

Growth in surface coal mining was accompanied by a second trend: the accelerated application of surface mining technology in large-scale area mines in the western region, characterized in optimal locations by box cut pits a mile or greater in length and about 200 ft wide, concentrated in the western states of Wyoming, Montana, North Dakota, Texas, Arizona, and New Mexico. In 1973, these six states accounted for 52 mmst out of a total of 599 mmst of U.S. coal mined, representing 9% of the total. By 2003, coal produced in these six western states accounted for 49% of all U.S. coal mined. No surface mines operating anywhere in the United States in 1973 had an annual output exceeding 5 mmst. By 2003, 64% of surface-mined coal was mined in the six western states in area mines exceeding 5 mmst per year of output[5]:

Surface production in western U.S. mines exceeding 5 mmst annually

Year	Percentage of U.S. total surface production (%)	Million short tons (mmst)
1973	0	0
1983	29	141
1993	50	299
2003	64	458

The third technological trend for the 1973–2003 period was the shift within underground mining from conventional room-and-pillar mining to longwall underground mining. Coal from longwall mining grew from 10 mmst in 1973 to 184 mmst in 2003, representing 52% of total U.S. underground production by 2003[6]:

Longwall production

Year	Percentage of U.S. total underground production (%)	Million short tons (mmst)
1973	3	10[a]
1983	27	80
1993	40	139
2003	52	184

[a]The 10 mmst of production in 1973 is based on 9.4 mmst of reported longwall machine coal recovery and 0.6 mmst estimated recovery by continuous mining of longwall entries.

States with substantial longwall production in 2003 included Alabama, Colorado, Pennsylvania, Utah, and West Virginia.

Due to superior productivity, large-scale surface and longwall technologies expanded faster than other mining methods. In 1983, large surface mines (greater than 5 mmst per year) had productivity higher than other surface mines, and in the next 20 years, they experienced higher rates of productivity growth. In 1983, longwall mines had about the same productivity as other underground mines; however, their productivity growth far outpaced other underground mines in the next 20 years:

Average productivity growth rates, 1983–2003

Surface Mines 5 mmst or greater	Other surface mines	Longwall mines	Other undergound mines
5.0%/year	3.1%/year	5.7%/year	2.9%/year

In the periods 1983–1993 and 1993–2003, large-scale surface technology and longwall technology saw about equal gains in productivity, decade over decade (Fig. 2). In contrast, other technologies saw decelerating gains in productivity. For a broader discussion, see the "Coal Mining Productivity" section below.

The fourth important trend was improvement in mining equipment durability and capability. Improvements to equipment, like the broad technology shifts described above, continue to raise productivity and keep coal mining costs low. For more on mining equipment, see "Mining Innovations" below.

COAL MINING PRODUCTIVITY

General production output and trends in productivity by type of coal mining by region are shown in Table 2. Effects of external and operational changes are described below. (For detailed, annual statistics, see the Energy Information Administration Web site: http://www.eia.doe.gov/emeu/aer/coal.html).

Productivity is calculated by dividing total coal production by the total direct labor hours worked by all employees engaged in production, preparation, processing, development, reclamation, repair shop, or yard work at mining operations, including office workers. In 1973, the average employee in the United States produced 2.16 short tons per hour (tph; Table 2, productivity section). In 1983, productivity had increased by 16%, to 2.50 tph; and by 1993 another 88%, to 4.70 tph. By 2003, average U.S. coal mining productivity was 6.95 short tph, an increase of 48% over 1993, and 222% over the 1973 level. Annual percentage increases in productivity during the 30-year span averaged 4.0%.

In 1973, productivity was in decline. It had fallen 10% since 1969, when the Coal Mine Safety and Health Act initiated or strengthened nationwide mine safety standards and their enforcement. This Act increased mine permitting and design requirements, added new safety and health standards in existing mines, and imposed new permitting and Black Lung fees on existing operations. The Mine Safety and Health Act of 1977 added additional safety, dust control, and mine ventilation requirements. Further, the federal government imposed strict new regulations on pollution and disruptions from mining through the Federal Water Pollution Control Act of 1972 and the Surface Mining Control and Reclamation Act of 1977. It can be argued that eventually these regulations improved productivity through safer, better-planned mines. The increasingly stringent controls of sulfur dioxide emissions under the Clean Air Act of 1970 and its amendments in 1977 and 1990 stimulated mining in low-sulfur coal regions, resulting in changes in mining techniques.

Underground coal mine productivity continued to decline through 1978, before starting a slow recovery. Underground productivity in 1973 was 1.45 tph. It fell to 1.04 tph in 1978, and then recovered to 1.61 tph by 1983. Productivity increased another 83% by 1993. By 2003, underground productivity had increased another 37%, to 4.04 tph. The annual average percentage increase in underground mining productivity for the last 30 years is nearly 4%.

Surface coal mining is less labor intensive, and its productivity is inherently higher, than underground mining. Surface productivity in 1973 was 4.56 tph. It decreased in 1983 by 16% to a level of 3.81 tph—a temporary result of the Federal Surface Mining Control and Reclamation Act of 1977, which required restoration of mined land, diverting some employees and equipment and increasing nonproduction labor hours per ton of mined coal. By 1993, surface productivity had recouped the earlier loss, increasing by 90%, to 7.23 tph. By 2003, surface productivity had increased another 49%, to 10.76 tph. The average annual percentage increase in surface mining productivity for the last 30 years is 3%.

REGIONAL PRODUCTIVITY

Regional geology, together with the type of mining, influences productivity. Appalachia (the mountainous bituminous coalfields of Pennsylvania, Maryland, Ohio, West Virginia, Virginia, eastern Kentucky, Tennessee, and Alabama, plus anthracite in Pennsylvania) has the highest number of mines, while the West has the least. As

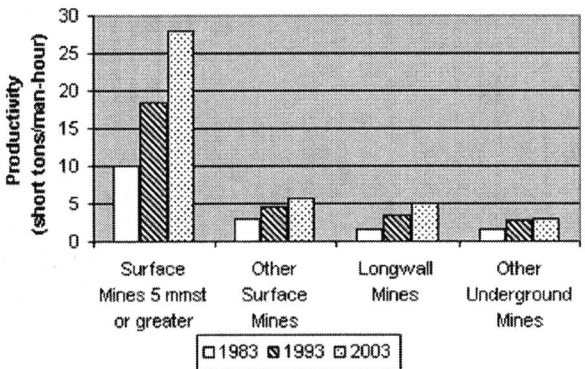

Fig. 2 U.S. Coal mining productivity.
Sources: From Energy Information Administration and Bureau of Mines and U.S. Geological Survey (see Refs. 7 and 22).

Table 2 Production and productivity at U.S. coal mines, selected years

Item	1973	1983	1993	2003
Production(thousand short tons)				
United States	598,568	782,091[a]	945,424	1,071,753
Underground	300,080	300,379[a]	351,053	352,785
Surface	298,491	481,713[a]	594,371	718,968
Appalachian region	381,629	377,952	409,718	376,775
Underground	239,636	230,191	257,433	244,468
Surface	141,993	147,761	152,285	132,307
Interior region	156,412	173,407	167,174	146,276
Underground	56,060	49,437	56,065	52,173
Surface	100,352	123,970	111,109	94,103
Western region	60,530	225,276	368,532	548,701
Underground	10,036	18,691	37,555	56,144
Surface	50,494	206,584	330,977	492,557
Number of employees				
United States	152,204	175,642	101,322	71,023
Underground	111,799	111,888	64,604	40,123
Surface	40,405	63,754	36,718	30,900
Appalachian	124,000	126,111	71,321	46,507
Underground	96,302	90,360	50,956	30,744
Surface	27,698	35,751	20,365	15,763
Interior	22,343	34,590	18,555	11,638
Underground	12,243	16,889	10,246	6,076
Surface	10,100	17,701	8,309	5,562
Western	5,861	14,941	11,446	12,878
Underground	3,254	4,639	3,402	3,303
Surface	2,607	10,302	8,044	9,575
Number of mines				
United States	4,744	3,405	2,475	1,316
Underground	1,737	1,638	1,196	580
Surface	3,007	1,767	1,279	736
Appalachian region	4,423	2,971	2,163	1,143
Underground	1,637	1,526	1,108	521
Surface	2,786	1,445	1,055	622
Interior region	226	311	219	109
Underground	55	64	54	36
Surface	171	247	165	73
Western region	95	123	93	64
Underground	45	48	34	23
Surface	50	75	59	41
Productivity (short tons per miner-hour)				
United States	2.16	2.50	4.70	6.95
Underground	1.45	1.61	2.95	4.04
Surface	4.56	3.81	7.23	10.76
Appalachian region	1.74	1.75	3.00	3.71
Underground	1.33	1.53	2.75	3.64
Surface	3.79	2.23	3.55	3.82
Interior region	3.43	2.69	4.43	5.56
Underground	2.27	1.87	3.06	3.83
Surface	4.80	3.26	5.71	7.43

(Continued)

Table 2 Production and productivity at U.S. coal mines, selected years *(Continued)*

Item	1973	1983	1993	2003
Western region	6.64	7.60	13.53	20.82
Underground	2.59	2.28	5.23	8.42
Surface	9.64	9.63	16.49	25.01

Note: Coal-Producing Regions: Appalachian includes Alabama, eastern Kentucky, Maryland, Ohio, Pennsylvania, Tennessee, Virginia, and West Virginia; Interior includes Arkansas, Illinois, Indiana, Iowa, Kansas, western Kentucky, Louisiana, Mississippi, Missouri, Oklahoma, and Texas; Western includes Alaska, Arizona, Colorado, Montana, New Mexico, North Dakota, Utah, Washington, and Wyoming.
All statistics incorporate data for Pennsylvania anthracite coal, except for number of mines in 1973, which was not reported. Anthracite statistics were collected and published separately from other U.S. coal in 1973.
[a] Production by regions does not total production for United States, underground, and surface (first three rows). The U.S. production for 1983, by surface and underground, represent *all* mines. Details such as regional production, employment, mine count, and productivity statistics were collected that year only from mines that produced 10,000 short tons or more. The small, excluded mines represented 5.5 mmst of coal, or only 0.7 percent of total U.S. production in 1983.
Source: From Energy Information Administration (see Ref. 7).

discussed earlier, coal production and mine size grew in tandem with shifts to more surface mining and toward the West. Keen price competition motivated productivity improvements accomplished through increased mine size and production.

Appalachian productivity (Table 2) did not increase between 1973 and 1983, primarily because of productivity declines in surface mining. Between 1983 and 2003, both surface and underground mining in Appalachia improved. The annual average percentage increase in productivity in Appalachia over those 30 years was 2.6%. The backsliding in surface productivity from 1973 to 1983 corresponds with closure of more than a thousand small contour mines (many were inefficient or seasonal operations), tightening of surface mine permitting and reclamation requirements, and greater public resistance to surface mining. Some production shifted to larger surface mines and MTR operations, but their costs and workforce requirements have been relatively high in Appalachia.

Productivity in the Interior region also declined between 1973 and 1983, before picking up over the next 20 years (Table 2). Active mining in the Interior region is primarily in bituminous coals in Illinois, western Indiana, and western Kentucky and the lignite deposits of Texas, Louisiana, and Mississippi; additionally, small bituminous mines are opened from time to time in Oklahoma, Arkansas, Missouri, Kansas, and Iowa. Most surface mines in the Interior are medium or large box cut area mines. Underground mines tend to be either shaft mines or drift mines entering the coal seam beneath the final highwall. The Interior region has never supported thousands of small surface mines as had Appalachian topography. In 1983, 247 Interior region surface mines produced almost as much coal (84%) as was mined in Appalachia's 1445 mostly contour mines. From 1973 to 2003, the annual average percentage increase in surface productivity was 1.6%; in underground productivity, it was 1.8%.

The increased productivity in the West between 1973 and 1983 can all be attributed to increased surface mining. Surface productivity from 1973 to 1983 did not in itself improve, but new mines opened, and the tonnage mined from the surface quadrupled (Table 2). Those gains in surface mining production share boosted overall productivity. During that time, western underground mining, working in reserves that tended to be thick-bedded, was slightly more efficient than in thinner-bedded eastern coal. In all regions, limited longwall experience and early mine development were insufficient to significantly boost productivity or increase average mine size until after 1983. Overall, productivity of the huge surface mines, principally in Wyoming and Montana, led the productivity growth between 1983 and 2003. The average annual increase in western mine productivity from 1973 to 2003 was 3.9%. Underground productivity gains averaged 4.0% annually. Little changed in surface productivity during the 1980s and early 1990s, but by 2003, ten mines were producing more than 15 million tons per year, with great economies of scale.[7] Productivity of western surface mining improved from 9.64 short tph in 1973 to 25.01 short tph in 2003, an average annual increase of 3.2%.

CHANGES IN REGIONAL COAL PRODUCTION

The relative importance of coal production regionally has changed over the past 30 years, primarily because of the increased size and productivity of western surface mines. Production in the western region increased by more than 800% from 1973 to 2003, dominated by surface production. In 1973, western surface production was 50.5 mmst. It rose to 206.6 mmst by 1983, 331.0 mmst by 1993, and to 492.6 mmst by 2003.

Appalachian coal production varied little during the same period. Without new growth, the Appalachian region dropped from the top U.S. coal-producing region to second place, as western coal filled rising coal demand. In 1973, Appalachian production was 381.6 mmst. Production was 378.0 mmst in 1983, 409.7 mmst in 1993, and only 376.8 by 2003. The split between surface and underground mining in the Appalachian region also has been relatively stable.

Coal production in the Interior region also has changed little over the course of 30 years. Interior production was 156.4 mmst in 1973, 173.4 mmst in 1983, 167.2 mmst in 1993, and 146.3 mmst in 2003. The production split between surface and underground mines was also stable.

COAL MINING EMPLOYMENT

Employment in the coal industry from 1973 to 2003 ties in with the factors discussed above: regional coal production levels, shifts in the type of mining, and changes in productivity within regions and by mining technology. Coal mining employment includes workers at preparation plants that process the mined coal prior to sale. They are allocated to underground or surface mining proportionately based on how the coal was extracted. The average number of mine employees working daily in the United States in 1973 was 152,204 (Table 2). By 1983, daily employment had increased to a total of 175,642 (although 1983 data on employment, number of mines, and productivity covered only mines producing 10,000 short tons or more during the year). The increase resulted from surface mines hiring workers to handle reclamation and from an increase in the number of small, less efficient mines. The average number of daily employees declined to 101,322 by 1993 and to 71,023 by 2003.

Daily underground employment in 1973 averaged 111,799. That figure increased slightly by 1983, to 111,888 employees working daily. Underground employment fell to 64,604 by 1993, and to 40,123 by 2003. This trend reflects the fact that by 1983, a significant number of low productivity mines were in operation—slightly fewer underground mines produced slightly less coal using slightly more employees. Over the next 20 years, declining coal prices and increasing competition forced many of those mines out of business.

Daily surface mine employment in 1973 was 40,405. In 1983, surface employment increased to 63,754 employees, reflecting increased reclamation requirements. By 1993, surface employment was down to 36,718 employees; by 2003, it was down to 30,900.

Regional mining employment reflects the trends in production and productivity discussed above and is outlined in Table 2.

MINING INNOVATIONS

The notable improvements in mining equipment from 1983 to 2003 include the following:

- Bigger and stronger longwall face coal belt conveyors.
- Conversion to belt conveyors to move coal out of underground mines.
- Better roof-bolting equipment, (Roof bolting is a technique to secure an underground mine roof and avoid rock falls by drilling 4–12 ft up into the overlying rock layers and inserting high-tensile-strength bolts and support plates to bind together weak layers with strong layers. Bolts hold via mechanical anchors, epoxy resin, or grout.) including combination continuous-miner/bolters.
- More powerful and durable longwall cutting bits.
- Better sensors for and automation of longwall roof shields.
- More powerful and more durable electric drive motors used in many applications.
- Continuous scale-up of haul trucks, loaders, and excavators for surface mining.

A feature of the improvements listed above is that significant benefits resulted from advances in materials and technology applied to existing mining techniques, not from pioneering entirely new mining machinery. That process continues. Roof bolting was a seminal change in underground coal mining. It allowed passageways to be secured with substantially fewer timbers and "cribs" (the pillars constructed of stacked short beams used to shore up million-pound roof loads). Roof bolting—a safety standard, mandated in the 1969 Coal Mine Safety and Health Act—resulted in safer, more open mine passages and led to single-operator roof-bolting machinery far more productive than the previous labor-intensive manual timbering and cribbing.[8] For areas subject to tangential forces, steel cable roof bolts, with higher tensile strength and resistance to shear failure, give superior results. Those same qualities, along with new flexible, sprayed rock coatings, are expected to attract more proponents as mines go deeper. Though cable bolts and coatings add cost, some mines have found that fewer are needed per unit area.[9]

Examples of other recent improvements in longwall mining include variable slip clutches in the power drives for coal face belt conveyors to accommodate surges in power demand due to irregular loading and to drag from oversized coal. Ceramic facings on belt drums now give better traction and wear. Stronger materials are being marketed in roof shields to extend usable life and reduce maintenance costs. With automated operation of longwall face-shearing drums and roof-shield positioning, operators can now monitor and control mining remotely from "outby" passages at the ends of panel cuts and away from some of the noise and moving machinery. New roof bolters are highly automated and shield the operator. Advances since the 1980s in distancing the operator from the working coal face also came with the accelerated use of highwall miners, which employ video or sensor-aided monitors to give the operator effective remote control of mining for distances approaching 1000 ft. Robotic mining is expected to grow. Scaled-down longwall machines are now being tested thinner than 4 ft, and robotic cutting tools have been bench tested that can extract coal as thin as 6 in.

Computerized control systems now monitor and coordinate belt speeds in some mines from the longwall face through the face belt conveyors and all downstream belt systems, out the portal, and to storage piles. Mine operators are adopting new machinery to reduce the downtime when a longwall system is moved to a new panel. With specially designed trailers that can haul several roof shields at 13–15 MPH vs the 2–3 MPH for single shields on the common fork-lift type shield mover, one operator recently cut 2 days out of a 14-day move.[10]

Underground mines of all kinds can now take advantage of manufactured crib materials that are stronger under load than timber and more impervious to water and oil, that interlock for greater solidity, and that can be assembled quickly by machine, reducing the injury potential of personnel handling heavy materials underground. Similarly, corrugated and "pumpable" supports save time and are safer to use. Portable roof shields are now available to improve safety and coal extraction in room and pillar "retreat mining," when piers of coal that had supported the roof are removed in final mining stages.

In surface mining, new improvements include innovative use of computerized process control, which are currently being used at progressive operations along with global positioning systems to schedule and dispatch haul trucks, and to control positioning and depths of cuts by bulldozers and scrapers preparing pits and exposing the coal seam. Sensors on dozer or loader blades are guiding operators in distinguishing and recovering coal vs black shales at a growing number of mines.

The opportunities for larger surface mining equipment do have physical and practical limitations, including wheel size and tire construction, but new configurations for haul trucks are on the drawing boards that may produce a 1000-ton haul truck by 2020. At the same time, in-pit excavators will be increasing bucket size from 50 to 150 cubic yards.[11]

COAL PREPARATION

Coal preparation is processing of run-of-mine coal—the raw coal coming out of the mine—in order to enhance its characteristics for shipping and ultimate utilization.

Benefits of coal preparation may include removal of noncombustible material, whose weight raises shipping costs and which can increase wear in coal grinding equipment and boilers; enhancement of deliverable heat content; removal of unwanted minerals that can foul boilers or damage the environment if entrained in boiler emissions or ash; suppression of dust; and improvement of handling and shipping qualities. These processes are carried out at preparation plants—also known as "prep" plants or wash plants—which may be located either at coal mines or at separate facilities serving numerous associated or independent mines in a mining region.

Preparation begins with crushing and screening freshly mined coal, which normally results in removal of some of the noncoal material. Some coal, especially coal from thick-bedded surface mines, is merely crushed and screened before shipping. Additional cleaning, known as mechanical cleaning, may entail separating out noncoal material in a liquid medium, which led to the widely used term "washing." The washing medium is an aqueous chemical solution prepared to enhance wettability and dissociation of the coal and noncoal materials or to produce specific gravities calibrated higher than water alone. The liquid medium may be combined with finely ground heavier minerals such as magnetite in a dense medium fluid, better to effect separation of unwanted rock and mineral matter from coal particles. Wet or "hydraulic" cleaning techniques may also include particle agitation by aeration of the coal-liquid feed, materials sorting via relative density in hydro-cyclonic chambers, and froth flotation to capture fine coal particles. To meet environmental regulations, technically advanced wash plants can remove as much as 40% of the inorganic sulfur in coal. Dry techniques, rarely used alone, include prewash segregation by vigorous shaking and pneumatic air-flow separation for crushed feed coal.

Prepared coal is commonly dewatered to some degree because excess moisture degrades deliverable heat content in the coal, and the added weight increases handling and shipping costs. Dewatering techniques range from inexpensive vibrating screens, filters, or centrifuges to the more costly use of heated rotary kilns or dryer units. Before burning, almost all coal for electric power and industrial boilers is either pulverized or crushed and sized. Precombustion coal washing is usually less costly than downstream options for removing ash and sulfur.

Two trends affect the amount of coal washed in the United States. First, production of western U.S. coal has outpaced the production of eastern U.S. coal. Most western coal is crushed and sized for market but rarely washed. Second, to meet environmental regulations, greater percentages of eastern coal are washed. In 1973, 28 and 69% of surface- and underground-mined coal, respectively, was washed. In addition, washed anthracite production, for which the type of mining was not identified, equated to 1% of U.S. production.[12] By 1983, the shares were at least 21 and 63%.[13,14] The term "at least" acknowledges that the 1973 statistics covered all mines with at least 1000 short tons of annual coal production; whereas the 1983 survey "supplement" covering prep plants was limited to larger mines, with at least 100,000 short tons of production.

Coal washing can produce large volumes of waste. In 2002, about 25% of the raw coal processed through preparation plants was discharged to waste ponds as "refuse," mixtures composed of shale, clay, coal, low-grade shaley coal, and preparation chemicals.[15] Like mining, coal-washing operations have undergone consolidation. Over the period 1983–2003, the number of U.S. wash plants fell from 362 to 132, and employment dropped from 7300 to about 2500 employees.

COAL PRICES

Except for price inflation generated following the energy crisis of 1973, U.S. coal prices were relatively stable from 1973 to 2003. When coal prices did rise, external factors like the 1973 oil embargo or burgeoning demand for coal and oil in China in 2003–2004 have been largely responsible. When real coal prices declined, however, as they did from 1975 to 2000, it was largely owing to improved labor productivity. That trend reflects the effects of "marked shifts in coal production to regions with high levels of productivity, the exit of less productive mines, and productivity improvements in each region resulting from improved technology, better planning and management, and improved labor relations."[16]

Adjusted for inflation, coal prices in year 2000 dollars decreased from $31.40 to $16.84 per short ton between 1950 and 2003 (see Table 3). The average price in nominal dollars went from $5.19 per short ton in 1950 to $17.85 in 2003. In energy terms—dollars per million Btu—coal has long been the lowest-cost fossil fuel. Petroleum products became more expensive than coal around the 1890s, when the first practical diesel and gasoline internal combustion engines were used in vehicles. Natural gas prices surpassed coal in 1979, in the first phase of natural gas price deregulation under the Natural Gas Policy Act of 1978. In 2003, one million Btu of coal sold for $0.87 on average, compared to $4.41 for natural gas and $4.75 for crude oil.[17]

The 1973 oil embargo spurred immediate and dramatic increases in coal prices, but in the long term it may have depressed prices through the long-lived excess productive capacity it generated. Between 1980 and 2000, coal prices remained under the overhang of excess capacity. Other suppliers often underbid contract coal prices considered reasonable, or even low, by mine operators.

Coal prices in Table 3 illustrate the changes that began in 1973. The average price in 1973 for U.S. coal was $8.59 per short ton, priced at the mine or original loading point. That price was unaffected by the oil embargo because in 1973 the federal government had not yet initiated policies to steer electricity producers away from petroleum as a fuel and to promote increased use of coal. Already, between 1968 and 1973, nominal coal prices had increased from $4.75 to $8.59 per short ton because coal producers passed through some of the increased costs of Black Lung taxes and the new mine safety regulations of the 1969 Coal Mine Safety and Health Act. Building on that beginning, historic real coal prices peaked in 1975, at $50.92 per short ton ($19.35 nominal).[18]

Coal prices are also commonly influenced by the end use. In the United States, the principal end uses are steam coal, metallurgical coal, and industrial coal. Steam coal, also known as "thermal" coal in international markets, is used to create steam or heat to power industrial processes. It is priced primarily on its deliverable heat content, and, because environmental regulations started in the 1970s, its value may be rated down for high sulfur content. Any rank of coal may be used as a steam coal. The average nominal price at the mine or origin of all U.S. coal produced in 2003 was $17.85 per short ton. By comparison, the average price of all coal delivered to electric power plants—which generally accounts for 90% of U.S. production—was $25.91. The $8.06 difference is roughly the average cost of handling and transporting the coal to the final consumers. Examples of real delivered prices of steam coal for electricity production[19] appear below:

Steam coal at electric power plants

Year	Delivered real price per short ton (in year 2000 dollars)
1973	$28.29
1983	$53.66
1993	$32.34
2003	$24.44

Metallurgical or coking coal is used to produce metallurgical coke. The coke is produced in sealed, oxygen-free ovens and used in blast furnaces in standard iron smelting for steel production. Coke is made from bituminous coal (sometimes blended with up to 1% anthracite). It must be low in sulfur and must "agglomerate," or fuse, incorporating ash-forming minerals in the coal, to produce a strong, porous, and carbon-rich fuel that can support the load of iron ore in a blast furnace. Coal for metallurgical use requires more thorough cleaning than for steam uses and it is priced higher[20]:

Table 3 Historical U.S. coal prices at the mine or source, by coal rank, selected years (prices in dollars per short ton, expressed in nominal dollars and in inflation-adjusted year-2000 dollars)

Year	Average price U.S. coal sales		Average price of bituminous coal[a]		Average price of subbituminous coal		Average price of lignite		Average price of anthracite	
	Nominal ($)	Real ($)	Nominal ($)	Real ($)	Nominal ($)	Real ($)	Nominal ($)	Real ($)	Nominal ($)	Real ($)
2003	17.85	16.84	26.73	25.22	7.73	7.29	11.20	10.57	49.55	46.75
2002	17.98	17.27	26.57	25.53	7.34	7.05	11.07	10.63	47.78	45.90
2001	17.38	16.97	25.36	24.77	6.67	6.51	11.52	11.25	47.67	46.55
2000	16.78	16.78	24.15	24.15	7.12	7.12	11.41	11.41	40.90	40.90
1999	16.63	16.99	23.92	24.44	6.87	7.02	11.04	11.28	35.13	35.90
1998	17.67	18.32	24.87	25.78	6.96	7.21	11.08	11.49	42.91	44.48
1997	18.14	19.01	24.64	25.82	7.42	7.78	10.91	11.43	35.12	36.81
1996	18.50	19.71	25.17	26.82	7.87	8.39	10.92	11.64	36.78	39.19
1995	18.83	20.44	25.56	27.75	8.10	8.79	10.83	11.76	39.78	43.19
1994	19.41	21.50	25.68	28.45	8.37	9.27	10.77	11.93	36.07	39.96
1993	19.85	22.46	26.15	29.59	9.33	10.56	11.11	12.57	32.94	37.27
1988	22.07	29.16	27.66	36.54	10.45	13.81	10.06	13.29	44.16	58.34
1983	25.98	39.84	31.11	47.71	13.03	19.98	9.91	15.20	52.29	80.19
1978	21.86	47.77	22.64	49.48	—[a]	—[a]	5.68	12.41	35.25	77.04
1973	8.59	26.97	8.71	27.35	—[a]	—[a]	2.09	6.56	13.65	42.86
1968	4.75	19.07	4.70	18.87	—[a]	—[a]	1.79	7.19	8.78	35.24
1963	4.55	20.87	4.40	20.19	—[a]	—[a]	2.17	9.96	8.64	39.64
1958	5.07	24.73	4.87	23.76	—[a]	—[a]	2.35	11.46	9.14	44.59
1953	5.23	28.67	4.94	27.08	—[a]	—[a]	2.38	13.05	9.87	54.10
1950	5.19	31.40	4.86	29.40	—[a]	—[a]	2.41	14.58	9.34	56.50

[a]Through 1978, subbituminous coal is included in "Bituminous Coal".
Source: From Energy Information Administration (see Ref. 7).

Coal at metallurgical coke plants

Year	Delivered real price per short ton (in year 2000 dollars)
1973	$62.07
1983	$90.94
1993	$53.68
2003	$47.77

Industrial coal can be of any rank. It is coal used to produce heat for steam or industrial processes. Typical industrial coal consumers include manufacturing plants, paper mills, food processors, and cement and limestone products. Prices[21] tend to be higher than for coal received at electricity producers, primarily because average tonnages purchased by industrial consumers are smaller and, in some cases, because the plant processes require specific or less-common coal characteristics:

Coal at other industrial facilities

Year	Delivered real price per short ton (in year 2000 dollars)
1973	NA
1983	$60.30
1993	$36.47
2003	$32.74

CONCLUSION

In recent years, about 90% of coal production in the United States has been consumed at domestic electric power plants. Coal use grew because of secure, abundant domestic reserves and relatively low prices. Demand has been met through increasing mine productivity, which in turn has been supported by operation of larger and larger mines, the use of larger, more efficient mining machinery, advances in technology and control systems, and the employment of fewer mine personnel.

REFERENCES

1. *Annual Energy Review 2004*, DOE/EIA-0384(2004); Energy Information Administration: Washington, DC, 2005; 207.
2. Definitions in the section are adapted from: *Coal Glossary*, Internet Website, Energy Information Administration: Washington, DC, 2005, http://www.eia.doe.gov/cneaf/coal/page/gloss.html (accessed August 2005); *A Dictionary of Mining, Mineral, and Related Terms*: U.S. Department of the Interior, Bureau of Mines: Washington, DC, 1968, various.
3. McDaniel, J.; Kitts, E. *A West Virginia Case Study*, Electronic slide presentation. Mining and Reclamation Technology Symposium: Morgantown, WV, 1999, http://www.epa.gov/region3/mtntop/appendix.htm "Case Study," (accessed August 2005).
4. *Draft Programmatic Environmental Impact Statement, Mountaintop Mining/Valley Fills in Appalachia*, http://www.epa.gov/region3/mtntop/eis.htm, Chapter 3, U.S. Environmental Protection Agency, Mid-Atlantic Region: Philadelphia, PA, 2003; III.J-1.
5. Mine Safety and Health Administration, Form 7000-2, *Quarterly Mine Employment and Coal Production Report*, file data, 1983–2003; Coal—bituminous and lignite chapter. *Minerals Yearbook 1973*; U.S. Department of the Interior, Bureau of Mines: Washington, DC; 326.
6. DOE/EIA-TR-0588. *Longwall Mining*; Energy Information Administration: Washington, DC, 1995; 34–35; Coal—bituminous and lignite chapter. *Minerals Yearbook 1973*; U.S. Department of the Interior, Bureau of Mines: Washington, DC; 353.
7. DOE/EIA-0584(2003). *Annual Coal Report*; Energy Information Administration: Washington, DC; 22; http://www.eia.doe.gov/cneaf/coal/page/acr/acr_sum.html (accessed September 2005).
8. Barczak, T. Q&A—research. Am. Longwall Mag. **2005**, *August*, 45.
9. Peterson, D.; LaTourrette, T.; Bartis, J. *New Forces at Work in Mining: Industry Views of Critical Technologies*; RAND Corporation: Arlington, VA, 2001; 26, [http://www.rand.org/publications/MR/MR1324/MR1324.ch3.pdf (accessed August 2005)].
10. Bahr, A. Saving time and money at elk creek. Am. Longwall Mag. **2005**, *August*, 43.
11. Peterson, D.; LaTourrette, T.; Bartis, J. *New Forces at Work in Mining: Industry Views of Critical Technologies*; RAND Corporation: Arlington, VA, 2001; 28, [http://www.rand.org/publications/MR/MR1324/MR1324.ch3.pdf (accessed August 2005)].
12. Coal—bituminous and lignite chapter. *Minerals Yearbook 1973*; U.S. Department of the Interior, Bureau of Mines: Washington, DC; 358–359, 391.
13. *Coal Production 1984*, Appendix C, DOE-EIA-0118(84), Energy Information Administration: Washington, DC; 118 (for quantities cleaned, by type of mining).
14. *Coal Production 1983*, DOE-EIA-0118(83), Energy Information Administration: Washington, DC; 4 (for quantities mined, by type of mining).
15. *Bituminous Coal Underground Mining: 2002*, U.S. Department of Commerce, Economics and Statistics Administration, U.S. Census Bureau, 2004, Tables 6a and 7.
16. Flynn, E. *Impact of Technological Change and Productivity on the Coal Market, Issues in Midterm Analysis and Forecasting*, DOE-EIA-0607(2000); Energy Information Administration: Washington, DC; 1; http://www.eia.doe.gov/oiaf/analysispaper/pdf/coal.pdf (accessed October 2005).
17. *Annual Energy Review 2004*, DOE/EIA-0384(2004), Energy Information Administration: Washington, DC, 2005; 67.
18. *Annual Energy Review 2004*, DOE/EIA-0384(2004), Energy Information Administration: Washington, DC, 2005; 219.

19. *Annual Summary of Cost and Quality of Electric Utility Plant Fuel 1976*; Bureau of Power, Federal Power Commission: Washington, DC, 1977; 12; *Cost and Quality of Fuels for Electric Utility Plants1983*, DOE/EIA-0191(83); Energy Information Administration: Washington, DC, 1984; 27; *Electric Power Monthly*, Energy Information Administration: Washington, DC, 68; http://www.eia.doe.gov/cneaf/electricity/epm/table4_1.html (accessed as Tables 4.1, October 2005).
20. *Quarterly Coal Report*, October–December 1984, 1994, and, 2004, DOE/EIA-0121(84) and (94) and (2004), Energy Information Administration: Washington, DC, 17, 53, 26; Coal—bituminous and lignite chapter. *Minerals Yearbook 1973*, U.S. Department of the Interior, Bureau of Mines: Washington, DC; 320.
21. *Quarterly Coal Report*, October-December 1984, 1994, and, 2004, DOE/EIA-0121(84) and (94) and (2004), Energy Information Administration: Washington, DC, 17, 53, 28.
22. U.S. Bureau of Mines, Annual Coal Production Surveys; Bureau of Mines and U.S. Geological Survey: Washington, DC, 1975.

BIBLIOGRAPHY

1. *Annual Coal Report*, DOE/EIA-0118 (years 1976–1992), Energy Information Administration: Washington, DC.

Coal Supply in the U.S.

Jill S. Tietjen
Technically Speaking, Inc., Greenwood Village, Colorado, U.S.A.

Abstract

The United States has approximately 275 billion tons of coal resources in the ground, enough to last more than 200 years at current rates of consumption. Coal ranks in the United States are (from lowest to highest) lignite, subbituminous, bituminous, and anthracite. Estimated recoverable reserves of coal are located in 32 states, and mining currently takes place in 26 of them. Coal is mined in underground and surface mines. Coal is transported from the mines to the ultimate users by conveyer belt, truck, barge, ship, train, or coal slurry pipeline and sometimes by multiple methods. More than 50% of the electricity generated in the United States comes from coal-fired power plants. Electric utilities and industrial users are the largest consumers of coal.

INTRODUCTION

The United States has approximately 275 billion tons of coal resources in the ground, enough to last more than 200 years at current rates of consumption. Coal constitutes 95% of the fossil energy reserves in the United States. Half of the recoverable coal reserves in the world are in China, the United States, and the former Soviet Union.[1]

These estimated coal reserves are generally characterized by coal rank. The coal ranks in the United States are (from lowest to highest) lignite, subbituminous, bituminous, and anthracite. Estimated recoverable reserves of coal are located in 32 states, and mining currently takes place in 26 of them. Coal deposits occur in additional states, but either reserve tonnages have not been estimated or physical conditions are not conducive to mining. Coal is mined in underground mines and in surface mines. Coal is transported from the mines to the ultimate users by conveyer belt, truck, barge, ship, train or coal slurry pipeline, and sometimes by multiple methods. More than 50% of the electricity generated in the United States comes from coal-fired power plants. Electric utilities and industrial users are the largest consumers of coal.[2]

RANKS OF COAL

Coal results from geologic forces having altered plant materials in different ways. The four coal types are derived from peat, which is the first stage in the formation of coal. Peat is partially decomposed plant material. The four types or ranks of coal are lignite, subbituminous, bituminous, and anthracite.[3] The rank of a coal refers to the degree of metamorphosis it has undergone. The longer the organic materials comprising the coal have been buried, along with the amounts of pressure and heat imposed, the greater is its conversion to coal.

Lignite is the lowest rank of coal and is often referred to as brown coal. It has the lowest heating value of any of the four categories of coal [4,000–8,300 Btu/lb (A British thermal unit (Btu) is the amount of heat needed to raise the temperature of 1 lb of water 1°F)]. It is brownish-black and has a very high moisture (sometimes as high as 55%) and ash content, and cannot be transported very long distances economically. About 8% of the coal produced in the United States is lignite, and most of it is in Texas and North Dakota. Lignite is primarily used as a fuel for electricity generation.

Subbituminous coal ranges from dull dark brown, soft and crumbly, to bright jet black, hard and relatively strong. Although it has a heating value higher only than lignite, averaging 8,400–8,800 Btu/lb in the Powder River Basin (PRB) of Wyoming, there are plentiful reserves of subbituminous coal in the West and in Alaska. Although this coal has a moderately high moisture content (20%–30%), the PRB coal resources in Wyoming and Montana have a lower sulfur content than many bituminous coal reserves and, thus, burn more cleanly. More than 90% of subbituminous coal production comes from the PRB. More than 40% of the coal produced in the United States is subbituminous.

Bituminous coal, often called soft coal in Europe, is the most common type of coal used for the generation of electricity in the United States. It is a dense coal, usually black and sometimes dark brown. This coal has a heating value from 10,500 to 15,000 Btu/lb. The primary use of bituminous coal—and all U.S. coal—is as "thermal" or steam coal, consumed mostly for electricity generation. Bituminous coal also has properties that allow it to be used as metallurgical coal for the steel and iron industries. Bituminous coal accounts for about one-half of U.S. coal production.

Keywords: Anthracite; Bituminous; Subbituminous; Lignite; Coal reserves; Underground mining; Surface mining; Barge; Railroad; Truck; Coal basin.

Table 1 U.S. coal regions and coal fields

Coal region	Coal field	States
Appalachia	Northern Appalachia	MD, OH, PA. Northern WV
	Central Appalachia	Eastern KY, VA, Southern WV, Northern TN
	Southern Appalachia	AL, Southern TN
Interior	Illinois Basin	Western KY, IL, IN
	Gulf Coast lignite	TX, LA, MS
	Other Western Interior	AR, IA, KS, MO, OK
West	Powder River Basin	WY, MT
	North Dakota lignite	ND
	Southwest	AZ, NM
	Rockies	CO, UT
	Northwest	AK, WA

Source: From Energy Information Administration (see Ref. 6).

Anthracite is the hardest coal (often referred to as hard coal), is brittle, has a high luster, and gives off the second greatest amount of heat when it burns (averaging 12,500 Btu/lb). It is low in volatile matter and has a high percentage of fixed carbon. Anthracite accounts for a small amount of the total coal resources in the United States. It is found mainly in Pennsylvania and is generally used for space heating. Since the 1980s, anthracite refuse or mine waste has been used to generate electricity.[4]

COAL BASINS IN THE UNITED STATES

The large majority of coal in the contiguous United States is found in the Appalachian, Interior, and Western coal regions. These regions are identified in Table 1 and Fig. 1. The characteristics of some typical kind of coals are shown in Table 2. Coal varies significantly from mine to mine and from region to region, even within the same classification (e.g., bituminous).[5]

The Appalachian coal region contains the largest deposit of high-grade bituminous coals in the United States. It is generally divided into three parts. Anthracite and bituminous coals are found in northern Appalachia, which includes the bituminous coal deposits found in the states of Pennsylvania, Maryland, Ohio, and northern West Virginia, and the anthracite fields of eastern Pennsylvania. The bituminous coals found in the central Appalachian area of southern West Virginia, Virginia, northern Tennessee, and eastern Kentucky include deposits that are low in sulfur and highly desirable as a fuel for electricity generation. The Alabama and southern Tennessee coal deposits that are characterized as southern Appalachia have been used primarily in the steel industry through history.

The Interior Basin coals, in general, contain a lower heating value than Appalachian coals, with higher sulfur

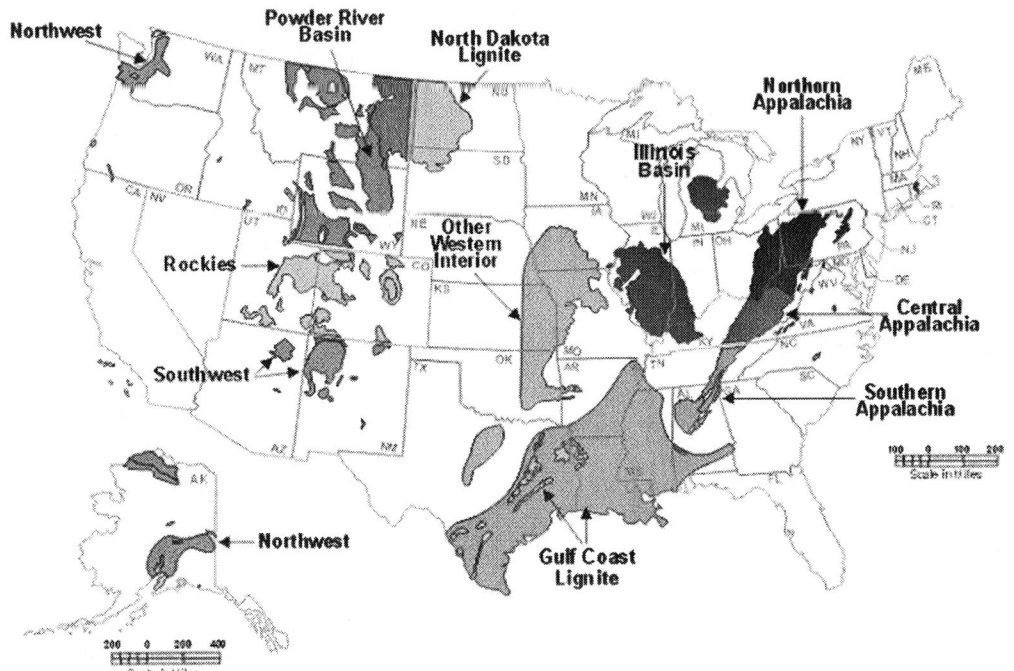

Fig. 1 Coal regions and coal fields.

Table 2 Characteristics of some typical U.S. coals

Coal region	Type of coal	Heating value (Btu/lb)	% Moisture	% Sulfur
Appalachia	Anthracite	12,440	2.4	0.5
Appalachia	Bituminous	12,790	3.3	1.0
Illinois Basin	Bituminous	11,440	8.5	4.3
Western	Bituminous	12,320	2.2	0.6
PRB	Subbituminous	8,400–8,800	28.0	0.2
Gulf Coast	Lignite	6,490	34.8	0.5

Source: From Combustion Engineering, Inc. (see Ref. 7).

content. Mining primarily is found today in the Illinois Basin states of Illinois, Indiana, and western Kentucky. In the Illinois Basin, most coal with lower chlorine content has already been mined, and most of the desirable remaining coal will need to be mined in deep underground mines in the future. Bituminous coal deposits are found in other states in the region, including Michigan, Iowa, Missouri, Nebraska, Kansas, Arkansas, Oklahoma, and part of Texas.

Lignites in the Gulf Coast region are found in Alabama, Mississippi, Louisiana, Texas, and Arkansas. Some of these lignites have a moisture content as high as 55% and heating values sometimes lower than 4,000 Btu/lb, which limits their marketability. Some of the Louisiana lignites are similar to those in North Dakota, with a lower moisture content of around 36% and sodium levels of 5%–8%.

Very large deposits of various kinds of coal are found in the Great Plains. Immense deposits of lignite are found in North and South Dakota. The PRB of Wyoming and Montana contains a very large deposit of subbituminous coal.

The Rocky Mountain states of Colorado and Utah contain deposits of bituminous coal, although other coals can be found throughout the region. Coal from the Southwest, primarily from New Mexico, is both bituminous and subbituminous, and is used for electricity generation. All four ranks of coal exist in Washington state, but the majority is subbituminous. Subbituminous coal is produced at the state's only active mine and is used for electricity generation.

Coal is distributed widely throughout Alaska, differing greatly in rank and geologic environment. Reserves are estimated to be 15% bituminous and 85% subbituminous and lignite. Developed reserves in Alaska are primarily located near the main lines of transportation.[8]

METHODS OF MINING

Coal is removed from the earth either through underground mining or surface mining. About two-thirds of current U.S. coal production comes from surface mines.

Underground mining, also referred to as deep mining, is used when the coal is more than several hundred feet underneath the surface of the earth. There are three types of underground mines: drift mines, slope mines, and shaft mines. The type of mine that will be constructed is dependent on the depth of the coal seam and the terrain. Drift mines have horizontal entries into the coal seam from a hillside. Slope mines, which are usually not very deep, are inclined from the surface to the coal seam. Shaft mines are generally the deepest type of mine and have vertical access to the coal seam via elevators that carry workers and equipment into the mine. Some underground mines are more than 2,000 ft underground. These types of mines are shown in Fig. 2.[9]

Surface mining can usually be used when the coal is buried less than 200 ft under the surface of the earth. Large machines remove the overburden—the layers of soil, rock, and other materials that are between the surface of the earth and the highest minable coal seam (see Fig. 3). Then the exposed coal seam is blasted, using explosives, and the coal is removed. Except in the PRB, where a single minable seam may be as thick as 100 ft, most surface mines produce from multiple coal seams, layered within the rock strata. When that is done, the process outlined above is repeated, removing successive layers of rock (called interburden) between mined coal and coal. For a surface mine, the ratio of overburden to the amount of coal removed is called the overburden ratio. Lower ratios mean that the mine is more productive, and the ratios may be lowered by recovering additional coal seams.

In most cases, after the coal has been mined, the materials that had been on the top of the coal are restored,

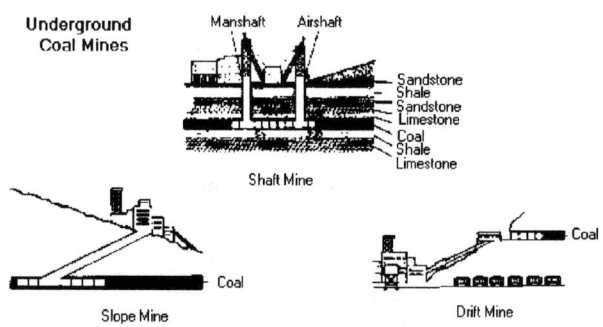

Fig. 2 Underground mining.

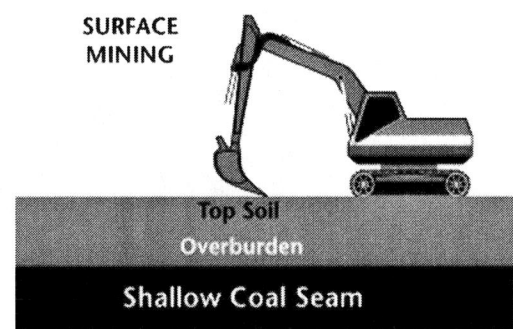

Fig. 3 Surface mining.

so that the surface on top of the pits can be reclaimed. When the area has been replanted, it can often be used for any purpose that could have occurred before the land was mined: cropland; wildlife habitat; recreation; or offices, homes, and stores. In the case of mountaintop removal mining, which is used primarily in Appalachia, the surface on top of the coal is not restored to its original contour.

There are several types of surface mines: area, contour, mountaintop removal, auger, and open pit. Area surface mines are usually found in flat terrain and consist of a series of cuts 100–200 ft wide. The overburden from one cut is used to fill in the mined-out area of the previous cut. Contour mining, which occurs in sloping and mountainous terrain, follows a coal seam along the side of a hill. Open pit mining is generally found where the coal seams are thick and the mines can reach depths of several hundred feet.[10]

COAL PRODUCTION

Coal is mined from surface and underground mines in 26 states. More coal is mined in Wyoming than in any other state. Other states with high production include West Virginia, Kentucky, Pennsylvania, and Texas. Table 3 presents coal production by state for 2001 through 2005. The total 2005 production just over 1 billion short tons of coal—represents more than one-fifth of the world's coal production.

The Appalachian coal region annually produces about 35% of total U.S. coal production from large underground and small surface mines. West Virginia is the largest coal-producing state in the Appalachian coal region and the second largest in the United States. Coal from this region is used as a power plant fuel to produce electricity, for production of metals, and to export to other countries.

Texas is the largest coal-producing state in the Interior region, almost all lignite for power plants. The Illinois Basin states are the next more significant producers: Indiana, Illinois, and Western Kentucky, respectively. Almost 13% of U.S. coal production is from the Interior region, most produced in midsize surface mines by midsize to large companies.

Wyoming is the largest producing state in the nation, a position it has held for 19 consecutive years as of 2005. Its 2005 level of production was just 11 million tons short of the production of the next five largest coal-producing states combined (West Virginia, Kentucky, Pennsylvania, Texas, and Montana). About 52% of the coal mined in the United States comes from the Western region, with more than 30% from Wyoming alone. The large surface mines in the West are the largest coal mines in the world.[12]

COAL TRANSPORTATION

Coal in the United States moves from the coal basin in which it was mined to its final use by rail, water, truck, tramway, conveyor, or slurry pipeline, or more than one of those modes (so-called multimodal transportation). Power plants that are located near or at a mine and burn coal from that mine are called minemouth power plants. The coal for these plants usually moves from the mine to the power plant by truck, tramway, or conveyor. Coal that has been mixed with a liquid, usually water, is called a slurry and is moved through a pipeline that is called a slurry pipeline. The longest coal slurry pipeline in operation in the United States moved coal 273 mi from Arizona to Nevada until January 1, 2006. (It was closed at that time pending the completion of environmental upgrades at the power plant where it is consumed). More than 65% of coal in the United States is transported for at least part of its trip to market by train.[13]

The primary manner in which coal was transported in 2002 is shown in Table 4. Fig. 4 shows the methods of transportation graphically, demonstrating the strong predominance of rail transportation for moving coal in the United States.

USES OF COAL

The overwhelming majority of coal in the United States is used by the electric utility sector, as shown in Table 5—almost 92% of all of the coal consumed in the United

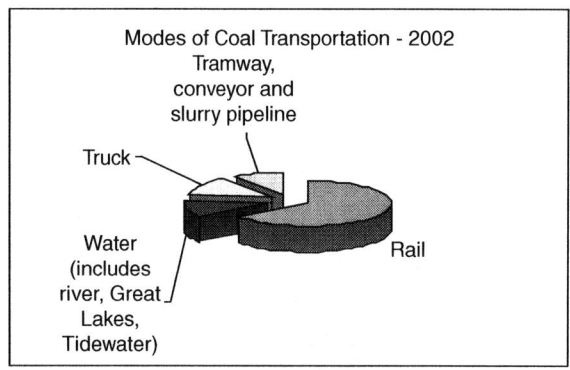

Fig. 4 Modes of coal transportation 2002.

Table 3 U.S. coal production by coal-producing region and state 2001–2005 (million short tons)

Coal producing region and state	2001	2002	2003	2004	2005
Appalachian total	**431.2**	**396.2**	**376.1**	**389.9**	**396.4**
Alabama	19.4	18.9	20.1	22.3	21.3
Kentucky, Eastern	109.1	99.4	91.3	90.9	93.4
Maryland	4.6	5.1	5.1	5.2	5.2
Ohio	25.4	21.2	22.0	23.2	24.7
Pennsylvania total	74.1	68.4	63.7	66.0	67.3
Anthracite	1.5	1.3	1.2	1.7	1.6
Bituminous	72.7	67.1	62.5	64.3	65.6
Tennessee	3.3	3.2	2.6	2.9	3.2
Virginia	32.8	30.0	31.6	31.4	27.7
West Virginia	162.4	150.1	139.7	148.0	153.6
Northern	38.2	34.0	34.9	40.6	42.6
Southern	124.5	116.0	104.8	107.3	110.9
Interior total	**146.9**	**146.6**	**146.0**	**146.0**	**149.2**
Arkansas	—	—	—	—	—
Illinois	33.8	33.3	31.6	31.9	32.1
Indiana	36.7	35.3	35.4	35.1	34.4
Kansas	0.2	0.2	0.2	0.1	0.2
Kentucky, Western	24.7	24.7	21.5	23.4	26.4
Louisiana	3.7	3.8	4.0	3.8	4.2
Mississippi	0.6	2.3	3.7	3.6	3.6
Missouri	0.4	0.2	0.5	0.6	0.6
Oklahoma	1.7	1.4	1.6	1.8	1.8
Texas	45.0	45.2	47.5	45.9	45.9
Western total	**547.9**	**550.4**	**548.7**	**575.2**	**587.0**
Alaska	1.5	1.1	1.1	1.5	1.5
Arizona	13.4	12.8	12.1	12.7	12.1
Colorado	33.4	35.1	35.8	39.9	38.5
Montana	39.1	37.4	37.0	40.0	40.4
New Mexico	29.6	28.9	26.4	27.2	28.5
North Dakota	30.5	30.8	30.8	29.9	30.0
Utah	27.0	25.3	23.1	21.7	24.5
Washington	4.6	5.8	6.2	5.7	5.3
Wyoming	368.7	373.2	376.3	396.5	406.4
Refuse recovery	**1.8**	**1.0**	**1.0**	**1.0**	**0.7**
U.S. total	**1127.7**	**1094.3**	**1071.8**	**1112.1**	**1133.3**

Source: From Energy Information Administration (see Ref. 11).

States. Industries and businesses also burn coal in their own power plants to produce electricity. The coal is burned to heat water to produce steam that turns turbines and generators to produce electricity.

Industries across the United States use coal for heat and as a chemical feedstock. Derivatives of coal, including methanol and ethylene, are used to make plastics, tar, synthetic fibers, fertilizers, and medicine. The concrete and paper industries also burn coal. Altogether, industrial customers consume more than 7% of the coal mined in the United States.

In the steel industry, coal is baked in hot furnaces to make coke, which is used to smelt iron ore into the pig iron or hot iron needed for making steel. The very high

Table 4 Coal transportation in the United States (millions of short tons, 2002)

Mode of transportation	Tonnage moved by this means
Rail	685,086
Water (includes river, Great Lakes, tidewater)	126,870
Truck	138,222
Tramway, conveyor and slurry pipeline	99,986
Total 2002	1,051,406

Source: From Energy Information Administration (see Ref. 14).

Table 5 Coal consumption by sector

Sector	2001	2002	2003	2004	2005
Electric power	964.4	977.5	1005.1	1016.3	1039.0
Coke plants	26.1	23.7	24.2	23.7	23.4
Other industrial plants	65.3	60.7	61.3	62.2	60.8
Combined heat and power (CHP)	25.8	26.2	24.8	26.6	20.6
Non-CHP	39.5	34.5	36.4	35.6	40.2
Residential and commercial users	4.4	4.4	4.2	5.1	5.1
Residential	0.5	0.5	0.5	0.6	0.6
Commercial	3.9	4.0	3.8	4.6	4.6
Total	1060.1	1066.4	1094.9	1107.3	1128.3

Source: From Energy Information Administration (see Ref. 15).

temperatures and the fluxing properties to isolate impurities made possible by using coke give steel the strength and flexibility required for bridges, buildings, and automobiles.

Coal provides more than half of the electricity in this country, as shown in Fig. 5, and in certain areas of the country accounts for about two-thirds of the fuel mix for electric power generation. Natural gas use has increased significantly in recent years, as newer generating facilities over the past decade were almost exclusively natural gas-fired because of formerly lower gas prices, lower emissions, and lower investment costs and lead times.[16] With the passage of recent legislation, including the Clean Air Interstate Rules and the Energy Policy Act of 2005, the federal government has put in place a number of incentives to encourage more use of clean, coal-fired power plants and the use of coal for synthetic natural gas and liquid transportation fuels.

CONCLUSION

Coal is a fossil fuel resource in the United States that serves an important role as a feedstock for providing electricity and fuel and/or feedstock for a variety of industries, including steel and plastics. Railroads move more than half of the coal from the mine to its final destination. Although natural gas usage has increased significantly for providing electricity over the past decade, coal remains the primary fuel for electricity production.

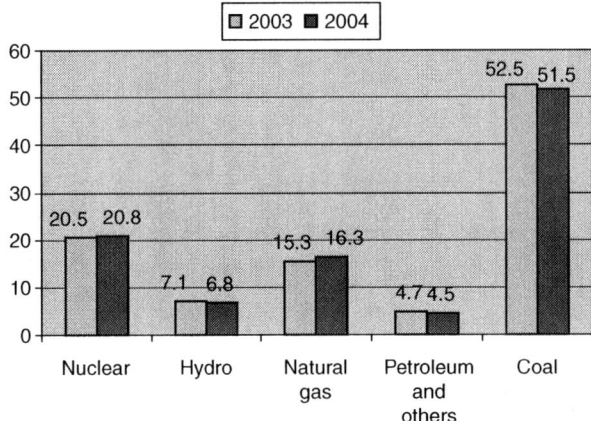

Fig. 5 Share of electric power sector net generation by energy sources 2003 and 2004.

REFERENCES

1. Energy Information Administration. U.S. Coal Reserves: 1997 Update. DOE/EIA-0529(97). Available at: http://tonto.eia.doe.gov/FTPROOT/coal/052997.pdf (accessed August 2006); Energy Information Administration. Energy Kid's Page. Coal—A Fossil Fuel. Available at: www.eia.doe.gov/kids/energyfacts/sources/non-renewable/coal.html (accessed April 2006); The Environmental Literacy Council. Coal. Available at: www.enviroliteracy.org/article.php?id=18&print=1 (accessed April 2006).
2. Energy Information Administration. Energy Kid's Page. Coal—A Fossil Fuel. Available at: www.eia.doe.gov/kids/energyfacts/sources/non-renewable/coal.html (accessed April 2006); The Environmental Literacy Council. Coal. Available at: www.enviroliteracy.org/article.php?id=18&print=1 (accessed April 2006).
3. Utah Mining Association. Types of Coal. Available at: www.utahmining.org/coaltype.htm (accessed April 2006).
4. Energy Information Administration. Coal: Glossary. Available at: www.eia.doe.gov/cnear/coal/page/gloss.html (accessed August 2006); The Environmental Literacy Council. Coal. Available at: www.enviroliteracy.org/article.php?id=18&print=1 (accessed April 2006); Texas Railroad Commission's Kids World School House. Types of Coal. Available at: http://kids.rrc.state.tx.us/school/classroom/coal/types.html (accessed April 2006); Utah Mining Association. Types of Coal. Available at: www.utahmining.org/coaltype.htm (accessed April 2006); Powder River Coal Company. Types of Coal. Available at: www.ccsd.k12.wy.us/mines/PR/CoalTypes.html (accessed April 2006); Energy Information Administration. Energy Kid's Page. Coal—A Fossil Fuel. Available at: www.eia.doe.gov/kids/energyfacts/sources/non-renewable/coal.html (accessed April 2006); Lignite. Available at: http://enl.wikipedia.org/wiki/Lignite (accessed April 2006); Sub-bituminous Coal. Available at: http://en.wikipedia.org/wiki/Sub-bituminous_coal (accessed April 2006); Bituminous Coal. Available at: http://en.wikipedia.org/wiki/Bituminous_coal (accessed April 2006); Anthracite Coal. Available at: http://en.wikipedia.org/wiki/Anthracite (accessed April 2006).
5. Energy Information Administration. Energy Kid's Page. Coal—A Fossil Fuel. Available at: www.eia.doe.gov/kids/energyfacts/sources/non-renewable/coal.html (accessed April 2006).
6. Energy Information Administration. Coal Transportation: Rates and Trends—1979–2001 (partial 2002). Available at: www.eia.doe.gov/cneaf/coal/page/trans/ratesntrends.html (accessed March 2006).
7. Singer, J.G. Ed., *Combustion: Fossil Power: A Reference Book on Fuel Burning and Steam Generation*; Combustion Engineering, Inc.: Windsor, CT, 1991; Accounting for Changes in Coal Properties When Optimizing Combustion. *Lehigh Energy Update*. 2002; 20(1). Available at: www.lehigh.edu/energy/leu/march_leu_2002.pdf (accessed August 2006); Mercury in U.S. Coal. U.S. Geological Survey Open-File Report 98-0772. Available at: http://pubs.usgs.gov/of/1998/of98-772 (accessed August 2006).
8. Energy Information Administration. Energy Kid's Page. Coal—A Fossil Fuel. Available at: www.eia.doe.gov/kids/energyfacts/sources/non-renewable/coal.html (accessed April 2006); Freme F. U.S. Coal Supply and Demand: 2004 Review. Energy Information Administration. Available at: www.eia.doe.gov/cneaf/coal/page/special/feature.html (accessed March 2006).
9. United Mine Workers of America. Underground Mining: Types of Underground Coal Mines. Available at: www.umwa.org/mining/ugtype.shtml (accessed April 2006); Energy Information Administration. Energy Kid's Page. Coal—A Fossil Fuel. Available at: www.eia.doe.gov/kids/energyfacts/sources/non-renewable/coal.html (accessed April 2006).
10. Energy Information Administration. Energy Kid's Page. Coal—A Fossil Fuel. Available at: www.eia.doe.gov/kids/energyfacts/sources/non-renewable/coal.html (accessed April 2006); United Mine Workers of America. Surface Coal Mining. Available at: www.umwa.org/mining/surmine.shtml (accessed April 2006); University of Kentucky and Kentucky Geological Survey. Methods of Mining. Available at: www.uky.edu/KGS/coal/coal_mining.htm (accessed August 2006).
11. Freme, F. *U.S. Coal Supply and Demand: 2005 Review*. Energy Information Administration. Available at: www.eia.doe.gov/cneaf/coal/page/special/feature.html (accessed August 2006). (The data for 2005 are preliminary).
12. Freme, F. *U.S. Coal Supply and Demand: 2004 Review*, Energy Information Administration. Available at: www.eia.doe.gov/cneaf/coal/page/special/feature.html (accessed March 2006).
13. Energy Information Administration. Energy Kid's Page. Coal—A Fossil Fuel. Available at: www.eia.doe.gov/kids/energyfacts/sources/non-renewable/coal.html (accessed April 2006); The Center for Land Use Interpretation. Black Mesa Coal Mine and Pipeline. Available at: http://ludb.clui.org/ex/i/AZ3134 (accessed April 2006).
14. Energy Information Administration. Coal Transportation: Rates and Trends in the United States, 1979–2001 (with supplementary data to 2002). Available at: www.eia.doe.gov/cneaf/page/trans/ratesntrends.html (accessed March 2006).
15. Freme, F. *U.S. Coal Supply and Demand: 2004 Review*. Energy Information Administration. Available at: www.eia.doe.gov/cneaf/coal/page/special/feature.html (accessed August 2006). (The data for 2005 are preliminary).
16. Energy Information Administration. Energy Kid's Page. Coal—A Fossil Fuel. Available at: www.eia.doe.gov/kids/energyfacts/sources/non-renewable/coal.html (accessed April 2006).

Coal-to-Liquid Fuels

Graham Parker
Battelle Pacific Northwest National Laboratory, U.S. Department of Energy, Richland, Washington, U.S.A.

Abstract
A chemical process used for turning coal into liquid fuels that has the potential for producing hundreds of thousands of barrels per day of hydrocarbon liquids and other byproducts—including electricity—is described. The key to converting coal to liquids is the Fischer–Tropsch (FT) process, which was invented in Germany in the 1920s. This process is used today in full-scale production plants in South Africa and it is being planned for use in plants in many other parts of world. A coal-to-liquids (CTL) industry is highly valued because of the security in using domestic sources of supply (coal) to produce hydrocarbons, in an environmentally acceptable process, that can be blended and refined into liquid fuels and transported to the end-user. In particular, FT fuels can play a significant role in providing a fuel currently used in the transportation industry and thus reducing dependence on imported petroleum and other refined transportation fuel products. This is of particular importance to the United States, which has an abundance of coal.

INTRODUCTION

A coal-to-liquids (CTL) plant is a chemical process plant that converts conventional pulverized coal to carbonaceous liquid fuels and byproduct hydrogen-based gases. These fuels are produced through a process that first converts the coal to coal–gas (or synthetic gas [syngas]) via conventional coal gasification and then converts liquids from the gas via the Fischer–Tropsch (FT) process. Depending on the coal quality and the way the plant is configured and operated, the CTL plant using the FT process can produce significant quantities of light- to mid-grade, high-value hydrocarbons along with other products such as naphtha, waxes, ammonia, hydrogen, and methane. Coal-to-liquids plants are often designed to produce ~2/3 liquid fuels and ~1/3 chemicals such as naphtha and ammonia. One of the key products from a CTL plant (that includes post-processing or refining of FT liquid products) is high-quality/low-sulfur diesel fuel.

The critical components of a CTL plant are the coal gasifier, the enrichment of the synthetic gas to increase the hydrogen/carbon monoxide ratio (H_2/CO ratio), and the selected FT process reactor. There are many options for the critical components and component configuration of a CTL plant. In particular, there are at least eight industry-proven gasifiers, primarily used for production of only pipeline-quality natural gas, and at least three commercial production FT processes.

Keywords: Fischer–Tropsch; Coal; Petroleum; Gasification; Diesel; Hydrocarbons; Catalyst; Cetane; Carbon dioxide; Sulfur.

PRODUCING LIQUIDS FROM COAL WITH THE FISCHER–TROPSCH PROCESS

The FT process was developed in the 1920s in Germany. Inventors Franz Fischer and Hans Tropsch developed a process to convert carbon monoxide (CO) and hydrogen (H) to liquid hydrocarbons using iron (Fe) and cobalt (Co) catalysts. The temperature, pressure, and catalyst determine whether a light or heavy liquid fuel is produced. During World War II, petroleum-poor but coal-rich Germany used the FT process to supply its war machine with diesel and aviation fuel after allied forces cut off petroleum imports. Germany's yearly synthetic oil production reached more than 90 million tons in 1944.

The FT process was (and still is) used to produce most of South Africa's diesel fuel during that country's isolation under apartheid. The South African company Sasol Ltd. has produced about 1.5 billion barrels of synthetic fuel from about 800 million tons of coal since 1955 and continues to supply about 28% of that nation's fuel needs from coal.[1]

A typical CTL plant configuration using the FT process is shown in Fig. 1. The FT process is comparable with a polymerization process, resulting in a distribution of chain-lengths of the products from the process. In general, the product range includes the light hydrocarbons methane (CH_4) and ethane (C_2H_6), propane (C_3H_8), butane (C_4H_{10}), gasoline (C_5H_{12}–$C_{12}H_{26}$), diesel fuel ($C_{10}H_{22}$–$C_{15}H_{32}$), and other long-chained hydrocarbons/waxes (>C_{15}). The distribution of the products depends on the FT catalyst used and the process operation conditions (temperature, pressure, and residence time).[2]

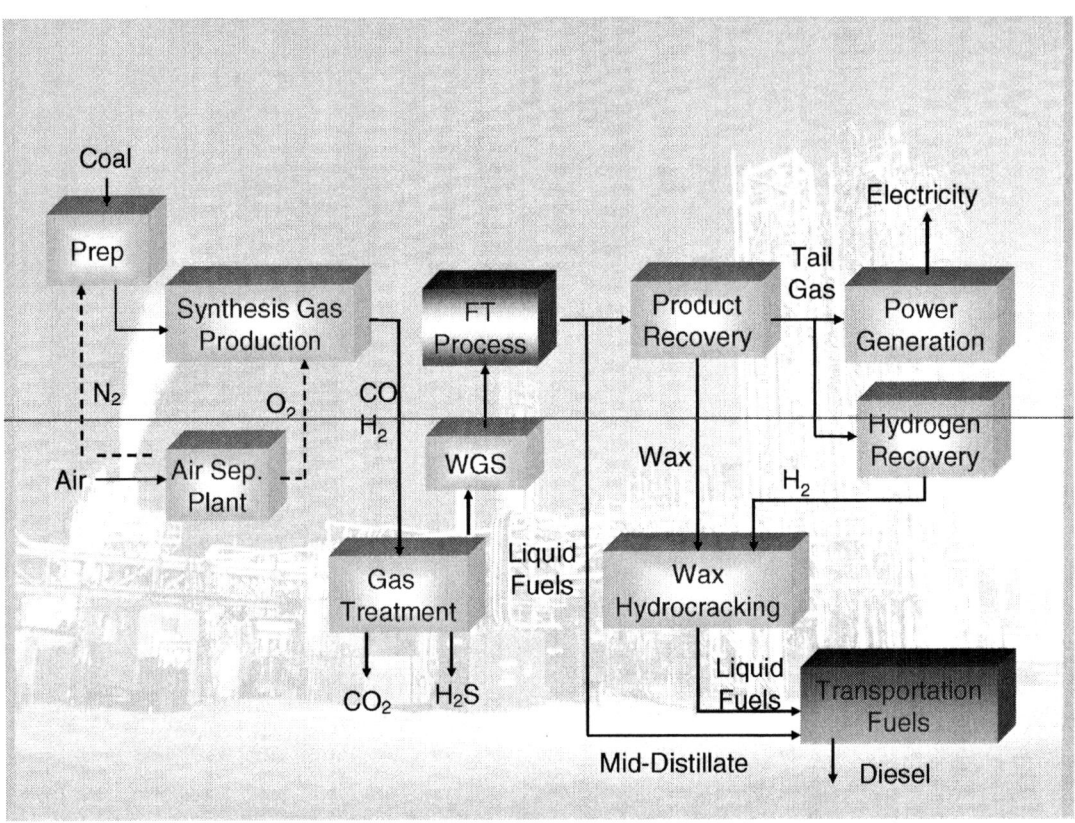

Fig. 1 Process flow diagram of a typical coal-to-liquids (CTL) plant using the Fischer–Tropsch (FT) process.

The FT process involves the use of slurry-bubble-column (slurry phase) systems using either cobalt-based (Co) or iron-based (Fe) catalysts. With these catalysts, two operating temperatures are available: low and high. For either catalyst, the FT process is exothermic. The FT reaction vessels must be actively cooled to keep them at optimal temperature for the reaction. The heat energy released by the FT process is not at a high enough temperature (200°C–300°C) to drive the production of syngas (the upstream gasification process); but it is adequate for downstream power generation if polygeneration is included in a CTL plant. (Note: polygeneration refers to the production of electricity in combustion turbines from the CTL process waste heat [primarily from the FT process] and tail gas [primarily methane] from the FT process not used to produce desired hydrocarbon products.)

Iron-based catalysts are the preferred catalysts for FT when using low CO/H_2 ratio synthesis gases derived from modern coal gasifiers. This is because in addition to reasonable FT activity, FT catalysts also possess high-water gas shift (WGS) activity. In the low temperature range, the iron catalyst can produce 50%–75% wax products. When operating at high temperatures (above 220°C), the gaseous and liquid products are highly olefinic. Another byproduct stream of the FT process is hydrogen gas. Currently, the low-temperature iron FT catalyst produces smaller quantities of products that can be used as chemicals than the high-temperature operation. A drawback with the use of Fe catalysts is their tendency to undergo attrition. This can cause fouling/plugging of downstream filters and equipment, making the separation of catalysts from the oil/wax product very difficult, if not impossible, and resulting in a steady loss of catalysts from the reactor. Iron catalysts have a higher tolerance for sulfur, are less expensive than Co catalysts, and produce more olefin products and alcohols. The lifetime of the Fe catalyst is short and in commercial installations it is generally limited to eight weeks. The Fe catalyst must then be replaced and the spent catalyst disposed of.[3]

A Co-based catalyst gives the benefit of greater management of throughput and output product selection. Co-based catalysts are often utilized because of their high FT activity, C^{+5} hydrocarbon selectivity, low WGS activity, and relatively low cost compared to Fe catalysts. The Co catalyst produces high-boiling, waxy products when operated in a low-temperature range, but attempts to operate in a high-temperature range (above about 220°C) result in the production of too much methane to be a viable option for producing liquid fuels.

Cobalt catalysts have the advantage of a higher conversion rate and a longer life (over five years); however, Co catalysts are less tolerant to sulfur and thus the upstream cleaning processes after gasification must remove most of the sulfur from the syngas. In general, the Co catalysts are more reactive for hydrogenation and therefore produce less unsaturated hydrocarbons and alcohols compared to iron catalysts. Processes have been developed to efficiently and cost effectively reactivate/regenerate and reuse a Co catalyst.

CRITICAL DRIVERS FOR A CTL INDUSTRY

There is considerable data—much of it from the oil industry—that indicates the proven reserves of crude oil and natural gas will sustain the world demand for ~100 years at current and future predicted consumption rates. There are, however, sufficient coal reserves to sustain world demand for liquid fuels for nearly 200 years. Coal is predicted to last twice as long as the combined proven crude petroleum and natural gas reserves at current usage rates. This is under a projected annual worldwide consumption of coal of 2.2 billion tons/year, or a ~1.5%/year increase.[4]

The current and future price of coal is relatively stable compared to other fossil fuels because of coal's abundance combined with the regional distribution of very large coal reserves. Currently there are about 273 billion short tons (bst) of coal reserves in North America, 173 bst in Russia and the former Soviet Union, 126 bst in China, 93 bst in India, and 90 bst in Australia.[4]

The resurgence in the interest in coal as a viable fossil fuel for direct combustion to produce electricity as well as for the production of liquid fuels is a result of the following energy dynamics:

- Steep increases and unpredictable spikes in crude oil and natural gas prices.
- A decline in domestic oil and natural gas production in those economies with high and increasing energy demands (particularly the United States, China, and India).
- Limitation in current domestic petroleum refining capacity and ability to site and build new refineries.
- The unstable political situation in the oil- and gas-rich Middle East.
- New technology developments in clean coal technologies, including coal combustion and CTL.

In addition, the production of liquid fuels from alternative and indigenous sources is addressed in the Energy Policy Act of 2005 Pub. L. 109-58, and is the linchpin of President Bush's "Advanced Energy Initiative" goal of new fuel-production technologies to replace more than 75% of our oil imports from the Middle East by 2025.

CTL PRODUCTS: DIESEL FUEL

The primary product from a CTL plant is high-quality diesel fuel. The diesel fuel from the FT process will require post-processing to make it compatible and comparable with diesel fuel derived from conventional oil refining. Typical diesel fuel characteristics from several sources are shown in Table 1.

With respect to the production of diesel fuel from a CTL plant, process conditions can be selected to produce maximum amounts of products in the diesel range. However, an even higher yield of diesel fuel

Table 1 Diesel product specifications

Property	Typical 2005 U.S. No. 2 diesel fuel	Petroleum-derived	Tar sands-derived	Fischer–Tropsch	Biodiesel
Debsity, g/cm^3	0.85	0.84	0.85	0.77	0.88
T10/T90°F	400/600	400/600	400/600	400/600	400/600
Cetane	45	48	42 (45[a])	75	56
Sulfur, ppm	400	12	12 (400 today)	3	12 (25 today)
Aromatics, %	35	32	20	0	0
Paraffins, %	35	37	20	100	0
	No. 2 diesel fuel	No. 2 diesel fuel	No. 2 diesel fuel	No. 2 diesel fuel or component	No. 2 diesel fuel or component
		Commercially available today	Commercially available today	Commercially available today	Commercially available today

[a]With Cetane improver.

can be achieved when the FT synthesis is optimized toward production of wax. Subsequently, the wax can be selectively post-processed (hydrocracking similar to processes in a refinery) to yield predominantly diesel fuel.

The resulting FT fuels from either a Fe or Co catalyst-based process are cleaner-burning than similar fuels from crude oil refining because many of the impurities are removed during the synthesis. The resulting fuels are colorless, odorless, and low in toxicity. Fischer–Tropsch fuels can be used in conventional diesel engines and have improved combustion, which reduces emissions. Fischer–Tropsch fuels have a higher cetane index and less sulfur (<5 ppm), nitrogen oxide, carbon monoxide and particulate matter emissions than petroleum fuels. In addition, the entire coal-to-liquid process is designed to remove other containments typical in fossil fuel combustion, including sulfur and carbon dioxide.

Diesel fuels from a CTL plant are being used as both a neat (a neat CTL diesel fuel is one that is introduced into the distribution system and used directly in a combustion engine without further processing or blending) fuel and as a blending fuel with conventional diesel fuel produced from petroleum processing in a refinery. The blended fuel can help refiners meet current and future sulfur standards as well as stretch the diesel fuel manufactured from conventional petroleum sources. (Note: from 2004 to 2006, the governments in some parts of Japan, Australia, and the European Union [EU] have limited sulfur in highway diesel to no more than 50 ppm. In 2006, the maximum falls to 15 ppm in some parts of the United States. Japan and the EU are expected to further restrict sulfur content to as low as 10 ppm. The cost for refiners to meet these sulfur limitations can range from $1.00 to as much as $3.00 per barrel, thus making the low sulfur diesel from a CTL process a valuable product for blending.)

The market for such fuel would include the domestic transportation industry, the department of defense (DOD), and potentially the agriculture industry. Additional post-FT processing and blending would be required and could produce other fuels, such as JP-4 and JP-8, which could be marketed to the domestic airline industry and DOD. The DOD alone requires over 3,00,000 barrels per day (bpd) of domestic diesel fuel for its operations and desires to supply this requirement with a domestic and secure resource. The Rocky Mountain (Western) states' usage will require nearly 2,00,000 bpd by 2010 and thus this is another large market. Still another large market for high-quality (low sulfur California air resources board [CARB]) diesel fuel will be California, where predictions show the state consuming over 3 billion gallons/year of diesel by 2010. (Note: current specifications for CARB diesel for use in California will be more stringent than the 2007 U.S. Environmental Protection Agency [EPA] diesel. Fischer–Tropsch diesel will be able to meet these specifications.)

Other products from a CTL plant that can be used in a combustion turbine to produce electricity and steam include elemental sulfur, naphtha, waxes, and tail gas (primarily C_4–C_6).

FT PROCESSES PROVIDERS

The Sasol Ltd. South Africa FT technology is the most mature full-scale technology in operation today, having been used in South Africa since 1955 and proposed to be used in several international CTL projects. The Sasol II and III CTL plants currently have the capacity to produce 1,50,000 bpd crude oil equivalent liquids.

There are currently three U.S.-based companies with FT technology that can be used in a CTL plant. Each company uses a proprietary technology. However, because there are no full-scale CTL plants in operation in the United States today, none of these three FT technologies have been deployed in a production-scale plant. These include:

Syntroleum Corporation
Tulsa, Oklahoma 74107
www.syntroleum.com

Syntroleum develops, owns, and licenses a proprietary process for converting natural gas to synthetic liquid hydrocarbons, known as gas-to-liquids (GTL) technology, as well as CTL technologies. For Syntroleum's GTL projects, they have executed an agreement with ExxonMobile to use ExxonMobile GTL patents to produce and sell fuels from natural gas, coal, or other carbonaceous substances.

Syntroleum is currently developing CTL projects with a recently announced project in Queensland, Australia. The CTL project will include removal and sequestration of CO_2 and the production of FT diesel.

Renergy Technologies Ltd. (Rentech)
Denver, Colorado 80202
www.rentech.com

Rentech began developing FT-fuel technology in 1981. It has designed FT plants in the range of 2000–40,000 bpd for potential projects in Bolivia, Indonesia, and the United States.

In June 2004, Rentech announced that it had entered into a contract with the Wyoming Business Council to perform engineering design and economic study using Rentech's patented and proprietary FT technology. The analysis evaluates the economic viability of constructing a mine-mouth plant capable of producing

10,000–12,000 bpd of ultra-low sulfur FT diesel for distribution in Wyoming, California, and other Western states. It is estimated that the facility will require about 3 million tons per year of Wyoming Powder River Basin coal for every 10,000 bpd of fuels production. The study also considered various levels of cogeneration of electric power for sale to the local transmission grid.[5]

Fischer–Tropsch projects based on coal are currently under development in Illinois, Kentucky, and Mississippi. The CTL plant in Kentucky is targeted at 57,000 bpd. In Illinois, Rentech has completed a study to convert an existing natural-gas-based ammonia plant into an integrated plant producing ammonia (900–950 tons/day), 1800–2000 bpd FT fuels, and ~10 MW exported electric power using high-sulfur Illinois coal as its feedstock. In Mississippi, Rentech has entered into an agreement with the Adams County Board of Supervisors to negotiate a contract under which Rentech would lease a site for a 10,000-bpd CTL plant producing primarily diesel fuel by 2010.

Headwaters technology innovation group (HTIG), Inc.
Lawrenceville, NJ 08648
www.htigrp.com

Headwaters technology innovation group is a wholly-owned subsidiary of Headwaters Incorporated (www.headwaters.com) that promotes and licenses technology for CTL projects using FT technology. Headwaters technology innovation group has developed an iron-based catalyst that is ideally suited for processing coal-derived syngas (synthetic gas) into ultra-clean liquid fuels. Headwaters technology innovation group has patents covering catalyst manufacturing, slurry phase reactor design and operation, production of FT liquids for fuel and chemical feedstocks, and the co-production of ammonia and FT liquids.

CURRENT AND PLANNED CTL INDUSTRY

To date, there is limited experience in full-scale (tens of thousands of bpd) production of liquid fuels from coal. Most experience in full-scale CTL plant operation is with the Sasol Ltd. operation.

A number of pilot-scale (up to hundreds of bpd) plants have been constructed and operated in the United States. These include a 35-bpd plant operating from 1975 to 1985 in Pennsylvania, a 35-bpd pilot plant operated by Air Products and Chemicals in 1984, a 230-bpd plant operated in Colorado by Rentech, a 70-bpd plant in Washington operated in 1998 by Syntroleum/ARCO, and a 2-bpd pilot plant in Oklahoma operated by Syntroleum from 1989 to 1990.

The technology challenge in a CTL plant focuses on the FT process and particularly the FT process fed with gasified U.S. coals. Experience in South Africa is with South African coal with unique coal qualities and with natural gas as a feedstock GTL. There is a dearth of experience in full-scale operation of a CTL plant in the United States with any FT process, and with the pilot/demonstration CTL plants, there has been limited testing of the qualities and performance of the fuels.

In addition to the projects noted above, there are a number of planned CTL projects under development around the world, as discussed below.

North America

A number of states are currently either considering or actively developing CTL plants using the FT process, including Kentucky, North Dakota, Mississippi, Missouri, Montana, Ohio, Pennsylvania, West Virginia, and Wyoming. Most of these projects are being developed as consortiums or partnerships of coal companies, gasification/FT suppliers, architect/engineering firms, universities, energy technology centers, and state governments.

Australia

The Australian Power and Energy Limited/Victorian Power and Liquids Project (APEL/VPLP) CTL project was planned in Australia as a joint venture between Australian Power and Energy Limited and Syntroleum. This project includes coproduction of power and hydrocarbon liquids from brown coal in the Labtrobe Valley in the state of Victoria. The initial phase of development envisions a 52,000-bpd plant with CO_2 capture and sequestration via subsurface injection.

Asia

Sasol is under discussions with China to build several CTL plants and could also take equity stakes of up to 50% in two proposed Chinese CTL projects. The Chinese proposal is part of a joint venture between Foster-Wheeler/Sasol and Huanqui for two facilities producing 80,000 bbl of liquids per day per site at the Ningxia autonomous region and the Shaanxi province, both in the coal-rich western part of China. An additional CTL/FT demonstration plant of unknown size is being planned by HTIG in Mongolia, China. There are also plans for an 80,000-bpd plant in Indonesia in partnership with Sasol.

India and Pakistan

There is considerable interest from coal companies in India and Pakistan in the Sasol CTL technology. Significant coal reserves in India and Pakistan could help to reduce dependence on imported crude oil.

SITING AND OPERATING CTL PLANTS

The development pathway for CTL plants is uncertain, given the myriad of choices of proven full-scale gasification processes that must be integrated with unproven (on a full-scale) FT processes other than the Sasol Ltd. Fischer–Tropsch process—which has only been used in a fullscale plant with South African coal. Although a CTL plant has the potential for producing tens of thousands of barrels of clean fuel (and other byproducts) in an environmentally-friendly process, there are a number of engineering, infrastructure, and institutional issues that need to be resolved before a full-scale plant is viable in the United States. These include but are not limited to improved materials/catalyst performance and reaction mechanisms for the FT process; permitting, siting, and regulating a first-of-a-kind plant that is neither a conventional coal-fired power plant nor a conventional chemical plant nor a conventional oil refinery; water use and water treatment requirements; securing a coal supply contract under high-demand conditions for coal for power plants; securing an FT-based fuel outtake contract with both a floor and ceiling price; cost-effective carbon capture and other emissions treatment strategies; and optimizing the plant output products that include liquid fuels, naphtha (feedstock for chemical production), ammonia (for fertilizer production), and electricity.

COAL FUEL SUPPLY

In the United States, there are vast deposits of coal—deposits more extensive than those of natural gas and petroleum, the other major fossil fuels. Identified resources include the demonstrated reserve base (DRB), which is comprised of coal resources that have been mapped within specified levels of reliability and accuracy and that occur in coal beds meeting minimum criteria for thickness and depth from the surface that may support economic mining under current technologies.

A typical CTL plant with the capacity of 10,000 bpd would require 10,000–15,000 tons of coal/day. Such a plant operating at 90+% capacity would consume 3.5–5 million tons of coal/year and produce over 3 million barrels of diesel fuels plus other marketable byproducts. It is anticipated that multiple plants could increase production capacity to 1,00,000 bpd by 2012, 1 million bpd by 2025, and 2–3 million bpd by 2035, and that the coal mining and coal transportation industry would be able to accommodate such production levels.

There are three major coal-producing regions in the United States: Appalachian (primarily Ohio, West Virginia, Kentucky, Tennessee, and Pennsylvania), Interior (which includes the Gulf Region and the Illinois Basin), and Western (which includes the Powder River Basin/Colorado Plateau and Northern Great Plains).[6]

U.S. coal production in 2004 totaled 1112.1 million short tons and was divided among the regions as follows:

- Appalachian: 389.9 short tons
- Western: 575.2 short tons
- Interior: 146.0 short tons

Of the total coal produced in 2004, 1.016 short tons (91%) were used to generate electricity.[4]

The actual proportion of coal resources that can be mined and recovered economically from undisturbed deposits varies from less than 40% in some underground mines to more than 90% at some surface mines. In some underground mines, much of the coal is left untouched as pillars, required to prevent surface collapse. Adverse geologic features such as folding, faulting, and inter-layered rock strata limit the amount of coal that can be recovered at some underground and surface mines.

COAL MINING AND TRANSPORT

There will be a significant increase in coal mining once the CTL industry matures and production reaches estimated full-scale operation by 2035. An estimated additional 2.5–4 million tons of coal per day will be required to supply CTL plants operational by 2035. A mature CTL industry would require an additional 2%–3% production increase over 2004 levels and this can be adequately handled by the coal industry. For example, current production in the Western Region's Powder River Basin is over 350 million tons/year, thus this increased demand would add 8%–12% additional demand if all the additional demand for coal were supplied by this region.[4] This additional demand can readily be provided by the existing mines, thus no new mines would need to be permitted.

There is, however, a current limitation to the amount of coal that can be transported over existing rail lines. Transportation of coal from the mine to the consumer continues to be an issue for the industry. The majority of coal in the United States is moved by railroads exclusively or in tandem with another method of transportation.

A nearby high-speed rail line (and connecting rail spur to the plant) would be required for coal transport capable of transporting ~100 coal cars/day for a 10,000–15,000-bpd plant. Petroleum pipelines are the preferred mode of moving FT diesel fuel; however, depending on the location of the CTL plant, rail transportation may also be the best alternative for transporting the diesel and possibly other plant byproducts to a refinery or end-user. A train with 40–45 tank cars/day would be required to transport the diesel fuel from a 10,000-bpd plant. Additional transportation modes (tanker truck, tanker car, or pipeline) would be required to haul away other products such as sulfur (truck transport), carbon for sequestration (pipeline transport), ammonia (road or rail transport), and possibly naphtha as a blending stock for gasoline refining (road or pipeline transport). Environmental issues related to increased coal transport (noise, dust) would need to be addressed by each state and community through which the trains would pass.

AIR QUALITY

A CTL plant emits far fewer criteria pollutants into the atmosphere than even the best-controlled and most efficient coal combustion power plant. Sulfur (as H_2S) and mercury (as elemental mercury and captured in impregnated activated carbon absorbent) are removed during the gasification. More than 99% of the sulfur and mercury are removed prior to producing liquids in the FT process.

A significant amount of carbon can also be captured as CO_2, with the percentage of carbon captured depending on how the plant is operated as well as the economics. Carbon can be further processed for sequestration and the sulfur is converted to elemental sulfur for dry disposal or sale. Carbon dioxide can also be used for enhanced oil recovery or for coal-bed methane extraction.

One of the key features of a CTL plant is the potential for substantial carbon capture and sequestration. This can make CTL plants environmentally preferable to combustion plants. Typical strategies for sequestering CO_2 include physical trapping, hydrodynamic trapping, solubility trapping, and mineral trapping. These sequestering and use of CO_2 strategies are illustrated in Fig. 2.

There are additional emissions of oxides of nitrogen from the combustion process used to generate electricity, chlorine, and particulates from combustion and coal transport, handling, and processing (pulverization). All of these emissions can be treated using the best available control technology (BACT). The goal of control technology for a CTL plant is to reduce the emissions to at least the level of those emitted by a conventional coal-fired power plant using the BACT to meet current air quality standards. Overall, a CTL plant allows for easier and more effective control of criteria pollutants—and additionally CO_2—compared to today's most efficient and controlled coal-fired power plants.

The only significant air quality issue in the siting and operation of a CTL plant is the potential impact the plant may have on Class I air-sheds such as those in national parks and other designated Class I areas. This would be addressed during the permitting process via air quality dispersion modeling.

WATER QUALITY

A CTL plant using high moisture content coal such as lignite (30%–50% water content) or sub-bituminous (10%–30% moisture content) will likely be a net water producer, depending on whether or not a dry or slurry feed is used for the gasifier (upstream of the FT process) and whether or not a significant amount of (excess) power is produced requiring a cooling tower. Under the plant design scenario using 30% moisture content coal, a net production of 100–200 gal/min would be discharged for a 10,000-bpd liquids plant. This water would require treatment and disposal in surface or underground wells.

Use of bituminous coal with a water content of 5%–10% or anthracite coal with a water content of <5% would likely require a water supply of 400–600 gal/min for a 10,000-bpd plant.

One achievable goal of a CTL plant is to design the plant to be "water-neutral," that is, near zero discharge and without substantial process water supply required. This will depend on the coal and specific processes used as well as the product mix (liquids and electricity generated).

LAND

The land requirement for a CTL plant is approximately 8–10 acres for a 10,000-bpd plant. This includes land for a rail spur, six days of coal storage, coal handling, the CTL plant itself, chemical and water treatment, post-FT fuel processing, electrical generation, water treatment, storage tanks, and auxiliary support equipment.

PERMITTING

The CTL plant is expected to produce discharges at the same or lower levels as the best coal-fired integrated (coal) gasification combined cycle (IGCC) power plant. Thus,

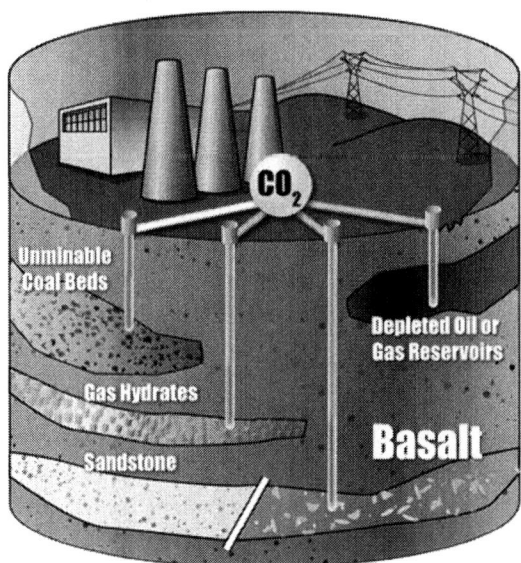

Fig. 2 Illustration of methods of sequestering and using CO_2 captured from a coal-to-liquids (CTL) plant.
Source: From Pacific Northwest National Laboratory.

the permitting process should not be onerous. Permitting would include addressing state and federal (Environmental Protection Agency) requirements for both a combustion power plant (power plant siting)—given electricity is generated on-site in a combustion turbine, and a chemical plant (industrial siting)—given the gasification and FT process are chemical processes. This includes new source performance standards and Clean Air Act requirements.

The significant solid waste to be disposed of is nontoxic, nonleachable slag (~140 kg/bbl) waste from the gasification process, and all Resource Conservation and Recovery Act (RCRA) requirements would need to be met. There are strategies that have been developed and are in use in the coal gasification industry to convert the slag into a usable and thus marketable product as well as to reuse a portion of the higher heating content slag in the gasification process. Thus, the total amount of slag can be reduced considerably if these processes are deployed.

Additional permitting would be required for (1) the construction of a rail spur; (2) coal handling/pulverization processes (fugitive dust); (3) electrical permitting for electrical supply requirements; (4) potable water supply; (5) water/wastewater discharge during construction and wastewater from the process; (6) construction and operating permits; and (7) permits related to cultural resources, native lands, National Environmental Policy Act of 1969 (NEPA), and the Federal Land and Policy Management Act.

EMPLOYMENT

A typical 10,000-bpd CTL plant would require ~1000 construction jobs with 150–200 permanent staff to operate the plant. There would be some marginal increase in employment in the coal mining and coal/diesel transportation industry, but not nearly as significant as in the plant operations. A rough estimate is an increase of 5–10 coal and transportation industry jobs for every 10,000-bpd plant.

NATURAL GAS

The CTL plant would not require significant quantities of natural gas, and this would primarily be during startup. At full production, the plant would be a net hydrocarbon producer including tail gas and naphtha from the FT process and possibly bypass gas from the gasifier. Most of the tail gas would be used to feed a combustion turbine to produce electricity for use by the plant and for the sale of any excess electricity.

ELECTRICITY

A transmission line and substation would be required to supply electricity for plant startup and an emergency generator (diesel-fueled) would be required for unanticipated plant shutdown. The CTL plant would be a net electricity generator (after use by the plant itself), using an efficient combined cycle combustion turbine generation plant ranging from 100 to 200+ MW, depending on how the plant is configured.

CONCLUSIONS

A CTL plant would be viable for countries with large reserves of relatively low-cost and readily accessible coal in grades that can be gasifiable. Stranded coal (due to location of coal quality) that cannot be easily monetized in other ways is also a viable source. The market for the liquid products should be nearby, and thus CTL plants would be most economically sited in major energy-consuming countries. In addition, there needs to be an ability to capture and sequester or otherwise use the CO_2 in order to minimize the environmental impact of the siting and operation of the plant.

Other conditions require crude oil prices above $40–$50/bbl (2005 dollars) combined with coal prices at $20–$50/ton (2005 dollars), favorable financing terms, and a short permitting process. The resulting FT fuels (diesel) must be compatible with the current diesel fuels used in engines and with a cost that is able to compete in the marketplace with diesel fuels produced from petroleum (or natural gas via a GTL process) feedstocks. If a domestic energy supply and security from using that supply are critical, there is a potential premium that could be applied to FT diesel fuel.

There remain significant challenges to a CTL industry, including the following:

- A CTL plant is highly capital-intensive and considered a risk to potential investors due to the few large-scale CTL plants operating worldwide.
- General inexperience in industry with the design, construction, and operation of a CTL plant.
- Few choices of large-scale demonstrated FT technologies and suppliers. Currently there is only one proven full-scale FT technology (Sasol) that has been used with syngas produced from coal.
- The historic volatility and uncertainty in the oil and natural gas industry makes it difficult to predict future oil and natural gas prices, and thus the economics of a CTL plant.
- The "true" cost of producing FT diesel fuel from a CTL plant and whether or not industry will pay a premium for that fuel without government guarantees or subsidies.
- The competition for coal (supply and transport) used for power generation by the utility industry potentially driving up the price of delivered coal for a CTL plant to the point of it not being economically viable.

- The dearth of industry/independent data on the long-term use of CTL/FT diesel fuel used in conventional (today's) diesel engines, particularly in the U.S. transportation industry.
- Environmental issues related to carbon capture and sequestration as well as inexperience with siting and operating a full-scale CTL plant, including issues related to "dirty coal" and the difficulty of siting of refineries.

REFERENCES

1. www.sasol.com (accessed on March 2006).
2. http://www.ecn.nl/biomassa/research/poly/ftsynthesis.en.html (accessed on March 2006).
3. www.clean-energy.us (accessed on March 2006).
4. www.eia.gov (accessed on March 2006).
5. Company mulls coal-to-diesel plant. *Casper Star Tribune*; 12 June 2004.
6. http://energy.er.usgs.gov (accessed on March 2006).

Cold Air Retrofit: Case Study*

James P. Waltz
Energy Resource Associates, Inc., Livermore, California, U.S.A.

Abstract

The first-cost focus of new construction can often result in buildings that do not work—particularly the Heating Ventilation and Air Conditioning (HVAC) systems. This article presents an innovative cold air retrofit that corrected designed-in HVAC inadequacies in an office building. By utilizing cold air, the retrofit overcame insufficient airflow and insufficient cooling capacity at the air-handling units, thereby avoiding a massive—and very disruptive—retrofit of a fully occupied office building. The resulting retrofit cost less than half of the conventional alternatives and was easily implemented during weekend hours. It resulted in a building that actually worked for the first time since its original construction some 15 years ago. The author's more than two decades of experience in restoration and remediation of existing buildings provides some valuable insight into how to creatively and cost effectively fix nagging comfort problems in existing buildings. The author additionally provides some reflection on the energy efficiency impact of the project and energy related implications for new building designers.

INTRODUCTION

In reading this article, it is important to realize that the world of retrofit is a poorly understood and completely unique niche of the building construction industry. We find that most building owners and most design professionals don't understand this. By and large, traditional design professionals, who grew up in the world of new construction, are ill equipped to face the constraints of fixing problems in existing buildings—short of wholesale replacement of systems (which gets us right back to "clean sheet" or new construction design, right?). In the world of retrofit, all those little problems that were resolved in the field by the original builders, and all those remodels and modifications need to be identified and dealt with by the retrofit engineer. Frequently, whole building testing is necessary to identify the cause of the complaints. In addition, retrofit frequently requires that the building be modified while the building is fully occupied, meaning that working conditions are difficult, and major disruption to the occupants cannot be allowed. We frequently liken it to performing a heart transplant on a marathon runner—during a marathon. The project described herein is just such a project.

THE SITUATION

We were called into the project by the service contractor of the building, who was trying to figure out how to help the

---
*This entry originally appeared as "Making Buildings Work: A Case Study of an Innovative Cold Air Retrofit" in *Energy Engineering*, Vol. 100, No. 6, 2003. Reprinted with permission from AEE/Fairmont Press.

Keywords: Cooling capacity; Retrofits; Case studies; Air flow; Redesign.

owners to keep their tenants happy. They weren't very happy, as the building was uncomfortably warm nearly all year long in its northern California climate. Only during the coldest months of the year was the building comfortable. Just as in our expert testimony work, we set about to ferret out the source of the problem. What we learned was that the original designer made some fundamental conceptual errors in determining the operating parameters of the air handling equipment, which effectively resulted in undersizing of both the airflow and the cooling coils. This was in spite of the fact that he had done a good job of estimating the cooling needs of the building. The problem wasn't in the capacity of the chiller, or in the apparent capacity of the air-handling units. There was a problem, though, in the performance of the air-handling units.

There are a couple of aspects of load calculations that, as the carton character Dilbert says about nuclear energy, "can be used for good or evil." Those aspects have to do with space loads versus system loads. Astute HVAC system designers are very careful when considering these loads, realizing that any cooling load that can be kept out of the occupied space allows the designer to reduce the supply air quantity needed to cool the space, and in turn allows the use of a smaller air-handling unit. Seems pretty obvious, right?

Well, in this case the designer assumed that 100% of the heat from the lights would go into the return air instead of the space. After all, he was assuming that return-air-troffer lighting fixtures would be used, and therefore all the heat from the lights would go into the return air passing through the fixtures. On the surface this seems plausible. However, certain fixture manufacturers actually document the percent of the total heat from a fluorescent fixture that is transferred into the air stream. The highest value we've seen so far is about 30%, and we think even that is a bit optimistic. When you consider that the lighting system heat gain can contribute as much as 40%–60% of the total heat gain in an occupied space, this had a dramatic effect on the calculated supply air cfm. Add to this the fact that the HVAC system was designed for a "shell" building and the eventual tenant build-out did not employ return air troffer lighting fixtures, you can start to get an idea of how much trouble this building was in. But there's more.

The "more" is the other effect of the designer's assumption about the space loads—its effect on the load system experiences. You see if you assume that 100% of the heat from the lights goes into the return air, instead of a return air temperature of, say 76°C, you will calculate a return air temperature of more like 86°C. This means that when combined with a fairly high ambient design temperature, the mixed air temperature will calculate out to about 88°C, instead of a more correct value like 78°C. This only becomes a problem when selecting a cooling coil for your (already undersized!) air-handling unit. Since there will be more heat transferred from 88°C air to 45°C chilled water than from 78°C air (total temperature difference is now thought to be 88–45 or 43°C rather than 78–45 or 33°C), it will appear that you can get all the cooling done that you need with a pretty small coil (i.e., fewer rows and/or fewer fins per inch). Indeed, such an undersized coil was selected by the system designer

The net–net of all the above is that by making one "fatally" wrong assumption, the system designer put into the building an air-side system that could never cool the building—and indeed it didn't.

SOLVING THE "PUZZLE"

Once we understood the root of the problem, we had to face the question of what to do about it.

The immediately obvious solution was to yank out the air handling units and replace them—what we would call the "traditional" approach. After all, this would correct the fundamental error that was made in the first place. The problems with this approach were, not surprisingly, many fold, and included:

- The air handling units were located in interior mechanical rooms in the "core" area of each floor, and replacing them would require knocking out walls and seriously disrupting the occupants and operations of the building.

- Increasing the horsepower of the air handling units would require significant cost for electrical work as all the air handling units were fed electrical power from the basement and the entire conduit and conductor riser would need to be replaced, as it had no excess capacity.
- The mechanical rooms were very cramped and there really was no room at all in them for larger air handling units. This would require re-configuring the floor plan layout of the "core," another very expensive proposition (and likely not really feasible).

Recognizing that a more traditional approach really did not constitute a suitable solution for this problem (and was likely the reason the problem had gone unresolved for 15 years), Energy Resonance Associates (ERA) set about to "re-engineer" the Heating, Ventilation and Air Conditioning (HVAC) system from the inside out, assuming that the air handling units themselves could not be replaced, nor could their fan horsepower be increased (due to the limitations of the building's power distribution system). Grinding away with a computerized coil selection program and rethinking other parts of the HVAC system, we determined that the air handling units could be made to perform by:

- Replacing the existing 4-row chilled water coils with 8-row coils of equal air pressure drop (examination of factory certified dimension drawings confirmed that they would fit in the air handling units).
- Increasing the chilled water flow through the coils (feasible with a much higher horsepower pump, and within the allowable flow rate for the chiller—and requiring more than twice the original horsepower to achieve a 30% increase in flow).
- Reducing the chilled water supply temperature (from 45°C to 40°C, also with the allowable operating parameters for the chiller).
- Installing new air handling unit temperature controls (to reset the planned very-low supply air temperature upwards during cool weather, else "cold" complaints would replace the prior "hot" complaints).

Without negatively impacting the air handling systems' air supply rate, the new system would be capable of supplying 46°C- air, thereby produce the actual cooling needed to satisfy the occupied space. As shown in the table below, some pretty interesting results can be achieved by optimizing the coil selection in particular!

Description	ROWS/FPI	EDB/EWB	LDB/LWB	EWT/LWT	GPM	APD	MBH
Existing	4/12	74/60	53.5/51.8	45/58.5	30	0.69	203
Retrofit	8/9	74/57	45.8/44.6	40/54.4	39	0.70	281

MAKING THE FIX

Upon completion of the study, ERA was engaged to prepare final installation documents. This work was performed in collaboration with the owner's selected contractor so as to achieve maximum integration of design concepts and the contractor's working knowledge of the building (the contractor had the service contract for the building). Final selection of equipment was made, simplified installation drawings were prepared, and the project installed and put into operation over a 90-day period, including startup. No tenant disruption was caused during the installation (which would have been the case had the conventional approach of replacing the air handling units been followed). Upon completion of the project, the building's HVAC systems provided comfort for the first time in the 15-year life of the building! The utterly prosaic business of HVAC engineering doesn't get any more exciting than this.

SOME INTERESTING CONCLUSIONS

One of the lessons that can be learned from this project is that the age-old tradition of linking engineering fees to construction cost—our traditional way of paying design professionals—would not have allowed this project to take place. After all, it took a lot of engineering to avoid spending money. So engineering fees went up, and construction costs went down, making the engineer's fees look "large" as a percent of construction costs. Many building owners would insist that less money be spent on engineering—with the result that the engineer is forced to get his eraser out to create a "clean sheet of paper" and do a very simple design, that doubles or triples construction costs. Voila! The engineer's fees look "small" as a percent of construction. Building owners, take heed.

Another, perhaps more technical lesson to be learned is that by understanding the essential nature of the engineering problem being faced, it is often possible to re-engineer a system from the inside out and make it work, even when it seems impossible. Design engineers, take heed.

A final lesson for new building HVAC designers is that if you want to build a little "safety" into your HVAC system, selecting a cooling coil with more rows (and perhaps a few less fins per inch) is really, really cheap "insurance."

Readers may have noticed that this retrofit would likely have the effect of increasing the energy use of this building. Our charter from the building owner on this project was to make the building work. They were not at all interested in energy conservation—much to the contrary. The truth is, even in today's energy sensitive environment, making buildings work, i.e., having them provide the function they were intended to provide (a comfortable and productive work environment), is equally, if not more important, than saving a few dollars on the utility bill (and this coming from an award-winning energy engineer).

For energy engineers, the lesson is that cold air works—and offers some interesting energy saving opportunities. If you were designing this building from scratch, using a conventional design would have required a larger (and probably more powerful) air handling unit, so cold air would have saved a lot of air circulation energy. Since the fans run whenever the building is occupied (or even longer), these savings would be dramatic. While we ran the chiller at a colder evaporator temperature, in a new building design the cooling tower could have been oversized (at relatively minimal cost) to compensate and keep the total chiller "lift" (which is what you pay for in terms of chiller power) the same as a conventional design, or even better. In this retrofit we had to increase pump power—rather dramatically. In a new design, a nominal (and relatively cheap) oversizing of the piping could have been done, and the system configured for variable flow (see our contracting business article on our web site for this) and the pumping power kept to a minimum as well. Finally, an energy engineer "worth his salt" would include variable speed drives on the fans and a digital control system to precisely reset supply air temperature (to minimize reheating) and manage the operation of the chilled water system and optimize run hours of the entire HVAC system.

Combined Heat and Power (CHP): Integration with Industrial Processes

James A. Clark
Clark Energy, Inc., Broomall, Pennsylvania, U.S.A.

Abstract
This entry discusses the integration of combined heat and power (CHP), otherwise known as cogeneration, with industrial processes. It builds on other entries in this encyclopedia that discuss the basics of CHP or cogeneration.

INTRODUCTION

This section discusses the integration of CHP, also referred to as cogeneration, with industrial processes. Topics discussed in this section include:

- Cogeneration unit location
- Industrial electric power systems
- Generator voltage selection
- Synchronous vs. induction generators
- Industrial thermal considerations
- Industrial process examples utilizing unique solutions
- Exhaust gas condensation solutions
- Cogeneration and interruptible fuel rates
- Future considerations

COGENERATION UNIT LOCATION

The location of the cogeneration unit is usually a function of the location of the following:

- Main electric switchgear
- Large electric branch circuit switchgear
- Natural gas lines or oil storage tank
- Thermal processes
- Exhaust stack
- Ancillary services, such as power, water, and sewer
- Utility relay and control location

A location that minimizes the cost of electric lines, piping, and other components required to supply the above services is the optimal solution. Industrial processes and branch circuit locations differentiate the location decision from nonindustrial sites. The cost of these components can have a substantial impact on the project cost.

Keywords: CHP; Cogeneration; Industrial; Processes; Heat recovery; Energy recovery; Interruptible gas.

INDUSTRIAL ELECTRIC POWER SYSTEMS

Industrial electric power systems typically involve multiple levels of voltage reduction through multiple levels of electric transformers. Most of these transformers are owned by the customer. By contrast, small commercial and residential customers are typically provided with one voltage supplied by a utility-owned transformer.

GENERATOR VOLTAGE

The generator voltage depends on the voltage of the industrial customer and the generator size. Manufacturers supply cogeneration units in multiple voltage sizes; however, the size range is generally limited to practical voltages. Because the size of the generator is inversely proportional to the voltage, there is a limitation to the minimum voltage based on the size and cost of the resultant generator.

Generator voltage should generally be matched to either the voltage after the main transformer or the voltage of the branch circuit, where most of the generator output will be utilized. In this manner, the electricity produced will travel through the least amount of transformers. Each time electricity travels through an electric transformer, approximately 2% or more of the electricity is given to transformer losses. When calculating the benefits of electricity production, these losses (or avoidance of these losses) should be considered. In nonindustrial applications, the cogeneration unit is typically installed at the electric service entrance. In industrial applications, the cogeneration unit can be installed in one of the branch circuits.

SYNCHRONOUS VS. INDUCTION GENERATORS

Generators are essentially electric motors in reverse; rather than using electricity, a generator produces electricity. The technology behind both can easily be described as an electric magnet spinning in a casing surrounded by wires.

A prime difference between synchronous vs. induction generators is that the synchronous generator is self-excited. This means that the power for the magnet is supplied by the generator and its control system. By contrast, an induction generator is excited by the utility. When utility power is lost, the loss of excitation power will theoretically shut down the induction generator. However, there is a chance that other electric system components could provide the excitation required to start the generator. Therefore, in addition to the electric utility engineers, a competent professional engineer should approve the design of the cogeneration installation.

Cogeneration systems require protective relays on the power system that will shut down the cogeneration installation in the event of a power interruption. They are not designed to run in power outages. They must be shut down or utility personnel will be endangered when they try to restore power to a nonpowered electric line. Many utilities have less stringent protective relay requirements for induction systems for the reasons explained above. The extra costs, however, can be prohibitive. It is important for owners to determine the costs of protective relays and other interconnection requirements before commencing construction. Many owners have been surprised after the fact and had to spend additional money to correct deficiencies.

Power factor is another consideratcion in the selection of induction vs. synchronous generators. Because induction generators use utility reactive power to excite the generator, the site power factor can degrade significantly. By using more reactive power while simultaneously reducing real (kW) power, there is a double effect on power factor. Why should an owner be concerned about power factor? Because many utilities measure and charge industrial users for reactive power and/or low power factor.

INDUSTRIAL THERMAL CONSIDERATIONS

Most cogeneration installations depend on almost full utilization of the thermal output of the plant. In order to utilize the full thermal output, the thermal usages and processes must be considered carefully. The thermal output each and every hour, as well as the temperature of the thermal output as it compares to the temperature of the industrial processes, must be examined.

Industrial thermal considerations include:

- Steam vs. hot water
- The temperature of cogeneration thermal output and industrial processes

Steam vs. Hot Water

The thermal output of gas turbine installations is usually steam, whereas the output of reciprocating engines is generally hot water in the 200°F range. There are exceptions to this general rule. For industrial processes that require higher temperatures for thermal processes, a gas turbine selection may be necessary. As described below, care should be taken to ensure that the thermal output can be utilized. In other words, if 240°F thermal energy is needed, a reciprocating engine may not be able to meet the requirements unless properly designed.

Temperature of Cogeneration Thermal Output and Industrial Processes

As mentioned above, the temperature of the thermal output of the cogeneration system may not be utilized in all thermal industrial processes. Simply put, the cogeneration thermal output should be at a higher temperature that the industrial process temperature. Otherwise, the owner may realize less savings than anticipated.

The engineer must perform a thermal balance to ensure that the thermal output of the cogeneration system is less than the industrial process requirements for all hours. The author likes to ask what the thermal needs are in July at midnight. At that time, comfort heating requirements are nonexistent and many industrial thermal processes may be shut down.

A common error occurs when thermal balance is made on an annual, rather than an hourly, basis. Even if the annual thermal output of the cogeneration system matches the annual thermal input of the industrial thermal processes, there are usually hours when excess thermal output must be discarded.

Thermal storage is a potential solution for the hourly ups and downs of thermal energy requirements. The author has found, however, that this solution often falls short; if more than an hour or two of storage is needed, the amount of storage required is both too costly and impractical.

INDUSTRIAL PROCESS EXAMPLES UTILIZING UNIQUE SOLUTIONS

The author has been involved in many installations where finding a unique thermal application has rescued a financially challenged project. A few of these applications are explained below.

Paper Plant

In a paper plant, three thermal industrial processes were added to existing systems and a steam generation system was designed and added to the reciprocating exhaust gas stream. In general, a better solution is to reduce the cogeneration system size by 10%–20%, rather than to add a complicated steam generation system. A hot water recovery system on the exhaust gas stream is generally more cost effective. However, in this installation, a steam generation system was added. Exhaust gas was fed into a steam generator to produce steam directly at 100 psig. A secondary heat exchanger was added after the steam generators in order to recover additional energy.

The second unique application was a heat exchanger being added to a water pulping tank. Water and paper pulp are added in one of the first steps of the paper-making process. This water/pulp slurry was heated by direct injected steam originally. Water at 200°F was introduced to the tank via a piping serpentine installed in the tank. The water in the tank was heated to approximately 140°F. The large temperature difference allowed for the use of nonfinned piping of a reasonable length. It is very important to note that an energy balance needed to be considered. The steam used in this tank was low pressure steam that had been recovered by a flash steam energy recovery tank. Essentially, this steam was free, and if it could not be utilized elsewhere, the value of the cogeneration hot water would be zero. Because the steam was utilized elsewhere, the energy was useful.

The third unique thermal energy recovery system added to this paper plant was a hot air blower system. In the paper-making process, the water/paper slurry is eventually dried on a steam drum by being drawn over the drum by a continuous paper sheet. Steam in the drum dries off the water in the slurry, leaving only paper. By blowing hot dry air over the paper sheet, the paper dries faster. The first benefit is that less steam is needed because the hot air dries the paper. An even more important benefit is that the speed of the paper sheet in the process can be increased and more paper is produced with the same overhead and labor costs. This financial benefit can dwarf the cogeneration system savings. The physical limitations of the existing system were overcome by the addition of this production enhancing system.

Car Parts Plant

A solution considered in a car parts plant was to direct the engine exhaust directly in a parts drying process. The 1200°F exhaust gas temperature was to be directed into a large process heater, where a 300°F temperature was maintained. One ironic outcome of this solution was that project economics were actually impaired because condensation of the exhaust gas could not be implemented. The benefits of exhaust gas condensation will be described below.

Boiler Plant Air and Water Reheat

Opportunities exist at the central boiler plant at many industrial plants. The boiler water is preheated in a makeup tank and/or condensate tank. Both returning condensate water at 180°F (or less) and makeup water at 60°F represent opportunities to use cogeneration thermal output. At both of these temperatures, an added opportunity to condense the exhaust gas, as explained later, also exists.

A less practical but potential opportunity exists to preheat the combustion air to the boiler. Each 40°F increase in the boiler air temperature equates to approximately a 1% drop in boiler fuel usage. The boiler air-fuel ratio and the combustion air fan speed must often be reset. This solution is often impractical, costly, and it provides marginal returns.

Fuel Switch on an Heating, Ventilating, and Air Conditioning System

A unique and profitable solution can be to utilize excess cogeneration thermal output for a new usage. At one facility, the author's company replaced an electric resistance coil in a heating system with a hot water coil. The electric rates were approximately four times the gas rates. In effect, the thermal output had four times the value and effectively boosted the electrical efficiency of the plant. The economic benefit of the displaced electric energy was almost equal to the fueling costs of the cogeneration plant, even though it only amounted to about a quarter of the thermal output of the cogeneration plant.

EXHAUST GAS CONDENSATION SOLUTIONS

The energy that is contained in the input fuel is generally not fully recovered. The more energy that has been recovered from the exhaust gas stream, the lower the temperature of the exiting exhaust gas stream. Approximately 10% of the entire energy that is contained in the fuel can only be recovered if the water in the exhaust is condensed. The same amount of energy that it takes to boil water is available if the reverse process of condensation is conducted. Converting the water vapor in the exhaust gas stream into water releases 10% additional energy. To condense the water in the exhaust, water at a temperature less than 200°F is generally needed.

If exhaust gas is condensed, plastic or stainless steel exhaust stack materials must be used. Care must be taken in design because an exhaust temperature that is too high can melt plastic exhaust stacks. Regular steel cannot be used because sulfur in the fuel and nitrogen in the air can cause the production of sulfuric and nitric acid in the condensate. Both of these compounds are extremely corrosive to normal steel.

COGENERATION AND INTERRUPTIBLE FUEL RATES

In the 1980s and 1990s, when natural gas prices at the well head were low, cogeneration was viewed as a natural way to increase gas companies' market share. A cogeneration system with a 40% thermal efficiency would result in twice as much, or more, gas usage as an 80% efficient boiler. Further discounts were offered for interruptible fuel rates, wherein a customer agrees to reduce their gas usage upon notification from the gas company. Gas companies were able to offer these interruptible discounts because they were using spare distribution pipe capacity during nonwinter months. In the late 1990s, natural gas fuel prices began to abruptly increase. Many discount gas rates have disappeared since then.

FUTURE CONSIDERATIONS

In the future, gas rates may decrease, making cogeneration more cost effective. Cogeneration gas rates and interruptible gas rates may return, even if for only a few months each year.

Increases in the efficiencies of reciprocating engines, gas turbines, and microturbines are already taking place and should continue. The cost of technologies, such as fuel cells and alternative energy technologies, are also decreasing. Some of these technologies may become cost competitive in the near future. Fuel cells already have higher electrical efficiencies than most present cogeneration technologies, but the cost is presently much higher than turbine and reciprocating engine technology.

CONCLUSION

The integration of CHP or cogeneration into industrial processes offers unique opportunities to optimize revenue and energy savings. The installation of new industrial thermal processes can make a financially challenged cogeneration project feasible. Condensation of the exhaust gas stream can both produce more revenue and increase overall system efficiency by more than 10%. Future advances in technologies, such as fuel cells, may offer additional options.

BIBLIOGRAPHY

1. Capehart, B.L.; Turner, W.C.; Kennedy, W.J. *The Guide to Energy Management (Chapter 14)*, 5th Ed.; Fairmont Press: Lilburn, GA, 2006.
2. Turner, W.C., Ed. *Energy Management Handbook (Chapter 7)*, 6th Ed., Fairmont Press: Lilburn, GA, 2006.
3. Petchers, N. *Handbook Combined Heating, cooling and Power*; Fairmont Press: Lilburn, GA, 2002.
4. Limaye, D.R., Ed. *Industrial Cogeneration Applications;* Fairmont Press: Lilburn, GA, 1987.
5. Payne, F.W., Ed. *Cogeneration Management Reference Guide;* Fairmont Press: Lilburn, GA, 1997.

Commissioning: Existing Buildings

David E. Claridge
Department of Mechanical Engineering, Associate Director, Energy Systems Laboratory, Texas A & M University, College Station, Texas, U.S.A.

Mingsheng Liu
Architectural Engineering Program, Director, Energy Systems Laboratory, Peter Kiewit Institute, University of Nebraska—Lincoln, Omaha, Nebraska, U.S.A.

W. D. Turner
Director, Energy Systems Laboratory, Texas A & M University, College Station, Texas, U.S.A.

Abstract

Commissioning an existing building is referred to by various terms, including recommissioning, retrocommissioning, and continuous commissioning® (CC®). A comprehensive study of 182 existing buildings totaling over 22,000,000 ft^2 in floor area reported average energy savings of 18% at an average cost of $0.41/ft^2 after they were commissioned, producing an average simple payback of 2.1 years. The commissioning process for an existing building involves steps that should include building screening, a commissioning assessment to estimate savings potential and cost, plan development and team formation, development of performance baselines, detailed measurements and commissioning measure development, implementation, and follow-up to maintain persistence. Existing building commissioning has been successfully used in energy management programs as a standalone measure, as a follow-up to the retrofit process, as a rapid payback Energy Conservation Measure (ECM) in a retrofit program, and as a means to ensure that a building meets or exceeds its energy performance goals. Very often, it is the most cost-effective single energy management option available in a large building.

INTRODUCTION

Commissioning an existing building has been shown to be a key energy management activity over the last decade, often resulting in energy savings of 10, 20 or sometimes 30% without significant capital investment. It generally provides an energy payback of less than three years. In addition, building comfort is improved, systems operate better, and maintenance cost is reduced. Commissioning measures typically require no capital investment, though the process often identifies maintenance that is required before the commissioning can be completed. Potential capital upgrades or retrofits are often identified during the commissioning activities, and knowledge gained during the process permits more accurate quantification of benefits than is possible with a typical audit. Involvement of facilities personnel in the process can also lead to improved staff technical skills.

This entry is intended to provide the reader with an overview of the costs, benefits, and process of commissioning an existing building. There is no single definition of commissioning for an existing building, so several widely used commissioning definitions are given. A short case study illustrates the changes made when an existing building is commissioned, along with its impact. This is followed by a short summary of published information on the range of costs and benefits. The major portion of the article describes the commissioning process used by the authors in existing buildings so the reader can determine whether and how to implement a commissioning program. Monitoring and verification (M&V) may be very important to a successful commissioning program. Some commissioning-specific M&V issues are discussed, particularly the role of M&V in identifying the need for follow-up commissioning activities.

COMMISSIONING DEFINITIONS

The commissioning of a navy ship is the order or process that makes it completely ready for active duty. Over the last two decades, the term has come to refer to the process that makes a building or some of its systems completely ready for use. In the case of existing buildings, it generally refers to a restoration or improvement in the operation or function of the building systems. A widely used short definition of new building commissioning is the process of ensuring systems are designed, installed, functionally

Keywords: Commissioning; Retrocommissioning; Recommissioning; Continuous commissioning®; Commissioning existing buildings; Monitoring and verification; Energy conservation measure.

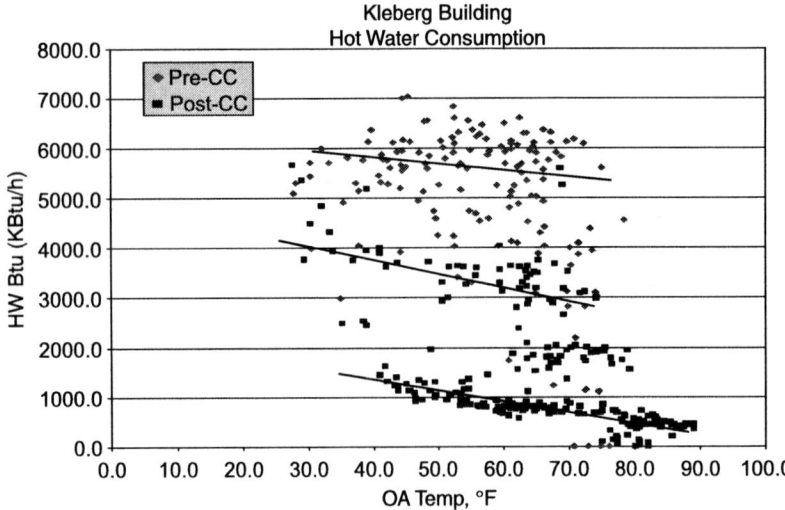

Fig. 1 Pre-CC and post-CC heating water consumption at the Kleberg building vs daily average outdoor temperature.
Source: From Proceedings of the 25th WEEC, Atlanta, GA, October 9–11 (see Ref. 4).

tested, and operated in conformance with the design intent. Commissioning begins with planning and includes design, construction, start-up, acceptance, and training and can be applied throughout the life of the building. Furthermore, the commissioning process encompasses and coordinates the traditionally separate functions of systems documentation, equipment start-up, control system calibration, testing and balancing, and performance testing.[1]

Recommissioning

Recommissioning refers to commissioning a building that has already been commissioned at least once. After a building has been commissioned during the construction process, recommissioning ensures that the building continues to operate effectively and efficiently. Buildings, even if perfectly commissioned, will normally drift away from optimum performance over time, due to system degradation, usage changes, or failure to correctly diagnose the root cause of comfort complaints. Therefore, recommissioning normally reapplies the original commissioning procedures in order to keep the building operating according to design intent, or it may modify them for current operating needs.

Optimally, recommissioning becomes part of a facility's continuing operations and maintenance (O&M) program. There is not a consensus on recommissioning frequency, but some consider that it should occur every 3–5 years. If there are frequent build-outs or changes in building use, recommissioning may need to be repeated more often.[2]

Retrocommissioning

Retrocommissioning is the first-time commissioning of an existing building. Many of the steps in the retrocommissioning process are similar to those for commissioning. Retrocommissioning, however, occurs after construction, as an independent process, and its focus is usually on energy-using equipment such as mechanical equipment and related controls. Retrocommissioning may or may not bring the building back to its original design intent, since the usage may have changed or the original design documentation may no longer exist.[2]

Continuous Commissioning

Continuous Commissioning (CC®)[3] is an ongoing process to resolve operating problems, improve comfort, optimize energy use, and identify retrofits for existing commercial and institutional buildings and central plant facilities. Continuous commissioning focuses on improving overall system control and operations for the building as it is currently utilized, and on meeting existing facility needs. Continuous commissioning is much more than an O&M program. It is not intended to ensure that a building's systems function as originally designed, but it ensures that the building and its systems operate optimally to meet the current uses of the building. As part of the CC process, a comprehensive engineering evaluation is conducted for both building functionality and system functions. Optimal operational parameters and schedules are developed based on actual building conditions and current occupancy requirements.

COMMISSIONING CASE STUDY—KLEBERG BUILDING

The Kleberg Building is a teaching/research facility on the Texas A&M campus consisting of classrooms, offices, and laboratories, with a total floor area of approximately 165,030 ft^2. A CC investigation was initiated in the summer of 1996 due to the extremely high level of simultaneous heating and cooling observed in the building.[4] Figs. 1 and 2 show daily heating and cooling

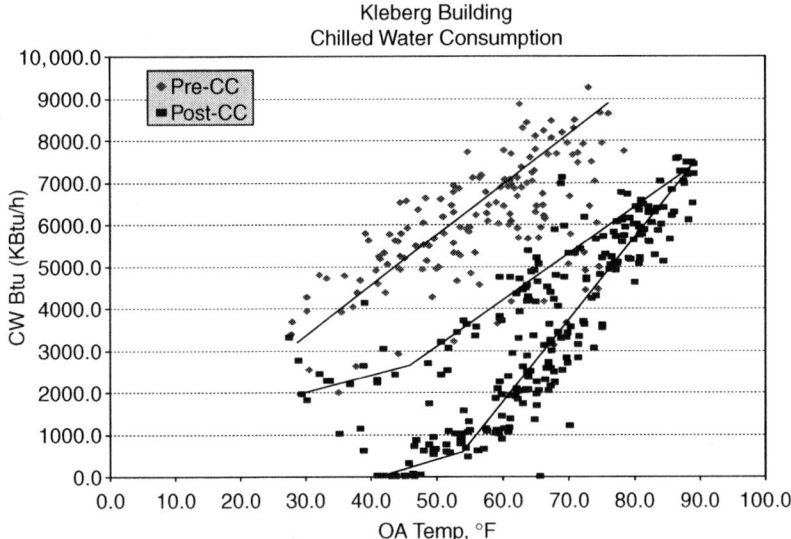

Fig. 2 Pre-CC and post-CC chilled water consumption at the Kleberg building vs daily average outdoor temperature.
Source: From Proceedings of the 25th WEEC, Atlanta, GA, October 9–11 (see Ref. 4).

consumption (expressed in average kBtu/h) as functions of daily average temperature. The pre-CC heating consumption data given in Fig. 1 show very little temperature dependence as indicated by the regression line derived from the data. Data values were typically between 5 and 6 MMBtu/h with occasional lower values. The cooling consumption is even higher (Fig. 2), though it shows more temperature dependence.

It was soon found that the preheat was operating continuously, heating the mixed air entering the cooling coil to approximately 105°F. The preheat was turned off, and heating and cooling consumption both dropped by about 2 MMBtu/h as shown by the middle clouds of data in Figs. 1 and 2. Subsequently, the building was thoroughly examined, and a comprehensive list of commissioning measures was developed and implemented. The principal measures implemented that led to reduced heating and cooling consumption were as follows:

- "Preheat to 105°F" was changed to "Preheat to 40°F."
- The cold deck schedule was changed from "55°F fixed" to "Vary from 62 to 57°F as ambient temperature varies from 40 to 60°F."
- The economizer was set to maintain mixed air at 57°F whenever the outside air was below 60°F.
- Static pressure control was reduced from 1.5 inH2O to 1.0 inH2O, and a nighttime set-back to 0.5 inH2O was implemented.
- A number of broken variable air volume terminal (VFD) boxes were replaced or repaired.
- Chilled water pump variable frequency drives (VFDs) were turned on.

These changes further reduced chilled water and heating hot water use as shown in Figs. 1 and 2 for a total annualized reduction of 63% in chilled-water use and 84% in hot-water use.

COSTS AND BENEFITS OF COMMISSIONING EXISTING BUILDINGS

The most comprehensive study of the costs and benefits of commissioning existing buildings was conducted by Mills et al.[5,6] This study examined the impact of commissioning 182 existing buildings with over 22,000,000 ft^2. The commissioning cost of these projects ranged from below \$0.10/ft^2–\$3.86/ft^2, but most were less than \$0.50/ft^2 with an average cost of \$0.41/ft^2. Savings ranged from essentially zero to 54% of total energy use, with an average of 18%. This range reflects not only differences among buildings in the potential for commissioning savings, but doubtless also includes differences in the level of commissioning applied and the skill of the commissioning providers. Simple payback times ranged from less than a month to over 20 years, with an average of 2.1 years. Fig. 3 illustrates the average payback as a function of building type and the precommissioning energy cost intensity. The sample sizes for office buildings and higher education are large enough that these averages for payback and energy savings may be representative, but the other sample sizes are so small that they may be significantly skewed by building specific and/or other factors.

Mills et al. concluded, "We find that commissioning is one of the most cost-effective means of improving energy efficiency in commercial buildings. While not a panacea, it can play a major and strategically important role in achieving national energy savings goals—with cost-effective savings potential of \$18 billion per year or more in commercial buildings across the United States."

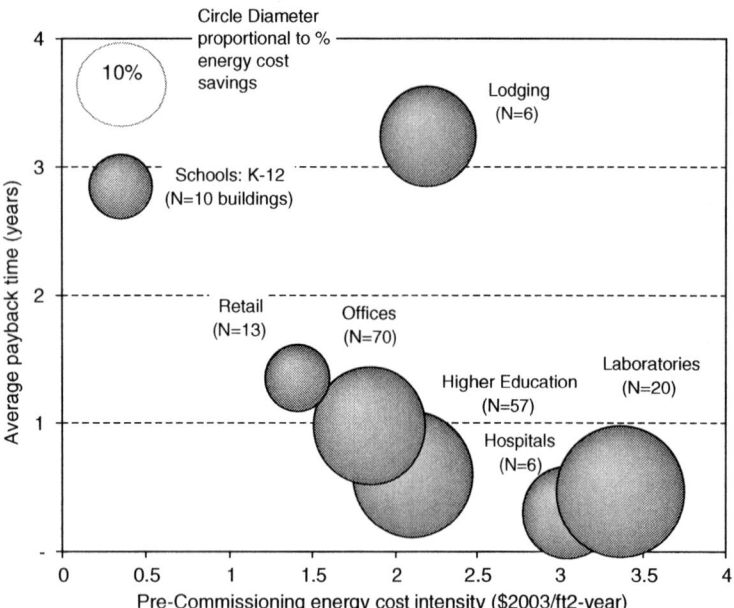

Fig. 3 Average simple payback time and percent energy savings from commissioning of existing buildings by building type.
Source: From Lawrence Berkeley National Laboratory Report No. 56637 (see Ref. 5).

THE COMMISSIONING PROCESS IN EXISTING BUILDINGS

There are multiple terms that describe the commissioning process for existing buildings, as noted in the previous section. Likewise, there are many adaptations of the process itself. The same practitioner will implement the process differently in different buildings, based on the budget and the owner requirements. The process described here is the process used by the chapter authors when the owner wants a thorough commissioning job. The terminology used will refer to the CC process, but many of the steps are the same for retrocommissioning or recommissioning. The model described assumes that a commissioning provider is involved, since that is normally the case. Some (or all) of the steps may be implemented by the facility staff if they have the expertise and adequate staffing levels to take on the work.

Continuous commissioning focuses on improving overall system control and operations for the building as it is currently utilized, and on meeting existing facility needs. It does not ensure that the systems function as originally designed, but ensures that the building and systems operate optimally to meet the current requirements. During the CC process, a comprehensive engineering evaluation is conducted for both building functionality and system functions. The optimal operational parameters and schedules are developed based on actual building conditions and current occupancy requirements. An integrated approach is used to implement these optimal schedules to ensure practical local and global

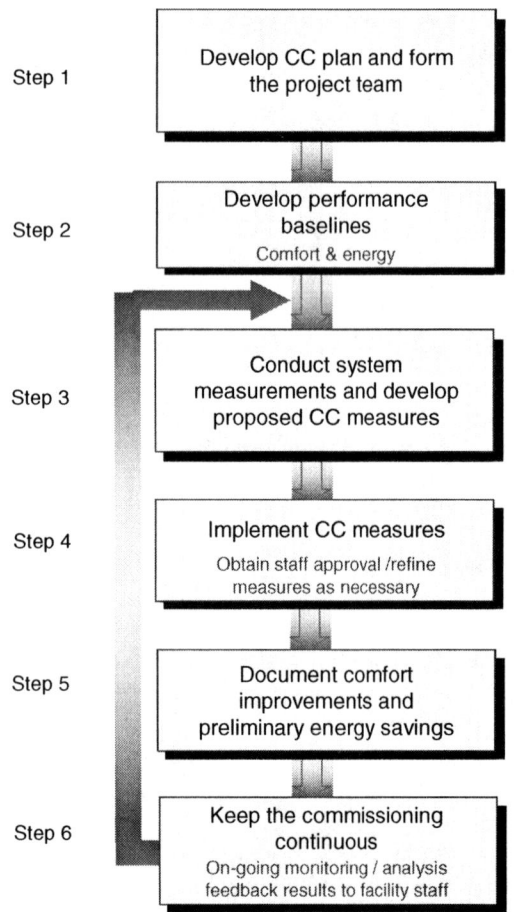

Fig. 4 Outline of phase II of the CC process: implementation and verification.
Source: From Energy Systems Laboratory.

system optimization and persistence of the improved operation schedules.

Commissioning Team

The CC team consists of a project manager, one or more CC engineers and CC technicians, and one or more designated members of the facility operating team. The primary responsibilities of the team members are shown in Table 1. The project manager can be an owner representative or a CC provider representative. It is essential that the engineers have the qualifications and experience to perform the work specified in the table. The designated facility team members generally include at least one lead heating, ventilating and air conditioning (HVAC) technician and an energy management control system (EMCS) operator or engineer. It is essential that the designated members of the facility operating team actively participate in the process and be convinced of the value of the measures proposed and implemented, or operation will rapidly revert to old practices.

Continuous Commissioning Process

The CC process consists of two phases. The first phase is the project development phase that identifies the buildings to be included in the project and develops the project scope. At the end of this phase, the CC scope is clearly defined and a CC contract is signed, as described in "Phase 1: Project Development." The second phase implements CC and verifies project performance through the six steps outlined in Fig. 4 and described in "Phase 2: CC Implementation and Verification."

Phase 1: Project Development

Step 1: Identify Candidate Buildings. Buildings are screened to identify those that will receive a CC assessment. Buildings that provide poor thermal comfort, consume excessive energy, or have design features of the HVAC systems that are not fully used are typically good candidates for a CC assessment. Continuous commissioning can be effectively implemented in buildings that have received energy efficiency retrofits, in newer buildings, and in existing buildings that have not received energy efficiency upgrades. In other words, virtually any building can be a potential CC candidate. The CC provider should perform a preliminary analysis to check the feasibility of using the CC process on candidate facilities before performing a CC assessment.

The following information is needed for the preliminary assessment:

- Monthly utility bills for at least 12 months.
- General building information—size, function, major equipment, and occupancy schedules.

Table 1 Commissioning team members and their primary responsibilities

Team member(s)	Primary responsibilities
Project manager	1. Coordinate the activities of building personnel and the commissioning team
	2. Schedule project activities
Continuous commissioning (CC) engineer(s)	1. Develop metering and field measurement plans
	2. Develop improved operational and control schedules
	3. Work with building staff to develop mutually acceptable implementation plans
	4. Make necessary programming changes to the building automation system
	5. Supervise technicians implementing mechanical systems changes
	6. Project potential performance changes and energy savings
	7. Conduct an engineering analysis of the system changes
	8. Write the project report
Designated facility staff	1. Participate in the initial facility survey
	2. Provide information about problems with facility operation
	3. Suggest commissioning measures for evaluation
	4. Approve all CC measures before implementation
	5. Actively participate in the implementation process
CC Technicians	1. Conduct field measurements
	2. Implement mechanical, electrical, and control system program modifications and changes, under the direction of the project engineer

- O&M records, if available.
- Description of any problems in the building, such as thermal comfort, indoor air quality, moisture, or mildew.

An experienced engineer should review this information and determine the potential of the CC process to improve comfort and reduce energy cost. If the CC potential is good, a CC assessment should be performed.

Step 2: Perform CC Assessment and Develop Project Scope. The CC assessment involves a site visit by an experienced commissioning engineer who examines EMCS screens, conducts spot measurements throughout the building systems, and identifies major CC measures suitable for the building. The CC assessment report lists and describes the preliminary CC measures identified, the estimated energy savings from implementation, and the cost of carrying out the CC process on the building(s) evaluated in the assessment. Once a commissioning contract is signed, the process moves to Phase 2.

Phase 2: CC Implementation and Verification

Step 1: Develop CC Plan and Form the Project Team. The CC project manager and project engineer develop a detailed work plan for the project that includes major tasks, their sequence, time requirements, and technical requirements. The work plan is then presented to the building owner or representative(s) at a meeting attended by any additional CC engineers and technicians on the project team. Owner contact personnel and in-house technicians who will work on the project are identified.

Step 2: Develop Performance Baselines. This step should document all known comfort problems in individual rooms resulting from too much heating, cooling, noise, humidity, or odors (especially from mold or mildew), or lack of outside air. Also, identify and document any HVAC system problems.

Baseline energy models of building performance are necessary to document the energy savings after commissioning. The baseline energy models can be developed using one or more of the following types of data:

- Short-term measured data obtained from data loggers or the EMCS system.
- Long-term hourly or 15-min whole building energy data, such as whole-building electricity, cooling, and heating consumption.
- Utility bills for electricity, gas, or chilled or hot water.

The baselines developed should be consistent with the International Performance Measurement and Verification Protocol,[7] with ASHRAE Guideline 14, or with both.[8]

Step 3: Conduct System Measurements and Develop Proposed CC Measures. The CC team uses EMCS trend data complemented by site measurements to identify current operational schedules and problems. The CC engineer conducts an engineering analysis to develop solutions for the existing problems; establishing improved operation and control schedules and set points for terminal boxes, air handling units (AHUs), exhaust systems, water and steam distribution systems, heat exchangers, chillers, boilers, and other components or systems as appropriate. Cost-effective energy retrofit measures can also be identified and documented during this step, if desired by the building owner.

Step 4: Implement CC measures. The CC project manager and/or project engineer presents the engineering solutions to existing problems and the improved operational and control schedules to the designated operating staff members and the building owner's representative to get "buy-in" and approval. Measures may be approved, modified, or rejected. A detailed implementation schedule is then developed by the CC engineer in consultation with the operating staff.

Continuous commissioning implementation normally starts by solving existing problems. Implementation of the improved operation and control schedules starts at the end of the comfort delivery system, such as at the terminal boxes, and ends with the central plant. The CC engineer closely supervises the implementation and refines the operational and control schedules as necessary. Following implementation, the new operation and control sequences are documented in a way that helps the building staff understand why they were implemented.

Step 5: Document Comfort Improvements and Preliminary Energy Savings. The comfort measurements taken in Step 2 (Phase 2) should be repeated at the same locations under comparable conditions and compared with the earlier measurements. The M&V procedures adopted in Step 2 should be used to determine the early post-CC energy performance and weather normalized to provide a preliminary evaluation of savings.

Step 6: Keep the Commissioning Continuous. The CC engineer should review the system operation after 6–12 months to identify any operating problems and make any adjustments needed. One year after CC implementation is complete, the CC engineer should write a project follow-up report that documents the first-year savings, recommendations or changes resulting from any consultation or site visits provided, and any recommendations to further improve building operations. Subsequently, the consumption should be tracked and compared with the first-year post-CC consumption during this period. Any significant and persistent increases in consumption should be investigated by the staff and/or CC engineer.

USES OF COMMISSIONING IN THE ENERGY MANAGEMENT PROCESS

Commissioning can be used as a part of the energy management program in several different ways:

- As a standalone measure. Commissioning is probably most often implemented in existing buildings because it is the most cost-effective step the owner can take to increase the energy efficiency of the building, generally offering a payback under three years, and often 1–2 years.
- As a follow-up to the retrofit process. Continuous commissioning has often been used to provide additional savings after a successful retrofit and has also been used numerous times to make an underperforming retrofit meet or exceed the original expectations.
- As an ECM in a retrofit program. The rapid payback that generally results from CC may be used to lower the payback of a package of measures to enable inclusion of a desired equipment replacement that has a longer payback in a retrofit package. This is illustrated by a case study in the next section. In this approach, the CC engineers conduct the CC audit in parallel with the retrofit audit conducted by the design engineering firm. Because the two approaches are different and look at different opportunities, it is very important to closely coordinate these two audits.
- To ensure that a new building meets or exceeds its energy performance goals. It may be used to significantly improve the efficiency of a new building by optimizing operation to meet its actual loads and uses instead of working to design assumptions.

CASE STUDY WITH CC AS AN ECM

Prairie View A&M University is a 1.7-million square foot campus, with most buildings served by a central thermal plant. Electricity is purchased from a local electric co-op.

University staff identified the need for major plant equipment replacements on campus. They wished to finance the upgrades through the Texas LoanSTAR program, which requires that the aggregate energy payback of all ECMs financed be ten years or less. Replacement of items such as chillers, cooling towers, and building automation systems typically have paybacks of considerably more than ten years. Hence, they can only be included in a loan if packaged with low payback measures that bring the aggregate payback below ten years.[9]

The university administration wanted to maximize the loan amount to get as much equipment replacement as possible. They also wanted to ensure that the retrofits worked properly after they are installed. To maximize their loan dollars, they chose to include CC as an ECM.

The LoanSTAR Program provides a brief walkthrough audit of the candidate buildings and plants. This audit is performed to determine whether there is sufficient retrofit potential to justify a more thorough investment grade audit.

The CC assessment is conducted in parallel with the retrofit audit conducted by the engineering design firm, when CC is to be included as an ECM. The two approaches look at different opportunities, but there can be some overlap, so it is very important to closely coordinate both audits. It is particularly important that the savings estimated by the audit team are not "double counted." The area of greatest overlap in this case was the building automation system. Considerable care was taken not to mix improved EMCS operation with operational improvements determined by the CC engineer, so both measures received proper credit.

The CC measures identified included the following:

- Hot and cold deck temperature resets.
- Extensive EMCS programming to avoid simultaneous heating and cooling.
- Air and water balancing.
- Duct static pressure resets.
- Sensor calibration and repair.
- Improved start, stop, warm-up, and shutdown schedules.

The CC engineers took the measurements required and collected adequate data on building operation during the CC assessment to perform a calibrated simulation on the major buildings. Available metered data and building EMCS data were also used. The CC energy savings were then written as an ECM and discussed with the design engineer. Any potential overlaps were removed. The combined ECMs were then listed and the total savings determined.

Table 2 summarizes the ECMs identified from the two audits:

The CC savings were calculated to be $204,563, as determined by conducting calibrated simulation of 16 campus buildings and by engineering calculations of savings from improved loop pumping. No CC savings were claimed for central plant optimization. Those savings were all applied to ECM #7, although it seems likely that additional CC savings will accrue from this measure. The simple payback from CC is slightly under three years, making it by far the most cost effective of the ECMs to be implemented. The CC savings represent nearly 30% of the total project savings.

Perhaps more importantly, CC accounted for two-thirds of the "surplus" savings dollars available to buy down the payback of the chillers and EMCS upgrade. Without CC as an ECM, the University would have had

Table 2 Summary of energy cost measures (ECMs)

ECM #	ECM	Annual savings				Cost to implement	Simple payback
		Electric kWh/yr	Electric demand kW/yr	Gas MCF/yr	Cost savings		
#1	Lighting	1,565,342	5221	(820)	$94,669	$561,301	6.0
#2	Replace chiller #3	596,891	1250	-0-	$33,707	$668,549	19.8
#3	Repair steam system	-0-	-0-	13,251	$58,616	$422,693	7.2
#4	Install motion sensors	81,616	-0-	(44.6)	$3567	$26,087	7.3
#5	Add 2 bldgs. to CW loop	557,676	7050	-0-	$60,903	$508,565	8.4
#6	Add chiller #4	599,891	1250	-0-	$33,707	$668,549	19.8
#7	Primary/secondary pumping	1,070,207	-0-	-0-	$49,230	$441,880	9.0
#8	Replace DX systems	38,237	233	-0-	$2923	$37,929	13.0
#9	Replace DDC/EMCS	2,969,962	670	2736	$151,488	$2,071,932	13.7
#10	Continuous commissioning	2,129,855	-0-	25,318	$204,563	$605,000	3.0
	Assessment reports					$102,775	
	Metering					$157,700	
	M&V					$197,500	
		9,606,677	15,674	40,440	$693,373	$6,470,460	9.3

to delete one chiller and the EMCS upgrades, or some combination of chillers and a portion of the building EMCS upgrades from the project to meet the ten-year payback criteria—one chiller and the EMCS upgrades, or some combination of chillers and limited building EMCS upgrades. With CC, however, the university was able to include all these hardware items, and still meet the ten-year payback.

SUMMARY

Commissioning of existing buildings is emerging as one of the most cost-effective ways for an energy manager to lower operating costs, and typically does so with no capital investment, or with a very minimal amount. It has been successfully implemented in several hundred buildings and provides typical paybacks of one to three years.

It is much more than the typical O&M program. It does not ensure that the systems function as originally designed, but focuses on improving overall system control and operations for the building as it is currently utilized and on meeting existing facility needs. During the CC process, a comprehensive engineering evaluation is conducted for both building functionality and system functions. The optimal operational parameters and schedules are developed based on actual building conditions. An integrated approach is used to implement these optimal schedules to ensure practical local and global system optimization and to ensure persistence of the improved operational schedules.

The approach presented in this chapter begins by conducting a thorough examination of all problem areas or operating problems in the building, diagnoses these problems, and develops solutions that solve these problems while almost always reducing operating costs at the same time. Equipment upgrades or retrofits may be implemented as well, but have not been a factor in the case studies presented, except where the commissioning was used to finance equipment upgrades. This is in sharp contrast to the more usual approach to improving the efficiency of HVAC systems and cutting operating costs, which primarily emphasizes system upgrades or retrofits to improve efficiency.

Commissioning of new buildings is also an important option for the energy manager, offering an opportunity to help ensure that new buildings have the energy efficiency and operational features that are most needed.

REFERENCES

1. ASHRAE. *ASHRAE Guideline 1–1996: The HVAC Commissioning Process*; American Society of Heating, Refrigerating and Air-Conditioning Engineers: Atlanta, GA, 1996.
2. U.S. Department of Energy. Building Commissioning: The Key to Quality Assurance, Washington, DC, 1999.
3. Continuous Commissioning® and CC® are registered trademarks of the Texas Engineering Experiment Station (TEES). Contact TEES for further information.
4. Claridge, D.E.; Turner, W.D.; Liu, M. et al. Is Commissioning Once Enough? Solutions for Energy Security & Facility Management Challenges: Proceedings of the 25th WEEC, Atlanta, GA, October 9–11, 2002, 29–36.
5. Mills, E.; Friedman, H.; Powell, T. et al. *The Cost-Effectiveness of Commercial-Buildings Commissioning: A Meta-Analysis of Energy and Non-Energy Impacts in Existing Buildings and New Construction in the United States*, Lawrence Berkeley National Laboratory Report No. 56637, 2004; 98. http://eetd.lbl.gov/EA/mills/emills.
6. Mills, E.; Bourassa, N.; Piette, M.A. et al. The Cost-Effectiveness of Commissioning New and Existing Commercial Buildings: Lessons from 224 Buildings. Proceedings of the 2006 National Conference on Building Commissioning, Portland Energy Conservation, Inc., New York, 2005. http://www.peci.org/ncbc/proceedings/2005/19_Piette_NCBC2005.pdf.
7. IPMVP Committee. International Performance Measurement & Verification Protocol: Concepts and Options for Determining Energy and Water Savings. U.S. Department of Energy, Vol. 1, DOE/GO-102001-1187, Washington, D.C., 2001.
8. ASHRAE Guideline 14-2002: Measurement of Energy and Demand Savings. American Society of Heating, Refrigerating and Air-Conditioning Engineers; Atlanta, GA, 2002.
9. Turner, W.D.; Claridge, D.E.; Deng, S.; Wei, G. The Use of Continuous Commissioning^{SM} as an Energy Conservation Measure (ECM) for Energy Efficiency Retrofits. Proceedings of 11th National Conference on Building Commisioning, Palm Springs, CA, CD, May 20–22, 2003.

BIBLIOGRAPHY

1. Two major sources of information on commissioning existing buildings are the *Continuous Commissioning^{SM} Guidebook: Maximizing Building Energy Efficiency and Comfort* (Liu, M., Claridge, D.E. and Turner, W.D., Federal Energy Management Program, U.S. Dept. of Energy, 144 pp., 2002) and *A Practical Guide for Commissioning Existing Buildings* (Haasl, T. and Sharp, T., Portland Energy Conservation, Inc. and Oak Ridge National Laboratory for U.S. DOE, ORNL/TM-1999/34, 69 pp. + App., 1999). Much of this entry has been abridged and adapted from the CC Guidebook.
2. The case studies in this entry have been largely abridged and adapted from Refs. 4 and 9.

Commissioning: New Buildings

Janey Kaster
Yamas Controls, Inc., South San Francisco, California, U.S.A.

Abstract

Commissioning is the methodology for bringing to light design errors, equipment malfunctions, and improper control strategies at the most cost-effective time to implement corrective action. The primary goal of commissioning is to achieve optimal building systems performance. There are two types of commissioning: acceptance-based and process-based. Process-based commissioning is a comprehensive process that begins in the predesign phase and continues through postacceptance, while acceptance-based commissioning, which is perceived to be the cheaper method, basically examines whether an installation is compliant with the design and accordingly achieves more limited results.

Commissioning originated in the early 1980s in response to a large increase in construction litigation. Commissioning was the result of owners seeking other means to gain assurance that they were receiving systems compliant with the design intent and with the performance characteristics and quality specified. Learn how commissioning has evolved and the major initiatives that are driving its growing acceptance.

The general rule for including a system in the commissioning process is: the more complicated the system is, the more compelling is the need to include it in the commissioning process. Other criteria for determining which systems should be included are discussed. Discover the many benefits of commissioning, such as improved quality assurance, dispute avoidance, and contract compliance.

Selection of the commissioning agent is key to the success of the commissioning plan. Learn what traits are necessary and what approaches to use for the selection process.

The commissioning process occurs over a variety of clearly delineated phases. The phases of the commissioning process as defined below are discussed in detail: predesign, design, construction/installation, acceptance, and postacceptance.

Extensive studies analyzing the cost/benefit of commissioning justify its application. One study defines the median commissioning cost for new construction as $1 per square foot or 0.6% of the total construction cost. The median simple payback for new construction projects utilizing commissioning is 4.8 years.

Understand how to achieve the benefits of commissioning, including optimization of building performance, reduction of facility life-cycle cost, and increased occupant satisfaction.

INTRODUCTION

This entry provides an overview of commissioning—the processes one employs to optimize the performance characteristics of a new facility being constructed. Commissioning is important to achieve customer satisfaction, optimal performance of building systems, cost containment, and energy efficiency, and it should be understood by contractors and owners.

After providing an overview of commissioning and its history and prevalence, this entry discusses what systems should be part of the commissioning process, the benefits of commissioning, how commissioning is conducted, and the individuals and teams critical for successful commissioning. Then the entry provides a detailed discussion of each of the different phases of a successful commissioning process, followed by a discussion of the common mistakes to avoid and how one can measure the success of a commissioning effort, together with a cost–benefit analysis tool.

The purpose of this entry will be realized if its readers decide that successful commissioning is one of the most important aspects of construction projects and that commissioning should be managed carefully and deliberately throughout any project, from predesign to postacceptance. As an introduction to those unfamiliar with the process and as a refresher for those who are, the following section provides an overview of commissioning, how it developed, and its current prevalence today.

OVERVIEW OF COMMISSIONING

Commissioning Defined

Commissioning is the methodology for bringing to light design errors, equipment malfunctions, and improper control strategies at the most cost-effective time to implement corrective action. Commissioning facilitates a thorough understanding of a facility's intended use and

Keywords: Commissioning; New building system optimization; Construction process; Functional testing.

ensures that the design meets the intent through coordination, communication, and cooperation of the design and installation team. Commissioning ensures that individual components function as a cohesive system. For these reasons, commissioning is best when it begins in the predesign phase of a construction project and can in one sense be viewed as the most important form of quality assurance for construction projects.

Unfortunately, there are many misconceptions associated with commissioning, and perhaps for this reason, commissioning has been executed with varying degrees of success, depending on the level of understanding of what constitutes a "commissioned" project. American Society of Heating, Refrigerating and Air-Conditioning Engineers (AHSRAE) guidelines define commissioning as: the process of ensuring that systems are designed, installed, functionally tested, and capable of being operated and maintained to perform conformity with the design intent… [which] begins with planning and includes design, construction, startup, acceptance, and training, and is applied throughout the life of the building.[4] However, for many contractors and owners, this definition is simplified into the process of system startup and checkout or completing punch-list items.

Of course, a system startup and checkout process carried out by a qualified contractor is one important aspect of commissioning. Likewise, construction inspection and the generation and completion of punch-list items by a construction manager are other important aspects of commissioning. However, it takes much more than these standard installation activities to have a truly "commissioned" system. Commissioning is a comprehensive and methodical approach to the design and implementation of a cohesive system that culminates in the successful turnover of the facility to maintenance staff trained in the optimal operation of those systems.

Without commissioning, a contractor starts up the equipment but doesn't look beyond the startup to system operation. Assessing system operation requires the contractor to think about how the equipment will be used under different conditions. As one easily comprehended example, commissioning requires the contractor to think about how the equipment will operate as the seasons change. Analysis of the equipment and building systems under different load conditions due to seasonal conditions at the time of system startup will almost certainly result in some adjustments to the installed equipment for all but the most benign climates. However, addressing this common requirement of varying load due to seasonal changes most likely will not occur without commissioning. Instead, the maintenance staff is simply handed a building with minimal training and left to figure out how to achieve optimal operation on their own. In this seasonal example, one can just imagine how pleased the maintenance staff would be with the contractor when a varying load leads to equipment or system failure—often under very hot or very cold conditions!

Thus, the primary goal of commissioning is to achieve optimal building systems performance. For heating, ventilation, and air-conditioning (HVAC) systems, optimal performance can be measured by thermal comfort, indoor air quality, and energy savings. Energy savings, however, can result simply from successful commissioning targeted at achieving thermal comfort and excellent indoor air quality. Proper commissioning will prevent HVAC system malfunction—such as simultaneous heating and cooling, and overheating or overcooling—and successful malfunction prevention translates directly into energy savings. Accordingly, energy savings rise with increasing comprehensiveness of the commissioning plan. Commissioning enhances energy performance (savings) by ensuring and maximizing the performance of specific energy efficiency measures and correcting problems causing excessive energy use.[3] Commissioning, then, is the most cost-effective means of improving energy efficiency in commercial buildings. In the next section, the two main types of commissioning in use today—acceptance-based and processed-based—are compared and contrasted.

Acceptance-Based vs Process-Based Commissioning

Given the varied nature of construction projects, contractors, owners, buildings, and the needs of the diverse participants in any building projection, commissioning can of course take a variety of forms. Generally, however, there are two types of commissioning: acceptance-based and process-based. Process-based commissioning is a comprehensive process that begins in the predesign phase and continues through postacceptance, while acceptance-based commissioning, which is perceived to be the cheaper method, basically examines whether an installation is compliant with the design and accordingly achieves more limited results.

Acceptance-based commissioning is the most prevalent type due to budget constraints and the lack of hard cost/benefit data to justify the more extensive process-based commissioning. Acceptance-based commissioning does not involve the contractor in the design process but simply constitutes a process to ensure that the installation matches the design. In acceptance-based commissioning, confrontational relationships are more likely to develop between the commissioning agent and the contractor because the commissioning agent and the contractor, having been excluded from the design phase, have not "bought in" to the design and thus may be more likely to disagree in their interpretation of the design intent.

Because the acceptance-based commissioning process simply validates that the installation matches the design, installation issues are identified later in the cycle.

Construction inspection and regular commissioning meetings do not occur until late in the construction/installation phase with acceptance-based commissioning. As a result, there is no early opportunity to spot errors and omissions in the design, when remedial measures are less costly to undertake and less likely to cause embarrassment to the designer and additional costs to the contractor. As most contractors will readily agree, addressing issues spotted in the design or submittal stages of construction is typically much less costly than addressing them after installation, when correction often means tearing out work completed and typically delays the completion date.

Acceptance-based commissioning is cheaper, however, at least on its face, being approximately 80% of the cost of process-based commissioning.[2] If only the initial cost of commissioning services is considered, many owners will conclude that this is the most cost-effective commissioning approach. However, this 20% cost differential does not take into account the cost of correcting defects after the fact that process-based commissioning could have identified and corrected at earlier stages of the project. One need encounter only a single, expensive-to-correct project to become a devotee of process-based commissioning.

Process-based commissioning involves the commissioning agent in the predesign through the construction, functional testing, and owner training. The main purpose is quality assurance—assurance that the design intent is properly defined and followed through in all phases of the facility life cycle. It includes ensuring that the budget matches the standards that have been set forth for the project so that last-minute "value engineering" does not undermine the design intent, that the products furnished and installed meet the performance requirements and expectation compliant with the design intent, and that the training and documentation provided to the facility staff equip them to maintain facility systems true to the design intent.

As the reader will no doubt already appreciate, the author believes that process-based commissioning is far more valuable to contractors and owners than acceptance-based commissioning. Accordingly, the remainder of this entry will focus on process-based commissioning, after a brief review of the history of commissioning from inception to date, which demonstrates that our current, actively evolving construction market demands contractors and contracting professionals intimately familiar with and expert in conducting process-based commissioning.

History of Commissioning

Commissioning originated in the early 1980s in response to a large increase in construction litigation. Owners were dissatisfied with the results of their construction projects and had recourse only to the courts and litigation to resolve disputes that could not be resolved by meeting directly with their contractors. While litigation attorneys no doubt found this satisfactory approach to resolving construction project issues, owners did not, and they actively began looking for other means to gain assurance that they were receiving systems compliant with the design intent and with the performance characteristics and quality specified. Commissioning was the result.

While commissioning enjoyed early favor and wide acceptance, the recession of the mid-1980s placed increasing market pressure on costs, and by the mid-to late 1980s it forced building professionals to reduce fees and streamline services. As a result, acceptance-based commissioning became the norm, and process-based commissioning became very rare. This situation exists in most markets today; however, the increasing cost of energy, the growing awareness of the global threat of climate change and the need to reduce CO_2 emissions as a result, and the legal and regulatory changes resulting from both are creating a completely new market in which process-based commissioning will become ever more important, as discussed in the following section.

Prevalence of Commissioning Today

There are varying degrees of market acceptance of commissioning from state to state. Commissioning is in wide use in California and Texas, for example, but it is much less widely used in many other states. The factors that impact the level of market acceptance depend upon:

- The availability of commissioning service providers
- State codes and regulations
- Tax credits
- Strength of the state's economy[1]

State and federal policies with regard to commissioning are changing rapidly to increase the demand for commissioning. Also, technical assistance and funding are increasingly available for projects that can serve as demonstration projects for energy advocacy groups. The owner should investigate how each of these factors could benefit the decision to adopt commissioning in future construction projects.

Some of the major initiatives driving the growing market acceptance of commissioning are:

- Federal government's U.S. Energy Policy Act of 1992 and Executive Order 12902, mandating that federal agencies develop commissioning plans
- Portland Energy Conservation, Inc.; National Strategy for Building Commissioning; and their annual conferences
- ASHRAE HVAC Commissioning Guidelines (1989)
- Utilities establishing commissioning incentive programs
- Energy Star building program
- Leadership in Energy Environmental Design (LEED) certification for new construction

- Building codes
- State energy commission research programs

Currently, the LEED is having the largest impact in broadening the acceptance of commissioning. The Green Building Council is the sponsor of LEED and is focused on sustainable design—design and construction practices that significantly reduce or eliminate the cradle-to-grave negative impacts of buildings on the environment and building occupants. Leadership in energy efficient design encourages sustainable site planning, conservation of water and water efficiency, energy efficiency and renewable energy, conservation of materials and resources, and indoor environmental quality.

With this background on commissioning, the various components of the commissioning process can be explored, beginning with an evaluation of what building systems should be subject to the commissioning process.[5]

COMMISSIONING PROCESS

Systems to Include in the Commissioning Process

The general rule for including a system in the commissioning process is: the more complicated the system is the more compelling is the need to include it in the commissioning process. Systems that are required to integrate or interact with other systems should be included. Systems that require specialized trades working independently to create a cohesive system should be included, as well as systems that are critical to the operation of the building. Without a commissioning plan on the design and construction of these systems, installation deficiencies are likely to create improper interaction and operation of system components.

For example, in designing a lab, the doors should be included in the commissioning process because determining the amount of leakage through the doorways could prove critical to the ability to maintain critical room pressures to ensure proper containment of hazardous material. Another common example is an energy retrofit project. Such projects generally incorporate commissioning as part of the measurement and verification plan to ensure that energy savings result from the retrofit process.

For any project, the owner must be able to answer the question of why commissioning is important.

Why Commissioning?

A strong commissioning plan provides quality assurance, prevents disputes, and ensures contract compliance to deliver the intended system performance. Commissioning is especially important for HVAC systems that are present in virtually all buildings because commissioned HVAC systems are more energy efficient.

The infusion of electronics into almost every aspect of modern building systems creates increasingly complex systems requiring many specialty contractors. Commissioning ensures that these complex subsystems will interact as a cohesive system.

Commissioning identifies design or construction issues and, if done correctly, identifies them at the earliest stage in which they can be addressed most cost effectively. The number of deficiencies in new construction exceeds existing building retrofit by a factor of 3.[3] Common issues that can be identified by commissioning that might otherwise be overlooked in the construction and acceptance phase are: air distribution problems (these occur frequently in new buildings due to design capacities, change of space utilization, or improper installation), energy problems, and moisture problems.

Despite the advantages of commissioning, the current marketplace still exhibits many barriers to adopting commissioning in its most comprehensive and valuable forms.

Barriers to Commissioning

The general misperception that creates a barrier to the adoption of commissioning is that it adds extra, unjustified costs to a construction project. Until recently, this has been a difficult perception to combat because there are no energy-use baselines for assessing the efficiency of a new building. As the cost of energy continues to rise, however, it becomes increasingly less difficult to convince owners that commissioning is cost effective. Likewise, many owners and contractors do not appreciate that commissioning can reduce the number and cost of change orders through early problem identification. However, once the contractor and owner have a basis on which to compare the benefit of resolving a construction issue earlier as opposed to later, in the construction process, commissioning becomes easier to sell as a win–win proposal.

Finding qualified commissioning service providers can also be a barrier, especially in states where commissioning is not prevalent today. The references cited in this entry provide a variety of sources for identifying associations promulgating commissioning that can provide referrals to qualified commissioning agents.

For any owner adopting commissioning, it is critical to ensure acceptance of commissioning by all of the design construction team members. Enthusiastic acceptance of commissioning by the design team will have a very positive influence on the cost and success of your project. An objective of this entry is to provide a source of information to help gain such acceptance by design construction team members and the participants in the construction market.

Selecting the Commissioning Agent

Contracting an independent agent to act on behalf of the owner to perform the commissioning process is the best way to ensure successful commissioning. Most equipment vendors are not qualified and are likely to be biased against discovering design and installation problems—a critical function of the commissioning agent—with potentially costly remedies. Likewise, systems integrators have the background in control systems and data exchange required for commissioning but may not be strong in mechanical design, which is an important skill for the commissioning agent. Fortunately, most large mechanical consulting firms offer comprehensive commissioning services, although the desire to be competitive in the selection processes sometimes forces these firms to streamline their scope on commissioning.

Owners need to look closely at the commissioning scope being offered. An owner may want to solicit commissioning services independently from the selection of the architect/mechanical/electrical/plumbing design team or, minimally, to request specific details on the design team's approach to commissioning. If an owner chooses the same mechanical, electrical, and plumbing (MEP) firm for design and commissioning, the owner should ensure that there is physical separation between the designer and commissioner to ensure that objectivity is maintained in the design review stages. An owner should consider taking on the role of the commissioning agent directly, especially if qualified personnel exist in-house. This approach can be very cost effective. The largest obstacles to success with an in-house commissioning agent are the required qualifications and the need to dedicate a valuable resource to the commissioning effort. Many times, other priorities may interfere with the execution of the commissioning process by an in-house owner's agent.

There are three basic approaches to selecting the commissioning agent:

Negotiated—best approach for ensuring a true partnership
Selective bid list—preapproved list of bidders
Competitive—open bid list

Regardless of the approach, the owner should clearly define the responsibilities of the commissioning agent at the start of the selection process. Fixed-cost budgets should be provided by the commissioning agent to the owner for the predesign and design phases of the project, with not-to-exceed budgets submitted for the construction and acceptance phases. Firm service fees should be agreed upon as the design is finalized.

Skills of a Qualified Commissioning Agent

A commissioning agent needs to be a good communicator, both in writing and verbally. Writing skills are important because documentation is critical to the success of the commissioning plan. Likewise, oral communication skills are important because communicating issues uncovered in a factual and nonaccusatory manner is most likely to resolve those issues efficiently and effectively. The commissioning agent should have practical field experience in MEP controls design and startup to be able to identify potential issues early. The commissioning agent likewise needs a thorough understanding of how building structural design impacts building systems. The commissioning agent must be an effective facilitator and must be able to decrease the stress in stressful situations. In sum, the commissioning agent is the cornerstone of the commissioning team and the primary determinant of success in the commissioning process.

At least ten organizations offer certifications for commissioning agents. However, there currently is no industry standard for certifying a commissioning agent. Regardless of certification, the owner should carefully evaluate the individuals to be performing the work from the commissioning firm selected. Individual experience and reputation should be investigated. References for the lead commissioning agent are far more valuable than references for the executive members of a commissioning firm in evaluating potential commissioning agents. The commissioning agent selected will, however, only be one member of a commissioning team, and the membership of the commissioning team is critical to successful commissioning.

Commissioning Team

The commissioning team is composed of representatives from all members of the project delivery team: the commissioning agent, representatives of the owner's maintenance team, the architect, the MEP designer, the construction manager, and systems contractors. Each team member is responsible for a particular area of expertise, and one important function of the commissioning agent is to act as a facilitator of intrateam communication.

The maintenance team representatives bring to the commissioning team the knowledge of current operations, and they should be involved in the commissioning process at the earliest stage, defining the design intent in the predesign phase, as described below. Early involvement of maintenance team representatives ensures a smooth transition from construction to a fully operational facility, and aids in the acceptance and full use of the technologies and strategies that have been developed during the commissioning process. Involvement of the maintenance team representatives also shortens the building turnover transition period.

The other members of the commissioning team have defined and important functions. The architect leads the development of the design intent document (DID). The MEP designer's responsibilities are to develop the mechanical systems that support the design intent of the facility and comply with the owner's current operating standards. The MEP schematic design is the basis for the systems installed and is discussed further below. The construction manager ensures that the project installation meets the criteria defined in the specifications, the budget requirements, and the predefined schedule. The systems contractors' responsibilities are to furnish and install a fully functional system that meets the design specifications. There are generally several contractors whose work must be coordinated to ensure that the end product is a cohesive system.

Once the commissioning team is in place, commissioning can take place, and it occurs in defined and delineated phases—the subject of the following section.

COMMISSIONING PHASES

The commissioning process occurs over a variety of clearly delineated phases. The commission plan is the set of documents and events that defines the commissioning process over all phases. The commissioning plan needs to reflect a systematic, proactive approach that facilitates communication and cooperation of the entire design and construction team.

The phases of the commissioning process are:

Predesign
Design
Construction/installation
Acceptance
Postacceptance

These phases and the commissioning activities associated with them are described in the following sections.

Predesign Phase

The predesign phase is the phase in which the design intent is established in the form of the DID. In this phase of a construction project, the role of commissioning in the project is established if process-based commissioning is followed. Initiation of the commissioning process in the predesign phase increases acceptance of the commissioning process by all design team members. Predesign discussions about commissioning allow all team members involved in the project to assess and accept the importance of commissioning to a successful project. In addition, these discussions give team members more time to assimilate the impact of commissioning on their individual roles and responsibilities in the project. A successful project is more likely to result when the predesign phase is built around the concept of commissioning instead of commissioning's being imposed on a project after it has been designed.

Once an owner has decided to adopt commissioning as an integral part of the design and construction of a project, the owner should be urged to follow the LEED certification process, as discussed above. The commissioning agent can assist in the documentation preparation required for the LEED certification, which occurs in the postacceptance phase.

The predesign phase is the ideal time for an owner to select and retain the commissioning agent. The design team member should, if possible, be involved in the selection of the commissioning agent because that member's involvement will typically ensure a more cohesive commissioning team. Once the commissioning agent is selected and retained, the commissioning-approach outline is developed. The commissioning-approach outline defines the scope and depth of the commissioning process to be employed for the project. Critical commissioning questions are addressed in this outline. The outline will include, for most projects, answers to the following questions:

What equipment is to be included?
What procedures are to be followed?
What is the budget for the process?

As the above questions suggest, the commissioning budget is developed from the choices made in this phase. Also, if the owner has a commissioning policy, it needs to be applied to the specifics of the particular project in this phase.

The key event in the predesign phase is the creation of the DID, which defines the technical criteria for meeting the requirements of the intended use of the facilities. The DID document is often created based in part upon the information received from interviews with the intended building occupants and maintenance staff. Critical information such as the hours of operation, occupancy levels, special environmental considerations (such as pressure and humidity), applicable codes, and budgetary considerations and limitations—is identified in this document. The owner's preference, if any, for certain equipment or contactors should also be identified at this time. Together, the answers to the critical questions above and the information in the DID are used to develop the commissioning approach outline. A thorough review of the DID by the commissioning agent ensures that the commissioning-approach outline will be aligned with the design intent.

With the commissioning agent selected, the DID document created, and the commissioning approach outline in place, the design phase is ready to commence.

Design Phase

The design phase is the phase in which the schematics and specifications for all components of a project are prepared.

One key schematic and set of specifications relevant to the commissioning plan is the MEP schematic design, which specifies installation requirements for the MEP systems. As noted, the DID is the basis for creating the commissioning approach outline in the predesign phase. The DID also serves as the basis for creating the MEP schematic design in the design phase. The DID provides the MEP designer with the key concepts from which the MEP schematic design is developed.

The completed MEP schematic design is reviewed by the commissioning agent for completeness and conformance to the DID. At this stage, the commissioning agent and the other design team members should consider what current technologies, particularly those for energy efficiency, could be profitably included in the design. Many of the design enhancements currently incorporated into existing buildings during energy retrofitting for operational optimization are often not considered in new building construction. This can result in significant lost opportunity, so these design enhancements should be reviewed for incorporation into the base design during this phase of the commissioning process. This point illustrates the important principle that technologies important to retrocommissioning should be applied to new building construction—a point that is surprisingly often overlooked in the industry today.

For example, the following design improvements and technologies should always be considered for applicability to a particular project:

- Variable-speed fan and pumps installed
- Chilled water cooling (instead of DX cooling)
- Utility meters for gas, electric, hot water, chilled water, and steam at both the building and system level
- CO_2 implementation for minimum indoor air requirements

This list of design improvements is not exhaustive; the skilled commissioning agent will create and expand personalized lists as experience warrants and as the demands of particular projects suggest.

In addition to assisting in the evaluation of potential design improvements, the commissioning agent further inspects the MEP schematic design for:

- Proper sizing of equipment capacities
- Clearly defined and optimized operating sequences
- Equipment accessibility for ease of servicing

Once the commissioning agent's review is complete, the feedback is discussed with the design team to determine whether its incorporation into the MEP schematic design is warranted. The agreed-upon changes or enhancements are incorporated, thus completing the MEP schematic design.

The completed MEP schematic design serves as the basis on which the commissioning agent will transform the commissioning-approach outline into the commissioning specification.

The commissioning specification is the mechanism for binding contractually the contractors to the commissioning process. Expectations are clearly defined, including:

- Responsibilities of each contractor
- Site meeting requirements
- List of the equipment, systems, and interfaces
- Preliminary verification checklists
- Preliminary functional-performance testing checklists
- Training requirements and who is to participate
- Documentation requirements
- Postconstruction documentation requirements
- Commissioning schedule
- Definition for system acceptance
- Impact of failed results

Completion of the commissioning specification is required to select the systems contractor in a competitive solicitation. Alternatively, however, owners with strong, preexisting relationships with systems contractors may enter into a negotiated bid with those contractors, who can then be instrumental in finalizing the commissioning specification.

Owners frequently select systems contractors early in the design cycle to ensure that the contractors are involved in the design process. As noted above, if there are strong, preexisting relationships with systems contractors, early selection without a competitive selection process (described in the following paragraph) can be very beneficial. However, if there is no competitive selection process, steps should be taken to ensure that the owner gets the best value. For example, unit pricing should be negotiated in advance to ensure that the owner is getting fair and reasonable pricing. The commissioning agent and the MEP designer can be good sources for validating the unit pricing. The final contract price should be justified with the unit pricing information.

If the system selection process is competitive, technical proposals should be requested with the submission of the bid price. The systems contractors need to demonstrate a complete understanding of the project requirements to ensure that major components have not been overlooked. Information such as the project schedule and manpower loading for the project provide a good basis from which to measure the contractor's level of understanding. If the solicitation does not have a preselected list of contractors, the technical proposal should include the contractor's financial information, capabilities, and reference lists. As in the negotiated process described above, unit pricing should be requested to ensure the proper pricing of project additions and deletions. The review of the technical proposals should be included in the commissioning agent's scope of work.

A mandatory prebid conference should be held to walk the potential contractors through the requirements and to reinforce expectations. This conference should be held regardless of the approach—negotiated or competitive—used for contractor selection. The contractor who is to bear the financial burden for failed verification tests and subsequent functional-performance tests should be reminded of these responsibilities to reinforce their importance in the prebid meeting. The prebid conference sets the tone of the project and emphasizes the importance of the commissioning process to a successful project.

Once the MEP schematic design and commissioning specification are complete, and the systems contractors have been selected, the construction/installation phase begins.

Construction/Installation Phase

Coordination, communication, and cooperation are the keys to success in the construction and installation phase. The commissioning agent is the catalyst for ensuring that these critical activities occur throughout the construction and installation phase.

Frequently, value engineering options are proposed by the contractors prior to commencing the installation. The commissioning agent should be actively involved in the assessment of any options proposed. Many times, what appears to be a good idea in construction can have a disastrous effect on a facility's long-term operation. For example, automatic controls are often value engineered out of the design, yet the cost of their inclusion is incurred many times over in the labor required to perform their function manually over the life of the building. The commissioning agent can ensure that the design intent is preserved, the life-cycle costs are considered, and the impact on all systems of any value engineering modification proposed is thoroughly evaluated.

Once the design aspects are complete and value engineering ideas have been incorporated or rejected, the submittals, including verification checklists, need to be finalized. The submittals documentation is prepared by the systems contractors and reviewed by the commissioning agent. There are two types of submittals: technical submittals and commissioning submittals. Both types of submittals are discussed below.

Technical submittals are provided to document the systems contractors' interpretation of the design documents. The commissioning agent reviews the technical submittals for compliance and completeness. It is in this submittal review process that potential issues are identified prior to installation, reducing the need for rework and minimizing schedule delays. The technical submittals should include:

- Detailed schematics
- Equipment data sheets
- Sequence of operation
- Bill of material

A key technical submittal is the testing, adjusting, and balancing submittal (TAB). The TAB should include:

- TAB procedures
- Instrumentation
- Format for results
- Data sheets with equipment design parameters
- Operational readiness requirements
- Schedule

In addition to the TAB, other technical submittals, such as building automation control submittals, will be obtained from the systems contractors and reviewed by the commissioning agent.

The commissioning submittal generally follows the technical submittal in time and includes:

- Verification checklists
- Startup requirements
- Test and balance plan
- Training plan

The commissioning information in the commissioning submittal is customized for each element of the system.

These submittals, together with the commissioning specification, are incorporated into the commissioning plan, which becomes a living document codifying the results of the construction commissioning activities. This plan should be inspected in regular site meetings. Emphasis on the documentation aspect of the commissioning process early in the construction phase increases the contractors' awareness of the importance of commissioning to a successful project.

In addition to the submittals, the contractors are responsible for updating the design documents with submitted and approved equipment data and field changes on an ongoing basis. This update design document should be utilized during the testing and acceptance phase.

The commissioning agent also performs periodic site visits during the installation to observe the quality of workmanship and compliance with the specifications. Observed deficiencies should be discussed with the contractor and documented to ensure future compliance. Further inspections should be conducted to ensure that appropriate corrective action has been taken.

The best way to ensure that the items discussed above are addressed in a timely manner is to hold regularly scheduled commissioning meetings that require the participation of all systems contractors. This is the mechanism for ensuring that communication occurs. Meeting minutes prepared by the commissioning agent document the discussions and decisions reached. Commissioning meetings should be coordinated with the regular project meetings because many participants in a construction project need to attend both meetings.

Typical elements of a commissioning meeting include:

- Discussing field installation issues to facilitate rapid response to field questions
- Updating design documents with field changes
- Reviewing the commissioning agent's field observations
- Reviewing progress against schedule
- Coordinating multicontractor activities

Once familiar with the meeting process, an agenda will be helpful but not necessary. Meeting minutes should be kept and distributed to all participants.

With approved technical and commissioning plan submittals, as installation progresses, the contractor is ready to begin the system verification testing. The systems contractor generally executes the system verification independently of the commissioning agent. Contractor system verification includes:

- Point-to-point wiring checked out
- Sensor accuracy validated
- Control loops exercised

Each of the activities should be documented for each control or system element, and signed and dated by the verification technician.

The documentation expected from these activities should be clearly defined in the commissioning specification to ensure its availability to the commissioning agent for inspection of the verification process. The commissioning agent's role in the system verification testing is to ensure that the tests are completed and that the results reflect that the system is ready for the functional-performance tests. Because the commissioning agent is typically not present during the verification testing, the documentation controls how successfully the commissioning agent performs this aspect of commissioning.

In addition to system verification testing, equipment startup is an important activity during this phase. Equipment startup occurs at different time frames relative to the system verification testing, depending on the equipment and system involved. There may be instances when the system verification needs to occur prior to equipment startup to prevent a catastrophic event that could lead to equipment failure. The commissioning agent reviews the startup procedures prior to the startup to ensure that equipment startup is coordinated properly with the system verification. Unlike in verification testing, the commissioning agent should be present during HVAC equipment startup to document the results. These results are memorialized in the final commissioning report, so their documentation ultimately is the responsibility of the commissioning agent.

Once system verification testing and equipment startup have been completed, the acceptance phase begins.

Acceptance Phase

The acceptance phase of the project is the phase in which the owner accepts the project as complete and delivered in accordance with the specifications, and concludes with acceptance of the project in its entirety. An effective commissioning process during the installation phase should reduce the time and labor associated with the functional-performance tests of the acceptance phase.

Statistical sampling is often used instead of 100% functional-performance testing to make the process more efficient. A 20% random sample with a failure rate less than 1% indicates that the entire system was properly installed. If the failure rate exceeds 1%, a complete testing of every system may need to be completed to correct inadequacies in the initial checkout and verification testing. This random-sampling statistical approach holds the contractor accountable for the initial checkout and test, with the performance testing serving only to confirm the quality and thoroughness of the installation. This approach saves time and money for all involved. It is critical, however, that the ramifications of not meeting the desired results of the random tests are clearly defined in the commissioning specifications.

The commissioning agent witnesses and documents the results of the functional-performance tests, using specific forms and procedures developed for the system being tested. These forms are created with the input of the contractor in the installation phase. Involvement of the maintenance staff in the functional-performance testing is important. The maintenance team is often not included in the design process, so they may not fully understand the design intent. The functional-performance testing can provide the maintenance team an opportunity to learn and appreciate the design intent. If the design intent is to be preserved, the maintenance team must fully understand the design intent. This involvement of the maintenance team increases their knowledge of the system going into the training and will increase the effectiveness of the training.

Training of the maintenance team is critical to a successful operational handover once a facility is ready for occupancy. This training should include:

- Operations and maintenance (O&M) manual overview
- Hardware component review
- Software component review
- Operations review
- Interdependencies discussion
- Limitations discussion
- Maintenance review
- Troubleshooting procedures review
- Emergency shutdown procedures review

The support level purchased from the systems contractor determines the areas of most importance in the training and therefore should be determined prior to the training process. Training should be videotaped for later use by new maintenance team members and in refresher courses, and for general reference by the existing maintenance team. Using the O&M manuals as a training manual increases the maintenance team's awareness of the information contained in them, making the O&M manuals more likely to be referenced when appropriate in the future.

The O&M manuals should be prepared by the contractor in an organized and easy-to-use manner. The commissioning agent is sometimes engaged to organize them all into an easily referenced set of documents. The manuals should be provided in both hard-copy and electronic formats, and should include:

- System diagrams
- Input/output lists
- Sequence of operations
- Alarm points list
- Trend points list
- Testing documentation
- Emergency procedures

These services—including functional-performance testing, training, and preparing O&M manuals—should be included in the commissioning plan to ensure the project's successful acceptance. The long-term success of the project, however, is determined by the activities that occur in the postacceptance phase.

Postacceptance Phase

The postacceptance phase is the phase in which the owner takes beneficial occupancy and forms an opinion about future work with the design team, contractors, and the commissioning agent who completed the project. This is also the phase in which LEED certification, if adopted, is completed. Activities that usually occur in the acceptance phase should instead occur in the postacceptance phase. This is due to constraints that are not controllable by the contractor or owner. For example, seasonal changes may make functional-performance testing of some HVAC systems impractical during the acceptance phase for certain load conditions. This generally means that in locations that experience significant seasonal climate change, some of the functional-performance testing is deferred until suitable weather conditions exist. The commissioning agent determines which functional-performance tests need to be deferred and hence carried out in the postacceptance phase.

During the postacceptance phase, the commissioning agent prepares a final commissioning report that is provided to the owner and design team. The executive summary of this report provides an overall assessment of the design intent conformance. The report details whether the commissioned equipment and systems meet the commissioning requirements. Problems encountered and corrective actions taken are documented in this report. The report also includes the signed and dated startup and functional-performance testing checklists.

The final commissioning report can be used profitably as the basis of a "lessons learned" meeting involving the design team so that the commissioning process can be continuously improved and adaptations can be made to the owner's commissioning policy for future projects. The owner should use the experience of the first commissioned project to develop the protocols and standards for future projects. The documentation of this experience is the owner's commissioning policy. Providing this policy and the information it contains to the design and construction team for the next project can help the owner reduce budget overruns by eliminating any need to reinvent protocols and standards and by setting the right expectations earlier in the process.

Commissioning therefore should not be viewed as a one-time event but should instead be viewed as an operational philosophy. A recommissioning or continuous commissioning plan should be adopted for any building to sustain the benefits delivered from a commissioning plan. The commissioning agent can add great value to the creation of the recommissioning plan and can do so most effectively in the postacceptance phase of the project.

Fig. 1 depicts the information development that occurs in the evolution of a commissioning plan and summarizes the information presented in the preceding sections by outlining the various phases of the commissioning process.

With this background, the reader is better positioned for success in future commissioning projects and better prepared to learn the key success factors in commissioning and how to avoid common mistakes in the commissioning process.

COMMISSIONING SUCCESS FACTORS

Ultimately, the owner will be the sole judge of whether a commissioning process has been successful. Thus, second only to the need for a competent, professional commissioning agent, keeping the owner or the owner's senior representative actively involved in and informed at all steps of the commissioning process is a key success factor. The commissioning agent should report directly to the owner or the owner's most senior representative on the project, not only to ensure that this involvement and information transfer occur, but also to ensure the objective implementation of the commissioning plan—a third key success factor.

Another key success factor is an owner appreciation—which can be enhanced by the commissioning agent—that commissioning must be an ongoing process to get full

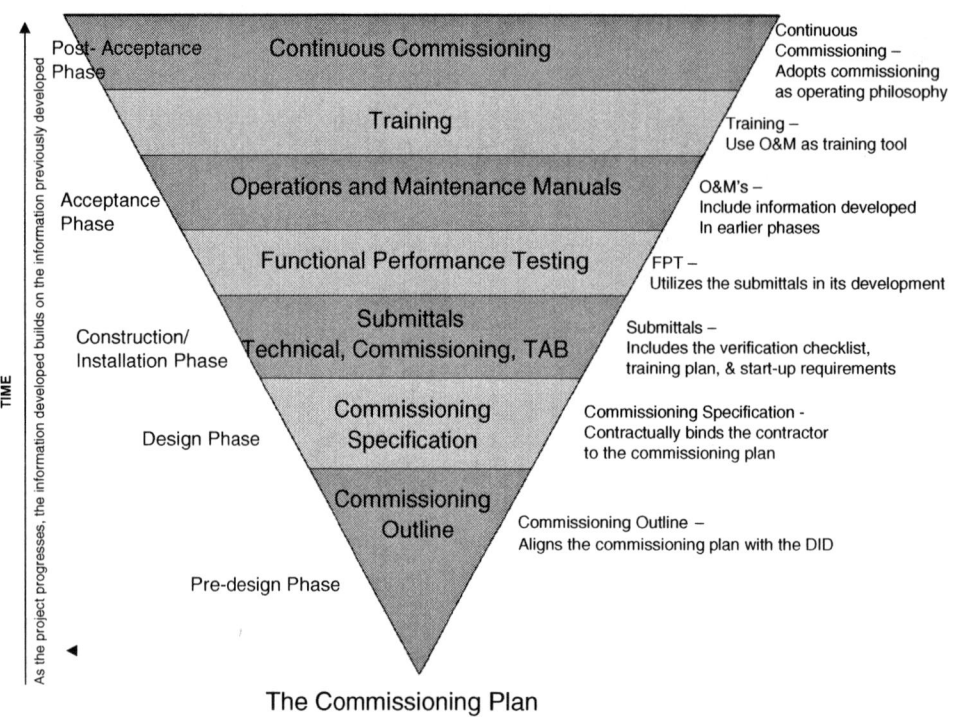

Fig. 1 The commissioning plan.

benefit. For example, major systems should undergo periodic modified functional testing to ensure that the original design intent is being maintained or to make system modification if the design intent has changed. If an owner appreciates that commissioning is a continuous process that lasts for the entire life of the facility, the commissioning process will be a success.

Most owners will agree that the commissioning process is successful if success can be measured in a cost/benefit analysis. Cost/benefit or return on equity is the most widely used approach to judge the success of any project. Unfortunately, the misapplication of cost/benefit analyses has been the single largest barrier to the widespread adoption of commissioning. For example, because new construction does not have an energy baseline from which to judge energy savings, improper application of a cost/benefit analysis can lead to failure to include energy savings technologies—technologies the commissioning agent can identify—in the construction process. Similarly, unless one can appreciate how commissioning can prevent schedule delays and rework by spotting issues and resolving them early in the construction process, one cannot properly offset the costs of commissioning with the benefits.

Fortunately, there are now extensive studies analyzing the cost/benefit of commissioning that justify its application. A study performed jointly by Lawrence Berkeley National Laboratory; Portland Energy Conservation, Inc.; and the Energy Systems Laboratory at Texas A&M University provides compelling analytical data on the cost/benefit of commissioning. The study defines the median commissioning cost for new construction as $1 per square foot or 0.6% of the total construction cost. The median simple payback for new construction projects utilizing commissioning is 4.8 years. This simple payback calculation does not take into account the quantified nonenergy impacts, such as the reduction in the cost and frequency of change orders or premature equipment failure due to improper installation practices. The study quantifies the median nonenergy benefits for new construction at $1.24 per square foot per year.[3]

While the primary cost component of assessing the cost/benefit of commissioning lies in whether there was a successful negotiation of the cost of services with the commissioning service provider, the more important aspect of the analysis relates to the outcomes of the process. For example, after a commissioning process is complete, what are the answers to these questions?

Are the systems functioning to the design intent?
Has the owner's staff been trained to operate the facility?
How many of the systems are operated manually a year after installation?

Positive answers to these and similar questions will ensure that any cost/benefit analysis will demonstrate the value of commissioning.

To ensure that a commissioning process is successful, one must avoid common mistakes. A commissioning plan is a customized approach to ensuring that all the systems operate in the most effective and efficient manner. A poor commissioning plan will deliver poor results. A common mistake is to use an existing commissioning plan and simply insert it into a specification to address commissioning. Each commissioning plan should be specifically tailored to the project to be commissioned.

Also, perhaps due to ill-conceived budget constraints, commissioning is implemented only in the construction phase. Such constraints are ill conceived because the cost of early involvement of the commissioning agent in the design phases is insignificant compared with the cost of correcting design defects in the construction phase. Significant cost savings can arise from identifying design issues prior to construction. Studies have shown that 80% of the cost of commissioning occurs in the construction phase.[2] Also, the later the commissioning process starts, the more confrontational commissioning becomes, making it more expensive to implement later in the process.[2] Therefore, adopting commissioning early in the project is a key success factor.

Value engineering often results in ill-informed, last-minute design changes that have an adverse and unintended impact on the overall building performance and energy use.[3] By ensuring that the commissioning process includes careful evaluation of all value engineering proposals, the commissioning agent and owner can avoid such costly mistakes.

Finally, the commissioning agent's incentive structure should not be tied to the number of issues brought to light during the commissioning process, as this can create an antagonistic environment that may create more problems than it solves. Instead, the incentive structure should be outcome based and the questions outlined above regarding compliance with design intent, training results, and postacceptance performance provide excellent bases for a positive incentive structure.

CONCLUSION

Commissioning should be performed on all but the most simplistic of new construction projects. The benefits of commissioning include:

- Optimization of building performance
  — Enhanced operation of building systems
  — Better-prepared maintenance staff
  — Comprehensive documentation of systems
  — Increased energy efficiency
  — Improved quality of construction
- Reduced facility life-cycle cost
  — Reduced impact of design changes
  — Fewer change orders
  — Fewer project delays
  — Less rework or postconstruction corrective work
  — Reduced energy costs
- Increased occupant satisfaction
  — Shortened turnover transition period
  — Improved system operation
  — Improved system reliability

With these benefits, owners and contractors alike should adopt the commissioning process as the best way to ensure cost-efficient construction and the surest way to a successful construction project.

REFERENCES

1. Quantum Consulting, Inc. *Commissioning in Public Buildings: Market Progress Evaluation Report*, June 2005. Available at: www.nwalliance.org/resources/reports/141.pdf (accessed).
2. ACG AABC Commissioning Group. ACG Commissioning Guidelines 2005. Available at: www.commissioning.org/commissioningguideline/ACGCommissioningGuideline.pdf (accessed).
3. Mills, E.; Friedman, H.; Powell, T. et al. *The Cost-Effectiveness of Commercial-Buildings Commissioning*, December 2004. Available at: http://eetd.lbl.gov/Emills/PUBS/Cx-Costs-Benefits.html (accessed).
4. American Society of Heating. *Refrigerating and Air-Conditioning Engineers*, ASHRAE Guidelines 1–1996.
5. Green Building Council Web site. Available at: www.usgbc.org/leed (accessed October 2005).

BIBLIOGRAPHY

1. SBW Consulting, Inc. *Cost-Benefit Analysis for the Commissioning in Public Buildings Project*, May 2004. Available at: www.nwalliance.org/resources/documents/CPBReport.pdf (accessed).
2. Turner, W.C. *Energy Management Handbook*, 5th Ed.; Fairmont Press: Lilburn, GA, 2005.
3. Research News Berkeley Lab. *Berkeley Lab Will Develop Energy-Efficient Building Operation Curriculum for Community Colleges*, December 2004. http://www.lbl.gov/Science-Articles/Archive/EETD-college-curriculum.html.
4. Interview with Richard Holman. *Director of Commissioning and Field Services for Affiliated Engineers, Inc.* Walnut Creek, CA.

Commissioning: Retrocommissioning

Stephany L. Cull
RetroCom Energy Strategies, Inc., Elk Grove, California, U.S.A.

Abstract
This entry examines the practice of commissioning in existing buildings, or retrocommissioning. It provides a definition and a practical understanding of the retrocommissioning process, outlines the energy and nonenergy benefits that result, and examines the link between retrocommissioning and maintenance activities. It also explains the relationship between retrocommissioning and the U.S. Green Building Council's Leadership in Energy and Environmental Design (LEED) certification process and the energy conservation outcomes of retrocommissioning application.

INTRODUCTION

Retrocommissioning is a popular method for reducing energy and operating costs in all types of large, existing buildings. This popularity is due largely to the fact that retrocommissioning pays for itself quickly through the energy and operating cost reductions it produces. Aside from its abundant energy conservation potential, it also offers additional benefits. Depending upon the nature of its application, retrocommissioning can decrease demand maintenance frequency and occupant comfort complaints, improve indoor air quality, and enhance building maintenance staff productivity.

Retrocommissioning is not a one-time event in the life of a building. Rather, its long-term benefits are better realized through a continuous or ongoing approach that is supported by appropriate maintenance activities, throughout the life of a building.

This entry provides a definition of retrocommissioning and how it differs from commissioning and recommissioning. It offers an overview of the retrocommissioning process, the energy and nonenergy benefits, methods used to maintain those benefits, and some typical results of this application. Although a comprehensive retrocommissioning project could extend to all of a building's components, such as operating equipment and systems, core, shell, envelope, and finishes, this discussion focuses on those issues that impact energy utilization, operating costs, and the U.S. Green Building Council's Leadership in Energy and Environmental Design (LEED) certification program.

Keywords: Retrocommissioning; Recommissioning; Commissioning; Continuous Commissioning®; Supported Retro Commissioning℠; Building tune-up; Building maintenance; Maintenance management; Indoor air quality; Energy management; Energy efficiency; Energy conservation; LEED; Green buildings.

DEFINING RETROCOMMISSIONING

Where commissioning refers to a function that occurs during the construction of a building, to ensure that a new construction project meets design intent, and recommissioning is the periodic reapplication of that same function during the life of a previously commissioned facility, retrocommissioning is essentially the commissioning of an existing building that was never commissioned. All three terms refer to a systematic quality management process designed to ensure that building systems and equipment, whether or not they consume energy, function as intended by design, or based upon current use requirements.

Historically, commissioning has not been widely embraced in the new construction industry and its application is currently limited. This is due largely to the lack of hard data concerning the benefits of commissioning and a perception that its application adds unnecessary costs to a construction project. The result is an inventory of buildings that were never commissioned, many of which suffer from undiscovered deficiencies left over from the construction phase. This unfortunate circumstance is becoming more evident to building owners, developers, designers, and contractors and the practice of commissioning, recommissioning, and retrocommissioning is now receiving increased attention.

Retrocommissioning is widely regarded as a more challenging activity than commissioning or recommissioning. This is the case because buildings that were not commissioned during the construction phase generally lack comprehensive or updated reference documentation that is critical to understanding the design and operating intent of building components and systems. Without this up-to-date documentation, the retrocommissioning provider must gather information and perform functional testing "retroactively" to determine the existing performance characteristics for building systems, and then make adjustments to compensate for changes in building configuration and usage requirements. When complete, a retrocommissioning project should provide new and updated documentation sufficient for future recommissioning.

Understandably, and depending upon the age of the building, much could have changed since original construction. This can include changes to the physical configuration of occupied spaces, the addition of equipment and fixtures, new operating sequences, and altered performance characteristics of installed equipment. In large facilities with complex building systems, understanding and documenting these changes can be a time consuming and expensive process. For this reason, many retrocommissioning projects tend to focus only on building systems that have a history of chronic performance problems or those systems that have an impact on energy or operating costs. Projects such as these are often referred to as building tune-ups and they focus more on the effective and efficient operation of a building, rather than static design issues and equipment retrofits. Heating, ventilating and air conditioning equipment, related mechanical and electrical systems, building management and control systems, lighting systems, domestic water systems, and the building envelope are all good candidates for this approach.

THE RETROCOMMISSIONING PROCESS

Retrocommissioning normally includes four phases of work. These phases are used to determine the available opportunities on a broad scale, put a plan in place to address those opportunities, implement that plan, and document results. Although each phase in the retrocommissioning process may require additional components of work depending upon project objectives or complexity of the tasks, most projects follow similar sequences. For the purposes of this discussion, it is generally accepted that a retrocommissioning project will include the following steps:

- A preliminary evaluation of building systems and operating characteristics designed to gather initial data and determine potential project requirements and project scope
- A detailed assessment and project development phase where comprehensive observation, monitoring, testing, and project design takes place
- A project implementation phase where chosen recommendations are implemented
- A documentation and training phase where post-project building characteristics, operating procedures, and equipment conditions are documented, and building operator training is conducted

During the preliminary evaluation stage, the provider will determine the general nature of the project and attempt to come to some preliminary conclusions regarding potential solutions. This phase includes discussions concerning the owner's project requirements, project goals, and the analysis of available data concerning building design and design intent. Current usage requirements, equipment inventories, energy consumption and costs, and occupant complaint histories are examined. Demand maintenance evaluation, equipment performance issues, operating staff interviews, and first hand observation of building system conditions are also carried out. At the conclusion of the preliminary evaluation the retrocommissioning provider should have sufficient information to approximate energy and operating cost reductions the project could produce, and the costs to deliver those reductions. This information is normally summated in a preliminary evaluation report.

The detailed assessment and project development stage of a retrocommissioning project is normally the phase where more definitive conclusions are reached regarding the current operating performance of equipment and systems within the building. At this time, monitoring devices are installed on various pieces of equipment and in occupied spaces, in order to collect and verify current operating data. Information about occupancy patterns, equipment scheduling, control sequences, temperatures, pressures, flows and loads can provide valuable insights into building and systems performance.

In many cases the building will be controlled by a building management system (BMS) or direct digital control (DDC) system capable of collecting data on equipment operation. This data can be extremely useful for populating trend logs that provide real time information about building operation. The degree to which a BMS is capable of collecting this data will determine the need for independent or standalone data collection devices.

The detailed assessment and project development phase is also the time when the condition and efficiency ratings of building equipment is evaluated, and utility cost, consumption and demand profiles are confirmed. At this point equipment demand maintenance histories are reviewed for trends, maintenance and operation procedures are examined, and building operator skills assessed. At the completion of this phase, the provider should have a detailed understanding of the current operating parameters of building systems and equipment, and will have identified the specific strategies and measures to be used in mitigating equipment performance issues and reducing energy and operating costs. The provider will typically provide a detailed assessment and project development report at this stage. This report will usually contain all the relevant information required to make a decision on the financial merits of the project.

During the third phase of a retrocommissioning project the recommended measures that the owner has selected are implemented. This stage of the project is guided by the implementation plan developed during the project development phase. Although it is typical for the provider to manage the quality of the implementation process, individual components of the work are usually performed by third party providers who have specific expertise related to each of the measure requirements. The implementation stage is an excellent time for building maintenance and operations staff to become more familiar with the operational and equipment improvements taking place in their building. Their participation in this phase of the work can be a valuable learning opportunity because it offers a hands-on understanding of the process and a foundation for ongoing commissioning activities in the future. At the conclusion of project implementation, the retrocommissioning provider will have completed the recommended measures outlined in the implementation plan, and will have put in place any project monitoring requirements that were included in that plan.

The final stage of retrocommissioning includes the preparation of as-built documentation concerning design and performance changes made as part of the implementation plan. This should provide the owner with all documentation associated with functional testing and load calculations, along with updates to drawings, specifications, and changes to equipment operation. Information concerning system set-points, operating schedules, operations and maintenance manuals, and any additional changes or alterations made during project implementation should also be provided.

It is also normal at this time to implement any project outcome monitoring requirements and to initiate maintenance and operations training that was identified during the project development phase. Aside from contractual obligations that may require return visits to perform follow-up training or provide project monitoring reports, the retrocommissioning provider's duties are generally complete at this point.

Although the four phases of work described here are widely accepted as standard for core retrocommissioning applications, variations are not uncommon. Unique site conditions, project outcome expectations, the skill level of existing building operations staff, and other factors can all have an impact on sequence and content. In some cases, the relationship past the point of project completion continues because the owner perceives a value in retaining the provider in a project-monitoring capacity. Some providers offer a comprehensive package of services that "support" the owner's operations staff in their efforts to continuously commission their building beyond project completion. Still others will provide a periodic review and reporting function designed to validate projected energy and operational improvements that occurred as a result of the retrocommissioning project. The value and selection of these additional services will always be unique to the circumstances of the project.

RETROCOMMISSIONING AS AN ENERGY CONSERVATION TOOL

Retrocommissioning has the potential to produce large reductions in energy use at a relatively low cost. For this reason, it is becoming a popular cost avoidance tool among building owners and operators. To understand why these energy savings are available it is must be understood that most existing buildings were not commissioned when they were constructed. As a consequence, many buildings do not meet design intent at completion. The result is a large inventory of buildings that are likely to be operating inefficiently. That is not to say that construction projects lack quality control or specification compliance. Simply, it suggests that the historical effectiveness of the quality control mechanism during the construction process is insufficient to fully ensure that a new building operates in accordance with design intent. When factors such as building age, traditionally low investments in maintenance staff training, changes in space configuration, and multiple adjustments to building systems over time are added to that equation, the result is often a building that fails to operate in an energy efficient manner.

Unfortunately, buildings that were not commissioned during construction will have the greatest degree of problems in the very systems that are the most responsible for energy consumption. Heating, ventilating and air conditioning systems; supporting mechanical and electrical equipment; and the controls that govern the operation of these systems play a critical role in the energy profile of any building. In the absence of a comprehensive quality management process during the construction, much can be overlooked. It is not unusual to discover:

- Pumps installed backwards
- Fans and terminal boxes that produce incorrect air volumes
- Missing dampers and damper motors
- Economizers that are incorrectly sequenced
- Disconnected valve actuators
- Systems or equipment that is undersized or oversized
- Inappropriate building management and control strategies
- Simultaneous heating and cooling
- Lighting systems that remain active past occupancy
- Voids in the building envelope
- Equipment or devices that are missing altogether

Although these sort of static deficiencies are not uncommon, they are not the only source for concern. Another, and perhaps more important factor, is the impact that a lack of effective documentation and training at building turnover can have. Recognizing that the cost to operate a building is significantly greater over its lifetime than the cost of original construction, it must be accepted that operations and maintenance plays a critical role in a lifecycle cost analysis. Projects that lack sufficient documentation and allowances for training at completion simply have a greater likelihood for inefficient operation. Unfortunately poor documentation and ineffective training is more the rule than the exception.

Overall, retrocommissioning is a powerful tool for improving energy efficiency and building operator effectiveness. In a 2004 study sponsored by the U.S. Department of Energy[1] that focused on the results of retrocommissioning applications across the country, it was found that, on average, each building had a total of 32 deficiencies. In one building, a total of 640 deficiencies were discovered. Of all the deficiencies found as a result of this study, 85% were related to heating, ventilating, and air conditioning systems. Often, deficiencies like these go undetected for years, causing repeated comfort complaints from building occupants, wasting energy, and increasing the rates of equipment failure and repair costs. Given the frequency of these sorts of deficiencies in new buildings, it should not be surprising that building owners are reaping large energy conservation and operating cost rewards through the application of retrocommissioning.

Although the magnitude of savings that retrocommissioning can produce is dependent on the types of deficiencies discovered, the research sponsored by the U.S. Department of Energy[1] suggests that energy savings can be very significant. In that same 2004 study, which took into account the results from 106 separate projects, energy savings of between 7 and 29% were reported, with paybacks ranging from 0.2 to 2.1 years. Additional data from that study is shown in Fig. 1 below.

NONENERGY BENEFITS OF RETROCOMMISSIONING

In addition to the reasonably quantifiable energy conservation benefits of retrocommissioning, there exists a group of nonenergy benefits that are more difficult to quantify. Although these benefits will have differing values depending upon the type of building and owner motivations, they are none the less important considerations in a decision to proceed with a retrocommissioning project. For example, in the 2004 U.S. Department of Energy[1] study of existing building commissioning, it was discovered that although energy conservation was a primary motivator for 94% of the projects reviewed, a surprising number of projects were motivated by nonenergy benefits. See Fig. 2 below.

As evidenced by this study, issues such as systems performance, improved thermal comfort, and indoor air quality were all deemed to be important factors in choosing to proceed with a retrocommissioning project.

These benefits are difficult to quantify in monetary terms because they are the outcomes of equipment operation or building performance. However, when they are viewed in the converse, their value becomes clearer. For instance, in an office building where the loss of a major tenant can be a costly event to the landlord, maintaining a healthy, comfortable, and productive indoor environment is critical from both a business perspective and a liability standpoint. The value of the benefit in this case would accrue to the avoided costs of finding a new tenant for a vacated space and the avoidance of potential litigation costs associated with an indoor air quality problem.

In certain types of buildings, energy conservation investments can produce other types of financial benefits. An example of this would be the relationship between reductions in operating costs and asset valuation in the commercial real estate industry. According to the ENERGY STAR® Program, every dollar invested in energy efficiency, at a 20%–30% saving rate, is equivalent to increasing net operating income by 3%–4%, and Net Asset Value by $2.50–$3.75. Comparisons like these can be made for all kinds of buildings in all major industries.

An additional nonenergy benefit of retrocommissioning can be derived from improvements in productivity for maintenance and operations staff. This benefit can be sourced from the reductions in occupant comfort complaints that result from improvements in building systems performance. With this reduction comes improvement in productivity or the amount of time that existing staff can expend on deferred maintenance items, other specific activities that impact energy use, or training opportunities that support those functions. Once again, these improvements are difficult to quantify in monetary terms and studies to date have failed to produce definitive findings. It is not difficult to accept, however, that improvements such as these can only have a positive impact on a building operator's time and on the overall cost of operations for any building.

RETROCOMMISSIONING AND MAINTENANCE ACTIVITIES

Retrocommissioning can provide a host of benefits in existing buildings, and studies conducted by many respected individuals and organizations supports that observation. Maintaining those benefits through the life of a building is, however, largely a function of how well the building is operated and maintained during the post-retrocommissioning period. It has been revealed through several studies that the quality of maintenance activities will impact the persistence of energy savings, and equally, the nonenergy benefits that are dependent upon equipment performance.

The relationship between maintenance activities and the persistence of the energy and non-energy benefits of retrocommissioning has received considerable attention. In a 2003 study[2] jointly sponsored by the Department of Energy and the California Energy Commission, the researchers noted that energy savings that averaged 41% of total energy used in a building decreased by 17% over

	Units	Number of projects	Min	Bottom 25%	Median	Average	Top 25%	Max
Commissioned floor area	ft^2	106	5,690	95,101	151,000	209,729	271,650	1,014,133
Commissioning Costs								
Total	$2003/building	102	3,214	26,112	33,696	46,442	45,862	476,554
Normalized - excluding non-energy impacts, NEIs*	$2003/ft^2	102	0.03	0.13	0.27	0.41	0.45	3.86
Normalized - only for cases including non-energy impacts, NEIs*	$2003/ft^2	11	-0.27	0.04	0.17	0.41	0.45	1.88
Cx agent fee as percentage of total commissioning fee	%	9	32%	35%	67%	57%	71%	76%
Costs paid by:								
Building owner	%	31	0%	32%	50%	47%	50%	100%
Utility (e.g. as rebate)	%	48	20%	50%	84%	75%	100%	100%
Other (e.g. research grant)	%	7	33%	100%	100%	90%	100%	100%
Utility rebates (included in above costs)	$2003/building	48	917	11,932	20,500	23,685	25,000	76,725
as % of total costs	%	48	20%	50%	84%	75%	100%	100%
Deficiencies								
Per building	Number/building	85	0.7	5.0	11	32	21.0	640.0
Per 100kft2	Number/100kft^2	85	0.1	2.8	6	24	18.3	225.6
Measures								
Per building	Number/building	75	1.0	4.5	9.0	20.3	18.0	481.0
Per 100kft2	Number/100kft^2	66	0.1	2.5	5.9	8.6	12.7	218.6
Total Energy Cost Saving								
Raw data (mixed energy prices and years)	nominal $/building-yr	100	-25,752	11,739	33,629	66,489	75,940	879,101
Local energy prices	$2003/building-yr	100	-26,595	13,351	37,376	75,393	80,615	1,034,667
Standardized US-average energy prices	$2003/building-yr	57	-39,043	14,646	44,629	105,156	98,708	1,776,371
Percent energy bill savings	%	74	-3%	7%	15%	18%	28%	54%
Normalized Energy Cost Savings								
Raw data (mixed energy prices and years)	nominal $/ft^2-yr	100	-0.09	0.11	0.24	0.42	0.46	3.83
Local energy prices	$2003/ft^2-yr	100	-0.09	0.11	0.27	0.47	0.52	4.33
Standardized US-average energy price	$2003/ft^2-yr	56	-0.13	0.11	0.26	0.54	0.72	3.23
Monetized non-energy Impacts (one-time)								
Per project	$2003/project (1000s)	10	-281	-31	-17	-45	-11	-1
Normalized by floor area	$2003/ft2-yr	10	-0.55	-0.45	-0.18	-0.26	-0.10	0.00
Energy Savings								
Electricity	kWh/ft^2-yr	57	-0.70	0.64	1.7	2.2	2.76	9.72
Percent savings	%	46	-5%	5%	9%	11%	15%	36%
Peak electrical power**	W/ft^2	6	0.1	0.4	0.6	0.7	0.8	1.6
Percent savings	%	3	1%	2%	2%	7%	9%	17%
Fuel	kBTU/ft^2-yr	29	-14.2	2.3	6.5	15.6	13.5	209.5
Percent savings	%	19	-16%	1%	6%	13%	23%	67%
Thermal (chilled water, hot water, steam)	kBTU/ft^2-yr	19	6	32	64	94	122	356
Percent savings	%	16	13%	23%	36%	37%	48%	63%
Total	kBTU/ft^2-yr	57	-15	7	17.0	49.3	56	357
Percent savings	%	46	-7%	7%	15%	19%	29%	57%
Payback Times [undiscounted]								
Raw data (mixed energy prices and years)	years	99	-1.5	0.4	1.0	2.1	2.0	20.7
Local energy prices and inflation-corrected cx costs	years	99	-1.5	0.3	1.0	2.1	2.4	26.1
Standardized U.S. energy prices and inflation-corrected cx costs	years	59	-1.0	0.2	0.7	1.7	2.1	10.4

* Non-energy impacts (NEIs) include increases or decreases in first or operating costs due to changes in maintenance costs, contractor callbacks, equipment life, and
** Most are averaged over the entire year, hence true "peak" savings are significantly higher than shown here.

Fig. 1 Result summary with quartile analysis—existing buildings.
Source: Reprinted from "The Cost Effectiveness of Commercial Buildings Commissioning" LBNL Publication No. 56637 (see Ref. 1).

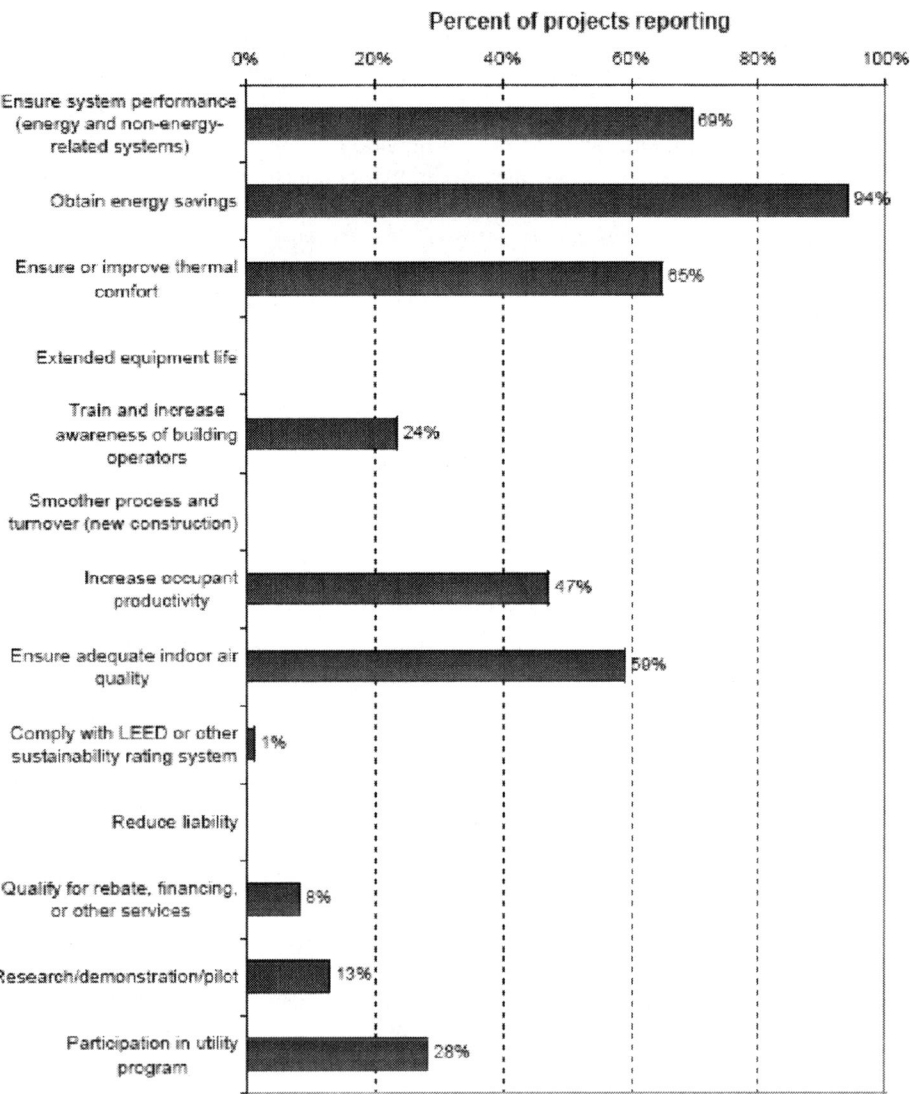

Fig. 2 Reasons for existing building commissioning.
Source: Reprinted from "The Cost Effectiveness of Commercial Buildings Commissioning" LBNL Publication No. 56637 (see Ref. 1).

two years. These findings also indicated that the long-term persistence of energy savings hinged on the abilities of the building operators to troubleshoot and understand how the systems in the building were supposed to operate.

In concluding why retrocommissioning benefits would persist in some applications and not in others, the study observed that persistence was influenced by a group of factors. The most notable of these appeared to be the working environment and operator training. Successful projects, where savings tended to persist, were those with working environments that provided high quality operator training, time to study and optimize building operation, and a management group that was focused on optimizing the performance of the building and reducing energy costs. In addition to the working environment and training, the study concluded that performance tracking and adequate documentation could impact benefit persistence as well. In the case of performance tracking it was suggested that energy use tracking and trend data analysis were important factors for persistence. Proper documentation concerning building equipment and its operating intent could also provide building operators with information on how to effectively operate the building.

Creating benefit persistence through building operator training can have a profound affect on how valuable a retrocommissioning project can be through the life of any building. This factor has been particularly well documented through the experiences at Texas A&M University over a number of years. Since 1993 the school has deployed a process called Continuous Commissioning® in more than 130 large campus buildings and has made training for building operators in this process a cornerstone of its maintenance efforts. The results of this focus have been dramatic, producing maintained energy savings for the school of between 15 and 25%. In a 2001 study conducted under the California Energy Commission's

Public Interest Energy Research Program and involving ten buildings at Texas A&M, the Continuous Commissioning process was projected to deliver $4,255,000 in energy savings over a four-year projection.[3]

Given the magnitude of energy savings and other benefits available from retrocommissioning, and the impact that maintenance functions and training have on the persistence of those savings, it is not difficult to conclude that a well-trained, effective, and supported maintenance program is essential for any building.

RETROCOMMISSIONING AND LEED CERTIFICATION

Commissioning and retrocommissioning have both found their way into the U.S. Green Building Council's LEED certification and point award process. This clearly illustrates the value that the USGBC places on these processes and provides further evidence that their application can have a positive impact on environmental and energy issues for any building.

The LEED certification process provides award points that encourage "green" building design, construction, and operation. Depending upon award points attained, a building can be ranked as LEED Certified, LEED Silver, LEED Gold, or LEED Platinum. It is important to point out that these award rankings can be achieved only if a new construction project is commissioned or an existing building is retrocommissioned. It is also important to note that in all rankings, LEED places a great deal of emphasis on the very systems in a building that account for the greatest impact on energy costs. This emphasis is similar for commissioning and retrocommissioning, and in many ways these processes, and the intent of the LEED rating system, are functionally complementary. Both have a positive affect on energy consumption, and therefore, a similar impact on the environmental footprint a building will make.

In existing buildings, retrocommissioning is regarded as a cornerstone of the LEED certification process. The LEED points available for existing buildings focus attention on building operations, testing, monitoring, repeated commissioning, and continuous improvement. Points are available for such things as the development of a Building Operations Plan, a commitment to 24 hours of training for building operators each year, performance monitoring of equipment operation and maintenance best practices. It is not coincidental that the most successful retrocommissioning projects encourage and support similar activities.

CONCLUSIONS

Retrocommissioning offers excellent energy conservation and operational improvement opportunities for existing buildings, with minimal investment. It can improve the effectiveness of operations and maintenance staff, provide a group of nonenergy benefits and can assist in the attainment of a LEED certification. However, the most important factor the reader can take away from this section is that commissioning during construction can eliminate many of the problems that owners inherit at the acceptance of their project. Ultimately, as commissioning is more widely adopted in new construction projects, the need for retrocommissioning could be eliminated altogether.

In the meantime, however, retrocommissioning is gaining an impressive following of supporters.

It is encouraging that the Department of Energy, ENERGY STAR®, American Society of Heating, Refrigerating and Air-Conditioning Engineers (ASHRAE), the Association of Energy Engineers, the California Department of General Services, the California Energy Commission, The U.S. Green Building Council, and many others now recognize the impact that this application brings to energy conservation and environmental protection. In California, retrocommissioning has found its way into the State's Green Building Action Plan by requiring that all buildings 50,000 ft^2 and over be retrocommissioned and that periodic recommissioning take place in the following years. In another signal of the value of this application, utilities in California are now offering rebates and incentives to encourage building operators and owners to implement retrocommissioning in their buildings.

Given the numbers of buildings in North America that have not been commissioned, and the ever increasing pressures to reduce energy consumption, it is not difficult to envision that retrocommissioning will become a very popular energy conservation measure in the years to come.

REFERENCES

1. Mills, E.; Friedman, H.; Powell, T.; et al. *The Cost Effectiveness of Commercial Buildings Commissioning-A Meta Analysis of Energy and Non-Energy Impacts in Existing Buildings and New Construction in the United States*; Lawrence Berkeley National Laboratory, Portland Energy Conservation Inc., Energy Systems Laboratory-Texas A&M University, Building Technologies Program, U.S. Department of Energy, 2004, Lawrence Berkeley National Laboratory Report No. 56637, 31,34,36, http://eetd.lbl.gov/emills/PUBS/Cx-Costs-Benefits.html.

2. Friedman, H.; Potter, A.; Haasl, T.; Claridge, D. *Strategies for Improving Persistence of Commissioning Benefits*; Portland Energy Conservation Inc., Texas A&M University Energy Systems Laboratory, U.S. Department of Energy, California Energy Commission, 2003; 2,4.

3. Turner, D.; Claridge, D.; Deng, S.; et al. *Persistence of Savings Obtained from Continuous Commissioning*; Texas A&M University Energy Systems Laboratory, Texas A&M University Physical Plant, University of Nebraska, Omaha, California Energy Commission, 2001; 13, HPCBS No. E5P2.2T5a2.

Compressed Air Control Systems

Bill Allemon
North American Energy Efficiency, Ford Land, Dearborn, Michigan, U.S.A.

Rick Avery
Sam Prudhomme
Bay Controls, LLC, Maumee, Ohio, U.S.A.

Abstract
This Compressor Control Systems entry provides an outline of the strategies used to manage compressed air generation at a typical manufacturing facility and discusses the components of a compressed air system, typical control methodology, and energy saving strategies. Compressed air distribution, metering, and monitoring is also reviewed. This entry is meant to give the reader a basic understanding of compressed air generation and management fundamentals.

INTRODUCTION

Compressor systems are benefiting from increasingly advanced control technologies to realize improved efficiency, safety, and operational benefits. Compressor controls now use microprocessors, computer networking, sophisticated control algorithms, and Web-based monitoring to provide superior control capabilities and features.

Today's control systems do more than just operate the air compressor. Using sophisticated technologies, advanced controls can significantly reduce energy expense and lower maintenance costs. Automation is another important feature, allowing much of the standard compressor system operation to be managed by software. By monitoring and controlling compressor auxiliary equipment, modern controls can help ensure that high-quality air at the lowest cost is reaching the end user. Zone management provides the ability to regulate air use in various plant departments. Remote and Web-based integration systems improve and streamline system control and management. Extensive data gathering and reporting provides real-time information to help drive business decisions, such as evaluating system expansion, identifying additional savings opportunities, and managing predictive maintenance.

Modern compressor control systems address the entire compressed air infrastructure. Today, a complete control approach can include the following facets of a compressor system: air production controls for air compressors and the motors that power them; air quality monitoring for dryers and other air conditioning equipment; distribution control in the form of zone management; and integration, management, and metering products to improve operations and decision making.

COMPRESSOR CONTROL SYSTEM COMPONENTS

Air Production Control

The supply side of a compressor system consists primarily of the compressor and the motor. Control systems have several important tasks: to ensure that the compressor system produces enough air to meet plant demand; to keep the compressor and motor running without costly shutdowns; to operate the system as efficiently as possible; and to prevent damage to the major pieces of equipment.

Older compressors use pneumatic and electro-mechanical control systems. Modern microprocessor-based controls improve performance by offering precise pressure regulation, networked capacity control, and additional features. For centrifugal compressors, advanced surge control can dramatically increase efficiency by reducing wasted air blowoff.

Electronic controls are available for all makes and models of compressors. To provide maximum efficiency and automation, a control system should electronically network the compressors, regardless of type or make. The primary goal of a control system is to ensure that the compressors produce enough air to meet plant demand.

Motor Control

Compressor systems begin with the prime mover, often an electric motor. A well-functioning motor is obviously crucial to the operation of a compressed air system.

Keywords: Compressor; Control; Air; Dryer; Centrifugal; Pressure; Surge; Blowoff; Leak; Automation; Regulate; System; Efficiency.

Effective electric motor control prevents damage and helps alleviate common electric power concerns.

Primary Motor Control

Modern electric motor controls offer a number of features that improve motor reliability. Extensive motor protection is built into these products, preventing motor overload and burnout that can result from drawing too much power. User-friendly, man–machine interfaces, with Liquid Crystal Display (LCD) screens and micro keypads, are now common features. These allow for easy initial configuration and modification of the control's operating parameters. Some motor controls are based on solid state components, which require less maintenance and replacement of costly electrical parts that otherwise wear out over time.

An important element of modern motor controls is the soft start function, wherein the motor is gradually brought up to its maximum power. Compared to a typical high torque start, a slow start greatly reduces mechanical and electrical system shock to the motor and attached compressor. This results in less wear and tear from starts and stops and reduces long-term maintenance of expensive motors. A soft start also reduces the initial electrical current inrush, placing less stress on the facility's electrical system and alleviating related low power problems.

Ride Through Motor Control

An important new auxiliary motor control is the ride through controller. This device addresses the problem of momentary interruptions to the motor's power supply, which, no matter how short, can often cause motor shutdown. Motor shutdown stops the supply of compressed air, halts the facility's production, and results in lost productivity and increased costs. A ride through device has the ability to keep the motor operating during momentary power interruptions and voltage sags. Because these are lower cost auxiliary controllers, they can often quickly justify installation costs.

Compressor Control

The most fundamental part of a compressor control system is the controller on the individual compressor. These devices are designed to protect the compressor from damage, operate it as efficiently as possible, and automate recurring control actions. Modern compressor controls offer features far beyond those of their predecessors; powerful microprocessor-based controls are becoming increasingly popular as their significant advantages become more widely known.

Compressor Control Types

Pneumatic

Many existing air compressors utilize electro-pneumatic control systems. These systems use electric and mechanically activated devices such as pressure switches, solenoid valves, and metering pins. To address monitoring and compressor protection, electro-pneumatic systems typically feature a series of mechanical trip switches which shut down the compressor when pressures or temperatures reach critical levels. With their mechanically limited control logic, pneumatic controls offer only a basic set of operating functions—usually simple compressor modulation and monitoring. In addition, the response time of pneumatic controls is typically inferior to other types, reducing the effectiveness and efficiency of control actions.

Programmable Logic Controllers

Programmable logic controller (PLC) systems are in common use today, and they represent a significant improvement over pneumatic-based systems. Programmable logic controller control systems utilize digital and analog control and monitoring instruments. The responsiveness and accuracy of these devices enable greater compressor efficiency. Additionally, because PLCs are electronically based, they can offer more functions, such as sequencing and advanced control and monitoring. Fundamental speed and performance limitations of PLC hardware, however, still leave room for improvement.

Microprocessor

The market for microprocessor-based controls is growing rapidly due to their comprehensive features and superior performance. The control and monitoring accuracy of these systems is excellent, allowing for tight pressure regulation and advanced protection strategies. Because these controls are based on hardware similar to modern computers, they can offer additional features such as sophisticated networking, multiple compressor control, and extensive operational record keeping. Many of the control benefits and features listed in this document are best implemented by these types of control systems.

Protection and Safety

On the most basic level, compressor controls must keep the machine running safely and reliably, preventing both serious compressor failure and more common mechanical damage.

Compressed Air Control Systems

Plant Safety

Maintaining plant safety by preventing catastrophic compressor failure is one of the most fundamental responsibilities of a compressor control. One example of severe compressor failure is when incompressible water builds up in a reciprocating compressor's compression chambers; if enough accumulates, the compression action can cause the equipment to physically fail, sometimes explosively. Other safety concerns include overheated and potentially combustible oil entering the air distribution system or violent vibration caused by a severe mechanical failure. Most modern control systems have the ability to shut down the compressor if a serious malfunction is detected.

Machine Protection

On a less extreme scale, compressor controls should guard the compressor from common wear and damage. Advanced control systems use analog and digital monitoring instruments to check the compressor for abnormal operating conditions. When the control system detects an unsafe measurement, it can either trigger a warning alarm or shut down the unit, depending on the nature of the situation.

Modern control systems examine every relevant operational value of a compressor and continuously check these values against standard ranges. Today's systems can visually display a list of the current values and alarm parameters of every monitoring point, allowing an operator to know precisely what the compressor is doing at any moment. Another useful feature is a recorded history of past operation and alarm events. If a problem occurs, such a record can provide critical diagnostic information to help resolve the issue quickly.

Centrifugal Compressor Surge

On centrifugal compressors, a key control feature is the ability to reduce or eliminate surge. Surge is a phenomenon where compressed air rapidly oscillates backwards then forwards through the compressor; unabated, this can cause severe damage. From a maintenance and protection standpoint, modern controllers offer far superior surge prevention compared to older systems. Advanced systems, with mathematical models of the compressor's surge line and fast control responses, work to keep the compressor out of surge. If surge does occur, these systems can detect the event and adapt the controller's operation to avoid any recurrence.

ENERGY SAVINGS WITH MODERN CONTROLS

One of the primary benefits of an advanced compressor controller is reduced energy expenses. Energy savings can result from any or all of these factors:

- Precise pressure regulation reduces the average system pressure output.
- Networked capacity control coordinates production among multiple compressors for maximum efficiency.
- Advanced centrifugal control can reduce wasted air from blowoff.
- Leak loss reduction is a byproduct of a lower average system pressure.
- Automated load scheduling can shut down or offload compressors when plant demand is lower.
- Proper intercooler control ensures better compressor efficiency.

Precise Pressure Regulation

Significant energy savings can be realized by lowering compressor discharge output pressure. Older systems are slow and inaccurate, resulting in large plant pressure swings. In order to keep pressure from swinging below the minimum, these older controls commonly maintain an average system pressure much higher than necessary. Modern controllers, with faster, more precise abilities, and sophisticated control strategies, can greatly reduce pressure swings. The smaller pressure range, typically within 2 psi of target pressure, allows a subsequent drop in average pressure setpoint. A 2-psi reduction in system pressure results in an approximately 1% drop in energy use; thus, the potential energy savings are substantial.

Networked Capacity Control

In plants with multiple compressors supplying a common system, uncoordinated compressor operations often offer opportunities for energy savings. Compressors that do not communicate with each other act independently, raising or lowering output as they detect changes in plant demand. This can often result in competing compressors; one compressor may be lowering its output while another is increasing its capacity. Unstable plant pressure levels are one result of this competition. Additionally, the compressors operate at inefficient part load capacity levels.

Sequencers, which were the first solution to this problem, assign compressors different fixed pressure levels at which point they come on or off line. Although this method does prevent competition between the individual compressors, it causes plant pressure to fluctuate along the range of assigned pressure levels (a four-compressor system with pressure intervals of 5 psi will have a 20 psi operating window) and tends to maintain an average system pressure higher than needed.

Advanced control systems with networked capacity control capabilities produce an efficient compressor system operation. Through network communication, the controls automatically operate as many individual compressors at their most efficient (full load) capacity levels as possible. Instead of multiple compressors operating at part load capacities, a single compressor is modulated to meet plant demand. The whole system is coordinated to maintain a single pressure setpoint, providing precise pressure regulation. Should demand fall, compressors are automatically shut down or unloaded, further saving energy. Rising demand will cause another compressor to come online, ensuring stable plant pressure. With such a system, multiple compressor installations can maintain plant pressure while operating individual compressors as efficiently as possible.

System Controllers

Built-in networked capacity control is limited to the more advanced control systems. When installation of these systems is not feasible, similar capabilities can be achieved with a system controller. In order to achieve the benefits of networked control with less capable, mismatched, or incompatible control devices, a system controller is sometimes used to coordinate the individual compressors. These master controllers offer many of the features of networked control, and can operate several compressors at a common plant pressure setpoint.

Surge Control and Blowoff Reduction

For centrifugal compressors, blowoff at minimum capacity is a significant energy waste. Modern controllers can reduce this waste using several methods. Advanced antisurge algorithms provide greater turndown, allowing a larger modulation range before blowoff begins. Additionally, when the minimum capacity point has been reached, the controller can switch the compressor into an unloaded state, where it produces little air and thus blows off little air. In a networked capacity control system, other compressors can modulate output, allowing the centrifugal compressor to operate at higher and more efficient capacity levels. These combined features offer great energy savings potential.

Leak Loss Reduction

Reducing the average pressure setpoint also reduces the amount of air that escapes from existing leaks. Leakage easily can be the largest energy problem in a compressed air system, ranging from 2 to 50% of compressor system capacity. An average plant has a leak rate of about 20% of total air production.

Lower pressure air has less air mass in the same volume. Because the volume of leaking air remains constant at a given pressure, a lower pressure results in less air escaping from existing leaks. Precise pressure control is one way to lower average system pressure.

Load Scheduling

Load scheduling automatically matches the compressor's output pressure with predetermined plant demand. During breaks and off-production periods, a significantly lower pressure often can be maintained in the plant. Any pressure reduction will save notable amounts of energy.

Compressor Intercooler Control

Effective intercooler control will help a compressor operate at top efficiency. Air that has not been cooled adequately by the intercooler will enter the next stage with a larger volume, reducing total compressor output. Alternately, air that is cooled too much can form liquid condensate, which can damage compressor components and increase maintenance costs. Modern control systems usually include intercooler control as a standard feature.

Automation

Automation is a key element of advanced control systems. Automated machine protection, data collection, and start/stop, combined with capacity regulation and load scheduling capabilities, give modern control systems the ability to automate nearly every operation of a compressor system. Remote monitor and control capabilities further reduce dependence on at-the-controller compressor supervision.

Start/Stop

Modern control systems can automate a compressor's start and stop procedures. With monitor and control connections to essential compressor subsystems (motor, lubricators, coolant, etc.), the controller can start and stop the complete compressor station while ensuring safe operation.

In a networked control system, automatic start capability can add additional system reliability. When one compressor is shut down because of a problem, the system can automatically compensate for reduced air supply by bringing additional compressor capacity online. Thus, plant pressure is maintained and production continues with limited interruption.

Scheduling

When start/stop capabilities are combined with a schedule, much of the day-to-day compressor operation can be automated with controls. Once an effective schedule is established, the system can essentially run on autopilot, with little need for immediate operator adjustments.

AIR QUALITY

An often overlooked element of compressor system controls is the equipment used to condition air, which primarily includes dryers, aftercoolers, and filters. These pieces of equipment have the essential task of ensuring that hot, wet air leaving the compressor is converted to high-quality, cool, dry air for use in the facility. The two largest air conditioning problems are insufficient drying and excessive differential pressure drops.

Wet air that enters the distribution system can eventually cause rust formation, leading to clogs and extra wear on end use equipment. If the problem becomes severe enough, it can require a shutdown of portions of the compressor system while corroded piping and failed components are replaced. These humidity problems can be caused by inadequate drying or poor aftercooling.

An improperly sized, poorly maintained, or outdated piece of equipment can cause an excessive pressure drop as air passes through it. Pressure drop across dryers alone can be greater than 6 psi; because a 2-psi pressure change roughly equates to 1% of energy used to compress the air, the opportunities for energy savings are substantial. Excessive pressure drop is a factor that can affect dryers, aftercoolers, and filters.

Monitoring Air Quality

Given that high quality air is so important, there are opportunities for mitigating these issues with monitoring techniques. By monitoring the pressure and quality of the air, both before and after conditioning has taken place, a facility can identify existing and potential problems. With continued monitoring, the results of equipment upgrades or maintenance actions can be verified for effectiveness. This approach is also able to flag when maintenance is necessary to upkeep the quality of air, reduce unnecessary pressure drops, and maintain system efficiency. Monitoring capabilities of this type are most often integrated into an overall compressor management system, where the metered values are displayed on a networked compressor management system workstation.

AIR DISTRIBUTION

Zone Management

An emerging control strategy for the distribution side of a compressed air system is known as zone management, which provides additional opportunities for energy savings and operational improvements.

Zone management technology gives the capability to monitor and control the demand side of a compressed air system. This strategy involves separating an air system into different distribution zones and regulating and metering the air supply to each one. The control of each zone can be scheduled for automatic operation, making complex zone management relatively easy. Zone management opens up new system operation options, such as running zones at different levels of pressure and shutting off zones when they are not in use.

With the ability to monitor the air consumption of each zone, users can gain much greater insight into the operational dynamics of a compressed air system. This advanced metering ability provides an accurate determination of the compressed air energy costs of different plant operations and can provide incentives and justifications for initiatives to reduce those costs.

On a design level, zone management involves logically separating the operations that use compressed air into different zones based upon factors such as concurrent air use, pressure setpoints, and air quality. The air supply to each zone is then individually controlled, allowing the air flow to each zone to be modulated according to the current use in that zone.

With a scheduling function, the system can be configured to automatically raise and lower pressures or to turn the air supply completely off to each individual zone.

Implementing zone management requires a combination of metering and control instruments for each zone, and a master control device to provide the monitoring, control, and scheduling functions. This master control functionality is often an add-on capability of a compressor management system workstation.

Benefits of Zone Management

A facility can realize one or more benefits from a zone management system: reduced compressed air energy consumption, zone cost regulation, increased data collection, air quality monitoring, and zone air leakage measurement.

Reduced Air Use

Zone management often results in a reduction in air demand due to its ability to lower the pressure or completely stop the flow of air to zones that are not in use. Less air is then used to maintain pressure in nonproduction areas and total demand is lowered. Less demand means less compressed air production, which directly reduces energy costs.

Regulate Air Costs

The ability to establish a separate cost center for each air use zone is another important benefit that comes from the capability of monitoring and metering the air use of each zone. With a comprehensive metering program, the facility has the ability to regulate the costs of compressed air for different segments of their production operations. Better information regarding air use and the costs of that use allows for better management decisions to be made.

End to End System Information

When zone management is paired with modern controls for other sections of the compressor system, a facility can have complete end to end system information and management. This system-wide approach enables better understanding of the operational dynamics of the entire compressed air system.

Monitor Air Quality

Air quality can be just as important as energy management. Correct pressure and humidity levels ensure the proper functioning of end use equipment and lower the lifelong maintenance costs of equipment based on wear and tear. Zone management allows constant monitoring of the pressure and dew point of the air entering each zone, ensuring that high-quality air reaches the end use point.

Measure Zone Leakage

Leakage is a frustrating aspect of compressed air systems and one that is difficult to measure. Zone management makes it easier to determine how much air is used and thus wasted by a particular zone when it is not in use. A majority of this wasted air comes from air leaks. By measuring the amount of air leaking from a zone, the facility can make effective decisions regarding leak control programs and the potential return from such efforts.

AIR MANAGEMENT, MONITORING, AND METERING

The final control layer for a compressed air system is an integration strategy that pulls all of the different control and monitoring features into one centralized location. This is usually accomplished through a compressor network, which is then routed to a human machine interface (HMI) workstation. This workstation can be a local monitor and control computer, or a Web-based management and analysis client server.

Facility Intranet Monitoring and Control

When connected to a network of individual compressor controllers, a remote computer station allows for monitor and control of an entire compressed air system from a single convenient location. With access to the entire range of monitored compressor data, the remote station can display the condition of the complete system at any given moment. A remote operator with security access to control capabilities can start, stop, and change the capacity of any compressor from the remote station.

Remote operation also allows an operator to pinpoint compressor problems as they occur. More immediate and precise information directs corrective actions, reducing compressor down time and associated costs from loss of adequate plant air supply.

A workstation connected to a network of compressor controls can also record and store data for the entire compressor system. This wealth of information is then available for analysis, which can quantify compressor system performance over time. Individual compressor problems can be identified and corrected before they become serious. This preventive maintenance reduces operating costs resulting from both inefficient operation and lost production due to compressed air equipment failure.

Below are some of the specific features of a typical HMI compressor workstation:

- A total system view, which displays the operating conditions of the complete compressed air system, including pressure, motor status, and alarm status.
- Complete remote control over every compressor on the network, including starting and stopping, capacity changes, and pressure setpoint changes.
- An individual compressor view, which displays the current readings of the major monitoring points, while graphically showing the user where the points are located on the unit.
- A complete list of all monitoring points for the individual compressor, with alarm and trip values and current status.
- Recorded history of all alarm and trip readings that occur.
- Recorded history of all operator events for the compressor.
- An ability to generate analytical performance reports based on the collected data. These reports can track compressor performance over time and spot trends before they become problems.

Web-Based Management Systems

A Web-based compressor management system offers many similar features to a local compressor network.

One of the central capabilities of a Web system, however, is the ability to provide access to multiple facilities from a standard internet connection.

Usually, a gateway device of some kind is used to connect the facility compressor network to a dedicated communication connection, such as a phone line or broadband connection. An off-site server then collects data from the system and a Web portal provides access to the information. From the site, a user can access real-time monitoring of plant air compressor systems, as well as the operating parameters of each individual compressor. For security purposes, a Web system often does not allow control actions; this prevents unauthorized tampering with the facility's compressor operation. If a company operates multiple plants, the Web site can provide centralized monitoring and metering capabilities for the entire enterprise. The Web-based nature also means that a user can access compressor system information from anywhere there is an internet connection. Additionally, Web systems frequently incorporate built-in efficiency and operations analysis and reporting, giving management powerful tools to measure the performance of their air compressor systems.

Benefits from Whole System Integration

These centralized management, monitoring, and metering systems offer many benefits to the operation of a compressed air system. Some of the more common benefits are described in the following sections.

Internet-Based Access

A Web-based system is accessible from any standard internet connection. This provides operators with monitoring capabilities from office, home, or even Web-enabled phones or PDAs (personal digital assistants, such as Treo or Blackberry).

Cost Metering

An integrated management solution offers the ability to meter the costs of compressed air operations. Comprehensive whole system metering provides operators and management with improved understanding of the costs and operational dynamics of the compressed air system. Real-world data can then be used to identify opportunities and drive business and management decisions.

System Efficiency

The ability of a management system to accurately and consistently calculate the cost of compressed air for single or multiple facilities makes it easier to measure any changes in efficiency and compressed air production. As future investments are made in the compressed air system, a management and metering system can compare system efficiencies before and after improvements are made to validate savings. Using centralized monitoring and analysis capabilities, a management system can often help identify further efficiency gains from operational changes. For example, these systems can help find the most optimal, most efficient mix of compressors to use during different shifts at a facility. Once the optimal mix is found, automatic scheduling ensures that the most energy efficient solution is used.

Measure Air Leakage

Leakage is a notoriously hard quantity to measure in normal circumstances. A system management tool can easily measure the air flow rate of a facility during nonproduction hours; this flow will be a close approximation of a system's air leakage.

Preventive Service Monitoring

The constant monitoring performed can be helpful in identifying and resolving problems with compressors before they become serious. Problems with sensors or other issues can be identified during periodic system reviews and resolved before the issue becomes serious.

Real-Time, Data Driven Troubleshooting

When a problem with a compressor does occur, the data recording and real-time monitoring capabilities of a whole system management solution can provide faster resolution to the problem. With a Web-based solution, experts from outside facility locations can view the current system status and help with troubleshooting.

Comprehensive Overview

Companies with multiple facilities find a Web system very useful in providing a centralized overview of all compressor networks. These tools provide an easy and accessible method to see the current status of every compressor on the system.

CONCLUSION

Modern compressor control systems now offer increased levels of sophistication and opportunities for control of the entire compressor system. A comprehensive, system-wide solution cannot only control and protect individual compressors, but it can also provide powerful new operational and management tools. This new control strategy provides powerful, efficient, and effective control of compressed air systems.

Compressed Air Energy Storage (CAES)*

David E. Perkins
Active Power, Inc., Austin, Texas, U.S.A.

Abstract
Compressed-air energy storage (CAES) is currently being deployed as an alternative to lead-acid batteries for uninterruptible power supplies. These systems use compressed air supplied from either transport cylinders delivered by local gas services, or from stationary cylinders refilled from on-site compressors to drive a variety of economical expansion engine topologies. Several factors make these systems feasible for use in small-scale CAES systems for load leveling in conjunction with wind or solar energy generation while opportunities exist for improving cycle efficiency.

INTRODUCTION

Large-scale CAES has been successfully used as a means of peak shaving as an alternative to peaking gas turbines. Two such systems are the 290-MW unit in Huntorf, Germany[1] and the 110-MW unit in McIntosh, Alabama.[2] These systems allow independent operation of the compression and expansion processes commonly found in conventional gas turbines. These systems have been called hybrid CAES; in that they continue to use fuel in the expansion process, the benefit being that the turbine need not produce power to drive the compressor when operating from the compressed air reserve. A disadvantage of these systems is that they require the use of fuel, which results in CO_2 emissions, and the heat of compression is discarded thus compromising the cycle efficiency.

One solution to these issues is to use adiabatic compression and thermal energy storage (TES) in place of the combustion process associated with hybrid CAES. Systems employing adiabatic compression and energy storage were explored in the 1970s and have received renewed interest. One program titled Advanced Adiabatic Compressed Air Energy Storage (AA-CAES)[3] is underway in Europe, which is focused on zero-emission storage technology for centralized storage as well as modular products for distributed storage.

Several companies are now offering CAES systems in various configurations as environmentally friendly alternatives to the lead-acid batteries found in uninterruptible power supply (UPS) systems. Earlier this year, active power-introduced products for the UPS market based on its Thermal and Compressed Air Storage (TACAS) technology.[4] Several companies, including a major supplier of photovoltaic cells have expressed interest in the use of this technology for electricity storage generated by renewable sources. At this time, however, systems based on TACAS technology achieve cycle efficiencies between 10 and 15%, since it currently relies on oil-lubricated reciprocating compressors, which are nearly isothermal. This paper presents the current embodiment of the TACAS technology for UPS and discusses proposed enhancements to improve cycle efficiency for electricity storage.

THERMAL AND COMPRESSED AIR STORAGE FOR UPS

A CAES system with thermal energy supplied by the grid has been developed for UPS applications. The output of this system is 85 kW and is capable of delivering power for up to 15-min (21 kWh). A schematic of the system architecture is shown in Fig. 1.

The TACAS technology system uses high-pressure gas cylinders for air storage since volumetric energy density should be compatible with batteries for UPS applications and the use of caverns is impractical. The use of high temperature TES heated by the grid allows higher turbine inlet temperature than would, otherwise, be available from direct expansion of the air from cylinders.

The specific energy available from compressed air is a function of the turbine pressure ratio and inlet temperature. System sizing then becomes a tradeoff between the mass of gas stored in the cylinders and the mass of TES required to heat the gas being delivered to the turbine to achieve a desired inlet temperature. Cost optimization was performed to balance the size of the cylinder banks and TES as a function of the turbine pressure ratio, inlet temperature, and discharge temperature. The turbine pressure ratio was constrained by manufacturing capabilities and the discharge temperature was constrained to be

---

*This entry was originally presented as "Compressed-Air Energy Storage for Renewable Energy Sources" at the World Energy Engineering Conference (WEEC), 13–15 September 2006, Washington DC, U.S.A. Reprinted with permission from AEE/Fairmont Press.

Keywords: Energy storage; CAES; Uninterruptible power supplies; UPS; Air turbine generators; Thermal energy storage; Renewable energy.

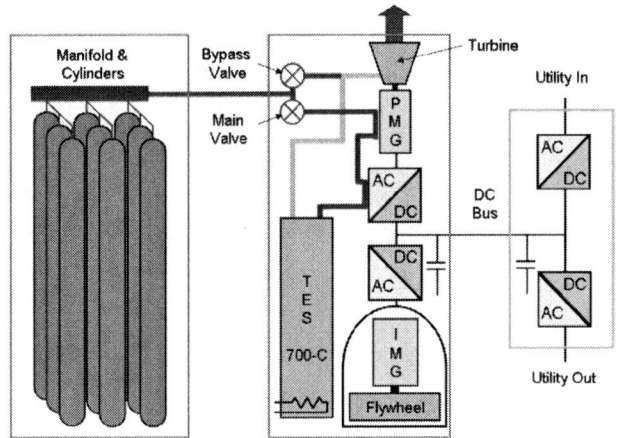

Fig. 1 Thermal and compressed air storage (TACAS) technology for UPS.

suitable for discharge into an interior building space. Maximum storage pressure was constrained by the most economical cylinders and compressors as well as diminishing returns due to significant effects of gas compressibility effects (>10%) above 310 bar [4500 psi].

Small-scale electric power and storage systems are generally more expensive to produce per kW and/or kWh output, so the optimization described above is essential. Furthermore, in order to meet system-cost targets, the simplest designs must be adopted.

EXPANSION TURBINE AND CONTROL

For small-scale systems, single-stage axial-flow impulse turbines offer an efficient and cost effective solution for high-pressure ratio expansion applications, particularly when direct-coupled to a high-speed permanent-magnet generator (PMG). A design speed of 70,000 rpm minimizes the PMG size helping to reduce the mass of expensive permanent magnet material and is a compromise between several machine design factors.

The system-optimized turbine shown in Fig. 2 operates at 385-psig and 230°C inlet temperature and delivers 100-kW shaft power at an efficiency of approximately 72% (small turbines typically have efficiencies between 70 and 80% whereas larger designs can approach 90%). Nearly the entire temperature drop occurs during expansion across the supersonic converging-diverging nozzles, so the temperature reaching the bladed disk approaches ambient. This allows the use of inexpensive rotor materials with integrally machined blades.

The single-stage turbine readily adapts to overhung-rotor architecture. The short overhang distance and short overall rotor length considerably simplify rotor-dynamic issues. In this particular configuration, rolling-element bearings with elastomer dampers are employed since the operational duty cycle for UPS is low. The design is easily configurable to foil-type gas bearings for higher duty service. Blade reaction is close to zero so axial bearing loads are very low and easily managed. Cooling of the bearing closest to the turbine inlet plenum is a challenge, but effective routing of air released from the control regulators helps to mitigate this. In fact, cooling of the turbine-PMG and its power converter is also provided by the process airflow.

Because the system is presently designed for UPS, transient response is of paramount importance. It takes about 1 s for the turbine-generator to reach full speed while carrying increasing load during the acceleration event. A small flywheel is used to "bridge" the gap between utility outage and full-load turbine output. This flywheel is configured with a bi-directional converter, which allows the flywheel to absorb step unloads and eliminates the need for unloading valves. During discharge, the flywheel also manages small power fluctuations so that turbine speed is held constant.

Simple turbines based on fixed nozzle geometry achieve power regulation through inlet pressure and temperature control. For this system, turbine inlet pressure and temperature are controlled by a pair of dome-loaded pressure regulators using a unique control scheme,[4,5] which routes air through or around the TES. Referring to Fig. 2, Main Regulator 340 controls airflow going to the Heat Exchanging System 350 (in this case, the TES). Bypass Regulator 320 controls airflow around the TES. Orifice 330 in the bypass path provides control stability. By combining flows through regulators 320 and 340 in varying proportions, constant fluid discharge temperature and pressure can be achieved throughout the sliding temperature range of the TES.

Regulators 320 and 340 identified in Fig. 3 are called dome-loaded regulators. This type of regulator is often employed when a high flow coefficient is needed. A dome-loaded regulator provides a discharge pressure equal to the pressure signal applied to the dome. Dome pressure

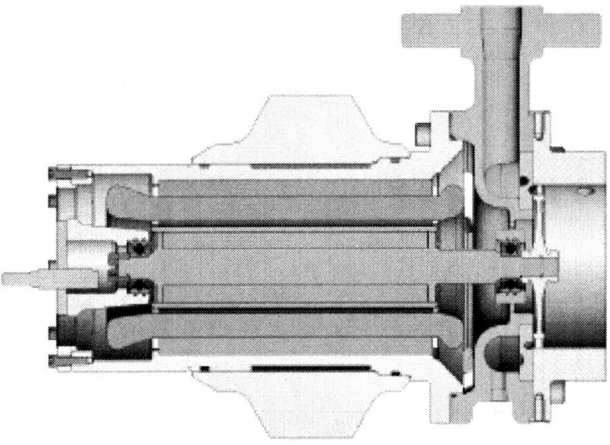

Fig. 2 100-kW single-stage axial-flow air turbine generator.

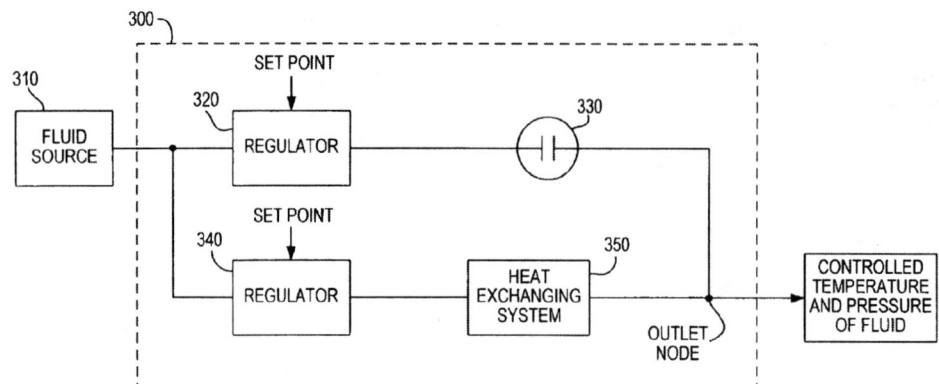

Fig. 3 Method of controlling constant turbine inlet temperature and pressure.
Source: From Ref. 5.

control is provided by a combination of solenoid valves 1050 and 1052 in Fig. 3 that, respectively, either pressurizes the dome with air supplied by an upstream regulator and accumulator, or depressurizes the dome by venting the dome to atmosphere. Operation of these two valves is provided by a digital controller with feedback from downstream sensors. For reliable termination of dome pressure in emergency situations, a redundant NO solenoid valve 1060 is in line with an Emergency Power Off (EPO) circuit and is powered closed during normal operation. Faults such as sensed turbine overspeed or manual EPO will de-power the valve and vent the dome causing the pressure regulator to close (Fig. 4).

THERMAL ENERGY STORAGE

High-temperature thermal storage using 304 stainless steel provides extremely compact, robust, and low-risk TES when designed in accordance with American Society of Mechanical Engineers (ASME) rules. Although material cost is relatively expensive, manufacturing processes are simplified by integrating the thermal storage and pressure retention functions and an annular channel configuration[7] achieves very high heat transfer coefficients and heat extraction efficiencies (Fig. 5).

Advanced micro-porous insulation is used to prevent excess heat loss and provides a temperature gradient of over 600°C with approximately 25-mm thickness. Maximum operating temperature of 700°C is chosen based on life considerations for replaceable cartridge heating elements and creep considerations for the storage material based on estimated time at pressure and temperature. In the case of UPS, the expected cycling is low, but low-cycle fatigue due to dilatation stresses in the inlet piping when cold air from the regulators is introduced were investigated. These were found to be non-issues for UPS. For more extensive cycling that would be required for wind or solar storage, further investigation is needed. In addition, work is ongoing to identify more economical TES designs.

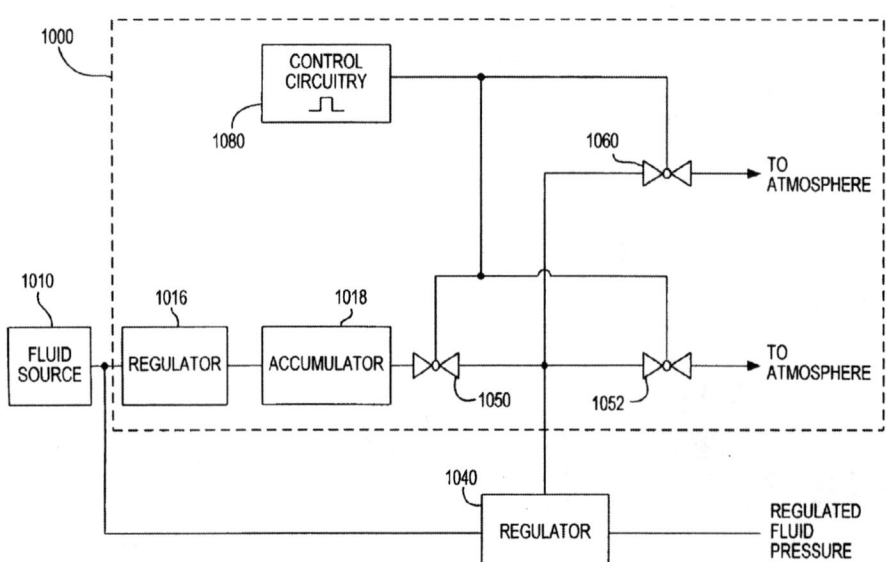

Fig. 4 Method of controlling constant fluid pressure.
Source: From Ref. 6.

Fig. 5 High energy thermal energy storage for UPS.

COMPRESSED AIR STORAGE IN VESSELS

Most large-scale CAES systems will probably continue to use underground caverns for pressurized air storage where available. However, for small-scale systems or where storage caverns are unavailable, small-storage vessels offer economies of scale due to mass production techniques employed. These cylinders are produced on automated process lines by backward extrusion of billet as opposed to larger vessels produced using seamless pipe with integrally forged heads. One manufacturer claims that the production rate for billet-formed cylinders is less than 30-s per cylinder.

In the United States, transport cylinders having Department of Transportation (DOT) exemption certification for 310-bar [4500-psi] service have been found most economical. Specific costs of less than $0.015 per bar-liter [$0.034 per psi-cu ft] can be obtained, whereas the cost of larger vessels has been found to be twice that or more. For the turbine inlet conditions being employed for UPS, this leads to energy cost for stored air of around $350/kWh. Energy storage costs for compressed air will decrease with increasing turbine inlet temperature. This must be balanced with the cost of the TES needed to achieve higher discharge temperatures.

In the United States, disagreement abounds on regulatory issues surrounding the on-site generation and storage of compressed air. Historically, stationary pressure vessels are designed to ASME standards, and installation and operation are regulated by the individual states. Transport vessels fall under the jurisdiction of the U.S. DOT and are exempted from state control. Many states allow the use of DOT vessels for stationary storage of air, but some do not and some have no regulations on pressure vessels. Attempts to qualify the most economical DOT designs through standard ASME channels have been fruitless since the two organizations' design rules are not harmonized, even though the service seen by the cylinders is comparable. Furthermore, some jurisdictions assess permitting fees on a per vessel basis, whereas others will consider a bank of vessels as one installation and asses a flat fee. Therefore, some users in the United States must pay twice or more for air storage than others.

The regulatory picture for compressed air storage in vessels in Europe is much more favorable with the recent harmonization of the European Union and introduction of the Pressure Equipment Directive (PED). Through appropriate Notified body channels, it is a straightforward matter to re-qualify designs originally intended for one application into another application so long as the types of service are similar.

MODIFICATIONS TO TACAS TECHNOLOGY

One of the most significant adaptations needed to improve TACAS technology for electricity storage is the development of a high-pressure adiabatic compressor and a slight modification to the system architecture to allow heat recovery. A schematic of the proposed architecture is shown in Fig. 6.

The exact form of the high-pressure compressor is under consideration. The high discharge temperatures imposed by adiabatic compression eliminates reciprocating compressors. Large-scale multi-stage centrifugal compressors with direct-drive induction motors have

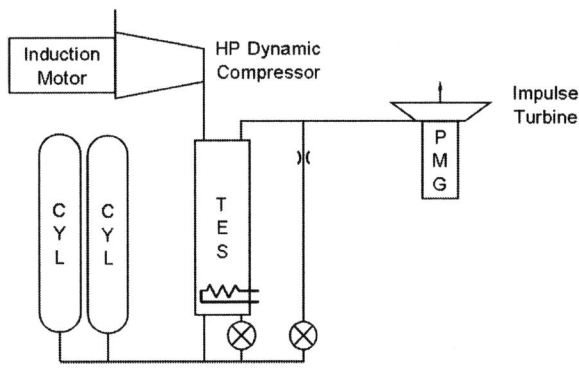

Fig. 6 Schematic arrangement of modifications to thermal and compressed air storage technology architecture for improved cycle efficiency.

been used for high-pressure applications in the oil and gas industry. Scaling dynamic compressor designs for small-scale applications and adapting for high-discharge temperatures poses significant technical challenges, but will be crucial if high cycle efficiency is to be achieved.

A further enhancement to the TACAS technology would allow higher turbine inlet temperatures since limitations on turbine-exhaust temperatures imposed by indoor discharge requirements of UPS could be relaxed. Increasing the discharge temperature reduces the TES discharge temperature range and so must be compensated by higher storage temperatures or additional TES mass for a given output. Further advances in TES material selection and design will be needed to achieve economic viability.

Finally, in applications for solar and wind energy storage, it may be possible to eliminate the bridging flywheel since the turbine has such rapid response. However, the excellent load regulation and speed control attributes of the flywheel in tandem with the turbine make this element of the architecture favorable if power quality enhancements are desired.

SUMMARY

Compressed-air energy storage and TES system have been successfully integrated for UPS applications. Selected features of the TACAS technology are presented and enhancements for improving storage cycle efficiency are discussed. These enhancements include:

- Development of a small-scale high-pressure adiabatic compressor with high discharge temperature capability.
- Modification to thermal storage charge and discharge flow path.
- Turbine bearing upgrades for longer operational life.
- Turbine modifications for increased inlet and discharge temperatures.
- Thermal energy storage cost reduction and or increase in allowable operating temperature.

Focus on these initiatives will allow consideration of TACAS technology for near-term electricity storage in conjunction with wind or solar power generating systems with lower upfront capital expenditure compared to large-scale systems.

REFERENCES

1. Crotogino, F.; Mohmeyer, K.; Scharf, R. *Huntorf CAES: More than 20 Years of Successful Operation*, Proceeding SMRI Spring Meeting, Orlando, FL, U.S.A., 15–18 April 2001.
2. Daly, J.; Loughlin, R.M.; DeCorso, M.; Moen, D.; Davis, L. *CAES—Reduced to Practice*, ASME 2001-GT-397, 2001.
3. Bullogh, C.; Gatzen, C.; Jakiel, C.; Koller, M.; Nowi, A.; Zunft, S. *Advanced Adiabatic Compressed Air Energy Storage for Integration of Wind Energy*, Proceeding of European Wind Energy Conference (EWEC), 22–25 November 2004.
4. Pinkerton, J.F. Thermal and Compressed Air Storage System. U.S. Patent 7086231, 8 August 2006.
5. Schuetze, K.T.; Hudson, R.S. Systems and Methods for Controlling Temperature and Pressure of Fluids. U.S. Patent Application 2006/0076426A1, Filed 5 February 2003.
6. Schuetze, K.T.; Weaver, M.D. Systems and Methods for Controlling Pressure of Fluids. U.S. Patent Application 2006/0060246A1, Filed 17 September 2004.
7. Perkins, D.E.; Hudson, R.S. Thermal Storage Unit and Methods of Using the Same to Heat a Fluid. U.S. Patent 6955050, 18 October 2005.

BIBLIOGRAPHY

1. Vosburgh, K. *Compressed Air Energy Storage*, AIAA Conference on New Options in Energy Technology, San Francisco, CA, 2–4 August, 1977.

Compressed Air Leak Detection and Repair

Robert E. Wilson
ConservAIR Technologies Co., LLP, Kenosha, Wisconsin, U.S.A.

Abstract

Compressed air is a major cost component in manufacturing. As such, it offers one of the largest savings opportunities. The investment in compressing air to energize it and then letting it escape from the system through leaks, without doing any useful work, is a complete waste. This waste can be minimized by implementing a program of leak detection and repairs. This entry covers the subject of how to use a handheld ultrasonic leak detector to locate leaks and the procedures required to implement repairs. The documentation and informational database required to ensure that leak waste is controlled and that new leaks are prevented is summarized. Different application technologies for controlling leaks are presented, and time and cost controls reviewed.

INTRODUCTION

This entry covers the basic methods of finding and repairing leakages.

The section on Leak Detection presents the most commonly used procedures to identify leakages. The sound signature of air leaks and the use of a handheld ultrasonic acoustic detector to locate leaks are explained. Suggested procedures for tagging and documenting air leaks are presented. The benefit of entering information into a database for historical trending is reviewed.

The section on leak repairs covers the most commonly found air leaks and the procedures to repair them. The logistical problems that create obstacles to expeditious repairs are discussed. The need for after-control, rechecks, and establishing standards are reviewed.

The section on leak control and prevention presents various approaches for managing leaks to minimize waste and ensure that the savings realized are ongoing. The application of flow monitoring and pressure regulation is presented.

Leak costs are discussed and a summary of the entry is presented.

LEAK DETECTION

In a compressed air system, the pressurized air confined by the pipes and vessels escapes from the system through openings as it expands back to atmospheric pressure. Ideally, all of these openings are created intentionally to extract energy from the compressed air in performing a desired task. In reality, however, many of the openings are unintentional, wasteful leaks.

The leak volume is directly proportionate to the area of the opening, the resistance to flow, and the applied pressure differential. The larger the area of the unrestricted opening and the higher the supply air pressure, the greater the leak flow. A chart showing the discharge of air through an orifice is included as an appendix. Note that the values listed are based on a 100% coefficient of flow and should be adjusted for other orifice configurations as suggested (Table 1).

When the air expands back to atmospheric pressure, it transitions from a high-pressure laminar flow to a low-pressure turbulent flow. The escape velocities become extreme as the air volume expands. This results in a full sound spectrum of noise, ranging from audible to high frequency inaudible.

One common method of detecting leakages is to use a soap-like liquid that forms bubbles. Products specifically formulated for high viscosity and film strength exaggerate the bubble effect to enhance the detection capabilities. The liquid is poured, sprayed, squirted, or brushed on a suspect area, and the formation of the bubbles is visually observed. This method allows the detection of leaks that cannot otherwise be heard or felt in the normal operating production environment, but bubble detection is time consuming and messy. It requires the inspection of every connection to the air system, and the foaming agent may require material approval before it can be used in a particular facility. It also is not practical for checking overhead ceiling pipes or under, behind, or inside operating machinery.

The more commonly accepted method for detecting leaks is to use a handheld, ultrasonic acoustic detector that can register the high frequency sound signature associated with gas leaks and translate it into an audible signal. Air leaks have a definitive ultrasonic sound component in their

Keywords: Industrial compressed air system leak detection; Repair; Control and prevention.

Table 1 Discharge of air through an orifice

Area sq. in. / press	1/64″ .00019	1/32″ .00077	3/64″ .00173	1/16″ .00307	5/64″ .00479	3/32″ .00690	7/64″ .0094	1/8″ .01227	9/64″ .01553	5/32″ .01973	3/16″ .02761	7/32″ .03758	1/4″ .04909	9/32″ .06213	5/16″ .07670	3/8″ .11045	7/16″ .15033	1/2″ .19635	9/16″ .24850	5/8″ .30680	3/4″ .44179	7/8″ .60132	1″ .78540
1	0.028	0.112	0.253	0.450	0.700	1.06	1.48	1.80	2.27	2.80	4.0	5.5	7.2	9.1	11.2	16.2	22.0	28.7	36.3	44.8	64.7	88	115
2	0.040	0.158	0.356	0.633	0.989	1.42	1.94	2.53	3.20	3.95	5.7	7.7	10.1	12.8	15.8	22.8	31.0	40.5	51.0	63.4	91.2	124	162
3	0.048	0.194	0.436	0.775	1.25	1.74	2.37	3.10	3.92	4.82	6.9	9.5	12.4	15.7	19.2	27.8	37.8	49.5	62.5	77.0	111.0	152	198
4	0.056	0.223	0.502	0.892	1.39	2.00	2.73	3.56	4.50	5.55	8.0	10.9	14.3	18.1	22.2	32.1	43.5	57.0	72.0	88.9	128.0	175	228
5	0.062	0.248	0.560	0.993	1.55	2.23	3.04	3.97	5.02	6.19	8.9	12.2	15.9	20.1	24.7	35.7	48.5	63.5	80.1	99.3	143.0	195	254
6	0.068	0.272	0.612	1.09	1.70	2.45	3.32	4.34	5.49	6.75	9.8	13.3	17.4	22.0	27.1	39.1	53.0	69.5	87.9	108.0	156.0	213	278
7	0.073	0.293	0.695	1.17	1.82	2.63	3.58	4.68	5.90	7.29	10.5	14.3	18.7	23.6	29.2	42.2	57.3	75.0	94.7	116.0	168.0	230	300
8	0.083	0.331	0.741	1.32	2.06	2.96	4.05	5.30	6.70	8.24	11.9	16.2	21.2	26.9	33.0	47.7	64.7	84.7	106.0	132.0	191.0	260	339
12	0.095	0.379	0.856	1.52	2.37	3.41	4.65	6.07	7.66	9.42	13.6	18.6	24.3	30.7	37.8	54.6	74.1	97.0	122.0	151.0	218.0	297	388
15	0.105	0.420	0.945	1.68	2.62	3.78	5.15	6.72	8.50	10.48	15.1	20.5	26.9	34.0	41.9	60.5	82.5	108.0	136.0	168.0	242.0	329	430
20	0.123	0.491	1.100	1.96	3.05	4.40	6.00	7.86	9.92	12.12	17.6	24.0	31.4	39.8	48.8	70.7	96.0	126.0	159.0	196.0	283.0	385	503
25	0.140	0.562	1.26	2.25	3.50	5.05	6.88	8.98	11.38	13.99	20.2	27.4	35.9	44.5	56.0	80.9	110.0	144.0	182.0	224.0	323.0	440	575
30	0.158	0.633	1.42	2.53	3.94	5.68	7.7	10.1	12.77	15.70	22.7	31.0	40.5	51.3	63.0	91.1	124.0	162.0	205.0	253.0	365.0	496	618
35	0.176	0.703	1.58	2.81	4.38	6.31	8.6	11.3	14.26	17.60	25.3	34.5	45.0	57.0	70.0	101.0	137.0	180.0	227.0	281.0	405.0	551	720
40	0.194	0.774	1.74	3.10	4.84	6.97	9.5	12.4	15.65	19.31	27.9	38.0	49.6	63.0	77.0	112.0	151.0	198.0	250.0	310.0	446.0	607	793
45	0.211	0.845	1.90	3.38	5.27	7.60	10.3	13.5	17.05	21.00	30.4	41.4	54.1	68.0	84.0	122.0	165.0	216.0	273.0	338.0	487.0	662	865
50	0.229	0.916	2.06	3.66	5.71	8.22	11.2	14.7	18.60	22.90	32.9	44.9	58.6	74.0	91.0	132.0	180.0	235.0	296.0	365.0	528.0	718	938
60	0.264	1.06	2.38	4.23	6.60	9.50	12.9	16.9	21.40	26.35	37.9	50.8	67.6	85.0	105.0	152.0	207.0	271.0	342.0	422.0	609.0	828	1,082
70	0.300	1.20	2.69	4.79	7.45	10.53	14.7	19.2	24.25	29.90	43.0	58.6	76.7	97.0	120.0	173.0	235.0	307.0	388.0	479.0	690.0	939	1,227
80	0.335	1.34	3.01	5.36	8.33	12.04	16.4	21.4	27.10	33.33	48.1	65.5	85.7	108.0	131.0	193.0	262.0	343.0	433.0	537.0	771.0	1,050	1,371
90	0.370	1.48	3.33	5.92	9.25	13.34	18.2	23.7	30.00	36.90	53.0	72.3	94.8	120.0	147.0	213.0	289.0	379.0	478.0	592.0	853.0	1,161	1,516
100	0.406	1.62	3.65	6.49	10.50	14.58	19.9	26.0	32.80	40.50	58.0	79.0	104.0	132.0	162.0	234.0	316.0	415.0	523.0	649.0	934.0	1,272	1,661
110	0.441	1.76	3.96	7.05	11.00	15.82	21.5	28.2	35.60	43.90	63.0	86.0	113.0	143.0	176.0	254.0	345.0	452.0	570.0	702.0	1,016.0	1,383	1,806
120	0.476	1.91	4.29	7.62	11.40	17.15	23.4	30.5	38.51	47.50	68.0	93.0	122.0	154.0	190.0	274.0	373.0	488.0	616.0	712.0	1,097.0	1,494	1,951
125	0.494	1.98	4.45	7.90	12.30	17.79	24.2	31.6	40.00	49.25	70.0	96.0	126.0	160.0	196.0	284.0	386.0	506.0	638.0	789.0	1,138.0	1,549	2,023
150	0.582	2.37	5.31	9.45	14.75	21.20	28.7	37.5	47.45	58.25	84.0	115.0	150.0	190.0	234.0	338.0	459.0	600.0	758.0	910.0	1,315.0	1,789	2,338
200	0.761	3.10	6.94	12.35	19.15	27.50	37.5	49.0	62.00	76.2	110.0	150.0	196.0	248.0	305.0	441.0	600.0	784.0	990.0	1,225.0	1,764.0	2,401	3,136
250	0.935	3.80	8.51	15.18	23.55	34.00	46.2	60.3	76.15	94.0	136.0	184.0	241.0	305.0	376.0	542.0	738.0	964.0	1,218.0	1,508.0	2,169.0	2,952	3,856
300	0.995	4.88	10.95	18.08	28.25	40.55	55.0	71.8	90.6	111.7	161.0	220.0	287.0	364.0	446.0	646.0	880.0	1,148.0	1,454.0	1,795.0	2,583.0	3,515	4,592
400	1.220	5.98	13.40	23.81	37.10	53.45	72.4	94.5	119.4	147.0	213.0	289.0	378.0	479.0	590.0	851.0	1,155.0	1,512.0	1,915.0	2,360.0	3,402.0	4,630	6,048
500	1.519	7.41	16.62	29.55	46.00	66.5	90.0	117.3	148.0	182.5	264.0	358.0	469.0	593.0	730.0	1,055.0	1,430.0	1,876.0	2,360.0	2,930.0	4,221.0	5,745	7,504
750	2.240	10.98	24.60	43.85	66.15	98.5	133.0	174.0	220.0	271.0	392.0	531.0	696.0	881.0	1,084.0	1,566.0	2,125.0	2,784.0	3,510.0	4,350.0	6,264.0	8,525	11,136
1000	2.985	14.60	32.80	58.21	91.00	130.5	177.0	231.0	291.5	360.0	520.0	708.0	924.0	1,171.0	1,440.0	2,079.0	2,820.0	3,696.0	4,650.0	5,790.0	8,316.0	11,318	14,784

Table is based on 100% coefficient of flow. For well-rounded orifice, multiply by 0.97. For a sharp-edged orifice, a multiplier of 0.65 will give approximate results. Values calculated by approximate formula proposed by S.A. Moss. $W = 0.5303(ACP/\sqrt{T})$; where: W, discharge (lb/s); A, area of orifice (in.^{2}); C, coefficient of flow; P, upstream pressure (PSI, abs.); T, upstream temperature (°F, abs.); Values used in calculating table: $C = 1$; $T = 530°R(70°F)$; $P =$ Gage pressure plus 14.7 psi; weights converted to volumes using density factor of 0.07494 lb/ft^3 (correct for dry air at 14.7 psi abs. and 70°F); values from 150 to 1000 psi calculated by Compressed Air Magazine and checked by Test Engineering Dept. of Ingersoll-Rand Co.

noise signature that is beyond the hearing threshold of the human ear. The ultrasonic leak detector translates the ultrasonic noise of the leak signature into an audible sound heard in the earphones worn by the leak surveyor. Some instruments are also equipped with display meters and indicator lights that visually register the magnitude of the air leak. A distinctive, loud rushing sound is produced in the earphones when the leak detector sensor probe is aligned with a leak. With the production background noise suppressed and filtered out by the headphone set, the leakage hissing is heard.

The sound wave generated by an air leak is directional in transmission. The intensity of the leak noise is based upon the shape of the orifice opening, the distance to the sensor probe, and the differential expansion pressure. The sound level is loudest at the actual point of the leakage exit. The procedure for detecting leaks ultrasonically uses this characteristic to locate the actual leaks. Initially, the leak detector is set at the maximum practical sensitivity consistent with the specific environment of the area being inspected. A sweep of the general area is performed as the surveyor walks the system. When a leak is detected, the direction of the leak is determined by scanning the area until the loudest noise level registers. With the probe pointing in the direction of the noise source, the surveyor moves towards the leak, adjusting the sensitivity of the leak detector accordingly. The intensity of the sound increases in the proximity of the leak and is loudest at the actual point of air exit. Extension tubes or cones attached to the sensor probe focus the sound and pinpoint the location of smaller leaks. The bigger, more serious leaks can be felt. A further test using a bubble solution can augment the process by visual enhancement of the exact location. One such product is formulated to produce an ultrasound shockwave as the bubbles burst, so the surveyor gains the benefits of both the visual observation and ultrasonic detection.

Competing sounds often mask a leak or otherwise distort the directional transmission. If possible, the best way to eliminate a competing sound is to shut the system off. If that is not possible, shielding techniques can be applied. The angle of the probe extension can be changed. The competing sound can be blocked using the body or other solid barrier like a piece of cardboard or clipboard. Cupping the hand over the leak, or using a rag, can often isolate the true source of the sound. Bubble tests can pinpoint the location regardless of the competing sound. It is imperative when working in and around operational machinery that safety be most important. Common sense dictates the extent of effort that should be expended to identify and quantify a specific leak.

The first step in the preparation for performing a leak survey is to establish a pattern for surveying the facility to ensure that all the piping, connected use points, and workstations in an area are inspected. Detected leaks are identified and tagged during the surveillance of the system. Different color tags can be used to visually indicate the severity of leaks and establish priorities. Typical classifications might include three levels:

Level 1: Not audible in any environment without an ultrasonic detector.
Level 2: Audible in a quiet environment but not in an operating facility.
Level 3: Serious leaks requiring immediate attention.

Level 1 leaks cannot be felt or heard under any conditions and require the use of the previously described procedures to detect. They are less than 1 scfm and are assigned no value, since the cost of the associated logistics and labor do not economically justify the repair, unless it is very simple, such as the ubiquitous push lock fitting on plastic tubing. Level 1 leaks are tagged and documented for future recheck, since air leaks never fix themselves and only grow larger over time. The cumulative effect of the Level 1 leaks on the compressed air system can be better controlled by maintaining a stable delivered air pressure at the lowest optimum level through the applications of pressure/flow control and regulating use points.

Level 2 leaks are in the 2 scfm range and can typically be felt but not heard without the use of an ultrasonic leak detector. Repairs are economically justifiable and should be performed within a 60-day period.

Level 3 leaks in excess of 2 scfm can typically be felt and sometimes heard by the human ear. These require immediate attention, since they not only waste air but impact the operational efficiency of the compressed air system. Leak flow is a real demand that adds to the filter/dryer loading, increases the pressure drop throughout the system, and creates pressure fluctuations that impact production.

While the true flow for any specific leak cannot be measured practically, the surveyor can assign values based upon the chosen leak volume associated with the various leak levels. These can then be totaled at the end of the survey to estimate the cumulative system leak waste. The surveyor will typically overestimate about the same amount of leakages that are underestimated, so the final figure gives a good portrayal of the total leak waste. As long as the survey procedures are replicated during the re-check, the comparative value for trending becomes an accurate measure for evaluating the remedial repair actions taken. A cost figure can be assigned for use in the financial analysis. Take into account power cost and associated compressor maintenance and repair costs, plus the costs to operate and maintain all the auxiliary equipment, when determining the real value of the leak waste.

Efforts have been made to estimate the actual volume of an air leak based upon pressure and the decibel level registered at a specific distance. People have assembled test stands using the most common orifice configurations

GUESS-TIMATOR CHART FOR UP9000/10,000
dB vs CFM

DIGITAL READING	100 PSIG	75 PSIG	50 PSIG	25 PSIG	10 PSIG
10 dB	0.5	0.3	0.2	0.1	0.05
20 dB	0.8	0.9	0.5	0.3	0.15
30 dB	1.4	1.1	0.8	0.5	0.4
40 dB	1.7	1.4	1.1	0.8	0.5
50 dB	2.0	2.8	2.2	2.0	1.9
60 dB	3.6	3.0	2.8	2.6	2.3
70 dB	5.2	4.9	3.9	3.4	3.0
80 dB	7.7	6.8	5.6	5.1	3.6
90 dB	8.4	7.7	7.1	6.8	5.3
100 dB	10.6	10.0	9.6	7.3	6.0

NOTES:
ALL READINGS ARE COMPENSATED FOR ATMOSPHERIC PRESSSURE.
All readings were taken at 40 kHz.

PROCEEDURE:

Use the Scanning Module to conduct the broad scanning to pinpoint the air leaks. The Scanning Module with the Rubber Focusing Probe (RFP) is used to determine air losses. The tip of the RFP on the UP9000 should be fifteen (15) inches away from the leak location for determination of the leak rate.

Notice: The values presented in this table are not stated as factual CFM measurement. This table is provided solely for convenience and should only be used as a general guideline.

Factors such as turbulence, leak orifice configuration, pressure, moisture and instrument sensitivity can significantly effect your results.

Fig. 1 Noise vs leak loss at various pressures.

found in compressed air systems, and then have measured air flow and decibel noise at different pressures and distances. One such Chart, published by UE Systems of Elmsford, NY, is presented in Fig. 1. Note the disclaimer that the values are not stated as "factual CFM" and are provided as a "general guideline." A leak signature is affected by many factors, and the loudness of the noise generated is by itself not the sole measure of the volume of the leakage. For example, a high-pitched whistle will sound a lot louder than a low-level whoosh sound, but the whistle will consume less air. At best, the leak detection process will provide an estimate for use in planning the priorities of the remedial repair procedures and a value for evaluating trends.

LEAK REPAIRS

Detected leaks must be visually tagged for future repair. Some tags are configured to enable you to tear off a copy to give to the maintenance supervisor responsible for the leak repairs. Regardless of the configuration of the tag, information sufficient to allow revisiting an individual leak for repair, even if the tag falls off or is missing, should be recorded on a separate worksheet. This typically consists of:

- Recording the unique, sequential tag number assigned to the specific leak.
- Defining its workplace location in a way that is meaningful to the air user.
- Identifying the specific item that is leaking.
- Identifying the actual point of air exit on the leaking item.
- Classifying the degree of leakage so priorities for remedial action are established.

The tag can be used for after-control by providing a place to enter the date and repairperson's name. The supervisor should check to ensure that the repair has been properly completed before signing off and removing the tag. The repair actions and associated time should be recorded on the original worksheet and entered into a database to establish time and cost control accounting procedures.

The surveyor should record complete information on the worksheet to describe the leak. The probable cause of the leak, such as aging, wear, damage, looseness, mishandling, breakage, or other reasons, should be noted with an explanatory note if required. Determine whether the leak should be repaired or a part replaced, and note it on the worksheet. If replacement is recommended, the surveyor should collect enough information about the item to allow for purchasing the repair part or replacement unit. Someone will have to do this if the leak is going to be fixed, so the surveyor should make the extra effort to record the information at the same time the leak is identified. The air user will also need to know if the leak is repairable without having to shut down the associated machinery. The worksheet should have areas for helpful comments and field notes to facilitate remedial actions or to alert people about other issues and opportunities that come to the attention of the surveyor.

Detected leaks must be repaired in order to realize any savings. Since most leaks occur at the operating machinery in the production area, repair procedures tend to be repetitive. Stresses are applied to all the various hoses and couplings, tubing connections, and pipe joints because of machinery vibration and movement of the connected tools and pneumatically driven devices. Over time, leaks develop at sealing areas. These are easily fixed by reconnecting the hose or reinstalling the pipe fitting after inspection and cleaning. Worn couplings or quick disconnects are replaced. The plastic components of point-of-use devises, such as filters, regulators, and lubrications, tend to age and crack over time. These must be replaced. Gaskets and seals dry out and become brittle, so they no longer seal effectively. Valve stem packing and sealing rings, manifold gaskets, hose reel rotary joints, and cylinder shaft seals wear over time and need to be replaced. Clamps, pipe unions, flanges, and pipe groove seals often require re-tightening. Leaks in the compressor room are found around air treatment equipment, condensate drains, receiver manholes, and control tubing.

Leaks on pipe joints are relatively easy to fix by either tightening or reinstalling a connection. Clean all surfaces before reassembly. Use a non-hardening sealing paste for threaded connections to prevent the possible contamination of the air system from torn or frayed Teflon™ tape.[1] Leaks in main headers and branch lines often require lifts or special rigging equipment to gain access, and may require special plumbing skills to repair. Advance planning and scheduling will be necessary for coordinating the repairs on machinery not accessible during production.

The largest obstacle to repairing leaks is the logistics involved in planning and implementing the repair procedures. These logistical problems often take months to resolve and sometimes impede the process entirely. A typical scenario follows.

LOGISTICAL PROCEDURES AND OVERHEAD ASSOCIATED WITH LEAK REPAIRS

1. Meetings and Planning
2. Maintenance requisitions
3. Purchaser—product and supplier identification
4. Order costs—cost per placed order
5. Transportation
6. Control of receipt—administration
7. Storage—space and logistics costs
8. Labor schedule—days/weeks
9. Leakage cost per week/month
10. Time control—verification and administration

LEAK MANAGEMENT

Air leaks grow bigger over time, and repaired leakages usually reappear within six months to one year after they are fixed. Steps must be taken to control the growth rate of leaks and to prevent reoccurrences after repairs are completed. The key to managed leakage control and prevention is rechecks and documentation. Periodic rechecks at predetermined intervals ensure that the leak rate is stabilized at a low level. Through documentation, the trends become obvious and developing patterns, both good and bad, are identified. Problems are recognized before creating issues that are more serious. Taking appropriate actions drives the leak trend downward until it reaches the target established by management, typically 5%–10% of the total air demand. This historical information is used to institute leak prevention measures and for calculating the most economical interval for rechecks to ensure that the gains realized are maintained in the future. Establishing standards and good practices minimizes future leakage. With the time and costs documented, controls can be put in place to properly administer a leak management program.

An alternative approach to implementing a full leak management program is to simply fix the leaks immediately upon discovery, assuming a system is checked for air leaks on a regular basis. The technician brings along a tool tote with the appropriate equipment needed to fix the most commonly found leaks. Usually, only the more serious leaks are addressed in the simple seek and fix approach. Little, if anything, is documented.

The total leakage for a facility can be estimated using techniques that measure pressure degradation over time when there are no production demands on the system.

One such method is to measure the load/unload cycle time of compressors when production is shut down and the only air demand on the system is leakage. Start the compressor(s) and record the on-load time and off-load time over a sampling period long enough to provide a representative average. Calculate the leakage lost as a total percentage of compressor capacity using the formula:

Leakage (%) = $[(T \times 100)/(T + t)]$

where: T = average on-load time, and t = average off-load time.

In systems configured with compressor controls other than load/unload, leakage can be estimated based upon the total system capacitance. The total estimated volume (V) of all air receivers, the main piping distribution, and other significant air containment vessels must be calculated in cubic feet. Pressure in the main header must be measured at the start and end of the evaluation test period. Production must be shut down so that the only demand on the system is leakage. The compressors are then started in order to pressurize the system to its normal operating pressure (P_1). The compressors are turned off, and the time (T) it takes the system to drop to a pressure equal to half the normal start pressure (P_2) is measured. The leakage is estimated using the formula:

Leakage (cfm of free air)

$= (V \times (P_1 - P_2)/T \times 14.7) \times 1.25$

where: V is the volume in cubic feet, P_1 and P_2 are in psig, and T is the time in minutes.

Because air escapes from the system at a rate proportional to the supply pressure, the leak volume rate at the normal start pressure will be much greater than the leak volume rate at the end of the timed cycle when the pressure is half. A 1.25 correction factor is applied to compensate for the difference in the leak rate and to provide a more accurate estimation of the loss.

Installing a flow meter to measure and record the actual flow improves the accuracy of the air leak estimate over using a calculated capacitance based upon estimated system volume. A properly configured flow meter can also be used to monitor the system consumption in order to (1) establish a baseline for evaluating the performance improvements realized from any remedial actions taken, (2) verify trends, and (3) verify that the gains continue to return the investment into the future.

Many of the leaks in an industrial compressed air system are intentional or planned. Condensate drainage and disposal, spot cooling, fume venting and exhausting, material conveying and blowing off, and drying are examples of intentional leaks. Devices are available to eliminate or mitigate the air used to perform these types of assigned tasks.

No air loss condensate drains collect the condensation in a vessel, until it fills with water. A float or sensor detects the high liquid level and opens a drain port, allowing compressed air to displace the water and forcing it to discharge from the vessel. The sensor shuts off the drain port before the vessel is completely drained, so that no air is lost with the water discharge. Some designs are entirely pneumatic, so no electric power is required at the use point. Electrically activated designs require a power source. While this is sometimes inconvenient, electric units have the advantages of (1) indicator lights that show the operational status of the drain and (2) contacts that can be interfaced with a building management system for remote monitoring.

High efficiency blowing devices are available to entrain surrounding ambient air in the primary air stream to increase the impingement force, so that less compressed air is required to perform the equivalent task. Air knives, nozzles, and jets are offered with a variety of different airflow patterns to better suit a specific task. Air amplification ratios as high as 40:1 over open blowing are achievable.[2] Supply pressure at the point of use can often be regulated to a lower level to further reduce compressed air consumption and the associated noise.

Air volume amplifiers are available to create directional air motion in their surroundings and to efficiently move air and light materials. A small amount of compressed air is used as a power source to amplify the flow of entrained ambient air. Airflow is directional, with an inlet and outlet, to exhaust and/or sweep an area in a shaped pattern. Air volume amplifiers can create output flows up to 25 times the compressed air consumption.[2]

LEAK CONTROL AND PREVENTION

Leaks of all sizes, both intentional and unintentional, can be controlled by supplying them, at minimum, an acceptable, delivered air pressure. An air system that has a cumulative equivalent of a 5/16" leakage orifice, for example, is illustrated in Fig. 2.

The application of Pressure/Flow Control in the compressed air system primary is a good method for minimizing leak waste. The smaller leaks, determined to be uneconomically repairable, leak less at the lower delivered pressure. The Pressure/Flow Control also prevents the

A Compressed Air Demand that consumes 80 CFM at 80 psig, will consume 100 CFM if the Upstream Pressure is increased to 100 psig.

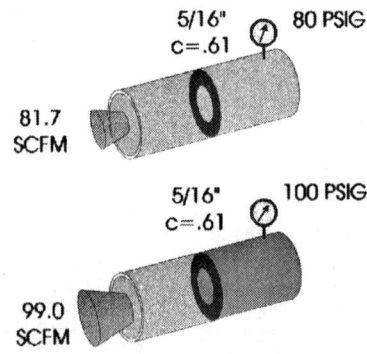

Fig. 2 Cumulative system leak demand at different pressures.

supply pressure from rising because of the lower demand that stems from leak repairs. Without some method of supply side pressure control, the system pressure increases inversely with demand, forcing leaks and other unregulated use points to consume more air. The savings achieved by lowering the leak demand are offset by air that is shunted out elsewhere in the system because of the rising pressure.

The application of point-of-use pressure regulation is another method for minimizing leakages. Setting and securing pressure regulators to supply air at the lowest minimal acceptable pressure maintains the respective leak losses at their lowest possible level.

Shutting off the air to non-productive workstations and assembly lines is another good method of leak control. Lock out valves and isolation valves can be installed to completely shut off the air to machinery that is shut down. The procedure can be manual or automated through the installation of actuated shut off valves that are activated by an external signal. Automation eliminates the dependency on a human action to stop the waste.

LEAK COSTS

The cost of leaks must be determined to allow management to make proper decisions about the compressed air system. The Chart in Fig. 3 illustrates the cost of air consumed by leaks.

In addition to the power cost shown in Fig. 3, consideration should be given to other associated costs, such as labor to log daily operations, scheduled maintenance, repair services, and periodic major overhauls. Leaks are a real demand that require real airflow to satisfy. There is an added cost burden that results from treating the leak air, removing condensation, additional compressor wear, and increased power to compensate for the greater pressure drop because of the higher flow. The final cost figures can more than double the cost based solely upon electrical power.

An effective leak control and prevention program requires continuous monitoring and verification that the gains realized are ongoing into the future. At a minimum, a leak survey should be performed several times a year in a system recheck. Results should be entered into a database and analyzed. Flow monitoring systems are available to measure actual flow. These can interface with management information systems that have remote access. Measuring real flow allows the true cost of the delivered air to be calculated in $/mmcf (Dollars per million cubic feet). Charting the savings in reports for management ensures continued support for the program.

SUMMARY

In summary, a good leak control and prevention program for a compressed air system starts with a leak survey. Ultrasonic leak detectors are the best tool to find air leaks and pinpoint their location. The information about the air leaks is recorded in a worksheet and documented in a database for use in generating reports and identifying trends. Savings are only realized if leaks are repaired. Repair procedures must be established and an investment made in satisfying all of the logistical obstacles before the actual remedial actions can be taken. Consideration should be given to contracting out the repairs, along with the logistical requirements, to expedite the process and realize the savings as soon as possible. Rechecks and monitoring are necessary to drive the leak trend down and keep it at the targeted rate.

REFERENCES

1. Ultraprobe™ Instruction Manual Volume 3. UE Systems, Inc., Elmsford, NY.
2. Exair Corporation Catalog No. 20, 11510 Gold Coast Drive, Cincinnati, OH.

BIBLIOGRAPHY

1. ConservAIR Technologies Company, LLP, Kenosha, WI, Industrial Seminar Series, Compressed Air Management.
2. *Improving Compressed Air System Performance a Sourcebook for Industry*, U.S. Department of Energy, Energy Efficiency and Renewable Energy, Prepared for the Compressed Air Challenge® and the United States Department of Energy, Prepared by Lawrence Berkeley National Laboratory, Washington, DC, and Resource Dynamics Corporation, Vienna VA, Section 2–3 *Compressed Air System Leaks*. Nov 2003.
3. Taranto, T. Data Power Inc., *Energy Efficiency in Compressed Air Systems*, 1998, Chapter 6-Compressed Air Systems, Performance.
4. Teflon™ is a trademark of DuPont.
5. Wilson, R. PEMCO Services, St. Petersburg, FL, Presentation, 2005.
6. Winkler, D. LeekSeek International Ltd., Training Presentation, 2005.

Air consumed by leaks at 100 psig

Diameter in.	SCFM Leakage	Annual volume		Cost per year*
1/64(.016)	0.41	215,496	cf/yr	$47.41
1/32(.032)	1.62	851,472	cf/yr	$187.32
1/16(.063)	6.49	3,411,144	cf/yr	$750.45
3/32(.094)	14.6	7,673,760	cf/yr	$1,688.23
1/8(.125)	26.0	13,665,600	cf/yr	$3,006.43
5/32(.156)	40.5	21,286,800	cf/yr	$4,683.10
1/4(.250)	113.0	59,392,080	cf/yr	$13,066.42

*Based upon rate used by US Dept. of Energy, EERE, A Sourcebook for Industry. C = 1.0

Fig. 3 The typical cost of air leaks.

Compressed Air Storage and Distribution

Thomas F. Taranto
ConservAIR Technologies, LLP, Baldwinsville, New York, U.S.A.

Abstract
Consistent stable operation of an industrial compressed air system is achieved when compressed air flow supplied to the system equals compressed air demand. Energy distributed to the system is available from two sources; rotating energy of the air compressors, and energy from compressed air storage. Optimum system energy efficiency is possible when the proper amount of energy is available from compressed air storage. Presented here are the physical and mathematical relationships that may be used to assess system performance and determine compressed air storage requirements. These relationships are also applied to design the air storage volume and distribution pressure profile necessary for effective compressed air storage.

INTRODUCTION

Compressed air systems have historically used on-line air compressor capacity sufficient to supply peak air demands. Where air receivers have been installed, the system's pressure profile and lack of storage control limit the effectiveness of compressed air energy storage. Today's energy costs and competitive world economy require that inefficient, wasteful practices of the past must end. This entry develops the necessary calculations to assess compressed air energy demand and calculate the usable compressed air energy in a properly engineered storage system. Methods of design, application, control, and optimization of compressed air energy storage are introduced.

ENERGY FLOW IN COMPRESSED AIR SYSTEMS

Compressed air systems can be viewed as having two parts: supply and demand. The supply system includes air compressors, dryers, filters, and equipment found in the powerhouse/compressor room. The demand side includes perhaps hundreds of use points, including tools, actuators, process use, blowing, cooling, material transport, and sparging (a process whereby air is injected into a tank of liquid, resulting in a bubbling of the solution to provide a desired action). The amount of energy necessary to drive productive air demands is changing constantly as equipment and processes start and stop. During normal machine cycles, compressed air use is often cyclical rather than continuous.

Normal demand variations and diversity of applications result in an airflow profile with peaks and valleys in airflow rates that are at times significantly more or less than the average air demand. For reliable operation of the system, peak air demands must be supplied as they occur. If air supply falls short of demand, system pressure will decrease. When peak air demands are not supplied, system pressure can fall below the minimum acceptable operating pressure. This often leads to lost productivity.

The supply of air is available from two sources; rotating on-line compressed air generation capacity and compressed air energy storage. Operating excessive air compressor capacity to supply peak airflow is inefficient and expensive. Properly engineered compressed air energy storage will supply peak air demand and reduce rotating on-line energy. The result is improved overall system efficiency and reduced power cost.

Compressed Air Generation Efficiency

Generation efficiency is highest when an air compressor is operating at full-load capacity. It is common for compressors to operate at less than full-load capacity, a condition referred to as part-load operation. During part-load operation, a compressor consumes less than full-load power as its compressed air output is reduced. "Specific power" measured as power per unit of air produced (kW/100 scfm), however, is greater during part-load operation. The result is that part-load operation is less efficient than full-load operation. Compressors may also operate in the no-load or unloaded condition. When unloaded, the air compressor produces no compressed air at all yet consumes 25%–30% of full-load power. Running compressors unloaded for a long period greatly reduces a system's overall generation efficiency.

Compressed Air System Efficiency

Supplying peak airflow demand with rotating on-line generation requires one or more compressors to operate in

---

Keywords: Air; Compressed air; Storage; Receiver; Pneumatic; Supply; Demand; Air storage; Energy; Compressor.

Compressed Air Storage and Distribution

the part-load or unloaded condition. As the peak demand occurs, the compressor(s) will load for a short time during the demand event and then return to part-load or unloaded operation. The result is poor overall system efficiency.

For a compressed air system to achieve maximum operating efficiency, the compressed air supply should incorporate both compressed air generation and storage. The goal is to supply average air demand with generation (on-line rotating energy) and to supply peak airflow requirements from storage (stored compressed air energy).

Compressed Air System Energy Balance

The energy delivered to and consumed in a compressed air system is a function of the weight or mass flow of air moving through the system. The mass of compressed air depends on pressure and temperature. Increasing pressure increases the density and, therefore, the mass of air. Increasing air temperature will decrease the air's density, decreasing the mass of air. This relationship is stated in the Ideal Gas Law.[1]

$$\frac{pV}{T} = a \text{ constant} \quad \text{(for a fixed mass of gas)} \quad (1)$$

Eq. 1 Ideal Gas Law for fixed mass of gas.

Compressed air is often measured in terms of its volume—ft^3, for example. The volumetric measure of air is irrelevant with respect to the air mass unless the temperature and pressure of the air volume are also known. Therefore, standards are adopted to express the mass of air under "Standard" conditions, resulting in the definition for a Standard Cubic Foot of air (scf). Standard conditions adopted by CAGI (Compressed Air and Gas Institute) and Compressed Air Challenge® (CAC) are 14.5 psia, 68°F, and 0% relative humidity.

Compressed air energy transfer can be expressed as the mass flow rate of air at a given operating pressure in standard cubic feet per minute (scfm). This is both a measure of volumetric flow rate and the mass or weight flow rate of compressed air. Higher-flow-rate scfm delivers a higher power rate, and the time duration of flow determines the energy transferred.

Compressed air power enters the system from the air compressors and exits the system through air demands, including productive demand, leaks, and all points where compressed air leaves the system, expanding back into the atmosphere. Power delivered from the compressors is measured as mass flow rate Q (scfm) from generation, or Q_{gen}, and power leaving the system also is measured as mass flow rate of air demand, or Q_{dmnd}.

Definition: Q_{gen}. Airflow rate of generation is the compressed air mass flow rate (scfm) produced by the rotating on-line compressor capacity at any moment.

Definition: Q_{dmnd}. Airflow rate of demand is the compressed air mass flow rate (scfm) escaping from the compressed air system to the atmosphere at any moment.

The ideal balance between generation and demand is achieved when $Q_{gen} = Q_{dmnd}$.

Ideal Air System Energy Balance

$$Q_{gen} = Q_{dmnd} \quad (2)$$

Only when system pressure = constant

Eq. 2 Ideal compressed air system energy balance.

The first law of thermodynamics for a change in state of a system,[2] or the law of conservation of energy, states that energy is not created or destroyed. This implies that the airflow rate of generation must equal the airflow rate of demand. From practical experience, it is observed that generation and demand are not always equal, resulting in changing system pressure. When $Q_{gen} > Q_{dmnd}$, system pressure increases, and when $Q_{gen} < Q_{dmnd}$, system pressure decreases. The energy imbalance between generation and demand is either absorbed into or released from storage (Q_{sto}).

With Compressed Air Entering and Exiting Storage

$$Q_{gen} \pm Q_{sto} = Q_{dmnd}$$

Entering storage pressure increases

Exiting storage pressure decreases

(3)

Eq. 3 Actual compressed air system energy balance.

Definition: Q_{sys}. Airflow rate of the system is the compressed air mass flow rate (scfm) produced by the rotating on-line compressor capacity (Q_{gen}) at any moment, minus the airflow absorbed into storage ($-Q_{sto}$ for increasing pressure) or plus airflow released from storage ($+Q_{sto}$ for decreasing pressure) (Fig. 1).

Practical Air System Energy Balance

$$Q_{sys} = Q_{dmnd}$$
$$Q_{sys} = Q_{gen} \pm Q_{sto} = Q_{dmnd} \quad (4)$$

accounts for changing system pressure

Eq. 4 Energy balance of compressed air systems.

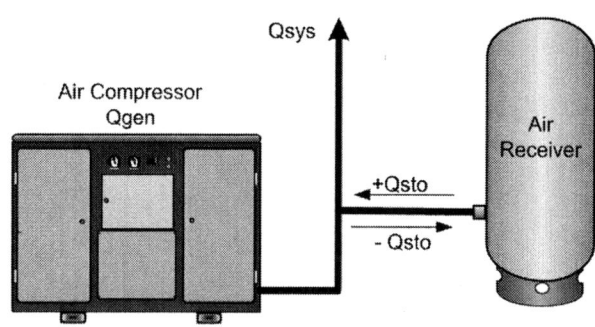

Fig. 1 Airflow relationship: generation, storage, and system flows.

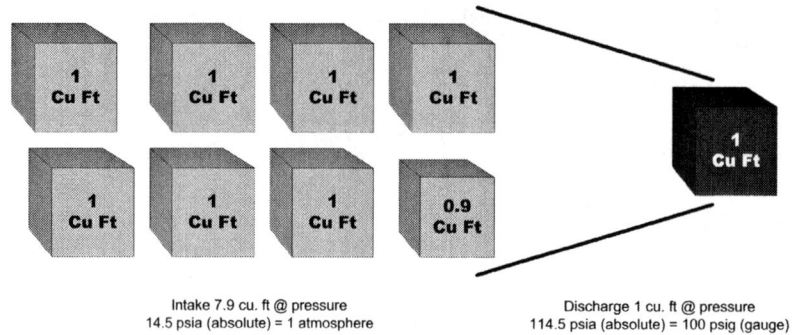

Fig. 2 Compression of air from atmospheric pressure to 100 psi gauge (114.5 psi).

COMPRESSED AIR ENERGY STORAGE

The compressed air system engineer must analyze the dynamics of compressed air energy storage. This section develops the mathematical expressions necessary to study the relationship among air system generation, storage, and demand. To optimize system operation, system energy supply must remain balanced with air demand. Furthermore, system energy supply must be optimized between generation (rotating on-line energy) and storage (stored compressed air energy).

Application of the Combined Gas Law

The combined gas law states that the pressure of an ideal gas multiplied by volume divided by temperature is a constant.

$$P_1 \left(\frac{V_1}{T_1}\right) = P_2 \left(\frac{V_2}{T_2}\right) \qquad (5)$$

Eq. 5 Combined Gas Law.

The compression process usually begins with air at ambient conditions of pressure and temperature (assuming no moisture content). A larger volume of air is compressed to a reduced volume, increasing the pressure of the air. During the compression process, the air's temperature increases. Assuming the air's temperature is ultimately returned to ambient, as a result it can be said that the end pressure of the compression process is equal to the initial pressure times the ratio of beginning volume to ending volume (compression ratio).

$$P_1 \frac{V_1}{V_2} = P_2, \quad \text{where } \frac{V_1}{V_2} = r(\text{Compression Ratio})$$

$$P_1 r = P_2 \qquad (6)$$

Eq. 6 Compression ratio.

If during compression, 7.9 ft^3 of air are reduced to a volume of 1 ft^3, the resultant pressure is 7.9 times the initial pressure or the ratio ($r = 7.9$ times).

Assume that the compressor intakes 7.9 ft^3 of dry atmospheric air at 14.5 psia. Multiplying by $r = 7.9$ results in a pressure of 114.5 psi absolute or 100 psi gauge (Fig. 2).

The application of the combined gas law is the basis of compressed air storage calculation. More advanced forms of the calculation can be used to assess many aspects of compressed air system performance.

If an air receiver of 1 ft^3 volume is pressurized to 114.5 psi absolute, and its discharge valve is opened to the atmosphere, how many ft^3 of air (at atmosphere) will be discharged from the receiver? The form of the ideal gas law used to solve this problem is

$$V_{\text{gas}} = V_{\text{rec}} \frac{\Delta P_{\text{rec}}}{P_{\text{atm}}} \qquad (7)$$

Eq. 7 Gas volume-receiver volume relationship (assuming temperature = constant).

The volume (at atmosphere) of air (V_{gas}) released from the receiver is equal to the air receiver's volume (V_{rec}) times the pressure change of the receiver (ΔP_{rec}) divided by atmospheric pressure (P_{atm}). Furthermore, the receiver's pressure change (ΔP_{rec}) is equal to the final pressure (P_{f}) minus initial pressure (P_{i}) giving ($\Delta P_{\text{rec}} = P_{\text{f}} - P_{\text{i}}$). Substituting:

$$V_{\text{gas}} = V_{\text{rec}} \frac{(P_{\text{f}} - P_{\text{i}})}{P_{\text{atm}}}$$

$$V_{\text{gas}} = 1 \text{ ft}^3 \frac{(14.5 \text{ psia} - 114.5 \text{ psia})}{14.5 \text{ psia}} \qquad (8)$$

$$= V_{\text{gas}} = -6.89 \text{ ft}^3$$

Eq. 8 Gas volume released from a receiver.

In the previous discussion of air compression, it was shown that reaching a pressure of 114.5 psia requires compressing 7.9 ft^3 of air. The receiver calculations above, however, show that only -6.89 ft^3 of air are released from the 1 ft^3 receiver above.

What happened to the other cubic foot of air? The other cubic foot of air is still inside the air receiver because the receiver's pressure remains at 14.5 psia (1 atm).

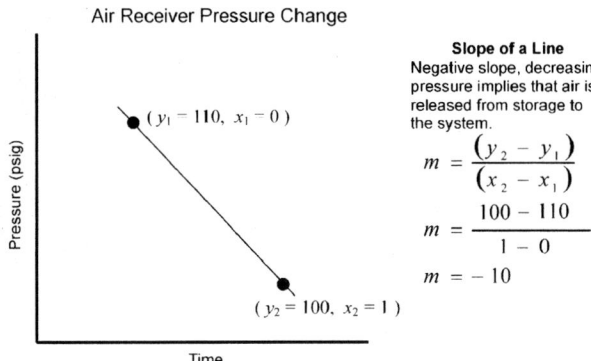

Fig. 3 Air receiver pressure change.

It is important to note that the receiver's pressure change is the slope of the line calculated from final pressure minus initial pressure. Looking at the receiver pressure throughout time, when pressure is falling (negative slope), air is flowing from the receiver to the system. The xy plot in Fig. 3 shows pressure (y) and time (x). Storage airflow Q_{sto} has an inverse relationship between the storage receiver and system—that is to say, air leaving the receiver ($-Q_{sto}$) is air entering the system ($+Q_{sto}$).

Pneumatic Capacitance

Pneumatic capacitance of a compressed air system (C_{pn}) represents the compressed air energy absorbed into or released by a compressed air system as its pressure increases or decreases. It is expressed in terms of the mass of air/unit change in pressure—for example, Standard Cubic Foot/Atmosphere (scf/atm) (Table 1).

$$C_{pn} = \frac{V_{sys}}{P_a} = \frac{V_{rec} + V_{pipe}}{P_a} \quad (9)$$

Eq. 9 Pneumatic capacitance.

For every 1 atm change of system pressure, increase or decrease, the system will absorb or release one times its volume (ft^3) of compressed air. Eq. 9 above gives the capacitance of the system in terms of ft^3 per atmosphere of pressure. Assume that the atmosphere is at standard conditions of pressure and temperature, and that the piping volume is negligible.

$$V_{gas} = C_{pn}\Delta P = \frac{V_{sys}}{P_a}\Delta P$$

$$C_{pn} = \frac{100 \text{ cu ft}}{1 \text{ atm}} = 100\frac{\text{scf}}{\text{atm}} \quad (10)$$

$$V_{gas} = 100\frac{\text{scf}}{\text{atm}} \times 1 \text{ atm} = 100 \text{ scf}$$

Eq. 10 Volume of gas as a function of pneumatic capacitance (C_{pn}).

Eq. 10 shows that a 100 ft^3-volume air receiver changing pressure y 1 atm will displace 100 scf of air into or out of the vessel. The pneumatic capacitance of the system is 100 scf/atm. If the ΔP is positive (i.e., initial pressure is lower than the final pressure), the air displaced is 100 scf, and air is absorbed into the air receiver tank. If the ΔP is negative (i.e., initial pressure is higher than the final pressure), the air displaced (100 scf) is delivered from the air receiver tank.

Because compressed air system pressure is often measured in psi, it is desirable to express the capacitance of compressed air systems in terms of scf per psig (scf/psi).

Table 1 Definition of variables and units of measure

C_{pn}	Pneumatic capacitance (scf/atm) or (scf/psia)
V_{rec}	Receiver volume (cu ft)
V_{pipe}	Piping volume (cu ft)
V_{sys}	System volume (cu ft)
P_a	Atmospheric pressure (psia)
P_i	Initial receiver pressure (psig)
P_f	Final receiver pressure (psig)
ΔP	Storage pressure delta ($P_f - P_i$)
r_s	Storage pressure ratio ($P_f - P_i$)/P_a
V_{gas}	Compressed air volume (scf) standard cubic feet
P_{load}	Compressor load pressure (psig)
P_{unload}	Compressor unload pressure (psig)
Q_{sys}	Airflow rate for the system (scfm)
Q_{gen}	Airflow rate from generation compressor(s) (scfm)
Q_{sto}	Airflow rate of storage (scfm)

Considering that $(P_f\,\text{psig} - P_i\,\text{psig})$ yields absolute pressure difference ΔP (psia): Substituting $P_f - P_i$ (psia) for the Storage Pressure Delta (ΔP atm).

$$V_{gas} = \frac{V_{sys}(\text{cu ft})}{P_a(\text{atm})} \Delta P(\text{atm})$$

$$V_{gas} = \frac{V_{sys}(\text{cu ft})}{P_a(\text{psia})} (P_f - P_i)(\text{psia}) \qquad (11)$$

Therefore: $\quad C_{pn} = \dfrac{V_{sys}(\text{cu ft})}{P_a(\text{psia})}$

Eq. 11 Capacitance and storage pressure delta (scf/psia).

Therefore, pneumatic capacitance of a compressed air system (C_{pn}) is a function of the total volume of the system and atmospheric pressure, which can be expressed as:

$$C_{pn} = \frac{V_{sys}}{P_a} = \frac{V_{rec} + V_{pipe}}{P_a} \qquad (12)$$

Eq. 12 Capacitance (C_{pn}) = scf/psia.

The V_{gas} of the system and the C_{pn} in units of scf/psia are directly related by the change in system pressure (ΔP) in units of psia.

$$V_{gas} = C_{pn}\Delta P \qquad (13)$$

Eq. 13 Gas volume as a function of capacitance and delta P.

For the system above with $V_{sys} = 100\,\text{ft}^3$, the capacitance is 6.896 scf/psia. If the system pressure delta is 14.5 psia (1 atm), the stored V_{gas} is 100 ft^3 (see Eq. 14), because atmosphere (14.5 psia) is the condition defined for Standard Gas Conditions $V_{gas} = 100$ scf.

$$C_{pn} = \frac{V_{sys}}{P_a} = \frac{100\,\text{cu ft}}{14.5\,\text{psia}} = 6.896\,\text{cf/psia}$$

$$V_{gas} = C_{pn}\Delta P$$

For $\Delta P = 14.5$ psia(1 atm) the Gas Volume is:

$$V_{gas} = C_{pn}\Delta P = 6.896 \times 14.5$$

$$V_{gas} = 100\,\text{scf}$$

(14)

Eq. 14 Calculating gas volume.

Usable Compressed Air Energy in Storage

In the previous discussion, it is apparent that two factors determine the amount of compressed air energy storage: the receiver volume and pressure delta (initial minus final pressure). The volume of a compressed air system is determined primarily by the number and size of air receivers in the system. Piping volume adds to the total, but unless there are several hundred feet of large-diameter pipe, it is often insignificant. For example, 1000 ft of 2-in. schedule 40 pipe has a volume of 23.3 ft^3 or 174 gal (7.48 gal/ft^3), and 1 mi of 1-in. schedule 40 pipe is only 31.7 ft^3 (237 gal). The available pressure delta for storage is determined by the system pressure profile.

Air System Pressure Profile and Storage Delta

The highest pressure available in the system is usually determined by the maximum working pressure of the air compressors. The lowest acceptable operating pressure is determined by manufacturing requirements. With an air compressor rated at 125 psig maximum working pressure and a required use-point pressure of 75 psig, for example, a maximum 50 psig pressure differential is available. Only a portion of this differential can be used for storage, as there are unrecoverable pressure losses as compressed airflows through the system. The pressure profile in Fig. 4 allows for 15-psig control pressure band, 5-psig treatment pressure drop, 2-psig loss through distribution piping, and 8-psig differential in the point-of-use connection piping.

Given a minimum demand-side use-point pressure of 75 psig, and including the imposed pressure delta (10 psi) through distribution plus point-of-use piping, the pressure profile in Fig. 4 shows that the lowest optimum target pressure of the supply-side header is 85 psig. The normal supply-side header pressure is 105 psig. This profile allows for primary storage pressure differential of 20 psig available to the system.

There are costs associated with compressed air energy storage. The discharge pressure at air compressors must be increased to provide storage pressure differential. Increased energy is about 1% for each 2-psig increase in compressor discharge pressure (for positive displacement-type compressors). Also, increasing the compressed air system pressure increases the air demand of the system. Compressed air leaves the system through various openings to the atmosphere, such as the open port of a control valve, a blowing nozzle or open blowing tube, or a leak in the piping. Any opening in the system that does not have a pressure regulator controlling the applied pressure will blow an increased amount of airflow as the system's applied pressure is increased. This increased airflow is called artificial demand. System air demand is increased by approximately 1.0%–1.3% for every 2-psig increase in system pressure. For systems with little effective use-point pressure regulation, artificial demand will be greater.

Definition: Artificial demand. Artificial demand is the additional compressed airflow demand consumed by the system due to actual applied air pressure being greater than the minimum required target pressure.

For the pressure profile shown in Fig. 4, the storage pressure differential is 20–35 psig as compressor controls cycle between their load and unload set points. If the average storage pressure differential is 28 psig, the

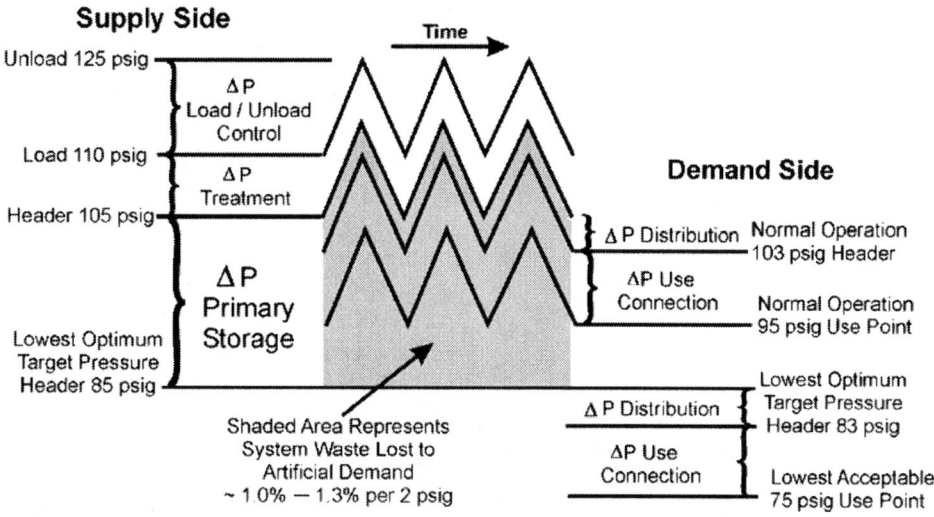

Fig. 4 Compressed air system pressure profile with uncontrolled storage.

resulting power increase is 14%, and waste to artificial demand is between 14 and 18% of the system's airflow.

As shown above, creating compressed air energy storage increases the system's energy requirement and power cost. Therefore, it is unwise to create more storage than the system requires. Proper design will minimize the increased compressor discharge pressure and power requirement. Also, proper control of compressed air energy in storage can virtually eliminate artificial demand. This topic is discussed in "Maximize and Control Compressed Air Energy Storage," presented later in this entry.

But first, the compressed air system engineer must consider various system requirements for stored energy and engineer the storage system appropriately. The following sections demonstrate common applications of the pneumatic capacitance calculation to solve various system energy storage requirements.

Calculate Useable Compressed Air in Storage

Compressed air energy in storage is of use to the air system only if storage pressure is greater than the minimum system supply pressure. Given the pressure profile shown in Fig. 4, and assuming that the air receiver volume of the system is 1000 gal, the amount of usable compressed air energy in storage can be calculated.

For an air receiver of 1000-gal volume with 20 psia primary storage delta P, see Fig. 4; solve for the usable compressed air energy storage V_{gas} (scf) (assuming that the air is at standard temperature and relative humidity).

$$C_{pn} = \frac{V_{sys}}{P_a} = \frac{V_{rec} + V_{pipe}}{P_a} = \frac{1000(gal)/7.48(gal/cu\,ft)}{14.5(psia)}$$

$$C_{pn} = \frac{133.67(cu\,ft)}{14.5(psia)} = 9.2(cu\,ft/psia)$$

$$V_{gas} = C_{pn}\Delta P = 9.2\left(\frac{cu\,ft}{psia}\right)20(psia) = 184(scf)$$

(15)

Eq. 15 Pneumatic capacitance (C_{pn}) and usable storage (V_{gas}).

Calculating Peak Air Demand

During a large demand event, the supply pressure in the system above is observed to draw down from 105 to 80 psig in 30 s. What is the airflow rate for Q_{sto} (scfm) during the demand event?

Adding time to Eq. 12 for gas volume (scf) allows solving for gas flow rate Q_{sto} (scfm).

$$V_{gas} = C_{pn}\Delta P$$

$$Q_{sto} = C_{pn}\frac{dP}{dT} \quad \text{Where time } T = \text{minutes}$$

(16)

Eq. 16 Solve for storage airflow rate Q_{sto} (scfm).

Solving for the peak airflow rate from storage (P_f, final pressure; P_i, initial pressure),

$$Q_{sto} = C_{pn}\frac{dP}{dT} = C_{pn}\frac{P_f - P_i}{dT}$$

$$Q_{sto} = 9.2\left(\frac{cf\ ft}{psia}\right)\frac{80(psig) - 105(psig)}{0.5(min)}$$

$$Q_{sto} = -460(scfm)$$

The storage airflow rate is 460 scfm, which is equal to approximately 100 hp of rotating on-line compressor capacity.

Calculating Required Receiver Volume for Demand Events

How much air receiver volume should be added to support the demand event while maintaining supply pressure at 85 psig minimum? Solve Eq. 15 for pneumatic capacitance (C_{pn}); then convert to gal and solve for additional receiver volume.

$$Q_{sto} = C_{pn}\frac{dP}{dT}$$

$$C_{pn} = Q_{sto}\frac{dT}{dP} = Q_{sto}\frac{dT}{P_f - P_i}$$

$$C_{pn} = -460(scfm)\frac{0.5(min)}{(85 - 105)(psia)}$$

$$C_{pn} = 11.5(scf/psia)$$

$$V_{rec} = C_{pn}P_a$$

$$V_{rec} = 11.5\left(\frac{scf}{psia}\right)14.5(psia) = 166.8(cu\ ft)$$

Convert to gallons : $166.8(cu\ ft) \times 7.48\left(\frac{gal}{cu\ ft}\right)$

$= 1248(gal)$

Additional Receiver Volume: $1248 - 1000$

$= 248(gal)$ \hfill (17)

Eq. 17 Solve for additional air receiver volume (gal).

Calculating Air Storage for Compressor Permissive Start-up Time

Storage of compressed air energy is also necessary to support compressed air demand during various supply-side events. One common supply-side event is the unanticipated shutdown of an air compressor (due to a motor overload or a high-temperature condition, for example). The startup of reserve compressor capacity requires a period that might range from many seconds to minutes depending on the type of compressors and controls. Most air compressors must start in an unloaded state to allow the electric motor to accelerate to normal running speed. For a typical lubricant-injected rotary screw compressor with part winding or Y-Delta, for example, starting might require 5–10 s for transition to full running torque. When the permissive time is past, the compressor's controls must open the inlet to begin compressing air. Then the internal piping, oil sump receiver, and possibly after-cooler must be pressurized before the compressor's internal pressure exceeds the system pressure and the first cubic foot of air is forced through the compressor's discharge check valve into the air system.

Consider the air system shown in Fig. 5, including three compressors operating in a baseload, trim capacity, and standby control configuration. The system air demand is 700 scfm required at 85 psig minimum pressure. The base-load and standby compressors are fixed-speed load/unload compressors with rated capacity of 400 scfm. The trim compressor is a variable-speed drive (VSD) compressor rated at 500 scfm capacity. The VSD trim compressor is set to maintain a target pressure of 90 psig. Assume that the standby compressor is set to start automatically at a pressure of 88 psig and requires a 15 s permissive startup time to deliver its first ft^3 of air into the system.

With the unanticipated shutdown of the base-load compressor, what size air receiver (gallons) is necessary to ensure that the system pressure does not fall below 85 psig during the permissive startup time of the standby compressor?

First, calculate the airflow required from storage after shutdown of the base-load compressor. System air demand is 700 scfm with shutdown of the base-load compressor; the VSD trim compressor will increase its air delivery to full capacity of 500 scfm. The remaining air deficit of 200 scfm must be supplied from storage (Q_{sto}). The pressure profile for the event will result in a fall of pressure to 88 psig before the standby compressor is signaled to start. The minimum pressure for the receiver is 85 psig. Therefore, the initial receiver pressure (P_i) is 88 psig, and the final receiver pressure (P_f) is 88 psig. The permissive startup event duration is 15 s or 0.25 min.

$$C_{pn} = Q_{sto}\frac{dT}{dP} = Q_{sto}\frac{dT}{P_f - P_i}$$

$$C_{pn} = -200(scfm)\frac{0.25(minutes)}{85 - 88(psia)}$$

$$C_{pn} = 16.7(scf/psia)$$

$$V_{rec} = C_{pn}P_a$$

Compressed Air Storage and Distribution

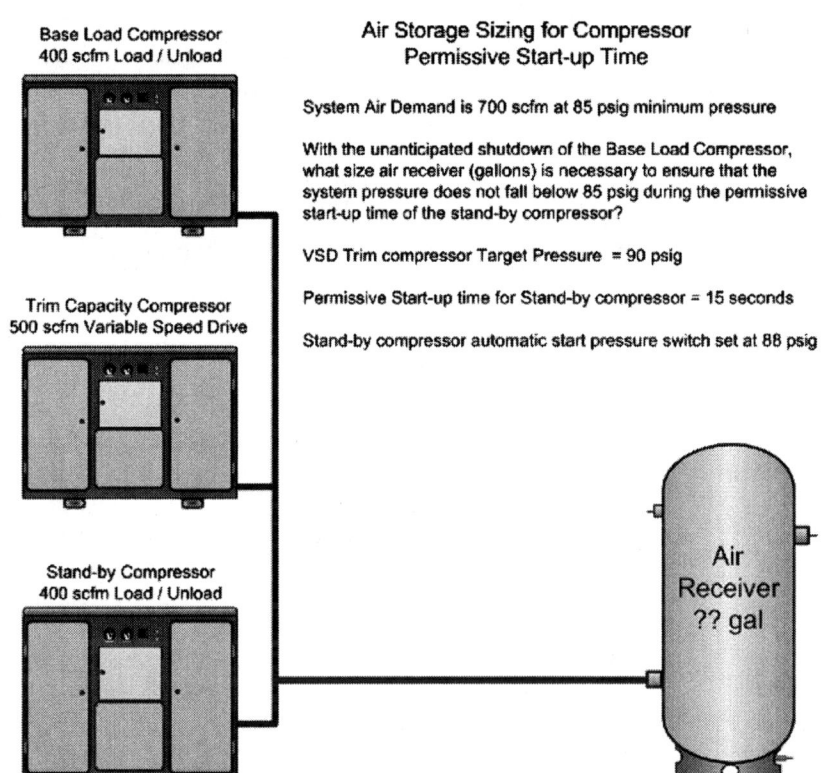

Fig. 5 Base-load, trim capacity, and standby compressed air system.

$$V_{rec} = 16.7 \left(\frac{scf}{psia}\right) 14.5(psia) = 242.2(cu\ ft)$$

$$\text{Convert to gallons}: 242.2(cu\ ft) \times 7.48 \left(\frac{gal}{cu\ ft}\right)$$

$$= 1812(gal) \tag{18}$$

Eq. 18 Permissive startup—solve for pneumatic capacitance and receiver volume.

Pneumatic capacitance calculations can be applied to solve a variety of compressed air storage requirements.

MAXIMIZE AND CONTROL COMPRESSED AIR ENERGY STORAGE

For compressed air energy in storage to be effective, the storage pressure must be higher than the demand-side target pressure. As supply-side pressure increases, the power required by positive displacement compressors also increases. The air compressor's power increase is approximately 1% for every 2-psig increase in discharge pressure. Increased storage pressure or increased air receiver volume increases usable air in storage. The economic tradeoff is the capital cost of increased air receiver volume vs the increased compressor supply-side energy cost of higher storage pressure.

Higher pressure also adds energy cost to the system's air demand. Increasing supply-side pressure creates a corresponding demand-side pressure increase. The result is additional energy consumption of the system through an air system loss called artificial demand. Simply stated, if the compressed air pressure applied to leaks and unregulated air use points is increased, the airflow consumed will also increase. Artificial demand is the additional compressed airflow demand consumed by the system when the actual applied air pressure is higher than the minimum required target pressure. Artificial demand in a system without any effective point-of-use pressure regulation will increase the system's energy demand by 2% for each 2-psig increase in pressure. In a "typical" compressed air system, it is common to find that 35%–50% of all air demands have effective pressure regulation. Therefore, artificial demand typically increases by 1.0%–1.3% for every 2-psig increase in applied system pressure.

Artificial demand can be eliminated by controlling the demand-side target pressure at an intermediate point separating the supply and demand sides of the system. An intermediate flow control valve is installed as shown in Fig. 6, downstream of the primary storage air receiver at the beginning of the distribution piping. This separates the supply side from the demand side of the system.

Flow control is used to control the energy (airflow) entering the system while maintaining a real-time energy balance between supply and demand. An intermediate

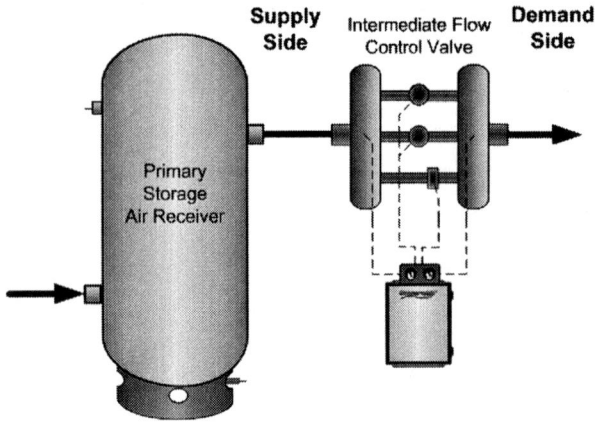

Fig. 6 Intermediate flow control separates supply and demand.

flow control is a packaged assembly of one or more flow control valves in a manifold arrangement with an automatic bypass or fail-safe open override device. It is normally installed in the compressor room at the beginning of the main piping distribution system. Adequate-size receiver(s) installed upstream of the flow control provide compressed air energy storage for controlled release into the system. The flow control senses air pressure at its discharge. Changes in the demand-side energy requirements cause fluctuating pressures. The flow control senses these changes and increases or decreases the airflow from storage as needed to maintain the system's energy balance between supply and demand. The result is stable system pressure set at the lowest optimum target pressure—normally, to within ±1 psi (Fig. 7).

The upstream compressed air energy storage is crucial to the satisfactory application of an intermediate flow control. The valve package responds immediately to the fluctuating pressure, so compressed air energy from storage must be available for instantaneous expansion into the system to maintain the supply/demand energy balance. The immediate energy supply cannot be dependent on rotating on-line generation. Air must be stored during dwells in the demand cycle when excess compressor capacity is available. Then stored energy is released by the intermediate flow control to satisfy the peak demands.

The intermediate flow control, like all components of a compressed air system, has some unrecoverable pressure loss, which represents an energy cost to the system. Intermediate flow controls are designed to operate with low unrecoverable pressure loss—typically, less than 5 psig. For the pressure profile shown in Fig. 6, the intermediate flow control has an unrecoverable pressure loss of 3 psig, which represents 1.5% increased energy cost at the compressor. In Fig. 4, it is shown that waste to artificial demand is between 14 and 18% of the system's air demand. The net savings achieved by eliminating artificial demand with application of intermediate flow control is 12.5%–16.5% of the system's energy input.

SUMMARY

Compressed air systems constantly undergo dynamic changes in their energy demand. For reliable, consistent, and efficient system operation, energy supply and demand must be balanced in real-time performance. Operating a system without proper energy storage results in part-load or no-load operation of compressors, which decreases system efficiency and increases energy cost.

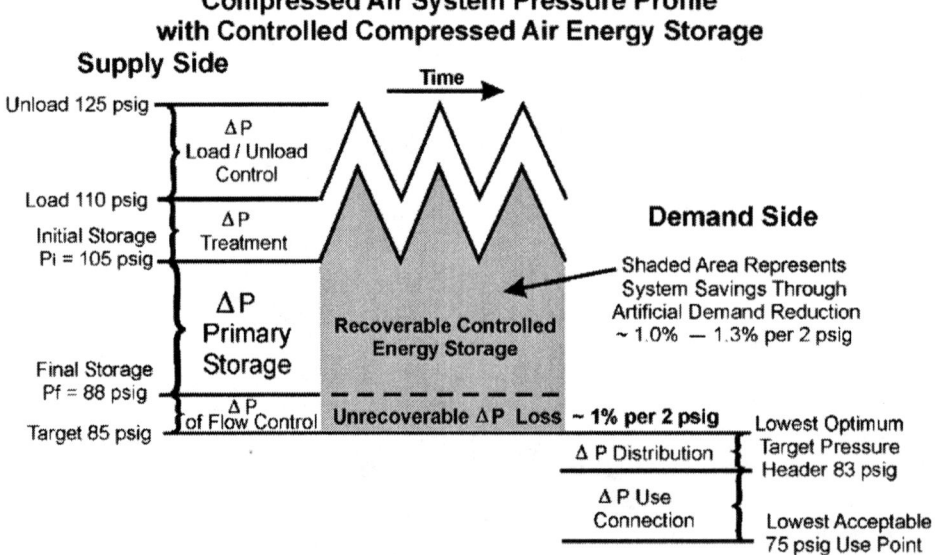

Fig. 7 Compressed air system pressure profile with controlled storage.

Compressed air energy storage can provide the necessary energy to meet peak demands. Pneumatic capacitance calculations derived from the Ideal Gas Law and First Law of Thermodynamics for systems allow mathematical modeling of compressed air energy storage. Usable air storage is a function of two factors: available storage volume, and the pressure difference between storage pressure and demand-side target pressure.

Demand-side target pressure should be the lowest optimum pressure required to support productive air demands. Increasing storage pressure increases the compressor's energy use by 1% per 2 psig. Uncontrolled compressed air storage pressure also increases the applied demand-side pressure, resulting in waste to artificial demand. Artificial demand typically wastes 1.0%–1.3% of the system's airflow for every 2-psig increase in system pressure.

Artificial demand can be eliminated through the application of intermediate flow control to separate supply and demand. The intermediate flow control paces the energy flow from supply to demand, maintaining a real-time energy balance. The result is reliable, consistent, and stable operation at an appropriate demand-side target pressure.

CONCLUSION

The compressed air system engineer must assess the dynamic energy characteristics of the air system. Compressed air energy storage requirements to support normal system events must be calculated. Energy storage must be optimized with an appropriate system pressure profile allowing the necessary storage pressure differential and adequate air receiver storage volume. Proper control of stored energy and the resultant control response of the system's air compressors, when optimized, will provide the best possible system operating efficiency.

REFERENCES

1. Resnick, R.; Halliday, D. *Kinetic Theory of Gasses: Physics for Students of Science and Engineering*; Wiley: New York, 1966; pp. 57–574.
2. Van Wylen, G.; Sonntag, R. *The First Law of Thermodynamics: Fundamentals of Classical Thermodynamics*, 2nd Ed.; Wiley: New York, 1973; pp. 90–97.

Compressed Air Systems

Diane Schaub
Industrial and Systems Engineering, University of Florida, Gainesville, Florida, U.S.A.

Abstract
Compressed air is a valuable resource for manufacturers, allowing the use of pneumatic-driven hand tools, which can be an ergonomic boon to employees. This resource comes with a price, however, in the form of higher energy costs. This article describes the use of compressed air and the creation and delivery of compressed air from both a supply side and demand side approach. A major focus of this article is on the costs associated with the generation of compressed air and ways to reduce the waste of this resource.

INTRODUCTION

The first section of this entry focuses on the use of compressed air and how it is generated. The section on generation is then separated into a discussion of the supply side and demand side components of a compressed air system. Finally, the costs associated with compressed air, as well as sources of further information, are found at the end of the entry.

OVERVIEW

Compressed air systems could be considered a unique source of energy despite the fact that they are actually powered by electricity. This similarity stems from the fact that compressed air lines can be designed to allow modular tools to plug into the air lines, just like electrical devices can be powered by tapping into electrical outlets.

By far, the most common use of compressed air is to drive pneumatic tools, ranging from nail guns to jackhammers to large drill presses. Pneumatic tools are favored over electric motor driven models because:

- They're smaller, lighter, and more maneuverable.
- They deliver smooth power and are not damaged by overloading.
- They have the ability for infinite, variable speed and torque control, and can reach these very quickly.
- They can be safer because they do not pose the potential hazards of electrical devices, particularly where water and gases are present.

Additional uses for compressed air in the manufacturing sector may include: filtration or control systems, driving conveyors, dehydration, aeration, or refrigeration.

Keywords: Compressor; Pneumatic; Supply side; Demand side; End-use applications; Manufacturing.

As these latter applications do not have the need for portability, and can be performed more cheaply without the additional process step of compressing air, their use is fairly limited. The economics of compressed air will be discussed later in this article.

COMPRESSED AIR SYSTEMS

Although a relatively simple-looking, self-contained air compressor can be purchased at a hardware store, they are limited in size and these small units (battery-powered, gas-powered, or plug-in models) are typically only to be used to fill tires or inflate rafts. Our discussion from this point onward will focus on larger, commercial compressed air systems.

The typical compressed air system is composed of:

- One or more in-series compressors
- An air dryer and air filters
- A receiving tank (for storage)
- Piping
- End uses

Compressed air systems should be perceived as possessing both a supply side and a demand side. Fig. 1 shows a typical block diagram of a industrial compressed air system, with both the supply and demand side noted. These block diagrams are a very helpful first step in understanding how to better manage compressed air systems, as recommended by Ref. 1.

- Improving and maintaining peak compressed air system performance requires addressing both the supply and demand sides, as well as how the two interact in order to have dependable, clean, dry, stable air delivered at the proper pressure. A well-planned balanced system will yield the cheapest and most energy efficient results.

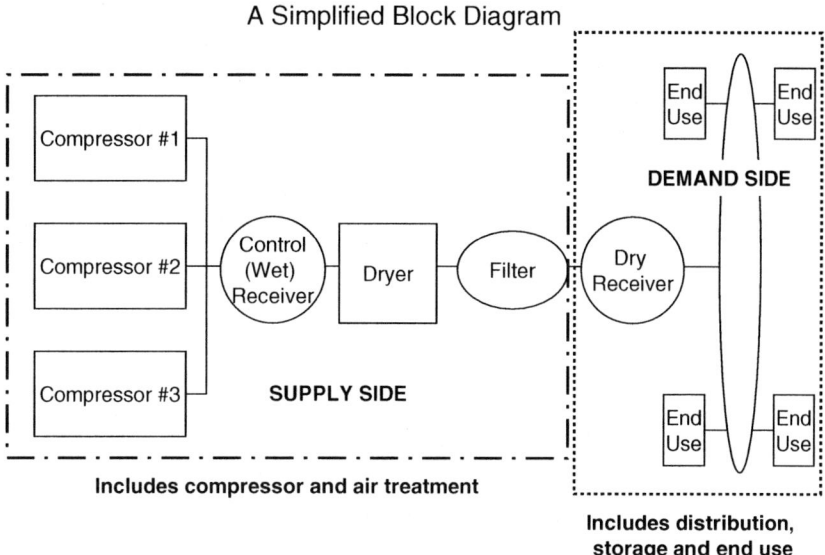

Fig. 1 Schematic of a compressed air system.

COMPONENTS OF A COMPRESSED AIR SYSTEM

Supply Side

A thorough understanding of the end-use compressed air needs, from both a volume and usage profile perspective, is necessary in order to select the appropriate number and size of air compressors. It is rare to find a manufacturing plant that has a constant, uniform use of compressed air throughout the day. Most manufacturing plants have cyclical flow and volume demands due to production schedules, and also desire back-up supply, so engineers typically plan for more than one air compressor to meet a facility's needs. A good strategy is to size a compressor for a base load, and have one or more compressors staged to come online to meet additional compressed air demand. In designing a compressed air system, altitude, inlet air temperature, and relative humidity should be considered, as they impact compressor capacity. More information on how to calculate the influence of these design considerations can be found in Ref. 2, pp. 9–10. It may also be helpful to have different size compressors, so that they can be tailored to fit the operating conditions. Additionally, a small compressor or separate booster may be appropriate for off-shift operations or a special high pressure, periodic application.

The vast majority of industrial compressors are of the rotary screw variety, but double-acting reciprocating or centrifugal compressors are also available for specific applications. Rotary screw compressors come in two configurations: lubricant-injected or lubricant-free. Both have various pros and cons associated with their use. Lubricant-free rotary screw compressors require higher electrical demand, but assure no lubricant carryover. This may be crucial when ultra-clean air is required. On the other hand, lubricant-injected rotary screw compressors have the ability to trim to partial loads to meet usage needs, which can further save on their already lower power costs.

Another issue that can greatly impact the energy efficiency of air compressors is their control strategy. Start/stop, load/unload, and modulating (or throttling) control strategies can be used, depending on the facility's compressed air usage profile.

In order to deliver clean compressed air, filters are installed downstream from the air compressors. The filters remove particulates, some condensate, and lubricant. Regular replacement of filters is necessary to prevent pressure drop, which results in a throttling effect. To illustrate the filter's importance, see the following example:

Example (Replacement of a Compressed Air Filter Element)

Assume a 100 hp compressor that operates continuously with an energy cost of seven cents/kWh, resulting in an annual energy cost of $55,328. As the filter becomes clogged, assume the pressure drop increases to six psi across the filter (as compared to a two psi pressure drop for a new filter). Consider that this four psi increase can cost two percent of the annual required energy, or $1100, as compared to $375 for a new filter element.

Another component of a compressed air system is the dryer(s). The compressing of air will condense out the moisture from the natural water vapor found in atmospheric air. This liquid water can cause rust problems in the lines or, should compressed air supply lines connect between buildings, freeze in the winter. Compressed air should be dried to a dew point at least 18°F below the lowest ambient temperature of the demand side.

The various types of dryers are:

- Refrigerated: This is the most common type, with both low initial and operating costs. It can be subject to freezing if operating at low capacities.
- Regenerative desiccant: Typically operated in tandem between two twin dryers, with one operating and the other regenerating. The required volume of purge air needed to regenerate can increase the load or even cause an idle compressor to be started. Heaters can be used in place of purge air, but present their own energy penalty.
- Heat of compression: Similar to the regenerative desiccant dryer, this type of dryer is available for lubricant-free rotary screw compressors and utilizes the hot discharge compressed air to regenerate the desiccant. Their efficiency is affected by changing air temperatures and additional heat may be required for low load situations.
- Deliquescent desiccant: A dissolvable desiccant is used. Regular replacement of this resource is necessary, requiring labor and material costs.
- Membrane-type: A porous membrane separates water vapor from the air and suppresses the dew point. Although there is a low initial cost, these dryers are appropriate only for low-volume applications.

Air receivers can be found on either the supply side (immediately after the compressor or the dryer) or on the demand side, close to the application end use. Air receivers store compressed air and help cover peak events of short duration. If sized properly, they can greatly reduce the frequent loading and unloading of the compressor, saving both energy and maintenance costs. They also stabilize system pressure, which improves performance of the end use.

Other components associated with the supply side may include aftercoolers or intercoolers (for lubricant-free systems), moisture separators, and condensate drains. Depending on the manufacturer, these latter items may be packaged in a single housing with the compressor itself.

Demand Side

Besides a downstream air receiver, the demand side consists of the distribution system or piping, and the end-use applications. Correct sizing of the distribution piping is a critical feature in compressed air system design in order to minimize energy costs.

The piping typically consists of rigid metal or plastic piping from the air compressor room to the general area of the end-use equipment. From this point, flexible rubber or plastic tubing is used, which may be plumbed directly to the end use, or have a shut-off valve with quick-connect attachment points. This flexible tubing may be subject to being run over by foot or equipment traffic and can wear out over time. As a result, air leaks can grow to epidemic proportions, and greatly increase the demand on the compressor. In fact, it isn't unusual to find a poorly maintained system running a compressor that is only feeding leaks. Some facilities will bury large portions of their distribution piping, which make finding and repairing leaks an expensive proposition. A 3/16" in. hole in a system operating at 100 psig can cost over $5000 a year.

Another operating consideration associated with the demand side is the cost of "normal production." Decisions to add additional applications should undergo a realistic cost evaluation. Consider the following example of an end-use application:

Example (Addition of an End-Use Application)

A quarter inch orifice required to operate a pneumatic hand tools at a recommended pressure of 100 psig was found to have a flow rate of 63.3 scfm (standard cubic feet per minute). After a year of constant use, this equates to 33.3 MMcf (million cubic feet) of compressed air. If compressed air generation costs $300/MMcf, then the power cost for this application will be approximately $10,000/year. If we add additional operating costs of $170/MMcf to account for the operator maintaining the compressed air equipment and the maintenance, lubricant, and repair costs for the system, we find that the cost of this new application use is over $15,000/year. Compare this with less than $2000/year to operate a comparable electrical tool.

High costs can also be incurred through the artificial demand associated with setting the compressor pressure level higher than needed. According to Ref. 3, p. 56, supplying 20% extra psig will force the system to consume 20% more air flow, resulting in 20% waste. Poor applications, such as stuck condensate drains, personnel use of compressed air for cooling or drying, or sparging (aerating of liquids), also use up precious compressed air.

ESTIMATING NECESSARY PRESSURE SET POINT

The determination of the pressure set point for the air compressors needs to be equated. Because of natural pressure drops associated with the components of a compressed air system, as well as unrepaired air leaks, the final point is more difficult to find than just dialing in the pressure recommended by the end-use equipment manufacturer. In fact, it is not unusual for plant personnel to reach the desired pressure by trial and error, increasing the set point until equipment operators stop complaining about low pressure. When possible, pressure measurements should be made after each component of the compressed air system to monitor system performance. Flow or electrical readings can also provide useful performance data. More information on how to calculate optimum compressed air system settings can be found in Ref. 2, p. 205. Fig. 2 provides an example of the pressure drops that can occur along the line.

Estimating Pressure Drop

Measurements can be taken at various points in a compressed air system to monitor the associated pressure drop from each component. The pressure profile shows the lowest pressure seen by the end-uses.

Compressor operating range:	115-105 psig	Air/Lubricant Separator	5 psid
FRL (Filter, regulator, lubricator)	7 psid	Hose and Disconnects	4 psid
Aftercooler	3 psid	Dryer	4 psid
Filter	3 psid	Distribution System	3 psid

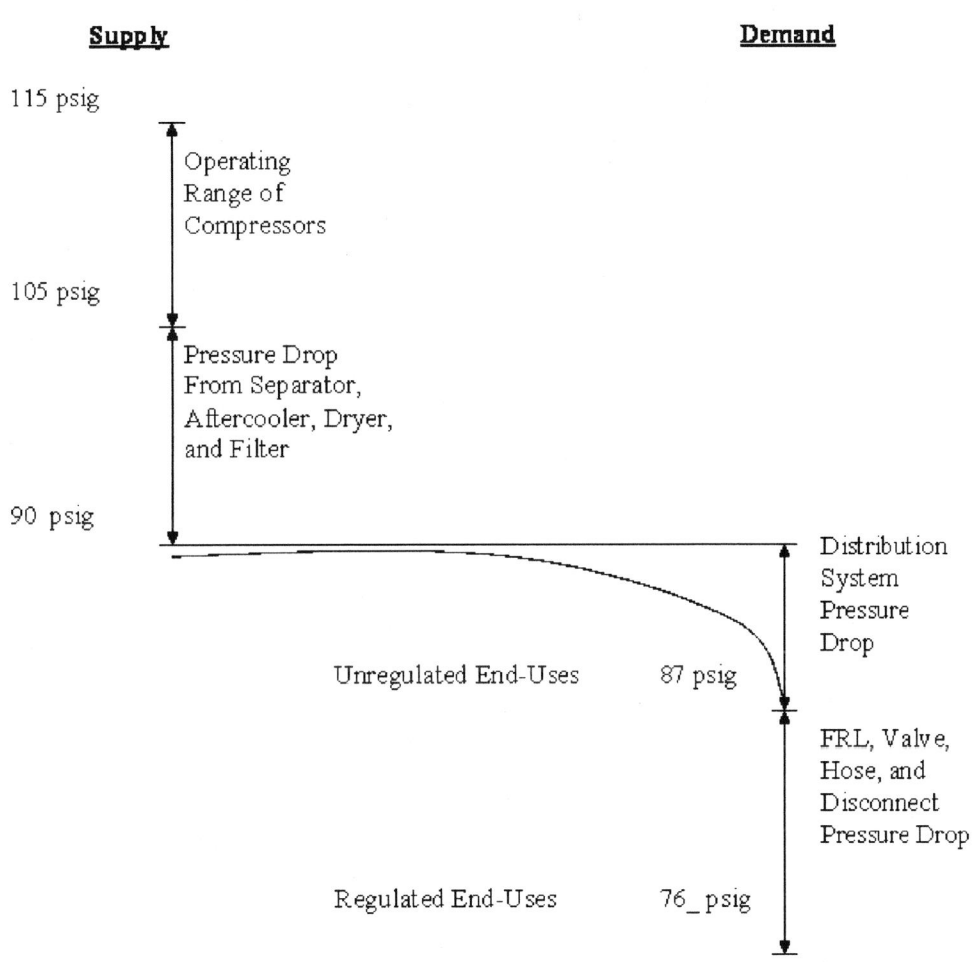

Fig. 2 Estimating compressed air system pressure drop.

COSTS OF COMPRESSED AIR

To operate a one hp air motor, seven to eight hp of electrical energy are required. This large energy penalty, along with the common employee perception that compressed air is essentially a free resource, makes it a challenge to control the costs of compressed air. Inadequate compressor control schemes can cause multiple compressors to run at partial loads, rather than turning them off. Problems with poor maintenance can increase consumption or cause pressure variability. In fact, it isn't unusual to find that compressed air can be the largest end user of electricity.

ARTICLES OF FURTHER INTEREST

The U.S. Department of Energy's Industrial Technologies Program sponsors compressed air training and Air-Master+tools through their Best Practices programs. See http://www.oit.doe.gov/bestpractices for more information. The organization charged with actually delivering the compressed air training can be found at: http://www.compressedairchallenge.org.

REFERENCES

1. *Fundamentals of Compressed Air Systems*; Training Manual for the Compressed Air Challenge, Prepared by Laurel and Associates, Ltd and Resource Dynamics Corporation and presented by the trade association known as the Compressed Air Challenge, (www.compressedairchallenge.org) 1999.
2. Scales, W.; McCulloch, D. *Best Practices for Compressed Air Systems*; The Compressed Air Challenge, 2003; 9–10.
3. Taranto, T. *Compressed Air Management*; ConservAIR Technologies Industrial Seminar Series, Conservair Technologies Company, LLP, Kenosha, WI, (www.conservair.com) 1998; 56.

BIBLIOGRAPHY

1. *Advanced Management of Compressed Air Systems*; Training Manual for the Compressed Air Challenge, 1999; 205.

Compressed Air Systems: Optimization

R. Scot Foss
IR Air Solutions, Davidson, North Carolina, U.S.A.

Abstract
This article will provide you with an complete action plan to optimize your compressed air system including compressor optimization, demand management, density management, and storage in a variety of different applications.

Compressed air represents one of the most critical utilities in most production and process environments. The efficiency of a compressed air system is 100% energy in and, when perfect, produces 11% useful work out. Understanding this, it will cost more to operate a compressor in the first year than it costs to buy and install. Despite this harsh information, power is thrown at symptoms of undefined problems every day. The opportunities of reducing operating cost and energy in air systems is typically more than 50%. This session will carve out a plan of attack to optimize the supply and demand systemically and yield the lowest demand at the highest rate of standard cubic feet per kilowatt of energy.

There are a number of essential actions that need to be taken to optimize the compressed air systems. You need to minimize demand, control the expansion of the air, distribute it while minimizing energy loss, store potential energy, and compress the air efficiently. Other than operating the compressors, as efficiently as possible, everything else seems to elude most everyone. This work cannot be done theoretically on each piece of equipment only in the compressor room. It must be done systemically. More efficient compressors make more air with the same amount of power. They cost more and can be an important part of a well-operated system. On the other hand, if you throw a more efficient compressor at a highly inefficient system, you will waste more air at the same operating cost and save nothing.

CONTROLLING DEMAND IN THE SYSTEM

1. Control the expansion of the compressed air to the point of use. You must control 100% of users with regulation, which is adjusted lower than the lowest supply pressure. If it is not possible to achieve this with operator discipline, then you must use a demand controller or expander at a central location adjusted in the same manner.
2. Reduce the pressure differentials on installation components such as filters, regulators, lubricators, tube, hose, and disconnects on the demand side of the system. The intent is to operate demand at the lowest possible supply pressure on critical high-pressure applications.
3. Flat line the high rate of flow, intermittent applications with dedicated storage and metered recovery. This is much like a battery charger or water tower application. This can also be a pressure driver for the operating protocol. You will slightly increase the base usage and eliminate peaks.
4. Review and add as necessary general and control storage to slow the rate of change in the system. This will allow you to maintain a higher point of use pressure if necessary without increasing the supply pressure. If there is any diligence used, you can normally reduce the supply pressure simultaneously.
5. Upgrade the quality of information to track progress and improve decision making. This should include a flow meter and demand pressure monitor at the discharge of the demand controller or the expander. If you do not use a demand controller, recognize that demand is only accurately displayed when the demand exceeds the supply. This is referred to as a negative rate of change. When supply exceeds demand, which is a positive rate of change, you are measuring supply response to demand. The system will take whatever supply power you throw at it. A 450 scfm negative rate of change will recover to the original pressure in 1 min, if we respond with a 200 hp compressor. If we throw a 400 hp compressor at the event, it will recover in 15 s at a more rapid rise in pressure. The inefficiency is the part load energy of the larger compressor for the balance of the 45 s. If we match

---

* This entry originally appeared as "Optimizing the Compressed Air System" in *Energy Engineering*, Vol. 102, No. 4, 2005. Reprinted with permission from AEE/Fairmont Press.

Keywords: Compressors; Compressed air; Storage; Demand controls; Expanders; Metered storage; Leaks; Transient events; Air dryers and filters; Potential energy.

the event with a 100 hp compressor, the pressure will hold at the load pressure of the compressor until the event stops, at which time, the pressure will recover at the same rate of rise as the initial rate of decay.

6. Review and add as necessary general and control storage to slow the rate of change in the system. This will allow you to maintain a higher point of use pressure, if necessary, without increasing the supply pressure. If there is any diligence used, you can normally reduce the supply pressure simultaneously.

REDUCING DEMAND IN THE SYSTEM

1. Develop a leak benchmarking program on a gradual reduction of the tolerance volume. Select a level at a known low load, and repair your way to that level. Every several weeks, check the low load and scan the system using an ultra sonic leak detector. Find and repair the largest leaks found to bring the system back into benchmark. When you are comfortable with this level, lower the level and begin again. You will reach a point where there are so many small leaks to fix during the benchmarking period, the labor hours cannot be justified. At this point return to the previous higher tolerance value.

   Record the types and nature of the leaks that you are fixing, so that you can leverage this information into buying more leak resistant components and improving best practices installations. Note that it is important that the reduction of demand does not cause the demand pressure to rise. If it does, then other unregulated users will increase at the elevated pressure. That is why it is so important to have demand controls installed before you become aggressive in demand reduction. It is also important to off load a linear amount of supply energy for the demand reductions.

2. Eliminate all open compressed air blowing applications and replace with low pressure centrifugal or positive displacement blowers, if at all possible. If it is not possible to use blowers, apply specialty air volume reducing nozzles for the application. Take your time with these applications developing the thrust per square inch as close as possible to the open blowing application. You will also need to filter the air for specialty nozzles, as they will easily plug up with pipe debris. Whenever possible, use a solenoid valve to shut of the air on cyclical applications.

3. Replace all applications, which are poor users of compressed air. Focus on operating cost alternatives. Use electricity whenever possible for its better wire to work energy relationship.

4. Reduce the size of demand events as seen by the system including high ramp applications. This can be accomplished by slowing down the introduction of these events into the system. This can be done by opening the demand valve slower manually or automatically. This reduces the "ramp in" rate of flow, so that the supply including control storage can match the event limiting the ultimate pressure drop, which would result.

5. Regulate all points of use, even if you have installed a demand controller or expander in the main supply system's piping. Make sure that the set points on the regulators are equal to the minimum supply pressure minus the point of use filter and regulator pressure drop or less. If you allow for a 2–3 psig margin below this value, small leaks and filter dirt loading will not cause frequent changes in process performance.

6. Limit the coincidence of events that cause peak demands in the system. This includes minimizing the blow duration on timer drains and adjusting intervals seasonally for relative humidity. Move large events to low load times where possible.

7. Shut off all air using equipment when not in use. Make sure that the shut off valves are ergonomically installed, so that operators can easily reach them. If this does not work, install solenoid shut off valves that are tied into the electrical shut off on the machine, work station, or process.

STORE POTENTIAL ENERGY TO SUPPORT TRANSIENT EVENTS INCLUDING A COMPRESSOR FAILURE IN THE SUPPLY SYSTEM

1. Convert enough kinetic energy to potential energy so that you can handle largest event without turning on another compressor during normal operation. If you do this, you will also handle all of the smaller transient events that are not controlled from downstream. This can include the coincidental impact of a third to first shift startup. Remember that storage is a function of the capacity to store air times the useful differential across it. If you are operating constant pressure compressor controls and they operate correctly, no amount of capacitance will generate any useful storage.

2. Store enough air on the supply side of the system to manage a desired pressure drop, while bringing up a backup compressor to replace a failed one. The intent would be that the event will have no impact

on the process or production serviced by the system. The intent is to operate only the supply that is required at any time with everything else off.

Example: largest compressor = 1600 scfm, maximum allowable pressure drop from the load pressure on the back up compressor = 10 psid, permissive time to load the compressor from a cold start signal to full load = 15 s, atmospheric pressure = 14.3 psia, gallons per standard cubic feet = 7.48 gal

$1600 \times (15/60) \times (14.3 \times /10) \times 7.48$

$= 4278.6$ gal

3. Create enough storage to control the maximum load cycles per time period on any trim compressor. It is safe to say that 3 min load–unload cycles or longer would be desirable on any positive displacement compressor. This can get trickier on large dynamic compressors, but it is not impossible.
4. If the size of any event or compressor is too large to handle with control storage or you want to protect the system and production against an electrical outage, single phase, or brown out, offline high-pressure peak shaving would be the most desirable approach to minimize on board power. It would not be unusual to store 30–40,000 ft^3 of air in a 100 psig differential supported by a 20 hp compressor offline. You would then introduce the air back into the system on variety of different cues or logic patterns to support the various events.

Note that it is the intent of all potential energy applications to either prevent the normal operation of an additional compressor, extend the mechanical life of a compressor or compressors, or both. Well applied storage will increase the base load in the system slightly, and eliminate the requirement for added compressors during peak plus the inefficient part load in between peaks.

DISTRIBUTE THE COMPRESSED AIR, WHILE MINIMIZING ENERGY LOSSES

1. The concept of design or redesign should be to minimize the highest amount of air mass or volume of air and the distance that the air must flow to support any part of the system from supply to demand.
2. Resistance to flow is necessary in the system. Without resistance to flow there is no flow. As the system is open on both ends of the system all of the time to a larger or lesser degree, resistance to flow and storage keeps it functioning. Mass flow restrictions are differential pressures in the system, which change as a square function of flow change. It is important to design or retrofit your system for a maximum differential at highest flow, highest temperature, and lowest inlet pressure. This will produce the highest differential pressure across the components being evaluated. Although we are recommending a conservative approach towards this process, the piping distribution system should not be made intentionally oversized or all the same size for convenience. Oversized piping will not provide economical storage and will make it difficult for supply to see demand efficiently. A reasonable differential pressure would be 1–2 psid from the discharge of the cleanup equipment at the supply or the discharge of the demand controller, as it applies to your system, to the farthest point in the demand system at the previously discussed design conditions.
3. In most systems that have distribution problems, you should minimize waste and flat line transient users with dedicated storage and metered recovery at the point of use before considering making changes in the piping distribution system.

As little as a 10%–20% demand reduction at the peak condition can be sufficient to eliminate the most distribution losses and the requirement for piping retrofits.

REDUCE SUPPLY ENERGY WHEREVER POSSIBLE

1. When 100% of demand is at a lower pressure than the lowest supply pressure, set up the supply pressure to optimize the pound per kilowatt of compressed air energy for the on board compressors. Operate all compressors that need to be on flat out and optimized except one compressor trimming and all other compressors off regardless of inlet conditions or relative demand load. You must optimize the compressor and the motor simultaneously. Optimal means the most pounds or standard cubic feet at the optimal density (pressure and temperature), while managing the highest power factor and motor efficiency simultaneously. In this scenario, the trim compressor is the only compromise to "optimal" assuming you can maintain a range of supply pressure across the range of load conditions that relates to optimal on the base load compressors. Another option is to trim with variable frequency drive compressors using storage continuously, while adding and subtracting base load compressors. The Variable

frequency drive (VFD) compressor or compressors will displace or fill in the removal or addition of a base. In this case, you will optimize both the base load compressors and the trim compressors at the same time.

Note that this is called a "Bellows Effect" operating protocol.

2. Base compressors should always be selected based on the best energy efficiency. Trim compressors should be selected first on operating speed to cold or hot start and shut off capabilities, and secondly on their flexibility for automation interface. If you are trimming with VFD/s, the same requirements are applicable. This typically translates into smaller, less permissive compressors. You must be certain that the total trim capacity (one, two, or three trim compressors) is equal to or larger in capacity than the largest base compressor in the supply arrangement. This will assure that there are no gaping holes in supply, so that you can make smooth transitions from one power level to the next. Supply systems that do not have this capability end up running too much power part loaded all of the time to support the transitions. Remember that bigger is more expensive.

3. Develop an operating profile for the supply system, which optimizes the compressors based on a full range of usage and conditions. In most systems, the only time the system is remotely efficient is during peak load. It generally goes down hill during lighter or low load. Also evaluate the full range of system's usage against the full range of ambient inlet and cooling conditions to determine how the system will work before you make any final plans on equipment selection. Make every attempt to manage peaks with potential energy instead of on line power. You must also evaluate the risk of a unit failure in order to have a solid curtailment plan. If brown outs or black outs are common, you must include this in your plan.

4. Unload all unnecessary ancillary power, such as dryers, pumps, fans, etc. through the use of more efficient controls and motor drivers. Size all filtration and dryer equipment for a total differential of 3.5–5 psid. The differential should be at the highest inlet flow, highest inlet temperature, and the lowest inlet pressure. The differential on the filtration should be in a wet and clean condition. Plan the additional differential from dirt loading when selecting the compression equipment, so that you do not overload the motor drives as you will absorb the added differential at the air end discharge. We would recommend no more than an additional 1.5–2 psid on the total filters. There are filters available to accomplish this with a change every 5–6 years at this dirt loading rate. The total differential across all cleanup equipment should not influence the total connected horsepower on the compressors by more than 4% at the worst case maintenance condition.

5. Use a master signal for the compressors located in the dry clean storage downstream of the contaminant control equipment. If the signals are in the compressors upstream of the cleanup equipment, the compressors controls will respond to the demand interpreted through the differential pressure, which changes as a square function of flow change. This causes the compressors to over shot and under shot, which results in hunting. This requires excess energy to compensate.

Please note in the illustration that we have installed a three-way valve so that you can return to local control signals when you wish to isolate the compressor from the system. It is also important to note that the adjustment of the compressor controls, with a master control signal, should be based on controlling downstream of the cleanup. If the pressure across the cleanup equipment is 10 psid, when you moved the signal, you would also want to reduce the control set points on the compressor/s an additional 10 psid. This is because you will absorb the differential when you move the signal and without adjusting the operating set points for the compressor/s, you may overload the motor.

6. Develop an operating profile which takes control storage, set points of the compressors, signal locations, and differentials into account. Put it down on paper prior to implementing it and check the range of conditions to make sure it will work. Do not put fudge factors into the profile. This is not an art form. It is a science. If you are not sure of what you are doing, contact a technology firm who can assist you. Literally, 95% of all compressor profiles are not set up correctly. Most engineering firms that design systems select the equipment and never think through the operating protocol or profile prior to installation.

7. Finally, you must get the system to operate effectively and efficiently before you automate it. More than 90% of the time, users try to apply automation to a system to get it to work properly. If you automate a system that does not work, you will have an automated mess. You must be able to get it to work correctly on the local controls first. When and if you automate, keep in mind that their purpose is to refine the operating cost and reliability issues across all conditions unattended. Automate the operation based on at least rate of change, storage, time, and pressure. You may even wish to add a selective rate of change protocol, which chooses the correct compressor for the situation. Take your time and test your concept

prior to making the decision by preparing algorithms including transitions of power and demand including failure scenarios. Keep in mind that you do not have to match the event in the system. You only need to slow it down so you can wait longer. The essence of a masterfully designed system is the ability to control demand by matching transient events as quickly as possible with an expander or demand controller serviced with potential energy.

Once this is accomplished, the compressors' control job is managing control storage by replenishing it as slowly as possible. The longer you can take, the less energy you will use.

SUMMARY

A compressed air system is a highly interactive configuration with all aspects affecting all other aspects. Developing an action plan to improve the efficiency and reduce the operating cost can be rewarding, but must be done in the correct order to enjoy the success and avoid production inconvenience. It is a process of black and white with a lot of gray in between. Far too many owners want to buy a solution, rather than apply one. Problem definition, metrology, and carefully planning are all essential. When you have completed the action plan, do not forget to measure the results. Validation is necessary to support you return on investment strategy.

Cooling Towers

Ruth Mossad
Faculty of Engineering and Surveying, University of Southern Queensland, Toowoomba, Queensland, Australia

Abstract
Cooling is necessary to many industrial processes, such as power generation units; refrigeration and air conditioning plants; and the manufacturing, chemical, petrochemical, and petroleum industries. As recently as 20 years ago, cooling towers were more the exception than the rule in the industry because of their high operating cost and the large capital required for their construction. Due to the recent stringent environmental protections, cooling towers became more common. Cooling towers range in sizes and types. Wet, dry, and hybrid are the main types, and each type has many variations in design according to the way the fluids are moved through the system. Some of the advantages and disadvantages of these types, methods of determining their performance, and some terminology common to the cooling industry are presented in this entry.

INTRODUCTION

Most industrial production processes need cooling of the working fluid to operate efficiently and safely. Refineries, steel mills, petrochemical manufacturing plants, electric utilities, and paper mills all rely heavily on equipment or processes that require efficient temperature control. Cooling water systems control these temperatures by transferring heat from hot process fluids into cooling water. Through this process, the cooling water itself gets hot, and before it can be used again, it must either be cooled or be replaced by a fresh supply of cool water.

A cooling system in which the water used in cooling processes or equipment is discharged to waste is called once-through cooling. Characteristically, it involves large volumes of water and small increases in water temperature. Once-through cooling is usually employed when water is readily available in large volume at low cost. Common sources are rivers, lakes, and wells, where the only cost involved is that of pumping. But with today's need for water conservation and minimal environmental impact, industry is turning more and more to recycling water in what are called cooling towers.

Recently, cooling towers are becoming widely used in most industrial power generation units; refrigeration and air conditioning plants; and the manufacturing, chemical, petrochemical, and petroleum industries to discard waste heat to the environment. They range in sizes—the smallest cooling towers are designed to handle water streams of only a few litres of water per minute supplied in small pipes like those in a residence, whereas the largest cool hundreds of thousands of litres per minute supplied in pipes as huge as 5 m in diameter in a large power plant.

Cooling towers are believed to be only the direct contact type heat exchangers. They can be direct or indirect, however, and they are also characterized in many other different ways based on the type of fluid being used in the cooling process, the means by which the fluids are moved, and the way the two fluids (hot and cold) move with respect to each other. Description of these types, some of their advantages and disadvantages, and methods of estimation of their performance are discussed in the following sections. Some useful terms common to cooling towers industry are also given. This information is a collection of materials published in the list of references and Web sites at the end of this article.

TYPES OF COOLING TOWERS

Cooling towers are classified mainly on the basis of the type of fluid used in the cooling process—water or air. These are three types: wet (evaporative), dry (nonevaporative), and wet–dry (called hybrid).

Wet Cooling Towers

The wet type is more common in large cooling towers, such as in electrical power generation. It is a direct contact heat exchanger, in which hot water from the condenser and cooling air come into direct contact. The water flows in either open circuit or closed circuit. In open circuit, cooling water is pumped into a system of pipes, nozzles, and sprayers within the tower, and is drawn by gravity into a pond below (Fig. 1). Air from the atmosphere enters the tower from the bottom of the tower and flows upward

---

Keywords: Cooling; Towers; Wet cooling; Dry cooling; Hybrid cooling; Plume; Performance; Counter-flow; Cross-flow.

Cooling Towers

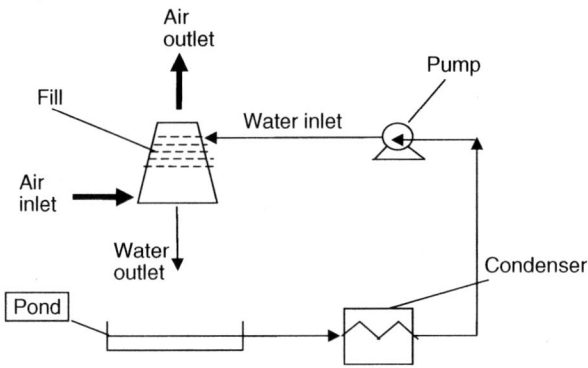

Fig. 1 Open-circuit cooling tower.

through the falling water. The two fluids go through a material that is provided to increase the surface area of contact between them, which is called packing (or fill). The heated and moisture-laden air leaving the fill is discharged to the atmosphere at a point remote enough from the air inlets to prevent it from being drawn back into the cooling tower. The water is collected at the bottom of the tower and then recirculated to remove more heat from the condenser. The temperature of the cold water entering the condenser will determine the steam condensate temperature and, hence, the backpressure, which impacts the efficiency of the whole power generation system.

The closed-circuit cooling tower (Fig. 2) involves no direct contact of the air and the liquid—usually, water or a glycol mixture—that is being cooled. This cooling tower has two separate fluid circuits. Water is recirculated in an external circuit outside a closed circuit made of tube bundles or coils containing the hot fluid being cooled. Air is drawn through the recirculating water cascading over the outside of the hot tubes, providing evaporative cooling similar to an open cooling tower. In operation, the heat flows from the internal fluid circuit, through the tube walls of the coils, to the external circuit and then (by heating of

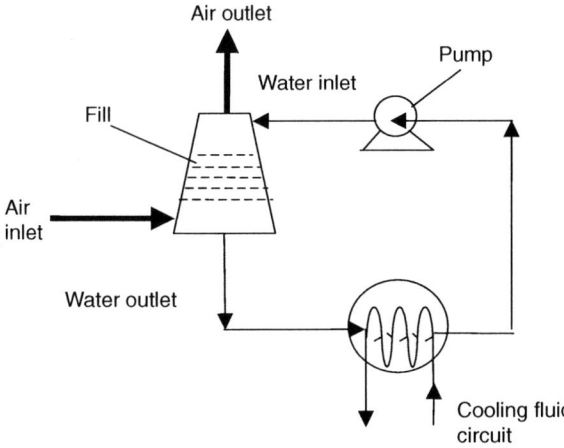

Fig. 2 Closed-circuit cooling tower.

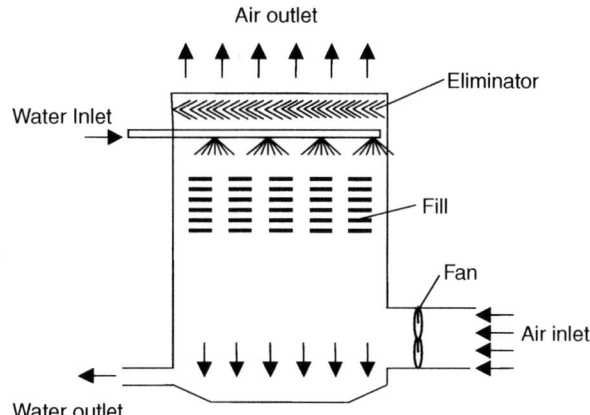

Fig. 3 Mechanical-draft counter-flow tower.

the air and evaporation of some of the water) to the atmosphere. Operation of the closed cooling towers is very similar to the open cooling tower, therefore, with one exception: the process fluid being cooled is contained in a "closed" circuit and is not exposed directly to the atmosphere or the recirculated external water.

Closed systems offer the advantages of precise temperature control (which is critical in many process applications) and low treatment cost. Because a secondary cooling system and heat exchangers are needed to cool the closed system, higher capital and operating costs are disadvantages of this design.

Mechanical and Natural Draft

In wet cooling towers, there are two types, based on the mechanism by which air is being circulated: mechanical draft and natural draft. The mechanical draft uses fans (one or more) to move large quantities of air through the tower. The mechanical draft is again divided into two types, based on the location of the air fan: forced and induced. In the case of the forced draft, the fan is located at the air entry at the base of the tower; in the induced draft, the fan is located at the air exit at the top of the tower. The induced draft produces more uniform airflow, which enhances its effectiveness over the forced draft and reduces the possibility of exhaust air recirculation.

There are many configurations of mechanical draft cooling towers that depend on the way the two fluids flow with respect to each other, such as counter flow, cross flow, and mixed flow. In a counter-flow cooling tower, air travels upward through the fill opposite to the downward motion of the water (Figs. 3 and 4).

In a cross-flow cooling tower, air moves horizontally through the fill as the water moves downward (Fig. 5 and 6). In a mixed-flow tower, air moves in a direction that is a combination of a counter flow and a cross flow. Cross-flow towers have greater air intake area, which results in considerably lower towers. This means that they have low

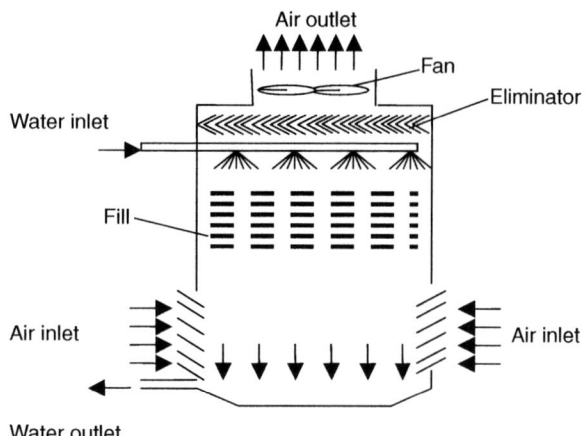

Fig. 4 Induced-draft counter-flow tower.

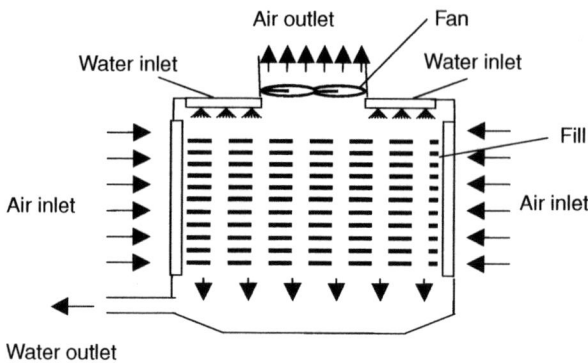

Fig. 6 Induced-draft cross-flow tower.

pressure drop in relation to their capacity and lower fan power requirement, leading to lower energy costs, but the risk of recirculation increases in tower exhaust air. On the other hand, counter-flow arrangements occupy less floor space than cross-flow towers but are taller for a given capacity, so they require higher pump heads. It shows better tower performance, since, the driest air contacts the coldest water, producing higher driving force to the heat.

A natural-draft tower is a large chimney and typically has a hyperbolic profile, which is chosen for its structural capability of withstanding wind-induced stresses and vibration; also, it requires less material. The design creates a chimney effect that causes air to move by natural convection through the fill region, which is located inside the base of the chimney (Fig. 7). As the air gets warmer from the contact with the cooling water, it gets lighter; buoyancy forces drive the air to the top of the tower and into the atmosphere, and draw fresh air into the bottom of the tower. The major economical advantage of natural-draft cooling towers is the extremely low auxiliary power consumption. Because there are no rotating parts, operational safety and low maintenance costs are maintained. The great distance between air inlet and air exit in the cooling tower prevents any hot-air recirculation back to the chimney that would otherwise reduce the performance of the cooling tower. The immediate vicinity is not affected by plumes from the cooling tower, because the hot-air exit is situated at a very high elevation. The only drawback to natural-draft towers is that they are large.

There is also the assisted-draft tower, which is a natural-draft tower with some fans added at the air entry that help reduce the size of the tower. Natural-draft towers are also divided into counter flow and cross flow, defined in a similar fashion to the mechanical-draft towers.

Heat Exchange in Wet Cooling Towers

The type of heat rejection in a wet cooling tower is termed evaporative, in that it allows a small portion of the water being cooled to evaporate into a moving air stream to provide significant cooling to the rest of that water stream. The heat from the water stream transferred to the air stream raises the air's temperature and its relative humidity to 100%, and then this air is discharged to the atmosphere. The ambient air wet-bulb temperature is the controlling factor in recirculated systems and will determine the steam condensate temperature. Evaporative-heat rejection devices such as cooling towers are commonly used to provide significantly lower water temperatures than are achievable with air-cooled or "dry" heat rejection devices. The evaporative process enhances the performance of wet cooling towers over dry cooling towers severalfold due to the change in both sensible and latent heats.

Consequences and Concerns for Wet Cooling Towers

Wet cooling towers are the most common type due to their high effectiveness, but there are some drawbacks. If cooled water is returned from the cooling tower to be reused, as in the circulating systems, some water must be added to replace, or make up, the amount of the water that evaporates. Because evaporation consists of pure water, the concentration of dissolved minerals and other solids in

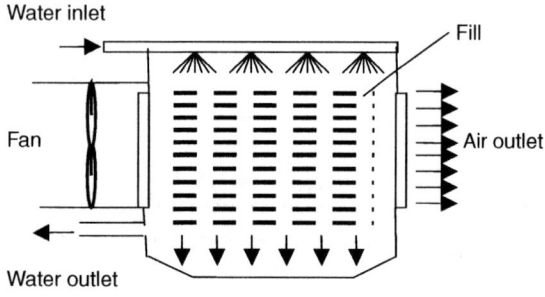

Fig. 5 Mechanical-draft cross-flow tower.

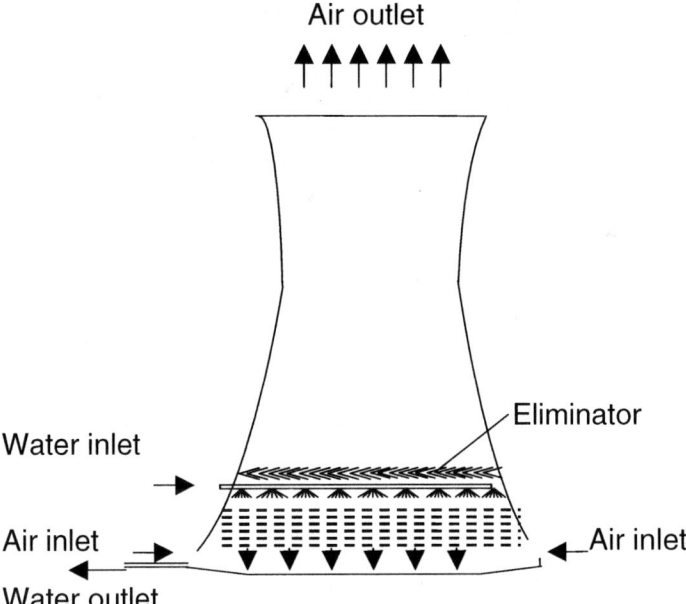

Fig. 7 Natural-draft cooling tower.

circulating water will tend to increase unless some means of dissolved-solids control (such as blow-down) is provided. Blow-down is the amount of the circulating water that is removed to maintain the quantity of dissolved solids and other impurities at an acceptable level. Some water is also lost from droplets being carried out with the exhaust air (drift). The makeup amount must equal the total of the evaporation, blow-down, drift, and other water losses (such as wind blow-out and leakage) to maintain a steady water level. Devices such as wind screens, louvers, splash deflectors, and water diverters are used to limit these losses.

The magnitude of drift loss is influenced by the number and size of droplets produced within the cooling tower, which in turn are determined by the fill design, the air and water patterns, and other interrelated factors. Drift is typically reduced by installing bafflelike devices, called drift eliminators, through which the air must travel after leaving the fill and spray zones of the tower to collect the droplets. Tower maintenance and operation levels can also influence the formation of drift, such as excessive water flow, excessive airflow, or bypassing the tower drift eliminators can increase drift emissions. Types of drift-eliminator configurations include herringbone (blade-type), wave form, and cellular (or honeycomb). The cellular units generally are the most efficient. Drift eliminators are made of various materials, such as ceramics, fibre-reinforced cement, fibreglass, metal, plastic, and wood.

Other unfavourable environmental impacts of wet cooling are pollutant discharge—e.g., zinc, chlorine, and chromium (chromium is used to protect cooling-system equipment from corrosion)—to the atmosphere. The spread of Legionnaires' disease is due to the bacteria that thrive at temperatures typical in wet cooling systems and that can be transported through air aerosols formed in cooling towers. Other impacts are mineral drift and the formation of visual plumes. Under certain conditions, a cooling-tower plume may present fogging or icing hazards to its surroundings (Fig. 8). Some interesting pictures of cooling towers can be found at The Virtual Nuclear Tourist: Nuclear Power Plants Around the World (www.nucleartourist.com/systems/ct.htm).

Types of Fill (Packing)

The fill may consist of multiple vertical, wetted surfaces upon which a thin film of water spreads (film fill); several levels of horizontal splash elements, which create a cascade of many small droplets that have a large combined surface area (splash fill); or trickle, which is a combination of the film- and splash-type fills. A wide variety of materials and geometries have been used for packing, such

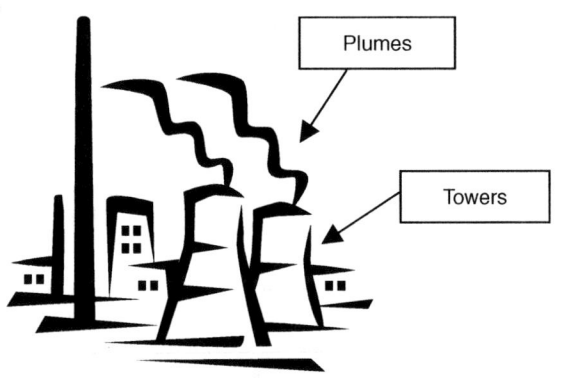

Fig. 8 Visual plume from cooling towers.

as corrugated roofing sheets made of cement-based or plastic material, timber laths of triangular or rectangular cross section, plastic-impregnated paper honeycomb, and complex cellular geometries made of thin plastic material.

Thermo-Fluid Dynamic Efficiency in Cooling Towers

To choose the most convenient fill, you need to find the one that produces the maximum heat transfer with the minimum pressure drop. Other factors to be considered are the physical and chemical characteristics required for the water to cool, fouling properties, suspended materials, etc. Sirena[1] has suggested a Thermo-Fluid Dynamic Efficiency that can be used to compare and select a fill material for a particular application. This efficiency is defined as the ratio of the number of units of diffusion to the pressure drop coefficient. In this paper, the pressure drop coefficient is given for some commercial fill materials. Al-Nimr[2] has studied the dynamic thermal behaviour of cooling towers containing packing material and was able to predict closed form solutions for the transient and steady performance of a counter-flow cooling tower.

Dry Cooling

Dry cooling towers transfer heat to the atmosphere without the evaporative loss of water. Dry cooling is capable of only smaller temperature variations (around 10°C), unlike wet cooling. Similar to wet cooling, there are two types of dry cooling: direct and indirect. Direct dry cooling systems utilize air directly to condense steam, which is exhausted from the turbine into ducts and headers for distribution into rows of small-diameter finned tubes (i.e., heat exchanger). Indirect dry cooling, which is also known as the Heller System, utilises a closed-cycle water cooling system as the primary coolant to condense steam, and the heated water is air cooled. The cooling water flows through bundles of tubes with air flowing over them, but the cooling air never comes into contact with the cooling water. In both systems, direct and indirect, the flow rate of air required to achieve the same cooling capacity will have to be three or more times greater than in a wet cooling tower, so the tower will have to be much larger and more expensive.

Cooling towers are also characterized by the means by which air is moved. Similar to wet cooling towers, dry cooling towers can be mechanical draft, natural draft, or fan-assisted draft. Mechanical-draft towers rely on power-driven fans to draw or force the air through the tower. Natural-draft cooling towers use the buoyancy of the exhaust air rising in a tall chimney to provide the draft. A fan-assisted natural-draft cooling tower employs mechanical draft to augment the buoyancy effect. Many early cooling towers relied only on prevailing wind to generate the draft of air.

In power generation applications, the heat transfer between air and cooling water is achieved by convection, and the driving force of this cooling process is the approach temperature (which is defined as cooling water temperature at outlet and air temperature at inlet)—not by evaporation, as in a wet cooling tower. Some of the advantages of dry cooling towers are that they do not need any makeup cooling water or water treatment and they do not generate plume, fog, mineral drift, and disposal issues associated with wet cooling. Size is a concern in dry cooling systems, however, because dry cooling is not as effective as wet cooling; the towers have to be much larger to achieve comparable heat rejection. Another challenge to direct dry cooling is operational control in regard to how to balance the steam flow to keep the desired steam condensation temperature (and, hence, turbine backpressure), which varies with loading. Other disadvantages of dry cooling are increase in noise, plume recirculation, maintenance of many components, and energy penalties caused by the variations of daily temperature and increases in air emissions.

Dry-cooling performance depends on the ambient air dry-bulb temperature (i.e., the sensible heat) instead of the wet-bulb temperature in the case of a wet cooling tower. Dry cooling becomes more economical when the approach temperature becomes considerably high. Other factors that affect the performance of dry cooling towers are the crosswind speed and the way that the heat exchanger bundles are arranged. The effect of crosswinds at different speeds and the effect of adding windbreak walls on the thermal performance of natural-draft dry cooling towers is given by Al-Waked and Behnia.[3] The effect of arranging the heat exchanger bundles—either vertically around the circumference of the tower or horizontally in the inlet cross section of the tower—is given by du Preez and Kröger.[4]

Wet–Dry Cooling (Hybrid)

Combined wet–dry cooling towers were introduced due to the recent stringent environmental protection laws. These towers effectively suppress detrimental plume formation at an efficiency level comparable to that of wet cooling towers. In hybrid wet–dry systems, the hot water from the power station condenser is cooled to the design discharge temperature as it passes in series first through the dry section and then through the wet section of the tower. The low-humidity hot air stream from the dry system is mixed with the moist warm air, leaving the tower at humidity levels sufficiently low to prevent the formation of visible plumes. The wet and dry components can be used separately or simultaneously for either water conservation or plume abatement purposes. At low ambient temperatures, the cooling tower can be operated as a dry cooling tower only, whereas at high temperatures, it can be used as a wet cooling tower only to achieve the required cooling without the risk of plume formation; the dry air is not put into operation until the ambient temperature starts to fall.

The design and construction of hybrid cooling towers are more complicated, and according to Streng,[5] the following data need to be specified for winter and summer operation. These data are thermal performance; cooling water flow or cooling range, which is the difference between the water temperature at inlet and the water temperature at outlet; ambient temperature; criteria for operating without plume; sound attenuation regulations; and limitations with respect to the erection area or overall height and operating weight and water analysis of the makeup water. In his work,[5] the construction, including material selection and automatic operation of the cooling system, is discussed in detail.

When a combination of wet and dry cooling technology is used, depending on system configuration, water consumption can approach that of recirculating wet systems or can be much lower. Design studies have ranged from 30 to 98% reduction in water use compared with all wet recirculating systems. As the hybrid cooling towers conform well to the stringent environmental protection requirements and to the standard operation reliability set for cooling systems, it is expected, therefore, that they will become more widespread.[5]

PERFORMANCE AND RATING OF COOLING TOWERS

In power generation, lower turbine backpressures are achieved when steam condensate temperatures are lower. Designing and operating a cooling system that can remove the heat of condensation consistently and continually at those low temperatures is essential. Therefore, the cooling system should be considered to be an integral part of the power generation process that can have a major influence on overall power plant performance and availability.

The choice of an appropriate cooling tower for a special application depends on many factors, such as capacity, availability, reliability, cost, and effectiveness. The effectiveness of a cooling tower is defined as the ratio of the actual energy that is exchanged to the maximum energy that could possibly be exchanged. The number of transfer units (NTU) is another parameter that measures the heat transfer size of the cooling tower. The higher the NTU value, the closer the cooling tower is to its thermodynamic limit. To estimate cooling-tower effectiveness and NTU, different analyses are used (similar to the ones used for analysing heat exchangers) that depend on the particular type under consideration.

The thermodynamic performance of any wet cooling tower is a function of the geometry and the ratio of the water flow rate (L) to the gas flow rate (G)—i.e., L/G. This value is quantified by means of a parameter known as the tower characteristic or number of diffusion units η ($\eta = K a V/L$), where K is the average mass transfer coefficient of condensed steam, a is the area of transfer surface per unit volume, and V is the effective volume. Manufactures supply charts for their cooling towers that present the tower characteristics as a function of L/G and the difference between the (outlet cooling water temperature (CWT) and air wet-bulb temperature (WBT).

Cooling-tower performance can be specified from the following parameters: water mass flow rate, inlet and exit temperature of water, and atmospheric wet- and dry-bulb temperatures of air. Many researchers have attempted to analyze wet cooling systems to estimate their performance. A basic theory of wet-cooling-tower operation was first proposed by Walker.[6] The practical use of basic differential equations, however, was first presented by Merkel,[7] who combined the equations for heat and water vapor transfer. He showed the utility of total heat or enthalpy difference as a driving force to allow for both sensible and latent heats. The basic approximations in Merkel's theory are

- The resistance for heat transfer in the liquid film is negligible.
- The mass flow rate of water per unit of cross-sectional area of the tower is constant (i.e., there is no loss of water due to evaporation).
- The specific heat of the air-stream mixture at constant pressure is the same as that of the dry air.
- The Lewis number (which relates heat transfer to mass transfer) for humid air is unity.
- The air exiting the tower is saturated with water vapor.

It is important to note that the formulation and implementation of Merkel's theory in cooling-tower design and rating are presented and discussed in most textbooks on unit operations and process heat transfer.

A summary of some of the methods that attempt to evaluate wet cooling towers' performance has been published by Kloppers and Kröger.[8] They compared cooling-tower performance obtained by Merkel, Poppe, and e-NTU methods. Merkel applied the mass and energy conservation laws to a differential control volume that includes the air and the water in a counter-flow cooling-tower arrangement, and derived the following differential relationships:

$$\frac{dh_a}{dz} = \frac{h_D a_{fi} A}{\dot{m}_a} = (h_{as,w} - h_a)$$

$$\frac{dT_w}{dz} = \frac{\dot{m}_a}{\dot{m}_w} \frac{1}{c_{pw}} \frac{dh_a}{dz}$$

where h, enthalpy; h_D, mass transfer coefficient; a, surface area per unit volume; A, frontal area; m, mass flow rate; z, vertical direction; c_p, specific heat at constant pressure;

subscript: a, air; w, water; s, saturated; fi, fill; (w), evaluated at water temperature T_w.

After the above equations were combined and integrated over the whole length of the tower, the Merkel equation was derived:

$$Me_M = \frac{h_D a_{fi} A L_{fi}}{\dot{m}_a} = \frac{h_D a_{fi} L_{fi}}{G_w} = \int_{T_{wo}}^{T_{wi}} \frac{c_{pw} dT_w}{(h_{as,w} - h_a)}$$

where Me, Merkel number; L, length; G, mass velocity; subscript: M, according to Merkel approach; i, inlet; o, outlet.

The term on the right side is a measure of the cooling requirement whereas the term on the left side is a measure of the performance of the packing.

Poppe included the effect of Lewis factor Le (defined as $h_c/h_D c_{pa}$) and the reduction in water flow due to evaporation. He derived two equations for the Merkel number based on the state of the air at exit—unsaturated or supersaturated. If the air is exiting as unsaturated, the Merkel number can be obtained by an iterative procedure of integrating the following equation:

$$\frac{dMe_P}{dT_w} = c_{pw}/\Big[h_{as,w} - h_a + (Le - 1)\{h_{as,w} - h_a - (w_{s,w} - w)h_v\} - (w_{s,w} - w)c_{pw}T_w\Big]$$

In the above equation, w, humidity ratio; the subscript P, Poppe approach; v, vapor; a, air; s, saturated, and w, water.

The Merkel number for air exits as supersaturated can be obtained by an iterative procedure of integrating the following equation:

$$\frac{dMe_P}{dT_w} = c_{pw}/\Big[h_{as,w} - h_{ss} + (Le - 1)\{h_{as,w} - h_{ss} - (w_{s,w} - w_{s,a})h_v\} - (w - w_{s,a})c_{pw}T_w\Big] + (w - w_{s,w})c_{pw}T_w$$

The subscript ss = supersaturated. Details can be found in Poppe and Rögener[9] and in Bourillot.[10]

According to the e-NTU method, in which the same simplification of Merkel is used, the Merkel number for the case where dry air mass flow rate $m_a > m_w c_{pw}/(dh_{asw}/dT_w)$ can be obtained by:

$$Me_e = \frac{c_{pw}}{dh_{as,w}/dT_w} \text{NTU}$$

The subscript e = the e-NTU approach.

The Merkel number for the case where dry air mass flow rate $m_a < m_w c_{pw}/(dh_{asw}/dT_w)$ can be obtained by:

$$Me_e = \frac{\dot{m}_a \text{NTU}}{\dot{m}_w},$$

where NTU is given by:

$$\text{NTU} = \frac{1}{1-C} \ln \frac{1-e^C}{1-e}.$$

C is the fluid capacity rate ratio and is defined as C_{min}/C_{max}. It is to be noted that the e-NTU method is applicable to the cross-flow arrangements, provided that the air and water streams should be defined, whether they are mixed, unmixed, or a combination.

Khan and Zubair[11] presented an analysis to estimate the effectiveness and NTU of a counter-flow wet cooling tower that matched the experimental data closely. They included in their model the effect of the Lewis number, defined in a similar fashion to Poppe's as the ratio of the convective heat transfer coefficient to the convective mass transfer coefficient times the specific heat at constant pressure of moist air, the heat resistance in the air-water interface, and the effect of water evaporation on the air states along the vertical length of the tower. In their analysis, they assumed constant convective heat and mass transfer coefficient, and ignored the heat lost through the tower walls, variation in specific heat properties, and water lost by drift. They applied the mass and energy conservation equations to a differential volume to relate the change in enthalpy of moist air to its humidity ratio, in terms of Lewis number Le and other properties of moist and saturated air. The outlet properties are obtained by numerically integrating the set of differential equations of conservation of mass and energy on an increment volume of the cooling tower.

They also gave the following definition of NTU and effectiveness:

$$\text{NTU} = \frac{h_D A_V V}{\dot{m}_a} = \int_{W_i}^{W_o} \frac{dw}{w_{s,w} - w}$$

$$\varepsilon = \frac{h_o - h_i}{h_{s,w,i} - h_i}.$$

A_V, surface area of water droplet per unit volume of the tower; V, tower volume; subscript: o, outlet, and i, inlet.

They also gave an empirical equation for NTU_{em}:

$$\text{NTU}_{em} = c\left(\frac{\dot{m}_w}{\dot{m}_a}\right)^{n+1}$$

c and n, empirical constants specific to a particular tower design, and subscript em, empirical.

Another approach is to return to the fundamental equations of fluid mechanics and heat and mass transfer, and arrive at numerical solutions with the aid of computational fluid dynamics technique (CFD). Some examples are the work by Al-Waked et al.[12] and Hasan et al.[13] among many others. These solutions can in principle be used as the sole basis of design or they can be used to examine, modify, and improve existing simpler methods—such as work by Kloppers and Kröger,[14] who used the finite difference method to compare the three approaches of Merkel, Poppe, and e-NTU.

CONCLUSION

The different types of cooling towers—wet, dry, and hybrid—have been presented. Research and experience show that the hybrid cooling towers conform well to the stringent environmental protection requirements and to the standard operation reliability set for cooling systems; it is expected, therefore, that they will become more widespread. Different methods to estimate cooling-tower performance are presented, based on some assumptions that simplify the problem. As the systems get more complicated, however, CFD is capable of predicting performance and can be used as the sole basis of design, or it can be used to modify and improve existing simpler methods to make them closer to reality.

Glossary

Approach temperature: The difference between the temperature of the condenser water leaving the tower and the wet-bulb temperate of the air entering the tower in the case of the wet tower and the dry-bulb temperature of the air entering the tower in the case of the dry tower.

Blow-down: The quantity of the circulating water that is removed to maintain the amount of dissolved solids and other impurities at an acceptable level.

Blow-out: Water droplets blown out of the cooling tower by wind—generally, at the air inlet openings. In the absence of wind, water may also be lost through splashing or misting.

Drift: Water droplets that are carried out of the cooling tower with the exhaust air.

Drift eliminator: Equipment containing a complex system of baffles designed to remove water droplets from cooling-tower air passing through it.

Noise: The sound generated by the impact of falling water; the movement of air by fans; the fan blades moving in the structure; and the motors, gearboxes, and drive belts.

Plume: The stream of saturated exhaust air leaving the cooling tower. The plume is visible when the water vapor it contains condenses in contact with cooler ambient air.

Range: The difference between the cooling water temperature entering the tower and the cooling water temperature leaving the tower.
More glossary words are available at the Cooling Technology Institute (CTI) Web site (www.cti.org/whatis/coolingtowerdetail.shtml).

REFERENCES

1. Sirena, J.A. The use of a thermo-fluid dynamic efficiency in cooling towers. Heat Trans. Eng. **2002**, *23* (2), 22–30.
2. Al-Nimr, M.A. Modelling the dynamic thermal behaviour of cooling towers containing packing materials. Heat Trans. Eng. **1999**, *20* (1), 91–96.
3. Al-Waked, R.; Behnia, M. The effect of windbreak walls on the thermal performance of natural draft dry cooling towers. Heat Trans. Eng. **2005**, *26* (8), 50–62.
4. du Preez, A.F.; Kröger, D.G. Effect of the heat exchanger arrangement and wind-break walls on the performance of natural draft dry-cooling towers subjected to cross-winds. J. Wind Eng. Ind. Aero. **1995**, *58* (3), 293–303.
5. Streng, A. Combined wet/dry cooling towers of cell-type construction. J. Energ. Eng. **1998**, *124* (3), 104–121.
6. Walker, W.H.; Lewis, W.K.; McAdams, W.H.,; Gilliland, E.R. *Principles of Chemical Engineering*; McGraw-Hill, Inc.: New York, NY, 1923.
7. Merkel, F. Verdunstungshuhlung. Zeitschrift des Vereines Deutscher Ingenieure (V.D.I) **1925**, *70*, 123–128.
8. Kloppers, J.C.; Kröger, D.G. Cooling tower performance evaluation: Merkel, Poppe, and e-NTU methods of analysis. J. Eng. Gas Turbines Power **2005**, *127* (1), 1–7.
9. Poppe, M.; Rögener, H. *Berechnung von Rückkühlwerken, VDI-Wärmeatlas*; 6. Auflage, VDI Verlag GmbH : Berlin, Germany, 1991.
10. Bourillot, C. TEFERI: Numerical Model for Calculating the Performance of an Evaporative Cooling Tower. EPRI Report CS-3212-SR, Electric Power Research Institute: Palo Alto, CA, 1983.
11. Khan, J-U-R.; Zubair, S.M. An improved design and rating analyses of counter flow wet cooling towers. J. Heat Trans. **2001**, *123* (4), 770–778.
12. Al-Waked, R.; Behnia, M. The performance of natural draft dry cooling towers under crosswind: CFD study. Int. J. Energ. Res. **2004**, *28* (2), 147–161.
13. Hasan, A.; Guohui, G. Simplification of analytical models and incorporation with CFD for the performance prediction of closed wet cooling towers. Int. J. Energ. Res. **2002**, *26*, 1161–1174.
14. Kloppers, J.C.; Kröger, D.G. A critical investigation into the heat and mass transfer analysis of counter flow wet-cooling towers. Int. J. Heat Mass Trans. **2005**, *48* (3–4), 765–777.

BIBLIOGRAPHY

Some useful sites and references that were used to collect some of the information in this article are:

1. Hewitt, G.F., Shires, G.L., Polezhaev, Y.V., Eds., *International Encyclopaedia of Heat and Mass Transfer*, CRC Press: New York, NY, 1997.
2. Shan, K.W. *Handbook of Air Conditioning and Refrigeration*, 2nd Ed.; McGraw-Hill, Inc.: New York, NY, 2001.
3. Cooling Technology Institute. Available at: www.cti.org.
4. Gulf Coast Chemical Commercial, Inc. Available at: www.gc3.com/techdb/manual/coolfs.htm

5. U.S. Environmental Protection Agency. Available at: www.epa.gov/waterscience/316b/technical/ch4.pdf#search = 'wetdry%20cooling%20towers
6. Online Chemical Engineering Information. Available at: www.cheresources.com/ctowerszz.shtml
7. GEA Cooling Tower Technologies. Available at: www.bgrcorp.com/default-gct.htm
8. Lenntech. Cooling Towers. Available at: www.lenntech.com/cooling%20towers.htm
9. Wikipedia. Available at: www.en.wikipedia.org/wiki/Cooling_tower
10. Legionella Control. Available at: www.legionellacontrol.com/legionella-glossary.htm

Data Collection: Preparing Energy Managers and Technicians

Athula Kulatunga
Department of Electrical and Computer Engineering Technology, Purdue University, West Lafayette, Indiana, U.S.A.

Abstract

Energy audits can be used to provide hands-on activities related to an energy management course. After learning the necessary background concepts, students need to be aware of what measurements must be taken to evaluate an existing energy system. In industry and universities, one may find apprentices and students from different educational backgrounds, such as electricians with no exposure to newer measuring instruments and students with no ability to take electrical and/or mechanical measurements. By studying the capabilities and limitations of measuring instruments, newcomers to the energy auditing may collect reliable data. This article introduces several hands-on activities that could be replicated to teach students how to take accurate measurements of electrical, light, and heat flow parameters, ultrasonic leak detection, electronic combustion analysis, and simple data acquisition before conducting energy audits. A sample laboratory activity includes a description of the measuring instrument, factors that contribute to inaccurate readings, safety concerns, and several practice measurements useful to energy audits.

INTRODUCTION

It has been accepted that engineering technology courses should have some hands-on activities such as labs, projects, and other practical experiences. In the field of energy management, energy audits have been used effectively to provide hands-on experiences.[1] An energy audit, also known as energy survey, energy analysis, or energy evaluation, is a process that examines the current energy consumption of a process or facility and proposes alternative ways to cut down energy consumption or costs. One aspect of the energy auditing process is to collect specific data of a process or a facility. Measuring temperature, flow rates (heat, liquid, and air), intensity of light, electrical current, voltage, power, power factor (PF), humidity, pressure, or vibration may be required to determine the energy consumption and waste. New measuring equipment is pouring into the measurement world making data collection easier, more accurate, and safer.

Accurate data collection is paramount not only to analyze energy consumption, but also to evaluate the effectiveness of proposed changes suggested in an energy audit report. Some energy-saving electrical retrofits may introduce electrical power quality problems that may not be accounted for by traditional meters, causing erroneous data. With inaccurate data, the conclusion of an energy improvement project holds no validity.[2]

Preliminary or walk-through energy audits are the most suitable for beginners. A preliminary energy audit is a process during which an auditor examines an existing energy consuming system according to a predetermined set of procedures. The procedures are outlined as a result of a historical data analysis of the targeted system and conversations the auditor had with the owner or the operator of the system. These procedures include taking electrical and other measurements under certain conditions.

ELECTRICAL MEASUREMENTS

Data collection of any system that consumes electrical energy requires at least three basic measurements—voltage, current, and PF—for energy analysis calculations. Utility meters collect all these data at a building service entrance point or at any other sub-metering location, if such meters are installed. By contacting the utility provider, one can easily obtain the historical data related to the above parameters and more for a given facility. However, when it comes to individual systems within a facility, these data may not be available, unless additional utility meters are installed at each of the service entrance points to individual systems. In a modern power distribution system, harmonics of the fundamental frequency, 50 or 60 Hz, appear due to non-linear devices connected into the system. Power line harmonics cause erroneous readings, if a meter is not capable of measuring

---

*This entry originally appeared as "Preparing Energy Managers and Technicians for Energy Data Collection" in *Energy Engineering*, Vol. 102, No. 4, 2005. Reprinted with permission from AEE/Fairmont Press.

Keywords: Energy audits; Energy data; Energy measurements.

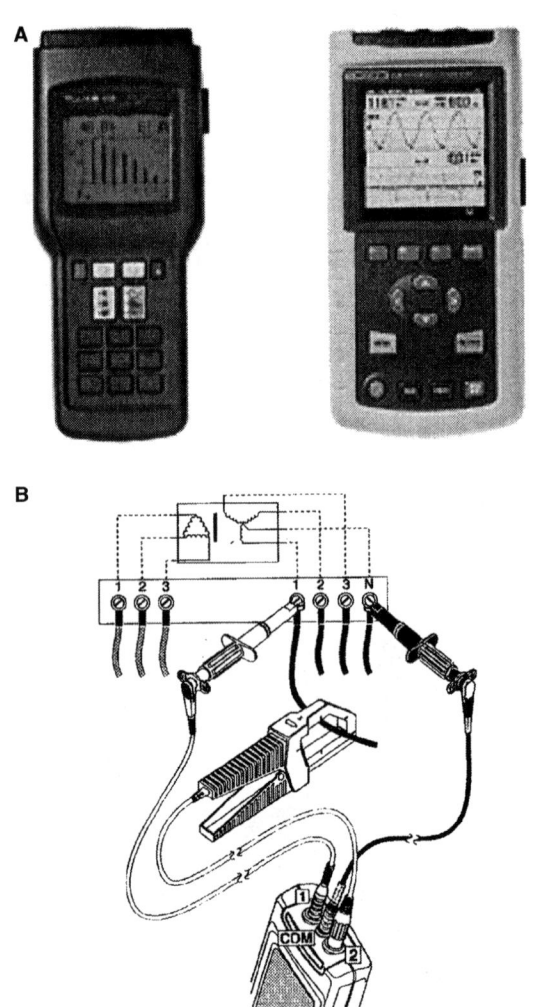

Fig. 1 (A) Fluke 41B and 43B power analyzer, (B) Using meter probes for three-phase. (Courtesy of Fluke Corporation).

and current probes that come with the meter. The meters are capable of measuring true RMS, peak, and total harmonic distortion (THD) for voltage and current. They also display true power (Watts), reactive power volt ampere reactance (VAR), PF, displacement power factor (DPF), crest factor, K-factor, and harmonics individually up to the 31st.

The meters are capable of displaying data in three views—waveforms, barographs showing harmonic levels, and numeric values for voltage, current, and power—as depicted in Fig. 2.

Each of the data displayed on Fig. 2 gives some clue about the nature of power quality. For example, flat-topped voltage waveform is an indication of the presence of current harmonics. One can use the THD levels to determine if they are within the specified limits. If not, the same data can be used to determine sizing of transformers and harmonic filters. The bar-graph display of the current reveals the percentage comparison of odd-numbered harmonics with respect to the fundamental frequency. Even-numbered harmonics cancel out in a power system; therefore their effects are not a concern in a power system. Finally, the meter displays numerical values such as total PF, DPF, kilovolt ampere, and kilowatt useful in determining power-factor correction methods. The data can be further analyzed by downloading FlukeView® software, which comes with the meter. Fig. 3 depicts current waveform of a nonlinear load and downloaded data in a tabular form.

Fluke 43B power quality analyzer has all the features of a 41B as well as sample and storage capabilities. These additional features allow a user to detect voltage/current sags and transients in a power system.

Practice Activity Outline

A typical electrical power system may have one of the following problems: (1) voltage sags, (2) current balance and loading, (3) harmonics, (4) grounding, or (5) loose connections. Taking measurements at the electrical service panel during an energy audit, the investigator can determine the sources of these problems. In a laboratory environment (or in a workshop), each of the above problems can be replicated to demonstrate the effects.

true root-mean-square (RMS) values of the fundamental frequency and the harmonics at a given instant.[3]

Two power quality analyzers made by Fluke Corporation are shown in Fig. 1A. One of these meters can be easily used to obtain all necessary data for an electrical system. Fig. 1B shows how to obtain measurements for a three-phase balance (5% or less imbalance) using voltage

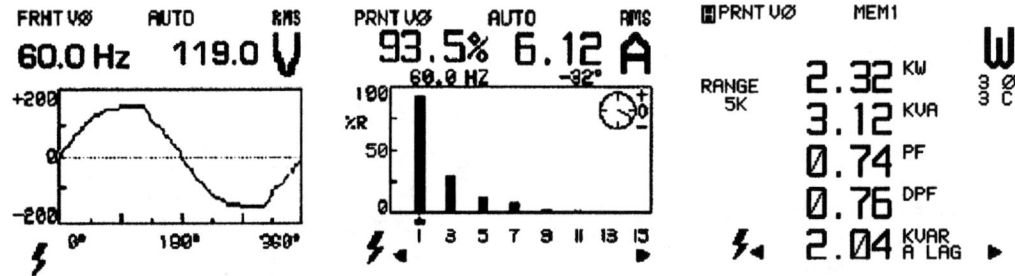

Fig. 2 Waveform, bar graph, and numerical displays of data presented by Fluke 41B. (Courtesy of Fluke Corporation).

Data Collection: Preparing Energy Managers and Technicians

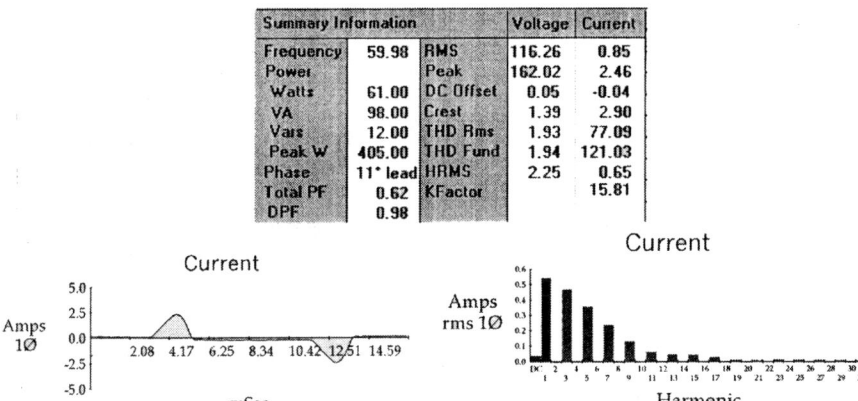

Fig. 3 Data table, current waveform, and harmonic representation of a non-linear load taken by Fluke View® software.

For example, the circuit shown in Fig. 4 may help students learn how to detect harmonics and locate the harmonic producing sources in a power system. The three-phase power supply should be taken from a Y-connected transformer secondary where the common point is taken as neutral. It is also necessary to ground the common point, if concepts related to overloading are introduced. A double-pole double-throw (DPDT) switch allows switching between the lamp-only circuit and the lamp with a dimmer circuit. The lamp-only path would allow the student to study the waveform characteristics and harmonic content of a linear load. The lamp with a dimmer can be used to observe the voltage and current waveform variation when harmonics are present to the right of the point, where the Fluke 43B meter is connected. Using the sag and swell mode of the Fluke 43B meter, a sample can be taken when DPDT is in position A and the Variable Speed Drive (VSD) is off. The waveform would look similar to the one shown in Fig. 5A. Note that the voltage sag coincides with a current swell indicating that the disturbance has occurred downstream of the measurement point.

The concept of upstream disturbance can be demonstrated by switching the DPDT switch to position B and starting the drive. A sample display taken under this condition is shown in Fig. 5B. Note that the voltage sag occurs simultaneously with the current sag, indicating that the source of the disturbance is upstream of the measurement point.[4]

Precautions

First of all, students must have a thorough understanding of the instrument's limitations. They should make voltage measurements after a circuit breaker and wear safety gloves and eyewear when taking measurements of live panels.

LIGHT MEASUREMENTS

On average, lighting consumes 35% of energy used in commercial buildings and 25% in industrial facilities. Lighting levels directly affect the productivity of

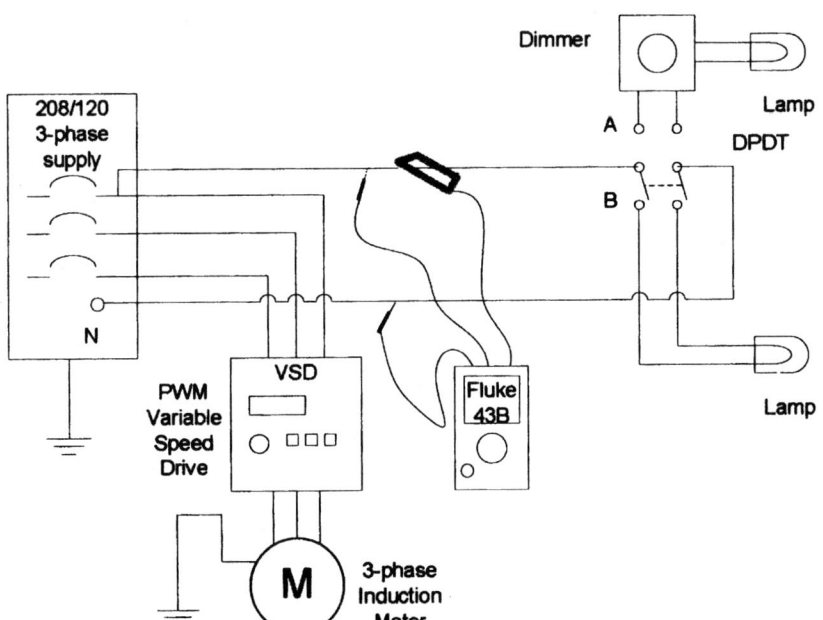

Fig. 4 A circuit for learning the sources of harmonics in an electrical power system.

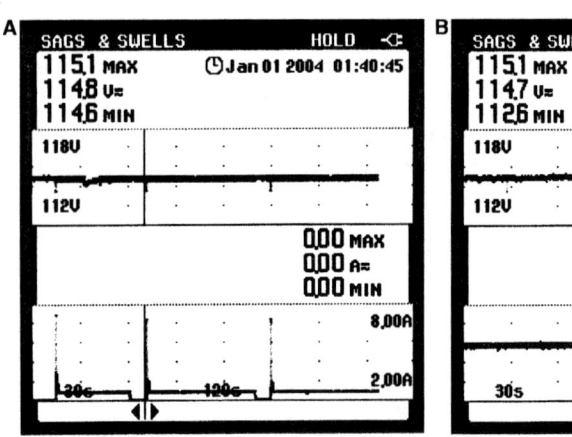
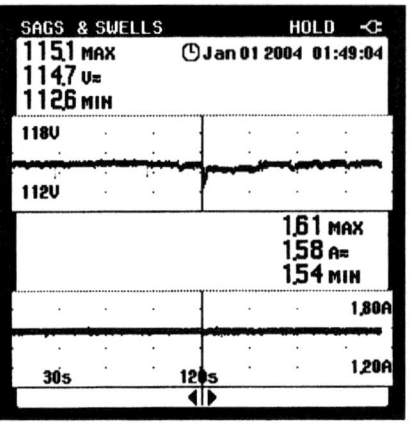

Fig. 5 (A) Downstream disturbance, (B) Upstream disturbance.

employees. However, many lighting systems are improperly designed and unattended over the years. Lighting is one of the areas where companies can save energy with the least amount of capital investments. Fig. 6 illustrates the before and after appearance of a warehouse, where 50% more light was obtained.

Light meters are very easy to use. The meter in Fig. 7 measures light intensity in Lumen and light density in foot-candles (fc). The purpose of measuring light is to determine the incident light on a horizontal or vertical surface. Illumination Engineering Society of America (IESA) publishes recommended light levels for specific tasks.

Practice Activity Outline

Select a location where students can move around with a ladder for at least 2 h. First, students draw an accurate layout of the space being audited including the location of furniture, windows, cabinets, and lighting fixtures. The location could be a classroom, a lab, a corridor, or a gym.

Using a light meter, students record the available light intensity in fc on horizontal and vertical surfaces with and without artificial lights. This would yield a light distribution map. Now compare the available light with the recommended light to determine levels and locate poorly illuminated or over-illuminated areas. Most classroom and labs are illuminated by fluorescent light fixtures. The amount of light generated by fluorescent light deteriorates with time. Students may replace existing fluorescent tubes with new ones to examine the available light.

Precautions

Students need to be aware that only the incident light on horizontal surfaces can be added. Light intensity must be measured at the recommended height as specified by IESA or any other standard. It is important to stay away from the detector to minimize the effect of body-reflected light entering into the detector.

BUILDING ENVELOP MEASUREMENTS

The heat loss or gains of a building depend on many factors: R-value of walls and roof material, window characteristics, ambient temperature, humidity level, and heating and cooling degree days of a given location where

Fig. 6 Fifty percent more light obtained through lighting retrofits in a warehouse.
Source: From Business News Publishing Company (see Ref. 5).

Fig. 7 A stick-type light meter with a retractable sensor. (Courtesy of Omega Inc.).

the building is allocated. When *R*-value and *U*-values are known, a detailed analysis of a building envelop would yield British thermal unit lost or gained through walls, windows, and other heating and cooling sources. It is very difficult to determine the *R*-values of even a several-year old facility due to poor record keeping and later add-ons to the structures. The OMEGA® OS-650 energy conservation and plant maintenance kit (see Fig. 8) is very useful in energy audits and general plant maintenance. The kit consists of an infrared (IR) thermometer capable of measuring temperatures from −2 to 200°F and a heat flow meter, which is a specially designed IR radiometer capable of displaying heat flow though a scanned wall in terms of British thermal unit per square feet hour.[6] The kit is priced around $1600. If purchased separately, either the thermometer or the heat flow meter would cost about $800.

Practice Activity Outline

A laboratory activity can be developed, so students will be able to (1) estimate *R*-value value of an unknown insulator, (2) measure heat flow through walls and windows, and (3) make energy cost analyzes at the end of the activity. To estimate *R*-value, net heat flow should be determined by using the heat flow meter. This is done by taking two measurements across a wall at the same height—one from inside and the other from outside of the room. The meter shows a + or − number, indicating the direction of heat flow.

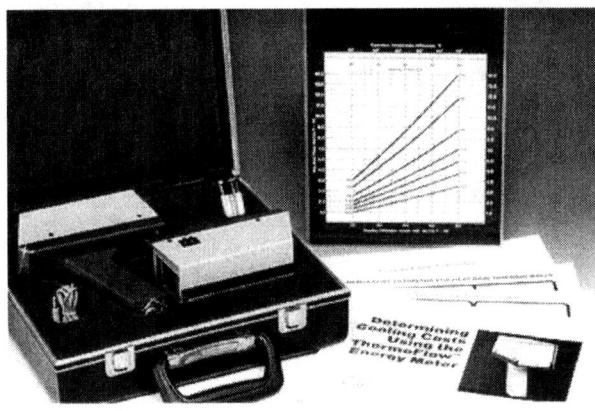

Fig. 8 OMEGA OS-650 Energy Conservation and Plant Maintenance Kit. (Courtesy of Omega Corporation).

THERMAL IMAGING

During any energy conversion process, heat is produced as a by-product. Poor insulation of buildings and pipes carrying hot or cold liquids causes energy wastes. New retrofits and devices containing switching power electronic circuits may introduce harmonics, which may overheat the neutral conductors. Poor connections of an electrical distribution system are the major contributor to system inefficiencies and may lead to a catastrophic fire. All of the above could be avoided, if one could detect them in advance. To detect such abnormalities, one requires a thermal imager. A thermal imager, compared to an IR thermometer, is capable of measuring temperature variation between two adjacent points. In the past, the cost of thermal imagers prevented widespread use in energy management and plant-maintenance activities. Industries hired a consultant to survey all electrical distribution panels and other critical locations in a facility to determine the thermal profile annually. But prices have come down significantly, under $10,000, over the past few years. A specially designed, hand-held thermal imager made for energy audits and plant maintenance is shown in Fig. 9.

The imager can hold up to 100 images, which are stamped with time and date as they are taken. Through an Universal Serial Bus (USB) port, the images can be downloaded to the accompanying software installed on a PC for further analysis. Images can be further analyzed by assigning a single color to a temperature range and creating a thermal profile. A thermal profile represents the temperature at the *x* and *y* axes as the cursor moves around the image.[7]

Practice Activity Outline

Many interesting activities can be developed around this device. Students could measure the temperature of different light sources and compare the power consumption of each light bulb and the surface temperature. The internal heat distribution of a room could be detected and documented by hanging black-painted aluminum sheets. Since this device measures the reflected thermal energy, the manufacturer recommends painting highly reflective surfaces with some dark color to minimize reading errors. Students could develop thermal images of buildings on campus and analyze the heat losses. In a power lab, students could measure the temperature of conductors when motors are driven with variable speed drives (Fig. 10).

Precautions

When taking an image, the focus is paramount. As with any digital camera, lack of focus blurs the image, thus minimizing the device's ability to distinguish the temperature difference between adjacent pixels. Some

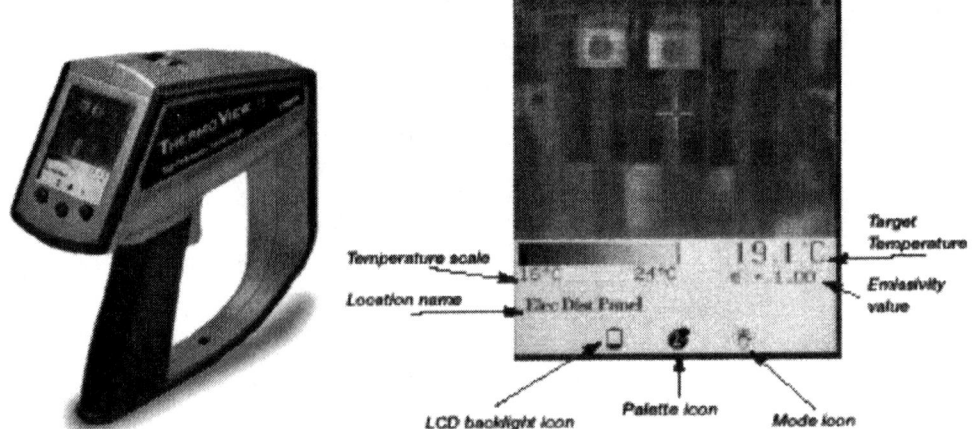

Fig. 9 Raytek ThermoView™ Ti30 portable imager and its display. (Courtesy of Raytek Corporation).

imagers have a minimum focusing distance that must be adhered to. Most current thermal imagers allow users to view the object using different color palettes such as rainbow, ironbow, and grayscale. Despite the popularity of color palettes, it is recommended to use grayscale for most applications, because the human eye can detect variations of grayscale better when thermal changes are subtle. In addition, students must be aware of concepts such as qualitative vs quantitative temperature measurements, distance to target ratio, field of view, effects of environmental conditions (steam, dust, smoke, etc.), and effects of emissivity.

COMBUSTION ANALYSIS

The effectiveness of fossil-fuel burning receives little attention during an energy audit process unless the auditor is well trained to perform mechanical tests. The mechanical test, a tedious process, requires a set of data from several different pieces of equipment such as draft gauges, thermometers, carbon monoxide (CO) stain length tubes, wet chemical absorption instruments, etc. Then the data are entered into a slide ruler calculator to determine the combustion efficiency.[8]

New electronic combustion analyzers provide more reliable and provable data than a traditional "eyeballing the flame" analysis. It also produces faster professional analysis than a mechanical test. Fyrite® Tec 50 and 60 residential combustion analyzers by Bacharach Inc. shown below (Fig. 11) falls under $500 and comes with many features that make this instrument ideal for technicians and students alike.

A user first selects the type of fuel being burnt. The fuel choices are natural gas, #2 oil, propane, and kerosene. The meter can measure flue gas oxygen content 0.0%–20.95% O_2, flue gas temperature up to 999°F, ambient temperature 32°F–104°F, and flue gas CO content 0–2000 ppm CO. Based on these measurements, the meter is capable of calculating the following: combustion efficiency 0.1%–99.9%, flue gas carbon dioxide content 0.0 to a fuel dependent maximum (in percent), flue gas CO air-free content 0–9999 ppm, and excess air 0%–400%.

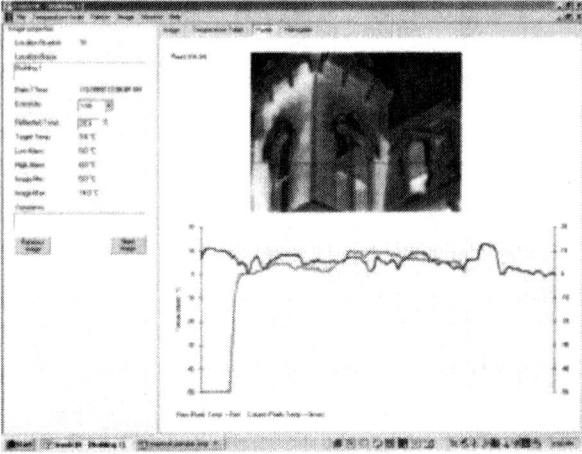

Fig. 10 Multi-image display and thermal profile using ThermoView™ Ti30 imager. (Courtesy of Raytek Corporation).

	Eyeball Analysis	Mechanical Tests	Electronic Analyzers (Fyrite Pro)
Reliability	poor	good	excellent
Accuracy	poor	good	excellent
Accurate/Safety Measurements (CO)	no	no	yes
Ease of use	n/a	no	yes
Method of Test	blind	snapshot	continuous
Speed	moderate	slow	fast
Documentation Capability	no	no	yes
Cost of Ownership (3 years)	n/a	moderate	low

Fig. 11 Fyrite® Tec 50 and 60 residential combustion analyzers and a manufacturer's comparison of combustion analysis techniques. (Courtesy of Bacharach Inc.).

Practice Activity Outline

Practice activities for this meter may include: (1) testing flames produced by the four different type flames while varying the air intake into the flame, (2) experimenting heat transfer characteristics of a jar containing water (or any other liquid applicable to a certain industrial process) under different insulations and ambient temperatures, and (3) simulating a specific type of burner used in an industrial facility to investigate efficiency improvement opportunities. Flames for activity-1 and activity-2 above can be set up by using off-the-shelf burners or commercial water heaters.

Precautions

When working with flammable gases in closed environments, one must adhere to Occupational Safety and Health Administration (OSHA) safety guidelines. Students must be aware of every detail of the gas being used and wear appropriate protection apparatus, which include burn-proof gloves and eye protection.

ULTRASONIC LEAK DETECTION

A leak, whether it is a compressed air, steam, or conditioned air, is a waste. Hidden costs due to leaks can be significant in very competitive market environments. Some leaks are very easy to detect, while others are not. Most leaks go undetected due to ambient noises. To detect some types of leaks, one has to use ultraviolet (UV) dyes, special lamps, bubble solutions, etc. Leak detection takes time and money. Any instrument that detects different types of leaks within a very short time would be the most practical solution.

Ultrasonic leak detectors minimize traditional problems associated with leak detection. Fig. 12 depicts a leak detector made by Superior Signal Company Inc., which comes with a price of less than $250.

According to the manufacturer's specifications, the meter can detect leaks of any type of refrigerant gas, vacuum, and pressure leaks. The reading is not affected by saturated gas areas or windy environments. An energy auditor may even detect leaks around freezer and cooler doors with this meter.[9]

Fig. 12 AccuTrak VPE by Superior Signal Company, Inc. (Courtesy of Superior Signal Company, Inc.).

Practice Activity Outline

Several simple activities can be developed to familiarize this equipment in a laboratory or workshop environment where compressed air is available. Several holes, at least one foot apart with different diameters, can be drilled in a copper piping tube. A connector needs to be soldered to one end and the other end must be capped. Once connected to a regulated compressed air outlet, learners may trace air leaks of each hole under different pressure to learn the nature of the leaks and the corresponding meter responses. The activity can be further expanded by moving away from the leaks and/or walking around the pipe. Similarly, to study the nature of leaks due to doors and windows, learners may take measurements around commercial soft drink coolers located in any facility.

Precautions

People have a natural tendency to clear the area around a suspected leak, especially painted or corroded surfaces, to examine the leak more closely. This may cause the hole to burst without much warning, spewing dust and loose particles all over. Therefore, students must wear safety glasses and dusk masks as appropriate when testing leaks of any form.

LOW COST DATA ACQUISITION

Energy managers have to rely on historical data when energy analyzes are performed. A trend of a measured parameter such as current, voltage, temperature, etc. presents a better picture of energy-saving opportunities than instantaneous readings. Most data loggers are expensive, because they are designed to minimize potential damages to the front-end electronics of the data acquisition

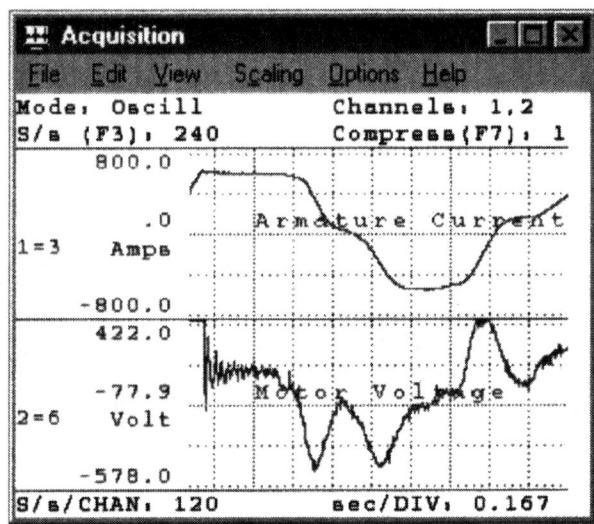

Fig. 13 DATAQ 4-channel data acquisition and chart recorder starter kit model. (Courtesy of DATAQ Instruments, Inc.).

system and the computers used to record the data. For laboratory applications and short-term data logging, one may use the D1-194RS low-cost (less than $25) data acquisition starter kit shown in Fig. 13. The kit includes hardware, recorder software, and a serial port cable.

The hardware is self-powered through the serial port of PC. Analog signals connected to any of the four channels are digitized and saved into the hard drive while showing on the screen as a strip-chart recorder. Each channel can be sampled up to 240 samples per second with 12-bit resolution, which is adequate for most analog signals in the energy field. Each channel accepts ± 10 V, which is large enough for most commercially available transducer outputs. The software includes Active-X control libraries that allow the user to program the kit from any Windows environment. The recoded data may be play backed for later analysis.[10] The input signals are not optically isolated and measured with respect to a common ground point.

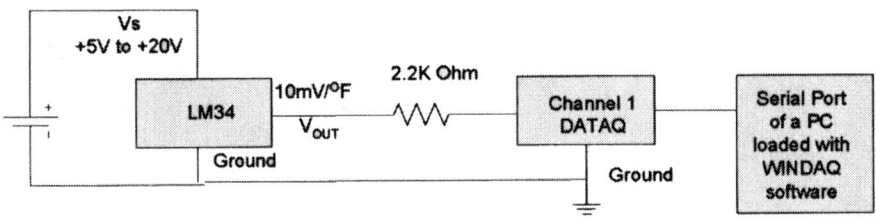

Fig. 14 Schematic diagram to measure ambient temperature directly with DATAQ starter kit.

Practice Activity Outline

Any transducer that produces 0 to +10 V or 0 to −10 V output proportional to the measured variable can be connected directly. As a precaution, measure the voltage between two grounds to verify that the two grounds are at the same electrical potential. When commercial transducers are not available, a very inexpensive ambient room temperature sensor could be developed by using a LM34 temperature sensor integrated circuits (IC) made by National Semiconductors Inc., as shown in Fig. 14.

LM34 IC produces 10 mV for every degree of Fahrenheit. For example, when it senses 72°F, the output would be 720 mV. The computer display would show this reading as a 720 mV, or it can be calibrated on the software to display in Fahrenheit. The software can be set up to save time-stamped signals for later analysis. The activity can be expanded by adding a second channel that tracks the turn-on–off signal of rooms' heating ventilating and air conditioning (HVAC) unit. Once the graphs are plotted, the learners may be able to study process characteristics of the room by measuring dead time, time constant, etc.

Precautions

Users of this data acquisition system should be aware that this unit does not provide electrical isolation. As long as the transducer's electrical ground and the unit hardware ground, which is the PC ground, are at the same potential, the unit will work accurately. A dc voltmeter reading between these two points reveals any potential problems. The unit measures only voltage signals. Current signals should be converted into a proportional voltage signal using a current shunt placed in series with the load. However, common mode voltage applied to the unit may destroy the unit hardware unless necessary precautions are taken. There are many useful literatures available in www.dataq.com website that would help users in making measurements safe and easy.

CONCLUSIONS

New meters appearing on the market make data collection for energy audits fun and instructive. With this new measuring equipment, an auditor can take measurements that were not economical or even possible several years ago. Collected data can be easily analyzed with software and sent to another individual conveniently over the internet. Seven measurement instruments have been presented. Each meter has its own unique capabilities compared to the traditional equipment. However, they have their own limitations as well. When taking sophisticated measurements, one should be aware of the conditions under which those measurements must be taken. One can understand the measuring equipment potentials and limitations by performing a set of controlled activities in a laboratory environment. This will allow the learner to change one variable at a time. Under no circumstances should one bypass safety requirements specified by the equipment manufacturers. Some meters require periodical calibrations.

REFERENCES

1. Kulatunga, A. *Integrating the Courses in Electrical and Mechanical Engineering Technologies to Fulfill the Needs of Energy Professionals*, Proceedings of ASEE Conference, Montreal, Canada, June 16–19, 2002.
2. Capehart, B.L.; Turner, W.C.; Kennedy, W.J. *Guide to Energy Management*, 3rd Ed.; The Fairmont Press: Lilburn, GA, 2000; [ISBN0-88173-336-9].
3. Kulatunga, A. *Peak Current Behavior of Nonlinear Loads in Residential and Commercial Settings*, Proceedings of International Association of Science and Technology for Development Conference (IASTED): Power and Energy Systems; Domijan, A., Ed.; ACTA Press: Calgary, AB, 2001; ISBN: 0-88986-317-2.
4. Power Quality Troubleshooting at the Service Panel: Application Note, Fluke Corporation, www.fluke.com/library, retrieved on December 12, 2003.
5. Hucal, M.C. Illuminating energy savings with efficient lighting. In *Environment Design + Construction*; Business News Publishing Company: Troy, MI, 2003; [retrieved on 11/4/2003 from www.edcmag.com].
6. OS-650 Series Energy Conservation and Plant Maintenance Kits: Operator's Manual. Omega Engineering, Inc., 1995. Available from: www.omega.com
7. ThermoView™ Ti30 Manual. Fluke Corporation (Formally Raytek), 2004. Available from: http://us.fluke.com/usen/products/Fluke_Ti30.htm?trck=ti30
8. Bacharach Inc. Why Test?, 1999; retrieved on February 21, 2004 from www.bacharach_inc.com
9. Superior Signal Company, Inc. retrieved on February 21, 2004 from www.SuperiorSignal.com
10. DI-154RS and DI-194RS Starter Kits, DATAQ Instruments, Inc., 2003; retrieved on February 22, 2004 from www.dataq.com

Daylighting

William Ross McCluney
Florida Solar Energy Center, University of Central Florida, Cocoa, Florida, U.S.A.

Abstract
Daylight illumination of building interiors is an ancient art now benefitting from relatively recent engineering advances. The benefits are numerous and include energy savings and enhanced visual and thermal comfort. The design must avoid overheating and the discomfort and reduced productivity resulting from glare. Good daylighting design can displace electric lighting and reduce air pollution, global warming, and dependence on dwindling supplies of fossil-fuel energy.

INTRODUCTION

Daylighting has been a primary means of interior illumination since the first buildings were constructed. Full use of natural daylight can displace daytime electric lighting energy consumption in buildings, resulting in considerable avoidance of building energy operating costs. Designing for daylight illumination is an ancient art, enhanced considerably in the modern era by improved materials, construction techniques, and computerized design and performance evaluation tools. Artists have known for centuries that natural daylight offers the best illumination for good color rendering of paintings and other objects. Now it is possible in many climates to use larger window areas than in the recent past—for the aesthetic, view, illumination, and health benefits these possess—without the adverse energy consequences large windows once represented.

Strong direct illumination from the solar disk, however, presents an important challenge to the daylighting designer. The sun's motion through the sky means that orientation and shading strategies for minimizing direct beam glare must incorporate the known paths of the sun through the sky each day. Fortunately, several design strategies are available to ease the design process. The hourly energy and illumination performances of windows and other daylighting systems now can be assessed quickly and with modest precision.

ILLUMINATION BASICS

Radiometry and Photometry

Radiometry is a system of language, mathematical formulations, and instrumental methodologies used to describe and measure the propagation of radiation through space and materials. Photometry is a subset of radiometry dealing with radiation in the visible portion of the spectrum. Only radiation within the visible portion of the spectrum, ranging from approximately 380 nm to approximately 720 nm, should be called light. Photometric quantities are defined in such a way that they incorporate the variations in spectral sensitivity of the human eye over the visible spectrum—as a spectral weighting function built into their definition. Though daylight illumination of building interiors deals primarily with photometric quantities, radiometric ones are important in assessing the energy-performance features of daylighting systems.

Photometric quantities may be derived from their spectral radiometric counterparts using the equation below. Let Q_λ be any of the four spectral radiometric quantities (flux, irradiance, intensity, or radiance), and let $V(\lambda)$ be the human photopic spectral luminous efficiency function. The photometric equivalent, Q_v, subscripted with the letter v (for visual) of the radiometric quantity Q_e, subscripted with the letter e (for energy) is the weighted integral of the spectral radiometric quantity over the visible portion of the spectrum.[1]

$$Q_v = 683 \int_{380 \text{ nm}}^{760 \text{ nm}} V(\lambda) Q_\lambda(\lambda) d\lambda$$

The Paths of the Sun Through the Sky

The sun moves in a predictable way through the sky each day. In the Northern Hemisphere at middle latitudes in the winter, it rises south of due east (azimuth 90) and sets south of due west (azimuth 270). In summer, it rises north of due east and sets north of due west. A plot of solar position vs time on a chart of solar coordinates is called a sunpath chart. Knowledge of solar movement is important in designing daylighting systems. Such systems perform best when they minimize glare and overheating from solar radiant heat gain. The World Wide Web offers several tools for determining the position of the sun in the sky.

Keywords: Daylight; Glazing; Aperture; Illumination; View; Windows; Energy.

A sunpath-chart drawing program is available for free download from the Web site of the Florida Solar Energy Center (www.fsec.ucf.edu).

Proper orientation of buildings and spaces with glazed apertures relative to solar movement is very important in building design for daylighting. Shading devices—including overhangs, side fins, awnings, window reveals, and a variety of exterior and interior shades and shutters—are important tools for the daylighting designer. In addition, new materials and design strategies permit the use of concentrated and piped daylighting systems, using the strong flux from direct beam sunlight to minimize aperture areas while delivering sunlight without glare to spaces somewhat remote from the building envelope.

The Solar Spectrum

Spectra for direct beam and diffuse sky radiation on a vertical wall, with the sun 60° above the horizon on a clear day, are plotted in Fig. 1, along with a scaled plot of the human photopic visibility function. The sky is blue as a result of spectrally selective scattering of light from air molecules. This is seen in the shift of the diffuse sky spectrum in Fig. 1 toward shorter (bluer) wavelengths. Light through windows from blue sky alone does not appear strongly colored to the human observer. One reason is the adaptability of the human visual system. Another reason is the presence of a wide range of colors in sun and sky light, except for the case of sunlight at sunrise and sunset (which is reddish by virtue of the greater mass of intervening atmosphere, which removes some of the blue light in the spectrum). The solar spectrum under an overcast sky has a shape approximately the same as the sum of the direct and diffuse spectra, because scattering of light from water droplets in the atmosphere—the principal mechanism producing light from clouds and overcast skies—has weak spectral selectivity.

Electric lighting is approximately constant in output, in the absence of dimming systems. Daylight is changing constantly. The variability in sun and sky light is both a problem and a benefit in daylighting design. The benefit stems from the positive responses most people feel when experiencing the natural changes in daylight illumination in the absence of glare. The problem comes from the need to design the daylighting system to respond well to the changes. The design process begins with an understanding of the daylight availability for the building site, including the blocking effects of trees, buildings, and other nearby objects. Daylighting systems intended for an area experiencing predominantly overcast skies will be different from those designed for mostly clear-sky conditions. The National Renewable Energy Laboratory in Golden, Colorado in the United States and other national laboratories around the world, including weather bureaus and other such services, can be consulted for information about daylight availability for sites in their jurisdictions. In 1984 the Illuminating Engineering Society of North America published a guide to daylight availability.[2]

Glare

There are two kinds of glare, illustrated in Figs. 2 and 3. With disability glare, light reflecting from the surface of a visual task masks the contrast in that task and degrades the ability to see it. Examples include light reflected from a

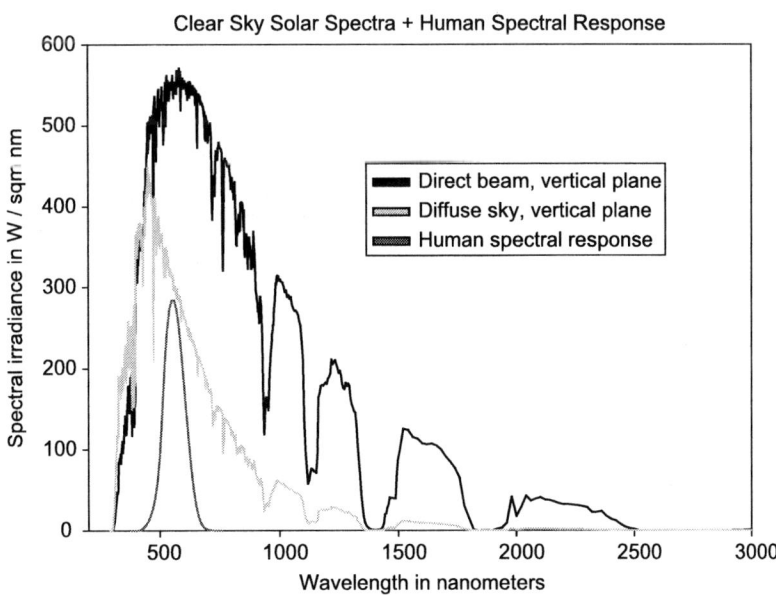

Fig. 1 Direct and diffuse solar spectra on a vertical plane. The sky is clear, and the solar-altitude angle is 60°. Also shown is the photopic spectral luminous efficiency function, whose peak value is 1.0, scaled up for clarity. (Credit: Author.)

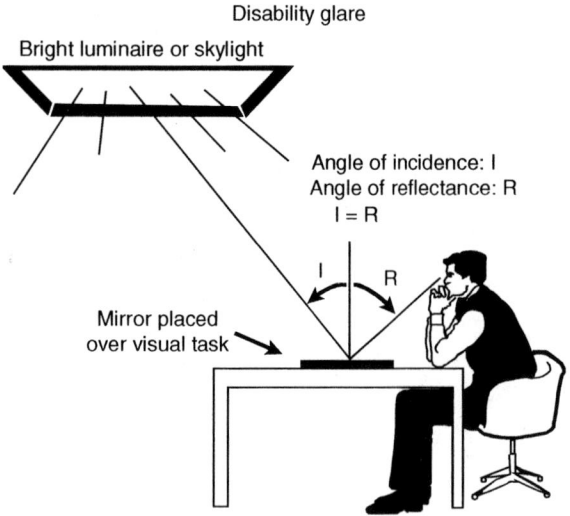

Fig. 2 Disability glare is produced with a strong source of light that masks a visual task, usually by reflection, reducing its contrast and disabling the person's ability to see that task well. Reorienting the task and the glare source often can ameliorate the problem, as can reducing the brightness of the source. (Credit: Florida Solar Energy Center.)

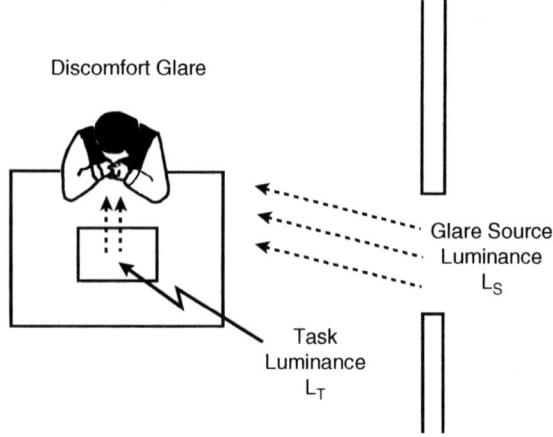

Fig. 3 Discomfort glare is produced by strong light from the side entering the eye without masking the target directly. (Credit: Florida Solar Energy Center.)

computer screen or television set and light reflected from a glossy magazine page. The reflected light reduces the contrast of the image. With a glossy magazine, for example, the reflected glare light can be as strong from the black ink as from the white paper, washing out the text and making it difficult or impossible to read.

Discomfort glare usually results when light entering the eye from the side is much brighter than that coming from the visual task. This extra-bright light is mentally and physically confusing, and can result in visual fatigue, discomfort, and even headaches.

Buildings have several potential sources of glare. Beam sunlight entering the eye directly perhaps is the worst, due to its extreme brightness. Both specularly and diffusely reflected beam sunlight also can produce both kinds of glare. A bright window surrounded by dark walls and furnishings nearly always produces discomfort glare unless the light from the window normally does not enter the eyes, due to the orientation of the visual task. Bare electric lamps, either incandescent or fluorescent, can produce glare, as can poorly designed electric luminaires. A successful lighting design will reduce or minimize the system's potential contribution to both kinds of glare, whether it be a daylighting or an electric lighting system.

A variety of metrics have been devised in attempting to quantify visual comfort in the presence of discomfort-glare sources. Most incorporate terms for the angles of the brighter source from the direct line of sight into the eye.[3] A rule of thumb for reducing discomfort glare is to keep the brightest light source in the visual field from being stronger than a few times the general surround luminance.

Fig. 4 illustrates two contrasting daylighting designs: one that promotes discomfort glare and one that ameliorates it. If the window in both cases looks out on the same uniform sky, it will have the same luminance in both cases. The larger window, therefore, will admit greater overall flux into the room. With moderately high room surface reflectances, this means that the room generally will be brighter and the room surface luminances will be closer to that of the window, producing less tendency toward glare. The same effect can be achieved by reducing the visible transmittance of the window while increasing the electric lighting in the room, reducing the contrast in room brightnesses, but this "solution" calls for more purchased energy.

A final caveat is offered regarding glare. One person's "killer glare" is another person's "sparkle." People living in a region with persistent clouds and long, dark winters may have a much higher tolerance for direct beam admission than those living in a hot region with

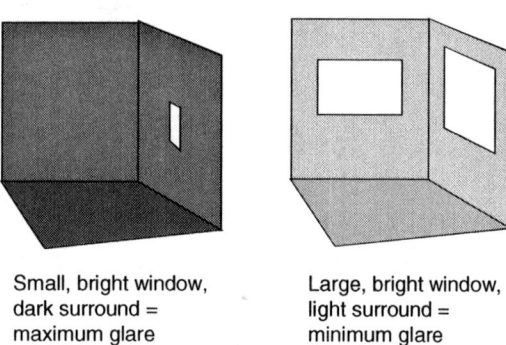

Fig. 4 Comparison of two daylighting designs. The one on the left has a relatively small window in a room with low surface reflectances, resulting in inadequate illumination, a gloomy appearance, and a propensity for discomfort glare. The one on the right, having larger window areas and higher surface reflectances, is brighter and less prone to glare. (Credit: Author.)

predominantly clear skies and high direct beam solar flux levels. It is important to use direct sunlight entry judiciously and with care to prevent the adverse impacts that can result. Some daylighting systems rely almost exclusively on direct beam sunlight, but these are designed to distribute the concentrated beams widely and diffusely, with minimal glare impact.

DAYLIGHTING DESIGN

The goals of good daylighting include the provision of good-quantity and good-quality daytime interior illumination and view, coupled with high visual comfort for occupants. Happy people are productive people. It is necessary to prevent common problems such as unwanted glare and overheating from excessive direct beam illumination. In the process of providing good-quality illumination of adequate quantity, it is desirable to design the system so as to displace as much daytime electric lighting as possible, to minimize the energy costs of building operation.

Some traditional means of admitting daylight into building spaces are drawn schematically in Fig. 5.

Additional approaches using light pipes and other beam-manipulation strategies can also be utilized. Several of these approaches are illustrated schematically in Fig. 6.

BUILDING OCCUPANCY

The most efficient buildings are unoccupied ones, with all the building's energy services turned off. Consequently, the more a building is occupied and using energy, the greater are the opportunities for energy savings. Daylighting design saves energy through averted electrical energy costs. If sufficient daylighting already is available in a building, however, additional daylighting could increase energy costs. On the other hand, health and comfort benefits often are sufficient to justify the introduction of some additional daylight if the extra energy costs are modest, limited by the use of high-performance windows.

If people are seldom in a building during the daylight hours, the need for illumination is minimal, and daylighting can't save much energy. This typically is the case for most residences. Exceptions include residences occupied by retirees and others not working outside the home in the daylight hours, and those daytime-occupied homes suffering inadequate daylighting. In these cases, added daylighting makes good energy sense, and offers additional visual comfort and psychological benefits. For offices and other buildings fully occupied during daylight hours, daylight illumination often saves more building energy than any other single strategy, and the attendant increases in worker productivity add further to the benefits.

DAYLIGHTING SIMULATION AND MODELING

A flow chart showing the connections among various aspects of daylighting system energy and illumination performance is shown in Fig. 7.

Over the past two centuries, a variety of methodologies has been developed for predicting both the energy and the illumination performances of daylighting systems. With the advent of the fast personal computer, most of these methods have given way to sophisticated new computer

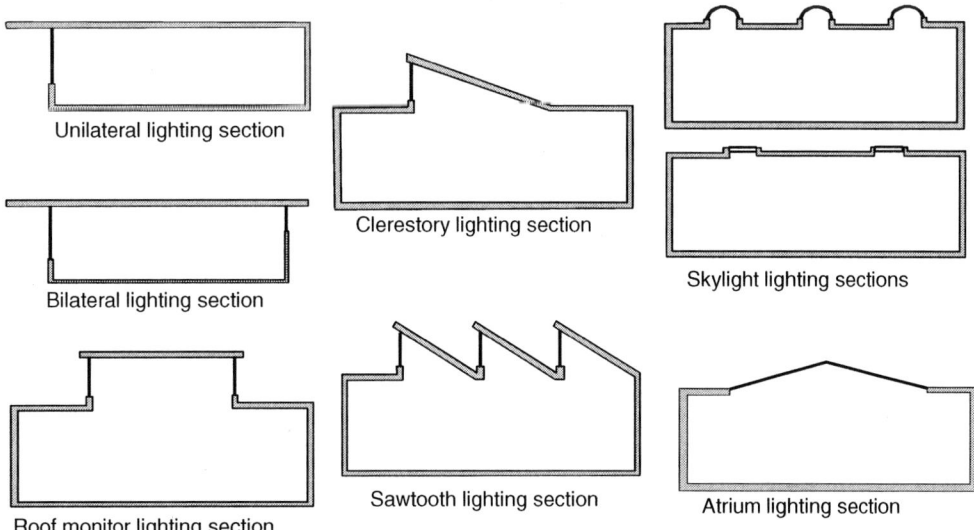

Fig. 5 Section views of several ways daylight can be admitted into buildings. (Credit: *IESNA Lighting Handbook*. 9th ed. Illuminating Engineering Society of North America, 2000.)

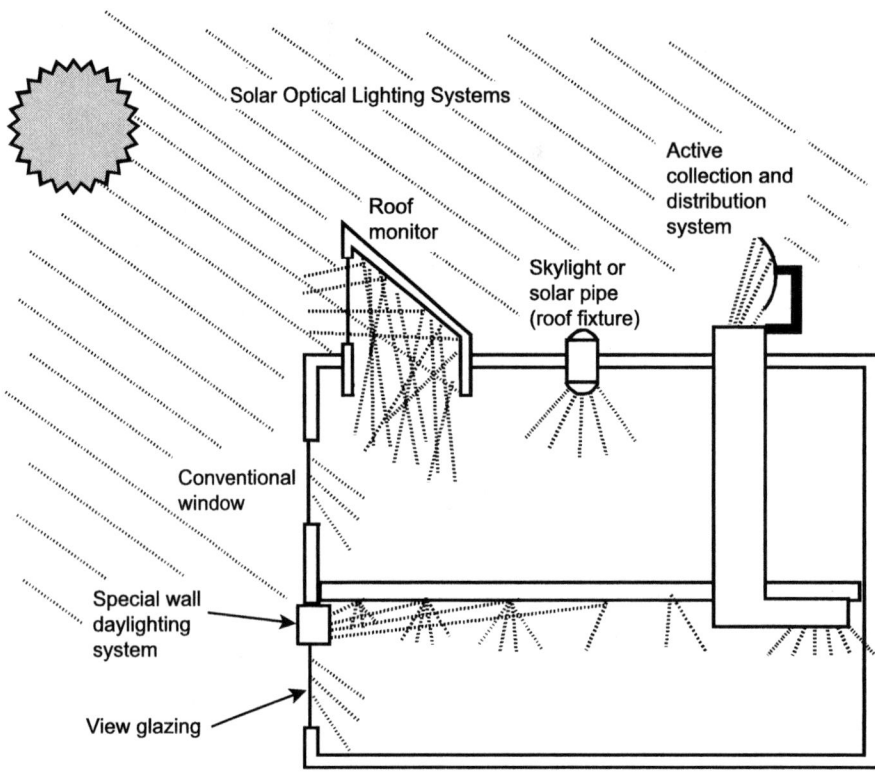

Fig. 6 Illustration of the variety of ways daylight can be admitted into building spaces for controlled illumination of the interior. (Credit: Author.)

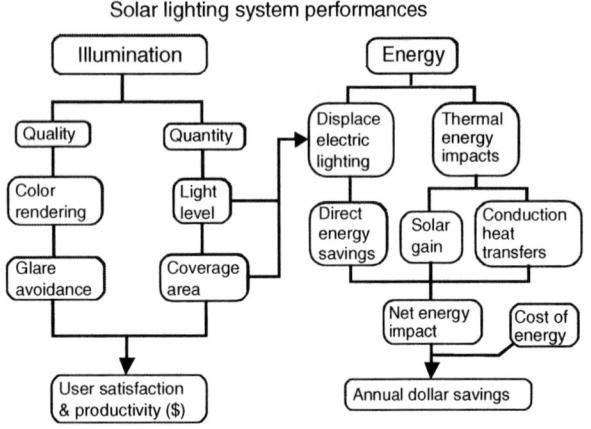

Fig. 7 Connection pathways for the energy and illumination performance components of daylighting systems. (Credit: Author.)

tools for determining the performance of both electric and daylighting systems. The new tools include commercially available computer programs and public-domain software. A Web search on such terms as daylighting design software and architectural lighting design should provide links to many sites describing these computer programs.

Another important tool is scale-model testing. Because illumination scales upward and downward well, small-scale models of buildings can be constructed of inexpensive materials. Light sensors can be placed inside these models to measure light levels when the models are placed in either simulated or natural daylight. Cameras also can be employed to image the light distributions inside the building's daylit rooms, under varied sun and sky conditions, to assess performance visually.

CONCLUSION

Humans need connections with the outdoors. This need is built into our genetic makeup; it promotes health and a sense of well being, and makes us happier and more productive. Even photographs of nature on the wall have proved to be helpful.

According to Judith Heerwagen, principal of J.H. Heerwagen and Associates, and senior scientist at the Pacific Northwest National Laboratory in Seattle, in our evolutionary past, information about our environment had a pronounced influence on survival and health. Changes in daylight provided time cues and assessment of cloud formations for information about future weather conditions. These events influenced our ancestors' daily decisions, such as where to sleep at night and where to look for food next week. Loss of illumination from and the view to the outdoors have been implicated in the poor recovery of patients in windowless intensive care units.

"Once you start thinking about it, [daylighting] design makes perfect sense," Heerwagen has written. "We didn't evolve in a sea of gray cubicles."[4]

Daylighting offers a number of benefits to building owners and occupants. Cool, natural daylight has good color rendering; it is healthy and offers clear psychological benefits. Daylighting can displace electric lighting, saving energy and reducing air pollution, global warming, and our dependence on dwindling supplies of fossil-fuel energy.

Glossary

Brightness: Brightness is a subjective term with no universally acceptable quantitative definition. It refers to a perception of the strength of illumination received by the human eye. It may be used in general characterizations of the appearances of different sources and objects. Luminance approximately characterizes the strength of illumination emanating from an object or source, as perceived by the eye.

Illuminance, E_v: The area density of luminous flux; the luminous flux per unit area at a specified point in a specified surface that is incident on, passing through, or emerging from that point in the surface (units: lm m^{-2} = lux).

Irradiance, E_e: The area density of radiant flux; the radiant flux per unit area at a specified point in a specified surface that is incident on, passing through, or emerging from that point in the surface (units: W m^{-2}).

Luminance, L_v: The area and solid angle density of luminous flux; the luminous flux per unit projected area and per unit solid angle incident on, passing through, or emerging from a specified point in a specified surface, and in a specified direction in space (units: lm m^{-2} sr^{-1} = cd m^{-2}).

Luminous flux, Φ_v: The $V(\lambda)$-weighted integral of the spectral flux Φ_λ over the visible spectrum (units: lumen or lm).

Luminous intensity, I_v: The solid angle density of luminous flux; the luminous flux per unit solid angle incident on, passing through, or emerging from a point in space and propagating in a specified direction (units: lm sr^{-1} = cd).

Photopic spectral luminous efficiency function, $V(\lambda)$: The standardized relative spectral response of a human observer under photopic (cone vision) conditions over the wavelength range of visible radiation.

Radiance, L_e: The area and solid angle density of radiant flux; the radiant flux per unit projected area and per unit solid angle incident on, passing through, or emerging from a specified point in a specified surface, and in a specified direction in space (units: W m^{-2} sr^{-1}).

Radiant flux, Φ_e: The time rate of flow of radiant energy (units: watt).

Radiant intensity, I_e: The solid angle density of radiant flux; the radiant flux per unit solid angle incident on, passing through, or emerging from a point in space and propagating in a specified direction (units: W sr^{-1}).

Radiation luminous efficacy, K_r: The ratio of luminous flux in lumens to radiant flux (total radiation) in watts in a beam of radiation (units: lumen/watt).

Spectral radiometric quantities: The spectral "concentration" of quantity Q, denoted Q_λ, is the derivative $dQ/d\lambda$ of the quantity with respect to wavelength λ, where Q is any one of: radiant flux, irradiance, radiant intensity, or radiance (units: same as that of quantity Q per nm).

System luminous efficacy, K_s: The ratio of luminous flux in lumens delivered in a space to the electrical consumption of the lighting system delivering that flux (units: lumen/watt).

Visible transmittance, T_v or VT: The ratio of transmitted to incident illuminance on a glazing system; a unitless quantity.

REFERENCES

1. McCluney, R. *Introduction to Radiometry and Photometry*, Artech House: Boston, MA, 1994.
2. IES Calculation Procedures Committee Recommended practice for the calculation of daylight availability. J. IES **1984**, July, 3-11–3-13.
3. Kaufman, J.E.; Christensen, J.F., Eds. In *IES Lighting Handbook, 1984 Reference Volume*, Illuminating Engineering Society of North America: New York, 1984.
4. Heerwagen, J.H. The Psychological Aspects of Windows and Window Design. EDRA 21 Symposium, Environmental Design Research Association, 1990.

Demand Response: Commercial Building Strategies*

David S. Watson
Sila Kiliccote
Naoya Motegi
Mary Ann Piette
Commercial Building Systems Group, Lawrence Berkeley National Laboratory, Berkeley, California, U.S.A.

Abstract

This paper describes strategies that can be used in commercial buildings to temporarily reduce electric load in response to electric grid emergencies in which supplies are limited or in response to high prices that would be incurred if these strategies were not employed. The DR strategies discussed herein are based on the results of three years of automated DR field tests in which 28 commercial facilities with an occupied area totaling over 11 million ft^2 were tested. Although the DR events in the field tests were initiated remotely and performed automatically, the strategies used could also be initiated by on-site building operators and performed manually, if desired. While energy efficiency measures can be used during normal building operations, DR measures are transient; they are employed to produce a temporary reduction in demand. Demand response strategies achieve reductions in electric demand by temporarily reducing the level of service in facilities. Heating, ventilating and air conditioning (HVAC), and lighting are the systems most commonly adjusted for DR in commercial buildings. The goal of DR strategies is, to meet the electric shed savings targets while minimizing any negative impacts on the occupants of the buildings, or the processes that they perform. Occupant complaints were minimal in the field tests. In some cases, "reductions" in service level actually improved occupant comfort or productivity. In other cases, permanent improvements in efficiency were discovered through the planning and implementation of "temporary" DR strategies. The DR strategies that are available to a given facility are based on factors such as the type of HVAC, lighting, and energy management and control systems (EMCS) installed at the site.

BACKGROUND

Power requirements on the electric grid are in constant flux, based on the demand of the devices connected to it. This demand varies based on time of day, weather, and many other factors. Traditionally, the supply is varied to meet the demand by increasing or decreasing electric generation capacity. Conversely, demand response (DR) can be defined as short-term modifications in customer end-use electric loads in response to dynamic price and reliability information.

As electric demand increases, generation costs increase in a non-linear fashion. A price spike caused by high demand on a hot summer afternoon would be an example of price information that might be used to initiate short-term modifications in customer end-use electric loads. A scenario in which a power plant failed unexpectedly would be an example of where short-term modifications in customer end-use electric loads could help other on-line plants manage the demand thereby increasing system reliability and avoiding blackouts.

Many electric utilities across the United States have implemented programs that offer financial incentives to ratepayers who agree to make their electric loads more responsive to pricing and/or reliability information. These programs are most prevalent for commercial and industrial customers in utility districts with known capacity or transmission constraints.

Recent studies have shown that customers have limited knowledge of how to develop and implement DR control strategies in their facilities.[2] Another barrier to participation in DR programs is the lack of systems that help automate the short-term modifications or strategies required during DR events.

This paper focuses on strategies that can be used to enable DR in commercial buildings (i.e., to make short-term modifications to their end-use equipment).

RESULTS OF FIELD TESTS

The strategies discussed herein are based on the results of a series of field tests conducted by the PIER Demand

---

* This paper was originally presented as "Strategies for Demand Response in Commercial Buildings" at the 2006 ACEEE Summer Study on Energy Efficiency in Buildings Conference, August 13–18, 2006 in Pacific Grove, California. Reprinted with Permission. ACEEE is the American Council for an Energy Efficient Economy, located in Washington, DC.

Keywords: Demand response; Electric load reduction; Load control; Peak load control; Peak load reduction; HVAC load reduction; Lighting load reduction; Global temperature adjustment.

Demand Response: Commercial Building Strategies

Table 1 Average and maximum peak electric demand savings during automated demand response (DR) tests

Results by year	Number of sites	Duration of shed (h)	Average savings (%)	Maximum savings (%)
2003	5	3	8	28
2004	18	3	7	56
2005	12	6	9	38

Response Research Center. While the tests focused on fully automated electric DR, some manual and semi-automated DR was also observed. The field tests included 28 facilities, 22 of which were in Pacific Gas and Electric territory. The other sites were located in territories served by Sacramento Municipal Utility District, Southern California Edison, City of Palo Alto Utilities and Wisconsin Public Service. The average demand reductions were about 8% for DR events ranging from 3 to 6 h.

Table 1 shows the number of sites that participated in the 2003, 2004, and 2005 field tests along with the average and maximum peak demand savings. The electricity savings data are based on weather sensitive baseline models that predict how much electricity each site would have used without the DR strategies. Further details about these sites and the automated DR research are available in previous reports.[4,5]

Fig. 1 shows the various DR strategies that were used in field tests and the frequency of each. The tests included building types such as office buildings, a high school, a museum, laboratories, a cafeteria, data centers, a postal facility, a library, retail chains, and a supermarket. The buildings range from large campuses, to small research and laboratory facilities.

Fig. 2 shows the various DR strategies that were used in field tests and the Demand Saving Intensity (W/ft^2) by Shed Strategy. The values shown are average savings over 1 h. Though the sample size is not large enough to generalize shed savings by strategy, it is clear that each of the three shed categories listed has the potential to shed about 0.5 W/ft^2. Most of the DR heating, ventilating and air conditioning (HVAC) strategies we have examined provide considerably greater savings on hotter days and the data in Fig. 2 were from a mild day. Lighting strategies are not weather dependent.

CONCEPTS AND TERMINOLOGY

Energy Efficiency

Energy efficiency can lower energy use without reducing the level of service. Energy efficiency measures are part of normal operations to permanently reduce usage during peak and off-peak periods. In buildings, energy efficiency is typically achieved through efficient building designs, the use of energy efficient equipment, and through efficient building operations. Since energy efficiency measures are a permanent part of normal operations, they are typically considered separate from DR which involves short term modifications to normal operations. However, some energy efficiency measures such as the use of variable frequency drives (VFDs) on electric motors can enable both energy efficiency and temporary DR modes when called to do so.

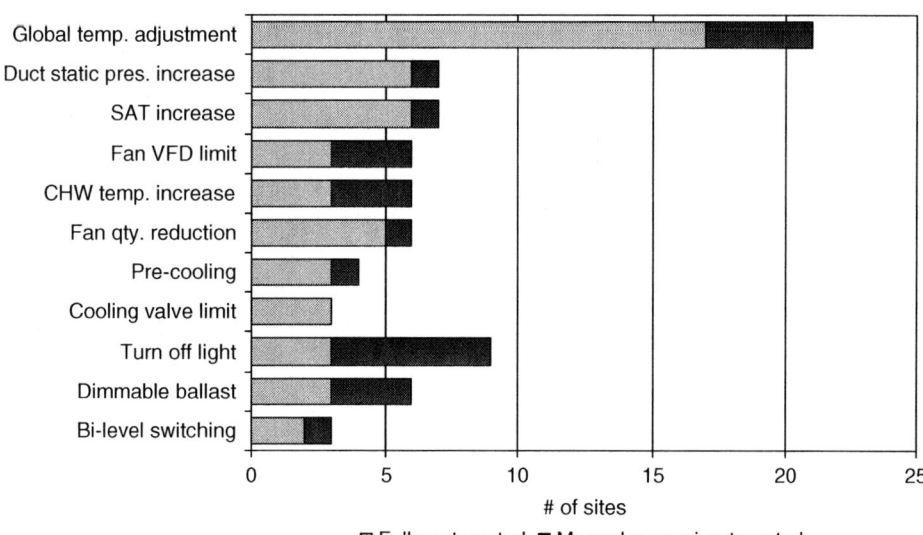

Fig. 1 Frequency of various demand response (DR) strategy usage.

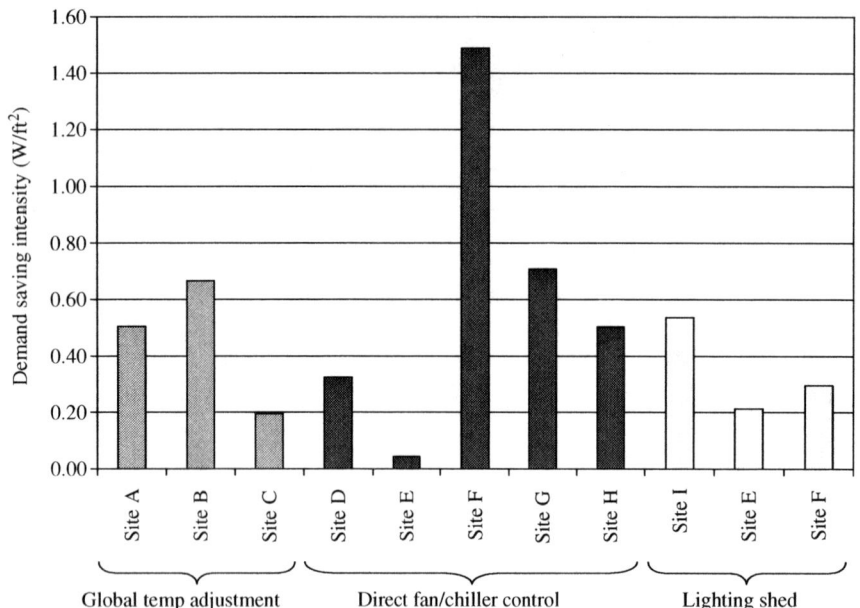

Fig. 2 Demand saving intensity (W/ft^2) by shed strategy on November 5, 2004.

Daily Peak Load Management

Daily peak load management is done in many buildings to minimize peak demand charges and time-of-use rates. Strategies that temporarily modify the operation of HVAC or lighting systems are often used to implement daily peak load management. Decisions about when to initiate daily peak load management are typically made by on-site staff or on-site automated equipment.

Demand Shifting

Demand shifting is achieved by changing the time that electricity is used. Thermal energy storage is an example of a demand shifting technology. Thermal storage can be achieved with active systems such as chilled water or ice storage, or with passive systems such as pre-cooling the mass of a building.[6] Both daily peak load management and demand shifting are typically done to minimize peak demand and time-of-use rate charges.

Demand Response

Demand response can be defined as short-term modifications in customer end-use electric loads in response to dynamic price and reliability information. Demand response events are dynamic and temporary. They are driven by factors such as low electricity reserves, warm weather and grid conditions.

One of the key components of DR is that the pricing and reliability information known at the grid system or utility level must be transmitted and translated into load reducing actions at the end-use sites. Signaling methods used to inform facility operators of upcoming DR events include: phone calls, pagers, text messages, and e-mail messages. Control signals are also used in some systems for direct signaling to energy management and control systems (EMCS) and control of electric loads. These digital control signals are broadcast using radio transmissions, power-line communications and the Internet.

Demand response can be implemented using various levels of automation. *Manual Demand Response* is performed by facilities staff physically turning off electric equipment after receiving notification of an upcoming DR event. *Semi-Automated Demand Response* is similar, but reduces facilities staff labor through use of a centralized control system with pre-programmed DR strategies. *Fully-Automated Demand Response* enables remotely generated event initiation signals to control loads directly or to initiate pre-programmed DR strategies at the site. Though Fully-Automated DR is capable of functioning without human intervention, it is recommended that facility operators are kept informed of the process and have the ability to "opt-out" of a DR event, if desired.

Reduction in Service

Demand response strategies achieve reductions in electric demand by temporarily reducing the level of service in facilities. Heating, ventilating and air conditioning, and lighting are the systems most commonly adjusted to achieve DR savings in commercial buildings. The goal of DR strategies is to meet the electric shed savings targets while minimizing any negative impacts on the occupants of the buildings or the processes that they perform. Occupant complaints were minimal in the field tests. In some cases, "reductions" in service level actually improved occupant comfort or productivity. Such cases can be caused by over-cooling that occurs in some buildings during normal operation. In other cases, permanent improvements in efficiency were discovered through the planning and implementation of "temporary" DR strategies. The DR strategies that are available to a given facility are based on factors such as the type of HVAC, lighting, and EMCS installed at the site.

Shared Burden

Demand response strategies that share the burden evenly throughout the facility are least likely to have negative effects on building occupants. For example, if it were possible to reduce lighting levels evenly throughout an entire facility by 25% during a DR event, impact to occupants may be minimal. However, turning off all of the lights in one quadrant of an occupied space would not be acceptable. In HVAC systems, strategies that reduce load evenly throughout all zones of a facility are superior to those that allow certain areas (such as those with high solar gains) to substantially deviate from normal temperature ranges.

By combining savings from sheds in HVAC and lighting (and other loads, if available), the impact on each system is minimized and the savings potential is increased.

Closed Loop Control

Comfort is maintained in modern buildings through the use of closed loop control of HVAC systems. Sensors are used to measure important parameters such as temperature and pressure. Controllers adjust actuators such as dampers or valves to maintain the desired setpoints for those parameters. The effect of the actuators on the controlled zone or system is measured by the sensor, hence "closing the control loop." Control sub-systems for which there is no feedback from sensors are known as "open loop" controls.

In order to maintain predictable and managed reductions of service during DR events, strategies should maintain the use of closed loop controls in HVAC systems.

Granularity of Control

For the purpose of DR control in buildings, the concept of granularity refers to how much floor area is covered by each controlled parameter (e.g., temperature). In HVAC systems, the ability to easily adjust the temperature setpoint of each occupied space is a highly granular way to distribute the DR shed burden throughout the facility. Less granular strategies such as making adjustments to chillers and other central HVAC equipment can provide effective shed savings, but can cause temperature in some zones to drift out of control. Granularity of control can also allow building operators to create DR shed behaviors that are customized for their facility. An example of this would be to slightly increase all office zone temperature setpoints, but leave computer server room setpoints unchanged.

Resolution of Control

In HVAC systems, parameters are controlled with great resolution. In many systems temperature setpoints can be adjusted by as little as 0.1°F. Although some modern lighting ballasts can adjust individual lamps in less than 1% increments, most commercial lights are only capable of being turned on or off. Additional information is provided in the "Lighting Based DR Strategies" section below.

Rebound

At the end of each DR event, the effected systems must return to normal operation. When lighting strategies are used for DR, normal operation is regained by simply re-enabling all lighting systems to their normal operation. Lights will come back on as commanded by time clocks, occupancy sensors or manual switches. There is no reason for lighting power to jump to levels that are higher than normal for that period.

However, without special planning HVAC systems tend to use extra energy following DR events in order to bring systems back to normal conditions. Extra energy is used to remove heat that is typically gained during the reduced service levels of the DR event. This post DR event spike in demand is known as "rebound." To minimize high demand charges and to reduce negative effects to the electric grid, rebound should be reduced or minimized through use of a strategy that provides a graceful return to normal operation. The simplest case is where the DR event ends or can be postponed until the building is unoccupied. If this is not possible, strategies that allow HVAC equipment to slowly ramp up or otherwise limit power usage during the return to normal period should be used.

HVAC BASED DR STRATEGIES

Heating, ventilating and air conditioning systems can be an excellent resource for DR shed savings for several reasons: (1) HVAC systems create a substantial electric load in commercial buildings, often more than one-third of the total; (2) the "thermal flywheel" effect of indoor environments allows HVAC systems to be temporarily unloaded without immediate impact to the building occupants; and (3) it is common for HVAC systems to be at least partially automated with EMCSs.

However, there are technical challenges to using commercial HVAC systems to provide DR sheds. These systems are designed to provide ventilation and thermal comfort to the occupied spaces. Operational modes that provide reduced levels of service or comfort are rarely included in the original design of these facilities. To provide reliable, repeatable DR sheds it is best to pre-plan and automate operational modes that will provide DR savings. The use of automation will reduce labor required to implement DR operational modes when they are called. In addition, timeliness of the response will typically be improved.

Heating, ventilating and air conditioning based DR strategies recommended for a given facility, vary based on the type and condition of the building, mechanical equipment and EMCS. Based on these factors, the best DR strategies are those that achieve the aforementioned goals of meeting electric shed savings targets while minimizing negative impacts on the occupants of the buildings or the processes that they perform. The following DR strategies are prioritized so as to achieve these goals:

1. Global temperature adjustment (GTA) of zones
2. Centralized adjustments to the air distribution and/or cooling systems.

All HVAC based DR strategies outlined in this paper allowed zone temperatures to drift outside of normal ranges. However, the rate at which the temperatures drifted was well below the rate of Acceptable Temperature Change defined in ASHRAE Standard 55-2004. Demand response strategies used to return the HVAC system to normal operation should be designed for a similarly gradual rate of change. In addition to the comfort benefits outlined in the ASHRAE standard, strategies that slowly return the system to normal have the additional benefit of limiting rebound spikes as described previously.

GLOBAL TEMPERATURE ADJUSTMENT OF ZONES

Description

Global temperature adjustment of occupied zones is a feature that allows commercial building operators to easily adjust the space temperature setpoints for an entire facility from one command from one location. Typically, this is done from a screen on the human machine interface (HMI) to the EMCS. In field tests, GTA was shown to be the most effective and least objectionable strategy of the five HVAC shed strategies tested.[4] It is most effective because it reduces the load of all associated air handling and cooling equipment. It is least objectionable because it shares the burden of reduced service level evenly between all zones. Global temperature adjustment based DR strategies can be implemented either manually by building operators or automatically based on remote signals.

Typical Implementation

Global temperature adjustment is typically implemented by broadcasting a signal from the central EMCS HMI server to the all final space temperature control devices distributed throughout the facility. Upon receipt of a global signal from the central EMCS server, the final space temperature control devices interprets the signal and reacts accordingly (e.g., DR Mode Stage-1 means increase space cooling setpoints 3°F and decrease space heating setpoints 3°F).

Final space temperature control devices suitable for GTA include:

- Space temperature controllers that adjust variable air volume (VAV) terminal box dampers (all types) (e.g., VAV boxes).
- Space temperature controllers that adjust hot water heating coil valves or chilled water cooling coils (e.g., fan coil units, CAV multi-zone heating and cooling coil valves).
- Space temperature controllers that adjust capacity of heat pumps or direct expansion (DX) units.

To avoid an unwanted increase in heating energy, heating setpoints should remain the same or be reduced during GTA mode.

Mode Transitions

In the most basic implementation, upon receipt of a DR signal the GTA enabled system will increase space cooling setpoints in one or two steps (two step increase shown in Table 2). Upon entering a DR mode (e.g., moderate shed), the global temperature setpoints will be increased and load on the air distribution and cooling systems will decrease.

More advanced implementations can adjust setpoints to follow linear or exponential curves.[6] Though more difficult to program, these strategies can provide added flexibility in creating shed profiles that are customized to provide optimal consistency or duration for a given facility.

Decay of Shed Savings

Over time, internal and external heat gains will increase zone temperatures until they exceed the new DR setpoints, causing fan and cooling systems to ramp back up. This phenomenon, known as "decay" of shed savings, can be prevented by further increasing the zone cooling setpoints to new levels (e.g., high shed). After a certain time duration, which varies by building type, weather and other factors, the shed savings will decay to the point where additional setpoint increases are not viable in an occupied building. In field tests, successful sheds of up to 6 h have been performed without substantial impact on commercial building occupants.

Table 2 Global temperature adjustment (GTA) setpoint adjustment—example of absolute and relative implementations

Demand response (DR) mode	Absolute space temperature cooling setpoints (°F)	Relative space temperature cooling setpoints
Normal	74 (globally)	Varies per zone
Moderate shed	76	Normal +2°F
High shed	78	Normal +4°F

Absolute vs Relative Implementation

Global temperature adjustment may be implemented on either an absolute or relative basis (Table 2). An absolute implementation of GTA allows the operator to set the space temperature setpoints for the entire facility to absolute values (e.g., heating setpoints at all final space temperature control devices = 68°F and cooling setpoints at all final space temperature control devices = 76°F). A relative implementation of GTA allows the operator to adjust the space temperature setpoints for the entire facility to new values that are offset from the current values by a relative amount (e.g., heating setpoints at all final space temperature control devices should decrease 2°F from current values and cooling setpoints should increase 2°F from current values). A relative implementation of GTA is best suited for sites where "normal" setpoints vary throughout the facility. It ensures that temperature will not deviate more than a fixed amount from the customized normal setpoint for each zone.

Factory vs Field Implementations of GTA

Several manufacturers offer GTA as a standard feature in their EMCS products. In field tests, sites that used EMCS products from these vendors provided some of the largest sheds and required the least amount of set-up labor. For sites that have EMCS controlled space temperature zones, but lack GTA, it can typically be added in the field. To add GTA to an existing site, each EMCS zone controller must be programmed to "listen" for global GTA commands from the central EMCS system. In addition, the central system must be programmed to send GTA commands to all relevant zone controllers on the EMCS digital network. Typically GTA commands are sent in a global broadcast to all controllers simultaneously.

Impediments to Using GTA Strategy

In field tests, sites that used HVAC shed strategies other than GTA usually did so because that feature was not available at their site. Reasons that GTA is not available include:

- Space temperature not controlled by EMCS (e.g., use of pneumatic controls in occupant zones).

- Space temperature is controlled by EMCS, but space temperature controllers do not include the GTA feature. (i.e., EMCS can adjust space temperature setpoints in each zone individually, but not globally). Adjusting each zone individually is more time consuming and error prone to use for DR purposes.

Evaluation of Global Temperature Adjustment of Zones

While the GTA DR strategy reduces the service level of the occupied spaces, it does so using a closed-loop control strategy in a highly granular fashion. This causes the DR shed burden to be evenly shared between all building occupants and keeps all zones under control. Since none of the zones are starved for airflow, there is no risk of ventilation rates dropping below specified design levels. If GTA of zones is available, the HVAC DR shed strategy recommended for commercial buildings.

AIR DISTRIBUTION AND COOLING SYSTEM ADJUSTMENT

In systems for which the aforementioned GTA of zones is not an option, strategies that make temporary adjustments to the air distribution and/or mechanical cooling systems can be employed to enable DR. Depending on the mechanical systems in place at a given facility, the following DR strategies may be used.

Duct Static Pressure Setpoint Reduction

For variable air volume systems, duct static pressure (DSP) is typically measured in the supply duct. The EMCS modulates the speed of the fan or the position of inlet guide vanes (IGV) to maintain a defined DSP setpoint at the measured location. The "normal" DSP SP at the measured point should be high enough to provide enough pressure for each terminal VAV box to function properly. In an ideal system, the DSP SP would be set just high enough to meet the pressure requirements of the VAV terminal box of greatest demand. But since the box is of greatest demand, and its associated pressure requirement are in constant flux, sub-optimal, yet substantially simpler strategies are usually used to control DSP. Typically DSP is measured at a single location about two-third of the way down the duct system. The DSP SP is set to a fixed value that is high enough to meet the needs of the box of greatest demand during design load conditions. During less demanding conditions energy is wasted due to losses associated with the DSP SP being higher than necessary to meet the demands of the VAV terminal boxes.

Fan energy and cooling energy can be reduced during DR events by reducing the DSP setpoint. This strategy is effective for three reasons:

1. The "normal" DSP SP is often higher than necessary. By reducing the DSP SP, some shed savings is provided without any reduction in comfort or service to the occupants.
2. Additional shed savings occurs when the DSP SP is set low enough to cause some VAV terminal boxes to "starve" from lack of air pressure. This reduction in service causes less air flow through the fans. There is some risk of ventilation rates dropping below specified design levels in some areas using this strategy.
3. When airflow drops below levels necessary to cool the space, electric load on the cooling system also drops.

Fan Speed Limit

Like DSP setpoint reduction mentioned above, this DR strategy is relevant to fans with VFD. During the DR event, the speed of the VFD is limited to a fixed value. To be effective, the fixed value must be lower than if it were allowed to operate under normal closed loop conditions. Fan speed limiting saves energy for the same reasons as DSP setpoint reduction. Its effect on the air distribution systems and associated occupied zones is somewhat less predictable because of the open-loop nature of the control. Fan speed limits may be useful as part of other DR strategies such as cooling system adjustments described below. This strategy may also be used on fans with IGV.

Fan Quantity Reduction

For constant air volume fan systems, the only way to reduce fan energy is by turning fans off completely. This is obviously a severe reduction in service, although it may be of some use in common areas served by multiple fans. If such a strategy is used, it should be noted that cooling energy in the fans that remain on will increase to make up for those that are off.

Increase Supply Air temperature

This strategy saves mechanical cooling energy. In packaged DX units and heat pumps, the savings will be achieved at each unit. For air handlers with cooling coils, the savings will occur at the central cooling plant. In either case, care must be taken to avoid increased fan energy in VAV systems due to increased air flow. This effect can be prevented by limiting fan speeds to levels in use prior to the increase in supply air temperature.

Central Chiller Plants

Most modern centrifugal, screw and reciprocating chillers have the capability of reducing their demand for power. This can be done by raising the chilled water supply temperature setpoint or by limiting the speed, capacity, the number of stages or current draw of the chiller. The quantity of chillers running can also be reduced in some plants.

Evaluation of Air Distribution and Cooling System Adjustment Strategies

While effective in terms of the ability to achieve load reductions, the use of centralized adjustments to air distribution systems and/or mechanical cooling systems for DR purposes have some fundamental drawbacks. In these strategies, the DR burden is not shared evenly between all the zones. Centralized, changes to the air distribution System and/or mechanical cooling systems allow zones with low demand or those that are closer to the main supply fan to continue to operate normally and hence not contribute toward load reduction in the facility. Zones with high demand, such the sunny side of the building or zones at the ends of long duct runs can become starved for air or otherwise go completely out of control. Centralized HVAC DR shed strategies can allow substantial deviations in temperature, airflow and ventilation rates in some areas of a facility. Increased monitoring of occupied areas should be conducted when using these strategies.

LIGHTING BASED DR STRATEGIES

Lighting systems offer great promise as a resource for DR shed savings for several reasons: (1) Lighting systems create a substantial electric load in commercial buildings, often more than 30% of the total; (2) lighting has no rebound effect during the transition from DR events to normal operations; and (3) the lighting systems in many California commercial buildings already have bi-level switching in place. Usually, this enables one-third or two-third or the lights in a given office to be turn off, leaving sufficient light for egress and many common office tasks.

However, there are major impediments to the use of lighting systems for DR: (1) Few office buildings have centralized control of lighting systems;[3] (2) even buildings with centralized lighting controls are not necessarily zoned in a way that would allow a reduction in lighting service that is adequate for occupancy.

Granularity of control is a very important factor in determining the usefulness of lighting systems for DR. The following lists five types of lighting systems from most coarse to most fine granularity: Zone Switching, Fixture Switching, Lamp Switching, Stepped Dimming, Continuous Dimming.

Zone Switching

In areas that are unoccupied or are illuminated by windows or other sources, entire lighting zones can be switched off for DR purposes. In some cases, this strategy can be applied to common spaces such as lobbies, corridors, and cafeterias.

Fixture/Lamp Switching

Fixture or lamp switching can be done by bi-level switching. California's Title 24 Energy Efficiency Building Standard, requires multiple lighting level controls in all individual offices built since 1983. With bi-level switching, each office occupant is provided with two wall switches near the doorway to control their lights. In a typical installation, one switch would control one-third of the fluorescent lamps in the ceiling lighting system, while the other switch would control the remaining two-third of the lamps. This allows four possible light levels: OFF, one-third, two-third and FULL lighting. The 2001 standards state that bi-level switching can be achieved in a variety of ways such as:

- Switching the middle lamps of three lamp fixtures independently of outer lamps (lamp switching)
- Separately switching "on" alternative rows of fixtures (fixture switching)
- Separately switching "on" every other fixture in each row (fixture switching)
- Separately switching lamps in each fixture (lamp switching).

Step Dimming

Through the use of ON/OFF switches, controls to regulate the level of electrical light, step dimming is a popular energy-saving retrofit solution for applications where existing fixtures are not equipped with dimming ballasts. Stepped dimming is often called bi-level dimming because the strategy often involves two levels of light output, usually 100 and 50%. However, if more flexibility is required, stepped dimming can involve three levels of light output.

Continuous Dimming

Continuous dimming ballasts allow light output to be gradually dimmed over the full range, from 100 to 10% (fluorescent) or 100 to 50% (HID). These lighting systems provide an excellent resource for DR purposes. These systems allow the lighting load to be reduced so gradually that modest changes may not even be noticed by building occupants.[1] Since the amount of reduction is continuously variable, specific DR shed goals can be achieved using straightforward strategies. As with GTA, shed strategies using continuously dimming lighting can be implemented in an absolute (building-wide) or relative fashion.

In addition to their use for DR, dimmable ballasts can be used in the design of energy efficient systems that reduce electric light requirements when daylight is available. Also, when dimming is available, for many tasks occupants often prefer light levels that are less than 100%.

Evaluation of Lighting for DR

The great potential for widespread use of lighting for DR will only be realized if more lighting systems are installed or upgraded to have the following features:

1. Centralized controls.
2. Zoning that allows light levels to be reduced with some degree of resolution that is minimally disruptive to building occupants.
3. Flexibility for various end-use scenarios.

SUMMARY AND FUTURE DIRECTIONS

This paper has presented a review of DR control strategies in commercial buildings based on a combination of results from field studies in 30 buildings over a three year period. The field studies have shown that there is a significant opportunity to enable DR capabilities in many existing buildings using existing EMCS and lighting controls. Further research is needed to understand the prevalence of controls in existing buildings to support a broad based deployment of these strategies. Newer, more advanced controls provide greater capability than older systems. Future work in this project will explore the applicability of these strategies to various building types, sizes, and climates.

ACKNOWLEDGMENTS

The authors are grateful for the extensive support from numerous individuals who assisted in this project. Many thanks to the engineers and staff at each building site. Special thanks to Ron Hofmann for his conceptualization of this project and ongoing technical support. Thanks also to Laurie ten Hope, Mark Rawson, and Dave Michel at the California Energy Commission. Thanks also to the Pacific Gas and Electric Company who funded the Automated CPP research. This work described in this report was coordinated by the Demand Response Research Center and funded by the California Energy Commission, Public Interest Energy Research Program, under work for others contract No. 150-99-003, Am #1 and by the U.S. Department of Energy under contract No. DE-AC03-76SF00098.

REFERENCES

1. Akashi, Y.; Neches, J. Detectability and acceptability of illuminance reduction for load shedding, IESNA Annual Conference Proceedings, August 3–6, 2003.
2. Goldman, C.; Hopper, N.; Sezgen, O. et al.; *Does Real-Time Pricing Deliver Demand Response? A Case Study of Niagara Mohawk's Large Customer RTP Tariff LBNL-54974*, 2004.
3. Kiliccote, S.; Piette, M.A. Control technologies and strategies characterizing demand response and energy efficiency. In *Proceedings of the Fifth Annual International Conference on Enhanced Building Operations*, Pittsburgh, PA, October 2005; LBNL #58179.
4. Piette, M.A.; Sezgen, O.; Watson, D.; Motegi, N.; Shockman, C. *Development and evaluation of fully automated demand response in large facilities*, CEC-500-2005-013. LBNL-55085. January 2005a. Available at http://drrc.lbl.gov/drrc-pubs1.html
5. Piette, M.A.; Watson, D.; Motegi, N.; Bourassa, N.; Shockman, C. Findings from the 2004 Fully Automated Demand Response Tests in Large Facilities, CEC-500-03-026. LBNL-58178, September 2005b. Available at http://drrc.lbl.gov/drrc-pubs1.html
6. Xu, P.; Haves, P. Case study of demand shifting with thermal mass in two large commercial buildings. ASHRAE Trans. **2005**, *LBNL-58649*.

BIBLIOGRAPHY

1. Kiliccote, S.; Piette, M.A.; Hansen, D. Advanced control and communication technologies for energy efficiency and demand response. In *Proceedings of Second Carnegie Mellon Conference in Electric Power Systems: Monitoring, Sensing, Software and Its Valuation for the Changing Electric Power Industry*, Pittsburgh, PA, January 2006; LBNL #59337.
2. Piette, M.A.; Watson, D.; Motegi, N.; Kiliccote, S.; Xu, P. *Automated critical peak pricing field tests: program description and results*, Report LBNL-59351, January 2006.
3. Xu, P.; Haves, P.; Piette, M.A.; Braun, J. Peak demand reduction from pre-cooling with zone temperature reset of HVAC in an office. In *Proceedings of ACEEE Summer Study on Energy Efficiency in Buildings*, Pacific Grove, CA, 2004; LBNL-55800.

Demand Response: Load Response Resources and Programs

Larry B. Barrett
Barrett Consulting Associates, Colorado Springs, Colorado, U.S.A.

Abstract
"Demand response" is a relatively new term to the electric utility industry for an old concept called peak load management. Demand response has gained currency, as the historically cumbersome peak load management programs have been transformed by real-time monitoring, digital controls, and robust communications. The costs of managing demand response resources have become more competitive relative to the costs of old load management techniques such as central station power plants and electric transmission system upgrades. Accordingly, demand response has joined the energy management lexicon as a more refined and flexible alternative to the old term of peak load management.

DEFINING DEMAND RESPONSE

Demand response in electricity markets is defined as "...load response called for by others and price response managed by end-use customers."[1] The definition of demand response conveniently divides activities into two categories: load response and price response, also called economic demand response.

Load response occurs when end users react to requests for reducing electric demand. Examples of load response programs are interruptible programs, curtailable programs, and cycling programs.

Price response occurs when end users react to price signals. Examples of these economic demand response programs include time-of-use rates, real-time pricing, and critical peak pricing.

The scope of this entry is focused on load response programs within the broader category of demand response. Material is presented elsewhere in this publication on the economic demand response programs such as real-time pricing.

The essential difference between load response and price response is who initiates and who follows in the short-term. In load response, an energy supplier initiates the call for load management and the customer acts under the terms of some agreement. In price response, the energy supplier sets the rates and the customer is responsible for initiating usage limitation actions, if any.

BENEFITS OF DEMAND RESPONSE

The potential for demand response is significant. For example, the largest grid operator in the United States, known as the PJM, counted demand response resources as equivalent to 5% of its peak demand in its Mid-Atlantic and Great Lakes regions.[2]

The benefits of demand response are varied and numerous, particularly when considering the many parties such as power grid operators, local electric distribution companies, facility managers, and all the customers who would have to pay higher rates otherwise. The benefits include the following:

- *Power system reliability*—Local and regional electric grids achieve greater reliability while avoiding blackouts and voltage reductions in emergency situations when customers are able to reduce loads on the grid. It has been estimated that "Power interruptions and inadequate power quality already cause economic losses to the nation conservatively estimated at more than $100 billion a year."[3]
- *Market efficiency*—Costs of power production and distribution can vary dramatically during the course of a day, a week, and a year as some plants operate only for the hours of peak demand. Yet standard rates present constant cost signals to consumers rather than the actual fluctuations in costs. Demand response programs provide incentives for customers that are able to reduce and shift loads, which not only benefit the program participants but all customers. The result is a more efficient use of the power generation, transmission, and distribution systems.
- *Bill savings*—Energy savings achieved during demand response periods typically translate to bill savings and are reinforced when load reductions are further rewarded with incentive payments. Electric bill savings

Keywords: Demand response; Peak load management; Load response; Curtailable load programs; Building automation; Air conditioner cycling; Ancillary services.

could reach about $7.5 billion per year in the United States according to a study for the Federal Energy Regulatory Commission (FERC).[4]
- *Cost reduction*—While similar to bill savings, cost reduction is more of a long-term benefit. Demand response reduces the long run costs of new power plants, defers new and expanded high voltage transmission systems, and mitigates the overloading of low voltage distribution systems.
- *Environmental quality*—To the extent that peaking power plants are old and inefficient, their reduced use can improve the environment. Even where peaking power plants are relatively clean in terms of emissions, demand response that results in a net reduction in energy use can improve environmental quality.
- *Customer service*—Demand response provides something that customers want—a choice. With demand response programs, customers that are able and willing to adjust their electricity usage for a few hours have another way to manage energy besides simply using less.
- *Risk management*—Providers of retail electricity must cover the risks of price volatility in wholesale markets if retail prices are less volatile. Demand response programs help cover those risks through greater energy availability, reliability, modularity, and dispensability.
- *Market power mitigation*—Market power refers to concentrating the central generation capacity into a few organizations. Demand response programs with hundreds, and indeed thousands, of owners of distributed assets can be called upon to help mitigate market power.
- *Complements to energy efficiency*—Demand response resources are discontinuous or occasional since they are only called upon for a few hours at a time. Energy efficiency actions are usually more permanent or continuous. Even where energy efficiency investments are made, there is opportunity for demand response participation.

LOAD RESPONSE PROGRAMS

Load response programs include those where the customer is responding to requests for short-term peak load reduction. They are sometimes referred to as "reliability-driven" programs in contrast to "market-based pricing" programs.[4]

Load response programs may be divided into two classes. The first is where virtually the entire facility or operation is interrupted. The second is where part of the facility reduces demand on the power grid so that the load is curtailed.

Interruptible Load Response Programs

Interruptible load response programs operate, as the name suggests, where the customer's entire facility or operations must be interrupted or shut down. A few circuits may be exempted to support lighting, HVAC, communications, and computer services in the administrative portions of the facilities. However, the major proportion of the electrical load is made up of much larger uses of power (e.g., production lines) which may be reduced through power interruptions.

A feature of interruptible programs is the ability of the utility to control the power flow. Thus, a feeder may be opened to prevent power to flow to the participating facility.

Another feature is that participation is voluntary and mandatory at the same time. Customers are free to join interruptible programs; however, once in the program, they must agree to power interruptions.

The incentive has historically been quite attractive to certain types of customers. Typically customers are rewarded with a lower rate that can prove to be a significant discount over the course of a year. In exchange, the customer agrees to interruptions. In most jurisdictions, the customer rarely has power interrupted. Thus, over decades of operation, these customers took advantage of the rate discount without any particular cost or inconvenience. And utilities offered the rate more as an inducement for economic development purposes to lure new facilities to their service territory.

Failure to perform or comply can be expensive. Penalties may be applied to facilities found not to be interrupting. These penalties can be quite substantial and may in fact negate the savings from the year-round rate reductions. Among other features, warning times of no more than 30 min may be imposed before interruption is required.

Participants in interruptible load response programs are typically industrial customers. One reason is that the utilities prefer larger operations with significant load reduction potential. A second is that industrial facilities are more likely to be subject to competing economic development offers, compared to commercial facilities such as office buildings or hospitals.

Load Curtailment Programs

These programs allow facility owners to curtail parts of their operations rather than entire facilities. Another feature of load curtailment programs is that smaller facilities may be eligible since lower thresholds of load reduction potential may still qualify. For example, many program participants must be able to provide only 100 kW of load reduction to qualify.

Another attraction of load curtailment programs is the voluntary nature of participation. Not only is selection into the program voluntary, each curtailment event may be voluntary. That is, when the utility calls for a load reduction, the customer may choose to curtail and get paid for the reduction or to continue operations and pay the price per the agreement. The customer may not be mandated to reduce their load, but failure to meet target load reductions can result in penalties. However, the penalties are not usually as severe as with interruptible load response programs.

Another feature is advance warning. Some programs may provide a 24-h notice before the curtailment event. Others offer a 2-h notice and others still a 30 min notice. Payments to customers for load curtailment may increase as the length of advance notice decreases.

Total power reduction may take place for certain facilities, such as those with standby generators sufficient to carry an entire facility or operation. While load curtailment implies a partial as opposed to total load reduction in most programs, customers may elect to provide the maximum load reduction possible by disconnecting from the grid either figuratively or in fact. In this case, the customer's load is generally met by standby generators designed to carry the entire facility.

Demand Buyback Programs—Pay for Performance

One variation on the operation of load curtailment programs is the demand buyback program. Also known as pay for performance, customers curtail loads in a two-step process.

First, the customer is notified by the energy supplier that a curtailment event is likely and bids will be accepted for peak load reduction. The customer decides whether to exercise the option of offering a certain amount of load reduction. The customer may also be required to suggest the price it wants to be paid in order to participate.

Then, the energy supplier has the option to accept the offer from the customer. Once accepted, the customer is typically obliged to meet the load reduction target in exchange for the customer's requested incentive.

The notice that a customer may receive before the possible curtailment event may vary from an hour to a couple of days. Once the customer bids and the energy supplier accepts the bid, the transaction should go through. If the customer does not shed load, a penalty may be imposed by the utility. The size of the penalty may be determined by various factors: the penalty may be based on just the load reduction that failed to materialize, or it could be based on the total load reduction promised under the buyback arrangement.

The amount of the buyback incentive may vary from event to event. Similarly the penalty may vary by event. The incentive may be based on kilowatt-hour reductions, kilowatt reductions or some other measure of performance. The incentive amount plus the performance achieved should determine the payment for customer participation.

COMMERCIAL BUILDINGS AND DEMAND RESPONSE SOLUTIONS

There are many ways to operate facilities for peak demand management.[5] Many of these solutions not only reduce peak demand but also save energy. Of course, any combination of solutions will vary by geography as well as from facility to facility and industry to industry.

Before participating in a demand response program, it is helpful for a facility to conduct an audit of the potential for peak load reduction. The utility that sponsors the demand response program may even provide an audit at no charge to inventory the facilities and equipment for their potential load reduction. An added benefit may be that in addition to obtaining a report on its peak load management opportunities, the audit may also suggest how the customer could increase general savings from energy efficiency upgrades.

The asset management options or resource solutions for demand response may be divided into practices requiring little investment and measures defined by significant investments. The asset options may also be divided according to energy end-uses as presented below.

Lighting represents a significant opportunity for load reduction. During peak periods, turn off unnecessary lights, including storerooms, mechanical rooms, wall washers, and spotlights. For retail facilities, this includes display lights on low-value or on-sale merchandise. In office settings, turn off lobby lights and a portion of hallway lights. In all cases, its advised to turn off exterior lights that happen to be on during the day. If the facility is wired for demand reduction potential, turn off selected lights on circuits with separate controls. For example, some fixtures allow users to turn off half the lights in a luminaire or turn off some fixtures while others continue to operate. For example, some circuits allow one row of fixtures to be turned off, while the adjacent row operates. Other lighting reduction options for peak periods include turning off lighting next to windows. Alternatively, dim lights where turning off lights is not feasible.

Air conditioning represents another large opportunity for reducing electricity usage. Options to reduce peak demand include increasing temperature set points on air conditioning thermostats and relying on outside air for cooling using economizers when weather conditions permit. With advance notice of a few hours or a day before an event, there is time to pre-cool space below normal temperatures prior to peak load conditions. Also, rotate the operation of chillers and packaged rooftop units and turn off condensing units while maintaining fan operations to continue air movement. If there are two-speed compressors on rooftop units, use the lower speed. Institute soft start procedures for multiple air conditioning units and let them coast during peak periods. Thermal energy storage is an air conditioning option that may be worth the investment, particularly in new facilities. This allows air conditioning to be supplied from chilled water or ice stored in tanks during peak hours. All of the air conditioning load can be moved off-peak with large thermal energy storage systems. Even partial systems allow substantial loads to be moved off peak.

Heating systems offer some potential as well for alleviating peaking electric systems in the winter. Options include reducing temperature set points on heating systems and pre-heating space above normal temperatures prior to peak load conditions. Electric thermal energy storage systems allow for off-peak charging at night with heat provided during peak daytime hours.

Ventilation options include installing carbon dioxide sensors that allow air intake to be reduced during peak hours, when levels of indoor air quality are acceptable. Reduced ventilation with outdoor air means less air conditioning load and reduced use of space conditioning systems. Separate carbon monoxide sensors for garages may prevent the operation of supply and exhaust ventilation systems that operate continuously, even though traffic patterns may only warrant operation for a few hours each day. Where permitted by codes, the carbon monoxide sensors can be tied to energy management systems and significantly reduce fan operation during peak load hours with little traffic.

Finally, building automation systems allow peak load strategies to be introduced reliably and consistently for the aforementioned air conditioning, heating, ventilation, and lighting systems. Automated controls also allow for the consolidation of multiple sites on a real-time basis to enable a single point of operation.

Options for commercial refrigeration including turning off some units for a few hours, postponing defrost cycles, staging operations to gain load diversity, and turning off electric strip heaters designed to remove moisture from glass covers. Water heaters are excellent candidates for peak load management, where storage capacity allows a facility to cycle units off for several hours.

Any facility with standby generators for emergency operation may be a candidate for operation during economic peak load management events. If cost-effective for such economic load management programs, facility owners may want to make upgrades to operate fully powered by the generators while in parallel to the electric grid. Another generic strategy for peak load reduction is to take some of the elevators and down escalators out of service for a few hours each day.

INDUSTRIAL DEMAND RESPONSE SOLUTIONS

Industrial facilities also present numerous opportunities to reduce and shift loads for several hours at a time. Some facilities can act within seconds of notification and most within the half hour. However, many facilities prefer advance notice of 2 h or a full day. The greater the advance notice the more consumers can do to reduce electric loads.

Prime candidates for demand response are facilities with storage capabilities in terms of their production materials and product shipping inventories. Air separation plants that produce oxygen, nitrogen, and other gases are examples where products can be stored and inventories managed. These plants operate automatically, and often remotely, allowing few complications with labor and other management considerations to accommodate requests for peak load reduction.

Water storage operations provide other options to reduce peak load. Manufacturing facilities with significant demands for water can coast through load curtailments by relying on stored water. Or, storage tanks may be filled at a slower rate during curtailment events, allowing some net drawdown of water supplies, and then the tanks can be replenished to capacity during off-peak conditions.

Public water delivery systems present a complex mix of pumping, gravity, and storage. With proper configuration, the pumping systems can be shut down or slowed to reduce peak loads. In wastewater treatment plants with multiple aerators, several options are available. One is to turn off the aerators for a few hours. Another is to cycle them so not all units are running at the same time. Still another option is to slow the aerators operation with variable frequency drives on the motors.

Refrigeration systems can assist in reducing peak loads with sufficient notice. A day's advance notice allows operators to pre-cool refrigeration cases to a lower than normal level and then coast through the curtailment period. Or, refrigeration temperatures may be permitted to migrate upward by a few degrees during curtailments and be restored afterwards.

Batch processing operations also lend themselves to curtailment. Once a batch is completed, the process may be halted and the load reduced until the curtailment event is over. In some processes, batches may be interrupted and then restarted without complications.

Continuous processing operations are also candidates for peak load management if they can be slowed down. Plants with variable frequency drives on pumps and motors can slow their operations for a few hours and still maintain some production.

Compressors are a large consumer of energy and present many options to save energy and peak demand. Plants with sufficient compressed air storage may be able to turn off the compressors for short periods of time. Plants with multiple compressors may be able to rotate their operation and reduce peak demand from simultaneous operation.[6]

Industrial processes with standby generators may not need to change their production operations at all. Generators may pick up significant portions of the internal plant load and free up capacity on the power grid for other purposes.

RESIDENTIAL DEMAND RESPONSE SOLUTIONS

Residential programs for demand response are in place with tens of thousands, and in some utilities, hundreds of thousands of homes participating. Three resources may be targeted for peak load reduction: central air conditioners, electric water heaters, and swimming pool pumps.

The large majority of demand response reduction comes from central air conditioners. While an average load reduction of 1 kW per residential air conditioner during peak summer periods may seem small, when aggregated over numerous households the available load response resource is significant.

Improved technologies are driving the growth of demand response with residential appliances such as air conditioners. In the 1980s, control switches were typically installed on outdoor condensing units of air conditioning systems. The switches were conspicuous and subject to damage. Now, special controls can be installed internally and linked with smart thermostats to cycle climate control systems. Also, new systems allow customers to override the controls on an exception basis and therefore increase load reduction participation rates.

Another improvement in load cycling has been with communications systems. Legacy products received signals to interrupt operations without an ability to acknowledge the request. Now, two-way communications are possible, allowing for more sophisticated control strategies and incentive plans to encourage wider participation.

Another improvement is in the measurement and verification of electrical usage. With better communications, it is possible to determine if systems are being cycled and for how long. Also, improved database systems and management software foster more robust and dependable measurement and verification protocols.

New technologies also allow alternative programs including different cycling strategies that can be varied according to a participant's energy settings and energy use. When controls are sensitive to natural duty cycles there is greater reliability in achieving optimum reductions. Also, there are more alternatives for different types of incentive plans such as those based on the number of cycling events, override frequencies, temperature adjustments, and the temporary shutdown of the compressor system for several hours under 100% cycling.

Load management strategies range from 25% cycling, where units are cycled off only seven and a half minutes per 30-min period, to 100% cycling where the compressor may be turned off for the duration of the event. Cycling can also be exercised from less than 2 h to over 6 h per day. A risk of cycling too short in hours and too low in percentage of time off is that the natural duty cycle is being replicated with little appreciable savings in peak demand.

An important advantage of load cycling is the ability to target certain neighborhoods and areas within a utility's service territory. This can be particularly beneficial where specific parts of the electrical system are in danger of being overloaded from rapid growth or aging infrastructure.

METERING, COMMUNICATIONS, AND CONTROL

Metering

Extra metering is typically required for commercial and industrial facilities. The most common application is with interval meters. These meters record usage at least hourly and for many applications, on 15 min intervals. As demand response expands into more time-sensitive operations, 5-min intervals may become more common. Interval meters are also called smart, automated, or advanced meters.

Interval meters record electricity consumption in kilowatt-hours, while demand is recorded in kilowatts and time of use. Some meters calculate the maximum or peak demand over a specified time period such as a day, week, or month. When connected to a personal computer with a modem, interval meters offer the utility of a rich database to analyze energy use levels and patterns.

Interval meters are essential to estimating the load reductions achieved by demand response programs. Load profiles may be developed for each time interval on the designated curtailment days. The load profiles can then be compared with normal days to calculate load reductions on which to base payments.

Communications

Communicating and advance notice of curtailment periods are becoming more sophisticated. In the early days of load response, communications were made manually through a telephone call or by an automated notification arrangement over dedicated telephone lines. Today, there is a trend toward wireless communications where signals are quite economical in short bursts of airtime.

Often multiple forms of communication are employed to insure that facility managers and operators receive notice of planned curtailments. Thus, a request for curtailment may be communicated by some combination of FAX, telephone, email, and pager.

Two-way communications are another feature of many demand response programs. The end-use facility is configured to receive a signal and request for curtailment. In addition, communications are sent back acknowledging the receipt of the request and, just as important, the real-time load levels. Such communications based on the interval meter recordings and stored on a web-enabled personal computer may be sent to the utility, the grid operation, or an additional third party that monitors performance.

The availability of load performance is an attractive feature of demand response programs. This performance information may be of value to the corporate energy engineer, the store manager, the plant superintendent, and others in the customer organization. A facility's finance and accounting departments may find the information of value, and, if participation may affect shipping schedules and sales, even those in marketing and sales.

Control

With data comes information, and with information comes control. Data from interval meters may be integrated with building management systems and energy management systems. These systems can control lighting, thermostat settings, equipment cycling, and other operations. Certain systems may be programmed to recognize a curtailment request and automatically shift into a different operating protocol to accommodate the event.

Residential applications are amenable to more sophisticated controls for load response. Programmable thermostats can be enhanced to accept signals for curtailment or cycling. Some thermostat models may also be configured to adjust the set-point by some specified amount, such as four degrees, with the effect of reducing air conditioning usage during the curtailment event.

Metering and communications are important in another way, namely by applying credit for a customer's load response performance. Credit may be issued in the form of a check or electric bill reduction. Done manually, settlement or payment for load response can take months where multiple parties are involved in requesting and managing the load reductions. Advances in metering and communications help accelerate the settlement process.

ANCILLARY SERVICES

Ancillary services refer to such functions as regulation, spinning reserve, supplemental reserve, and replacement reserve on the bulk power grid.[7] Regulation services operate in fractions of a second to maintain the balance between power generation and customer loads while still maintaining voltages within required ranges.

Spinning reserve services are called upon in the event of a generator outage or transmission interruption. Spinning reserves need to be synchronized to the grid and meet capacity within 15 min. Supplemental reserves are similar to spinning reserves but do not need to be synchronized immediately, as long as they can reach capacity within 15 min. Replacement reserves are similar to supplemental reserves but have a 30–60 min window to reach capacity. The supplemental and replacement reserves may be called upon to replace spinning reserves, allowing the spinning reserves to stand down and be ready for another contingency.

Traditionally, ancillary services have been provided by central generation plants. Plants are kept in spinning reserve, but not under load, in the event that there is a system failure on the grid and additional load is needed suddenly. The reserves must be able to supply capacity and regulate power within minutes.

However, many demand response resources perform within the short time deadlines needed to supply spinning reserves. Some demand response resources can respond within seconds. For example, standby generators can respond within 10–15 s.

As another example, air conditioner cycling programs for residential and small commercial buildings can be signaled within a minute. If called upon for spinning reserves, cycling programs are even more advantageous, since they can easily operate for the minimum of 30 min required by ancillary services.

When air conditioner loads are interrupted to meet needs for spinning reserves, the available capacity can be triple the amount of capacity that is available compared to cycling.[8] The reason is that cycling disrupts, but does not necessarily eliminate, the natural diversity of air conditioner operations over each 30 min cycle. Interrupting loads for a spinning reserve can prevent even this part-time operation over 30 min.

The 30 min time period allows central station generation resources to come on line and make up for the capacity lost in spinning reserves. Then, the cycling schedule can be discontinued and the resources once again are available in standby to provide spinning reserves. Multiple parties are satisfied since power plants can operate for hours to meet system needs, while if cycling programs operate for too many hours, the customers may start to experience more discomfort and inconvenience than they are not likely to notice in a 30 min event.

Another benefit is that the aggregation of many demand response resources, while small individually, makes it highly probable that the assigned ancillary services will be achieved in total. This means that demand response resources are more reliable, when compared to a central generation resource, where the failure of one turbine could cause large losses of spinning reserves.

Compared to central station power plants, demand response resources can help "level the playing field" by providing an equivalent capacity, with high reliability, in an economic manner.

ECONOMICS OF DEMAND RESPONSE

By balancing financial benefits with costs, the primary motivation for end users to manage peak loads is economic. Whether the customer is residential, commercial, or industrial, incentives must be sufficient to cover whatever costs, inconvenience, and perhaps discomfort that may arise with load reductions. Furthermore, electric bill savings may be realized through reduced usage.

Another motivation to participate in demand response programs is for community reasons such as helping to improve the environment. A third motivation is to gain information about energy use and operating conditions associated with the more detailed monitoring protocols attendant to demand response programs.

Incentives may be paid from multiple parties. Grid operations are willing to pay for demand response resources at many times the normal electric rates. For example, in 2004, the New York Independent System Operator and the Independent System Operator of New England offered $500 per megawatt-hour or 50 cents per kilowatt-hour for load response resources during peak hours on the grid.

Other parties that may pay for demand response resources are traditional utilities with operations that are vertically integrated from the power plant to the meter, or utilities with only distribution businesses attempting to reduce high demand charges. There can be third parties such as curtailment service providers that make a business of aggregating customer loads to bid into demand response programs.

A key consideration in the economics of demand response is the cost of metering, communication, and control. In some cases, the cost may be underwritten by utilities, grid operators, and even government agencies with energy responsibilities.

There may be other costs to consider, such as modifications to production equipment to increase capacity or improve controls. Of course, labor costs associated with interrupting operations are a factor, particularly if overtime wages must be paid to accommodate the shifts in production schedules.

In general, it is advantageous to anticipate demand response opportunities when building new facilities or upgrading load capacity and operations. As demand response programs continue to expand in size and scope, the relationship of benefits and costs for participation should continue to improve.

CONCLUSION

Load response programs, as one of the principal forms of demand response, are an important way to achieve peak load management in electricity markets. There are many benefits including increased reliability of the power grid, higher efficiency of energy markets, and increased savings in consumer energy bills. Load response can be achieved in commercial, industrial, and residential buildings from numerous assets including air conditioning, water heating, and refrigeration. Standby generation equipment is a common asset deployed for load response in commercial and industrial facilities.

Load response programs are enabled by improved technologies in metering, communications, and control. The economics of load response favor utility systems with high costs, such as those associated with capacity shortages, operating inefficient units at peak loads, and transmission bottlenecks. Financial rewards from load response are expanding beyond the traditional markets for generation and transmission capacity shortages that may be anticipated hours or days in advance. Load response is gaining acceptance in ancillary services markets to provide stability on the power grid with only a few seconds or minutes notice.

REFERENCES

1. Peak Load Management Alliance. Demand Response: Principles of Regulatory Guidance. February 2002, 1.
2. Barrett, L.B. Demand Response Resources Ready and Able, *Barrett Consulting Associates Report*, January, 2005.
3. EPRI Technology action plan addresses western power crisis. EPRI J. **2001**, 5.
4. U.S. Government Accounting Office. Electricity Markets: Consumers Could Benefit from Demand Programs, but Challenges Remain. August 2004, 4.
5. California Energy Commission. Enhanced Automation: Technical Options Guidebook. 2002.
6. U.S. Department of Energy, Office of Industrial Technologies. Best Practices: Compressed Air System Renovation Project Improves Product at a Food Processing Facility. June 2001.
7. Hirst, Eric. Price-Response Demand as Reliability Resources. Oak Ridge, Tennessee. April 2002, 7.
8. Kirby, B.J. Spinning Reserve from Response Loads. Oak Ridge National Laboratory, ORNL/TM 0-19, March 2003, 35.

Demand-Side Management Programs

Clark W. Gellings
Electric Power Research Institute (EPRI), Palo Alto, California, U.S.A.

Abstract
Demand-side management is the active planning and implementation of programs that will change consumers' use of electricity. These programs may encourage the adoption of more efficient appliances, the use of new technologies, and the time these and other devices are used.

INTRODUCTION

Demand-side management programs result from the planning and implementation of those activities designed to influence consumer use of energy in ways that will produce desired changes in the time pattern and magnitude of energy demand. Programs and initiatives falling under the umbrella of demand-side management include load management, new uses, strategic conservation, electrification, and adjustments in the market share of energy-consuming devices and appliances.

Demand-side management includes only those activities that involve a deliberate intervention in the marketplace. This intervention has often been affected by electric or natural gas utilities. Examples of demand-side management programs include those that encourage consumers to install energy-efficient refrigerators, through either incentives or advertising.

Demand-side management extends beyond conservation and load management to include programs designed specifically to modify energy use in all periods. Thus, demand-side management alternatives warrant consideration by entities such as energy and energy service suppliers with ambitious construction programs, those with high reserve margins, and those facing high marginal costs.

Key definitions for demand-side management are given below.

WHY CONSIDER DEMAND-SIDE MANAGEMENT

Since the early 1970s, economic, political, social, technological, and resource supply factors have combined to change the electricity sector's outlook for the future. Many utilities face significant fluctuations in demand and energy growth rates, declining financial performance, and political or regulatory and consumer concern about rising prices. Although demand-side management is not a cure-all for these difficulties, it does provide additional alternatives.

For utilities facing strong load growth, load management and strategic conservation can provide an effective means to reduce or postpone acquisition of power contracts or construction of new generating facilities; for others, load growth can improve the utility load characteristics and optimize asset utilization. Changing the purchase pattern or the amount of consumer demand on an energy system can reduce operating costs.

Implementing demand-side management can lead to greater flexibility in facing rapid change in today's business environment. Although demand-side management will not solve all the problems facing the energy industry and its stakeholders, it does provide additional alternatives for meeting the challenges of the future (Fig. 1).

SELECTING THE RIGHT ALTERNATIVE

Demand-side management encompasses planning, evaluation, implementation, and monitoring of activities selected from among a wide variety of programmatic and technical alternatives. Due to the large number of alternatives, assessing which alternative is best suited is not a trivial task. The choice is complicated by the fact that the attractiveness of alternatives is influenced strongly by specific local and regional factors, such as industry structure, generating mix, expected load growth, capacity expansion plans, load factor, load shapes for average and extreme days, regulatory climate, and reserve margins. Therefore, it is often inappropriate to transfer these varying specific factors from one area to another without appropriate adjustments. In addition, the success of any alternative depends on the specific combination of promotional activities selected to aid in promoting customer acceptance.

Keywords: Demand-side management; Energy efficiency; Conservation; Load management; Consumer; Load shape; Load control; Marketing; End use.

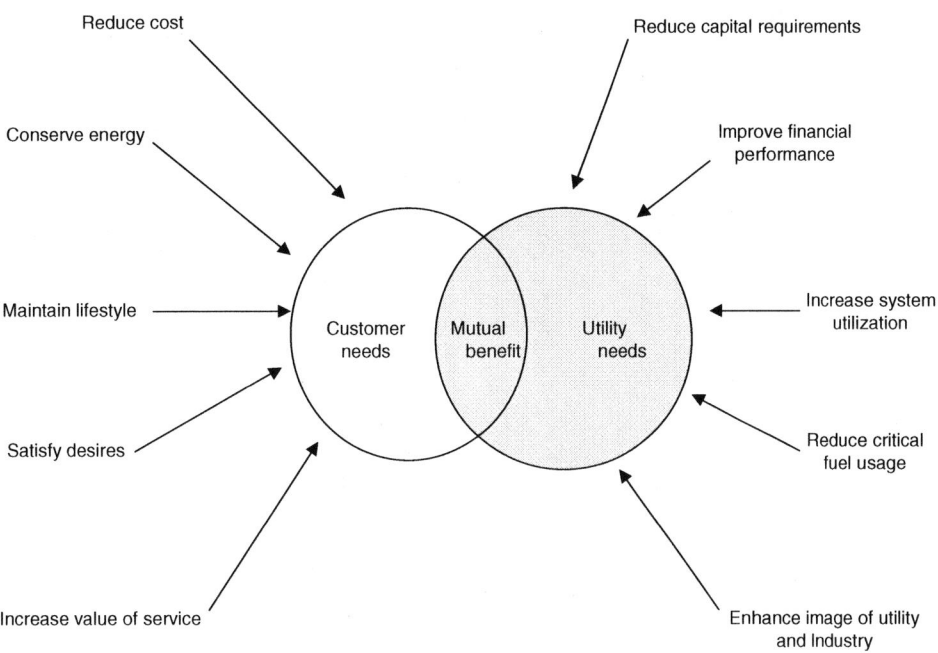

Fig. 1 Goals of demand-side management programs.

LOAD-SHAPE CHANGES

Although there is an infinite combination of load-shape changing possibilities, six can illustrate the range of possibilities: peak clipping, valley filling, load shifting, strategic conservation, strategic load growth, and flexible load shape. These six are not mutually exclusive and may frequently be employed in combinations.

The demand-side management planning approach does provide policymakers with a whole new set of alternatives with which to meet energy needs. The concept that the load shape is not fixed but can be altered deliberately opens a new dimension in planning and operation.

DEMAND-SIDE MANAGEMENT PRACTICE

Demand-side management can help achieve a broad range of operational objectives merely by changing the system's load shape. Numerous industries have found that changing the pattern of the demand for their product can be profitable. Telephone utilities, for example, have long offered reduced evening rates to shift demand and to encourage off-peak use. Airlines offer night coach fares, and movie theaters offer reduced matinee prices— examples of deliberate attempts to change the demand pattern for a product or service.

In an electric utility, the physical plant is designed to serve the projected demand for electricity in the least-cost manner, given a specified level of desired quality and reliability. If the load shape is not fixed but may be altered, the cost of serving the load can be reduced still further.

Cost reductions due to changes in the load shape arise primarily from three attributes:

- Reduction in the requirements for new assets or energy
- Higher utilization of facilities
- More efficient operation of facilities

System expansion can be delayed or eliminated and the use of critical resources reduced by significantly reducing energy and peak consumption through the use of demand-side management.

Higher utilization of existing and planned facilities can also be achieved through a program of load growth. Although such programs increase total costs due to higher fuel and other operating expenses, they reduce unit costs by spreading the fixed costs (debt service and dividends) over more units of energy sold.

Six generic load-shape objectives can be considered to be part of demand-side management (Fig. 2).

SELECTING DEMAND-SIDE MANAGEMENT ACTIVITIES

Although customers and suppliers act independently, resulting in a pattern of demand, the concept of demand-side management implies a supplier/customer relationship that produces mutually beneficial results. To achieve these mutual benefits, suppliers must carefully consider the manner in which the activity will affect the patterns and amount of demand (load shape). In addition, suppliers must assess the methods available for obtaining customer

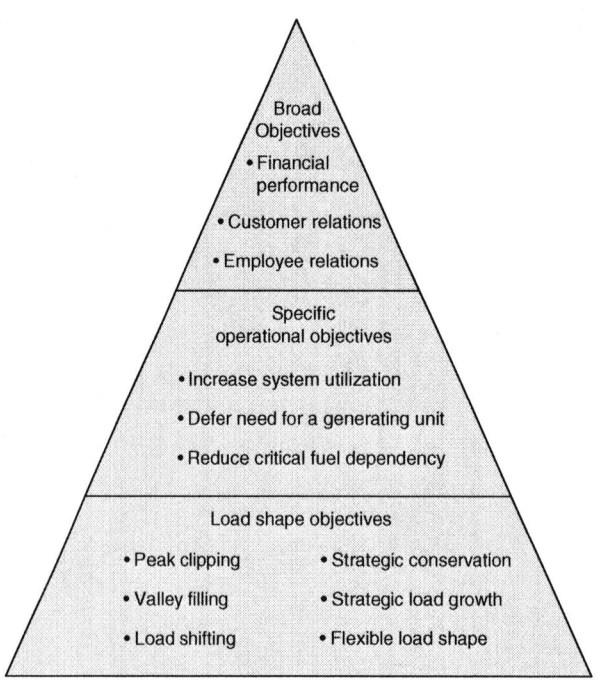

Fig. 2 Hierarchy of energy planning objectives.

participation, and the likely magnitudes of costs and benefits to both supplier and customer prior to attempting implementation.

Because there are so many demand-side management alternatives, the process of identifying potential candidates can best be carried out by considering key aspects of the alternatives in an orderly fashion. Demand-side management activities can be categorized in a two-step process:

- Step 1: load-shape objectives
- Step 2: end-use technology activities

LOAD-SHAPE OBJECTIVES

The first step in identifying demand-side alternatives is the selection of a load-shape objective to ensure that the desired result is consistent with goals and constraints.

END USE

When the load-shape objective has been established, it is necessary to find end uses that can achieve it. This identification process involves three dimensions. The first dimension involves identifying the appropriate end uses whose consumption characteristics generally match the requirements of the load-shape objectives. In general, each end use exhibits predictable demand or load patterns.

Nine major residential end-uses of electricity are examples of having the most potential for electric demand-side management. They are space heating, space cooling, water heating, lighting, refrigeration, cooking, laundry, swimming pools, and miscellaneous other uses. Each of these end uses provides a different set of opportunities to meet electric load-shape modification objectives. Some of the end uses can serve successfully as the focus of programs to meet any of the load-shape objectives; others realistically can be useful for meeting only one or two objectives. In general, space heating, space cooling, and water heating are the residential end uses with the greatest potential applicability for achieving objectives. These end uses tend to be among the most energy intensive and among the most adaptable.

TECHNOLOGY ALTERNATIVES

The second dimension of demand-side management alternatives involves choosing appropriate technology alternatives for each end use. This process should consider the suitability of the technology.

Residential demand-side management technologies can be grouped into four general categories:

- Building-envelope technologies
- Efficient equipment and appliances
- Thermal storage technologies
- Energy and demand control technologies.

These four main categories cover most of the available residential options. Many of the individual options can be considered to be components of an overall program.

HOW TO SELECT ALTERNATIVES

Selection of the most appropriate demand-side management alternatives is the most crucial question. The relative attractiveness of alternatives depends on specific characteristics, such as load shape, summer and winter peaks, generation or product system mix, customer mix, and the projected rate of load growth.

MARKET IMPLEMENTATION METHODS

Among the most important dimensions in the characterization of demand-side management alternatives is the selection of the appropriate market implementation methods (Fig. 3).

Planners and policymakers can choose among a wide range of methods designed to influence customer adoption, which can be broadly classified in six categories:

- Customer education
- Direct customer contact
- Trade-ally cooperation
- Advertising and promotion
- Alternative pricing
- Direct incentives

Energy suppliers, utilities, and government entities have used many of these strategies successfully. Typically, multiple marketing methods are used to promote demand-side management programs. The selection of the individual market implementation method or mix of methods depends on a number of factors, including:

- Experience with similar programs
- Existing market penetration
- The receptivity of policymakers and regulatory authorities
- The estimated program benefits and costs to suppliers and customers
- Stage of buyer readiness
- Barriers to implementation

The objectives of deploying market implementation methods are to influence the marketplace and to change customer behavior. The key question is the selection of the market implementation method(s) to obtain the desired customer acceptance and response. Customer acceptance refers to customer willingness to participate in an implementation program, customer decisions to adopt the desired technology, and behavior change as encouraged by the supplier. Customer response is the actual load-shape change that results from customer action, combined with the characteristics of the devices.

Customer acceptance and responses are influenced by the demographic characteristics of the customer, income, knowledge, and awareness of the technologies and programs available, as well as attitudes and motivations. Customer acceptance and response are also influenced by

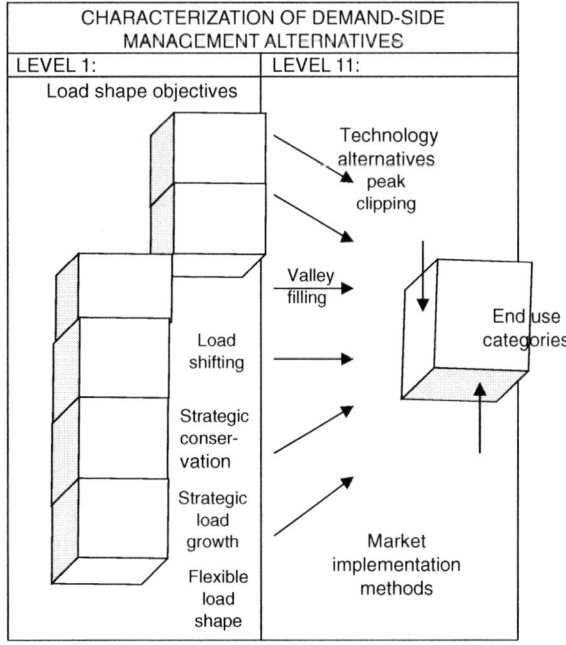

Fig. 3 Characterization of demand-side management alternatives.

other external factors, such as economic conditions, prices, technology, regulation, and tax credits.

CUSTOMER EDUCATION

Many energy suppliers and governments have relied on some form of customer education to promote general customer awareness of demand-side management programs. Brochures, bill inserts, information packets, clearinghouses, educational curricula, and direct mailings are widely used. Customer education is the most basic of the market implementation methods available and can be used to:

- Inform customers about products/services being offered and their benefits, and influence customer decisions to participate in a program
- Increase the perceived value of service to customers
- Inform customers of the eligibility requirements for program participation
- Increase customers' knowledge of factors influencing energy purchase decisions
- Provide customers other information of general interest
- Generally improve customer relations.

DIRECT CUSTOMER CONTACT

Direct customer contact techniques refer to face-to-face communication between the customer and an energy supplier to encourage greater customer acceptance of programs. Energy suppliers have for some time employed marketing and customer service representatives to provide advice on appliance choice and operation, sizing of heating/cooling systems, lighting design, and even home economics.

- Energy audits are particularly useful for identifying heating/air conditioning system improvements, building-envelope improvements, water heating improvements, and the applicability of renewable resource measures.
- Program services involve activities undertaken to support specific demand-side management measures, including heat pumps, weatherization, and renewable energy resources. Examples of such programs include equipment servicing and analyses of customer options.
- Storefronts are business areas where energy information is made available, and appliances and devices are displayed to citizens and consumers.
- Workshops and energy clinics are special sessions that may cover a variety of topics, including home energy conservation, third-party financing, energy-efficient appliances, and other demand-side technologies.
- Exhibits and displays are useful for large public showings, including conferences, fairs, and large showrooms.

TRADE-ALLY COOPERATION

Trade-ally cooperation can contribute significantly to the success of many demand-side management programs. A trade ally is defined as any organization that can influence the transactions between the supplier and its customers. Key trade-ally groups include home builders and contractors, local chapters of professional societies (e.g., the U.S. American Society of Heating, Refrigeration and Air Conditioning Engineers; the Illuminating Engineering Society of North America; and the Institute of Electrical and Electronic Engineers), trade associations (e.g., local plumbing and electrical contractor associations), and associations representing wholesalers and retailers of appliances and energy-consuming devices.

Depending on the type of trade-ally organization, a wide range of services is performed, including:

- Development of standards and procedures
- Technology transfer
- Training
- Certification
- Marketing/sales
- Installation, maintenance, and repair

In performing these diverse services, trade allies may significantly influence the consumer's technology choice. Trade allies can assist substantially in developing and implementing demand-side management programs.

ADVERTISING AND PROMOTION

Energy suppliers have used a variety of advertising and promotional techniques to influence customers. Advertising uses various media to communicate a message to consumers so as to persuade them. Advertising media applicable to demand-side management programs include radio, television, magazines, newspapers, outdoor advertising, and point-of-purchase advertising. Promotion usually includes activities to support advertising, such as press releases, displays, demonstrations, coupons, and contests/awards. Some prefer the use of newspapers, based on consumer research that found this medium to be the major source of customer awareness; others have found radio and television advertising to be more effective.

Advertising and promotion also have widespread applicability. A number of radio and television spots have been developed to promote demand-side management measures. Other promotional techniques used have been awards, energy-efficient logos, and residential home energy rating systems, to name a few.

ALTERNATIVE PRICING

Pricing of energy as a market-influencing factor generally performs three functions:

- Transfers to producers and consumers information regarding the implied value of products and services being provided
- Provides incentives to use the most efficient technologies
- Allocates supply and demand

Alternative pricing through innovative schemes can be an important implementation technique for utilities promoting demand-side management options. Rate incentives for encouraging specific patterns of utilization of electricity can often be combined with other strategies (e.g., direct incentives) to achieve electric utility demand-side management goals.

Various pricing structures are more suited to certain types of demand-side management options. For utilities, time-of-use rates may generally be offered or tied to specific technologies (e.g., storage heating and cooling, or off-peak water heating). They can be useful for thermal storage, energy and demand control, and some efficient equipment options.

DIRECT INCENTIVES

Direct incentives are used to increase market penetration of an option by reducing the net cash outlay required for equipment purchase. Incentives also reduce customer resistance to options without proven performance histories.

The individual categories of direct incentives include

- Cash grants
- Rebates
- Buyback programs
- Billing credits
- Low-interest or no-interest loans

CONCLUSION

Demand-side management offers key advantages to conventional alternatives of supply. Demand-side management is a viable and cost-effective resource that should be part of all electricity supply portfolio planning.

BIBLIOGRAPHY

1. *Customer's Attitudes and Customers' Response to Load Management, Electric Utility Rate Design Study*, Report No. EPRI SIA82-419-6; Published by the Electric Power Research Institute, December 1983.
2. Decision Focus, Inc. *Cost/Benefits Analysis of Demand-Side Planning Alternatives*, EPRI EURDS 94 (RP 1613); Published by Electric Power Research Institute, October 1983.
3. Decision Focus, Inc. *Demand-Side Planning Cost/Benefit Analysis*, Report No. EPRI RDS 94 (RP 1613); Published by the Electric Power Research Institute, November 1983.
4. Decision Focus, Inc. *Integrated Analysis of Load Shapes and Energy Storage*, Report No. EA-970 (RP 1108); Published by the Electric Power Research Institute, March 1979.
5. Demand-Side Management Vol. 1: *Overview of Key Issues*, EPRI EA/EM-3597, Vol. 1 Project 2381-4 Final Report.
6. Demand-Side Management Vol. 2: *Evaluation of Alternatives*, EPRI EA/EM-3597, Vol. 2 Project 2381-4 Final Report.
7. Demand-Side Management Vol. 3: *Technology Alternatives and Market Implementation Methods*, EPRI EA/EM-3597, Vol. 3 Project 2381-4 Final Report
8. Demand-Side Management Vol. 4: *Commercial Markets and Programs*, EPRI EA/EM-3597, Vol. 4 Project 2381-4 Final Report.
9. EPRI Project, RP 2547, *Consumer Selection of End-Use Devices and Systems*.
10. EPRI Reports prepared by Synergic Resources Corporation. Electric Utility Conservation Programs: *Assessment of Implementation Experience* (RP 2050-11) and 1983 *Survey of Utility End-Use Projects* (EPRI Report No. EM 3529).

Desiccant Dehumidification: Case Study

Michael K. West
Building Systems Scientists, Advantek Consulting, Inc., Melbourne, Florida, U.S.A.

Glenn C. Haynes
RLW Analytics, Middletown, Connecticut, U.S.A.

Abstract

Desiccant dehumidification is primarily a nonresidential end-use technology that can be important to certain commercial businesses such as restaurants, hotels, grocery stores, and hospitals; in public buildings such as courthouses, jails, and auditoriums; and in manufacturing sectors such as pharmaceuticals and microelectronics. This rigorous case study presents results of the field test and performance evaluation of a typical, commercially available, two-wheel gas-fired desiccant air conditioner. Field-measured performance is compared with the manufacturer's specifications, with predictions made using DOE-2 hourly modeling, with all-electric technologies, and with the theoretical limits of the technology. Comparisons were made between the manufacturer's published data, the manufacturer's site test data taken at the time of installation, the collected field data, the computer model, and the theoretical best-case performance. The desiccant unit as installed delivers less cooling and dehumidification capacity than the manufacturer's rating, and much less than it would if the equipment design were optimized and the installation were commissioned. While the measured energy efficiency at peak load conditions is better than the rated efficiency, the data clearly show this rating is not representative of long-term field performance.

BACKGROUND

A field test was initiated to demonstrate and evaluate natural gas desiccant technology in the commercial market segment as a means of controlling weather-sensitive kilowatt electric demand. The serving Florida investor-owned utility, in cooperation with the local gas company, randomly selected and then recruited a commercial customer. The customer installed a new gas desiccant dehumidification system as an alternate technology to the existing electric-DX overcooling and electric reheating system. Advantek Consulting, Inc. was tasked, as an independent third party, with collecting and analyzing field performance data in light of the manufacturer's published data and the results of computer modeling. The customer paid for purchase and installation.

FIELD TESTING

The dehumidification equipment, as well as key components of the building's heating, ventilation, and air-conditioning (HVAC) system, was fitted with a comprehensive instrumentation package to continuously monitor both overall system and sublevel component performance. The field monitoring system collects averaged 1-hour interval data for 35 data points. The most current set of data includes electric kilowatt-hour and natural gas cubic feet (CF) consumption as well as ambient, space, and system temperatures and humidities. The customer integrated the operation of the unit into the existing building management system and is responsible for all maintenance and repairs.

The collected data was screened and used in the calculation of secondary quantities such as the amount of dehumidification capacity delivered, the quantity of moisture removed from the air, and the energy efficiency of the equipment. These quantities were used to assess the performance of the unit as compared with the manufacturer's published performance data. The manufacturer's rated cooling capacity at the peak load condition is 248 MBH (MBH = 1000 Btuh = 0.083 tn); however, the average as-installed capacity was measured to be considerably lower at 155 MBH.

The manufacturer's rated efficiency at the peak load condition (93°F dry bulb, or 78°F wet bulb) is COP 0.73 [COP = (Btuh Capacity)/(Btuh Gas and Electric Input)]; the measured efficiency at this condition was COP 0.83. However, the average as-installed efficiency was measured to be considerably lower than the rated efficiency at COP 0.53. The cooling capacity at peak load was measured to be 19% less than the manufacturer's rating. The heat input at peak load was measured to be 12% less than rated. In comparison, the optimized efficiency of this type of equipment is much higher at COP 1.0–1.2 (Fig. 1).

As designed and installed, the gas dehumidification unit is not optimized nor does it represent the maximum

Keywords: Dehumidification; Humidity; Desiccant; Evaporative; Regeneration; Reactivation; Process; COP (Coefficient of Performance); Preconditioner.

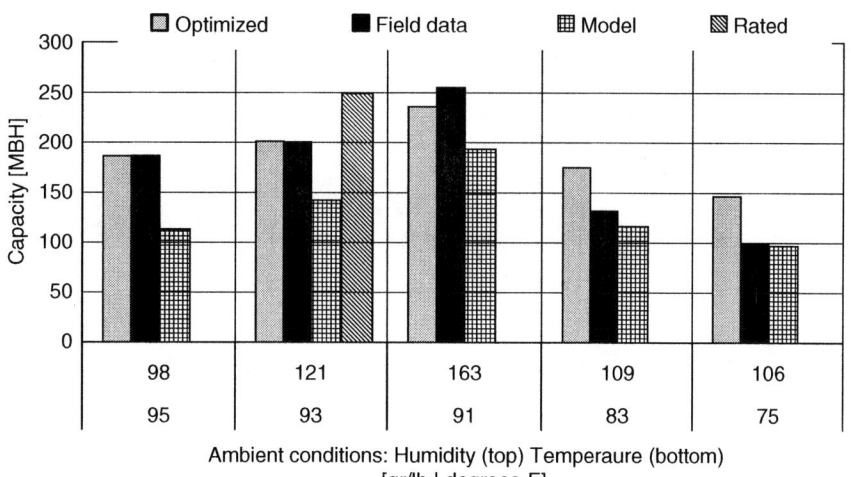

Fig. 1 Comparison of cooling capacity.

efficiency potential of desiccant equipment. Even so, it does (in our opinion) represent a "typical" commercial installation. The simple gas boiler control does not have the ability to vary heat output or gas consumption according to the need for dehumidification. The boiler is either full on or shut off, and the data clearly shows it unnecessarily operates full-on almost constantly. Our data indicates that less than 60% of the natural gas energy consumed by the unit is actually utilized. Likewise, the evaporative cooler is not nearly as effective as it should be. The analysis also indicates the possibility of moisture carry-over from the regeneration side of the evaporative cooler to the process side via the heat wheel.

Control of the unit is based simply on supply air temperature, and to a lesser degree, humidity. The data clearly shows that the control sequence does not take into account the cooling needs of the building; it aims merely to supply air at a fixed temperature regardless of whether additional mechanical cooling or reheating is necessary downstream.

COMPUTER MODELING

The most complete, representative, accurate, and reliable contiguous sets of data were used to develop, calibrate, and validate an hourly computer model. These sets included 55 days of hourly data from various periods of the project—a total of some 46,000 data points. Performance was evaluated using results from these sets of screened field data and a full-year set of computer model results as driven by the serving utility company's typical 30-year hourly weather data.

The hourly computer model consists of a set of submodels for each of the components of the system. These component submodels, such as the evaporative cooler and the desiccant wheel, are assembled together to

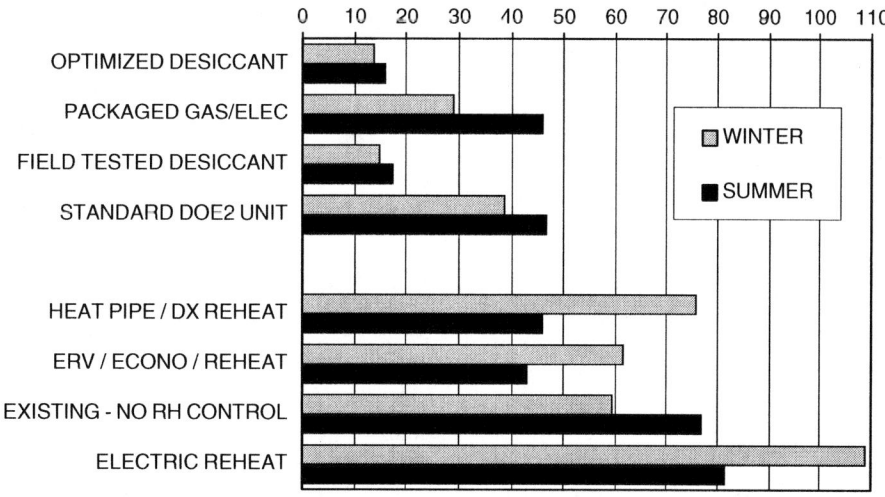

Fig. 2 Comparison of peak electric use (kW).

simulate the performance of the equipment as a whole. Each submodel was used to simulate the performance of a single component for each of the 8760 h in the typical weather year.

As a final check, the results of the model were compared against the standard DOE-2.1e hourly simulation software developed by the U.S. Department of Energy. A static comparison and validation was also performed at the outdoor temperature and humidity conditions published in the manufacturer's equipment performance specifications. Comparisons were made between the manufacturer's published data, the manufacturer's site test data taken at the time of installation, the collected field data, the computer model, and the theoretical best-case performance.

Two baseline options were developed to simulate comparable all-electric dehumidification equipment commonly used in the commercial sector. The baseline computer model simulates the existing electric-DX overcooling and electric reheating system; and alternatively, two all-electric packaged roof-top system configurations that satisfy the Florida Energy Code criteria of minimizing or avoiding the use of new energy for reheat. The first unit incorporates an energy recovery wheel (ERV) and an economizer function, and the second is equipped with wrap-around heat pipes and condenser waste heat recovery. The results of these models were also checked against the standard DOE-2.1e hourly simulation software (Fig. 2).

RESULTS

The desiccant unit as installed delivers less cooling and dehumidification capacity than the manufacturer's rating, and much less than it would if the equipment designs were optimized and the installation was commissioned. The unit consumes less energy than rated; however, it consumes considerably more than it would with optimization and commissioning. While the measured energy efficiency at peak load conditions is better than rated, the data clearly shows that this rating is not representative of long-term field performance. In contrast to all-electric cooling equipment, the efficiency of this type of unit tends to decrease as conditions become less humid and cooler. Because peak load conditions are experienced only a fraction of the time, the average efficiency is considerably lower than the rated efficiency. Furthermore, the measured decline in performance with decreasing cooling load—when dehumidification is most critical—is more severe than would be expected.

On the plus side, the primary benefit to the customer of installing the unit has been decreased humidity and increased ventilation for building occupants. The desiccant unit has provided this improvement at annual energy and maintenance savings of about 30% per year, as compared with achieving a similar improvement with the existing all-electric overcool and reheat equipment. The desiccant unit could provide the same level of comfort as existed before its installation (no improvement in humidity or ventilation) at annual energy and maintenance savings of about 16%. The peak demand of the desiccant unit is 15 kW, as compared with 77 kW for the existing equipment. The incremental cost of the desiccant installation will pay back in roughly 7 years.

Two all-electric packaged rooftop system alternatives that satisfy the Florida Energy Code criteria of minimizing or avoiding the use of new energy for reheat were also compared. Peak demand during cooling mode would be about 45 kW. A gas/electric package unit (not desiccant) would have provided an annual savings of about 30%, and a peak demand reduction from 77 to 46 kW. Any of these three options would pay back in about 5 years.

The potential savings available from optimization and field commissioning of the existing desiccant unit is an additional 25% per year, increasing the total savings to about 42% as compared with the baseline.

ACCURACY

Minor inaccuracies in the results of the computer modeling arise mostly from the assumption of linear behavior and use of linear equations in the model. Unlike most other HVAC components, the combined heat and mass transfer occurring in the desiccant wheel experiences hysteresis and nonlinear transients. For example, during relatively humid conditions, the wheel can remove significantly more humidity from the process air than it expels in regeneration. The wheel "stores" moisture in this manner typically over a period that can last hours, and sometimes days. Nonetheless, the average error between the measured field data and the computer predictions is just 7%.

Minor errors in the field data propagated from a number of sources: temperature sensor calibration error of ± 0.8 to $\pm 1.3°F$, plus airflow measurement error of ± 50 fpm, plus dimensional measurement error of ± 0.5 in., plus relative humidity measurement error of $\pm 4\%$ rh. These field data errors result in a sensible cooling capacity error of 11%, a dehumidification capacity error of 25%, a total unit cooling capacity error of 15%, and an energy efficiency error of 18%. These errors were inherent to the sensors and equipment used, the use of "point" rather than "averaging" RTD sensors, the sometimes large differential between point sensor reading and bulk flow conditions, and the different data averaging and sampling rates of the K20 and CS data loggers.

CONCLUSIONS

1. The long-term as-installed performance of typical desiccant HVAC equipment may be less than expected in terms of both delivered capacity and energy efficiency.
2. Engineered improvements to the design and installation of typical desiccant HVAC equipment can provide large performance and cost benefits.
3. Field monitoring and computer analysis of HVAC equipment performance can reveal many cost-effective energy-saving measures.

Distributed Generation

Paul M. Sotkiewicz
Public Utility Research Center, University of Florida, Warrington College of Business, Gainesville, Florida, U.S.A.

Ing. Jesús Mario Vignolo
Instituto Ingeniería Eléctrica, Universidad de la República, Montevideo, Uruguay

Abstract

Distributed generation (DG) is generally thought of as small-scale generation that is used on site and/or connected to the distribution network. Distributed generation development has been driven by technological changes, the availability of inexpensive natural gas, the evolution of electricity competition, and perhaps most of all by the need for extremely reliable electricity supply and combined heat and power (CHP) applications. Distributed generation technologies range from small microturbines and fuel cells to larger reciprocating engines and simple cycle gas turbines. Although DG is not competitive in most applications with grid-supplied power, it has benefits such as increased thermal efficiency in co-generation applications, enhanced reliability, the potential to reduce system losses, the potential to delay or avert new infrastructure investment, and lower emissions compared with traditional coal- and oil-fired technologies.

INTRODUCTION: SOME HISTORY AND EVOLUTION TOWARD DISTRIBUTED GENERATION

Distributed generation (DG) is generally thought of as small-scale generation that is used on site and/or connected to the distribution network. Historically, the type of technologies employed has varied but generally is limited to small engines or combustion turbines fueled by diesel, gasoline, or natural gas and expensive to run relative to grid-supplied power.[1] More recently, intermittent renewable resources such as solar photovoltaic (PV), small hydro, and wind have been thought of as DG that is seen as being deployed to reduce overall emissions. Consequently, small-scale, fossil-fired generation was seen, and still is seen, as primarily providing reliable backup generation in the event of grid-supplied power interruptions, with an estimated 70% of diesel distributed generators in the United States being used for emergency purposes.[2] In contrast, the electricity industry was historically seen as possessing economies of scale in the production and delivery of power. Such economies of scale necessitated larger and larger generating facilities to meet the increasing demand. This brings us to the power system of today, where we have large, centrally dispatched power stations that are connected to one another and to consumers by the high-voltage transmission system, eventually leading to lower distribution voltages and consumers of power.

Several developments, however, have made the idea of DG not only possible, but potentially desirable. The first development is the technological change relating to costs and economies scale that came to fruition in the 1990s: combined cycle, natural gas technology. In Fig. 1, the change in economies of scale compared with historical trends is a fundamental shift toward smaller, lower-cost generating units.

The second development, as indicated by proponents of DG in Conseil International des Grands Résaux Électriques (CIGRE) (in English the International Council on Large Electric Systems) Working Group 37.23,[3] was the availability of relatively inexpensive natural gas supplies, which made potential DG technologies more affordable to operate. Consequently, then it would be possible, as argued in International Energy Agency (IEA)[2] and CIGRE Working Group 37.23,[3] for DG to operate at costs competitive to that of traditional central-station power while averting, deferring, or reducing network costs.

The third development, aided by the previous two developments and described by Hunt and Shuttleworth,[4] is the policy change around the world, moving from vertically integrated monopolies toward more competitive market structures in the generation sector, allowing for more diverse ownership of generating assets that would compete to drive the price of electricity down.

The last development, driven by ongoing environmental policy, is the idea that DG can help countries reduce emissions, especially carbon emissions.[5] Natural gas-fired technologies have lower carbon emissions than traditional coal-fired technologies but higher emissions than renewable technologies, which have zero carbon emissions.

The remainder of the entry is organized as follows. First, we will discuss how DG is defined and contrast that

Keywords: Distributed generation; Reliability; Distribution networks; Distributed generation policies.

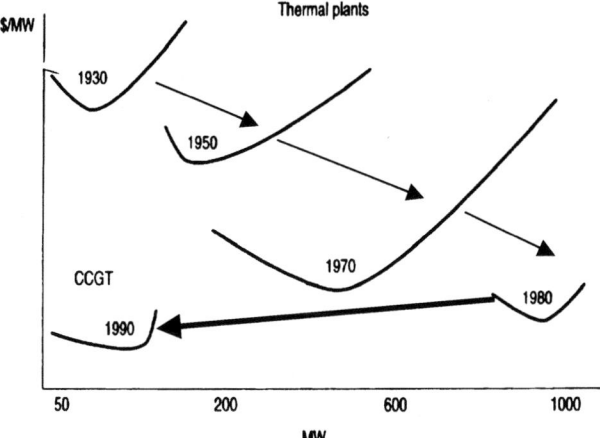

Fig. 1 Generating plants costs curves concerning power (1930–1990).
Source: From Wiley (see Ref. 4).

with other notions of distributed resources. Next, we will briefly outline the types of technologies that are deployed as DG and summarize their cost characteristics. Following that, we will discuss the potential benefits attributed to DG and provide some cautions about overstating the benefits. Finally, we will discuss policies affecting DG and provide concluding remarks.

WHAT IS DISTRIBUTED GENERATION?

Many terms have emerged to describe power that comes from sources other than from large, centrally dispatched generating units connected to a high-voltage transmission system or network. In fact, there is no clear consensus as to what constitutes DG.[2,6]

Congrès International des Réseaux Electriques de Distribution (CIRED) (in English the International Conference on Electricity Distribution) Working Group No. 4[6] created a questionnaire that sought to identify the current state of dispersed generation in the various CIRED member countries. Response showed no agreement on a definition, with some countries using a voltage level definition while others considered direct connection to consumer loads. Other definitions relied on the type of prime mover (e.g., renewable or co-generation), while others were based on non-centrally dispatched generation.

This diversity is also reflected in the CIGRE Working Group 37.23[3] definition, which characterizes dispersed generation as not centrally planned or dispatched, connected to lower-voltage distribution networks, and less than 50–100 MW.

The World Alliance for Decentralized Energy (WADE)[7] defines decentralized energy (DE) as:

- Electricity production at or near the point of use, irrespective of size

- Technology or fuel used, both off grid and on grid, including (1) high-efficiency cogeneration on any scale, (2) on-site renewable energy, and (3) energy recycling systems powered by waste gases, waste heat, and pressure drops to generate electricity and/or useful thermal energy on site.

The IEA[2] defines DG as the following:

Distributed generation is a generating plant serving a customer on site or providing support to a distribution network, connected to the grid at distribution level voltages. The technologies include engines, small (and micro) turbines, fuel cells, and PV systems.

The IEA definition excludes wind power, arguing that it is mostly produced on wind farms usually connected to transmission, rather than for on-site power requirements. In addition to providing a definition for DG, the IEA[2] has provided nomenclature for other dispersed, distributed, or DE resources that we outline below for completeness and to alert the reader of the different terms that are often used with respect to DG. It should be noted that in each of the bulleted definitions below, DG is a subset of the defined category:

- Dispersed generation includes DG plus wind power and other generation, either connected to a distribution network or completely independent of the grid.
- Distributed power includes DG plus energy storage technologies such as flywheels, large regenerative fuel cells, or compressed air storage.
- Distributed energy resources include DG plus demand-side measures.
- Decentralized power refers to a system of distributed energy resources connected to a distribution network.

For the purpose of this work, we will consider DG as generation used on site (and possibly unconnected to the distribution network) and/or connected to the lower-voltage distribution network irrespective of size, technology, or fuel used. This nomenclature encompasses the definitions of IEA[2] and WADE.[7]

DG Technologies

Reciprocating Engines

Reciprocating engines, according to IEA,[2] are the most common form of DG. This is a mature technology that can be fueled by either diesel or natural gas, though the majority of applications are diesel fired. The technology is capable of thermal efficiencies of just over 40% for electricity generation, relatively low capital costs, but relatively high running costs, as shown in Table 1. The technology is also suitable for backup generation, as it can be started quickly and without the need for grid-supplied power. When fueled by diesel, this technology has the

Table 1 Cost and thermal efficiencies of distributed generation (DG) technologies, inclusive of grid connection costs and without combined heat and power (CHP) capability

	Installed cost ($/kW)	O&M (c/kWh)	Efficiency (%)	Levelized cost (c/kWh)
Simple cycle gas turbine	650–900	0.3–0.8	21–40	6–9
Microturbines	1000–1300	0.5–1.0	25–30	7–9
Diesel engines	350–500	0.5–1.0	36–43	7–11
Gas engines	600–1000	0.7–1.5	28–42	6–9
Fuel cells	1900–3500	0.5–1.0	37–42	11–14
Solar photovoltaic (PV)	5000–7000	0.1–0.4	NA	34.5–46.0
Small hydro	1450–5600	0.7	NA	3.5–8
Wind	790	NA	NA	7.6

Source: From OECD/IEA (see Ref. 2) except for the wind which is from AWEA (see Ref. 8) and Small Hydro from WADE (see Ref. 7). Levelized cost numbers assume 60% capacity factor except for Solar PV from WADE (see Ref. 7) at 1850 h/year, Small Hydro from WADE (see Ref. 7) at 8000 h/year, and wind at 39% capacity factor.

highest nitrogen oxide (NO_x) and carbon dioxide (CO_2) emissions of any of the DG technologies considered in this entry, as shown in Table 2.

Simple Cycle Gas Turbines

This technology is also mature, deriving from the use of turbines as jet engines. The electric utility industry uses simple cycle gas turbines as units to serve peak load, and they generally tend to be larger. Simple cycle gas turbines have the same operating characteristics as reciprocating engines in terms of startup and the ability to start independently of grid-supplied power, making them suitable as well for backup power needs. This technology is often run in combination for combined heat and power (CHP) applications, which can increase overall thermal efficiency. Capital costs are on par with those of natural gas engines, as shown in Table 1, with a similar operating and levelized cost profile. The technology tends to be cleaner, as it is designed to run on natural gas, as shown in Table 2.

Microturbines

This technology takes simple cycle gas technology and scales it down to capacities of 50–100 kW. The installed costs are greater than for gas turbines, and the efficiencies are lower as well, as shown in Table 1. Microturbines are much quieter than gas turbines, however, and have a much lower emissions profile than gas turbines, as shown in Table 2. The possibility also exists for microturbines to be used in CHP applications to improve overall thermal efficiencies.

Fuel Cells

Fuel cell technology is also fairly new and can run at electrical efficiencies comparable to those of other mature technologies. Fuels cells have the highest capital cost

Table 2 Emission profiles of distributed generation (DG) technologies

Technology	lbs. NO_x/MWh	lbs. NO_x/mmBtu	lbs. CO_2/MWh	lbs. CO_2/mmBtu
Average coal boiler 1998	5.6	0.54	2115	205
Combined cycle gas turbine 500 MW	0.06	0.009	776	117
Simple cycle gas turbine	0.32–1.15	0.032–0.09	1154–1494	117
Microturbines	0.44	0.032	1596	117
Diesel engines	21.8	2.43	1432	159
Gas engines	2.2	0.23	1108	117
Fuel cells	0.01–0.03	0.0012–0.0036	950–1078	117
Solar photovoltaic (PV)	0	0	0	0
Small hydro	0	0	0	0
Wind	0	0	0	0

Source: From RAP (see Ref. 9).

among fossil-fired technologies and consequently have the highest levelized costs, as shown in Table 1. Offsetting that, the emission footprint of fuel cells is much lower than that of the other technologies, as shown in Table 2.

Renewable Technologies

We discuss three major types of renewable energy technologies here: solar PV, small hydro, and wind. Each of these technologies is intermittent, in that it is dependent upon the sun, river flows, or wind. Consequently, these technologies are not suitable for backup power, but also have no fuel costs and have a zero emissions profile, as shown in Table 2. The capital costs vary significantly among the technologies, however, and operating conditions over the year affect their respective levelized costs. Solar PV is by far the most expensive in both capital costs and levelized costs, as shown in Table 1. Capital costs for small wind are much lower, but levelized costs are in the range of more traditional technologies, as shown in Table 1. Small hydro capital costs can vary widely, with levelized costs reflecting the same variation.

The Role of Natural Gas and Petroleum Prices in Cost Estimates

The levelized cost figures in Table 1 make assumptions about the price of natural gas and diesel. As shown in U.S. Energy Information Administration (USEIA),[10] the prices of natural gas and petroleum products have risen substantially in recent years relative to the time the levelized cost estimates have been calculated. Consequently, if the forecasts in USEIA[10] turn out to be relatively accurate, the levelized cost of all the fossil technologies will be greater than the revelized costs stated here, all else equal.

Potential Benefits of Distributed Generation

Distributed generation has many potential benefits. One of the potential benefits is to operate DG in conjunction with CHP applications, which improves overall thermal efficiency. On a stand-alone electricity basis, DG is most often used as backup power for reliability purposes but can also defer investment in the transmission and distribution network, avert network charges, reduce line losses, defer the construction of large generation facilities, displace more expensive grid-supplied power, provide additional sources of supply in markets, and provide environmental benefits.[11] Although these are all potential benefits, however, one must be cautious not to overstate the benefits, as we will discuss as well.

Combined Heat and Power (CHP) Applications

Combined heat and power, also called cogeneration, is the simultaneous production of electrical power and useful heat for industrial processes, as defined by Jenkins et al.[12] The heat generated is used for industrial processes and/or for space heating inside the host premises or is transported to the local area for district heating. Thermal efficiencies of centrally dispatched, large generation facilities are no greater than 50% on average over a year, and these are natural gas combined cycle facilities.[9] By contrast, cogeneration plants, by recycling normally wasted heat, can achieve overall thermal efficiencies in excess of 80%.[7] Applications of CHP range from small plants installed in buildings (hotels, hospitals, etc.) up to big plants on chemical works and oil refineries, although in industrialized countries, the vast majority of CHP is large industrial CHP connected to the high-voltage transmission system.[2] According to CIGRE Working Group 37.23,[3] the use of CHP applications is one of the reasons for increased DG deployment.

Table 3 shows the costs of DG with CHP applications and their levelized costs. Compared with the levelized costs of stand-alone electricity applications, these costs are lower, especially at high capacity factors (8000 h), showing evidence of lower costs along with greater efficiency in spite of the higher capital cost requirements.

Impact of DG on Reliability (Security of Supply)

It seems quite clear that the presence of DG tends to increase the level of system security. To confirm this idea, the following example is presented.

Table 3 Distributed generation (DG) technology costs, inclusive of combined heat and power (CHP) infrastructure

			Levelized cost (c/kWh)	
	Installed capital cost ($/kW)	O&M (c/kWh)	8000 h/year	4000 h/year
Simple cycle gas turbine	800–1800	0.3–1.0	4.0–5.5	5.5–8.5
Combined cycle gas turbine	800–1200	0.3–1.0	4.0–4.5	5.5–6.5
Microturbines	1300–2500	0.5–1.6	5.0–7.0	7.0–11.0
Reciprocating engines	900–1500	0.5–2.0	4.5–5.5	6.0–8.0
Fuel cells	3500–5000	0.5–5.0	9.0–11.5	14.5–19.5

Sources: From OECD/IEA (see Ref. 2) and WADE (see Ref. 7).

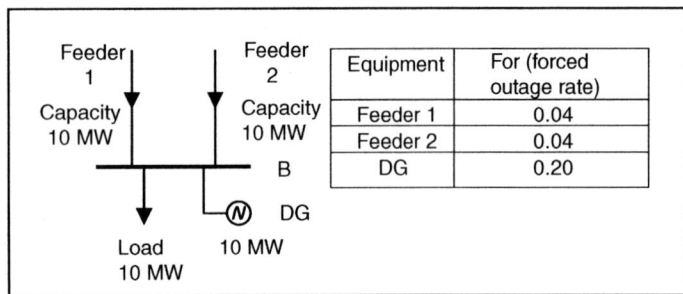

Fig. 2 Security of supply example with distributed generation (DG).

Fig. 2 shows a very simple distribution network. It consists of two radial feeders, each with 10 MW of capacity, that feed busbar B. A constant load of 10 MW is connected to B. The Forced outage rate (FOR) of the two feeders is given in the table in Fig. 2. Additionally, consider a 10 MW DG source with an availability factor of 80%.

To begin with, let us consider only the two feeders and assume that there is no distributed resource connected to busbar B. The loss of load probability (LOLP)—the probability that load is not served—is simply the probability that both feeders will be out of service at the same time, which can be calculated by multiplying the two probabilities of failure. Consequently, LOLP = (0.04 × 0.04) = 0.0016. The expected number of days in which the load experiences troubles can also be calculated by multiplying the LOLP by 365, which results in 0.584 days/year. This number can be expressed in hour/year by multiplying by 24, resulting in 14 h/year.

Now let us consider including the DG source. It has an outage rate greater than the two feeders at 0.20, but it also adds a triple redundancy to the system. Thus, we would expect the addition of the DG source to decrease the LOLP. The new LOLP is the probability that both feeders will fail and that the DG source is not available. Therefore, LOLP = (0.04 × 0.04 × 0.20) = 0.00032—that is, the probability of being unable to serve load is 5 times less than before. This translates to an expected number of hours per year unable to serve load at just less than 3 h/year in our example.

Impact of Distributed Generation on Network Losses, Usage, and Investment

The presence of DG in the network alters the power flows (usage patterns) and, thus, the amount of losses. Depending on the location and demand profile in the distribution network where DG is connected and DG operation, losses can either decrease or increase in the network. A simple example derived from Mutale et al.[13] can easily show these concepts.

Fig. 3 shows a simple distribution network consisting of a radial feeder that has 2 loads (D1 and D2 at points A and B, respectively) and a generator (G) embedded at point C. The power demanded by the loads is supposed to be constant and equal to 200 kW. The power delivered by the G is 400 kW. The distance between A and B is the same as the distance between B and C. In addition, the distance between T and A is twice the distance between A and B. Moreover, we assume that the capacity of each of the sections is equal to 1000 kW. Impedances for sections AB and BC are assumed to be equal, as are the distances. The impedance on TA is assumed to be twice that of AB and BC, as the distance is double. We also assume that voltages are constant and that losses have a negligible effect on flows.

From this hypothesis, it is easy to demonstrate that the line losses (l) can be calculated by multiplying the value of line resistance (proxy for impedance) (r) by the square of the active power flow (p) through the line: $l = rp^2$.

If distributed G is not present in the network (disconnected in Fig. 4), the loads must be served from point T, with the resulting power flows, assuming no losses for the ease of illustration, of Fig. 4.

Losses in the network are $l = 4^2(2 \times 0.001) + 2^2 \times 0.001 = 0.036$ p.u., or 3.6 kW. Additionally, the usage of the network is such that the section TA is used to 40% of its capacity (400 kW/1000 kW) and section AB is used to 20% of its capacity (200 kW/1000 kW).

Now assume that DG G is connected at point C, as shown in Fig. 5.

The resulting power flows, assuming no losses again for ease of illustration, are the following:

The losses are $l = 0.001[2^2 + 4^2] = 0.02$ p.u., or 2 kW, which is a 44% reduction in losses in the case without DG. The reduction from losses comes from transferring flows from the longer-circuit TA to the shorter-circuit BC. Moreover, because less power must travel over the transmission network to serve the loads D1

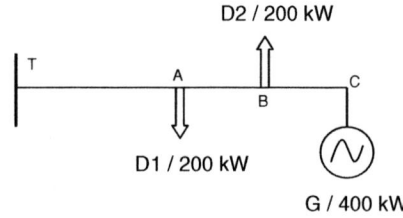

Fig. 3 A simple distribution network.

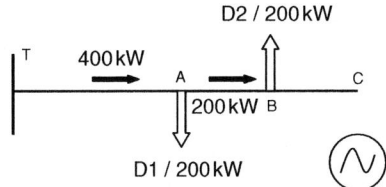

Fig. 4 Power flows without distributed generation (DG).

and D2, losses on the transmission system are reduced, all else being equal.

Additionally, the pattern of usage has changed. The usage on AB is still 200 kW, but the flow is in the opposite direction from the situation without DG. The flow on TA has been reduced from 400 to 0 kW. In effect, the DG source at C has created an additional 400 kW of capacity on TA to serve growing loads at A and B. Suppose that the loads D1 and D2 increased to 700 kW each. Without DG, this would require extra distribution capacity to be added over TA, but with DG, no additional distribution capacity is needed to serve the increased load. In short, DG has the ability to defer investments in the network if it is sited in the right location.

Finally, depending on the distribution and transmission tariff design, DG can avert paying for network system costs. This is especially true in tariff designs, in which all network costs are recovered through kWh charges rather than as fixed demand charges. This is another reason, according to CIGRE Working Group 37.23,[3] for DG deployment.

It is important to emphasize that the potential benefits from DG are contingent upon patterns of generation and end use. For different generation and end-use patterns, losses and usage would be different. In fact, losses may increase in the distribution network as a result of DG. Let G produce 600 kW, for example. In this case, losses are 6 kW—greater than the 3.6 kW losses without DG. Moreover, although DG effectively created additional distribution capacity in one part of the network, it also increased usage in other parts of the network over circuit BC. Consequently, one must be cautious when evaluating the potential for DG to reduce losses and circuit usage.

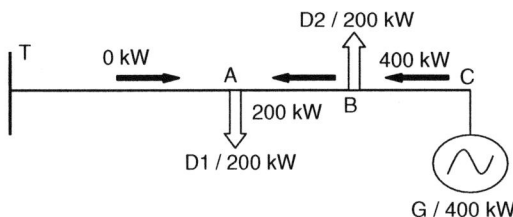

Fig. 5 Power flows and usage with generator (G) producing 400 kW.

Potential to Postpone Generation Investment

In addition to the potential network benefits and reliability (security of supply benefits), DG may bring other benefits to power systems. The first is the ability to add generating capacity in a modular fashion, which does not require building large power plants that will have excess capacity for some time and that, because of size, may be easier to site and permit and faster to complete. In this vein, Hadley et al.[14] modeled DG in the Pennsylvania-New Jersey-Maryland Interconnection (PJM) market and found the potential to displace some existing units, as well as postpone new combined cycle gas units. One must be cautious with this potential benefit, however, as the overall costs of DG may be more than central-station power.

Potential Electricity Market Benefits

In an electricity market environment, DG can offer additional supply options to capacity markets and the ancillary-services market, thereby leading to lower costs and more competition.[15] In the same vein, the owner of DG has a physical hedge against price spikes in electricity markets, which not only benefits the owner of DG but also should help dampen the volatility in the market.[2]

Potential Environmental Benefits

Finally, DG resources may have lower emissions than traditional fossil-fired power plants for the same level of generation, as shown in Table 2, depending on technology and fuel source. This is true for renewable DG technologies, of course. The benefits are potentially large in systems in which coal dominates electricity generation, as shown in Table 2. Hadley et al.[14] models DG in the PJM market and finds that DG displacing generation on the system led to lower emission levels. These reasons were cited in CIGRE Working Group 37.23[3] as determining factors for some DG deployment. Moreover, because losses may also be reduced, DG may reduce emissions from traditional generation sources as well. Additionally, customer demand for renewable energy may be driving renewable energy deployment.[16]

POLICIES AND CONCLUDING REMARKS

Distributed generation as defined in IEA[2] and WADE[7] can provide many benefits, though it is not yet quite competitive with grid-supplied power on its own. Current policies to induce DG additions to the system generally consist of tax credits and favorable pricing for DG-provided energy and services that are subsidized by government.[2] Although such policies may be effective to capture some potential benefits from DG, such as environmental benefits, they do not address the network or market benefits of DG. Only recently has serious consideration been given to considering locational pricing of network services as a way to provide better incentives without subsidies,[17,18] as recommended by IEA.[2] Moreover, only recently has DG been recognized as a potential player in wholesale power markets to provide marketwide benefits.[15] Finally, any barriers that prevent the efficient entry of DG should be reconsidered.[1,2]

REFERENCES

1. U.S. Congressional Budget Office (USCBO). *Prospects for Distributed Electricity Generation*; Congressional Budget Office: Washington, DC, 2003.
2. International Energy Agency (IEA). *Distributed Generation in Liberalised Electricity Markets*; OECD/IEA: Paris, 2002.
3. CIGRE Working Group 37.23. *Impact of Increasing Contribution of Dispersed Generation on the Power System*; Paris, France, 1999.
4. Hunt, S.; Shuttleworth, G. *Competition and Choice in Electricity*; Wiley: London, 1996.
5. World Alliance for Decentralized Energy (WADE). *World Survey of Decentralized Energy 2005*. Available at: www.localpower.org/documents_pub/report_worldsurvey05.pdf (accessed on November 2006).
6. CIRED Working Group No. 4 on Dispersed Generation (CIRED). *Preliminary Report Discussion at CIRED 1999*; Nice: France, June 2, 1999.
7. World Alliance for Decentralized Energy (WADE). *WADE Guide to DE Technologies: Cogeneration Technologies*, 2003; Available at: www.localpower.org/documents_pub/report_de_technologies.pdf (accessed on November 2006).
8. American Wind Energy Association (AWEA). *Economics of Wind Energy*, Washington, DC, AWEA Available at: www.awea.org/pubs/factsheets/EconomicsofWind-March2002.pdf (accessed on November 2006).
9. Regulatory Assistance Project (RAP). *Expected Emissions Output from Various Distributed Energy Technologies*. Available at: www.raponline.org/ProjDocs/DREmsRul/Collfile/DGEmissionsMay2001.pdf (accessed on November 2006).
10. U.S. Energy Information Administration (USEIA). Annual Energy Outlook 2006 with Projections to 2030. Available at: www.eia.doe.gov/oiaf/aeo/pdf/0383(2006).pdf (accessed on November 2006).
11. Ianucci, J.J.; Eyer, J.M.; Pupp, R.L. *DER Benefits Analysis Study: Final Report*, NREL/SR-620-34636; National Renewable Energy Laboratory: Golden, CO, September 2003.
12. Jenkins, N.; Allan, R.; Crossly, P.; Kirschen, D.; Strbac, G. Embedded Generation. In *IEE Power and Energy Series 31*; IEE: London, 2000.
13. Mutale, J.; Strbac, G.; Curcic, S.; Jenkins, N. Allocation of losses in distribution systems with embedded generation. IEE Proc.-Gener. Transm. Distrib. **2000**, *147* (1), 1–8.
14. Hadley, S.W.; Van Dyke, J.W.; Stovall, T.K. *The Effect of Distributed Energy Resource Competition with Central Generation*, ORNL/TM-2003/236; Oak Ridge National Laboratory, U.S. Department of Energy, Office of Energy Efficiency and Renewable Energy: Oak Ridge, TN, October 2003.
15. Sotkiewicz, P.M. *Nodal Pricing and MW-Mile Methods for Distribution: Have We Uncovered Markets or Elements for Wholesale Markets*, Presented at: Harvard Electricity Policy Group 42nd Plenary Session, La Jolla, CA, March 2, 2006; Available at: www.ksg.harvard.edu/hepg.
16. Hyde, D. Will consumers choose renewable energy, and should you care one way or the other?. *IEEE Power Engineering Review*; IEEE: London, May 1998.
17. Sotkiewicz, P.M.; Vignolo, J.M. Nodal pricing for distribution networks: efficient pricing for efficiency enhancing DG. *IEEE Transactions on Power Systems*; IEEE: London, 2006; Vol. 21, No. 2, 1013–1014.
18. Sotkiewicz, P.M.; Vignolo, J.M. Allocation of fixed costs in distribution networks with distributed generation. *IEEE Transactions on Power Systems*; IEEE: London, 2006; Vol. 21, No. 2, 639–652.

Distributed Generation: Combined Heat and Power

Barney L. Capehart
Department of Industrial and Systems Engineering, University of Florida College of Engineering, Gainesville, Florida, U.S.A.

D. Paul Mehta
Department of Mechanical Engineering, Bradley University, Peoria, Illinois, U.S.A.

Wayne C. Turner
Industrial Engineering and Management, Oklahoma State University, Stillwater, Oklahoma, U.S.A.

Abstract

Distributed generation (DG) is electric or shaft power generation at or near the site of use as opposed to central power station generation. Combined heat and power (CHP) takes advantage of this site location to recover the normally wasted thermal energy from power generation and utilizes it beneficially to increase the total system efficiency. This article explores the rapidly developing world of DG and associated CHP. First the article shows why DG is necessary in the U.S. power future and that DG is going to happen. Then, the article briefly looks at the different technologies that might be employed and their relative advantages and disadvantages. The article then explores who should be the major designers and implementers of DG and CHP technologies, and develops a strong argument that in many cases this should be an Energy Service Company (ESCO). Finally, the reasons for selecting either an independent ESCO or a local utility-affiliated ESCO are discussed, and in particular, opportunities for the local utility ESCO (the local grid) to be a major moving force in this effort are examined in depth.

INTRODUCTION

Distributed generation (DG) is electric or shaft power generation at or near the user's facility as opposed to the normal mode of centralized power generation and utilization of large transmission and distribution lines. Since DG is at or near the user's site, combined heat and power (CHP) becomes not only possible, but advantageous for many facilities. The CHP is the simultaneous production of electric or shaft power, and the utilization of the thermal energy that is "left over" and normally wasted at the central station generating site. Since DG means the power is generated at the user's site, CHP can be used to beneficially recover "waste" heat, and provide the facility with hot water, hot air, or steam, and also cooling through the use of absorption chillers.

Normal power generation using a steam Rankine cycle (steam turbine) is around 35% efficient for electric power production and delivery to the using site. The DG with its associated CHP potential means, the total system efficiency can be improved dramatically and sometimes even doubled. Thus, even though DG cannot usually beat the electrical generation efficiency of 35–40% (at the central station), it can save the user substantial amounts of money through recovery of the thermal energy for beneficial use at the site. In addition, there are many other potential benefits from DG/CHP, discussed below.

Thus, the user can choose the objective he/she desires and will likely find a technology in this list that meets that objective. Objectives might include power production that is environmentally friendly, cost effective, more reliable, or yields better power quality. Each of these candidate technologies will be briefly explored in the sections that follow.

Why Distributed Generation?

To explore why DG should be (and is becoming) more popular, the question of why DG needs to be addressed from the perspective of the user, the utility, and society in general. Each of these perspectives is examined below.

The user might desire more reliable power. This can occur with DG, especially, when it is connected to and backed-up by the grid. The user might desire better quality power which can result, because there will be fewer momentary interruptions and possibly better voltage consistency. Often, the user desires better economics (cheaper power), which is quite possible with DG when CHP is employed. Finally, there could be a competitive

---

[*] This entry originally appeared as "Distributed Generation and Your Energy Future" in *Cogeneration and Distributed Generation Journal*, Vol. 18, No. 4, Fall 2003. Reprinted with permission from AEE/Fairmont Press.

Keywords: Distributed generation (DG); Combined heat and power (CHP) systems; Cogeneration; Self-generation; On-site generation; Utility-affiliated ESCOs; Reciprocating engines; Gas turbines; Fuel cells; Photovoltaic cells; Wind turbines.

advantage during a utility power outage when the user has power and the competition does not.

The utility might desire less grid congestion and less future grid construction, both of which DG definitely yields. The utility may be able to hold on to a customer better, if the customer has DG and CHP at their site. Certainly, this is true if the utility constructs and runs the local power facility. Today's technology is capable of allowing the utility to remotely dispatch literally hundreds of DG units scattered in the grid, which could dramatically improve their ability to handle peak load and grid congestion problems.

Society in general likes DG/CHP, which can provide strong environmental advantages (when the appropriate technology is utilized). For example, CHP means less total fuel will be consumed placing less strain on the environment. Also, DG means less grid construction, and "green technology" (wind, photovoltaics, etc.) can be used if desired. Since less total fuel is consumed, there will be reduced reliance on imported oil and gas, and an improved balance of payments for the United States.

Basic Philosophies

In the development to follow, some basic objectives are assumed. They are:

- Any expenditure of funds should be cost effective
- Any change should be good for all parties, or at least be of minimal harm to any party
- Existing partnerships that are working should be maintained if at all possible.

These assumptions should be clear as stated above; but more explanation will follow as the arguments are developed.

Existing and Future Markets for DG

Resource dynamics corporation has estimated that the installed base of DG units greater than 300 kW in size is at least 34 GW (34×10^9 W). The Gas Research Institute (GRI) estimates that the installed DG capacity in the United States of all sizes is 75 GW. Just over 90% of the installed units are under 2 MW (2×10^6 W) and well over 90% of the installed units are reciprocating engines. The use of DG power plants for back-up purposes is growing steadily at 7% per year while other DG applications for baseload and peaking requirements are growing at 11 and 17%, respectively. Resource Dynamics Corporation says there is the potential to double the installed DG capacity by adding as much as 72 GW by 2010. All these sources confirm that, there is already a large base of DG units nationally, and that the growth will be significant. Distributed generation is happening and will continue to happen.

Without trying to stratify it with exact numbers, the potential market could be broken into three components. They are (1) large and medium, (2) small, and (3) smaller. Each of these categories is examined briefly below.

The large and medium market is often 25 MW and larger (sometimes hundreds of megawatts) and is a mature market because there have been plants operating for many years. Typically these are in the larger process industries such as petroleum refining, pulp and paper, and chemical plants. Steam production may be in the range of hundreds of thousands of pounds per hour. While there are many such operating plants today, this mature market probably still offers the largest immediate growth potential. There is much more that could be done.

The small market will range somewhere between 50 (or 100) kW and 25 MW. These might be plants that need significant steam and could easily add a topping or bottoming steam turbine to become DG–CHP. Important to their success is the need for thermal energy and electricity (or shaft power) and the relative sizes of those needs might dictate which technology is appropriate. This market is virtually untouched today and the management/maintenance talents in these facilities might easily support this DG/CHP technological addition to their needs. Some facilities of this size will not have the management backing and maintenance talent that it takes to make these DG/CHP systems operate successfully. Those facilities would likely seek "outsourcing" for the power plant. The growth possibility here is extremely large, but will likely take a few years to realize its full potential.

The smaller market would include those small manufacturing plants or commercial facilities that need less than 50 (or 100) kW and do not have large thermal needs. These plants and facilities likely do not possess the management backing and desire or maintenance talent it takes to run them. The market potential here is tremendous in numbers of applications, but small in numbers of total megawatts. Finally, there is a significant drop in economies of scale somewhere around 200–500 kW, so the economics here would not be as exciting as the other two markets. Thus, the authors' opinion is that this market will not be as robust in the near to immediate future. Note that this could change overnight, if a local Energy service company (ESCO) such as the utility ESCO offered to design, install, run, and dispatch these units.

DG Technologies

There is a wide range of technologies possible for DG. They include:

- Reciprocating engines—diesel, natural gas, and dual fuel
- Gas turbines—microturbines, miniturbines, and large turbines
- Steam turbines

Distributed Generation: Combined Heat and Power

Table 1 Overview of distributed generation (DG) technology

Technology	Pros	Cons
Fuel cell	Very low emissions	High initial investment
	Exempt from air permitting in some areas	Only one manufacturer producing commercially available units today
	Comes in a complete "ready to connect" package	
Gas turbine	Excellent service contracts	Requires air permit
	Steam generation capabilities	The size and shape of the generator package is relatively large
	Mature technology	
Micro turbine	Low initial investment	Relatively new technology
	High redundancy with small units	Requires an air permit
	Low maintenance cost	Possible synchronization problems at large installations
	Relatively small size	
	Installation flexibility	
Engine	Low initial investment	High maintenance cost
	Mature technology	Low redundancy in large sizes
	Relatively small size	Needs air permit
Photovoltaics	Low operations and maintenance (O&M) costs	Very expensive initially
	Environmentally friendly	Very large footprint
		Sun must shine
		Battery storage usually needed
Wind	Low to medium O&M costs	Large footprint
	Environmentally friendly	Wind must blow

- Fuel cells
- Photovoltaic cells
- Wind turbines
- Storage devices (batteries or flywheels).

The following table briefly summarizes the pros and cons of these different DG/CHP technologies (Table 1). If more detailed information is needed, the authors recommend Capehart, et al[1]; Turner[2]; or Petchers[3].

The table above demonstrates that there is a wide range of technologies available. Some are environmentally friendly, some are not. Some are more economically feasible, while others are extremely expensive. Some use mature technologies, while others are still somewhat of a gamble. What is badly needed for this market to mature are more ESCOs that are broadly experienced in DG/CHP applications and that are good at all of the above technologies, including the nontraditional approaches. Their tool sack contains all of these technologies and they know when and how to apply each of these technologies. To our knowledge, today only a few ESCOs can claim this broad a talent base.

Who?

Thus far, we have demonstrated that there is a significant market projected for DG/CHP systems and that this market needs to be satisfied. We have also shown there is a wide range of technologies that is available. What is needed is someone to "make this happen." Rather obviously, there are about three groups that could make this happen. They are

The users themselves
Energy service companies (ESCOs)
— Independent ESCOs (consultants)
— Utility affiliated ESCOs.

This section examines each of these groups, and shows how they might contribute to the expanded need for DG/CHP. One consideration in evaluating the potential success of a DG/CHP project is the goal alignment of the participants, where the goals of the user or the facility are compared to the goals of the organization that is implementing the project. The closer these goals match up, or align, the more likely the DG/CHP project is to succeed.

The Group of Users Themselves

The users' goals are to have a DG/CHP project that provides an appropriate solution to their needs for electric or shaft power, and probably heat; works well for them both in the short term and the long term; and maximizes their economic benefit from this investment.

The user knows its process better than anyone else. This is a real advantage of doing DG/CHP projects in house and leads to the best economics if it works. Finally, the goal alignment for this group is the best of the three groups, as it is the user itself doing the job.

For this to work, the user must have a staff of technically qualified people who can analyze potential technologies, evaluate the options, select the best technology for their application, permit and install the equipment, and operate the DG/CHP system in a manner which produces the desired results. In addition, management and maintenance must both commit to the project. This often will not occur, if they wish to devote their time and efforts to building better products, delivering better services, or expanding into new products or services. Another disadvantage is that a very large capital investment is normally required and many plants and facilities simply do not have the necessary capital. Finally, these projects would involve grid interconnections and environmental permitting. Many plants and facilities are very unfamiliar with these requirements.

However, if the facility or plant does have a committed and skilled management and staff that can select, permit, install, finance and operate the DG/CHP system, this approach will most likely provide them with the highest rate of return for this kind of project.

The ESCO Group

For facilities that cannot or do not want to initiate and implement their own DG/CHP projects, the involvement of an ESCO, is probably their most appropriate alternative. Energy Service Companies bring a very interesting set of talents to DG/CHP projects. The right ESCO knows how to connect to the grid, what permits are required, and how to obtain those permits. The right ESCO knows all of the technologies available and how to choose the best type and size to utilize for this application. The right ESCO is a financial expert that knows all of the financing options available and which might be the best. Often, this means they have partnered with a financing source and have the money available with payback based on some mutually agreeable terms (interest bearing loan, shared savings, capital lease, true lease, etc.).

One of the disadvantages of using an ESCO is, they are sometimes "in a hurry." When this happens, the project design is not as well done as it should be, and they may leave before the equipment is running properly. Commissioning becomes extremely important here. Another disadvantage is that some ESCOs choose the same technology (cookie cutting) for all projects. A certain type turbine made by a particular company is always chosen, when this may not always be the best solution. If the ESCO approach is to work, all technologies must be considered and the best one chosen. For this group, goal alignment is not the best as the user no longer is in charge and the relationship is likely to be of limited duration (outsourcing being a possible exception).

However, if an "ideal" ESCO can be found and utilized, this approach offers a very satisfactory arrangement.

Independent ESCOs (Consultants)

The goals of the independent ESCO are typically to sell the customer a technology solution that the ESCO is familiar with; get the equipment installed and checked out quickly; maximize their profit on the project; and in the absence of a long term contract to provide maintenance or operating assistance to get out as quickly as they can. Sometimes the independent ESCOs goals do not line up that closely with the customer's. The independent ESCO may try to sell the customer a particular piece of equipment that they are most familiar with, and may be the one that gives them the largest profit. If there is not going to be any long-term contract for the ESCO, then they want to get the project completed as quickly as possible, and then get out. This may leave the user with a DG/CHP system that is not thoroughly checked out and tested, and leaves the user to figure out how to operate the system and how to maintain it. The project ESCO team may then depart the facility, and return to their distant office, which may be in a very different part of the country.

However, as long as the user is willing to pay the ESCO for continuing their support, the ESCO is almost always willing to do that. Unless the user is willing to pay for a part time or full time person to remain at their facility, they will have to deal with the ESCO by phone, FAX, FedEx, or Email.

One of the other potential problems with an independent ESCO is the question of its permanence. Will it be around for the long term? Historically, making the comparison of current DG/CHP ESCOs to the solar water heating companies of the 1970s and 1980s, leads to the concern that some of the DG/CHP ESCOs may not be around for the long term. Very few of the companies that manufactured, sold, or installed solar water heating systems at that time are still around today. Many of these solar water heating companies were gone within a few years of the customers purchasing the systems. Most of these companies were actually gone long before the useful lives of the solar systems had been reached. Repair services, parts, and operating advice were often no longer available, so many solar water heating system users simply stopped using them, or removed the systems. Based on this history, selection of an ESCO that is likely to be around for the long term is an important consideration.

Utility-Based ESCOs

Next, consider a utility-based ESCO. Utility-affiliated ESCOs have goals similar to the independent ESCOs' goals in many respects, but the big difference is that the utility is a permanent organization that is local is there for the long term, and is interested in seeing the user succeed, so that they will be an even better customer in the long run. Also, since the utility and the affiliated ESCO are local, they can send someone out periodically to check on the facility and the DG/CHP project to help answer questions and make sure the project is continuing to operate successfully. The utility is financially secure, stable, and, in most instances, is regarded by the community as an honest and trustworthy institution.

This ESCO now is an independent branch of the local utility. If they have the full set of tools (knowledge of all the technologies) then their advantages include all those listed above. In addition, there is much better goal alignment. They will be there as a partner as long as the wires are connected and that likely is almost forever. Thus, both parties want this to work. They are the grid, so the grid interconnection is not as much of a problem. The user and the utility have been partners for years. This would change the relationship; not destroy it. (The devil you know vs the one you do not.) Finally, this is what they do (almost).

One limitation of the utility-based ESCO is that they must change their mindset of "sell as much electricity as possible," and recognize that there is a lot of business and income to be captured from becoming an energy service organization. Someone is going to do these DG/CHP projects; the utility revenue base is most enhanced when they do it and when the project is successful. Their services could involve design, installation, start up, commissioning, and passing of the baton to the user or they might run it themselves (outsourcing).

Now, if the local utility company ESCO can take advantage of the opportunities they have, then they have a lot to offer to facilities and plants that are interested in working with them to put in DG and CHP systems. The old Pogo adage "We are surrounded by insurmountable opportunities" is always around in these situations. Another old saying would describe this DG/CHP opportunity for utility-affiliated ESCOs as "the business that is there for them to lose." The utility-affiliated ESCOs need to aggressively pursue these opportunities.

Some Local Utility ESCO Successes

A very good example of a local utility ESCO success story comes from the experiences shared by AmerenCILCO, a utility company in central Illinois, This company has experiences with both DG and CHP projects.

DG Only Projects

AmerenCILCO has extensive experience in using reciprocating engine generator sets as DG to meet peak load conditions on their system. The specifications of some of their DG projects are as follows:

- Hallock Substation, 18704 N. Krause Rd., Chillicothe, IL
- Eight reciprocating diesel engine generator sets
- Nominal capacity 1.6 MWe each, 12.8 MWe total
- Owned by AmerenCILCO
- Kickapoo Substation, 1321 Hickox Dr., West Lincoln, IL
- Eight reciprocating diesel engine generator sets
- Nominal capacity 1.6 MWe each, 12.8 MWe total
- Owned by Altorfer Inc.; power purchase agreement with AmerenCILCO, operating agreement provides for operations and maintenance (O&M)
- Tazewell Substation, 18704 N. Krause Rd., Chillicothe, IL
- Fourteen reciprocating diesel engine generator sets
- Nominal capacity 1.825 MWe each, 25.55 MWe total
- Owned by Altorfer Inc.; power purchase agreement with AmerenCILCO; operating agreement provides for O&M.

Although these DG power module facilities are primarily used as peaking facilities, they are also used to maintain system integrity in the event of an unanticipated outage at another AmerenCILCO generating station. They are unmanned and remotely operated from the company's Energy Control Center. They have proven to be a reliable and low cost option for the company to meet its peaking requirements. The power module sites were constructed at a cost of approximately $400/kW and have an operating cost of $75/MWh using diesel fuel at $0.85/gallon.

DG/CHP Projects

AmerenCILCO also has some successful CHP projects. A summary of two such DG/CHP projects are given below.

Indian Trails Cogeneration Plant

The Indian Trails Cogeneration Plant is owned and operated by AmerenCILCO. It is located on the property of MGP Ingredients of Illinois (MGP) in Pekin, Illinois and provides process steam to MGP and electricity to AmerenCILCO. The plant was constructed at a cost of $19,000,000 and went into full commercial operation in June 1995.

The plant consists of three ABB/Combustion Engineering natural gas-fired package steam boilers and one ABB STAL backpressure turbine-generator. Two of the boilers, boilers 1 and 2, are high-pressure superheat boilers rated at 185,000 lb/h of steam at 1250 psig and 900°F. Boiler 3 is a low-pressure boiler rated at 175,000 lb/h of steam at 175 psig and saturated temperature. Boilers 1 and 2 are normally in operation, with Boiler 3 on standby to insure maximum steam production reliability for MGP.

The high-pressure steam from boilers 1 and 2 passes through the ABB backpressure turbine-generator, which is rated at 21 MW. The steam leaving the turbine is at 175 psig and is desuperheated to 410°F to meet MGP's process steam requirements. The electricity produced goes to the AmerenCILCO grid to be used to meet utility system requirements.

The plant configuration provides significant operating efficiencies that benefit both MGP and AmerenCILCO. The Indian trails has an overall plant efficiency in excess of 80% and an electric heat rate of less than 5200 Btu/kWh. The construction of the Indian Trails by AmerenCILCO created an energy partnership with a valued customer. It allowed MGP to concentrate its financial and personnel resources on its core business. In turn, AmerenCILCO used its core business of producing energy to become an integral part of MGP's business, making AmerenCILCO more than just another vendor selling a product.

Medina Valley Cogen Plant

The Medina Valley Cogeneration Plant is owned and operated by AmerenCILCO. It is located on the property of Caterpillar and provides process steam and chilled water to Caterpillar, and electricity to AmerenCILCO. The plant was constructed at a cost of $64,000,000 and went into full commercial operation in September 2001.

The 40 MW electric generating plant consists of three natural gas-fired Solar Titan 130 model 18001S combustion turbines equipped with SoloNO$_x$ (low NO$_x$) combustion systems manufactured by Caterpillar driving electric generators rated at 12.2 MW (gross generating capacity) each. There are also two Dresser-Rand steam turbine-generators with a total rated capacity of 8.9 MW.

The 410,000 #/h steam plant consists of three Energy Recovery International (ERI) VC-5-4816SH heat recovery steam generators (HRSGs) equipped with Coen low NO$_x$ natural gas-fired duct burners and catalytic converters to reduce carbon monoxide (CO), rated at 109,000 lb/h at 600 psig each. There is also one Nebraska natural gas-fired steam generation boiler equipped with low NO$_x$ burners, rated at 100,000 lb/h at 250 psig.

The plant configuration provides significant operating efficiencies that benefit both Caterpillar and AmerenCILCO. Medina Valley has an overall plant efficiency in excess of 70% and an electric heat rate of less than 6400 Btu/kWh. The construction of Medina Valley by AmerenCILCO created an energy partnership with a valued customer, whereby competitive electricity and steam prices were provided as well as greater operational flexibility, improved quality control in manufacturing, and improved steam reliability. It also allowed Caterpillar to concentrate its financial and personnel resources on its core business. In turn, AmerenCILCO used its core business of producing energy to become an integral part of Caterpillars business, strengthening its ties with a major customer as well as adding additional efficient-generating capacity, and improving air quality (399 fewer tons pollutants/year).

CONCLUSIONS

Distributed generation and DG/CHP should, must, and will happen. The benefits to all parties when CHP is utilized are too much to ignore. Therefore, the question becomes who should do it, not should it be done.

If management is behind the project, the engineering and maintenance staff is capable, and financing is available, the project should be done in-house. Maximum economic benefits would result. However, the user must commit to this project.

If any of the above is not true, the best approach for the facility or plant is to seek the help of an ESCO. It is important that the ESCO chosen must be fully equipped with knowledge and experience in all of the technologies and be able to provide the financing package. Such ESCOs do exist today, but some of them need a better understanding of the different technologies required, as well as insuring that the DG/CHP project is successfully completed and turned over to a facility that can operate it and maintain it. Commissioning and baton passing must be part of the contract.

The authors believe there is a tremendous opportunity for local utility ESCOs to successfully participate in this movement to DG, and particularly to the use of CHP. A utility-affiliated ESCO, properly equipped as we have defined it, can do these projects, and can do them successfully. The local utility ESCO has entries with local facilities and plants that few other ESCOs have. If the utility can exploit this opportunity, they have the chance to help many facilities and to help themselves in the process. This is a true win–win opportunity for the utility affiliated ESCO. All utilities should be ready to fill this need or recognize that they will likely lose market share.

REFERENCES

1. Capehart, B.L., Turner, W.C., Kennedy, W.J., Eds.; *The Guide to Energy Management*, 5th Ed.; Fairmont Press: Lilburn, GA, 2006 (Chapter 14).
2. Turner, W.C. Ed.; *Energy Management Handbook*, 6th Ed.; Fairmont Press: Lilburn, GA, 2006 (Chapter 7).
3. Petchers, N. Ed.; *Combined Heating, Cooling and Power Handbook*, Fairmont Press: Lilburn, GA, 2003.

District Cooling Systems

Susanna S. Hanson
Global Applied Systems, Trane Commercial Systems, La Crosse, Wisconsin, U.S.A.

Abstract

Seemingly small efficiency improvements to traditionally designed chiller plants multiply to create impressive savings in district cooling plants. The size and scope of these larger projects make the benefits more obvious and very valuable. For example, the United Arab Emirates is fully embracing district cooling for accommodating its tremendous growth, rather than using individual cooling systems in each building. It has been estimated that 75% of the energy used in Dubai, U.A.E. is for cooling. By using district cooling, Dubai planners expect to reduce the amount of electrical energy used by 40%. More than 30% of the Dubai market is selecting pre-engineered, packaged chilled water systems, many of which employ series chillers, variable-primary distribution design, and other concepts discussed in this article.

INTRODUCTION

District cooling was inspired by district heating plants, which became popular in the 19th century for providing steam to a number of buildings and often to entire portions of cities. Recent district cooling projects exceed 100,000 tons, typically in urban areas with dense, simultaneous development. One project of this size can, all by itself, magnify an incremental efficiency improvement into nearly 10 MW and offset $10 million or more in power generation capacity.

These ideas can naturally be used on smaller projects, where the benefits are less immediately obvious. To illustrate the energy and cost savings potential, this entry includes an analysis from a 10,500-ton convention center. The key design elements used on these projects violate traditional rules of thumb and allow the system designer to focus on the energy consumed by the entire chilled water system.

MOVING AWAY FROM CHILLER-ENERGY-ONLY ANALYSES

In the last 40 years, water-cooled chiller efficiency expressed in coefficient of performance (COP) has improved from 4.0 to over 7.0. Cooling tower and pump energy has remained largely flat over the same period. The result is that towers and pumps account for a much higher percentage of system energy use.

Because of these dramatic improvements, it is tempting to go after chiller savings in new plants and chiller replacements. After all, the chiller usually has the biggest motor in a facility. Many chiller plant designs continue to use flow rates and temperatures selected to maximize chiller efficiency and ignore the system effects of those decisions. Chilled water system designs frequently default to using the flow rates and temperatures used for the rating tests developed by the Air Conditioning and Refrigeration Institute (ARI) standards 550/590 for vapor compression chillers[1] and ARI 560 for absorption chillers.

While these benchmarks provide requirements for testing and rating chillers under multiple rating conditions, they are not intended to prescribe the proper or optimal flow rates or temperature differentials for any particular system. As component efficiency and customer requirements change, these standard rating conditions are seldom the optimal conditions for a real system. There is great latitude in selecting flow rates, temperatures, and temperature differences.

Today, an equal price centrifugal chiller can be selected for less condenser flow, with no loss in efficiency. At the same time, chilled water temperatures are dropping. Pump savings usually exceed chiller efficiency losses. Larger chilled water distribution systems will realize higher savings. Analyze the entire system to define the right design conditions.

MARKET TRENDING TOWARD HIGHER LIFT

Chiller efficiency is dependent on several variables—capacity (tons) and lift (chiller internal differential temperature) are two of them. Lift is the difference between the refrigerant pressures in the evaporator and condenser, and it can be approximated using the difference between the water temperatures leaving the evaporator and condenser.

Designers are converging on higher-lift systems for a variety of reasons; e.g., more extreme climates, system

Keywords: Chillers; Series chillers; Counterflow; Variable-primary pumping; District cooling; Efficient chiller plant design; Chiller lift; Part load efficiency.

optimization, replacement considerations, thermal storage, and heat recovery. Higher lift creates solutions for engineering problems and leverages chiller capabilities.

Chiller plant design for U.S. conditions usually calls for a maximum tower-leaving temperature of 85°F. Conventional designs used 44°F/54°F evaporators and 85°F/95°F condensers—easy on the chiller (and the system designer). But higher ambient wet bulb conditions are common in the Middle East and China. Many parts of China call for 89.6°F design condenser/tower water, and parts of the Middle East design for 94°F. As the cooling markets grow in areas with extreme weather, technology providers must find better ways to deliver cooling at the higher lift conditions.

A reduced flow rate in the condenser saves tower energy and condenser pump energy. In the replacement and renovation market, the resulting increase in delta-T (tower range) delivers the same capacity with a smaller cooling tower, or more capacity with the existing cooling tower.

Chilled water temperatures below 40°F allow system designers to implement a larger chilled water delta-T. The result is less costly chilled water distribution and more effective cooling and dehumidification. Besides reducing water flow rates, colder chilled water gives higher delta-Ts at the chiller and better utilization of the chillers.

Thermal storage is also becoming more prevalent. For stratified chilled water storage tanks, 39°F water is the magic number, because it corresponds to the maximum density of water and keeps the charged section below the thermocline.[2]

Higher condensing temperatures provide more useful heat to recover from the condenser water. Some commercial building codes require condenser heat recovery for applications with simultaneous heating and cooling.

A proper chiller selection can deliver these conditions with little or no extra cost. It is possible to increase lift while increasing overall chiller plant efficiency by using a series-counterflow chiller arrangement.

SERIES CHILLERS

In their heyday of the early 1960s, series chillers were widely used in government buildings in Washington, D.C. Series arrangements were necessary for perimeter induction cooling systems, which supply cold primary air to the space and require colder water from the chiller. As previously discussed, chillers in the 1960s had a COP of about 4.0, with high-flow (velocity) smooth-bore tubes, low tube counts, and one-pass evaporators to reduce pressure drop. The most efficient centrifugal chillers today have a COP of more than 7.0—more than 75% higher than chillers used in these early series chiller plants.

Designers virtually stopped using series arrangements in the 1970s because variable air volume (VAV) systems made the colder chilled water used for induction systems unnecessary. Given the chiller's relatively low efficiency by today's standards, it made sense to raise temperatures and save chiller energy. Variable air volume systems were widely adopted because they saved energy and adapted to unknown cooling loads. Variable air volume systems are still the most popular choice for comfort cooling applications.

Induction systems are virtually nonexistent in 2006, but the dramatically improved centrifugal chiller efficiencies are driving resurgence in series chiller arrangements. Series chiller plants offer energy and first cost savings, even in VAV systems.

Using multiple-stage centrifugal chillers and putting chillers in series creates higher lift more easily and more efficiently, with a more rigorous, stable operating profile. In contrast, chillers lined up in parallel must each create the coldest water required for the entire system while rejecting heat to the warmest condensing temperature.

While series arrangements of chiller evaporators have been used in many applications,[3,4] series condenser arrangements are less common. A series-counterflow chilled water plant design arranges the evaporators in series, but also arranges the condensers in series using a counterflow configuration. Counterflow means that the condenser water and the evaporator water flow in opposite directions.

Chiller plants with series evaporators and parallel condensers aren't recognizing the highest efficiency gains because the chiller producing the coldest chilled water must create more lift and therefore do more work. By arranging the condensers in series counterflow, the lift of each compressor is nearly the same. The result is a pair of chillers working together to create high lift while increasing overall plant efficiency.

VARIABLE FLOW

A relatively new concept, the variable-primary system, has removed one barrier to series-chiller plant design. Series chillers often have a higher pressure drop. In traditional primary–secondary systems, constant flow through the chillers equals constant pressure drop and constant pump energy, so the series chillers' higher pressure drop results in higher pump energy all the time. Varying the flow through series chillers eliminates the penalty for much of the operating hours.

Variable-primary systems send variable amounts of water flow through the chillers to reduce the pumping energy and enable the delta-T seen by the chiller to remain equal to the system delta-T. Because pump energy is approximately proportional to the cube of the flow (subject to losses through the distribution system), even small flow reductions are valuable.

Consider the following ideal relationship between flow and energy in a pumping system:

Pump Energy $\propto$ Flow^3

For a system using 80% of the design flow—a modest 20% reduction—the energy required is reduced by nearly 50%. In turn, a more aggressive 50% reduction in flow is an 87% reduction in power.

Varying the flow through series chillers reduces the total operating cost, despite an increased pressure drop at design conditions. Improved tube designs and extensive testing in manufacturers' testing labs have cut minimum water velocities in half, leading to better turndown for all chillers. The additional implicit turndown capability of the series configuration enables further pumping energy savings. Single-pass evaporators and condensers, when practical, reduce water pressure drop and pumping costs compared to two- or three-pass configurations.

VARIABLE FLOW COMBATS THE INCREDIBLE SHRINKING CHILLER

So far, we have only discussed efficiency and energy savings. But what about the installed costs associated with chiller-water system design? Any chiller, big or small, is a significant capital investment. This is one reason most chillers are factory tested for capacity. Cooling capacity is the product of chilled water flow and chilled water delta-T. For the same capacity, as gallons per minute goes up, delta-T goes down.

Chilled water system design has a unique and sometimes dramatic effect on chiller capacity. Primary-secondary chilled water systems shrink chiller capacities, while variable-primary chilled water systems help chiller capacities expand. At times, this chiller capacity expansion can exceed nominal chiller capacity.

Due to this variation in chiller capacity, primary-secondary chilled water systems have excessive chiller starts as well as higher chiller run hours, because more chillers will be operating than necessary. Variable-primary chilled water systems have fewer chiller starts and lower chiller run hours, by squeezing more capacity out of the operating chillers. The plant controller can delay the operation of an additional chiller by increasing the flow through the operating chillers, thus increasing the chiller capacity. Operating the fewest number of chillers possible is a well-known energy optimization strategy.

It is the design of the chilled water system that causes this change in net chiller capacity, not the chiller itself. Chillers are constant flow devices when employed in a primary–secondary chilled water system. A bypass pipe, commonly called a decoupler, serves as a bridge between the primary loop serving only the chillers and the secondary loop distributing chilled water throughout the building to all the air handling units. Because the chiller is a constant flow device, we must subject the chiller to a smaller chilled water delta-T in order to unload the chiller. The decoupler pipe mixes surplus-chilled water with water returning from the load to produce cooler return water for the chiller. But what if some event other than the decoupler lowers the temperature of the returning chilled water? This reduction of chiller-entering water temperature unloads the chiller—even when we don't want the chiller unloaded.

In a primary–secondary system, additional chillers are sequenced on when the demand for chilled water exceeds the constant flow capacity of each chiller. To illustrate, consider three equally sized chillers in a primary-secondary system. Each chiller is sized for 500 tons and 750 gpm of chilled water (1.5 gpm per ton, or a 16°F chilled water delta-T). When the chilled water distribution system demands more than 750 gpm, two chillers must operate. When the chilled water system demand exceeds 1500 gpm, all three chillers must be on. These chillers will produce their full 500-ton cooling capacity only when the cooling coils create a full 16°F water-temperature rise. If the cooling coils collectively can only produce a 12°F water-temperature rise, the 500-ton chillers can only produce 375 tons. 125 tons of installed cooling capacity is lost. This inability of the chilled water distribution system to achieve the design chilled-water-temperature rise is called "low delta-T syndrome."

There are several contributors to low delta-T syndrome, but the three greatest offenders are excess distribution pump head, three-way control valves, and chilled water reset. Three-way control valves allow cold chilled water to bypass the cooling coil. The bypassed water is dumped into the return chilled water line, diluting the return chilled water. Any high water-temperature rise created by the cooling coil is destroyed.

The chilled water distribution pump is variable speed, producing more pressure when more chilled water flow is required, and producing less pressure when less chilled water flow is required. The speed of the pump is controlled by sensing the available pressure at the end of the chilled water distribution loop. If this pump creates excess pressure, even two-way control valves will have a difficult time reducing chilled water flow through the cooling coil. When the cooling coil receives excess flow, there is not enough heat in the air to adequately warm the water. Again, water-temperature rise is hampered.

Perhaps the most insidious destroyer of water-temperature rise is chilled water reset. The chiller will consume less energy when it produces warmer chilled water, but chiller energy savings may be dwarfed by the additional pump energy consumption required for delivering warmer chilled water. Coils use less chilled water when it is delivered at a colder temperature. As the chilled water

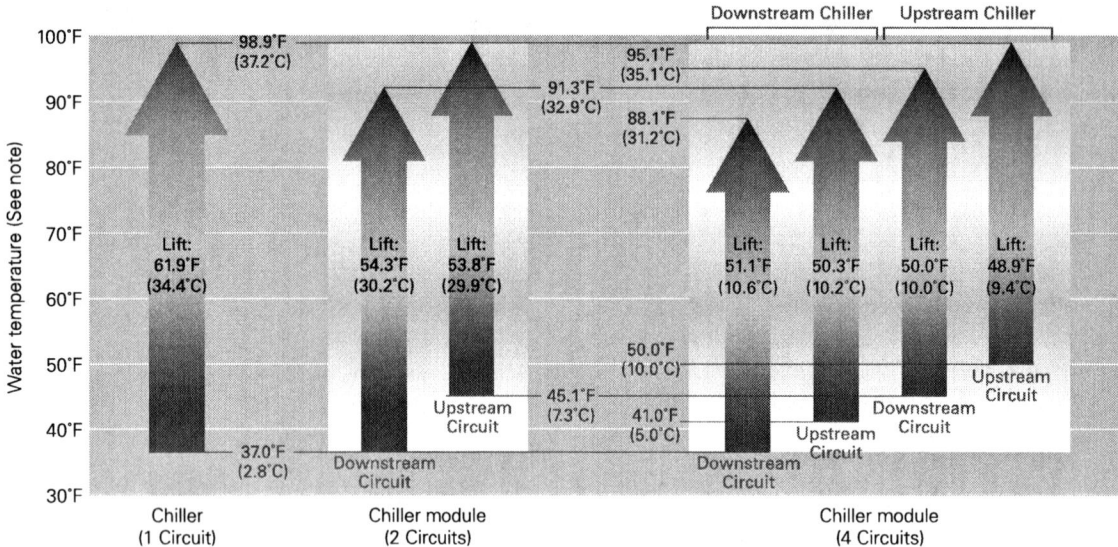

Fig. 1 Series counterflow chiller arrangement equalizes lift performed by each compressor, minimizing the energy needed to create high lift.

temperature is set upwards, each coil will demand more water to meet the same cooling load.

CHILLER PLANT DESIGN ENERGY COMPARISON

The following example is for a design created initially for the Washington DC Convention Center. Each pair of series-counterflow chillers (assuming multiple-stage compressors on each circuit) has eight to twelve stages of compression equally sharing the load (Fig. 1). The chiller module depicted in Fig. 2 created series-pair efficiencies of 0.445 kW per ton (7.8 COP) at standard ARI rating conditions.

Fig. 3 shows the component and system energy use of various parallel and series chiller configurations using variable evaporator flow with reduced condenser water flow. The series–series counterflow arrangement for the chillers reduces the chiller energy to compensate for additional pump energy. In the case of this particular installation, series–series counterflow saved $1.4 million in lifecycle costs over the parallel–parallel alternative. The low-cost alternative used six electric centrifugal chillers with dual refrigeration circuits, 2 gpm of condenser water per ton of cooling, piped in a "series evaporator-series condenser" arrangement. It also used 1040 kW less than the parallel–parallel configuration.[5]

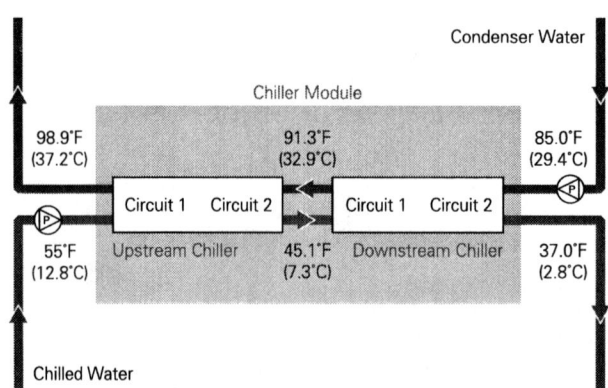

Fig. 2 Module with dual-circuit chillers in series provides 8–12 stages of compression and uses 0.445 kW per ton for the chillers at standard ARI rating conditions.

DESIGN PARAMETERS

A larger-than-conventional difference between the entering and leaving chilled water temperatures permits a lower flow rate, reducing the initial costs for distributing the chilled water (pumps, piping) in central chilled water plants. Smaller pipes and pumps can then be used to satisfy the same capacity.

Because supplying colder chilled water requires more power from the chillers, the cost savings from reducing the pumping power and pipe size and installation must offset the chiller power increase. Chiller designs and controls have improved to the point at which producing 37°F water no longer causes concern for freezing evaporator tubes. Experience shows that fast, accurate chiller controls and algorithms can safely accommodate temperatures as low as 34°F without the addition of antifreeze.

- Entering-chiller water temperature: 55°F.
- Leaving-chiller water temperature: 37°F.
- Evaporator flow rate/capacity: 1.33 gpm/ton.

Many plant configurations are possible:

- Both evaporators and condensers in parallel.
- Evaporators in series and condensers in parallel.
- Both evaporators and condensers in series.

At design conditions:

- Chilled water enters the upstream chiller at 55°F and exits at 45.1°F.
- Chilled water enters the downstream chiller at 45.1°F and exits at 37°F.
- Condenser water enters the downstream chiller at 85°F and exits at 91.3°F.
- Condenser water enters the upstream chiller at 91.3°F and exits at 98.9°F.

FULL-LOAD EFFICIENCY IMPROVEMENT

The series–series counterflow arrangement yields the lowest full-load chiller power (about 14% lower than the parallel–parallel configuration) (Fig. 3). The dramatic reduction in chiller power occurs because the upstream chiller in the series–series counterflow arrangement operates at a higher chilled water temperature, which means that the refrigerant temperature and refrigerant pressure in the evaporator are also higher in the upstream machine. The downstream chiller "sees" a lower condenser-leaving water temperature—and therefore has a lower condenser refrigerant pressure—than it would in a plant with the chiller condensers arranged in parallel. Fig. 1 illustrates the concept of reduced lift using the design

Arrangement		Chillers*		Evaporator				Condenser				Cooling Towers		System	
Evaporator	Condenser	Units/ modules	Compressor efficiency kW/ton	Flow gpm	ΔP Feet of Water	Number of pumps	Power per pump kW	Flow gpm	ΔP Feet of Water	Number of Pumps	Power per pump kW	Number of cells	Power per cell kW	Total power kW	Life-cycle cost $USD
Parallel	Parallel	5/5	0.649	2,800	3.26	5	2.18	4,200	3.66	5	3.67	8	60	7324	18,836,302
Parallel	Parallel	6/6	0.618	2,333	4.18	6	2.33	3,500	3.53	6	2.95	8	60	7001	18,076,391
Series	Series-Counterflow (1.5 gpm/ton)	6/3	0.560	4,667	17.96	3	19.99	5,250	14.8	3	18.54	8	48	6379	16,819,167
Series	Series-Counterflow (2.0 gpm/ton)	6/3	0.535	4,667	17.96	3	19.99	7,000	25.2	3	42.08	8	60	6284	16,656,947
Series	Parallel (2.0 gpm/ton)	6/3	0.555	4,667	17.96	3	19.99	3,500	3.53	6	2.95	8	60	6385	16,888,493

*The chillers represented in this table all have dual refrigerant circuits. The full analysis included single refrigerant circuit chillers at various flow rates and efficiencies.

Fig. 3 Projected energy-use and lifecycle costs for series and parallel chiller configurations.

parameters for this chilled water plant. Chiller power can be reduced by decreasing compressor "lift." In this example, the difference in average lift at design is nearly 13%.

$$1 - \frac{(54.3 + 53.8)/2}{61.9} = 0.126$$

Now, consider a series–series counterflow arrangement of two dual-circuited chillers. Because each of the chillers in this design has two refrigeration circuits, the reduced lift effect is multiplied. Instead of two lifts, there are four. The difference in average lift at design for the system with four independent refrigeration circuits in a series–series counterflow arrangement exceeds 19%.

$$1 - \frac{(51.1 + 50.3 + 50.0 + 48.9)/4}{61.9} = 0.191$$

At the design conditions defined for the system, chiller performance is well above the 6.1 COP requirement set by ANSI/ASHRAE/IESNA Standard 90.1-2004, Energy Standard for Buildings Except Low-Rise Residential Buildings. At standard ARI rating conditions, each chiller module would operate with an efficiency of 0.445 kW/ton.

PART-LOAD EFFICIENCY IMPROVEMENT

The reduction in lift provided by the series–series counterflow arrangement also occurs at part-load conditions. The temperature of the water leaving the evaporator of the upstream chiller is always warmer than the system water, and the temperature of the water leaving the condenser of the downstream chiller is always cooler than the system water. The upstream chiller does not need to perform the same amount of cooling as the downstream chiller. The benefit comes from the upstream chiller's ability to produce chilled water at an elevated temperature.

SAVINGS AMPLIFIED BY POWER INFRASTRUCTURE

While the chiller performance is remarkable with this design, the performance conditions for this application were carefully selected to optimize the overall energy consumption of the entire chilled water plant. Series chiller configurations are not just about the chiller or the pump savings, but reduced electrical infrastructure requirements and environmental impact. The previous example was a 10,500-ton plant. Consider the reduction in power generation requirements when multiplied tenfold. In a 100,000-ton cooling and power infrastructure project, increasingly common in the Middle East and China, our example's 1040-kW reduction blossoms into nearly 10 MW. A conservative estimate for the cost of the generation equipment is $1000 per kW. Using that round budget number, series chiller configuration and low-flow, low-temperature conditions could save $10 million or more in power generation equipment on a 100,000-ton project.

PACKAGED SOLUTIONS MINIMIZE COMPLEXITY AND RISK

Frequent users of these concepts are packaged chiller plant manufacturers that are pre-engineering and packaging series-counterflow chillers with built-in optimization controls. Packaged chiller plants utilizing series-counterflow chiller arrangements are currently available in sizes up to 8000 tons. The chillers in these packaged chiller plants can be factory-performance tested in accordance with ARI procedures prior to shipment from the chiller manufacturers' facility.

Packaging companies and astute engineers have put two series-counterflow chillers in series with each other—essentially creating a 4-chiller series module (Fig. 2). These solutions enhance the thermodynamic benefit created by series-counterflow chillers while minimizing complexity and risk for the system owner and operator.

CONTROL STRATEGIES

How the plant should respond to varying system conditions is a topic for discussion with the design engineer, plant owner, and plant operators. For example, if the entering-chiller water temperature is not reaching design conditions, the operators could:

1. Increase pump speed or turn on more pumps to increase flow rates and more fully load the active chillers.
2. Reset the setpoints of the upstream chillers to 55% of the total temperature difference. Lowering the setpoint of the upstream chillers as the result of a drop in entering-chiller water temperature lessens the benefit of reduced lift. However, the upstream chillers will always run at a higher evaporator pressure than the downstream chillers, which saves energy consumption and costs.
3. Address chiller sequencing in the context of the system options, variable or constant chiller flow, extra pumps, or other concerns. It might be most cost effective to use a startup strategy that fully loads one chiller module and then activates the remaining chillers in modules (pairs). Activating the upstream chiller and operating it at the higher water temperature takes advantage of all of the available heat transfer surface area without increasing the energy consumed by ancillary equipment.

USEFUL REDUNDANCY

Large chiller plants can be more adaptive and efficient with multiple chillers rather than fewer large, field-erected chillers. In plants with more chillers, redundancy is easily created through parallel banks of upstream and downstream chillers. Different combinations of upstream and downstream chillers can meet the load, so if one chiller is being serviced, its duty can be spread out to the other chillers. The same is true for pumps, which do not have to be sequenced with the chillers.

EFFICIENCY, FLEXIBILITY IN SMALLER PLANTS AND RETROFITS

The benefits of low flow, low temperature, and high efficiency apply to other types of chillers as well. Smaller, noncentrifugal chillers can benefit proportionately more under these conditions when placed in series.

Helical-rotary chillers are sensitive to increased lift and decreased condenser water flow. Absorption chillers struggle to make water colder than 40°F, and their cooling capacity increases when placed upstream. Both can be put upstream in the sidestream position for reduced first cost and higher efficiency. Reusing existing, older, less efficient chillers upstream is also an interesting option to explore. These sidestream configurations combine the benefits of series and parallel chillers while isolating some chillers from water flow variations.

CONCLUSION

As chiller efficiencies continue to improve, district energy and central plant designers can optimize the entire system to achieve even lower costs of ownership. Owners can expect more first cost and energy savings from low-flow, low-temperature and highly efficient chiller configurations. The unique benefits and flexibility of series chiller plant designs with variable-primary pumping arrangements include lower overall chilled water system operating costs, reduced emissions, and improved environmental responsibility.

REFERENCES

1. *Standard for water chilling packages using the vapor compressor cycle*, ARI Standard 550/590-2003, Air-Conditioning and Refrigeration Institute, Arlington, VA, 2003.
2. *ASHRAE Handbook-HVAC Systems and Equipment*, Chapter 11, District Heating and Cooling, 2000; 11.1-11.34.
3. Coad, W.J. A fundamental perspective on chilled water systems. Heat Piping Air Cond. **1998**, *August*, 59–66.
4. Trane, *Multiple-Chiller-System Design and Control*, SYS-APM001-EN, American Standard Inc., 2001; 58–59.
5. Groenke, S.; Schwedler, M. ASHRAE J. **2002**, *44* (6), 23–31.

BIBLIOGRAPHY

1. *ASHRAE Handbook-HVAC Systems and Equipment Handbook*, Chapter 12, Hydronic Heating and Cooling System Design, 12.1-12.18.
2. Avery, G. Improving the efficiency of chilled-water plants. ASHRAE J. **2001**, *43* (5), 14–18.
3. Avery, G. Controlling chillers in variable flow systems. ASHRAE J. **1998**, *40* (2), 42–45.
4. Bahnfleth, W.P.; Peyer, E. Comparative analysis of variable and constant primary-flow chilled-water-plant performance. HPAC Eng. **2001**, *April*, 50.
5. Demirchian, G.H.; Maragareci, M.A. *The Benefits of Higher Condenser Water DT at Logan International Airport central chilled water plant*, IDEA 88th Annual conference Proceeding, 1997; 291–300.
6. Hartman, T. All-variable speed centrifugal chiller plants. ASHRAE J. **2001**, *43* (9), 43–51.
7. Kelly, D.W.; Chan, T. Optimizing chilled water plants. Heat Piping Air Cond. **1999**, *January*, 145–147.
8. Kirsner, W. The demise of the primary–secondary pumping paradigm for chilled water plant design. Heat Piping Air Cond. **1996**, *November*, 73–78.
9. Luther, K. Applying variable volume pumping. Heat Piping Air Cond. **1998**, *October*, 53–58.
10. Rishel, J.B. *HVAC Pump Handbook*; McGraw Hill: New York, 1996; 109–112.
11. Schwedler, M.; Bradley, D. Variable-primary-flow systems: An idea for chilled-water plants the time of which has come. Heat Piping Air Cond. **2000**, *April*, 41–44.
12. Taylor, S. Primary-only vs primary–secondary variable flow chilled water systems. ASHRAE J. **2002**, *44* (2), 25–29.
13. Waltz, J.P. Don't ignore variable flow. Contracting Bus. **1997**, *July*, 133–144.

District Energy Systems

Ibrahim Dincer
Faculty of Engineering and Applied Science, University of Ontario Institute of Technology (UOIT), Oshawa, Ontario, Canada

Arif Hepbasli
Department of Mechanical Engineering, Faculty of Engineering, Ege University, Izmir, Turkey

Abstract
This entry presents some historical background on district heating and cooling along with cogeneration and geothermal applications, and discusses some technical, economical, environmental, and sustainability aspects of geothermal energy and performance evaluation tools in terms of energy and exergy analyses for district heating systems. Case studies are also presented to highlight the importance of exergy use as a potential tool for system analysis, design, and improvement.

NOMENCLATURE

COP	coefficient of performance
E	energy
Ex	exergy
f	figure of merit
h	specific enthalpy
Q	heat interaction
R	energy grade function
s	specific entropy
T	temperature
W	shaft work
η	energy efficiency
τ	exergetic temperature factor
ψ	exergy efficiency

Subscripts

C	cooling
ch	chiller
d	natural direct discharge
dest	destruction
DH	district heating
elec	electrical
equiv	equivalent
f	fuel
gen	generation
H	heating
HE	heat exchanger
heat	heat
in	inlet
net	net
o	environmental state
out	outlet
r	reinjected
sys	system
tot	total
UC	user cooling
UH	user heating

Superscripts

·	rate with respect to time
CHP	combined heat and power (cogeneration)
r	room
s	space
u	user
w	water

Abbreviations

CHP	combined heat and power
DC	district cooling
DES	district energy system
DH	district heating
DHC	district heating and cooling
GDHS	geothermal district heating system
IBGF	Izmir–Balcova geothermal field
IEA	International Energy Agency
IPCC	Intergovernmental Panel on Climate Change

INTRODUCTION

District energy systems (DESs) can utilize a wide range of energy resources, ranging from fossil fuels to renewable energy to waste heat. They are sometimes called community energy systems because, by linking a

Keywords: District heating; District cooling; Cogeneration; Efficiency; Energy; Exergy; Environment; Sustainability; Performance; System.

community's energy users, DESs maximize efficiency and provide opportunities to connect generators of waste energy (e.g., electric power plants or industrial facilities) with consumers who can use that energy. The heat recovered through district energy can be used for heating or can be converted to cooling using absorption chillers or steam turbine drive chillers.

District energy system cover both district heating (DH) and district cooling (DC), and distribute steam, hot water, and chilled water from a central plant to individual buildings through a network of pipes. District energy systems provide space heating, air conditioning, domestic hot water, and/or industrial process energy, and often also cogenerate electricity. With district energy, boilers and chillers in individual buildings are no longer being required. District energy is considered an attractive, more efficient, and more environmentally friendly way to reduce energy consumption.

A basic DH system consists of three main parts, as shown in Fig. 1a, and for DC the flow chart becomes somewhat different, as shown in Fig. 1b. An example is that the DC system uses hot water produced from the DH system to operate an absorption refrigerator installed in the substation of the consumers' building instead of the conventional vapor-compression refrigeration system operated by electricity.

Note that storage of chilled water or ice is an integral part of many DC systems. Storage allows cooling energy to be generated at night for use during the hottest part of the day, thereby helping manage the demand for electricity and reducing the need to build power plants.[1]

District heating and cooling (DHC) systems can provide other environmental and economic benefits, including:

- Reduced local and regional air pollution
- Increased opportunities to use ozone-friendly cooling technologies
- Infrastructure upgrades and development that will provide new jobs
- Enhanced opportunities for electric peak reduction through chilled water or ice storage
- Increased fuel flexibility
- Better energy security

In fact, the Intergovernmental Panel on Climate Change (IPCC) has identified cogeneration/DHC as a key greenhouse-gas reduction measure, and the European Commission has been developing a European Union cogeneration/DHC strategy.

District heating and cooling potential can be realized through policies and measures to increase awareness and knowledge; recognize the environmental benefits of district energy in air-quality regulation; encourage investment; and facilitate increased use of district energy in government, public, commercial, industrial, and residential buildings.

During the past few decades there have been various key initiatives taken by major energy organizations (e.g., International Energy Agency (IEA), U.S. Department of Energy, Natural Resources Canada, etc.) on the implementation of DHC all over the world as one of the most significant ways to:

- Maximize the efficiency of the electricity generation process by providing a means to use the waste heat, saving energy while displacing the need for further heat-generating plants
- Share heat loads, thereby using plants more effectively and efficiently
- Achieve fuel flexibility and provide opportunities for introduction of renewable sources of energy as well as cogeneration and industrial waste heat

Furthermore, the IEA recently developed a strategic document[2] as an implementing agreement on DHC including the integration of combined heat and power (CHP), focusing on:

- Integration of energy-efficient and renewable energy systems for limited emissions of greenhouse gases

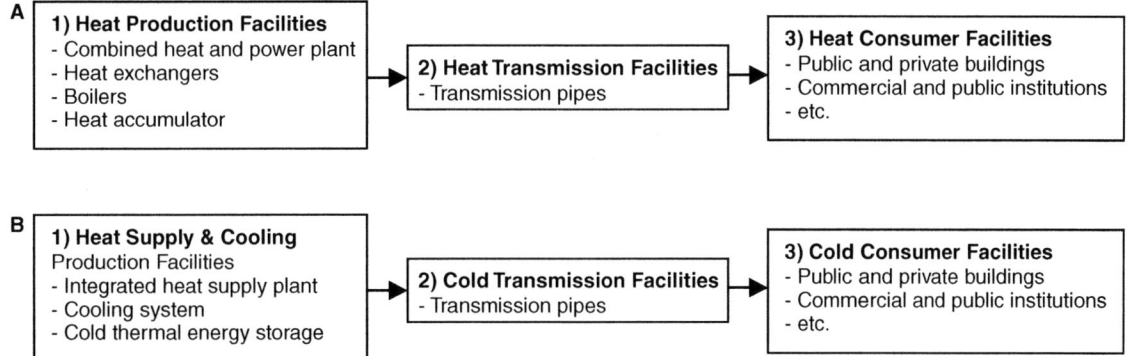

Fig. 1 A basic flow chart for (A) DH application and (B) DC application.

- Community system integration and optimization, and use of waste thermal energy, renewable energy and CHP, for a better environment and sustainability
- Reliability, robustness, and energy security for effective maintenance and management of buildings
- Advanced technologies for improved system integration, including information systems and controls
- Dissemination and deployment for rapid change toward energy efficiency and sustainability

In this entry we present some historical background on DHC systems and applications, and discuss some technical, economical, environmental, and sustainability aspects of these systems and their performance evaluations tools in terms of energy and exergy efficiencies. A case study is also presented to highlight the importance of exergy use as a potential tool for system analysis, design, and improvement.

COGENERATION AS A KEY PART OF DISTRICT HEATING AND COOLING

Cogeneration, also referred to as CHP, is the simultaneous sequential production of electrical and thermal energy from a single fuel. During the past couple of decades, cogeneration has become an attractive and practical proposition for a wide range of thermal applications, including DHC. Some examples are the process industries (pharmaceuticals, paper and board, cement, food, textile, etc.); commercial, government, and public-sector buildings (hotels, hospitals, swimming pools, universities, airports, offices, etc.); and DHC schemes. Fig. 2 shows a comparison of conventional power systems and cogeneration systems. The main drawback in the conventional system is the amount of intensive heat losses, resulting in a drastic drop in efficiency. The key question is how to overcome this and make the system more efficient. The answer is clear: by cogeneration. In this regard, we minimize the heat losses and increase the efficiency, and provide the opportunity to supply heat to various applications and facilities. The overall thermal efficiency of the system is the percentage of the fuel converted to electricity plus the percent of fuel converted to useful thermal energy. Typically, cogeneration systems have overall efficiencies ranging from 65 up to 90%, respectively.

The key point here is that the heat rejected from one process is used for another process, which makes the system more efficient compared with the independent production of both electricity and thermal energy. Here, the thermal energy can be used in DH and/or DC applications. Heating applications basically include generation of steam or hot water. Cooling applications basically require the use of absorption chillers that convert heat to cooling. Numerous advanced technologies are available to achieve cogeneration, but the system requires an electricity generator and a heat-recovery system for full functioning.

Cogeneration has been widely adopted in many European countries for use in industrial, commercial/institutional, and residential applications. It currently represents 10% of all European electricity production and more than 30% of electricity production in Finland, Denmark, and the Netherlands. Within Canada, however, cogeneration represents just over 6% of national electricity production.[3] This relatively lower penetration is attributed to Canada's historically low energy prices and electric-utility policies on the provision of backup power and the sale of surplus electricity. Despite these conditions, cogeneration has been adopted in some industrial applications, notably the pulp-and-paper and chemical products sectors, where a large demand for both heat and electricity exists. Several classical technologies currently are available for cogeneration, such as steam turbines, gas turbines, combined cycle (both steam and gas) turbines, and reciprocating engines (gas and diesel). In addition, there has been increasing interest in using some new technologies—namely, fuel cells, microturbines, and Stirling engines. Note that heat output from

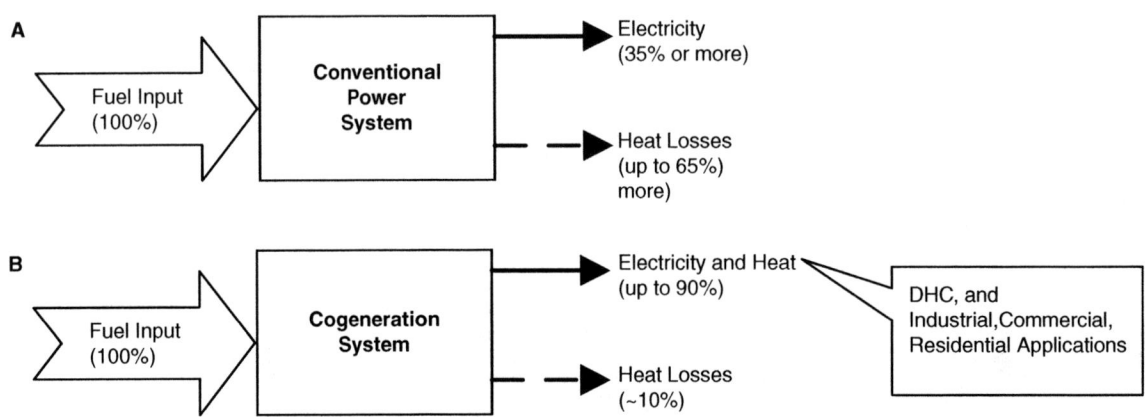

Fig. 2 Illustration of (A) conventional power system and (B) cogeneration system.

Table 1 Main characteristics and technical aspects of cogeneration systems

Technology	Fuel type	Capacity (MW$_e$)	Electrical efficiency (%)	Overall efficiency (%)	Average capital cost (US$/kW$_e$)	Average maintenance cost (US$/kWh)
Steam turbine	Any	0.5–500	7–20	60–80	900–1800	0.0027
Gas turbine	Gaseous and liquid fuels	0.25–50 or more	25–42	65–87	400–850	0.004–0.009
Combined cycle	Gaseous and liquid fuels	3–300 or more	35–55	73–90	400–850	0.004–0.009
Reciprocating engines	Gaseous and liquid fuels	0.003–20	25–45	65–92	300–1450	0.007–0.014
Micro turbines	Gaseous and liquid fuels	-	15–30	60–85	600–850	<0.006–0.01
Fuel cells	Gaseous and liquid fuels	0.003–3 or more	35–50	80–90	—	—
Stirling engines	Gaseous and liquid fuels	0.003–1.5	~40	65–85	—	—

Source: From United Nations Environment Programme (see Ref. 4).

the system varies greatly depending on the system type. The output can range from high-pressure, high-temperature (e.g., 500°C–600°C) steam to hot water (e.g., 90°C). High-pressure, high-temperature steam is considered to be high-quality thermal output because it can meet most industrial-process needs. Hot water is considered to be a low-quality thermal output because it can be used for only a limited number of DHC applications.

Cogeneration can be based on a wide variety of fuels, and individual installations may be designed to accept more than one fuel. Although solid, liquid, or gaseous fossil fuels currently dominate, cogeneration from biomass fuels is becoming increasingly important. Sometimes, fuels are used that otherwise would constitute waste (e.g., refinery gases, landfill gas, agricultural waste, and forest residues). These substances increase the cost efficiency of cogeneration.[4] Table 1 gives a comprehensive summary of cogeneration technologies in terms of fuel type, capacity, efficiency, average capital cost, maintenance cost, etc.

HISTORY OF DISTRICT HEATING AND COOLING

The oldest DH system was created in the early 14th century in the village of Chaudes-Aigues Cantal in France. This system distributed warm water through wooden pipes and is still in use today. The first commercial DH system was created by Birdsill Holly in Lockport, New York in 1877. In this system the boiler was used as the central heat source; it supplied a loop consisting of steam pipes, radiators, and even condensate-return lines. Later, the system attracted a dozen customers. Only three years later, it served several factories as well as residential customers and had extended to a ~5 km loop. The roots of DC go back to the 19th century. It was initially introduced as a scheme to distribute clean, cool air to houses through underground pipes. The first known DC system began operations at Denver's Colorado Automatic Refrigerator Company in late 1889. In the 1930s, large DC systems were created for Rockefeller Center in New York City and for the U.S. Capitol in Washington, DC.

It is believed that district energy in Canada began in London, Ontario in 1880. The London system was built in the form of a group of systems serving the university, hospital, and government complexes. The University of Toronto is known to have developed a DH system in 1911 that served the needs of the university. The first commercial DH system in Canada was established in 1924, in the city of Winnipeg's commercial core. Canada boasts the site of one of the northernmost DESs in North America: Fort McPherson, located in the Northwest Territories. The Canadian District Energy Association (CDEA) was created in 1993 in recognition of the fact that the emerging Canadian district energy industry needed to create a common voice to promote DHC applications. It aims to exchange and share information and experience for its stakeholders. It has also been instrumental in helping provide a forum for the exchange of ideas and information, and in identifying and addressing key technical and policy issues to advance the use of district energy in Canada.[5]

TECHNICAL ASPECTS OF DISTRICT HEATING AND COOLING

District heating and cooling is the distribution of heating (hot water, steam) and cooling (cold water) energy transfer mediums from a central energy production source to meet the diverse thermal energy needs of residential, commercial, and industrial users. Thermal energy needs or demands include space heating and cooling systems for maintaining human comfort, domestic hot water requirements, manufacturing-plant process heating and cooling system requirements, etc. In many of the systems that have been established around the world, both district heating and district cooling have not been provided. In Europe, for example, where moderate summer temperatures prevail, most DESs provide heating capability only. DC has only recently become more widespread, with the most prevalent application being in North America, where summer temperatures can reach extremes of 30°C–40°C over extended periods.

To implement a DH, DC, or DHC in a community, one should weigh several factors to conduct a feasibility study for determining whether a DH, DC, or DHC system is suitable for application. Essential factors include energetic, environmental, economic, and social criteria; operating conditions; fuel availability; efficiency considerations; local benefits; viability of competing systems; local climatic conditions; user characteristics such as load density; total load requirements; characteristics of the heating and cooling systems currently in place; developer's perspectives; and local utility considerations.

As shown in Fig. 1, a basic DHC consists of three subsystems—namely, energy generation, energy transmission, and energy use. These subsystems are described below:

1. *Energy generation.* In this section, steam or hot water (in the case of DH) and chilled or cold water (in the case of DC) are produced.
2. *Energy transmission.* In this part, the thermal energy medium (steam or water) is distributed via pipelines from the production sources to the network of users.
3. *Energy use.* In this section, energy (either heat or cold) is consumed for either heating or cooling purposes in the facilities. A combination of residential, public, commercial, and industrial users may be involved with varying uses of the thermal energy, including space heating and cooling, domestic water heating, and plant process heating and cooling.

Note that a DH and/or DC system differs fundamentally from a conventional system in that in the latter system, thermal energy is produced and distributed at the location of use. Examples of conventional systems include residential heating and cooling with, respectively, furnaces and air conditioners; electric heating of offices; package boilers/chillers providing heating/cooling of apartment complexes; and a dedicated boiler plant providing heat to an industrial facility.

DISTRICT HEATING AND COOLING AND ENVIRONMENTAL IMPACT

Problems with energy supply and use are related not only to global warming, but also to such environmental concerns as air pollution, acid precipitation, ozone depletion, forest destruction, and emission of radioactive substances. These issues must be taken into consideration simultaneously if humanity is to achieve a bright energy future with minimal environmental impact. Much evidence exists to suggest that the future will be negatively impacted if humans keep degrading the environment.

One solution to both energy and environmental problems is to enhance much more use of DHC applications.

Numerous fuels are used at DHC plants, including various grades of oil and coal, natural gas, refuse, and other biofuels (e.g., wood chips, peat, and straw). The combustion of such fuels may produce environmentally hazardous products of combustion; thus, flue-gas cleaning devices and other emission reduction measures are often incorporated. Some measures are usually required under increasingly strict legislation before approval to operate a facility is granted. Examples of pollution control equipment used at DHC plants include acid–gas scrubbers. These systems typically utilize hydrated lime to react with the moisture, SO, and other acid gases in the flue gases discharged from the combustion system. With such systems, the lime–acid, gas–water vapor reaction products are efficiently collected by electrostatic precipitators as particulate matter. Bag filters are also utilized in many applications to capture the particulate matter as well as the acid–gas scrubbing reaction products. Conventional oil-/gas-fired boilers utilizing low NO_x, burners to reduce NO_x emissions dramatically are also becoming more common. Flue-gas recirculation to reduce NO_x emissions has also proved to be effective. Other emission control or reduction techniques can be introduced with DHC systems, including optimization of combustion efficiency (i.e., reduced CO_2, CO, and hydrocarbon emissions) through the use of modern computerized combustion control systems, and utilization of higher-quality, lower-emission-producing fuels. With the above, it is apparent that heating and cooling systems that minimize the quantity of fuel and electrical power required to meet users' needs will result in reduced impact on the environment.

In addition, DHC systems that comprise several types of thermal energy generation plants can optimize plant and system efficiency by utilizing, whenever possible, the thermal energy sources with the highest energy conversion efficiencies for base and other partial load conditions. Then the sources with the poorer conversion efficiencies can be utilized only to meet peak loads. Essentially, improved efficiency means use of less fuel for the same amount of energy produced, which in turn results in the conservation of fossil fuels, reduced emissions of pollutants, improved air quality, and reduced use of CFC refrigerants (if any) in DC applications.

District heating and cooling systems are well suited to combine with electric power production facilities as cogeneration plants. The amalgamation of these two energy production/utilization schemes results in a substantial improvement in overall energy conversion efficiency, because DH systems can effectively utilize the otherwise-wasted heat associated with the electric power production process. A district system meeting much or all of its load requirements with waste heat from power generation facilities will have a positive environmental impact, as fuel consumption within the community is reduced considerably. Conservation of fossil fuels and a reduction of combustion-related emissions are resultant direct benefits of such a DHC system.

The centralized nature of DHC energy production plants results in a reduced number of emissions sources in a community. This introduces the potential for several direct benefits.

The higher operating efficiency afforded to larger, well-maintained facilities translates directly to reduced fuel consumption, which in turn results in conservation of fossil fuels and reduced emissions. Higher operating efficiency of the combustion process (where parameters such as temperature, combustion air and fuel input levels, and residence time are closely monitored) also impacts emission production in that the concentration of certain pollutants produced—particularly CO_2 and NO_x—is reduced.

Furthermore, measures to increase energy efficiency can reduce environmental impact by reducing energy losses. From an exergy viewpoint, such activities lead to increased exergy efficiency and reduced exergy losses (both waste exergy emissions and internal exergy consumption).

A deeper understanding of the relations between exergy and the environment may reveal the underlying fundamental patterns and forces affecting changes in the environment, and help researchers deal better with environmental damage.

The second law of thermodynamics is instrumental in providing insights into environmental impact. The most appropriate link between the second law and environmental impact has been suggested to be exergy, in part because it is a measure of the departure of the state of a system from that of the environment. The magnitude of the exergy of a system depends on the states of both the system and the environment. This departure is zero only when the system is in equilibrium with its environment.

To achieve the energetic, economic, and environmental benefits that DHCs offer, the following integrated set of activities should be carried out[6]:

- *Research and development.* Research and development priorities should be set in close consultation with industry to reflect its needs. Most research is conducted through cost-shared agreements and falls within the short-to-medium term. Partners in these activities should include a variety of stakeholders in the energy industry, such as private-sector firms, utilities across the country, provincial governments, and other federal departments.
- *Technology assessment.* Appropriate technical data should be gathered in the lab and through field trials on factors such as cost benefit, reliability, environmental impact, safety, and opportunities for improvement. These data should also assist the preparation of technology status overviews and strategic plans for further research and development.
- *Standards development.* The development of technical and safety standards is needed to encourage the acceptance of proven technologies in the marketplace. Standards development should be conducted in cooperation with national and international standards-writing organizations, as well as with other national and provincial regulatory bodies.
- *Technology transfer.* Research and development results should be transferred through sponsorship of technical workshops, seminars, and conferences, as well as through the development of training manuals and design tools, Web tools, and the publication of technical reports.

Such activities will also encourage potential users to consider the benefits of adopting DHC applications, using renewable energy resources. In support of developing near-term markets, a key technology transfer area is to accelerate the use of cogeneration and DHC applications, particularly for better efficiency, cost effectiveness, and environmental considerations.

DISTRICT HEATING AND COOLING AND SUSTAINABLE DEVELOPMENT

Sustainable development requires a sustainable supply of clean and affordable energy resources that do not cause negative societal impacts.[7] Supplies of such energy resources as fossil fuels and uranium are finite. Green energy resources (e.g., solar and wind) are generally considered to be renewable and, therefore, sustainable over the relatively long term.

Sustainability often leads local and national authorities to incorporate environmental considerations into energy planning. The need to satisfy basic human needs and aspirations, combined with increasing world population, will make the need for successful implementation of sustainable development increasingly apparent. Various criteria that are essential to achieving sustainable development in a society follow:

- Information about and public awareness of the benefits of sustainability investments

- Environmental education and training
- Appropriate energy and exergy strategies
- The availability of renewable energy sources and cleaner technologies
- A reasonable supply of financing
- Monitoring and evaluation tools

The key point here is to use renewable energy resources in DHC systems. As is known, not all renewable energy resources are inherently clean, in that they cause no burden on the environment in terms of waste emissions, resource extraction, or other environmental disruptions. Nevertheless, the use of DHC almost certainly can provide a cleaner and more sustainable energy system than increased controls on conventional energy systems.

To seize the opportunities, it is essential to establish a DHC market and gradually build up the experience with cutting-edge technologies. The barriers and constraints to the diffusion of DHC use should be removed. The legal, administrative, and financing infrastructure should be established to facilitate planning and application of geothermal energy projects. Government could and should play a useful role in promoting geothermal energy technologies through funding and incentives to encourage research and development, as well as commercialization and implementation in both urban and rural areas.

Environmental concerns are significantly linked to sustainable development. Activities that continually degrade the environment are not sustainable. The cumulative impact on the environment of such activities often leads over time to a variety of health, ecological, and other problems. Clearly, a strong relationship exists between efficiency and environmental impact, because for the same services or products, less resource utilization and pollution are normally associated with increased efficiency.[8]

Improved energy efficiency leads to reduced energy losses. Most efficiency improvements produce direct environmental benefits in two ways. First, operating energy input requirements are reduced per unit output, and the pollutants generated are reduced correspondingly. Second, consideration of the entire life cycle for energy resources and technologies suggests that improved efficiency reduces environmental impact during most stages of the life cycle.

In recent years, the increased acknowledgment of humankind's interdependence with the environment has been embraced in the concept of sustainable development. With energy constituting a basic necessity for maintaining and improving standards of living throughout the world, the widespread use of fossil fuels may have impacted the planet in ways far more significant than first thought. In addition to the manageable impacts of mining and drilling for fossil fuels, and discharging wastes from processing and refining operations, the "greenhouse" gases created by burning these fuels is regarded as a major contributor to a global-warming threat. Global warming and large-scale climate change have implications for food-chain disruption, flooding, and severe weather events.

Use of renewable energy sources in DHC systems with cogeneration can help reduce environmental damage and achieve sustainability.

Sustainable development requires not just that sustainable energy resources be used, but also that the resources be used efficiently. The authors and others feel that exergy methods can be used to evaluate and improve efficiency, and thus to improve sustainability. Because energy can never be "lost," as it is conserved according to the first law of thermodynamics, whereas exergy can be lost due to internal irreversibilities, this study suggests that exergy losses that represent potential not used, particularly from the use of nonrenewable energy forms, should be minimized when striving for sustainable development. In the next section, the authors discuss the exergetic aspects of thermal systems and present an efficiency analysis for performance improvement.

Furthermore, this study shows that some environmental effects associated with emissions and resource depletion can be expressed based on physical principles in terms of an exergy-based indicator. It may be possible to generalize this indicator to cover a comprehensive range of environmental effects, and research in line with that objective is ongoing.

Although this work discusses the benefits of using thermodynamic principles—especially exergy—to assess the sustainability and environmental impact of energy systems, this area of work is relatively new. Further research is needed to gain a better understanding of the potential role of exergy in such a comprehensive perspective. This includes the need for research to (1) better define the role of exergy in environmental impact and design; (2) identify how exergy can be better used as an indicator of potential environmental impact; and (3) develop holistic exergy-based methods that simultaneously account for technical, economic, environmental, and other factors.

PERFORMANCE EVALUATION

From the thermodynamics point of view, exergy is defined as the maximum amount of work that can be produced by a system or a flow of matter or energy as it comes to equilibrium with a reference environment. Exergy is a measure of the potential of the system or flow to cause change, as a consequence of not being completely in stable equilibrium relative to the reference environment. Unlike energy, exergy is not subject to a conservation law (except for ideal, or reversible, processes). Rather, exergy is consumed or destroyed, due to irreversibilities in any real process. The exergy consumption during a process is proportional to the entropy created due to irreversibilities associated with the process.

Exergy analysis is a technique that uses the conservation of mass and conservation of energy principles together with the second law of thermodynamics for the analysis, design, and improvement of DHC systems, as well as others. It is also useful for improving the efficiency of energy-resource use, for it quantifies the locations, types, and magnitudes of waste and loss. In general, more meaningful efficiencies are evaluated with exergy analysis rather than energy analysis, because exergy efficiencies are always a measure of how nearly the efficiency of a process approaches the ideal. Therefore, exergy analysis identifies accurately the margin available to design more efficient energy systems by reducing inefficiencies. We can suggest that thermodynamic performance is best evaluated using exergy analysis because it provides more insights and is more useful in efficiency-improvement efforts than energy analysis. For exergy analysis, the characteristics of a reference environment must be specified. This is commonly done by specifying the temperature, pressure, and chemical composition of the reference environment. The results of exergy analyses, consequently, are relative to the specified reference environment, which in most applications is modeled after the actual local environment. The exergy of a system is zero when it is in equilibrium with the reference environment.

In exergy analysis, the temperatures at different points in the system are important, and the equivalent temperature T_{equiv} between the supply (Eq. 1) and return (Eq. 2) temperatures can be written as

$$T_{equiv} = \frac{h_1 - h_2}{s_1 - s_2} \quad (1)$$

where h and s denote specific enthalpy and specific entropy, respectively.

Here in the cogeneration section, the electricity production rate $\dot{W}$ can be expressed for a cogeneration-based system using electric chillers as a function of the product-heat generation rate $\dot{Q}_H$ as

$$\dot{W} = \left(\frac{\eta_{elec}^{CHP}}{\eta_{heat}^{CHP}}\right)\dot{Q}_H \quad (2)$$

and for a cogeneration-based system using absorption chillers as a function of the product-heat generation rates, $\dot{Q}_H$ and $\dot{Q}_{gen}$, as

$$\dot{W} = \left(\frac{\eta_{elec}^{CHP}}{\eta_{heat}^{CHP}}\right)(\dot{Q}_H + \dot{Q}_{gen}) \quad (3)$$

where η_{elec}^{CHP} and η_{heat}^{CHP} denote, respectively, the electrical and heat efficiencies of the cogeneration, or CHP, plant.

The total energy efficiency can be written for the cogeneration plant using electric chillers as

$$\eta^{CHP} = \frac{\dot{W} + \dot{Q}_H}{\dot{E}_f} \quad (4)$$

and for the cogeneration plant using absorption chillers as

$$\eta^{CHP} = \frac{\dot{W} + \dot{Q}_H + \dot{Q}_{gen}}{\dot{E}_f} \quad (5)$$

where $\dot{E}_f$ denotes the fuel energy input rate. The corresponding total exergy efficiency can be expressed for the cogeneration plant using electric chillers as

$$\psi^{CHP} = \frac{\dot{W} + \tau_{Q_H}\dot{Q}_H}{R\dot{E}_f} \quad (6)$$

and for the cogeneration plant using absorption chillers as

$$\psi^{CHP} = \frac{\dot{W} + \tau_{Q_H}\dot{Q}_H + \tau_{Q_{gen}}\dot{Q}_{gen}}{R\dot{E}_f} \quad (7)$$

where τ_{Q_H} and $\tau_{Q_{gen}}$ are the exergetic temperature factors for $\dot{Q}_H$ and $\dot{Q}_{gen}$, respectively.

For heat transfer at a temperature T, the exergetic temperature factor can be written as

$$\tau \equiv 1 - \frac{T_0}{T} \quad (8)$$

The fuel exergy flow rate, which is all chemical exergy, is evaluated as $R\dot{E}_f$ here, where R and $\dot{E}_f$ denote, respectively, the energy grade function and energy flow rate of the fuel.

In the chilling process, the exergy efficiency can be written for the chilling operation using electric chillers as

$$\psi_{ch} = \frac{-\tau_{Q_C}\dot{Q}_C}{\dot{W}_{ch}} \quad (9)$$

and using absorption chillers as

$$\psi_{ch}' = \frac{-\tau_{Q_C}\dot{Q}_C}{\tau_{Q_{gen}}\dot{Q}_{gen}} \quad (10)$$

In this part, we deal with DHC. District heating utilizes hot-water supply and warm-water return pipes, whereas DC utilizes cold-water supply and cool-water return pipes. The pipes are assumed to be perfectly insulated so that heat loss or infiltration during fluid transport can be minimized. Hence, the energy efficiencies of the DHC portions of the system are both 100%. The exergy efficiency can be evaluated for DH as

$$\psi_{DH} = \frac{\tau_{Q_H^u}\dot{Q}_H^u}{\tau_{Q_H}\dot{Q}_H} \quad (11)$$

and for DC as

$$\psi_{DC} = \frac{-\tau_{Q_C^u}\dot{Q}_C^u}{-\tau_{Q_C}\dot{Q}_C} \quad (12)$$

Heat loss and infiltration for the user heating and cooling subsystems are assumed to be negligible, so that their energy efficiencies are assumed to be 100%. The exergy efficiency can be expressed for the user-heating subsystem as

$$\psi_{UH} = \frac{\tau_{Q_H^{u,s}} \dot{Q}_H^{u,s} + \tau_{Q_H^{u,w}} \dot{Q}_H^{u,w}}{\tau_{Q_H^u} \dot{Q}_H^u} \qquad (13)$$

and for the user-cooling subsystem as

$$\psi_{UC} = \frac{-\tau_{Q_C^{u,r}} \dot{Q}_C^{u,r}}{-\tau_{Q_C^u} \dot{Q}_C^u} \qquad (14)$$

The left and right terms in the numerator of Eq. 13 represent the thermal exergy supply rates for space and water heating, respectively.

For the overall process, because three different products (electricity, heat, and cooling) are generated, application of the term energy efficiency here is prone to be misleading, in part for the same reason that the term energy efficiency is misleading for a chiller. Here, an overall-system "figure of merit" f_{sys} is used, calculated as follows:

$$f_{sys} = \frac{\dot{W}_{net} + \dot{Q}_H^{u,s} + \dot{Q}_H^{u,w} + \dot{Q}_C^{u,r}}{\dot{E}_f} \qquad (15)$$

The corresponding exergy-based measure of efficiency is simply an exergy efficiency and is evaluated as

$$\psi_{sys} = \frac{\dot{W}_{net} + \tau_{Q_H^{u,s}} \dot{Q}_H^{u,s} + \tau_{Q_H^{u,w}} \dot{Q}_H^{u,w} - \tau_{Q_C^{u,r}} \dot{Q}_C^{u,r}}{R\dot{E}_f} \qquad (16)$$

For further details about the energy and exergy analysis of the systems.[9,10]

CASE STUDIES

Here, we present an efficiency analysis, accounting for both energy and exergy considerations, for two case studies—namely, a cogeneration-based DES and a geothermal district heating system (GDHS).

Case Study I

The case considered here for analysis is a major cogeneration-based DHC project in downtown Edmonton, Alberta,[11,12] having (1) an initial supply capacity of 230 MW (thermal) for heating and 100 MW (thermal) for cooling; (2) the capacity to displace about 15 MW of electrical power used for electric chillers through DC; and (3) the potential to increase the efficiency of the Rossdale power plant that would cogenerate to provide the steam for DHC from about 30 to 70%, respectively. The design includes the potential to expand the supply capacity for heating to about 400 MW (thermal). The design incorporated central chillers and a DC network. Screw chillers were to be used originally and absorption chillers in the future. Central chillers are often favored because (1) the seasonal efficiency of the chillers can increase due to the ability to operate at peak efficiency more often in a large central plant; and (2) lower chiller condenser temperatures (e.g., 20°C) can be used if cooling water from the environment was available to the central plant, relative to the condenser temperatures of approximately 35°C needed for air-cooled building chillers. These two effects can lead to large central chillers having almost double the efficiencies of distributed small chillers.

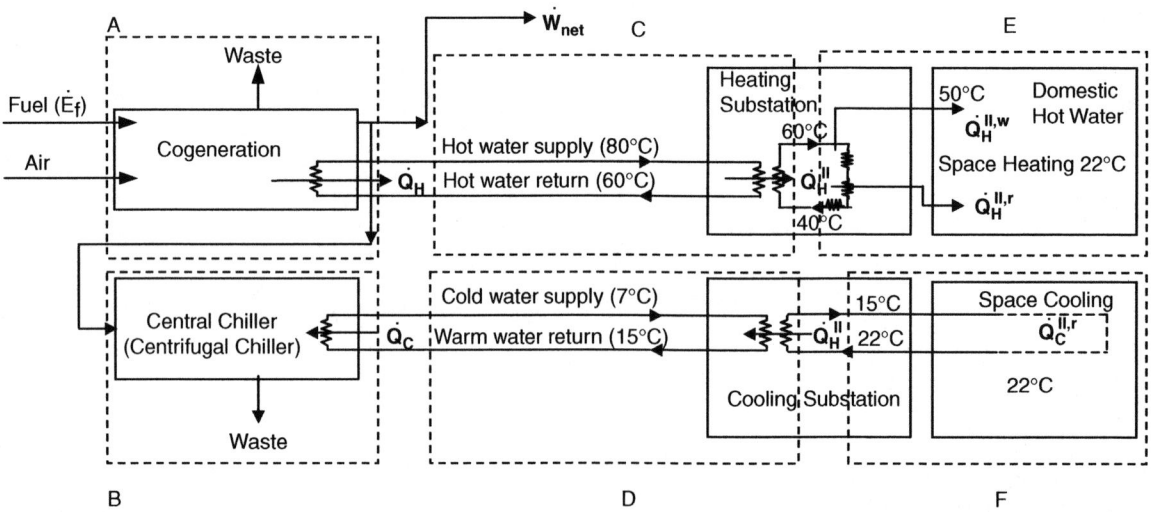

Fig. 3 Simplified diagram of the cogeneration-based district energy system (DES) in the Edmonton Power. The system, which uses electric chillers, is divided into six subsections within three categories. On the left are production processes, including cogeneration of electricity and heat (A) and chilling (B). In the middle are district-energy transport processes, including district heating (DH) (C) and DC (D). On the right are end-user processes, including user heating (E) and user cooling (F).
Source: Adapted from Refs. [9,10].

Table 2 Monthly heating and cooling load breakdown (in %) in the design area of Edmonton, Alberta

	Period 1 (Winter)								Period 2 (Summer)					
	Oct	Nov	Dec	Jan	Feb	Mar	Apr	Total	May	June	July	Aug	Sep	Total
Heating	6.90	12.73	16.83	18.67	14.05	12.95	7.34	89.46	2.39	1.56	1.34	1.92	3.33	10.54
Cooling	0.0	0.0	0.0	0.0	0.0	0.0	0.0	0.0	10.62	22.06	32.00	26.80	8.52	100

Source: Adapted from Refs. [9–11].

There are two main stages in this case study, taken from.[9,10] First, the design for cogeneration-based DHC[11,12] is evaluated thermodynamically. Then the design is modified by replacing the electric centrifugal chillers with heat-driven absorption chillers (first single- and then double-effect types) and reevaluating it thermodynamically.

The cogeneration-based DES considered here (see Fig. 3) includes a cogeneration plant for heat and electricity, and a central electric chiller that produces a chilled fluid. Hot water was produced to satisfy all heating requirements of the users, at a temperature and pressure of 120°C and 2 bar, respectively. The heat was distributed to the users via heat exchangers, DH grids, and user heat-exchanger substations. A portion of the cogenerated electricity was used to drive a central centrifugal chiller, and the remaining electricity was used for other purposes (export, driving other electrical devices, etc.). The central chiller produces cold water at 7°C, which was distributed to users via DC grids.

For the cogeneration-based DES using absorption chillers, the design was modified by replacing the electric chiller with single-effect absorption chillers. Hot water was produced at 120°C and 2 bar to satisfy all heating requirements of the users and to drive the central absorption chillers. A small portion of the cogenerated electricity was used to drive the absorption solution and refrigeration pumps, and the remaining electricity was used for purposes other than space cooling. Then this cogeneration-based DES was further modified by replacing the electric centrifugal chillers with double-effect absorption chillers. The system was similar to the cogeneration-based DES using single-effect absorption chillers, except that higher-quality heat (170°C and 8 bar) was produced to drive the double-effect absorption chillers.

For the analysis, the year was divided into two seasonal periods (see Table 2). Period 1 (October to April) has an environmental temperature of 0°C and was considered to be a winter period with only a heating demand. Period 2 (May to September) has an environmental temperature of 30°C and was considered to be a summer period with a cooling demand and a small demand for water heating. The small variations in plant efficiency that occur with changes in environmental temperature are neglected here.

The overall energy efficiency of the proposed cogeneration plant was 85%; the electrical efficiency (i.e., the efficiency of producing electricity via cogeneration) was 25%; and the heat production efficiency was 60%. Also, the total heating requirement of the buildings in the design region was $\dot{Q}_H = 1040$ GWh/yr for space and water heating, and the cooling requirement was $\dot{Q}_C = 202$ GWh/yr for space cooling. The total fuel energy input rate can be evaluated for cogeneration plant using electric chillers as $\dot{E}_f = 1040/0.6 = 1733$ GWh/yr. Because 33 GWh/yr of this cooling was provided through free cooling, the cooling requirement of the chilling plant was 169 GWh/yr.[11] The COP of the single-effect absorption chiller used here was taken to be 0.67, a typical representative value. Therefore, the annual heat required to drive the single-effect absorption machine was $\dot{Q}_{gen} = 169/0.67 = 252$ GWh/yr. Thus, the total fuel energy input rate to the cogeneration plant can be evaluated as $\dot{E}_f = (1040 + 252)/0.6 = 2153$ GWh/yr.[9,10]

As mentioned above, steam was required at higher temperatures and pressures to drive the double-effect absorption chillers, and more electricity was curtailed as higher quality of heat or more heat was produced. The overall energy efficiency of the proposed cogeneration plant was unchanged (85%) in Period 2. Only the electrical and heat efficiencies changed due to more heat being produced in this period, when the absorption chiller was in operation. Thus, the electrical efficiency (i.e., the efficiency of producing electricity via cogeneration) was 25 and 21%, respectively, in Periods 1 and 2, respectively, and the heat production efficiency was 60 and 64%, respectively, in Periods 1 and 2, respectively. The COP of the double-effect absorption chiller used here was taken to be 1.2, a typical representative value. Therefore, the annual heat required to drive the double-effect absorption machine was $\dot{Q}_{gen} = 169/1.2 = 141$ GWh/yr. The total fuel energy input rate to the cogeneration plant can be evaluated as the sum of the fuel energy input rate to the plant in two periods. Thus, $\dot{E}_f = 1942$ GWh/yr.[9,10]

The average supply and return temperatures, respectively, were taken as 80 and 60°C for DH, and 7 and 15°C for DC. The supply and return temperatures, respectively, were taken as 60 and 40°C for the user-heating substation, and 15 and 22°C for the user-cooling substation. Furthermore, the user room temperature was considered constant throughout the year at 22°C. For DH the equivalent temperature was 70°C for the supply system and 50°C for the user substation, whereas for DC the equivalent temperature was 11°C for the supply system and 19°C for the user substation.

Table 2 shows that 89.46 and 10.54% of the total annual heat loads occur in Periods 1 and 2, respectively. Because there was assumed to be no space heating demand in Period 2, the 10.54% quantity was taken to be the heat needs for water heating (which was assumed to be constant throughout the year). Table 2 also presents the space cooling breakdown in Period 2. Annual energy transfer rates for the cogeneration-based DES are shown in Table 3, with details distinguished where appropriate for the three chiller options considered. The data in Table 3 are used to calculate exergy efficiencies for the systems for each period and for the year.

Edmonton Power had annual free cooling of 33 GWh/yr; the cooling requirement of the chilling plant was 169 GWh/yr. The COP of the centrifugal chiller in the design was 4.5. Thus, the annual electricity supply rate to the chiller was $\dot{W}_{ch} = 169/4.5 = 38$ GWh/yr. For the chilling operation, including free cooling and electrical cooling, COP $= (169 + 33)/38 = 5.32$. The net electricity output ($\dot{W}_{net}$) of the combined cogeneration/chiller portion of the system was $433 - 38 = 395$ GWh/yr, where the electrical generation rate of the cogeneration plant was 433 GWh/yr. Similarly, for the chilling operation, including free cooling and single-effect absorption cooling, the coefficient of performance was COP $= 202/252 = 0.80$,

Table 3 Annual Energy Transfer Rates (in GWh/yr) for the Cogeneration-Based DHC System in Edmonton, Alberta

Type of Energy	Period 1, $T_o = 0°C$	Period 2, $T_o = 30°C$
District heating, $\dot{Q}_H$	$0.8946 \times 1040 = 930$	$0.1054 \times 1040 = 110$
Water heating, $\dot{Q}_H^{u,w}$	(22 GWh/yr/mo.) × 7 mo. = 154	$0.1054 \times 1040 = 110$ (or 22 GWh/yr/mo.)
Space heating, $\dot{Q}_H^{u,s}$	$930 - 154 = 776$	0
Space cooling, $\dot{Q}_C$	0	$1.00 \times 202 = 202$
Electric chiller case		
Total electricity, $\dot{W}$	$0.8946 \times 433 = 388$	$0.1054 \times 433 = 45.6$
Input energy, $\dot{E}_f$	$0.8946 \times 1733 = 1551$	$0.1054 \times 1733 = 183$
Single-effect absorption chiller case		
Heat to drive absorption chiller, $\dot{Q}_{gen}$	0	$1.00 \times 252 = 252$
Total electricity, $\dot{W}$	$0.8946 \times 433 = 388$	$25/60 \, (110 + 252) = 151$
Input energy, $\dot{E}_f$	$0.8946 \times 1733 = 1551$	$(110 + 252)/0.6 = 603$
Double-effect absorption chiller case		
Heat to drive absorption chiller, $\dot{Q}_{gen}$	0	$1.00 \times 141 = 141$
Total electricity, $\dot{W}$	$0.8946 \times 433 = 388$	$21/64 \, (110 + 141) = 82$
Input energy, $\dot{E}_f$	$0.8946 \times 1733 = 1551$	$(110 + 141)/0.64 = 391$

Source: Adapted from Refs. [9,10].

Table 4 System and subsystem efficiencies for the cogeneration-based DES for several types of chillers

	Efficiency (%)					
	Energy (η)			Exergy (ψ)		
System	Centrifugal chiller	1-Stage absorption chiller	2-Stage absorption chiller	Centrifugal chiller	1-Stage absorption chiller	2-Stage absorption chiller
Individual Subsystems						
Cogeneration	85	85	85	37	37	37
Chilling	450[a]	67[a]	120[a]	36	23	30
District heating (DH)	100	100	100	74	74	74
District cooling (DC)	100	100	100	58	58	58
User heating (UH)	100	100	100	54	54	54
User cooling (UC)	100	100	100	69	69	69
Combination subsystems[b]						
Cogeneration + chilling	94	83	88	35	35	35
District energy (DE)	100	100	100	73	73	73
User energy (UE)	100	100	100	53	53	53
Cogeneration + DH	85	85	85	34	35	34
Cogeneration + DH + UH	85	85	85	30	31	31
Chilling + DC	532[a]	80[a]	143[a]	21	14	18
Chilling + DC + UC	532[a]	80[a]	143[a]	14	9	12
DH + UH	100	100	100	40	40	40
DC + UC	100	100	100	41	41	41
Cogeneration + chilling + DE	94	83	88	32	32	32
DE + UE	100	100	100	40	40	40
Overall process	94	83	88	28	29	29

[a] These are coefficient of performance (*COP*) values when divided by 100.
[b] DE = DH + DC and UE = UH + UC.
Source: Adapted from Refs. [9,10].

and for double-effect absorption cooling, it was COP = 202/141 = 1.43. It should be noted that the work required to drive the solution and refrigeration pumps was very small relative to the heat input to the absorption chiller (often less than 0.1%); this work was thus neglected here.

Table 4 lists the energy and exergy efficiencies evaluated for the individual subsystems, several subsystems comprised of selected combinations of the individual subsystems, and the overall system for cogeneration-based DES using electric chillers, single-effect absorption chillers, and double-effect absorption chillers. Overall energy efficiencies are seen to vary for the three system alternatives considered, from 83 to 94%, respectively, and exergy efficiencies vary from 28 to 29%, respectively. Table 4 demonstrates that energy efficiencies do not provide meaningful and comparable results relative to exergy efficiencies when the energy products are in different forms. The energy efficiency of the overall process using electric chillers, for example, is 94%, which could lead one to believe that the system is very efficient. The exergy efficiency of the overall process, however, is 28%, indicating that the process is far from ideal thermodynamically. The exergy efficiency is much lower than energy efficiency because the heat is being produced at a temperature (120°C) much higher than the temperatures actually needed (22°C for space heating and 40°C for water heating). The low exergy efficiency of the chillers is largely responsible for the low exergy efficiency for the overall process. The exergy-based efficiencies in Table 4 are generally lower than the energy-based ones because the energy efficiencies utilize energy quantities that are in different forms, whereas the exergy efficiencies provide more meaningful and useful results by evaluating the performance and behavior of the systems using electrical equivalents for all energy forms. The results for cogeneration-based DESs using absorption chillers (single-effect and double-effect absorption chillers) and those using electric chillers are, in general, found to be similar.[11,12]

For cogeneration-based district energy, in which electricity, heating, and cooling are produced simultaneously, exergy analysis provides important insights into the performance and efficiency for an overall system and its separate components. This thermodynamic analysis technique provides more meaningful efficiencies than energy analysis, and pinpoints the locations and causes of inefficiencies more accurately. The present results indicate that the complex array of energy forms involved in cogeneration-based DESs make them difficult to assess and compare thermodynamically without exergy analysis. This difficulty is attributable primarily to the different nature and quality of the three product energy forms: electricity, heat, and cooling. The results are expected to aid designers of such systems in development and optimization activities, and in selecting the proper type of system for different applications and situations.

Case Study II

Geothermal district heating has been given increasing attention in many countries during the past decade, and many successful geothermal district heating projects have been reported. For district heating to become a serious alternative to existing or future individual heating and/or cooling systems, it must provide significant benefits to both the community in which it is operated and the consumers who purchase energy from the system. Further, it must provide major societal benefits if federal, state, or local governments are to offer the financial and/or institutional support that is required for successful development.[13]

The case study here is the Izmir–Balcova GDHS, which is one example of the high-temperature district heating applications in Turkey. The Balcova region is about 7 km from the Centrum of the Izmir province, located in western Turkey, and is endowed with considerably rich geothermal resources. The Izmir–Balcova geothermal field (IBGF) covers a total area of about 3.5 km^2 with an average thickness of the aquifer horizon of 150 m. In the district heating system investigated, there are two systems—namely, the Izmir–Balcova GDHS and the Izmir–Narlidere GDHS. The design heating capacity of the Izmir–Balcova GDHS is equivalent to 7500 residences. The INGDHS was designed for 1500 residence equivalence but has a sufficient infrastructure to allow capacity growth to 5000 residence equivalence. The outdoor and indoor design temperatures for the two systems are 0 and 22°C, respectively. Fig. 4 illustrates a schematic of the IBGF, where the Izmir–Balcova GDHS, the Izmir–Narlidere GDHS, and hotels and official buildings heated by geothermal energy were included. The Izmir–Balcova GDHS consists mainly of three cycles, such as (a) energy production cycle (geothermal well loop and geothermal heating center loop), (b) energy distribution cycle (district heating distribution network), and (c) energy consumption cycle (building substations). As of the end of 2001, there are 14 wells ranging in depth from 48 to 1100 m in the IBGF. Of those, seven and six wells are production and reinjection wells, respectively, while one well is out of operation. The wellhead temperatures of the production wells vary from 95 to 140°C, with an average value of 118°C, and the volumetric flow rates of the wells range from 30 to 150 m^3/h. Geothermal fluid, collected from the seven production wells at an average wellhead temperature of 118°C, is pumped to a mixing chamber, where it is mixed with the reinjection fluid at an average temperature of 60°C–62°C, cooling the mixture to 98°C–99°C. Then this geothermal fluid is sent to two primary-plate-type heat exchangers and cooled to about 60°C–62°C as its heat is transferred to the secondary fluid. The geothermal fluid whose heat is taken at the geothermal center is reinjected into the reinjection wells, while the secondary fluid (clean hot water) is transferred to the heating circulation water of

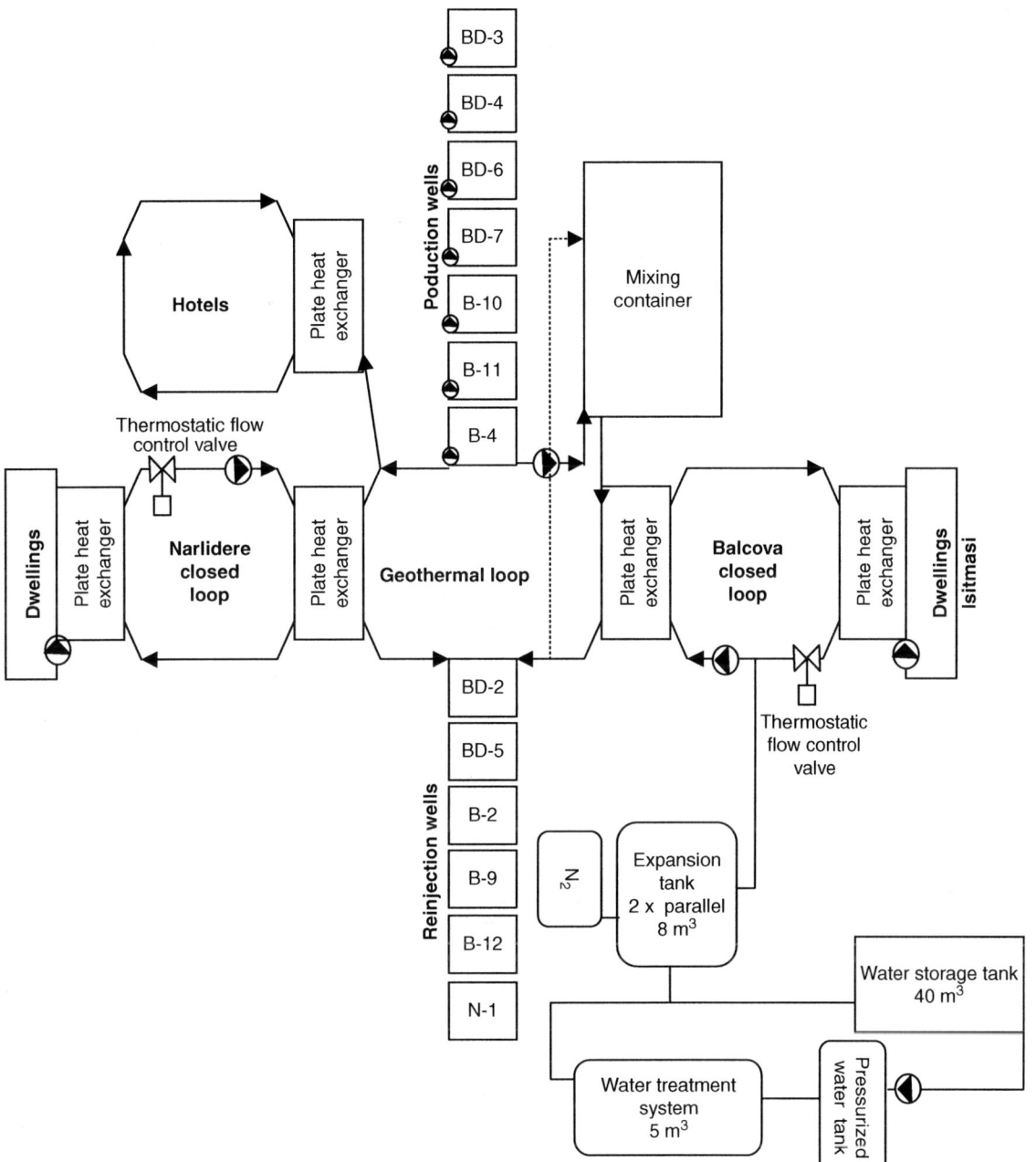

Fig. 4 A schematic of the Izmir–Balcova–Narlıdere geothermal district heating system.

the building by the heat exchangers of the substations. The average conversion temperatures obtained during the operation of the IBGDHS are, on average, 80°C/57°C for the district heating distribution network and 65°C/45°C for the building circuit. By using the control valves for flow rate and temperature at the building substations, the needed amount of water is sent to each housing unit, and the heat balance of the system is achieved.[14]

In the following paragraphs, we give main relations for mass, energy, and exergy flows, along with the energy and exergy efficiencies for the Izmir–Balcova GDHS.[15]

The mass balance equation is written as follows:

$$\sum_{i=1}^{n} \dot{m}_{w,tot} - \dot{m}_r - \dot{m}_d = 0 \qquad (17)$$

where $\dot{m}_{w,tot}$ is the total mass flow rate at the wellhead, $\dot{m}_r$ is the flow rate of the reinjected thermal water, and $\dot{m}_d$ is the mass flow rate of the natural direct discharge.

We define the energy efficiency as follows:

$$\eta_{system} = \frac{\dot{E}_{useful,HE}}{\dot{E}_{brine}} \qquad (18)$$

The geothermal brine exergy input from the production field is calculated as follows:

$$\dot{Ex}_{brine} = \dot{m}_w[(h_{brine} - h_0) - T_0(s_{brine} - s_0)] \quad (19)$$

The exergy destructions in the heat exchanger, pump, and the system itself are calculated using the following:

$$\dot{Ex}_{dest,HE} = \dot{Ex}_{in} - \dot{Ex}_{out} = \dot{Ex}_{dest}, \quad (20)$$

$$\dot{Ex}_{dest,pump} = \dot{W}_{pump} - (\dot{Ex}_{out} - \dot{Ex}_{in}), \text{ and} \quad (21)$$

$$\dot{Ex}_{dest,system} = \sum \dot{Ex}_{dest,HE} + \sum \dot{Ex}_{dest,pump} \quad (22)$$

We define the exergy efficiency as follows:

$$\psi_{sys} = \frac{\dot{Ex}_{useful,HE}}{\dot{Ex}_{brine}}$$

$$= 1 - \frac{\dot{Ex}_{dest,sys} + \dot{Ex}_{reinjected} + \dot{Ex}_{natural\ discharged}}{\dot{Ex}_{brine}} \quad (23)$$

In this study, the reference environment was taken to be the state of environment at which the temperature and the atmospheric pressure are 13.1°C and 101.325 kPa, respectively, which were the values measured at the time when the GDHS data were obtained. For analysis purposes, the actual data were taken from the BGDHS on January 1, 2003, and the respective thermodynamic properties were obtained based upon these data. It is important to note that the number of the wells in operation in the IBGF may vary depending on the heating days and operating strategy.

Using Eq. 17, the total geothermal reinjection fluid mass flow rate is 111.02 kg/s at an average temperature of 66.1°C, and the production well total mass flow rate is 148.19 kg/s, and then the natural direct discharge of the system is calculated to be 37.17 kg/s on January 1, 2003. This clearly indicates that in the BGDHS, a significant amount of hot water is lost through leaks in the hot-water distribution network.

The exergy destructions in the system particularly occur in terms of the exergy of the fluid lost in the pumps, the heat-exchanger losses, the exergy of the thermal water (geothermal fluid) reinjected, and the natural direct discharge of the system, accounting for 3.06, 7.24, 22.66, and 24.1%, respectively, of the total exergy input to the BGDHS. Both the energy and the exergy efficiencies of the overall BGDHS are investigated for system performance analysis and improvement, and are determined to be 37.60 and 42.94%, respectively.

In the GDHSs, the temperature difference between the geothermal resource and the supply temperature of the district heating distribution network plays a key role in terms of exergy loss. In fact, the district heating supply temperature is determined after the optimization calculation. In this calculation, it should be taken into account that increasing the supply temperature will result in a reduction of investment cost for the distribution system and the electrical energy required for pumping stations, while it causes an increase of heat losses in the distribution network. Unless there is a specific reason, the district heating supply temperature should be higher to increase the exergy efficiency of the heat exchangers and, hence, the entire system. Besides this, in the design and operating condition of the primary heat exchangers, a temperature approach of about 3°C is desired. On the other hand, dropping the district heating supply temperature increases the amount of building heating equipment to be oversized. Oversizing does not mean only cost, but also more exergy production due to unnecessarily inflated pumping, pipe frictions, etc. In this regard, there is an optimum district flow rate and the minimum possible exergy loss (mainly due to pumping), of which determination is planned as a further future work to be conducted.

CONCLUSIONS

We have presented some historical background on DHC, along with cogeneration and GDHS applications, and discussed some technical, economical, environmental, and sustainability aspects of geothermal energy and performance evaluations tools in terms of energy and exergy analyses for such DHC systems. We also presented two case studies to highlight the importance of exergy use as a potential tool for system analysis, design, and improvement. The benefits have been demonstrated of using the principles of thermodynamics via exergy to evaluate energy systems and technologies as well as environmental impact. Thus, thermodynamic principles—particularly the concepts encompassing exergy—can be seen to have a significant role to play in evaluating energy and environmental technologies.

For societies to attain or try to attain sustainable development, effort should be devoted to developing DHC applications and technologies, which can provide an important solution to current environmental problems, particularly if renewable energy resources are used. Advanced renewable energy technologies can provide environmentally responsible alternatives to conventional energy systems, as well as more flexibility and decentralization.

REFERENCES

1. Spurr, M. District Energy/Cogeneration Systems in U.S. Climate Change Strategy. Climate Change Analysis Workshop, June 6–7, Springfield: Virginia, 1996.
2. IEA. Implementing Agreement on District Heating and Cooling Including the Integration of CHP. Strategy document, International Energy Agency: Paris, 2004.
3. Strickland, C.; Nyboer, J. Cogeneration Potential in Canada-Phase II. Report for Natural Resources Canada, 2002; 40, 6.

4. UNEP. Cogeneration, Energy Technology Fact Sheet, Division of Technology, Industry and Economics-Energy and Ozone Action Unit. United Nations Environment Programme, Nairobi, Kenya. Available at: www.uneptie.org/energy. Accessed on 14th November 2006.
5. Enwave. History of District Energy, 2006. Available at: www.enwave.com/enwave/view.asp?/solutions/heating/history. Accessed on 14th November 2006.
6. Dincer, I. Renewable energy and sustainable development: a crucial review. Renew. Sust. Energy Rev. **2000**, *4* (2), 157–175.
7. Dincer, I.; Rosen, M.A. Thermodynamic aspects of renewables and sustainable development. Renew. Sust. Energy Rev. **2005**, *9* (2), 169–189.
8. Dincer, I. The role of exergy in energy policy making. Energy Policy **2002**, *30* (2), 137–149.
9. Rosen, M.A.; Le, M.N.; Dincer, I. Thermodynamic assessment of an integrated system for cogeneration and district heating and cooling. Int. J. Exergy **2004**, *1* (1), 94–110.
10. Rosen, M.A.; Le, M.N.; Dincer, I. Efficiency analysis of a cogeneration and district energy system. Appl. Therm. Eng. **2005**, *25* (1), 147–159.
11. Edmonton Power. City of Edmonton District Energy Development Phase. Section 2: Engineering Report, 1991.
12. MacRae, K.M. *Realizing the Benefits of Community Integrated Energy Systems*; Canadian Energy Research Institute: Calgary, Alberta, 1992.
13. Bloomquist, R.; Nimmons, J. Geothermal district energy. In *Course on Heating with Geothermal Energy: Conventional and New Schemes* (Conveyor: Lienau, P.J.). WGC2000 Short Courses, June 8–10, 2000, Kazuno, Tohoku District, Japan, 2000; 85–136.
14. Hepbasli, A.; Canakci, C. Geothermal district heating applications in Turkey: a case study of Izmir–Balcova. Energy Convers. Manage. **2003**, *44*, 1285–1301.
15. Ozgener, L.; Hepbasli, A.; Dincer, I. Thermo-mechanical exergy analysis of Balcova geothermal district heating system in Izmir–Turkey. ASME J. Energy Res. Tech. **2004**, *126* (4), 293–301.

Drying Operations: Agricultural and Forestry Products

Guangnan Chen
Faculty of Engineering and Surveying, University of Southern Queensland, Toowoomba, Queensland, Australia

Abstract
This entry presents an overview of the methods for drying agricultural and forestry products. The need for the drying of agricultural and forestry products is presented, and the principles of drying operations are described. It is shown that a diversity of drying systems are currently used. The performance of these dryers varies significantly. Significant progress has been made in the research and development of drying technology to improve product quality, reduce cost, and improve environmental performance.

INTRODUCTION

Drying, the removal of water from products, is a necessary operation in many industries for the purpose of preserving product quality or adding value to the products. This entry focuses on the drying of agricultural and forestry products; in particular, grain drying and (solid) wood drying, which form the bulk of drying operations in these two industries.

THE NEED FOR DRYING OF AGRICULTURAL AND FORESTRY PRODUCTS

The main purpose of agricultural drying is often to reduce field loss and weather damage, and to maintain product quality during subsequent storage and delivery. This is in contrast to the drying of wood, whose main purpose is to have all the shrinkage and distortion take place before the wood is put into use. The reduction of insect and fungal attack is another reason for the need for wood drying.

In addition to the above difference, agricultural drying is also typically a seasonal activity, while wood drying is normally a year-long operation. This difference can have a significant impact on a number of aspects of drying operation and dryer design.

Overall, drying may be regarded as a risk management tool for the agricultural industry, and a value-adding tool for the forestry industry.

METHOD OF THERMAL DRYING

In most cases, drying involves the application of thermal energy. This is achieved by heating up the product and forcing hot air through it, therefore vaporizing and removing the moisture inside the product. In addition to the promotion of heat and mass transfer, the circulation of air also helps to carry the heat to and the moisture away from the product.

Significant energy is required in the drying process for several reasons:

- Raising the temperatures of air, the product, and water
- Vaporizing the water
- Compensating heat loss through radiation, convection, and operational losses (e.g., leaks)
- Compensating heat loss through the venting of heated humid air

Thermal drying, which involves water phase change, is a very energy intensive activity. For example, evaporating one cup (250 mL) of water would require approximately the same amount of energy as it would to heat a big pan of soup (3 L) from 25 to 70°C. Thus, the efficiency of energy use in drying processes is significant in the context of energy, economic, and environmental policy goals.

Because artificial drying normally offers the advantages of better control over product quality and higher productivity, this method has been widely used in the agricultural and forestry industries. Most artificial dryers also use the direct heat and vent method to drive the drying process.

HOW DRYING TAKE PLACE

During drying, evaporation may take place in two stages.[1,2] At first, there may be sufficient moisture within the product to replenish the moisture lost at the surface, until the critical point is reached and a dried surface forms. Evaporation is then principally dependent upon the rate of internal moisture diffusion. This is called the falling rate period or second period of drying, and is often a diffusion

Keywords: Grain drying; Wood drying; Drying performance; Drying process; Dryer design; Agricultural and forestry products; Product quality; Process energy efficiency; Drying technology.

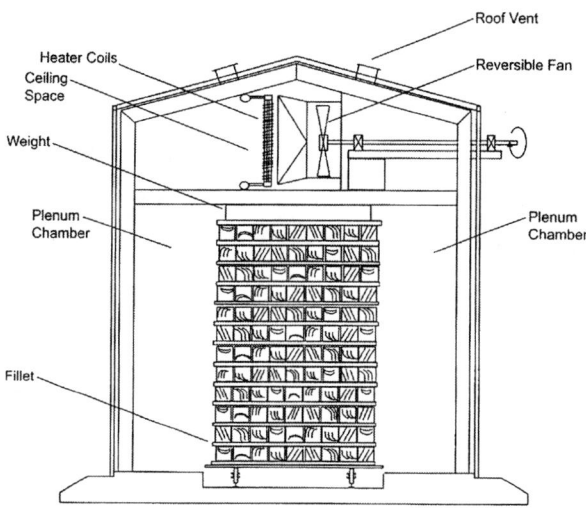

Fig. 1 Typical components and arrangement of a batch wood drying kiln.

process. Compared with the first period of constant drying at the rate of liquid water evaporation, diffusion is typically a slow process, and is mainly controlled by internal moisture transport of the product. Diffusion processes may be considerably accelerated with increased temperatures. External mass transfer plays a relatively small role at this stage.

Corresponding to the above process, initially, as the product surface is dried, it is restrained by the wet core so that it is subjected to a tensile stress. Later, as the core dries, it is in turn restrained by the drier surface, so that the stress profile inverts. At the end of drying, the product surface may be left with a residual compressive stress, whereas the core is subjected to a tensile stress. This is called case-hardening.[3] A drying schedule will therefore need to ensure that the stresses developed during any period of the drying process not exceed the strength of the material, so that stress damage of the product does not take place. At the end of drying, stress relief for the residual stresses may also be carried out.[4,5] This is particularly important when a product needs further processing or when a high-temperature fast drying schedule is employed. Cooling of agricultural products also minimizes the water condensation on the product surface.

TYPES OF DRYERS

A dryer generally consists of a chamber, a heating and air circulation system, and a control system.

Dryers may be classified in many ways, such as by modes of operation (e.g., batch or continuous dyers), by fuel sources, by drying temperature ranges, or by dryer throughputs. Further classifications are also possible, including heat transfer methods (e.g., direct or indirect heat transfer) and relative directions of the flows of product and air (e.g., cross-flow, counter-flow, and concurrent-flow).

In the agriculture and forestry sector, the most common type of dryer is still the fixed-bed batch (bin, shed, or compartment) dryer (Fig. 1). In these dryers, the product remains stationary while the drying environments are successively varied. This drying mode is particularly suitable for small to medium operators, with the advantage of low capital cost requirement. However, this method is also generally of lower capacity and more labor intensive. In comparison, a continuous dryer would typically use higher temperatures, have much larger capacity, and be more suitable for large operations. This method, however, requires large capital outlay. It is also sometimes more difficult to achieve accurate product specification with continuous driers, because of the potential impact of process air leakage and ambient conditions.

In the past decade, there has been an increasing interest in the use of various drying facilities, particularly low-cost, low-temperature dryers, as more and more farmers have begun to appreciate the importance of drying in the total harvesting system. The rapid growth of the forestry plantation industry also promotes the widespread installation of various timber-drying kilns.

Other specialist drying methods are also available. These include fluidized-bed drying for moist particulate products (such as grains, peas, and sliced vegetables) and drying by the application of energy from microwave or dielectric sources. However, many of these methods may only be cost effective for particular high-value products or for obtaining specific attributes for specific products. Freeze drying is reported to be able to achieve smaller product shrinkage, longer product storage life, and better retention of biological activity, so it is popular with the food industry.

DRYING SCHEDULES

A drying schedule may be described as a series of temperature, humidity, and air velocity settings used to dry the product to a specific moisture content, and to produce consistent, defect-free dry products in as short a time as possible with the least amount of energy use. Many agricultural and forestry products are required to be dried to a final moisture content of around 10%–12%.

There are two ways to define moisture content.[6] In general, the moisture content of forestry products is often expressed in dry basis (MC_{db}), which is the fraction of the mass of water in comparison with the mass of the oven-dry product. By contrast, the moisture content of agricultural products is normally expressed in wet basis (MC_{wb}), which is the amount of water in the product divided by the total product weight. These two definitions of MC_{db} and MC_{wb} can be converted to each other by the relationship

$MC_{db} = MC_{wb}/(1 - MC_{wb})$.

During drying, agricultural and forestry products usually undergo considerable changes, including shrinkage, cracking (both externally and internally), nutrient loss, and color changes. By imposing harsh drying conditions, a high-temperature regime may bring in the benefit of shortened drying schedules. However, such a process may also increase the risk of quality degradation, so a right balance needs to be achieved between these two competing factors.

The required drying time varies greatly for different agricultural and forestry products, ranging from a few hours to several days or even several weeks, depending on the product characteristics and the specified quality grades. Many agricultural products also have to be dried within a certain time frame to avoid significant quality deterioration.[7]

At present, most commercial dryers operate at comparatively moderate drying schedules to avoid the risk of quality loss. Low-value and more permeable materials may be dried more rapidly. In general, it may be categorically found that a high-temperature, high-humidity schedule may be suitable for fruit products, while a low-temperature, low-humidity schedule would be good for high-value seed products. In the middle, grain and timber are reasonably robust and may be suitable for a high-temperature, low-humidity schedule.

Currently, simple staged temperature controls are preferred for drying agricultural and forestry products, particularly for grain drying. To minimize degrades, some additional pre- or post-treatments such as initial air drying and post-drying cooling and conditioning may also be employed. Additional quality control may be attained by presorting the material prior to kiln drying so that the properties of batches are relatively homogenous. Grain inverter or airflow reversal may also be adopted.

Current drying schedules have been largely derived from trial-and-error experiments over a number of years. However, this method could be lengthy and expensive. Recently, a number of theoretical models have been developed to simulate and optimize the drying process and to reduce the number of laboratory and field experiments.[8,9]

Typical drying schedules for several agricultural and forestry products are as follows:

Permeable softwoods (such as radiate pine) for structural uses:
- Dry straight through from green to an average moisture content of 4% at a dry-bulb temperature of 120°C, a wet-bulb temperature of 70°C, and an air velocity of 5 m/s.
- Cool outside under cover for 90 min.
- Steam for 2 h.
- Cool with weight on and de-fillet within 24 h of steaming.

Permeable softwoods (such as radiate pine) for furniture uses (appearance grades):
- Dry at a dry-bulb temperature of 90°C and a wet-bulb temperature of 60°C, with a total duration of 2–3 days. Final steaming is also required, in order to remove the residual stress generated during the drying process. This is necessary, as the dried timber will be further reprocessed during furniture making.

Different from the softwoods, hardwoods are generally less permeable and more difficult to dry, so they are usually kiln dried by moisture content schedules.[10] This means that the dry- and wet-bulb temperatures are changed when the timber (lumber) reaches certain moisture contents. Hardwood is also often air dried first, before the kiln drying. Depending on the species and thickness of the lumber, the drying times may vary from one to a few weeks, with the final temperature being gradually raised from the ambient temperature to between 45 and 65°C. The air velocity is typically maintained at 1–1.5 m/s.

For grain drying (milling grade), it is generally recommended that the maximum drying temperature be limited to 70°C to minimize the heat damage, particularly at initial period of high moisture content. Feedstock grades can be dried at much higher temperatures. Seed drying is usually limited to 30°C–40°C. The common airflow rate for grain dryers is in the range of 200–1000 L/s/t of grain.

DRYER PERFORMANCE AND ENERGY EFFICIENCY

A dryer may be regarded as an energy system. Various energy sources may be used in the drying process, including electricity and various primary fuels such as coal, diesel, and gas. These energy sources all have different heating values, costs, and environmental impacts. Although electricity is a convenient and "clean" energy source, it is a high-grade energy because the typical efficiency of thermal generation of electricity is only 35%–50%. Overall, electricity is generally more expensive, particularly after taking account of the associated supply and transmission charges.

Typically, the energy cost for a small or medium drying operation may range from a few thousand dollars to over twenty thousand dollars; depending on and significantly influenced by the quantity and initial moisture content of the product, and operation practice such as drying schedules and controls. Assuming a 5% moisture removal, the total amount of water being removed from one ton of product is about 50 kg. Currently, in a commercial dryer, the energy required to evaporate 1 kg of moisture from a product ranges from 3.5 to 7.0 MJ.[11,12]

In many cases, it has been found that there is little correlation between the dryer energy performance and product process requirement. Lower process requirements do not necessarily lead to higher energy efficiency. This indicates that there is a significant potential to improve the dryer energy performance.

Although the technology is currently available, it is noted that there are significant barriers for the uptake of energy-efficient technology in the drying industry. This is because present production methods have historically been based on considerations of process throughput, reliability, and capital cost. Energy costs, although comprising a significant part of total operating costs in the drying of agricultural and forestry products, typically represent only 2%–5% of product value, and are therefore often of low priority. This is further reinforced by the factor that drying may be a secondary activity for many farmers and operators. This is particularly the case for agricultural dryers, as agricultural drying is typically a highly seasonal activity. Most agricultural dryers are only utilized for one to three months.

DRYER DESIGN AND SELECTION

Established procedures are now available for the design of agricultural and forestry product dryers. In spite of this, the actual performance of different industries and different dryers still varies considerably. Poor drying systems can lead to significant penalties in terms of lost production and lost income, including increased energy cost and degraded or non-uniform product. In comparison with the "seasonal" agricultural drying industry, the timber industry is typically a year-long operation, and hence more expensive technology and personnel training may be justified. For example, in the timber drying industry, automatic kiln monitoring, management, and control systems have now been routinely implemented to improve the dryer performance. This is relatively rare for the drying of agricultural products. Because of the short period used, most agricultural crop dryers are also not insulated.

To obtain the maximum performance and the desired product quality, it is important that a suitable drying system be selected, with correct system sizing, matching of subsystems, and operating procedures. These factors are often interlinked, so an integrated and holistic approach is required. For example, when employing a high-temperature regime, more water vapor is produced, and higher airflow rates will be required to carry away this vapor. When contemplating increasing the dryer capacity or adopting a new dryer, it is also important to consider the impact on all the other components of crop harvesting and storage systems. A decision should not be made on the basis of the effects of that particular facility alone.

To save energy, some large sites may have several dryers so that a cascade arrangement for exhaust energy recovery is justified. For a large production, it may be possible to carry out electricity tendering or form an electricity user "club" to reduce the electricity tariff. Optimization of fan sizing and operation is also important, as fan laws stipulate that a 50% fan speed reduction can result in a reduction of fan power to only one-eighth of the original power requirement. Less fan power may be needed during the later stages of the diffusion drying process.

At present, a number of computer models have been developed and used in the dryer design. The main advantage of this method is to achieve a precision sizing of the equipment and to produce a predictable design to reduce the customer's business risk. For agricultural drying, climate-based models[13] have also been developed to ensure optimal design and integration between various agricultural machinery, crop performance, and perceived weather risk. Together with local historical weather data and future climate forecasts, these models have been used to assist in the decisions of both long-term investment in drying facilities and short-term tactical operation decisions (e.g., by adjusting the crop planting and harvesting schedule, by crop choice and crop diversification, or by early negotiation with harvest and drying contractors).

RECENT RESEARCH AND DEVELOPMENT

Significant research has been carried out in the area of agricultural and forestry product drying. This has included research on dryer design, the impact of drying on product quality, improvements of drying energy efficiency, new methods of drying, and applications of new technologies.

Dryer Design and Operation

Because most of the current agricultural dryers are of small to medium throughputs and are operated by rural family businesses, the main constraints for agricultural drying are often the capital expenditure and the technical competence and skills of farmers and local dryer manufacturers. It has been found that the common reasons for poor dryer performance are poor dryer design, inappropriate equipment selection and installation, and bad control and operating practice. Practical information on the best operating practice also tends to be fragmented and not readily available in a useable form.

One of the current research priorities is therefore to demonstrate and establish appropriate technologies to overcome the above barriers and to improve the integration of heat and mass transfer processes and the matching of subsystems.[14]

For commercial operators, uneven drying is also a significant problem. This may be difficult to eliminate, because drying environment is inherently dynamic and varies with locations inside the dryer. Furthermore, dryers may also be required to handle variable resources, including feed materials of different species, non-uniform initial moisture contents, and different sizes. Computational fluid dynamics (CFD) has now been widely used to improve the dryer design and to minimize air recirculation loss. The latest research is also focusing on the development of new sensor techniques and enhanced machine vision tools for collecting quality control information and developing expert systems for rapid problem diagnosis.

Considerable effort is also being made to investigate the effect of drying conditions on the shrinkage, stress development, and quality of products.

New Methods of Drying and New Technologies

A number of innovative methods and drying technologies are being developed. Among them, drying with a modified atmosphere has shown significant promise. Drying and storing fruit and foods in a controlled atmosphere (CA) can lead to improved product quality, because displacing oxygen with other gases such as N_2 and CO_2 retards the oxidation process. Considerable commercial success has been achieved in the area of CA fruit storage and transport. Similar opportunities have also been identified in the area of fruit drying, particularly in terms of eliminating the use of chemical preservatives or other additives. Recent experiments have shown that the use of innovative CA, oxygen-free drying for apples can significantly improve the product attributes, in particular reducing brown staining and avoiding the requirement of using sulfur chemical pretreatment.[15] This can lead to more healthy products, with the additional benefits of improved taste (no acid) and better texture.

Since drying is an energy-intensive operation, much attention is also given to the development of an energy-efficient drying process. Heat-pump drying is one such technology, because the heat normally vented to the atmosphere is recovered. A heat-pump dryer (HPD) is essentially an industrial adaptation of a normal air conditioning system. Energy (electricity) inputs to the dryer include those to the compressor and the fans. For each unit of electrical energy used by the heat pump, generally three to four units of energy are available for drying the product. In a heat-pump dehumidifier, most of the moisture is also removed from the kiln as liquid rather than moist warm air.

Due to the limits of currently applied working fluids (refrigerants), the HPD normally has to operate at low to medium temperatures, so the drying rates are also slower. This is suitable for a number of heat-sensitive products, but also makes it difficult to compete with alternative mainstream technologies, where the emphasis is often put on the fast drying rate and quick return of plant capital costs.[16] Solar-assisted HPDs are also being developed[17] to accelerate the drying process.

CONCLUSION

Drying is a significant operation in the agricultural and forestry industries. Considerable progress has been made in the research and development of drying technology to improve product quality, reduce cost, and improve environmental performance. It has been shown that a diversity of drying systems are used. The efficiency and performance of different driers vary significantly. Energy consumption is strongly influenced by the dryer design, the particular operation practice, and the individual skills of the operator.

A number of innovative methods and drying technologies are being developed. Among them, drying with a modified atmosphere and improvements in sensor and control technology have shown significant promise.

REFERENCES

1. Mujumdar, A.S., Ed. *Handbook of Industrial Drying*; Marcel Dekker: New York, 1989.
2. Keey, R.B. *Drying of Loose and Particulate Materials*; Hemisphere Publishing Corporation: New York, 1992.
3. Keey, R.B.; Langrish, T.A.G.; Walker, J.C.F. *Kiln-Drying of Lumber*; Springer: Berlin, 2000.
4. Pratt, G.H. *Timber Drying Manual*; Building Research Establishment: London, 1974.
5. Haslett, A.N.; Simpson, I. Final steaming critical to HT drying of appearance-grade timber. NZ Forest Industries **1991**, (9), 40–41.
6. Broker, D.B.; Bakker-Arkema, F.W.; Hall, C.W. *Drying and Storage of Grains and Oilseeds*; AVI Publishing Co.: New York, 1992.
7. Raghavan, G.S.V. Drying of agricultural products. In *Handbook of Industrial Drying*; Mujumdar, A.S., Ed.; Marcel Dekker: New York, 1987; 555–570.
8. Stanish, M.A.; Schajer, G.S.; Kayihan, F. A mathematical model of drying for hygroscopic porous media. AIChE J. **1986**, *32* (8), 1301–1311.
9. Wu, Q.; Milota, M.R. Rheological behaviour of Douglas-fir at elevated temperature. Wood Fibre **1995**, *27* (3), 285–295.
10. Simpson, W.T., Ed. *Dry Kiln Operator's Manual, Agriculture Handbook No. 188*; U.S. Department of Agriculture: Washington, DC, 1991.
11. Chen, G.; Anderson, J.A.; Bannister, P.; Carrington, C.G. Monitoring and performance of a commercial grain dryer. J. Agric. Eng. Res. **2001**, *81* (1), 73–83.
12. Hill, M., Ed. *Drying and Storage of Grains and Herbage Seeds*; Foundation of Arable Research: Lincoln, New Zealand, 1999.
13. Abawi, G.Y. A simulation model of wheat harvesting and drying in northern Australia. J. Agric. Eng. Res. **1993**, *54*, 141–158.

14. Carrington, C.G.; Sun, Z.F.; Sun, Q.; Bannister, P.; Chen, G. Optimising efficiency and productivity of a dehumidifier batch dryer. Part 1—Capacity and Airflow. Int. J. Energy Res. **2000**, *24*, 187–204.
15. Perera, C.O. Modified atmosphere drying. Chapter 6. In *Drying of Products of Biological Origin*; Mujumdar, A.S., Ed.; Oxford and IBH Publishers: New Delhi, India, 2004; 153–163.
16. Bannister, P.; Carrington, C.G.; Chen, G. Heat pump dehumidifier drying technology—status, potential and prospects. Proceedings of 7th IEA (International Energy Agency) Heat Pump Conference, Vol. 1, 219–230, China Architecture and Building Press: Beijing, 2002.
17. Chen, P.Y.S.; Helmer, W.A.; Rosen, H.N.; Barton, D.J. Experimental solar-dehumidification kiln for drying lumber. Forest Prod. J. **1982**, *32* (9), 35–41.

Drying Operations: Industrial

Christopher G. J. Baker
Chemical Engineering Department, Kuwait University, Safat, Kuwait

Abstract
Dryers are widely used on an industrial scale and are major consumers of energy. This article first briefly reviews some of the more common types of drying equipment. It then goes on to discuss the energy consumption of dryers, with particular emphasis on an ideal adiabatic dryer model against which the performance of drying equipment in the field can be benchmarked. It will then describe the use of energy audits to quantify potential energy savings. Finally, it will discuss low-cost and capital-intensive schemes for reducing the energy consumption of dryers.

NOMENCLATURE

C_{pg}	Specific heat of dry air (kJ/kgK)
E_s	Specific energy consumption (GJ/t)
$E_{s,a}$	Specific energy consumption of adiabatic dryer (GJ/t)
E_s^*	Measured specific energy consumption (GJ/t)
F	Feedrate (dry solids basis) (kg/s)
Q_{ev}	Heat required to provide the latent heat of evaporation (kW)
Q_{htr}	Heat supplied to dryer by heater (kW)
T	Temperature (°C)
W_{ev}	Evaporation rate (kg/s)
X	Moisture content (mass water per unit mass of dry solids) (kg/kg)
Y	Humidity (mass of water vapor per unit mass of dry air) (kg/kg)

Greek letters

η	Thermal efficiency of dryer (defined by Eq. 1) (%)
λ_{ref}	Latent heat of evaporation of water at 0°C (GJ/t)
ξ	$(T_o - T_a)/(Y_o - Y_a)$ (°C)

Subscripts

a	Ambient
i	Inlet
o	Outlet

INTRODUCTION

Drying can be defined as that unit operation which converts a liquid, solid, or semi-solid feed material into a solid product that has a significantly lower moisture content. Although there are some notable exceptions, drying is normally achieved through the application of thermal energy, which is used in part to supply the latent heat of the evaporation of water. In certain industrial processes, drying may also involve the removal of organic solvents, either alone or in combination with water.

Drying forms an integral part of many industrial manufacturing processes. Examples can be found in the following sectors: chemicals, petrochemicals, polymers, food, agriculture, pharmaceuticals, ceramics, minerals, paper and board, textiles, etc. Dryers come in a wide variety of configurations with throughputs ranging from 50 kg/h or less to tens of tonnes per hour. As would be anticipated, batch dryers are employed at relatively low throughputs and continuous dryers at higher throughputs. The exact demarcation between these two categories of dryer is based on a number of technical and economic factors.

The principal types of batch and continuous dryers are classified into different categories as shown in Figs. 1 and 2, respectively. In these figures, the term "layer dryer" is used to describe those devices in which a surface within the dryer is employed to support and/or heat the coherent mass of drying solids. Conversely, in dispersion dryers, the solids are freely suspended in the hot air flow. The feedstock may be heated directly by convection with hot air, as in dispersion dryers, or indirectly by conduction through a heat-exchange surface, as in contact dryers. In the latter case, operation is possible in a vacuum as well as at atmospheric pressure. Special dryers include those in which energy is wholly or partially supplied by means of dielectric (microwave and radiofrequency) heating or water is removed by sublimation, as in freeze dryers. Both types are relatively expensive in terms of capital and operating costs. Therefore, the use of dielectric dryers is largely restricted to the removal of relatively small traces of moisture, which is expensive by conventional means. Freeze drying can be employed when low-temperature processing enhances a product's quality and value, as is the case with freeze-dried coffee.

A selection of typical and widely used industrial dryers is described in Table 1 and illustrated schematically in

Keywords: Dryers; Dryer selection; Dryer operation; Energy consumption; Energy auditing; Energy conservation.

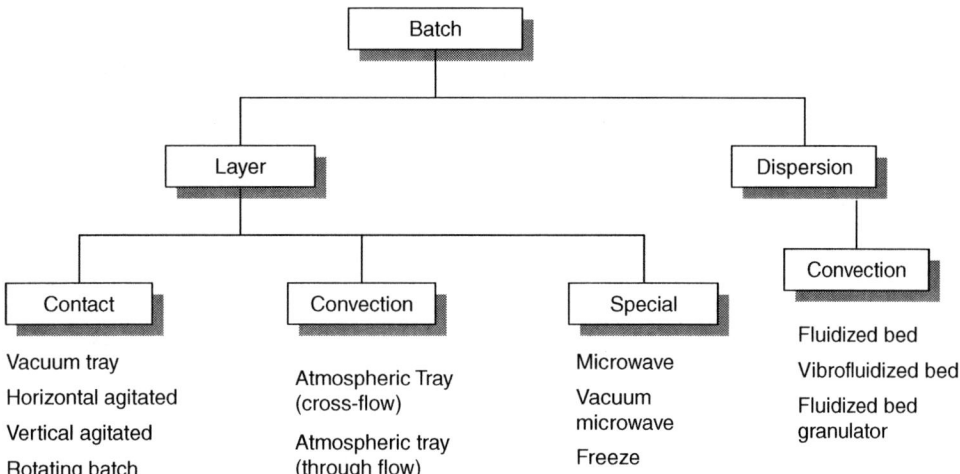

Fig. 1 Simplified classification of batch dryers.

Figs. 3–7. Because of the complexity of drying operations, many factors have to be considered and weighed when selecting an appropriate dryer for a given application.[1] Often there is no one "right" answer as several options may be both technically and economically viable. The optimal choice can be defined as that dryer which satisfies all process requirements at minimum cost. Process requirements may include the specification of designated quality parameters in addition to exit moisture content. Minimum cost is often taken to mean minimum capital cost, but this ignores the expenditures on fuel, which dominate the cost of the drying operation over its economic life.

Drying processes involve simultaneous heat and mass transfer, the underlying theory of which is relatively complex and falls beyond the scope of this article. The interested reader is therefore referred to one of a number of specialized handbooks on drying[2,3] or to the appropriate chapter in a more general Chemical Engineering textbook.[4,5] This article concentrates primarily on those aspects of dryer operation that impact on their energy use.

ENERGY CONSUMPTION OF DRYERS

Drying processes consume very large quantities of energy. There are several reasons for this. The first, as noted above, is their widespread use throughout industry. Secondly, by their very nature, dryers are highly energy intensive. As a rule of thumb, a typical convective dryer

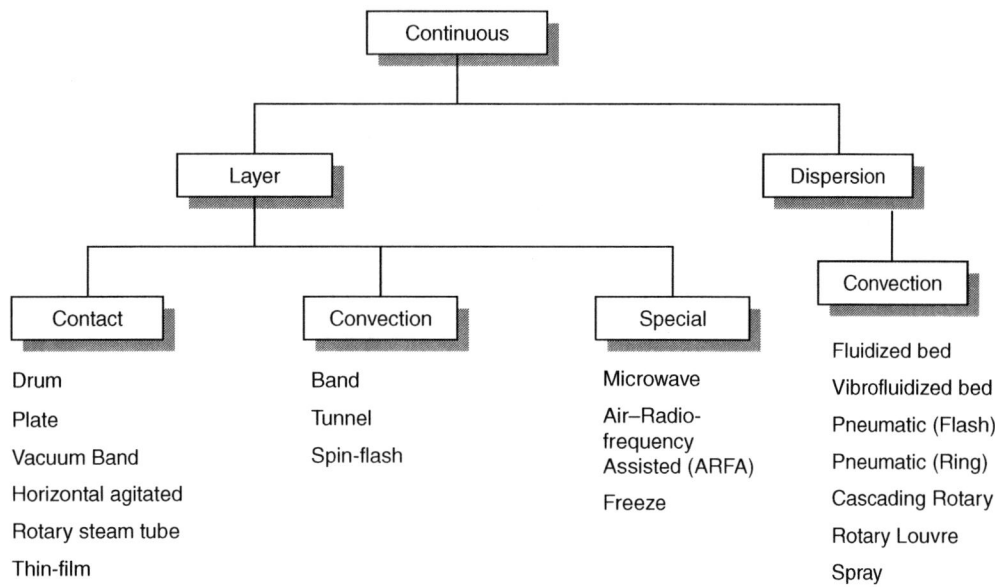

Fig. 2 Simplified classification of continuous dryers.

Table 1 Examples of different dryer types

Type	Description	Feedstocks and products
Horizontal agitated dryer (Several configurations, batch or continuous operation)	Horizontal cylindrical heated shell with axial paddles. Atmospheric or vacuum operation	Solids, pastes, or solutions. Used to dry, e.g., products ranging from foodstuffs to pigments. Low throughputs
Drum dryer (Several configurations, continuous operation)	Slowly rotating hollow drum heated internally by steam. Continuous operation. Most models operate at atmospheric pressure; vacuum model available	Solutions and pastes. Products range from instant potatoes to salts. Low throughputs
Spray dryer (Several configurations, continuous operation)	Used for drying liquid feeds, which are atomized into small droplets, and contacted with hot gas in a large drying chamber	A large number of products including liquid foods (e.g., coffee, milk), chemicals, plastics, etc. Capable of processing very high throughputs
Fluidized bed dryers (Several configurations, batch or continuous operation)	Floating bed of solids supported by stream of hot drying air	Widely used to dry small, free-flowing solids—foods, minerals, chemicals, plastics, etc. Can be used to dry/agglomerate particles. High throughputs
Rotary dryers (Several configurations, continuous operation)	Large, slowly rotating sloping drum. In one version, solids are picked up by flights on the periphery and showered through a hot air stream	Very high throughputs of solid products ranging from minerals to chemicals, fertilizers and food by-products

can be expected to consume at least 1 MW of thermal power per t/h of evaporation. Finally, as dryers are frequently operated very inefficiently, this figure may in practice be considerably higher than it needs to be. As a result, many companies view dryers as popular targets in their energy conservation programs. The recent escalation in the price of oil and natural gas has naturally provided an added incentive. Additionally, the Kyoto Protocol has focused the need for governments of signatory countries to take drastic actions to curb

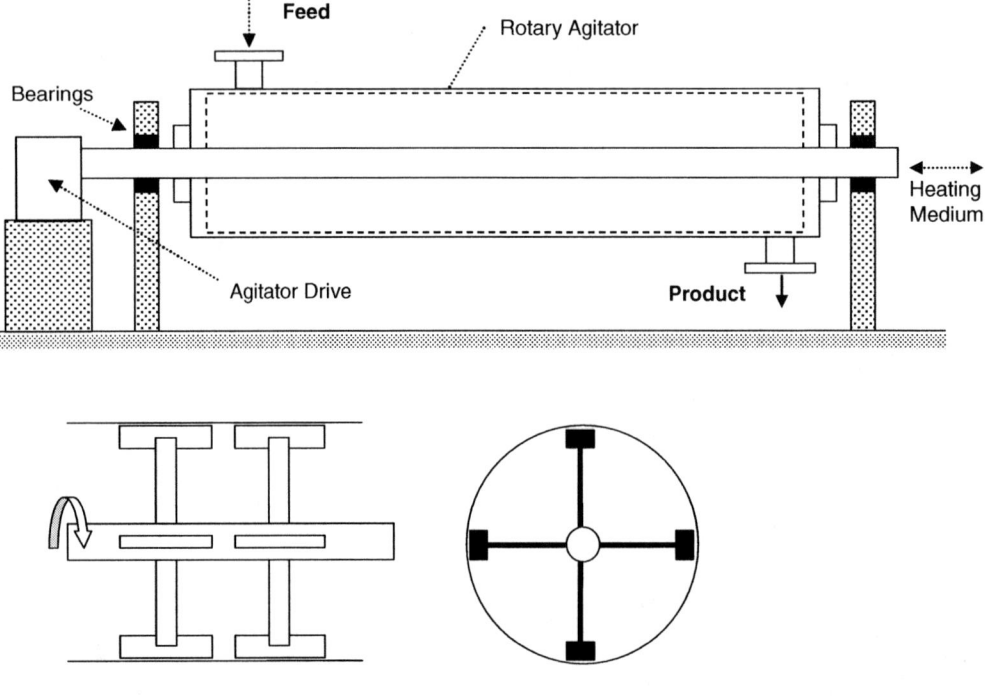

Fig. 3 Horizontal agitated dryer.

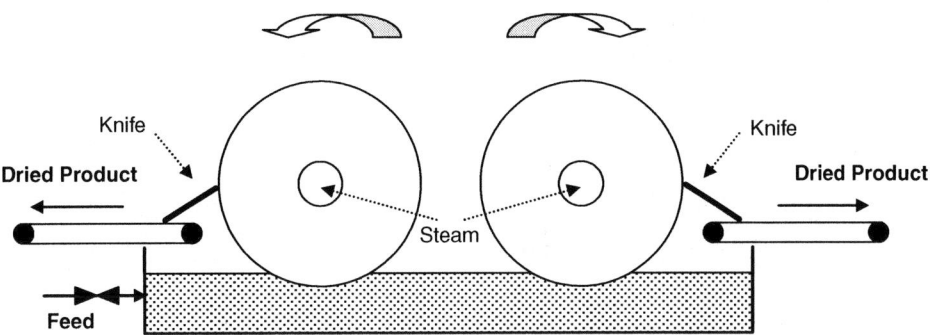

Fig. 4 Drum dryer (Double drum, dip feed).

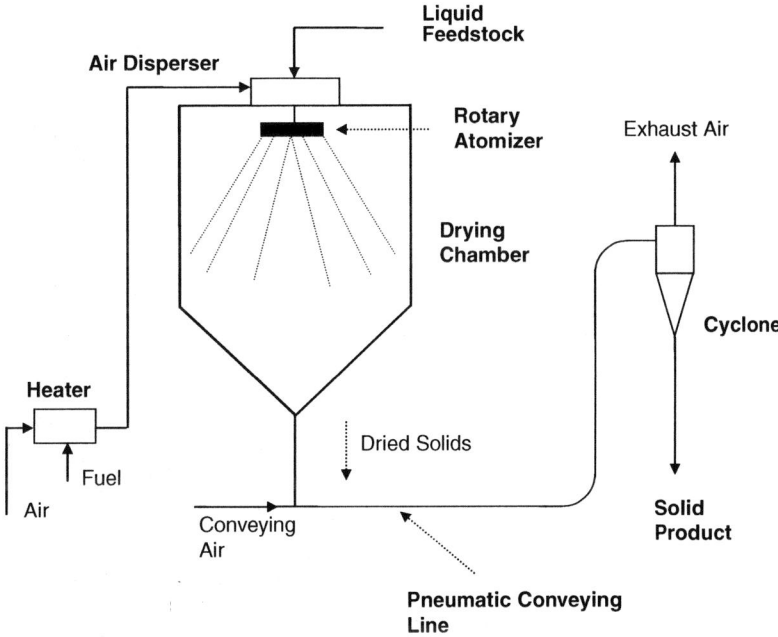

Fig. 5 Cocurrent spray dryer with rotary atomizer and pneumatic conveying of dried powder.

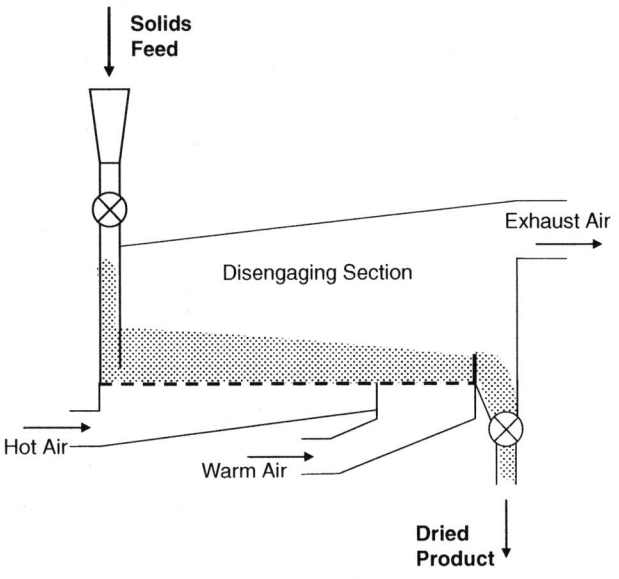

Fig. 6 Plug flow fluidized bed dryer.

greenhouse gas emissions. In order to achieve this, a body of enabling legislation has been introduced, particularly within the European Union. This legislation and its likely impact on both dryer manufacturers and operators is reviewed elsewhere.[6]

The most recent analysis of dryer energy consumption within the United Kingdom was published by Gilmour et al.[7] and includes a number of interesting statistics. For example, in 1994, estimates of the energy consumed in drying ranged from 348.6 to 379.5 PJ, depending on the method of calculation employed; the corresponding figure for the total industrial energy consumption in that year was 1969 PJ. Moreover, their analysis suggests that the proportion of energy consumed in drying progressively increased from around 11.6% in 1982 to 17.7%–19.3% in 1994. Another interesting statistic cited by Gilmour et al. is that the cost of fuel consumed by a typical convective dryer over its lifetime will likely exceed five times its initial capital cost. Given the recent worldwide

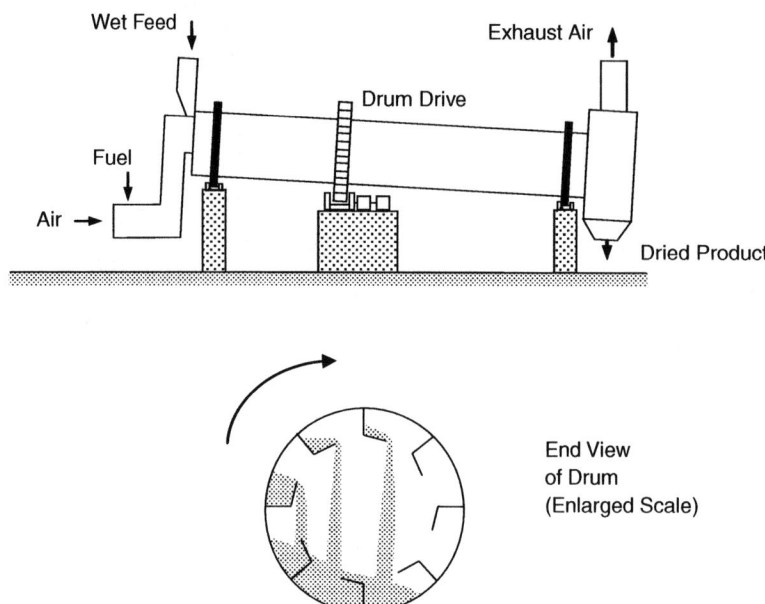

Fig. 7 Cascading rotary dryer with cocurrent solids-air flow.

escalation in oil and natural gas prices, this figure is probably a significant underestimate in today's terms.

The thermal efficiency of a dryer can be expressed in several ways. A typical measure is:

$$\eta = 100 \frac{Q_{ev}}{Q_{htr}} \qquad (1)$$

where Q_{htr} is the total rate at which thermal energy is supplied to the dryer. Of this, Q_{ev} is required to provide the latent heat of evaporation. Alternatively, the specific energy consumption E_s of the dryer is defined as the thermal energy required to evaporate unit mass of water:

$$E_s = 0.001 \frac{Q_{htr}}{W_{ev}} \qquad (2)$$

where W_{ev} is the evaporation rate. One might expect E_s to be approximately equal to the latent heat of evaporation of water, namely 2.5 GJ/t of water evaporated. In practice however, much higher values are observed (see below).

Gilmour et al.[7] cited previously published values of η for different industry sectors in the United Kingdom. As can be seen in Table 2, these differed quite widely, presumably reflecting the mix of dryer types employed and products dried, amongst other factors. Assuming that these figures are typical of those for other industrialized countries, the variation in the drying efficiencies cited suggests that there should be considerable scope for reducing dryer energy consumption. However, as will be discussed below, this statement has to be viewed with caution as many theoretical improvements may not be achievable in practice.

Baker and McKenzie[8] undertook a survey of the energy consumption of 32 industrial spray dryers in the ceramics, chemicals, and food industries. The survey, which was commissioned by the U.K. Government's Energy Efficiency Best Practice Programme, included dryers evaporating a total of 67.8 t/h of water. The thermal energy input to these dryers was 92.6 MW. The results of this survey, which included dryers having evaporation rates ranging from 0.1 to 12 t/h, revealed values of E_s varying from around 3–20 GJ/t of water evaporated. The average for all dryers included in the survey was 4.87 GJ/t.

The data obtained in the above survey were interpreted with the aid of a model that enabled the performance of a particular dryer to be compared with that of its ideal

Table 2 Dryer performance in selected industrial sectors

Industrial sector	Average drying plant efficiency, %	Total drying energy, PJ
Paper and board	50.0	91.8
Ceramics and building materials	69.3	79.0
Food and agriculture	47.1	123.0
Plaster and plasterboard	60.0	2.5
Textiles	57.3	38.7
Timber	55.0	9.6
Chemicals	58.0	3.3
Pharmaceuticals	70.0	0.02
Laundry	53.0	0.7

Source: From Gilmour et al. (see Ref. 7).

adiabatic counterpart. Baker and McKenzie showed that the specific energy consumption $E_{s,a}$ of such a dryer is not fixed in the absolute sense, but rather that it depended on the temperature and humidity of the outlet air:

$$E_{s,a} = 0.001 \left[C_{pg} \left(\frac{T_o - T_a}{Y_o - Y_a} \right) + \lambda_{ref} \right]$$
$$= C_{pg} \xi + \lambda_{ref} \qquad (3)$$

In this equation, C_{pg} is the specific heat of dry air, and λ_{ref} is the latent heat of water at 0°C. T_a and Y_a denote the temperature and humidity of the ambient air, and T_o and Y_o denote the corresponding values for the outlet air leaving the dryer. The latter are set so as to achieve the product's desired moisture content.

For an adiabatic dryer, it follows from Eq. 3 that a plot of $E_{s,a}$ against $\xi = (T_o - T_a)/(Y_o - Y_a)$ should be linear with a slope of 0.001 C_{pg} and an intercept of $0.001\lambda_{ref}$. This plot can be used as a baseline against which the performance of non-adiabatic dryers can be judged. Fig. 8 shows the data obtained in Baker and McKenzie's survey[8] plotted as E_s against $(T_o - T_a)/(Y_o - Y_a)$, in which T_a was taken as 25°C and Y_a as 0.005 kg/kg. As would be expected, all of the points scattered on or above Eq. 3, which depicts the performance of an ideal adiabatic dryer.

Eq. 3 can be used to interpret the performance of dryers in the field. As illustrated schematically in Fig. 9, a dryer exhibiting a given specific energy consumption may either be wasting considerable quantities of energy or operating close to peak efficiency. Here, the actual energy consumption of Dryer 1, E_s^*, is significantly larger than the value $(E_{s,a})_1$ required for an adiabatic dryer for which $\xi = \xi_1$. The efficiency of this dryer is relatively low. In contrast, for Dryer 2 ($\xi = \xi_2$), $E_s^* - (E_{s,a})_2 \ll E_s^* - (E_{s,a})_1$. This indicates that Dryer 2 is much more efficient than Dryer 1. Baker and McKenzie[8] found that the efficiency of the 32 spray dryers included in their survey varied widely. On average, though, almost 30% of the energy supplied to the dryers was wasted.

Specific energy consumption therefore provides a useful guide as to how efficiently (or not) a particular dryer is operating. The value of E_s must be measured by means of an energy audit on the dryer. Guidance on undertaking such audits is provided in the following section.

DRYER ENERGY AUDITS

Any attempt at reducing the fuel consumption of an industrial dryer should begin with assessments of how much energy is currently being consumed and whether this is being expended usefully or is being wasted. This establishes the benchmark against which future energy savings can be judged. In order to make the above assessments, it is first necessary to carry out an energy audit. This consists of a series of measurements designed to establish pertinent energy and mass flows into and out of the dryer. As each dryer is different, it follows that each audit should be customized to fit it.

There are essentially two types of audit, namely a basic audit and a detailed audit. In both cases, the dryer should be properly instrumented. Fig. 10 illustrates an ideal arrangement of measuring points. The primary purpose of a basic audit is to establish the specific energy consumption of the dryer and to assess the potential energy savings that are possible. If appropriate, a detailed audit can subsequently be undertaken to determine where the energy losses are

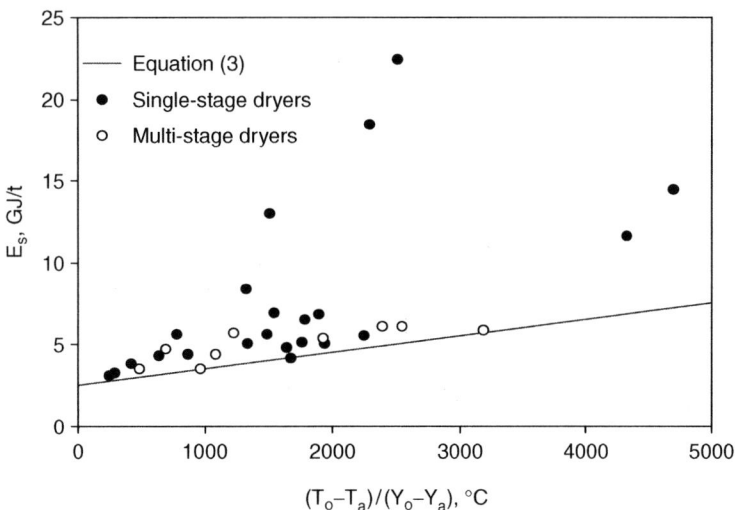

Fig. 8 Plot of specific energy consumption against $(T_o - T_a)/(Y_o - Y_a)$ for spray dryers (after Baker and Mckenzie[8]).

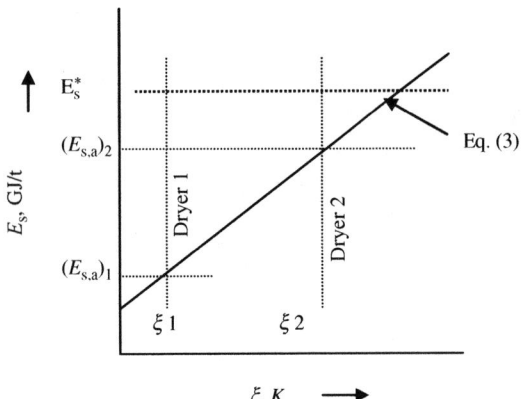

Fig. 9 Schematic representation of dryer efficiency.

occurring and to identify the corrective actions necessary to improve the performance of the dryer.

Process measurements on operating dryers are difficult to perform and the following precautions should be taken to minimize errors. All measurements should be made in duplicate or triplicate in order to ensure their accuracy and to confirm that the dryer is operating at steady-state. Triangulation, in which the value of a particular parameter is arrived at by more than one approach, is always advisable as it provides added confidence in the data. It is also important to ensure that all the measuring instruments employed are properly calibrated and used in accordance with recognized (e.g., ISO) standards to ensure accurate results.

Basic Audits

Table 3 lists the measurements that are required for a basic audit, the purpose of which is to determine E_s and ξ so that the performance of the dryer relative to its adiabatic counterpart can be assessed. The specific energy consumption can be calculated from the following equation:

$$E_s = \frac{0.001 \; Q_{htr}}{F(X_i - X_o)} \quad (4)$$

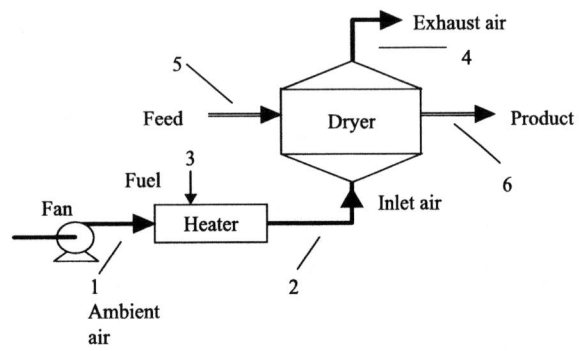

Fig. 10 Location of measurement points (after Baker[6]).

Table 3 Measurements required for a basic audit

Number	Measurement point	Measurements made
1	Ambient air	Air flowrate, temperature, and humidity
2	Inlet air	Air temperature
3	Fuel	Fuel flowrate[a]
4	Exhaust air	Air temperature and humidity[b]
5	Feed	Solids moisture content
6	Product	Solids flowrate, moisture content

[a]If steam is used as the heating medium, its temperature and pressure should be measured as well. Where possible, the condensate temperature and flow should also be recorded.
[b]It may be more appropriate to calculate this—see text.

in which F is the dry-basis solid feed rate, X_i and X_o are the moisture contents of the feed material and product, respectively, and Q_{htr} is as defined above. The measurement of F, X_i, and X_o is normally straightforward. Assuming that the dryer is operating at steady-state and that solid losses in, for example, the exhaust air, are minimal, F can be calculated from the production rate, which will undoubtedly be recorded on an ongoing basis for commercial reasons. Two methods are commonly used to determine the heat load Q_{htr}:

1. Direct measurement—This naturally requires that a fuel or steam meter be fitted.
2. Indirect measurement—This is useful in cases where a fuel meter is not fitted. One commonly used method is to carry out an energy balance over the heater.

The use of Method 2 presupposes that the air flowrate is known. This is normally measured using, for example, a Pitot tube. If this is not possible, the original dryer design specifications may, in some circumstances, give a reasonable estimate. However, there is clearly some uncertainty in the latter approach as the operating conditions may have changed considerably since the dryer was commissioned.

Methods 1 and 2 can of course both be used in the same audit. This is a good example of triangulation and increases confidence in the measured values of the energy supplied by the heater and the mass flowrate of the drying air.

In order to determine ξ, it is necessary to measure T_a, T_o, Y_a, and Y_o. The first three of these variables are straightforward. However, the measurement of Y_o is more problematical as the exhaust air is often dusty and its temperature and humidity are relatively high. Under these circumstances, the most accurate technique to determine

Table 4 Results of hypothetical basic audit on dryers having an evaporative load of 5 t/h

Dryer	ξ, °C	$E_{s,a}$, GJ/t	E_s^*, GJ/t	$E_s^* - E_{s,a}$, GJ/t	Potential saving, MW	Potential saving, %
A	6545	9.04	11.90	2.86	4.0	24.0
B	3333	5.83	9.47	3.64	5.1	38.6
C	2870	5.37	6.20	0.83	1.2	13.4

Y_o is often to calculate it from a mass balance over the dryer.

Table 4 shows some typical results that might be obtained through basic audits on three hypothetical dryers, each evaporating 5 t/h of moisture. In this table, the potential energy saving in MW is $W_{ev}(E_s^* - E_{s,a})$, where W_{ev} is the evaporation rate in kg/s and E_s^* and $E_{s,a}$ are in GJ/t. The numbers cited represent the maximum possible energy savings that can be achieved when the dryer is operating at the same exhaust air temperature and humidity.

Detailed Audits

Detailed audits are much wider in their scope than basic audits and may involve considerably more measurements. Their principal purpose is to identify the causes of inefficiencies in the dryer. This is accomplished by constructing detailed mass and energy balances around both the dryer and its air heater, as shown in Fig. 11. From these balances, it should be possible, for example, to determine the following:

- Magnitude of air leaks (if any) into or out of the dryer
- Magnitude of internal steam leaks (if any) within the heater
- Magnitude of the heat losses from the heater and dryer

Although interpretation of the data obtained in detailed audits requires a specialist's knowledge, they can, as discussed below, provide useful guidance on possible measures that can be employed to reduce a dryer's energy consumption.

PRACTICAL MEASURES FOR REDUCING DRYER ENERGY CONSUMPTION

Data obtained from detailed audits can be used to evaluate the effectiveness of various options for reducing the energy consumption of the dryer. These can be divided into (i) schemes involving little or no capital expenditure and (ii) schemes involving significant capital expenditure. A detailed quantitative analysis of the results of a typical dryer audit has been described by Baker.[6]

It should be stressed that any alterations to the dryer itself or to its operating conditions may involve an element of risk and should not be undertaken without a proper evaluation of all the factors involved. Where appropriate, expert guidance should be sought.

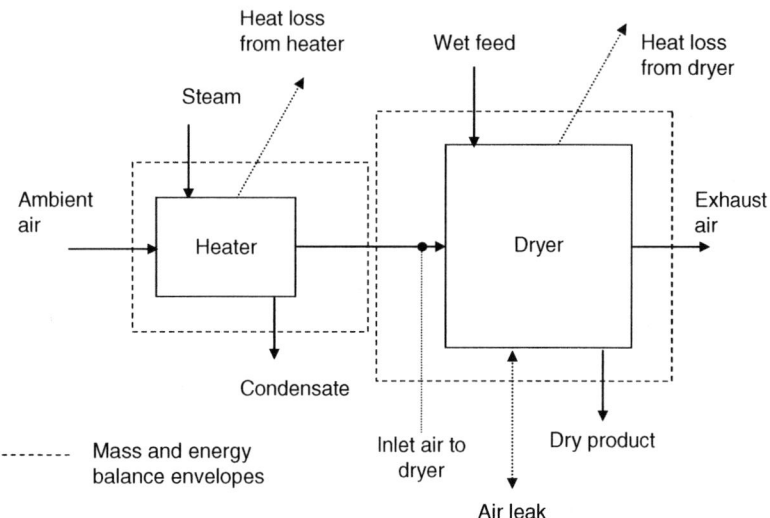

Fig. 11 Energy balances around dryer and heater.

Schemes Involving Little or No Capital Expenditure

Examples of schemes that require little or no capital expenditure include those which can be described as housekeeping measures, 1–4 below, and those that are based exclusively on a beneficial change in the dryer's operating conditions, 5–7.

1. Reducing air leaks
2. Eliminating steam leaks
3. Improving dryer insulation
4. Improving heater insulation
5. Reducing the air mass flowrate
6. Increasing the inlet air temperature
7. Eliminating over-drying

Brief outlines of each of these measures are as follows. Leakage of hot air out of the system or cold air into it clearly represent a waste of energy which can normally be reduced or eliminated by appropriate maintenance procedures. Air in-leakage through, for example, a warped access door, can occur when the dryer is fitted with an exhaust fan and is operated under a slight vacuum. It has the effect of reducing the temperature of the air in the drying chamber below the desired value. As a result, the inlet air has to be heated to a somewhat higher temperature to compensate. The primary effect of steam leaks in the heater is to increase the humidity of the inlet air to the dryer, which, in turn, may reduce the drying rate. To compensate, the inlet air temperature has to be raised above the design value in order to maintain the desired product moisture content.

Adequate insulation of the dryer and the air heater is vital in order to maintain heat losses within acceptable levels. As a result, damaged or water-laden insulation needs to be replaced as soon as possible. It is also sensible to periodically review the optimum insulation thickness, as this will naturally increase as the price of fuel escalates. Repairing and upgrading the insulation, however, will impact on the dryer's operation. Assuming that the control settings remain unchanged, the temperature of the drying air will increase as a result of the reduced heat loss and the product will therefore be over-dried. There are three possible ways in which the dryer operating conditions can be changed to overcome this problem:

1. Maintain constant solids feedrate and air flow and reduce the inlet air temperature to the dryer.
2. Maintain constant solids feedrate and inlet air temperature and reduce the air flowrate to the dryer.
3. Maintain constant inlet air temperature and flowrate and increase the solids feedrate.

Each of the above approaches needs to be evaluated in order to determine the best strategy to adopt.

It may also be possible to reduce the dryer energy consumption simply by modifying its operating conditions. Most of the heat leaves the dryer via the exhaust air stream. It can be shown that if the air flowrate is reduced, its humidity will be increased and energy will be saved. In practice, there are limits to this approach. For example, it is normally necessary to maintain a difference of at least 10°C between the temperature and dewpoint of the exhaust air in order to prevent condensation in the downstream ductwork and gas-cleaning equipment. Moreover, in a fluidized bed dryer, a minimum air velocity is required to fluidize the particles. Similar limitations exist on other types of dryer. It can also be shown that operating a dryer at as high an inlet air temperature as possible also reduces its energy consumption. Naturally, other adjustments will have to be made in order to maintain a constant product moisture content and care should be taken to avoid damage to the product if it is heat sensitive.

Based on 1978 figures for the United Kingdom, Baker and Reay[9] estimated that, on a national scale, around 11×10^6 GJ/y could be saved by implementing the low-cost measures described above. This represents about 8.6% of the energy that was then expended on drying in the United Kingdom.

Schemes Involving Significant Capital Expenditure

Only when the low-cost measures described above have been exhausted and existing dryer performance optimized should capital-intensive schemes even be considered. These possibilities include:

1. Recovering heat from the exhaust air
2. Partially recirculating the exhaust air
3. Utilizing waste heat
4. Monitoring and advanced control
5. Switching from an indirect to a direct heater
6. Prior dewatering of the dryer feedstock

Most of these techniques have been discussed by Mercer,[10] who also summarized the results of several industrial case studies. In practice, with the possible exception of monitoring and control, such schemes are rarely economically viable as retrofits because of the associated plant modifications (e.g., to the ductwork) that are required. However, when considering the purchase of a new dryer, it is worthwhile considering all the above options, which are described briefly below.

1. As noted above, most of heat leaving a dryer is contained in its exhaust air. If this can be at least partially recovered and used to preheat the incoming air to the dryer, significant energy

savings, typically 17%–40%, may result.[10] However, industrial experience has produced mixed results. Fouling of the heat exchangers by entrained particles in the dryer exhaust and, to a lesser extent, corrosion, are commonly encountered and result in poor performance and severe operating problems. These can be overcome by employing glass heat exchangers fitted with clean-in-place washing systems; however, such equipment is expensive. Other possible devices include heat wheels, heat pipes, and run-around coils.

2. An alternative means of heat recovery, which avoids the need for heat exchangers and their attendant problems, is to partially recycle, or recirculate, the exhaust air. Typically, 10%–50% of this stream is mixed with the hot inlet air from the heater before entering the dryer. Under appropriate conditions (e.g., an exhaust air temperature in excess of 120°C and a heat-resistant product), fuel savings of up to 20% can be achieved using this technique.[11]

3. In some cases, it may be possible to use process waste heat (flue gases, low-pressure steam, etc.) as a full or partial replacement for conventional energy sources. Clearly, each case must be judged on its own technical and economic merits. Principal factors to be considered include: the quantity of "free" heat available, its reliability of supply, its compatibility with the product (temperature, feasibility of direct contact, etc.), incremental cost of engineering work required to recover the heat, etc. Solar dryers,[12] which are frequently used in tropical and sub-tropical countries to dry agricultural produce in particular, represent an extreme example of the use of free energy.

4. Many older dryers are fitted with, at best, only very basic open-loop controls. Plant trials have frequently shown that effective control systems not only reduce the energy consumption of the dryer, but may also improve product quality, reduce the amount of off-spec product, and increase through put. Effective controllers range from relatively simple devices (e.g., proportional-integral-derivative (PID) controllers) to more sophisticated types (e.g., adaptive, model-based, and fuzzy logic). The optimal choice should be based on process conditions and the type of dryer employed. Mercer[10] reported energy savings ranging from 0 to 50%, together with associated benefits of typically 0.5–1 times the direct energy savings.

5. Provided that the product being dried is compatible with the combustion gases, a direct-fired dryer should be used in preference to an indirectly heated dryer. In the former case, most (95%–98%) of the energy produced in the burner is transferred to the dryer's inlet air stream. This compares with 85% for a typical steam heater. In practice, if we also take boiler efficiency into account, a primary fuel savings of around 30% can be achieved by switching from an indirect to a direct heater.

6. Finally, given the poor thermal efficiency of many dryers, every effort should be made to minimize the moisture content of the feed by using more energy-efficient processes. In the case of liquid feedstocks, evaporation is inevitably employed to partially dewater the feedstock prior to spray drying. With solid feeds, it may be possible to use various mechanical dewatering techniques such as vacuum or compression filtration, centrifugation, etc.

Reducing Electrical Energy Consumption

To date, the emphasis in this article has been directed towards reducing thermal energy consumption. However, the electrical energy consumed by ancillary equipment such as fans, pumps, atomizers, and conveyors should not be neglected. Baker and McKenzie[8] reported that the average thermal-to-electric power ratio for the spray dryers included in their survey was around 27; this ratio is equivalent to around nine on a primary fuel basis. Therefore, consideration should always be given to implementing measures that reduce the consumption of electricity. These include installing high-efficiency electric motors, reducing the pressure drop across the system by, for example, fitting variable-speed fans rather than dampers in the ductwork, and employing bag filters rather than cyclones for dust collection.

THE FUTURE

High fuel prices and recognition of the need to cut greenhouse gas emissions are undoubtedly encouraging companies to reduce their energy consumption. In order to adapt to this new environment, dryer operators in particular will need to make some adjustments to the ways in which they purchase and operate their equipment. For example:

1. When buying a new dryer, more emphasis should be placed on lifetime cost rather than initial capital cost, which is often the case at present. This type of long term thinking can be expected to favour those options that have a lower specific energy consumption while, naturally, fulfilling all other process requirements. Contact dryers, for instance, are more efficient ($E_s \sim 2.8$ GJ/t of water evaporated) than their convective counterparts. Superheated-steam dryers are claimed to consume less than 2 GJ/t when heat recovery is employed. Other advantages include faster drying, reduced emissions, and the possibility of recovering volatile organic

compounds (VOCs). Pulsed fluidized bed and vibrofluidized bed dryers are more efficient than their conventional counterparts because they use less air. Finally, the efficiency of two-stage dryers is normally higher than that of their single-stage counterparts.
2. Despite their added cost, the heat recovery options and other energy-saving measures described above become more attractive as fuel prices rise. However, a thorough technical evaluation needs to be undertaken to ensure that any potential benefits are not negated, e.g., by fouling of the heat exchanger surfaces.
3. More effort needs to be taken to monitor the fuel consumption of individual dryers as useful energy savings can frequently be achieved by means of low-cost measures. Fitting dryers with advanced control systems can also result in quality improvements.

Collectively, these measures can help to offset rising fuel bills, combat inflation, and conserve our planet's natural resources.

CONCLUSIONS

By their very nature, dryers are major consumers of energy and are, in addition, often operated inefficiently. The techniques described in this article outline a rigorous approach to benchmarking dryer performance and suggest a hierarchy of methods for reducing their energy consumption.

REFERENCES

1. Lababidi, H.M.S.; Baker, C.G.J. Web-based expert system for food dryer selection. Comput. Chem. Eng. **2003**, *27*, 997–1009.
2. Mujumdar, A.S., Ed., *Handbook of Industrial Drying*, 2nd Ed., Marcel Dekker: New York, 1995.
3. Baker, C.G.J., Ed., *Industrial Drying of Foods*, Blackie Academic and Professional: London, 1997.
4. Richardson, J.F.; Harker, J.H.; Backhurst, J.R., 5th Ed. In *Chemical Engineering*; Butterworth Heinemann: New York, 2003, Vol. 2; 901–969 [Chapter 16].
5. Geankoplis, C.J. *Transport Processes and Unit Operations*, 3rd Ed.; Prentice-Hall: London, 1993; 520–583 [Chapter 9].
6. Baker, C.G.J. Energy efficient dryer operation—an update on developments. Drying Technol. **2005**, *23*, 2071–2087.
7. Gilmour, J.E.; Oliver, T.N.; Jay, S. Energy use for drying processes: the potential benefits of airless drying. In *Proceedings 11th International Drying Symposium*, Halkidiki, Greece, August 19–22, 1998; Vol. A, 573–580.
8. Baker, C.G.J.; McKenzie, K.A. Energy consumption of industrial spray dryers. Drying Technol. **2005**, *23*, 365–386.
9. Baker, C.G.J.; Reay, D. Energy usage for drying in selected U.K. industrial sectors. In *Proceedings 3rd International Drying Symposium*, Vol. 2, Drying Research Ltd: Birmingham, U.K., 1982, 201–209.
10. Mercer, A. Learning from experiences with industrial drying technologies. In *CADDET Energy Efficiency Analyses Series No. 12*; Sittard: Netherlands, 1994.
11. Masters, K. *Spray Drying Handbook*; Longman: London, 1985; 106–108.
12. Imre, L. Solar dryers. Baker, C.G.J., Ed. In *Industrial Drying of Foods*, Blackie Academic and Professional: London, 1997; [Chapter 10], 210–241.

Electric Motors

H. A. Ingley III
Department of Mechanical Engineering, University of Florida, Gainesville, Florida, U.S.A.

Abstract
This entry on electric motors is written from the perspective of a mechanical engineer involved in the design and specification of mechanical systems powered by electric motors. The entry describes several motor types and their intended applications. Specific components and performance criteria for motor selection are illustrated. The entry concludes with a discussion on motor efficiencies and the use of adjustable speed drives (ASDs) to enhance electric motor performance.

INTRODUCTION

This chapter on electric motors is written from the perspective of a mechanical engineer involved in the design and specification of mechanical systems powered by electric motors. The reader may have the impression that electrical engineers have the prime responsibility to specify electric motors, and it may come as a surprise that in most cases, it is the mechanical engineer specifying the electric motor. This is primarily because most of the mechanical equipment specified by mechanical engineers is packaged with electric motors. However, careful coordination with an electrical engineer on the electrical service requirements for the motor is essential in any design.

Electric motors find wide use in the industrial, commercial, and residential sectors of the United States. The following table summarizes the extent of these applications (Table 1).

Electric motors are responsible for a significant fraction of this nation's electrical consumption (Table 2). As an example, electric motor energy use represents 60% of the total energy consumption in the United States' industrial sector. A basic knowledge of electric motor design and applications equips the engineer with the skills to provide motor selections that insure reliable service with minimal energy consumption.

As might be expected, there are about as many electric motor classifications as there are applications. The criteria listed in Table 3 should be considered when selecting motors for a given application.

A general classification of common motor types with comments on their applications and special characteristics is given in Table 4. The three basic types of motors are alternating current induction/asynchronous, alternating current synchronous and direct current. Over 90% of the electrical energy input to electric motors serves to power alternating current induction motors.[2]

Because induction motors account for the highest percentage of energy use among motors, it is important to have a basic understanding of how these motors work and what technologies are available to enhance the performance of these motors. Induction motors are constructed to run on single-phase or three-phase power. Most homes in the United States are supplied with single-phase power; therefore, the majority of motors used in these homes are single-phase induction motors. Even though numerically the majority of motors fall in this category (see Table 2), the highest percentage of energy use results from the use of larger three-phase induction motors.

One other important distinction between the smaller single-phase motors and the three-phase motors is the drive method used in the application. Generally, single-phase motors have their shafts directly connected to the device; e.g., a motor driving a small hermetically sealed compressor or a small bathroom exhaust fan. When the motor fails, the whole device generally requires replacement. Larger three-phase motors are coupled to their devices with belts or couplings that allow for relatively easy motor replacement.

Another distinction between single-phase and three-phase motors is the method used to start the motor. In a three-phase motor, there are three stator (the fixed coil) windings spaced at 120° intervals around the rotor (the rotating member of the motor). These motor windings result in a rotating field around the rotor, thus starting and maintaining rotation of the rotor. Single-phase motors do not have this magnetic field arrangement; therefore, they require an auxiliary method to start the rotation. Generally, this is accomplished by a separate winding set at an angle of 90° from the main winding. This auxiliary winding is connected in series with a capacitor to provide a starting torque to initiate rotor rotation. After the motor starts, the auxiliary winding is disconnected by an internal switch. This type of motor is referred to as a capacitor-start motor.

Keywords: Electric motors; Induction motors; Slip; Motor efficiency; Power factor; Adjustable speed drives; Energy estimating.

Table 1 Electric motor application statistics

14% of the population of motors power centrifugal fans
33% are used for material handling
34% are used for positive displacement pumps
18% are used for centrifugal pumps
32.5% are used in variable torque applications
67.5% are used in constant torque applications

Source: From American Council for an Energy Efficient Economy (see Ref. 1).

In permanent split-capacitor induction motors, the capacitor is not disconnected after motor startup.

The difference between a synchronous and an asynchronous induction motor is dictated by the amount of full-load motor slip resulting from the motor design.

$$\text{Synchronous speed in rpm} = \frac{\text{applied voltage frequency(Hz)} \times 60}{\text{number of pole pairs in the motor}} \quad (1)$$

Referring to Eq. 1 above, a motor having one pole pair (two poles) would have a synchronous speed in the United States (where the applied voltage frequency is 60 Hz) of 3600 rpm. Four poles would result in a synchronous speed of 1800 rpm, and six poles would result in a synchronous speed of 1200 rpm. However, induction motors experience "slip," which results in a lower operating speed. These motors, referred to as asynchronous induction motors, can experience full-load slip in the range of 4% for small motors and 1% for large motors. Thus, a four-pole asynchronous motor would operate at 1750 rpm.

Induction motors can be further classified according to the design of their rotors. When these rotor components are formed into a cylindrical shape resembling a squirrel cage, the motor is referred to as a "squirrel cage" motor. This type of motor is relatively inexpensive and reliable; therefore, it finds wide use in commercial and industrial applications. In motors where starting current, speed, and torque require close control, the rotor is comprised of copper windings much like the stator. This requires an external source of power for the rotor, which can be accomplished with slip

Table 2 Electric motor facts

It is estimated that 60% of all electric power produced in the United States is used by electric motors
90% of all motors are under 1 hp
8% of all motors are in the 1–5 hp range
2% of all motors are over 5 hp
70% of the electric use by motors is by motors over 5 hp; 22% is by motors in the 1–5 hp range; and 8% is by motors under 1 hp

Source: From American Council for an Energy Efficient Economy (see Ref. 2).

Table 3 Electric motor selection criteria

Brake horsepower output required
Torque required for the application
Operating cycles (frequency of starts and stops)
Speed requirements
Operating orientation (horizontal, vertical, or tilted)
Direction of rotation
Endplay and thrust limitations
Ambient temperature
Environmental conditions (water, gasoline, natural gas, corrosives, dirt and dust, outdoors)
Power supply (voltage, frequency, number of phases)
Limitations on starting current
Electric utility billing rates (demand, time of day)
Potential application with variable frequency drives

rings and brushes. As might be expected, this increases the first cost of the motor. Maintenance costs are also higher. These wound-rotor motors are generally sold in sizes of 20 hp and up.

When an engineer is considering an electric motor for a given application, it is important to match the enclosure of the motor to the type of operating environment involved. There are tradeoffs to be considered in this selection. A more open motor will stay cooler, which will improve its efficiency and service life. A closed motor will be less subject to contamination by a wet or dirty environment, but it may be less efficient. National Electrical Manufacturers Association (NEMA) Standard MG-1-1978 describes 20 different types of enclosures which address these two tradeoffs. Generally, in commercial and industrial applications, the enclosures fall in one of three categories: open drip proof (ODP), totally enclosed fan cooled (TEFC); and explosion proof (EXP). Fig. 1 is an example of a TEFC motor enclosure. Table 5 describes some of the other basic nomenclature used to describe motor enclosures.

Proper motor selection also includes an understanding of temperature ratings, insulation classifications, and service factors. National Electrical Manufacturers Association Standard MG 1-1998 relates the allowable temperature rises for each insulation classification. The most common insulation class is class B, which allows for a temperature rise of 80°C for Open or TEFC motors with service factors of 1.0, and 90°C for motors operating at a service factor condition of 115% rated load. When applying adjustable speed drives (ASDs) to motors, it is important to ensure that the insulation class is rated for ASD operations. In general, Class F will meet this requirement. Most premium-efficiency motors should also have this classification of insulation. At times, consideration may be given to selecting motors

Electric Motors

Table 4 Classification of common motor types

a.c.	Induction	Squirrel cage	Three-phase, general purpose, >0.5 hp, low cost, high reliability
			Single-phase, <0.5 hp, high reliability
		Wound rotor	>20 hp, special purpose for torque and starting current regulation, higher maintenance requirement than for squirrel cage
			Very large sizes, high efficiency and reliability, higher maintenance requirement than for squirrel cage
a.c.	Synchronous	Reluctance	Standard, small motors, reliable, synchronous speed
			Switched, rugged high efficiency, good speed control, high cost
		Brushless permanent magnet (overlaps d.c. also)	High efficiency, high performance applications, high reliability
d.c.	Wound rotor	Limited reliability, relatively high maintenance requirements	Series, traction, and high torque applications
			Shunt, good speed control
			Compound, high torque with good speed control
			Separated, high performance drives; e.g., servos

Source: From American Council for an Energy Efficient Economy (see Ref. 2).

(especially large motors, 200 hp and up) with heaters in the motor windings. These heaters are energized when the motor is de-energized and they keep the motor windings warm to inhibit moisture wicking into the winding interstitial areas.

The service factor rating for a motor addresses the capacity at which an electric motor can operate for extended periods of time at overload conditions. As an example, if the service factor is 1.0, the motor cannot operate above full-load capacity for a significant period of time without damaging the motor. Service factors of 1.15, 1.25, or 1.35 indicate that a motor can operate at 1.15, 1.25, or 1.35 times its rated full load, respectively, for extended periods of time without failure. It should be noted, however, that this does not mean that the motor's service life is not affected. Insulation life can be reduced by as much as 50% operating under these conditions.

When evaluating motor replacement options, the frame size for a motor is also a matter of consideration. U-frame and T-frame designations are typical with electrical motors. New high-efficiency motors with U-frames are not always interchangeable with older style U-frame motors. It is important to check the frame size and determine if a conversion kit is required. The method of attachment or integration of the frame with the motor is also important. The author has often experienced owner preferences in selecting motors that will be used to drive centrifugal fans with a belt drive system. In one case, the owner had experienced structural failures with motors in which the motor housing was welded to the base frame.

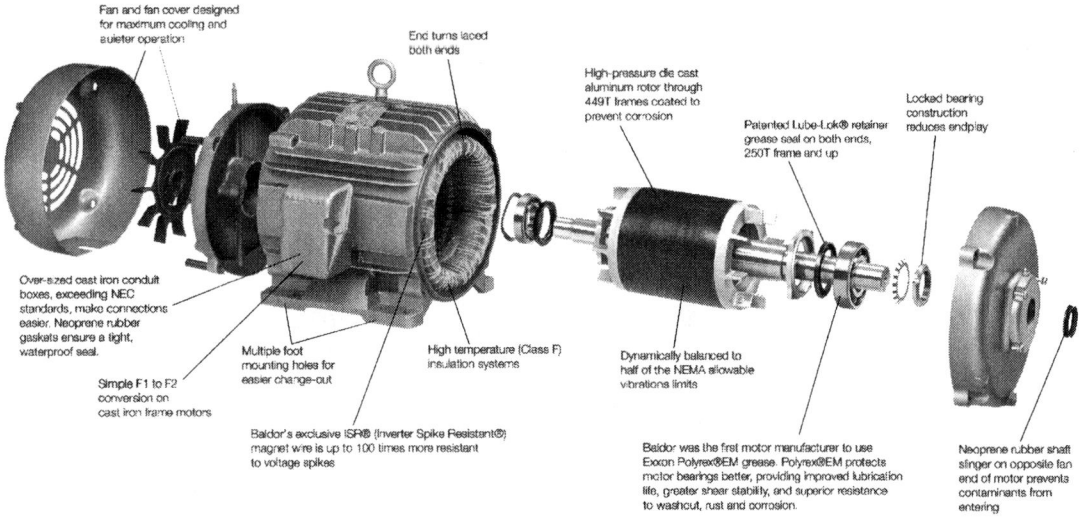

Fig. 1 Premium-efficiency motor (courtesy of Baldor).

Table 5 Examples of motor enclosure nomenclature

Open-type—full openings in frame and endbells for maximum ventilation, low cost

Semi-protected—screens on top openings to keep falling debris out, protected with screens on all openings (older style motors)

Drip-proof—upper parts are covered to keep drippings out falling at an angle not over 15° from vertical

Splash-proof—baffled at bottom to keep particles out coming from an angle of 100° from vertical

Totally enclosed—explosion proof, nonventilated, or separately ventilated for hazardous atmospheres

Fan-cooled—totally enclosed motor with fan built in to ventilate motor

This particular owner preferred motors in which the base and motor housing was a single integrated piece.

In motor replacement or in selection of new motors, the supply voltage should be coordinated with the selection. Most three-phase motors are designed to operate at 460 V and 60 Hz in the United States. Many commercial or institutional facilities are served with 208- or 230-V services. Residential motors and small fractional horsepower industrial motors would be served with 120-V power. It should be noted that even though a motor might be rated to operate at either 208 or 230 V, the operation at the lower voltage will result in a lower efficiency and shorter service life. Off-voltage operation can adversely affect electric motor performance and should be avoided.

Associated with the power supply to the motor is the motor's contribution to the overall power factor for the facility using the motor. Utilities generally penalize customers with poor power factors, because this translates into reduced availability of transformer capacity and inefficient use of power. The typical threshold level for power factors may be in the range of 85%–95% for some utilities. When specifying electric motors, the power factor should be included, as well as the other factors indicated above.

Now that we have covered some of the basics in motor types and selection, the discussion will turn to addressing three questions that consulting engineers often have to consider when evaluating options for motor replacement: what motor efficiency is most cost effective? Should motor rewinding be considered if an existing motor has failed and requires replacement? Are there other options to consider that might save energy?

The importance of motor efficiency is certainly obvious considering the energy consumption trends referenced at the introduction of this chapter. Improvements in electric motor efficiency have been recognized as a means of reducing energy consumption in all sectors of the economy.

The efficiency for motors is generally defined by Eq. 2:

$$\text{Efficiency} = \frac{746 \times \text{output horsepower}}{\text{input watts}} \quad (2)$$

In October of 1992, the U.S. Congress signed into law the Energy Policy Act (EPAct) that established energy efficiency standards for general-purpose, three-phase alternating current industrial motors ranging in size from 1 to 200 hp. In October of 1997, the EPAct became effective.[3] Table 6 illustrates a sample of required full-load nominal efficiencies for general purpose motors. Note that the power factors are not given in this table. Consideration should be given to selecting the highest possible power factor when selecting any motor. There are several standards available for testing motors. However, for the same motor, the tests performed using these standards may result in differing efficiencies. The generally recognized standards for testing are the IEEE 112 Method B and CSA C-390-93 standards. These two testing standards typically provide the same results when applied to the same motor.

A motor that meets or exceeds the minimum efficiencies specified by the Consortium for Energy Efficiency (CEE) is referred to as a CEE premium-efficiency motor. These motors generally have efficiencies

Table 6 Required full-load nominal efficiencies for general purpose motors.

Motor hp	Open motors (%)			Enclosed motors (%)		
	6-pole	4-pole	2-pole	6-pole	4-pole	2-pole
1	80.0	82.5		80.0	82.5	75.5
5	87.5	87.5	85.5	87.5	87.5	87.5
7.5	88.5	88.5	87.5	89.5	89.5	88.5
10	90.2	89.5	88.5	89.5	89.5	89.5
15	90.2	91.0	89.5	90.2	91.0	90.2
20	91.0	91.0	90.2	90.2	91.0	90.2
100	94.1	94.1	93.0	94.1	94.5	93.6

Courtesy of Department of Energy

Table 7 Comparison of Energy Policy Act (EPAct) to premium-efficiency motors

Horsepower	Average EPAct efficiency at 75% load (%)	Average premium efficiency at 75% load (%)	Ratio of premium-efficiency motor cost to EPAct motor cost
1	82.4	85.2	1.20
5	88.5	90.5	1.20
10	91.1	91.9	1.09
20	92.3	93.5	1.08
50	93.9	94.8	1.14
75	94.5	95.7	1.09
100	94.9	95.7	1.17
150	96.1	95.9	1.24
200	95.3	96.3	1.14

Source: From American Council for an Energy Efficient Economy (see Ref. 2, Table Aa-1).

that exceed those of EPAct motors. Table 7 illustrates a comparison between EPAct motor efficiencies and CEE premium motor efficiencies for several motor sizes. Even though the price of CEE premium-efficiency motors exceeds EPAct motors by as much as 20%, energy and demand savings can result in simple paybacks of 2 to 3 years.

The EPAct of 2005 required that federal agencies select and purchase only premium efficient motors that meet a specification set by the Secretary of Energy. On August 18, 2006, the DOE set forth the specifications developed by the Federal Energy Management Program to be used for purchasing. These standards are consistent with those recommended by the NEMA and the CEE. Tables 8 and 9 illustrate these new standards. In order to meet Energy Star requirements, the efficiencies in Tables 8 and 9 must also be met.

In evaluating motors in the field for general performance; in particular, to determine if the motor is overloaded or underloaded, measurements of applied voltage and amperes can provide valuable information. This data can also be used to evaluate annual energy consumption for a motor if an estimate of operating periods can be made. Eq. 3 illustrates a method for estimating the power consumption of a motor in kilowatts[4]:

Three-Phase Power, kW

$$= \frac{\text{Volts} \times \text{Amperes} \times \text{Power Factor} \times 1.732}{1000} \quad (3)$$

When replacing or rewinding motors—premium-efficiency motors (motors whose efficiencies exceed EPAct requirements)—the economics for the decision making will obviously depend on several factors such as motor type, operating regime (hours/year), and cost. Consortium for Energy Efficiency premium-efficiency motors[2] with operating regimes in the range of 4000 h per year will experience simple paybacks of 6 years or less compared to EPAct motors. Totally enclosed fan cooled motors in the size range of 2–25 hp will pay back in 3 years or less. Motors in sizes of 10 hp and below are generally replaced when they fail; larger motors are often repaired. When the rewinding or repair option of a standard-efficiency motor is compared to replacement with a premium-efficiency motor, the simple payback can be less than 2 years (for ODP motors in sizes up to 200 hp and for TEFC motors in sizes up to 40 hp). For single-phase and fractional horsepower motors, the simple payback varies significantly (2.5–10 years) when replacement with a premium-efficiency motor is evaluated.

Another area of concern consulting engineers face is in the selection of motors for pumps. Many engineers base their selection of centrifugal pump motor size on nonoverloading criteria. In order to ensure that the pump motor never overloads, the engineer checks the box on the pump selection software that will pick nonoverloading motor sizes. The resulting pump curve will always be to the left of the motor hp curve. The significance of this is that a pump with a duty horsepower of 2.3 hp could have a 5-hp motor selected. This results in the motor operating many hours of the year at part load and consequently, at lower than full-load efficiencies. Consulting engineers should evaluate their decisions to select nonoverloading motors with energy efficiency in mind. With the advent of reliable electronic ASDs, significant energy savings are now available for many applications involving electrically driven pumps and fans. It is very common today to design commercial building air conditioning systems with ASD technology. Variable-air-volume air conditioning systems use ASDs to vary fan speed as the demand for supply air varies with cooling and heating loads. Similarly, in large chilled-water or heating-hot-water distribution systems, ASDs are used to vary the flow rate of water in response to varying loads. In one such application involving two 400-hp circulating pumps on a chilled water system, the

Table 8 Nominal efficiencies for induction motors rated 600 V or less

| Horsepower | Random wound ||||||
| | Open drip-proof (%) ||| Totally enclosed fan-cooled (%) |||
	6-pole	4-pole	2-pole	6-pole	4-pole	2-pole
1	82.5	85.5	77.0	82.5	85.5	77.0
1.5	86.5	86.5	84.0	87.5	86.5	84.0
2	87.5	86.5	85.5	88.5	86.5	85.5
3	88.5	89.5	85.5	89.5	89.5	86.5
5	89.5	89.5	86.5	89.5	89.5	88.5
7.5	90.2	91.0	88.5	91.0	91.7	89.5
10	91.7	91.7	89.5	91.0	91.7	90.2
15	91.7	93.0	90.2	91.7	92.4	91.0
20	92.4	93.0	91.0	91.7	93.0	91.0
25	93.0	93.6	91.7	93.0	93.6	91.7
30	93.6	94.1	91.7	93.0	93.6	91.7
40	94.1	94.1	92.4	94.1	94.1	92.4
50	94.1	94.5	93.0	94.1	94.5	93.0
60	94.5	95.0	93.6	94.5	95.0	93.6
75	94.5	95.0	93.6	94.5	95.4	93.6
100	95.0	95.4	93.6	95.0	95.4	94.1
125	95.0	95.4	94.1	95.0	95.4	95.0
150	95.4	95.8	94.1	95.8	95.8	95.0
200	95.4	95.8	95.0	95.8	96.2	95.4
250	95.4	95.8	95.0	95.8	96.2	95.8
300	95.4	95.8	95.4	95.8	96.2	95.8
350	95.4	95.8	95.4	95.8	96.2	95.8
400	95.8	95.8	95.8	95.8	96.2	95.8
450	96.2	96.2	95.8	95.8	96.2	95.8
500	96.2	96.2	95.8	95.8	96.2	95.8

Table 9 Nominal efficiencies for induction motors rated 5 kV or less

| Horsepower | Form wound ||||||
| | Open drip-proof (%) ||| Totally enclosed fan-cooled (%) |||
	6-pole	4-pole	2-pole	6-pole	4-pole	2-pole
250	95.0	95.0	94.5	95.0	95.0	95.0
300	95.0	95.0	94.5	95.0	95.0	95.0
350	95.0	95.0	94.5	95.0	95.0	95.0
400	95.0	95.0	94.5	95.0	95.0	95.0
450	95.0	95.0	94.5	95.0	95.0	95.0
500	95.0	95.0	94.5	95.0	95.0	95.0

author observed a payback of under 3 years for a major retrofit to variable speed pumping. This project also included modifications to the control valves to accommodate the variable flow strategy.

The application of ASD technology does require care. Consider the effect on power factor when selecting an ASD. Some ASD applications can actually result in an installed system with a power factor better than the original motor power factor.

However, improper selection of an ASD can also result in a lower overall power factor. Another concern deals with the harmonics that an ASD can superimpose on the system. From a consulting mechanical engineer's perspective, this is an electrical engineering problem! Consulting with the project's electrical engineer on the effects of harmonics is important. Applying too many ASDs on an electrical system can cause harmful harmonics that can affect the overall system performance. Many ASD suppliers offer a service for evaluating these harmonics and can assist in dealing with this issue.

Another significant issue in the selection of an ASD for an application is having a reasonable estimate of the operating profile for the system.[1] Without this knowledge, energy savings associated with the ASD application can be over- or even underestimated. Once this issue is resolved, and assuming the estimate is favorable, the resulting energy savings are still in question if the ASD system is not maintained properly. The total success of the ASD system hinges on the accuracy and reliability of sensing pressure differentials or flow rates. If the sensor goes out of calibration or ceases to function, the ASD will be rendered useless.

Because of the complex electronics in an ASD system (at least, complex to the consulting mechanical engineer!) and the various sensors and controllers that might be connected to the ASD, lightning and surge protection is another consideration in the overall system design. Again, the consulting electrical engineer should be able to provide insight into the appropriate lightning- and surge-protection systems for the project. The use of fiber optics in the communication system between the sensors and the ASD can eliminate some of the problems from lightning. Surge protection and optical relays can provide additional protection.

One last consideration in the control of electrical motors addresses the method of starting electric motors. There are many technologies available that can be used to start motors (such as mechanical motor starters, ASDs, and soft-start starters) and improve the life expectancy of these motors. The selection of the motor starter sometimes falls on the mechanical engineer, and sometimes the electrical engineer. Whichever the case, consideration should be given to coordinating the starter selection with the motor selection to ensure the most efficient combination results. The electrical engineer's role at this point is also significant. Careful coordination is required to verify and make necessary adjustments to the electrical service for the motor, to verify branch circuit wiring capacity, and to coordinate branch circuit protection as well as overload protection and disconnect requirements. Then it would be prudent to check the motor rotation (or else pumps will run backwards!) and alignment.

CONCLUSION

In summary, electric motor selection, replacement, and repairs require careful evaluation in order to provide an efficient, reliable, and easy-to-maintain motor drive system whether in the residential, commercial, or industrial sector. With over 60% of our nation's industrial energy use represented by the use of electric motors, it is irresponsible not to apply the utmost of care in our practice of motor selection and replacement.

REFERENCES

1. Elliott, R.N.; Nadel, S. *Realizing energy efficiency opportunities in industrial fan and pump systems*, Report A034; American Council for an Energy Efficient Economy: Washington, DC, April 2003.
2. Nadel, S.; Elliott, R.N.; Shepard, M.; Greenberg, S.; Katz, G.; de Almeida, A.T. *Energy-Efficient Motor Systems: A Handbook on Technology, Program, and Policy Opportunities*, 2nd Ed.; American Council for an Energy Efficient Economy: Washington, DC, 2002.
3. *Engineering Cookbook*, 2nd Ed.; Loren Cook Company: Springfield, MO, 1999.
4. Turner, W.C., Ed. *Energy Management Handbook* 5th Ed.; Fairmont Press: Atlanta, GA, 2005.

Electric Power Transmission Systems

Jack Casazza
American Education Institute, Springfield, Virginia, U.S.A.

Abstract
This entry discusses electric power transmission system functions, the benefits produced, components, ratings and capacity, alternating current (AC) and direct current (DC) transmission, transmission system operations, the need for coordination, control areas, NERC and reliability councils, reliability standards, transmission access, and new technology.

TRANSMISSION FUNCTIONS

Basic Role

The transmission system provides the means by which large amounts of power are delivered from the generating stations where it is produced to other companies or to locations where voltage is reduced, to supply subtransmission systems or substations where it is distributed to consumers. Transmission in the United States is mostly by three phase, 60 Hz (cycles per seconds) alternating current (AC) at voltages between 115,000 and 765,000 V. Direct current (DC) is used at a few locations but its potential role is increasing.

Benefits of Interconnection[1]

Electric power must be produced at the instant it is used. Needed supplies cannot be produced in advance and stored for future use. It was soon recognized that peak use for one system often occurred at a different time than peak use in other systems. It was also recognized that equipment failures occurred at different times in various systems. Engineering analyses showed significant economic benefits from interconnecting systems to provide mutual assistance. The investment required for generating capacity could be reduced. Reliability could be improved. Differences in the cost of producing electricity in the individual companies and regions often resulted in one company or geographic area producing some of the electric power sold to another company in another area for distribution. This lead to the development of local, then regional, and subsequently, five grids in North America with three United States transmission grids, as shown in Fig. 1. Fig. 2 shows the key stages of the evolution of these transmission grids.

Keywords: Electric power; Transmission systems; Interconnections; Voltage control; Synchronous operation; Capacity; Losses; Reliability.

Summarizing the transmission system[2]:

- Delivers electric power from generating plants to large consumers and distribution systems.
- Interconnects systems and generating plants to reduce overall required generating capacity requirements by taking advantage of:
  — The diversity of generator outages, i.e., when outages of units occur in one plant, units in another plant can provide an alternative supply.
  — The diversity of peak loads because peak loads occur at different times in different systems.
- Minimizes fuel costs in the production of electricity by allowing its production at all times at the available sources having the lowest incremental production costs.
- Facilitates the location and use of the lowest cost additional generating units available.
- Makes possible the buying and selling of electric energy and capacity in the marketplace.
- Helps provide for major emergencies such as hurricanes, tornadoes, floods, strikes, fuel supply disruptions, etc.

Transmission of Real and Reactive Power[3]

The delivery of electric energy to perform desired functions requires the delivery of "real" and "reactive power." The real power is produced in the power plant from fuels or energy sources and provides the energy that is used. This real power must be accompanied by reactive power that provides the electric fields required by various devices for the utilization of this energy. This reactive power does not include any energy. It is produced by the fields of the generators, by capacitors installed for that purpose, and by the "charging current" of the transmission system.

Electric Power Transmission Systems

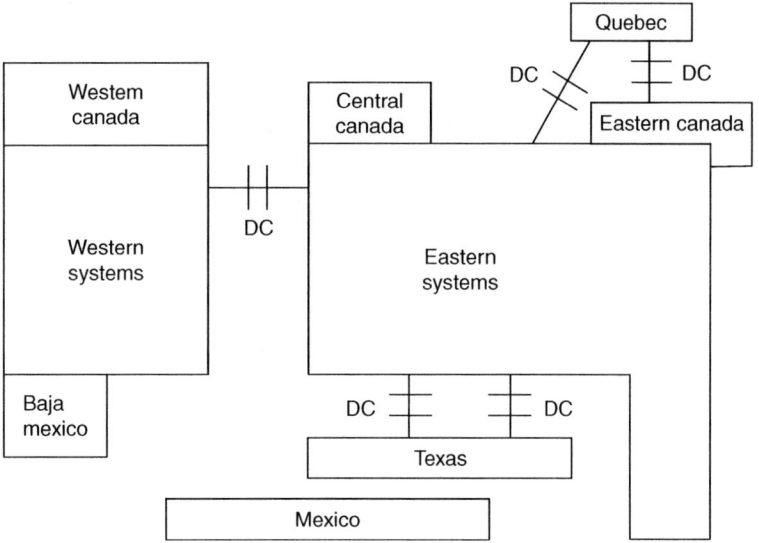

Fig. 1 The five synchronous systems of North America.

Electrical Characteristics[4]

Transmission systems have resistance (R), which causes the heating of conductors and the loss of energy when current (I) flows and reactance (X), which causes voltage drops when current flows. Reactance can be positive or negative; positive when it is an inductive reactance and negative when it is a capacitive reactance.

TRANSMISSION COMPONENTS

Transmission Lines

The transmission system consists of three-phase transmission lines and their terminals, called substations or switching stations. Transmission lines can be either overhead or underground (cable). High-voltage alternating current (HVAC) lines predominate, with high-voltage direct current lines (HVDC) used for special applications. Overhead transmission, subtransmission, and primary distribution lines are strung between towers or poles. In urban settings, underground cables are used primarily because of the impracticality of running overhead lines along city streets. While underground cables are more reliable than overhead lines (because they have less exposure to climatological conditions such as hurricanes, ice storms, tornadoes, etc.), they are also much more expensive than overhead lines to construct per unit of capacity and they take much longer to repair because of the difficulty in finding the location of a cable failure and replacement.

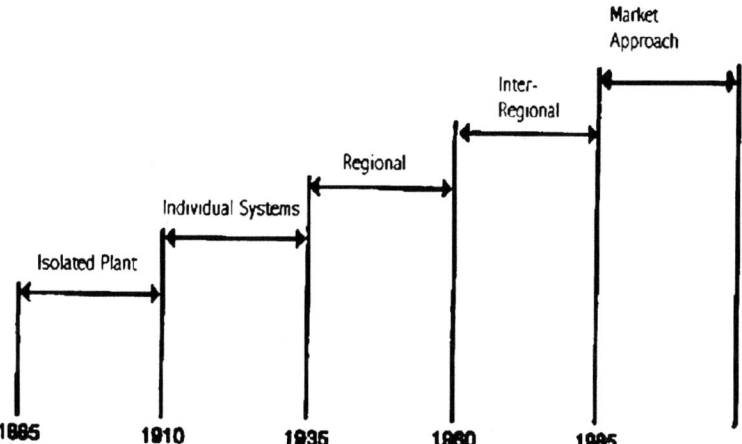

Fig. 2. Stages of transmission system development.

The primary components of an overhead transmission line are:

- Conductors (three, one per phase)
- Ground or shield wires
- Insulators
- Support structures
- Land or right-of-way (ROW)

Conductors consist of stranded aluminum woven around a core of stranded steel that provides structural strength. When there are two or more of these wires per phase, they are called bundled conductors.

Ground or shield wires are wires strung from the top of one transmission tower to the next, over the transmission line. Their function is to shield the transmission line from lightning strokes. Insulators are used to attach the energized conductors to the supporting structures, which are grounded. The higher the voltage at which the line operates, the longer the insulator strings.

The most common form of support structure for transmission lines is a steel lattice tower, although wood H frames (so named because of their shape) are also used. In recent years, as concern about the visual impact of these structures has increased, tubular steel poles also have come into use. The primary purpose of the support structure is to maintain the electricity carrying conductors at a safe distance from the ground and from each other. Higher-voltage transmission lines require greater distances between phases and from the conductors to the ground than lower-voltage lines and therefore they require bigger towers. There has been some concern about the biological effects of transmission lines with the general conclusion being that there are no serious effects.[5]

Ratings

The capability of an individual overhead transmission line or its rating is usually determined by the requirement that the line does not exceed code clearances with the ground. As power flows through the transmission line, heat is produced. This heat will cause an expansion of the metal in the conductor, and as a result increase the amount of its sag. The amount of sag will also be impacted by the ambient temperature, wind speed, and sunlight conditions. The heating can also affect the characteristics of the metals in the conductors, reducing their strength.

Ratings are usually of two types—normal and emergency—and are usually determined for both summer and winter conditions. Some companies average the summer and winter ratings for the fall and spring. Ratings are also specified for various time periods. A normal rating is the level of power flow that the line can carry continuously. An emergency rating is the level of power flow the line can carry for various periods of time, for example, 15 and 30 min, 2, 4, and 24 h, and so forth.

In recent years, there has been a trend in calculating ratings for critical transmission lines on a real-time basis, reflecting actual ambient temperatures as well as the recent loading (and therefore heating) patterns.

Cables

The majority of the transmission cable systems in the United States are high-pressure fluid filled (HPFF) or high-pressure liquid filled (HPLF) pipe-type cable systems. Each phase of a high-voltage power cable usually consists of stranded copper wire with oil-impregnated paper insulation. All three phases are enclosed in a steel pipe. The insulation is maintained by constantly applying a hydraulic pressure through an external oil adjustment tank to compensate for any expansion or shrinkage of the cable caused by temperature variations.

Cable capacity is determined by the effect of heat on the cable insulation. Because the cable is in a pipe that is buried in a trench, dissipation of the heat is a major issue in cable design and operation. Cable capacity can be increased by surrounding the pipe with thermal sand, which helps dissipate heat.

A limitation on the application of ac cables is their high capacitance, which increases with their length. Cable capacitance causes a charging current to flow equal to the voltage divided by the capacitive reactance. This can limit the length of cable that can be used without some intermediate location where shunt reactor compensation can be installed to absorb the charging current.

Substations

Substations are locations where transmission lines are tied together. They fulfill a number of functions:

- Allow power from different generating stations to be fed into the main transmission grid.
- Provide for interconnections with other systems.
- Provide transformers to be connected to feed power into the subtransmission or distribution systems. Transformers are generally equipped to change voltage ratios, with fixed taps that require de-energization to change and tap changing equipment that can change taps while the transformer is in operation.
- Allow transmission lines to be independently switched to isolate faulty circuits or for maintenance.
- Provide a location where compensation devices such as shunt or series reactors or capacitors can be connected to the transmission system.
- Provide a location for protection, control, and metering equipment.

Substation equipment includes:

- Bus work through which lines, transformers, etc., are connected.
- Protective relays that monitor voltages and currents and initiate disconnection of lines and equipment in the event of failures or malfunctions.
- Circuit breakers that interrupt the flow of electricity to de-energize facilities.
- Shunt capacitors to help provide needed reactive power.
- Disconnect switches.
- Lightning arrestors.
- Metering equipment.
- System control and data acquisition (SCADA) systems.
- Shunt reactors to limit high voltages.
- Series reactors to increase the impedance of lines.
- Phase angle regulating transformers and other devices to control power flow and voltage in specific circuits.

The bus/circuit breaker connection arrangements used in substations affect substation costs, ease of maintenance, and transmission reliability.[6]

Direct Current Transmission

An alternate means of transmitting electricity is to use HVDC technology. Direct current facilities are connected to HVAC systems by means of rectifiers, which convert alternating current to direct current, and inverters, which convert direct current to alternating current.

Starting in the 1970s, thyristors became the valve type of choice for the rectifiers and inverters. Thyristors are controllable semiconductors that can carry very high currents and can operate at very high voltages. They are connected in series to form a valve, which allows electricity to flow during the positive half of the alternating current voltage cycle but not during the negative half. Because all three phases of the HVAC system are connected to the valves, the resultant voltage is unidirectional but with some residual oscillation. Smoothing reactors are provided to dampen this oscillation.

High-voltage direct current lines transmission lines can either be single pole or bipolar, although most are bipolar—they use two conductors operating at different polarities, such as $+/-500\,kV$. There have been a number of applications for DC transmission:

- To transmit large amounts of power over long distance is not feasible with AC.
- To transmit power across water where the capacitance of AC cables would limit circuit capacity.
- The need to connect two AC systems in a manner that prevents malfunctions or failures in one system from causing problems in another system.
- To provide direct control of power flow in a circuit.
- To limit short-circuit duties.
- To increase the ability to transfer power over existing right-of-ways because DC requires two conductors versus three for AC.

The difficulty of DC is its higher costs and the lack of reliable DC current breakers.

HOW TRANSMISSION SYSTEMS WORK

Synchronous Operation

Because of the synchronous operation of all generators, an interconnected electric power system functions as a single large machine that can extend over thousands of miles. One of this machine's characteristics is that changes in any one portion of it instantly affect the operation of all other portions. Future plans for any one part of it can affect transmission conditions in other parts of it.

In system operation, the effects of contingencies in one system can be felt throughout a large geographic area. For example, if a large generating unit is lost in New York city, the interconnected power system (the machine) suddenly has less power input than power output and slows down. As the individual rotors of every generator in this system slow down in unison and system frequency declines, each rotor gives up a certain amount of its rotating energy ($\omega^2 R$) to compensate for the lost input from the unit that has tripped off.

Instantaneously, with the loss of a large unit or power plant, there is an inrush of power from units all over the synchronous network, feeding into the system or region that has lost the generator unit. While these inrushes of power from individual units long distances away are not large, they accumulate and build up like water flowing from creeks into a river, increasingly loading the transmission lines near the system that has lost generation. This surge of power exists until automatic generator controls cause them to increase their output to compensate for the lost generating capacity and restore system frequency.

The very reason that systems have been tied together and operate in synchronism causes this effect. By having the various generator units throughout the region assist with the loss of a large generator unit in a specific system, the total amount of spare or reserve generating capacity required can be reduced. This is similar to the insurance business. The larger the number of policy holders, the better able the insurance company is to cope with any specific major disaster and the less percentage reserves it requires. The great strength of operating in synchronism and being tied together in an integrated system is the ability of one system to be helped by others. Its greatest weakness, however, is that the inrush of power into any one system can cause transmission system overloads in other systems.

Because of these characteristics of modern electric power systems, the design and operation of the key elements in a synchronous network must be coordinated. Business decisions, government legislation and regulations, and other institutional processes must be compatible with the technical characteristics. Many problems can be solved by technical solutions, some can be solved by institutional solutions, and in some cases, problems can be solved by coordinating both.

Kirchhoff's Laws[7]

In AC transmission networks, the flow of power in the various circuits is determined by Kirchhoff's Laws. Power flow is determined by the impedance of the individual circuits, the location in the network of the sources of power, and the location of the substations to which power is delivered. The flow of electrical power in the synchronous grid does not respect company boundaries, contracts or ownership of facilities. Power cannot be scheduled to flow over a specific line or lines, but will divide in accordance with Kirchhoff's Laws.

This sometimes results in two phenomena known as "parallel path flow" and "loop flow." Parallel path flow results when power to be delivered from system A to system B goes through systems in parallel. Such flows can increase transmission loadings in these other systems, reducing their ability to be used by the owners for their own purposes. Loop flows are circulating flows that occur with all systems supplying their own loads from their own sources. They are the result of network characteristics and often the result of deliberate network designs to limit total transmission investment requirements.

Transmission Capacity

Transmission limits can be determined by a number of factors, a common one being the maximum allowable thermal loading on circuits. Current flowing through conductors causes heating, which causes expansion and "sagging," reducing clearance to ground, possibly contacting trees or other obstacles, and resulting in a circuit trip out.

Potential stability disturbances can limit the amount of power that can be transferred. If stability limits are exceeded, the occurrence of a critical fault can cause generators to oscillate wildly and trip out. Voltage conditions can limit the ability to transfer power. An inadequate supply of reactive power can cause transmission voltages to become too low, causing excessive transmission currents and voltage instability, resulting in circuit trip outs.

While recognizing these various causes of transmission limits, it is also essential to recognize that the ability to deliver power is the ability of the interconnected network, i.e., the system that forms the grid, to transfer power, not the sum of the capacities of the individual circuits involved. Often, one circuit can become overloaded while another has unused capacity because of Kirchoff's Laws.

Transmission Access

The restructuring and deregulation of the electric power industry has lead to significant changes in the use of transmission systems. The owners of transmission lines must make them available to everyone who wants to use them on the same basis as they are used for their own customers. This has increased the complexity and difficulty of planning and operating transmissions systems because there are many more potential users whose decisions affect the entire system.[8]

Short Circuit Duties

An often overlooked factor is that transmission systems must have the ability to interrupt the very high currents that result when short circuits occur. This is done by circuit breakers that must have an interrupting capacity sufficient to interrupt the magnitudes of the fault currents involved. These fault currents are increased as generation and new transmission lines are added to the system. Transmission system design requires studies of the ability of circuit breakers to interrupt expected faults. When breaker capability is not adequate, expensive replacements or changes in substation or system designs may be needed.

Transmission System Losses

Power losses in the transmission system consume 3%–5% of the electric power produced. Because they are supplied by the highest cost generation available, they are responsible for more than 5% of the cost of energy produced. Many believe that the deregulation of the electric power industry has increased transmission losses.

There are two basic types of power (MW) losses in transmission systems:

- Core losses are dissipated in the steel cores of transformers. These losses typically vary with at least the third, and often higher, powers of the voltage variations at their terminals. If transmission voltages are fairly constant, core losses are also constant. An increase in the power carried by a transmission system does not substantially affect the core losses, but a variation in transmission voltage can.
- Conductor losses are dissipated in transmission lines and cables and transformer windings. As these losses depend on the resistance of the circuit (R) and vary with the square of the current and therefore

approximately with the square of the power carried by each component, they vary greatly between light load and heavy-load times and are also affected by increases in the power carried by the transmission system.

There are also reactive power (MVAR) losses in the transmission system. These losses depend on the reactance of the system (X) and again vary as the square of the current.

When electrical energy is transported across large distances through the transmission system, a portion of the energy is "lost." For a given amount of electric power, the higher the operating voltage of the transmission line, the lower the current flowing through the line. Therefore, use of higher transmission voltages permits the transmission of electric power with lower currents and a resulting reduction in energy losses.

The losses in the transmission system at a specific moment, typically at the time of system peak, are measured in megawatts and are referred to as "capacity losses." The energy expended in transmission losses over a given period is given in megawatt-hours and is referred to as "energy loss." Capacity losses require the installation of additional generation and transmission equipment; energy losses require the consumption of fuel or equivalent energy sources. The increase in a system's total losses due to a specific action is referred to as an "incremental loss."

In addition to real power losses, significant reactive power losses occur in the transmission systems, and these reactive losses are typically about 4 or 5 times the real power capacity losses.

OPERATION OF ELECTRIC BULK POWER SYSTEMS[9,10]

Need for Coordination[11,12]

The operation of the bulk power system in the United States involves the interdependency of the various entities involved in supplying electricity to the ultimate consumers.[13] These interdependencies have evolved as the utility industry grew and expanded over the century.

In a power system, the coordination of all elements of the system and all participants are required from an economic and a reliability perspective. Generation, transmission, and distribution facilities must function as a coordinated whole. The scheduling of generation must recognize the capability of the transmission system. Voltage control must involve the coordination of reactive power supplies from all sources, including generators, the transmission system, and distribution facilities. Actions and decisions by one participant, including decisions not to act, affect all participants.

In parallel with its early growth, the industry recognized it was essential that operations and planning of the system be coordinated and organizations were formed to facilitate the joint operation and planning of the nation's electric grid. Initially, holding companies and then power pools were established to coordinate the operation of groups of companies. Recently, new organizations, independent system operators (ISOs) and regional transmission operators (RTOs), have been formed to provide this coordination.

Reliability Councils and Nerc[14]

After the Northeast Blackout of 1965, regional electric reliability councils were formed to promote the reliability and efficiency of the interconnected power systems within their geographic areas. These regional councils joined together shortly afterwards to form a national umbrella group, NERC—the North American electricity reliability council. At present, there are ten regional councils. Each Council has a security coordinator who oversees the operation of the grid in their region.

The members of NERC and these regional councils come from all segments of the electric industry; investor-owned utilities; federal power agencies; rural electric cooperatives; state, municipal and provincial utilities; independent power producers; power marketers; and end-use customers. These entities account for virtually all the electricity supplied in the United States, Canada, and a portion of Baja California North, Mexico.

When formed in 1968, the NERC operated as a voluntary organization to promote bulk electric system reliability and security—one that was dependent on reciprocity, peer pressure, and the mutual self-interest of all those involved.

The growth of competition and the structural changes taking place in the industry have significantly altered the incentives and responsibilities of market participants to the point that a system of voluntary compliance is no longer adequate. New federal legislation in the United States has required formation of an electric reliability organization (ERO) to monitor and enforce national reliability standards under FERC oversight. In response to these changes, NERC is transforming itself into an industry-led self-regulatory reliability organization ERO that will develop and enforce reliability standards for the North American bulk electric system.

Control Areas

While overall system control is, in some cases, the responsibility of newly formed ISOs and RTOs, more than 140 "control areas" still perform needed functions.

A control area can consist of a generator or group of generators, an individual company, or a portion of a company or a group of companies providing it meets certain certification criteria specified by NERC. It may be a specific geographic area with set boundaries or it may be scattered generation and load.

The control centers require real-time information about the status of the system. This information includes power line flows, substation voltages, the output of all generators, the status of all transmission lines and substation breakers (in-service or out-of-service), and transformer tap settings. Some areas are implementing real-time transmission line rating systems requiring additional information such as weather conditions, conductor temperatures, and so forth.

Each control area monitors on an on-going basis the power flow on all of its interties (in some cases delivery points) and the output of each generator within its control. The sum of the internal generation and the net flow on the interties is equal to the consumer load and all transmission losses within the area.

The various commercial interests that are involved within the area are required to notify the control area personnel of their contractual arrangements on an ongoing basis for either sales or purchases of electricity with entities outside the area's boundaries.

Oasis and Transmission Capacity

The open access same-time information system (OASIS) is an Internet-based bulletin board that gives energy marketers, utilities, and other wholesale energy customers real-time access to information regarding the availability of transmission capacity. OASIS provides the ability to schedule firm and nonfirm transactions.

The North American electricity reliability council[15] has defined transmission capacity as follows:

Available transfer capacity (ATC) = Total transfer capability (TTC) − Existing commitments − transmission reliability margin (TRM) − Capacity benefit margin (CBM),

where:

- Available transfer capability is a measure of the transfer capability remaining in the physical transmission network for further commercial activity over and above already committed uses.
- Total transfer capability is the amount of electric power that can be transferred over the interconnected transmission network in a reliable manner while meeting all of a specific set of defined pre- and post-contingency system conditions.
- Transmission reliability margin is the amount of transmission transfer capability necessary to ensure that the interconnected transmission network is secure under a reasonable range of uncertainties in system conditions.
- Capacity benefit margin is the amount of transmission transfer capability reserved by load serving entities to ensure access to generation from interconnected systems to meet generation requirements in emergencies.

With this information, the control area operators can compare the total scheduled interchange into or out of the control area with the actual interchange. If the receipt of electricity exceeds the schedule, the control area must increase generation levels. If the receipt is too low, generation within the control area is reduced. These schedules are typically made a day ahead and then adjusted in real time. Because these adjustments are ongoing simultaneously by all control areas, the adjustments balance out.

The process where individual contracts scheduled within OASIS are identified as to source and customer is known as tagging. This information, while it may be commercially sensitive, is critical if system operators are to adjust system power flows to maintain reliable levels.

Concurrently, the system operators can also evaluate the expected power flows internal to the control area to determine if adjustments are required in the generation pattern to insure that all internal transmission facilities are operated within the capabilities.

Each control area also participates in maintaining the average system frequency at 60 Hz. The system frequency can deviate from normal when a large generating unit or block of load is lost. In addition to adjustments made because of variations of tie flows from schedule, another adjustment is made to correct frequency deviations.

Reliability standards have been developed by the regional reliability councils and NERC for many years. They define the reliability aspect of the interconnected bulk electric systems in two dimensions:

- Adequacy—the ability of the electric systems to supply the aggregate electrical demand and energy requirements of their customers at all times, taking into account scheduled and reasonably expected unscheduled outages of system elements.
- Security—the ability of the electric systems to withstand sudden disturbances such as electric short circuits or unanticipated loss of system elements.

Detailed reliability standards exist that specify allowable system voltage and loading conditions for various system contingencies. These are developed by the various regional reliability councils and must meet the minimum standards established by NERC. There are standards for various single contingencies and for various combinations of outages.

Meeting reliability standards in the planning of the transmission system is difficult because the time required to install new transmission is longer than the time required to install new generation. Attempting to meet reliability standards in planning for future transmission needs involves considerable uncertainties because future generation locations are not known. The general industry consensus is that the restructuring and deregulation of the electric power industry has resulted in a decrease in reliability. How will this affect future transmission policies is uncertain.

NEW TECHNOLOGIES

Future developments may have long-range effects on transmission requirements, transmission system characteristics, and the capacity of the various networks. New transmission technologies, some involving "power electronics," are under study. These include:

- The development of methods to control the division of flow of power in AC networks.
- The development of "smart systems" or "self-healing" systems[16] that may involve a redesign and upgrade of presently electromechanically controlled transmission systems.
- The subdivision of huge synchronous AC networks into smaller synchronous networks interconnected by DC.

There are other possible developments that may have a significant effect on transmission systems, including:

- The development of significant amounts of small distributed generation, including the use of solar energy, wind power, micro turbines, etc.
- A major national shift to large nuclear or coal units to reduce dependence on foreign oil and gas.
- The development of low-cost energy storage devices to allow power to be produced at one time for use at another.
- Increasing use of hydrogen as a mechanism for transferring energy from one location to another, including possible linking of hydrogen production with off peak generating capacity.

The future holds many uncertainties and requires analyses similar to post national power surveys[17] to determine how to develop transmission systems to meet potential future developments. Failure to make such analyses will result in wasteful transmission additions and design of a poor system.

REFERENCES

1. Casazza, J.A. The development of electric power transmission—the role played by technology, institutions, and people, *IEEE Case Histories of Achievement in Science and Technology*; Institute of Electrical and Electronic Engineers: New York, 1993, www.lulu.com.
2. Casazza, J.A.; Delea, F. *Understanding Electric Power Systems: An Overview of the Technology and the Marketplace*; Wiley: New York, 2003; 17.
3. Casazza, J.A.; Delea, F. *Understanding Electric Power Systems: An Overview of the Technology and the Marketplace*; Wiley: New York, 2003; 34, 35.
4. Casazza and Delea *Understanding Electric Power Systems: An Overview of the Technology and the Marketplace*; Wiley: New York, 2003; 25.
5. *Biological Effects of Power Frequency Electric and Magnetic Fields*, U.S. Congress Office of Technology Assessment, 1989.
6. Casazza, J.A.; Delea, F. *Understanding Electric Power Systems: An Overview of the Technology and the Marketplace*; Wiley: New York, 2003; 81.
7. Named after Gustav Robert Kirchoff, a German physicist, (1824–1887).
8. Non-Technical Impediments to Power Transfers, National Regulatory Research Institute, 1987.
9. Nagel, T.J. Operation of a major electric utility today. Science **1978**, *2* (4360).
10. Rustebakke, H.M. *Electric Utility Systems and Practices*, 4th Ed.; Wiley: New York, 1983.
11. Casazza, J.A. *Coordinated Regional EHV Planning in the Middle Atlantic States—U.S.A.*; CIGRE paper No. 315: Paris, France, 1964.
12. Casazza, J.A. The Development of Electric Power Transmission—The Role Played by Technology, Institutions, and People. IEEE Case Histories of Achievement in Science and Technology, Institute of Electrical and Electronic Engineers, www.lulu.com, 1993; 81.
13. Hughes, T.P. Systems builders technology's master craftsmen. CIGRE Electra **1986**, *109*, 21–30.
14. See www.NERC.com.
15. See www.NERC.com.
16. Stahlkoph, K.; Sharp, P.R. Where technology and politics meet. Electric Power Transmission Under Deregulation, IEEE, 2000.
17. Federal power commission *National Power Survey—A Report by the Federal Power Commission*; U.S. government printing office: Washington, DC, 1964.

Electric Power Transmission Systems: Asymmetric Operation

Richard J. Marceau
University of Ontario Institute of Technology, Oshawa, Ontario, Canada

Abdou-R. Sana
Montreal, Québec, Canada

Donald T. McGillis
Pointe Claire, Québec, Canada

Abstract

Asymmetric operation of an electric power transmission corridor is an operating strategy that enables a 3-phase line to be operated with one or two phases out of service in the case of single-line transmission corridors, or with one, two, or three phases out of service in the case of multiple-line corridors, while preserving 3-phase symmetrical operation at corridor extremities. This article shows how asymmetric operation is implemented, how much it costs, and how it can improve the reliability and economics of electric power transmission systems.

INTRODUCTION

Existing electric power transmission systems are operated with three physically different systems of conductors referred to as "phases," where the sinusoidal voltages and currents in each phase are offset with respect to one another to take advantage of Nicolas Tesla's groundbreaking invention of the 3-phase alternating-current (AC) generator and motor. The 3-phase power generator is a remarkably robust and economical technology that creates the three sets of voltages and currents, the so-called phases, in 3-phase transmission lines. This results in a steady torque characteristic, which in turn translates into high generator reliability. The 3-phase AC motor, which requires an infeed of the three different phases to operate, is also a highly reliable and economic piece of machinery, and is a workhorse of industry.

Because 3-phase generators and motors are so widely used, they are connected to 3-phase high-voltage (HV), extra-high-voltage (EHV), or ultra-high-voltage (UHV) power transmission systems that are symmetrically operated, meaning that if any problem develops on one or more phases, all three phases are taken out of service. This requirement is due to the fact that if symmetric operation is not enforced, undesirable voltages and currents are generated that are harmful not only to rotating machinery, but also to many other types of loads. There are exceptions to this rule: For economical reasons, single-phase systems equipped with appropriate mitigating measures have been used in railway electrification and for distributing power at the household level. Throughout the world, however, standard practice is to operate bulk 3-phase power transmission systems symmetrically.

Symmetric Operation of Power Transmission Systems

When at least one phase of a power transmission line touches an object, causing the current of that phase to be redirected either to the ground (i.e., through a tree) or to another phase, such short-circuit conditions are generally referred to as "faults." Under normal operation, when a fault condition occurs, circuit breakers at both ends of the line interrupt the current in all three phases, effectively removing the line from operation until such time as the fault condition has been removed, even though one or two healthy phases remain that conceivably could carry useful power. Such a strategy has several disadvantages. First, a problem on 33% of a line automatically deprives the network of 100% of this same line. Second, the operation of three phases as a single organic whole rather than three independent conductors augments the probability of loss of 3-phase transmission by nearly a factor of three with respect to single-phase transmission. Finally, after a fault is cleared, the power system operator must apply remedial measures, such as changing the amount of generation from different power plants, to redirect power flows and ensure that the system is capable of sustaining further contingencies with no impact on load power delivery. This exposes power system operation to the possibility of human error and, therefore, increases operating risks.

Though such a conservative operating strategy served the industry well while it was economically possible to do so, it would seem that the electric power industry's prevalent transmission strategy today is wasteful of

Keywords: Electric power transmission; Reliability; Symmetric operation; Asymmetric operation.

expensive transmission equipment, costly in terms of loss of potential revenue, and stressful to both power system equipment and operators. In the light of the considerable pressures on electric utilities due to deregulation, greater environmental awareness, normal load growth, and (as shown later in this article) symmetrical operation expose the system to numerous risks while wasting valuable transmission capacity. This was particularly shown to be true in the August 2003 blackout in the northeastern United States, and Canada in which many lines were switched out due to single-phase faults.[1]

Asymmetric Operation of Power Transmission Systems

Asymmetric operation is essentially defined as the operation of a 3-phase transmission line as three independently operated entities.[2] In this approach, a 3-phase line can be operated with one or two phases out of service for single-line transmission corridors, or with one, two, or even three phases out of service in the case of multiple-line corridors. For this to occur, the strategy implements three operational objectives:

1. Upon entering asymmetric operation, undesirable voltages and currents are "contained" within the affected corridor.
2. From the system perspective, the corridor appears to operate symmetrically at both extremities.
3. From the system perspective, the faulted corridor returns to its precontingency electrical state (i.e., in terms of impedance, voltages, and currents) and maintains precontingency power flows.

Because the post-contingency system is electrically indistinguishable from its precontingency state, this eliminates the need for operator-driven postfault remedial measures. Postfault remedial measures are, therefore, built into the strategy.

To enable asymmetric operation as described above, compensating equipment must be introduced into the corridor just as the faulted phase(s) is(are) switched out by means of circuit breaker action. Such compensating equipment can either be conventional, inexpensive passive devices, such as capacitors and reactors equipped with appropriate switching equipment, or more complex and expensive power-electronic devices such as Flexible AC Transmission System (FACTS) controllers. Both approaches have their strengths and weaknesses, and the final choice will often depend on system-specific constraints imposed by the system planner.

The consequence of asymmetric operation is to add flexibility to transmission system operation, increase corridor reliability, and increase security limits (i.e., the amount of power transferred under normal conditions) and thereby improve transmission economics.

BENEFIT OF ASYMMETRIC OPERATION

Statistics of Transmission Line Failures

Most transmission line faults are single-phase faults (varying from 60 to 97% with increasing voltage level).[3,4] This alone is sufficient motivation for studying asymmetric operation, even if the practice of single pole reclosure is successful up to 50% of the time at voltages up to 765 kV. Because it is so simple to define and to employ, the 3-phase fault—often used concurrently with the subsequent loss of major transmission equipment—has long been the industry norm for estimating system performance under difficult conditions. Even so, one must occasionally be reminded of the fact that 3-phase faults have no more than a 1% probability of occurrence and that their use as a criterion is clearly limited in terms of physical significance. Indeed, given such a statistic, one could argue that a symmetrical 3-phase response to normally occurring events and contingencies is an inappropriate response 99% of the time and, therefore, is far from optimal from an operations perspective.

In the past, this criterion served a useful purpose as an umbrella contingency, covering a large number of contingencies and accounting for lack of knowledge of either a specific operating context or of power system dynamics more generally. Asymmetric operation, however, enables power systems to respond surgically to contingencies while subsuming symmetrical response capability when required.

Reliability Analysis

The most frequently used reliability index in transmission planning is the loss of load expectation (LOLE), which is the expected mean of energy not supplied due to the failure of network components.[5] Here, the LOLE is used for estimating and comparing the risk of operating a single- or multiple-line transmission corridor under symmetric or asymmetric operation.

Fig. 1 shows the logic circuits for the reliability analysis of a 3-phase transmission line under symmetrical or asymmetrical operation. In the symmetrical approach, events leading to the loss of any phase results in the loss of all three phases; in this case, the equivalent logical circuit presents the three phases in series. In the asymmetrical approach, the three phases function independently; thus, the equivalent logical circuit presents the three phases in parallel.

If one considers a corridor of N 3-phase lines transmitting a total power value of T, where the probability of successful transmission of each phase of each 3-phase line is p, the expected mean nontransmitted power according to symmetric operation $LOLE_{sym}$ and the expected mean nontransmitted power according to asymmetric operation $LOLE_{asym}$ are[2]

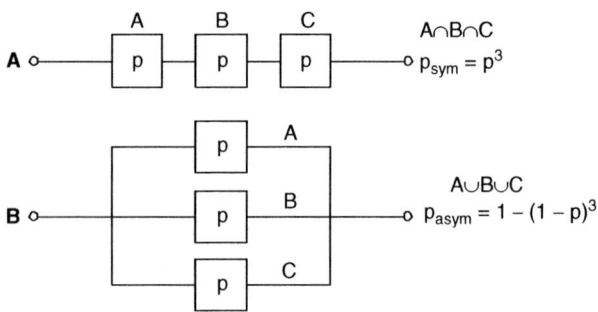

Fig. 1 Equivalent logical circuits of a 3-phase line under (A) symmetric operation and (B) asymmetric operation.

$$\begin{aligned} \text{LOLE}_{\text{sym}} &= f_{\text{ch}} T(1 - p^3) \\ \text{LOLE}_{\text{asym}} &= f_{\text{ch}} T(1 - p)^3 \end{aligned} \quad (1)$$

where f_{ch} is a load factor that takes average load variations into account. An arbitrary load factor between 50 and 75% is considered to be acceptable in the industry.

The difference between the LOLE of symmetric and asymmetric operation, ΔLOLE, yields the benefit of asymmetric operation, which can be evaluated at the energy generation cost

$$\begin{aligned} \text{LOLE}_{\text{sym}} &- \text{LOLE}_{\text{asym}} \\ &= [(1 - p^3) - (1 - p)^3] f_{\text{ch}} T \\ &= 3p(1 - p) f_{\text{ch}} T \end{aligned} \quad (2)$$

Eq. 2 shows that ΔLOLE is always greater than zero (ΔLOLE > 0) for $0 < p < 1$. This means that if the probability of successful transmission of the energy from the sending end to the receiving end is the same for all phases, the risk of nontransmission is always higher in the case of symmetric operation.

Example 1 Asymmetric Operation of a 2-Line, 400 kV, 300 km Corridor

The probability of nontransmission can be evaluated as 0.133% per 100 km of line.[6] Thus, for each phase with an equivalent length of 300 km, one has: $q = 1 - p = 0.004$ and $p = 0.996$. The benefit of asymmetric operation is obtained from Eq. 2 as $\Delta\text{LOLE} = 0.01195 f_{\text{ch}} T$. Assuming a generation cost of 2500 $/kW and a load factor of 75%, the benefit of the asymmetric approach over the symmetric approach is approximately 23 $/kW.

IMPLEMENTATION

Two distinct cases must be addressed for the purpose of implementing asymmetric operation: (1) the multiple-line corridor and (2) the single-line corridor. As previously pointed out, either of two implementation strategies can be employed: (1) conventional devices (i.e., passive LC elements with electromechanical switches or circuit breakers) or (2) power-electronic devices.[7] Though the latter incorporate significant advantages, including rapid response and precise control, the following sections focus on the use of conventional devices due to their lower cost.

Multiple-Line Corridor

A lossless, uncoupled, lumped-parameter transmission line model is used as a starting point for quantifying the compensation strategy; this simplifies the analysis and focuses on the underlying concepts while leading to a reasonable estimate of the capacity and cost of the required compensating equipment. Though more complex distributed parameter line models enable such factors as line resistance to be thoroughly accounted for, the results presented in the following sections are remarkably precise because line resistance is typically very low.[2]

Compensating Impedances

The design of the compensation strategy begins by considering a corridor of N parallel lines ($N \geq 2$). The corridor, therefore, includes N instances of each phase, where a_i, b_i, and c_i, respectively, refer to the A, B, and C phases of the ith line. Let us consider the case with L of the individual a_i-phases out of service. The problem is to determine the conditions for which the power transmitted on all of the N individual a_i-phases in symmetric mode is equal to that of $N - L$ compensated remaining a_i-phases in asymmetric mode and then deduce the values of the compensating impedances.

Fig. 2A shows the equivalent circuit of the N parallel a_i-phases working in symmetric mode, all individual phases being in service. V_S and V_R are, respectively, the line-to-line rms voltages of the sending and the receiving ends of the transmission system, and X_p and B_p are, respectively, the series impedance and the shunt susceptance of each a_i-phase.

With L of the a_i-phases out of service due to faults, the equivalent circuit of the $N - L$ remaining sound a_i-phases in parallel in asymmetric mode is shown in Fig. 2B. X_S and B_C are, respectively, the series impedance and the shunt susceptance of the compensating devices for each a_i-phase, defined as follows:

$$\begin{aligned} X_S &= -\frac{L}{N} X_p \\ B_C &= \frac{L}{N - L} B_p \end{aligned} \quad (3)$$

With these compensating elements, the asymmetrically operated compensated corridor A-phase with L of

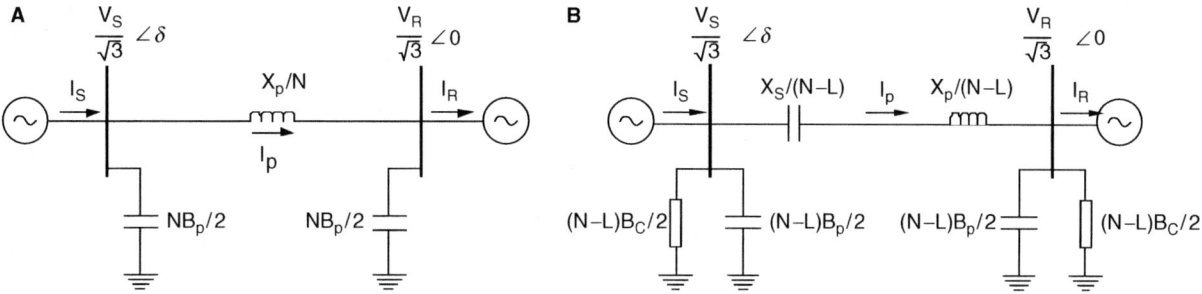

Fig. 2 Equivalent circuit of one phase of a corridor in symmetric and asymmetric operation: (A) corridor A-phase in symmetric operation, consisting of N individual a_i-phases; (B) corridor A-phase in asymmetric operation, consisting of $N-L$ individual a_i-phases.

the a_i-phases out of service has the same electrical characteristics and carries the same power as the original symmetrically operated corridor A-phase with N operational a_i-phases.

Installed Reactive Power

The total installed reactive power for series compensation Q_{Tseries} is calculated assuming that series-connected reactive power is available to every phase of every line. The total installed series reactive power in the $3N$ phases of the corridor is

$$Q_{\text{Tseries}} = 3NX_S \left(\frac{I_p}{N-L}\right)^2$$
$$= -3\frac{L}{(N-L)^2} X_p I_p^2 \quad (4)$$

where I_p is the rms value of the current in the equivalent line, X_S is the impedance of the series compensating device, and X_p is the series impedance of each a_i-phase.

The value of the total installed shunt reactive power Q_{Tshunt} is again calculated based on the fact that any phase of any line can be lost. There is no need to compensate for any particular a_i-phase, however, as shunt compensation can be installed on the sending-end and receiving-end buses. Consequently, the total reactive power installed for shunt compensation is given by:

$$Q_{\text{Tshunt}} = 3(N-L)B_C V_N^2 = 3LB_p V_N^2 \quad (5)$$

where B_C is the shunt susceptance of the compensating devices for each a_i-phase; B_p is the shunt susceptance of each a_i-phase; and V_N is the rms phase-to-ground voltage, assumed to be the same at the two ends of the corridor.

Example 2 Two-Line, 400 kV, 300 km Corridor with Lossless Lines (Fig. 3)

In this example, the methodology is applied for the particular case of two 400 kV, 300 km, lossless, transposed lines loaded to their surge impedance loading

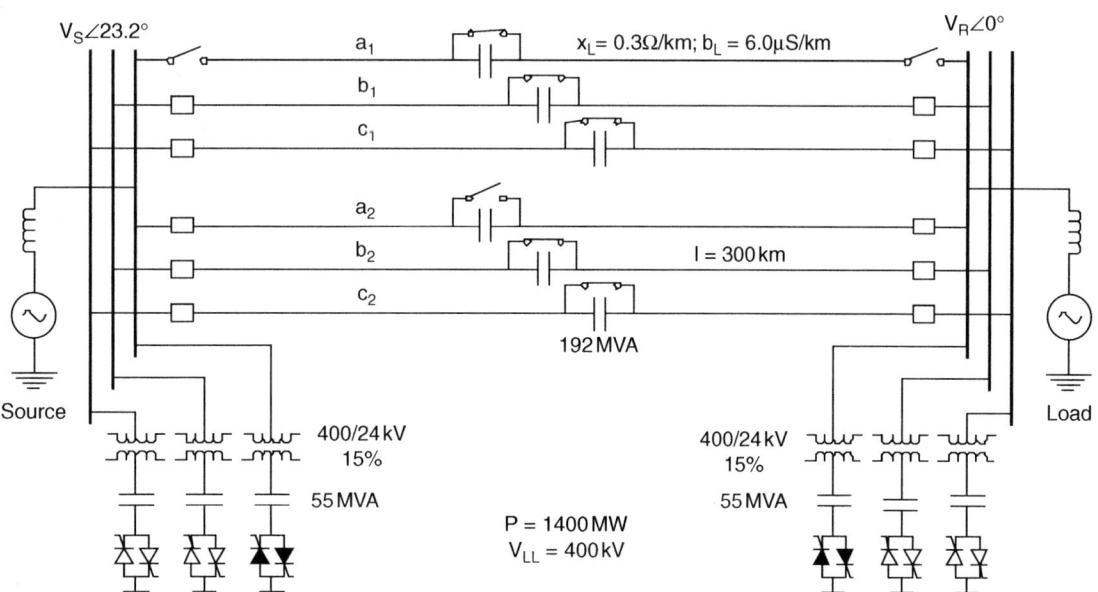

Fig. 3 Compensation scheme for asymmetric operation of a 2-line, 400 kV, 300 km corridor (assuming a lossless conductor model).

(SIL = 1400 MW) and equipped to sustain the loss of three different phases on either of the two 3-phase lines. Fig. 3 shows this corridor with one phase out of service on one line. The unit length parameters are $x_L = 0.3$ Ω/km and $b_L = 6.0$ μS/km. Thus, for 300 km, $X_L = 90$ Ω, and $B_L = 1.80 \times 10^{-3}$ S.

For $L = 1$, Table 1 gives the required reactive power resulting from the application of Eqs. 4 and 5. For series compensation, a total 1150 Mvar is required. The total need for shunt compensation, including 15% reactive impedance representing Static Var System (SVS) transformer losses, is 330 Mvar.

Remarks

The case developed above for N lines with L of the a_i, b_i, or c_i phase out of service is theoretical. Indeed, the case of $L > 1$ has such a low probability of occurrence that it is questionable whether such a contingency need ever be considered. Additionally, as L goes from 1 to 2, compensation requirements and operational complexity are far greater, as one must provide suitable reactive power sources and associated switching for all values of L.

For these reasons, it seems practical to design for the loss of only one a_i-, b_i-, or c_i-phase. Even so, the case of $L = 1$ covers the loss of 1 a_i-phase, 1 b_i-phase, or 1 c_i-phase for any combination of lines, or all three phases of a single line. In the case of a 2-line corridor, the maximum corridor power transfer will be maintained after sustaining up to three single-phase contingencies for any combination of phases on the two different lines. After having sustained a third contingency, however, power transfer would normally be reduced for security considerations. As can be seen, this goes beyond the 3-phase $N-1$ criterion traditionally used to establish security limits, where N is the number of 3-phase lines in a corridor. Asymmetric operation thus maintains full corridor capacity under challenging circumstances while providing the system operator the precious time required to restore it to its precontingency physical state.

Single-Line Corridor

The solution proposed for the single-line case is based on the application of symmetrical components to balance the resulting 2-phase or 1-phase transmission after the loss of one or two phases, respectively. The currents of the negative and zero sequences resulting from a 1-phase-open or 2-phase-open situation are not negligible and must be filtered or compensated so as to guarantee an adequate asymmetric operation. As the single-phase fault has the highest probability of occurrence, the loss of two phases is not considered here.

Fig. 4 shows the principle of asymmetric operation of a 3-phase line with phase a out of service. In normal symmetric operation, the two connected networks and the line are perfectly balanced. During asymmetric operation with one phase open, three basic compensating elements must be introduced to rebalance the voltages and currents:

1. Series compensation of the sound phases (phases b and c in Fig. 4) to lower the series reactance, maintain the same angular spread, and thus ensure the flow of the precontingency power transfer. Series compensation can be supplied by means of either conventional capacitors or series-controlled voltage sources employing power-electronic devices. The compensating elements can be placed at the sending end, the receiving end, or the center of the line.
2. Zero-sequence filters, at each end of the line, to afford a low-impedance path for the zero sequence current. Many designs can be used to implement such filters:
   - a single T, zigzag, or Δ-Y transformer
   - combinations of passive LC elements that may be variable to reflect large load variations
   - controlled shunt current sources made up of power-electronic devices.
3. Negative sequence compensators at each end of the line to eliminate the negative-sequence current by injecting an opposite current of the same magnitude. Here again, the negative sequence compensators can be constructed of passive LC elements connected in delta or star, or of controlled shunt current sources employing power electronic converters. Depending on the planning criteria, the negative-sequence compensator and zero-sequence filter at each end of the line can be grouped together.[2]

Table 1 Total reactive power requirements for the asymmetric operation of a 2-line, 400 kV, 300 km corridor equipped to sustain the loss of any three phases in the corridor (assuming a lossless conductor model)

Total transmitted power $P = 1400$ MW	
Reactive power Q (Mvar)	
Series compensation	1150 Mvar
Shunt compensation	330 Mvar

Example 3 Asymmetric Operation of a Single-Line, 120 kV, 100 km Corridor (Fig. 5)

A 3-phase, 120 kV, 100 km line with parameters $r = 0.061$ Ω/km, $x_L = 0.3644$ Ω/km, $b_L = 4.54 \times 10^{-6}$ S/km is considered. The load power is 50 MW. Fig. 5 and Table 2 show the calculation results and the reactive power needed for asymmetric operation of the line with a phase open. Compensation of the positive sequence requires three

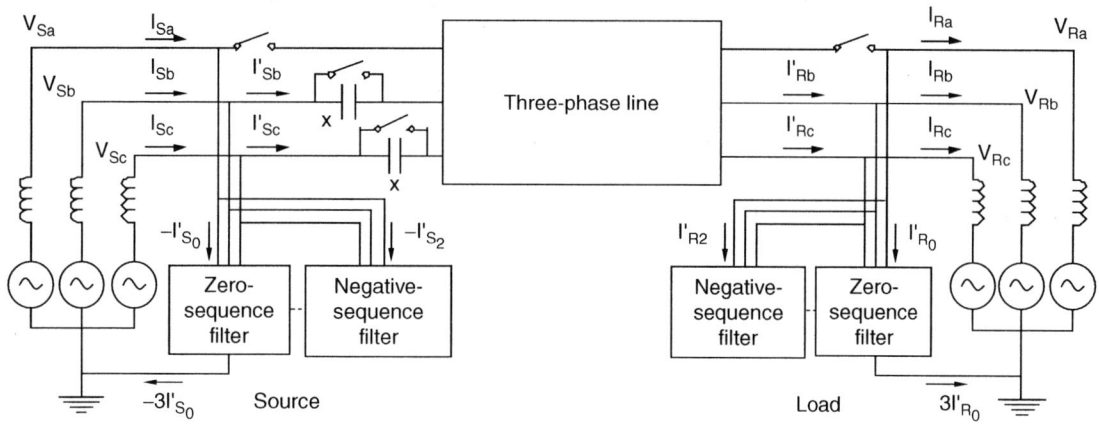

Fig. 4 Concept of asymmetric operation of a single-line corridor.

capacitors of 2 Mvar each for a total 6 Mvar. The compensation of the negative sequence requires 68 Mvar for the two compensators at the ends of the line. Filtering the zero sequence requires 16 Mvar.

Remarks

The values of the passive LC elements of the negative-sequence compensator depend on the actual load, and it may be necessary to adjust them for large variations of the load from SIL (design load). Adjusting the grounding transformers of the zero-sequence filters according to the actual load is not necessary because impedances are designed for maximum load conditions.

In a sense, this solution is an extension of the principle of load compensation.[8] As pointed out earlier, FACTS controller-based implementations can provide better and more rapid control but cost more than conventional solutions.

FINANCIAL ANALYSIS

The LOLE can also be expressed as the loss of revenue related to the energy not delivered in 1 year; such a

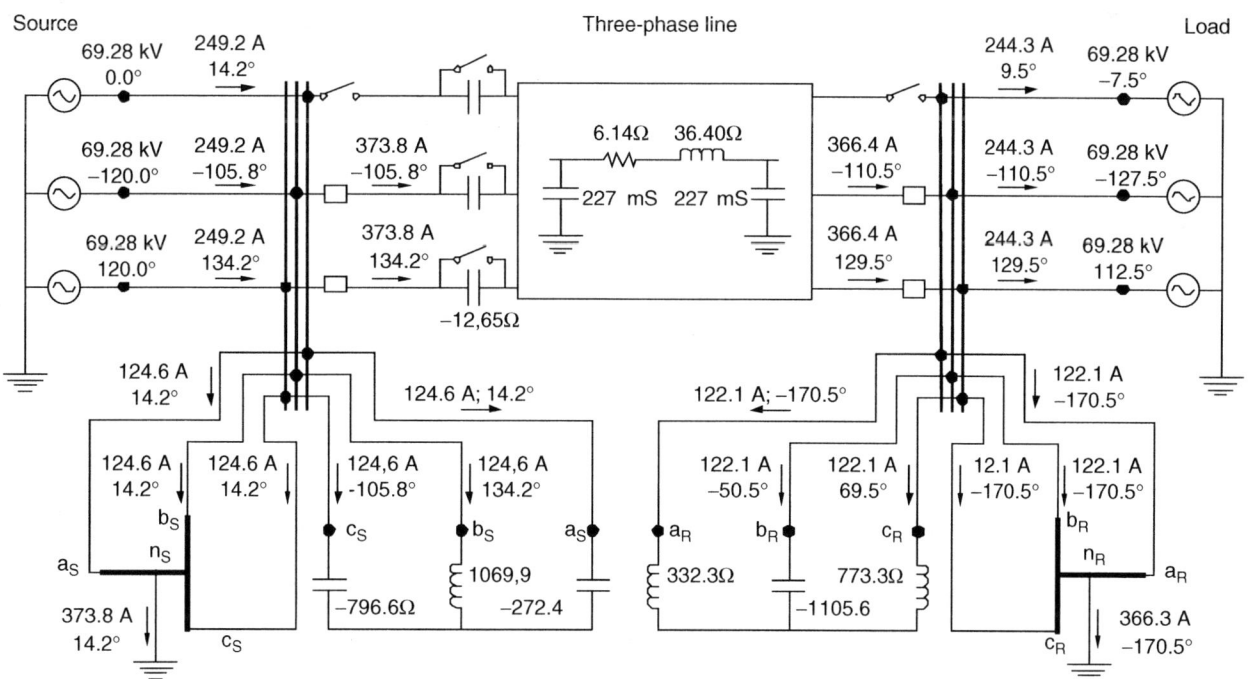

Fig. 5 Asymmetric operation of a single-line, 120 kV, 100 km corridor with one phase open (assuming the use of conventional compensating devices).

Table 2 Calculated results of the operation of a single-line, 120 kV, 100 km corridor with one phase open, using conventional compensating devices

		Sending end	Receiving end 50 MW 16.4 Mvar
Characteristics of the line: $R_L = 6.14\ \Omega$; $X_L = 36.4\ \Omega$; $B_c = 4.54 \times 10^{-4}$ S; $X_c = 12.65\ \Omega$ (3×2 Mvar)			
System voltages	E_{ab} (kV)	120.0 ∠ 30.0°	120.0 ∠ 22.6°
Line currents before asymmetric operation	I_a (kA)	0.249 ∠ 9.3°	0.244 ∠ 2.1°
Line currents during asymmetric operation	I'_b (kA)	0.374 ∠ −106°	0.366 ∠ −110.5°
	I'_c (kA)	0.374 ∠ 134°	0.366 ∠ 129.5°
Negative-sequence compensator			
Compensating impedances	X_a (Ω)	−272.4	332.3
	X_b (Ω)	1069.9	−1105.6
	X_c (Ω)	−797.6	773.3
Reactive power of the compensator	S_a (Mvar)	−4.2	5.2
	S_b (Mvar)	16.6	−17.2
	S_c (Mvar)	−12.4	12.0
Total	S_T (Mvar)	33.2	34.4
Zero sequence filter			
Transformer capacity	S_{Tg} (MVA)	8.1	8.1

figure can then be used to determine how long it takes to pay for the equipment required to implement either asymmetric operation or some other alternative, such as adding another line. The purpose of this section is to show that asymmetric operation is a practical and economical alternative for improving transmission system capacity.

Retrofit of Existing Corridors

The benefit of asymmetric operation is obtained by comparing (1) the total investment cost of retrofitting an existing 3-phase corridor for asymmetric operation and (2) the LOLE of 3-phase symmetrical operation—expressed in dollars—which is offset by means of asymmetric operation. Initially, the annual savings generated by asymmetric operation can be used to reimburse the equipment investment; after this investment has been repaid, this represents additional revenue to the utility. Table 3 summarizes the financial analysis for three 400 kV, 150 km asymmetrically operated corridors involving, respectively, one, two, and three transmission lines. For purposes of comparison, one also finds both the cost of a new line on this table and the cost of implementing a typical FACTS-based compensation strategy.[2]

For the single-line case, the cost of the LOLE is lower than the investment cost of asymmetric operation (and much lower than the cost of building a new line). The payback period, therefore, is greater than 1 year. The cost of LOLE for this case, however, does not take into account the larger social and economic costs associated with the total loss of the power supply in the symmetric operation mode, which are not addressed here.

For two or three lines, the investment cost of asymmetric operation using conventional devices is lower than the cost of the LOLE: in this case, the payback time of asymmetric operation is less than 1 year.

In a general way, one sees that the cost of implementing asymmetric operation by retrofitting multiple lines is always lower than in the single-line case because one exploits the transmission equipment already in place. In other words, part of the equipment required to implement asymmetric operation is already there!

As a final note, all three cases provide remarkable performance in relation to the standard 3-phase $N-1$ security criterion.[9] Case 1, when operated asymmetrically, respects the single-phase $N-1$ criterion, whereas symmetrical operation of this same line is incapable of respecting any $N-1$ criterion, either 1-phase or 3-phase. Cases 2 and 3, when operated asymmetrically, are capable of respecting the full 3-phase $N-1$ criterion in addition to a single-phase $N-1$ criterion. All this translates into high transmission reliability and improved transmission economics.

Table 3 Comparing the cost of three options at 400 kV for reducing the risk of nontransmitted energy (costs are expressed in millions of dollars Canadian)

	Case 1	Case 2	Case 3
Number of lines	1	2	3
Voltage (kV)	400	400	400
Line length (km)	150	150	150
Transmitted power (MW)	700	1400	2000
Probability of failure of one phase[6]	0.002	0.002	0.002
Load factor (%)	75	75	75
Cost of power not supplied: LOLE (M$CA)	8	16	23
Asymmetric operation with conventional devices			
Investment Cost (M$CA)	48	15.0	12.1
Payback time (years)	6.0	0.93	0.53
Asymmetric operation with FACTS devices			
Investment cost (M$CA)	70	42.8	34.7
Payback time (years)	8.8	2.7	1.5
New line			
Investment cost (M$CA)	60	60	60
Payback time (years)	7.5	3.8	2.6

New Corridors

Three voltage scenarios (345, 500, and 735 kV; see Table 4) have been selected to compare the performance of different transmission options under symmetric and asymmetric operation for planning new transmission capacity from the point of view of the $N-1$ criterion. Each scenario compares the cost of a 2-line, symmetrically operated transmission corridor to a 1-line, asymmetrically operated corridor of the same capacity, both of which respect an $N-1$ criterion. In each case, the $N-1$ criterion is interpreted as follows: For a symmetrically operated system, $N-1$ represents the loss of a 3-phase line; for an asymmetrically operated system, $N-1$ represents the loss of a single phase.

In each voltage scenario, both cases have the same capacity, as normal planning and operating criteria generally load a double-circuit corridor to no more than the SIL of a single line for reliability purposes. This ensures that the corridor respects the 3-phase $N-1$ criterion—in this case, the loss of one of the two lines without loss of load.

Because the operational reliability of these asymmetric and symmetric scenarios is essentially identical (as they respect their respective $N-1$ criterion), LOLE is not an appropriate basis for comparison. To compare the three scenarios, one must consider their respective investment costs.

Cost Analysis

Table 5 presents a summary of the costs associated with the construction of a 300 km, symmetrically operated 2-line corridor at 345 kV, 500 kV, and 735 kV; it also presents a summary of those associated with the construction of a 300 km, asymmetrically operated single line corridor at each of these respective voltage levels. These estimates are based on the use of conventional elements and include rights of way.

In all three scenarios, the cost of a single, asymmetrically operated 300 km line is less than that of two symmetrically operated 300 km lines. This is because the cost of the reactive power for implementing asymmetric operation of a 300 km line is less than the cost of the additional line required in symmetric operation.

Table 4 Transmission system scenarios for comparing symmetric and asymmetric operation

	Scenario 1	Scenario 2	Scenario 3
Transmitted power	450 MW	1000 MW	2200 MW
Symmetric operation	345 kV; 2×300 km lines	500 kV; 2×300 km lines	735 kV; 2×300 km lines
Asymmetric operation	345 kV; 1×300 km line	500 kV; 1×300 km line	735 kV; 1×300 km line

Table 5 Costs associated with symmetric and asymmetric operation for the three scenarios of Table 4 (costs are expressed in millions of dollars Canadian)

	Scenario 1	Scenario 2	Scenario 3
Transmitted power (MW)	450	1000	2200
Symmetric operation			
Voltage level; Nb. lines	345 kV; 2 lines	500 kV; 2 lines	735 kV; 2 lines
Total cost (lines) (M$CA)	120	240	360
Asymmetric operation			
Voltage level; Nb. Lines	345 kV; 1 line	500 kV; 1 line	735 kV; 1 line
Cost of the lines (M$CA)[10]	60	120	180
Cost of compensation (M$CA)	26.8	59.2	128.6
Total cost (M$CA)	86.8	179.2	306.6

Effect of Line Length

In the above example, line length was set somewhat arbitrarily at 300 km even though all scenarios are technically realistic. There is considerable merit, however, in comparing the costs of a 1-line, asymmetrically operated corridor as a function of distance with respect to those of a 2-line symmetrically-operated corridor. This was examined, therefore, in the case of the three voltage scenarios considered above and for line lengths ranging from 1 to 400 km.

As Fig. 6 shows, the costs of both symmetric and asymmetric operation vary linearly with line length but at different rates. For short lines, the cost of asymmetric operation is higher because the cost of shunt compensation is predominant. For long lines, the cost of symmetric operation is higher because the cost of the lines increases more quickly than the cost of compensation. As shown, there exists a point for each scenario at which, between 100 and 200 km, an asymmetrically operated 1-line corridor costs less to build than a symmetrically operated 2-line corridor of the same voltage. Such cost behavior is similar to that found in comparisons of AC and DC transmission corridors, where, beyond a certain point, DC transmission is less costly than the equivalent AC solution.

CONCLUSION

Asymmetric operation transforms power system planning and operation by virtue of the greater flexibility available in finding solutions to specific challenges. Though actual implementations will generally require detailed simulation and engineering of components and systems for such purposes as insulation coordination, protection, control, security, and reliability, the numerous examples presented here show that the concepts are applicable to any voltage level.

In the case of multiple-line corridors, it has been shown that conversion to asymmetric operation increases the corridor availability, reduces the loss of load expectation, and increases the secure power transfer limit (as each and every line can be operated

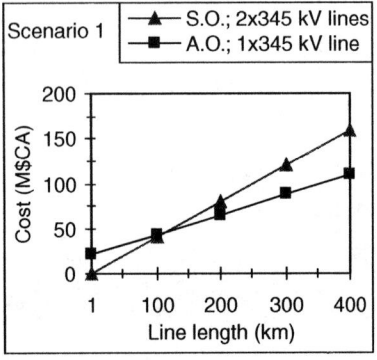
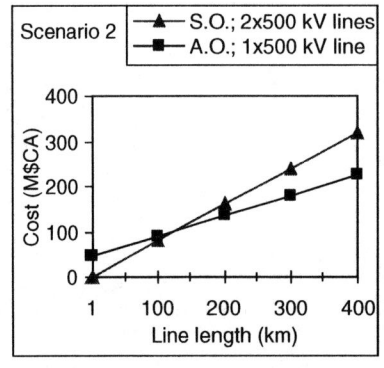
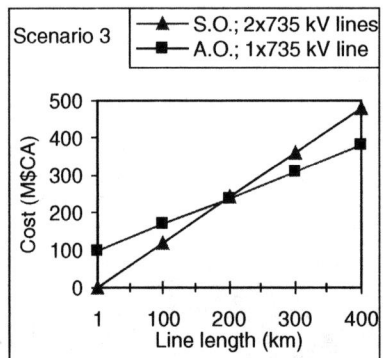

Fig. 6 Cost comparison of symmetric (S.O.) and asymmetric (A.O.) operation for three scenarios as a function of line length (costs are expressed in millions of dollars Canadian).

at its SIL) while respecting existing 3-phase $N-1$ security criteria if need be. In the case of new transmission corridors, point-to-point transmission, interarea network interconnections, or ring-type transmission grids feeding large metropolitan areas, asymmetric operation has such a positive impact on the reliability of single-line transmission systems that it redefines one's outlook on such fundamental issues as choice of voltage, number of lines, the contingency that defines one's security criterion, the amount of power transferred securely, and the amount of land required for transmission rights of way. In short, asymmetric operation redefines the reliability, environmental impact, and economics of the planning and operation of AC power transmission.

REFERENCES

1. U.S.-Canada Power System Outage Task Force. Final Report on the August 14, Blackout in the United States and Canada: Causes and Recommendations. April 2004. Available at: www.nrcan-rncan.gc.ca/media/docs/final/finalrep_e.htm (accessed on October 2005).
2. Marceau, R.J.; Sana, A-R.; McGillis, D.T. *Asymmetric Operation of AC Power Transmission Systems*; Presses Internationales Polytechnique: Montreal, Quebec, Canada, 2006.
3. Blyakov, N.N.; et al. Utilisation du Réenclenchement Monophasé Automatique dans un Grand Réseau EHT Comportant des Lignes 1200 kV. CIGRE **1990**, 34–207.
4. Automatic reclosing of transmission lines. In *IEEE Transactions on Power Apparatus and Systems*, IEEE Power System Relaying Committee Report, 1984; PAS-103(2).
5. Anders, G.J. *Probability Concepts in Electric Power Systems*; Wiley: New York, 1990.
6. Canadian Electrical Association. Forced Outage Performance of Transmission Equipment for the Period of January 1, 1994 to December 31, 1998. February 2000.
7. Hingorani, N.G.; Gyugyi, L. Understanding FACTS. IEEE Press: New York, Power Systems Engineering, 2000.
8. Gyugyi, L.; Otto, R.A.; Putman, T.H. Principles and applications of static, thyristor-controlled shunt compensators. IEEE Trans. Power Apparatus Systems **1978**, *PAS-97* (5), 1935–1945.
9. Marceau R.J. et al. Power system security assessment: A position paper. CIGRE Task Force 38.3.12 Final Report. June 30, 1997.
10. Picard, B.; Galiana F.D.; McGillis, D. A knowledge-based system for the structural design of high-voltage lines. Proceedings of IEEE-CCECE'1999, Edmonton, Alberta, Canada, 1999.

Electric Supply System: Generation

Jill S. Tietjen
Technically Speaking, Inc., Greenwood Village, Colorado, U.S.A.

Abstract

The electric utility system is comprised of three major components: the generation system, the transmission system, and the distribution system. The generation system, where the electricity is produced, is comprised of power plants, also called generating units. Generation in the United States is produced at facilities categorized as conventional and renewable resources. Conventional resources are those in which the fuel is burned, and include coal, nuclear, natural gas, oil, and diesel power plants. Renewable resources are those in which the "fuel" consumed to produce electricity is replenished by nature or can be replenished by humankind. Generation resources in the United States are predominantly conventional resources, although renewable resources are in operation and providing more and more electricity every year.

INTRODUCTION

The electric utility system is comprised of three major components: the generation system, the transmission system, and the distribution system. The generation system, where the electricity is produced, is comprised of power plants, also called generating units. The transmission and distribution systems both consist of wires and other equipment that carry the electricity from the power plants (or generation sources) to the homes and businesses where we consume the power. The transmission lines and towers move large amounts of power from the power plants to large population areas (cities and towns), where it is converted at substations to lower voltages. The distribution system carries the power at lower voltages from those substations to our actual houses and businesses.

Generation in the United States is produced at facilities categorized as conventional and renewable resources. Conventional resources are those in which the fuel is typically burned, and include coal, nuclear, natural gas, oil, and diesel power plants. Renewable resources are those in which the "fuel" consumed to produce electricity is replenished naturally by nature or can be replenished by humankind; these resources include hydroelectric, geothermal, solar, wind, and biomass. Generation resources in the United States are predominantly conventional resources, although renewable resources are in operation and are providing more and more electricity every year in the United States.

CONVENTIONAL RESOURCES

The fuels used in conventional resources to generate electricity in the United States include coal, nuclear, natural gas, oil, and diesel. Coal, natural gas, and oil are burned in a boiler that heats water to produce steam. Nuclear power plants generate steam from the heat given off by nuclear fission. Natural gas is burned to generate electricity in combustion turbines and combined cycle power plants. Oil can also be burned to produce electricity in combustion turbines. Diesel generators tend to have an internal combustion engine that directly turns a generator to produce electricity.

A typical coal-fired power plant works as demonstrated in Fig. 1. The coal is fed into the boiler, where it is burned to heat water to convert the water to steam. At that point in the cycle, the steam is high temperature and high pressure. This steam moves into one or more turbines mounted on the same or on separate shafts; the steam turns the turbine blades. The turbine blades are connected to the turbine shaft, which is connected directly or indirectly to the generator rotor (shaft). As the generator shaft rotates, it produces electricity (see Fig. 2). The steam that exits the turbine is now low pressure and lower temperature, because its pressure and temperature have been reduced in the process of its working to turn the turbine blades. The steam is sent to a cooling tower or condenser, where additional heat is removed so that the steam can be converted back to water. Water cooled through the cooling tower or condenser then can be discharged back into the body of water from which it came—usually, a river or a lake. Or as shown in Fig. 1, water can be pumped directly back to the boiler from the cooling tower or condenser.

Coal-fired generating units currently produce about half of the energy in the United States.[1] Many smaller, older units are still in operation. Newer units tend to be in the size range of 250 MW to more than 1,000 MW.

Keywords: Power plants; Conventional resources; Renewable resources; Coal; Nuclear; Natural gas; Oil; Diesel; Wind; Solar; Geothermal; Biomass; Turbine; Generator.

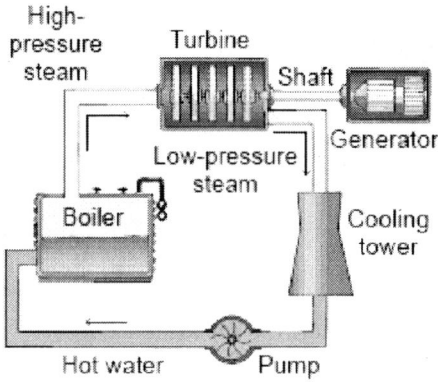

Fig. 1 Schematic diagram of a conventional fossil fueled generating station.
Source: From Wiley (see Ref. 2).

Coal is mined and shipped to the power plants from a broad range of states. The vast majority of coal is mined in Wyoming, West Virginia, Kentucky, Pennsylvania, and Texas.

Coal-fired power plants are very reliable and operate in a manner that is called dispatchable. This means that the amount of power that the facility provides minute to minute is controllable by a human operator and can be increased or decreased at any time, depending on the amount of electricity required by a utility's customers. Typically, the output is controlled by an integrated control system based on automatic inputs and feedback. New coal-fired power plants have a significant amount of pollution control equipment installed that can include some or all of the following: electrostatic precipitators (to control particulate matter), flue-gas desulfurization (to reduce sulfur dioxide emissions), and selective catalytic reduction (SCR) (to reduce nitrogen oxide (NO_x) emissions). Coal-fired power plants also produce carbon dioxide (CO_2) in the combustion process.

Nuclear power plants are in some ways similar to coal-fired power plants and in some ways different. Similarities include that the nuclear fuel heats water to turn it into steam and that turbines turn generators to produce electricity.

In boiling water reactors, there is one water system, as with coal-fired power plants. For pressurized water reactors, however, there are two water circulation systems. As shown in Fig. 3, these are called the primary water loop and the secondary water loop. The water in the primary water loop stays within the containment building, where it is heated by nuclear fission in the reactor and in turn heats the water in the secondary water loop, causing it to turn into steam. The secondary water loop is very similar to the water system in a coal-fired power plant, as it goes through the turbine, is cooled, and then is pumped back through the cycle to be reheated.

Nuclear power plants produce no greenhouse gases in the process of combustion. The fuel and other portions of the power plant are radioactive, however, and require special handling and storage. Nuclear power plants currently account for about 10% of the installed generating capacity in the United States.[1] Because of their high reliability and low fuel cost, nuclear units generally run at full capacity when they are online, and utility dispatchers rarely lower or raise the amount of power being generated from them on a minute-to-minute basis. No new nuclear units have been built in the United States for many years, but as of 2006, several consortia are planning new units that could be constructed and providing electricity by 2015 or later.

Combustion turbine technology burns natural gas (or some other type of liquid fuel) in an internal combustion chamber to heat compressed air.[4] The heated air turns the turbine blades, which turn the generator to produce electricity (see Fig. 4). Combustion turbines are primarily used by utilities to provide energy during the peak period (when the load is at its highest—generally, on hot summer

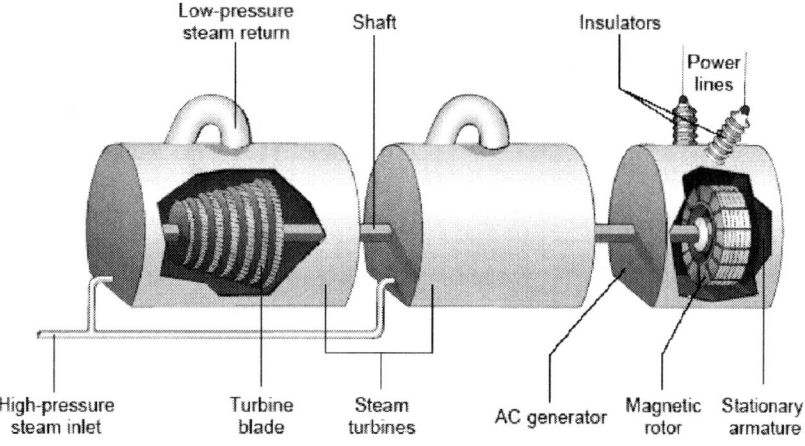

Fig. 2 Turbine-generator configuration.
Source: From Wiley (see Ref. 2).

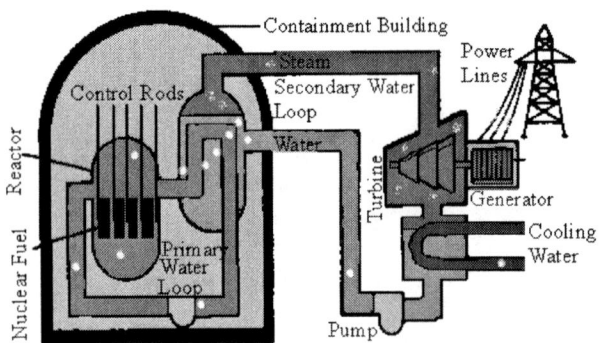

Fig. 3 Diagram of a nuclear power plant: pressurized water reactor.
Source: From Nuclear Reactors (see Ref. 3).

afternoons and cold winter mornings) or during emergencies (such as the loss of a major generating unit or transmission line).

Combustion turbines range in size from about 1 MW to more than 300 MW. New units use selective catalytic reduction in addition to water or steam injection to control NO_x emitted during the combustion process. Combustion turbines also emit CO_2.

In special combined cycle units, much of the heat that is exhausted (not used) in the process of generation from a combustion turbine is captured using a heat recovery steam generator (HRSG). This HRSG uses the waste heat to turn water into steam and produce additional electricity. This combination of combustion turbines and an HRSG is much more efficient than the standard combustion turbine power plant in a stand-alone mode.

The combined cycle unit can be used throughout the day by the electric utility dispatcher, and the amount of generation that it produces can be dispatched (raised and lowered as the electric utility customers' load increases and decreases). Because combined cycle units include a combination of combustion turbines and HRSGs, the total installed capacity for this type of unit ranges from 10 MW to more than 950 MW. Emissions from these units include NO_x and CO_2, and SCR is generally used. Installed capacity in the United States fueled by natural gas—including combustion turbines, combined cycle units, and steam units—comprises about one-quarter of all capacity.[1]

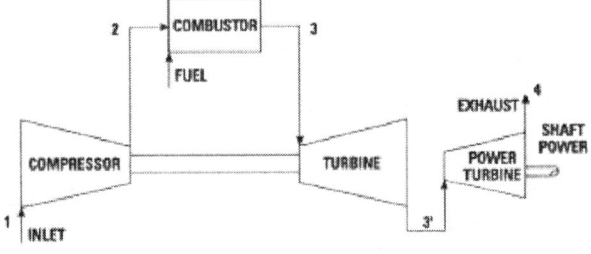

Fig. 4 Combustion turbine schematic.
Source: From Energy Solutions Center (see Ref. 4).

Diesel generators are a reciprocating engine technology that burns diesel fuel. In these engines, an air-and-fuel mixture is burned. These smaller generators are often used as the source of electricity in remote villages and, thus, run 24 h per day. In addition, they are used as emergency backup, particularly at nuclear power plants, and run only for very limited hours in a year, if at all. Diesel generators emit particulates, sulfur dioxide, NO_x, carbon monoxide and CO_2. A small percentage of total U.S. installed generating capacity consists of diesel generators.

RENEWABLE RESOURCES

Renewable resources in the United States have received added attention in the past few years as a means of producing power without the environmental effects associated with the burning of fossil fuels (coal, oil, natural gas, and diesel). Renewable resources in the United States currently used for electricity generation include hydroelectric (conventional and pumped storage), geothermal, wind, solar, and biomass.

Hydroelectric energy (energy that comes from water) has been in use in the electric utility industry since that industry's infancy. Conventional hydroelectric resources use flowing water to move turbines that turn generators, which produce electricity. These facilities can either be run-of-the-river or storage hydro facilities. Run-of-the-river means that as the water flows in the river, it passes through the dam and generates electricity. This type of capacity is not within the control of the utility dispatcher. Storage hydro means that water can be stored behind a dam for some period and then released to produce electricity at the command of the electric utility dispatcher. The dispatcher normally would want to use the water to produce electricity during system peaks. About 8% of the current installed electric generating capacity in the United States is provided by hydroelectric power plants.[1]

Pumped storage hydro takes advantage of significant height differences between an upper reservoir and a lower reservoir. During the day and at the control of the operating utility, water flows downhill and produces electricity. At night, water is pumped back uphill so that it can be used the next day or otherwise in the future. Pumped storage hydro actually requires energy to perform the pumping, and because it is 70%–85% efficient, it actually consumes more energy than it produces. The cost differential between the energy used to pump the water uphill at night and the energy that would be generated by other fuel sources during peak hours that instead is replaced with the water from the pumped storage facility allows this form of electric generating capacity to be cost effective. Pumped storage capacity amounts to just over 2% of installed generating capacity in the United States.[1]

Neither conventional hydroelectric generation nor pumped storage hydro generation emits any greenhouse gases. Environmental concerns related to hydroelectric

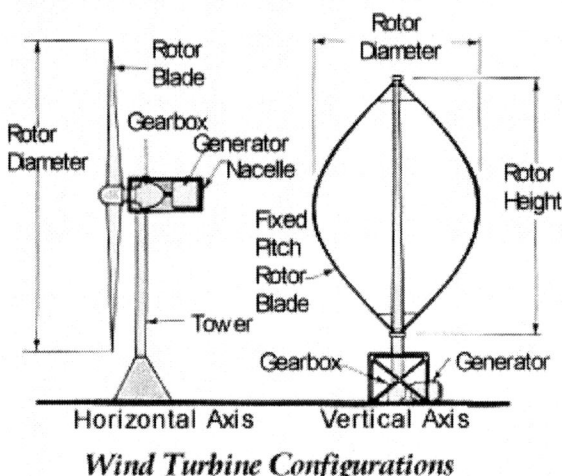

Fig. 5 Wind turbine configurations.
Source: From AWEA (see Ref. 7).

facilities include changes to stream flows, oxygen content of the water, and the disruption of wildlife habitat.[5]

The word geothermal comes from Latin words that mean "heat from the earth." Geothermal resources range from shallow ground to hot water and rock several miles below the Earth's surface, to even farther down, to molten rock known as magma. In the United States, most geothermal resources are located in the Western states, Alaska, and Hawaii.

The three types of geothermal power plants operating today are dry steam plants, flash steam plants, and binary-cycle plants. Dry steam plants directly use geothermal steam to turn turbines. Flash steam plants pull deep, high-pressure hot water into lower-pressure tanks and use the resulting flashed steam to drive turbines. Binary-cycle plants pass moderately hot geothermal water by a secondary fluid with a much lower boiling point than water. This process causes the secondary fluid to flash to vapor, and this vapor then drives the turbines. Geothermal energy emits little to no greenhouse gases and is very reliable (with an average system availability of 95%). Geothermal heat pumps use the nearly constant temperature of the upper 10 ft of the Earth's surface to both heat and cool residences.[6]

Wind energy transforms the kinetic energy of the wind into electrical energy. Wind turbines come in 2 types: vertical-axis (eggbeater style), and horizontal-axis (propeller-style) machines (see Fig. 5). The turbine subsystems include a rotor or blades to convert the wind energy to rotational shaft energy, a nacelle (an enclosure) to cover the drive train (which usually includes a gearbox and a generator), a tower to support the rotor and drive train, and electronic equipment to connect the facility to the utility's transmission grid.

Wind energy produces no greenhouse gases but does raise concerns with respect to noise and presenting a danger to migratory birds and bats. Another concern about wind energy generation is that power is produced only when the wind blows and, thus, is not dispatchable. The wind may not be blowing at the times when it is most needed (i.e., on hot summer afternoons). Unlike fossil fuel-fired power plants, the maximum usage expected from wind-energy turbines would be that they generate energy at about 30%–40% of the maximum amount of energy that would be possible if the wind were blowing every hour.[7]

Installed wind energy capacity reached 9149 MW at the end of 2005 (out of total installed U.S. generating capacity of more than 960,000 MW). The states with the most wind turbines installed, in decreasing order, are California, Texas, Iowa, Minnesota, Oklahoma, New Mexico, Washington, Oregon, Wyoming, and Kansas. Much new capacity is expected to come online during 2006 and 2007 due to the availability of the Federal Production Tax Credit.[1,8]

The rays from the sun can be harnessed as what is called solar power. A variety of forms are possible for utility application in the form of power plants. Concentrating solar power (CSP) technologies include dish/engine systems, trough systems, and power towers. CSP systems use reflective materials such as mirrors to concentrate the sun's energy. The concentrated heat energy then is converted to electricity. Photovoltaic (PV) systems (so-called solar cells that are familiar from America's space program) are built into arrays that directly convert sunlight into electricity. To date, solar facilities that generate electricity are primarily located in California. Much research and development is ongoing to make solar energy more cost effective for a wide range of applications.

Solar generation facilities do not emit any greenhouse gases, but current facilities use up many more acres of land compared with a conventional resource for the same amount of electricity output. Solar generation typically produces electricity only when the sun is shining unless some form of energy storage device has been installed.[9]

Biomass electric generation is the largest source of renewable energy that is not hydroelectric. Biomass means any plant-derived organic matter available on a renewable basis, including dedicated-energy crops and trees, agricultural food and feed crops, agricultural crop wastes and residues, wood wastes and residues, aquatic plants, animal wastes, and municipal wastes. Waste energy consumption generally falls into categories that include municipal solid waste, landfill gas, other biomass, and other. Other biomass includes agriculture byproducts and crops; sludge waste; tires; and other biomass solids, liquids, and gases. Biofuels being developed from biomass resources include ethanol, methanol, biodiesel, Fischer–Tropsch diesel, and gaseous fuels such as hydrogen and methane.[10]

Biomass power systems fall into 4 categories: direct-fired, co-fired, gasification, and modular systems. Most biomass systems today are direct-fired systems that are quite similar to most fossil fuel-fired power plants where

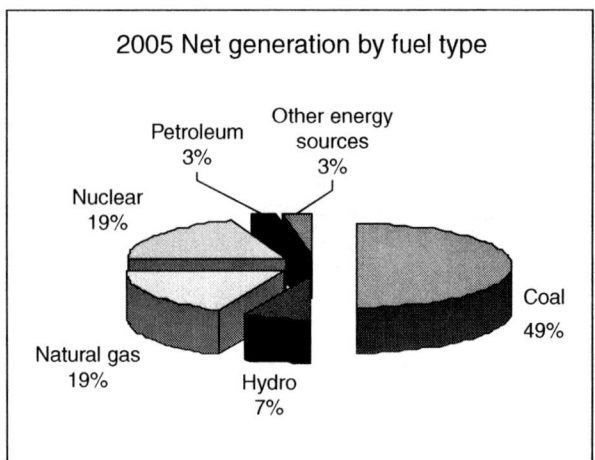

Fig. 6 2005 net generation by fuel type.
Source: From Energy Information Administration, Electric Power Monthly, March 2006: With Data for December for 2005, DOE/EIA-0226 (2006/03), www.eia.doe.gov/cneaf/electricity/epm/epm_sum.html, accessed April 2006.

the biomass is burned, providing heat to turn water into steam, which turns blades in a turbine that turns a generator and produces electricity. The direct-fired systems produce many of the same products of combustion as do coal-fired generating units and require much of the same equipment for cleaning up the byproducts. Direct-fired systems tend to be dispatchable, although some are constrained by the amount of fuel available to the facility.

Co-firing involves substituting biomass for a portion of coal in an existing power plant boiler. Co-firing is a much less expensive way to burn biomass than building a new biomass power plant. Biomass gasifiers heat the biomass in equipment in which the solid biomass breaks down to a flammable gas. The gas can be cleaned and filtered to remove problem elements and then burned much like natural gas in a combined cycle unit. Modular systems are either direct-fired or use gasifiers in small-scale situations, such as for villages, farms, and small industry.[11]

GENERATION IN THE UNITED STATES

Coal is the predominant fuel used for electricity generation in the United States. Fig. 6 shows the actual generation for 2005, with coal producing 49.9% of the electricity consumed during the year. The large majority of the remainder of electricity produced in 2005 came from nuclear and natural gas. Hydroelectric is large enough to be identified separately, but the remaining renewable resources are included in the other energy sources.[12]

CONCLUSION

The generation system is one of the three components of an electric utility's system. That generation (the power plants) is comprised of both conventional and renewable resources. Conventional generation resources including coal, nuclear, natural gas, and oil continue to provide the large majority of electricity produced in the United States. Renewable resources are growing but represent a small slice of total electricity production.

REFERENCES

1. Energy Information Administration. Electric Power Annual (with date for 2004), released November 2005. Available at: www.eia.doe.gov/cneaf/electricity/epa/epa_sum.html (accessed April 2006).
2. Bloomfield, L.A. *How Things Work: The Physics of Everyday Life*; Wiley: New York. Available at: www.howthingswork.virginia.edu/supplements/electric_power_generation.pdf (accessed April 2006).
3. Nuclear Reactors. University of Wisconsin Nuclear Reactor Tour, College of Engineering, University of Wisconsin, Madison. Available at: http://reactor.engr.wisc.edu/power.html (accessed April 2006).
4. Energy Solutions Center (ESC). Combustion Turbine. Available at: www.energysolutionscenter.org/DistGen/Tutorial/CombTurbine.htm (accessed April 2006).
5. California Energy Commission. Hydroelectric Power in California. Available at: www.energy.ca.gov/electricity/hydro.html (accessed April 2006).
6. U.S. Department of Energy. Geothermal Basics Overview. Energy Efficiency and Renewable Energy Geothermal Technologies Program. Available at: www1.eere.energy.gov/geothermal/printable_versions/overview.html (accessed April 2006).
7. AWEA. Wind Energy Basics. Available at: www.awea.org/faq/tutorial/wwt_basics.html (accessed April 2006).
8. AWEA. Wind Energy Projects Throughout the United States of America. Available at: www.awea.org/projects (accessed April 2006).
9. U.S. Department of Energy. Concentrating Solar Power. Energy Efficiency and Renewable Energy, Solar Energy Technologies Program. Available at: www1.eere.energy.gov/solar/printable_versions/csp.html (accessed April 2006). U.S. Department of Energy. Solar Energy Topics. Energy Efficiency and Renewable Energy. Available at: www.eere.energy.gov/RE/solar.html (accessed April 2006).
10. U.S. Department of Energy. Biomass Topics. Energy Efficiency and Renewable Energy. Available at: www.eerer.energy.gov/RE/biomass.html (accessed April 2006). Energy Information Administration. Landfill Gas. Available at: www.eia.doe.gov/cneaf/solar.renewables/page/landfillgas/landfillgas.html (accessed April 2006).
11. U.S. Department of Energy—Energy Efficiency and Renewable Energy Biomass Program. Electrical Power Generation. Available at: http://eereweb.ee.doe.gov/biomass/printable_versions/electrical_power.html (accessed April 2006).
12. Energy Information Administration. Electric Power Monthly, March 2006: With Data for December for 2005. DOE/EIA-0226 (2006/03). Available at: www.eia.doe.gov/cneaf/electricity/epm/epm_sum.html (accessed April 2006).

Electricity Deregulation for Customers

Norm Campbell
Energy Systems Group, Newburgh, Indiana, U.S.A.

Abstract

Deregulation of the United States electric industry has been a roller coaster ride with roots in the 1920's and continuing as a work in progress to present day. From the birth of retail electricity markets and the initial regulations provided through PUCA in 1935, the electric industry has fueled the growth of the United States. Since PURPA in the 1970's, the Energy Policy Act of 1992 and most recent Energy Policy Act of 2005, regulators have been attempting to balance the yoke of regulation with that of open competition while continuing the history of high reliability and reasonably stable costs to the consumers. This article frames the history of the electric industry with the attempts to deregulate and provide for the benefits of a competitive environment into the 21st century.

INTRODUCTION

Accessible and reliable energy—in particular, electricity—is one of the driving forces of a successful economy. Since the early stages of the industrial revolution in the United States, abundant energy has been a cornerstone of domestic economic growth. Initially this growth was fueled by the creation of an inconsistent and unregulated utility system. The system energy supply business gave rise to the unregulated monopoly structures of the 1920s, which created ineffective and anticapitalistic business structures. These aggressive monopolies forced the regulated energy supply structure, encapsulated by the Public Utility Holding Act (PUCA) of 1935, and set in place the framework for today's regulated utility companies with the means of delivery and regulation of electric supply. For over 40 years, the resulting system helped build a foundation of manufacturing, service, and technology achievement. However, as with any system, changes are inevitable and the business of energy is no exception. Today, some customers vie for choice in electric suppliers and the potential for lower energy costs, but the structures of old do not allow for these opportunities. Starting with the Public Utility Regulatory Policy Act (PURPA) of 1978 and continuing with the Energy Policy Act of 1992, customer choice at the wholesale and the retail level were viable options but not without some major challenges. In the next few pages, the origins of competition in the electric industry to today will be traced and some of the results and current activities will be reviewed. Customer choice is a concept that strikes a chord at the heart of American business, and that chord is competition. The final tally has not been written because there are examples of success and failure in the U.S. market; however, activity continues in attempting to find the balance between reliability, quality, cost, and adequate supply that will drive the next century of economic growth.

THE BEGINNINGS

Electric supply systems were born in the city that never sleeps, New York, when Thomas Edison installed the first generation and distribution system to serve customers in the metropolitan area searching for a way to use a new idea called an electric light. From these early halcyon days, competition was fierce and fully unstructured as the race was started to see who could serve the needs of an interested and power hungry populace. This race gave way to multiple providers in close proximity and required oversight by government. As early as 1905, the city of Chicago had granted 21 municipal franchises to provide electricity including three franchises that covered the entire city.[1] However, this system of supply had too many choices with too few rules, which created the first natural monopolies based on economies of scale. The 1920s yielded the unregulated electric monopoly with stable costs but the need for growth beyond affluent customers, which were the primary customers of the time. In Pennsylvania, the state government proposed centralization of electric generation and a transmission system to connect existing customers and new customers and this was decried as un-American. This initial step was the first movement towards centralized generation with associated transmission system to deliver the fuel to power America's ascendancy to the top of the industrial peak. By 1932, nearly all investor owned utilities (IOU) moved towards the same centralized generation and transmission model proposed in Pennsylvania, but in an ominous turn, 50% of these IOUs were controlled by three companies. This ownership structure created a worry that unregulated

Keywords: Electric competition; Customer choice; Open access; Reduced costs; Reliability; Wheeling; Wholesale energy trading.

monopolies would be able to control the electric system without any limitations. With unfettered control on the horizon, the federal government passed the Public Utilities Holding Act of 1935 (PUCHA) to limit electric monopolies and offer a framework for a structured generation and transmission system to supply electricity to a growing nation. This system and its evolution served well for nearly 40 years until the 1973 oil crisis shocked the nation into a new era of energy awareness and regulation.

GROWING PAINS

1973 was a year of long lines at gas stations and a collective shock to the economy based on the new limitations of energy. With this painful ordeal seemingly without solution, politicians and regulators alike searched for a way to help increase the opportunity for energy production from nontraditional central power stations. This "out of the box" thought gave birth to the Public Utilities Regulatory Policy Act of 1978 (PURPA), which had many components but for this discussion yielded the mainstreaming of cogeneration and of nonutility generation (power production) and direct interconnection if specific requirements were met by the energy producer. The two key components of PURPA in this regard were (1) that certain generators of electricity would be able to produce power and not fall under the utility regulation formats and (2) that there is a requirement of the local regulated utility to purchase excess generation from the nonutility generator at the utility's avoided cost.[2] This requirement opened the crack in the door for third-party generator interconnection to the electric grid and sale/purchase of this power. PURPAs intent was very clear, but the results were mixed. Because the state utility regulators oversaw the implementation of the program, calculations and processes surrounding avoided costs, interconnection standards, and local regulations created a patchwork of activity in the Unites States. California and New York were among the most ardent supporters of PURPA and the implementation of programs, but even today the rules in place for PURPA have directly impacted the efforts to restructure the electric markets.

In 1992, after many years of activity surrounding deregulation of the interstate natural gas supply industry, the Energy Policy Act of 1992 (EPAct '92) was passed to create the advent of electric market restructuring on a large scale. Based on the lesson learned from the natural gas market unbundling, the Federal Energy Regulatory Commission (FERC) provided the option of "wheeling" or open access across transmission systems based on the petition of buyers or generators of electricity. The effect of EPAct '92 was the initial opening of the wholesale transmission market, public information regarding transmission facilities, and expansion of the exemptions from utility regulation granted to certain generators started with PURPA. The main limitation of this law was that FERC was unable to extend open access to "retail" customers, but this right was reserved for the states to address on an individual basis.[3] These laws and events set into motion the start of the race for deregulation of the last true monopoly in American business.

EVENTS AFTER PASSING THE LAWS

After the passage of EPAct '92, the world of electricity was facing a new frontier. Wholesale energy trading was now an approved business and the race was on to see exactly what kind of business could be developed. As with any new business, the rules did not cover every possible permutation or scenario and so some of the practices had to be determined through experience. FERC used Order 888 to create the framework for movement of energy over the national electric grid in a way that promoted competition between generators to supply wholesale electric customers (for this discussion, wholesale energy customer includes regulated utilities, municipal utilities, rural electric cooperatives, and the like) based on more capitalistic principles. Order 888 (1996) required that all transmission owners and operators file tariffs stating rates, terms, and conditions for moving power over their transmission lines. Additionally, the order required that the transmission owners and operators charge comparative rates to what they would charge their utility or affiliates, thus offering a level playing field for moving power from one side of the transmission system to the other at standard prices.

Order 888 also set the foundation for the development of regional transmission operators (RTO) to control flow and access to the system. Prior to Order 888, the United States electric system was arranged into three major interconnections (East, West, and the majority of Texas, known as ERCOT). After the blackouts of 1965, the North American Electric Reliability Council (NERC) was created with ten regions dividing the country to address the causes of the blackout and to help prevent future occurrences. The regions developed operating, design, and communication procedures between member utilities and fostered cooperation between utilities to maintain electric grid reliability. These regions were created for traditional utilities and their operation and helped to provide processes and standards of support among utilities using the transmission grid. Under these rules and procedures, each utility owned, maintained, and controlled transmission systems in their franchise area or from their generation source to the load. However, the arrangement did not adequately address nonutility generators, which gave rise to the need for RTOs to help maintain reliability working with the existing utilities and enabling nonutility generators to have access to provide their product to the grid. RTOs would now be the hub for access and delivery of electricity to wholesale customers. The theory of the RTO was that pricing, maintenance, scheduling, development, expansion, and access would be most effectively controlled by one entity rather than allow for the multiple structures of franchise utility controls, as historically was in place. There was also a desire to prevent discriminatory practices to prevent nonutility generators from using the grid to supply alternatives to wholesale customers.

Is the RTO concept successful? The final tally has yet to be determined simply based on the fluid nature of the system. What was developed in the late 1990s is not the same system seen today based on the changes in regulations to adapt to business practices and precedents developed through regulator rulings on disputes and contentions. What is available for review is a comparison to the original goals of EPAct and Order 888, which included increased access to transmission systems by nontraditional generators, transparency in costs, nonpreferential treatment in transmission system use, and maintenance of existing reliability of service. With the exception of price spikes in the Northeast and the Midwest based on market forces and one reliability failure in 2003 for the Northeast, it appears that the wholesale side of electric deregulation has been mostly successful to this point based on these limited benchmarks. Is it a perfect system for "open" access and "full competition"? Not yet. There is much work remaining to make the process more effective and provide increased access, reliability, and control without dramatically increasing costs for wholesale and, in turn, retail consumers. Some of the major issues remaining include interconnection standards and processes, transmission planning, and capacity limitations

WHOLESALE REASONING

Why was there a push to open the wholesale markets? Was the existing system broken? Were customers clamoring for full choice? These seem like simple questions, but in reality, the answers are very complex. The existing system was not broken. Utilities provided reliable and consistent energy supplies to wholesale and retail customers. However, the deregulation of other industries including natural gas, trucking, airlines, and telecommunications fully placed the spotlight on the one major remaining regulated industry. Additionally, international electric industry restructuring, especially in the United Kingdom, offered apparent support for the economic benefits of open markets. There were significant pricing disparities between the coastal regions and the central United States, and conventional wisdom held that a competitive market was, in the long run, the most efficient model for the consumer. These thoughts and others pushed regulators, lawmakers, and customer towards the opening of the markets.

At the time of PURPA in 1978, the country's leaders were searching for ways to reduce the cost of living and operating in the United States. Faced with increasing interest rates, a deepening recession, and a general increase in the cost of goods or services, politicians, regulators, and citizens were looking for ways to reduce costs (and lessen dependence on foreign energy sources—oil for starters). Increasing energy production from sources using combined heat and power looked like an ideal solution to address energy costs while allowing access to the electric grid and not increasing utility regulation or generating companies. It certainly seemed like a perfect first start. However, as with any first concept, there were many hurdles. What was initiated in 1978 needed many years and many lawsuits to find a balance when national events once again pushed legislators to action. This time, the events surrounded the turmoil in the Middle East via armed conflict. Once again, the focus was on the United States' apparent economic exposure to imported crude oil. This coupled with a heightened urgency to address environmental concerns led to the passage of the Clean Air Act Amendments of 1990 and the Energy Policy Act of 1992. Both of these provided an impact to the energy markets: the Clean Air Act started to address emission issues while the Energy Policy Act attempted to build on PURPA and natural gas deregulation.

However, what the regulation or deregulation of the wholesale power market did not address was the creation of a liquid market structure. It was anticipated that the "market" would create a clearinghouse and develop liquidity of price, much like commodity markets for natural gas. The acceptance of electricity as a trade-worthy commodity like oil, wheat, cotton, and natural gas by the financial markets was and is a cornerstone of a deregulated marketplace. In effect, if standard financial tools and techniques are available for large-scale energy trading, the market of supply and demand would set the price of the commodity based on pure economic principles. As with any new commodity market, it takes some time for experience and the application of hedging, derivatives, and other generally accepted trading tools. These tools were needed to help contract negotiations between suppliers and consumers to help provide some risk mitigation. As much as farmers who grow corn can use the financial markets to bracket their financial risk and improve their returns, so too would a municipality or large energy purchaser be able to do the same when buying electricity from suppliers in adjoining regions. The only problem was that electricity did not respond like other established trading commodities. With the inability to store electricity and inconsistent rules across varying transmission systems, there was great potential for supply and demand imbalances. The difficulties with trading electricity became apparent in the summer of 1998 when prices in the Midwest skyrocketed to more than $10,000 per megawatt. As the prices moved higher, contractual defaults started to increase. Suppliers and users were not ready for this type of fast price movement after the years of steady wholesale supply and contractual prices. Other regions were also affected with similar supply and demand imbalance events, and some of these, such as the activities in California, have created court actions to determine if markets were artificially manipulated.

Clearly the wholesale market for electricity moved from infancy to adolescence in a big hurry with all of the usual "heartbreak" one would associate with this growth. On the wholesale side of deregulation, the march continues forward in strengthening RTOs and developing more liquidity in the financial markets for electricity. Owners and operators of transmission systems and system traffic controllers are adapting to the speed and technical concerns to balance supply and demand in an open market. The complexities of a deregulated wholesale electric market have raised the bar on the need for accurate information flow, speed of response, flexibility of resources, and a host of other concerns. In general, the opening of markets has seen an increase in participation via nonregulated generators. The largest question still remains—is the deregulated marketplace better than the regulated version that was replaced? At this time, the opinions on this question remain split and are fully dependent upon issues of personal concern including reliability, delivered cost, access, environmental impacts, economic impacts, security, and many more. Not since telephone deregulation have more people in all stages of the value chain felt an impact of the change of an industry's structure, and the changes will continue as the experience grows. Only time will help to determine the success or failure of wholesale electric deregulation. A return to the previous system seems very unlikely and nearly a physical improbability.

RETAIL CONSTRUCTION

With a wholesale deregulated market driven by federal regulations and laws, the call for retail deregulation of electricity was not far behind (some might even say that the two calls were nearly simultaneous). The initial structure of PURPA and EPAct '92 addressed the concerns of large generators and cogeneration facilities. These large entities had and now have a greater ability of electric self-determination than a small business owner or a residential consumer. This is where the theory of retail deregulation tries to match consumer choice with the end consumer. There are many arguments advanced for deregulation of electricity to the home meter, including lower delivered costs, enhanced customer service, increased quality, faster development of new associated products and services, the reduction in power of the incumbent monopoly, and much more. The reality of course is much different than the theory.

The opening of wholesale markets was dictated by FERC and federal regulations, however, these same regulators and legislators decided not to address the retail customer due to large disparities in geography, existing regulation infrastructure, complexity of rule-making, and the rights of states to make and enforce rules pertaining to their citizens. With these thoughts and the need to focus on wholesale market structure, the federal regulators provided each state the flexibility to address retail service as it deemed. Reasoning that the existing regulatory structure would best know how and when to provide retail electric choice, federal regulators opted to concentrate on transmission and generation rules.[4] What this created was the potential for 50 different scenarios of retail electric competition with varied rules for access, service, reliability, and business practices.

In reality, the resulting "void" of federal mandates on retail competition provided for local choice in participating in the deregulation of electric suppliers. Debates of the issue were very strong in most all state regulatory and legislative arenas, with leadership in opening retail markets to electric competition developing in the coastal states. California, Rhode Island, Massachusetts, and New York were among the first states to open the retail market for electric competition in 1998. Citing high retail energy costs and the potential for dramatic reductions in electricity costs for consumers, markets were opened to energy suppliers. In the ensuing several years, other states followed with regulatory, legislative, and combination open-access rules for the retail electric markets. All total, 18 states have adopted retail electricity restructuring initiatives (California suspended their activities due to major issues in 2003) and another two states have large customer open access.[5] However, more than half of the states have opted to delay or not take any major action in the retail deregulation of electricity. If the opportunities are so good, why did not all states open their electric markets?

This is a large and complex question, but looking at some of the major concerns helps to shed light on decisions that dramatically affect the daily lives of individual and business consumers. Begin by seeking to find the driving factors of electric deregulation in the areas where these actions have taken place. Initial analysis provided an indication that customers could save between 5 and up to 40% in electricity costs if retail open access was available (these price reductions did not account for stranded asset cost recovery). Anecdotal discussions regarding the increased advance of new generation technology, the decreased environmental impact, and the improved customer service were also noted as benefits of customer choice in electricity. However, these points lacked validation from existing electric restructuring

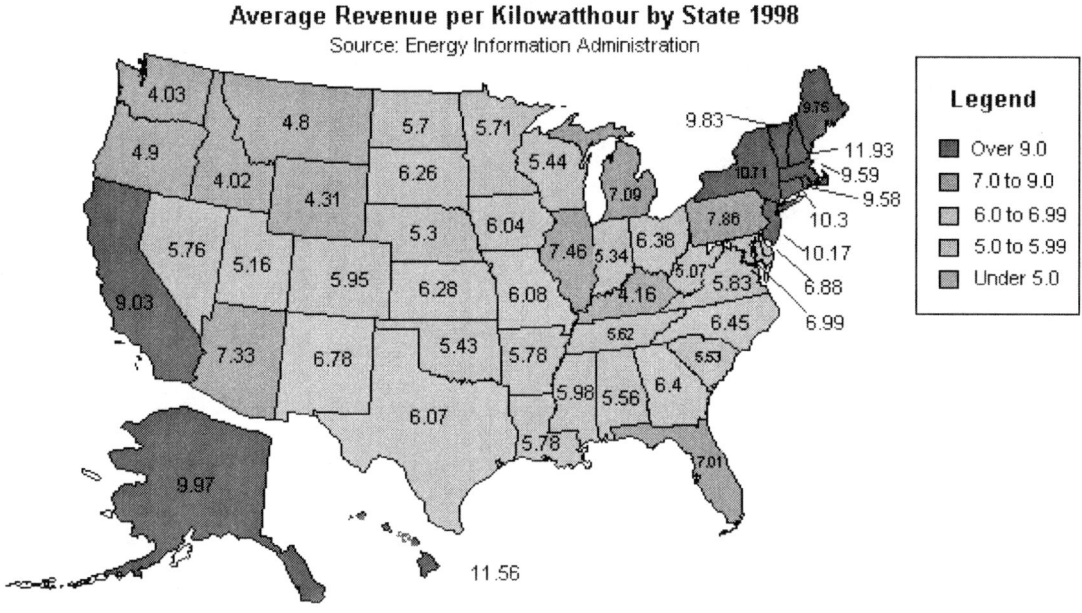

Fig. 1 Average revenue per kilowatthour by state 1998.
Source: From Energy Information Administration, 1998.

activities and appeared to have been interpolated or extracted from other industry's experiences with increased competition. So the remaining (and most prominent) driver of issues in United States business is economic improvement or lower consumer costs. Because this text is limited on space, there will only be a highlighting of each group's concerns, but this should help to define the overall image.

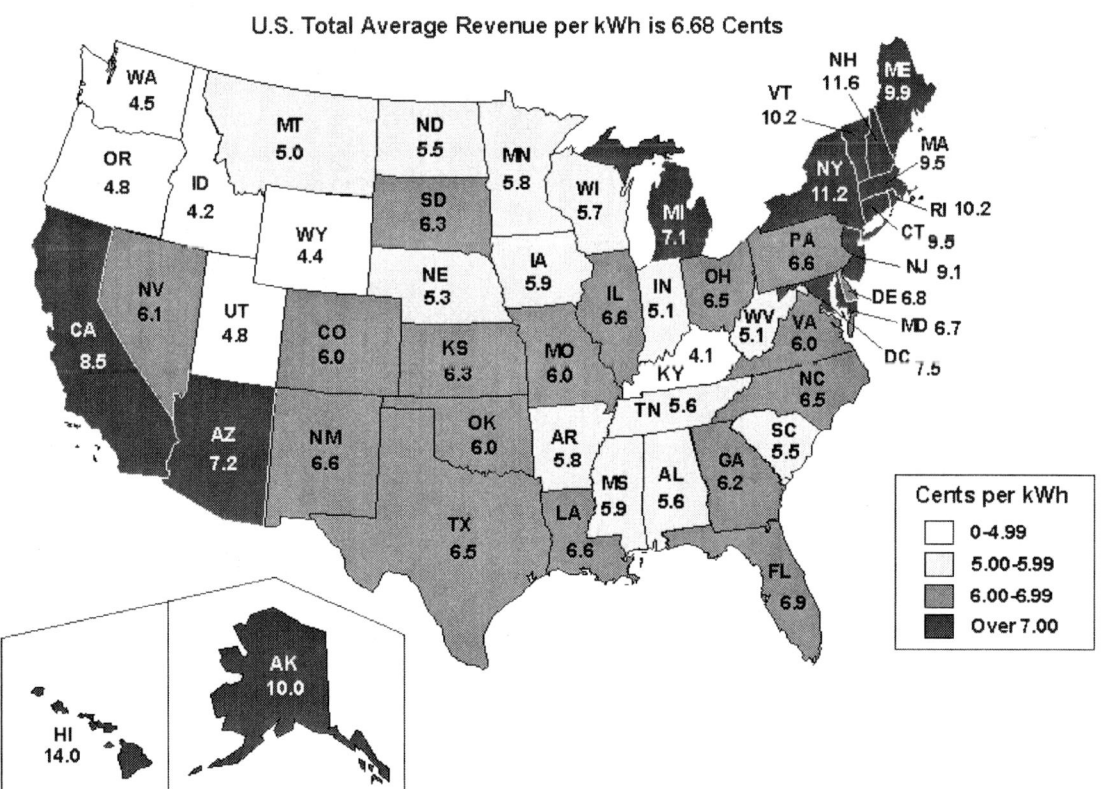

Fig. 2 U.S. total average revenue per kWh is 6.68 Cents.
Source: From Courtesy of Energy Information Administration, 2000.

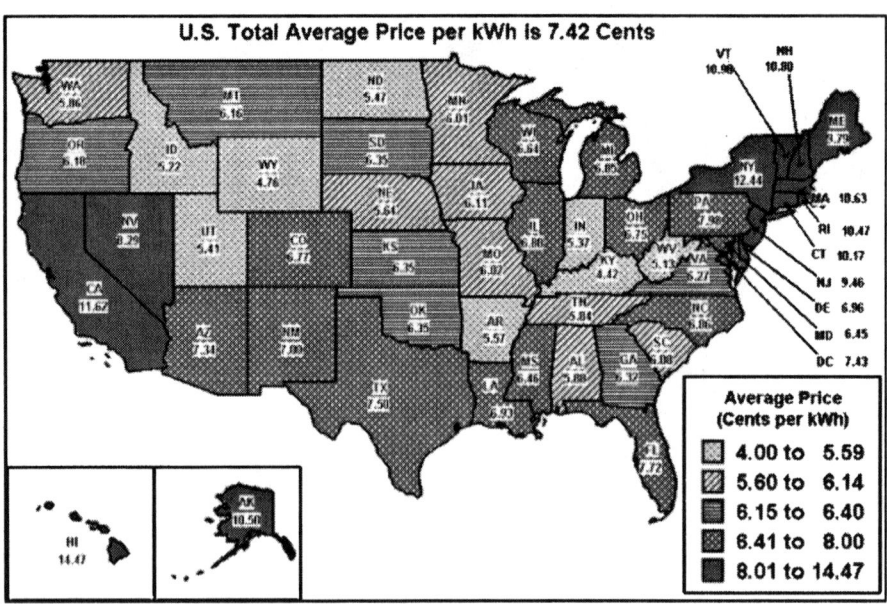

Fig. 3 U.S. total average price per kWh is 7.42 Cents.
Source: From Courtesy of Energy Information Administration, 2003.

THE CONSUMER'S POINT OF VIEW

The consumers of electricity are usually classified into residential, commercial, and industrial users. Of these, typically the industrial users are considered the largest consumers of electricity and thus they are the most prominent players in the economic situation. However, because electricity is also a regulatory/legislative issue, the residential and commercial interests are important in economic and political terms. With this in mind, one can generalize that industrial, commercial, and business clients are more apt to pursue cost reductions in energy to strengthen their bottom lines—they would view energy as any other commodity in their supply chain and use political pressure to advocate their business interests. From a residential economic perspective, lower costs of electricity would create availability of funds to meet basic needs or for discretionary spending and this group would use the court of public opinion to help open access. The majority of the costs savings certainly favor the larger users based on bulk purchasing capabilities; however, small consumers could have the chance to bundle purchasing power in some manner (the Sam's Club effect), but their impacts would be less than that of a single or multiple large business user for economic purposes if simply viewed in the increased complexity of supplying such an aggregate grouping. Additional consumer arguments of increased reliability and improved service based on ability to change suppliers (open market theory) are also very prevalent but not necessarily applicable based on the experience of open markets.

THE REGULATOR/POLITICAL POINT OF VIEW

Tip O'Neil once stated that "All Politics is local" and this is very evident in the actions regarding deregulation of electricity. Federal rules and regulations focused on the large picture of regional transmission, wholesale transactions, large-scale electric generation, and utility interaction. With a commitment in Order 888 that retail or customer choice was the domain of the states, FERC provided a Pandora's box and a golden fleece to each state utility regulator and lawmaker at the same time. The regulators had (and still have) complete control to develop programs that are beneficial to the local population. However, this also left the option of having 50 different programs and with many utilities covering multiple states, the potential for confusion on a grand scale. So what are some of the driving forces on the "local" level? In this case, the driving forces include (but are not limited to) responding to the call from customers to create a "free" market place, attempting to lower energy costs, increasing options in providers, increasing service, allowing once regulated companies to compete without going bankrupt, retaining property taxes from utilities, increasing the economic development capabilities for their areas, retaining the local energy marketplace, and improving the opportunities for the environment. This is a large list of opposing forces, and based on the states, the results and the main focus vary greatly. For example, in California, the main focus was on lowering consumer costs and potentially reducing environmental impacts. In Texas, the programs focus on customer choice and reliability, whereas in New York there is a focus to reduce customer costs (NY had the highest cost electricity provider in the continental United States) and offer options in the spirit of PURPA, of which they were a leading proponent. In other states such as Kentucky and Indiana, the focus is on maintaining the existing low electric rates and economic development opportunities for their communities.

So the regulators and legislators have multiple forces pulling them in many directions. Naturally, these opposing forces are not necessarily interested in the same outcome and the scales of balance on the issue move from one side to another, fully depending upon the individual regulators or political perspective. Looking at the various states and their positions regarding deregulation as well as the reviews of their implementations, one finds many interesting stories. For instance, California—which led the fray into deregulation under the concepts of lower electricity costs, increased open market activity, and reduced long-term environmental impacts—suspended their retail program in 2002 based on the results of the program, which included near bankrupt utilities, massive rolling blackouts, increased wholesale costs, and the like. Not all of these results stemmed from the regulators and political actions to create a deregulated marketplace, but the combination of the type of deregulation, market forces, market manipulation, and increased energy consumption helped to develop a very unstable system which tipped towards collapse. California is attempting to learn from their experience and at present is aggressively pursuing both supply and demand side activities to create balance, and perhaps at some point they might be able to advance their efforts to create an open market for retail customers as well as what exists for wholesale companies. In other states such as Pennsylvania, electric costs before deregulation (1996) and after deregulation have remained stable with reduction in cost for industrial customers of slightly more than 5% (based on EIA data comparing 1990–1995 vs 1996–2003 FSP costs). For Pennsylvania consumers, the regulators and legislators developed a program that appears to work and as the market matures, more benefits may be seen (Figs. 1–3). The following charts illustrate average costs for electricity in the United States. However, changes in costs cannot be solely attributed to deregulation due to the immense complexities of the mix of items that develop an average cost.

THE UTILITY POINT OF VIEW

Deregulation of retail electricity service was a shocking blow to the electric utility industry in that they had approached the business with a long-term (30-year view) based on the systems in place through PUCHA. PURPA created several new issues, such as qualified facility energy production, nonutility interconnects for wholesale power sales, and so on. The progression of FERC Order 888 and the opening of the wholesale market caused an increase in the angst at the utility company. In effect, FERC 888 created a new way to address wholesale concerns from a physical and financial approach for the utilities. Physically, the utilities were used to move power from point one to point two based on some economic activity but mostly based on reliability needs as developed by the NERC control program. Now, the same transmission system was to be used for economic dispatch. System models that were built for reliability now needed to act as a highway. To further the complications, new nonutility generators (NUG) were looking to establish facilities to sell power to areas in which they could potentially create a profit for their shareholders.

The physical challenges were intertwined with financial challenges and the existing structure. Utilities were accustomed to planning on long-term cycles and power plants, transmission systems, distribution systems, staffing, and so forth were determined through a rate-based process using integrated system planning. The infrastructure needs developed from this type of analysis and the allowable rates of returns guided investment decision for regulated utilities. NUGs did not develop or perform business in this way, which set the stage for competition

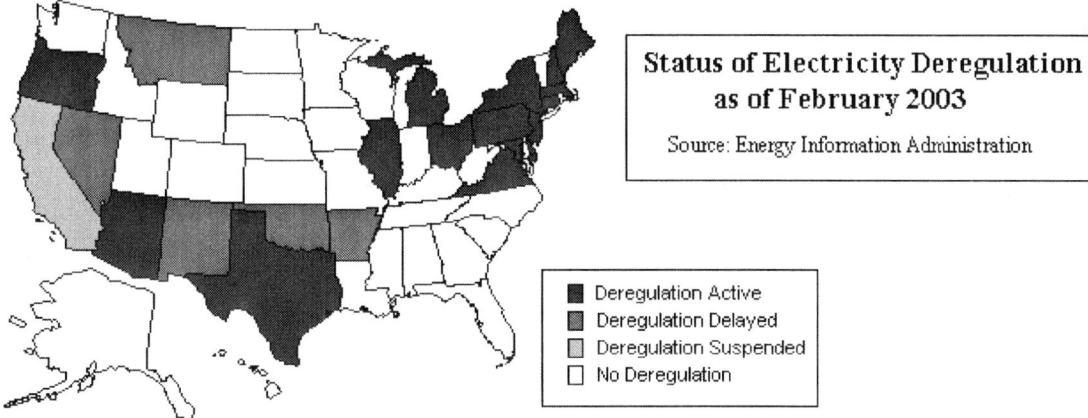

Fig. 4 Status of electricity deregulation as of February 2003. Source: From Energy Information Administration.

among various business models with varying rules. These rule variations lead to the discussion of stranded assets (those assets built by regulated utilities in past years with the understanding the assets would be recovered or paid by the customers over time). When deregulation became a distinct possibility, utilities naturally were worried that investments in infrastructure prudently made under the "old" rules needed to be addressed if customers began to select new suppliers. Some of these concerns were addressed on the wholesale side through FERC but because the majority of issues arose on the state level, each state had to determine the best way to balance the stranded asset issue without front loading prices in a deregulated market. This would in effect increase consumer prices in the short run if stranded assets would be recovered in a short period of time. Other issues concerned billing of customers as well as providing service to credit risk consumers. Electric utilities have a requirement to serve customers with highly regulated disconnect or refusal of service processes. These rules would not necessarily be applied to unregulated competitors and thus they would conceivably leave the highest risk customers with the incumbent utility.

These were just some of the concerns of each stakeholder in the deregulation story. Currently, there are 16 states with active programs and another five states that have delayed implementation but are ready to start once specific concerns are addressed. California has suspended their deregulation experiment based on the many issues and the existing public debt created by the original program. Their actions in attacking the demand side portion of the energy will solve one of the concerns to help balance the energy equation. The initial deregulation design did not offer any consumer triggers to reduce consumption based on price signals or supply/demand imbalances, which helped to drive up costs and, in part, increase the probability of demand out-stripping supply. In this area of demand control, California is showing national leadership and will increase the likelihood of future resumption of deregulation. The majority of states have decided to continue to study deregulation or take no action. Their existing price and supply structure shows little or no consumer benefits to pursue open markets. As energy prices increase and supply tightens, even these markets may open to a more competitive environment, but the shock of events surrounding other unsuccessful actions will increase the threshold for action.

CONCLUSION

Deregulation of electricity comes in wholesale and retail flavors with distinctly different goals and results. (Fig. 4) Wholesale open access started with PURPA in 1978 and marched forward with FERC 888 and EPAct 1992 and continues through state retail access programs. So is deregulation of electric markets beneficial to consumers? It depends on who is asked and their perspective of the situation. In most cases (based on information from EIA), the consumers have seen some benefits with increased complexity of the offering. This appears equivalent to the scenario of long distance telephone competition in which consumers are inundated with varying offers. Through scrutiny by state regulators, these electric offerings are more controlled and thus not as likely for confusion. The story of electric deregulation is far from complete and many chapters are yet to be written. Lessons learned from unsuccessful as well as successful programs will help guide regulators and legislators to find balance between consumers, utilities, and electric providers. Based on other previously competitively regulated industries, competition has proven some benefits. The stakes with electricity are very high because this country is so dependent on electricity and this major fact warrants extreme caution in program design and implementation. So far, caution has been applied and the future for competitive electricity is waiting for a technology break-through to help ease the transition. Only time will tell.

REFERENCES

1. Christensen, P.C. *Retail Wheeling—A Guide for End-Users*, 3rd Ed.; Pennwell Publishing: Tulsa, OK, 1998.
2. P.L. 95–617, Public Utilities Regulatory Policy Act of 1978, www.thomas.loc.gov.
3. P.L. 102–486, Energy Policy Act of 1992, www.thomas.loc.gov.
4. *Commission Orders Sweeping Changes for Electric Utility Industry, Requires Wholesale Market to Open to Competition*, Convergence Research, 1996, www.converger.com/FERC-NOPR/888_889.htm.
5. Status of Electricity Competition in the States, Edison Electric Institute, 2003, www.eei.org/industry_issues/electricity_policy/state_and_local_policies/state_restructuring_and_regulatory_policy/Competition_states_map.pdf.

BIBLIOGRAPHY

1. *Electricity Prices in a Competitive Environment: Marginal Cost Pricing of Generation Services and Financial Status of Electric Utilities. A Preliminary Analysis Through 2015*, Energy Information Administration, 1997, DOE/EIA—0614.

Electricity Enterprise: U.S., Past and Present

Kurt E. Yeager
Galvin Electricity Initiative, Palo Alto, California, U.S.A.

Abstract
This article provides a brief historical synopsis of the development of the electricity enterprise in the United States. The end of this article includes a reference list of more detailed accounts of the electricity enterprise, on which this summary is based.

INTRODUCTION

The organization of this synopsis corresponds with the stages in the Electricity Sector Life-Cycle shown in Fig. 1. The first three stages of the life-cycle are discussed in this article. A second article discusses the future of the U.S. Electricity Enterprise.

Electricity now powers American life to an unprecedented degree. In 1900, electricity's share of the nation's energy use was negligible. That share has now risen to nearly 40%. Electricity is produced from fuels through a costly conversion process so that its price per thermal unit has always been higher than that of the fuels themselves. So, something other than cost must account for the sustained growth in electricity's market share. Simply put, electricity can be used in ways that no other energy form can. Technological progress over the past century has led to radically improved ways of organizing productive activities as well as new products and new techniques of production, all of which have been heavily dependent on electricity. As a result, electricity has become the lifeblood of the nation's prosperity and quality of life. In fact, the U.S. National Academy of Engineers declared that "the vast networks of electrification are the greatest engineering achievement of the 20th Century."

Fig. 2 summarizes the historical trends in the energy sources for U.S. electricity generation. Coal is notable in its persistence of as the dominant fuel.

Electricity, despite its mystery and complexity, is simply the movement of electrons. Each of these tiny sub-atomic particles travels only a short distance as it displaces another electron around a circuit, but this transfer occurs at the speed of light (186,000 mi per second). This invisible wonder occurs virtually everywhere in nature. For example, it transmits signals from our brains to contract our muscles. What's relatively new is our ability to put electricity to work lighting and powering our world.

For example, steady advances during the course of the 20th century improved electric lighting efficiencies a great deal. Looking forward, full-spectrum light-emitting diodes (LEDs) may increase the efficiency of U.S. lighting by 50% within 50 years. At the same time, electric motors have revolutionized manufacturing through unprecedented gains in the reliability and localized control of power vis-à-vis steam engines. By 1932, electric motors provided over 70% of all installed mechanical power in U.S. industries. The proliferation of household appliances has also been primarily due to the use of small electric motors. Today, the ubiquitous electric motor in all its forms and sizes consumes two-thirds of all U.S. electricity production.

Electricity is indeed a superior energy form; however it is not a tangible substance, but rather a physical effect occurring throughout the wires that conduct it. Electricity must be produced and consumed in absolutely instantaneous balance and it can't be easily stored. Its delivery, therefore, today requires the ultimate just in time enterprise that balances supply and demand at literally the speed of light. Yet the status quo suffers numerous shortcomings. Efficiency, for instance, has not increased since the late 1950s, and U.S. generators throw away more energy than Japan consumes. Unreliable power—the result of blackouts or even just momentary surges and sags—costs America more than $100 billion annually. This is equivalent to about a 50¢ surcharge on every dollar of electricity purchased by consumers.

Moreover, the U.S. bulk electricity infrastructure is aging and becoming obsolescent. The average generating plant was built in 1964 using 1950s technology, whereas factories that construct computers have been replaced and updated five times over the same period. Today's high-voltage transmission lines were designed before planners ever imagined that enormous quantities of electricity would be sold across state lines in competitive transactions. Consequently, the wires are often overloaded and subject to blackouts. Yet demand is increasing at twice the rate of capacity expansion. Finally, the local distribution

Keywords: Electric enterprise; Restructuring; Electric system infrastructure; Electric system planning; Electric system history; Electrification; Electric transmission system planning; Grid system; Electric system transformation.

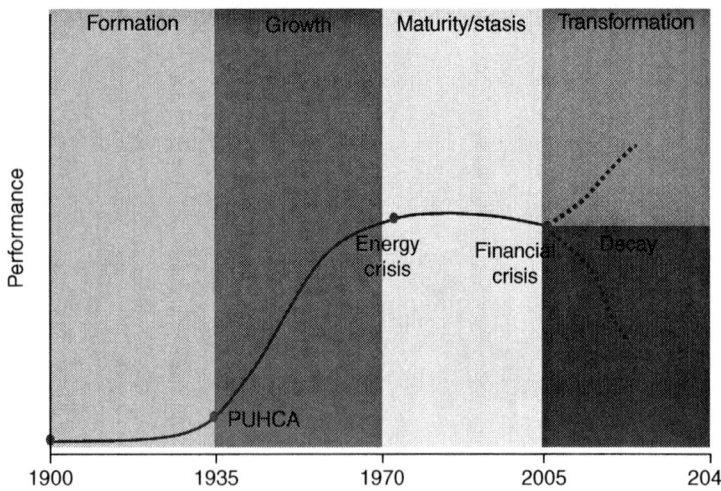

Fig. 1 Electricity sector life-cycle—a fork in the road.

systems that connect the power supply to each consumer are effectively a last bastion of analog, electromechanically controlled industry. This is a particularly notable paradox given the fact that the nation's electricity supply system powers the digital revolution on which much of the current and future value depends. Keeping the lights on 99.97% of the time is simply not good enough. That still means the average consumer doesn't have power for 2.5 h a year. In today's impatient, increasingly computerized world that is more than just a nuisance.

In spite of these deficiencies, the traditional producers and deliverers of electricity—the nation's electric utilities—hold assets exceeding $600 billion; with 70% invested in power plants, 20% in distribution facilities, and 10% in transmission. They form one of the largest industries in the United States—roughly twice the size of telecommunications and nearly 30% larger than the U.S. automobile industry in terms of annual sales revenues. The Achilles' heel here is the fact that supplying electricity is also extremely capital intensive, requiring far more investment per unit of revenue than the average manufacturing industry. This investment challenge is further intensified by the fact that the U.S. electricity enterprise is made up of over 5000 commercial entities, both public and private. The largest individual corporate market cap in the enterprise today is that of Exelon at $32 billion. This compares with Exxon-Mobil, for example, with a market cap of $365 billion. In fact, only about 17 electric utilities have market equity value greater than $10 billion. As a result, the decision to invest the billion or more dollars needed to construct a major new power plant or power delivery (T&D) line is effectively an uncertainty-laden, long-term, "all or nothing" decision that is constantly avoided by most electric enterprise corporations today.

Meanwhile, the market for portable electric devices continues to grow dramatically from the traditional flashlights and auto ignitions to a diverse array of

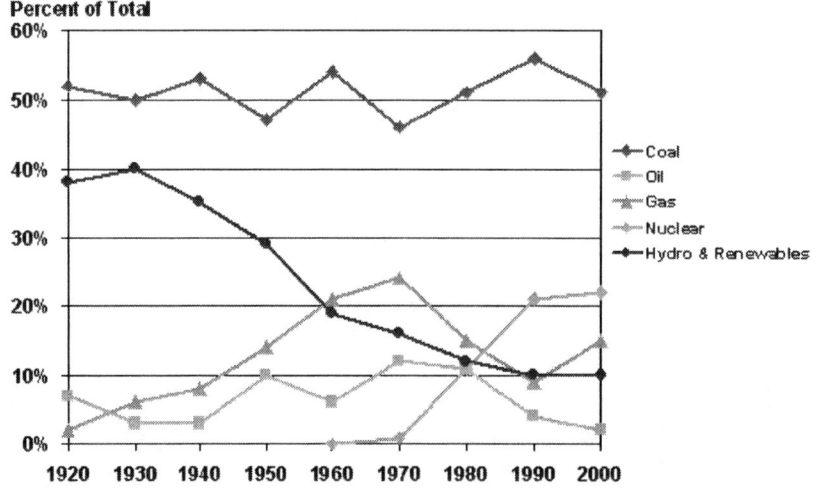

Fig. 2 Sources of energy for U.S. electricity generation.

computers, communications and entertainment products, cordless tools, medical devices, military products, etc. This innovative diversity has been accomplished by exploiting the synergy between the products themselves, the electricity storage devices they employ—including batteries, ultracapacitors, and fuel cells—and the power-management systems that charge these storage devices. Today, the global portable electricity storage market is about $50 billion per year, of which, $5 billion is allocated to rechargeable (secondary) batteries. Demand growth in this market is estimated at 6% annually, compared to grid-supplied electricity demand growth of about 1.5% per year.

A new generation of energy-hungry devices, such as digital cameras, camera phones, and high-performance portable computing devices, is expected to continue to drive this rapid growth in portable electrification. Notably, the kWh-equivalent price of portable electricity can be as much as $100, compared to about 10 cents for grid-supplied power. This is one indication of the potential for highly flexible, individualized electricity services to enhance electricity's value beyond the commoditized energy value proposition and business model of the traditional electricity enterprise.

The history of this magical energy form, electricity, provides keen perspective on the real-world interplay among technical progress, business struggles, and political debates. Electricity's ongoing evolution also suggests how the potential for renewed electricity-based innovation could curtail pollution and spur a wide array of electro-technical advances, while continuing to improve the U.S. quality of life, and that of billions more worldwide.

> Electrification is not an implacable force moving through history, but a social process that varies from one time period to another and from one culture to another. In the United States electrification was not a thing that came from outside society and had an impact; rather, it was an internal development shaped by its social context.

Table 1A and B summarize some recent basic statistics for the U.S. electricity enterprise.

FORMATION

Neither electricity nor electric lighting began with Edison. In 1808, Sir Humphrey Davey sent a battery-powered electric current between two carbon rods to produce an arc of light. In 1831, Michael Faraday invented the dynamo, which, when turned by a steam engine, supplied a cheap electric current by means of electromagnetic induction. Davey's arc lamp was used in a production of the Paris Opera in 1844 and was part of the Philadelphia Exposition in 1876. The development of the high vacuum Crookes tube served to rekindle interest in incandescent electric lamps that had begun in the 1820s.

Edison, with a characteristic vision that distinguished him from his competitors, worked not only on the incandescent lamp but on the entire system that powers the lamp. "The same wire that brings the light will also bring power and heat—with the power you can run an elevator, a sewing machine, or any mechanical contrivance, and by means of the heat you may cook your food." Scientists and rival investors predicted failure. In 1879, Edison had a working incandescent lamp and, within months, patented a direct current (DC) electric distribution system.

The next question was: who would buy the lights and equipment? Edison conceived the central power station that would distribute electricity through lines to the customers. On September 4, 1882, Edison's Pearl Street Station in New York City went into business to serve 85 customers with 400 lamps. This marked the beginning of the electric utility industry. Unfortunately, with Edison's low-voltage DC system, it was too expensive to distribute electricity more than a mile from the power plant. Transformers that could raise and lower voltage did not work with direct current.

In 1888, Nicola Tesla, who had been previously employed by Edison, announced his polyphase alternating current (AC) power system. That same year George Westinghouse bought the rights to Tesla's system. Westinghouse saw the potential for locating a central station at the source of water power or coal, shipping the power for great distances at high voltages, and then stepping down the power for distribution. But inventing a practical AC system created new problems. AC and DC systems could not be linked until Charles Bradley, another former Edison worker, invented the rotary converter in 1888 which converted DC to AC. Westinghouse also bought Bradley's idea.

The Westinghouse engineers developed a universal system in which the polyphase AC generator at the central station produced electricity that went to a local substation where it was transformed to the voltage required by the user. This system had many advantages. First was a realization of economies of scale in power generation. The second was the need for only one wiring grid. The third was that the generating stations could serve a wider area. The fourth was that the new system's productivity could benefit from load diversity; e.g., lighting in the evening, streetcars during rush hours, and factory motors during the periods in between. Interestingly, the introduction of practical electric streetcars in the late 1880s provided the major concentrated electricity demand that dramatically pushed the enterprise toward more powerful equipment and larger service areas favoring AC.

In little more than a decade, Edison had put more than a half century of research into practical applications, conceived and invented an entire industry, and then

Table 1A U.S. electricity industry statistics (2002)

Generating capacity and generators	Capacity (1000 MW)	Net generation (billion kWh)
Investor-owned utility	398	1740
Government and cooperatives	204	807
Non-utility (unregulated producers)	380	1294
Total	982	3841

Total net generation by fuel	(Billion kWh)	Percentage
Coal	1926	50
Nuclear	780	20
Natural gas	695	18
Hydroelectric and pumped storage	253	7
Fuel oil	92	2
Biomass	72	2
Geothermal	13	1
Wind	9	
Photovoltaic	1	
Total	3841	100

Electricity sales	Customers (million)	Energy (billion kWh)
Residential	115	1267
Commercial	15	1122
Industrial	0.6	973
Other	1	109
Total	132	3471

Revenues	Sales ($billion)	Percent total electricity energy (%)	Average/kWh (cents)
Residential	107	37	8.4
Commercial	89	32	7.9
Industrial	47	28	4.8
Other	7	3	6.6
Total	250		Average price 7.2

Financial	($Billion)
Total assets	598
Total operating revenues	250
Operating expenses	220
Operating income	30
Construction	25

Source: From Edison Electric Institute (see Ref. 1).

became a reactionary who threatened to stagnate the industry at in its primitive stage of development. In 1892, Edison's financier, J.P. Morgan, stepped in and forced a merger with Thompson-Houston and put their management in charge of Edison's General Electric Co. In 1896, General Electric and Westinghouse exchanged patents, a typical move in the age of trusts, so even General Electric used Westinghouse concepts. The age of Edison had ended.

By 1892, Samuel Insull of Chicago Edison, another former Edison associate, had formulated an understanding of the economics of the electric utility business that was

Table 1B U.S. consumption of electricity (2001)

A. Residential	Billion kWh	Percentage
Air conditioning	183	16
Refrigerators	156	14
Space heating	116	10
Water heating	104	9
Lighting	101	9
Ranges and ovens	80	7
Laundry	76	7
Color TV, VCR/DVD, Stereos	55	5
Freezers	39	3
Furnace fans	38	3
Dishwashers	29	2
Personal computers and communication	28	2
Pools and hot tubs	17	2
Other[a]	118	10
Total	1140	100

B. Commercial	Billion kWh	Percentage
Space cooling	288	26
Lighting	255	23
Office equipment/computing	200	18
Refrigeration	100	9
Ventilation	78	7
Space heating	56	5
Cooking	22	2
Water heating	11	1
Other	100	9
Total	1110	100

C. Manufacturing	Billion kWh	Percentage
Machine drive	512	53
Process heating	104	11
Electro-chemical	86	9
HVAC	82	9
Process cooling and refrigeration	63	7
Lighting	62	6
Other	61	6
Total	970	100

[a] Composed of about 15 additional consumption categories, each representing less than 1% of residential electricity consumption.
Source: From U.S. Energy Information Agency (EIA), Energy Consumption Survey.

sustained through most of the 20th century. When Insull took over the Chicago Edison Co. in that year, it was just one of 20 electric companies in the city. Although Chicago had a population of more than one million, only 5000 had electric lights. He vowed to serve the entire population. Insull and other leaders of the Association of Edison Illuminating Companies (AEIC) realized that the industry had high fixed costs because of the infrastructure investment needed. At the same time, the cost of operating the plants was fairly low. The question was how to translate that into profits, especially in an industry that had concluded it was selling a luxury item.

Insull began a sales campaign, cut prices as necessary to get customers, and wrote long-term contracts for large customers. He utilized a demand meter (invented in England) and set the price of electricity to cover both fixed and operating costs. Insull also concluded that profits were maximized by keeping the power plant running as much as possible to exploit the diversity of load. As a result, the U.S. led the world in the rates of electrification. Insull in Chicago sold more electricity per capita, ran larger power stations, kept the plants running longer each day, and charged customers less. Insull also discovered that there were clear advantages to tying together urban and rural loads. For example, Chicago had a winter peak and the farm towns a summer peak.

By 1911, thanks to the development of the ductile metal filament lamp, electric lighting ceased to be a luxury, manufacturers developed new uses (e.g., refrigerators and sewing machines), and the demand for electricity skyrocketed. Also, Charles Parsons had recognized the limits of the reciprocating steam engine in 1884, and developed the steam turbine that produced rotary motion directly as high-pressure steam pushed against blades attached to a shaft. This elegantly simple machine occupied one-tenth the space, and cost one-third as much as the reciprocating engine of equivalent capacity. By 1911, 12,000 kW turbine generators also became the norm. Thus, the keys to the success of the traditional, declining cost commodity, grow-and-build electric utility business model were established, i.e., economies of increasing scale, rapidly rising consumer demand, and consumer diversity for load stabilization and higher capacity factors.

However, during much of this era of rapid sales growth and technological progress, electric utilities managed to earn unspectacular profits. This was addressed by consolidating the over-fragmented industry into everlarger holding companies. Centralized ownership served to facilitate raising money and engineering the best systems. Also non-utility, industrial, electricity generation declined from more than 50% of the U.S. total as late as 1914 to 20% by 1932. Although states regulated the operating subsidiaries that sold electricity, none regulated the holding companies. By 1932 the eight largest holding companies controlled 73% of the investor-owned electric businesses. The Insull empire alone operated in 32 states and controlled at least a half billion dollars in assets with an investment of only $27 million. As a result of these excesses committed, the electricity holding companies were condemned in the wake of the Depression, and controlling legislation was passed that created the present structure of the electric utility industry.

Under this legislation, interstate holding companies had to register with the SEC. This included any company that owned 10% or more of the voting securities of an electric or gas utility. The Act also broke up holding company systems that were not contiguous and eliminated intermediate holding companies from the financial structure.

Table 2 Electrification of the U.S. economy

	1902	1932
Percentage of population in electric-lighted dwellings	2	70
Percentage power in industry (horsepower equivalent)	5	73
Average power plant size (MW)	0.5	8.5
Electricity generation (10^9 kWh)	6	100
Residential service price (¢ per kWh—1983$)	~40	15

This Public Utility Holding Company Act of 1935 (PUHCA) also effectively marked the end of the formative period of the U.S. electricity enterprise. Table 2 summarizes the rapid progress of the electricity enterprise during this formation period.

GROWTH

In 1932, Franklin Roosevelt denounced the "Insull monstrosity" and proposed four Federal hydropower projects—The St. Lawrence, Muscle Shoals, Boulder Dam, and the Columbia. "Each of these in each of the four quarters of the United States will be forever a national yardstick to prevent extortion against the public and to encourage the wider use of that servant of the people—electric power." In the same general time frame, Lenin also underscored the universal impact of electricity by declaring that "Communism equals the Soviet power plus electrification."

Even during the Depression and through World War II, the U.S. electric utility industry continued to expand and to cut its costs. Government-supported entities, such as the Rural Electrification Administration, brought electricity to the farms. Although investor-owned utilities lost territory to governmentally owned (public power) utilities, the most significant change was the devolution of operating control from holding companies to locally operated utilities. These were incented to concentrate on customer service rather than on complex financial frameworks. Between 1935 and 1950, 759 companies were separated from holding company systems, and the number of registered holding companies declined from over 200 to 18.

During this period of industry consolidation and growth, utilities desired three features from new technology: reliability, greater power at lower costs, and higher thermal efficiency. As a result of this demanding market, manufacturers initially developed their new machines using a "design-by-experience" incremental technique. As with many other engineering endeavors then, the people who built these complex machines learned as they went along. This was reflected by steady increases in steam pressure and temperature in boilers and generators, providing corresponding improvements in thermal efficiency. Steam temperature and pressure in 1903 were typically 530°F and 180 PSI respectively. By 1930, water-cooled furnace walls permitted the production of steam at 750°F and upto 1400 PSI. By 1960, these parameters had increased to 1000°F and 3000 PSI, turning water into dry, unsaturated, supercritical steam, effectively exploiting the full potential of the Rankine steam cycle.

Improvements in transmission systems also occurred incrementally during the power industry's first several decades. While comprising a relatively small portion of a power system's total capital cost, transmission systems nevertheless contributed significantly to providing lower costs and more reliable service. They did this by operating at ever-increasing voltages and by permitting interconnections among different power plants owned by contiguous power companies. Transmission voltage increased from 60,000 V in 1900 to 240,000 V in 1930 and upto 760,000 V by 1960. Increased voltage, like higher water pressure in a pipe, allows more electricity to pass through a transmission wire. Doubling the voltage, for example, increased a line's volt-ampere capacity by a factor of four. In short, the development of high-voltage transmission systems contributed as much to the steady increase in capacity of power production units as did advances in turbine speed or generator-cooling techniques.

U.S. energy consumption grew in lock-step with the economy after World War II, but electricity sales rose at double that rate until about 1970. The result, over the period of 1935–1970, was an 18-fold increase in electricity sales to end users with a corresponding 12-fold increase in electric utility revenues. This growth was stimulated by the dramatic and continuing drop in the real price of electricity, compared to other fuels. Much of the success in reducing costs was due to these continued improvements in the generating process and in higher voltages and longer distance transmissions, which together more than offset the impact of the break up of the holding companies. Over the post-war (1945–1965) period, the average size of a steam power plant rose fivefold, providing significant economy-of-scale advantages.

Efficiency improvements (heat-rate) did not, however, keep pace after the late 1950s, even with higher operating steam temperatures and pressures. The inherent limitations of the Rankine steam cycle coupled with metallurgical constraints caused this efficiency plateau. Electricity distribution system expense per customer year also increased over 100% during this period, from $8 to $17.

After World War II, the accelerated growth of the industry caused manufacturers to modify their incremental, design-by-experience approach to one of "design-by-extrapolation." This enabled manufacturers to produce larger technologies more rapidly. The push for larger unit sizes reflected the postwar U.S. economic prosperity and the introduction of major new electricity uses including air conditioning, electrical space-heating, and television. The all-electric home loomed large on the horizon. The biggest concerted promotional push began in 1956 with the "Live Better Electrically" campaign employing celebrities such as Ronald Reagan, on the heels of the very successful "Reddy Kilowatt" mascot for modern electric living.

While the best steam turbine generating units only improved in thermal efficiency from 32 to a 40% plateau in the postwar period, turbine unit sizes jumped from about 160 MW in 1945 to over 1000 MW in 1965 through the design-by-extrapolation approach. Correspondingly, new plant construction costs declined from $173/kW in 1950 to $101/kW in 1965. Between 1956 and 1970, utilities operated 58 fewer plants to produce 179% more electricity. As regulated monopolies, electric utilities could not compete with each other for market share, but competition existed during this period as engineer-managers strived for technical leadership among their peers. This type of competitive environment contributed to rapid technological advances and production efficiencies. Utility managers encouraged manufacturers to build more elegant technology so they could get credit for using the "best" machines. The risks to gain customized technological supremacy often meant an economic tradeoff, but this was a price readily paid in the 1950s and 1960s by utility managers who retained their engineering values and goals as they became leaders of large business enterprises.

During this period of rapid expansion and success, a third participant—in addition to electric utilities and manufacturers—played a largely invisible supporting role. This third party consisted of the state regulatory bodies, which performed two tasks relative to the electricity enterprise. First, they protected the public from abusive monopoly practices while assuring reasonably priced, reliable utility service. Second, they guaranteed the financial integrity of the utility companies. Conflicts rarely arose because utilities were steadily reducing their marginal costs of producing power, and they passed along some of these savings to consumers. Thus, few people complained about a service where declining costs countered the general trend toward cost-of-living increases. Regulatory actions also tended to reinforce the industry's grow-and-build strategy by permitting utilities to earn a return only on capital expenditures. This "social contract" served the industry and its stakeholders well for more than half a century providing a robust, state-of-the-art infrastructure. Although not articulated at the time, these stakeholders had forged an implicit consensus concerning the design, management, and regulation of a national technological system. As long as benefits continued to accrue to everyone, the consensus remained intact.

For electric utilities, this consolidation and growth period was, in summary, one of reorganization out of the holding companies, minimal need for rate relief, declining costs and prices, an average doubling in electricity demand

Table 3 Electricity enterprise growth

	1932	1950	1968
Ultimate customers (million)	24	43	70
Net generation (10^9 kWh)	100	389	1,436
Installed generating capacity (10^3 MW)	43	83	310
Average power plant size (MW)	8.5	18	85
Circuit miles of hi-voltage line[a] (10^3 mi)	NA	236	425
Residential service price (¢ per kWh—1983$)	15	10	7.1

[a] 22,000 V and above.

every decade, incentives to add to the rate base, satisfied customers and investors, and acceptable returns for owners. That environment of few operating problems and little need to question the prevailing regulatory structure left the electricity enterprise and its stakeholders unprepared to either anticipate, or respond quickly to, the challenges that rapidly followed.

Table 3 summarizes the progress of the electricity enterprise during this period of growth and consolidation.

MATURITY AND STASIS

By the mid 1960s, the electricity enterprise and its stakeholders were beginning to experience the first cracks in the traditional business model and its associated regulatory compact. The fundamental concepts of the enterprise began to be challenged and investment started to erode. The most notable initial event was the November 9, 1965 Northeast Blackout that spread over 80,000 mi² affecting 30 million people. This, and other outages that followed, forced utilities to redirect expenditures from building new facilities to improving the existing ones. Specifically, they had to upgrade the fragile transmission and distribution (T&D) system in order to handle larger power pools and more frequent sales among utilities. These new costs led to higher rates—for the first time, literally, in decades—and despite expensive public relations efforts, the public grew increasingly critical of utility monopolies.

1967 marked a second major turning point for the U.S. electricity enterprise—generation efficiency peaked. Rather than lower the average cost of electricity, a new steam power plant would henceforth increase it. Economies of scale ceased to apply (bigger was no longer necessarily better or cheaper) and continued expansion in the traditional manner no longer held the same consumer benefits. The grow-and-build strategy had seemingly reached the end of the line. A third turning point was Earth Day in 1970. This launched environmental activism and focused fresh attention on electric utilities, ultimately leading to further investment redirection for environmental control equipment, most notably for sulfur dioxide scrubbing on the industry's fleet of coal-fired power plants. The Clean Air Act of 1970 made environmental concerns an integral part of the utility planning process while planning for growth became more difficult.

The fourth major turning point event for the electricity enterprise was the Oil Embargo of 1973. OPEC's actions led to a rapid rise in the cost of all fuels, including coal. Accelerated inflation and interest rates resulting from the Vietnam War economy also led to higher borrowing rates for utilities. The sum of these turning point issues led to ever higher electricity prices, reducing the growth of U.S. electricity sales in 1974 for the first time since World War II. Consolidated Edison missed its dividend and utility stock prices fell by 36%, the greatest drop since the Depression.

In spite of these troubling events, the electricity enterprise's commitment to growth was slow to respond. In 1973 electric utilities issued $4.7 billion in new stock, almost seven times that sold by all U.S. manufacturing companies combined. Finally in 1975, capital expenditures declined for the first time since 1962. The traumatic decade of the 1970s concluded with perhaps the most strategically serious turning point issue of all—three mile Island. On March 28, 1979, a cooling system malfunction at the 3 mi Island nuclear plant in Pennsylvania destroyed public and political confidence in nuclear power, which had been seen as a technological solution to restoring the declining commodity cost and financial strength of the electricity enterprise. This event fell immediately on the heels of the nuclear accident-themed movie, *The China Syndrome,* and seemed to validate nuclear power plant risks in the public mind. Although the lack of any core meltdown or even radiation leakage was testament to the quality and integrity of nuclear power plant design and construction, the demand for stricter safety regulations led to rapid cost escalation.

The first commercial nuclear power unit built in 1957 had a rating of 60 MW. By 1966, utilities were ordering units larger than 1000 MW, even though manufacturers had no experience with units larger than 200 MW at the time. This arguably over-aggressive design-by-extrapolation, plus uneven utility operations and maintenance (O&M) training and management, led to reactor cost overruns that were sending power companies to the brink of bankruptcy while average power prices soared 60% between 1969 and 1984. Utilities and manufacturers in 1965 predicted that 1000 reactors would be operating by 2000 and providing electricity "too cheap to meter." The reality was that only 82 nuclear plants were operating in

2000, and no new U.S. orders had been placed in two decades.

These issues were profoundly impacting the electricity enterprise in the 1980s. The very ways electricity was generated and priced were being challenged for the first time in nearly a century. No longer could planners count on a steady rise in electricity demand. No longer could utilities count on low-cost fuels or the promise of the atom. They could no longer construct larger and more efficient generation, nor could they avoid the costs associated with environmental emissions. Competition from innovative technologies and hustling entrepreneurs could no longer be blocked, and the long-standing social contract consensus among the stakeholders of the electricity enterprise began to unravel.

The push for open power markets started when energy-intensive businesses began demanding the right to choose their suppliers in the face of rising electricity prices. Recognizing this pressure, the Energy Policy Act of 1992 also greased the skids for greater competition. The Act let new unregulated players enter the electricity generation market and opened up the use of utilities' transmission lines to support wholesale competition. The deregulation of natural gas in the 1980s made gas a more available, affordable, and cleaner fuel for electricity generation. This, coupled with rapid advancements in aircraft-derivative combustion turbines, provided an attractive vehicle for new, independent power producers to enter the market with low capital investment. Notably, non-utility sources as late as 1983 supplied only 3% of the U.S. generation market. By 2003, however, unregulated non-utility generators had captured nearly 30% of the U.S. generation market, exceeding the combined share from rural coops, the federal government, and municipal utilities. Wholesale electricity trading also soared—from approximately 100 million kWh in 1996 to 4.5 billion kWh in 2000.

Between 2000 and 2003, about 200,000 MW of new natural gas-fired combustion turbine capacity were added to the U.S. electricity generating fleet. Sixty-five percentage of this new deregulated generating capacity utilizes a combined-cycle technology. However, the rate of new combustion turbine-based capacity addition has dropped off dramatically since then. In addition, the performance of this new capacity has suffered in terms of both heat-rate and capacity factor. These are all symptoms of the boom-bust cycle in power generation that now exists in the restructured industry. For example, the average capacity factor for the fleet of new combined-cycle power plants dropped from 50% in 2001 to below 30% in 2004 as electricity supply capability significantly exceeded demand, market access was physically limited by transmission constraints, and natural gas prices rose dramatically.

The 21st century has not begun well for the performance and integrity of the U.S. electricity enterprise. The biggest power marketer, Enron, collapsed amid scandal while facing a slew of lawsuits. Pacific Gas and Electric, one of the largest investor-owned utilities, filed for bankruptcy amid the chaotic power markets in California. 50 million people in the Northeast and Midwest lost electricity because of a cascading power failure in 2003 that could have been prevented by better coordination among utility operators. High natural gas prices and the U.S. economy's overall slowdown caused electricity demand to falter and wholesale prices to fall. As noted, the natural gas-fired combustion turbine boom collapsed, carrying many independent generation companies with it.

The end result was a $238 billion loss in market valuation for the electricity enterprise by early 2003 and the worst credit environment in more than 70 years. Those companies able to maintain good credit ratings and stable stock prices bought nearly $100 billion in assets from weaker firms. Since the Energy Policy Act of 1992, competitive electricity generators have been able to charge market rates, while the transmission and distribution sides of the enterprise have remained regulated with relatively low investment returns. As a result, more power is being produced, but it is being sent over virtually a frozen grid system. The U.S. Department of Energy predicts that transmission investment is only likely to expand 6% compared to the 20% growth in electricity demand expected over the coming decade. Another rising controversy pits residential against business customers as electricity rates increase.

In summary, as shown in Fig. 3, the past 30 years have focused on efforts to restore the electricity enterprise's declining cost commodity tradition. All have failed to meet this challenge and there are no "silver bullets" on the horizon that are likely to change this reality within the context of today's aging electricity supply infrastructure. At the same time, electricity has become increasingly politicized as an essential retail entitlement where market price volatility is effectively allowed to

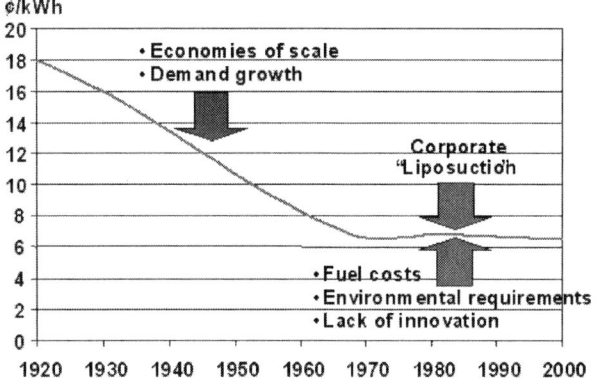

Fig. 3 Average U.S. price of residential electricity service (in 1984 $).

operate only in one direction—downward. Thus, the essential foundation for restoring vitality to the electricity enterprise rests first and foremost on innovation, principally in the consumer/delivery interface and in end-use electro-technologies. This represents a profound transformational challenge for the enterprise, which, throughout its history following the Edisonian beginning, has focused on supply-side technology as the wellspring of progress.

This combination of rising costs and artificially constrained price creates an economic vise on electricity supply that is squeezing out more and more value from the enterprise and the nation. Unfortunately, the dominant financial imperative has been to contain immediate costs at the expense of infrastructure development and investment. Unless the resulting standstill is ended and the assets of the enterprise are urgently reinvented, they risk being left behind as industrial relics of the 20th century. For example, the total capital expenditure rate of the electricity enterprise, both regulated and unregulated, as a fraction of its electricity revenues is now about 10%, less than one-half of the historic minimum levels and, in fact, a percentage only briefly approached during the depths of the Depression (refer to Fig. 4).

Unfortunately, this emphasis on controlling costs at all cost has also resulted in a period of profound technological stasis throughout the grid-based electricity enterprise. This has not been for lack of innovative opportunity but rather the lack of financial incentives. Every aspect of the enterprise has both the need and opportunity for technological renewal. For example, although coal remains the backbone of the power generation fleet producing over half the nation's electricity, the outdated technology being used is both inefficient and unable to keep pace with rising environmental demands, including carbon control. Integrated coal gasification-combined cycle (IGCC) technology which, in effect, refines coal into a clean synthesis gas for both electric power generation and synthetic petroleum production, could fundamentally address these constraints. Similarly, advanced nuclear power cycles could resolve the waste management, proliferation, and efficiency problems limiting the use of this essential clean energy source. Without considerably greater R&D emphasis and modernization of the power delivery system, renewable energy will also remain a theoretically attractive but commercially limited resource opportunity. The same is true of electrical storage.

Throughout the history of commercial electrification, large-scale storage, the long-sought after "philosopher's stone" of electricity, has remained elusive and thus a fundamental constraint on addressing optimal load management and asset utilization. Pumped storage, the use of off-peak electricity to pump water behind dedicated hydroelectric dams, has gained acceptance where feasible within geologic and environmental constraints. Demonstrations of compressed air storage using evacuated salt domes and aquifers were also successful, although this technology has not yet achieved significant commercial acceptance. At the other end of the spectrum of electricity storage, small-scale devices, including batteries and capacitors, are used for short-term load stability purposes and to dampen current fluctuations that affect reliability. A variety of advances in storage, including super conducting magnetic energy storage (SMES) and flywheels are being explored, but all have suffered from the general technological malaise constraining the grid-based electricity enterprise.

Only in applications for power portability has innovation in electricity storage made significant technical and commercial progress during this period. This progress also represents quite a different set of players than those of the traditional electric utility-based industry. Also indicative

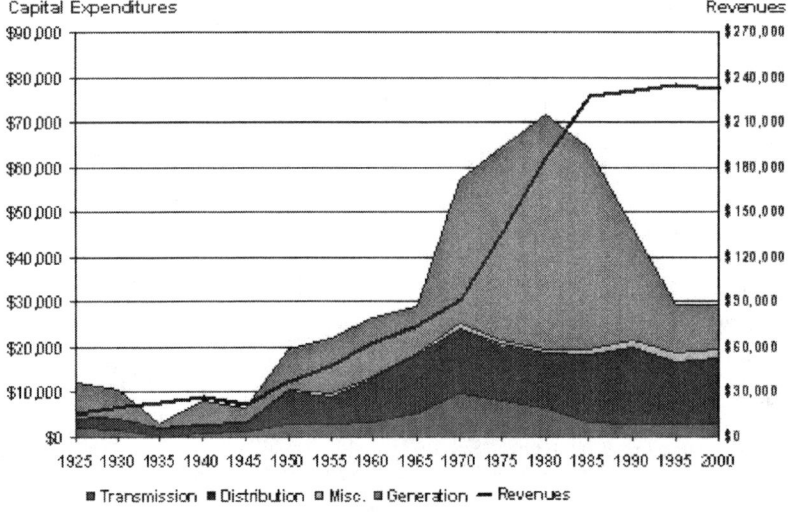

Fig. 4 Electric utility revenues and capital expenditures (in 2003 millions of $), 1925–1999.

of new players taking advantage of the growing value gaps in the nation's traditional electricity supply capability is the trend toward distributed power resources. Consumers with urgent needs for high quality power are increasingly taking advantage of the emergence of practical on-site generation technologies. These include diesel power sets, microturbines utilizing natural gas or landfill methane, fuel cells, and hybrid power systems incorporating photovoltaics. All of these are, in effect, competitors with the power grid, although ideally they could be integrated as grid assets within a truly modernized national electricity service system.

Similarly, in terms of power delivery technology, a wide array of thyristor-based digital control systems (e.g., FACTS), wide-area monitoring and communications, and highly sensitive anticipatory condition monitors, have been demonstrated and could revolutionize the reliability, capacity, and operability of the nation's electricity transmission and distribution network. Superconductivity represents another potential breakthrough technology that could fundamentally improve the efficiency of both power delivery and end use. This has been enabled by the recent development of so-called "high-temperature" superconductive materials operating at relatively modest liquid nitrogen temperatures. These materials, in effect, have no electrical resistance, thus eliminating transmission distance limitations, and are capable of significantly increasing the electrical capacity of existing distribution systems. The primary constraint today is the brittle ceramic nature of these superconducting materials and the resulting difficulties in manufacturing durable wiring, etc.

However, unless and until the electricity system advances from its current "life-support" level of infrastructure investment, all these potential advances remain, at best, on-the-shelf novelties. (These and others will be addressed further in the following entry on Transformation.) This investment gap is exacting a significant cost that is just the tip of the iceberg in terms of the electricity infrastructure's growing vulnerabilities to reliability, capacity, security, and service challenges. In fact, since the mid-1990s, the electric utility industry's annual depreciation expenses have exceeded construction expenditures. This is typical of an industry in a "harvest the assets" rather than an "invest in the future of the business" mode.

An even more dramatic measure of stasis is the minimal R&D investment by the electricity enterprise. In the wake of restructuring in the early 1990s, the enterprise's R&D investment rate has declined to about 0.2% of annual net revenues. This compares to a U.S. industry-wide average of about 4%. Even with inclusion of federal electricity-related R&D, the total is still only equivalent to a fraction of one percent of annual electricity sales revenues. The bill for this mortgaging of the future has come due and will, unless promptly paid, impose a heavy price on the nation's productivity, economy, and the welfare of its citizens.

Fig. 5 compares recent relative price trends among a variety of essential retail consumer goods and services. On the surface, the trend for electricity looks quite favorable but, after factoring in the rapidly growing cost of service unreliability (the shaded area), the real cost of electricity service has been increasing significantly and continues to do so at an escalating rate. This is an indelible reminder that it is only the lack of quality that adds to cost.

There is also a significant and growing stakeholder concern that the electricity enterprise, as currently constituted, is out of step with the nation at large. As economic growth resumes, will the enterprise be able to keep pace with energy quantity and quality needs, and can it also satisfy investor expectations without again resorting to questionable, high-risk financial schemes?

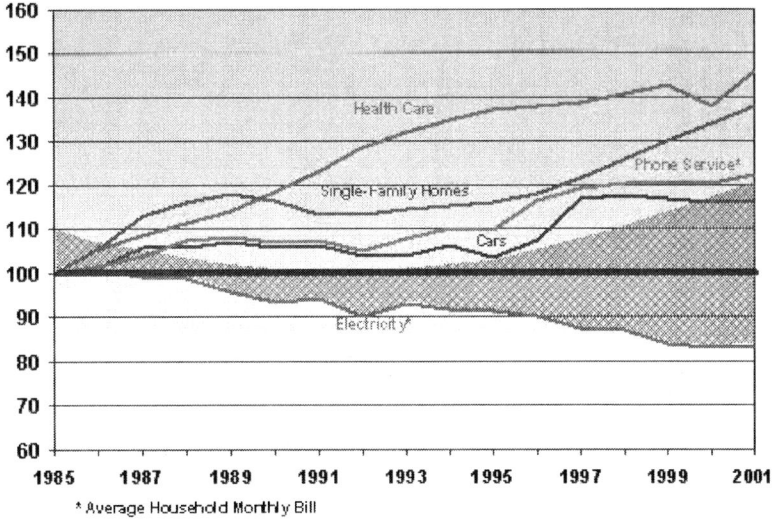

Fig. 5 Relative consumer prices, 1985–2001.

Table 4 Maturity and stasis

	1968	1985	2002
Ultimate customers (in millions)	70	101	132
Net generation (10^9 kWh)	1,436	2,545	3,841
Installed generating capacity (10^3 MW)	310	712	981
Average power plant size (MW)	85	227	300
Circuit miles of high-voltage line[a] (10^3 mi)	425	605	730
Residential service price (¢ per kWh—1983$)	7.1	6.8	6.8

[a] 22,000 V and above.

Table 4 summarizes the course of the enterprise during the maturity and stasis period of the last 35 years.

CONCLUSION

The past century has witnessed the creation of the electric utility industry, and the profound growth in electric use. Electricity is now consumed by virtually every residential, commercial, industrial, and institutional facility in the United States. For the first half century of this new technology, costs of electricity continually declined; and by the 50 year mark, a unified electric grid system reached the far corners of the nation. The electric power system in the U.S. had reached its zenith. But, by the mid-1970s, the electric industry was hit by several price shocks—the most damaging being the end of ever cheaper power plants and ever cheaper electric power. The more recent trend toward deregulation in the electric industry has had a significant impact on electricity customers, as well as on electric utilities. Cost increases for customers, economic problems for utilities, and reliability problems for the electric grid have all become serious problems for the electricity enterprise. A "reinvention" of the electricity enterprise is needed to control costs, increase the economic health of electric utilities, and prepare for the future uses of electricity. A modernized electricity enterprise would provide widespread benefits for the U.S. economy and society. The need for a transformed electricity enterprise in the U.S. is the topic of a second article in this area.

REFERENCE

1. *Statistical Yearbook of the Electric Utility Industry*; Edison Electric Institute: Washington, DC, 2003.

BIBLIOGRAPHY

1. Brennan, T.J. et al. *Alternating Currents—Electricity Markets and Public Policy*; Resources for the Future: Wahington, DC, 2002.
2. *Electricity Sector Framework for the Future*; Electric Power Research Institute (EPRI): Palo Alto, CA, 2003, www.epri.com/corporate/esff/default.asp.
3. *Electricity Technology Roadmap*; Electric Power Research Institute (EPRI): Palo Alto, CA, 2003.
4. Gellings, C.; Yeager, K. *Transforming the Electricity Infrastructure, Physics Today*; American Institute of Physics: College Park, MD, 2004; 45–51.
5. Hirsch, R.F. *Technology and Transformation in the American Electric Utility Industry*; Cambridge University press,: Cambridge, UK, 1989.
6. Hyman, L.S. *America's Electric Utilities—Past, Present and Future*, 6th Ed.; Public Utilities Reports, Inc.: Arlington, VA, 1997.
7. Nye, D.E. *Electrifying America*; The MIT Press: Cambridge, MA, 1990.
8. Patterson, W. *Transforming Electricity*; The Royal Institute of International Affairs/Earthbeam Publications Ltd: London, 1999.
9. *Reviving the Electricity Sector*; National Commission on Energy Policy: Washington, DC, 2003, www.energycommission.org.
10. Schurr, S.H. et al. *Electricity in the American Economy*. Contributions in Economies and Economic History-117, Greenwood Press: New York, 1990.
11. Smi, V. *Energy at the Crossroads*; The Royal Institute of International Affairs/Earthbeam Publications Ltd: London, 1999.
12. Yeager, K. *The Widening Technological Divide*; Public Utilities Fortnightly: Washington, DC, April 2005; 54.

Electricity Enterprise: U.S., Prospects

Kurt E. Yeager
Galvin Electricity Initiative, Palo Alto, California, U.S.A.

Abstract
This entry examines the present status of the U.S. electricity enterprise, and seeks to identify future opportunities for technological innovation in the electric power systems (as broadly defined) that will best serve the changing needs of consumers and businesses over at least the next 20 years. Of paramount importance will be insuring that the electricity system provides absolutely reliable and robust electric energy service in the context of ever-changing consumer needs.

INTRODUCTION

The stages in the Electricity Sector Life-Cycle shown in Fig. 1. The first three stages of the life-cycle are discussed in another entry. This entry discusses the fourth stage of Transformation, and suggests a path to the future of the U.S. Electricity Enterprise.

Electricity can be used in ways that no other energy form can be used. In addition to new products and new techniques of production, technological progress over the past century has led to radically improved ways of organizing productive activities as well, and all of these innovations have been heavily dependent on electricity. As a result, electricity has become the lifeblood of the nation's prosperity and quality of life. In fact, the U.S. National Academy of Engineers declared that "the vast networks of electrification are the greatest engineering achievement of the 20th Century."

Looking to the future, this entry discusses the performance changes within the U.S. electricity enterprise needed to respond to the rapidly growing reliability and service value expectations of 21st century consumers and society. One possible approach to such a transformation is described here as an example. Ultimately, a broad range of potential technological innovations bearing on the future of the electricity enterprise, from both the supply and utilization perspective, will need to be examined. Based on that broad objective examination, a comprehensive blueprint for elevating the reliability and value proposition of electric energy service will need to be formulated.

Above all, such a transformed system must be able to remain robust in the face of future complications of all sorts. It should therefore incorporate mechanisms for learning, innovation, and creative problem solving well beyond the capabilities of the current electric energy service system. The robustness of the system will also require that the initiative and its participants consider the role of the system's evolutionary history in determining its current and future state, and the set of strategic options open to the system (*Robust Design: A Repertoire of Biological, Ecological and Engineering Case Studies*, edited by Erica Jen, Santa Fe Institute Studies in the Science of Complexity, Oxford University Press (2005). This book uses robustness as follows: "Robustness is an approach to feature persistence in systems that compels us to focus on perturbations, and assemblages of perturbations, to the system that are different from those considered in its design, or from those encountered in its prior history.")

TRANSFORMATION

The first step in restoring the integrity and building the value of the electricity enterprise, in the context of 21st century needs, is to focus on the fundamentals that stakeholder input has helped to highlight:

1. Electricity is more than a form of commoditized energy; it is the underpinning of the modern quality of life, and the nation's indispensable engine of prosperity and growth.
2. Electricity is a service-based enterprise whose value to consumers depends on the most technically complex machines ever built.
3. The opportunities for technology to relieve cost pressures are principally in its ability to increase the service value of electricity. Building service value, over and above electricity's historic commodity energy value, is essential to every element of, and participant in, the electricity value chain.
4. The ability to capitalize on these new value opportunities requires a transformation of today's electricity infrastructure. This transformation must

---

Keywords: Electric enterprise; Restructuring; Electric system infrastructure; Grid system; Electric system transformation; Electric system stakeholders; Modern grid initiative; Galvin electricity initiative; Power quality; Digital quality power; Digital revolution; Portable power; Intelligrid.

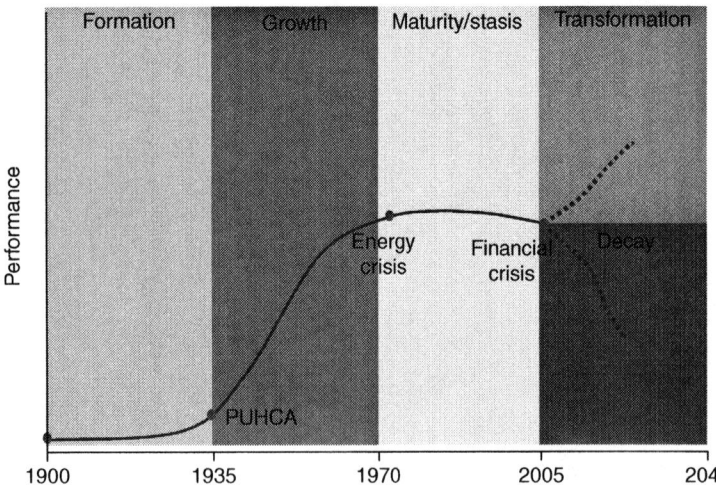

Fig. 1 Electricity sector life-cycle—a fork in the road.

enable all consumers to become active participants in, and benefactors of, the electricity enterprise, rather than remaining captive to the historic model of energy as strictly a commodity. Consumers want more choice and control.

The stakeholders' feedback also underscores that this "forward-to-fundamentals" transformation of the electricity enterprise is a process, not an event. This process should develop and proceed based on local benefits and costs—just as the nation's highway system was developed in the 20th century. It will also be a process because it requires a transformation of the institutional as well as physical infrastructure.

This "reinvention" of the electricity enterprise is likely to be as daunting as it is essential. Indeed, it is more typical for incumbents to consider innovative transformation as a greater threat to their status than as an opportunity. However, the current conditions prevailing in the enterprise and the growing service-value gap created by these conditions provide a situation in which transformative change is rapidly becoming a matter of survival for many incumbents, in addition to being a national imperative for productivity.

Unfortunately, there are very few examples of successful transformative reinvention of systems by established incumbents in any endeavor. The issue is even more challenging in the case of the electricity enterprise because of the significant barriers to entry by new players. Although the Public Utility Holding Company Act (PUHCA) was repealed in the just-passed Energy Policy Act of 2005, this is unlikely to attract significant new entrants. That is, unless the prevailing regulatory conflicts, uncertainties, and other disincentives are resolved that effectively create an impenetrable "iron curtain" around the enterprise, and transformational change leadership is mobilized.

Transformation requires that both the industry and its regulators move beyond their traditional cost-plus (or minus) commodity culture in terms of electricity's value proposition to consumers and society. A compelling, unified leadership vision is needed that conveys the fact that electricity, through innovative technology, provides a service value to consumers and society that is significantly greater than its basic commodity value. In order to realize this vision, regulatory policy must facilitate true consumer participation by urgently resolving the tension in the changing role of regulation from one primarily of protecting to one of enabling.

The foundation for confidence in this enterprise transformation stems from the revolution underway in the enabling technologies affecting all network infrastructures. A portfolio of innovative technologies can comprehensively resolve the vulnerability of today's power supply system in terms of its capacity, reliability, security, and consumer service value. These "smart technologies" will also open the door to fully integrating distributed resources and central station power into a single network, in a manner than can reduce system vulnerability rather than add to it—as is typically the case today—while also steadily improving the efficiency and environmental performance of the system.

Since 1980, the electricity intensity (kilowatt-hours per dollar GDP) of the U.S. economy has declined about 10%, leaving the intensity today about where it was three and a half decades ago. Is this the result of real efficiency improvement or just another measure of stasis? The electricity enterprise stands at a critical fork in its road of progress. Today its stakeholders have the necessity, the opportunity, and the means to make a clear choice about the future value of the enterprise. The decisions made and the path taken will make a profound difference, not only to the destiny of the electricity enterprise, but also to the nation, and ultimately to the world.

A primary rationale behind the restructuring of the electric utility industry ten years ago was that competitive markets manage supply and demand, provide incentives for innovation, and allocate investments more effectively than centrally regulated monopolies. While fundamentally sound in principle, the policy implementation of this rationale has not adequately reflected either the unique physics or the public entitlement characteristics of electricity. The consequence has been a breakdown in the traditional public/private partnership built around the obligation to reliably serve, and upon which the value and reputation of the electricity enterprise was built. The decision to begin the competitive market transformation of electricity with the wholesale supply function rather than with retail service has proven to be particularly counterproductive. As a result, most of the potential consumer benefits of innovation have been left on the table thus seriously compromising the consumer value of a transformation.

In short, the electricity enterprise has tended, through restructuring, to become a victim of its historic success in maintaining universal service reliability at ever-lower costs. The essential foundation for restoring enterprise vitality in the coming decade is rebuilding this fundamental public/private partnership, based on technology innovations that can increase the quality and value of electricity service, particularly providing higher levels of reliability and security. This transformation of the traditional electricity supply network into tailored, multifunctional service networks should also result in significant new business growth opportunities, reinforced by greater consumer satisfaction.

A modernized electricity enterprise would provide widespread benefits for the economy and society. In this enhanced scenario, productivity growth rates are higher and the economy expands more rapidly, while energy consumption, intensity, and carbon emissions are reduced relative to business as usual. Higher productivity rates can be sustained because a more reliable digital power delivery infrastructure would enable workers to perform existing and new functions more accurately and efficiently. This accelerated productivity growth has been demonstrated and established selectively in the economy, but its potential could be expanded dramatically through a smart power supply system. In effect, improved reliability and quality of electricity would enable the digital economy to expand at a broader and faster rate—an essential factor for successful U.S. competition in a global economy.

The Digital Revolution is the third major economic transformation in the last 100 years, and each has increasingly depended on electricity. Each also has created substantial new levels of wealth, as well as winners and losers at the scale of individuals, corporations, and nations—all depending on the effectiveness with which the innovative technology underpinning the economic transformation is exploited. In this new electricity business environment, it is the quality of customer connectivity and responsiveness that increasingly will differentiate the winners from the losers. The most precious business asset becomes the customer access portal. Every electricity supply function preceding it will be under relentless cost pressure, only moderated by the value that each consumer ultimately receives.

This transformed electricity infrastructure and business model will also serve to catalyze entirely new capabilities for consumer participation in the electricity marketplace, while significantly reducing the parasitic costs of power disturbances that are characteristic of today's system. This technological innovation will finally break open the commodity box currently constraining both the electricity enterprise and consumers, and will usher in an era of ever-higher valued energy/information services even beyond our imaginations. The payoff from this economic progress could easily exceed $1 trillion per year in additional U.S. GDP within a decade. This accelerated economic expansion is essential to meeting the nation's growing debt, security, and aging population costs.

Above all, this modernized electricity system would provide much greater functionality and service value for consumers. From a business perspective, this additional value is increasingly necessary to compensate for the expected significant rise in the cost of electricity. A number of upward trends are likely to increase electricity cost by at least 30% during the coming decade. These upward cost pressures largely occur because of forces beyond the control of the electricity enterprise, including fuel prices environmental protection regulation, and protection of the physical and cyber infrastructures against potential incursions.

The prices of fuels—natural gas, coal, and uranium—are rising, and are likely to remain significantly elevated. As a result, fuel costs are expected to account for at least half of the net increase in the price of electricity over this period.

Similarly, the cost of compliance with steadily tightening environmental regulations also continues to increase. In addition, there is a growing possibility that mandatory carbon control requirements will be instituted, possibly in the coming decade. This would have a significant additional impact on electricity costs, including the need for major strategic investments in cleaner replacement generation facilities using natural gas and nuclear, Integrated Gasification Combined Cycle (IGCC) for coal, and renewable resources.

Security improvements, both to discourage terrorist attacks and to recover from them, may further add to the cost of electricity. A fully functional security program would, for example, combine enhanced physical security systems with self-healing grid capabilities and cyber security advances. Cyber security would focus on increasing the security of communications, data monitoring, and automated system control functions, plus provide vulnerability assessments to support the self-correcting grid and related adaptive islanding capabilities. Rising fears of terrorist attacks may also fuel political pressure to further escalate infrastructure security measures, regardless of their cost. On the other hand, a modern, digitally monitored and controlled power supply system would comprehensively and confidently address these concerns as part of the process of technical modernization.

The need to improve the reliability and quality of power systems is another important reason the cost of electricity likely will increase through 2015. Mandatory reliability standards have been endorsed widely and were moving through Congress until stalled with the rest of the 2003 Energy Policy Act legislation. However, to be relevant to the needs of the new century, these standards will need to move beyond the traditional "keep the lights on" level of reliability to reflect the more stringent requirements of the digital economy.

Probably the greatest long-term challenge to the electricity sector is the fact that even as the demand for power is growing, the nature of electricity demand is undergoing a profound shift due to digital technology. Twenty years ago when the personal computer was introduced, few foresaw the widespread proliferation of "smart" devices. Today, for every microprocessor inside a computer, there are 30 more in stand-alone applications, resulting in the digitization of society. In applications ranging from industrial sensors to home appliances, microprocessors now number more than 12 billion in the United States alone.

These digital devices are highly sensitive to even the slightest disruption in power (an outage of less than a fraction of a single cycle can disrupt performance), as well as to variations in power quality due to transients, harmonics, and voltage surges and sags. "Digital quality power," with sufficient reliability and quality to serve these growing digital loads, now represents about 10% of the total electrical load in the United States, for example. It is expected to reach 30% by 2020 under business-as-usual conditions, and as much as 50% in a scenario where the power system is revitalized to provide universal digital-grade service.

However, the current electricity infrastructure in the United States, designed decades ago to serve analog (continuously varying) electric loads, is unable to consistently provide the level of digital quality power required by our digital manufacturing assembly lines, information systems, and soon even our home appliances. The economic loss of power disturbances mentioned earlier is attributable in part to the sensitivity of these new digital technologies.

Advanced technology now under development or on the drawing boards sustains the promise of fully meeting the electricity needs of a robust digital economy. The architecture for this new technology framework is becoming clear through early research on concepts and the necessary enabling platforms. In broad strokes, the architectural framework envisions an integrated, self-healing, electronically controlled electricity supply system of extreme resiliency and responsiveness—one that is fully capable of responding in real time to the billions of decisions made by consumers and their increasingly sophisticated microprocessor agents. In short, the potential exists to create an "IntelliGrid1" electricity supply system that provides the same reliability, efficiency, precision, and interconnectivity as the billions—ultimately trillions—of microprocessors that it will power. The following summarizes a potential set of key steps in this performance transformation:

- *Digitally controlling the power delivery network* by replacing today's relatively slow electro-mechanical switching with real-time, power-electronic controls. This will become the foundation of a new "smart, self-healing power delivery system" that will enable innovative productivity advances throughout the economy to flourish. Digital control is the essential step needed to most cost-effectively address the combined reliability, capacity, security, and market-service vulnerabilities of today's power delivery system. As a practical matter, this technical expansion is the only way that these vulnerabilities can be comprehensively resolved.
- *Integrating communications* to create a dynamic, interactive power system as a new "mega-infrastructure" for real-time information and power exchange. This is the capability needed to enable retail energy markets; power interactive, microprocessor-based service networks; and fundamentally raise the value proposition for electricity. Through advanced information technology, the system would be "self healing" in the sense that it is constantly self-monitoring and self-correcting to keep high-quality, reliable power flowing. It would sense disturbances and instantaneously counteract them, or reconfigure the flow of power to isolate any damage before it can propagate. To realize the vision of the smart power delivery system, standardized communications architecture must first be developed and overlaid on today's power delivery system. This "integrated energy and communications system architecture" should reflect an open, standards-based architecture for data communications and distributed computing infrastructure.
- *Automating the distribution system* to meet changing consumer needs. The value of electricity distribution system transformation—fully automated and integrated with communications—derives from four basic functionality advantages:
  — Reduced number and duration of consumer interruptions, system fault anticipation, and faster restoration.
  — Increased ability to deliver varying "octane" levels of reliable, digital-grade power.

— Increased functional value for all consumers in terms of metering, billing, energy management, demand-response, and security monitoring, among others.
— Access to selective consumer services including energy-smart appliances, power-market participation, security monitoring, and distributed generation.

To a power system operator, automation means a self-healing, self-optimizing smart power delivery system that automatically anticipates and quickly responds to disturbances, thus minimizing, if not ultimately eliminating, power disruptions altogether.

- *Transforming the meter* into a consumer gateway that allows price signals, decisions, communications, and network intelligence to flow back and forth through the two-way energy/information portal. This will be the linchpin technology that leads to a fully functioning marketplace with consumers responding (through microprocessor agents) to price signals. For consumers and providers alike, this gateway or portal through today's opaque electric service "iron curtain" provides the tool for moving beyond the commodity paradigm of 20th century electricity service. The result will quite possibly usher in a new set of energy/information services at least as diverse as those in today's telecommunications. This portal would sit between consumers' "in-building" communications network and wide-area "access" networks, enabling two-way, secure, and managed communications between consumers' equipment and energy service and/or communications entities.
- *Integrating distributed energy resources.* The new system would also be able to seamlessly integrate an array of locally installed, distributed power generation units (such as fuel cells and renewable resources) as power system assets addressing a variety of challenges. These challenges include the needs to increase the resiliency and reliability of the power delivery infrastructure, provide high-quality power, facilitate the provision of a range of services to consumers, and provide consumers with lower-cost, higher-quality power. These distributed power sources could be deployed on both the supply and consumer side of the energy/information portal as essential assets dispatching reliability, capacity, and efficiency. Unfortunately, today's electrical distribution system, architecture, and mechanical control limitations, in effect, prohibit this enhanced system functionality.
- *Accelerating end-use efficiency* through digital technology advances. The growing trend toward the digital control of processes can enable sustained improvements in efficiency and worker productivity for nearly all industrial and commercial operations. Similarly, the growth in end-use electrotechnologies, networked with system controls, will afford continuous improvements in user productivity and efficiency.
- *Expanding portable power.* The added value of individualized consumer electricity services is exemplified by the rapidly proliferating market for portable electricity devices and power supplies. These are likely to capture more and more of the higher-value electricity uses. More efficient power usage is reducing the overall power demands of these devices while new cell chemistries are expected to double usable power densities of these power supplies.

Lithium-ion battery technology is rapidly leading the demand for powering portable devices, while the market for nickel–cadmium (NiCad) is shrinking under environmental pressures. The price of lithium-ion batteries has also dropped by 20%–50% during the last few years, while NiCad and nickel-metal hydride battery prices have declined by 10%–20%. Primary batteries can be stored up to 10 years and have much higher energy densities than rechargeable secondary batteries.

Fuel cells are a particularly attractive alternative for greater power portability. However, until major cost, size, and performance breakthroughs are achieved, fuel cell use will remain limited in portable applications. Fortunately, this is an area of technology that is receiving substantial developmental investment to resolve these constraints.

With the switch to electronic automobile braking and steering by wire, the 3 kW capability of the rechargeable lead-acid and the single 12-V battery will no longer be sufficient, and is likely to usher in the 42-V system for automobiles. Hybrid vehicles require a high-voltage battery of about 150 V. This is currently provided by connecting nickel-metal-hydride cells in series. Battery life in this application is crucial since replacement costs as much as a new motor.

ACHIEVING THE TRANSFORMATION

There is growing recognition by the electricity sector of the need to modernize its aging infrastructure and business model. This is reflected in a series of complementary initiatives being pursued by various public and private organizations. These initiatives include the Modern Grid Initiative sponsored by the U.S. Department of Energy and its National Energy Technology Laboratory; the Gridwise Architectural Council sponsored by the Battelle Pacific Northwest National Laboratory; and the IntelliGrid Consortium sponsored by the Electric Power Research Institute. Each of these initiatives is intended to comprehensively modernize the nation's electric energy supply system within the current regulated utility structure.

All of these initiatives have been developed over the last several years in response to the concerns of the broad electricity stakeholder community about the nation's aging electricity supply system. This reflects the fact that the nation and its economy are dependent on the integrity of a complex web of digital networks for which the electric power system is the foundation. These initiatives reflect the broad-based collaboration of leaders in energy, technology, and government, working to address the looming electricity supply industry issues and set the United States on a migration path toward the intelligent, self-healing power system of the future.

The foundation of this new system is an open-systems-based, comprehensive reference architecture. This architecture enables the sustainable integration of intelligent equipment and data communications networks on an industry-wide basis. This intelligent system will enable real-time energy information and power exchanges, in addition to fundamentally enhanced system reliability and security. These initiatives apply the latest system engineering methods and computational tools to model the advanced energy enterprise of the future. In so doing, they cut across traditional operating boundaries, promoting greater interoperability, and enabling unprecedented improvement in performance and asset utilization.

The result will be an electricity and information mega-infrastructure that enables technological advancement to continue to flourish. This system will be "always on and active," and interconnected into a national network of real-time information and power exchanges. This system will also have the intelligence to seamlessly integrate traditional central power generation with distributed energy resources.

The foundational reference architecture of this transformed electricity supply system also eliminates the constraints on the consumer gateway to the system now imposed by the meter. The result will allow price signals, communications, and network intelligence to flow back and forth through a two-way portal. This is the linchpin enabling a fully functioning retail power marketplace with consumers responding (through microprocessor agents) to price signals. Specific capabilities can include the following:

- Pricing and billing processes that support real-time pricing.
- Value added services such as billing inquiries, service calls, emergency services, and diagnostics.
- Improved building and appliance standards.
- Consumer energy management through sophisticated on-site systems.
- Easy "plug and play" connection of distributed energy resources.
- Improved real-time system operations.
- Improved short-term local forecasting.
- Improved long-term planning.

These grid modernization initiatives are being designed to accommodate widespread deployment of distributed energy resources (DER). DER includes microturbines, fuel cells, photovoltaics, and energy storage devices installed close to the point of use—all of which can reduce the need for intensive power delivery system investment. Particularly suitable for a variety of industry and commercial applications, DER reinforces the bulk power system by providing consumers with lower cost peak power, higher reliability, and improved power quality. Many DER devices also provide recoverable heat suitable for cogeneration.

In summary, these various grid modernization initiatives are intended to ultimately provide the technological and engineering means to resolve the large and growing gap between the performance capability of today's bulk electric energy supply system and the needs and expectations of modern society. Also fundamental to resolving this gap is a corresponding commitment to restore the level of infrastructure investment needed to fully develop and deploy the essential technological advancements.

Recognizing the deeply entrenched resistance to transformative change that exists in established enterprises—particularly such as the regulated monopoly structure of the electricity sector, another complementary initiative is taking a fundamentally different approach to achieving electricity system transformation.

In contrast to the various grid modernization initiatives described above, the Galvin Electricity Initiative seeks to transform the electric energy supply system starting from the consumer rather than from the supplier. This recognizes that the most important asset in resolving the growing electricity cost/quality dilemma and its negative reliability, productivity, and value implications is consumer-focused innovation that disrupts the performance status-quo by targeting and quickly demonstrating the advantages of system modernization.

The goal of the Galvin Electricity Initiative is the "Perfect Power System". That is, a system that never fails to meet, under all conditions, each consumer's expectations for electricity service confidence, convenience, and choice. In the context of this Initiative, the electric energy supply system includes all elements in the chain of technologies and processes for electricity production, delivery, and use across the spectrum of industrial, commercial, and residential applications. This focus on the consumer also reflects the relatively intractable nature of the highly regulated monopoly of bulk power infrastructure that dominates U.S. electric energy supply and service.

In addition to absolute quality, a second principle guiding the Galvin Electricity Initiative is enabling self-organizing entrepreneurs to engage in the nation's electricity enterprise. Innovative entrepreneurial leadership guided by consumer service opportunities is seen by the Initiative as providing the most confident engine for quality transformation and sustainable system improvement. The emphasis here is on creative, "outside the box" thinking that focuses on achieving maximum consumer value through innovation with as short a turnaround as possible.

The basic approach to developing the Perfect Power System is to increase the independence, flexibility, and intelligence of electric energy management and use at the device level and then integrate these capabilities at larger scales as necessary to achieve the delivery of perfect power. In this context, the Initiative is developing three levels of generic electric energy system configurations that have the potential to achieve early perfection, together with the corresponding technology innovation opportunities that are essential to their success. These configurations are: (a) Device-Level (portable) systems serving a highly mobile digital society; (b) Building Integrated Systems which focus on modular facilities serving individual consumer premises; and (c) Distributed Microgrid Systems including interconnection with local bulk power distribution networks. It is at this largest configuration level that the Galvin Electricity Initiative seamlessly interfaces with the modernization initiatives of the bulk power system described previously. Thus, each level of Galvin Perfect Power configuration reflects a consumer-guided step on the path to ultimately transforming the nation's entire bulk electric energy supply system.

In the context of each system configuration level, the Galvin Electricity Initiative is also performing comprehensive evaluations of eight target innovation nodes and their enabling technologies. The innovation nodes being considered are: communications; computational ability; distributed generation; power electronics and controls; storage; building systems; efficient appliances and devices; and sensors. These evaluations include, in each case, quantifying the performance gaps that must be filled to achieve perfection, and defining the opportunities and risks involved in resolving these gaps within a 10–20 year period. Business opportunity templates plus a quality leadership/management guidebook and associated training courses are also being developed. In total, the results will provide a comprehensive and confident roadmap for achieving prompt, successful commercial implementation of this consumer-focused electric energy supply and service modernization Initiative.

Modern society is increasingly dependent on electric energy, and it expects individualized service that enables the intelligent control of energy consumption with a premium on reliability, efficiency, and environmental performance. Transformation to an interactive, consumer-directed, service capability will change the business dynamics of the electricity enterprise from one of simply providing electric energy as a bulk commodity to offering a portfolio of individualized energy services of far greater value. This consumer-focused quality transformation will also serve to restore innovation and investment, and resolve the growing vulnerabilities facing the regulated bulk electric energy infrastructure. Prototype commercial Perfect Power System installations using state-of-the-art enabling technologies are also being implemented to immediately demonstrate these performance advantages.

CONCLUSIONS

The overarching priority of the electricity enterprise and all its stakeholders is to pursue the policies and actions needed to stimulate system modernization. The current rate of investment in the power delivery system alone, some $18–$20 billion per year, is barely enough to replace failed equipment. To correct deficiencies in the existing system and to enable the smart power delivery system of the future will require double this amount annually. The result would be an electricity supply system fully capable of meeting the escalating needs and aspirations of the 21st century society. In summary, these needs and aspirations dictate a future where:

- The electricity enterprise confidently provides the nation's most essential platform for technical innovation, productivity growth, and continued economic prosperity. Eventually, it is expected that this platform will enable every end-use electrical process and device to be linked, in real-time, with the open marketplace for goods and services, including, but not limited to, electric power.
- Economic productivity increases substantially as a result of the transformation of the electricity enterprise, generating additional wealth to deal with the large societal, security, and environmental challenges of the 21st century.
- The roles, responsibilities, and rules governing the electricity enterprise have been clarified, enabling a revitalized public/private partnership that maintains confidence and stability in electricity sector financing. As a result, the rate of investment in the essential electricity infrastructure is substantially increased.
- The role of regulation has evolved from oversight of company operations and "protection" of ratepayers to oversight of markets, as well as enabling and guiding market transparency and specific public-good services (i.e., reliability standards, provider-of-last-resort, market transformation, etc.).
- National security and energy policies emphasize U.S. fuel diversity, placing electricity at the center of a strategic thrust to: (1) create a clean, robust

portfolio of domestic energy options including fossil, nuclear, and renewable energy sources, along with enhanced end-use efficiency; (2) develop a sustainable electric energy system providing the highest value to all consumers with perfect reliability; and (3) electrify transportation to reduce dependence on foreign oil.

The cost/benefits of electricity system modernization will be profoundly positive. For example, the cost to the average household would be less than $5 per month before taking any credit for the considerable energy-savings opportunities that each consumer would be empowered to achieve through a modernized, more functional power system. In return, as power reliability and quality improve, each consumer would save hundreds of dollars per year in the price of purchased goods and services. Even more financially significant would be the income potential added to each household as the nation's productivity, security, and competitiveness increase. Thus, the service quality improvements from electricity system modernization would not result in higher costs, but are the genesis of cost savings for all consumers. Again, improving quality always costs less.

In support of this more positive future, a proactive technology development program will be necessary. This should emphasize power system reliability and functionality, management of greenhouse gases, and the development of higher-value, more efficient, smart electricity end-use devices and services. Technological innovation will remain, as it has been throughout the history of commercial electricity enterprise, the essential asset determining the destiny of the electricity sector and its value to society.

The result is likely to be a profoundly transformed, multi-dimensional electricity service capability incorporating an array of distributed, stored, and portable power resources as assets. At the same time, this infrastructure transformation will enable the convergence of electricity, telecommunications, and sensors into a smart, sustainably robust, mega-infrastructure powering a universal digital society with absolute reliability. Most importantly, this transformation will enable an array of innovations in electricity service that is only limited by our imaginations.

In closing, the Galvin Electricity Initiative seeks to urgently catalyze this performance and value transformation. It will do so through a two-phase effort that first explores and evaluates the most promising opportunities for technological innovation throughout the electricity value chain, as seen from the broad consumer perspective. The goal here is a demonstrable and compelling perfection in electricity service quality that will mobilize the broad community of stakeholders to demand the necessary performance and value transformation. In its second phase, the Initiative will develop the comprehensive change in leadership plan for most effectively stimulating this electricity service and supply transformation. This plan will broadly consider and specifically recommend the essential policies, institutions, standards, and incentives, etc. needed to break today's pervasive innovation/investment logjam in the electricity enterprise. The results will truly electrify the nation and the world in the full meaning of the word.

A message of enduring vision:

So long as there remains a single task being done by men or women which electricity could do as well, so long will that development (of electrification) be incomplete. What this development will mean in comfort, leisure, and in the opportunity for the larger life of the spirit, we have only begun to realize.

—Thomas Edison, 1928

BIBILIOGRAPHY

1. Brennan, T.J.; Palmer, K.L.; Martinez, S. Alternating Currents—Electricity Markets and Public Policy, Resources for the Future, Washington, DC, 2002.
2. *Electricity Sector Framework for the Future*, Electric Power Research Institute (EPRI), Palo Alto, CA, www.epri.com/corporate/esff/default.asp; 2003.
3. *Electricity Technology Roadmap*, Electric Power Research Institute (EPRI), Palo Alto, CA , 2003.
4. Gellings, C.; Yeager, K. *Transforming the Electricity Infrastructure, Physics Today*; American Institute of Physics: College Park, MD, 2004; PP. 45–51.
5. Hirsch, R.F. *Technology and Transformation in the American Electric Utility Industry*; Cambridge University Press: Cambridge, U.K., 1989.
6. Hyman, L.S. *America's Electric Utilities—Past, Present and Future*, 6th Ed.; Public Utilities Reports, Inc.: Arlington, VA, 1997.
7. Nye, D.E. *Electrifying America*; The MIT Press: Cambridge, MA, 1990.
8. Patterson, W. *Transforming Electricity,* The Royal Institute of International Affairs; Earthbeam Publications Ltd.: London, 1999.
9. Reviving the Electricity Sector, National Commission on Energy Policy, Washington, DC, www.energycommission.org; 2003.
10. Schurr, S.H.; Burwell, C.C.; Devine, W.D., Jr.; Sonenblum, S. *Electricity in the American Economy, Contributions in Economies and Economic History-117*; Greenwood Press: New York, 1990.
11. Smil, V. *Energy at the Crossroads*; The Royal Institute of International Affairs, Earthbeam Publications Ltd.: London, 1999.
12. *Statistical Yearbook of the Electric Utility Industry*, Edison Electric Institute, Washington, DC, 2003.
13. Yeager, K. The Widening Technological Divide, *Public Utilities Fortnightly*; Washington, DC, 2005; 54 ff.

Electronic Control Systems: Basic

Eric Peterschmidt
Honeywell Building Solutions, Honeywell, Golden Valley, Minnesota, U.S.A.

Michael Taylor
Honeywell Building Solutions, Honeywell, St. Louis Park, Minnesota, U.S.A.

Abstract
Every piece of energy-consuming equipment has some form of control system associated with it. This article provides information about electronic control systems primarily used to control HVAC equipment. The same principles are used to control other equipment, such as lighting, compressed air systems, process equipment, and production equipment.

INTRODUCTION

Every piece of energy-consuming equipment has some form of control system associated with it. The controls can be as simple as a snap switch or as complicated as a dedicated microcomputer chip system. Larger pieces of equipment, along with buildings and industrial processes, typically use complex computer-based control systems to optimally control and operate them. This article provides information about electronic control systems primarily used to control HVAC equipment. However, the same technologies and principles are used to control other equipment, such as lighting, compressed air systems, process equipment and production equipment.

An electronic control system comprises a sensor, controller and final control element. The sensors used in electronic control systems are simple, low-mass devices that provide stable, wide-range, linear and fast response. The electronic controller is a solid-state device that provides control over a discrete portion of the sensor range and generates an amplified correction signal to control the final control element.

Features of electronic control systems include the following:

- Controllers can be remotely located from sensors and actuators.
- Controllers can accept a variety of inputs.
- Remote adjustments for multiple controls can be located together, even though sensors and actuators are not.
- Electronic control systems can accommodate complex control and override schemes.
- Universal type outputs can interface to many different actuators.
- Display meters can indicate input or output values.

The sensors and output devices (e.g., actuators, relays) used for electronic control systems are usually the same ones used on microprocessor-based systems. The distinction between electronic control systems and microprocessor-based systems is in the handling of the input signals. In an electronic control system, the analog sensor signal is amplified, then compared to a setpoint or override signal through voltage or current comparison and control circuits. In a microprocessor-based system, the sensor input is converted to a digital form, in which discrete instructions (algorithms) perform the process of comparison and control.

Fig. 1 shows a simple electronic control system with a controller that regulates supply water temperature by mixing return water with water from the boiler. The main temperature sensor is located in the hot water supply from the valve. To increase efficiency and energy savings, the controller resets the supply water temperature setpoint as a function of the OA (outdoor air) temperature. The controller analyzes the sensor data and sends a signal to the valve actuator to regulate the mixture of hot water to the unit heaters. These components are described in the section titled "Components."

A glossary of control system terms is given in the last section of this article.

Electronic control systems usually have the following characteristics:

Controller. Low voltage, solid state.
Inputs. 0–1 V d.c., 0–10 V d.c., 4–20 mA, resistance element, thermistor, thermocouple.
Outputs. 2–10 V d.c. or 4–20 mA device.
Control Mode. Two-position, proportional, proportional plus integral (PI) or step.

Keywords: Electronic control; HVAC controls; Automatic temperature controls; Sensors; Actuators; Controllers; Output devices; Indicating devices; Energy efficiency.

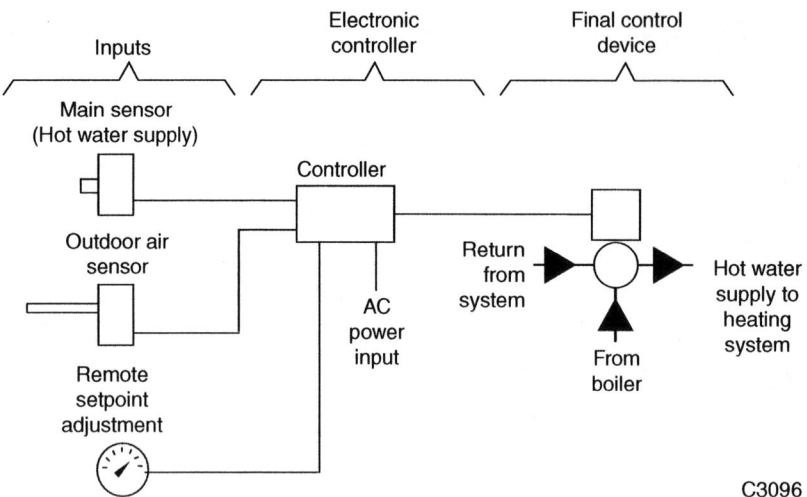

Fig. 1 Basic electronic control system.

Circuit diagrams in this article are basic and fairly general. A resistance-temperature input and a 2–10 V d.c. output are used for purposes of discussion. A detailed discussion on control modes can be found in the "Control Fundamentals" section of the Engineering Manual of Automatic Controls.[1]

ELECTRONIC CONTROL SYSTEM COMPONENTS

An electronic control system includes sensors, controllers, output devices such as actuators and relays; final control elements such as valves and dampers; and indicating, interfacing, and accessory devices. Fig. 2 provides a system overview for many electronic system components.

Sensors

A sensing element provides a controller with information concerning changing conditions. Analog sensors are used to monitor continuously changing conditions such as temperature or pressure. The analog sensor provides the controller with a varying signal such as 0–10 V. A digital (two-position) sensor is used if the conditions represent a fixed state such as a pump that is on or off. The digital

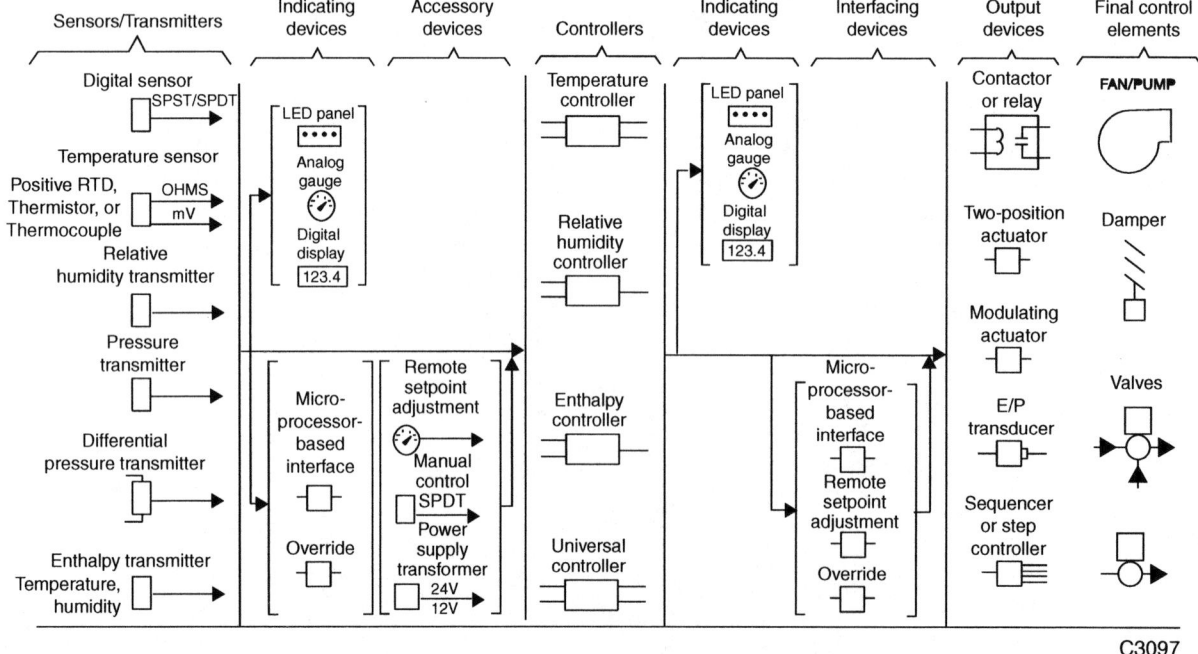

Fig. 2 Typical electronic control system components.

sensor provides the controller with a discrete signal such as open or closed contacts.

Some electronic sensors use an inherent attribute of their material (e.g., wire resistance) to provide a signal and can be directly connected to the electronic controller. Other sensors require conversion of the sensor signal to a type or level that can be used by the electronic controller. For example, a sensor that detects pressure requires a transducer or transmitter to convert the pressure signal to a voltage that can be used by the electronic controller. Typical sensors used in electronic control systems are included in Fig. 2. A sensor-transducer assembly is called a transmitter.

Temperature Sensors

For electronic control, temperature sensors are classified as follows:

- Resistance temperature devices (RTDs) change resistance with varying temperature. RTDs have a positive temperature coefficient (resistance increases with temperature).
- Thermistors are solid-state resistance-temperature sensors with a negative temperature coefficient.
- Thermocouples directly generate a voltage as a function of temperature.

Resistance Temperature Devices

In general, all RTDs have some common attributes and limitations:

- The resistance of RTD elements varies as a function of temperature. Some elements exhibit large resistance changes, linear changes, or both over wide temperature ranges.
- The controller must provide some power to the sensor and measure the varying voltage across the element to determine the resistance of the sensor. This action can cause the element to heat slightly—called self-heating—and can create an inaccuracy in the temperature measurement. By reducing the supply current or by using elements with higher nominal resistances, the self-heating effect can be minimized.
- Some RTD element resistances are as low as 100 Ω. In these cases, the resistance of the lead wires connecting the RTD to the controller may add significantly to the total resistance of the connected RTD, and can create an offset error in the measurement of the temperature. Fig. 3 shows a sensor and controller in relation to wire lead lengths. In this figure, a sensor 25 ft from the controller requires 50 ft of wire. If 18 AWG solid copper wire with a d.c. resistance of 6.39 Ω/Mft is used, the 50 ft of wire has a total d.c. resistance of 0.319 Ω. If the sensor is a 100-ohm platinum sensor with a

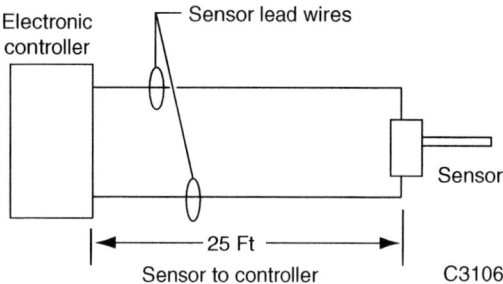

Fig. 3 Lead wire length.

temperature coefficient of 0.69 Ω/°F, the 50 ft of wire will introduce an error of 0.46°F. If the sensor is a 3000-ohm platinum sensor with a temperature coefficient of 4.8 Ω/°F, the 50 ft of wire will introduce an error of 0.066°F.

Significant errors can be removed by adjusting a calibration setting on the controller, or—if the controller is designed for it—a third wire can be run to the sensor and connected to a special compensating circuit designed to remove the lead length effect on the measurement. In early electronic controllers, this three-wire circuit was connected to a Wheatstone Bridge configured for lead wire compensation. In digital controllers, lead wire compensation on low resistance sensors may be handled by software offset.

- The usable temperature range for a given RTD sensor may be limited by non-linearity at very high or low temperatures.
- RTD elements that provide large resistance changes per degree of temperature reduce the sensitivity and complexity of any electronic input circuit. (Linearity may be a concern, however.)

A sensor constructed using a BALCO wire is a commonly used RTD sensor. BALCO is an annealed resistance alloy with a nominal composition of 70 percent nickel and 30 percent iron. A BALCO 500-ohm resistance element provides a relatively linear resistance variation from −40 to 250°F. The sensor is a low-mass device and responds quickly to changes in temperature.

Another material used in RTD sensors is platinum. It is linear in response and stable over time. In some applications, a short length of wire is used to provide a nominal resistance of 100 Ω. However, with a low resistance value, the element can be affected by self-heating and sensor-leadwire resistance. Additionally, due to the small amount of resistance change of the element, additional amplification must be used to increase the signal level.

To use the desirable characteristics of platinum and minimize any offset, one manufacturing technique deposits a film of platinum in a ladder pattern on an

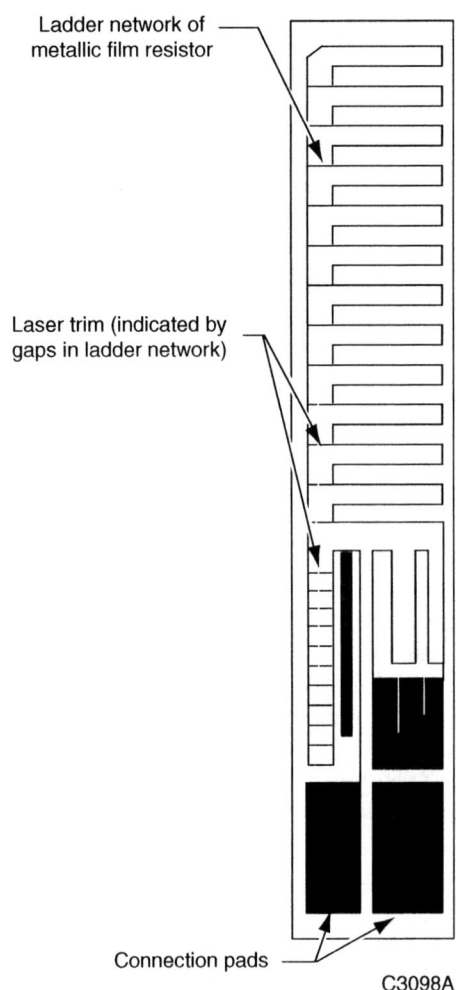

Fig. 4 Platinum element RTD sensor.

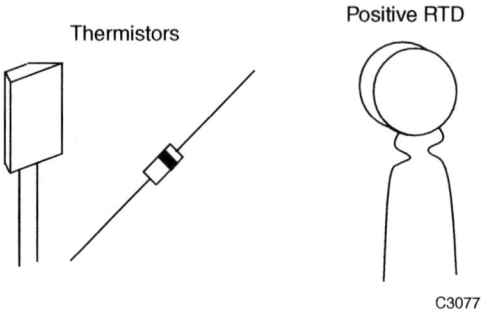

Fig. 5 Solid-state temperature sensors.

Positive temperature coefficient solid-state temperature sensors may have relatively high resistance values at room temperature. As the temperature increases, the resistance of the sensor increases (Fig. 6). Some solid-state sensors

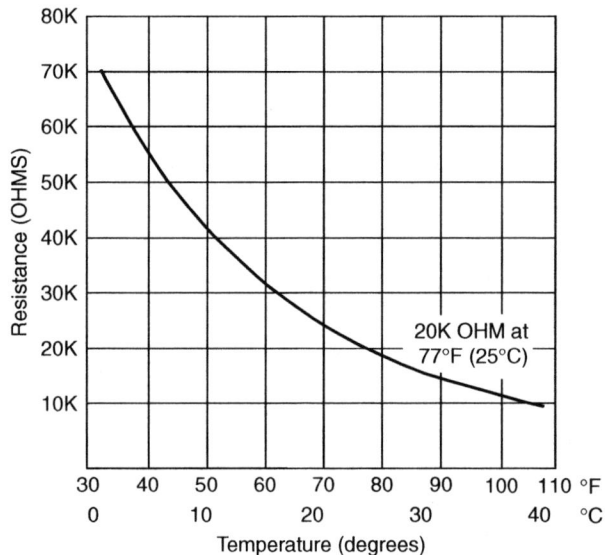

20K OHM NTC Thermistor

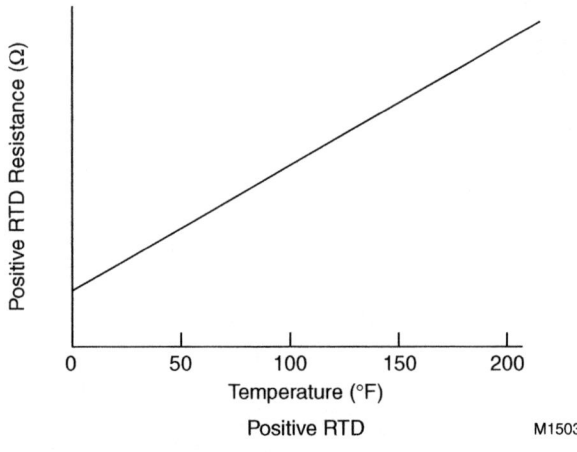

Positive RTD

Fig. 6 Resistance vs temperature relationship for solid-state sensors.

insulating base. A laser trimming method (Fig. 4) then burns away a portion of the metal to calibrate the sensor, providing a resistance of 1000 Ω at 74°F. This platinum film sensor provides a high resistance-to-temperature relationship. With its high resistance, the sensor is relatively immune to self-heating and sensor-leadwire resistance offsets. In addition, the sensor is an extremely low-mass device and responds quickly to changes in temperature. RTD elements of this type are common.

Solid-State Resistance Temperature Devices

Fig. 5 shows examples of solid-state resistance temperature sensors having negative and positive temperature coefficients. Thermistors are negative temperature coefficient sensors typically enclosed in very small cases (similar to glass diodes or small transistors) that provide quick response. As the temperature increases, the resistance of a thermistor decreases (Fig. 6). Selection of a thermistor sensor must consider the highly nonlinear temperature-resistance characteristic.

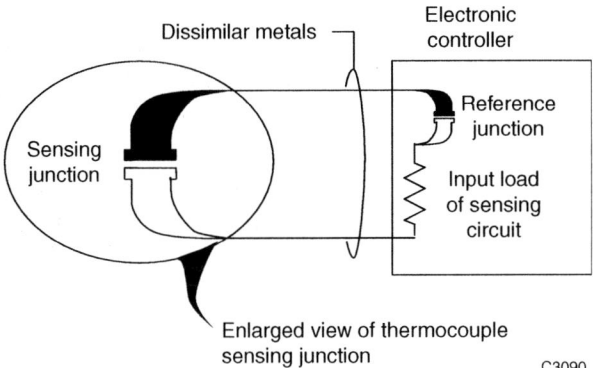

Fig. 7 Basic thermocouple circuit.

have near-perfect linear characteristics over their usable temperature range.

Thermocouples

In a thermocouple, two dissimilar metals such as iron and constantan are welded together to form a thermocouple junction (Fig. 7). When this junction is exposed to heat, a voltage in the millivolt range is generated and can be measured by the input circuits of an electronic controller. The amount of voltage generated is directly proportional to the temperature (Fig. 8). At room temperatures for typical HVAC applications, these voltage levels are often too small to be used, but are more usable at higher temperatures of 200°F–1600°F. Consequently, thermocouples are most common in high-temperature process applications.

Transmitter/Transducer

The input circuits for many electronic controllers can accept a voltage range of 0–10 V d.c. or a current range of 4–20 mA. The inputs to these controllers are classified as universal inputs because they accept any sensor having the correct output. These sensors are often referred to as transmitters as their outputs are an amplified or conditioned signal. The primary requirement of these transmitters is that they produce the required voltage or current level for an input to a controller over the desired sensing range.

Transmitters measure various conditions such as temperature, relative humidity, airflow, water flow, power consumption, air velocity and light intensity. An example of a transmitter would be a sensor that measures the level of carbon dioxide (CO_2) in the return air of an air handling unit. The sensor provides a 4–20 mA signal to a controller input, which can then modulate outdoor/exhaust dampers to maintain acceptable air quality levels. Since electronic controllers are capable of handling voltage, amperage or resistance inputs, temperature transmitters are not usually used as controller inputs within the ranges of HVAC systems due to their high cost.

Relative Humidity Sensor

Various sensing methods are used to determine the percentage of relative humidity, including the measurement of changes of resistance, capacitance, impedance and frequency.

An older method that used resistance to determine relative humidity depended on a layer of hygroscopic salt, such as lithium chloride or carbon powder, deposited between two electrodes (Fig. 9). Both materials absorb and release moisture as a function of the relative humidity, causing a change in resistance of the sensor. An electronic controller connected to this sensor detects the changes in resistance, which it can use to provide control of relative humidity.

A method that uses changes in capacitance to determine relative humidity measures the capacitance between two conductive plates separated by a moisture-sensitive material such as polymer plastic (Fig. 10A). As the material absorbs water, the capacitance between the plates decreases, and the change can be detected by an electronic circuit. To overcome any hindrance of the material's ability to absorb and release moisture, the two plates and their electric leadwires can be on one side of the polymer plastic, with a third sheet of extremely thin conductive material on the other side of the polymer plastic forming the capacitor (Fig. 10B). This third plate, too thin for attachment of leadwires, allows moisture to penetrate and be absorbed by the polymer, thus increasing sensitivity and response.

A relative humidity sensor that generates changes in both resistance and capacitance to measure moisture level is constructed by anodizing an aluminum strip and then applying a thin layer of gold or aluminum (Fig. 11). The anodized aluminum has a layer of porous oxide on its surface. Moisture can penetrate through the gold layer and fill the pores of the oxide coating, causing changes in both

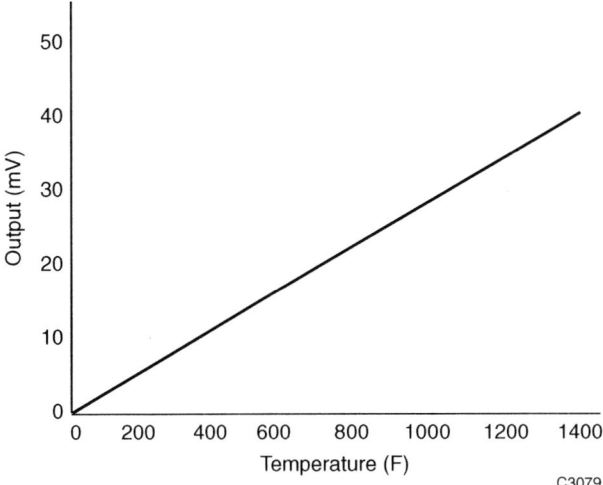

Fig. 8 Voltage vs temperature for iron-constantan thermocouple.

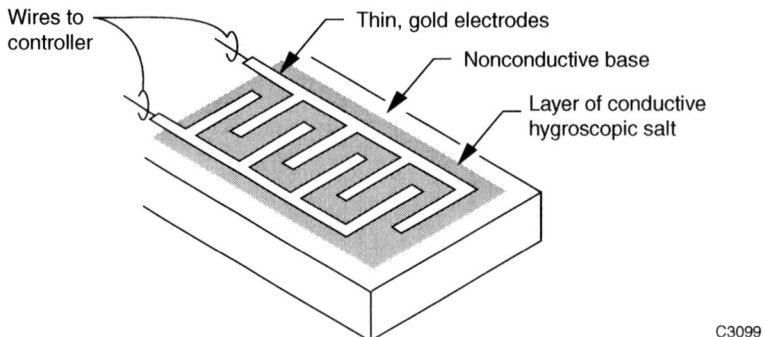

Fig. 9 Resistive-type relative humidity sensor.

resistance and capacitance that can be measured by an electronic circuit.

Sensors that use changes in frequency to measure relative humidity (Fig. 12) can use a quartz crystal coated with a hygroscopic material such as polymer plastic. When the quartz crystal is energized by an oscillating circuit, it generates a constant frequency. As the polymer material absorbs moisture and changes the mass of the quartz crystal, the frequency of oscillation varies and can be measured by an electronic circuit.

Most relative humidity sensors require electronics at the sensor to modify and amplify the weak signal and are referred to as transmitters. The electronic circuit compensates for the effects of temperature and both amplifies and linearizes the measured level of relative humidity. The transmitters typically provide a voltage or current output that can be used as an input to the electronic controller.

Pressure Sensors

An electronic pressure sensor converts pressure changes into a signal such as voltage, current or resistance that can be used by an electronic controller.

A method that measures pressure by detecting changes in resistance uses a small, flexible diaphragm and a strain gage assembly (Fig. 13). The strain gage assembly includes very fine (serpentine) wire or a thin metallic film deposited on a nonconductive base. The strain gage assembly is stretched or compressed as the diaphragm flexes with pressure variations. The stretching or compressing of the strain gage (shown by a dotted line in Fig. 13) changes the length of its fine wire or thin film metal, which changes the total resistance. The resistance can then be detected and amplified. These changes in resistance are small. Therefore, an amplifier is provided in the sensor assembly to amplify and condition the signal so the level

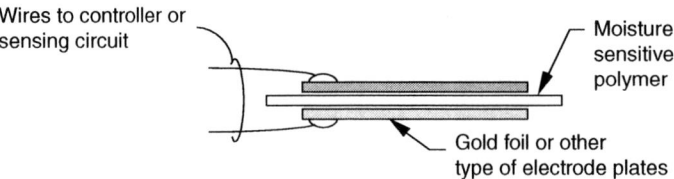

A. Moisture sensitive material between electrode plates.

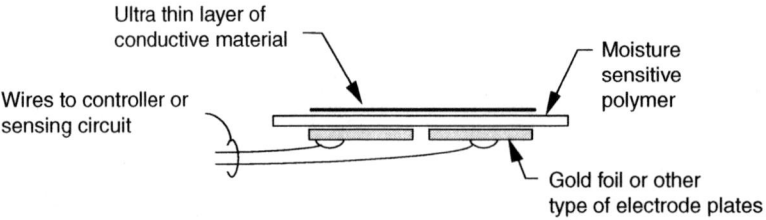

B. Moisture sensitive material between electrode plates and third conductive plate.

Fig. 10 Capacitance-type relative humidity sensor.

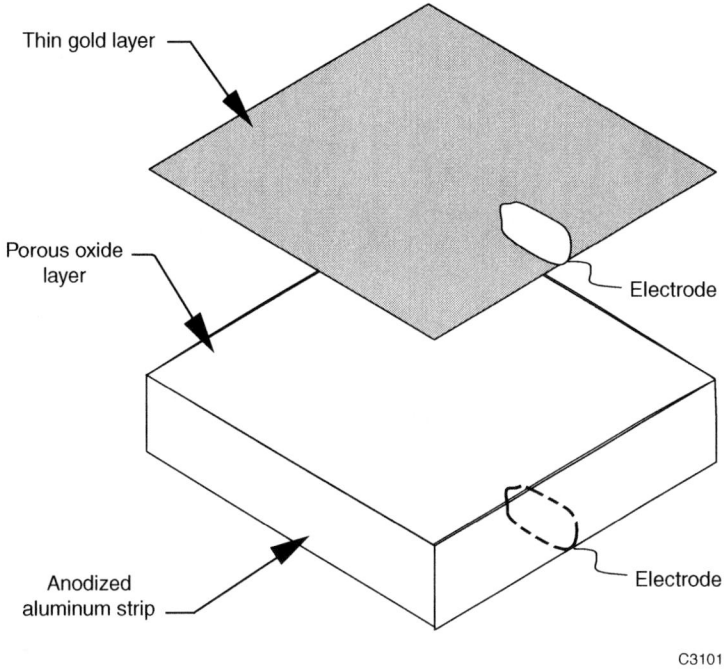

Fig. 11 Impedance-type relative humidity sensor.

sent to the controller is less susceptible to external noise interference. The sensor thus becomes a transmitter.

Another pressure sensing method measures capacitance (Fig. 14). A fixed plate forms one part of the capacitor assembly, and a flexible plate is the other part of the capacitor assembly. As the diaphragm flexes with pressure variations, the flexible plate of the capacitor assembly moves closer to the fixed plate (shown by a dotted line in Fig. 14) and changes the capacitance.

A variation of pressure sensors is one that measures differential pressure using dual pressure chambers (Fig. 15). The force from each chamber acts in an opposite direction with respect to the strain gage. This type of sensor can measure small differential pressure changes even with high static pressure.

Controllers, Output Devices and Indicating Devices

Controller

The electronic controller receives a sensor signal, amplifies and/or conditions it, compares it with the setpoint, and derives a correction if necessary. The output signal typically positions an actuator. Electronic controller circuits allow a wide variety of control functions and sequences, from very simple arrangements to multiple-input circuits with several sequential outputs. Controller circuits use solid-state components, such as transistors, diodes and integrated circuits, and include the power supply and all the adjustments required for proper control.

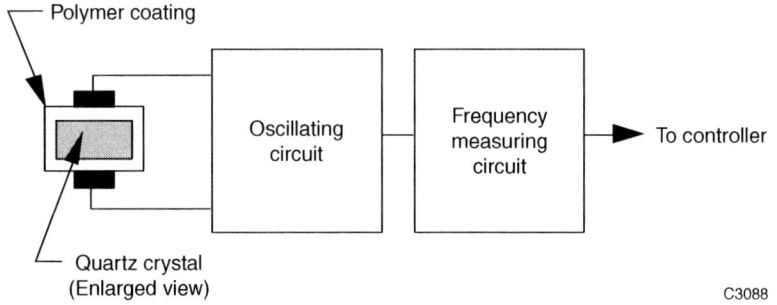

Fig. 12 Quartz crystal relative humidity sensor.

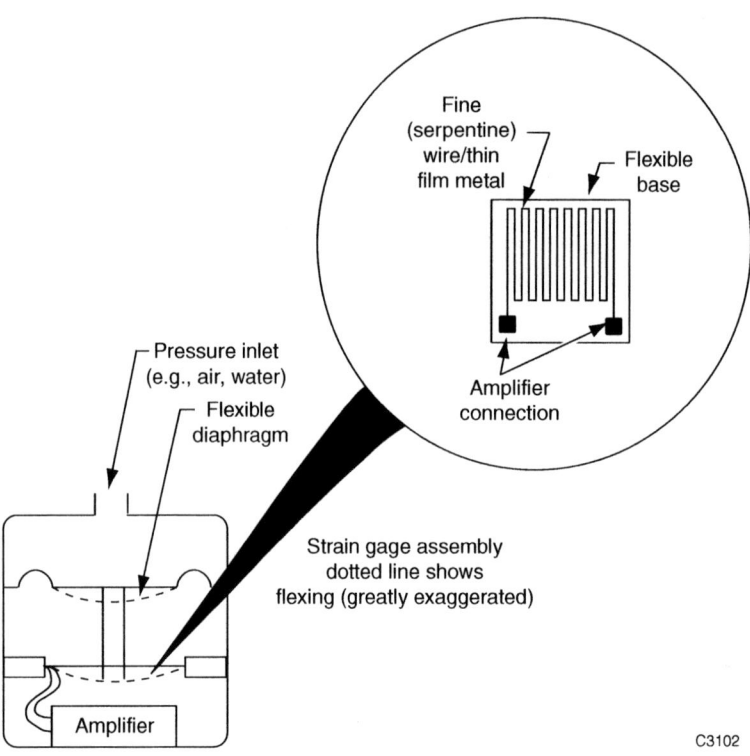

Fig. 13 Resistance-type pressure sensor.

Input Types

Electronic controllers are categorized by the type or types of inputs they accept, such as temperature, humidity, enthalpy or universal.

Temperature Controllers

Temperature controllers typically require a specific type or category of input sensors. Some have input circuits to accept RTD sensors such as BALCO or platinum elements, while others contain input circuits for thermistor sensors. These controllers have setpoint and throttling range scales labeled in degrees Fahrenheit or Celsius.

Relative Humidity Controllers

The input circuits for relative humidity controllers typically receive the sensed relative humidity signal already converted to a 0–10 V d.c. voltage, or a

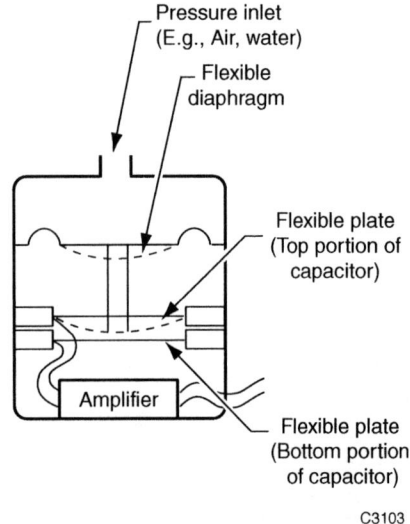

Fig. 14 Capacitance-type pressure transmitters.

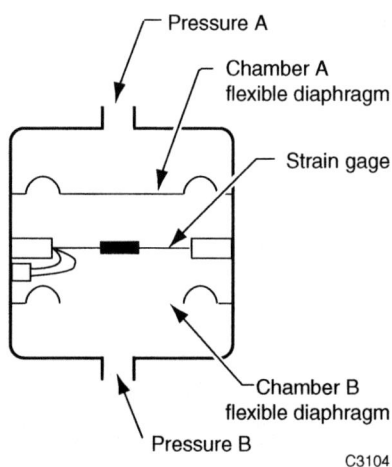

Fig. 15 Differential pressure sensor.

4–20 mA current signal. Setpoint and scales for these controllers are in percent relative humidity.

Enthalpy Controllers

Enthalpy controllers are specialized devices that use specific sensors for inputs. In some cases, the sensor may combine temperature and humidity measurements and convert them to a single voltage to represent enthalpy of the sensed air. In other cases, individual dry-bulb temperature sensors and separate wet-bulb or relative humidity sensors provide inputs, and the controller calculates enthalpy. In typical applications, the enthalpy controller provides an output signal based on a comparison of two enthalpy measurements, indoor and outdoor, rather than on the actual enthalpy value. In other cases, the return air enthalpy is assumed constant so that only OA enthalpy is measured. It is compared against the assumed nominal return air value.

Universal Controllers

The input circuits of universal controllers can accept one or more of the standard transmitter or transducer signals. The most common input ranges are 0–10 V d.c. and 4–20 mA. Other input variations in this category include a 2–10 V d.c. and a 0–20 mA signal. Because these inputs can represent a variety of sensed variables, such as a current of 0–15 A or pressure of 0–3000 psi, the settings and scales are often expressed in percent of full scale only.

Control Modes

The control modes of some electronic controllers can be selected to suit the application requirements. Control modes include two-position, proportional and proportional-integral. Other control features include remote setpoint, the addition of a compensation sensor for reset capability, and override or limit control.

Output Control

Electronic controllers provide outputs to a relay or actuator for the final control element. The output is not dependent on the input types or control method. The simplest form of output is two-position, in which the final control element can be in one of two states. For example, an exhaust fan in a mechanical room can be turned either on or off. The most common output form, however, provides a modulating output signal which can adjust the final control device (actuator) between 0 and 100 %, such as in the control of a chilled water valve.

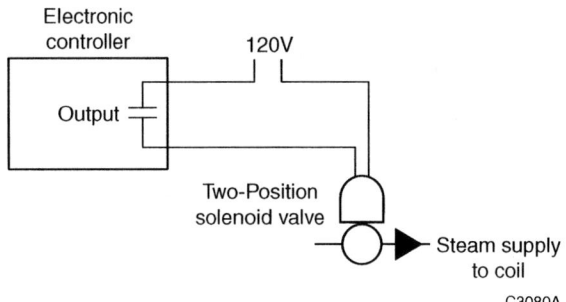

Fig. 16 Two-position control.

Output Devices

Actuators, relays, and transducers (Fig. 2) are output devices which use the controller output signal (voltage, current, or relay contact) to perform a physical function on the final control element such as starting a fan or modulating a valve. Actuators can be categorized as devices that provide two-position action or as those that provide modulating action.

Two-Position

Two-position devices such as relays, motor starters, and solenoid valves have only two discrete states. These devices interface between the controller and the final control element. For example, when a solenoid valve is energized, it allows steam to enter a coil that heats a room (Fig. 16). The solenoid valve provides the final action on the controlled media, steam. Damper actuators can also be designed to be two-position devices.

Modulating

Modulating actuators use a varying control signal to adjust the final control element. For example, a modulating valve controls the amount of chilled water entering a coil so that cool supply air is just sufficient to match the load at a desired setpoint (Fig. 17). The most common modulating actuators accept a varying voltage input of 0–10 V, or 2–10 V d.c., or a current input of 4–20 mA. Another form of actuator requires a pulsating (intermittent) or duty cycling signal to

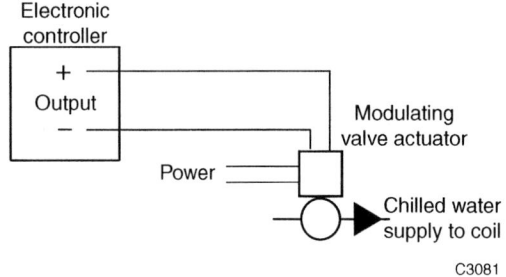

Fig. 17 Modulating control.

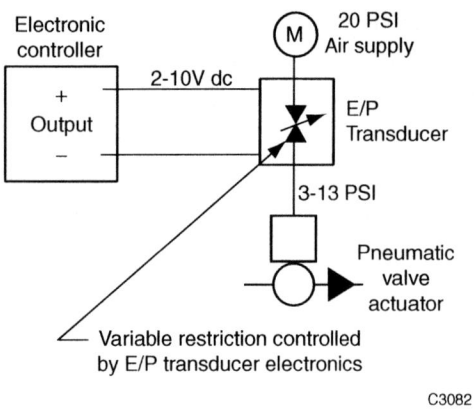

Fig. 18 Electric-to-pneumatic transducer.

perform modulating functions. One form of pulsating signal is a Pulse Width Modulation (PWM) signal.

Tranducer

In some applications, a transducer converts a controller output to a signal that is usable by the actuator. For example, Fig. 18 shows an Electronic-to-Pneumatic (E/P) transducer that converts a modulating 2–10 V d.c. signal from the electronic controller to a pneumatic proportional modulating 3–13 psi signal for a pneumatic actuator.

Indicating Devices

An electronic control system can be enhanced with visual displays that show system status and operation. Many electronic controllers have built-in indicators that show power, input signal, deviation signal and output signal. Fig. 19 shows some types of visual displays. An indicator light can show on/off status or, if driven by controller circuits, the brightness of a light can show the relative strength of a signal. If a system requires an analog or digital indicating device and the electronic controller does not include this type of display, separate indicating devices can be provided.

Interface with Other Systems

It is often necessary to interface an electronic control device to a microprocessor-based building management system or other related system. An example is an interface that allows a building management system to adjust the setpoint or amount of reset (compensation) for a specific controller. Compatibility of the two systems must be verified before they are interconnected.

ELECTRONIC CONTROLLER FUNDAMENTALS

General

The electronic controller is the basis for an electronic control system. Fig. 20 shows the basic circuits of an electronic controller including power supply, input, control and output. For greater stability and control, internal feedback correction circuits also can be included, but these are not discussed here. The circuits described provide an overview of the types and methods of electronic controllers.

Power Supply Circuit

The power supply circuit of an electronic controller provides the required voltages to the input, control, and output circuits. Most voltages are regulated DC voltages. The controller design dictates the voltages and current levels required.

All power supply circuits are designed to optimize both line and load regulation requirements within the needs and constraints of the system. Load regulation refers to the ability of the power supply to maintain the voltage output at a constant value even as the current demand (load) changes. Similarly, line regulation refers to the ability of the power supply to maintain the output load voltage at a constant value when the input (AC) power varies. The line regulation abilities or limitations of a controller are usually part of the controller specifications such as 120 V AC +10%, –15%. The degree of load regulation involves the end-to-end accuracy and repeatability, and is usually not explicitly stated as a specification for controllers.

TYPICAL SYSTEM APPLICATIONS

Fig. 21 shows a typical air-handling system controlled by two electronic controllers, C1 and C2; sequencer S; multi-compensator M; temperature sensors T1 through T4; modulating hot- and chilled-water valves V1 and V2; and outdoor, return, and exhaust air damper actuators. The control sequence is as follows:

- Controller C1 provides outdoor compensated, summer/winter control of space temperature for a heating/cooling system which requires PI control with a low limit. Sensor T4 provides the compensation signal through multi-compensator M, which allows one

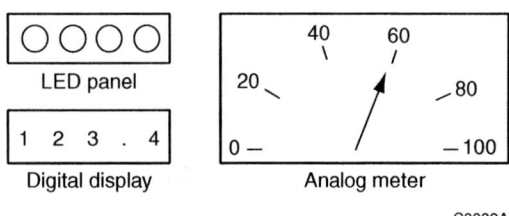

Fig. 19 Indicating devices.

Electronic Control Systems: Basic

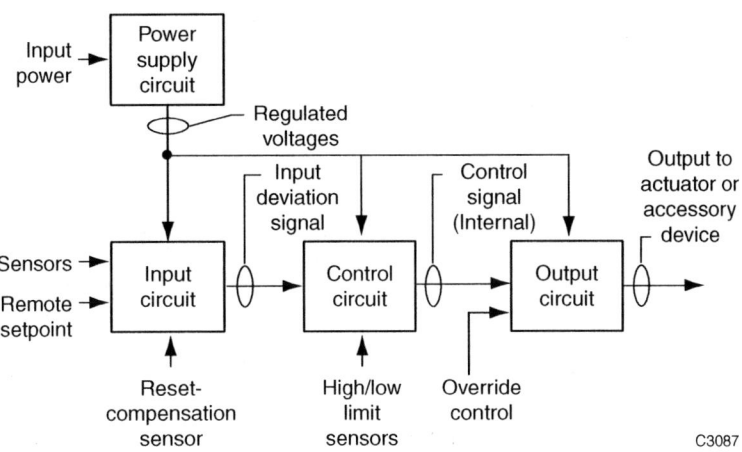

Fig. 20 Electronic controller circuits.

outdoor temperature sensor to provide a common input to several controllers. Controller C1 modulates the hot- and chilled-water valves V1 and V2 in sequence to maintain space temperature measured by sensor T1 at a pre-selected setpoint. Sequencer S allows sequencing the two valve actuators from a single controller. Low-limit sensor T2 assumes control when the discharge air temperature drops to the control range of the low-limit setpoint. A minimum discharge air temperature is maintained regardless of space temperature.

When the outdoor temperature is below the selected reset changeover point set on C1, the controller is in the winter compensation mode. As the outdoor air temperature falls, the space temperature setpoint is raised. When the outdoor temperature is above the reset changeover point, the controller is in the summer compensation mode. As the outdoor temperature rises, the space temperature setpoint is raised.

- Controller C2 provides PI mixed air temperature control with economizer operation. When the OA temperature measured by sensor T4 is below the setting of the economizer startpoint setting, the controller provides proportional control of the dampers to maintain mixed air temperature measured by

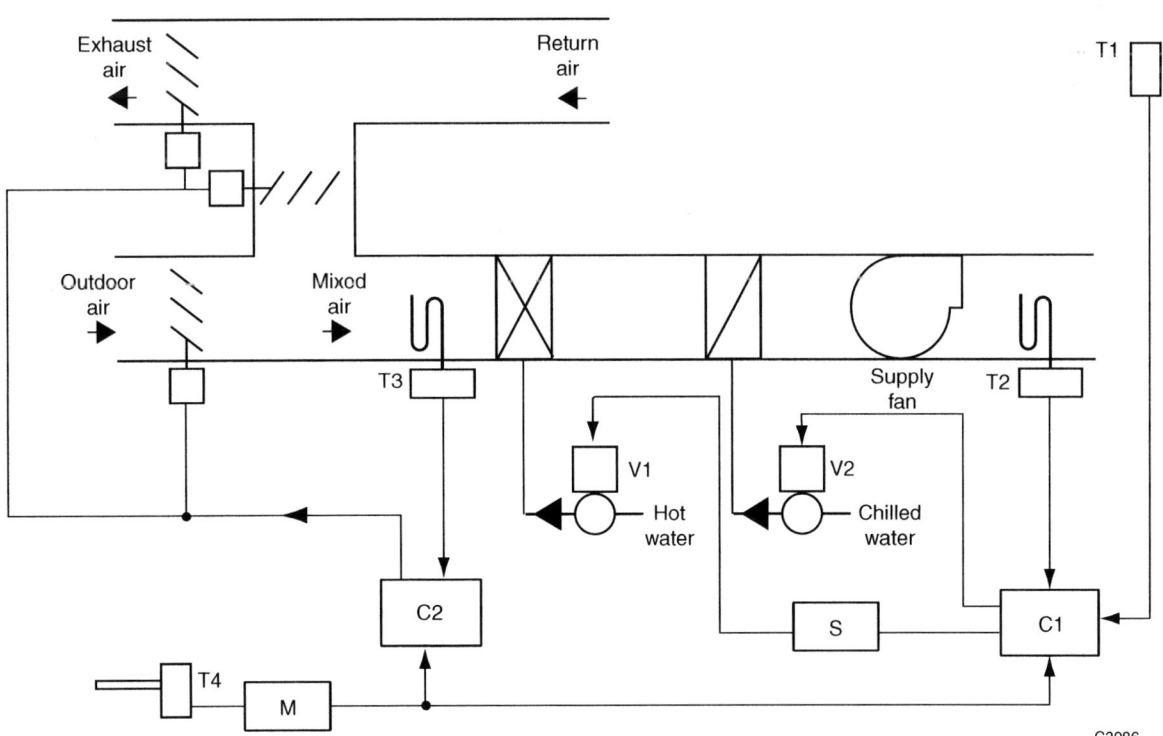

Fig. 21 Typical application with electronic controllers.

sensor T3 at the selected setpoint. When the OA temperature is above the economizer startpoint setting, the controller closes the OA dampers to a preset minimum.

ADDITIONAL DEFINITIONS

Authority (reset authority or compensation authority). A setting that indicates the relative effect a compensation sensor input has on the main setpoint (expressed in percent).

Compensation change-over. The point at which the compensation effect is reversed in action and changes from summer to winter or vice versa. The percent of compensation effect (authority) may also be changed at the same time.

Compensation control. See *Reset Control.*

Compensation sensor. See *Reset Sensor.*

Control Point. The actual value of a controlled variable (setpoint plus or minus offset).

Deviation. The difference between the setpoint and the value of the controlled variable at any moment. Also called "offset."

Direct acting. A direct-acting controller increases its output signal on an increase in input signal.

Electric control. A control circuit that operates on line or low voltage, and uses a mechanical means, such as a temperature-sensitive bimetal or bellows, to perform control functions, such as actuating a switch or positioning a potentiometer. The controller signal usually operates or positions an electric actuator, although relays and switches are often controlled.

Electronic control. A control circuit that operates on low voltage and uses solid-state components to amplify input signals and perform control functions, such as operating a relay or providing an output signal to position an actuator. Electronic devices are primarily used as sensors. The controller usually furnishes fixed control routines based on the logic of the solid-state components.

Electronic controller. A solid-state device usually consisting of a power supply, a sensor amplification circuit, a process/comparing circuit, an output driver section, and various components that sense changes in the controlled variable and derive a control output which provides a specific control function. In general, adjustments such as setpoint and throttling range necessary for the process can be done at the controller via potentiometers and/or switches.

Final control element. A device such as a valve or damper that changes the value of the manipulated variable. The final control element is positioned by an actuator.

Integral action (I). An action in which there is a continuous linear relationship between the amount of increase (or decrease) on the output to the final control element and the deviation of the controlled variable to reduce or eliminate the deviation or offset.

Limit sensor. A device which senses a variable that may be other than the controlled variable and overrides the main sensor at a preset limit.

Main sensor. A device or component that measures the variable to be controlled.

Negative (reverse) reset. A compensating action in which a decrease in the compensation variable has the same effect as an increase in the controlled variable. For example, in a heating application, as the outdoor air temperature decreases, the control point of the controlled variable increases. Also called "winter reset or compensation."

Offset. A sustained deviation between the control point and the setpoint of a proportional control system under stable operating conditions. Also called "deviation."

Positive (direct) reset. A compensating action in which an increase in the compensation variable has the same effect as an increase in the controlled variable. For example, in a cooling application, as the OA temperature increases, the control point of the controlled variable increases. Also called "summer reset or compensation."

Proportional band (throttling range). In a proportional controller, the control point range through which the controlled variable must pass to drive the final control element through its full operating range. Proportional band is expressed in percent of the main sensor span. A commonly used equivalent is "throttling range," which is expressed in values of the controlled variable.

Proportional control (P). A control algorithm or method in which the final control element moves to a position proportional to the deviation of the value of the controlled variable from the setpoint.

Proportional-integral (PI) control. A control algorithm that combines the proportional (proportional response) and integral or deviation control algorithms. Integral action tends to correct the offset resulting from proportional control. Also called "proportional plus reset" or "two-mode" control.

Remote setpoint. A means for adjusting the controller setpoint from a remote location, in lieu of adjusting it at the controller itself. The means of adjustment may be manual with a panel or space mounted potentiometer, or automatic when a separate device provides a signal (voltage or resistive) to the controller.

Reset control. A process of automatically adjusting the control point of a given controller to compensate for changes in a second measured variable such as outdoor air temperature. For example, the hot deck control point is reset upward as the outdoor air temperature decreases. Also known as "compensation control."

Reset sensor. The system element which senses a variable other than the controlled variable and resets the main sensor control point. The amount of this effect is established by the authority setting.

Reverse acting. A reverse-acting controller decreases its output signal on an increase in input signal.

Setpoint. The value on the controller scale at which the controller is set, such as the desired room temperature set on a thermostat. The setpoint is always referenced to the main sensor (not the reset sensor).

Throttling range. In a proportional controller, the control point range through which the controlled variable must pass to move the final control element through its full operating range. Throttling range is expressed in values of the controlled variable such as temperature in degrees Fahrenheit, relative humidity in percent, or pressure in pounds per square inch. A commonly used equivalent is "proportional band," which is expressed in percent of sensor span for electronic controls.

Transducer. A device that converts one energy form to another. It amplifies (or reduces) a signal so that the output of a sensor or transducer is usable as an input to a controller or actuator. A transducer can convert a pneumatic signal to an electric signal (P/E transducer) or vice versa (E/P transducer), or it can convert a change in capacitance to an electrical signal.

Transmitter. A device that converts a sensor signal to an input signal usable by a controller or display device.

CONCLUSION

Basic automatic electronic control systems are extremely important to provide desirable operational features of energy-using equipment and systems. Proper control is critical to achieving functional performance, as well as energy-efficient performance in equipment, buildings and processes. Electronic control systems hold a large share of the control technologies used in most of our modern energy control applications.

REFERENCES

1. Engineering Manual of Automatic Controls, Honeywell International, Golden Valley, MN, 2005; available online at www.buildingcontrolworkbench.com.

BIBLIOGRAPHY

1. Control Systems. In *Chapter 12 in Energy Management Handbook*; 5th Ed.; Turner, W.C., Ed.; Fairmont Press Atlanta, GA, 2005.
2. Control Systems and Computers. In *Chapter 9 in Guide to Energy Management*; 5th Ed.; Capehart, B.L., Turner, W.C., and Kennedy, W.J., Eds.; Fairmont Press Atlanta, GA, 2006.
3. HVAC Control Systems, ASHRAE Handbook of Fundamentas In *Chapter 15, American Society of Heating, Refrigerating, and Air Conditioning Engineers*, Atlanta, GA.
4. Wolovich, W.A. In *Automatic Control Systems: Basic Analysis and Design*; Oxford University Press: U.S.A., 1994.
5. Bradshaw, V. In *Building Control Systems*; Wiley: New York, 1993.
6. Kuo, B.C. *Automatic Control Systems*, 7th Ed.; Wiley: New York, 1995.
7. Underwood, C. *HVAC Control Systems*; Taylor and Francis Books, Ltd.: Washington, DC, 1998.
8. Bolton, W. *Control Engineering*, 2nd Ed.; Pearson Higher Education: 1998.
9. Bishop, R.H.; Dorf, R.C. *Modern Control Systems*, 9th Ed.; Pearson Higher Education: 2000.

Emergy Accounting

Mark T. Brown
Department of Environmental Engineering Sciences, University of Florida, Gainesville, Florida, U.S.A.

Sergio Ulgiati
Department of Sciences for the Environment, Parthenope, University of Naples, Napoli, Italy

Abstract
In this chapter, we briefly review H.T. Odum's concepts and principles of emergy and related quantities.[1–5] The concept of energy quality is introduced and defined by transformity and specific emergy. Tables are given of data on global emergy flows, from which the emergy and transformities of most products and processes of the biosphere are calculated. Tables of transformity and specific emergy for many secondary products are provided. Finally, the concept of net emergy yield is introduced and defined using an Emergy Yield Ratio (EYR).

DEFINITIONS

Energy is sometimes referred to as the ability to do work. Energy is a property of all things that can be turned into heat and is measured in heat units (Btus, calories, or joules).

Emergy is the availability of energy [exergy (See the entry "Exergy" in this same encyclopedia.) For high-quality flows, such as fuels and electricity, the energy content and the available energy do not differ significantly. For this reason, the energy of a flow instead of its exergy is sometimes used for the sake of simplicity.] of one kind that is used up in transformations directly and indirectly to make a product or service. The unit of emergy is the emjoule, a unit referring to the available energy of one kind consumed in transformations. For example, sunlight, fuel, electricity, and human service can be put on a common basis by expressing them all in the emjoules of solar energy that are required to produce each. In this case, the value is a unit of solar emergy expressed in solar emjoules (abbreviated sej). Although other units have been used, such as coal emjoules or electrical emjoules, in most cases all emergy data are given in solar emjoules.

The emjoule, short for "emergy joule," is the unit of measure of emergy. It is expressed in the units of the energy previously used to generate the product; the solar emergy of wood, for example, is expressed in the joules of solar energy that were required to produce the wood.

The emdollar (abbreviated em$) is a measure of the money that circulates in an economy as the result of some process. In practice, to obtain the emdollar value of an emergy flow or storage, the emergy is multiplied by the ratio of total emergy to the Gross National Product for the national economy.

Unit emergy values are calculated based on the emergy required to produce them. There are three types of unit emergy values, as follows:

Transformity is defined as the emergy per unit of available energy (exergy). For example, if 4000 solar emjoules are required to generate a joule of wood, the solar transformity of that wood is 4000 solar emjoules per joule (abbreviated sej/J). Solar energy is the largest, but most dispersed, energy input to the Earth. The solar transformity of the sunlight absorbed by the Earth is 1.0 by definition.

Specific emergy is the unit emergy value of matter defined as the emergy per mass, usually expressed as solar emergy per gram (abbreviated sej/g). Solids may be evaluated best with data on emergy per unit mass for its concentration. Because energy is required to concentrate materials, the unit emergy value of any substance increases with concentration. Elements and compounds not abundant in nature therefore have higher emergy/mass ratios when found in concentrated form, because more work was required to concentrate them, both spatially and chemically.

Emergy per unit money is a unit emergy value used to convert money payments to emergy units. The amount of resources that money buys depends on the amount of emergy supporting the economy and the amount of money circulating. An average emergy/money ratio in solar emjoules per dollar can be calculated by dividing the total emergy use of a state or nation by its gross economic product. It varies by country and has been shown to decrease each year. This emergy/money ratio is useful for evaluating service inputs given in money units when an average wage rate is appropriate.

Keywords: Energy quality; Emergy; Emergy yield ratio; Specific emergy; Transformity.

Empower is a flow of emergy (i.e., emergy per unit of time). Emergy flows are usually expressed in units of solar empower (solar emjoules per unit of time).

ENERGY, QUALITY, AND EMERGY

Probably the least understood and most criticized parts of H.T. Odum's body of work[1–5] are his concepts and theories of energy quality, which are embodied in the 35 year development of the emergy concept. The development of emergy and its theoretical base cannot be separated from the development of the concept of energy quality. Beginning in the early 1970s, Odum suggested that different forms of energy have different capacities to do work. He reasoned that whereas energy is measured in units of heat (Btus, joules, calories), not all calories are the same when it comes to work processes, especially complex work processes. All energies can be converted to heat at 100% efficiency; thus, it is relatively easy and accurate to express energies in their heat equivalents.

Although heat-equivalent energy is a good measure of the ability to raise the temperature of water, it is not a good measure of more complex work processes. Processes outside the window defined by heat engine technology do not use energies that lend themselves to thermodynamic heat transfers. As a result, converting all energies of the biosphere to their heat equivalents reduces all work processes of the biosphere to heat engines. Human beings, then, become heat engines, and the value of their services and information is nothing more than a few thousand calories per day. Different forms of energy have different abilities to do work. A calorie of sunlight is not the same as a calorie of fossil fuel or a calorie of food unless it is being used to power a steam engine. A system organized to use concentrated energies like fossil fuels (or food) cannot process a more dilute energy form like sunlight, calorie for calorie. By the same token, a system organized to process dilute energy like sunlight (a plant, for instance) cannot process more concentrated energy like fossil fuels.

In this way, the use and transformation of energy sources is system dependent; the appropriateness of an energy source in a particular system is dictated by its form and is related to its concentration. The processes of the biosphere are infinitely varied and are more than just thermodynamic heat engines. As a result, the use of heat measures of energy that can recognize only one aspect of energy—its ability to raise the temperature of things—cannot quantify adequately the work potential of energies used in more complex processes of the biosphere. In the larger biosphere system as a whole, energies should be converted to units that span this greater realm, accounting for multiple levels of system processes, ranging from the smallest scale to the largest scales of the biosphere, and accounting for processes other than heat engine technology.

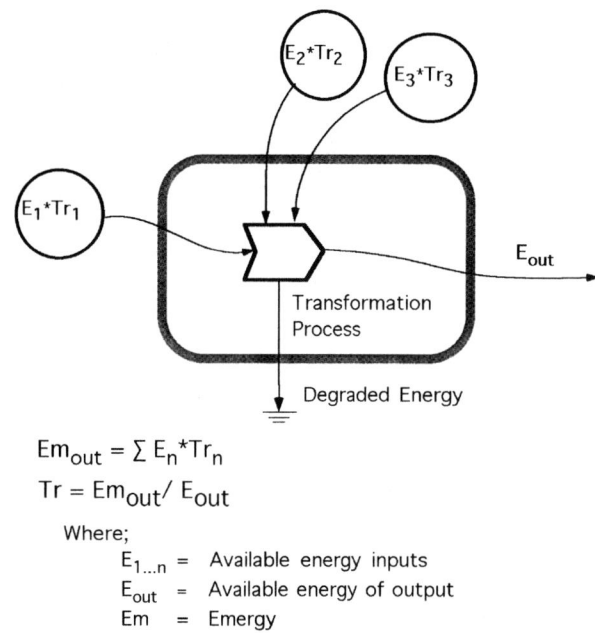

$$Em_{out} = \Sigma E_n * Tr_n$$
$$Tr = Em_{out} / E_{out}$$

Where;
$E_{1...n}$ = Available energy inputs
E_{out} = Available energy of output
Em = Emergy
Tr = Transformity

Fig. 1 In all processes, some energy is degraded and some is transformed into higher quality energy. The energy out is equal to the sum of the input energies minus the degraded energy. The emergy out is equal to the sum of the input emergies. The equations at the bottom of the figure show the general calculation of emergy of a product.

TRANSFORMITY AND SPECIFIC EMERGY

Transformity and specific emergy are unit emergy values calculated as the total amount of emergy required to make a product or service divided by the available energy of the product (resulting in a transformity) or divided by the mass of the product (resulting in a specific emergy). Figs. 1 and 2 illustrate the method of calculating a transformity, first in equation form (Fig. 1) and then with example numbers (Fig. 2). The transformity of the product is the emergy of the product divided by the energy of the product (in sej/J). If the output flow is in mass, the specific emergy of the product is the emergy of the output divided by the mass (in sej/g).

TRANSFORMITY AND QUALITY

Quality is a system property, which means that an "absolute" scale of quality cannot be made; neither can the usefulness of a measure of quality be assessed without first defining the structure and boundaries of the system. Self-organizing systems (be they the biosphere or a national economy) are organized with hierarchical levels (Fig. 3), and each level is composed of many parallel processes. This leads to two possible definitions of quality: parallel quality and cross quality. The first, parallel

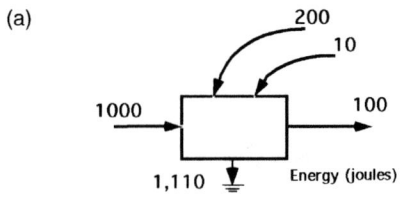
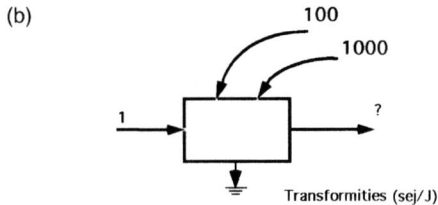
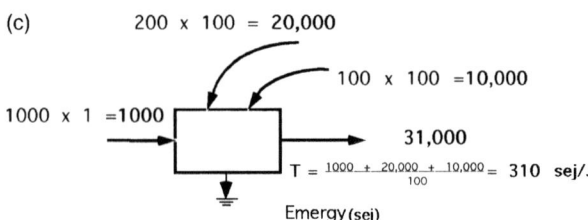

Fig. 2 Method of calculating transformity. (a) energy flows; (b) transformity of the output is calculated by dividing the emergy of the output in c by the energy of the output in a.

quality, is related to the efficiency of a process that produces a given flow of energy or matter within the same hierarchical level. (See Fig. 3 for an example of comparison among units in the same hierarchical level.) For any given output—say, electricity—there is almost an infinite number of ways to produce it, including all the generators, chemical processes, solar voltaic cells, and hydroelectric dams presently in service. A recent compilation of transformities for electricity from various production systems has yielded transformities from 6.23×10^4 sej/J, for a 2.5-MW wind generator, to 2.0×10^5 sej/J,

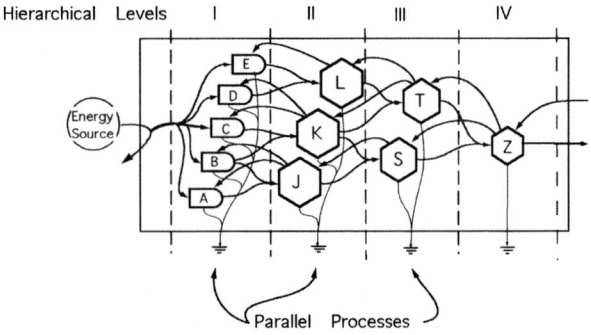

Fig. 3 Complex systems are organized hierarchically where it takes many small components to support the next level in the hierarchy which in turn supports even fewer components at the next level, and so on. Comparison between components of the same level is a comparison of parallel quality while comparison of components from different levels is a comparison of cross quality.

for a 1280-MW oil-fired power plant in Italy.[6] A mean value of 1.66×10^5 sej/J from these and other plants was suggested by Odum for electricity when the source is assumed to come from "average" conditions.[4] The same rationale can be used for any energy or material flow as long as the flow is believed to represent the average. Each individual process has its own efficiency, and as a result, the output has a distinct transformity. Quality as measured by transformity in this case relates to the emergy required to make like products under differing conditions and processes. For the most part, transformities of like products are within the same order of magnitude.

The second definition of quality, cross quality, is related to the hierarchical organization of systems. In this case, transformity is used to compare components or outputs from different levels of the hierarchy, accounting for the convergence of emergy at higher and higher levels. (See Fig. 3 for an example of comparison of transformity between different hierarchical levels.) At higher levels, a larger convergence of inputs is required to support the component: many villages are necessary to support a city, many kilograms of grass to support a cow, etc. Also, higher feedback and control ability characterize components at higher hierarchical levels. Therefore, higher transformity, as equated with a higher level in the hierarchy, often means greater flexibility and is accompanied by greater spatial and temporal effects. In this definition of quality, the higher the transformity, the higher the quality of the process or product. Transformities of products from different hierarchical levels usually differ by at least one order of magnitude.

EMERGY OF THE GEOBIOSPHERE: THE BASIS FOR COMPUTING UNIT EMERGY VALUES

Annual Budget of Emergy Supporting the Geobiosphere

An emergy evaluation table of the main inputs to the geobiosphere of the Earth (omitting, for the moment, the emergy use from nonrenewable resources) is given in Table 1. The annual budget of emergy flow (empower) supporting the geobiosphere (the atmosphere, ocean, and Earth's crust) includes solar energy, tidal energy, and heat energy from the deep Earth. These contributions to the geobiosphere total about 15.83×10^{24} sej/year.

Average Emergy Unit Values for Main Global Processes

Table 2 calculates unit emergy values for some main flows of the Earth. The total emergy input to the geobiosphere in solar emergy (15.83×10^{24} sej/year from Table 1) is divided by each global product's ordinary measure (the number of joules or grams). The unit values that result are

Table 1 Annual emergy contributions to global processes

Input	Inflow (J/year)	Emergy (sej/J)	Empower $\times 10^{24}$ (sej/year)
Solar insolation[a]	3.93×10^{24}	1.0	3.93
Deep earth heat[b]	6.72×10^{20}	1.20×10^4	8.06
Tidal energy[c]	0.52×10^{20}	7.39×10^4	3.84
Total	—	—	15.83

Not including non-renewable resources.
Abbreviations: sej, solar emjoules.
[a]Sunlight: solar constant 2 gcal/cm^2/min = 2 Langley per minute; 70% absorbed; earth cross section facing sun 1.27×10^{14} m^2.
[b]Heat release by crustal radioactivity 1.98×10^{20} J/year plus 4.74×10^{20} J/year heat flowing up from the mantle Ref. 16. Solar transformity 1.2×10^4 sej/J based on an emergy equation for crustal heat as the sum of emergy from earth heat, solar input to earth cycles, and tide Ref. 8.
[c]Tidal contribution to oceanic geopotential flux is 0.52×10^{20} J/year Ref. 17. Solar transformity of 7.4×10^4 sej/J is based on an emergy equation for oceanic geopotential as the sum of emergy from earth heat, solar input to the ocean, and tide following Refs. 8 and 18.
Source: From University of Florida (see Ref. 7).

useful for other emergy evaluations for which global averages can be used.

Temporary Emergy Inputs to the Geobiosphere

In the past two centuries, the production and consumption processes of human civilization that are using the large emergy in the geologic stores of fuels and minerals have reached a scale with global impact. Because these storages are being used much faster than they are being generated in geologic cycles, they are often called nonrenewable resources. They are actually very slowly renewed resources. Table 3 summarizes these additional components of the global emergy budget.

At present, the emergy from nonrenewable energy use that is contributed to the geobiosphere by human civilization is greater than the inputs from renewable sources. The result of this "temporary surge" of emergy is the accumulation of carbon dioxide in the atmosphere, adding to the greenhouse effects that may be altering ocean temperatures and, ultimately, the pattern and intensity of weather. The total renewable and nonrenewable emergy contributions to the global systems, including those released by humans, are 50.1×10^{24} sej/year. (Fig. 4).

UNIT EMERGY VALUES FOR FUELS AND SOME COMMON PRODUCTS

Unit emergy values result from emergy evaluations. Following are several tables of unit emergy values for some common materials and energy sources. In Table 4,

Table 2 Emergy of products of the global energy system

Product	Emergy[a] $\times 10^{24}$ (sej/year)	Production	Emergy
Global latent heat[b]	15.83	1.26×10^{24} J/year	12.6 sej/J
Global wind circulation[c]	15.83	6.45×10^{21} J/year	2.5×10^3 sej/J
Global precipitation on land[d]	15.83	1.09×10^{20} g/year	1.5×10^5 sej/g
Global precipitation on land[e]	15.83	5.19×10^{20} J/year	3.1×10^4 sej/J
Average river flow[f]	15.83	3.96×10^{19} g/year	4.0×10^5 sej/g
Average river geopotential[g]	15.83	3.4×10^{20} J/year	4.7×10^4 sej/J
Average river chem. energy[h]	15.83	1.96×10^{20} J/year	8.1×10^4 sej/J
Average waves at the shore[i]	15.83	3.1×10^{20} J/year	5.1×10^4 sej/J
Average ocean current[j]	15.83	8.6×10^{17} J/year	1.8×10^7 sej/J

[a]Main empower of inputs to the geobiospheric system from Table 1 not including non-renewable consumption (fossil fuel and mineral use).
[b]Global latent heat = latent heat of evapotranspiration 1020 mm/year, (1020 mm/year)(1000 g/m^2/mm)(0.58 Cal/g)(4186 J/Cal)(5.1×10^{14} m^2) = 1.26×10^{24} J/year.
[c]Global wind circulation, 0.4 watts/m^2 Ref. 19 (0.4 J/m^2/sec)(3.15×10^7 sec/year)(5.12×10^{14} m^2/earth) = 6.45×10^{21} J/year.
[d]Global precipitation on land = 1.09×10^{11} m^3/year Ref. 20 (1.09×10^{14} m^3)(1×10^6 kg/m^3) = 1.09×10^{20} g/year.
[e]Chemical potential energy of rain water relative to sea water salinity (1.09×10^{20} g/year)(4.94 J Gibbs free energy/g) = 5.19×10^{20} J/year.
[f]Global runoff, 39.6×10^3 km^3/year (Todd 1970) (39.6×10^{12} m^3/year)(1×10^6 g/m^3) = 3.96×10^{19} g/year.
[g]Average river geopotential work; average elevation of land = 875 m (39.6×10^{12} m^3/year)(1000 kg/m^3)(9.8 m/sec^2)(875 m) = 3.4×10^{20} J/year.
[h]Chemical potential energy of river water relative to sea water salinity (3.96×10^{19} g/year)(4.94 J Gibbs free energy/g) = 1.96×10^{20} J/year.
[i]Average wave energy reaching shores, Ref. 21 (1.68×10^8 Cal/m/year)(4.39×10^8 m shore front)(4186 J/Cal) = 3.1×10^{20} J/year.
[j]Average ocean current: 5 cm/sec Oort et al. 1989; 2 year turnover time (0.5)(1.37×10^{21} kg water)(0.050 m/sec)(0.050 m/sec)/(2 year) = 8.56×10^{17} J/year.
Source: From University of Florida (see Ref. 7).

Table 3 Annual emergy contributions to global processes including use of resource reserves

Inputs	Inflow (J/year)	Emergy[a]	Empower $\times 10^{24}$ (sej/year)
Renewable inputs[b]	—	—	15.8
Nonrenewable energies released by society:			
Oil[c]	1.38×10^{20}	9.06×10^4 sej/J	12.5
Natural gas (oil eq.)[d]	7.89×10^{19}	8.05×10^4 sej/J	6.4
Coal (oil eq.)[e]	1.09×10^{20}	6.71×10^4 sej/J	7.3
Nuclear power[f]	8.60×10^{18}	3.35×10^5 sej/J	2.9
Wood[g]	5.86×10^{19}	1.84×10^4 sej/J	1.1
Soils[h]	1.38×10^{19}	1.24×10^5 sej/J	1.7
Phosphate[i]	4.77×10^{16}	1.29×10^7 sej/J	0.6
Limestone[j]	7.33×10^{16}	2.72×10^6 sej/J	0.2
Metal ores[k]	9.93×10^{14}	1.68×10^9 sej/g	1.7
Total non-renewable empower			34.3
Total global empower			50.1

Abbreviations: sej, solar emjoules; t, metric ton; oil eq., oil equivalents.
[a]Values of solar emergy/unit from Ref. 4 and modified to reflect a global resource base of 15.83×10^{24} sej/year.
[b]Renewable Inputs: Total of solar, tidal, and deep heat empower inputs from Ref. 4.
[c]Total oil production = 3.3×10^9 t oil equivalent Ref. 8. Energy flux = $(3.3 \times 10^9$ t oil eq.$)(4.186 \times 10^{10}$ J/t oil eq.$) = 1.38 \times 10^{20}$ J/year oil equivalent.
[d]Total natural gas production = 2.093×10^9 m^3 Ref. 8. Energy flux = $(2.093 \times 10^{12}$ m$^3)(3.77 \times 10^7$ J m$^3) = 7.89 \times 10^{19}$ J/year.
[e]Total soft coal production = 1.224×10^9 t/year Ref. 8. Total hard coal production = 3.297×10^9 t/year Ref. 8. Energy flux = $(1.224 \times 10^9$ t/year$)(13.9 \times 10^9$ J/t$) + (3.297 \times 10^9$ t/year$)(27.9 \times 10^9$ J/t$) = 1.09 \times 10^{20}$ J/year.
[f]Total nuclear power production = 2.39×10^{12} kwh/year Ref. 8. Energy flux = $(2.39 \times 10^{12}$ kwh/year$)(3.6 \times 10^6$ J/kwh$) = 8.6 \times 10^{18}$ J/year electrical equivalent.
[g]Annual net loss of forest area = 11.27×10^6 ha/year Ref. 23. Biomass = 40 kg m^2; 30% moisture Ref. 24. Energy flux = $(11.27 \times 10^6$ ha/year$)(1 \times 10^4$ m^2/ha$)(40$ kg m$^2)(1.3 \times 10^7$ J/kg$)(0.7) = 5.86 \times 10^{19}$ J/year.
[h]Total soil erosion = 6.1×10^{10} t/year Refs. 25 and 26. Assume soil loss 10 t/ha/year and 6.1×10^9 ha agricultural land = 6.1×10^{16}/g/year (assume 1.0% organic matter), 5.4 Cal/g. Energy flux = $(6.1 \times 10^{16}$ g$)(0.01)(5.4$ Cal/g$)(4186$ J/Cal$) = 1.38 \times 10^{19}$ J/year.
[i]Total global phosphate production = 137×10^6 t/year Ref. 27. Gibbs free energy of phosphate rock = 3.48×10^2 J/g. Energy flux = $(137 \times 10^{12}$ g$)(3.48 \times 10^2$ J/g$) = 4.77 \times 10^{16}$ J/year.
[j]Total limestone production = 120×10^6 t/year Ref. 27. Gibbs free energy phosphate rock = 611 J/g. Energy flux = $(120 \times 10^{12}$ g$)(6.11 \times 10^2$ J/g$) = 7.33 \times 10^{16}$ J/year.
[k]Total global production of metals 1994: Al, Cu, Pb, Fe, Zn Ref. 28: 992.9×10^6 t/year = 992.9×10^{12} g/year.
Source: After Elsevier (see Ref. 30).

unit emergy values are given for primary nonrenewable energy sources. In some cases, the unit emergy value is based on only one evaluation—plantation pine, for example. In other cases, several evaluations have been done of the same primary energy but from different sources, and presumably different technology, so unit emergy value is an average. Obviously, each primary energy source has a range of values depending on source and technology. Because they use data from typical production facilities (and actual operating facilities), the unit emergy values represent average conditions and can be used for evaluations when actual unit values are not known. If it is known that the conditions under which an evaluation is being conducted are quite different from the averages suggested here, detailed evaluations of the sources should be conducted.

Table 5 lists the unit emergy values for some common products in order of their transformity. Only a few products are given in this table, but many more evaluations leading to unit emergy values have been conducted and are presented in a set of Emergy Folios published by the Center for Environmental Policy at the University of Florida.[7,8]

NET ENERGY AND EMERGY YIELD RATIO

The concept of net energy has played an important role in the development of energy quality and emergy. The concept of net production in ecosystems is widely used and understood as a measure of fitness of ecological systems. When applied to the human economy, the concept suggests that an energy source must be able to provide a net contribution to the economy of the larger system in which it is embedded (i.e., it must provide more energy than it costs in extraction and processing). Because this principle must be applicable to all living systems—from ATP-providing energy to the biochemical reactions in

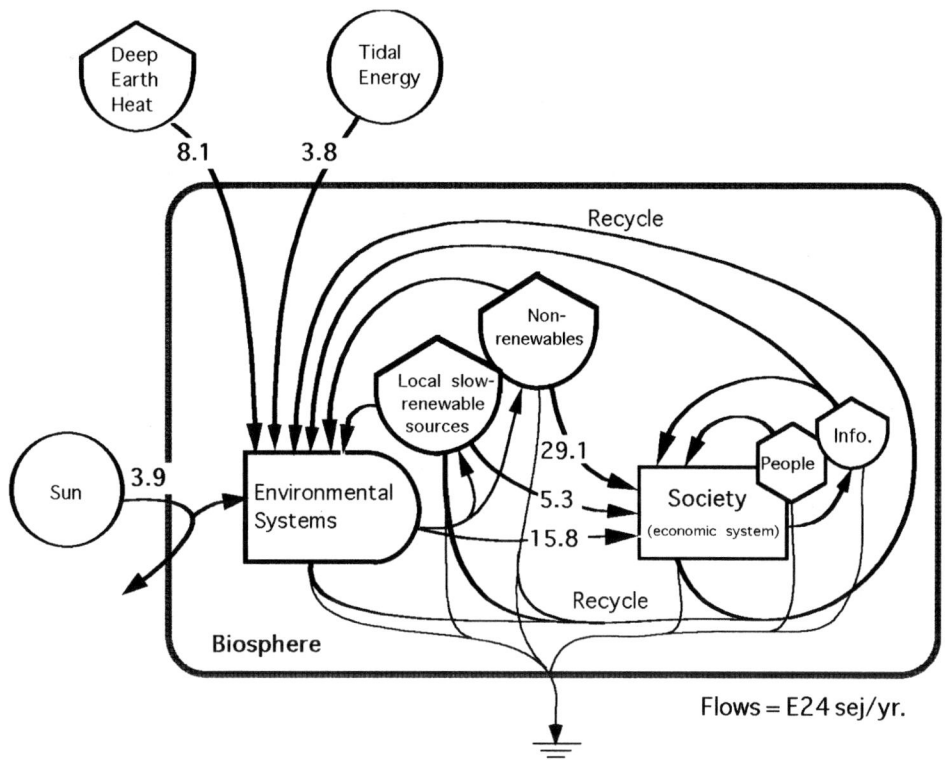

Fig. 4 Total emergy driving biosphere processes. For millions of years the driving emergy was dominated by the sources of sun, deep heat, and tidal momentum, now the dominant emergy flows are those associated with human society.

living systems, to photosynthesis, to the energy expended by animals as they graze or chase prey—it seems logical that it must also be applied to the processes of extracting fossil fuels from the Earth and to energy production of all sorts that drives economic sectors and human societies.

Table 4 Unit emergy values for primary non-renewable energy sources

	Transformity	
Item	sej/J	sej/g
Plantation pine (in situ)	1.1×10^4	9.4×10^7
Peat	3.2×10^4	6.7×10^8
Lignite	6.2×10^4	
Coal	6.7×10^4	
Rainforest wood (chipped, trans.)	6.9×10^4	4.1×10^8
Natural gas	8.1×10^4	
Crude oil	9.1×10^4	
Liquid motor fuel	1.1×10^5	
Electricity	3.4×10^5	

Source: From Wiley (see Ref. 4).

Odum suggested, "The true value of energy to society is the net energy, which is that after the costs of getting and concentrating that energy are subtracted."[9]

An Emergy Yield Ratio (EYR) is used to calculate the net contribution of energy sources to the economy. The EYR, as its name implies, is the ratio of the yield from a process (in emergy) to its costs (in emergy). The diagram in Fig. 5 illustrates the concept. The yield from this process is the sum of the input emergy from all sources: the environmental renewable source on the left (R), the nonrenewable storage (N), and the two purchased flows from the right (F). In the case of fossil fuels, there is little input from renewable sources, because the vast majority of the input comes from deep storages in the Earth. The EYR is the ratio of the yield (Y) to the costs (F) of retrieving it. The costs include energy, materials, and human service purchased from the economy, all expressed in emergy.

In practice, calculating the net energy of a process is far more complex than is shown in Fig. 5. Most processes, especially energy technologies, have many stages (or unit processes) and many inputs at each stage. As an example, Fig. 6 illustrates a series of processes beginning with an "energy crop"—fast-growing willow—and ending with wood chips that will be used in an electric power plant. At each stage in the process, an EYR can be calculated.

Table 5 Unit emergy values for some common products

Item	Transformity (sej/J)	Specific emergy (sej/g)
Corn stalks	6.6×10^4	
Rice, high energy[a]	7.4×10^4	1.4×10^9
Cotton	1.4×10^5	
Sugar (sugar cane)[b]	1.5×10^5	
Corn	1.6×10^5	2.4×10^9
Butter	2.2×10^6	
Ammonia fertilizer	3.1×10^6	
Mutton	5.7×10^6	
Silk	6.7×10^6	
Wool	7.4×10^6	
Phosphate fertilizer	1.7×10^7	
Shrimp (aquaculture)	2.2×10^7	
Steel[b]	8.7×10^7	7.8×10^9

[a]After Ref. 22.
[b]After Ref. 29.
Source: From Wiley (see Ref. 4).

Notice that the EYR decreases at each stage as more and more resources are used to process the wood into chips.

NET EMERGY AND TIME

Net energy is related to time, in that the longer a product has to develop, the higher its quality and the greater its net contribution. Doherty, for example, evaluated several processes that convert wood to higher-quality energy, such as ethanol.[10] Then he graphed the EYR vs the time it takes to grow the input wood. The graph in Fig. 7 resulted. The cycle time in this case is the time (in years) it takes to grow the wood from seeding to harvest. Some processes use fast-growing shrubby biomass (willow), whereas the longer cycle times are for climax rainforest species that take well over 100 years to mature. As the time increases, the quality of the wood (heat content) increases; thus, the EYR increases.

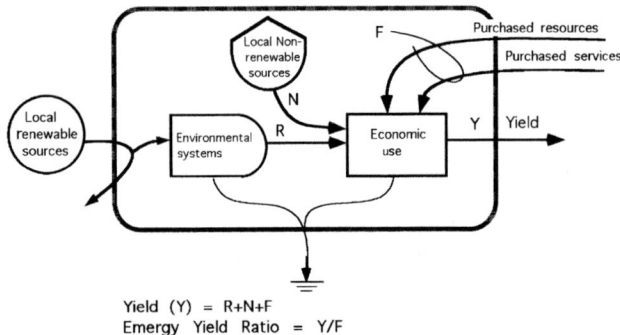

Fig. 5 Emergy Yield Ratio (EYR) is equal to the yield (expressed in emergy) divided by the purchased resources and services (also expressed in emergy).

Corn converted to ethanol is sometimes offered as an alternative fuel to fossil fuels. The impressive increases in yields per acre over time have been offered as evidence of the potential for corn to provide net energy and ultimately provide a renewable alternative to conventional fuels. The corn-to-ethanol process, however, uses a significant amount of fossil fuels directly and indirectly in growing, transporting, and conversion processes. In addition, the net yields are such that the result of using ethanol made from corn is that more energy will be consumed. Finally, there is no evidence that the net energy is increasing; therefore, the contribution to the economy is not increasing. Fig. 8 shows a graph of EYR and transformity for U.S. corn production from 1945 to 1993.[11–15] The EYR appears to be relatively constant at a value of around 1.7 over the past 55 years, which means that the contribution to the economy is not changing, even with increased yields. Transformities show declining values over this same period—probably the result of increased efficiency in the use of input resources and the fact that there have been increasing yields per hectare. As the EYR is constant, however, the net yield to society is not increasing.

NET EMERGY OF ENERGY SOURCES

Critical to continued prosperity, the net yields from fossil fuel energy sources that drive our economy are declining. As the richest and largest oil fields are tapped, and the remaining energy gets harder to find and even harder to drill for, the energy costs of obtaining oil and gas rise. As these limits are felt throughout modern economies, society looks to alternative sources: wind, waves, tides, solar, biomass, ethanol, and others. The graphs in Figs. 9 and 10 show the EYR for various energy sources used in modern economies. In Fig. 9, conventional nonrenewable sources are shown, and in Fig. 10, some of the so-called renewable energies are shown. It is imperative that the net contributions of proposed new energy sources be evaluated and all costs included. Many of the so-called renewable energy sources are actually guzzlers of fossil fuels. Take, for instance, proposed corn-to-ethanol programs. Our repeated evaluations over the past decade continue to show net yields of less than 2–1. Therefore, if ethanol is used to replace fossil fuels having yields of 8–1, the ethanol is actually using energy at four times the rate and increasing greenhouse-gas emissions over the burning of fossil fuels.

CONCLUSION

Emergy accounting—the process of evaluating the contributions of energy, material, and information inputs to processes—is intended to account for aspects that are

Emergy Accounting

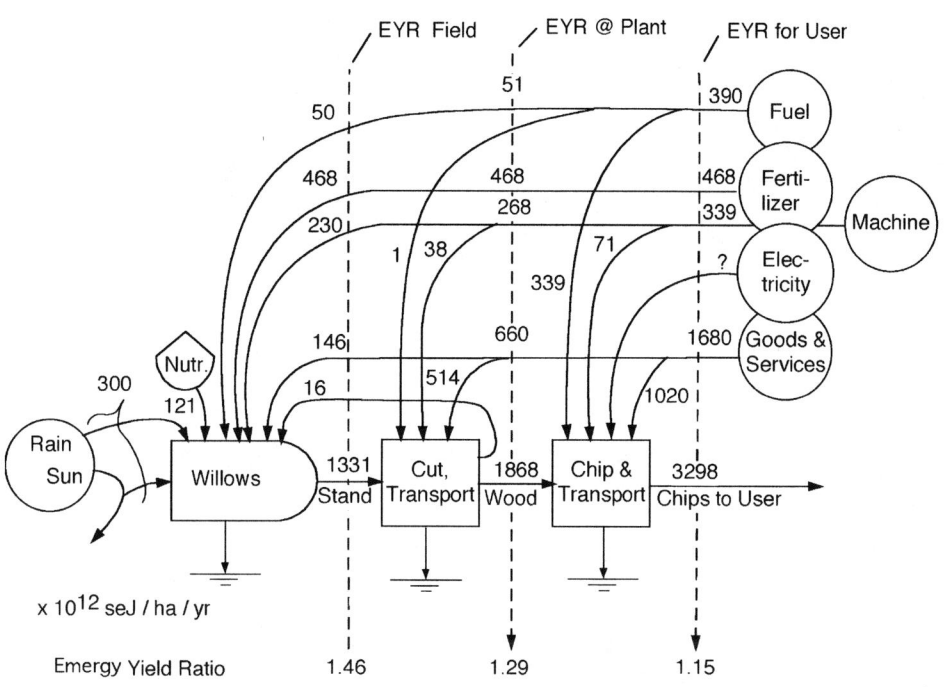

Fig. 6 EYR of fast growing willow crop used as a feed stock for electric production. At each step in the process an EYR can be calculated if all the inputs are known (dashed lines). The final EYR (at the right) is the ratio that the user receives.

usually not accounted for by other energy evaluation methods. Nonemergy approaches most often evaluate only nonrenewable resources and often do not account for the free services that a system receives from the environment (e.g., the photosynthetic activity driven by the solar radiation or the dilution of pollutants by the wind), which are just as much a requirement for the productive process as are fossil fuels. Finally, most nonemergy methods do not have an accounting procedure for human labor, societal services, and information. Emergy accounting includes all of these, perhaps not perfectly, but it places them in perspective and, thus, helps us understand the huge network of supporting energies necessary to support any particular economic activity in our culture.

The idea that a calorie of sunlight is not equivalent to a calorie of fossil fuel or electricity, or even to a calorie of human work, strikes many people as preposterous, because they believe that a calorie is a calorie is a calorie. Some have rejected the concept as being impracticable, because from their perspective, it is impossible to quantify the amount of sunlight that is required to produce a quantity of oil. Others reject it because emergy does not appear to conform to first-law accounting principles. The concept of energy quality has been most controversial, and emergy and transformity have been even more so. Although quality has been recognized somewhat in the energy literature, in which different forms of fossil energy are expressed in coal or oil equivalents (and some researchers have even expressed electricity in oil equivalents by using first-law efficiencies), there has been widespread rejection of quality corrections of other forms of energy.

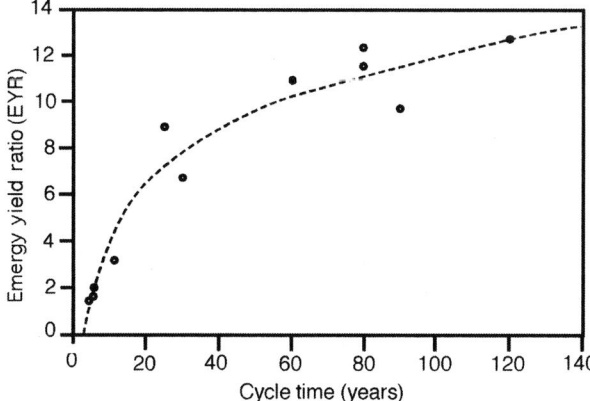

Fig. 7 The EYR is related to the cycle time of the resource. The graph shows the emergy yield ratio for 11 different forest cropping systems having very different cycle times, from willow systems of a couple of years to rainforest tress that require well over 100 years to mature. As the cycle time gets longer, the quality of the wood increases and the EYR increases.

The emergy approach represents a conceptual framework that is absolutely needed for a reliable investigation of the interplay of natural ecosystems and human-dominated systems and processes. The common thread is the ability to evaluate all forms of energy, materials, and human services on a common basis by converting them to equivalents of one form of energy: solar emergy, a measure of the past and present environmental support to

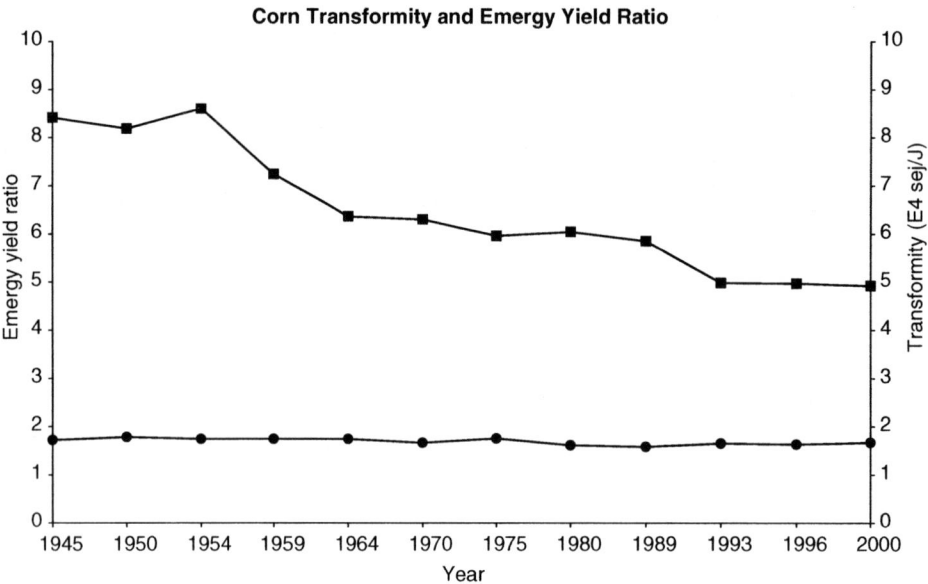

Fig. 8 Historical perspective on transformity and EYR of corn grown in the U.S.A. The transformity has declined over the years reflecting increased efficiencies and yields per acre. However, the EYR has remained essentially static during this time at about 1.7–1.

any process occurring in the biosphere. Through this quality correction, it is possible to evaluate all the inputs to processes and compute true net yields for processes, including potential energy sources. Without quality correction, net energy accounting can evaluate fossil energy return only for fossil energy invested; it cannot include human services, materials, and environmental services, in essence accounting for only a portion of the required inputs. The result can easily be a false assumption of the contributions from energy sources, but more important, this reasoning could lead to the wasteful use of energies in a futile pursuit of Maxwell's Demon.

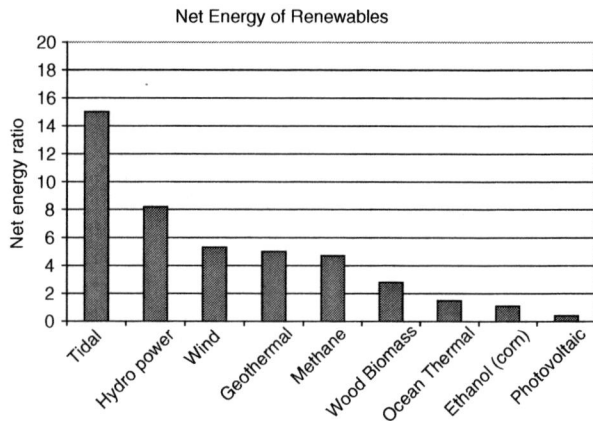

Fig. 10 Emergy yield ratios for renewable energy sources.

REFERENCES

1. Odum, H.T. *Environment, Power and Society*; Wiley: New York, 1971; 336.
2. Odum, H.T. Maximum power and efficiency: a rebuttal. Ecol. Model. **1983**, *20*, 71–82.
3. Odum, H.T. *Ecological and General Systems: An Introduction to Systems Ecology*; University Press of Colorado: Niwot, Colorado, 1994; [644, Revised edition of Systems Ecology, 1983, Wiley].
4. Odum, H.T. *Environmental Accounting. Emergy and Environmental Decision Making*; Wiley: New York, NY, 1996.
5. Odum, H.T.; Odum, E.C. *A Prosperous Way Down: Principles and Policies*; University Press of Colorado: Niwot, Colorado, 2001.

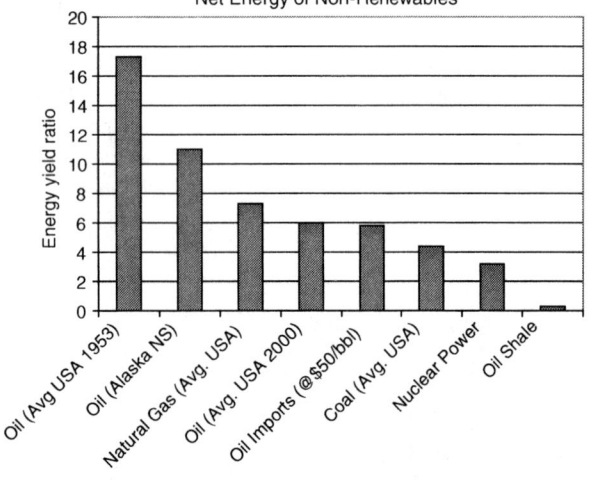

Fig. 9 Emergy yield ratios for major nonrenewable energy sources.

6. Brown, M.T.; Ulgiati, S. Emergy evaluation and environmental loading of electricity production systems. J. Clean. Prod. **2002**, *10*, 321–334.
7. Odum, H.T.; Brown, M.T.; Williams, S.B. *Handbook of Emergy Evaluation: A Compendium of Data for Emergy Computation Issued in a Series of Folios. Folio *1—Introduction and Global Budget*; Center for Environmental Policy, Environmental Engineering Sciences, University of Florida: Gainesville, FL, 2000; 16.
8. Odum, H.T. *Handbook of Emergy Evaluation: A Compendium of Data for Emergy Computation Issued in a Series of Folios. Folio #2—Emergy of Global Processes.* Center for Environmental Policy, Environmental Engineering Sciences, University of Florida: Gainesville, FL, 2000. British Petroleum *BP Statistical Review of World Energy*; The British Petroleum Company: London, 1997; 41.
9. Odum, H.T. Energy, ecology and economics. AMBIO **1973**, *2* (6), 220–227.
10. Doherty, S.J. Emergy Evaluations of And Limits to Forest Production. Ph.D. Dissertation, Environmental Engineering Sciences, University of Florida, Gainesville, 1995; 215.
11. Pimentel, D.; Hurd, L.E.; Bellotti, A.C.; et al. Food production and the energy crisis. Science **1973**, *182*, 443–449.
12. Pimentel, D.; Pimentel, M. *Food, Energy and Society*, Wiley: New York, 1979; 165.
13. Pimentel, D.; Warnaeke, A.F.; Teel, W.S.; et al. Food versus biomass fuel: socioeconomic and environmental impacts in the United States, Brazil, India and Kenya. Adv. Food Res. **1988**, *32*, 185–238.
14. Pimentel, D.; Wen, D. Technological changes in energy use in U.S. agricultural production. In *Agroecology*; Carrol, C.R., Vandermeer, J.H., Rosset, P.M., Eds., McGraw Hill: New York, NY, 1990; 147–164.
15. Giampietro, M.; Ulgiati, S.; Pimentel, D. Feasibility of large-scale biofuel production. Does an enlargement of scale change the picture? BioScience **1997**, *47* (9), 587–600.
16. Sclater, J.F.; Taupart, G.; Galson, I.D. The heat flow through the oceanic and continental crust and the heat loss of the earth. Rev. Geophys. Space Phys. **1980**, *18*, 269–311.
17. Miller, G.A. The flux of tidal energy out of the deep oceans. J. Geophys. Res. **1966**, *71*, 2485–2489 [Gainesville, 30].
18. Campbell, D. Emergy analysis of human carrying capacity and regional sustainability: an example using the state of Maine (Appendix). Environ. Monit. Assess. **1998**, *51*, 531–569.
19. Wiin-Nielsen, A.; Chen, T. *Fundamentals of Atmospheric Energetics*, Oxford Press: New York, 1993; 376.
20. Ryabchikov, A. *The Changing Face of the Earth*, Progress Publishers: Moscow, 1975; 203; [Translated by J. Williams].
21. Kinsman, B. *Wind, Waves, Their Generation and Propagation on the Ocean Surface*; Prentice-Hall: Englewood Cliffs, NJ, 1965; 676.
22. Brown, M.T.; McClanahan, T. Emergy analysis perspectives for Thailand and Mekong river dam proposals. Ecol. Model. **1996**, *91*, 105–130.
23. Brown, L.R.; Renner, M.; Flavin, C. *Vital Signs, 1997. The Environmental Trends That Are Shaping Our Future*; W.W. Morton & Co: New York, 1997; 165.
24. Leith, H.; Whittaker, R.H. *Primary Productivity of the Biosphere*; Springer: New York, 1975.
25. Oldeman, L.R. The global extent of soil degradation. In *Soil Resilience and Sustainable Land Use*; Greenland, D.J., Szabolcs, I., Eds., CAB International: Wallington, U.K., 1994; 561.
26. Mannion, A.M. *Agriculture and Environmental Change: Temporal and Spatial Dimensions*; Wiley: New York, 1995; 405.
27. USDI. *Mineral Commodity Summaries, January 1997*; U.S. Department of Interior: Washington, DC, 1996.
28. World Resources Institute *World Resources 1996–97*; Oxford University Press: New York, 1996.
29. Odum, E.C.; Odum, H.T. System of ethanol production from sugarcane in Brazil. Ciencia e Cultura **1984**, *37* (11), 1849–1855.
30. Brown, M.T.; Ulgiat, S. Emergy evaluation of the biosphere and natural capital. Ambio **1999**, *28* (6), 486–493.

BIBLIOGRAPHY

1. Oort, A. H., Ascher, S. C., Levitus, S., Peixoto, J. P. New estimates of the available potential energy in the World Ocean. J. Geophys. Res. **1989**, *94*, 3187–3200.
2. Todd, D. K.; ed. The Water Encyclopedia, 2nd Ed.; Lewis Publication: Chelsea, MI, 1970.

Emissions Trading

Paul M. Sotkiewicz
Public Utility Research Center, University of Florida, Warrington College of Business, Gainesville, Florida, U.S.A.

Abstract

Emission trading is a system of rights or permits that gives the holder the right to emit 1 unit of a designated pollutant. Permits or rights to pollute then can be considered an input to production and are priced like any other commodity. The idea behind emissions trading is to meet environmental goals at the lowest possible cost compared with other environmental policies, such as command-and-control (CAC) or emissions taxes. Simple in concept, emissions trading can become complex in practice, considering all elements that must be in place. The most widely used trading mechanism is cap-and-trade, wherein the overall emissions level is capped and permits can be traded between emissions sources. Other forms of trading, though not used as often as cap-and-trade, include offset or project trading and emission rate trading. Early experience with emissions trading via offset or project-based trading, as well as the current experience with cap-and-trade, have resulted in significant cost savings to participants vs traditional CAC methods but have not achieved all the cost savings available.

INTRODUCTION: IDEAS AND CONTEXT BEHIND EMISSIONS TRADING

The idea of emissions trading, popularized by Dales[1] and then formalized by Montgomery,[2] is to create a system of property rights or permits—or, as they are called in many trading programs, allowances—in the spirit of Coase[3] that would give the holders of the rights/permits the right to emit 1 unit of a pollutant. These rights/permits/allowances can be thought of as inputs to production much like any other input, such as coal, oil, or natural gas, and thus would have a market-determined price and be tradable like any other commodity. These rights have value because the number of rights available is limited (capped) either explicitly or implicitly. As shown by Montgomery[2] and reproduced in Baumol and Oates,[4] emissions trading has the property of meeting an aggregate emissions (reduction) target at the lowest possible cost, because trading provides the ultimate flexibility to polluting sources in how best to meet the emissions target. Sources not only have flexibility in choosing technologies or input mixes to minimize the cost of meeting emission targets at individual sources, but also can buy and sell the permits/rights/allowances to pollute among one another to allocate the burden of emissions reductions in such a way as to minimize the cost in aggregate across all sources. A cost-minimizing allocation of emissions reductions results in sources with low costs of abatement making greater reductions and those with higher costs of abatement making fewer reductions than they may otherwise make under command-and-control (CAC) policies. Thus, it can be said that the cost-saving benefits of emission trading vis-à-vis CAC policies is greater in proportion with the variability in emissions control costs.

As a policy option to achieve environmental compliance with pollution reduction goals, emissions trading is relatively new in its widespread application, though the first trading programs go back to the mid-1970s and have been used in a variety of contexts.[5] Prior to the launching of the first emissions trading schemes, the policy option to meet environmental objectives came in the form of CAC regulations that required emissions sources to meet a legislated emissions rate standards or to meet a stated technology standard. On a larger scale, the early policy for air pollution in the United States, beginning in 1970, mandated that specified concentration levels of pollutants be attained and then maintained at or below those levels going forward under the National Ambient Air Quality Standards (NAAQS). Many areas were in nonattainment of the standards, which would not permit the entry of new emission sources that would be associated with economic growth.[6] Consequently, the first emissions trading scheme, an offset policy or emission reduction credit (ERC) trading mechanism, was born out of the necessity to accommodate economic growth while moving toward attainment of the NAAQS in the middle 1970s.[7] The system was quite simple in concept. Existing sources in an area could reduce their emissions below an administratively defined baseline level and then could sell those offsets or ERCs to a new source entering the area at a price agreed upon by the parties. A variant of offset trading known as a bubble was introduced in 1979. The bubble provided flexibility to allocate emissions among multiple sources at the same facility (e.g., multiple generating units

Keywords: Emissions trading; Cap-and-trade; Emissions reduction credits; Offset trading; Least-cost emissions reduction.

at the same plant), so long as total facility emissions did not exceed a specified level.[8]

The movement to emissions trading as a policy option has also been driven by the cost of CAC policies relative to the least-cost way of meeting emissions standards. As shown in Portney,[6] a multitude of studies conducted during the 1980s showed ratios of CAC cost to least cost in a range from as low as 1.07 to as high as 22. The movement toward widespread application and acceptance of cap-and-trade programs led by the Title IV Sulfur Dioxide (SO_2) Trading Program (SO_2 Program) from the 1990 Clean Air Act Amendments (CAAA) can be seen as the meeting of environmental interests that want to see further emissions reductions with business and political interests that want to see market-driven policies.[9]

TYPES OF TRADING MECHANISMS

Cap-and-Trade

Under a cap-and-trade scheme, the aggregate level of emissions is capped, and property rights/permits/allowances are created such that the number of allowances available does not exceed the cap. Examples of cap-and-trade markets include the markets facilitated by the U.S. Environmental Protection Agency (USEPA), including the current SO_2 Program and NO_x SIP Call Program and the soon-to-be implemented Clean Air Interstate Rule (CAIR) and Clean Air Mercury Rule trading programs,[10] the Regional Clean Air Incentives Market (RECLAIM) in California,[11] and the European Union's Emissions Trading Scheme (EU ETS).[12] Cap-and-trade programs are perhaps the most used and visible of all emissions trading programs.

Offset or Project-Based Trading

In an offset or project-based trading scheme similar to that described above, potential emissions sources create credits by reducing emissions below their administratively determined baselines, so that credits can be sold to other sources that may be emitting more than their baselines. The ERC generated in this scheme is generally not a uniform commodity like the permit/property right/allowance that is defined under a cap-and-trade regime, but the number of ERCs created or needed is often determined on a project (case-by-case) basis. The spirit of an offset scheme is to cap emissions implicitly, though this is likely not the case in practice.[8] In-depth descriptions of such programs for the United States can be found in Hahn and Hester[13] and Environmental Law Institute.[14] An example in the context of carbon policy is the Clean Development Mechanism (CDM).[15]

Emissions Rate-Based Trading

In a rate-based trading environment, an emissions rate standard (e.g., lbs/mmBtu) is determined that must be met in aggregate, but sources can create credits by reducing emissions rates below the standard and sell these to sources with emissions rates above the standard. An example of this type of trading program exists for electric utility nitrogen oxide (NO_x) sources subject to Title IV of the 1990 CAAA.[16] Under this program, sources within the same company may trade credits to meet the NO_x emissions rate standard. Because credits are being traded to meet the standard, emission are in general not capped.[8]

ELEMENTS OF EMISSIONS TRADING PROGRAMS

As cap-and-trade emissions trading programs are the most prevalent, active, and visible, most of the elements in trading regimes are described with cap-and-trade in mind, though many of these elements also relate to other forms of trading in many cases. The format of this section closely follows U.S. EPA.[8]

Definition of Affected Sources

Determination of the emission sources to be included in the program (affected sources) is essential. Ideally, as many emissions sources as possible should be included in any trading program, but consideration must also be given to the size of the source, ability to monitor and report emissions from the source, and any other considerations that may be deemed important. Under the SO_2 Program, for example, existing simple cycle combustion turbines and steam units less than 25 MW in capacity were exempt from the program. One could surmise that such technologies were not large sources of SO_2 emissions or were too small to monitor in a cost-effective manner.

Measurement, Verification, and Emissions Inventory

Without the ability to measure emissions, emissions trading programs would not be workable. The measurement of emissions for the inventory can be done through a monitoring system or through the use of mass-balance equations. To verify emission monitoring, results can be checked against mass-balance equation derived emissions readings to ensure robust readings. The measurement of emissions prior to the commencement of a trading program can help provide a basis by which to set a cap and allocate permits/allowances in a cap-and-trade system, to set a baseline by which the emissions reductions can be measured in an offset system, or to determine emissions rates.

Determination of an Emissions Cap

In cap-and-trade systems, the element that makes emissions reductions valuable is the programwide limit on total emissions. The decision on the level of the emissions cap is as much political as it is scientific. In an ideal world with perfect information, the cap would be set so that the net benefits to society would be maximized (marginal costs of emissions reductions would equal the marginal benefits of reduction). Determining benefits is not as easy as determining costs of pollution reduction, however, though great strides have been made in recent years. As a matter of practice, although consideration is given to maximizing net benefits to society, the level of the cap is often determined through political means to gain wider stakeholder acceptance.[9]

Unit of Trade: Allowance/Permit/Emissions Reduction Credit

To facilitate trading among sources, it is crucial to define the units of trade between emissions sources. In the academic literature, these units are sometimes called permits. In the language of the U.S. EPA, these units are known as allowances in cap-and-trade systems, and as ERCs in offset and bubble systems in the United States. Regardless of the nomenclature, a permit/allowance/ERC gives the holder the right to emit 1 unit of pollutant where units can be defined in pounds, tons, kilograms, or any other accepted unit of measure. In effect, the allowance/permit/ERC is a property right to pollute and can be traded between sources at a price amenable to the parties, as any other commodity could be.

Compliance Period and True-Up

The time period for which emissions are to be controlled must be defined. For emissions in the SO_2 Trading Program, the compliance period is January 1–December 31, whereas in the NO_x OTC Market, the predecessor to the current NO_x SIP Call Market, it was May 1–September 31.[16] Sources must have allowances at least equal to their emissions during the compliance period. A trading regime may also allow a true-up period, during which sources may verify their actual emissions during the compliance period and then buy or sell allowances for the purposes of meeting the just-concluded compliance-period obligations.

Allowance/Permit Allocation or ERC Baseline

Under cap-and-trade, permits/allowances must be allocated to affected sources, or in the case of an offset system, the baseline must be established by which reductions are measured and ERCs are created.

With respect to cap-and-trade, there are three primary allocation methods: historical baseline, fixed; auction; and historical baseline with updating. Allocations may also be created for new units, or as a reward for undertaking certain actions to reduce emissions quickly or by other means. Under historical baseline, fixed methods, the allocation is *gratis* and is determined by a measure of performance for affected sources from the past. The performance measure could be based on output or input. Being based on the past, affected sources cannot engage in any behavior in an attempt to gain larger allowance allocations. For Phase I units in the SO_2 Program announced in 1990, for example, allocations were based on an emissions rate per unit of heat input from 1985 to 1987.

Under an auction allocation method, the allowances are sold directly to sources at a predetermined interval in advance of the time when affected sources will need the allowances to cover their emissions.

Under an updating methodology, allowance allocations beyond the first years of the program are determined based on updated performance measures such as heat input or output, rather than being permanently fixed to historic performance. Some countries in the EU ETS, for example, have decided to use an updating allocation method in which sources that are shut down permanently will have their allocations taken away.[17]

Choosing a baseline is crucial for offset programs, as the baseline determines how many ERCs are created through abatement. The determination of what the baseline might be varies across jurisdictions and is often open to negotiation in U.S.-based programs.[14]

Spatial and Temporal Trading Rules

The wider trading opportunities across space and time are, the greater is the potential for cost savings from trading. Still, political or environmental considerations may necessitate rules defining and restricting how trade can be made across space and time. If the pollutant being traded is seen to create greater damages where it is concentrated (such as mercury) or may become highly concentrated due to wind and weather patterns (NO_x and SO_2), it may be necessary to create spatial trading ratios that differ from a one-to-one exchange or to restrict trades from one zone to another, as has been done in the RECLAIM program.[18]

The ability to create ERCs or to save allowances for future use is known as banking. Banking ERCs or allowances is a way of trading between time periods and is allowed in many programs. Such a practice is warranted if concentration increases at a given point in time are not troublesome. But if pollutant concentrations increase at a given point in time, such as NO_x during summer ozone

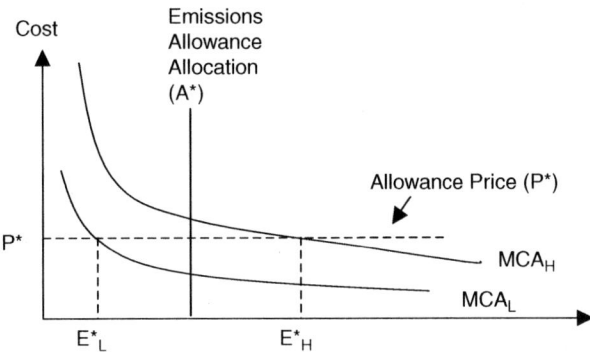

Fig. 1 Benefits from emissions trading.

season, banking may not be allowed such as in RECLAIM or by some states in the NO_x SIP Call Program.[18]

Penalties and Enforcement

All affected sources in a trading program must possess enough allowances to cover its emissions in a cap-and-trade program. Penalties and enforcement are necessary because without penalties or enforcement, there is no reason for sources to hold the necessary allowances to be in compliance. In cap-and-trade systems, a penalty per allowance not held, well in excess of the market price of allowances, for any shortfalls in allowances is necessary so that sources will participate in the market and maintain the emissions cap, and will not *de facto* opt out by paying no penalty or a small penalty.

FIRM INCENTIVES UNDER EMISSIONS TRADING

Consider a cap-and-trade system for electric generating units in which allowances have already been allocated, but keep in mind that the same logic applies to offset and emissions rate trading systems. If a generating unit has low abatement costs, that unit can reduce emissions below its allowance allocation and sell the remaining allowances or simply bank them for future use. As long as the marginal (incremental) cost of abatement (emissions reduction) is less than the allowance price, it pays the generating unit to reduce emissions further and sell the freed-up allowance, as shown in Fig. 1.

Marginal cost of abatement (MCA) MCA_L represents a low marginal abatement cost source. Being allocated A^* allowances, if the market price of allowances is P^*, it pays a generating unit that has low abatement costs to reduce emissions until it reaches E_L^*. The revenue from allowance sales is the rectangle with the width $A^* - E_L^*$ and height P^*. The cost to the utility company is the area under MCA_L between A^* and E_L^*. The net profit from the allowance sale is the area of the revenue rectangle above MCA_L. Conversely, a unit may have high abatement costs, represented by MCA_H. Rather than reduce pollution, that unit may find it less expensive to buy allowances in the open market and use the purchased allowances, along with the allowance allocation, to cover its emissions obligation. Units will continue buying allowances as long as the marginal (incremental) cost of abatement (emissions reduction) is greater than the allowance price. A more formal way of expressing this idea is that the unit with high abatement costs (Fig. 1) will buy $E_H^* - A^*$ allowances in the market at the price P^*. That unit's expenditure on allowances is the rectangle with width $E_H^* - A^*$ and height P^*. Because of its reduction in abatement costs, the area between A^* and E_H^* and below MCA_H, is greater than the expenditure on allowances, and the unit with high abatement costs will benefit. Also note that the allowance market leads to the equalization of the marginal costs of abatement across generating units.

COST-MINIMIZING POLLUTION ABATEMENT WITH EMISSIONS TRADING

Consider the following example of two firms with the objective of minimizing the cost of achieving the aggregate emissions restriction of 2000 tons. (Table 1). Let E_i in Table 1 represent the unrestricted or baseline emissions level for firm i. Let e_i be the emissions level of firm i after abatement, so that abatement for firm i is equal to $(E_i - e_i)$.

The least-cost solution for emissions abatement can be solved by minimizing the cost of abatement subject to the aggregate emissions restriction:

$$\text{Min}_{e_1, e_2} C_1(E_1 - e_1) + C_2(E_2 - e_2)$$

s.t. $e_1 + e_2 \leq A^*$

The solution to this problem requires that the MCA be equalized across the firms, as shown in the solution to this problem in Table 2. Also note in Table 2 that Firm 2 makes much larger reductions (2000 vs 400) than Firm 1, as its cost of abatement is only a fifth of that for Firm 1.

Table 1 Two-firm cost-minimizing example

	Firm 1	Firm 2
Unrestricted/baseline emissions (tons) E_i	2000	2400
Total cost of abatement function	$C_1(E_1-e_1)=0.5(E_1-e_1)^2$	$C_2(E_2-e_2)=0.1(E_2-e_2)^2$
Marginal cost of abatement function	$MCA_1=(E_1-e_1)$	$MCA_2=0.2(E_2-e_2)$
Aggregate emission restriction	$e_1+e_2\leq 2000$	

Now consider a cap-and-trade emissions trading program. Let X_i be the allowance allocation for firm i and x_i be the allowance purchase ($x_i>0$) or allowance sales ($x_i<0$) position of firm i. Let P be the price of allowances in the market. Each firm in the market minimizes its cost of pollution abatement and allowance purchases/sales subject to the restriction that emissions, e_i, are less than or equal to the allowance allocation plus the net position:

$$Min_{e_i,x_i} C_i(E_i - e_i) + Px_i$$

s.t. $e_i \leq X_i + x_i$

The solution to this problem for each firm requires that its MCA be equal to the allowance price P just as shown in Fig. 1, where the allowance price is the mechanism by which marginal costs of abatement are equalized across firms. Additionally, the aggregate emissions constraint must be satisfied $\Sigma_i\, e_i \leq \Sigma_i X_i$, and assuming no banking, the sum of allowance sales and purchases is equal to zero $\Sigma_i\, x_i = 0$.

Extending the example in Table 1, assume that each firm is initially allocated 1000 allowances signifying the right to emit 1000 ton. We know that each firm reduces emissions up to the point where MCA = P, and the MCAs are equal across firms. Consequently, we arrive at the same emissions outcome and MCA as the least-cost solution in Table 2. This results in Firm 2's having 600 surplus allowances, which it sells to Firm 1, which needs 600 allowances at a price of 400/ton (MCA). Table 3 shows the result.

It is important to note that the allowance purchases and sales cancel each other out in aggregate and that the actual abatement cost is the same as the least-solution found in Table 2.

An important lesson from the results in Table 3 is that emissions trading can achieve the least-cost solution without the need to collect detailed information on sources' abatement costs, and as we will see below, the method by which allowances are allocated does not change this result.

Allowance Allocation and Distribution of Costs

How allowances are allocated across firms, whether they are allocated gratis or by auction, the distribution of the initial allocation, once determined, does not change the aggregate abatement cost, although updating methods introduce other inefficiencies and effects, as discussed in Ahman et al.[17] and Burtraw et al.[19] Shifting allocations does change the distribution of the cost burden to meet the aggregate emissions constraint, however. In the previous example, we assumed that each firm was allocated 1000 allowances. Suppose instead that Firm 1 is allocated all 2000 allowances, and Firm 2 gets none. This does not change the optimizing behavior on how much is emitted; neither does it change the aggregate abatement cost. What is does do is change the allowance position of each of the firms: Firm 1 sells 400 ton, giving it allowance revenue of 160,000, and Firm 2 buys 400 ton, adding 160,000 in allowance costs. The allowance price, P, remains unchanged at 400. All that has changed is the distribution of the cost burden in meeting the aggregate emissions constraint.

Suppose that instead of there being a *gratis* allocation of allowances, the allowances were auctioned off, and the government kept the revenue for use elsewhere, such as offsetting other taxes. In this case, the allocations X_1 and X_2 are equal to zero, and the net allowance position for each firm is equal to the number of allowances they would need to satisfy their emissions constraints. Once again, the change in allocation method does not change the

Table 2 Least-cost solution to the two-firm example

	Firm 1	Firm 2
Emissions level, e_i	1,600	400
Abatement level, (E_i-e_i)	400	2,000
Total cost of abatement	80,000	400,000
Marginal cost of abatement	400	400
Aggregate abatement cost	480,000	

Table 3 Solution to the two-firm emission-trading example

	Firm 1	Firm 2
Allowance allocation, X_i	1,000	1,000
Emissions level, e_i	1,600	400
Abatement level, $(E_i - e_i)$	400	2,000
Allowance position, x_i	600	−600
Total cost of abatement	80,000	400,000
Marginal cost of abatement	400	400
Allowance price	400	
Allowance costs	240,000	−240,000
Aggregate abatement cost	480,000	

optimizing behavior of firms, as shown in Table 3, as they still produce the same emissions ($e_1 = 1600$, $e_2 = 400$); neither does it change the allowance price, which is still $P = 400$. What does change is the allowance cost for the firms. Under the gratis allocation firms are, in effect, being allocated a subsidy in the sense of not needing to pay for any costs associated with their emissions covered by the allocation as they would under an auction scheme. Under an auction, firms pay the government directly for their emissions through the purchase of allowances at auction. Given the optimal emissions levels and the allowance price, Firm 1 would pay 640,000 in allowance costs at auction, and Firm 2 would pay 160,000 in allowance costs, providing the government 800,000 in auction proceeds that were forgone with the gratis allocation scheme.

Emissions Trading Vs CAC

Suppose that the environmental regulator promulgated a CAC regime in which each firm had to reduce its emissions by 1200 tons, and an equal share of the reductions needed to get emissions down to 2000 tons. Such a regime leads to certainty regarding the emissions level, but mandating each firm to reduce by the same amount (in total quantity or percentage terms) is quite unlikely to lead to the least-cost solution. Table 4 shows the results of the above CAC scheme.

The aggregate abatement cost under this CAC regime is almost double the cost from emissions trading (864,000 vs 480,000). The MCAs are not equalized under CAC; the MCAs indicate that Firm 2 should engage in more abatement and Firm 1 should engage in less abatement activity in an effort to equalize the marginal costs across firms.

The only way in which the CAC regime could achieve the least-cost solution is to collect detailed information on the costs of abatement at the firm or source level so as to implement the least-cost outcome as the CAC target.

Emissions Trading Vs Emissions Taxes

Rather than using CAC or emissions trading to reduce emissions, the environmental authority wants to employ emissions taxes to reduce emissions. The incentives under emissions taxes are similar to those under emission trading, as shown in Fig. 1. Firms will want to reduce emissions until the marginal cost of abatement is equal to the tax rather than the allowance price. The difference between the two regimes involves the certainty with which an emissions target will be met. Under emissions trading, there is certainty about the emissions resulting from the program, assuming no banking, but the allowance price is uncertain, as it is determined endogenously. With emissions taxes, the price of emissions is certain, but the resulting emission level is determined endogenously.

Table 4 Command-and-control (CAC) costs

	Firm 1	Firm 2
Required reductions $(E_i - e_i)$	1,200	1,200
Emissions level, e_i	800	1,200
Total cost of abatement	720,000	144,000
Marginal cost of abatement	1,200	240
Aggregate abatement cost	864,000	

Table 5 Emissions tax of 300/ton results

	Firm 1	Firm 2
Emissions level, e_i	1,700	900
Reductions ($E_i - e_i$)	300	1,500
Total cost of abatement	45,000	81,000
Marginal cost of abatement	300	300
Aggregate emissions	2,600	
Aggregate abatement cost	126,000	

Suppose that the environmental regulator imposes an emission tax of 300 per ton. By design, the marginal costs of abatement are equalized across firms, thus minimizing the cost of meeting the uncertain emissions level. Table 5 shows the result for the tax of 300 per ton.

The resulting emissions of 2600 are greater than the target set forth under either emission trading or CAC, although this higher emissions level is achieved at least-cost. If the goal is to achieve the 2000-ton limit with emissions taxes, this would require a constant adjustment of the tax level until the goal is met. Such adjustments to the tax, however, would introduce uncertainty and increase risk for firms operating in their respective industries, and would likely be fought by the owners of the affected sources.

EXPERIENCE WITH EMISSIONS TRADING PROGRAMS AND CONCLUDING THOUGHTS

The early experiences with offset trading programs were that the programs achieved cost savings, but many opportunities for cost savings went unexploited due to administrative complexity and burden, and that the environmental improvements were not as great as was hoped.[7,13] More recent programs have seen little trading, as other environmental programs have resulted in greater reductions, reducing the demand for credits.[14]

Burtraw et al.[18] provides a survey of the performance of U.S. cap-and-trade programs. Ellerman et al.[9] offers a comprehensive analysis of the early years of the SO_2 Program. Burtraw and Evans[16] offer insight into federal NO_x trading programs. Overall, there is general agreement that the cap-and-trade programs in the United States have offered significant cost savings and technological innovation, and have resulted in significant emissions reductions. Moreover, no emissions hot spots or locally high concentrations have been found, as were feared by environmentalists. Still, there is a growing consensus that the existing programs have not achieved all the possible cost savings from trading, One possible explanation offered by Sotkiewicz and Holt[20] is that affected sources also face economic regulation by state public utility commissions which may provide incentives to affected sources that lead to deviations from the overall cost minimizing solution.

The EU ETS was only 18 months into operation at the time this article was written, and little can be said about its performance or the performance of the CDM to date. Still, the movement of the EU toward emissions trading, based on the U.S. experience, shows confidence in emissions trading; it also shows that the experience to date has been more positive than negative, and has delivered reduced emissions at lower cost than traditional CAC regimes.

REFERENCES

1. Dales, J.H. *Pollution, Property, and Prices*; University of Toronto Press: Toronto, Ont., Canada, 1968.
2. Montgomery, W.D. Markets in licenses and efficient pollution control programs. J. Econ. Theory **1972**, *5*, 395–418.
3. Coase, R.H. The problem of social cost. J. Law Econ. **1960**, *3*, 1–44.
4. Baumol, W.J.; Oates, W.E. *The Theory of Environmental Policy*, 2nd Ed.; Cambridge University Press: Cambridge, U.K., 1988.
5. Stavins, R.N. Market-based environmental policies. In *Public Policies for Environmental Protection*; 2nd Ed; Portney, P.R., Stavins, R.N., Eds.; Resources for the Future: Washington, DC, 2000.
6. Portney, P.R. Air pollution policy. In *Public Policies for Environmental Protection*; 2nd Ed; Portney, P.R., Stavins, R.N., Eds.; Resources for the Future: Washington, DC, 2000.
7. Hahn, R.W. Economic prescriptions for environmental problems: how the patient followed the doctor's orders. J. Econ. Perspect. **1988**, *3* (2), 95–114.
8. U.S. Environmental Protection Agency (USEPA). Tools of the trade: a guide to designing and operating a cap and trade program for pollution control, EPA430-B-03-002; Available at: www.epa.gov/airmarkets/international/tools.pdf (accessed on November 2006).
9. Ellerman, A.; et al. *Markets for Clean Air: The U.S. Acid Rain Program*; Cambridge University Press: Cambridge, U.K., 2000.
10. U.S. Environmental Protection Agency (USEPA). Clean Air Markets Program home page. Available at: www.epa.gov/airmarkets (accessed on November 2006).

11. South Coast Air Quality Management District (SCAQMD). RECLAIM home page. Available at: www.aqmd.gov/reclaim/index.htm (accessed on November 2006).
12. European Commission. Emission Trading Scheme (EU ETS) home page. Available at: http://ec.europa.eu/environment/climat/emission.htm (accessed on November 2006).
13. Hahn, R.W.; Hester, G.L. Where did all the markets go? An analysis of EPA's emissions trading program. Yale J. Regul. **1989**, *6* (109), 109–153.
14. Environmental Law Institute (ELI). *Emissions Reduction Credit Trading Systems: An Overview of Recent Results and an Assessment of Best Practices*, Washington, DC, (accessed on November 2006).
15. CDMWatch. *The Clean Development Mechanism (CDM) Toolkit: A Resource for Stakeholders, Activists, and NGOs.* Available at: www.cdmwatch.org/files/CDMToolkitVO19-02-04.pdf.
16. Burtraw, D.; Evans, D.A. NO_x emissions in the United States. In *Choosing Environmental Policy: Comparing Instruments and Outcomes in the United States and Europe*; Harrington, W., Morgenstern, R.D., Sterner, T., Eds.; Resources for the Future: Washington, DC, 2004.
17. Ahman, M. et al. *The Ten Year Rule: Allocation of Emission Allowances in the EU Emission Trading System*, Resources for the Future Discussion Paper 05–30, Washington, DC, June 2005.
18. Burtraw, D. et al. *Economics of Pollution Trading for SO_2 and NO_x*, Resources for the Future Discussion Paper 05–05, Washington, DC, January 2005.
19. Burtraw, D. et al. *The Effect of Allowance Allocation on the Cost of Carbon Emission Trading*, Resources for the Future Discussion Paper 01–30, Washington, DC, April 2001.
20. Sotkiewicz, P.M.; Holt, L. Public utility commission regulation and cost effectiveness of title IV: lessons for CAIR. Electricity J. **2005**, *18* (8), 68–80.

Energy Codes and Standards: Facilities

Rosemarie Bartlett
Pacific Northwest National Laboratory, Richland, Washington, U.S.A.
Mark A. Halverson
Pacific Northwest National Laboratory, West Halifax, Vermont, U.S.A.
Diana L. Shankle
Battelle Pacific Northwest National Laboratory, U.S. Department of Energy, Richland, Washington, U.S.A.

Abstract

Energy codes and standards play a vital role in the marketplace by setting minimum requirements for energy-efficient design and construction. They outline uniform requirements for new buildings as well as additions and renovations. This article covers basic knowledge of codes and standards; development processes of each; adoption, implementation, and enforcement of energy codes and standards; and voluntary energy efficiency programs.

INTRODUCTION

Energy-efficient buildings offer energy, economic, and environmental benefits. They reduce energy expenditures and environmental pollutants. They also create economic opportunities for business and industry by promoting new energy-efficient technologies.

Unfortunately, the marketplace does not guarantee energy-efficient design and construction. Owners of commercial buildings generally pass on energy costs to consumers or tenants, eliminating any incentive for energy-efficient design and construction. Homebuyers often are motivated more by up-front costs than operating costs.

Energy codes and standards play a vital role by setting minimum requirements for energy-efficient design and construction. They outline uniform requirements for new buildings as well as additions and renovations.

THE DIFFERENCE BETWEEN ENERGY CODES AND ENERGY STANDARDS, AND THE MODEL ENERGY CODE

Energy codes—specify how buildings must be constructed or perform and are written in mandatory, enforceable language. States or local governments adopt and enforce energy codes for their jurisdictions. Residential and commercial energy codes typically include requirements for building envelopes, mechanical systems, service water heating, and lighting and electrical power.

Energy standards—describe how buildings *should* be constructed to save energy cost-effectively. They are published by national organizations such as the American Society of Heating, Refrigerating, and Air-Conditioning Engineers (ASHRAE). They are not mandatory, but serve as national recommendations, with some variation for regional climate. States and local governments frequently use energy standards as the technical basis for developing their energy codes. Some energy standards are written in mandatory, enforceable language, making it easy for jurisdictions to incorporate the provisions of the energy standards directly into their laws or regulations. Residential and commercial energy standards typically include requirements for building envelopes, mechanical systems, service water heating, and lighting and electrical power.

The model energy code (MEC)[*]—The International Code Council (ICC) publishes and maintains the International Energy Conservation Code (IECC), which is an MEC that makes allowances for different climate zones. Because it is written in mandatory, enforceable language, state and local jurisdictions can easily adopt the model as their energy code. Before adopting the IECC, state and local governments often make changes to reflect regional building practices.

Table 1 provides an overview of energy standards and the MEC.

How are Energy Standards Developed and Revised?

Standards 90.1 and 90.2 are developed and revised through voluntary consensus and public hearing processes that are critical to widespread support for their adoption.

Keywords: Energy; Code; Standard; ASHRAE; IECC; ICC; Adoption; Enforcement.

[*] MEC in this article refers to any model energy code, not specifically to the predecessor to the IECC.

Table 1 Overview of national energy standards and the model energy code (MEC)

Title	Type	Sponsoring organization(s)	Description	Commonly used versions
International Energy Conservation Code (IECC)	MEC	International Code Council (ICC)	Applies to residential and commercial buildings. Written in mandatory, enforceable language	1998 IECC 2000 IECC 2003 IECC 2006 IECC
American Society of Heating, Refrigerating, and Air-Conditioning Engineers (ASHRAE)/IESNA/ANSI standard 90.1: Energy-Efficient Design of New Buildings Except Low-Rise Residential Buildings	Energy standard	ASHRAE, together with the Illuminating Engineering Society of North America (IESNA) and the American National Standards Institute (ANSI)	Applies to all buildings except residential buildings with three stories or less	90.1-1989 90.1-1999[a] 90.1-2001 90.1-2004
ASHRAE Standard 90.2 Energy-Efficient Design of New Low-Rise Residential Buildings	Energy standard	ASHRAE	Applies to residential buildings with three stories or less	90.2-1993 90.2-2001 90.2-2004

[a]This and subsequent versions written in mandatory, enforceable language.

Who is Involved?

ASHRAE works with other standards organizations, such as the Illuminating Engineering Society of North America (IESNA), American National Standards Institute (ANSI), American Society of Testing and Materials (ASTM), Air Conditioning and Refrigeration Institute (ARI), and Underwriters Laboratories (UL). The voluntary consensus process also includes representation from other groups:

- The design community, including architects, lighting, and mechanical designers
- Members of the enforcement community, including building code officials, representatives of code organizations, and state regulatory agencies
- Building owners and operators
- Industry and manufacturers
- Utility companies
- Representatives from the U.S. Department of Energy (DOE), energy advocacy groups, and the academic community.

DOE's Role

Federal law requires the DOE to determine whether revisions to the residential portion of the IECC would improve energy efficiency in the nation's residential buildings and whether revisions to ASHRAE/IESNA/ANSI Standard 90.1 would improve energy efficiency in the nation's commercial buildings.

When DOE determines that a revision would improve energy efficiency, each state has 2 yr to review the energy provisions of its residential or commercial building code. For residential buildings, a state has the option of revising its residential code to meet or exceed the residential portion of the IECC. For commercial buildings, a state is required to update its commercial code to meet or exceed the provisions of Standard 90.1.

How does the Process Work?

Standards 90.1 and 90.2 are both on continuous maintenance and are maintained by separate Standing Standards Project Committees. Committee membership varies from 10 to 60 voting members. Committee membership includes representatives from the list above to ensure balance among all interest categories.

After the committee proposes revisions to the standard, the revised version undergoes public review and comment. The committee usually incorporates nonsubstantive changes into the standard without another review. Substantive changes require additional public review. Occasionally, mediation is necessary to resolve differing views.

When a majority of the parties substantially agree (known as consensus), the revised standard is submitted for approval to the ASHRAE board of directors. Those not in agreement with the decision may appeal to the board. If an appeal is upheld, further revision, public comment, and resolution occur. If the board denies the appeal, publication of the revised standard proceeds.

What's the Timing of Revisions to Standards 90.1 and 90.2?

Standards 90.1 and 90.2 are automatically revised and published every 3 yr. However, anyone may propose a revision at any time. Approved interim revisions (called addenda) are posted on the ASHRAE Web site and are included in the next published version.

Key activities relating to revisions, including responding to public comments, typically occur during one of ASHRAE's annual (June) or midwinter (January) meetings. Public review of standards commonly occurs 2–4 months after one of these meetings.

HOW ARE MODEL ENERGY CODES DEVELOPED AND REVISED?

The most recent MECs are the 2003 IECC and the 2006 IECC. These are developed and published by the ICC through an open public-hearing process. Prior to 1998, the IECC was known as the Council of American Building Officials MEC.

Who is Involved?

The IECC Code Development Committee typically comprises 7–11 code, building science, and energy experts appointed by the ICC. Most, but not all, committee members are code officials. They may or may not be members of the ICC. The International Residential Code (IRC) Building and Energy Committee is approximately the same size, and includes builders, code officials, and industry representatives.

How does the Process Work?

Anyone may suggest a revision to the IECC or IRC by requesting a code change proposal from the committee and preparing a recommended change and substantiation. The committee publishes proposed changes and distributes them for review. This occurs about 6 weeks prior to an open public hearing, which is held in front of the code development committee.

At the public hearing, the committee receives testimony and then votes to approve, deny, or revise each change. The committee publishes its results.

Those wishing to have a proposed change reconsidered may submit a challenge to the committee's recommended action. Proponents and opponents present additional information at a second public hearing, followed by a vote by the full ICC membership. This outcome may be appealed to the ICC board of directors.

What's the Timing of the Process?

The IECC and IRC are revised on and 18-months cycle. However, full publication of the documents occurs every third yr, with supplements issued in the interim years. When developing and adopting their own energy codes, states and local governments typically adopt the fully published IECC or IRC. By specifically adopting the supplements as well, state and local governments ensure that their energy codes include important additions and clarifications to the IECC or IRC.

ADOPTION OF ENERGY CODES ON THE STATE AND LOCAL LEVEL

Before adopting or revising an energy code, states and local governments often assemble an advisory board comprising representatives of the design, building construction, and enforcement communities. This body determines which (if any) energy standards and MECs should be adopted. The group also considers the need to modify energy standards and MECs to account for local preferences and construction practices. The body also may serve as a source of information during the adoption process.

Overview of the Adoption Process

The adoption process generally includes the following steps:

- Change is initiated by a legislative or regulatory agency with authority to promulgate energy codes. Interested or affected parties also may initiate change. An advisory body typically is convened. The proposed energy code is developed.
- The proposal undergoes a legislative or public review process. Public review options include publishing a notice in key publications, filing notices of intent, and holding public hearings. Interested and affected parties are invited to submit written or oral comments.
- The results of the review process are incorporated into the proposal, and the final legislation or regulation is prepared for approval.
- The approving authority reviews the legislation or regulation. Revisions may be submitted to the designated authority for final approval or for filing.
- After being filed or approved, the code is put into effect, usually on some specified future date. This grace period allows those regulated to become familiar with any new requirements. The period between adoption and effective date typically varies from 30 days to 6 months.
- Details of the adoption process vary depending on whether the energy code is adopted by legislation, regulation, or a local government. Each is discussed below.

Adoption Through Legislation

State legislation rarely includes the complete text of an energy standard or MEC. More commonly, legislation references an energy standard or MEC that is already published. The legislation often adds administrative provisions addressing enforcement, updating, variances, and authority.

Another common approach is to use legislation to delegate authority to an agency, council, or committee. The delegated authority is empowered to develop and adopt regulations governing energy-related aspects of building design and construction. Such regulations are discussed in "Adoption through Regulation" later in this entry. Some states adopt the administrative provisions of the energy code by legislation and the technical provisions by regulation, or vice versa.

Adoption Through Regulation

A key factor in a state's ability to regulate the energy-related aspects of design and construction is the extent to which the state has authority over adoption, administration, implementation, and enforcement of building construction regulations. In most states, a single state agency has such authority. In some states, no such authority exists. If multiple state agencies, committees, or councils are involved, the authority is diluted.

When a state agency, council, or committee has authority to adopt regulations, it must follow requirements outlined in the legislation that enables development, revisions, and adoption of the regulations. The technical provisions of the regulation may be unique to the state, or the regulations may adopt, by reference, national energy standards or an MEC. When a state adopts regulations, it typically includes its own administrative provisions within the regulations.

Adoption by Local Government

If a state has limited authority to adopt an energy code [a "home rule" state (In the energy codes and standards arena, home rule means the state cannot interfere or control on the local level.)], units of local government have the option to assume that responsibility. Local governments also can adopt standards or codes that are more stringent than the state's.

A local government's municipal code typically includes a title or provision covering building construction, under which energy provisions can be adopted.

Most local governments adopt an MEC by reference. They apply administrative provisions from other building construction regulations to implementation and enforcement of the energy code.

Timing of the Adoption and Revision of State and Local Codes

Most states adopt or revise energy codes in concert with the publication of a new edition of a national energy standard or MEC. This may occur either through a regulatory process or automatically because state regulation or legislation refers to "the most recent edition."

Adoption also can be tied to the publication date of an energy standard or MEC; e.g., "This regulation shall take effect 1 month from publication of the adopted MEC."

IMPLEMENTATION OF ENERGY CODES ON THE STATE AND LOCAL LEVEL

During implementation, the adopting jurisdiction(s) must prepare building officials to enforce the energy code and prepare the building construction community to comply with it. It is important for all stakeholders to know that a new code is coming and understand what is required. Many states or jurisdictions start this education process several years in advance of an energy code change—often before adoption itself. The more publicity about and training on the new code there are, the more it will be accepted and used.

Communication and information exchange should occur in several contexts:

- Between the code-adopting bodies and the code-enforcing bodies
- Between the code-adopting bodies and the building construction community
- Between the code-enforcing bodies and the building construction community
- Within the building construction community and the code-enforcing bodies

Training is critical. To be effective, training must cater to the specific needs of building officials, architects, designers, engineers, manufacturers, builders and contractors, and building owners. Training for specific stakeholders can be provided or sponsored by the following:

- State energy offices and agencies
- Universities and community colleges
- Professional organizations and societies
- Utilities
- Trade associations
- National or regional code organizations
- Others, such as product distributors.

The DOE, the ICC, ASHRAE, and other codes organizations can supply tools and materials to make implementation and training easier for states and local jurisdictions.

ENFORCEMENT OF ENERGY CODES ON THE STATE AND LOCAL LEVELS

Enforcement ensures compliance with an energy code and is critical to securing energy savings. Enforcement strategies vary according to a state or local government's regulatory authority, resources, and manpower. Enforcement can include all or some of the following activities:

- Plan review
- Product, material, and equipment specifications review
- Testing and certification review
- Supporting calculation review
- Building inspection during construction
- Evaluation of materials substituted in the field
- Building inspection immediately prior to occupancy.

Sometimes a state or local government has no enforcement authority. The courts address enforcement if and when legal action is sought by a building owner against a designer or contractor.

State Enforcement

State enforcement is a common approach in smaller states, in rural jurisdictions that have no code officials, and for state-owned or financed construction. Enforcement by a single state agency usually is more uniform than enforcement conducted by several local agencies. Plan review is generally performed by one office. Although there may be numerous state field inspectors, they are bound under one organization. This arrangement benefits the building construction community by offering a single point of contact. However, if state resources are limited, plan reviews and construction inspections may not be performed as thoroughly as warranted.

Local Enforcement

Local enforcement agencies are closer to the construction site and in more direct contact with the design and construction community. This offers the potential for more regular enforcement during design and construction. However, local jurisdictions may lack sufficient resources to support enforcement. Because jurisdictions vary, local enforcement may lead to some noncompliance across a state. Compliance is enhanced when a state code agency actively supports local governments in their efforts to enforce the state code.

Some states allow local jurisdictions to petition to conduct enforcement activities that are usually the responsibility of the state. This strategy offers the advantages associated with state enforcement, recognizes those local governments with equivalent enforcement capabilities, and helps ensure comparable levels of compliance. Continued state oversight is necessary to ensure a consistent level of enforcement by local jurisdictions. A hybrid approach might involve the state conducting the plan review and the local authority conducting the construction inspection.

Third-Party Alternatives

Some states and local governments allow qualified third parties to conduct plan reviews. Often, these reviewers have more experience dealing with the complexities and subtleties of energy codes and standards; have better sources, references, and contacts because of affiliations with professional organizations; and can help ease heavy workloads.

VOLUNTARY ENERGY-EFFICIENCY PROGRAMS

Voluntary programs encourage a level of energy efficiency above code. They can help motivate consumers to recognize the value of energy efficiency. Examples include the following:

- Home energy rating systems—Also known as HERS, these compare the energy efficiency of a home with that of a computer-simulated reference house. The rating involves analysis of the home's construction plans and at least one onsite inspection. This information is used to estimate the home's annual energy costs and give the home a rating between 0 and 100. The lower the score, the more efficient the home.
- ENERGY STAR—The U.S. Environmental Protection Agency outlines criteria for ENERGY STAR certification of homes and commercial buildings. ENERGY STAR homes are typically 30% more energy efficient than average minimum energy codes. For more information, go to www.energystar.gov.
- Utility, government, and other programs—Utilities, state and local governments, and other organizations often sponsor programs that qualify buildings based on certain standards. Examples include the following:
  — The DOE's Building America Program, www.eere.energy.gov/buildings/building_america
  — The U.S. Green Building Council's Leadership in Energy and Environmental Design (LEED™), www.usgbc.org/DisplayPage.aspx?CategoryID=19
  — The New Buildings Institute Advanced Buildings Benchmark™ www.poweryourdesign.com/benchmark.htm
  — ASHRAE's Advanced Energy Design Guides, http://resourcecenter.ashrae.org/store/ashrae

DOE SUPPORT

The DOE's Building Energy Codes Program (BECP) supports state and local governments in their efforts to implement and enforce building energy codes. This support includes the following activities:

- Developing and distributing easy-to-use compliance tools and materials
- Providing financial and technical assistance to help adopt, implement, and enforce building energy codes
- Participating in the development of MECs and energy standards
- Providing information on compliance products and training, and energy-code-related news.

For more information on BECP products and services, visit the BECP Web site at www.energycodes.gov.

CONCLUSION

Energy codes and standards set baseline requirements for energy-efficient design and construction. Several organizations play a role in their development, and there is a long, intertwined development history. The development processes vary, but suggested code and standard changes may be submitted by any interested party. Development is only the first step, however; adoption and enforcement are critical next steps.

BIBLIOGRAPHY

1. American Society of Heating, Refrigerating, and Air-Conditioning Engineers (ASHRAE), www.ashrae.org.
2. Association of Energy Engineers, www.aeecenter.org.
3. Bartlett, R.; Halverson, M.A.; Shankle, D.L. *Understanding Building Energy Codes and Standards*; Pacific Northwest National Laboratory: Richland, WA, 2003; [PNNL-14235].
4. Bartlett, R.; Halverson, M.A. *Codes 101*; Pacific Northwest National Laboratory: Richland, WA, 2006; [PNNL-SA-51223].
5. Codes 101. Available at: http://energycode.pnl.gov/moodle/course/view.php?id=7.
6. International Code Council, www.iccsafe.org.
7. National Conference of States on Building Codes and Standards (NCSBCS), www.ncsbcs.org.
8. Residential Energy Services Network, www.natresnet.org.
9. U.S. Department of Energy's Building Energy Codes Program, www.energycodes.gov.

Energy Conservation

Ibrahim Dincer
Faculty of Engineering and Applied Science, University of Ontario Institute of Technology (UOIT), Oshawa, Ontario, Canada

Adnan Midilli[*]
Department of Mechanical Engineering, Faculty of Engineering, Nigde University, Nigde, Turkey

Abstract
This study highlights these issues and potential solutions to the current environmental issues; identifies the main steps for implementing energy conservation programs and the main barriers to such implementations; and provides assessments for energy conservation potentials for countries, as well as various practical and environmental aspects of energy conservation.

INTRODUCTION

Civilization began when people found out how to use fire extensively. They burned wood and obtained sufficiently high temperatures for melting metals, extracting chemicals, and converting heat into mechanical power, as well as for cooking and heating. During burning, the carbon in wood combines with O_2 to form carbon dioxide (CO_2), which then is absorbed by plants and converted back to carbon for use as a fuel again. Because wood was unable to meet the fuel demand, the Industrial Revolution began with the use of fossil fuels (e.g., oil, coal, and gas). Using such fuels has increased the CO_2 concentration in the air, leading to the beginning of global warming. Despite several warnings in the past about the risks of greenhouse-gas emissions, significant actions to reduce environmental pollution were not taken, and now many researchers have concluded that global warming is occurring. During the past two decades, the public has became more aware, and researchers and policymakers have focused on this and related issues by considering energy, the environment, and sustainable development.

Energy is considered to be a key catalyst in the generation of wealth and also a significant component in social, industrial, technological, economic, and sustainable development. This makes energy resources and their use extremely significant for every country. In fact, abundant and affordable energy is one of the great boons of modern industrial civilization and the basis of our living standard. It makes people's lives brighter, safer, more comfortable, and more mobile, depending on their energy demand and consumption. In recent years, however, energy use and associated greenhouse-gas emissions and their potential effects on the global climate change have been of worldwide concern.

Problems with energy utilization are related not only to global warming, but also to such environmental concerns as air pollution, acid rain, and stratospheric ozone depletion. These issues must be taken into consideration simultaneously if humanity is to achieve a bright energy future with minimal environmental impact. Because all energy resources lead to some environmental impact, it is reasonable to suggest that some (not all) of these concerns can be overcome in part through energy conservation efforts.

Energy conservation is a key element of energy policy and appears to be one of the most effective ways to improve end-use energy efficiency, and to reduce energy consumption and greenhouse-gas emissions in various sectors (industrial, residential, transportation, etc.). This is why many countries have recently started developing aggressive energy conservation programs to reduce the energy intensity of the their infrastructures, make businesses more competitive, and allow consumers to save money and to live more comfortably. In general, energy conservation programs aim to reduce the need for new generation or transmission capacity, to save energy, and to improve the environment. Furthermore, energy conservation is vital for sustainable development and should be implemented by all possible means, despite the fact that it has its own limitations. This is required not only for us, but for the next generation as well.

Considering these important contributions, the energy conservation phenomenon should be discussed in a comprehensive perspective. Therefore, the main objective of this article is to present and discuss the world's primary energy consumption and production; major environmental problems; potential solutions to these issues;

Keywords: Energy; Energy conservation; Environment; Sustainability; Policies; Strategies; Life-cycle costing.

[*] Currently on sabbatical leave at University of Ontario Institute of Technology (UOIT).

practical energy conservation aspects; research and development (R&D) in energy conservation, energy conservation, and sustainable development; energy conservation implementation plans; energy conservation measurements; and life-cycle costing (LCC) as an excellent tool in energy conservation. In this regard, this contribution aims to:

- Help explain main concepts and issues about energy conservation
- Develop relations between energy conservation and sustainability
- Encourage energy conservation strategies and policies
- Provide energy conservation methodologies
- Discuss relations between energy conservation and environmental impact
- Present some illustrative examples to state the importance of energy conservation and its practical benefits

In summary, this book contribution highlights the current environmental issues and potential solutions to these issues; identifies the main steps for implementing energy conservation programs and the main barriers to such implementations; and provides assessments for energy conservation potentials for countries, as well as various practical and environmental aspects of energy conservation.

WORLD ENERGY RESOURCES: PRODUCTION AND CONSUMPTION

World energy consumption and production are very important for energy conservation in the future. Economic activity and investment patterns in the global energy sector are still centered on fossil fuels, and fossil-fuel industries and energy-intensive industries generally have been skeptical about warnings of global warming and, in particular, about policies to combat it. The increase of energy consumption and energy demand indicates our dependence on fossil fuels. If the increase of fossil-fuel utilization continues in this manner, it is likely that the world will be affected by many problems due to fossil fuels. It follows from basic scientific laws that increasing amounts of CO_2 and other greenhouse gases will affect the global climate. The informed debate is not about the existence of such effects, but about their magnitudes and seriousness. At present, the concentration of CO_2 is approximately 30% higher than its preindustrial level, and scientists have already been able to observe a discernible human influence on the global climate.[1]

In the past, fossil fuels were a major alternative for overcoming world energy problems. Fossil fuels cannot continue indefinitely as the principal energy sources, however, due to the rapid increase of world energy demand and energy consumption. The utilization distribution of fossil-fuel types has changed significantly over the past 80 years. In 1925, 80% of the required energy was supplied from coal, whereas in the past few decades, 45% came from petroleum, 25% from natural gas, and 30% from coal. Due to world population growth and the advance of technologies that depend on fossil fuels, reserves of those fuels eventually will not be able to meet energy demand. Energy experts point out that reserves are less than 40 years for petroleum, 60 years for natural gas, and 250 years for coal.[2] Thus, fossil-fuel costs are likely to increase in the near future. This will allow the use of renewable energy sources such as solar, wind, and hydrogen. As an example, the actual data[3,4] and projections of world energy production and consumption from 1980 to 2030 are displayed in the following figures, and the curve equations for world energy production and consumption are derived as shown Table 1.

As presented in Figs. 1 and 2, and in Table 2, the quantities of world primary energy production and consumption are expected to reach 14,499.2 and 13,466.5 Mtoe, respectively, by 2030. World population is now over six billion, double that of 40 years ago, and it is likely to double again by the middle of the 21st century. The world's population is expected to rise to about seven billion by 2010. Even if birth rates fall so that the world population becomes stable by 2030, the population still would be about ten billion. The data presented in Figs. 1 and 2 are expected to cover current energy needs provided that the population remains constant. Because the population is expected to increase dramatically, however, conventional energy resource shortages are likely to occur, due to insufficient fossil-fuel resources. Therefore, energy

Table 1 World energy production and consumption models through statistical analysis

Energy	Production (Mtoe)	Correlation coefficient	Consumption (Mtoe)	Correlation coefficient
World primary	$=148.70 \times Year - 287,369$	0.998	$=139.62 \times Year - 269,953$	0.998
World oil	$=44.47 \times Year - 85,374$	0.997	$=42.18 \times Year - 80,840$	0.998
World coal	$=20.05 \times Year - 37,748$	0.946	$=23.18 \times Year - 43,973$	0.968
World NG	$=45.73 \times Year - 89,257$	0.999	$=46.27 \times Year - 90,347$	0.999

Mtoe, million tons of oil equivalent.

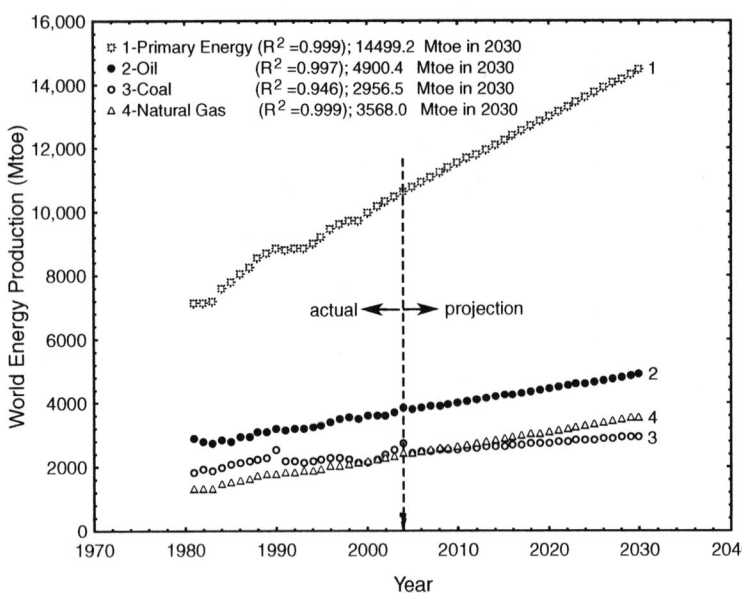

Fig. 1 Variation of actual data taken from Refs. 3,4, and projections of annual world energy production. Mtoe, million tons of oil equivalent.
Source: From Refs. 3,4.

conservation will become increasingly important to compensate for shortages of conventional resources.

MAJOR ENVIRONMENTAL PROBLEMS

One of the most important targets of modern industrial civilizations is to supply sustainable energy sources and to develop the basis of living standards based on these energy sources, as well as implementing energy conservation measures. In fact, affordable and abundant sustainable energy makes our lives brighter, safer, more comfortable, and more mobile because most industrialized and developing societies use various types of energy. Billions of people in undeveloped countries, however, still have limited access to energy. India's per-capita consumption of electricity, for example, is one-twentieth that of the United States. Hundreds of millions of Indians live "off the grid"-that is, without electricity-and cow dung is still a major fuel for household cooking. This continuing reliance on such preindustrial energy sources is also one of the major causes of environmental degradation.[5]

After many decades of using fossil fuels as a main energy source, significant environmental effects of fossil fuels became apparent. The essential pollutants were from greenhouse gases (e.g., CO_2, SO_2, and NO_2). Fossil fuels are used for many applications, including industry, residential, and commercial sectors. Increasing fossil-fuel utilization in transportation vehicles such as automobiles, ships, aircrafts, and spacecrafts has led to increasing pollution. Gas, particulate matter, and dust clouds in the atmosphere absorb a significant portion of the

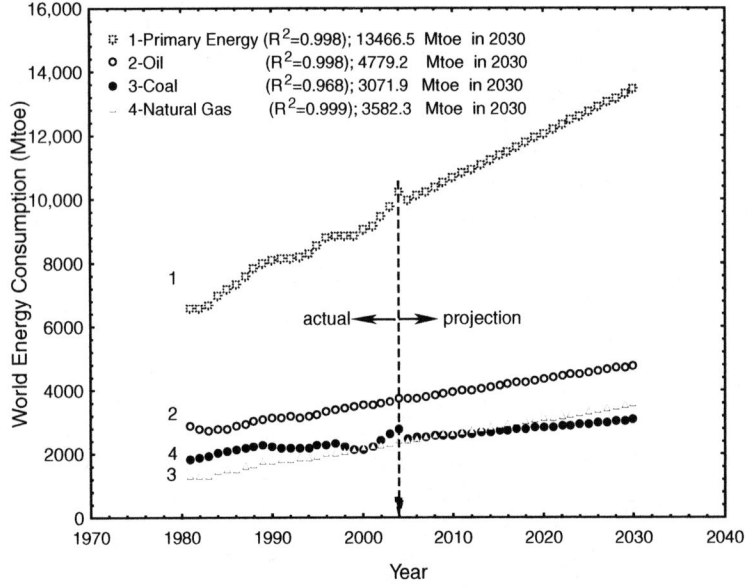

Fig. 2 Variation of actual data taken from Refs. 3,4, and projections of annual world energy consumption. Mtoe, million tons of oil equivalent.
Source: From Refs. 3,4.

Table 2 Some extracted values of world primary and fossil energy production and consumption

Year	Primary energy production (Mtoe)	Primary energy consumption (Mtoe)	Oil prod. (Mtoe)	Oil cons. (Mtoe)	Coal prod. (Mtoe)	Coal cons. (Mtoe)	NG prod. (Mtoe)	NG cons. (Mtoe)
1994	8,996.9	8,310.1	3237.1	3204.4	2178.1	2185.5	1891.2	1876.7
2000	9,981.9	9,079.8	3614.0	3538.7	2112.4	2148.1	2189.9	2194.5
2006	10,930.3	10,115.7	3833.1	3767.0	2475.3	2515.7	2470.5	2471.8
2012	11,822.5	10,953.4	4099.9	4020.0	2595.6	2654.8	2744.9	2749.5
2018	12,714.7	11,791.1	4366.7	4273.1	2715.9	2793.8	3019.2	3027.1
2024	13,606.9	12,128.8	4633.5	4526.1	2836.2	2932.9	3293.6	3304.7
2030	14,499.2	13,466.5	4900.4	4779.2	2956.5	3071.9	3568.0	3582.3

Mtoe, million tons of oil equivalent.

solar radiation directed at Earth and cause a decrease in the oxygen available for the living things. The threat of global warming has been attributed to fossil fuels.[2] In addition, the risk and reality of environmental degradation have become more apparent. Growing evidence of environmental problems is due to a combination of factors.

During the past two decades, environmental degradation has grown dramatically because of the sheer increase of world population, energy consumption, and industrial activities. Throughout the 1970s, most environmental analysis and legal control instruments concentrated on conventional pollutants such as SO_2, NO_x, particulates, and CO. Recently, environmental concern has extended to the control of micro or hazardous air pollutants, which are usually toxic chemical substances and harmful in small doses, as well as to that of globally significant pollutants such as CO_2. Aside from advances in environmental engineering science, developments in industrial processes and structures have led to new environmental problems.[6,7] In the energy sector, for example, major shifts to the road transport of industrial goods and to individual travel by cars has led to an increase in road traffic and, hence, to a shift in attention paid to the effects and sources of NO_x and to the emissions of volatile organic compounds (VOC). In fact, problems with energy supply and use are related not only to global warming, but also to such environmental concerns as air pollution, ozone depletion, forest destruction, and emission of radioactive substances. These issues must be taken into consideration simultaneously if humanity is to achieve a bright energy future with minimal environmental impact. Much evidence exists to suggest that the future will be negatively impacted if humans keep degrading the environment. Therefore, there is an intimate connection among energy conservation, the environment, and sustainable development. A society seeking sustainable development ideally must utilize only energy resources that cause no environmental impact (e.g., that release no emissions to the environment). Because all energy resources lead to some environmental impact, however, it is reasonable to suggest that some (not all) of the concerns regarding the limitations imposed on sustainable development by environmental emissions and their negative impacts can be overcome in part through energy conservation. A strong relation clearly exists between energy conservation and environmental impact, because for the same services or products, less resource utilization and pollution normally are associated with higher-efficiency processes.[8]

Table 3 summarizes the major environmental problems-such as acid rain, stratospheric ozone depletion, and global climate change (greenhouse effect)-and their main sources and effects.

As shown in Fig. 3, the world total CO_2 production is estimated to be 18,313.13 million tons in 1980, 25,586.7 million tons in 2006, 27,356.43 million tons in 2012, and 29,716.1 million tons in 2020 whereas fossil-fuel

Table 3 Major environmental issues and their consequences

Issues	Description	Main sources	Main effects
Acid precipitation	Transportation and deposition of acids produced by fossil-fuel combustion (e.g., industrial boilers, transportation vehicles) over great distances through the atmosphere via precipitation on the earth on ecosystems	Emissions of SO_2, NO_x, and volatile organic compounds (VOCs) (e.g., residential heating and industrial energy use account for 80% of SO_2 emissions)	Acidification of lakes, streams and ground waters, resulting in damage to fish and aquatic life; damage to forests and agricultural crops; and deterioration of materials, e.g., buildings, structures
Stratospheric ozone depletion	Distortion and regional depletion of stratospheric ozone layer though energy activities (e.g., refrigeration, fertilizers)	Emissions of CFCs, halons (chlorinated and brominated organic compounds) and N_2O (e.g., fossil fuel and biomass combustion account for 65%–75% of N_2O emissions)	Increased levels of damaging ultraviolet radiation reaching the ground, causing increased rates of skin cancer, eye damage and other harm to many biological species
Greenhouse effect	A rise in the earth's temperature as a result of the greenhouse gases	Emissions of carbon dioxide (CO_2), CH_4, CFCs, halons, N_2O, ozone and peroxyacetylnitrate (e.g., CO_2 releases from fossil fuel combustion ($\sim 50\%$ from CO_2), CH_4 emissions from increased human activity)	Increased the earth's surface temperature about 0.6°C over the last century and as a consequence risen sea level about 20 cm (in the next century by another 2°C–4°C and a rise between 30 and 60 cm); resulting in flooding of coastal settlements, a displacement of fertile zones for agriculture and food production toward higher latitudes, and a decreasing availability of fresh water for irrigation and other essential uses

Source: From Refs. 9–11.

consumption is found to be 6092.2 million tons in 1980, 8754.5 million tons in 2006, 9424.3 million tons in 2012, and 10,317.2 million tons in 2020. These values show that the CO_2 production will probably increase if we continue utilizing fossil fuel. Therefore, it is suggested that certain energy conversion strategies and technologies should be put into practice immediately to reduce future environmental problems.

The climate technology initiative (CTI) is a cooperative effort by 23 Organization for Economic Co-operation and

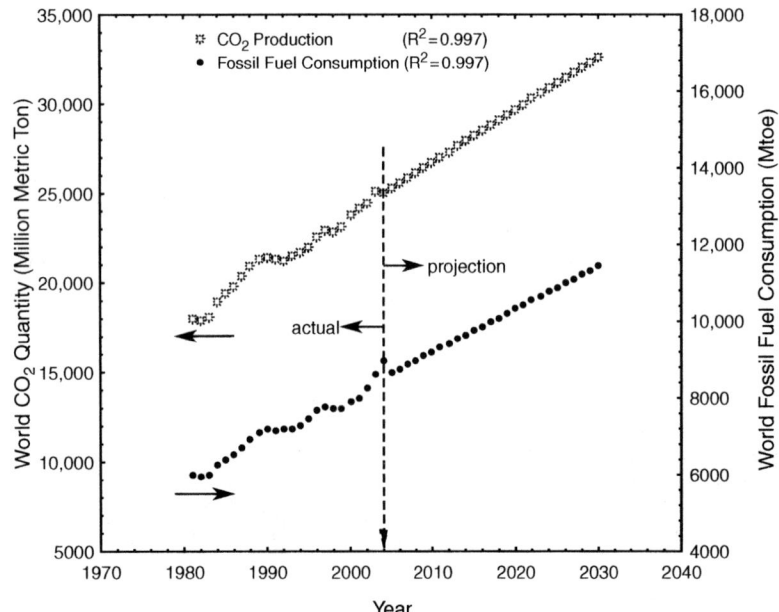

Fig. 3 Variation of world total fossil-fuel consumption and CO_2 production; actual data from Ref. 3 and projections. Mtoe, million tons of oil equivalent.
Source: From Ref. 3.

Development (OECD)/International Energy Agency (IEA) member countries and the European Commission to support the objectives of the united nations framework convention on climate change (UNFCCC). The CTI was launched at the 1995 Berlin Conference of the Parties to the UNFCCC. The CTI seeks to ensure that technologies to address climate change are available and can be deployed efficiently. The CTI includes activities directed at the achievement of seven broad objectives:

- To facilitate cooperative and voluntary actions among governments, quasigovernments, and private entities to help cost-effective technology diffusion and reduce the barriers to an enhanced use of climate-friendly technologies
- To promote the development of technology aspects of national plans and programs prepared under the UNFCCC
- To establish and strengthen the networks among renewable and energy efficiency centers in different regions
- To improve access to and enhance markets for emerging technologies
- To provide appropriate recognition of climate-friendly technologies through the creation of international technology awards
- To strengthen international collaboration on short-, medium-, and long-term research; development and demonstration; and systematic evaluation of technology options
- To assess the feasibility of developing longer-term technologies to capture, remove, or dispose of greenhouse gases; to produce hydrogen from fossil fuels; and to strengthen relevant basic and applied research

POTENTIAL SOLUTIONS TO ENVIRONMENTAL ISSUES

Although there are a large number of practical solutions to environmental problems, three potential solutions are given priority, as follows[11]:

- Energy conservation technologies (efficient energy utilization)
- Renewable energy technologies
- Cleaner technologies

In these technologies, we pay special attention to energy conservation technologies and their practical aspects and environmental impacts. Each of these technologies is of great importance, and requires careful treatment and program development. In this work, we deal with energy conservation technologies and strategies in depth. Considering the above priorities to environmental solutions, the important technologies shown in Fig. 4 should be put into practice.

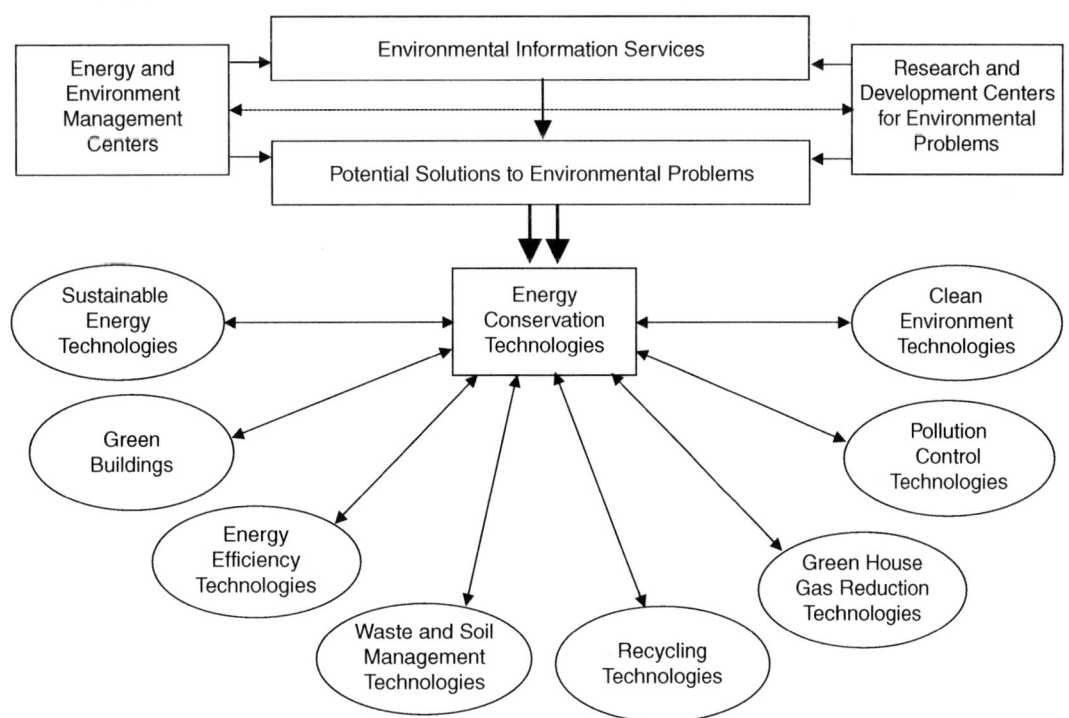

Fig. 4 Linkages between possible environmental and energy conservation technologies.

PRACTICAL ENERGY CONSERVATION ASPECTS

The energy-saving result of efficiency improvements is often called energy conservation. The terms efficiency and conservation contrast with curtailment, which decreases output (e.g., turning down the thermostat) or services (e.g., driving less) to curb energy use. That is, energy curtailment occurs when saving energy causes a reduction in services or sacrifice of comfort. Curtailment is often employed as an emergency measure. Energy efficiency is increased when an energy conversion device-such as a household appliance, automobile engine, or steam turbine-undergoes a technical change that enables it to provide the same service (lighting, heating, motor drive, etc.) while using less energy. Energy efficiency is often viewed as a resource option like coal, oil, or natural gas. In contrast to supply options, however, the downward pressure on energy prices created by energy efficiency comes from demand reductions instead of increased supply. As a result, energy efficiency can reduce resource use and environmental impacts.[12]

The quality of a country's energy supply and demand systems is increasingly evaluated today in terms of its environmental sustainability. Fossil-fuel resources will not last indefinitely, and the most convenient, versatile, and inexpensive of them have substantially been used up. The future role of nuclear energy is uncertain, and global environmental concerns call for immediate action. OECD countries account for almost 50% of total world energy consumption: Current use of oil per person averages 4.5 bbl a year worldwide, ranging from 24 bbl in the United States and 12 bbl in western Europe to less than 1 bbl in sub-Saharan Africa. More than 80% of worldwide CO_2 emissions originate in the OECD area. It is clear, then, that OECD countries should play a crucial role in indicating a sustainable pattern and in implementing innovative strategies.[11]

From an economic as well as an environmental perspective, energy conservation holds even greater promise than renewable energy, at least in the near-term future. Energy conservation is indisputably beneficial to the environment, as a unit of energy not consumed equates to a unit of resources saved and a unit of pollution not generated.

Furthermore, some technical limitations on energy conservation are associated with the laws of physics and thermodynamics. Other technical limitations are imposed by practical technical constraints related to the real-world devices that are used. The minimum amount of fuel theoretically needed to produce a specified quantity of electricity, for example, could be determined by considering a Carnot (ideal) heat engine. However, more than this theoretical minimum fuel may be needed due to practical technical matters such as the maximum temperatures and pressures that structures and materials in the power plant can withstand.

As environmental concerns such as pollution, ozone depletion, and global climate change became major issues in the 1980s, interest developed in the link between energy utilization and the environment. Since then, there has been increasing attention to this linkage. Many scientists and engineers suggest that the impact of energy-resource utilization on the environment is best addressed by considering exergy. The exergy of a quantity of energy or a substance is a measure of the usefulness or quality of the energy or substance, or a measure of its potential to cause change. Exergy appears to be an effective measure of the potential of a substance to impact the environment. In practice, the authors feel that a thorough understanding of exergy and of how exergy analysis can provide insights into the efficiency and performance of energy systems is required for the engineer or scientist working in the area of energy systems and the environment.[8] Considering the above explanations, the general aspects of energy conservation can be summarized as shown in Fig. 5.

RESEARCH AND DEVELOPMENT STATUS ON ENERGY CONSERVATION

Now we look at R&D expenditures in energy conservation to assess the importance attached to energy conservation in the long range. The share of energy R&D expenditures going into energy conservation, for example, has grown greatly since 1976, from 5.1% in 1976 to 40.1% in 1990 and 68.5% in 2002.[11] This indicates that within energy R&D, research on energy conservation is increasing in importance. When R&D expenditures on energy conservation are compared with expenditures for research leading to protection of the environment in the 2000s, the largest share was spent on environment research. In fact, it is not easy to interpret the current trends in R&D expenditures, because energy conservation is now part of every discipline from engineering to economics. A marked trend has been observed since the mid-1970s, in that expenditures for energy conservation research have grown significantly, both in absolute terms and as a share of total energy R&D. These expenditures also grew more rapidly than those for environmental protection research, surpassing it in the early 1980s. Therefore, if R&D expenditures reflect long-term concern, there seems to be relatively more importance attached to energy conservation as compared with environmental protection.

In addition to the general trends discussed above, consider the industrial sector and how it has tackled energy conservation.

The private sector clearly has an important role to play in providing finance that could be used for energy efficiency investments. In fact, governments can adjust

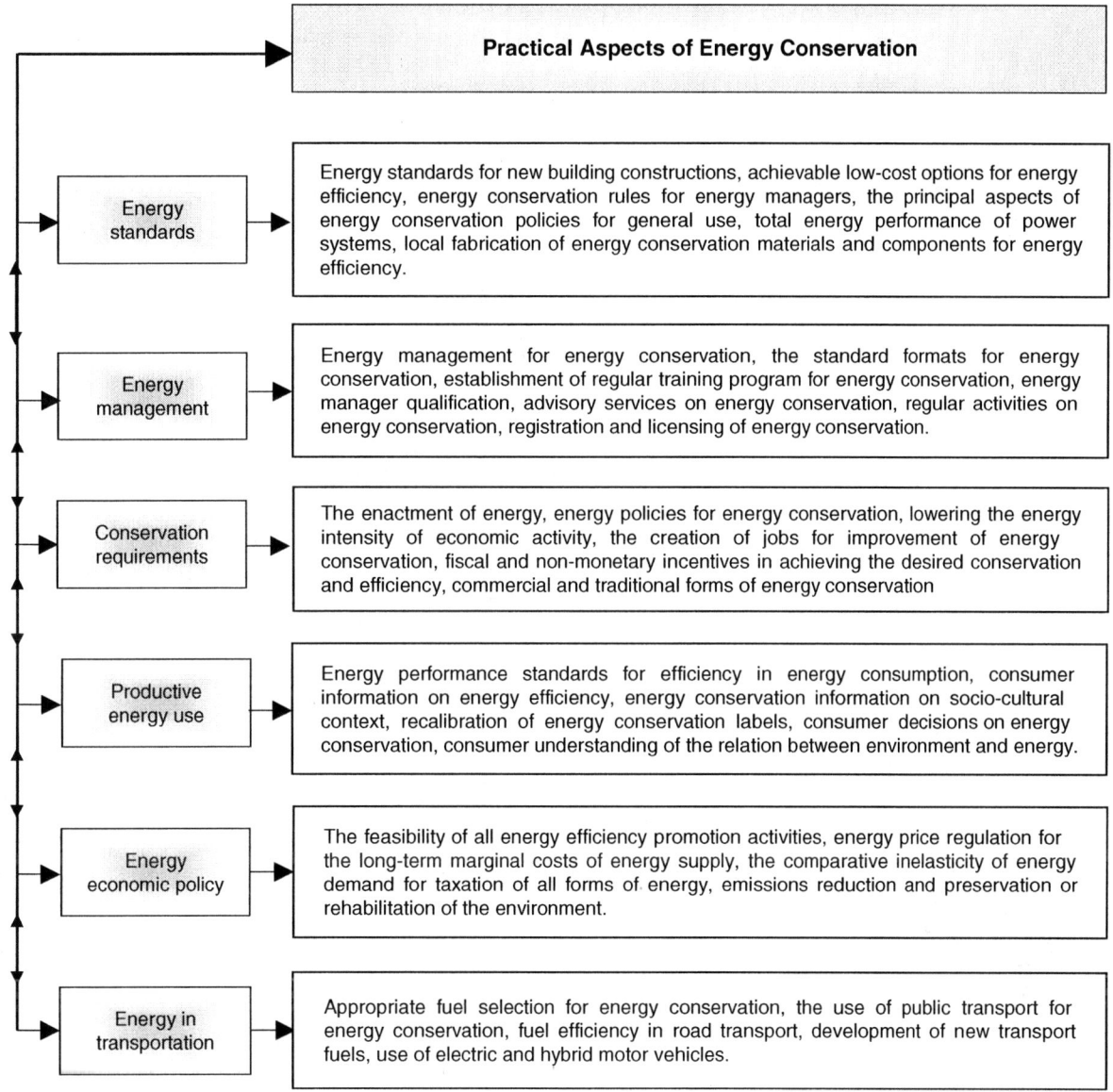

Fig. 5 A flow chart of practical energy conservation aspects.

their spending priorities in aid plans and through official support provided to their exporters, but they can influence the vast potential pool of private-sector finance only indirectly. Many of the most important measures to attract foreign investors include reforming macroeconomic policy frameworks, energy market structures and pricing, and banking; creating debt recovery programs; strengthening the commercial and legal framework for investment; and setting up judicial institutions and enforcement mechanisms. These are difficult tasks that often involve lengthy political processes.

Thus, the following important factors, which are adopted from a literature work[13] can contribute to improving energy conservation in real life. Fig. 6 presents the improvement factors of energy conservation.

ENERGY CONSERVATION AND SUSTAINABLE DEVELOPMENT

Energy conservation is vital for sustainable development and should be implemented by all possible means, despite the fact that it has its own limitations. This is required not only for us, but for the next generation as well.

A secure supply of energy resources is generally considered a necessity but not a sufficient requirement for development within a society. Furthermore, sustainable development demands a sustainable supply of energy resources that, in the long term, is readily and sustainably available at reasonable cost and can be utilized for all required tasks without causing negative societal impact. Supplies of such energy resources as fossil fuels (coal, oil,

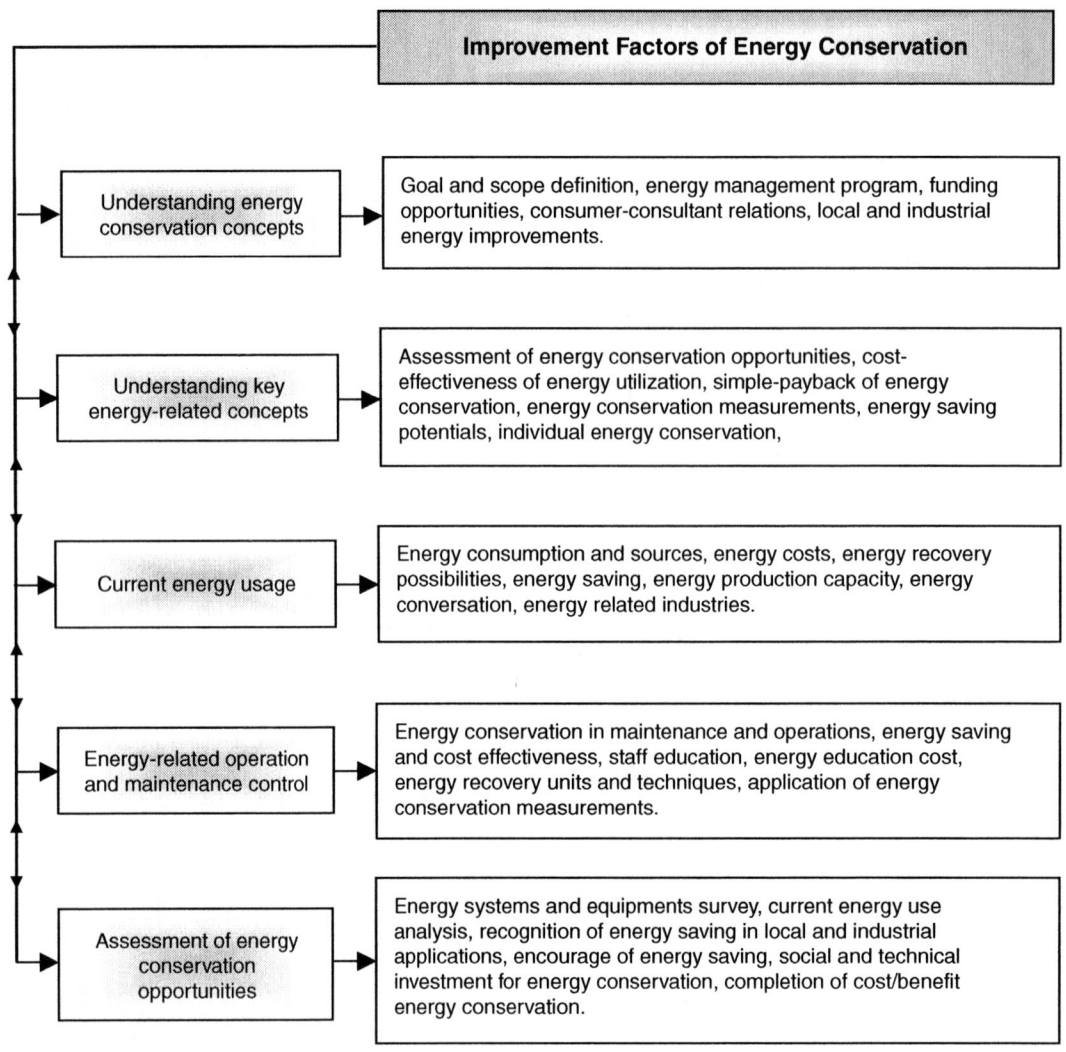

Fig. 6 Improvement factors of energy conservation.

and natural gas) and uranium are generally acknowledged to be finite. Other energy sources (such as sunlight, wind, and falling water) are generally considered to be renewable and, therefore, sustainable over the relatively long term. Wastes (convertible to useful energy forms through, for example, waste-to-energy incineration facilities) and biomass fuels usually also are viewed as being sustainable energy sources. In general, the implications of these statements are numerous and depend on how the term *sustainable* is defined.[14]

Energy resources and their utilization are intimately related to sustainable development. For societies to attain or try to attain sustainable development, much effort must be devoted not only to discovering sustainable energy resources, but also to increasing the energy efficiencies of processes utilizing these resources. Under these circumstances, increasing the efficiency of energy-utilizing devices is important. Due to increased awareness of the benefits of efficiency improvements, many institutes and agencies have started working along these lines. Many energy conservation and efficiency improvement programs have been developed and are being developed to reduce present levels of energy consumption. To implement these programs in a beneficial manner, an understanding is required of the patterns of "energy carrier" consumption-for example, the type of energy carrier used, factors that influence consumption, and types of end uses.[15]

Environmental concerns are an important factor in sustainable development. For a variety of reasons, activities that continually degrade the environment are not sustainable over time-that is, the cumulative impact on the environment of such activities often leads over time to a variety of health, ecological, and other problems. A large portion of the environmental impact in a society is associated with its utilization of energy resources. Ideally, a society seeking sustainable development utilizes only energy resources that cause no environmental impact (e.g., that release no emissions to the environment). Because all energy resources lead to some environmental impact, however, it is reasonable to suggest that some (not all) of the concerns regarding the limitations imposed on sustainable development by environmental emissions and their negative impacts can be overcome in part through increased energy efficiency. Clearly, a strong relationship exists between energy efficiency and environmental impact, because for the same services or products, less resource utilization and pollution normally are associated with increased energy efficiency.

Here, we look at renewable energy resources and compare them with energy conservation. Although not all renewable energy resources are inherently clean, there is such a diversity of choices that a shift to renewables carried out in the context of sustainable development could provide a far cleaner system than would be feasible by tightening controls on conventional energy. Furthermore, being by nature site-specific, they favor power system decentralization and locally applicable solutions more or less independently of the national network. It enables citizens to perceive positive and negative externalities of energy consumption. Consequently, the small scale of the equipment often makes the time required from initial design to operation short, providing greater adaptability in responding to unpredictable growth and/or changes in energy demand.

The exploitation of renewable energy resources and technologies is a key component of sustainable development.[11] There are three significant reasons for it:

- They have much less environmental impact compared with other sources of energy, because there are no energy sources with zero environmental impact. Such a variety of choices is available in practice that a shift to renewables could provide a far cleaner energy system than would be feasible by tightening controls on conventional energy.
- Renewable energy resources cannot be depleted, unlike fossil-fuel and uranium resources. If used wisely in appropriate and efficient applications, they can provide reliable and sustainable supply energy almost indefinitely. By contrast, fossil-fuel and uranium resources are finite and can be diminished by extraction and consumption.
- They favor power system decentralization and locally applicable solutions more or less independently of the national network, thus enhancing the flexibility of the system and the economic power supply to small, isolated settlements. That is why many different renewable energy technologies are potentially available for use in urban areas.

Taking into consideration these important reasons, the relationship between energy conservation and sustainability is finally presented as shown in Fig. 7.

ENERGY CONSERVATION IMPLEMENTATION PLAN

The following basic steps are the key points in implementing an energy conservation strategy plan[11]:

1. *Defining the main goals.* It is a systematic way to identify clear goals, leading to a simple goal-setting process. It is one of the crucial concerns and follows an organized framework to define goals, decide priorities, and identify the resources needed to meet those goals.
2. *Identifying the community goals.* It is a significant step to identify priorities and links among energy, energy conservation, the environment, and other primary local issues. Here, it is also important to identify the institutional and financial instruments.
3. *Performing an environmental scan.* The main objective in this step is to develop a clear picture of the community to identify the critical energy-use areas, the size and shape of the resource-related problems facing the city and its electrical and gas utilities, the organizational mechanisms, and the base data for evaluating the program's progress.
4. *Increasing public awareness.* Governments can increase other customers' awareness and acceptance of energy conservation programs by entering into performance contracts for government activities. They can also publicize the results of these programs and projects. In this regard, international workshops to share experiences on the operation would help overcome the initial barrier of unfamiliarity in countries.
5. *Performing information analysis.* This step carries out a wide range of telephone, fax, email, and Internet interviews with local and international financial institutions, project developers, and bilateral aid agencies to capture new initiatives, lessons learned, and viewpoints on problems and potential solutions.
6. *Building community support.* This step covers the participation and support of local industries and communities, and the understanding the nature of conflicts and barriers between given goals and local actors; improving information flows; activating education and advice surfaces; identifying institutional barriers; and involving a broad spectrum of citizen and government agencies, referring to the participation and support of local industrial and public communities.
7. *Analyzing information.* This step includes defining available options and comparing the possible options with various factors (e.g., program implementation costs, funding availability, utility

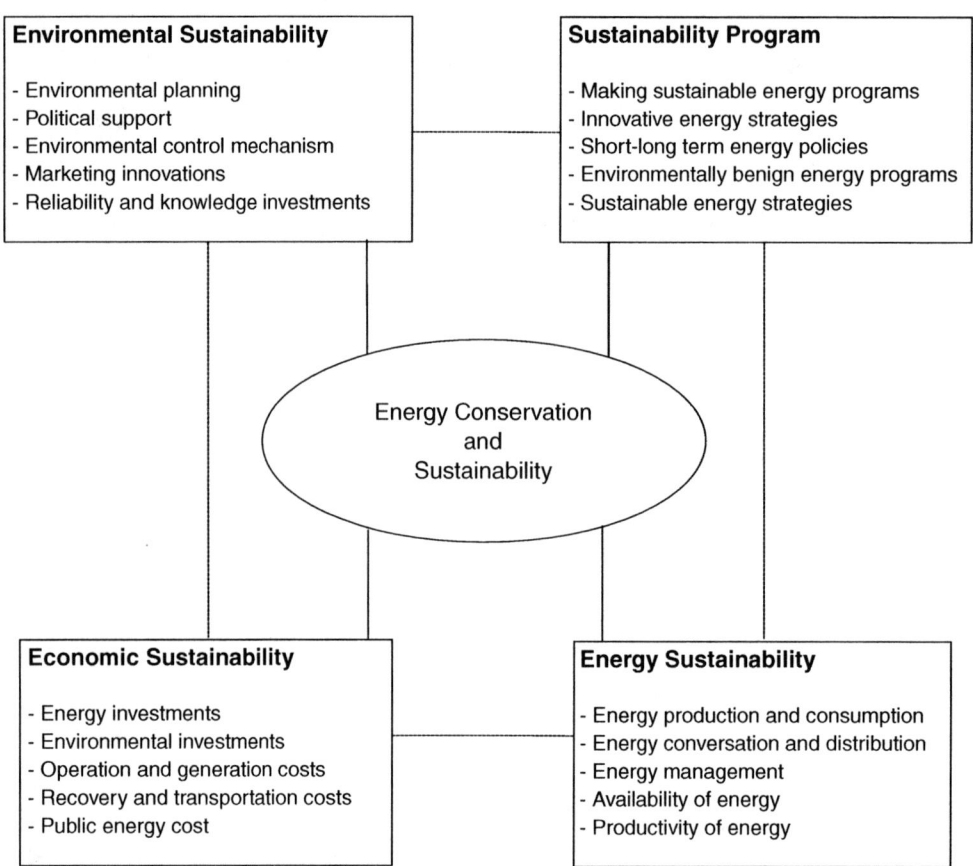

Fig. 7 Linkages between energy conservation and sustainable development.

capital deferral, potential for energy efficiency, compatibility with community goals, and environmental benefits).

8. *Adopting policies and strategies.* Priority projects need to be identified through a number of approaches that are best for the community. The decision-making process should evaluate the cost of the options in terms of savings in energy costs; generation of business and tax revenue; and the number of jobs created, as well as their contribution to energy sustainability and their benefit to other community and environmental goals.
9. *Developing the plan.* When a draft plan has been adopted, it is important for the community to review it and comment on it. The public consultation process may vary, but the aim should be a high level of agreement.
10. *Implementing new action programs.* This step involves deciding which programs to concentrate on, with long-term aims being preferred over short-term aims. The option that has the greatest impact should be focused on, and all details should be defined, no matter how difficult the task seems. Financial resources needed to implement the program should be identified.
11. *Evaluating the success.* The final stage is evaluating and assessing how well the plan performed, which helps identify its strengths and weaknesses and to determine who is benefiting from it.

ENERGY CONSERVATION MEASURES

For energy conservation measures, the information about the measure's applicability, cost range, maintenance issues, and additional points should be presented. Energy conservation involves efficiency improvements, formulation of pricing policies, good "housekeeping practices," and load management strategies, among other measures. A significant reduction in consumer energy costs can occur if conservation measures are adopted appropriately. The payback period for many conservation programs is less than 2 years.

In spite of the potentially significant benefits of such programs to the economy and their proven successes in several countries, conservation programs have not yet been undertaken on a significant scale in many developed and developing countries. Some reasons for this lack of energy conservation programs relate to the following factors:

- Technical (e.g., lack of availability, reliability, and knowledge of efficient technologies)

- Institutional (e.g., lack of appropriate technical input, financial support, and proper program design and monitoring expertise)
- Financial (e.g., lack of explicit financing mechanisms)
- Managerial (e.g., inappropriate program management practices and staff training)
- Pricing policy (e.g., inappropriate pricing of electricity and other energy commodities)
- Information diffusion (e.g., lack of appropriate information)

Reduced energy consumption through conservation programs can benefit not only consumers and utilities, but society as well. In particular, reduced energy consumption generally leads to reduced emissions of greenhouse gases and other pollutants into the environment.

Accelerated gains in energy efficiency in energy production and use, including those in the transportation sector, can help reduce emissions and promote energy security. Although there is a large technical potential for increased energy efficiency, there exist significant social and economic barriers to its achievement. Priority should be given to market forces in effecting efficiency gains. Reliance on market forces alone, however, is unlikely to overcome these barriers. For this reason, innovative and bold approaches are required by governments, in cooperation with industry, to realize the opportunities for energy efficiency improvements, and to accelerate the deployment of new and more efficient technologies.

Here, we look at energy conservation measures, which may be classified in six elements:

- Sectoral measures
- Energy conservation through systematic use of unused energy
- Energy conservation by changing social behavior
- International cooperation to promote energy conservation to counteract global warming
- Enhancing international and government-industry-university cooperation in developing technologies for energy conservation
- Promoting diffusion of information through publicity and education

The emphasis is on sectoral energy conservation. Table 4 presents some examples of such sectoral energy conservation measures. After determining which energy conservation measures are applicable, you should read the description of each of the applicable energy conservation measures. Information about the savings that can be expected from the measure, maintenance issues related to the measure, and other items to consider is provided for each energy conservation measure.

Table 4 Sectoral energy conservation measures

Sector	Measures
Industrial	Strengthening of financial and tax measures to enhance adoption and improvement of energy saving technologies through energy conservation equipment investments
	Re-use of waste energy in factories and/or in surrounding areas
	Enhancing recycling that reduces primary energy inputs such are iron scraps and used papers, and devising measures to facilitate recycling of manufactured products
	Retraining of energy managers and diffusion of new energy saving technologies through them
	Creating database on energy conservation technologies to facilitate diffusion of information
Residential and Commercial	Revising insulation standards provided in the energy conservation law, and introducing financial measures to enhance adoption of better insulation
	Development of better insulation materials and techniques
	Developing 'energy conservation' model homes and total energy use systems for homes
	Revising or adopting energy conservation standards for home and office appliances
	Developing more energy saving appliances
	Revising guidelines for managing energy use in buildings, and strengthening advisory services to improve energy management in buildings
Transportation	Because 80% of energy consumption of the sector is by automobiles, further improvement in reducing fuel consumption by automobiles is necessary together with improvement in transportation system to facilitate and reduce traffic flow
	Diffusion of information about energy efficient driving
	Adopting financial measures to enhance the use of energy saving transportation equipment such as wind powered boats

Source: Adapted from Ref. 16.

To evaluate the energy conservation measures, the following parameters should be taken into consideration[13]:

- *Cost estimation.* The first step is to estimate the cost of purchasing and installing the energy conservation measure. Cost estimates should be made for the entire development rather than for a single piece of equipment (e.g., obtain the cost for installing storm windows for an entire development or building, rather than the cost of one storm window). If you are planning to implement the energy conservation measure without the help of an outside contractor, you can obtain cost estimates by calling a vendor or distributor of the product. If, on the other hand, you will be using a contractor to install or implement the energy conservation measure, the contractor should provide estimates that include all labor costs and contract margins.
- *Data survey.* In this step, the questions on fuel consumption and cost should be listed for more than one possible fuel type (e.g., gas, oil, electric, or propane). The appropriate data for each fuel type should be selected and used accordingly for the cost estimation of each fuel.
- *Energy savings.* The amount of energy or fuel used should be estimated.
- *Cost savings.* This step determines the level of savings.
- *Payback period.* The last step in the cost/benefit analysis estimates the simple payback period. The payback period is found by dividing the cost of the measure by the annual cost savings.

LIFE-CYCLE COSTING

The term LCC for a project or product is quite broad and encompasses all those techniques that take into account both initial costs and future costs and benefits (savings) of a system or product over some period of time. The techniques differ, however, in their applications, which depend on various purposes of systems or products. Life-cycle costing is sometimes called a cradle-to-grave analysis. A life-cycle cost analysis calculates the cost of a system or product over its entire life span. Life-cycle costing is a process to determine the sum of all the costs associated with an asset or part thereof, including acquisition, installation, operation, maintenance, refurbishment, and disposal costs. Therefore, it is pivotal to the asset management process.

From the energy conservation point of view, LCC appears to be a potential tool in deciding which system or product is more cost effective and more energy efficient. It can provide information about how to evaluate options concerning design, sites, materials, etc., how to select the best energy conservation feature among various options; how much investment should be made in a single energy conservation feature; and which is the most desirable combination of various energy conservation features.

A choice can be made among various options of the energy conservation measure that produces maximum savings in the form of reduction in the life-cycle costs. A choice can be made between double-glazed and triple-glazed windows, for example. Similarly, a life-cycle cost comparison can be made between a solar heating system and a conventional heating system. The one that maximizes the life-cycle costs of providing a given level of comfort should be chosen. The application of such techniques to energy conservation is related to determining the optimum level of the chosen energy conservation measure. Sometimes, energy conservation measures involve the combination of several features. The best combination can be determined by evaluating the net LCC effects associated with successively increasing amounts of other energy conservation measures. The best combination is found by substituting the choices until each is used to the level at which its additional contribution to energy cost reduction per additional dollar is equal to that for all the other options.

Illustrative Example

Here, we present an illustrative example on LCC to highlight its importance from the energy conservation point of view. This example is a simple LCC analysis of lighting for both incandescent bulbs and compact fluorescent bulbs, comparing their life-cycle costs as detailed in Table 5. We know that incandescents are less expensive (95% to heat and 5% to usable light) and that compact fluorescent bulbs are more expensive but much more energy efficient. So the question is which type of lighting comes out on top in an LCC analysis.

This example clearly shows that LCC analysis helps in energy conservation and that we should make it part of our daily lives.

CONCLUSION

Energy conservation is a key element in sectoral (e.g., residential, industrial, and commercial) energy utilization and is vital for sustainable development. It should be implemented by all possible means, despite the fact that it sometime has its own limitations. This is required not only for us, but for the next generation as well. A secure supply of energy resources is generally considered a necessary but not a sufficient requirement for development within a society. Furthermore, sustainable development demands a sustainable supply of energy resources that, in the long term, is readily and sustainably available at reasonable cost and can be utilized for all required tasks without causing negative societal impact.

Table 5 An example of life-cycle costing (LCC) analysis

Cost of purchasing bulbs	Incandescent	Compact fluorescent
Lifetime of one bulb (hours)	1,000	10,000
Bulb price ($)	0.5	6.0
Number of bulbs for lighting 10,000 h	10	1
Cost for bulbs ($)	$10 \times 0.5 = 5.0$	$1 \times 6 = 6$
Energy cost		
Equivalent wattage (W)	75	12
Watt–hours (Wh) required for lighting for 10,000 h	$75 \times 10,000 = 750,000$ Wh $= 750$ kWh	$12 \times 10,000 = 120,000$ Wh $= 120$ kWh
Cost at 0.05 per kWh	750 kWh $\times$ $0.05 = $37.5	120 kWh $\times$ $0.05 = $6
Total cost ($)	$5 + 37.5 = 42.5$	$6 + 6 = 12$

An enhanced understanding of the environmental problems relating to energy conservation presents a high-priority need and an urgent challenge, both to allow the problems to be addressed and to ensure that the solutions are beneficial for the economy and the energy systems.

All policies should be sound and make sense in global terms-that is, become an integral part of the international process of energy system adaptation that will recognize the very strong linkage existing between energy requirements and emissions of pollutants (environmental impact).

In summary this study discusses the current environmental issues and potential solutions to these issues; identifies the main steps for implementing energy conservation programs and the main barriers to such implementations; and provides assessments for energy conservation potentials for countries, as well as various practical and environmental aspects of energy conservation.

REFERENCES

1. Azar, C.; Rodhe, H. Targets for stabilization of atmospheric CO_2. Science **1997**, *276*, 1818–1819.
2. Midilli, A.; Ay, M.; Dincer, I.; Rosen, M.A. On hydrogen and hydrogen energy strategies: I: Current status and needs. Renew. Sust. Energy Rev. **2005**, *9* (3), 255–271.
3. BP. Workbook 2005. *Statistical Review—Full Report of World Energy*, British Petroleum, 2005, Available at: www.bp.com/centres/energy (accessed July 25, 2005).
4. IEE. World Energy Outlook 2002. Head of Publication Services, International Energy, **2002**, Available at: www.worldenergyoutlook.org/weo/pubs/weo2002/WEO21.pdf.
5. Kazman, S. *Global Warming and Energy Policy*; CEI; 67–86, **2003**. Available at: www.cei.org.
6. Dincer, I. Energy and environmental impacts: Present and future perspectives. Energy Sources **1998**, *20* (4/5), 427–453.
7. Dincer, I. *Renewable Energy, Environment and Sustainable Development*. Proceedings of the World Renewable Energy Congress; September 20–25, 1998; Florence, Italy; 2559–2562.
8. Rosen, M.A.; Dincer, I. On exergy and environmental impact. Int. J. Energy Res. **1997**, *21* (7), 643–654.
9. Dincer, I. Renewable energy and sustainable development: A crucial review. Renew. Sust. Energy Rev. **2000**, *4* (2), 157–175.
10. Dincer, I. *Practical and Environmental Aspects of Energy Conservation Technologies*. Proceedings of Workshop on Energy Conservation in Industrial Applications; Dhahran, Saudi Arabia, February 12–14, 2000; 321–332.
11. Dincer, I. On energy conservation policies and implementation practices. Int. J. Energy Res. **2003**, *27* (7), 687–702.
12. Sissine, F. Energy efficiency: Budget, oil conservation, and electricity conservation issues. *CRS Issue Brief for Congress*. IB10020; 2005.
13. Nolden, S.; Morse, D.; Hebert, S. *Energy Conservation for Housing: A Workbook*, Contract-DU100C000018374; Abt Associates, Inc.: Cambridge, MA, 1998.
14. Dincer, I.; Rosen, M.A. A worldwide perspective on energy, environment and sustainable development. Int. J. Energy Res. **1998**, *22* (15), 1305–1321.
15. Painuly, J.P.; Reddy, B.S. Electricity conservation programs: Barriers to their implications. Energy Sources **1996**, *18*, 257–267.
16. Anon. *Energy Conservation Policies and Technologies in Japan: A Survey*, OECD/GD(94)32, Paris, 1994.

Energy Conservation: Industrial Processes

Harvey E. Diamond
Energy Management International, Conroe, Texas, U.S.A.

Abstract
Energy Conservation in Industrial Processes will focus on energy conservation in industrial processes, will distinguish industrial processes and characteristics that differentiate them, will outline the analytical procedures needed to address them, will identify and discuss the main industrial energy intensive processes and some common ways to save energy for each of them, and will address managerial methods of conservation and control for industrial processes.

INTRODUCTION AND SCOPE

Energy conservation is a broad subject with many applications in governmental, institutional, commercial, and industrial facilities, especially because energy costs have risen so high in the last few years and continue to rise even higher. Energy conservation in industrial processes may well be the most important application—not only due to the magnitude of the amount of potential energy and associated costs that can be saved, but also due to the potential positive environmental effects such as the reduction of greenhouse gases associated with many industrial processes and also due to the potential of the continued economic success of all of the industries that provide jobs for many people.

This article will focus on energy conservation in industrial processes—where energy is used to manufacture products by performing work to alter feedstocks into finished products. The feedstocks may be agricultural, forest products, minerals, chemicals, petroleum, metals, plastics, glass, or parts from other industries. The finished products may be food, beverages, paper, wood building products, refined minerals, refined metals, sophisticated chemicals, gasoline, oil, refined petroleum products, metal products, plastic products, glass products, and assembled products of any kind.

This article will distinguish industrial processes and the characteristics that differentiate them in order to provide insight into how to most effectively apply energy conservation within industries. The level of applied technology, the large amount of energy required in many cases to accomplish production, the extreme conditions (e.g., temperature, pressure, etc.) that are frequently required, and the level of controls that are utilized in most cases to maintain process control will be addressed in this article.

This article will outline the analytical procedures needed to address energy conservation within industrial processes and will comment on general analytical techniques that will be helpful in analyzing energy consumption in industrial facilities.

Many of the main energy intensive processes, systems, and equipment used in industries to manufacture products will be identified and discussed in this article and some common ways to save energy will be provided.

This article will cover main energy intensive processes, systems, and equipment in a general format. If more in-depth instruction is needed for explanation of a particular industrial process, system, or type of equipment or regarding the analytical procedures required for a specific process, then the reader should refer to the many other articles included in this Encyclopedia of Energy Engineering and Technology, to references at the Association of Energy Engineers, to the references contained in this article, and if further detail is still needed, then the reader should contact an applicable source of engineering or an equipment vendor who can provide in-depth technical assistance with a specific process, system, or type of equipment.

In addition to the analytical methods of energy conservation, managerial methods of energy conservation will be briefly discussed. The aspects of capital projects versus managerial and procedural projects will be discussed. The justification of managerial efforts in industrial processes will be presented.

INDUSTRIAL PROCESSES—DIFFERENTIATION

Industrial processes require large amounts of energy, sometimes the highest level of technology, and often require very accurate process controls for process specifications, safety, and environmental considerations.

Industrial processes utilize an enormous amount of energy in order to produce the tons of production that are

Keywords: Energy conservation; Industrial processes; Industrial equipment; Process analyses; Process heating; Chemical reactions; Process cooling; Distillation; Drying; Melting and fusion; Mechanical processes; Electrical processes; Energy management.

being produced within industrial facilities. Industrial processes utilize over one-third of the total energy consumed in America.[1] Consider the amount of energy that is required to melt all of the metals being manufactured, to vaporize all of the oil and gasoline being refined, to dry all of the finished products that are made wet, to heat all of the chemicals that react at a certain temperature, to vaporize all of the chemicals that must be distilled for purity, to vaporize all of the steam that is used to heat industrial processes, to mechanically form all of the metal objects that we use, etc.,—this list is too long to be fully included in this article. This is an enormous amount of energy that produces all of the things that humans need and use—food, clothes, homes, appliances, cars, municipal facilities, buildings, roads, etc.

The level of technology required by current industrial processes is the highest in many cases and it is always at a high level in most industrial processes. Most industrial processes are utilizing technology that has been developed in the last 100 years or so, and consequently it has been further improved in the most recent years. Industrial processes most often utilize aspects of chemistry and physics in a precise manner in order to produce the sophisticated products that benefit people in our culture today. Very often, industrial processes require a very high or low temperature or pressure. Often they require a very precise and sophisticated chemistry and commonly they require highly technical designed mechanical processes. The application of electrical equipment and facilities in industrial processes is the highest level of technology for electrical power systems.

Industrial processes often require the highest level and accuracy of controls in order to produce products that meet product specifications, keep processes operating in a safe manner, and maintain environmental constraints. Due to each of these requirements or due to a combination of these requirements, the process controls for the processes within industrial facilities are often real-time Distributed Control Systems (DCSs), that are of the most sophisticated nature. A typical DCS for industrial processes functions to control process variables instantaneously on a real-time basis, whereby each process variable is being measured constantly during every increment of time and a control signal is being sent to the control element constantly on a real-time basis. The accuracy of a DCS in an industrial facility today is comparable to that of the guidance systems that took the first men to the moon.

Most industrial facilities with DCS controls also utilize a process historian to store the value of most process variables within the facility for a certain increment or period of time. The stored values of these process variables are used for accounting purposes and technical studies to determine optimum operating conditions and maintenance activities.

ENERGY CONSERVATION ANALYSES FOR INDUSTRIAL PROCESSES

In any industrial facility, the first analysis that should be performed for the purpose of energy conservation should be that of determining a balance of the energy consumed for each form of energy. This balance is used to determine how much energy is consumed by each unit, area, or division of the plant or possibly by major items of equipment that are consuming major portions of the energy consumed by the plant. This balance should be determined for each form of energy, whether it is for natural gas, electricity, coal, fuel oil, steam, etc. (see Table 1 below for an example of an energy balance). It

Table 1 Energy balance

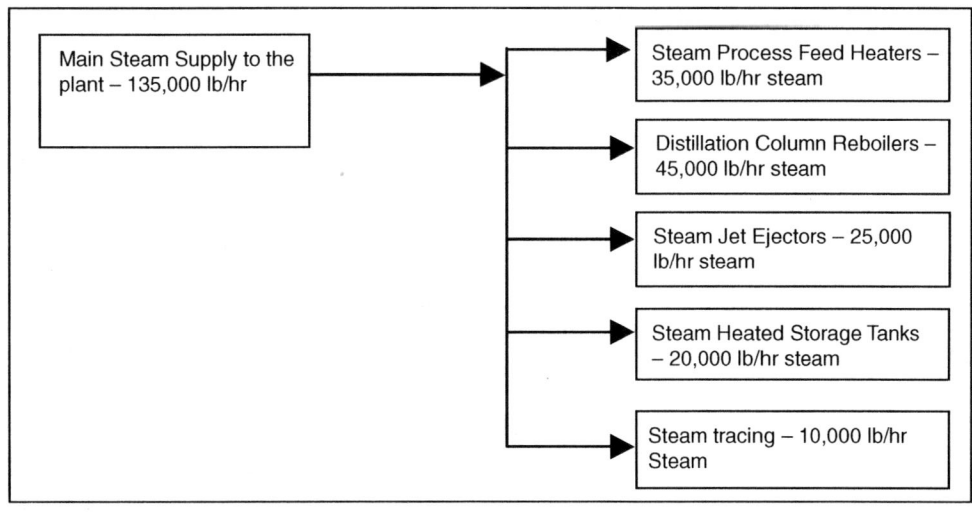

might be best that this determination not be called a balance (in that the numbers might not exactly come to a precise balance) but that it sufficiently quantifies the amount of energy consumed by each unit, area, or division of the plant. A better term for this determination might be an "Energy Consumption Allocation". The term balance is more usually applied to chemical and thermodynamic processes where heat and material balances are worked together mathematically to determine a calculated variable and the numbers have to exactly balance in order to arrive at the correct mathematical solution.

Once the amount of energy that is being actually consumed by each part of a plant has been determined, an energy consumption analysis should be performed for each item of energy consuming equipment and each major energy consuming facility in order to determine how much energy should realistically be consumed by each part of the plant. Notice that these calculations are called "realistic" as opposed to just theoretical because the object of these calculations is to determine as closely as possible how much actual energy each item of equipment or part of the plant should be consuming. By comparing these calculations with the actual energy consumption allocations mentioned above, it should be possible to obtain at least an initial indication of where energy is being wasted in a plant (see Table 2 below for an example of an energy consumption analysis for a process feed heater). During the course of obtaining the values of process variables that are required to make these energy consumption analyses, it is possible that indications will be observed of energy wastage due to the presence of an inordinate value of some process variable, such as too high or low of a temperature or pressure. When this type of indication is discovered, it usually also provides insight into what is operating in the wrong way to waste energy. There are numerous instances of energy wastage that can be discovered during these analyses, such as the inordinate manual control of a process, loose operational control of a process variable, or simply not shutting down a piece of equipment when it is not needed.

Table 2 Example of energy consumption analysis: process feed HTR

A process feed heater heats 1,199,520 lb/day of liquid feed material with a specific heat of 0.92 Btu/lb-°F, from 67 to 190°F. A realistic heater efficiency for this type of heater is determined to be 88%. The amount of realistic heat required for this heater is calculated to be: $Q = 1{,}199{,}520 \text{ lb/day} \times 0.92 \text{ Btu/lb-°F} \times (190°F - 67°F) \div 0.88 = 154{,}247{,}367.3$ Btu/day of realistic heat consumption.

It is observed that this feed heater is consuming 186,143,720 Btu/day

This feed heater is being operated in a wasteful way and is wasting over 20% of its heat

The analyses discussed in the above paragraph encompass all technical engineering science subjects, such as chemistry, thermal heat transfer, thermodynamics, fluid mechanics, mechanical mechanisms, and electrical engineering.

The next set of energy conservation analyses that should be performed are used to calculate the efficiencies of each item of equipment or facility to which efficiency calculations would be applicable, such as boilers, fired heaters, furnaces, dryers, calciners, and all other thermodynamic processes (efficiency calculations for boilers and other combustion equipment is available in the *Energy Management Handbook* by Wayne C. Turner and Steve Doty[2] and in the *Guide to Energy Management* by Barney L. Capehart, Wayne C. Turner, and William J. Kennedy)[3] and for electrical and mechanical equipment such as motors, pumps, compressors, vacuum pumps, and mechanical machinery. Once the actual efficiencies of any of the above have been determined, these numbers can be compared to the realistic efficiency for the type of equipment or facility that is prevalent throughout industry. These calculations and comparisons will also reveal wastage of energy and will frequently identify the causes of energy wastage and the possible issues to be corrected.

The next level of energy conservation analysis that may be performed is process analysis that can be conducted on a particular chemical, thermodynamic, thermal, fluid flow, mechanical, or electrical process. These analyses are usually performed by experienced engineers to examine the process itself and the process variables to determine if the process is being operated in the most effective and efficient manner. Here again, an indication will be provided as to whether or not energy is being wasted in the actual operation of the process. Chemical, thermodynamic, thermal, fluid flow, and other processes, as well as combinations of any of these processes can often require process simulation software such as PROMAX by Bryan Research & Engineering, Inc.,[4] in order to properly analyze these processes. The analysis of distillation columns, evaporators, and dryers can fall into this category. A good example would be the process analysis of a distillation column to determine if an effective and efficient level of reflux to the column and reboiler duty is being used.

Another analysis that has been very useful in the past few years in identifying energy conservation projects is Pinch analysis. This analysis is performed on thermodynamic and thermal processes in order to identify sources of energy within existing processes that can be used to supply heat for these processes instead of having to add additional heat to the entire process. The net effect is to reduce the amount of energy required for the overall process. The performance of a Pinch analysis on a particular process or facility will usually identify capital projects where revisions to the facility can be made to decrease the total amount of energy required. These are very often waste heat recovery projects. See "Use Pinch Analysis to Knock Down Capital Costs and Emissions" by Bodo Linnhoff, Chemical Engineering, August 1994[5] and "Pinch Technology: Basics for the Beginners".[6]

MAIN INDUSTRIAL ENERGY PROCESSES, SYSTEMS, AND EQUIPMENT

This section provides an overview and a list of the more common energy intensive industrial processes that are used to manufacture products in industrial facilities. Most energy intensive industrial processes can be classified into about eight general process categories—process heating, melting, chemical reactions, distillation-fractionation, drying, cooling, mechanical processes, and electrical processes. These processes are intended to be the main general energy intensive processes that are most commonly used and to which variations are made by different industries in order to make a specific product. In this regard, this is an overview—these processes are often not the specific process but a general category to which variations can be made to achieve the specific process.

In the following paragraphs, each process will be discussed by addressing its description, what systems it utilizes, what products are generally made, how it uses energy, and frequent ways that energy can be saved.

Common energy consuming systems and equipment that work to manufacture products in industrial facilities are also listed below and discussed in the same manner as the main industrial processes, as they are also common to industrial facilities and are related to these processes.

Process Heating

Description. The addition of heat to a target in order to raise its temperature. Temperatures can range from the hundreds to the thousands in industrial process heating.
Energy form. Heat must be generated and transferred to the intended object or medium.
Energy unit. Btu, calorie, joule, therm or watt-hour.
Examples. The application of heat in order to heat feed materials, to heat chemical processes, to heat metals for forming, to heat materials for drying, to heat materials in a kiln or calciner, to heat minerals and metals for melting.
Applied systems. Combustion systems, steam systems, thermal systems, hot oil systems, heating medium systems such as Dowtherm[7] or Paracymene,[8] and electrical resistance or induction heating systems.
Common equipment. Boilers, furnaces, fired heaters, kilns, calciners, heat exchangers, waste heat recovery exchangers, preheaters, electrical resistance heaters, and electrical induction heaters.
Common energy conservation issues. Keeping the heat targeted at the objective—proper insulation, seals on enclosures, eliminating leakage, and eliminating unwanted air infiltration. Control issues—maintaining sufficient control of the heating process, temperatures, and other process variables to avoid waste of heat. Management issues—shutting down and starting up heating processes at the proper times in order to avoid waste of heat and management of important process variables to reduce the amount of heat required to accomplish the proper process. Application of Pinch Technology—identify process areas where heat can be recovered, transferred, and utilized to reduce the overall process heat requirement. Waste heat recovery.

Melting and Fusion

Description. The addition of heat or electrical arc energy at a high temperature in order to melt metals, minerals, or glass. The melting process involves more than just process heating, it involves fluid motion, fluid density equilibrium, chemical equilibrium, cohesion, and sometimes electro-magnetic inductance. Reference: "The study showed that the fluid equations and the electromagnetic equations cannot be decoupled. This suggests that arc fluctuations are due to a combination of the interactions of the fluid and the electromagnetics, as well as the rapid change of the boundary conditions."[9]
Energy forms. Heat at high temperatures or electrical arc energy in the form of high voltage and high current flowing in an arc.
Energy units. Heat, Btu, calorie, joule, or therm.
Electrical arc. KWhrs.
Examples. Melting of ores in order to refine metals such as iron, aluminum, zinc, lead, copper, silver, etc. Melting of minerals in order to refine minerals such as silica compounds, glass, calcium compounds, potassium alum, etc.
Applied systems. Combustion systems, chemical reactions, and electrical systems.
Common equipment. Blast furnaces, arc furnaces, electrical resistance heaters, and electrical induction heaters.
Common energy conservation issues. Pre-condition of feed material—moisture content, temperature, etc. Feed method—efficiency of melting process effected by the feed method, feed combinations, and feed timing. Control of electromagnetics during melting and use of magnetic fields during separation. Overheating can waste energy without yielding positive process results. Heat losses are due to poor insulation, the failure of seals, or lack of shielding or enclosure.

Chemical Reactions

Description. Chemicals react to form a desired chemical, to remove an undesired chemical, or to break out a desired chemical. The chemical reaction can involve heat, electrolysis, catalysts, and fluid flow energy.
Energy forms. Heat, electrolysis, and fluid flow.
Energy units. Heat, Btu, calorie, joule, or therm.
Electrolysis. KWhrs.
Fluid flow. Ft-Lbs or Kg-M.
Examples. Reaction of chemical feed stocks into complex chemicals, petrochemical monomers into polymers, the oxidation of chemicals for removal, dissolving of chemicals to remove them, reaction of chemicals with other chemicals to remove them, the reaction of lignin with reactants in order to remove it from cellulose, the electro-plating of metals out of solution to refine them.
Applied systems. Feed systems, catalysts systems, heating systems, cooling systems, vacuum systems, run-down systems, separation systems, filtering systems, and electro-plating systems.
Common equipment. Reactors, digesters, kilns, calciners, smelters, roasters, feed heaters, chillers, pressure vessels, tanks, agitators, mixers, filters, electrolytic cells.
Common energy conservation issues. Close control of heating and cooling for chemical reactions. Close control of all reaction process variables—balance of all constituents, amount of catalyst, proper timing on addition of all components. Management of feedstocks, catalysts, and run-down systems for proper timing and correct balance for highest efficiency. Pinch analysis of feed heating, run-down products, cooling system, etc. Conservation of heating and cooling—proper insulation, sealing, and air infiltration. Waste heat recovery.

Distillation-Fractionation

Description. A thermo-dynamic and fluid flow equilibrium process where components of a mixture can be separated from the mix due to the fact that each component possesses a different flash point.
Energy form. Heat and fluid flow.
Energy units. Heat, Btu, calorie, joule, or therm.
Fluid flow. Ft-Lbs or KG-M.
Examples. Distillation-fractionation of hydrocarbons in oil and gas refineries and chemical plants. Distillation of heavy hydrocarbons in gas processing plants where natural gas is processed to remove water and heavy hydrocarbons.
Applied systems. Feed heating systems, over-head condensing systems, reflux systems, reboil systems, vacuum systems.
Common equipment. Distillation columns or towers, over-head condenser heat exchangers and accumulators—vessels, reflux pumps, reboiler heat exchangers, feed pumps, feed—effluent heat exchangers, vacuum steam jet ejectors.
Common energy conservation issues. Feed temperatures, reflux ratios, reboiler duty. Close control on pressures, temperatures, feed rates, reflux rates, and reboil duty. Management of overall operation timing—running only when producing properly. Concurrent use of vacuum systems—only when needed. Pinch analysis for feed and effluent streams and any process cooling systems. Proper insulation and elimination of lost heat for fired heater reboilers.

Drying

Description. The use of heat and fluid flow to remove water or other chemical components in order to form a more solid product.
Energy forms. Heat and fluid flow.
Energy units. Heat, Btu, calorie, joule, or therm.
Fluid flow. Ft-Lbs or KG-M.
Examples. Spray dryers that dry foods, sugar, fertilizers, minerals, solid components, and chemical products. Rotary dryers that dry various loose materials. Line dryers that dry boards, tiles, paper products, fiberglass products, etc. Other dryers that dry all kinds of products by flowing heated air over finished products in an enclosure.
Applied systems. Combustion systems, steam systems, thermal heating systems, cyclone systems, air filter systems, incinerator systems, Regenerative Thermal Oxidizer (RTO) systems.
Common equipment. Spray dyers, spray nozzles, natural gas heaters, steam heaters, electrical heaters, blowers, fans, conveyors, belts, ducts, dampers.
Common energy conservation issues. Efficient drying process for the components being eliminated. Proper amount of air flowing through dryer for drying. Proper insulation, seals, and elimination of lost heat due to infiltration. Waste heat recovery.

Process Cooling

Description. The removal of heat by a cooling medium such as cooling water, chilled water, ambient air, or direct refrigerant expansion.
Energy form. Heat.
Energy unit. Btu, calorie, joule, or therm.
Examples. Cooling water or chilled water circulated through a cooling heat exchanger, an air cooled heat exchanger, or a direct expansion evaporator that cools

air for process use.

Applied systems. Cooling water systems, chilled water systems, refrigerant systems, thermal systems.

Common equipment. Cooling towers, pumps, chillers, refrigeration compressors, condensers, evaporators, heat exchangers.

Common energy conservation issues. Use evaporative cooling as much as possible. Keep chillers properly loaded. Restrict chilled water flow rates to where 10°F temperature difference is maintained for chilled water. Limit cooling water pumps to the proper level of flow and operation. Apply Pinch analysis to achieve most efficient overall cooling configuration. Proper insulation, seals, and elimination of air infiltration.

Mechanical Processes

Description. Physical activities that involve force and motion that produce finished products. Physical activities can be discrete or can be by virtue of fluid motion.

Energy form. Physical work.

Energy unit. Ft-Lbs or KG-M.

Examples. Machining of metals, plastics, wood, etc.; forming or rolling or pressing of metals, minerals, plastics, etc.; assembly of parts into products; pumping of slurries thru screens or filters for separation; cyclone separation of solids from fluids; pneumatic conveyance systems that remove and convey materials or products and separate out solids with screens or filters.

Applied systems. Machinery, electrical motors, hydraulic systems, compressed air systems, forced draft or induced draft conveyance systems, steam systems, fluid flow systems.

Common equipment. Motors, engines, turbines, belts, chains, mechanical shafts, bearings, conveyors, pumps, compressors, blowers, fans, dampers, agitators, mixers, presses, moulds, rolls, pistons, centrifuges, cyclones, screens, filters, filter presses, etc.

Common energy conservation issues. Equipment efficiencies, lubrication, belt slippage, hydraulic system efficiency, compressed air system efficiency. Control of process variables. Application of variable speed drives and variable frequency drives. Management of system and equipment run times.

Electrical Processes

Description. The application of voltage, current, and electromagnetic fields in order to produce products.

Energy form. Voltage-current over time; electromagnetic fields under motion over time.

Energy units. KWh.

Examples. Arc welding, arc melting, electrolytic deposition, electrolytic fission, induction heating.

Applied systems. Power generator systems, power transmission systems, amplifier systems, rectifier systems, inverter systems, battery systems, magnetic systems, electrolytic systems, electronic systems.

Common equipment. Generators, transformers, relays, switches, breakers, fuses, plates, electrolytic cells, motors, capacitors, coils, rectifiers, inverters, batteries.

Common energy conservation issues. Proper voltage and current levels, time intervals for processes, electromagnetic interference, hysteresis, power factor, phase balance, proper insulation, grounding. Infrared scanning of all switchgear and inter-connections.

Combustion Systems

Combustion systems are found in almost all industries in boilers, furnaces, fired heaters, kilns, calciners, roasters, etc. Combustion efficiency is most usually a prime source of energy savings.

Boilers and Steam Systems

Boilers and steam systems may well be the most widely applied system for supplying process heat to industrial processes. "Over 45% of all the fuel burned by U.S. manufacturers is consumed to raise steam."[10] Boiler efficiencies, boiler balances (when more than one boiler is used), and steam system issues are usually a prime source of energy savings in industrial facilities.

Flare Systems and Incinerator Systems

Flare and incinerator systems are used in many industrial facilities to dispose of organic chemicals and to oxidize volatile organic compounds. Proper control of flares and incinerators is an issue that should always be reviewed for energy savings.

Vacuum Systems

Vacuum systems are used to evaporate water or other solvents from products and for pulling water from products in a mechanical fashion. Vacuum systems are also used to evacuate hydrocarbon components in the petroleum refining process. Vacuum systems are frequently used in the chemical industry to evacuate chemical solvents or other components from a chemical process. Steam jet ejectors and liquid ring vacuum pumps are commonly used to pull vacuums within these systems. The efficiencies of the ejectors and the liquid ring vacuum pumps can be a source of energy savings as well as the management of vacuum system application to production processes. Pneumatic conveyance systems that utilize a fan or blower to create a low-level vacuum are sometimes used to withdraw materials or products from a process and separate the matter within a screen or filter. For large conveyance systems, the efficiencies of the equipment and the management of their operation can be a source of energy savings.

Furnaces, Fired Heaters, Kilns, Calciners

The above comments on combustion systems are applicable to these equipment items and additional energy savings issues can be found relative to them.

Centrifugal Pumps

Centrifugal pumps are used widely in industries. The flow rate being pumped is a primary determining factor for the amount of power being consumed and it is sometimes higher than required. Good control of the pumping rate is an important factor in saving energy in centrifugal pumps. The application of variable frequency drives to the motor drivers can be a good energy saving solution for this issue.

Fans and Blowers

The flow rate for fans and blowers is analogous to the pumping rate above for centrifugal pumps. Good control of the flow rate and the possible application of Variable Frequency Drive (VFD) apply here as well for fans and blowers.

Centrifugal Compressors

Compressors are used widely in industry. The above discussions of flow rates, control of flow rates, and application of VFDs apply here as well. Centrifugal compressors frequently will have a recycle flow that is controlled in order to prevent the compressor from surging. Close control of this recycle flow at its minimum level is very important for compressor efficiency.

Liquid Ring Vacuum Pumps

As mentioned above in several places, liquid ring vacuum pumps are used widely in industry. The amount of sealing liquid that is recycled to the pump and the temperature of the sealing liquid are important determinates of the efficiency of the Liquid Ring Vacuum Pump (LRVP).

The above overview and list of industrial processes, systems, and equipment has been general in nature due to the limitations of this article. Greater and more specific familiarity with each of these industrial energy intensive processes, systems, and equipment will yield greater applicable and effective insight into ways to save energy related to each of these items.

CAPITAL PROJECTS VERSUS IMPROVED PROCEDURES

Energy conservation effort applied in industrial facilities can identify capital projects whereby the facilities can be changed in order to achieve greater overall energy efficiency or the efforts can identify changes to in day-to-day operating and maintenance procedures that can reduce waste of energy and also improve the overall efficiency of the facility. Frequently, energy-saving procedural changes to day-to-day operations and maintenance activities within an industrial plant can be identified by taking and recording operating data once the processes, systems, and equipment have been studied and analyzed for energy consumption. Procedural changes to operations and maintenance within an industrial plant can often amount to low costs or possibly no costs to the facility. This aspect of energy conservation is often overlooked by highly technical personnel that have worked hard to design industrial facilities because they have technically designed the facility very well for energy consumption considerations and the more mundane activities related to day-to-day operation and maintenance tend to not register in their highly technical perspective. None-the-less, a considerable amount of energy can usually be saved within most industrial processes, systems, and equipment due to changes in the way they are operated and maintained. A general tendency within industrial plants is that operations will often operate the processes and systems at a point that provides a comfortable separation between an operating variable and its limitation in order to understandably ensure no upsets occur within the process or system. However, with the cost of energy being what it is today, it is frequently found that a significant amount of energy can be saved by operating processes and systems more tightly and efficiently, even though it may require more attention, increased control, and the monitoring of process variables.

EFFECTIVE ENERGY MANAGEMENT SYSTEMS

Another aspect of energy conservation that can be very productive in saving energy within industrial processes is that of an effective energy management system. An effective energy management system is comprised of operational and maintenance managers functioning in conjunction with an accurate and concurrent data collection system in order to eliminate waste and improve overall efficiency of industrial processes. It is not possible to manage any activity unless the activity is being properly monitored and measured with key performance metrics (KPMs). The data collection system part of an effective energy management system within any industrial facility provides the accurate and concurrent measurement data (KPMs) that is required in order to identify actions that are needed to eliminate waste of energy and improve overall efficiency of the facility. An effective energy management system is first built upon acquiring total knowledge of the facility down into every level of operation and maintenance of the facility. Such a level of thoroughness and complete analysis of energy consumption within a facility is sometimes referred to as *Total Quality Energy* management.[11] Once an effective energy management system has been established and is effectively controlling energy consumption of an industrial facility, it should be maintained, in effect, so that it will continue to monitor KPMs to maintain energy conservation for the facility. An effective energy management system within an industrial manufacturing facility can eliminate as much as two to three percent of the energy costs by eliminating waste of energy on a day-to-day operational and maintenance basis. In most industrial facilities, this level of cost reduction is significant and will justify an effective energy management system.

THE CURRENT NEED FOR GREATER ENERGY CONSERVATION IN INDUSTRY

With the present cost of all forms of energy today, it would certainly seem logical that all of industry would be seeking greater energy conservation efforts within their facilities. Unfortunately, many corporate industrial managers are not aware of the true potential of conserving energy within their processes and facilities. Greater awareness of the ability to conserve energy on the basis of increased efficiencies of processes, systems, and equipment is needed; and also due to the application of an effective energy management system. For the good of society and environment, corporate industrial managers should be more open to the possibility of the improvement of industry that will work to sustain their business and improve the world that we live in. This is in opposition to corporate political thinking, which does not want to consider making changes and wants no one to interfere with their present activities. Human beings should be willing to examine themselves and make changes that will make things better. The same outlook should be applied to businesses and industry in order to make things better. Greater management support is needed in industry today to accomplish greater and very much needed increased energy conservation.

CURRENT APPLICATION OF INCREASED ENERGY MANGEMENT

With the recent technological advancements that have been made in digital computer and communications systems, data collection systems can be implemented in industrial facilities in a much more cost effective manner. Wireless communication systems for metering and data collection systems have advanced dramatically in the last few years and network-based computer communication has enabled whole new systems for measurement and control. With all of these new fields of configuration for data collection systems, with the increased technology, and with the lower costs to accomplish data collection systems, it is now possible to apply energy management systems to industry today with much greater applicability. Hopefully this will be recognized and result in greater applications of effective energy management systems.

From recent observations, it appears that most of industry today is a candidate for improved and more effective energy management systems. In conjunction with the increased technology and lower cost potentials, it seems that there is a definite match between supply and need for the application of increased energy management systems.

SUMMARY

Industrial processes have commonality in processes, systems, and equipment. There are logical and systematic analyses that can be performed in industrial processes that can identify ways to save energy. Effective energy management systems are needed in industry today and there are great possibilities to save energy in industrial processes. Energy can be conserved in industrial processes by analyses that will improve efficiencies, by implementation of procedures that eliminate waste, and by application of an effective energy management system.

REFERENCES

1. U.S. Department of Energy–Energy Efficiency and Renewable Energy, http://www.eere.energy.gov/EE/industry.html (accessed on 2006).
2. Turner, W.C.; Doty, S. *Energy Management Handbook*, 6th Ed.; Fairmont Press: Lilburn, GA, 2005.
3. Capehart, B.L.; Turner, W.C.; Kennedy, W.J. *Guide to Energy Management*, 5th Ed.; Fairmont Press: Lilburn, GA, 2004.
4. Bryan Research & Engineering, Inc. PROMAX; BRE, Bryan, TX; http://www.bre.com (accessed on 2006).
5. Linnhoff, B. Use pinch analysis to knock down capital costs and emissions. Chemical Engineering **August 1994** http://www.che.com.

6. Solar Places Technology. Pinch Technology: Basics for the Beginners; http://www.solarplaces.org/pinchtech.pdf (accessed on 2006).
7. Dowtherm, Dow Chemical http://www.dow.com/heattrans/index.html (accessed on 2006).
8. Paracymene, Orcas International, Flanders, NJ 07836 http://www.orcas-intl.com (accessed on 2006).
9. King, P.E. Magnetohydrodynamics in Electric Arc Furnace Steelmaking. Report of Investigations 9320; United States Department of the Interior, Bureau of Mines; http://www.doi.gov/pfm/ar4bom.html (accessed on 2006).
10. U.S. Department of Energy–Energy Efficiency and Renewable Energy, http://www.eere.energy.gov/EE/industry.html the common ways (accessed on 2006).
11. Energy Management International, Inc. Total Quality Energy; http://www.wesaveenergy.com (accessed on 2006).

Energy Conservation: Lean Manufacturing

Bohdan W. Oppenheim
U.S. Department of Energy Industrial Assessment Center, Loyola Marymount University,
Los Angeles, U.S.A.

Abstract

Productivity has a major impact on energy use and conservation in manufacturing plants—an impact often more significant than optimization of the equipment energy efficiency. This article describes Lean Manufacturing, which represents the current state-of-art in plant productivity. A significant opportunity for energy savings by transforming production into single-piece Lean Flow is demonstrated. The impact of major individual productivity elements on energy is discussed. Simple metrics and models are presented as tools for relating productivity to energy. Simple models are preferred because productivity is strongly influenced by intangible human factors such as work organization and management, learning and training, communications, culture, and motivation, which are difficult to quantify in factories.

INTRODUCTION

At the time of this writing (2005), the world is experiencing strong contradictory global trends of diminishing conventional energy resources and rapidly increasing global demands for these resources, resulting in substantial upwards pressures in energy prices. Because the energy used by industry represents a significant fraction of the overall national energy use, equal to 33% in the United States in the year 2005, a major national effort is underway to conserve industrial energy.[1] The rising energy prices place escalating demands on industrial plants to reduce energy consumption without reducing production or sales, but by increasing energy density.

Optimization of industrial hardware and its uses, including motors and drives, lights, heating, ventilation and cooling equipment, fuel-burning equipment, and buildings, are well understood, have been practiced for years,[2] and are important in practice. However, they offer only limited energy conservation opportunities, rarely exceeding a few percent of the preoptimization levels. In contrast, the impact of productivity on energy use and energy density offers dramatically higher savings opportunities in energy and in other costs. In the extreme case, when transforming a factory from the traditional "process village" batch-and-queue system to the state-of-the-art, so-called Lean system, the savings in energy can reach 50% or more.

The best organization of production known at this time is called Lean, developed at Toyota in Japan.[3] It is the flow of value-added work through all processes required to convert raw materials to the finished products with minimum waste. Major elements of Lean organization include: steady single-piece flow with minimum inventories and no idle states or backflow; flexible production with flexible equipment and operators and flexible floor layouts ready to execute the order of any size profitably and just-in-time; reliable and robust supplies of raw materials; minimized downtime due to excellent preventive maintenance and quick setups; first-pass quality; clean, uncluttered, and well-organized work space; optimized work procedures; and, most importantly, an excellent workforce–well trained, motivated, team-based and unified for the common goals of having market success, communicating efficiently, and being well-managed. The Lean organization of production is now well understood among productivity professionals, but it is not yet popular among the lower tier suppliers in the United States. Its implementation would save energy and benefit the suppliers in becoming more competitive.

The engineering knowledge of energy conservation by equipment improvements is well understood and can be quantified with engineering accuracy for practically any type of industrial equipment.[2] In contrast, industrial productivity is strongly influenced by intangible and complex human factors such as management, work organization, learning and training, communications, culture, and motivation. These work aspects are difficult to quantify in factory environments. For this reason, the accuracy of productivity gains and the related energy savings are typically much less accurate than the energy savings computed from equipment optimization. Simple quantitative models with a conservative bias are therefore recommended as tools for energy management in plants. This article includes some examples. They are presented in the form of energy savings or energy cost savings that would result from implementing a given productivity

Keywords: Productivity impact on energy; Productivity; Energy; Lean; Just-in-time; Energy savings; Energy conservation; Energy density; Industrial energy.

improvement, or eliminating a given productivity waste, or as simple metrics measuring energy density.

It is remarkable that in most cases, these types of energy savings occur as a natural byproduct of productivity improvements, without the need for a direct effort centered on energy. Thus, the management should focus on productivity improvements. In a traditional non-Lean plant intending to transform to Lean production, the first step should be to acquire the knowledge of the Lean system. It is easily available from industrial courses and workshops, books,[3,4] and video training materials.[6] The next step should be the actual transformation of production to Lean. Most of the related energy savings will then occur automatically. Implementation of individual productivity elements such as machine setup time reduction will yield some energy savings, but the result will not be as comprehensive as those yielded by the comprehensive implementation of Lean production.

TRADITIONAL VS LEAN PRODUCTION

The traditional organization of production still used frequently in most factories tends to suffer from the following characteristics:

- Supplier selection is based on minimum cost, resulting in a poor level of mutual trust and partnership, the need for receiving inspection, and often large inventories of raw materials (RM).
- Work-in-progress (WIP) is moving in large batches from process village to process village and staged in idle status in queues in front of each machine, while the machine moves one piece at a time. This work organization is given the nickname "batch-and-queue" (BAQ).[3]
- Finished goods (FG) are scheduled to complex forecasts rather than customer orders, resulting in large inventories.
- The floor is divided into "process villages" populated with large, complex, and fast similar machines selected for minimum unit cost.
- Minimum or no information is displayed at workstations, and the workers produce quotas.
- Work leveling is lacking, which results in a random mix of bottlenecks and idle processes.
- Unscheduled downtime of equipment occurs frequently.
- Quality problems with defects, rework, returns, and customer complaints are frequent.
- Quality assurance in the form of 100% final inspections attempts to compensate for poor production quality.
- The floor space is cluttered, which makes moving around and finding items difficult.
- The workforce has minimum or no training and single skills.
- The management tends to be authoritarian.
- A language barrier exists between the workers and management.
- There is a culture of high-stress troubleshooting rather than creative trouble prevention.

In such plants, the waste of materials, labor, time, space, and energy can be as much as 50%–90%.[3]

The Lean production method developed primarily at Toyota in Japan under the name Just-In-Time (JIT), and generalized in the seminal work[3] is the opposite of the traditional production in almost all respects, as follows:

- Raw materials are bought from reliable supplier–partners and delivered JIT in the amount needed, at the price agreed, and with the consistently perfect quality that obviates incoming inspection.
- Single-piece flow (SPF) of WIP is steadily moving at a common takt time (Takt time is the common rhythm time of the pieces moving from workstation to workstation on the production line. It is the amount of time spent on EACH operation. It precisely synchronizes the rate of all production operations to the rate of sales JIT.), from the first to the last process.
- The FG are produced to actual customer orders JIT resulting in minimum inventories.
- The floor is divided into flexible production lines with small simple machines on casters that can be pushed into position and setup in minutes.
- The labor is multiskilled, well motivated and well trained in optimized procedures.
- Quality and production status are displayed on large visible boards at each workstation, making the entire production transparent for all to see.
- Preventive maintenance assures no unscheduled downtime of equipment.
- All process operators are trained in in-line quality checks and variability reduction.
- No final inspection is needed, except for occasional sampled checks of FG.
- Defects, rework, returns, and customer complaints are practically eliminated.
- The floor space is clean and uncluttered.
- The workforce is trained in company culture and commonality of the plant mission, customer needs, workmanship, and quality.
- The culture promotes teamwork, multiple job skills, supportive mentoring management, and company loyalty.
- The management promotes trouble prevention and "stopping the line" at the first sign of imperfection so that no bad pieces flow downstream.

According to Womack et al. the transformation from traditional to Lean production can reduce overall cost,

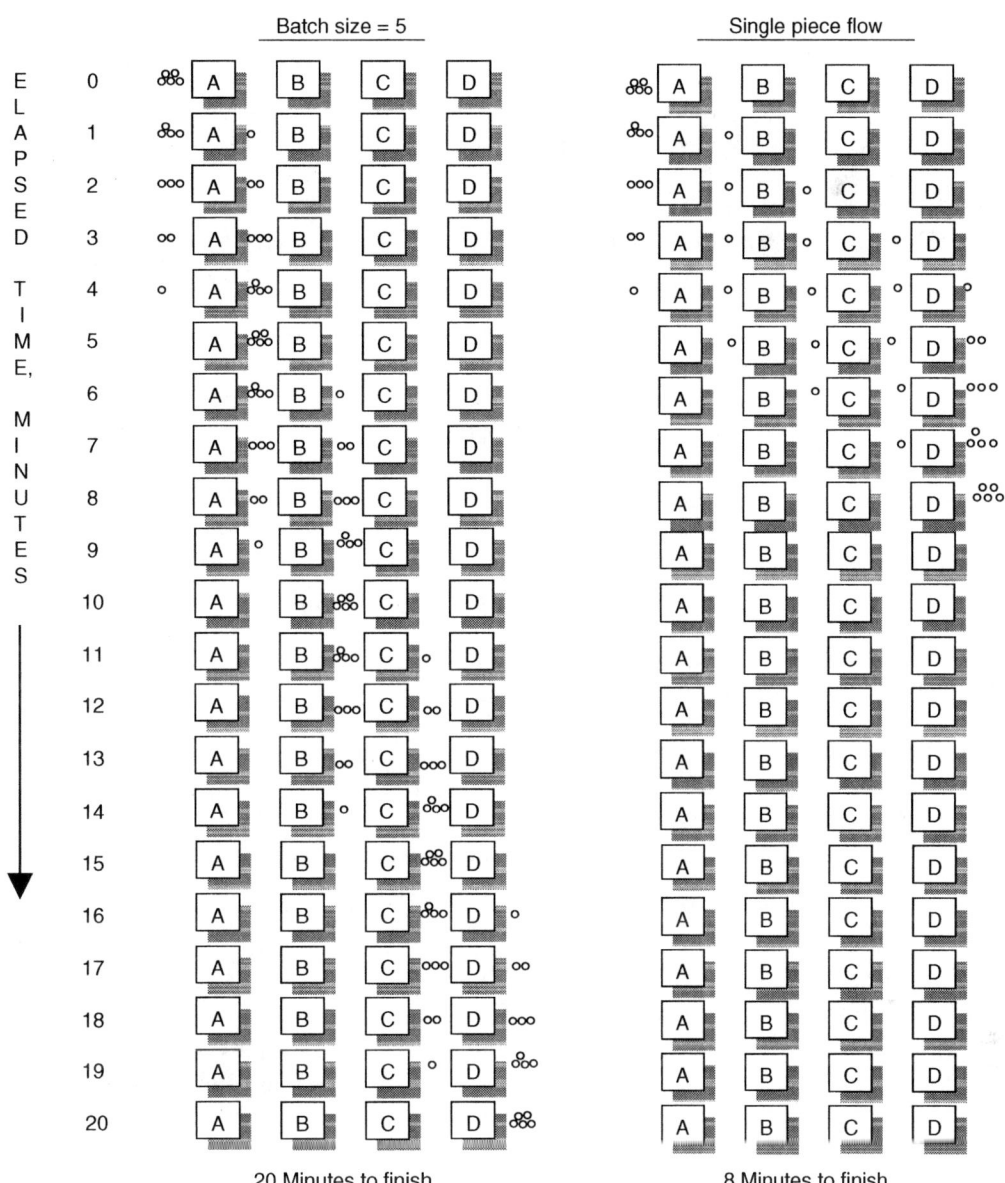

Fig. 1 BAQ with batch size of five vs SPF.

inventory, defects, lead times by 90%, and space by 50%, and vastly increase plant competitiveness, customer satisfaction, and workforce morale. The resultant energy savings can be equally dramatic. Ref. [4] contains interviews with industry leaders who have succeeded in this transformation.

IMPACT ON ENERGY

The impact of productivity on plant energy falls into the following two broad categories:

1. Productivity improvements that save infrastructure energy. These improvements reduce the energy consumed by all plant support systems, which tend to be energized regardless of the actual production activities, such as lights, space cooling and heating devices, cooling towers, combustion equipment (boilers, molten metal furnaces), air compressors, forklift battery chargers, conveyors, etc. To the first approximation, the infrastructure energy is reduced in proportion to the production time reductions, which can be huge in the Lean system. In order to perform more detailed estimates of the infrastructure energy savings, the management would have to conduct detailed energy accounting and understand how much energy is used by each support system under different production conditions. This knowledge is rarely available;

therefore the former simplistic approach, combined with conservative estimates, offer useful tools.

2. Process energy savings. In this category, the energy savings of process equipment are obtained by improving the process productivity. Examples include the reduction of unscheduled machine downtime or setup time and the elimination of process variability, defects, rework, scrap, excessive labor time, etc.

Single Piece Flow (SPF)

Changing the traditional BAQ production to Lean production is by far the most effective productivity transformation a plant can undertake, creating dramatic savings in the overall throughput time, cost, quality, and energy. The example shown in Fig. 1 compares just one aspect of the transformation—a reduction of batch size from five to one, i.e., the SPF. In both cases, four processes of equal one-minute takt time are assumed. The benefits of the SPF alone are dramatic, as follows:

1. In BAQ, the batch is completed in 20 min and in SPF in only 8 min, a 60% reduction.
2. In BAQ, only one machine at a time produces value, while three others are idle. If the idle machines remain energized, as is the case, e.g., with injection molding, three of the four machines (75%) would be wasting energy, and doing it for 16 min each, adding up to 64 min of machine energy wasted. In the SPF system, no machine energy is wasted as no machine would be idle, except for the lead and tail of each process of 4 min, adding up to 16 min of machine energy wasted, a savings of 75% from BAQ.
3. Reducing the batch throughput time by 60% reduces the infrastructure energy by the same amount, assuming the production is completed faster and the plant is de-energized. Alternatively, the freed 60% time and energy could be used for additional production and profits.
4. An important additional benefit is that in SPF, a defect can be detected on the first specimen–as soon as it reaches the next process, while in the BAQ, the entire batch may be wasted before the defect is discovered and a corrective action undertaken, with the energy used for making the batch wasted.

This simple example clearly illustrates the dramatic impact of SPF on both overall productivity and energy consumption. Typically, as the factories transform to the Lean system, their sales, production, and profits increase simultaneously and the energy used decreases. A convenient metric to track the overall benefit is the gross energy density, ED_1 or ED_2:

$$ED_1 = \frac{EC_T}{P} \tag{1a}$$

$$ED_2 = \frac{EC_T}{AC} \tag{1b}$$

where EC_T is the overall annual cost of energy in the plant, P is the number of products produced per year and AC is the total annual costs (sales minus net profit). ED_1 should be used if similar products are made most of the time, and ED_2 should be used if the plant has a wide menu of dissimilar products. The ED ratios will decrease as progress is made from BAQ to SPF. If the volume of production remains constant during the transformation, energy savings and energy cost savings alone may be more convenient metrics to track plant energy efficiency.

Inventory Reduction

All inventories, whether in RM, WIP, or FG, beyond the immediate safety buffers, are detrimental. Inventory means that company capital is "frozen" on the floor; cutting into the cash flow; wasting labor for inventory control, storage, and security; wasting infrastructure energy for lights, forklift energy, and possible cooling or heating of the inventory spaces if the goods require temperature or humidity control; wasting space and the associated lease/mortgage fees and taxes; and becoming scrap if not sold (a frequent waste in large inventories). Inventory and inventory space reductions lead to infrastructure energy savings. Process energy can also be saved by not making the FG that end up in inventory, cannot be sold, and become scrap. Refs. 3 and 4 contain case studies for, among others, inventory reductions. A convenient nondimensional metric to track the overall impact of all inventories on energy savings is

$$EC_T \times \frac{I_T}{AC} \tag{2}$$

where I_T is the number of inventory turns per year.

Workmanship, Training, and Quality Assurance

In the ideal Lean system, the processes, equipment, procedures, and training are perfected to the degree that guarantees consistent and robust production with predictable effort, timing, quality, and cost; with no variability, defects, or rework, and with maximum ergonomics and safety. This is accomplished by a consistent long-term strategy of continuous improvement of all the above elements, including intensive initial training of the workforce and subsequent retraining in new procedures. A procedure must be developed for each process until it is robust and predictable and optimized for

Example 1 Energy waste from poor workmanship

> A plant with \$20,000,000 in sales and \$2,000,000 in profits spends \$1,000,000 on energy per year. The typical order requires 10 processes of roughly equal energy consumption. The production equipment consumes 60% and the supportive infrastructure consumes 40% of the plant energy. Sequential process #5 has the defect rate of 10%. In order to compensate for the defects, the first 5 processes must produce 10% extra pieces. The annual waste of energy cost (and the energy cost savings, if the defective process is fixed) is then:
>
> $$(\$1,000,000/\text{yr})(5/10 \text{ processes})(60\% \text{ process energy})(10\% \text{ defect rate}) = \$30,000/\text{yr} \qquad (3)$$
>
> The additional production time of 10% wastes not only the cost of the process energy computed in (3) but also the infrastructure energy cost of:
>
> $$(\$1,000,000/\text{yr})(40\% \text{ infrastructure energy})(10\% \text{ defects}) = \$40,000/\text{yr} \qquad (4)$$
>
> Such delays also extend the promised delivery time and reduce customer satisfaction and factory competitiveness. Adding (1) and (2) together, (not counting the direct productivity losses), the wasted energy cost alone of \$70,000/yr represents 3.5% of the annual profits and 7% in annual energy costs. Based on the author's experience,[5] these numbers are not infrequent in industry. Fixing the productivity of process #5 would eliminate these wastes.

minimum overall cost, required quality, maximum ergonomics, and safety. Process operators must be trained in the procedures as well as in the process quality assurance, and they must be empowered to stop the process and take corrective action or call for help if unable to avoid a defect. Management culture must be supportive for such activities. Any departure from this ideal leads to costly penalties in quality, rework, delays, overtime or contract penalties, crew frustrations, and customer dissatisfaction. These, in turn, have negative impacts on energy as follows:

1. Defects require rework, which requires additional energy to remake or repair the part. The best metric to use here is the energy or energy cost per part used in the given defective process multiplied by the number of bad parts produced per year.
2. Variability in the process time or delays caused by defects mean that the production takes more time and more infrastructure and process energy for the same amount of value work and profits when compared with the ideal nonvariable process. Example 1 illustrates cases (1) and (2).
3. Defective processes usually require a massive final inspection to sort out the good products. Finding the finished goods defective is the most inefficient means of quality assurance because often the entire batch must then be remade, consuming the associated energy. The inspection space, labor, and energy represent a direct waste and should be replaced with in-line quality assurance (ILQA) that detects the first bad piece (Governmental, medical, etc. orders usually require a 100% final inspection. In the Lean system, this is performed as a formality because everybody in the plant knows that all pieces will be perfect because all imperfections have been removed in real time before the inspection process.) and immediately undertakes a corrective action. Typically, the ILQA can be implemented in few days of operators' training and has the simple payback period measured in days or weeks.[5]

Overage Reduction

Many a plant compensates for its notorious defects by routinely scheduling production in excess of what the customer orders. Some minimum overage is usually justified for machine setups, adjustments, and QA samples. In a Lean plant this rarely exceeds a fraction of one percent. In a traditional plant, the value of 5%–15% is not infrequent. A 5% overage means that the plant spends 105% of the necessary costs. If the profit margin is 5%, the overage alone may consume the entire profit. The overall energy waste (and the opportunity to save energy) is simply proportional to the overage amount. Overage is one of the most wasteful ways of compensating for defective processes. The best remedy is to simply identify the defective process with ILQA, find the root cause (typically the lack of training, excessive work quotas, or bad process or material), and repair it.

Unintentional overage can also be destructive to profits and energy use. Example: A worker is asked to cut only a few small pieces from a large sheet of metal, but instead he cuts the entire sheet, thinking, "my machine is already setup and soon they will ask me to cut the rest of the sheet anyway, so I may as well do it now". The excessive pieces then move through all processes, unknowingly to the management, consuming energy, labor and fixed costs, to end up as excessive FG inventory and, in the worst case, find no buyer and end up as scrap. Uncontrolled and careless overage can easily consume all profits, and, of course, waste energy proportionately to the overage amount.

Example 2 Energy savings from setup time reduction

A plant operates on two shifts, 260 days per year, performing on average of 20 two-hour setups per day on their electrically heated injection molding machines. Each machine consumes 20 kW when idle but energized. By a focused continuous improvement system and training, the crew reduces the routine setup time to 0.5 h, with few, if any expenses for additional hardware, thus saving:

(260 days/yr) (20 setups/day) (1.5 h saved/setup) = 7800 machine h/yr.

The resultant process energy saved will be:

(7800 h/year) (20 kW) = 156,000 kWh/yr (5)

In addition, infrastructure energy will be saved because of the reduced downtime. Using the data from Example 1, if the work is done in two shifts for 260 days per year (4160 h/yr), the plant infrastructure uses 40% of the plant energy, and each machine consumes 2% of the plant infrastructure energy during the setup, the additional energy cost savings due to the setup time reduction will be:

(7,800 hr/yr) (0.02) (0.04) ($1,000,000)/(4160 h/yr) = $15,000 (6)

Downtime

Equipment downtime and idleness may occur due to scheduled maintenance, unscheduled breakdowns, machine setups, and poor process scheduling. The downtime may cause proportional loss of both profits and energy. The downtime may have fourfold impact on energy use, as follows:

1. When a process stops for whatever reason during an active production shift, the plant infrastructure continues to use energy and loosing money, as in Eq. 4. A good plant manager should understand what fraction of the infrastructure energy is wasted during the specific equipment downtime. With this knowledge, the energy waste can be estimated as being proportional to the downtime.
2. Some machines continue using energy during maintenance, repair, or setup in proportion to the downtime (e.g., the crucible holding molten metal for a die casting machine remains heated by natural gas while the machine is being setup or repaired). Reducing the setup time or eliminating the repair time saves the gas energy in direct proportion to the downtime saved. In order to calculate energy savings in such situations, it is necessary to understand the energy consumption by the equipment per unit of time multiplied by the downtime reduction.
3. When a particular machine is down, additional equipment upstream or downstream of that machine may also be forced into an idle status but remain energized, thus wasting energy. In an ideal single-piece flow, the entire production line (As in the saying "In Lean either everything works or nothing works.") will stop. In order to estimate the energy-saving opportunity from reducing this cumulative downtime, the energy manager must understand which equipment is idled by the downtime of a given machine and how much energy it uses per unit time while being idle.
4. Lastly, energized equipment should be well managed. A high-powered machine may be left energized for hours at a time when not scheduled for production. A good practice is to assign each of these machines to an operator who will have the duty of turning the machine off when not needed for a longer time, if practical, and to turn it back on just in time to be ready for production exactly when it is needed.

Preventive maintenance and setup time reduction have a particularly critical impact on both productivity and related energy use, as follows:

Preventive Maintenance

Practical and routine preventive maintenance should be done during the hours free of scheduled production (e.g., during night shifts, on weekends, or during layover periods). The maintenance should be preventive rather than reactive (The term "preventive" tends to be replaced with "productive" in modern industrial parlance). Well-managed "total" preventive[6] maintenance involves not only oiling and checking the machines per schedule but also ongoing training of the mechanics; developing a comprehensive database containing information on the particular use and needs of various machines; preparing a schedule of part replacement and keeping inventory of frequently used spare parts; and a well-managed ordering system for other parts, including vendor data so that when a part is needed it can be ordered immediately and shipped using the fastest possible means. Industry leaders have demonstrated that affordable preventive maintenance can reduce the unscheduled downtime and associated energy waste to zero. This should be the practical goal of well-run factories.

Setups

Modern market trends push industry towards shorter series and smaller orders, requiring, in turn, more and shorter setups. Industry leaders have perfected routine setups to take no more than a few minutes. In poorly managed plants, routine setups can take as long as several hours. In all competitive modern plants, serious efforts should be devoted to setup time reductions. The effort includes both training and hardware improvements. The training alone, with only minimal additional equipment (such as carts), can yield dramatic setup time reductions (i.e., from hours to minutes). Further gains may require a change of the mounting and adjustment hardware and instrumentation. Some companies organize competitions between teams for developing robust procedures for the setup time reductions. In a plant performing many setups, the opportunity for energy savings may be significant, both in the process and infrastructure energy, as shown in Example 2.

Flexibility

Production flexibility, also called agility, is an important characteristic of competitive plants. A flexible plant prefers small machines (if possible, on casters) that are easy to roll into position and plug into adjustable quick-connect electrical and air lines and that are easy to setup and maintain over the large fixed machines selected with large batches and small unit costs in mind (such machines are called "monuments" in Ref. 3). Such an ideal plant will also have trained a flexible workforce in multiple skills, including quality assurance skills. This flexibility allows for the setup of new production lines in hours or even minutes, optimizing the flow and floor layout in response to short orders, and delivers the orders JIT. The energy may be saved in two important ways, as follows:

- Small machines processing one piece at a time use only as much energy as needed. In contrast, when excessively large automated machines are used, the typical management choice is between using small batches JIT, thus wasting the large machine energy, or staging the batches for the large machine, which optimizes machine utilization at the expense of throughput time, production flow, production planning effort, and the related infrastructure energy.
- Small machines are conducive to flexible cellular work layout, where 2–4 machines involved in the sequential processing of WIP are arranged into a U-shaped cell with 1–3 workers serving all processes in the cell in sequence, and the last process being quality assurance. This layout can be made very compact—occupying a much smaller footprint in the plant compared to traditional "process village" plants, roughly a reduction of 50%[3,4]—and is strongly preferred by workers because it saves walking and integrates well the work steps. Such a layout also saves forklift effort and energy and infrastructure energy due to the reduction of the footprint.

Other Productivity Elements

The complete list of productivity elements is beyond the scope of this article, and all elements have some leverage on energy use and conservation. In the remaining space, only the few most important remaining aspects are mentioned, with their leverage on energy. Descriptive details can be found in Ref. 7 and numerous other texts on Lean production.

- *Visual factory*: Modern factories place an increasing importance on making the entire production as transparent as possible in order to make any problem visible to all, which is motivational for immediate corrective actions and continuous improvements. Ideally, each process should have a white board displaying the short-term data, such as the current production status (quantity completed vs required); the rate of defects or rejects and their causes; control charts and information about the machine condition or maintenance needs; and a brief list and explanation of any issues, all frequently updated. The board should also display long-term information such as process capability history, quality trends, operator training, etc. Such information is most helpful in the optimization of, among other things, process time and quality, which leads to energy savings, as discussed above.
- *"Andon" signals*: The term refers to the visual signals (lights, flags, markers, etc.) displaying the process condition, as follows: "green = all OK," "yellow = minor problem being corrected," and "red = high alarm, stopped production, and immediate assistance needed." The signals are very useful in identifying the trouble-free and troubled processes, which is conducive to focusing the aid resources to the right places in real time, fixing problems immediately and not allowing defects to flow downstream on the line. These features, in turn, reduce defects, rework, delays, and wasted costs, which improve overall productivity and save energy, as described above. It is also useful to display the estimated downtime (Toyota and other modern plants have large centrally located Andon boards that display the Andon signal, the workstation number, and the estimated downtime.). Knowing the forecasted downtime frees other workers to perform their pending tasks which have waited for such an opportunity rather than wait idle. This leads to better utilization of the plant resources, including infrastructure energy.

Fig. 2 In this messy plant, the workers waste close to 20% of their time looking for items and scavenging for parts and tools, also wasting the plant energy.

- "5Ss": The term comes from five Japanese words that begin with the "s" sound and loosely translate into English as: sorting, simplification, sweeping, standardization, and self-discipline (many other translations of the words are popular in industry); and describes a simple but powerful workplace organization method.[8] The underlying principle of the method is that only the items needed for the immediate task (parts, containers, tools, instructions, materials) are kept at hand where they are needed at the moment, and everything else is kept in easily accessible and well-organized storage in perfect order, easy to locate without searching, and in just the right quantities. All items have their designated place, clearly labeled with signs, labels, part numbers, and possibly bar codes. The minimum and maximum levels of inventory of small parts are predefined and are based on actual consumption rather than the "just-in-case" philosophy. The parts, tools, and materials needed for the next shift of production are prepared by a person in charge of the storage during the previous shift and delivered to the workstation before the shift starts. The floor is uncluttered and marked with designated spaces for all equipment. The entire factory is spotlessly clean and uncluttered. Walls are empty except for the visual boards. In consequence of these changes, the searching for parts, tools, and instructions which can represent a significant waste of labor and time is reduced, and this, in turn, saves energy. Secondary effects are also important. In a well-organized place, fewer mistakes are made; fewer wrong parts are used; less inspection is needed; quality, throughput time, and customer satisfaction are increased; and costs and energy are decreased. Fig. 2 illustrates a fragment of a messy factory, where the average worker was estimated to waste 20% of his shift time looking for and scavenging for parts and tools. This percentage multiplied by the number of workers yields a significant amount of wasted production time, also wasting plant energy in the same proportion. Sorting, cleaning, and organizing the workplace is one of the simplest and most powerful starting points on the way to improved productivity and energy savings.

CONCLUSION

Large savings in energy are possible as an inherent byproduct of improving productivity. The state-of-the-art Lean productivity method can yield dramatic improvements in productivity. In the extreme case of converting from the traditional batch-and-queue and "process village" manufacturing system to Lean production, overall costs, lead times, and inventories can be reduced by as much as 50%–90%, floor space and energy by 50%, and energy density can be improved by 50%. The amount of energy that can be saved by productivity improvements often radically exceeds the savings from equipment optimization alone, thus providing a strong incentive to include productivity improvements in energy-reduction efforts.

Productivity strongly depends on human factors such as management, learning, and training, communications, culture, teamwork, etc. which are difficult to quantify, making accurate estimates of the cost, schedule, and quality benefits from various productivity improvements and the related energy savings difficult to estimate with engineering accuracy. For this reason, simple metrics and models are recommended, and some examples have been presented. If applied conservatively, they can become useful tools for energy management in a plant. The prerequisite knowledge includes an understanding of Lean Flow and its various productivity elements and a good accounting of energy use in the plant, including the knowledge of the energy used by individual machines and processes both when in productive use and in the idle but energized state, as well as the energy elements used by the infrastructure (various light combinations, air-compressors, cooling and heating devices, combusting systems, conveyers, forklifts, etc.). In the times of ferocious global competition and rising energy prices, every industrial plant should make every effort to improve both productivity and energy use.

ACKNOWLEDGMENTS

This work is a result of the studies of energy conservation using the Lean productivity method performed by the Industrial Assessment Center funded by the U.S. Department of Energy at Loyola Marymount University. The author is grateful to Mr. Rudolf Marloth, Assistant Director of the Center, for his help with various energy estimates included herein and his insightful comments, to the Center students for their enthusiastic work, and to his son Peter W. Oppenheim for his diligent editing.

REFERENCES

1. U.S. Department of Energy, Energy Efficiency and Renewable Energy, http://www.eere.energy.gov/industry/ (accessed on December 2005).
2. U.S. Department of Energy, http://eereweb.ee.doe.gov/industry/bestpractices/plant_assessments.html, (accessed on December 2005).
3. Womack, P.J.; Jones, D.T. *Lean Thinking. 2nd Ed.*; Lean Enterprise Institute: Boston, MA, 2005; (www.lean.org), ISBN: 0-7432-4927-5.
4. Liker, J. *Becoming Lean, Inside Stories of U.S. Manufacturers*; Productivity Inc.: Portland, OR, 1998; service@productivityinc.com.
5. Oppenheim, B.W. *Selected Assessment Recommendations*, Industrial Assessment Center, Loyola Marymount University: Los Angeles, (boppenheim@lmu.edu), unpublished 2004.
6. *Setup Reduction for Just-in-Time*, Video/CD, Society of Manufacturing Engineers, Product ID: VT90PUB2, http://www.sme.org/cgi-bin/get-item.pl?VT392&2&SME&1990, (accessed on December 2005).
7. Ohno, T. *Toyota Production System: Beyond Large Scale Production*; Productivity Press: New York, 1988; info@productivityinc.com.
8. Hiroyuki, H. *Five Pillars of the Visual Workplace, the Sourcebook for 5S Implementation*; Productivity Press: New York, NY, 1995; info@productivityinc.com.

Energy Conversion: Principles for Coal, Animal Waste, and Biomass Fuels

Kalyan Annamalai
Department of Mechanical Engineering, College Station, Texas, U.S.A.

Soyuz Priyadarsan
Texas A&M University, College Station, Texas, U.S.A.

Senthil Arumugam
Enerquip, Inc., Medford, Wisconsin, U.S.A.

John M. Sweeten
Texas Agricultural Experiment Station, Amarillo, Texas, U.S.A.

Abstract

A brief overview is presented of various energy units; terminology; and basic concepts in energy conversion including pyrolysis, gasification, ignition, and combustion. Detailed sets of fuel properties of coal, agricultural biomass, and animal waste are presented so that their suitability as a fuel for the energy conversion process can be determined. It is also found that the dry ash free (DAF) heat values of various biomass fuels, including animal waste, remain approximately constant, which leads to a presentation of generalized results for maximum flame temperature as a function of ash and moisture contents. The cofiring technology is emerging as a cost-effective method of firing a smaller percentage of biomass fuels, with coal as the major fuel. Various techniques of cofiring are summarized. Gasification approaches, including FutureGen and reburn technologies for reduction of pollutants, are also briefly reviewed.

INTRODUCTION AND OBJECTIVES

The overall objective of this entry is to provide the basics of energy conversion processes and to present thermochemical data for coal and biomass fuels. Energy represents the capacity for doing work. It can be converted from one form to another as long as the total energy remains the same. Common fuels like natural gas, gasoline, and coal possess energy as chemical energy (or bond energy) between atoms in molecules. In a reaction of the carbon and hydrogen in the fuel with oxygen, called an oxidation reaction (or more commonly called combustion), carbon dioxide (CO_2) and water (H_2O) are produced, releasing energy as heat measured in units of kJ or Btu (see Table 1 for energy units). Combustion processes are used to deliver (i) work, using external combustion (EC) systems by generating hot gases and producing steam to drive electric generators as in coal fired power plans, or internal combustion (IC) engines by using the hot gases directly as in automobiles or gas turbines; and (ii) thermal energy, for applications to manufacturing processes in metallurgical and chemical industries or agricultural product processing.

Fuels can be naturally occurring (e.g., fossil fuels such as coal, oil, and gas, which are residues of ancient plant or animal deposits) or synthesized (e.g., synthetic fuels). Fuels are classified according to the phase or state in which they exist: as gaseous (e.g., natural gas), liquid (e.g., gasoline or ethanol), or solid (e.g., coal, wood, or plant residues). Gaseous fuels are used mainly in residential applications (such as water heaters, home heating, or kitchen ranges), in industrial furnaces, and in boilers. Liquid fuels are used in gas turbines, automotive engines, and oil burners. Solid fuels are used mainly in boilers and steelmaking furnaces.

During combustion of fossil fuels, nitrogen or sulfur in the fuel is released as NO, NO_2 (termed generally as NO_x) and SO_2 or SO_3 (termed as SO_x). They lead to acid rain (when SO_x or NO_x combine with H_2O and fall as rain) and ozone depletion. In addition, greenhouse gas emissions (CO_2, CH_4, N_2O, CFCs, SF_6, etc.) are becoming a global concern due to warming of the atmosphere, as shown in Fig. 1 for CO_2 emissions. Global surface temperature has increased by 0.6°C over the past 100 years. About 30%–40% of the world's CO_2 is from fossil fuels. The Kyoto protocol, signed by countries that account for 54% of the world's fossil based CO_2 emissions, calls for reduction of greenhouse gases by 5% from 1990 levels over the period from 2008 to 2012.

The total worldwide energy consumption is 421.5 quads of energy in 2003 and is projected to be 600 quads in 2020, while U.S. consumption in 2004 is about 100 quads and is projected to be 126 quads in 2020. The split

Keywords: Energy units; Fuels; Biomass; Pyrolysis; Gasification; Cofiring; Reburn; Heating value; Pollutants.

Table 1 Energy units and terminology

The section on energy Units and Conversion factors in Energy is condensed from Chapter 01 of Combustion Engineering by Annamalai and Puri [2005] and Tables.

Energy Units

1 Btu (British thermal unit) = 778.14 ft lb$_f$ = 1.0551 kJ, 1 kJ = 0.94782 Btu = 25,037 lbmft/s^2

1 mBtu = 1 k Btu = 1000 Btu, 1 mmBtu = 1000 k Btu = 10^6 Btu, 1 trillion Btu = 10^9 Btu or 1 giga Btu

1 quad = 10^{15} Btu or 1.05×10^{15} kJ or 2.93×10^{11} kW h,

1 Peta J = 10^{15} J = 10^{12} kJ » 0.00095 Quads

1 kilowatt-hour of electricity = 3,412 BTU = 3.6 Mj,

1 cal: 4.1868 J, One (food) calorie = 1000 cal or 1 Cal,

1 kJ/kg = 0.43 Btu/lb, 1 Btu/lb = 2.326 kJ/kg

1 kg/GJ = 1 g/MJ = 2.326 lb/mmBtu; 1 lb/mmBtu = 0.430 kg/GJ = 0.430 g/MJ

1 Btu/SCF = 37 kJ/m^3

1 Therm = 10^5 Btu = 1.055×10^5 kJ

1 m^3/GJ = 37.2596 ft^3/mmBTU

1 hp = 0.7064 Btu/s = 0.7457 kW = 745.7 W = 550 lbf ft/s = 42.41 Btu/min

1 boiler HP = 33475 Btu/h, 1 Btu/h = 1.0551 kJ/h

1 barrel (42 gal) of crude oil = 5,800,000 Btu = 6120 MJ

1 gal of gasoline = 124,000 Btu = 131 MJ

1 gal of heating oil = 139,000 Btu = 146.7 MJ, 1 gal of diesel fuel = 139,000 Btu = 146.7 MJ

1 barrel of residual fuel oil = 6,287,000 Btu = 6633 MJ

1 cubic foot of natural gas = 1,026 Btu = 1.082 MJ, 1 Ton of Trash = 150 kWh

1 gal of propane = 91,000 Btu = 96 MJ, 1 short ton of coal = 20,681,000 Btu = 21821 MJ

Emission reporting for pollutants: (i) parts per million (ppm), (ii) normalized ppm, (iii) emission Index (EI) in g/kg fuel, (iv) g/GJ, v) mg/m^3 of flue gas:

Conversions in emissions reporting: (ii) normalized ppm = ppm $\times$ (21-O_2% std)/(21-O_2% measured); (iii) EI of species k: C % by mass in fuel $\times$ mol Wt of $k \times$ ppm of species $k \times 10^{-3}/\{12.01(CO_2\% + CO\%)\}$, (iv) g/GJ = EI/ {HHV in GJ/kg}; (v) mg/m^3 = ppm of species $k \times$ Mol Wt of k/24.5

Volume of 1 kmol (SI) and 1 lb mole (English) of an ideal gas at STP conditions defined below:
Pressure at 101.3 kPa (1 atm, 14.7 psia, 29.92 in.Hg, 760 Torr) fixed; T changes depending upon type of standard adopted

Scientific (or SATP, standard ambient T and P)	US standard (1976) or ISA (International standard atmosphere)	NTP (gas industry reference base)	Chemists-standard-atmosphere (CSA)
25°C (77°F)	15°C (60°F)	20°C (68°F), 101.3 kPa	0°C (32°F)
24.5 m^3/kmol (392 ft^3/lb mole); $\rho_{air,SATP}$ = 1.188 kg/m^3 = 0.0698 lb$_m$/ft^3	23.7 m^3/kmol (375.6 ft^3/lb mole); $\rho_{air,ISA}$ = 1.229 kg/m^3 = 0.0767 lb$_m$/ft^3	24.06 m^3/kmole or 385 ft^3/lb mole; $\rho_{air,NTP}$ = 1.208 kg/m^3 = 0.0754 lb$_m$/ft^3	22.4 m^3/kmol (359.2 ft^3/lb mole), $\rho_{air,CSA}$ = 1.297 kg/m^3 = 0.0810 lb$_m$/ft^3

---

is as follows: 40 quads for petroleum, 23 for natural gas, 23 for coal, 8 for nuclear power, and 6 for renewables (where energy is renewed or replaced using natural processes) and others sources. Currently, the United States relies on fossil fuels for 85% of its energy needs. Soon, the U.S. energy consumption rate which distributed as electrical power (40%), transportation (30%), and heat (30%), will outpace the growth in the energy production rate, increasing reliance on imported oil. The Hubbert peak theory (named after Marion King Hubbert, a geophysicist with Shell Research Lab in Houston, Texas) is based upon the rate of extraction and depletion of conventional fossil fuels, and predicts that fossil-based oil would peak at about 12.5 billion barrels per year worldwide some time around 2000. The power cost and percentage use of coal in various U.S. states varies from

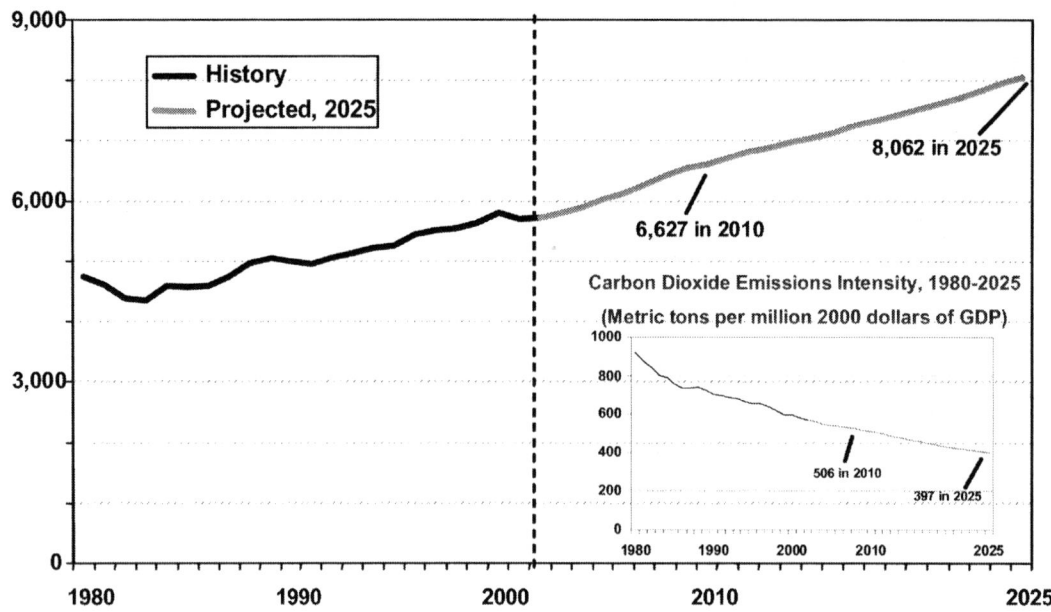

Fig. 1 Total CO_2 emission in million metric tons per year: History and Projected 1980–2025.
Source: From DOE-EIA (see Ref. 1).

10 cents (price per kWh) at 1% coal use for power generation in California to 48 cents at 94% use of coal in Utah.

Biomass is defined as "any organic material from living organisms that contains stored sunlight (solar energy) in the form of chemical energy."[1] These include agro-based materials (vegetation, trees, and plants); industrial wastes (sawdust, wood chips, and crop residues); municipal solid wastes (MSWs), which contain more than 70% biomass (including landfill gases, containing almost 50% CH_4); and animal waste. Biomass is a solid fuel in which hydrogen is locked with carbon atoms. Biomass production worldwide is 145 billion metric tons. Biomass now supplies 3% of U.S. energy, and it could be increased to as high as 20%. Renewable energy sources (RES) include biomass, wind, hydro, solar, flowing water or hydropower, anaerobic digestion, ocean thermal (20°C temperature difference), tidal energies, and geothermal (a nonsolar source of energy), and these supply 14% of the world demand. The RES constitute only 6%, while coal, petroleum, and natural gas account for 23%, 40%, and 24%, respectively. About 9% of the world's electricity is from RES, and 65% of the electricity contributed by biomass. About 97% of energy conversion from biomass is by combustion. Many U.S. states have encouraged the use of renewables by offering REC (Renewable Energy Credits). One REC = 1 MW/h = 3.412 mmBtu; hence the use of 1 REC is equivalent to replacing approximately 1500 lb of coal, reducing emission of NO_x and SO_x by 1.5 lb for every 1 REC, assuming that emissions of NO_x and SO_x are 0.45 lb per mmBtu generated by coal. Several emission-reporting methods and conversions are summarized in Table 1. Recently, H_2 is being promoted as a clean-burning, non-global-warming, and pollution-free fuel for both power generation and transportation.

Fig. 2 shows a comparison between biomass and hydrogen energy cycles. In the biomass cycle, photosynthesis is used to split CO_2 into C and O_2, and H_2O into H_2 and O_2, producing Hydrocarbons (HC) fuel (e.g., leaves) and releasing O_2. The O_2 released is used to combust the HC and produce CO_2 and H_2O, which are returned to produce plant biomass (e.g., leaves) and O_2. On the other hand, in the hydrogen cycle, H_2O is disassociated using the photo-splitting process to produce H_2 and O_2, which are then used for the combustion process. The hydrogen fuel can be used in fuel cells to obtain an efficient conversion. Photosynthesis is water intensive; most of the water supplied to plants evaporates through leaves into the atmosphere, where it re-enters the hydrology cycle.

This entry is organized in the following format: (i) coal and biosolid properties; (ii) coal and biosolid pyrolysis (a process of thermal decomposition in the absence of oxygen), combustion, and gasification; (iii) combustion by cofiring coal with biosolids; (iv) gasification of coal and biosolids (a process that includes pyrolysis, partial oxidation due to the presence of oxygen, and hydrogenation); and (v) reburn for NO_x reduction.

FUEL PROPERTIES

Fuel properties play a major role in the selection, design, and operation of energy conversion systems.

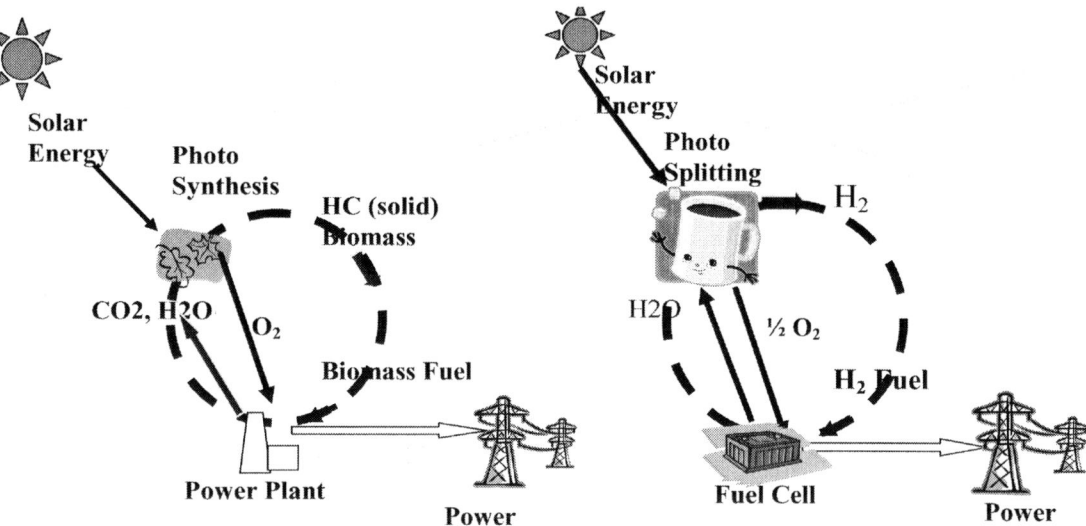

Fig. 2 Comparison of biomass energy and H_2 energy cycles.

Solid Fuels

The primary solid fuel widely used in power plants is coal containing combustibles, moisture, and intrinsic mineral matter originating from dissolved salts in water. During the "coalification" process, lignite, the lowest rank of coal (low C/O ratio), is produced first from peat, followed by sub-bituminous (black lignite, typically low sulfur, noncaking), bituminous (soft coal that tends to stick when heated and is typically high in S), and finally anthracite (dense coal; has the highest carbon content, >90%, low volatile <15%) with a gradual increase in the coal C/O ratio. The older the coal, the higher its rank. Anthracite (almost carbon) is the highest-ranked coal, with a high heating value. To classify coals and ascertain the quality of coal, it is essential to perform proximate and ultimate analyses according to American Society of Testing Materials (ASTM) standards.

Proximate Analysis (ASTM D3172)

A solid fuel consists of combustibles, ash, and moisture. Combustibles together with ash are called the solid content of fuel. A proximate analysis provides the following information: surface moisture (SM) or dry loss (DL), i.e., moisture in air-dried coal; the inherent moisture in the coal (M); volatile matter (VM; produced by pyrolysis, a thermal decomposition process resulting in release of water, gases, oil and tar); fixed carbon (FC; skeletal matter left after release of volatiles); mineral matter (MM; inert collected with solid fuel); and heating value (HV). On combustion, the MM may be partially oxidized or reduced, and the material left after combustion of C and H in the fuel is called ash (CaO, $CaCO_3$, Fe_2O_3, FeO, etc).

Table 2 shows comparative proximate analyses of coal, advanced feedlot biomass (FB, low-ash cattle manure; see "Coal and Bio-Solids Cofiring"), and litter biomass (LB, chicken manure).[2] Feedlot manure has higher moisture, nitrogen, chlorine, and ash content than coal. With aging or composting, the VM in manure decreases as a result of the gradual release of hydrocarbon gases or dehydrogenation, but fuel becomes more homogeneous.

Ultimate/Elemental Analysis (ASTM D3176)

Ultimate analysis is used to determine the chemical composition of fuels in terms of either the mass percent of their various elements or the number of atoms of each element. The elements of interest are C, H, N, O, S, Cl, P, and others. It can be expressed on an "as received" basis, on a dry basis (with the moisture in the solid fuel removed), or on a dry ash free (DAF) basis (also known as the moisture ash free basis MAF). Tables 3 and 4 show the ultimate analyses of various types of coal and biomass fuels.[3] While nitrogen is not normally present in natural gas, coal has 1%–1.5%; cattle manure and chicken waste contain high amounts of N (Table 2).

Heating Value (ASTM D3286)

The gross or higher heating value (HHV) of a fuel is the amount of heat released when a unit (mass or volume) of the fuel is burned. The HHV of solid fuel is determined using ASTM D3286 with an isothermal jacket bomb calorimeter. For rations fed to animals and animal waste fuels, the HHV for DAF roughly remains constant at about 19,500 kJ/kg (8400 Btu/lb),[4] irrespective of stage of decomposition of animal waste. The HHV can also be

Table 2 Coal, advanced feedlot biomass (FB) and litter biomass (LB)

Parameter	Wyoming coal	Cattle manure (FB)	Chicken manure (LB)[a]	Advanced Feedlot biomass (AFB)[b]	High-ash Feedlot biomass (HFB)[b]
Dry loss (DL)	22.8	6.8	7.5	10.88	7.57
Ash	5.4	42.3	43.8	14.83	43.88
FC	37.25	40.4	8.4	17.33	10.28
VM	34.5	10.5	40.3	56.97	38.2
C	54.1	23.9	39.1	50.08	49.27
H	3.4	3.6	6.7	5.98	6.13
N	0.81	2.3	4.7	38.49	38.7
O	13.1	20.3	48.3	4.58	4.76
S	0.39	0.9	1.2	0.87	0.99
Cl	<0.01%	1.2			
HHV-as received (kJ/kg)	21385	9560	9250	14983	9353
$T_{adiab,\ Equil}$[c]	2200 K (3500°F)	2012 K (3161°F)			
DAF formula	$CH_{0.76}O_{0.18}N_{0.013}S_{0.0027}$	$CH_{1.78}O_{.64}N_{.083}S_{.014}$	$CH_{2.04}O_{0.93}N_{0.10}S_{0..012}$	$CH_{1.4184}O_{0.5764}N_{0.078}S_{0.0066}$	$CH_{1.4775}O_{0.5892}N_{0.083}S_{0.0076}$
HHV-DAF (kJ/kg)	29785	18785	18995	20168	19265
CO_2, g/GJ					
N, g/GJ					
S, g/GJ					

[a] Ref. 2.
[b] Ref. 37.
[c] Equilibrium temperature for stoichiometric mixture from THERMOLAB Spreadsheet software for any given fuel of known composition (Ref. 36. website http://www.crcpress.com/e_products/downloads/download.asp?cat_no=2553)

Table 3 Coal composition (DAF basis)

ASTM Rank	State (U.S.A.)	Ash, % (dry)	C	H	N	S*	O**	HHV$_{Est}$ kJ/kg	CO$_2$ kg/GJ	N kg/GJ	S kg/GJ
Lignite	ND	11.6	63.3	4.7	0.48	0.98	30.5	24,469	94.8	0.196	0.401
Lignite	MT	7.7	70.7	4.9	0.8	4.9	22.3	28,643	90.4	0.279	1.711
Lignite	ND	8.2	71.2	5.3	0.56	0.46	22.5	28,782	90.7	0.195	0.160
Lignite	TX	9.4	71.7	5.2	1.3	0.72	21.1	29,070	90.4	0.447	0.248
Lignite	TX	10.3	74.3	5	0.37	0.51	19.8	29,816	91.3	0.124	0.171
Sbb. A	WY	8.4	74.3	5.8	1.2	1.1	17.7	31,092	87.6	0.386	0.354
Sbb. C	WY	6.1	74.8	5.1	0.89	0.3	18.9	30,218	90.7	0.295	0.099
HVB	IL	10.8	77.3	5.6	1.1	2.3	13.6	32,489	87.2	0.339	0.708
HVC	IL	10.1	78.8	5.8	1.6	1.8	12.1	33,394	86.5	0.479	0.539
HVB	IL	11.8	80.1	5.5	1.1	2.3	11.1	33,634	87.3	0.327	0.684
HVB	UT	4.8	80.4	6.1	1.3	0.38	11.9	34,160	86.2	0.381	0.111
HVA	WV	7.6	82.3	5.7	1.4	1.8	8.9	34,851	86.5	0.402	0.516
HVA	KY	2.1	83.8	5.8	1.6	0.66	8.2	35,465	86.6	0.451	0.186
MV	AL	7.1	87	4.8	1.5	0.81	5.9	35,693	89.3	0.420	0.227
LV	PA	9.8	88.2	4.8	1.2	0.62	5.2	36,153	89.4	0.332	0.171
Anthracite	PA	7.8	91.9	2.6	0.78	0.54	4.2	34,974	96.3	0.223	0.154
Anthracite	PA	4.3	93.5	2.7	0.24	0.64	2.9	35,773	95.8	0.067	0.179

HHV$_{est}$: Boie Equation. CO$_2$ in g/MJ or kg/GJ = C content in % × 36645/(HHV in kJ/kg). CO$_2$ in lb per mmBtu = Multiply CO$_2$ in (g /MJ) or (g/MJ) by 2.32. N in g/MJ or kg/GJ = N% × 10000/(HHV in kJ/kg).
For NO$_x$ estimation, multiply N content in g/MJ by 1.15 to get NO$_x$ in g/MJ which assumes 35% conversion of fuel N.
For SO$_2$ estimation, multiply S content in g/MJ by 2 to get SO$_2$ in g/MJ assuming 100% conversion of fuel S (Multiply HHV in kJ/kg by 0.430 to get Btu/lb).
*Organic sulfur; **by difference.

Table 4 Ultimate analyses and heating values of biomass fuels

Biomass	C	H	O	N	S	Residue	Measured HHV$_M$	[a]Estimated HHV	CO2 g/MJ	N, g/MJ	S, g/MJ
Field crops											
Alfalfa seed straw	46.76	5.40	40.72	1.00	0.02	6.07	18.45	18.27	92.9	0.542	0.011
Bean straw	42.97	5.59	44.93	0.83	0.01	5.54	17.46	16.68	90.2	0.475	0.006
Corn cobs	46.58	5.87	45.46	0.47	0.01	1.40	18.77	18.19	90.9	0.250	0.005
Corn stover	43.65	5.56	43.31	0.61	0.01	6.26	17.65	17.05	90.6	0.346	0.006
Cotton stalks	39.47	5.07	39.14	1.20	0.02	15.10	15.83	15.51	91.4	0.758	0.013
Rice straw (fall)	41.78	4.63	36.57	0.70	0.08	15.90	16.28	16.07	94.0	0.430	0.049
Rice straw (weathered)	34.60	3.93	35.38	0.93	0.16	25.00	14.56	12.89	87.1	0.639	0.110
Wheat straw	43.20	5.00	39.40	0.61	0.11	11.40	17.51	16.68	90.4	0.348	0.063
Switchgrass[b]	42.02	6.30	46.10	0.77	0.18	4.61	15.99	15.97	96.3	0.482	0.113
Orchard prunings											
Almond prunings	51.30	5.29	40.90	0.66	0.01	1.80	20.01	19.69	93.9	0.330	0.005
Black Walnut	49.80	5.82	43.25	0.22	0.01	0.85	19.83	19.50	92.0	0.111	0.005
English Walnut	49.72	5.63	43.14	0.37	0.01	1.07	19.63	19.27	92.8	0.188	0.005
Vineyard prunings											
Cabernet Sauvignon	46.59	5.85	43.90	0.83	0.04	2.71	19.03	18.37	89.7	0.436	0.021
Chenin Blanc	48.02	5.89	41.93	0.86	0.07	3.13	19.13	19.14	92.0	0.450	0.037
Pinot Noir	47.14	5.82	43.03	0.86	0.01	3.01	19.05	18.62	90.7	0.451	0.005
Thompson seedless	47.35	5.77	43.32	0.77	0.01	2.71	19.35	18.60	89.7	0.398	0.005
Tokay	47.77	5.82	42.63	0.75	0.03	2.93	19.31	18.88	90.7	0.388	0.016
Energy Crops											
Eucalyptus Camaldulensis	49.00	5.87	43.97	0.30	0.01	0.72	19.42	19.19	92.5	0.154	0.005
Globulus	48.18	5.92	44.18	0.39	0.01	1.12	19.23	18.95	91.8	0.203	0.005
Grandis	48.33	5.89	45.13	0.15	0.01	0.41	19.35	18.84	91.5	0.078	0.005
Casuarina	48.61	5.83	43.36	0.59	0.02	1.43	19.44	19.10	91.6	0.303	0.010
Cattails	42.99	5.25	42.47	0.74	0.04	8.13	17.81	16.56	88.5	0.415	0.022
Popular	48.45	5.85	43.69	0.47	0.01	1.43	19.38	19.02	91.6	0.243	0.005
Sudan grass	44.58	5.35	39.18	1.21	0.08	9.47	17.39	17.63	93.9	0.696	0.046

Fuel											
Forest residue											
Black Locust	50.73	5.71	0.57	0.01	41.93	0.97	19.71	19.86	94.3	0.289	0.005
Chaparral	46.9	5.08	0.54	0.03	40.17	7.26	18.61	17.98	92.3	0.290	0.016
Madrone	48	5.96	0.06	0.02	44.95	1	19.41	18.82	90.6	0.031	0.010
Manzanita	48.18	5.94	0.17	0.02	44.68	1	19.3	18.9	91.5	0.088	0.010
Ponderosa Pine	49.25	5.99	0.06	0.03	44.36	0.3	20.02	19.37	90.1	0.030	0.015
Ten Oak	47.81	5.93	0.12	0.01	44.12	2	18.93	18.82	92.6	0.063	0.005
Redwood	50.64	5.98	0.05	0.03	42.88	0.4	20.72	20.01	89.6	0.024	0.014
White Fur	49	5.98	0.05	0.01	44.75	0.2	19.95	19.22	90.0	0.025	0.005
Food and fiber processing wastes											
Almond hulls	45.79	5.36	0.96	0.01	40.6	7.2	18.22	17.89	92.1	0.527	0.005
Almond shells	44.98	5.97	1.16	0.02	42.27	5.6	19.38	18.14	85.0	0.599	0.010
Babassu husks	50.31	5.37	0.26	0.04	42.29	1.73	19.92	19.26	92.5	0.131	0.020
Sugarcane bagasse	44.8	5.35	0.38	0.01	39.55	9.79	17.33	17.61	94.7	0.219	0.006
Coconut fiber dust	50.29	5.05	0.45	0.16	39.63	4.14	20.05	19.2	91.9	0.224	0.080
Cocoa hulls	48.23	5.23	2.98	0.12	33.09	10.25	19.04	19.56	92.8	1.565	0.063
Cotton gin trash	39.59	5.26	2.09		36.33	16.68	16.42	16.13	88.4	1.273	0.000
Macadamia shells	54.41	4.99	0.36	0.01	39.69	0.56	21.01	20.55	94.9	0.171	0.005
Olive pits	48.81	6.23	0.36	0.02	43.48	1.1	21.39	19.61	83.6	0.168	0.009
Peach pits	53	5.9	0.32	0.05	39.14	1.59	20.82	21.18	93.3	0.154	0.024
Peanut hulls	45.77	5.46	1.63	0.12	39.56	7.46	18.64	18.82	90.0	0.874	0.064
Pistachio shells	48.79	5.91	0.56	0.01	43.41	1.28	19.26	19.25	92.8	0.291	0.005
Rice hulls	40.96	4.3	0.4	0.02	35.86	18.34	16.14	15.45	93.0	0.248	0.012
Walnut shells	49.98	5.71	0.21	0.01	43.35	0.71	20.18	19.45	90.8	0.104	0.005
Wheat dust	41.38	5.1	3.04	0.19	35.19	15.1	16.2	16.78	93.6	1.877	0.117

[a] HHV based on Boie equation.
[b] Ref. 20; [Adapted from Refs. 3 and 17] See foot note of Table 3.2 for conversions to English units and estimation of NO_x and SO_2 emissions.

estimated using the ultimate analysis of the fuel and the following empirical relation from Boie[5]:

HHV_{fuel}(kJ/kg fuel)

$$= 35,160\ Y_C + 116,225\ Y_H - 11,090\ Y_O$$
$$+ 6280\ Y_N + 10465\ Y_S \qquad (1)$$

HHV_{fuel}(BTU/lb fuel)

$$= 15,199\ Y_C + 49,965\ Y_H - 4768\ Y_O$$
$$+ 2700\ Y_N + 4499\ Y_S, \qquad (2)$$

where Y denotes the mass fraction of an element C, H, O, N, or S in the fuel. The higher the oxygen content, the lower the HV, as seen in biomass fuels.

Annamalai et al. used the Boie equation for 62 kinds of biosolids with good agreement.[6] For most biomass fuels and alcohols, the HHV in kilojoules per unit mass of stoichiometric oxygen is constant at 14,360–14,730 kJ/kg of O_2 (6165–6320 Btu/lb of O_2).[7]

Estimate of CO_2 Emission

Using the Boie-based HVs for any fuel of known elemental composition, one can plot the CO_2 emission in g/MJ (Fig. 3) as a function of H/C and O/C ratios.[8] Comparisons for selected fuels with known experimental HVs are also shown in the same figure. Coal, with H/C ratio ≈ 0.5, releases the highest CO_2, while natural gas (mainly CH_4) emits the lowest CO_2. Because the United States uses fossil fuels for 86% of its energy needs (100 quads), the estimated CO_2 emission is 6350 million ton/year, assuming that the average CO_2 emission from fossil fuels is 70 kg/GJ (methane: 50 kg/GJ vs coal: 90 kg/GJ). Fig. 1 seems to confirm such estimation within a 10% error.

Flame Temperature

Fig. 4 shows a plot of maximum possible flame temperature vs moisture percentage with combustion for biomass fuels. The result can be correlated as follows[4]:

$$T(K) = 2290 - 1.89\ H_2O + 5.06\ Ash$$
$$- 0.309\ H_2O\ Ash - 0.180\ H_2O^2$$
$$- 0.108\ Ash^2 \qquad (3)$$

$$T(°F) = 3650 - 3.40\ H_2O + 9.10\ Ash$$
$$- 0.556\ H_2O\ Ash - 0.324\ H_2O^2$$
$$- 0.194\ Ash^2 \qquad (4)$$

The adiabatic flame temperature decreases if the ash and moisture contents increase.

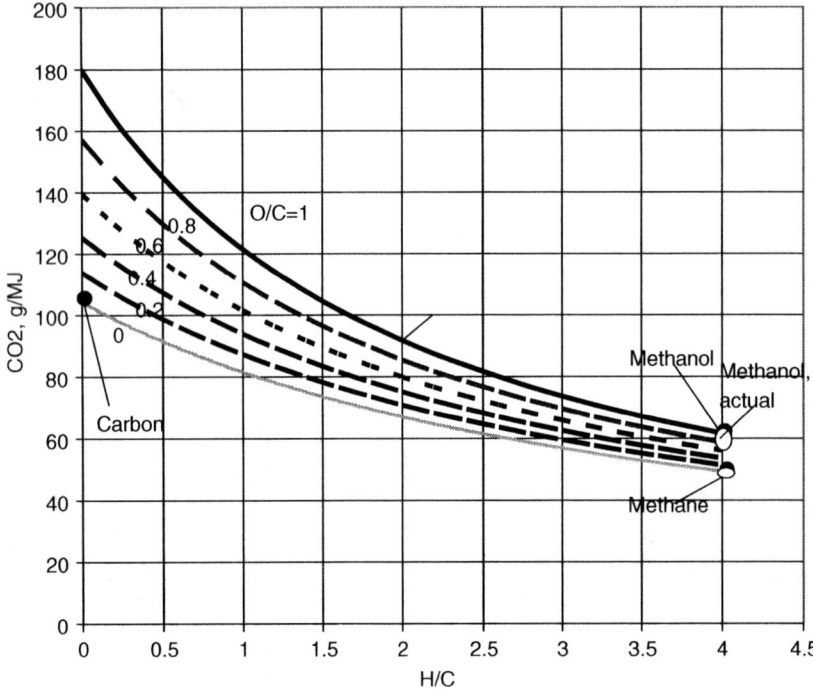

Fig. 3 Emission of CO_2 as a function of H/C and O/C atom ratios in hydrocarbon fuels.
Source: Adapted from Taylor and Francis (see Ref. 8).

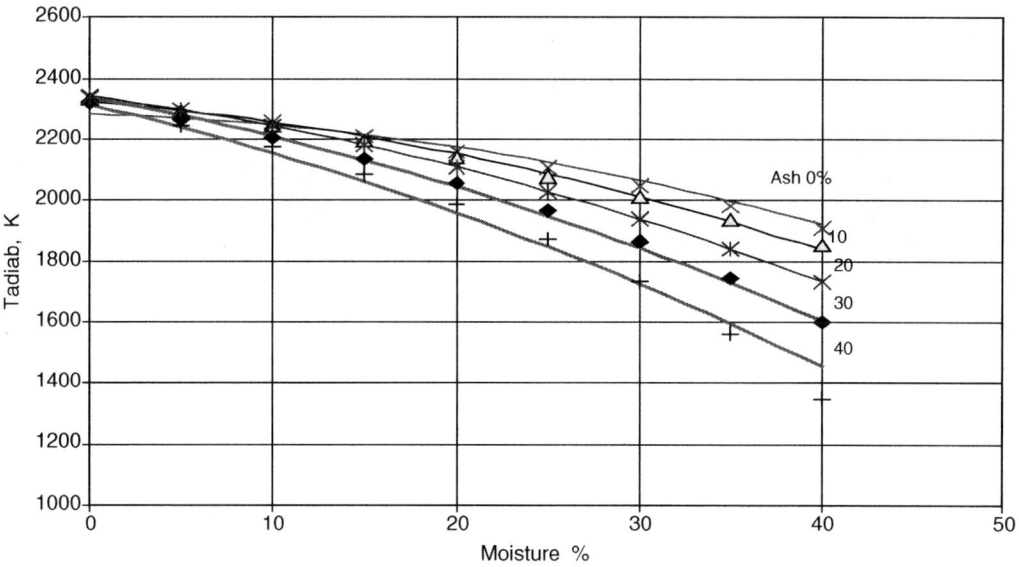

Fig. 4 Correlation of adiabatic flame temperature with moisture and ash contents.

Flue Gas Volume

The flue gas volume for C–H–O is almost independent of O/C ratios. The fit at 6% O_2 in products gives the following empirical equation for flue gas volume (m^3/GJ) at SATP[8]:

$$\text{Flue gas}_{vol}(m^3/GJ) = 4.96\left(\frac{H}{C}\right)^2 - 38.628\left(\frac{H}{C}\right) + 389.72 \quad (5)$$

$$\text{Flue gas}_{vol}(ft^3/mmBtu) = 184.68\left(\frac{H}{C}\right)^2 - 1439.28\left(\frac{H}{C}\right) + 14520.96 \quad (6)$$

Liquid Fuels

Liquid fuels, used mainly in the transportation sector, are derived from crude oil, which occurs naturally as a free-flowing liquid with a density of $\rho \approx 780$ kg/m^3–1000 kg/m^3, containing 0.1% ash and 0.15%–0.5% nitrogen. Crude oil normally contains a mixture of hydrocarbons, and as such, the "boiling" temperature keeps increasing as the oil is distilled. Most fuel oils contain 83%–88% carbon and 6%–12% (by mass) hydrogen.

Gaseous Fuels

The gaseous fuels are cleaner-burning fuels than liquid and solid fuels. They are a mixture of HC but dominated by highly volatile CH_4 with very little S and N. Natural gas is transported as liquefied natural gas (LNG) and compressed natural gas (CNG), typically at 150–250 bars. Liquefied petroleum gas (LPG) is a byproduct of petroleum refining, and it consists mainly of 90% propane. A low-Btu gas contains 0–7400 kJ/SCM (Standard Cubic Meter, 0–200 Btu/SCF, standards defined in Table 1); a medium-Btu gas. 7400–14,800 kJ/SCM (200–400 Btu/SCF); and a high-Btu gas, above 14,800 kJ/SCM (more than 400 Btu/SCF). Hydrogen is another gaseous fuel, with a heat value of 11,525 kJ/SCM (310 Btu/SCF). Because the fuel quality (heat value) may change when fuel is switched, the thermal output rate at a fixed gas-line pressure changes when fuels are changed.

COAL AND BIOMASS PYROLYSIS, GASIFICATION, AND COMBUSTION

Typically, coal densities range from 1100 kg/m^3 for low-rank coals to 2330 kg/m^3 for high-density pyrolytic graphite, while for biomass, density ranges from 100 kg/m^3 for straw to 500 kg/m^3 for forest wood.[9] The bulk density of cattle FB as harvested is 737 kg/m^3 (CF) for high ash (HA-FB) and 32 lbs/CF for low ash (LA-FB).[10] The processes during heating and combustion of coal are illustrated in Fig. 5, and they are similar for biomass except for high VM. The process of release of gases from solid fuels in the absence of oxygen is called pyrolysis, while the combined process of pyrolysis and

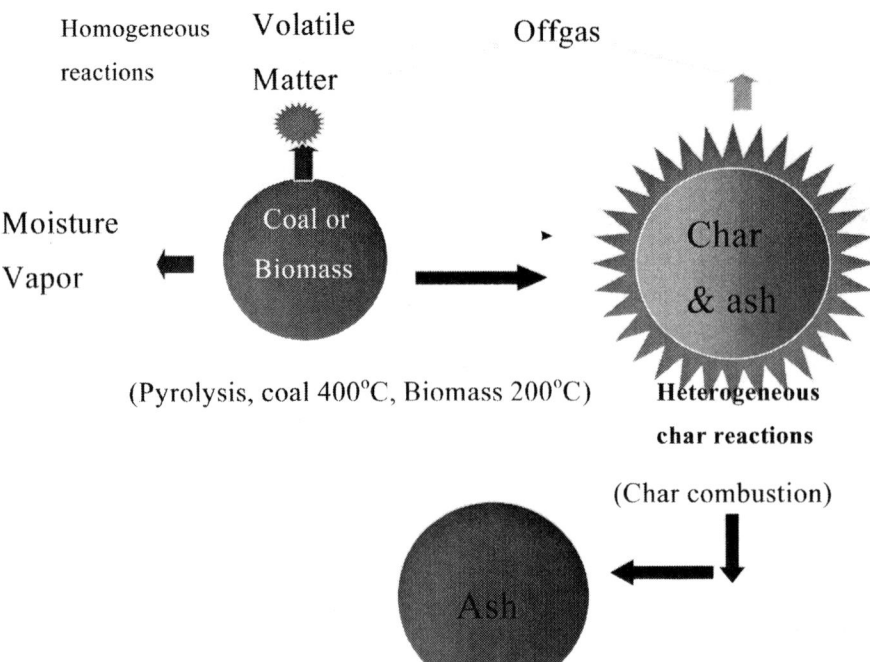

Fig. 5 Processes during coal pyrolysis, gasification, and combustion.

partial oxidation of fuel in the presence of oxygen is known as gasification. If all combustible gases and solid carbon are oxidized to CO_2 and H_2O, the process is known as combustion.

Pyrolysis

Solid fuels, like coal and biomass, can be pyrolyzed (thermally decomposed) in inert atmospheres to yield combustible gases or VM. While biomass typically releases about 70%–80% of its mass as VM (mainly from cellulose and hemicellulose) with the remainder being char, mainly from lignin content of biomass, coal releases 10%–50% of its mass as VM, depending upon its age or rank. Typically, a medium-rank coal consists of 40% VM and 60% FC, while a high-rank coal has about 10% VM. Bituminous coal pyrolyzes at about 700 K (with 1% mass loss for heating rates <100°C/s), as in the case of most plastics. Pyrolytic products range from lighter volatiles like CH_4, C_2H_4, C_2H_6, CO, CO_2, H_2, and H_2O to heavier molecular mass tars. Apart from volatiles, nitrogen is also evolved from the fuel during pyrolysis in the form of HCN, NH_3, and other compounds or, more generally, XN.

Sweeten et al. performed the thermogravimetric analysis (TGA) of feedlot manure.[4] The results are shown in Fig. 6. In the case of manure, drying occurred between 50 and 100°C, pyrolysis was initiated around 185°C–200°C for a heating rate of 80°C/min, and the minimum ignition temperature was approximately 528°C. The gases produced during biomass pyrolysis can also be converted into transportation fuels like biodiesel, methanol, and ethanol, which may be used either alone or blended with gasoline.

Volatile Oxidation

Once released, volatiles (HC, CO, H_2, etc.) undergo oxidation within a thin gas film surrounding the solid fuel particle. The oxidation for each HC involves several steps. The enveloping flame, due to volatile combustion, acts like a shroud by preventing oxygen from reaching the particle surface for heterogeneous oxidation of char. Following Dryer,[11] the one-step global oxidation of a given species can be written as

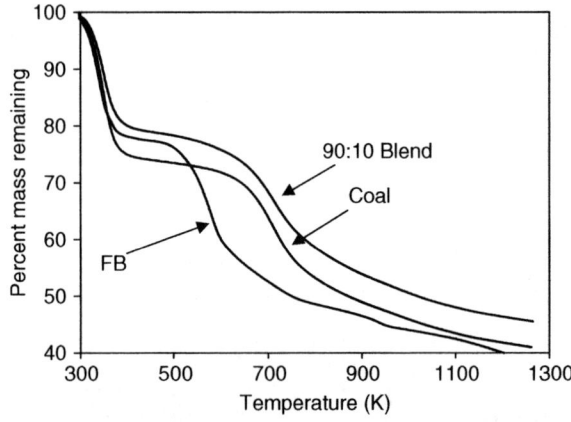

Fig. 6 Thermo-gravimetric analyses of Feedlot Biomass (FB or cattle manure), coal, and 90:10 coal: FB blends.
Source: From Elsevier (see Ref. 4).

$$\text{Fuel} + v_{O_2}O_2 \xrightarrow{\text{Oxidation}} v_{CO_2}CO_2 + v_{H_2O}H_2O \qquad (7)$$

$$-\frac{d[\text{Fuel}]}{dt}, \frac{\text{kg}}{\text{m}^3 \text{ sec}} = A \exp\left(\frac{-E}{RT}\right)[Y_{\text{fuel}}]^a [Y_{O_2}]^b, \qquad (8)$$

where [ ] represents the concentration of species in kg/m^3, Y the mass fraction, A the pre-exponential factor, E the activation energy in kJ/kmole, and a and b the order of reaction; they are tabulated in Bartok and Sarofim for alkanes, ethanol, methanol, benzene and toluene.[11]

Char Reactions

The skeletal char, essentially FC, undergoes heterogeneous reactions with gaseous species. The heterogeneous combustion of carbon or char occurs primarily via one or more of the following reactions:

Reaction I. $C + \frac{1}{2}O_2 \rightarrow CO$

Reaction II. $C + O_2 \rightarrow CO_2$

Reaction III. $C + CO_2 \rightarrow 2\,CO$

Reaction IV. $C + H_2O \rightarrow CO + H_2$

Assuming a first-order reaction for scheme I, the oxygen consumption rate is given as

$$\dot{m}_{O_2} \approx \pi d_p^2 B_I T^n \exp\left(-\frac{E}{R_v T_p}\right)\rho_\infty Y_{O_2,w}. \qquad (9)$$

The dominant oxygen transfer mechanism at high temperatures is via reaction I with an E/R (a ratio of activation energy to universal gas constant) of about 26,200 K, where $B_I = 2.3 \times 10^7$ m/s and $n = 0.5$ to 1. Reaction II has an E/R of 20,000 K, and $B_{II} = 1.6 \times 10^5$ m/s. Reaction III, the Boudouard reduction reaction, proceeds with an E/R of about 40,000 K. The reduction reactions, III and IV, may become significant, especially at high temperatures for combustion in boiler burners. Reaction with steam is found to be 50 times faster than CO_2 at temperatures up to 1800°C at 1 bar for 75–100 micron-sized Montana Rosebud char.[12] The combustible gases CO and H_2 undergo gas phase oxidation, producing CO_2 and H_2O.

Ignition and Combustion

Recently, Essenhigh et al. have reviewed the ignition of coal.[13] Volatiles from lignite are known to ignite at $T > 950$ K in fluidized beds. Coal may ignite homogeneously or heterogeneously depending upon size and volatile content.[14,15] A correlation for heterogeneous char ignition temperature is presented by Du and Annamalai, 1994.

Once ignited, the combustion of high volatile coal proceeds in two stages: combustion of VM and combustion of FC. Combustion of VM is similar to the combustion of vapors from an oil drop. The typical total combustion time of 100-micron solid coal particle is on the order of 1 s in boilers and is dominated by the time required for heterogeneous combustion of the residual char particle, while the pyrolysis time ($t_{\text{pyr}} = 10^6 (\text{s/m}^2) d_p^2$) is on the order of 1/10th–1/100th of the total burning time. Since bio-solid contains 70%–80% VM (coal contains 10% VM), most of the combustion of volatiles occurs within a short time (about 0.10 s).

For liquid drops and plastics of density ρ_c, simple relations exist for evaluating the combustion rates and times. If the transfer number B is defined as

$$B = \frac{c_p\{T_\infty - T_w\}}{L} + \frac{Y_{O_2,\infty}}{v_{O_2}}\frac{h_c}{L}, \qquad (10)$$

where $T_w \approx$ TBP for liquid fuels; $T_w = T_g$, the temperature of gasification for plastics; L is the latent heat for liquid fuel and $L = q_g$, heat of gasification for plastics; $Y_{O_2,\infty}$ is the free-stream oxygen mass fraction; v_{O_2} is the stoichiometric oxygen mass per unit mass of fuel (typically 3.5 for liquid fuels); and h_c is the lower heating value of fuel; then the burn rate ($\dot{m}$) and time (t_b) for spherical condensates (liquid drops and spherical particles of diameter d_p and density ρ_c) are given by the following expressions:

$$\dot{m} \approx 2\pi \frac{\lambda}{c_p} d_p \ln(1 + B) \qquad (11)$$

$$t_b = \frac{d_0^2}{\alpha_c}, \qquad (12)$$

where

$$\alpha_c = 8\frac{\lambda}{c_p}\frac{\ln(1 + B)}{\rho_c} \qquad (13)$$

and c_p and λ are the specific heat and thermal conductivity of gas mixture evaluated at a mean temperature (approximately 50% of the adiabatic flame temperature).

The higher the B value, the higher the mass loss rate, and the burn time will be lower. The value of B is about 1–2 for plastics (polymers), 2–3 for alcohols, and 6–8 for pentane to octane. The burn time of plastic waste particles will be about 3–4 times longer than single liquid drops of pentane to octane ($\approx$ gasoline) of similar diameter.

COMBUSTION IN PRACTICAL SYSTEMS

The time scales for combustion are on the order of 1000, 10, and 1 ms for coal burnt in boilers and liquid fuels burnt in gas turbines and diesel engines. Coal is burnt on grates in lumped form (larger-sized particles, 2.5 cm or greater with a volumetric intensity on the order of 500 kW/m^3), medium-sized particles in fluidized beds (1 cm or less, 500 kW/m^3), or as suspensions or pulverized fuel (pf; 75 micron or less, 200 kW/m^3) in boilers.

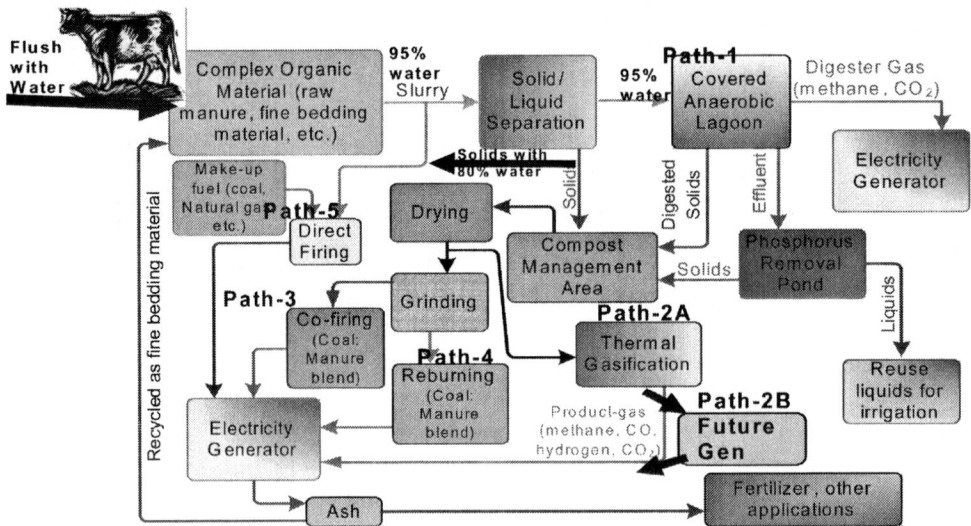

Fig. 7 Flow chart showing several energy conversion options for a typical dairy or cattle operation.

Apart from pyrolysis, gasification, and combustion, another option for energy conversion (particularly if solid fuel is in slurry firm, such as flushed dairy manure), is the anaerobic digestion (in absence of oxygen) to CH_4 (60%) and CO_2 (40%) using psychrophylic (ambient temperature), mesophyllic (95°F) and thermophyllic (135°F) bacteria in digesters.[10] Typical options of energy conversion, indicated in Fig. 7, include anaerobic digestion (path 1, the biological gasification process), thermal gasification with air to produce CO, HC, CO_2 (path 2A) or with steam to produce CO_2 and H_2 (path 2B), cofiring (path 3), reburn (path 4; see "Reburn with Bio-Solids"), and direct combustion (path 5).

Suspension Firing

In suspension-fired boilers, solid fuel is pulverized into smaller particles ($d_p = 75$ μm or less) so that more surface area per unit mass is exposed to the oxidant, resulting in improved ignition and combustion characteristics. Typical boiler burners use swirl burners for atomized oil and pulverized coal firing, while a gas turbine uses a swirl atomizer in highly swirling turbulent flow fields. A swirl burner for pf firing is shown in Fig. 8. The air is divided into a primary air stream which transports the coal (10%–20% of the total air, heated to 70°C–100°C to prevent condensation of vapors and injected at about 20 m/s to prevent settling of the dust, loading dust and gas at a ratio of 1:2) and a secondary air stream (250°C at 60–80 m/s) which is sent through swirl vanes, supplying the remaining oxygen for combustion and imparting a tangential momentum to the air. In wall-fired boilers, burners are stacked above each other on the wall; while in tangential-fired boilers, the burners are mounted at the corners of rectangular furnaces.

Stoker Firing

The uncrushed fuel [fusion temperature <1093°C (2000°F); volatile content >20%; sizes in equal proportions of 19 mm×12.5 mm (3/4 in.×½ in.), 6.3 mm×3.2 mm (½ in.×¼ in.), 3.2 mm×3.2 mm (¼ in.×¼ in.)][16] is fed onto a traveling chain grate below to which primary air is supplied (Fig. 9), which may be preheated to 177°C (350°F) if moisture exceeds 25%. The differential pressure is on the order of 5–8 mm (2–3 in.). The combustible gases are carried into an over-fire region into which secondary air (almost 35% of total air at three levels for low emissions) is fired to complete combustion.

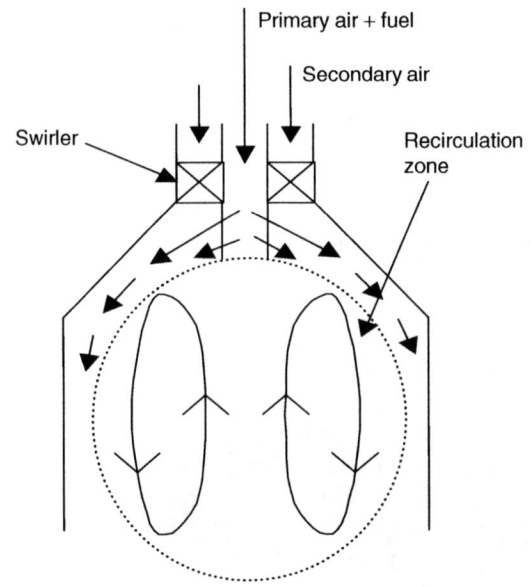

Fig. 8 Pulverized Fuel (pf) fired swirl burner.

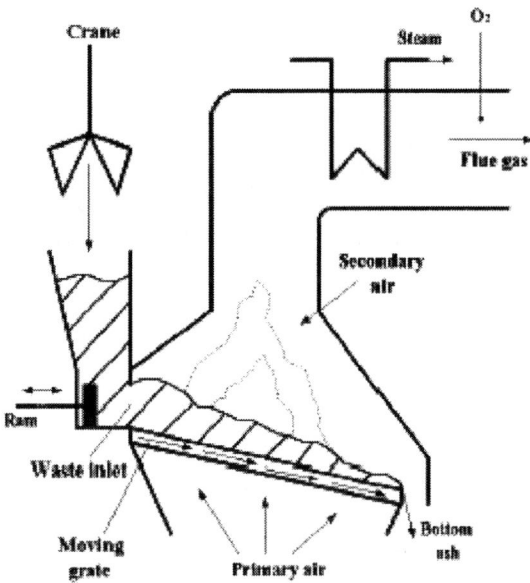

Fig. 9 Schematic of stoker firing.
Source: From Ref. 39.

The over-fire region acts like a perfectly stirred reactor (PSR). It is apparent that solid fuels need not be ground to finer size.

Fixed Bed Combustor

The bed contains uncrushed solid fuels, inert materials (including ash), and processing materials (e.g., limestone to remove SO_2 from gases as sulfates). It is fed with air moving against gravity for complete combustion, but the velocity is low enough that materials are not entrained into the gas streams. Large solid particles can be used.

Fluidized Bed Combustor

When air velocity (V) in fixed bed combustor (FXBC) is increased gradually to a velocity called minimum fluidization velocity V_{mf}, the upward drag force is almost equal to the weight of the particle, so that solids float upward. The bed behaves like a fluid (like liquid water in a tank), i.e., it becomes fluidized. If $V > V_{mf}$, then air escapes as bubbles and is called a bubbling fluidized bed combustor (BFBC). The bed has two phases: the bubble phase, containing gases (mostly oxygen), and the emulsion phase (dense phase, oxygen deficient), containing particles and gas. Many times gas velocity is so high that gaseous combustibles produced within the bed burn above the bed (called free board region), while solids (e.g., char and carbon) burn within the bed. Fluidized Bed Combustor (FBC) is suitable for fuels which are difficult to combust in pf-fired boilers.

Circulating Fluidized Bed Combustor (CFBC)

When air velocity in FBC is increased at velocity $V \gg V_{mf}$, particles are entrained into the gas stream. Since the residence time available to particles for combustion is shorter, unburned particles are captured using cyclones located downstream of the combustor and circulated back to the bed.

The residence time (t_{res}) varies from a low value for pf-fired burners to a long residence time for fixed-bed combustors. The reaction time (t_{reac}) should be shorter than t_{res} so that combustion is complete. The reaction time includes time to heat up to ignition temperature and combustion. The previous section on fuel properties and the homogenous (e.g., CH_4, CO oxidation) and heterogeneous (e.g., carbon oxidation) reaction kinetics can be used to predict t_{reac} or burn time t_b.

COAL AND BIO-SOLIDS COFIRING

General Schemes of Conversion

Most of the previously reviewed combustion systems typically use pure coal, oil, or gas. The same systems require redesign for use with pure biomass fuels. A few of the technologies, which utilize bio-solids as an energy source, are summarized in Annamalai et al.[17] These technologies include direct combustion (fluidized beds), circulating fluidized beds, liquefaction (mostly pyrolysis), onsite gasification for producing low to medium Btu gases, anaerobic digestion (bacterial conversion), and hydrolysis for fermentation to liquid fuels like ethanol.[18,19]

Cofiring

Although some bio-solids have been fired directly in industrial burners as sole-source fuels, limitations arose due to variable moisture and ash contents in bio-solid fuels, causing ignition and combustion problems for direct combustion technologies. To circumvent such problems, these fuels have been fired along with the primary fuels (cofiring) either by directly mixing with coal and firing (2%–15% of heat input basis) or by firing them in between coal-fired burners.[20–24]

Cofiring has the following advantages: improvement of flame stability problems, greater potential for commercialization, low capital costs, flexibility of adaptation of biomass fuels and cost effective power generation, mitigated NO_x emissions from coal-fired boilers, and reduced CO_2 emissions. However, a lower melting point of biomass ash could cause fouling and slagging problems.

Some of the bio-solid fuels used in cofiring with coal are cattle manure,[25,26] sawdust and sewage sludge,[21] switch grass,[20] wood chips,[24,27] straw,[22,28] and refuse-derived fuel (RDF).[21] See Sami et al. for a review of literature on cofiring.[7]

Coal and Agricultural Residues

Sampson et al. reported test burns of three different types of wood chips (20%, HHV from 8320 to 8420 Btu/lb) mixed with coal (10,600 Btu/lb) at a stoker (traveling grate) fired steam plant.[24] The particulate emission in grams per SCF ranged from 0.05 to 0.09. An economic study, conducted for the 125,000 lb/h steam power plant, concluded that energy derived from wood would be competitive with that from coal if more than 30,000 tons of wood chips were produced per year with hauling distances less than 60 mi. Aerts et al. carried out their experiments on cofiring switch grass with coal in a 50-MW, radiant, wall-fired, pulverized coal boiler with a capacity of 180 tons of steam at 85 bar and 510°C (Fig. 10). The NO_x emissions decreased by 20%, since switchgrass contains lesser nitrogen (Table 4).[20] It is the author's hypothesis that a higher VM content of bio-solids results in a porous char, thus accelerating the char combustion process. This is validated by the data from Fahlstedt et al. on the cofiring of wood chips, olive pips and palm nut shells with coal at the ABB Carbon 1 MW Process Test Facility; they found that blend combustion has a slightly higher efficiency than coal-only combustion.[27]

Coal and RDF

Municipal solid waste includes residential, commercial, and industrial wastes which could be used as fuel for production of steam and electric power. MSW is inherently a blended fuel, and its major components are paper (43%); yard waste, including grass clippings (10%); food (10%); glass and ceramics (9%); ferrous materials (6%); and plastics and rubber (5%). Refer to Tables 5 and 6 for analyses. When raw waste is processed to remove non-combustibles like glass and metals, it is called RDF. MSW can decompose in two ways, aerobic and anaerobic. Aerobic decomposition (or composting) occurs when O_2 is present. The composting produces CO_2 and water, but no usable energy products. The anaerobic decomposition occurs in the absence of O_2. It produces landfill gas of 55% CH_4 and 45% CO_2.

Coal and Manure

Frazzitta et al. and Annamalai et al. evaluated the performance of a small-scale pf-fired boiler burner facility (100,000 Btu/h) while using coal and premixed coal-manure blends with 20% manure. Three types of feedlot

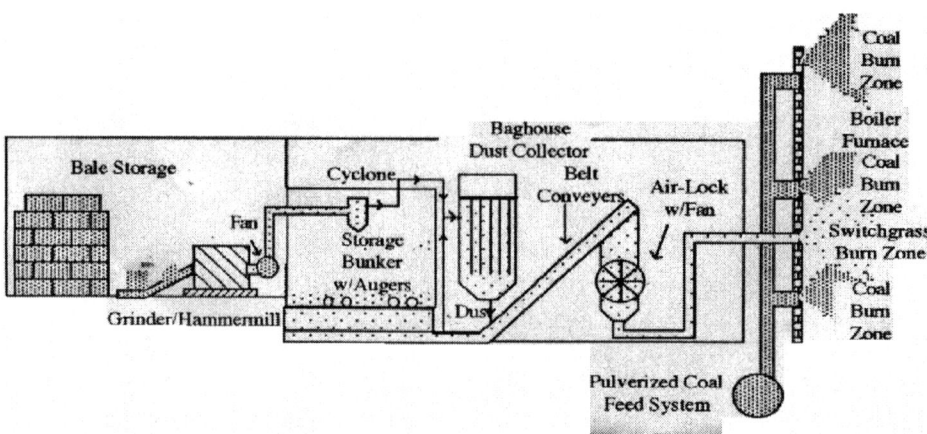

Fig. 10 A cofiring Scheme for coal and biomass (Alternate Fuel Handling Facility at Blount St. Generating Station). Source: From Ref. 20.

manure were used: raw, partially composted, and fully composted. The burnt fraction was recorded to be 97% for both coal and coal-manure blends.[25,26]

NO_x Emissions

During combustion, the nitrogen evolved from fuel undergoes oxidation to NO_x; and this is called fuel NO_x to distinguish it from thermal NO_x, which is produced by oxidation of atmospheric nitrogen. Unlike coal, most of the agricultural biomass being burned is very low in nitrogen content (i.e., wood or crops), but manure has a higher N content than fossil fuels. A less precise correlation exists between cofiring levels on a Btu basis and percent NO reduction under cofiring. The following Eq. (valid between 3 and 22% mass basis cofiring) describes NO_x reduction as a function of cofiring level on a heat input basis:

$$NO_x \text{ Reduction (\%)} = 0.0008 \, (COF\%)^2 + 0.0006 \, COF\% + 0.0752, \quad (14)$$

where COF% is the percentage of co-firing on a heat input basis. The mechanisms used to reduce NO_x emissions by cofiring vary between cyclone firing and PC firing.

Fig. 11 shows the percentage reduction in NO with percentage cofiring of low-N agricultural biomass fuels. This relationship does not apply to high-N biofuels such as animal manure.

Fouling in Cofiring

Hansen et al. investigated the ash deposition problem in a multi-circulating fluidized bed combustor (MCFBC) fired with fuel blends of coal and wood straw.[25] The Na and K lower the melting point of ash. For ash fusion characteristics see Table 7. Rasmussen and Clausen evaluated the performance of an 80-MW co-generation power plant at Grenaa, Denmark, fired with hard coal and bio-solids (surplus straw from farming). Large amounts of Na and K in straw caused superheater corrosion and combustor fouling.[29] Annamalai et al. evaluated fouling potential when feedlot manure biomass (FB) was cofired with coal under suspension firing.[30] The 90:10 Coal:FB blend resulted in almost twice the ash output compared to coal and ash deposits on heat exchanger tubes that were more difficult to remove than baseline coal ash deposits. The increased fouling behavior with blend is probably due to the higher ash loading and ash composition of FB.

Table 5 Chemical composition of solid waste

	Percent	
Proximate analysis	Range	Typical
Volatile matter (VM)	30–60	50
Fixed carbon (FC)	5–10	8
Moisture	10–45	25
Ash	10–30	25

	Percent by mass (dry basis)					
Ultimate analysis	C	H	O	N	S	Ash
Yard wastes	48	6	38	3	0.3	4.7
Wood	50	6	43	0.2	0.1	0.7
Food wastes	50	6	38	3	0.4	2.6
Paper	44	6	44	0.3	0.2	5.5
Cardboard	44	6	44	0.3	0.2	5.5
Plastics	60	7	23			10
Textiles	56	7	30	5	0.2	1.8
Rubber	76	10		2		12
Leather	60	9	12	10	0.4	8.6
Misc. organics	49	6	38	2	0.3	4.7
Dirt, ashes, etc.	25	3	1	0.5	0.2	70.3

Table 6 Heat of combustions of municipal solid waste components

Component	Inerts (%)		Heating values (kJ/kg)	
	Range	Typical	Range	Typical
Yard wastes	2–5	4	2,000–19,000	7,000
Wood	0.5–2	2	17,000–20,000	19,000
Food wastes	1–7	6	3,000–6,000	5,000
Paper	3–8	6	12,000–19,000	17,000
Cardboard	3–8	6	12,000–19,000	17,000
Plastics	5–20	10	30,000–37,000	33,000
Textiles	2–4	3	15,000–19,000	17,000
Rubber	5–20	10	20,000–28,000	23,000
Leather	8–20	10	15,000–20,000	17,000
Misc. organics	2–8	6	11,000–26,000	18,000
Glass	96–99	98	100–250	150
Tin cans	96–99	98	250–1,200	700
Nonferrous	90–99	96		
Ferrous metals	94–99	98	250–1,200	700
Dirt, ashes, etc.	60–80	70	2,000–11,600	7,000

GASIFICATION OF COAL AND BIO-SOLIDS

Gasification is a thermo-chemical process in which a solid fuel is converted into a gaseous fuel (primarily consisting of HC, H_2 and CO_2) with air or pure oxygen used for partial oxidation of FCs. The main products during gasification are CO and H_2, with some CO_2, N_2, CH_4, H_2O, char particles, and tar (heavy hydrocarbons). The oxidizers used for the gasification processes are oxygen, steam, or air. However, for air, the gasification yields a low-Btu gas, primarily caused by nitrogen dilution present in the supply air. Syngas ($CO+H_2$) is produced by reaction of biomass with steam. The combustible product, gas, can be used as fuel burned directly or with a gas turbine to produce electricity; or used to make chemical feedstock (petroleum refineries). However, gas needs to be cleaned to remove tar, NH_3, and sulfur compounds. The integrated gasification combined cycle (IGCC) (Fig. 12),

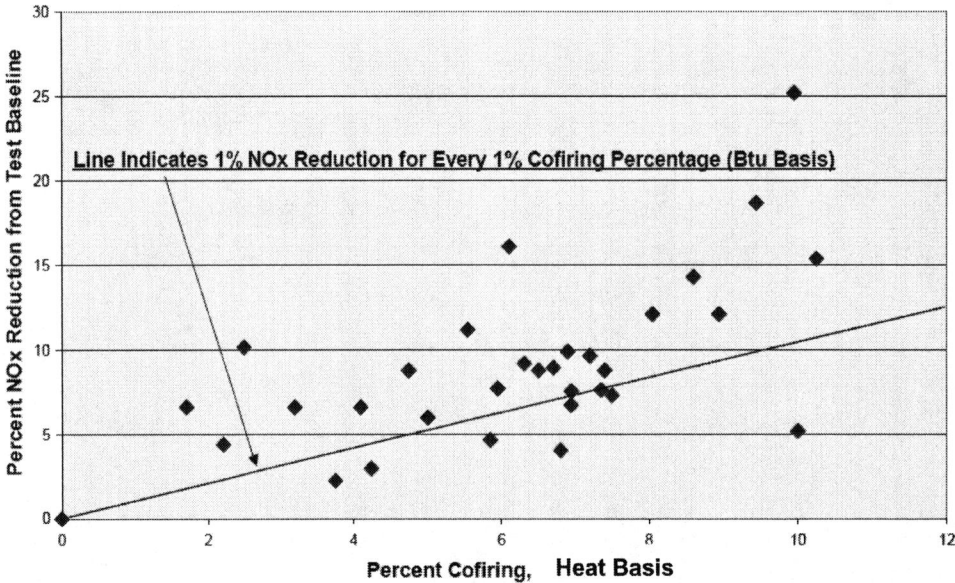

Fig. 11 NO_x reduction due to cofiring with low-N agricultural residues.
Source: From Technical Advisory Committee (see Ref. 40).

for combined heat and power (CHP), and traditional boilers use combustible gases from gasifiers for generation of electric power.

Typically in combined cycles, gaseous or liquid fuel is burnt in gas turbine combustors. High-temperature products are expanded in a gas turbine for producing electrical power; a low-temperature (but still hot) exhaust is then used as heat input in a boiler to produce low-temperature steam, which then drives a steam turbine for electrical power. Therefore, one may use gas as a topping cycle medium, while steam is used as fluid for the bottoming cycle. The efficiency of a combined cycle is on the order of 60%, while a conventional gas turbine cycle has an efficiency of 42%.[31] Commercial operations include a 250-MW IGCC plant at Tampa, Florida, operating since 1996; a 340-MW plant at Negishi, Japan, since 2003; and a 1200-MW GE-Bechtel plant under construction in Ohio for American Electric Power, to start in 2010.[31]

There are three basic gasification reactor types: (i) fixed-bed gasifiers (Fig. 13); (ii) fluidized-bed gasifiers, including circulating-bed (CFB) or bubbling-bed; and (iii) entrained-flow gasifiers. The principles of operation are similar to those of combustors except that the air supplied is much below stoichiometric amounts, and instead of a combination of steam, air and CO_2, air can also be used. The oxidant source could also include gases other than air, such as air combined with steam in Blasiak et al.[32]

Table 7 Ash fusion behavior and ash composition, fusion data: ASTM D-188

	FB	PRB Coal	Blend
Ash Fusion, (reducing)			
Initial deformation, IT, °C (°F)	1140 (2090)	1130 (2060)	NA
Softening, °C (°F)	1190 (2170)	1150 (2110)	NA
Hemispherical, HT, °C (°F)	1210 (2210)	1170 (2130)	NA
Fluid, °C (°F)	1230 (2240)	1200 (2190)	NA
Ash fusion, (oxidizing)			
Initial deformation, IT, °C (°F)	1170 (2130)	1190 (2180)	NA
Softening, °C (°F)	1190 (2180)	1200 (2190)	NA
Hemispherical, HT, °C (°F)	1220 (2230)	1210 (2210)	NA
Fluid, °C (°F)	1240 (2270)	1280 (2330)	NA
Slagging Index, Rs, °C (°F)	1160 (2120)	1140 (2090)	
Slagging classification	High	Severe	
Ash composition (wt%)			
SiO_2	53.63	36.45	43.56
Al_2O_3	5.08	18.36	12.87
Fe_2O_3	1.86	6.43	4.54
TiO_2	0.29	1.29	0.88
CaO^+	14.60	19.37	17.40
MgO^+	3.05	3.63	3.39
Na_2O^+	3.84	1.37	2.39
K_2O^+	7.76	0.63	3.58
P_2O_5	4.94	0.98	2.62
SO_3	3.71	10.50	7.69
MnO_2	0.09	0.09	0.09
Sum	98.84	99.11	99.00
Volatile Oxides	30.77	28.25	
Basic oxides	32.73	35.51	
Silica ratio	0.73	0.53	
$Na_2O + K_2O$	11.60	2.00	5.97
Inherent Ca/S Ratio	6.71	1.86	2.48
kg alkali ($Na_2O + K_2O$)/GJ	5.37	0.06	0.29

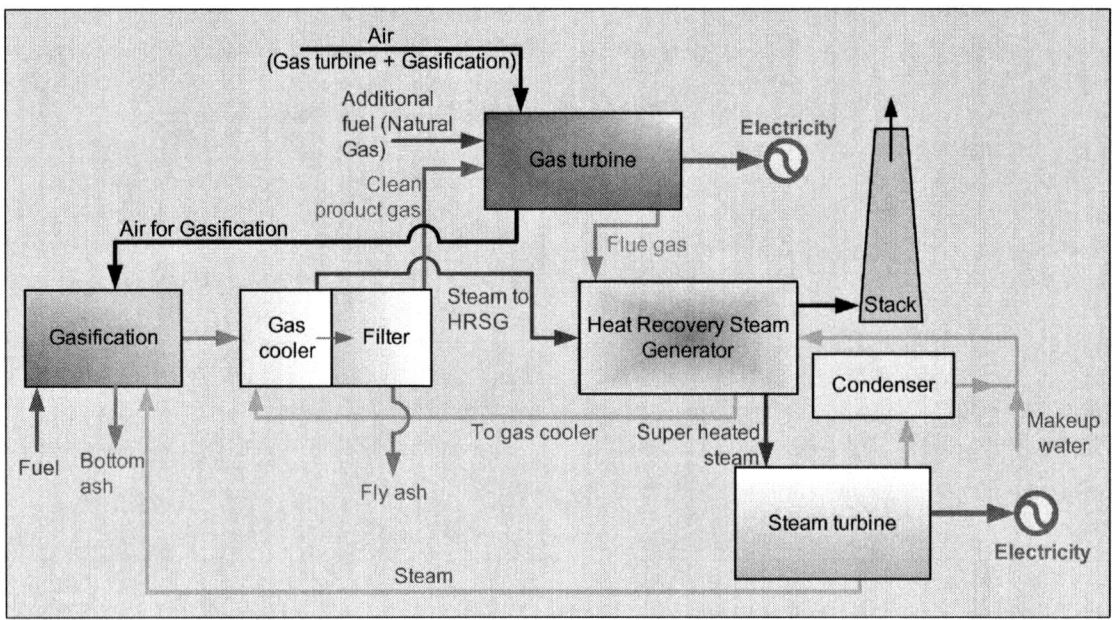

Fig. 12 Fluidized-bed gasification for Integrated Gasification Combined Cycle (IGCC) Process.

FUTUREGEN

FutureGen is a new U.S. initiative to build the world's first integrated CO_2 sequestration and H_2 production research power plant using coal as fuel. The technology shown in Fig. 14 employs modern coal gasification technology using pure oxygen, resulting in CO, C_nH_m (a hydrocarbon), H_2, HCN, NH_3, N_2, H_2S, SO_2, and other combinations which are further reacted with steam (reforming reactions) to produce CO_2 and H_2. The bed materials capture most of the harmful N and S compounds followed by gas-cleaning systems; the CO_2 is then sequestered and H_2 is used as fuel, using either combined cycle or fuel cells for electricity generation or sold as clean transportation fuel. With partial oxidation of gasification products and char supplying heat for pyrolysis and other endothermic reactions (i.e, net zero external heat supply in gasifier), the overall gasification reaction can be represented as follows for 100 kg of DAF Wyoming coal:

Reaction V $C_{6.3}H_{4.4}N_{0.076}O_{1.15}S_{0.014} + 8.14\ H_2O(\ell)$
$+ 1.67\ O_2 \rightarrow 6.3\ CO_2 + 10.34\ H_2$
$+ 0.038\ N_2 + 0.014\ SO_2$

It is apparent that the FutureGen process results in enhanced production of H_2, using coal as an energy source to strip H_2 from water. For C–H–O fuels, it can be shown that theoretical H_2 production (N_{H_2}) in moles for an empirical fuel CH_hO_o is given as $\{0.4115\ h - 0.6204\ o + 1.4776\}$ under the above conditions. For example, if glucose $C_6H_{12}O_6$ is the fuel, then empirical formulae is CH_2O; thus, with $h=2$, $o=1$, $N_{H_2} = 1.68$ kmol, using atom balance, $CH_2O + 0.68\ H_2O(\ell) + 0.16\ O_2 \rightarrow CO_2 + 1.68\ H_2$.

REBURN WITH BIO-SOLIDS

NO_x is produced when fuel is burned with air. The N in NO_x can come both from the nitrogen-containing fuel

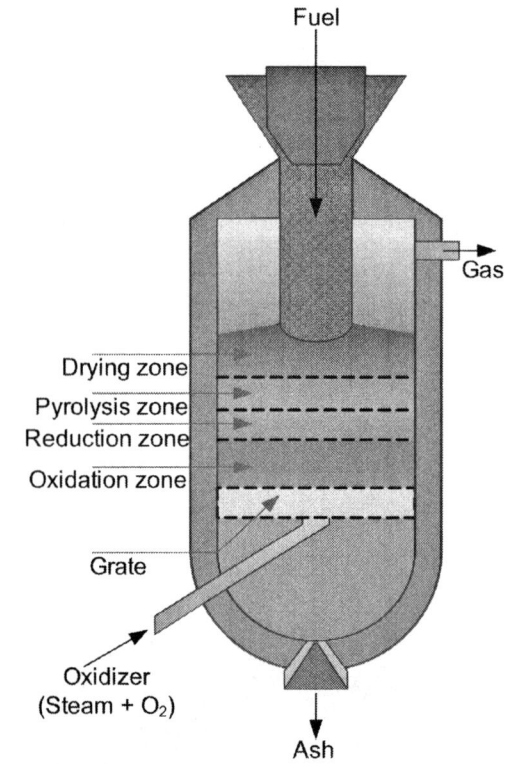

Fig. 13 Updraft fixed-bed gasifier.

Energy Conversion: Principles for Coal, Animal Waste, and Biomass Fuels

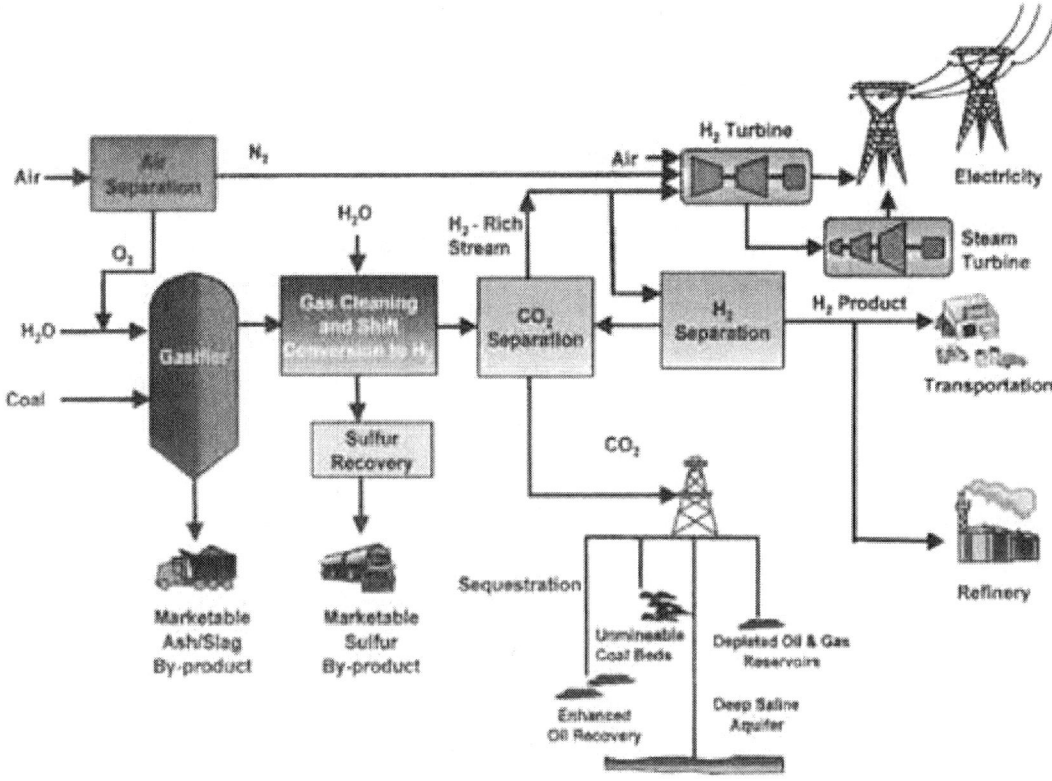

Fig. 14 FutureGen layout.
Source: From http://www.fe.doe.gov.

compounds (e.g., coal, biomass, plant residue, animal waste) and from the N in the air. The NO_x generated from fuel N is called fuel NO_x, and NO_x formed from the air is called thermal NO_x. Typically, 75% of NO_x in boiler burners is from fuel N. It is mandated that NO_x, a precursor of smog, be reduced to 0.40–0.46 lb/mmBtu for wall and tangentially fired units under the Clean Air Act Amendments (CAAA). The current technologies developed for reducing NO_x include combustion controls (e.g., staged combustion or low NO_x burners (LNB), reburn) and post-combustion controls (e.g., Selective Non-Catalytic Reduction, SNCR using urea).

In reburning, additional fuel (typically natural gas) is injected downstream from the primary combustion zone to create a fuel rich zone (optimum reburn stoichiometric ratio (SR), usually between SR 0.7 and 0.9), where NO_x is reduced up to 60% through reactions with hydrocarbons when reburn heat input with CH_4 is about 10%–20%. Downstream of the reburn zone, additional air is injected in the burnout zone to complete the combustion process. A diagram of the entire process with the different combustion zones is shown in Fig. 15. There have been numerous studies on reburn technology found in literature, with experiments conducted, and the important results summarized elsewhere.[33] Table 8 shows the percentages of reduction and emission obtained with coal or gas reburn in coal-fired installations and demonstration units.

The low cost of biomass and its availability make it an ideal source of pyrolysis gas, which is a more effective reburn fuel than the main source fuel, which is

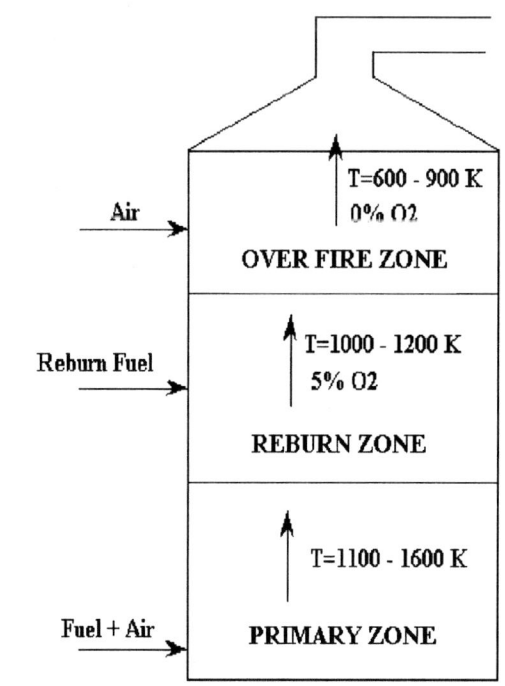

Fig. 15 Schematic of reburn process.

Table 8 Percentage reduction in NO_x: demonstration and/or operating reburn installations on coal-fired boilers in the United States

Type of Burner	% Reburn Heat in	% Reduction	NOx with Reburn lb/mmBtu[a]
Gas reburning			
Tangential	18	50–67	0.25
Cyclone	20–23	58–60	0.39–0.56
Wall without LNB	18	63	0.27
Coal reburn			
Cyclone (micronized)	30 (17)	52 (57)	0.39 (0.59)
Tangential(micron)with LNB	14	28	0.25

LNB: Low NO_x Burners.
[a] 1 lb per mmBtu = 0.430 kg/GJ.
Source: From U.S. Department of Energy (see Ref. 38).

typically coal. Recently, animal manure has been tested as a reburn fuel in laboratory scale experiments. A reduction of a maximum of 80% was achieved for pure biomass, while the coal experienced a reduction of between 10 and 40%, depending on the equivalence ratio.[34] It is believed that the greater effectiveness of the feedlot biomass is due to its greater volatile content on a DAF basis and its release of fuel nitrogen in the form of NH_3, instead of HCN.[35]

ACKNOWLEDGMENTS

Most of this work was supported in part by the U.S. Department of Energy of Pittsburgh, PA, the DOE of Golden, CO, the USDA, the Texas Advanced Technology Program, and the Texas Commission on Environmental Quality (TCEQ) of Austin, TX, through the Texas Engineering Experiment Station (TEES), Texas A&M University, and College Station, TX.

REFERENCES

1. U.S. DOE, Energy Information Administration (www.eia.doe.gov/, accessed April 12, 2005).
2. Priyadarsan, S.; Annamalai, K.; Sweeten, J.M.; Holtzapple, M.T.; Mukhtar, S. Co-gasification of blended coal with feedlot and chicken litter biomass, Proceedings of the 30th Symposium (International) on Combustion; The Combustion Institute: Pittsburgh, PA, 2005; Vol. 30, 2973–2980.
3. Ebeling, J.M.; Jenkins, B.M. Physical and chemical properties of biomass fuels. Trans. ASAE **1985**, *28* (3), 898–902.
4. Sweeten, J.M.; Annamalai, K.; Thien, B.; McDonald, L. Co-firing of coal and cattle feedlot biomass (FB) fuels, part I: feedlot biomass (cattle manure) fuel quality and characteristics. Fuel **2003**, *82* (10), 1167–1182.
5. Boie, W. Wiss z Tech. Hochsch. Dresden **1952/1953**, *2*, 687.
6. Annamalai, K.; Sweeten, J.M.; Ramalingam, S.C. Estimation of the gross HVs of biomass fuels. Trans. Soc. Agric. Eng. **1987**, *30*, 1205–1208.
7. Sami, M.; Annamalai, K.; Wooldridge, M. Co-firing of coal and biomass fuel blends. Prog. Energy Combust. Sci. **2001**, *27*, 171–214.
8. Annamalai, K.; Puri, I.K. 2006, *Combustion Science and Engineering*, Taylor and Francis: Orlando, FL, 2006.
9. Tillman, D.A.; Rossi, A.J.; Kitto, W.D. *Wood Combustion: Principles: Processes and Economics*; Academic press: New York, 1981.
10. Sweeten, J.M.; Heflin, K.; Annamalai, K.; Auvermann, B.; Collum, Mc; Parker, D.B. *Combustion-Fuel Properties of Manure Compost from Paved vs Un-paved Cattle feedlots*; ASABE 06-4143: Portland, OR, July 9–12, 2006.
11. Bartok, W., Sarofim, A.F. 1991, Fossil Fuel Combustion, chapter 3: FL Dryer, John Wiley, and Hobokan, NJ. pp. 121–214.
12. Howard, J.B.; Sarofim, A.F. Gasification of coal char with CO_2 and steam at 1200–1800°C. *Energy Lab Report*; Chemical Engineering; MIT: MA, 1978.
13. Essenhigh, R.H.; Misra, M.K.; Shaw, D.W. Ignition of coal particles: a review. Combust. Flame **1989**, *77*, 3–30.
14. Annamalai, K.; Durbetaki, P. A theory on transition of ignition phase of coal particles. Combust. Flame **1977**, *29*, 193–208.
15. Du, X.; Annamalai, K. The transient ignition of isolated coal particles. Combust. Flame **1994**, *97*, 339–354.
16. Johnson, N. Fundamentals of Stoker Fired Boiler Design and Operation, CIBO Emission Controls Technology Conference, July 15–17, 2002.
17. Annamalai, K.; Ibrahim, Y.M.; Sweeten, J.M. Experimental studies on combustion of cattle manure in a fluidized bed combustor. Trans. ASME, J. Energy Res. Technol. **1987**, *109*, 49–57.
18. Walawender, W.P.; Fan, L.T.; Engler, C.R.; Erickson, L.E. Feedlot manure and other agricultural wastes as future material and energy resources: II. Process descriptions. *Contrib.30*, Deptartment of Chemical

Engineering, Kansas Agricultural Experiment Station: Manhattan, KS, 30, 1973.
19. Raman, K.P.; Walawander, W.P.; Fan, L.T. Gasification of feedlot manure in a fluidized bed: effect of temperature. Ind. Eng. Chem. Proc. Des. Dev. **1980**, *10*, 623–629.
20. Aerts, D.J.; Bryden, K.M.; Hoerning, J.M.; Ragland, K.W. *Co-firing Switchgrass in a 50 MW Pulverized Coal Boiler*, Proceedings of the 59th Annual American Power Conference, Chicago, IL, 1997; Vol. 50(2), 1180–1185.
21. Abbas, T.; Costen, P.; Kandamby, N.H.; Lockwood, F.C.; Ou, J.J. The influence of burner injection mode on pulverized coal and biosolid co-fired flames. Combust. Flame **1994**, *99*, 617–625.
22. Siegel, V.; Schweitzer, B.; Spliethoff, H.; Hein, K.R.G. Preparation and co-combustion of cereals with hard coal in a 500 kW pulverized-fuel test unit. Biomass for energy and the environment, *Proceedings of the 9th European Bioenergy Conference*, Copenhagen, DK, 24–27 June, 1996; 2, 1027–1032.
23. Hansen, L.A.; Michelsen, H.P.; Dam-Johansen, K. Alkali metals in a coal and biosolid fired CFBC—measurements and thermodynamic modeling, *Proceedings of the 13th International Conference on Fluidized Bed Combustion*, Orlando, FL, May 7–10, 1995; 1, 39–48.
24. Sampson, G.R.; Richmond, A.P.; Brewster, G.A.; Gasbarro, A.F. Co-firing of wood chips with coal in interior Alaska. Forest Prod. J. **1991**, *41* (5), 53–56.
25. Frazzitta, S.; Annamalai, K.; Sweeten, J. Performance of a burner with coal and coal: feedlot manure blends. J. Propulsion Power **1999**, *15* (2), 181–186.
26. Annamalai, K.; Thien, B.; Sweeten, J.M. Co-firing of coal and cattle feedlot biomass (FB) fuels part II: performance results from 100,000 Btu/h laboratory scale boiler burner. Fuel **2003**, *82* (10), 1183–1193.
27. Fahlstdedt, I.; Lindman, E.; Lindberg, T.; Anderson, J. Co-firing of biomass and coal in a pressurized fluidized bed combined cycle. Results of pilot plant studies, *Proceedings of the 14th International conference on Fluidized Bed Combustion*, Vancouver, Canada, May 11–14, 1997; 1, 295–299.
28. Van Doorn, J., Bruyn, P., Vermeij, P. Combined combustion of biomass, fluidized sewage sludge and coal in an atmospheric Fluidized bed installation, Biomass for energy and the environment, *Proceedings of the 9th European Bioenergy Conference*, Copenhagen, DK, June 24–27, 2, 1996; 1007–1012.
29. Rasmussen, I.; Clausen, J.C. ELSAM strategy of firing biosolid in CFB power plants, *Proceedings of the 13th International Conference on Fluidized Bed Combustion*, Orlando, FL, May 7–10, 1, 1995; 557–563.
30. Annamalai, K.; Sweeten, J.; Freeman, M.; Mathur, M.; O'Dowd, W.; Walbert, G.; Jones, S. Co-firing of coal and cattle feedlot biomass (FB) fuels, part III: fouling results from a 500,000 Btu/hr pilot plant scale boiler burner. Fuel **v.82,10**, 82 (10), 1195–1200, 2003.
31. Langston, L.S. 2005 New Horizons. Power and Energy vol 2, 2, June 2005.
32. Blasiak, W.; Szewczyk, D.; Lucas, C.; Mochida, S. *Gasification of Biomass Wastes with High Temperature Air and Steam*, Twenty-First International Conference on Incineration and Thermal Treatment Technologies, New Orleans, LA, May 13–17, 2002.
33. Thien, B.; Annamalai, K. National Combustion Conference, Oakland, CA, March 25–27, 2001.
34. Arumugam, S.; Annamalai, K.; Thien, B.; Sweeten, J. Feedlot biomass co-firing: a renewable energy alternative for coal-fired utilities, Int. Natl J. Green Energy, vol. 2, No. 4, 409–419, 2005.
35. Zhou, J.; Masutani, S.; Ishimura, D.; Turn, S.; Kinoshita, C. Release of fuel-bound nitrogen during biomass gasification. Ind. Eng. Chem. Res. **2000**, *39*, 626–634.
36. Annamalai, K.; Puri, I.K. *Advanced Thermodynamics*; CRC Press: Boca Raton, FL, 2001.
37. Priyadarsan, S.; Annamalai, K.; Sweeten, J.M.; Holtzapple, M.T.; Mukhtar, S. Waste to energy: fixed bed gasification of feedlot and chicken litter biomass. Trans. ASAE **2004**, *47* (5), 1689–1696.
38. DOE, *Reburning technologies for the control of nitrogen oxides emissions from coal-fired boilers* Topical Report, No 14, U.S. Department of Energy, May, 1999.
39. Loo, S.V.; Kessel, R., 2003, Optimization of biomass fired grate stroker systems, EPRI/IEA Bioenergy Workshop, Salt Lake City, Utah, February 19, 2003.
40. Grabowski, P. *Biomass Cofiring, Office of the Biomass Program*, Technical Advisory Committee, Washington, DC, March 11, 2004.

Energy Efficiency: Developing Countries

U. Atikol
H.M. Güven
Energy Research Center, Eastern Mediterranean University, Magusa, Northern Cyprus

Abstract

Statistics and projections show that not only the rate of energy consumption, but also the carbon emissions of developing countries are rising very fast. The rate of increase in the capacity needs in developing countries can be decreased in two steps. First, the component associated with the old infrastructure should be dealt with; and second, the application of end-use energy efficiency measures should be put in effect. The energy efficiency technologies that are available in the industrialized countries may not always be feasible for transfer to developing countries. Readily available economic and social indicators, supported by more detailed end-use research, can be used to determine which technologies are suitable for a given developing country. Case studies performed in Northern Cyprus and Turkey show that in Northern Cyprus, transfer of these technologies would be more successful in the residential and commercial sectors, whereas in Turkey, they would be more feasible for the industrial sector.

INTRODUCTION

The quest for energy efficiency dates back to 1973, at which time the Oil Producing and Exporting Countries (OPEC) placed an embargo on crude oil during the war in the Middle East. Until that year, most developed countries had experienced decades of low energy prices and plentiful fuel supplies; consequently, high and growing per capita use of energy was of little concern to most governments. In 1979, there was another interruption in supplies when U.S. President Jimmy Carter announced a ban on oil imports from Iran after the hostage crisis began at the U.S. embassy in Tehran. Subsequently, rapidly rising energy costs and interruptions in supplies forced re-evaluation of existing policies. In most developed countries, conservation and efficiency improvements to energy systems became an important component of energy policy, due not only to the realization that the world is vulnerable to interference with its energy supplies, but also to awareness that excessive use of fossil fuels is causing vital damage to the world ecosystem.

In the United States before the 1973 crises, both primary energy and electricity consumption increased at almost the same rate as the Gross National Product (GNP). (The Gross National Product is the dollar value of all goods and services produced in a nation's economy, including goods and services produced abroad). After that year, however, high oil prices led to progressive energy policies, and the demand was halted (see Fig. 1). Between 1973 and 1986, the rate of growth of electricity use decreased, growing only 2.5% per year, or 3.2% per year less than projected by pre1973 trends.[1] In 1986, projected electricity use was 50% higher than actual electricity use, indicating a savings of 1160 TWh. The gap between the two trends is referred to as JAWS, coined by Rosenfeld.

Since 1970, the growth of electricity consumption has been around 3.5% for the countries in the Organization for Economic Cooperation and Development (OECD), (Organization for Economic Cooperation and Development countries are Australia, Austria, Belgium, Canada, Denmark, Finland, France, Germany, Greece, Iceland, Ireland, Italy, Japan, Luxembourg, The Netherlands, New Zealand, Norway, Portugal, Spain, Sweden, Switzerland, Turkey, the United Kingdom, and the United States.) while that of developing countries has been 8.2%. For the OECD countries, this growth is almost parallel with the growth of their Gross Domestic Product (GDP). (Gross Domestic Product is a measure of the value of all goods and services produced by the economy. Unlike GNPs, GDP includes only the values of goods and services earned by a nation within its boundaries). In the developing countries, however, electricity generation rose much faster than the GDP. While the total GDP (in U.S. dollars) increased slightly over twofold between 1970 and 1989, electricity generation grew more than fourfold.[2]

It was rather fortunate that there was ample scope for improvements in energy efficiency. Insulation and other measures, such as using more efficient air conditioners, were applied to reduce space heating and air conditioning requirements. The wide use of household appliances meant a large market for more energy-efficient electric motors, compressors, water heaters, and other equipment. In transportation, improvement in fuel use was also achieved for new vehicles. In industrialized countries,

Keywords: Developing countries; Energy efficiency; Demand-side management; Energy planning; Technology transfer.

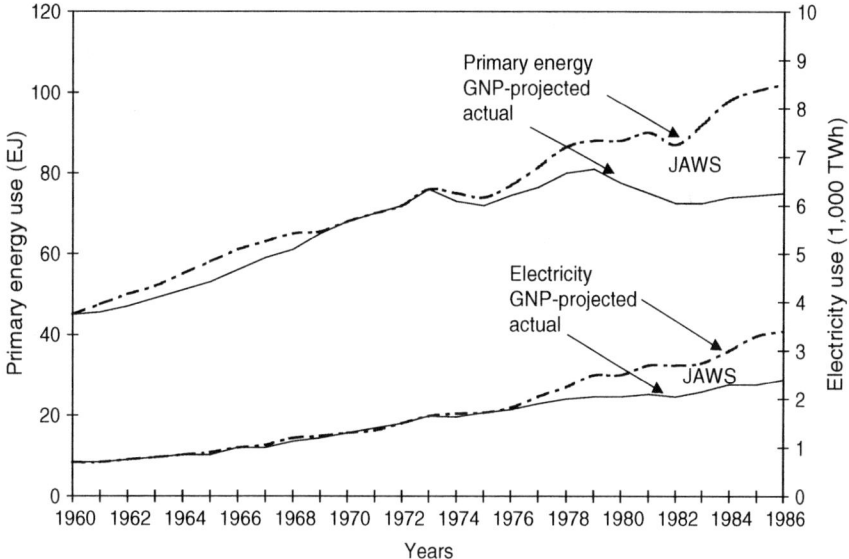

Fig. 1 U.S. total primary energy and electricity use immediately after 1973: actual and GNP projected. The graph shows the effect of energy efficiency applications between 1973 and 1986.
Source: From Island Press (see Ref. 1).

several mechanisms were developed to implement the changes; two important ones were the introduction of utility demand-side management (DSM) and the establishment of energy service companies (ESCOs) backed up with accurate data collection and improved codes and standards. Recently, distributed power and renewable energy (RE) technologies have been promoted by the introduction of new legislation.

In developing countries, these mechanisms do not exist, mainly due to lack of institutional formation, technical expertise, and sufficient infrastructure. In this article, there will be an attempt to evaluate these problems with suggestions for possible solutions.

TRENDS IN WORLDWIDE ENERGY DEMAND

In a report prepared by the Energy Information Administration,[3] the International Energy Outlook 2000 (IEO2000), much of the growth in worldwide energy use is projected for the developing world (Fig. 2a). In particular, energy demand in developing Asia and Central and South America is projected to more than double between 1997 and 2020. Both regions are expected to sustain energy demand growth of more than 3% annually throughout the forecast, accounting for more than 50% of the total projected increment in world energy consumption and 83% of the increment for the developing world alone. World carbon emissions are projected to rise from 6.2 billion metric tons in 1997 to 8.1 billion metric tons in 2010 and to 10 billion metric tons in 2020, according to the IEO2000. This analysis does not take into account the potential impact of the Kyoto Protocol. (The Kyoto Climate Change Protocol is a treaty signed by 83 countries and the European Union that requires reductions or limits to the growth of carbon emissions within the Annex I countries between 2008 and 2012. The Annex I countries under the protocol are Australia, Bulgaria, Canada, Croatia, the European Union, Iceland, Japan, Liechtenstein, Monaco, New Zealand, Norway, Romania, Russia, Switzerland, Ukraine, and the United States. Turkey and Belarus are Annex I countries that did not commit to quantifiable emissions targets under the protocol). In this forecast, world carbon emissions will exceed their 1990 levels by 40% in 2010 and by 72% in 2020. Emissions in the developing countries accounted for about 28% of the world total in 1990, but they are projected to make up 44% of the total by 2010 and nearly 50% by 2020. As a result, even if the Annex I countries were able to meet the emissions limits or reductions prescribed in the Kyoto Protocol, worldwide carbon emissions still would grow substantially. The increase is expected to be caused both by rapid economic expansion, accompanied by growing demand for energy, and by continued heavy reliance on coal (the most carbon intensive of the fossil fuels), particularly in developing Asia.

There is an expected increase in electricity consumption worldwide by 76% in the IEO2000 reference case, from 12 trillion kWh in 1997 to 22 trillion kWh in 2020. Long-term growth in electricity consumption is expected to be strongest in the developing countries of Asia, followed by those of Central and South America. Those two regions alone account for 52% of the world's net electricity consumption increment in the IEO2000 reference case (Fig. 2b). Rapid growth in population and income, along with greater industrialization and more

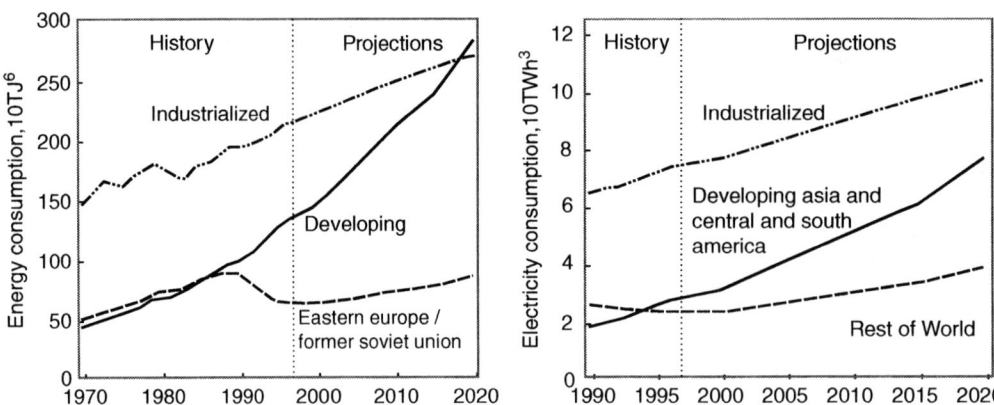

Fig. 2 World energy and net electricity consumption by region, 1970–2020.
Source: From U.S. Department of Energy (see Ref. 3).

household electrification, are responsible for the increase.

Developing countries need energy to raise productivity and improve the living standards of their populations. Traditionally, developing countries have addressed their energy needs by expanding their supply base, with little attention to the efficient use of energy. This approach has been raising serious financial, institutional, and environmental problems. The magnitude of these problems has been underlining the need for improving the efficiency with which energy is currently used and produced in developing countries.

Factors contributing to the rapidly rising energy consumption in developing countries include population growth, economic growth, and increased consumer demand.

Population Growth in Developing Countries

The world population is just over 6 billion and is projected by the U.S. Bureau of Census[4] to reach 9.1 billion in 2050. Over the next two decades, the population of the developing world is projected to increase by nearly 1 billion, to almost 6 billion total, while that of the industrialized countries will increase by only 50 million, to 1 billion total.[3] By 2050, the population of the less-developed regions could reach 7.75 billion or more.[5] Developing countries then could account for 85% of the global population. The increase in population between 1998 and 2050 for the developing countries would be 64%, and this alone would account for a large increase in their energy consumption by 2050, even if per capita consumption remained at current levels.

Economic Growth in Developing Countries

Securing higher living standards for this rising population requires rapid economic growth, further increasing the demand for energy services. The steady increase in per capita energy consumption in developing countries is due merely to economic growth, which includes urbanization, increased use of commercial fuels instead of traditional biomass fuels (such as wood, crop residues, and animal dung), and increased use of energy-intensive materials. When annual per capita energy consumptions of the industrialized and developing countries are compared, it is observed that the gap between them is more than 4.5 million tons of oil equivalent (TOE).[3] This shows that the developing countries have a long way to go to increase their living standards to those of industrialized countries. Closing this gap will lead to astronomic increases in the world energy consumption. This is the inevitable and unavoidable eventuality that may lead to catastrophic environmental consequences. Therefore, utilizing energy as efficiently as possible and trying to use cleaner sources of energy is absolutely essential.

Accelerated Consumer Demand

In recent years, modern manufacturing techniques and improved materials have sharply lowered the real costs of consumer goods such as radios, refrigerators, and televisions. In the 1990s, for example, a study[6] revealed that the real cost of refrigerators decreased by a factor of 5 between 1950 and 1990. Similarly, in Northern Cyprus, due to the decreasing cost of air conditioners by a factor of 2.5 between 1996 and 2006, not only the number of residences, but also the number of small to medium-size enterprises using them increased dramatically. Global distribution systems have also increased the accessibility of these appliances. Thus, people in developing countries can purchase these goods at a far earlier stage in the development cycle (as measured by per capita GDP) than did people in today's industrial countries. The rapidly increasing use of these consumer goods has a strong impact on residential electricity use, creating additional demand at peak times.

ENERGY AND DEVELOPING COUNTRIES

Developing countries need energy to raise productivity and improve the living standards of their populations. Traditionally, developing countries have addressed their energy needs by expanding their supply base, with little attention to the efficient use of energy, which led to the need to expand the supply base frequently. This approach, however, raises serious financial, institutional, and environmental problems. The magnitude of these problems underlines the need for improving the efficiency with which energy is currently used and produced in developing countries.

The magnitude of funding needed for the required power supply projects is growing so fast that these projects have little chance of being mobilized. The power sectors in developing countries frequently experience a wide range of institutional problems, including excessive staffing, inadequate management, weak planning, poor maintenance, deficient financial monitoring, and few incentives to improve efficiency of operation.[7] This raises questions about the ability of this key sector to continue expanding rapidly even if financial resources were available. The continuous increase in energy consumption (fossil-fuel combustion in modern industry, transport, and electricity generation) causes air pollution and contaminates water supplies and land. On the other hand, nonfossil energy sources may also cause environmental damage. Hydroelectric development often involves the flooding of vast areas of land, with a resulting loss of agricultural land, human settlements, fish production, forests, wildlife habitats, and species diversity. Therefore, using energy as efficiently as possible should be an essential part of the planning process.

ENERGY EFFICIENCY IMPLEMENTATION BARRIERS

As the demand for electricity increased and the global climate change issues caused governmental concern, energy efficiency attracted attention in some developing countries. Difficulties were experienced in the implementation of these energy efficiency measures and programs, however. The implementation barriers were identified as follows:[8,9]

- Lack of central coordination and institutional formation
- Lack of awareness and general misinformation
- Lack of technical information and expertise
- Inappropriate pricing policies
- Lack of capital
- Lack of appropriate laws
- Taxes and tariffs that discourage energy efficiency activities
- Poor infrastructure
- Customs and maintaining the status quo.

CHALLENGES AND REWARDS OF IMPLEMENTATION OF ENERGY EFFICIENCY IN DEVELOPING COUNTRIES: RESIDENTIAL, COMMERCIAL, AND INDUSTRIAL SECTOR CASE STUDIES AND EXAMPLES

How Economic Development Is Measured

The statistical information that indicates the level of economic development can be used in the DSM decision-making process. Economic development can be measured through a number of social, economic, and demographic indexes (indicators). The following indicators are often used to measure the economic development of a country:

- *Per capita income*. This is a statistic that is seldom readily available, and GNP or GDP per capita is often used instead. According to 1999 GNP per capita, calculated using the World Bank Atlas method,[9] the following grouping of countries is possible: low-income, $755 or less; lower-middle-income, $756–$2995; upper-middle-income, $2996–$9265; and high-income, $9266 or more.
- *Per capita purchasing power*. This is a more meaningful measure of actual income per person, because it includes not only income, but also the price of goods in a country. It is usually measured in per capita GDP. Per capita GDP can be expressed both by using the current exchange rates in USD and by using the current purchasing power parities (PPPs). Purchasing power parities are the number of currency units required to buy goods equivalent to what can be bought with 1 unit of the currency of the base country or with 1 unit of the common currency of a group of countries.
- *Economic structure of the labor force*. In the United States and Western Europe, fewer than 2% of the workers are engaged in agriculture, whereas in certain African nations, India, and China, more than 70% of the laborers are in this sector. More than 75% of U.S. laborers are engaged in wholesaling, retailing, professional and personal services (including medical, legal, and entertainment), and information processing (such as finance, insurance, real estate, and computer-related fields).[10]
- *Consumer goods purchased*. The quantity and quality of consumer goods purchased and distributed in a society

also provide a good measure of the level of economic development in that society. Televisions, automobiles, home electronics, jewelry, watches, refrigerators, and washing machines are some of the major consumer goods produced worldwide on varying scales. The ratio of people to television sets in developing countries is 150 to 1, for example, and the ratio of the population to automobiles is 400 to 1. In California, the ratio is almost 1 to 1 for these consumer items.[10] The number of consumer goods such as telephones and televisions per capita is a good indicator of a country's level of economic development.

- *Education and literacy of a population*. The more men and women who attend school, usually the higher the level of economic development in a country. The literacy rate of a country is the percentage of people in the society who can read and write.
- *Health and welfare of a population*. Measures of health and welfare, in general, are much higher in developed nations than in les- developed ones. One measure of health and welfare is diet. Most people in Africa do not receive the United Nations' daily recommended allowance of diet. People in less-developed countries also have poor access to doctors, hospitals, and medical specialists.
- *Per capita energy consumption*. The energy consumption of the population is a good indicator of a country's level of economic development.

Transfer of Energy Efficiency Technologies

There may be more indicators of economic development to list, but what is sought for purposes of making energy policy is (preferably) a list of readily available indicators for deciding on the feasibility of the transfer of energy efficiency technologies. In a previous study,[9,11] a similar analysis was carried out for DSM technology transfer to developing countries.

Indicators such as GDP, per capita energy consumption, installed capacity, capacity factor, growth rate, income distribution, and carbon emissions will only give an idea about the level of the economic development of the country in question. More information is needed for choosing the correct DSM options. Details on the characteristics of end uses, which are seldom available in developing countries, would be a very useful indicator. These data can be obtained by monitoring, auditing, or surveying. In a recent study,[9] it was found that surveying is the least time-consuming low-cost method to obtain such data. To generalize the method, the indicators can be grouped under two headings: macro- and micro-level indicators (see Fig. 3).

The macro indicators are those that will be useful in determining which of the following categories the developing country belongs to. The categories are as follows:

- Advanced developing countries—middle-income countries that are advanced industrially
- Developing countries—low-income and lower-middle-income countries that are developing
- Least developed countries—low-income countries that are either not developing or developing very slowly.

The determination of which category a country belongs to is based on a cross-check of macro- and micro-level indicators. The execution of the decision-making is carried

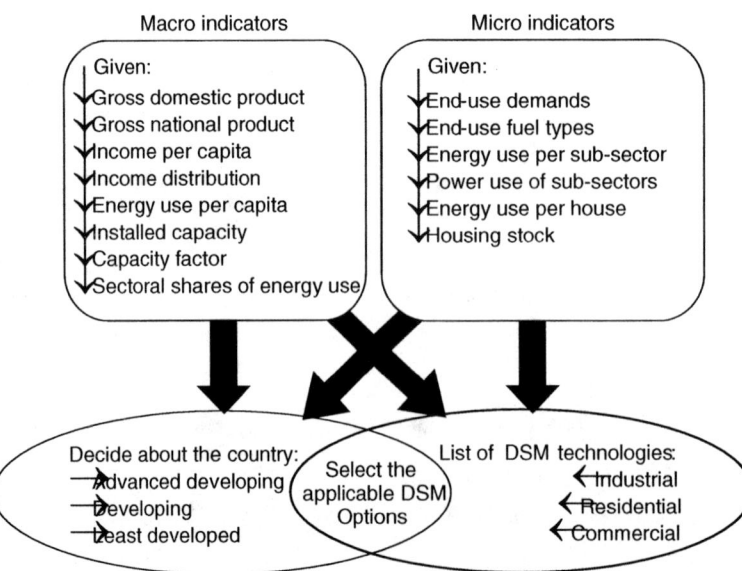

Fig. 3 Macro- and micro-level input variables used in decision-making for the most applicable DSM options in any developing country. Source: From Eastern Mediterranean University and Elsevier Science (see Refs. 9 and 11).

out through the consideration of the variables shown in Fig. 3. Using the end-use information, the available DSM options can be determined from the pool of DSM technologies. Then the applicability of the DSM technologies to the country in question is examined further by using information about the category of the country. The final decision is made by checking the cost-to-benefit ratio of the applicable DSM program.

Finally, the implementation of the DSM programs is evaluated, and if there are any doubts about their success, more surveys may be conducted to update the data at the micro level.

Reducing Power Capacity Needs

In developing countries, the slow-occurring technology transfer and, hence, delayed energy efficiency improvements in all sectors may lead to exaggerated power demand. In Turkey, the growth of installed power capacity has been greater than that of the GNP projected between 1968 and 2003 (see Fig. 4). In 1997, installed power was almost three times greater than it should have been with the GNP-projected levels; and in 2003, installed power was almost four-and-a-half times greater than those levels. The gap between the two trends (shaded area in Fig. 4) is a common feature of the developing countries. It is the reverse of JAWS, which was described for developed countries in the introduction. To close the gap between the two trends, in Fig. 4, the old infrastructures that are associated with the energy services should be renewed both at the power supply side and at the demand side. Therefore, this gap is referred to as an excess energy consumption (EXEC) gap.

The traditional method of meeting the demand with little attempt to rebuild the infrastructure and increase the efficiency of energy services needs to be revised in developing countries. Traditionally, the policy-makers in developing countries believed that the opportunity in power reduction lay in reducing the EXEC. The opportunity for power reduction is not limited to this amount, however. As the replacement of the old infrastructure continues, ways of improving energy efficiency at customer level and, hence, further reducing the power capacity requirements at the utilities and achieving the JAWS, a feature of developed countries, should also be considered. The gap between the maximum installed power and the minimum that can be achieved is called the energy-saving opportunity gap (ESOG).[12] The energy-saving opportunity in developing countries lies in reducing both EXEC and JAWS (i.e., ideally ESOG = EXEC + JAWS).

Case Studies of DSM Implementation

The two countries examined were Northern Cyprus and Turkey.[11] The flow chart in Fig. 5 shows a simplified approach that was applied for determining the best applicable DSM technologies for Northern Cyprus and Turkey. In this simple approach, the GNP per capita income of the country in question was used to determine the income level of the country. At the same time, a decision was made about the industrial development level of the country. For this reason, the shares of electricity consumption were used. This synthesis was carried out at the macro level, leading to setting priorities on considering

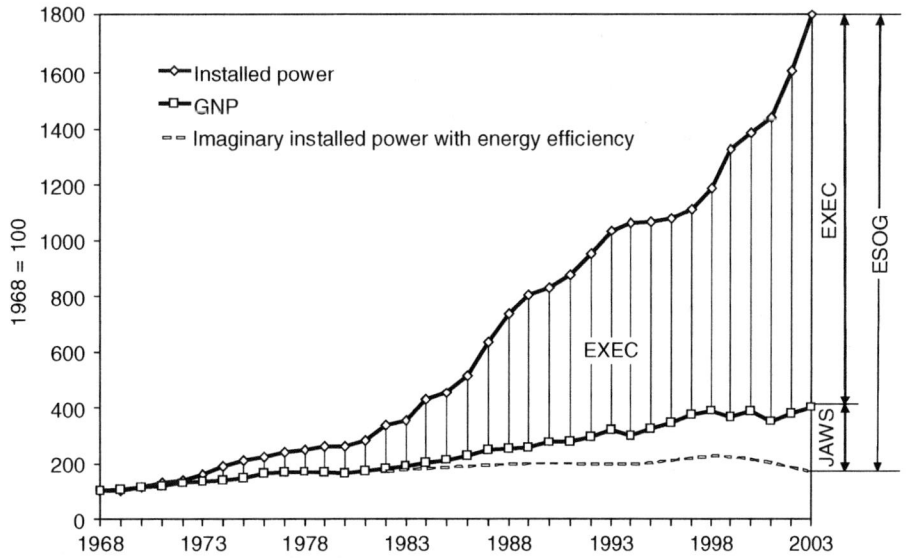

Fig. 4 Economic and power capacity developments of Turkey between 1968 and 2003.
Source: From Prime Ministry of Republic of Turkey, State Institute of Statistics Publication (see Ref. 13).

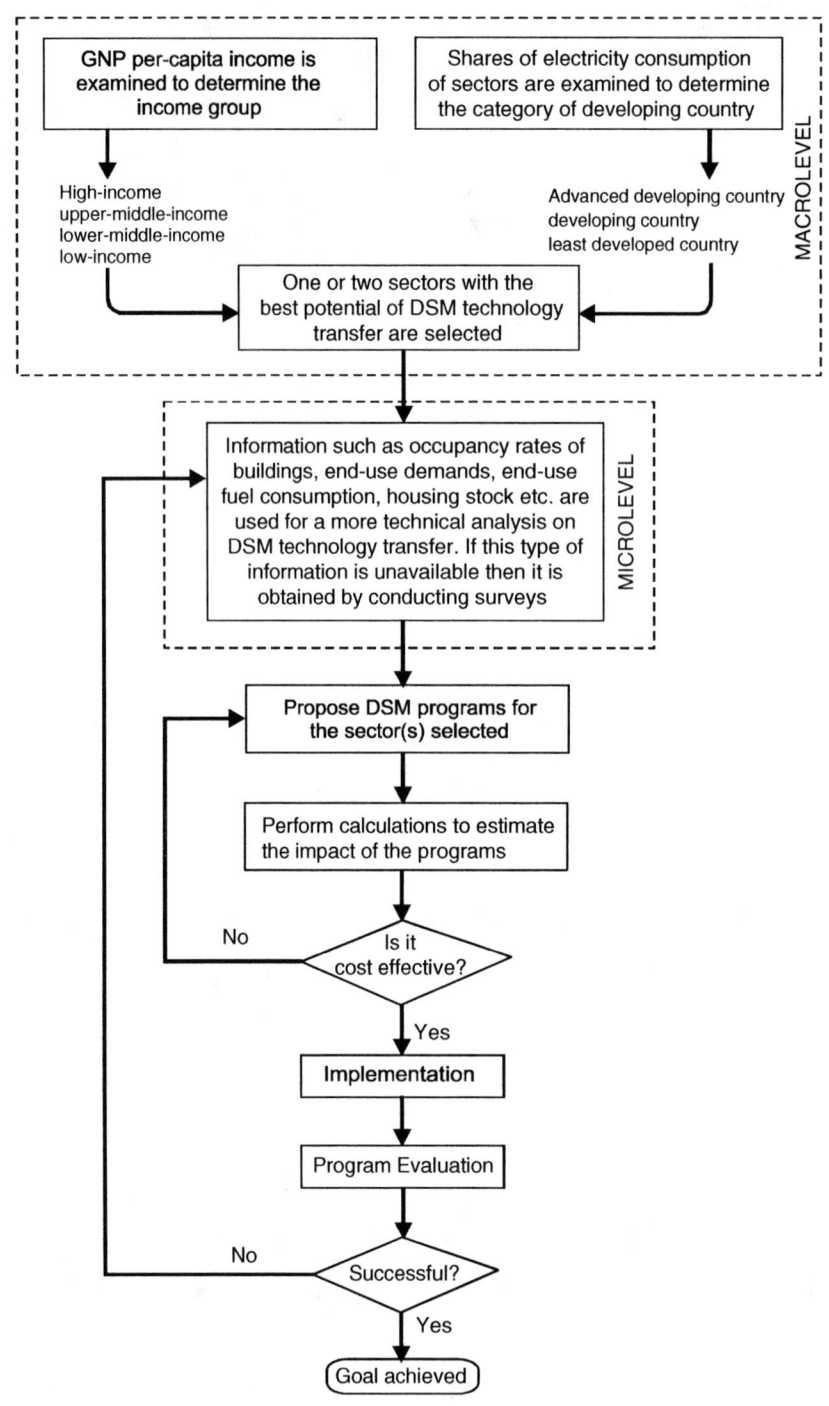

Fig. 5 A simplified approach for DSM technology transfer to developing countries.
Source: From Elsevier (see Ref. 11).

the sectors with the best applicability of DSM options. To determine the specific DSM options to apply, however, detailed information on the end uses was required. Therefore, a survey was conducted for this purpose, and DSM programs were proposed.

Northern Cyprus, being a small island, is not an industrially advanced country. Sector-based electricity consumption shares are given as 35% for residential, 18.74% for commercial, and 9.39% for industrial sectors.[11] The great majority of the population (approximately 70%) is employed in the services sector; hence, the DSM measures are best selected for the residential and commercial technologies. The 1996 GNP per capita being $4222 USD, Northern Cyprus is an upper-middle-income country according to the World Bank criterion. The population is highly educated, and the literacy rate is almost 100%. With these macro indicators, it was decided that the residential and commercial sectors would have the best potential for load reduction and that the DSM technology transfer should be considered for these sectors. Surveys were conducted in both sectors to determine the micro indicators leading to the selection of the DSM options to apply.

In the residential sector, statistical information was obtained on the number of electrical appliances and their time of use, and end-use load curves were obtained. It was discovered that electrical water heaters demanded 50 MW at the winter peak hour (19:00), which constituted 45% of the 110-MW peak. The other shares of end-use demands of the winter peak were estimated to be 28% for space heating, 9% for television sets, 7% for lighting, and 5% for refrigeration. Demand-side management programs concerning the water heating, space heating, and lighting activities were proposed. The programs proposed were estimated to defer the need for a new 60 MW power unit worth $100 million (U.S. dollars) for at least 19 years at an expense of $12 million (U.S. dollars).[14]

The diversity of the commercial buildings and their end uses in Northern Cyprus preclude the adoption of the approach used in the residential sector. Not only do the system sizes for end uses vary greatly, but also, their time of use may vary for different building types. The problem can be simplified by selecting a segment of the commercial sector. In Northern Cyprus, the tourism sector was selected, being the segment in the commercial sector that consumes the most electricity. According to the surveys conducted, the power demand of this sector is greater during the summer season due to the heavy use of air conditioners. Air-conditioner use in the hotels constituted 27% of the utility load in summer. The proposed DSM technology transfer to this sector could reduce the summer peak by 11% at a total cost of $5,60,000 (U.S. dollars).[15]

Turkey, on the other hand, is an advanced developing country due to its past two decades of industrial development. The primary energy consumption share of its industrial sector is currently 29% and is expected to reach 40% in 2020. Due to its nonuniform income distribution, however, the 1999 GDP per capita is $2807 USD, which makes Turkey a lower-middle-income country. The level of education is low, and a rebate program on replacing incandescent lamps with compact fluorescent lamps (CFLs) in the residential sector is not expected to be as successful as in Northern Cyprus, especially in the low-income homes. The macro indicators of Turkey imply that the transfer of DSM technology would be best selected from the industrial options. Micro indicators such as end-use fuel types, average grid losses at 20%, and the magnitude of power and rate of heat demands of the factories pointed to sizable energy gains from a cogeneration program. A more detailed study[16] showed that by replacing the nuclear program with the cogeneration program, Turkey not only would save $72.6 billion (U.S. dollars), but also would reduce total primary energy demand by 11% in 2020.

CONCLUSIVE COMMENTS

Statistics and projections show that not only the rate of energy consumption, but also the carbon emissions of developing countries are rising very fast. In 2001, carbon dioxide emissions from industrialized countries were 49% of the global total, followed by developing countries at 38% and Eastern Europe/Former Soviet Union (EE/FSU) at 13%. If business continues as usual, by 2025, developing countries are projected to account for the largest share of world carbon dioxide emissions, at 46%, followed by the industrialized world at 42% and the EE/FSU at 12%.[17] To save the world, it is essential not only to transfer the latest energy efficiency technologies to the developing countries to achieve substantial reductions in energy consumption, but also to explore ways of rebuilding their poor infrastructure.

On the other hand, energy efficiency technologies, which are successfully used in many industrialized countries, may not always be feasible to apply in developing countries. Some DSM technologies may be easily transferable to some developing countries but not applicable in others.

Compact fluorescent lamps, for example, which were offered as a DSM measure in some developing countries, may not be suitable for other developing countries. In Turkey, the squatters, who are low-income people, probably would find this program luxurious, and even if they were given a full rebate, they might prefer selling the lamps rather than using them for themselves. On the other hand, in the villages of Bangladesh, 90% of the energy required for lighting is supplied from petroleum products; only 10% comes from electricity. Therefore, CFLs cannot be worthy of any consideration as a DSM option.

It is possible that some countries, although having a lower per capita income than others, may be industrially advanced, and the DSM technologies devised for the industrial sector may be more suitable than those for the residential sector. Turkey, for example, falls into the category of "industrially advanced developing country," and applying DSM measures in the industrial sector would be more successful and have more impact on the utility load management of the country.

Statistical information on end uses is also very useful in determining the suitability of the proposed energy efficiency or DSM technology. In Northern Cyprus, it was discovered that the largest demand for electricity came from domestic water heating. In South Africa, figures as high as 40%–50% of the monthly electricity use of an average middle-to-upper-income household were quoted for water heating. DSM technologies regarding efficient water heating would be most appropriate for these countries, but in many other countries, this may not be the case. In Thailand, for example, the highest potential savings could come from improving the energy efficiency of the refrigerators, the reason being the intense use of old-technology, inefficient refrigerators.

Therefore, it can be concluded that to apply energy efficiency technology transfer to developing countries, the social and economic realities of these countries need to be taken into account, along with the technical applicability of the programs. The approach should be twofold: First, the economic and social state of the country should be determined; and second, the suitability of the energy efficiency options with the highest potential savings should be determined.

REFERENCES

1. Rosenfeld, A.H.; Ward, E. Energy use in buildings. In *The Energy-Environment Connection*; Hollander, J.M., Ed., Island Press: Washington, DC, 1992; 223–257.
2. Levine, M.D.; Koomey, J.G.; Price, L.; Geller, H.; Nadel, S. Electricity end-use efficiency: Experience with technologies, markets and policies throughout the world. Energy Int. J. **1995**, *20* (1), 37–61.
3. Energy Information Administration *International Energy Outlook*; U.S. Department of Energy: Washington, DC, 2000.
4. U.S. Bureau of Census *International Database*; U.S. Bureau of Census: Washington, DC, 2001.
5. United Nations *1997 Energy Statistics Yearbook*; United Nations: New York, NY, 1998.
6. Office of Technology Assessment. *Energy in Developing Countries*, U.S. Congress Publication, OTA-E-486. U.S. Congress: Washington, DC, 1991.
7. Office of Technology Assessment. *Fueling Development—Energy Technologies for Developing Countries*. U.S. Congress Publication, OTA-E-516. U.S. Congress: Washington, DC, 1992.
8. Painuly, J.P.; Reddy, B.S. Electricity conservation programs: Barriers to their implementation. Energy Sources **1996**, *18* (3), 257–267.
9. Atikol, U. Utility demand-side management technology-transfer and impact assessment methodologies for developing countries [PhD thesis]. Eastern Mediterranean University: Magusa, Northern Cyprus, 2001.
10. Stutz, F.P.; Souza, A.R. *The World Economy: Resources, Location, Trade and Development*, 3rd Ed.; Prentice Hall: Englewood Cliffs, NJ, 1998.
11. Atikol, U.; Guven, H. Feasibility of DSM-technology transfer to developing countries. Appl. Energy **2003**, *76* (1–3), 197–210.
12. Atikol, U. An algorithm for reducing the power capacity needs in developing countries. In: *Modelling in Energy Economics and Policy*. Sixth IAEE European Conference, Zurich, Switzerland, September 1–3, 2004; International Association for Energy Economics, Cleveland, Ohio, 2004.
13. State Institute of Statistics *Statistical Year Book 2004*; Prime Ministry of Republic of Turkey, State Institute of Statistics Publication: Ankara, Turkey, 2005.
14. Atikol, U.; Dagbasi, M.; Guven, H. Identification of residential end-use loads for demand-side planning in northern cyprus. Energy Int. J. **1999**, *24* (3), 231–238.
15. Atikol, U. A demand-side planning approach for the commercial sector of developing countries. Energy Int. J. **2004**, *29* (2), 257–266.
16. Atikol, U.; Guven, H. Impact of cogeneration on integrated resource planning of Turkey. Energy Int. J. **2003**, *28* (12), 1259–1277.
17. Energy Information Administration *International Energy Outlook*; U.S. Department of Energy: Washington, DC, 2004.

Energy Efficiency: Information Sources for New and Emerging Technologies

Steven A. Parker
Pacific Northwest National Laboratory, Richland, Washington, U.S.A.

Abstract
The purpose of this entry is to share a list of useful organizations that provide reliable information on new and emerging energy-efficient technologies based on research and experience. Experienced energy managers may use the information provided by these organizations to enhance their knowledge and understanding, thereby improving their energy management programs. The scope is limited to publicly available and open-membership organizations that deal with new and emerging energy-efficient technologies, strategies, and products. The sources identified should not be considered exhaustive, but rather a first step "go to" list suggested by the author when searching for useful information on new and emerging energy-efficient technologies.

BACKGROUND

There are many strategies that can be used to reduce energy consumption and costs. These strategies might include:

- Awareness programs to motivate users and occupants to change behaviors that result in saved energy;
- Operations and maintenance programs, including commissioning to keep equipment running efficiently and to optimize run times;
- Training programs to educate people on how to reduce energy consumption and costs;
- Equipment procurement programs to buy the most efficient equipment available (rather than the cheapest first cost) to minimize life-cycle costs, which include energy costs;
- Commodity procurement programs to buy the least expensive energy source available; and
- Energy-efficient technology programs to replace existing or conventional equipment with more efficient equipment. This strategy also includes new and emerging energy-efficient technologies.

An effective energy management program will likely use many, if not all, of these strategies.

Removing old, inefficient equipment and replacing it with newer, more energy-efficient equipment is one method of reducing a site's energy consumption and costs. Two questions that are often asked include: "Which energy-efficient technology should I select?" And, just as important, "From what list of alternatives should I consider?"

The answer to the first question should be "I should select the alternative with the lowest life-cycle cost." The answer to the second question may be more difficult. If you do not place a new technology on the list of alternatives, it can never be selected. One issue with new and emerging technologies is the concern about risk and reliability. When something is new, it is perceived to have more risk. "It claims to be more energy efficient, but I have not seen it in operation. Does it really work? Will it work here? Will it work in this application? Is my maintenance team ready to work with this new technology?" These are all important and relevant questions.

The time it takes for a new technology to be accepted to the point of being considered conventional can be surprising. Take, for example, the fluorescent T-8 lamp and electronic ballast. How many years did it take for this technology to be seen as conventional for standard office lighting configurations and to fully penetrate the market? Think of the energy savings that could have been realized if this technology would have fully penetrated the market in only 5–10 years instead of 15–20 years. The same is true for all new equipment end uses (cooling, heating, ventilation, water heating, battery chargers, etc.). New technologies take a long time to fully penetrate the market and considerable energy may be saved if this transitory period were reduced.

Of course, balanced against the potential energy reduction is the potential for risk. Not every new technology will be cost effective and survive the market process. The way an experienced energy manager manages the risk, or at least better understands the risk, is to evaluate the new technology on a limited basis before it is mass deployed. This is the purpose of research, evaluations, demonstrations, pilot projects, and case

Keywords: New technology; Energy-efficient technology; Emerging energy-efficient technology; Energy efficiency; Energy-efficient equipment; State-of-the-art technology.

studies. With these actions, experienced energy managers can look before they leap, and in some cases, benefit from the lessons learned from others.

INTRODUCTION

Finding useful and reliable information on new and emerging energy-efficient technologies can be daunting, even for the experienced energy manager. Today, more than ever, a lot of information is available. There are books, professional associations with journals, magazines, e-newsletters, other periodicals, and the gamut of Internet Web sites. A lot of the information available is even well intended. One of the problems is that there is too much information to sort through. Finding the right type of information on the right subject can be difficult.

There are several reasons why it is difficult to find good information on new and emerging energy-efficient technologies. First and foremost, there may be little information on the technology and its potential effectiveness because the product is new. Second, collecting, analyzing, and reporting information on technologies or equipment can be expensive and time consuming. Third, many tests may be performed in-house, and may not be publicly available. Even with these barriers, there is a lot of information out there. In addition to their regular, full-time duties, energy managers can spend years building up a network of sources and resources.

There are a number of periodicals that can be used as sources for information on new and emerging energy-efficient technologies. In addition to journals provided through membership organizations, there are a number of low-cost and no-cost periodicals, many of which can be very useful in learning about new technologies or new approaches to energy efficiency. Unfortunately, there are so many periodicals that it may be impossible for any energy manager to stay on top of their reading list.

The purpose of this entry is to share a list of useful sources. This is not a list of sources on energy management—that list would be too long—but a list of sources (organizations) that specifically deal with new and emerging, energy-efficient technologies, strategies, and products. The sources identified in this entry should not be considered exhaustive. In fact, to keep its length manageable, this list has purposely been limited to sources within the United States. The author acknowledges that there are many organizations outside the United States performing similar useful services and many organizations, both public and private, that perform a lot of good research and provide a host of useful services in the area of energy efficiency. The list of sources included in this article may be considered an initial or first step or a go-to list suggested by the author in your search for useful information on new and emerging energy-efficient technologies.

Sources are not one-size-fits-all. Not every source does everything. The reader will find that some sources are more research- and development-oriented and focus on pre-commercial technologies, while others are more focused on enhancing the deployment of commercially-available technologies. Some sources are research-oriented, while others are more applied. In addition, some sources simply document results, whereas other sources document more detailed analysis, expand on the lessons learned, or offer guidance and tools to support future applications. You may also find different definitions for the terms "new" and "emerging," as well as "technology."

It is the author's hope that the reader will benefit from the information made available on these proactive organizations. Furthermore, it is the author's hope that more organizations will take proactive steps and make even more (reliable) information on new and emerging energy-efficient technologies publicly available.

UNITED STATES GOVERNMENT AGENCIES AND PROGRAMS

While some believe we should be spending more, the United States federal government supports a lot of research and development in the area of energy efficiency and renewable energy. Because the federal government supports a lot of this work, it is also a good source for information on new and emerging energy-efficient technologies. The information, however, can be difficult to find. This section identifies several sources of information within the federal government.

U.S. Department of Energy

Obviously, one of the largest sources of information on energy-efficient technologies is the U.S. Department of Energy (DOE). Within the DOE, there are three major offices that the advanced energy manager should be aware of—the Office of Energy Efficiency and Renewable Energy (EERE), the Office of Fossil Energy (FE), and the Office of Electricity Delivery and Energy Reliability (OE).

Each of the DOE program offices plays a unique role in the larger energy picture. FE deals primarily with power generation; OE deals primarily with electric delivery and infrastructure; and EERE deals primarily with end use. However, you will find some overlap among the DOE program offices. For example, distributed generation (renewable energy, fuel cells, cogeneration, etc.) is one of those technology sets that can be found in several programs.

Energy Efficiency and Renewable Energy

Simply put, the EERE mission is to strengthen America's energy security, environmental quality, and economic vitality in public–private partnerships that:

- Enhance energy efficiency and productivity;
- Bring clean, reliable, and affordable energy technologies to the marketplace; and
- Make a difference in the everyday lives of Americans by enhancing their energy choices and quality of life.

EERE manages and supports a wide variety of energy research, development, and deployment activities. EERE (www.eere.energy.gov) consists of 10 major program offices:

- Biomass Program (www.eere.energy.gov/biomass)
- Building Technologies Program (www.eere.energy.gov/buildings)
- Federal Energy Management Program (www.eere.energy.gov/femp)
- FreedomCAR and Vehicle Technologies Program (www.eere.energy.gov/vehiclesandfuels)
- Geothermal Technologies Program (www.eere.energy.gov/geothermal)
- Hydrogen, Fuel Cells, and Infrastructure Technologies Program (www.eere.energy.gov/hydrogenandfuelcells)
- Industrial Technologies Program (www.eere.energy.gov/industry)
- Solar Energy Technologies Program (www.eere.energy.gov/solar)
- Weatherization and Intergovernmental Program (www.eere.energy.gov/wip)
- Wind and Hydropower Technologies Program (www.eere.energy.gov/windandhydro)

Each of these program offices publishes or supports the publication of several reports every year. Many of these publications are available through the DOE Web site. Because the Web site is rather large, use the search engine.

While a large part of DOE is actively involved in the research and development of new, energy-efficient technologies, there are three specific subprograms that the author would like to highlight: the Industrial Technologies Program's inventions and innovations, Building Technologies Program's emerging technologies, and the Federal Energy Management Program's new technologies.

Inventions and Innovation (I&I), part of the DOE Industrial Technologies Program, provides grants to independent inventors and small companies with sound ideas for energy-efficient technologies. Distinct from other federal grant programs, I&I provides grantees not only with funding, but with additional resources such as training, market assessments, technical assistance, access to promotional events and materials, and special contacts to aid in commercialization endeavors. Although I&I is part of the Industrial Technologies Program organization, their mission is cross cutting; I&I supports energy efficiency and renewable energy technology development in focus areas that align with 10 EERE programs. For more information on I&I, see www.eere.energy.gov/inventions.

Emerging Technologies, part of the DOE Building Technologies Program (BTP), advances the research and development of the next generation of energy-efficient components, materials, and equipment. To help near-term emerging technologies overcome market introduction barriers, the program uses a combination of strategies, including demonstration/evaluation and technology procurement. These strategies are designed to increase buyer confidence and build demand for new energy-efficient technologies. For more information on BTP's Emerging Technologies, see www.eere.energy.gov/buildings/emergingtech.

The DOE Federal Energy Management Program (FEMP) supports a limited number of technology demonstrations through their new technology activities, which are designed to provide independent performance data to federal decision-makers and support timely federal adoption of energy saving and environmentally beneficial technologies. While these publications are developed to benefit the federal energy and building manager, they are useful to any energy manager wishing to advance their energy management program. For more information on FEMP's new technology publications, see www.eere.energy.gov/femp/technologies/tech_demos.cfm.

Office of Fossil Energy

According to the FE Web site (www.fossil.energy.gov), the primary mission of DOE's Office of Fossil Energy is to ensure that we can continue to rely on clean, affordable energy from our traditional fuel resources. Fossil fuels supply 85% of the nation's energy, and FE is working on such priority projects as pollution free coal plants, carbon sequestration, FutureGen, more efficient power generation, more productive oil and gas fields, and the continuing readiness of federal emergency oil stockpiles.

While FE deals primarily with fossil fuels and source fuels for power generation, a significant amount of fuel-cell-related R&D is supported by this DOE office. If you are looking for information on fuel cells, do not overlook this important source.

Office of Electricity Delivery and Energy Reliability

According to the OE Web site (www.oe.energy.gov), their mission is to lead national efforts to modernize the electric grid, enhance the security and reliability of the energy infrastructure, and facilitate recovery from disruptions to the energy supply. OE is a relatively new office within the DOE that was formed when the Office of Energy Assurance and the Office of Electric Transmission and Distribution combined. Research and development programs within OE include: security and reliability of the transmission infrastructure, distributed energy, transmission grid modernization, energy storage, and superconductivity. New technologies related to the utility grid, including grid reliability and GridWise, are supported through this office. In addition, the distributed energy programs support distributed generation, combined heat and power, and thermally activated technologies.

DOE National Laboratories

To accomplish its mission, the DOE maintains several national laboratories and technology centers (www.energy.gov/organization/labs-techcenters.htm). Of the 13 DOE national laboratories, 11 are actively involved in EERE-supported research, development, and deployment activities (www.eere.energy.gov/site_administration/doe_labs.html). All publicly available reports published by DOE national laboratories are available through the National Technical Information Service (NTIS) at www.ntis.gov/search.

Each DOE national laboratory has specialty or focused areas of research, although there is considerable overlap in some areas. While the DOE national laboratories may be performing world-class research, their Web sites may not fully identify the breadth or depth of their knowledge and the information available. This is where nothing beats a good professional network or a (series of) well placed telephone calls.

Office of Scientific and Technical Information

At the DOE Office of Scientific and Technical Information (OSTI) (www.osti.gov), you can search for DOE research results and find out about ongoing research projects. OSTI makes R&D findings available to DOE researchers and the public.

U.S. Environmental Protection Agency

Under the U.S. Environmental Protection Agency (EPA), the Office of Air and Radiation, and the Office of Atmospheric Programs (OAP) is the Climate Protection Partnership Division (CPPD). The CPPD (www.epa.gov/cpd) promotes greater use of energy-efficient products and practices in the residential, commercial, and industrial sectors through the ENERGY STAR® partnership program (www.energystar.gov). The Division's programs, such as Climate Leaders (www.epa.gov/climateleaders), provide information, technical assistance, and recognition for environmental leadership to organizations as they develop strategies and take advantage of proven opportunities to reduce their greenhouse gas emissions. CPPD also encourages investments in efficient, clean technologies, such as combined heat and power (CHP) (www.epa.gov/CHP) and green power (www.epa.gov/greenpower) from renewable resources.

ENERGY STAR is a joint program of the EPA and the DOE. In 1992, the EPA introduced ENERGY STAR as a voluntary labeling program designed to identify and promote energy-efficient products to reduce greenhouse gas emissions. The ENERGY STAR label is now on major appliances, office equipment, lighting, home electronics, and more. EPA has also extended the label to cover new homes and commercial and industrial buildings.

[Note from the author: While I would trust the information on the ENERGY STAR Web site (www.energystar.gov) and information published by EPA and DOE; I would not trust any product or claim just because it had an ENERGY STAR label. ENERGY STAR is a voluntary program with a marketing foundation, and it is not difficult to find their logo being utilized by others to make unsubstantiated claims.]

U.S. Department of Commerce

The Department of Commerce is a very diverse organization within the federal government. Within the Department of Commerce, Office of Technology Administration, however, are two specific organizations an advanced energy manager should be aware of. These are the National Technical Information Service (better known as NTIS) and the National Institute of Standards and Technology (better known as NIST).

National Technical Information Service

The National Technical Information Service (NTIS) is part of the U.S. Department of Commerce's Technology Administration. NTIS' primary purpose is to assist U.S. industries to accelerate the development of new products and processes, as well as helping the United States maintain a leading worldwide economic competitive position. NTIS is the largest central resource for government-funded scientific, technical, engineering, and business-related information available today. They have information on more than 600,000 information products covering over 350 subject areas from over 200 federal agencies. To search the NTIS library for federal reports and documents published since 1990, go to www.ntis.gov/search/

National Institute of Standards and Technology

The National Institute of Standards and Technology (NIST) is a nonregulatory federal agency within the U.S. Commerce Department's Technology Administration. NIST's mission is to develop and promote measurement, standards, and technologies to enhance productivity, facilitate trade, and improve the quality of life. NIST consists of four primary programs. Of specific interest to energy efficiency are the NIST laboratories (www.nist.gov/public_affairs/labs2.htm) and the

Advanced Technology Program (www.atp.nist.gov). The NIST laboratories conduct research that advances the nation's technology infrastructure and is needed by U.S. industry to continually improve products and services. The Advanced Technology Program accelerates the development of innovative technologies for broad national benefit by cofunding R&D partnerships with the private sector.

Other Federal Agency Sources

As a result of legislation, every federal agency has an internal energy management organization and/or program working toward a series of energy-efficiency goals. In response to these demands, several federal agencies are performing in-house research on new and emerging energy-efficient technologies to identify those technologies that can best serve the goals of the agency. Some of this information is publicly available on agency Web sites (although it can be very difficult to find), while other agencies have found it necessary to keep the information on internal-access-only intranets.

Department of Defense, Fuel Cell Program

There are a number of energy-efficiency research and development programs within the U.S. Department of Defense (DOD), each serving specific DOD missions. While most of the energy-efficiency programs serve internal DOD objectives, the DOD fuel cell program (dodfuelcell.cecer.army.mil) has provided significant benefit to the overall advancement and commercialization of fuel cells. The DOD fuel cell program is managed by the U.S. Army, Corps of Engineers, Engineer Research and Development Center (ERDC), Construction Engineering Research Laboratories (CERL). The DOD fuel cell projects Web site provides information guides, demonstration results, and information on the climate change (fuel cell) rebate program.

STATE AGENCIES AND UNIVERSITY PROGRAMS

While the federal government has some very substantial research, development, and deployment programs for new and emerging energy-efficient technologies, several state agencies and university programs are also excellent resources and are often closer to technologies ready for commercial deployment.

While the resources are good, this presents a case where perspective, or bias, should be considered by the reader. State energy offices focus on the priorities of the state. Furthermore, climate factors can also skew results. Support infrastructure and product availability may also be regional. For example, a dehumidification technology demonstrated in Orlando will have significantly different findings in San Francisco. Similarly, sunny California is subject to both high electric energy and electric demand rates. Therefore, you may find considerable information supporting the technical appropriateness and cost effectiveness of a technology such as photovoltaic energy. The results will be significantly different for an application in the Pacific Northwest, where solar radiation has different technical factors and both energy and demand rates are notably lower. The author's point is that many factors can support the appropriateness of a new technology: end-use application, energy, peak demand, average demand, climate, location, and infrastructure are just some of these factors.

Almost every state in the United States has a state energy office. The experienced energy manager should become familiar with their local state energy office and the services it offers. In addition, there are several universities with energy-research programs. This section does not purport to be all inclusive; however, the following state and university programs are particularly noted for performing research and making publications available on new and emerging energy-efficient technologies.

California Energy Commission

The California Energy Commission (CEC) is one of the largest state energy programs. The CEC (www.energy.ca.gov) has program offices on energy efficiency, renewable energy, transportation energy, and significant research and development activities. Some CEC programs appear to be larger than their DOE counterparts.

Public Interest Energy Research

The California Energy Commission's Public Interest Energy Research (PIER) program creates state-wide environmental and economic benefits by supporting energy research, development, and demonstration (RD&D) projects that will help improve the quality of life in California by bringing environmentally safe, affordable, and reliable energy products and services to the marketplace. PIER includes a full range of RD&D activities that will advance science or technology not adequately provided by competitive and regulated markets. The PIER Web site is located at www.energy.ca.gov/pier.

While PIER is involved in RD&D, they are not involved—at least not directly—in the commercialization or deployment of technologies. In some cases, you may find that local utilities become involved with new technologies that achieve successful results from PIER demonstrations. For this reason, local utilities can be good sources for information. In California, the major utilities have collaborated to form the Emerging Technology Coordinating Council (ETCC). The ETCC is discussed in more detail later in this entry.

Energy Center of Wisconsin

The Energy Center of Wisconsin (www.ecw.org) is a private, nonprofit organization dedicated to improving energy sustainability, including support of energy efficiency, renewable energy, and environmental protection. The Energy Center of Wisconsin provides objective research, information, and education on energy issues to businesses, professionals, policymakers, and the public. The Energy Center's education department supports comprehensive energy education programs.

They offer hundreds of publications on a myriad of energy-efficiency topics and host a unique collection of industrial fact sheets and many other research materials. The Energy Center also has a comprehensive collection of energy-related library materials, many of which are available through their Web site.

Florida Solar Energy Center

The Florida Solar Energy Center (FSEC) is a state-supported renewable energy and energy-efficiency research, training, testing, and certification institute. FSEC (www.fsec.ucf.edu) is a research institute of the University of Central Florida (UCF) and functions as the state's energy research, training, and certification center. FSEC's mission is to research and develop energy technologies that enhance Florida's and the nation's economy and environment, and to educate the public, students, and practitioners on the results of the research. FSEC has gained national and international recognition for its wide range of research, education, training, and certification activities. FSEC maintains a wide variety of technical articles, research reports, newsletters, and public information documents. Their joint-use research library is reported to be one of the largest repositories of energy publications in the nation.

Iowa Energy Center

The Iowa Energy Center (www.energy.iastate.edu) invests its resources to create a stable energy future for the state of Iowa. The Iowa Energy Center works quietly and steadily, producing dividends that support Iowa communities, businesses, and individuals. Most of the Iowa Energy Center's activities are spent on energy-efficiency research, demonstrations, and education projects. These projects address energy use in agriculture, industry, commercial businesses, municipalities, and residential areas. The Iowa Energy Center is constantly at work seeking opportunities to improve the total energy picture of the state, its businesses, and communities.

The National Building Controls Information Program (www.buildingcontrols.org) was established by the Iowa Energy Center with support from the EPA to facilitate the adoption of energy-efficient building control products and strategies through testing, demonstration, education, and dissemination of product information. A special Web site, DDC Online (www.ddc-online.org), was developed by the Iowa Energy Center and is reported to be the most complete, unbiased listing available of direct-digital controls (DDC) vendors and information related to DDC systems.

New York State Energy Research and Development Authority

The New York State Energy Research and Development Authority (NYSERDA) is a public benefit corporation created by the New York state legislature. NYSERDA (www.nyserda.org) has been cited by the DOE as being among the best government research organizations in North America. NYSERDA's principal goal is to help all New York state utility customers solve their energy and environmental problems while developing new, innovative products and services that can be manufactured or commercialized by New York state firms, through which NYSERDA supports a wide assortment of research and development in energy-efficient technologies. Many of NYSERDA's technical reports are available through their Web site (www.nyserda.org/publications). In addition, many of NYSERDA's reports are available from NTIS.

Rensselaer Polytechnic Institute, Lighting Research Center

The Rensselaer Polytechnic Institute, Lighting Research Center (LRC) is a university-based research center devoted to lighting (www.lrc.rpi.edu). The LRC is a reliable source for objective information about lighting technologies, applications, and products. The LRC provides training programs for government agencies, utilities, contractors, lighting designers, and other lighting professionals. LRC programs cover a range of activities, including both laboratory testing of lighting products and real world demonstration and evaluation of lighting products and designs. They conduct research into energy efficiency, new products and technologies, lighting design, and human factors issues.

Washington State University, Energy Programs

The Washington State University (WSU) Energy Program (www.energy.wsu.edu) is a self-supported department within the university's extension service. The WSU Energy Program provides support to other national and local energy organizations, such as EnergyIdeas.org. The program has developed a significant library of materials on new and emerging technologies.

EnergyIdeas.org provides comprehensive information about commercial, industrial, and residential energy efficiency. You will find actual customer questions with detailed answers, case studies, reports, and articles on topics ranging from appliances and lighting to motors and solar energy. The EnergyIdeas Web site is sponsored by the nonprofit Northwest Energy Efficiency Alliance (www.nwalliance.org) and it is managed by the WSU Energy Program.

More on University Programs

There are several universities that operate energy management research programs. Each major university energy program will have specialty or focused areas of research. Unfortunately, there does not appear to be a composite index identifying the universities and their points of contact. Furthermore, while the university programs may be performing world-class research, their Web sites may not fully identify the breadth or depth of their knowledge. This is where nothing beats a good professional network or a well-placed telephone call.

MEMBERSHIP ORGANIZATIONS AND ASSOCIATIONS

There are a number of membership associations. Some membership organizations are associations of companies, while others allow individual membership. Both types of organizations add value to the field of energy management. In addition to supporting research, training, and publications, several of these organizations can also respond to special inquiries from members. The following is a nonexhaustive list of membership associations and other associations that may be useful to members in obtaining information on new and emerging energy-efficient technologies.

Alliance to Save Energy

The Alliance to Save Energy (ASE) (www.ase.org) is a nonprofit coalition of business, government, environmental, and consumer leaders. The Alliance to Save Energy supports energy efficiency as a cost-effective energy resource under existing market conditions and advocates energy-efficiency policies that minimize costs to society and individual consumers, and lessen greenhouse gas emissions and their impact on the global climate. To carry out its mission, the Alliance to Save Energy undertakes research, educational programs, and policy advocacy; designs and implements energy-efficiency projects; promotes technology development and deployment; and builds public–private partnerships in the United States and other countries. Although the ASE is very policy-oriented (and, therefore, very politically oriented), it does support several technical programs.

American Council for an Energy-Efficient Economy

The American Council for an Energy-Efficient Economy (ACEEE) (www.aceee.org) is a nonprofit organization dedicated to advancing energy efficiency as a means of promoting both economic prosperity and environmental protection. ACEEE conducts in-depth technical and policy assessments, organizes conferences, and publishes books and reports. While ACEEE is not a membership organization, they do send out notices of publications, conferences, and other activities. ACEEE sponsors several events, but they are probably most noted for the biennial Summer Study on Energy-Efficient Buildings. Recently, ACEEE has also been organizing a biennial Summit on Emerging Technologies in Energy Efficiency.

Consortium for Energy Efficiency

The Consortium for Energy Efficiency (CEE) (www.cee1.org) is a nonprofit, public benefits corporation that promotes energy-efficient products and services. CEE develops initiatives for its members to promote the manufacture and purchase of energy-efficient products and services. CEE's goal is to induce lasting structural and behavioral changes in the marketplace, resulting in the increased adoption of energy-efficient technologies.

E Source

E Source Companies LLC (www.esource.com) is a membership-based information services company. E Source information services provide member organizations with unbiased, independent analysis of retail energy markets, services, and technologies. According to their Web site, E Source serves as a high-value filter of the torrent of information on developments in the energy services marketplace, sorting through the hype and providing clients with concise strategic insights and in-depth technology assessments. The E Source core products and services are offered through an integrated package of membership benefits. Additional services for energy service providers, including multiclient studies and focused research services, may be purchased.

Professional and Trade Associations

There are a number of professional and trade membership associations. Professional and trade associations provide their members benefits such as journals, publications, and training, as well as a professional network. Some organizations also sponsor certification programs, conferences and trade shows, technical research, or other benefits. The following is a nonexhaustive list of professional and trade membership associations with significant ties to advancing the field of energy management through the research, development, deployment, assessment, and/or the use of new and emerging energy-efficient technologies.

- Association of Energy Engineers (AEE), www.aeecenter.org
- American Institute of Architects (AIA), www.aia.org
- American Society of Heating, Refrigerating, and Air-Conditioning Engineers (ASHRAE), www.ashrae.org
- American Society of Mechanical Engineers (ASME), www.asme.org
- California Commissioning Collaborative (CCC), www.cacx.org
- Fuel Cells 2000, www.fuelcells.org
- Gas Appliance Manufacturers Association (GAMA), www.gamanet.org
- Geothermal Heat Pump Consortium, www.geoexchange.org
- Geothermal Resources Council (GRC), www.geothermal.org
- Illuminating Engineering Society of North America (IESNA), www.iesna.org
- Institute of Electrical and Electronics Engineers (IEEE), www.ieee.org
- Institute of Industrial Engineers (IIE), www.iienet.org
- International Facility Management Association (IFMA), www.ifma.org
- International Ground-Source Heat Pump Association (IGSHPA), www.igshpa.okstate.edu
- National Council on Qualifications for the Lighting Professions (NCQLP), www.ncqlp.org

UTILITY ENERGY CENTERS AND RESEARCH PROGRAMS

Many if not most utilities perform research on new and emerging energy-efficient technologies. However, much of this research is designed to support in-house programs. On the other hand, some utilities work aggressively with their customers to research, demonstrate, and deploy these new technologies. Even if the information is not made publication-ready, the experienced energy manager will benefit from establishing an active and on-going relationship with local utility customer account-service representatives. Through the account representatives, the energy manager will learn of the utility's involvement in new and emerging technologies and can benefit from the lessons learned.

California Statewide Emerging Technology Program

The statewide Emerging Technologies (ET) program seeks to accelerate the introduction of "near market-ready" energy-efficiency innovations that are not widely adopted by utility customers in California. Four investor-owned utilities (Pacific Gas and Electric, San Diego Gas and Electric, Southern California Edison, and Southern California Gas) and the California Energy Commission (CEC) are working together cooperatively to pool resources and knowledge for project selection and results dissemination.

The ET program consists of two main components: (1) demonstration and information transfer and (2) participation in the ETCC. The demonstration component provides technology assessments and information to utility customers and industry, often in the form of technology demonstrations at customer facilities. The ETCC coordinates activities and information among the utilities. The ETCC also maintains a Web site listing projects and summary results in its database (www.etcc-ca.com).

The ETCC is charged with administrating California utility, ratepayer-funded programs for energy-related research and energy-efficient emerging technologies. The ETCC coordinates among its members to facilitate the application of energy-efficient emerging technologies that will transform the market and benefit California ratepayers.

The Emerging Technologies project database contains a compilation of technologies and applications that are part of either current or recent projects that council members have sponsored. The database also contains information about the technologies, their applications, and the resulting assessments from member projects.

Energy Design Resources

Energy Design Resources (www.energydesignresources.com) offers energy design tools, software, design guides, case studies, and more. Their goal is to educate architects, engineers, lighting designers, and developers about techniques and technologies, making it easier to design and build energy-efficient commercial and industrial buildings in California. Energy Design Resources is funded by California utility customers and administered by Pacific Gas and Electric Company, Sacramento Municipal Utility District, San Diego Gas and Electric, Southern California Edison, and Southern California Gas under the auspices of the California Public Utilities Commission.

Lighting Design Lab

The Lighting Design Lab (www.lightingdesignlab.com) works to transform the Northwest lighting market by promoting quality design and energy-efficient technologies. The Lighting Design Lab accomplishes its mission through education and training, consultations, technical assistance, and demonstrations. The Lighting Design Lab is sponsored by the Northwest Energy Efficiency Alliance, Seattle City Light, Puget Sound

Energy, Snohomish County Public Utility District, British Columbia Hydro, and Tacoma Power.

Sacramento Municipal Utility District, Customer Advanced Technologies Program

The Sacramento Municipal Utility District (SMUD) Customer Advanced Technologies (CAT) program is a research and development program designed to encourage customers to use and evaluate new or underutilized technologies. According to one presentation, CAT provides the following benefits: it helps customers sort fact from fiction, identify the most promising technologies through direct, first-hand experience, and avoid making major investments in technologies that do not work. Unlike many R&D programs, research is accomplished through implementing real world demonstration projects (instead of laboratory testing). Completed demonstration projects include lighting technologies, light emitting diodes (LEDs), building envelopes, heating ventilation, and air-conditioning (HVAC) systems, as well as a wide variety of other technologies. Reports describing the results for many of these projects are available through their Web site (www.smud.org/education/cat/index.html).

Southern California Edison, Energy Centers

Southern California Edison (SCE) operates two energy centers (www.sce.com/RebatesandSavings/EnergyCenters)—the Agricultural Technology Application Center (ATAC) and the Customer Technology Application Center (CTAC). Both facilities offer hands-on demonstrations of the latest state-of-the-art technologies, as well as workshops, classes, and interactive displays.

Southern California Gas Company, Energy Resource Center

The mission of the Energy Resource Center (ERC) is to serve as a one-stop "idea shop" where customers can find the most efficient, cost-effective, and environmentally sensitive solutions to their energy needs. The ERC showcases innovations in resource-efficient designs, materials, and equipment to help businesses make informed choices about energy consumption and conservation. For more information, see www.socalgas.com/business/resource_center/erc_home.shtml.

OTHER SOURCES OF INFORMATION

The following organizations are worthwhile sources that did not fit into any of the previous categories.

Centre for Analysis and Dissemination of Demonstrated Energy Technologies

The Centre for Analysis and Dissemination of Demonstrated Energy Technologies (CADDET) is an international information source that helps managers, engineers, architects, and researchers find out about renewable energy and energy-saving technologies that have worked in other countries. Along with its sister program, GREENTIE (www.greentie.org), CADDET ceased collecting new information at the end of March 2005. Nevertheless, the information currently remains available through the Web site (www.caddet.org). CADDET's objective was to enhance the exchange of information on new, cost-effective technologies that have been demonstrated in applications such as industry, buildings, transport, utilities, and agriculture. The information was not only collected and disseminated to a very wide audience; it was also analyzed to provide a better understanding of the benefits of the technologies.

Portland Energy Conservation, Inc.

Portland Energy Conservation, Inc. (PECI) considers itself passionate about energy efficiency. Their mission is to help everyone use energy more effectively. PECI assists clients in the promotion of energy-efficient practices and technologies that benefit both businesses and individual consumers. PECI helps clients deliver long-term energy savings by helping transform markets through education and incentive programs that build demand for more efficient products and services. PECI's Web-based library (www.peci.org) offers considerable information on commissioning, as well as operations and maintenance.

CAVEAT EMPTOR

It is important to remember that the Internet is a tool and not a source. Almost anyone has the capability to make information available through the Internet. Furthermore, not every article in a periodical or presentation at a conference undergoes rigorous peer review. Every source has an intent or purpose when making information available. It is important to understand the source to determine potential bias (or perspective) and the reliability of the information. Even if the intent is altruistic, the reader needs to be aware of the qualifications, or basis, of the source, as well as any potential bias.

While each of the organizations identified in this entry provide publications or other informational sources about new and emerging energy-efficient technologies, as with all things new, there are risks. Some risks are minimized by evaluating new technologies on a limited basis before they are deployed on a larger scale. Demonstrations and pilot projects allow users to "look before you leap" when applying new technologies.

Many of the organizations identified in this entry are doing their part to get information on new and emerging energy-efficient technologies into the hands of energy managers, design engineers, building owners, and others, and they are encouraging them to consider the new technologies so that they may save energy, reduce costs, reduce emissions, or achieve other objectives.

The reader should also understand that most of these organizations are not making guarantees and implying endorsements. For example, those who claim endorsements from federal agencies are misleading the public. Furthermore, publications provided by these organizations are not substitutes for sound engineering or due diligence on the part of the reader. These programs do strive to be accurate and responsible. Remember, their objectives are to help.

New and emerging technologies are also subject to change. It is important to note the date of the publication or other information source. Publications usually offer a snapshot of a technology at a given time. As time passes, technologies, costs, maintenance recommendations, and even manufacturers change. In many cases, this can be a good thing; manufacturers can make numerous improvements to the equipment as more is learned about the technology and its operation. However, there may be other changes. New manufacturers may develop a product line that may not have been in that business when the report was published. Other manufacturers, who were known at the time of publication, may relocate, merge, consolidate, drop the product line, or even go out of business. Readers must continue to do their homework.

Energy managers and facility staff need to be wary. *Caveat emptor*, let the buyer beware, definitely applies. Readers are encouraged to check the original sources and follow up with other users. Do not rely on third-hand sources (for example, a government report should be checked out from a government source and not from a second- or third-party distributor).

CONCLUSIONS

Energy management is a diverse and growing field. As a result, there are many more sources available to the energy manager beyond those identified in this article. The organizations identified in this article are simply the ones this author turns to first when seeking useful information about an emerging or unknown technology. Many organizations are involved in innovation, research, and development. Many other firms are involved in assessment, experimentation, validation, training, and other forms of deployment. The energy manager can benefit from each of these organizations' contributions.

Technologies are continuously evolving. Efficiencies are improving, controls are improving, and maintenance is being simplified. New and emerging energy-efficient technologies can be used to help the experienced energy manager achieve and surpass their goals. Trying something new may not be risk-free, but these programs are doing their part to assist experienced energy managers.

Disclaimer

The information contained in this article is based on the author's 25 meandering years in energy management. The information does not purport to be comprehensive or complete, but is provided in the hope that it may be of some use to the reader. Any source provided in the entry should not be interpreted as an endorsement, just as any source not provided in this entry should not be considered a negative endorsement. If you do not want to use the information contained in this entry in the spirit in which it is intended, then do not use it. The reader should also be aware that there is a difference between 25 years of experience and 1 year of experience 25 times over. The author is unsure where he fits in the spectrum.

ACKNOWLEDGMENTS

The Pacific Northwest National Laboratory is operated for the U.S. Department of Energy by Battelle Memorial Institute under contract DE-AC05-76RL01830.

Energy Efficiency: Low Cost Improvements

James Call
James Call Engineering, PLLC, Larchmont, New York, U.S.A.

Abstract
In most energy efficiency initiatives, one is blessed with a number of energy-saving projects, all competing for a company's limited corporate resources. The various projects are typically interrelated; one of them, if implemented, would impact the others. They have a wide range of implementation costs and related energy-saving impacts—that is, they have different payback periods. If the quickest payback projects are done first, their savings will be unavailable to "subsidize" long payback projects. Thus, the long payback projects will have to fend for themselves in their fight for capital budget allocations. The end result may free up capital for other company projects in completely different areas.

This is why low-cost/no-cost energy projects are so valuable: they not only save money in their own right, but also set up a level playing field for the ongoing prioritization of corporate resources generally. Fourteen low-cost/no-cost projects from actual field experience are described as examples of the kind of projects that can introduce big savings up front and, thus, change the business case for long payback projects subsequently considered.

BACKGROUND

Longer-Term Consequences of Implementing Low-Cost Options First

If low-cost, quick-payback solutions are bundled in with longer-payback projects, the quick-payback projects can inadvertently subsidize the longer-payback projects that overlap on some of the benefits.

Consider a case in which an $1800, 12-channel "smart" time clock (holidays, sun tracker, outside air temperature sensor, etc.) would reduce energy costs by 25% and have a 2-mo payback. Alternatively, an $80,000, full-featured Energy Management System (EMS), if implemented alone, would reduce energy costs by 30% and have a 5-year payback. The issue is that if the 12-channel smart time clock is installed first, it co opts the "savings" from the EMS project, stretching the EMS payback to perhaps 20 years. This is because the EMS also has a smart time clock inside it, as one of its many features, and that single feature was mostly why the EMS seemed attractive in the first place.

It generally will be impossible to develop a fair and accurate business case for a longer-payback system if low-cost, quick-payback solutions are "reserved" to bundle in with them.

Put another way, mixing in low-cost, quick-payback approaches to sweeten the case for longer-payback projects is analogous to improving a $5 bottle of wine by mixing it with a $90 bottle—a strategy perhaps to be avoided ambitiously.

Corporate Payback Analyses Should Include All Company Opportunities

In asking energy projects to compete for corporate resources (i.e., in comparing payback periods), companies will generally want to compare all projects, not just other energy projects. A proposed energy savings project, for example, should undergo comparison with projects in other departments, such as Advertising and New Product Development.

Maintaining or Improving Comfort and Safety Must be Addressed in All Analyses

In all projects, of course—low cost, no cost, or otherwise—employee comfort and safety, and compliance with applicable codes, must be attended to. Such requirements obviously should have the final say in all project decisions and implementations.

EXAMPLES FROM THE FIELD

Safety First

Any implementation project must put safety first. The engineering, design, and implementation should be done by qualified professionals only and with the greatest attention to safety.

Following are summaries of a number of low-cost/no-cost opportunities that have been encountered in actual

---

Keywords: ASHRAE 12; ASHRAE 62; Comfort and safety; Cost accounting feedback; Open loop; Payback period; Prioritize.

energy cost reduction initiatives. These are the kinds of opportunities that frame the issues in comparing payback periods and prioritizing competing projects.

Replace Lugs on Time Clocks to Reestablish Night Setback

Surprisingly, many buildings with installed time clocks have their OFF lugs removed and, thus, let the HVAC systems run 24/7. They were removed for reasons that probably seemed compelling, or at least expedient, at the time.

In a 100,000-ft^2. office building, for example, employees working late at night or on weekends had complained some years previously that the HVAC was not on and that it was too cold, or too hot. This was because the building was in the scheduled setback mode during off hours. As an expedient solution to this issue, the Building Services Department simply removed the OFF lugs on the 7-day time clock. Because energy expense was added in with numerous other overhead expenses and then charged back to about 20 departments on a square-foot basis, the sudden increase in energy costs went unnoticed for a number of years.

To reclaim the lost savings from the abandoned night and weekend setback, a low-cost solution was implemented in the following steps:

1. Replace the OFF lugs on the master time clock to run the HVAC 6:00 a.m. to 7:00 p.m. 5 days a week.
2. Install latching relays across the existing time clock's low-voltage outputs for the 12 building zones of interest, and wire their latch trigger circuits to user-accessible pushbuttons.
3. Provide a labeled "red button" in an accessible place (by the copy machine) in each of the 12 zones, and let employees know that if they come in to work during off hours, they should simply push the button, and their zone will come on for up to 6 h. Later, if the zone shuts OFF, employees should push the button again to turn on the zone for 6 more hours.
4. Provide a reset to the latch coil circuit by putting a $100 time clock in series with the latch current that turns OFF 1 min every 6 h.
5. Provide high-temperature and low-temperature "night setback" thermostats, which would ensure that the space did not get extremely hot in summer or so cold as to freeze pipes or fan coils in winter.

This solution cost under $1000, was implemented in 1 month, and saved more than $30,000 per year, giving it a nominal payback period of a couple of weeks.

Fig. 1 is a photo of the relay logic as actually constructed, and Fig. 2 shows the pushbutton setup that triggers the temporary comfort cycle.

Fig. 1 Relay logic to cycle certain zones, temporarily, with timed-out latch release.

Tighten Schedules When Possible

Building systems may be set on an automatic schedule that, upon review, can be substantially tightened without affecting employee comfort. A good way to determine the optimum schedule is to make a manual count of cars in the parking lot—say, every 15 min around the beginning and end of the business day. A rough hand-drawn plot of arrivals per quarter-hour period will indicate when the space should be ready to satisfy, say, 98% of the people. The optimum "be ready" time likely will not occur at an even quarter-hour (e.g., 7:30 a.m.), and if the existing schedule is in fact set at such even times (as most are), there is probably an opportunity to reduce energy usage.

Data loggers are invaluable in determining the warming and cooling curves of a building. The curves generally have the characteristic shape of the classical capacitor charge and discharge (i.e., an exponentially decaying rise or fall toward a final value).

A longer lead time may be needed on Monday mornings, of course, or any time when the building levels have drifted greatly during the off hours. Given the usual 30–90 min that most spaces need to achieve the target temperature, conditions would still be fairly comfortable, plus or minus 15 or 20 min of the actual target time.

Update Schedules to Reflect Current Building Use

During a 1:00 a.m. visit to a commercial office space as part of an energy audit, it was noticed that a large wing of the building was still running the HVAC system, even though the rest of the building had gone into night setback, as was expected, from the time clock.

Upon investigation, it was determined that the time clock did indeed have the zone still ON until 2:00 a.m. each weekday night, while all the other zones shut OFF at 7:00 p.m. It turned out that the zone still running formerly

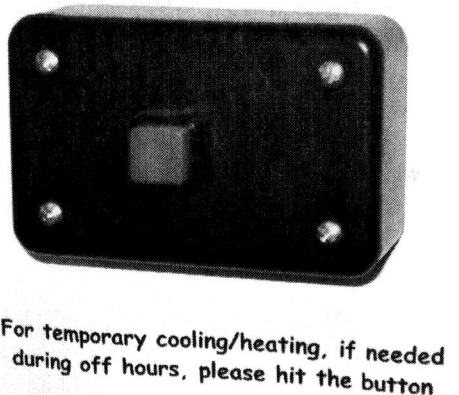

Fig. 2 User pushbutton to start temporary comfort period.

was used by a second shift in a data entry department that finished at 2:00 a.m. That was no longer the case, but the time clock had never been reset when the late shift was discontinued. It had been more than 4 years since the late shift had used the space.

Because the cooling tower and the entire chiller plant were interlocked to run when any zone called for cooling, resetting the time clock to reflect the current schedule saved an estimated $8000 per summer.

Shut Off Boiler in Summer by Installing a Dedicated Domestic Hot Water Heater

Domestic hot water for a company's 10 bathrooms was provided by a loop in the building's main boiler running through a heat exchanger. Thus, the main boiler was kept on all summer, at a cost of some $25,000 per year for gas.

The solution was to install a 60-gal dedicated hot water tank for domestic hot water and to shut off the main boiler during the summer. The dedicated tank was fitted with a mixing valve, expansion tank, and pump to address ASHRAE 12[1] requirements.

The improvement took 2 months to implement, cost some $4000, and had a payback period of less than 3 months.

Reduce Gross System Overcapacity by Shutting off Equipment as Appropriate

It is not unusual to find gross overcapacity in HVAC systems. The reasons behind this appear to include

- "Rounding up" in each serial design step when the building was originally built. That is, installed overcapacity may have resulted from the cumulative effect of what the building owner initially saw as the load, then the design engineer's safety factor, and finally installing 7.50-ton units when the nominal calculation showed 6.74 ton.
- Building use changes since the original design.
- Accumulation of ad hoc package systems additions, over the years, to meet special needs.

The following two examples illustrate low-cost/no-cost approaches that were used to reduce overcapacity:

a. *Retiring a Package Unit and Attaching the Served Ductwork to a Nearby Remaining Unit*: A company's office space had three package air conditioners, each with dual 5-ton compressors. Theoretical calculations, which were confirmed by actual tests using loggers, determined that the installed cooling capacity was more than 3 times the required amount for cooling comfort on the hottest day of the year. Moreover, one of the units had been scheduled for replacement due to its age and unreliability.

To eliminate the need for the unit's replacement and reduce electric demand at the same time, the following steps were implemented:

1. Confirmed by field measurements that there would be sufficient air volume from the other two good units if the old unit were retired.
2. Decommissioned the old unit and relocated its discharge ductwork over to one of the adjacent remaining good units.
3. Rebalanced the served space.

The result was a savings of more than $12,000 of averted costs to replace the old unit, as well as a 20% drop in electric demand.

b. *Run Only One Compressor in Two-Compressor Package Units*: A 90,000-ft^2. office building had 30 rooftop package units, each controlled by a local thermostat in the related zone. Calculations, confirmed by loggers, indicated that the building as a whole had well over twice the cooling capacity needed even for the hottest day of the year. Twenty of the 30 rooftop units had 2 (hermetically sealed) compressors, typically 5 or 7.5 ton each. In their related zones, two-stage T-stats would bring on the first and then the second compressor as needed. On morning startup or on very hot days, however, the T-stats would bring the second compressor on, even though the general cooling overcapacity in the building would have eventually (in about 15 min) taken care of the entire load without the second compressor's being needed. That is, the two-stage T-stat was unduly "impatient."

The overcapacity was corrected as follows:

- To reduce the excess capacity, the second-stage low-voltage control wire was simply cut, and the open ends were taped off. The cut was made about 1 in. back from the screw terminal connector, leaving an "audit trail" for future service personnel in case anyone ever wanted to reinstate the redundant disconnected compressor. Indeed, the "extra" compressor then became a potential spare unit that could easily replace the other compressor should it ever fail.
- It can be appreciated that each modified roof top unit would be left running more air across both the condenser coil and the DX coil than was actually needed, but even so, the improvement in occupant comfort and electric demand reduction was dramatic. Fig. 3 shows the two-compressor unit with the cover removed.

Fig. 3 Two-compressor HVAC package unit where one compressor was not needed.

Resolve Systems Fighting Each Other

Days in the spring and fall present probably the most challenging building-comfort control issue. Building managers sometimes cope by running both heating and cooling systems at the same time, sometimes for weeks. This situation almost always presents significant energy saving opportunities, which may be hidden if the heating and cooling systems fight each other to a comfortable standoff.

Approaches to break the standoff include automating the summer–winter cutover. Relatively inexpensive boiler controls use inputs from outside air temperature and/or the space ambient temperature to decide when to go into so-called warm-weather shutdown. Setting the chiller's startup temperature a few degrees higher than the boiler's warm-weather shutdown point will create a useful dead band (i.e., a gap in temperature between shutting down the boiler and starting the chiller).

The dead-band concept leads to significant energy savings, because neither active heating nor active cooling is going on.

Relocate Thermostats to the Areas That They Serve

A company renovated its space by building out new offices but leaving the dropped ceiling unchanged. After the renovation, HVAC diffusers in the ceiling served some areas totally walled off from their controlling thermostats. As a result, certain spaces became grossly overcooled, with their ambient temperature dropping as low as 62°F (in the summer) because there was no way to close the feedback loop from the T-stat to the HVAC unit.

A project to relocate T-stats to their controlled spaces was completed for less than $500. The payback was estimated at about 2 months, although the real payback was improved employee comfort.

Reduce Lag/Lead Triggered by Thermostat Cases

To prevent tampering, vented steel boxes (which are commonly sold for this purpose) had been installed over thermostats in a company's office space. Even so, there were occupant complaints of being too hot and too cold.

By placing small data loggers inside and outside the metal boxes, and then comparing their readings on the same chart, it was discovered that the space ambient temperature was oscillating wildly around the set point. The following scenario was identified:

1. On summer-morning startup, the HVAC unit would cool the space to the comfort-level set point of 73°F in 20 min.
2. Because the metal box cover was initially warm and had considerable thermal mass, however, it took about 40 min for the space temperature signal to reach 73°F.

3. This introduced a 20-min lag in the feedback loop, thus allowing the HVAC to overshoot grossly. The space temperature would drop to 62°F by the time the metal cover and the enclosed thermostat were down to 73°F.
4. At this point, the HVAC would cycle off, but the process would begin again in the opposite direction as the room heated. The metal box kept the thermostat cool until the room was nearly 80°F.

This energy-saving and comfort-improvement opportunity was addressed simply by removing the metal boxes and using stop lugs (which were found already available inside the thermostat) to limit manual tampering to plus or minus 2°F. Then the T-stat, with its quick time constant thus restored, could control the room's ambient temperature accurately.

Use Small Dedicated AC Units to Cool Server Closets, Rather Than Run the Whole Chiller Plant

A company scheduled an HVAC operation in its 90,000-sq.-ft. building during normal business hours. The HVAC system included a 300-ton chilled-water plant supplying some 14 air handlers.

After the building was wired for local area network (LAN) access, two network server closets (small rooms, actually) were provided for LAN infrastructure equipment; these closets required air conditioning continuously. At that point, the company began to run the two air handlers serving the equipment areas on a 24/7 basis. This also meant, of course that the 300-ton chiller plant and associated pumps would be on continuously. Moreover, because only 2 of the 14 air handlers were running, the 300-ton chiller plant tended to short cycle, even after the compressor controls had unloaded as many cylinders as they could.

To address the energy saving opportunity of shutting off the chiller plant and its associated large pumps, a dedicated 2.5-ton AC unit was installed in each server room. These were "split mini" units (air handlers wall mounted and cooled by a DX coil).

Total installed cost (the two units themselves, plus installation labor for both) was less than $12,000. Because the units removed the need to cool two entire building zones during weekends and off hours, however, the payback period was estimated at less than 6 months. Fig. 4 shows a typical wall-mounted air handler, the DX side, as installed.

Tighten Leaky Outside Air Dampers

Logger data from morning warmups taken in each of 7 zones (relating to the 7 air handlers in a company's building) showed that 5 of the zones reached comfortable levels in 40 min, whereas 2 of the zones took nearly twice as long. When the warmup curves were plotted on the same graph, distinctly different slopes were shown. It was appreciated that the original capacities of the air handlers might be different for each area, of course, but CO_2 monitoring also showed much lower levels of CO_2 concentrations in the slow-warming spaces even though occupancy density was about the same.

These findings suggested that the outside air dampers in the two long-warmup zones were not closing or were open way above the normal minimum for fresh air.

Subsequent inspection showed that in one of the suspect zones, the damper actuator linkage was slack, and in the other zone, the actuator shaft had broken off inside the air handler.

Repairs were made in less than 2 h, and the heating bill dropped by about 8%, giving the project a payback period of less than 1 week.

Reduce Fan Speeds by Adjusting Variable-Speed Pulleys When Appropriate

Reducing fan speeds (consistent with ASHRAE 62[2] and building comfort requirements) is a common energy saving opportunity. Although it can be done, when applicable, by retrofitting with variable-speed drives, a lower-cost approach is simply to adjust the variable-speed pulley that often already exists on the drive motor shaft.

Pulley adjustment produces a fixed reduction, as opposed to the wide range of variable speeds provided by new variable-speed drives, but the savings impact vs the difference in cost between the two approaches often suggests the low-cost pulley adjustment as at least a first step. Then the business case can be evaluated to see whether there are enough savings left to justify new drives.

Reset on Ambient Air Temperature When Appropriate

A company's HVAC VAV system was designed to provide a single fan discharge temperature of 60°F and

Fig. 4 Dedicated AC air-handler DX unit in computer server room.

Fig. 5 Use of T-fitting to pick off fan discharge temperature signal to control fan coil valves.

then rely on the VAV box in each zone to control the space by releasing more or less of this fixed-temperature air (with a preset maximum and minimum amount). This is very near to running the system "open loop" except for the (considerable) action of the VAV boxes.

There were days and conditions during occupied periods, however, when no zone in the entire building needed 60°F air—when all zones became cold on a winter morning, for example. (This discussion assumes that the building had already undergone a warmup cycle where all VAV boxes and fan coil hot water valves were 100% open.)

An energy-saving and comfort-improvement solution is to control fan discharge air temperature based on ambient air temperature. Under this approach, there are two complementary feedback loops in play; the return air T-stat informs fan discharge air temperature, and the zone T-stats control the VAV boxes. The savings result from the fact that the VAV boxes will, on average, not need to try so hard to control the space, because they will have more appropriate air. Thus, they will be more closed (i.e., expelling less air and, thus, reducing fan horsepower). This assumes that fan volume is reduced by throttling or by lowering fan speed.

Fig. 5 shows how the pneumatic signal for Fan Discharge Temperature was picked off simply by tapping into the line to the Fan Discharge Temperature panel gauge line. The signal was routed to the existing Receiver-Controller.

Optimize the Strategy for Manual Light Switches

Short of installing occupancy sensors, companies have used a number of low-cost approaches that can save energy on lighting. They include

- Relocate light switches to the spaces they serve. Light switches sometimes are located far from the spaces they control—in some cases, on other floors. Understandably, this leads users to turn lights on, just to be sure, even when they will not be working near the lights. Consider relocating switches to obvious positions close to where users will enter and leave the affected space.
- Discourage users from turning on every light in the place when the first one arrives. When switches are arranged in banks of 4, 6, or even 8, there is a strong tendency of the first person who arrives each morning to turn all the switches on, even lighting a whole floor for an extra 1 or 2 h per day. By labeling the switches with the zones they cover (and/or relocating them as just described) and by suggesting to users that they turn on just what they need, important savings can be found.
- Leave a few selected fluorescent units on 24/7 as night lights. Understandably, users leaving a lighted space at night would not turn off lights if that left them in total darkness. An energy-saving solution is to provide lighting levels sufficient for a safe exit by identifying intermittently spaced fluorescent (or other approved energy-efficient) lights to be wired on 24/7. Although some energy will be used all the time, on balance, there can be savings from the larger majority of lights that can then be turned off. When adequate night lights are implemented, place friendly signage by key light switches reminding the last person out to shut off the lights.

Provide Actionable Cost-Accounting Feedback to Users

When the company's cost-accounting system inadvertently hides energy costs from users and managers, savings opportunities will invariably be missed.

A cost-accounting system with the following features can go a long way toward addressing this situation:

- Show both dollars and usage (gals, therms, kWh, kW, etc.) on the same report.
- Separate the impact of varying prices from the impact of varying usage.
- In a column adjacent to actual dollars and usage, display dollars and usage adjusted for weather, number of days in the billing period, and/or changes in the space served. The impact of weather can be estimated by setting up a regression model in which at least one of the independent variables is heating degree days or cooling degree days, thus allowing an apples-to-apples comparison over different periods.
- Track the overall building impact of energy-saving improvements by comparing results to a base year (e.g., the year before a formal energy initiative began).

Use a realistic but easy-to-understand measure, such as Btu per square foot.

CONCLUSION

This paper explored the consequences of implementing low-cost/no-cost energy-saving projects before considering longer-payback projects. A number of examples of such low-cost/no-cost projects and approaches were reviewed.

When a company lets energy projects compete for funds in terms of the whole company's needs, as opposed to considering only the impact on the energy or building services department, low-cost/no-cost approaches, done first, create a reordering of project priorities. Indeed, the success of low-cost approaches may even make subsequent investment in long-payback projects unwarranted.

REFERENCES

1. ASHRAE Guideline 12-2000. *Minimizing the Risk of Legionellosis Associated with Building Water Systems*, American Society of Heating, Refrigerating and Air-Conditioning Engineers, Inc.: Atlanta, GA.
2. ASHRAE Standard 62.2-2004 *Ventilation for Acceptable Indoor Air Quality*, American Society of Heating, Refrigerating and Air-Conditioning Engineers, Inc.: Atlanta, GA.

Energy Efficiency: Strategic Facility Guidelines

Steve Doty
Colorado Springs Utilities, Colorado Springs, Colorado, U.S.A.

Abstract
The intent of this article is to offer a convenient list of strategic guidelines to help steer new building designs toward energy efficiency. It is hoped that the project owner, interested in achieving greater than average savings, will provide this list to the design team as part of their project intent instructions. To encourage its use and acceptance, emphasis is placed on keeping the list at a manageable size, with only proven solutions.

Through a top-down approach, using this document as a convenient tool, building owners can set the expectations of energy frugality, create positive changes in building practices, and reward themselves with operational savings.

INTRODUCTION

The old saying "an ounce of prevention is worth a pound of cure" is very applicable to creating energy efficient buildings. The economic realities of retrofitting existing buildings usually result in only the worst ones getting attention, with the average ones left to their life cycle. By contrast, evaluating improvements to designs that exist only on paper has a much smaller economic barrier to change, that being the differential cost from the standard designs. This simple fact makes it easier to create a new efficient building than forcing an older one to be more efficient.

A one-size-fits-all approach is not expected to work for every area of the country, nor for every business segment. Design practices, standard of care, and things that make practical sense will vary by region along with the climate differences and the business it serves. The content of this document was originally prepared for building owners in the Colorado Springs area, and will need local amending for use in other areas like Miami, Florida (all cooling, no heating), and Alaska (all heating, no cooling), and especially in areas with persistent high humidity. Still, many of the concepts presented will be applicable to many areas. One referenced text[1] provides recommendations by climate zone, acknowledging the regional differences. Local adjustments aligned with the eight climate zones are suggested, but beyond the scope of this text. As a starting point, basic allowances have been made to adapt climate-dependent criteria to other areas as follows:

- For most humidity-dependent heating ventilating and air conditioning (HVAC) operations, a distinction is made for climates that are either below 65°F or above 75°F wet bulb temperatures. Interpolation and judgment for points in between is required.
- Evaporative cooling constraints are unique and presented very conservatively, erring on the side of occupant comfort. The climate distinction presented is outdoor air consistently below 52°F wet bulb and below 42°F dew point. (This will no doubt be debated by some designers in semi-arid climates. However, experience has shown that at moisture levels much above these values it becomes increasingly hard to *guarantee* comfort all the time. Since the choice to use evaporative cooling is a large fork in the road for the customer, presenting conservative parameters is deliberate and are sure to work. If evaporative cooling is chosen in "fringe" climates, a supplemental conventional cooling system integrated into the first stage of evaporative cooling is suggested).
- Glazing U-value limits are relaxed in areas where HVAC heating is not used. Note that this suggests double pane windows anywhere HVAC heating is used.
- Glazing shading coefficients are relaxed in areas where HVAC cooling is not used, but are otherwise strict. Solar loads drive peak electrical demands if nothing else.

These climate-dependent items will no doubt be a work in progress, and so they are marked with an asterisk (*) with suggested criteria in *italics* for convenience in editing. I did my best!

To the owner who may use this: not all of the suggestions will apply to all buildings, but many of them will. Don't worry that a number of these won't make sense in your particular building—but do ask the design team to

---

*This entry originally appeared as "Strategic Facility Guidelines for Improved Energy Efficiency in New Buildings" in *Strategic Planning for Energy and the Environment*. Reprinted with permission from AEE/Fairmont Press.

Keywords: Commissioning; Strategic facility guidelines; Sustainability; Internal rate of return; Simple payback; Energy transport; Energy budget; Maintenance activities; Optimization; Evaporative cooling; Comfort envelope; Integrated design; Guide specification; Downsizing; Test and balance; Thermal breaks; Simultaneous heating and cooling.

explain which measures do not apply, and why. Doing so will reinforce your intentions to achieve operational savings by smart design decisions up front, and motivate the design team to make a good effort to accommodate.

THE FACILITY GUIDE SPECIFICATION AND THE TOP-DOWN APPROACH

The approach used for this article is to tabulate a list of suggested "Dos and Don'ts," suitable for use in a facility **guide specification**. A guide specification is a hand-out document given to a design team at the beginning of a project to provide general instructions and owner preferences. The owner handing out the guidelines has more of an effect on the end result than the same suggestions made from a lone team member, hence the term **top-down**. These instructions are then integrated into the other governing documents and codes that eventually form the design. Traditional guide specifications, used by national accounts, campuses, and large facilities, spell out preferred manufacturers, acceptable types of piping, valves, light fixtures, pavement, etc. The concept of energy efficient guidelines in the owner's guide specification is a natural and overdue extension of an existing document. Even if the owner does not have a guide specification to add this to, the listed items in this document can be used in stand-alone fashion to serve the same purpose and provide the same benefit to the owner.

More Top-Down Features

Beginning at the point of use is a common theme for all energy engineers. In the case of the office building, stipulating temperature values can have a pronounced effect on energy use. The indoor temperatures become design parameters that calculations and equipment sizing hinge upon. A design that requires 72°F summer indoor conditions will be larger and use more energy than one designed for 76°F, and yet with proper attire and humidity control, 76°F is a very reasonable temperature. A 72°F design can save energy by turning up the thermostat, but a 76°F design will additionally conserve first-cost dollars. Many designers are reluctant to push the comfort envelope for fear of reprisal against them, and that is where the top-down support concept comes in. With management support, and some education and encouragement, paradigms can slowly change.

Integrated Design

With more creative design solutions such as **integrated design**, more options are apparent to design teams now than in the past. This concept acknowledges cases where decisions in one trade (e.g., electrical or architectural) benefit another trade (e.g., mechanical). It may seem odd, but traditional design practices are compartmentalized by trade, and haven't always allowed the big-picture benefits to find their way to the owner. This is partly due to traditional fee structures where design fees are influenced by the cost of construction; the resistance to downsizing anything becomes obvious in this light. The integrated design process allows an extra expense in one trade to be considered if it produces a corresponding savings in another trade. To avoid redesign costs, this is best implemented with emphasis on group schematic designs (to air out the ideas early), as well as requiring good cost-estimating skills on the design team. A good architect is needed who can be receptive to new ideas, flexible, encourage the integration process, and also know enough to call a halt to the options at some point to maintain a reasonable schedule and finish the job without escalating design fees. By carefully selecting the design team members, constructive opportunities will come from this process which are usually well worth the effort. Examples of integrated design concepts follow, many of them revenue-neutral with sustained energy benefits for "free."

- The added cost of high performance suspended film windows can be offset in some climates by the elimination of perimeter fin-tube heating.
- Downsizing cooling systems in conjunction with reduced lighting power design.
- Downsizing cooling systems in conjunction with improved window shading coefficients, films, or exterior shading systems.
- Downsizing heating and cooling systems by using 1% or 2% ASHRAE design outdoor weather conditions[2] instead of 0.4%, allowing the temperature to drift up a few degrees for a few hours of the year.
- Downsizing a boiler or hot water unit by virtue of selecting higher efficiency equipment. Since output is the design driver, it is often possible to utilize the next smaller size unit, but at higher efficiency, to achieve the same result.
- Downsizing primary heating and cooling equipment in conjunction with upgrades in envelope elements like insulation or, especially, window shading.
- Downsizing primary heating and cooling equipment in conjunction with heat recovery systems.
- Downsizing fan and pump motors, and primary cooling equipment, by increasing duct and pipe sizes, filter areas, coil areas, etc. as the trade-off for using less transport energy. Note that the extra heat of transport energy elements, especially fans, often drives the equipment size up a notch.
- Downsizing overhead lighting and HVAC cooling via an owner commitment to a greater use of task lighting.

Table 1 Equivalent rate of return values for various simple payback

	ECM PROJECT LIFE			
SIMPLE PAYBACK	5 YEAR IRR	10 YEAR IRR	15 YEAR IRR	20 YEAR IRR
2	40.0%	49.0%	50.0%	50.0%
3	20.0%	31.0%	33.0%	33.0%
4	8.0%	21.0%	23.5%	24.6%
5		15.0%	18.0%	19.4%
6		10.6%	14.5%	15.8%
7		7.0%	11.5%	13.0%
8		4.3%	9.0%	11.0%
SPB typ 9		1.8%	7.1%	9.2%
10-yr max 10			5.5%	7.7%

Simple Payback vs. Internal Rate of Return
payback periods noted for various project lifespans that achieve greater than 15% rate of return

Derivation: Internal Rate of Return (IRR) is that interest rate where the present worth of the savings is equal to the initial investment.

$$P = A * (P/A, i, n)$$

$$P = A * \frac{(1+i)^n - 1}{i(1+i)^n}$$

so, for some value of i,

$$P/A \text{ (simple payback)} = \frac{(1+i)^n - 1}{i(1+i)^n}$$

In addition to downsizing, integrated design can be used as an investment tool, with utility use reduction as the return. If sufficient funding exists, the owner can allow upgrades with identifiable costs and annual savings to be proposed, with a stipulated payback period such as 2–5 yr, to capture the fruit that is just above the proverbial low-hanging level. For example, an upgrade with a 10 yr life and a 4 yr simple payback is an equivalent internal rate of return of 21%, which is an attractive return for investment dollars.

To the energy engineer: be *sure* of your calculations, and even derate them, before leading your owner out with borrowed money (Table 1).

Sustainability

The guideline includes instructions to the owner's staff as well as designers and contractors. This is because the construction of a well-performing building is not the end of the story. The energy and water use of the building will also depend upon user habits at the points of use, and interaction with the building occupants is a vital component for program success. While many endeavors begin a long slide in performance after inception, it is a further intent of this document to encourage **sustainability** of operations. To this end, the suggestions for thorough documentation, commissioning, maintenance, and occupant education are included, and require an ongoing commitment by the owner.

STRATEGIC FACILITY GUIDELINES FOR IMPROVED ENERGY EFFICIENCY IN NEW BUILDINGS

Purpose

- Implementation of the contents of this guideline will reduce facility energy use by 30–50% when compared with ASHRAE 90.1 Base Building and Minimum Local Energy Codes.
- Suggested use of this document is for the project owner interested in achieving the stated savings to provide to the design team as part of their project intent guidance.

General

- Design document submittals must include detailed narrative descriptions of system functionality, features, limitations, design assumptions, and parameters, for use by the owner. The narratives will be detailed enough to provide benefit to subsequent design teams, and will be written to be informative and useful to building operations personnel. The narrative will be provided as a deliverable with the schematic design, and will be updated with each subsequent design delivery including design development (DD) and construction documents (CD) phases. In its final form, this document shall be placed on the first sheet of the drawing set behind the title page, so that the information is retrievable years later when all that is available to facility operations are the drawings. Design assumptions include number of people, indoor and outdoor HVAC design conditions, foot-candles of illumination, hours of operation, provisions for future expansion (if any), roof snow load, rainfall rates, etc. that all define the capabilities of the building.
- All equipment schedules, including HVAC, plumbing, lighting, glazing, and insulation shall be put onto the drawings and shall not reside in the specification books, so that the information is retrievable years later when all that is available to facility operations are the drawings.
- Design thermal insulation values and glazing properties that affect energy use (U-value, shading coefficient, etc.) shall be clearly noted on the drawings.

Energy Efficiency: Strategic Facility Guidelines

- Project commissioning that includes identifying measurable energy savings goals and monitoring the design and construction activities with these as project intent items, with early detection and notification of any project changes that impact energy use or demand.
- Project final payment contingent upon:
  — Receipt of accepted accurate as-built drawings, with accuracy verified by owner and signed by the contractor.
  — Receipt of accurate and complete operations and maintenance (O/M) manuals, with certified factory performance data, repair parts data, and vendor contact information for all energy consuming equipment, including all HVAC and lighting equipment and controls.
  — Receipt of test and balance report that demonstrates design intent is met for air, water, and ventilation quantities, showing design quantities, final adjusted quantities, and percent variance. This would include all variable air volume (VAV) box minimum settings shown, including both heating and cooling balanced air quantities. This would also include any equipment performance testing that was specified for the project.
  — Verification by the owner that the test and balance settings include permanent markings and hence these settings can be preserved over time.
  — Receipt of on-site factory-authorized start-up testing for primary HVAC equipment including chillers and boilers, with efficiency and heat/cool performance figures and heat exchanger approach temperatures to serve as baseline. The submitted reports would include as a minimum heating/cooling output, gas/electric energy input, heat exchanger approach temperatures, water and air flows.
  — Receipt of control shop drawings with detailed descriptions of operation.
  — Acceptance testing of the automatic control system using the approved sequence of operation, and verification that the sequences are fully descriptive and accurate. Acceptance testing also includes review of the control system man-machine interface provisions to become familiar with each adjustable point in the system. Acceptance is by the owner, who will witness each sequence as part of the turnover training requirements.
- Building design must prevent negative pressure condition, unless safety considerations require it.
- Electric resistance space heating, air heating, and water heating are not allowed, unless there is no means to get natural gas to the site.
- Portable space heaters not allowed, unless required for an approved emergency measure.

Energy Use, Overall Performance

- Using ASHRAE 90.1 or local energy code as a baseline, demonstrate through computer modeling that the building energy use will be at least 30% less than this value.

Irrigation Water Use, Overall Performance

- Using standard Kentucky Bluegrass sod and average regional rainfall rates as a baseline, demonstrate that irrigation use for the property will be 50% or less of this value.

Test and Balance

- Balance using "proportional balancing," a technique that strives to reduce throttling losses, which permanently energy transport penalties (pump and fan power).
- Any motor over 5 hp found to be throttled with a resistance element (valve or damper) more than 25% must be altered by sheave change or impeller trim to eliminate lifelong energy waste from excessive throttling losses.
- All 3-phase motor loads, including HVAC equipment, must include voltage balance verification as part of the test and balance (TAB) work. Voltage imbalance of more than 1% indicates unbalanced electrical service in the building and unacceptable efficiency losses.
- Vertical return air shafts serving multiple floors require a balancing damper at each branch outlet to proportion the return air by floor.
- Air flow performance testing for all Air Conditioning and Refrigeration Institute (ARI) certified HVAC factory packaged unitary equipment greater than 5-tons capacity. Heating and cooling performance and efficiency verification is assumed via the ARI certification process.
- Heating efficiency, cooling efficiency, and air flow performance testing for all HVAC *split system* equipment greater than 5-tons capacity or 200,000 Btuh input heating capacity.
- Water flow performance testing for all ARI certified factory packaged water chillers. Cooling performance and efficiency verification is assumed via the ARI certification process.
- Water flow and combustion efficiency testing for all boiler equipment.
- Combustion efficiency testing for all boiler equipment unless factory startup is provided on site.
- Cooling tower thermal performance verification is assumed via the Cooling Tower Institute (CTI) certification process.

Electrical Service

- Provide separate utility metering for electric, gas, and water for the building, separate from other buildings.
- Electrical transformer conversion efficiency not less than 95% efficient at all loads from 25% to 100% capacity. Dry-type transformers National Electrical Manufacturers Association (NEMA) TP-1 compliant.
- Locate transformers in perimeter areas that do not require air conditioning for cooling.
- Power factor (PF) correction on large motor loads, for overall building PF of 90% or better. Large mechanical equipment can be provided with the correction equipment. If motor loads are segregated, this can be done at the switchgear.
- Arrange switchgear and distribution to allow metering of the following electrical loads (requires segregating loads):
  — Lighting.
  — Motors and Mechanical.
  — Plug Loads and Other.

Envelope

- Orient buildings long dimensions E–W where possible to reduce E–W exposure and associated solar load.
- Provide building entrance vestibule large enough to close one door before the next one opens (air lock).
- Where thermal breaks are used, the thermal break material must have thermal conductivity properties an order of magnitude better than the higher conductivity material it touches, and must be at least 1/2 in. thick.
- *Minimum wall insulation 25% *beyond ASHRAE 90.1 values, but not less than R-19.* Insulation is generally not expensive during new construction. Incorporate exterior insulation system (outboard of the studs) for at least one-half of the total R-value, to avoid thermal short circuits of standard metal stud walls, which derate simple batt insulation system by approximately 50%, e.g., a standard stud wall with R-19 batts between the studs yields an overall R-9.5.
- *Minimum roof insulation R-value 25% *beyond ASHRAE 90.1 values, but not less than R-30.* Insulation is generally not expensive during new construction. Select insulation that will retain its thermal properties if wet, e.g., closed cell material.
- Glazing meeting the following requirements:
  — Thermal breaks required.
  — *U-factor of 0.35 or less *where HVAC heating is provided.*
  — *Low-E coatings on east- and west-facing glass *where HVAC cooling is provided.*
  — *Max shading coefficient of 0.2 *where HVAC cooling is provided.* Note: any combination of tinting, coating, awnings, or other exterior shading can be used to achieve this. This is to say that no more than 20% of the heat energy from the sunlit glazing is to get into the building.
  — Glazing not more than 25% of gross wall area.
- Skylight/Clerestory elements must meet the following requirements:
  — Thermal breaks required.
  — Triple pane (layer) construction with sealed air space(s).
  — Overall U-value of 0.25 or less.
  — *Skylight shading coefficient must be 0.2 or less *where HVAC cooling is provided.*
  — *Low-E coating *where HVAC cooling is provided.*
- Skylight/Clerestory area not to exceed 5% of roof area.
- Return plenums and shafts designed with an air barrier for leakage not exceeding 0.25 CFM/ft^2 of building envelope surface area at 50 Pa (Energy Efficient Building Association (EBBA) criteria). Shaft construction requires field testing and verification.
- Building envelope devoid of thermal short circuits. Provide thermal break at all structural members between outside and inside surfaces.
- Building leakage testing required (new buildings), with no more than 0.25 CFM/ft^2 of building envelope surface area at 50 Pa (EBBA criteria).
- Utilize lower ceilings to reduce necessary light input power for equivalent light levels at the work surface.
- Utilize reflective (light) color interior colors for ceilings, walls, furniture, and floors, to allow reduced lighting power for comparable illumination. It can take up to 40% more light to illuminate a dark room than a light room with a direct lighting system.
- Good reflectance parameters to use when picking interior surfaces and colors follow. If these values are used and the lighting designer is informed of it, the integrated design process will allow reduced lighting power to achieve the desired light levels.
  — Min 80% reflective ceiling.
  — Min 50% reflective walls.
  — Min 25% reflective floor and furniture.
- Provide operable blinds for vision glass.

Lighting

- Follow ASHRAE 90.1 or local energy code requirements for lighting power budget guidelines, and verify that designs are lower than these limits while meeting current applicable Illuminating Engineers Society of North America (IESNA) lighting illumination requirements.
- Utilize task lighting and less on overhead lighting for desk work.
- Provide separate circuits for perimeter lights within 10 ft of the wall, to allow manual or automatic light harvesting.

Energy Efficiency: Strategic Facility Guidelines

- Use 1-2-3 switching for large open interior area spaces.
- Use ballast that will tolerate removing at least one bulb with no detriment.
- Where occupancy sensors are used, provide "switching ballast" that will tolerate large numbers of on–off cycles without bulb or ballast life span detriment.
- Use electronic ballast instead of magnetic ballast.
- Use ballast factor in the lighting design to improve lighting system efficiency. Because the ballast mostly determines how many watts are used, ballast choice is critical to achieving best energy efficiency.
- Coordinating light output with "ballast factor" is an excellent tool for providing optimum light levels and energy use.
- Use high power factor ballast, with minimum PF of 95% at all loads.
- Occupancy sensors in conference rooms, warehouses, and multifunction rooms. Also in locker rooms and restrooms, but with some continuous manual switched lighting in these areas.
- Photo-cell controlled lights in the vicinity of skylights.
- Do not use U-tube fluorescent lights, due to high replacement bulb costs.
- Do not use incandescent lights.
- Outdoor lighting on photocell or time switch.

Motors and Drives

- All motors meet or exceed EPACT-1992 efficiency standards.
- Variable frequency drive (VFD) on all HVAC motors larger than 10 hp that have variable load.
- Motor nameplate horsepower (hp) not more than 20% higher than actual brake horsepower served (i.e., do not grossly oversize motors).

HVAC

- Provide HVAC calculations and demonstrate equipment is not oversized. Equipment selection should not be more than 10% greater capacity than calculated values indicate.
- HVAC calculations will include both maximum and minimum heat/cool loads and equipment shall be designed to accommodate these load swings, maintaining heat/cool efficiency equal to or better than full load efficiency at reduced loads down to 25% of maximum load, e.g., equipment capacity will track load swings and energy efficiency will be maintained at all loads.
- Provide necessary outside air (OA), but no more than this. Excess ventilation represents a large and controllable energy use. Reduce exhaust to minimum levels and utilize variable exhaust when possible instead of continuous exhaust. Reduce "pressurization" air commensurate with building leakage characteristics. If the building is tested to low leakage as indicated herein, there should be little need for this extra air, or the heat/cool energy it requires. Design controls to dynamically vary outside air with occupancy.
- VAV box primary heating cubic feet per minute (CFM) shall be not higher than the cooling minimum CFM. This is to say the VAV box primary damper will *not* open up during heating mode.
- Zoning:
  — Design HVAC zoning to require heating *or* cooling, not both. This will improve comfort and also reduce the inherent need for simultaneous heating and cooling.
  — Do not zone any interior areas together with any exterior areas.
  — Do not zone more than three private offices together.
  — Do not zone more than one exposure (N, S, E, and W) together.
- Design and control settings for ASHRAE Standard 55 comfort envelope, which indicates 90% occupant comfort. Appropriate temperatures will vary depending on humidity levels. For example, in Colorado Springs (dry) the following space temperatures are appropriate:
  — 71°F heating.
  — 76°F cooling.
  — Facilities may institute a range 68–72°F heating and 74–78°F cooling, provided that a 5° dead band is kept between the heating and cooling settings.
- Do not heat warehouses above 60°F.
- Do not cool data centers below 72°F.
- Do not use electric resistance heat.
- Do not use perimeter fin-tube hydronic heating.
- *In cooler climates where HVAC economizers are used, designs should normally favor air-economizers over water-economizers since the efficiency kW/ton is better for the air system. The water economizer "free cooling" includes the pumping and cooling tower fan horsepower, as well as the air handler fan. If the air handler fan power is considered required regardless of cooling source, the air-side economizer is truly "free" cooling.
- *In very dry climates, with outdoor air wet bulb temperatures consistently less than 52°F and dew point consistently less than 42°F, evaporative cooling (direct, indirect, or direct-indirect) should be used in lieu of mechanical refrigeration cooling, as long as indoor humidity of 40% rH or less can be maintained. To the water consumption issue, it is this author's opinion that water is a renewable resource and does not disappear from the planet like fossil fuels do, and hence this technology should be used without environmental resources concern.
- Packaged HVAC cooling equipment not less than SEER-13 or EER-12, as applicable.

- *Air-side economizers for all rooftop equipment, regardless of size, *for climates with design wet bulb temperatures below 65°F.*
- Avoid duct liner and fiber-board ducts due to higher air friction and energy transport penalties.
- Insulate all outdoor ductwork to R-15 minimum.
- Use angled filters in lieu of flat filters, to reduce air friction loss.
- Reduce coil and filter velocities to a maximum of 400 fpm to lower permanent air system losses and fan power.
- Avoid series-fan-powered VAV boxes.
- For fan-powered VAV boxes, use energy conservation measure (ECM) motors to achieve minimum 80% efficiency. Although the motors are not large, when there are many of them this efficiency benefit is significant.
- Heat recovery for any 100% outside air intake point that is greater than 5000 CFM when the air is heated or cooled.
- Air filter requirements:
  — Terminal units (fan coils, fan powered boxes, unit vents): 20% (1-in. pleated). Note: this may require an oversize fan on small terminal equipment, and not all manufacturers can accommodate.
  — Air handlers with 25% or less OA: 30%—MERV-7
  — Air handlers with 25–50% OA: 45%—MERV-9
  — Air handlers with more than 50% OA: 85%—MERV-13
  — Provide manometers across filter banks for all air handlers over 20 tons capacity. Equip manometers with means to mark the "new-clean" filter condition, and change-out points.
- *Air-cooled condensing units over 25 tons, provide evaporative precooling, *for climates with design wet bulb temperatures below 65°F.*
- Make-up meter for all hydronic systems to log system leaks and maintain glycol mix.
- Separate systems for 24-7 loads to prevent running the whole building to serve a small load.
- *Direct evaporative post cooling for all chilled water systems, *for climates with design wet bulb temperatures below 65°F.*
- Require duct leakage testing for all ducts 2 in. w.c. design pressure class or greater.
- For process exhaust and fume hoods, design for variable exhaust and make-up.
- Utilize general exhaust air as make-up for toilet exhaust and other exhaust where possible.
- Dedicated outside air system (DOAS) for large office facilities (over 50,000 SF) with VAV systems, allowing ZERO minimum settings for all VAV boxes. This will eliminate the VAV reheat penalty, and the internal zone over-cooling effect from VAV minimums which often requires running the boilers throughout the year for comfort control.
- Separate interior and exterior VAV zoning for open-plan rooms to utilize zero-minimums in the interior spaces.
- Do not use grooved pipe fittings in hydronic heating or cooling piping systems to prevent operating central heating and cooling equipment year-round on account of these fittings.
- Verify that all manufacturer's recommended clearances are observed for air cooled equipment.
- Humidification:
  — Do not humidify any general occupancy buildings such as offices, warehouses, or service centers.
  — In data centers *only*, humidification should not exceed 30–35% rH.
  — Where humidification is used, humidifiers should be ultrasonic, mister, or pad type, and should not be electric resistance or infrared type.
  — Do not locate humidifiers upstream of cooling coils, to avoid simultaneous humidification—dehumidification.
  — Where humidification is used, provide for elevated apparatus dew point of cooling coils or other means to prevent simultaneous humidification—dehumidification.
- Dehumidification:
  — Do not dehumidify below 45% rH.
- Provide performance and efficiency testing of package heating and cooling equipment over 7000 CFM or 20 tons or 500,000 Btu input heating units with factory authorized equipment representatives. Test figures to include on-site gross heat/cool output, fuel and electrical input, and efficiency, compared to advertised values.

Energy Transport Systems—Energy Budget

- For HVAC air systems, the maximum energy transport budget will be:
  — No less than 10 Btu cooling and heating delivered to the space per Btu of fan energy spent at design conditions.

    This will generally steer the design toward generous sizing of sheet metal ducts, air handler cabinetry, coils and filters, higher efficiency fans (0.7 or better), and higher system differential temperatures to reduce air flow rates, but it will result in greatly reduced lifetime energy use since it lowers the bar of system pressure.
  — Fan hp limitation from:

  Cooling fan hp max input

  $$= \frac{\text{Cooling Btu gross output}}{(10 \times 3413 \times \text{motor eff})}.$$

Energy Efficiency: Strategic Facility Guidelines

— Air hp limitation from:

Cooling fan hp max budget × fan eff.

— TSP limitation from:

TSP = (air hp × fan eff × 6360)/CFM.

For example, a 100-ton HVAC air system using 80% e motor, 70% e fan, and 350 CFM per ton would be limited to 44 hp motor load and 3.9 in. w.c. TSP. NOTE: for systems with both supply and return fans, the transport energy considers both combined as the "fan."

- For HVAC water systems, the maximum energy transport budget will be:
— No less than 50 Btu cooling and heating delivered to the space per Btu of pump energy spent, at design conditions.

This will generally steer the design toward generous sizing of piping, strainers, coils, and heat exchangers, higher efficiency pumps (0.75 or better) and higher system differential temperatures to reduce water flow rates, but will result in reduced lifetime energy use since it lowers the bar of system pressure.

— *Pump hp limitation from*:

Cooling pump hp max input

$$= \frac{\text{Cooling Btu gross output}}{(50 \times 3413 \times \text{motor eff})}.$$

— *Water hp budget from*:

Pump max hp × pump eff

— *HEAD limitation from*:

HEAD = (water hp × pump eff × 3960)/GPM.

Hydronic Circulating Systems

- Heating: minimum 40° dT design, to reduce circulating flow rates and pump hp.
- Cooling: minimum 16° dT design, to reduce circulating flow rates and pump hp.

Boilers and Furnaces

- No atmospheric burners.
- No standing pilots.

- Design hydronic system coils to return water to the boiler at or below 140°F water with a minimum of 40°F temperature drop. This will reduce circulating pump energy and improve boiler efficiencies.
- Minimum efficiency of 85% at all loads down to 25% load.
- For heating load turn-down greater than 4:1, provide modular boilers or a jockey boiler.
- For multiple boilers sharing multiple pumps, provide motorized valves to cause water flow to occur *only* through the operating boiler.
- Provide stack dampers interlocked to burner fuel valve operation.

Chillers

- *Water-cooled centrifugal efficiency 0.5 kW/ton or less with 70°F condenser water and 45°F chilled water *for climates with design wet bulb 65°F and lower. 0.58 kW/ton or less with 85°F condenser water and 45°F chilled water in climates where design wet bulb temperatures are above 75°F.*
- *Water-cooled centrifugal units able to accept 55°F condenser water at 3 gpm per ton, all loads. *Beneficial in dry climates with design wet bulb temperatures less than 65°F and typical wet bulb temperatures less than 50°F.*
- *Water-cooled positive displacement units 0.7 kW/ton or less with 70°F entering condenser and 45°F chilled water *for climates with design wet bulb 65°F and lower 0.81 kW/ton or less with 85°F condenser water and 45°F chilled water in climates where design wet bulb temperatures are above 75°F.*
- Do not provide chilled water temperatures less than 45°F. Select cooling coils to provide necessary cooling with 45°F chilled water or higher.
- Air-cooled chiller efficiency 1.0 kW/ton or less with 95°F entering air.
- *Air-cooled chillers over 25 tons, provide evaporative precooling *where design wet bulb temperatures are less than 65°F.*

Cooling Towers

- Selected for 7°F approach at design wet bulb and 0.05 kW/ton or less fan input power. This will steer the design toward a larger free-breathing cooling tower box with a small fan, minimizing parasitic losses from the cooling tower fan. Cooling tower fan kW/ton should not be more than one-tenth of the chiller it serves.
- *Set condenser water temperature set point not higher than 70°F *for climates with design wet bulb 65°F and lower. For climates with higher wet bulb temperatures,*

design to 7°F above design wet bulb with reset controls to lower the setting whenever conditions permit.
- Water treatment control for minimum seven cycles of concentration to conserve water.
- Specify cooling tower thermal performance to be certified in accordance with CTI STD-201.

Air-Cooled Equipment and Cooling Towers in Enclosures

- Locate to prevent air short-circuiting and associated loss of thermal performance. Rule of thumb is the height of the vertical finned surface projected horizontally. The fan discharge must be at or above the top of the enclosure, the distance to the enclosure walls should be as indicated above, and there should be amply sized inlet air openings in the enclosure walls as low as possible.

Ground Source Heat Pumps

- Coefficient of performance (COP) 4.0 or higher at 40°F entering water.
- EER 17 or higher at 80°F entering water.
- No electric resistance heating.

Controls

- Design OUT all simultaneous heating and cooling through the use of proper zoning, interlocks, and dead bands. This includes all constant volume systems and terminal unit systems. VAV systems inherently have an overlap which should be minimized by water and air reset in heating season, prudent use of minimum VAV box settings, and consideration of systems that separate the outside air from the supply air (SA).
- Programmed start–stop for lighting and HVAC systems with option for temporary user overrides. Use these controls to prevent unnecessary operating hours.
- Lock out air flows for conference rooms and intermittent occupancy rooms by interlocking VAV box to close with occupancy sensors.
- Lock out chiller operation below 50°F, except for data centers or humidity-sensitive areas that cannot use outside air for cooling.
- Lock out boiler operation above 60°F, unless space temperatures cannot be maintained within the specified ranges any other way.
- *All cooling by air economizer below 55°F *for climates with design wet bulb 75°F and lower.*
- Night setback for heating. Suggested temperature for unoccupied time is 60°F.
- No night set-up for cooling—no cooling operation in unoccupied times for general occupancy buildings. If building temperature rise during unoccupied times can cause detriment, then limit off-hours cooling operation to 85°F indoor temperature.
- Reset boiler hot water temperature settings in mild weather.
- *Reset chilled water temperature settings in mild weather, *provided that outdoor air dew point is below indoor dew point levels.* Refrigeration savings generally exceeds increases in pump power.
- Provide appropriate interlock for all exhaust fans to prevent infiltration of outside air from uncontrolled exhaust fans that operate in unoccupied times.
- All analog instruments—temperature, pressure, etc. other than on–off devices—must be calibrated initially (or verified for non-adjustable devices). Merely accepting out-of-the-box performance without verification is not acceptable.
- Two-year guarantee on calibration, with 18-month re-calibration of all analog inputs.
- Air handler control valves with a residual positive seating mechanism for positive closure. Use of travel stops alone for this is not acceptable.
- For terminal units and heating/cooling hydronic water flow rates less than 10 gpm, use characterized ball valves for control valves instead of globe valves or flapper valves, for their inherent improved long-term close-off performance. This will reduce energy use from simultaneous heating and cooling.
- Valve and damper actuator close-off rating at least 150% of max system pressure at that point, but not less than 50 psid (water) and 4 in. w.c. (air).
- Dampers at system air intake and exhaust with leakage rating not more than 10 CFM/ft^2 at 4 in. water column gage when tested in accordance with Air Movement and Control Association (AMCA) Standard 300.
- Water coil control valve wide open pressure drop sizing not to exceed the full flow coil water-side pressure drop.
- Provide main electrical energy and demand metering, and main gas metering. Establish baseline and then trend; log kBtu/SF, kWh/SF-yr, and kW perpetually and generate alarm if energy use exceeds baseline.
- Implement demand-limiting or load leveling strategies to improve load factor and reduce demand charges. Stagger-start any large loads, e.g., morning warm-up or cool-down events. Use VFDs to limit fan, pump, and chiller loads to 90% during peak hours, etc.
- Independent heating and cooling set points for space control.
- Space temperature user adjustment locked out or, if provided, limited to ±2°F.

Energy Efficiency: Strategic Facility Guidelines

- 5°F dead band between space heating and cooling set points to prevent inadvertent overlap at zone heat/cool equipment, and from adjacent zones.
- 5°F dead band between air handler heating and cooling (or economizer) set points, e.g., preheat coil cannot share a single, sequenced, set point with the economizer or cooling control.
- Provide separate lighting and HVAC time schedules.
- For chillers (condenser) and hot water boilers, use temperature sensors to log heat exchanger approach values, to prompt predictive maintenance for cleaning fouled heat exchange surfaces. New-equipment approach will be the baseline value, and approach temperature increases of 50% will prompt servicing.
- Interlock heating and cooling equipment in warehouses serving doorway areas to shut off when roll-up doors are open to reduce waste.
- Optimization routines:
  — Automatically adjust ventilation rates for actual people count.
  — Optimal start to delay equipment operation as long as possible.
  — Demand limiting control point that will limit all VFD-driven air handler fans components to a maximum of 90% max output in summer. This will cause system temperatures to drift up slightly during extreme weather, but will reduce electrical demand for this equipment (and the cooling equipment it serves) compared to full output operation, during times when utility demand is highest. Do not oversize equipment capacity to compensate for this requirement.
  — Optimal static pressure setting based on VAV box demand, not a fixed set point. This is a polling routine.
  — *For areas with design wet bulb temperatures below 65°F only*, optimal SA reset that will reset the SA temperature set point upward from 55°F to 62°F for VAV systems during heating season, to reduce reheat energy. This can either be from two methods.
    * *Method 1.* Basic Optimization. When the main air handler fan is below 40% of capacity and OA temperature is below 40°F.
    * *Method 2.* Fully Optimized. Polling VAV boxes (at least 80% of the boxes served are at minimum air flows).
  — Do not reset SA temperature from return air. Do not reset SA temperature during cooling season.
  — Reset condenser water temperature downward when outdoor conditions permit, using the lowest allowable condenser water the chiller can accept.

Plumbing

- Max shower flow 1.5 gpm.
- Max bathtub volume 35 gal.
- Max urinal water flow 0.5 gpf, or waterless.
- Max lavatory water flow 0.5 gpf.
- Metering (self closing) or infrared lavatory faucets.
- Avoid single lever faucets since these encourage complacency for the use of hot water.
- All domestic hot water piping insulated.
- Heat trap in domestic hot water main outlet piping.
- If a circulating system is used, provide aquastat or timer to prevent continuous operation.
- Max domestic hot water temperature for hand washing 125°F.
- Gas water heaters in lieu of electric where natural gas is available.
- Domestic water heater equipment separate from the building boiler and heating system to prevent year-round operation of central heating equipment.
- Water fountains instead of chilled water coolers.
- Operate the building at reduced pressure (such as 50 psig) instead of 70 psig, to reduce overall usage. Verify that design maintains at least 10 psig over the required minimum pressure at all flush valves.

Management and Maintenance Activities to Sustain Efficiency

- Management support
  — Create buy-in from the building occupants. Distribute information to building occupants to raise awareness of energy consumption, especially communicating that the user's habits are an essential ingredient to overall success, and are useful and appreciated. This would be in the form of occasional friendly and encouraging reminders of how user participation is helping, fun facts, etc. along with estimated benefits from behavior changes. Provide measured results whenever available.
  — Enforce temperature setting limitations, including the explanation of why this is helpful and also why it is reasonable. Encourage seasonal dress habits to promote comfort and conservation together.
  — Prohibit space heaters.
  — For offices, utilize LCD monitors and the software-driven "monitor power-off" feature, since the monitor represents two-thirds of the whole personal computer (PC) station energy use.
  — Track monthly energy and water use and maintain annual graphing lines, comparing current and prior years. Establish new benchmark curves after major renovations, alterations, or energy conservation projects. Compare annual use with benchmark and

verify that building energy and water usage per SF is not increasing. Report results to the building occupants as an annual energy use report for their feedback.
— Escrow (save) approximately 5% of the replacement cost per year for the energy consuming equipment in the facility that has a normal life cycle, such as HVAC systems, lighting systems, and control systems. This will allow 20-year replacement work without "surprises" to sustain efficient building operations.
— For leased office space, show the tenants their utility costs to increase awareness and encourage conservation by the users. The typical industry arrangement is to build in utilities into the lease price, so the tenants do not see a separate utility bill. Although the customers are paying for the utilities, having those costs clearly shown will reduce the complacency in utility use.
- Chillers:
— Owner provides annual equipment "tune up," including cooling efficiency testing and heat exchanger approach measurements.
— Owner adjusts temperature settings or cleans heat exchangers, or adjusts water flows whenever cooling efficiency tests are less than 90% of new-equipment values. For example, if the new equipment benchmark is 0.5 kW/ton, then a measurement of 0.5/0.905 = 0.55 kW/ton would trigger corrective action.
- Boilers:
— Owner provides annual equipment "tune up," including combustion efficiency testing and heat exchanger approach measurements.
— Owner adjusts temperature settings, cleans heat exchangers, or adjusts air-fuel mixture whenever combustion efficiency tests are less than 95% of new-equipment values. For example, if new equipment benchmark is 80%, then a measurement of $0.8 \times 0.95 = 0.76$ would trigger corrective action.
- HVAC air coils:
— Owner changes filters at least quarterly, and verifies there are no air path short circuits allowing air to bypass the filters.
— Owner cleans HVAC coils whenever there is any sign of visible accumulation or if air pressure drop is found to be excessive.
- HVAC air-cooled condensers:
— Owner provides location free from debris, leaves, grass, etc. and adequate spacing for free "breathing" and no recirculation.
— Owner cleans heat exchange surfaces annually.

- Controls:
— Owner re-evaluates system occupancy several times each year, to reduce unnecessary HVAC and lighting operating hours.
— Owner re-evaluates control set points each year including space temperature settings, duct pressure settings, SA temperature settings, reset schedules, and heating and cooling equipment lock-out points.
— Owner re-calibrates control instruments each two years other than on–off devices.
— Owner cycles all motorized valves and dampers from open to closed annually, and verifies tight closure.
— Owner cycles all VAV box dampers from open to closed annually and verifies that the control system is responsive, since these often have a short life and can fail without the user knowing it.

CONCLUSIONS

The listed items in this document are intended to supplement traditional facility guide specifications, as a tool to help steer new building designs toward sustained low energy use.

It is easier to design efficiency into a new building than to retrofit an existing one, for practical and monetary reasons. While we continue to search for ways to upgrade existing buildings, we should influence the new buildings as much as possible for the long term benefits. Engineers and architects alone may understand the benefits and opportunities available, but may not be effective at altering the default course of events for new building designs. The most effective way to assure energy savings as a built-in feature is through a top-down approach where owner support conveys efficiency as a design team priority. The energy-efficient design commitment is made more effective through the use of integrated design, and commissioning can be an effective tool to be certain the design intentions are realized through construction. The sustained energy efficiency goal requires an ongoing commitment from the owner, maintenance staff, and building occupants, and includes training, appreciation, and feedback.

REFERENCES

1. Advanced Energy Design Guide for Small Office Buildings, ASHRAE, 2000.
2. ASHRAE Fundamentals Handbook, 2001.

Energy Information Systems

Paul J. Allen
Reedy Creek Energy Services, Walt Disney World Co., Lake Buena Vista, Florida, U.S.A.

David C. Green
Green Management Services, Inc., Fort Myers, Florida, U.S.A.

Abstract
Advances in new equipment, new processes, and new technology are the driving forces in improvements in energy management, energy efficiency, and energy-cost control. Of all the recent developments affecting energy management, the most powerful technology to come into use in the past several years has been information technology (IT). The combination of cheap, high-performance microcomputers and emerging high-capacity communication lines, networks, and the Internet has produced explosive growth in IT and its application throughout our economy. Energy Information Systems (EIS) have been no exception. Information technology and Internet-based systems are the wave of the future. This entry will introduce basic principles, structures, and definitions, and will also show examples for typical EISs.

ENERGY INFORMATION SYSTEMS

The philosophy "If you can measure it, you can manage it" is critical to a sustainable energy management program. Continuous feedback on utility performance is the backbone of an Energy Information Systems (EIS).[1] A basic definition of an EIS is equipment and computer programs that allow users to measure, monitor, and quantify the energy usage of their facilities and to help identify energy conservation opportunities.

Everyone has witnessed the growth and development of the Internet—the largest computer communications network in the world. Using a Web browser, one can access data around the world with a click of a mouse. An EIS should take full advantage of these new tools.

EIS PROCESS

There are two main parts to an EIS: (1) data collection and (2) Web publishing. Fig. 1 shows these two processes in a flow-chart format.

Data Collection

The first task in establishing an EIS is to determine the sources of the energy data. Utility meters monitored by an energy management system or other dedicated utility-monitoring systems are a good source. The metering equipment collects the raw utility data for electric, chilled and hot water, domestic water, natural gas, and compressed air. The utility meters communicate to local data storage devices by preprocessed pulse outputs; by 0–10 V or 4–20 mA analog connections; or by digital, network-based protocols.

Data gathered from all the local data storage devices at a predefined interval (usually on a daily basis) are stored on a server in a relational database (the data warehouse). Examples of relational databases are FoxPro, SQL, and Oracle. A variety of methods are used to retrieve these data:

Modem connection. A modem connection uploads the data from the local data storage device to the energy data server. Typically, the upload takes place on a daily basis, but it may occur more frequently if needed.

LAN or WAN network connection. A local area network (LAN) or a wide area network (WAN) connection established between computers transfers energy data files to the energy data server.

FTP network connection. File transfer protocol (FTP) is an Internet protocol used for transferring files from one computer to another. It moves the energy data files from the local data storage devices to the energy data server.

When the energy data have been transferred to the energy data server, an update program reads all the various data files and reformats them in a format that is used by the Web publishing program.

Web Publishing

To publish the energy data on an Intranet or on the Internet, client/server programming is used. The energy

---
* More information about energy information systems, as well as links to many of the tools discussed in this entry, can be found at www.utilityreporting.com

Keywords: Energy information system; Internet; Meters; Utility data; Web publishing.

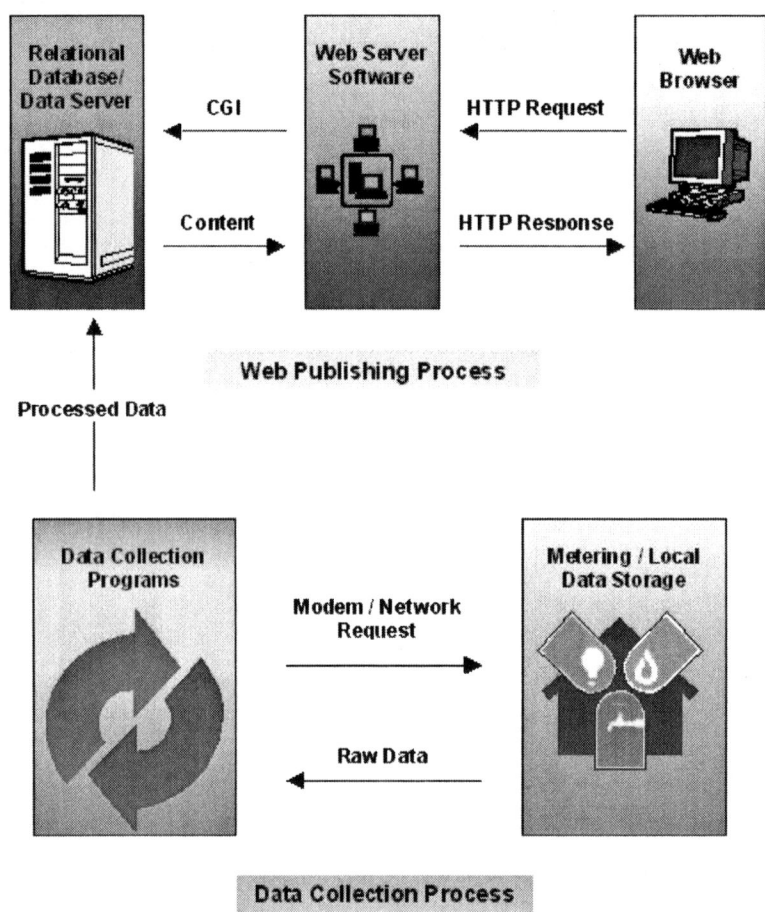

Fig. 1 EIS schematic.

data are stored on a central computer—the server—and wait passively until a user makes a request for information using a Web browser—the client. A Web-publishing program retrieves the information from a relational database and sends it to the Web server, which then sends it to the client Web browser that requested the information.

Many software choices are available for the Web-publishing process. One choice uses a server-side Common Gateway Interface (CGI) program to coordinate the activity between the Web server and the Web-publishing program. CGI is a method used to run conventional programs through a Web browser. The Web-publishing client/server process for an EIS uses the steps below (see Fig. 1):

1. A user requests energy information by using a Web browser (client) to send an hypertext transfer protocol (HTTP) request to the Web server.
2. The Web server activates the CGI interface program, which then starts the Web-publishing program.
3. The Web-publishing program retrieves the information from the relational database, formats the data in hypertext markup language (HTML), and returns it to the CGI interface program.
4. The CGI interface program sends the data as HTML browser content to the Web server, which sends the content to the Web browser that requested the information.

This entire process may take only seconds, depending on the speed of the client computer's connectoin to the Web.

PROGRAMMING CHOICES FOR EIS WEB PUBLISHING

Server-Side Programs

Server-side programs are programs that run on the network server. CGI programming is the most fundamental way to access relational database information over the Internet or an Intranet and to display dynamic content. This is important to any EIS because the amount of data involved will undoubtedly require processing with a relational database of some kind.

FoxWeb is a small CGI program that connects to a Visual FoxPro database application. The application then calls any number of custom-designed queries and procedures to return results to the browser as HTML. More information about FoxWeb is available at www.foxweb.com

Perl (practical extraction and report language) is an application used for CGI programming. Applications written in Perl will work on any operating system. It also has the ability to connect to many types of databases. A good source for information on Perl is www.perl.com

ColdFusion is a server-side application that uses a ColdFusion server. The server executes templates containing a mixture of HTML and ColdFusion instructions and then returns the results to the browser as pure HTML.

Active Server Pages (ASP) are also a popular choice recently. The ASP program on the Windows Web server will automatically interpret Web pages with the extension.asp. The Web pages are a mixture of ASP instructions, Visual Basic code, and HTML.

Java Servlets are Java programs that run on a Web server and build Web pages. Java is the latest of a long line of higher-level programming languages such as Fortran, Pascal, and C++. It is also portable across operating systems. Java Servlets written for one Web server on one operating system will run on virtually any Web server and on any operating system.

Java Server Pages (JSP) are similar to ASP except that the pages consist of Java code instead of Visual Basic code. This makes the code portable across operating systems. The Web pages typically have the extension.jsp. This tells the Web server to interpret the embedded Java code like any other Java Servlet. Information about Java Servlets and JSP is available on this Johns Hopkins University Web site: www.apl.jhu.edu/~hall/java

Hypertext Preprocessor (PHP). In an HTML document, PHP script (similar syntax to that of Perl or C) is enclosed within special PHP tags. Because PHP is embedded within tags, the author can jump between HTML and PHP (similar to ASP and ColdFusion) instead of having to rely on heavy amounts of code to output HTML.

Any organization wishing to develop an EIS should carefully consider which server-side applications to use. The decision should be a practical one rather than a popular one. All of the criteria below should be part of the evaluation process:

- What operating system is predominantly available to the facility?
- What programming languages are support personnel willing to work with?
- What applications are compatible with the existing database?
- How much of the budget is available to spend?

Client-Side Programs

Client-side applications can create a deeper level of interactivity within Web pages. Scripting languages such as JavaScript and VBScript are less complex versions of other languages, such as Java and Visual Basic, respectively. They reside within the HTML of a Web page and provide a great deal of functionality that HTML itself cannot. Scripts such as these validate input, control the cursor programmatically, and do much more.

Dynamic HTML (DHTML) is the result of scripting languages taking advantage of the extensions common to the latest browsers to make the pages change after they are loaded. A good example of this is a link that changes color when the user places the mouse pointer on it. Much more dramatic effects are possible using DHTML. However, the two most popular browsers—Internet Explorer and Netscape—interpret DHTML differently. Good information about DHTML is available at www.dynamicdrive.com

Cascading Style Sheets (CSS) are special HTML features that allow much more flexibility in formatting elements of a Web page. The ability of CSS to describe the style of an element only once rather than every time you display the element provides a separation of content and presentation. This makes Web pages less complex, much smaller, and therefore faster to load. Beware that CSS is fully supported only in the latest versions of browsers (4.0 and later).

Java applets are small Java programs that are stored on a Web server and called from the HTML in a Web page. Statements in the HTML pass to the applet parameters that affec its functionality. Unlike what it does with Java Servlets, the browser downloads the applet and runs it using the browser's own operating system rather than the operating system of the Web server. Free Java applets are widely available on the Internet. KavaChart applets are very useful for charting trends in data. KavaChart applets are available at www.ve.com

Extensible markup language (XML) is a metalanguage that has many uses. A metalanguage is a language used to explain another language. Extensible markup language organizes data in a predefined format for the main purpose of sharing between or within computer systems. Furthermore, its uses include data organization and transfer, data presentation, data caching, and some that probably have not been invented yet. More information about XML is available at http://xml.com

Developers can use any or all of these to enhance the content of an EIS. Three important points to remember about using client-side applications to enhance browser content are

- Many client-side applications require later versions of browsers to work correctly. Be sure that all your users are using the required browser versions.

- Many client-side applications are available free. Search the Internet before spending resources to develop custom client-side applications.
- Client-side applications will make your Web pages more complex, which adds to development and maintenance costs. Be sure to weigh the benefits of these enhancements against their costs.

Choosing EIS Web-Publishing Tools

We might define tools, in this case, as utility applications that require more configuration effort than programming effort. There are some exceptions, of course. In any case, tools fall into three categories:

- Open-source or free
- Purchased
- Developed

Tools perform such functions as batch emailing, charting, scheduling application run times, and enabling database-to-Web connectivity. The relational database and Web server are also tools.

Batch email applications can be of any category. There are good free ones and purchased ones. Some email servers have batch-processing capability, and some do not. Purchased batch email applications are relatively inexpensive. They have some variations in features, such as whether or not they will send HTML or attachments. They are also easy to develop.

Charting tools are really too complex to warrant developing. There are some very good free and open-source charting tools that have all the features of purchased ones.

The EIS needs scheduling programs to launch data collection applications at predefined times. Some operating systems have scheduling programs built in, but they may be difficult to configure. A purchased version is probably the best choice, because the cost will be quite low.

Some applications do most of the database-to-Web connectivity. These applications either require purchase or (in the case of the free applications) a great deal of programming. The purchased applications are a good choice because much of the error reporting is part of the application. The purchased versions can be very expensive or relatively inexpensive. Each database has its own connectivity options, so much of this decision rests on which relational database is used.

The database used is likely a purchased one or may actually be open-source. Open-source databases like MySQL are competing with the best of the others. Commercial relational database systems range in cost from very inexpensive (Microsoft Access) to very expensive (Oracle).

Open-source Web servers like Apache are available for some operating systems and are very widely used and reliable. Others are free with operating systems such as Microsoft Internet Information Server (IIS). Some are also commercially available for a few hundred dollars. The Web-server choice depends mostly on the operating system of the server itself.

The EIS tools will likely be a mix of open-source, purchased, and developed applications. It is important to consider budget constraints, operating systems, and support when deciding which tools and what types of tools to use. Always plan for compatibility among all the tools the EIS will use.

EIS Choice: Purchase or Do It Yourself

Up to this point in this entry, we have gone "under the hood" and provided details on the processes and software tools needed to create a custom EIS. However, users may not be willing to invest the time and effort required for this do-it-yourself approach.

Numerous companies provide an EIS for an ongoing monthly service fee. The services available range from monthly utility billing data processing and analysis to interval data recorded from submeters for utility cost allocation and load profiling. The advantage of this approach is that the user does not get involved with the details and operation of the EIS, but instead is able to work with the EIS service provider to develop the utility data reports most helpful to the user's operation. An ongoing monthly service fee is a function of the amount of data processed; the more meters or bills processed, the higher the monthly fee. There are additional costs for customizing any reporting from the standard reports already created by the EIS service provider. Ultimately, the monthly cost for the EIS service would have to be justified by the benefit from the EIS reports provided.

The do-it-yourself approach works well for users that either have sufficient IT software personnel in house or that can hire an IT consultant to develop and maintain their own custom EIS. However, having adequate IT resources is only half the battle. The user also needs to create an EIS requirements document that defines what data need to be collected and at what frequency, as well as a list of report outputs. As explained earlier in this entry, the software choices for developing the EIS are numerous, which can result in debates among IT personnel as to which system is best. The best solution will be the one that will result in the desired EIS for the least cost in the least amount of time. The energy manager can help facilitate this decision (and prevent IT gridlock) by estimating the cost, advantages, and disadvantages of each approach.

The decision between the EIS service approach and the do-it-yourself approach will be based on several factors, but it will ultimately be based on the ongoing cost to operate the EIS.

UTILITY REPORT CARDS: EIS EXAMPLE

The Utility Report Cards (URC) program[2] is a Web-based EIS that reports and graphs monthly utility data for schools. The URC was developed and prototyped by the Florida Solar Energy Center (FSEC)[3] using Orange County Public Schools (OCPS)[4] utility data. Each month, the EIS automatically generates a Web-based report and emails it to school staff to examine the school's electricity usage (energy efficiency) and to identify schools with high-energy consumption for further investigation. The easy-to-use Web-style report includes hyperlinks allowing users to (1) drill down into further meter details, (2) display graphs for a 12-month comparison with prior-year data, (3) filter the data to show selected schools, and (4) re-sort the data to rank schools based on the data selected. The URC is also for teachers and students to use as an instructional tool to learn about school energy use as a complement to the energy education materials available through the U.S. Department of Energy's EnergySmart Schools program (ESS).[5] To run the URC, go to www.utilityreportcards.com and click URC Live.

URC Data Collection

On a monthly basis, the OCPS utilities (Orlando Utilities Commission[6] and Progress Energy[7]) electronically transmit to FSEC the OCPS utility data, which then adds it to the URC relational database. Because there was no consistency between the utility's data format, FSEC created a custom program called URC_DPP (URC Data Processing Program) that processes each utility's data file separately and loads the data into a common Oracle database.

Florida Solar Energy Center sends an email to a designated email address at OCPS, with a copy of the current month's URC embedded in the email message. Then OCPS forwards this email to its internal email distribution list for the OCPS staff. This procedure makes it easy for users, because all they need to do is click the hyperlinks in the URC email to produce graphs and detailed reports.

URC Web-Publishing Program

The URC program is informative, intuitive, and flexible for all users. It takes advantage of extensive use of hyperlinks that create graphs and detailed reports from an overall summary report listing all schools. Users are able to view graphs and see the electric consumption patterns simply by clicking hyperlinks in the URC Web page.

When the user selects the URC Live hyperlink, a summary report (see the example in Fig. 2) shows totals for each school type in the entire school district—primary (elementary) schools, middle schools, high schools, etc.—as the rows in the report. The data presented in the columns include the electric consumption (kWh), the cost (dollars), and the efficiency (Btu/ft^2). To provide meaningful comparisons to the same time in the prior year, the URC program divides the data by the number of days in the billing period to produce per-day values. The URC also allows the user to change the values shown to per-month

	UTILITY REPORT CARDS								
Electric Per Day	ORANGE COUNTY PUBLIC SCHOOLS						<< February ▼ 2004 ▼ >>		
	Consumption (kWh/day)			Cost ($/day)			Efficiency (Btu/sq ft/day)		
School Type	Current Period	Previous Period	Percent Change	Current Period	Previous Period	Percent Change	Current Period	Previous Period	Percent Change
Regular Primary School	178,073	181,162	-2 %	$ 13,923	$ 12,012	16 %	146	149	-2 %
Regular Middle School	121,735	124,550	-2 %	$ 9,119	$ 7,984	14 %	139	142	-2 %
Regular High School	168,196	167,136	1 %	$ 12,135	$ 10,121	20 %	161	160	1 %
Special Other	6,351	6,092	4 %	$ 479	$ 404	19 %	143	137	4 %
Vocational High School	5,910	6,572	-10 %	$ 469	$ 448	5 %	106	118	-10 %
Grand Total ORANGE COUNTY PUBLIC SCHOOLS	480,266	485,511	-1 %	$ 36,125	$ 30,969	17 %	148	150	-1 %

On   Denotes Increase From Previous Year
On   Denotes Decrease From Previous Year by ◀ 0 % ▶ or more

Fig. 2 Overall school-district summary page.

figures (the per-day values multiplied by 30 days). Fig. 2 shows the overall school-district summary report for February 2004.

The URC program interface makes the program easy to use, considering the enormous amount of data available. The top-down approach lets users view different levels of data from the overall districtwide summary to individual meters in a single school. For example, the user can click a school type to display the details for each school. Fig. 3 shows the result of clicking the hyperlink Regular High School: A Report on All High Schools in the Database for February 2004. The report is sorted based on the percent change in kilowatt-hour usage from the prior-year levels. The schools that changed the most are at the top of the list. Re-sorting the schools is accomplished simply by clicking the column title. To produce the report shown in Fig. 3, the original report sorted by the efficiency percentage change was re-sorted by clicking the consumption percentage change column.

Graphing is accomplished by clicking any current period value in the report. To display a 12-month graph for Apopka Senior High School, clicking the number 22,399 would produce the graph shown in Fig. 4. Note that the consumption levels are significantly higher than prior-year levels for this school. Focusing on the reasons for this increase should be the next step for the OCPS facility personnel. When it is determined that the increase is not due to other factors, such as increased enrollment or extreme weather, personnel can consider making adjustments to the energy management system controls to turn the trend around.

URC Web-Program Details

Two Web programs make up the URC application interface. One is for reporting, and one is for graphing. The programs are written in PHP. "Hypertext preprocessor is a widely-used Open Source general-purpose scripting language that is especially suited for Web development and can be embedded into HTML," according to PHPBuilder.com. This means that the PHP application itself is not for sale. It is developed and supported solely by volunteers. What makes it an attractive choice is that the developers' code is part of the HTML page itself. Hypertext preprocessor is ideal for connecting to databases and running SQL queries return dynamic content to a Web page.[8]

Using PHP, the reports accommodate both novice and expert users. Hyperlinks in the reports pass values back into the same two programs in a recursive manner, using

UTILITY REPORT CARDS									
Electric Per Day			ORANGE COUNTY PUBLIC SCHOOLS			<< February 2004 >>		
Regular High School	Consumption (kWh/day)			Cost ($/day)			Efficiency (Btu/sq ft/day)		
SCHOOL	Current Period	Previous Period	Percent Change	Current Period	Previous Period	Percent Change	Current Period	Previous Period	Percent Change
APOPKA SENIOR HIGH SCHOOL	22,399	20,360	10 %	$ 1,777	$ 1,375	29 %	211	192	10 %
UNIVERSITY HIGH SCHOOL	20,713	19,056	9 %	$ 1,482	$ 1,131	31 %	163	150	9 %
WINTER PARK HIGH SCHOOL	22,631	21,157	7 %	$ 1,604	$ 1,247	29 %	168	157	7 %
EVANS HIGH SCHOOL	17,108	16,526	4 %	$ 1,229	$ 1,022	20 %	156	151	4 %
OLYMPIA HIGH SCHOOL (FORMERLY DR. PHILL	13,441	13,119	2 %	$ 980	$ 805	22 %	118	116	2 %
TIMBER CREEK HIGH SCHOOL	15,133	15,379	-2 %	$ 1,081	$ 924	17 %	134	136	-2 %
WEST ORANGE HIGH SCHOOL	21,375	21,898	-2 %	$ 1,473	$ 1,261	17 %	190	194	-2 %
ROBERT HUNGERFORD PREPARATORY HIGH SCHOOL (FORMERL	5,525	5,727	-4 %	$ 424	$ 393	8 %	192	199	-4 %
OAK RIDGE HIGH SCHOOL	12,940	14,669	-12 %	$ 908	$ 855	6 %	148	168	-12 %
CYPRESS CREEK SENIOR HIGH SCHOOL	16,932	19,245	-12 %	$ 1,176	$ 1,108	6 %	153	174	-12 %
Regular High School	168,196	167,136	1 %	$ 12,135	$ 10,121	20 %	161	160	1 %

Fig. 3 High-school summary report sorted by percentage change in consumption from prior year.

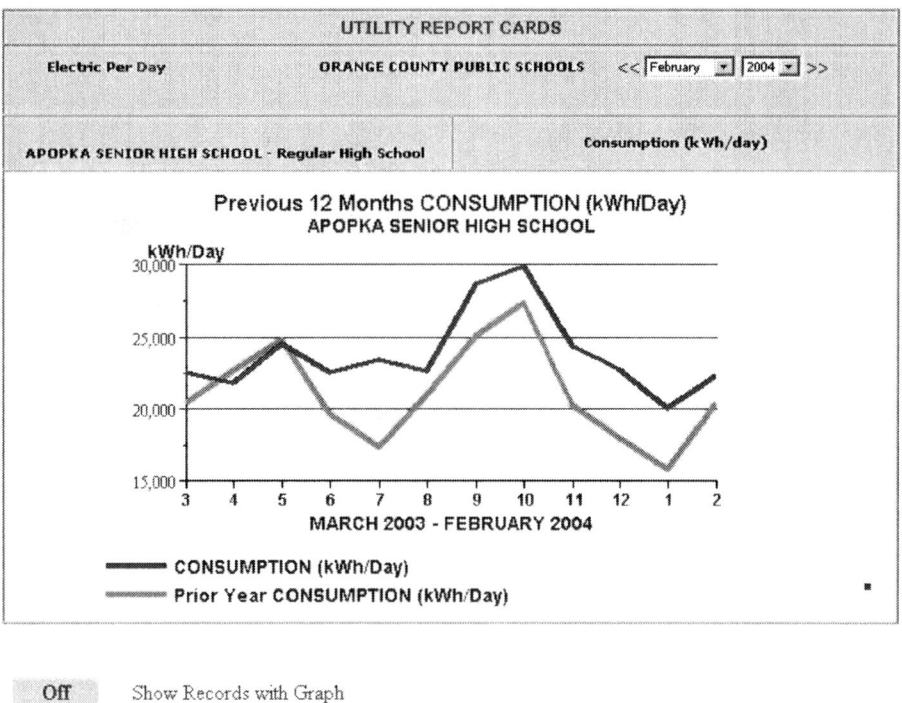

Fig. 4 Apopka Senior High School 12-month kWh graph.

the CGI query string. These values are processed by a set of predetermined rules built into the program by the developer to change the appearance of the reports incrementally to suit the user. The hyperlink construction includes messages using the onMouseOver event to explain the action of the hyperlink. Total and subtotal lines provide summary information. Data that show an increase or decrease from the previous year are flagged with a different background cell color. The user can define percentage criteria for marking decreases from the prior year because this is helpful in tracking progress toward a particular goal, such as a 5% decrease from the previous year. Graphs are created using KavaChart, a collection of Java applets available from Visual Engineering at www.ve.com.[9]

CONCLUSION

Today's energy manager needs to be knowledgeable about the basic principles and concepts of IT because it is a fast-growing area of new systems and services. Web-based EISs provide users feedback on how and where energy is consumed. Building automation systems control the devices that use energy. Together, these two systems provide the energy manager the tools needed for a successful energy management program.

REFERENCES

1. Capehart, B.; Allen, P.; Pawlik, K.; Green, D. How a web-based energy information system works. *Information Technology for Energy Managers*; The Fairmont Press Inc.: Lilburn, Georgia; 2004.
2. Allen, P.; Green, D.; Kalaghchy, S.; Kivler, B.; Sheinkopf, B. The utility report cards—an energy information system for orange county public schools *Information Technology for Energy Managers*; The Fairmont Press Inc.: Lilburn, Georgia; 2005; Vol. II.
3. Florida Solar Energy Center; A Research Institute of the University of Central Florida; http://www.fsec.ucf.edu/ (accessed April 2004).
4. Orange County Public Schools, Orlando, Florida; http://www.ocps.k12.fl.us/ (accessed 28th March 2007).
5. U.S. Department of Energy's EnergySmart Schools; Rebuild America, Helping Schools Make Smart Choices About Energy; http://www.rebuild.org/sectors/ess/index.asp (accessed April 2004).
6. Orlando Utilities Commission; OUC The Reliable One; http://www.ouc.com/ (accessed April 2004).
7. Progress Energy; People, Performance, Excellence; http://www.progress-energy.com (accessed April 2004).
8. The PHP Group; PHP Manual, Preface; http://www.phpbuilder.com/manual/preface.php (accessed April 2004); last updated, October 26, 2002.
9. Visual Engineering; KavaChart, The Complete Solution for Java-Based Charting; http://www.ve.com/ (accessed April 2004).

BIBLIOGRAPHY

1. Apache HTTP Server Project, Apache Software Foundation, www.apache.org
2. ColdFusion MX, Macromedia, Inc., www.macromedia.com/software/coldfusion
3. Acquisuite Energy Information Server, Obvius, Inc., Portland, Oregon, www.obvius.com
4. FoxWeb, Eon Technologies, Alameda, California, www.foxweb.com
5. Java, Sun Microsystems, Inc., http://java.sun.com
6. KavaChart (Java applets for graphing data), www.ve.com
7. MySQL, MySQL AB, www.mysql.com
8. Perl, Perl Mongers—The Perl Advocacy People, www.perl.org
9. Visual FoxPro, Microsoft Corporation, http://msdn.microsoft.com/vfoxpro
10. PHP, PHP Builder, http://phpbuilder.com
11. Utility Report Cards, Florida Solar Energy Center, www.utilityreportcards.com

Energy Management: Organizational Aptitude Self-Test

Christopher Russell
Energy Pathfinder Management Consulting, LLC, Baltimore, Maryland, U.S.A.

Abstract

Human, technical, and financial criteria all contribute to a manufacturer's ability to build wealth through energy management. Collectively, these attributes constitute a "culture" and receptiveness not only to energy management, but to operational efficiency in general. Manufacturers will enjoy a wider range of energy management options by nurturing several key organizational attributes, including staff awareness, competence, leadership, commitment, and removal of institutional barriers. This entry offers a typology and self test of corporate "aptitudes" for energy management. This discussion is based on the author's observation and research (The author served as director, industry sector, for the Alliance to Save Energy in Washington, DC from 1999 to 2006. His work included (1) over 40 workshops serving industrial energy users, each attracting 25 to over 120 participants; (2) presentations at 30 industry conferences; and (3) over 30 articles for trade press. All of these activities have generated communications with hundreds of individuals, all of whom add another dimension to the cumulative story of industrial energy management). Until these theories can be properly tested, readers are asked to merely consider this persuasive argument.

LIST OF ATTRIBUTES THAT FACILITATE CORPORATE-WIDE ENERGY MANAGEMENT

Fundamental business viability. The manufacturer's front office stability is important. Companies that are the subject of a merger or acquisition, labor disputes, bankruptcy, or severe retrenchment may have fundamental distractions that will interfere with the attention that energy management deserves. A preponderance of such conditions indicates management turmoil that makes energy management impractical.

Replication capacity. Manufacturers with multiple facilities should spread knowledge of energy efficient techniques and compare their ongoing results. The ability to cooperate across multiple sites and across departmental boundaries is required to maximize industrial energy management potential.

Energy leadership (or "champion"). Successful energy improvements are usually led by an "energy champion," a manager that (i) understands both engineering and financial principles (ii) communicates effectively both on the plant floor and in the boardroom, and (iii) is empowered to give direction and monitor results.

Energy market capability. This dimension is straightforward: does the corporation wish to purchase energy through open-market activity or procure as usual from the local utility? If open markets are the choice, the corporation should be prepared to maintain sophisticated search and verification procedures to support its contracting activities. Purchasing decisions should reflect the collaboration of procurement, production, and maintenance personnel.

Leadership intensity. Quality of operations should be demanded, facilitated, and recognized by top officers of the corporation. Adoption of professional and industry standards, such as ISO 9000, are helpful in attaining this attribute. Energy-smart operations will hold employees accountable for adherence to energy management goals and other quality standards.

Pride intensity. Energy efficiency is very much dependent on the behavior of line workers. Employees' awareness of their impact on energy costs must be achieved. A positive, can-do attitude on the part of staff is helpful in attaining potential energy savings. Rewards and recognition can be harnessed to good effect.

Fiscal protocol. Financial considerations involve far more than invoice quotes. Are purchasing decisions made on first-cost or lifecycle costs? Which department pays for improvements and which claims the savings? Do savings count only fuel bill impacts or do they include the value of material waste minimization and greater capacity utilization? What criteria determine adequate payback?

Engineering protocol. Successful energy management depends on an ability to understand energy consumption. This requires benchmarking, documenting, comparing, remediating, and duplicating success stories. Internal skills, procedures, and information services are engaged. The likelihood of building value through energy efficiency varies directly with the depth of these technical capabilities.

In the absence of an energy management process, energy expense control is reduced to one-dimensional efforts. Many manufacturers (either wittingly or not) settle for something less than full energy efficiency potential due to a lack of time, interest, or understanding. The approach

Keywords: Energy; Energy efficiency; Energy management; Industry; Manufacturing; Risk management.

taken by individual manufacturers is very much a function of their organizational attributes and business culture.

PREVAILING ENERGY MANAGEMENT STRATEGIES

The aim of this section is to present the range of typical energy management strategies practiced by industry. Every manufacturer employs some energy management strategy, even if the choice is to do nothing about energy consumption. Consequently, every manufacturing organization adopts one or more of these strategies:

- *Do Nothing*. Ignore energy improvement. Just pay the bill on time. Operations are business-as-usual or "that's the way we've always done it." The result is essentially "crisis management" in that energy solutions are induced by fire-drill emergencies and undertaken without proper consideration of the true costs and long-term impacts.

  Who does this? Companies that do not understand that energy management is a strategy for boosting productivity and creating value. Or, companies that are subject to merger, buy-out, bankruptcy, union disputes, relocation, or potential closure. Or, companies that are extremely profitable and don't consider energy costs to be a problem.
  PROs: you don't have to change behavior or put any time or money into energy management.
  CONs: you don't save anything. Income is increasingly lost to uncontrolled waste. Because you don't inventory your energy usage, you are exposed to volatility in energy markets. You are less prepared to adapt to evolving emissions compliance agendas and you are less capable of spotting opportunities presented by new technologies. Because you don't monitor anomalies in energy flow data, you are more susceptible to lapses in mechanical integrity and plant reliability.

- *Price Shopping*. Switch fuels and shop for lowest fuel prices. No effort to upgrade or improve equipment. No effort to add energy-smart behavior to standard operating procedures.

  Who does this? Companies that "don't have time" or "don't have the money" to pursue improvement projects. Or, these companies truly believe that fuel price is the only variable in controlling energy expense.
  PROs: you don't have to bother plant staff with behavioral changes or create any more work in the form of data collection and analysis.
  CONs: lack of energy consumption knowledge exposes the subject company to a variety of energy market risks. You don't know where your waste occurs nor do you identify opportunities to boost savings and productivity. You are also exposed to energy market volatility and emissions and safety compliance risks.

- *Occasional O&M Projects*. Make a one-time effort to tune-up current equipment, fix leaks, clean heat exchangers, etc. Unable/unwilling to make capital investments. Revert to business-as-usual O&M behavior after one-time projects are completed.

  Who does this? Companies that are insufficiently organized to initiate procedural changes or make nonprocess asset investments. They cannot assign roles and accountabilities for pursuing ongoing energy management.
  PROs: you spend very little money when just pursuing quick, easy projects.
  CONs: savings are modest and temporary because you don't develop procedures for sustaining and replicating your improvements. Familiar energy problems begin to reappear. Energy bills begin to creep back up.

- *Capital Projects*. Acquire big-ticket assets that bring strategic cost savings, but beyond that, day-to-day O&M procedures and behavior are business-as-usual.

  Who does this? This strategy is adopted by companies that believe that advanced hardware is the only way to obtain real, measurable savings. Similarly, they believe that operational and behavioral savings are "weak" and not measurable. However, they have the fiscal flexibility to acquire strategic assets that boost productivity and energy savings.
  PROs: obtain fair to good savings without having to change behavior or organize a lot of people.
  CONs: forfeit savings attributable to sustained procedural and behavioral efforts. Also, savings from the new assets may be at risk if adequate maintenance is not applied.

- *Sustained Energy Management*. Merge energy management with day-to-day O&M discipline. Diagnose improvement opportunities and pursue these in stages. Procedures and performance metrics drive improvement cycles over time.

  Who does this? Companies with corporate commitment to quality control and continual improvement, well-established engineering and internal communications protocol, and staff engagement through roles and accountabilities.
  PROs: maximize savings and capacity utilization. Increased knowledge of in-plant energy use is a hedge against operating risks. Greater use of operating metrics will also improve productivity and scrap rates while reducing idle resource costs.

CONs: you need a lot of in-house talent, cooperation, and a capable energy "champion" to do this.

It is beyond the scope of this entry to comment on which strategies are predominantly encountered in industry. Anecdotal evidence suggests that all industrial energy management strategies can be categorized per one of these five selections. It is also possible for firms to practice multiple strategies simultaneously—for example, price shopping for low-priced fuel commodities in concert with a capital-projects focus.

It should be noted that most of the ten of the experiences documented in the Alliance's corporate energy management case study series can be categorized as "sustained energy management." As such, these companies integrate energy management with day-to-day operating procedures and accountabilities.

ENERGY MANAGEMENT PATHFINDING: MATCHING STRATEGIES WITH CORPORATE ATTRIBUTES

This section will build on the theory of corporate receptiveness to energy management, as presented above. The energy management strategies available to a manufacturer are a function of its organizational attributes, as summarized in Table 1. Note that this is currently presented as theory.

> *Examples for interpreting this Table 1* manufacturer should have attained the attributes of "fundamental viability," "leadership intensity," "fiscal protocol," and "engineering protocol" in order to effectively pursue capital projects as a single-site energy reduction strategy.
>
> Alternatively, a manufacturer that has attained "fundamental viability," "replication capacity," "leadership intensity," "pride intensity," "engineering protocol," and has an "energy champion," should be capable of pursuing both the occasional O&M projects and sustained energy management strategies across multiple sites. In this instance, the company may wish to start with the lesser strategy (O&M projects) and evolve into the practice of sustained energy management.

This typology presumes that energy management for multisite organizations is more demanding than it is for single-site companies. Accordingly, adoption of a certain strategy by a multisite organization requires all the organizational attributes that a single-site organization would be expected to muster, plus the capacity to replicate.

Managers that are contemplating improved energy management are encouraged to consider the case study results and theory presented in this paper. To act on this information, the steps are to:

Table 1 Theory: matching corporate attributes to energy management strategies

	Organizational attributes							
	Fundamental viability	Replication capacity	Energy champion	Energy market capability	Leadership intensity	Pride intensity	Fiscal protocol	Engineering protocol
Strategies for single-site energy reduction								
Do nothing								
Price shop				Required				
Capital projects	Required				Required		Required	Required
Occasional O&M projects	Required				Required	Required		
Sustained energy management	Required		Required		Required	Required		Required
Strategies for replicating energy reduction at multiple sites								
Do nothing								
Price shop		Required		Required				
Capital projects	Required	Required	Required		Required		Required	Required
Occasional O&M projects	Required	Required	Required		Required	Required		
Sustained energy management	Required	Required	Required		Required	Required		Required

Source: From The alliance to save energy.

- Refer to Appendix A, "Determining an Organization's Aptitude for Energy Management." Note which organizational attributes have been substantially attained by the subject company.
- Compare the attained attributes to the information in Table 1. The presence (or absence) of certain attributes determines which energy management strategies are available to the subject company.
- Use these findings to understand what the subject organization can or cannot achieve in terms of energy management.

Keep in mind that this exercise indicates what a manufacturer can expect from energy management given its current organizational attributes and business culture. There may be a desire to evolve to a higher level of energy management than what the current organization allows. What if a manager wants to advance energy management in his or her organization? There are windows of opportunity. An obvious example is when energy market turmoil brings top management's attention to fuel costs. Also, take advantage of annual planning sessions or strategic reorganizations to propose the kind of organizational processes needed to practice sustained energy management. Remember that energy cost control is as much dependent on people as it is on technology.

CONCLUSION

Volatile energy markets are here to stay, and so are competitive and regulatory pressures. Energy price movements will put some manufacturers out of business, while others will decide to move offshore. Surviving manufacturers will not only provide superior products and service, but they will maximize value through operating efficiencies. Energy efficiency is an indispensable component of wealth creation.

Energy procurement strategies such as shopping for low energy prices and supply contracts are only partial solutions to soaring energy expenses. Management of consumption is an underappreciated opportunity. While technology is the foundation for managing consumption, it is the human dimension that makes technology work. Organizational procedures, priorities, and accountabilities are crucial to energy management.

A manufacturer's ability to manage energy consumption is ultimately a function of organizational attributes and corporate culture. This entry advances "energy management pathfinding" concepts. Appendix A presents the criteria that define seven distinct organizational attributes needed for energy management. While sustained, day-to-day energy management is recommended for providing the greatest and most durable value, and it is also the most demanding in terms of operational character. Many companies will find that they are suited for strategies that are less challenging but may also provide less value. The same management diagnostic presented in this entry serves as a pathfinder for matching organizational characteristics with appropriate energy management strategies.

APPENDIX A

Determining an Organization's Aptitude for Energy Management

This Appendix serves two purposes:

- To further define the organizational attributes that a manufacturer needs to pursue energy management as a continuous-improvement process, and
- To determine if a subject organization has substantially attained each of the organizational attributes listed ("Fundamental Viability," "Replication Capacity," etc.).

Please see below. For each attribute, a number of conditions are posed in a bulleted list. When considering a subject company, ask: are most or all of these conditions true? If the answer is yes, then the subject company has substantially attained that attribute. The degree of attainment for each attribute varies directly with the number of considerations that can be affirmed for each attribute. There are no scores, per se. If the subject company has attained a majority of the bulleted considerations listed under an attribute, consider that attribute to be substantially attained.

The range of topics covered by these conditions would be best answered by a high-level manager or perhaps a team of managers. After this exercise, note all of the attributes that have been substantially attained. Compare those results to Table 1. That table indicates which energy management strategies are available to the company, given its organizational attributes.

Fundamental Viability

Your plant capacity is generally stable or growing.
Your company is not currently experiencing excessive turnover of managerial and corporate personnel.
Strikes or other labor-related work stoppages are not considered an ongoing concern for management.
Your company is not the current subject of a merger or acquisition attempt.
Your company is not in receivership, Chapter 7, or Chapter 11 status.

Replication Capacity

Your company operates more than one manufacturing facility.

Energy Management: Organizational Aptitude Self-Test

Your manufacturing processes and products are mostly similar across all plants.

Your facilities are designed and operated per one standard; standards do not significantly vary by facility for asset selection, procedures, and management styles.

Staff members from different plants (or divisions) regularly collaborate to share their common issues and solutions.

Maintenance management is set up to serve multiple sites; individual sites adhere to centralized maintenance planning and procedures.

Your corporation currently uses (or is it willing to use) contract vendors for ongoing energy management.

Energy Champion (Note: All of These Conditions Must be Met to Have a True "Energy Champion")

Your lead energy person has thorough knowledge of technology and staff capabilities at the facility level.

Your lead energy person can prepare financial analyses to support engineering proposals and convincingly present these to top managers.

Your lead energy person applies more than 50% of his or her time to energy issues.

Your lead energy person can give direction or at least influence decision making by general managers.

Your lead energy person understands utility tariff structures and administers relations with utility providers.

Leadership Intensity

Your organization actively maintains disciplines of excellence such as Six Sigma, ISO 9000, or Total Quality Management.

Process technologies, procedures, or staff expertise are a selling point in marketing your products.

Current and future environmental impacts from manufacturing operations are a concern to your top management.

A corporate officer consistently reviews cost and quality performance data for all facilities.

To most of your corporate leaders, "energy efficiency" is perceived as an "opportunity" as opposed to a "hassle".

Staff compensation, raises, and rewards are impacted by their stewardship of energy, raw materials, and other inputs.

Production metrics are integral to performance evaluations for facility managers and staff.

Your facilities are subject to public scrutiny or "good citizenship" expectations.

Pride Intensity

All or most plants are consistently high performers with respect to health and safety compliance.

Most plant-floor staff members are well trained for their jobs.

Staff turnover is not considered to be a problem.

Your typical plant worker philosophy can be described as "do what's right" instead of "do what's easy."

You describe your plant equipment as "well maintained" as opposed to "poorly maintained."

To most of your facility staff, "energy efficiency" means "opportunity" as opposed to "hassle."

Key facility personnel maintain professional certifications.

Your organization prescribes and enforces technical training for facility personnel.

Fiscal Protocol Intensity

Asset purchases are judged primarily by life-cycle costs (acquisition plus life-time operating, maintenance, etc.), instead of first costs (cost of acquisition).

Your organization uses (or is willing to use) leases and other off-balance sheet methods to finance major acquisitions.

Your organization's investing strategy seeks large payback as opposed to fast payback.

Most of your facilities take utility tariffs into account when planning their operating times.

Facilities invest in plant improvements (as opposed to simply fixing what's broken).

Energy-related capital project proposals assigned a hurdle rate equal to or lower than other project proposals.

Any energy savings are returned to the facilities that successfully implement capital improvements.

Your facility managers understand utility tariffs and their role in determining energy expenses.

Energy Market Capability

Your company is willing to make an on-going effort to use energy marketing services to obtain lowest-cost energy commodities and risk-hedging securities.

Engineering Protocol Intensity

Your facilities maintain a scheduled maintenance routine for powerhouses, motor drives, pumps, compressed air, and similar utilities.

Your facilities maintain a protocol for responding to anomalies in operating performance data.

Your chief engineers are comfortable with using software to analyze engineering issues.

Plant managers develop (or help to develop) project proposals for capital budgeting purposes.

Your facilities maintain procedures for safety, health, and waste management.

Most or all of your facilities maintain an action plan for improving process efficiencies.

Your organization maintains a database or archive that documents engineering problems and solutions.

Your facilities track the volume of factor inputs required per unit of production.

Your facilities monitor scrap or error rates.

Your annual budgets include factor inputs and production targets as well as dollar figures.

Production, inputs, and cost-performance data are created and utilized at the facility level.

Your engineering problems and emergencies are generally unpredictable and unique as opposed to predictable and recurring.

Company-wide production stats are made available to all facility staff by publication, discussion, or graphic display.

BIBLIOGRAPHY

1. Thumann, A. *Plant Engineer & Managers Guide to Energy Conservation*, 6th Ed.; The Fairmont Press: Lilburn, GA.
2. Curl, R.S. *Successful Industrial Energy Reduction Programs*, The Fairmont Press: Lilburn, GA.
3. John, M.S. *Slashing Utility Costs Handbook*, The Fairmont Press: Lilburn, GA.

Energy Master Planning

Fredric S. Goldner
Energy Management & Research Associates, East Meadow, New York, U.S.A.

Abstract
Energy master planning (EMP) is the process of transitioning an organization's culture from the traditional "fixed cost, line item" view of energy to one in which energy is recognized as both an opportunity and a risk that that can be managed. An EMP can guide an organization in longer-range planning of energy cost reduction and control as part of facility maintenance, management, and design. Though an EMP encompasses traditional efforts to cut energy costs, it also includes many steps not usually taken under conventional, technically oriented energy management, including energy procurement, energy-related equipment purchasing, measurement and verification (M&V), staffing and training, communications, and setting energy consumption targets and tracking/feedback loop systems. A critical difference between energy master planning and traditional energy management is an orientation toward the future. An EMP deals with a longer timeframe than just simple payback periods and goes beyond merely reducing energy use. EMP requires sustained commitment from top levels of an organization down through the rank and file, a dedicated energy manager, and creation of an energy team with membership from across the organization.

INTRODUCTION

Developing an energy management program or more broadly energy master planning (EMP) is the process of transitioning an organization's culture from the traditional "fixed cost, line item" view of energy to one in which energy is recognized as the opportunity and risk that it has become.

An EMP can guide an organization in longer-range planning of energy cost reduction and control as part of their facility maintenance, management, and design. An EMP can even lead the energy budget to be recognized as a potential profit center and source of opportunity rather than just another business expense. An EMP moves beyond the confines of traditional engineering to include energy procurement, energy-related equipment purchasing, measurement and verification (M&V), staffing and training, communications, and setting energy consumption targets and tracking/feedback loop systems. The long-term perspective goes beyond simply cutting last year's energy use. It makes energy awareness part of the everyday operation and "mindset" of the organization.

If you're thinking this doesn't apply to your firm or clients because you're too small, think again. This approach works for an organization as small as a single site to an owner with half a dozen small buildings to Fortune 100 companies. The effort and level of detail vary, respectively, but the approach is basically the same. The good news is that there are resources available to assist professionals and the organizations they serve to understand the EMP process and get started on this path.

To be successful an organization must treat energy in the same business-like manner that they do all other major expenses, such as labor and materials. "If you can measure it, you can manage it"[1] is the catch phrase of Paul Allen, the energy manager at Disney World, who has been instrumental in implementing one of the most successful EMPs in the country. "Energy is a competitive opportunity... Winners manage it effectively!" is the driving force of another highly successful program at Owens Corning. Many other organizations throughout this country and across the globe have recognized that to achieve significant and sustained energy cost control, organization needs to make energy management an integrated part of their business/operations. There are various ways to pursue the process, but a key requirement is an interdisciplinary mix of engineering/technical, behavioral, and organizational or management components. The EMP must be integrated into the basic business operations.

UNEXPECTED BENEFITS

One of the most potent driving factors in many organizations' efforts to address energy issues is increased profitability that can be realized through reduced/optimized energy expenditures. Beyond the "bottom line" impacts, an EMP can also provide an organization with a more secure energy supply, reduced downtime of systems, improved equipment availability, reduction in maintenance costs/premature system replacement expenditures, and overall productivity gains. Additional benefits that

Keywords: Strategic; Energy management; Bottom line; Upper management; Profit center; Buy-in; Commitment; Long-term; Integrated; Energy accounting; Business planning; Road map.

have been documented include quality-of-life improvements, enhanced product quality, better operational safety, reduced raw material waste in industrial plants, and increased rentability in commercial facilities. An often overlooked outcome of an EMP is the reduction of emissions, among other environmental impacts that help organizations become perceived as better corporate citizens.

As more companies move toward an integrated corporate strategy that links environmental, economic, and social considerations, the results of an EMP can be used to considerable public relations advantage. Ratings in one of the sustainability indices and publication of an annual sustainability report (using, for example, the Global Reporting Initiative guidelines) can give an organization a higher standing in the business community and can result in a higher level of trust by stockholders.

A PROCESS FOR OPTIMIZATION

Though energy prices are volatile, and energy security is often far from reliable, facilities now face leaner operating funds and increased directives to do more with less. Optimizing a facility's operations budget frequently means cutting energy costs. But how do you do so without cutting occupant comfort or productivity? How do you know where to start and what steps to take? And how do you persuade upper management that energy costs can be controlled?

While the general goal of an EMP is the same as that of conventional energy management, the two disciplines are far from identical. Traditional energy management, which is technically oriented, is essentially centered around the boiler or mechanical room. Energy master planning, on the other hand, is a business management procedure for commercial, institutional, and industrial operations. With this approach, it is not enough simply to manage installations. The process involves:

- Developing strategies
- Creating processes to fulfill those strategies
- Identifying barriers and finding procedures to overcome them
- Creating accountability
- Providing feedback loops to monitor and report progress.

Clarification of the terminology for these disciplines is important, as the terms mean different things to different people. In other English-speaking nations, for example, the defining feature of "energy management" is the emphasis on integration with business practices to analyze, manage, and control energy. In the U.S., however, "energy management" has traditionally referred to developing technical and operational measures involving equipment handled by facility managers—not processes such as energy procurement and business planning usually handled by purchasing agents, production personnel, and corporate economists. A typical U.S. energy manager's responsibility rarely extends much beyond utility bill analysis, an occasional energy audit, or managing installation of system upgrades.

Though an EMP encompasses traditional efforts to cut energy costs, it also includes many steps not usually taken under standard energy management. Rather than being just equipment-oriented, an EMP starts long before a comprehensive energy audit and extends beyond commissioning of new systems. An EMP may be thought of as a road map to savings that starts before and continues after energy-efficiency measures are involved. Why do you need such a map? Because you can't get there if you don't know where "there" is. How many of us are willing to undertake a trip that will have costs and risks (like any business or personal decision) if we don't know where we are going? A map makes clear not just your final destination, but also how to get there—and it leaves no question about the starting point.

An EMP is a process to organize and improve your existing energy-related resources and capabilities. Resources, in this case, include standard operating procedures, institutional memory, and actual records (such as energy bills, plans and blueprints, and energy contracts). Capabilities include facilities staff familiar with mechanical room equipment, consultants for energy costs or usage, energy cost accounting and management systems, and meters and software that monitor them. Once organized and integrated, these resources and capabilities become powerful tools for managing energy and producing savings.

TODAY'S PRACTICES FOUND WANTING

Current thinking about managing energy often falls short of the EMP perspective. Today's energy management frequently reflects a short-term crisis mentality: A facility manager or energy manager concentrates on whatever immediate 'fires' have to be put out at his facility. By contrast, with an EMP mindset, the energy manager might first look to increase the efficiency of systems already in place and then move ahead to lay a solid foundation for improving performance via tight energy specs and training.

Current thinking is also often characterized by piecemeal 'solutions' that lead to short-sighted component replacement. When equipment breaks down, it gets replaced without anyone's asking whether this is the best option, long-term. Typically, business thinks in a quarterly mindset because of the short budget cycles of our economic system. That kind of "right now and right here" viewpoint creates situations in which life-cycle thinking is not possible because potentially higher initial costs are visible, but potential benefits tend to be invisible. As a result, when first cost becomes the main criterion for purchasing, such a focus distorts planning and decision-making. Too often, facilities choose the cheapest solution, based on the current quarter's budget, without realizing it could cost them more later.

A critical difference between energy master planning and conventional energy management is an orientation toward the future. With an EMP, you don't just look to increase the efficiency of systems already in place. Instead, you plan for new or changed loads based on your detailed information about the facility's long-term business strategy and projected growth.

Energy master planning deals with a longer timeframe than just simple payback periods. For example, typical financial constraints for a commercial building upgrade often dictate a 2- to 3-year timeframe. Energy master planning, however, looks deeper than simple payback and goes beyond merely reducing energy use. Therefore, an energy professional with an EMP considers life-cycle costs and views long-range planning of energy cost minimization/optimization as part of overall facility maintenance, management, and design. To sustain the savings over time, an EMP calls for hiring an energy manager/coordinator and setting up an energy team. But who is this energy leader? Not simply the facility manager wearing yet another hat, but rather, a highly trained and, often, certified specialist with sufficient acumen and expertise to understand, handle, and maintain whatever new energy systems and practices are to be put in place. Willingness to identify (from within the organization) or hire an individual with the required qualifications and to define his or her responsibility as managing energy rather than managing the facility is a requisite indication of senior management's true commitment to energy master planning. The policy guidance needs to come from senior management levels to convey "buy-in" throughout the management structure of the organization.

IMPROVING BUSINESS AS USUAL

Energy master planning is one way of creating a new norm of "business as usual," taking its cues from time-tested business management practices. For example, rather than carrying out upgrade projects with a defined start and end point, an energy manager uses processes that continue to turn up new sources of savings. This requires a level of creativity for identifying and capturing new opportunities. If a chiller replacement is needed in Building A, for instance, a better approach might be to expand the capacity of the existing central chilled water plant in Building B and run piping from Building B to Building A.

Without an EMP, the Purchasing department often buys equipment and energy, rather than the Facilities department. And Facilities is so busy handling emergencies that they have little interaction with other departments in the organization, let alone industry groups or other end users. An EMP avoids either departmental isolation and turf wars by including representatives from such departments on the energy team. The team consists of more than an energy manager, facilities personnel, and design and construction specialists. To be effective, it needs to incorporate representatives from every department in the organization impacted by energy. This may mean including purchasing, accounting, engineering, environmental affairs, maintenance, legal, health and safety, corporate relations, human resources and training, public relations and marketing, and members from the rank and file (hourly employees).

Energy master planning is a significant challenge. It often rejects the status quo and may question existing components of an organizational culture that do not support energy master planning. If you have always done something one way, you don't necessarily have to perpetuate what could be a costly mistake. For example, using outdated specs (e.g., calling for T12 lamps instead of T8) allows inefficiency to continue.

ORIGINS OF THE AMERICAN EMP APPROACH

A common set of energy master planning definitions and processes has taken root in English-speaking countries other than the US. In the United Kingdom, the energy master planning concept has been practiced for well over a decade and vigorously promoted by the government's Action Energy program. It has been so successful that Canada, New Zealand, and Australia have each adapted the process to their own conditions.

In the US, there are a few recent models cover some, but not all, aspects of energy master planning. management system for energy (MSE) 2000 is a specialized quality improvement standard (like ISO 14000), developed by Georgia Tech. It's an Amercian National Standards Institute (ANSI)-approved management system for energy that covers all sectors, not only buildings. Georgia Tech offers a certificate program to train energy professionals in this standard. The Association of Energy Engineers delivers the Developing an Energy Management Master Plan and Creating a Sustainable Energy Plan Workshops to both 'real' and 'virtual' end users. Live presentations and online seminars present an energy master planning approach with strong emphasis on integration with business strategy.

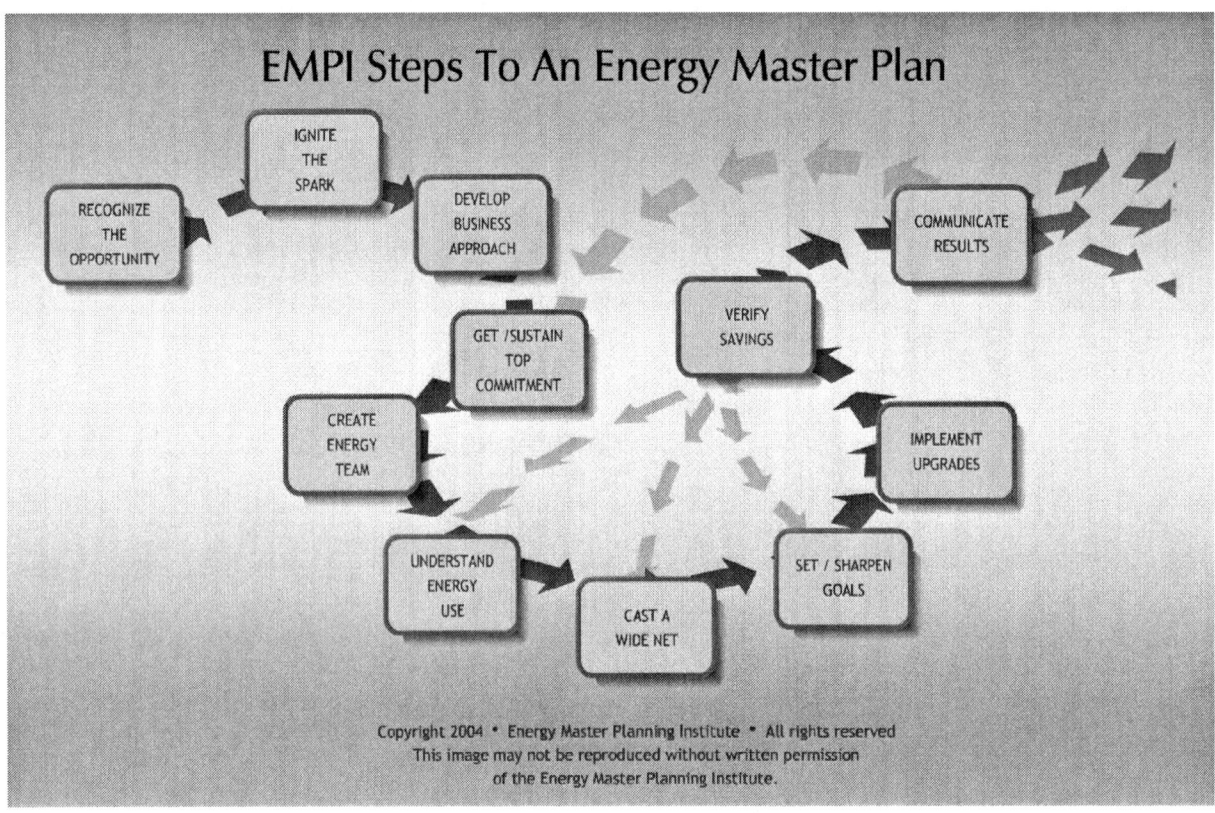

Fig. 1 Energy Master Planning Institute (EMPI) steps to an energy master plan (EMP).

STEPS TO AN ENERGY MASTER PLAN

To fill this gap in the U.S., the Energy Master Planning Institute (EMPI) was established and has developed a set of steps that lay out the process for applying energy master planning to a commercial, institutional, or industrial facility (See Figure 1). This model, which builds on accepted international approaches, offers US organizations a broad and integrated business approach for managing energy that is both strategic and sustainable.

The steps presented in Figure 1 appear sequential, but built into the energy master planning process are series of feedback loops, evident in figure 1. These should not be overlooked, as they guarantee the viability of the process and offer many points for input from internal and external stakeholders, from the Chief Executive Officer (CEO) to boiler room personnel, and from the local community to organizational peers across the country.

Recognize the Opportunity

This is where the process starts—becoming aware of the facility's major energy-related opportunities and challenges. Whether it's the facility manager or the energy manager, whoever takes the initial action must define the opportunities and challenges succinctly so they can be clearly communicated. When the leader creates and implements an EMP, that plan can actually generate a revenue stream and gain recognition for the leader's contribution to the company's bottom line. By integrating energy concerns into the overall corporate business strategy, the energy budget will come to be seen as a potential profit center and source of opportunity—and not just an uncontrollable expense.

Ignite the Spark

Since top executives in an organization are the ultimate decision-makers, particularly concerning funding, it's critical to spark their attention early in the process. It's unlikely that the facility manager (or the energy manager, if there is one), has direct access to upper management, so the right individual to make the pitch to the CEO or Chief Financial Officer CFO must be identified. This could be someone along your management chain, or a consultant or board member or a senior officer in another department who can sell an idea at the top. To find the right person, it's important to understand the decision-making structure of your organization as well as the vision, mission, and long-term business plan.

Develop a Business Approach

To get access to senior executives and persuade them to listen to a new idea, you must speak to them in their own language—dollars/ft^2/year savings, not kWh. Since most executives have not been introduced to the bottom-line value of energy master planning, your task is to change management culture so executives no longer view energy as an uncontrollable expense. Part of the marketing message is that managing energy is no different from tracking, controlling, and accounting for the costs of raw materials, IT, personnel, safety, or the corporate fleet.

Obtain and Sustain Top Commitment

Without question, this step is the most difficult in the energy master planning process. Serious commitment from senior management means providing ongoing financial resources and personnel with appropriate credentials. A one-time memo of support is not effective. To secure top executive commitment, you must show how an EMP can support key business goals, such as growth, customer satisfaction, or a sharper competitive advantage. You might, for instance, explain that lower operating costs and increased energy efficiency can bring a higher level of occupant comfort, which, in turn, can mean a lower worker absentee rate–and, possibly, a greater employee retention rate. To gain top executive commitment, make sure that CEO and CFO see the dollar savings highlighted in the energy team's regular reports.

Top management commitment is the single most important goal in an effective and lasting EMP. Not only is it crucial to have this commitment, it must also be obvious to everyone throughout the organization. Top management should participate in the program start-up and continue to reinforce that commitment periodically with both words and actions. Such organizations as Walt Disney have achieved this top-level buy-in and gained significant and lasting bottom-line results from energy master planning.

Create an Energy Team

A middle manager in your organization may declare that energy master planning is only the responsibility of the Facilities department. However, when an energy team represents the company's broad interests, energy concerns can be successfully integrated into the overall business plan. Every department that's impacted by the organization's energy use should be invited to participate on the team. This includes:

- Facilities
- Construction/engineering
- Purchasing
- Accounting
- Inventory
- Environmental, health, safety
- Legal
- Public affairs
- Property/asset management
- Leasing/real estate
- Risk management
- Security
- Financial service

This should result in the creation of an Energy Committee. Depending on the size of the organization, there may be separate Technical and Steering Energy Committees. In addition to representatives from the departments listed above it is important to include members of the rank & file (hourly employees or line workers). By including such folks in the planning process it allows not only those individuals, but their peers as well, to become aligned with the EMP objectives and particular initiatives during the early stages. Having these employees on the team helps the rest of the rank and file staff (who will be needed to carry out many of the activities) see the energy program as something other than 'just another management flavor of the month'. It also increases success, as feedback from these employees often helps address many of the nuts-and-bolts "bugs" in advance.

The energy team should be headed by an Energy Manager or coordinator. The Energy Manager needs to have a mix of technical, people, and communications skills, and must be enthusiastic. He or she should thoroughly understand the organization's operations and should report to someone as close to the top of the organization as possible, so he or she has the clout need to get things accomplished.

Understand the Organization's Energy Use

Conduct disciplined information analysis. Collect and use metered–not just billed–energy data, and analyze it with software designed to manage energy costs. "Fully understanding current energy use practices–the when, where and how much of energy consumption–in detailed qualitative terms is an essential precondition to energy management master planning."[2] For example, to determine exactly where the energy is going, specify and install sub-metering and a data acquisition system. Such tools will help you monitor and collect, as well as analyze, the meter data for electricity, fossil fuels, steam and condensate, and water. For loads of several hundred kW, it's advantageous to use interval meters that allow a close look at 15-minute interval or hour-by-hour use of energy, so you can see where peaks and valleys in usage occur. This kind of metering is also essential for internal billing. Besides establishing points of excessive usage, it can also help manage loads and pinpoint efficiency opportunities.

Cast a Wide Net

Casting a wide net means looking beyond the central plant for savings opportunities, such as lighting, office equipment, elevators, and localized process loads. It means making construction and equipment specs energy-conscious and creating ways to "enforce" such specs. For example, specifying T8 lamps for efficiency upgrades is not sufficient. The architecture, design, and construction staff should build T8s into their specs for non-energy-related upgrades, such as converting a library to training rooms. Casting a wide net also means revising inventory and purchasing practices to support energy efficiency. As in the example above, to ensure that the new installation continues to function properly, the purchasing specs need to list T8s, not T12s.

Set or Sharpen Goals

Once the organization's energy use is determined, strategic thinking linked to the long-term growth objectives will determine how much energy you should aim to save, where the savings should come from, and when those savings should occur. The energy manager along with the energy team, proposes a phase-one timetable with quantifiable goals to reduce energy use and operating and energy costs. Senior management, however, must mandate the goals and schedule or they carry no weight. Ideally, the CEO, CFO, or the Board issues a position calling for measurable reductions at the end of a 3-year fiscal cycle, based on current use: a percentage reduction in peak electrical demand across the entire facility, a reduction in annual kWh consumption of electricity, and a reduction in overall British Thermal Unit (BTU) for fuel per gross square foot. Lesser annual goals will help keep progress on track. A key is to set achievable goals, and then work to meet or beat them.

Implement Upgrades

Because energy managers generally have experience with efficiency upgrades, it is critical not to fall into old patterns of sporadic efforts and a piecemeal approach, with an eye on the quickest payback. To carry out an EMP, start the upgrade process with a comprehensive energy audit—not just a walk-through—to determine which upgrades will give the best results, not just the best rate-of-return. For example, in a commercial building operation, to avoid a rush upgrade for a new tenant, pre-audit all your buildings so you know what work needs to be done before that tenant signs their lease. Or, if you're considering recommissioning, look at it from the longer-term energy master planning perspective. Recommissioning may not be sufficient, as it means only the existing systems would operate more efficiently. But what if they need to be ripped out as part of the upgrade?

Don't fall into the common trap of focusing just on the high-profile, capital-intensive, projects. While a new chiller or micro-turbine cogeneration system may be a good photo opportunity, focusing on the less visible details can often provide the lion's share of savings through lower-cost measures that improve operations & maintenance (O&M) or maintenance and management practices.

Verify Savings

Once the upgrade process is underway, you must first validate the savings with a recognized technique such as the International Performance Measurement & Verification Protocol (IPMVP or MVP). However, instead of following the M&V protocol after the fact, build measurement and verification into the design—that is, install it as part of the upgrade—so you can identify the savings as soon as the device is turned on. Then, set up an energy accounting system that tracks usage and savings, thereby providing objective accountability. All too often energy management activities are not perceived as being of value because the energy team did not plan ahead and position themselves and theirs efforts for recognizable success. A North American floor coverings manufacturer reported how "in the past we would complete a project to find that we did not have the baseline or operational data to judge whether the project was successful."[3] How willing will management be to fund the next energy related project or initiative if the energy team cannot prove that prior efforts actually saved what was projected?

With an energy accounting system, everyone on the energy team will be working from the same data, and you can prepare regular progress reports, plus quarterly reports to the CFO and an annual report to the Board. This is critical: The energy team must account for the savings—or lack of savings–to senior management, using a feedback process so goals and targets can be reset if they aren't met. Use this process to verify energy use and to help identify any new opportunities for savings, and then to fine-tune your goals. If you fall short of your targets, you need to figure out why. Or perhaps you did meet your targets, but parts of the data were wrong. Perhaps the metering is off, or the energy accounting system needs readjustment. Regularly scheduled feedback from these reports will keep the energy master planning process up-to-date and realistic and will ensure accountability. One approach that has been successfully used in many organizations is the use of "energy report cards" and "intra-organizational listings/scorecards". Produced by the energy manager and sent out to all facilities/operating groups/departments on a monthly basis, these tools inform as well as motivate, based on the certainty that no one wants to be on the bottom of the list.

Communicate Results

Successful implementation of all stages of the EMP can be a useful vehicle for departmental—and personal—recognition. With a representative of the corporate PR office already on the energy team—enlist that individual's expertise for internal and external communications. Inform the organization's Board about the financial savings and improved asset value. Let the staff and community know about the environmental benefits. Apply for energy awards for national recognition in your sector. And use the good will generated by the success to keep the energy master planning process moving forward.

TIPS FOR SUCCESS

From these steps to an EMP, two points emerge as the most critical to success.

The first is ensuring buy-in from the top, with a long-term commitment and binding, formal statement of energy policy for the organization. Commitment to action also means that senior executives support the people in the middle—delegating authority to the energy manager or facility manager. If the EMP lacks commitment from the top down, no amount of effort by the energy manager will make it succeed. Once you've caught executives' attention, keep energy master planning on their radar screen by having the energy team build it into the organization's business plan. Such a commitment must also be apparent. Top management must continue to be seen (by all levels of staff) reaffirming their commitment, lest the EMP be perceived as just another short-term initiative of the organization.

The second critical component is line-management accountability, making specific individuals accountable for sustaining the savings. On the management side, executives need to mandate quantifiable goals and targets, strengthened by obligatory deadlines. Such requirements should even be built into job descriptions and evaluated as part of annual performance standards reviews. Likewise, incentives can be offered through these same personnel standards for individuals who meet their energy goals. Employee teams with day-to-day knowledge of the facility's operation can also be organized to identify additional opportunities for savings. At the upper level of the organization, corporate accountability for successful energy management can be communicated (and made public) through such mechanisms as the Global Reporting Initiative, especially in firms with sustainability principles that follow a "triple bottom-line".

CONCLUSION

Energy master planning is an effective and long-term shift in organizational cultures to address and adapt to the impact energy resources can have on their competitive posture and economic success. Energy is a universal raw material and is essential to the operation of almost every commercial, institutional, governmental, and even non-profit organization. Significant changes to the basic structure of the energy supply chain and significant energy price volatility, driven by a wide range of political and physical events (e.g., climate change, weather), have made energy both a competitive opportunity and a risk requiring active long-term planning to manage. Energy master planning requires buy-in from top levels of an organization down through the rank and file, selection/hiring of a dedicated energy manager, and creation of an energy team with membership from across the organization. This team must be given the resources and top-level access to develop and implement long-term planning for procurement and operations throughout the organization that places a premium on lasting energy-use reductions and cost optimization, rather than "burst" efforts that provide quick and quickly forgotten energy management efforts.

Successful organizations do not exist on a quarterly basis. They plan for the long-term in all aspects of their operations. Experience has shown that organization-wide adoption of EMP is effective in optimizing energy costs and needs to ensure competitive posture and success in the long-term business reality in which organizations exist. This is the new bottom line for energy.

ACKNOWLEDGMENTS

Permission has kindly been granted by the Energy Master Planning Institute for use of materials on its Web site and for materials published and presented by Coriolana Simon, founder of the EMPI.

REFERENCES

1. Allen, P.J.; Kivler, W.B. *Walt Disney World's Utility Efficiency Awards and Environmental Circles of Excellence*, Proceedings of WEEC 1995; Association of Energy Engineers: Atlanta, GA, 1995.
2. Tripp, D.E.; Dixon, S. *Making Energy Management "Business as Usual": Identifying and Responding to the Organizational Barriers*, Proceedings of the 2003 World Energy Engineering Congress; Association of Energy Engineers: Atlanta, GA, 2003.
3. Key, G.T.; Benson, K.E. *Collins and Aikman Floorcoverings: A Strategic Approach to Energy Management*, Proceedings of the 2003 World Energy Engineering Congress; Association of Energy Engineers: Atlanta, GA, 2003.

Energy Project Management

Lorrie B. Tietze
Interface Consulting, LLC, Castle Rock, Colorado, U.S.A.

Sandra B. McCardell
Current-C Energy Systems, Inc., Mills, Wyoming, U.S.A.

Abstract

Today's energy projects are becoming ever more complicated technically, beginning to involve more people, and beginning to include more integrated components and systems. As the complexity of an individual project increases, so does the importance of project management. Even if the responsibilities for project management are outside the business organization, good project management principles apply. To successfully manage an energy project, the project manager must have a plan by which he or she can direct project activities; the tools to document agreements and project work; the ability to facilitate communication; and the skills to manage the project's technical side (cost and schedule) as well as the project's human side (conflict resolution, team building, and gaining commitment from stakeholders). The process of energy project management is discussed here along with several tools that can assist managers in developing their own energy projects.

INTRODUCTION

Energy project management was once simple—meeting costs and schedules were considered sufficient. That is no longer true, and it is imperative that good project management principles are used and that the energy business owner remains intimately involved. The goal of this article, written for the energy business owner and the energy business project manager, is to discuss key requirements for successfully managing an energy project.

Today's energy projects are technically complicated, involve many people and disciplines, and consist of numerous integrated components and systems. In an energy project, the up-front sales and marketing, initial energy data collection, and data analysis take a tremendous amount of time and effort. After those efforts, there is often an assumption that project implementation will take care of itself. Of course that is not the case, and a poorly managed project will not deliver the intended results.

There are also time and cost advantages to good energy project management. Buildings last for 30–50 years and errors and omissions grow in importance and cost over time.

It is also less expensive to pay attention to project management than to ignore it. This allows potential problems to be identified and solved earlier and at a lower cost. The general rule is that if an item costs $1 to correct while the project is just an idea, it costs $10 during project definition, $100 during project design, $10,000 during project start-up, and $100,000 after the project is up and running.

To successfully manage an energy project, an energy project manager needs a plan by which he or she can direct project activities; the tools needed to document the agreements, the actual project work, and to facilitate communication; and the skills to manage the technical side (cost and schedule) and the human side (conflict resolution, gaining commitment) of the project. This article will introduce basic energy project definitions, an energy project management plan, and the tools and skills necessary for the successful energy project.

Keywords: Energy project management; Commissioning; Start-up; Project stages; Performance bid; Energy efficiency; Schedule risk analysis.

DEFINITIONS

Commissioning. The systematic process of assuring by verification and documentation—from the design stage to a minimum of one year after construction—that all building facility systems perform interactively in accordance with the design documentation and intent and with the owner's operational needs, including the preparation of operation personnel.

Energy project management plan. A set of organized activities, each having a definite beginning and end, that when completed by a project team, deliver a specific business objective.

Milestone. Key, high-level summary activities that must be completed by a set date.

Operation. Business operation for which the energy project was performed.

Energy Project Stage	Stage Deliverables
Conceptual Stage	<ul><li>Project's general description or project brief,</li><li>Needs basis,</li><li>Return on Investment (ROI) estimate, such as a Life Cycle Cost Analysis (LCCA),</li><li>Initial cost range,</li><li>Initial benefit range (including intangibles, where appropriate),</li><li>Initial project timing,</li><li>Project Charter,</li><li>Initial project management team.</li></ul>
Definition Stage	<ul><li>Detailed project scope,</li><li>Detailed project cost,</li><li>Initial project funding strategy,</li><li>Bid process strategy,</li><li>Updated Return on Investment (ROI) or Life Cycle Cost Analysis (LCCA),</li><li>Project Management strategy,</li><li>Final Project management team,</li><li>Initial project schedule,</li><li>Initial delivery plan,</li><li>Initial skilled resource personnel plan,</li><li>Initial communication plan,</li><li>Initial operational and maintenance training plan,</li><li>Major risk and challenges,</li><li>Initial contingency plans.</li></ul>
Design Stage	<ul><li>Detailed engineering designs,</li><li>Detailed project plan and schedule,</li><li>Detailed skilled personnel resource plan,</li><li>Risk analysis and contingency plan,</li><li>Financing plan,</li><li>Operations and maintenance plan,</li><li>Commissioning plan,</li><li>Operations plan,</li><li>Building modeling as appropriate,</li></ul>

Fig. 1. Energy project stages and deliverables.
Source: From Interface Consulting, LLC 2005.

	• Sustainability analysis and design plan, • Bid for funding.
Construction Stage	• Detailed construction schedule, • Detailed construction personnel resource plan, • Coordination plan for sub-contractors providing different system components, • Construction communication plan, • Initial start-up and ongoing operations plan, • Updated operational and maintenance training plan.
Delivery Stage	• Detailed start-up schedule, • Detailed operational and maintenance personnel resource plan, • Detailed start-up and ongoing operations plan, • Detailed turnover plan, • Updated operational and maintenance training plan, • Project Post mortem.

Fig. 1 (*Continued*)

Project stage. A logical sequence of activities, milestones, and deliverables.

Schedule risk analysis. Equipment and computer programs that let users measure, monitor, and quantify schedule impact on their energy project and help identify schedule risks and opportunities.

ENERGY PROJECT MANAGEMENT PLAN

The energy project management plan is the roadmap for the project. It is a set of organized activities, each having a definite beginning and end, intended to meet a specific business objective when completed by a project team. To make the management of the process easier, project activities are divided into separate structured stages.

Each project stage contains a logical sequence of activities, milestones, and deliverables required to achieve the project objectives. The milestones are key, high-level summary activities that must be completed by a set date, and are linked to the success criteria for each particular stage. The number of stages and milestones is dependent upon the energy project plan length and complexity. The deliverables ensure that all prerequisite work is completed prior to moving to the next stage of the project. Transitions from one stage to the next are times to assess and review both the stage just completed and the project plan.

Fig. 1 illustrates the typical stages and deliverables in an energy project plan.

The plan's main purpose is to provide the roadmap for specific project activities, but it is also a collaboration, alignment, and communication tool. It details which activities are happening at what time, who is involved, and what is happening next. It is also an acknowledgement tool—recognizing when the project team has achieved milestones and when it is on track to delivering the overall project objectives. When the energy project plan is well crafted, the plan followed, and the success criteria realized, an energy project manager can produce maximum results in delivering the overall project objectives.

Conceptual Stage

The conceptual stage is the beginning of an energy project. Its deliverables are broad and are intended to evaluate the project's appropriateness for the business and its needs by answering two key questions:

- "Is this project worth a level of investment?"
- "How does this project further our business plan and mission?"

Business stakeholders answer these questions and develop the initial cost estimate. At this point, there is uncertainty in the project scope and associated costs. While strategic level decisions can be made with these numbers, the conceptual stage costs should not be used as the project cost commitment.

If the project is a fit for the business, a project summary or project brief is written and then reviewed with key stakeholders. The project brief reviews have two purposes—to educate and to gain alignment and commitment to the proposed project. Once the project has commitment from all appropriate parties, a project charter is created. The project charter includes much of the same information as the project brief, as well as the names of the project team and project authorization signatures. It is the official document authorizing the project manager to begin work. Once the charter is approved, the project moves into the first tactical stage—definition.

Definition Stage

This is the stage at which the "how?" and the "when?" for an energy project are defined. "Definition" requires the largest amount of documentation and the most deliverables of any stage and it builds on information from the conceptual stage. Project work and scope is identified for each step in the project.

The scope detail includes electrical and mechanical layout, as well as civil and structural where appropriate, and the general layout of major pieces of equipment, piping, and electrical systems. A physical inspection of the planned location is recommended so that obstacles can be identified as early as possible. For LEED certification (U.S. Green Building Council's Leadership in Energy and Environmental Design), as well as for good business practice, all interactions between building systems should also be taken into consideration from this point forward. A third party "owner's agent" or "commissioning agent" is also a valuable team member who can ensure that the building is both designed and constructed in an energy efficient way and that the finished product meets the project objectives.

The initial personnel resource plan for the project should be completed as early as possible, including a decision on whether to have the project managed by in-house personnel or to use project management services provided by an outside vendor. Remaining personnel are identified iteratively during the scope refinement and activity identification stage and include core project team members, technical resource members, support members, operating department members, and outside vendor or resource team members. All personnel, titles, and general responsibilities should be summarized in a single contact document for wide distribution throughout the project and the business.

Once the project personnel are named and engaged, a communication plan is developed. The plan specifies how project information will be collected, what information will be disseminated, who the information will go to, and when the information should be communicated. A RACI chart (discussed in "Collaboration and Alignment Tools") provides structure to this communication plan throughout the project.

All project activities are planned for during the definition stage and it is critical to remember that the construction and training activities inherent in energy projects involve large numbers of people—so the human elements of the project must be considered at all times. Due diligence in definition provides a basis for all downstream work and provides important details for communication with all personnel involved in the project. This is the first stage where all people involved will have enough detail to see what the project will really look like, how it will impact them, and what steps should be taken to minimize the impact on existing business operations.

Another face of the critical human element in energy projects that carries heavy operational impact is training of the operational and maintenance personnel. Training should be planned during the definition stage and is developed by the following model:

- Step 1—Identify the new tasks and activities.
- Step 2—Identify the skills to support the Step 1 tasks.
- Step 3—Identify the knowledge and demonstration to show mastery of the Step 2 skills.
- Step 4—Identify the training required to accomplish Step 3.

Once the steps are identified, corresponding theoretical and practical project activities are incorporated into the project schedule.

Another key milestone document in this stage is the delivery plan, which includes success criteria for the project completion substages of start-up, full-going, and turnover. The success criteria are negotiated and mutually agreed to by both the project team and the business operation. The delivery plan documents what the project team is promising to provide for the operation and what the operation promises to provide for the project. When designed well, the delivery plan is a powerful communication tool that can engender agreement among all partners throughout the life of the project.

Start-up criteria are designed to demonstrate that equipment and processes can function at the engineered design rates, quality, and efficiency, usually for a normal business operating shift. Start-up is led and mainly staffed by the energy project team members with operations personnel assisting.

The full-going criteria are designed to demonstrate that the equipment, processes, and people are capable of running at the engineering design rates, efficiencies, and specified quality for 2–7 days. Full-going is led by either project team or operational personnel, but staffed by operational personnel.

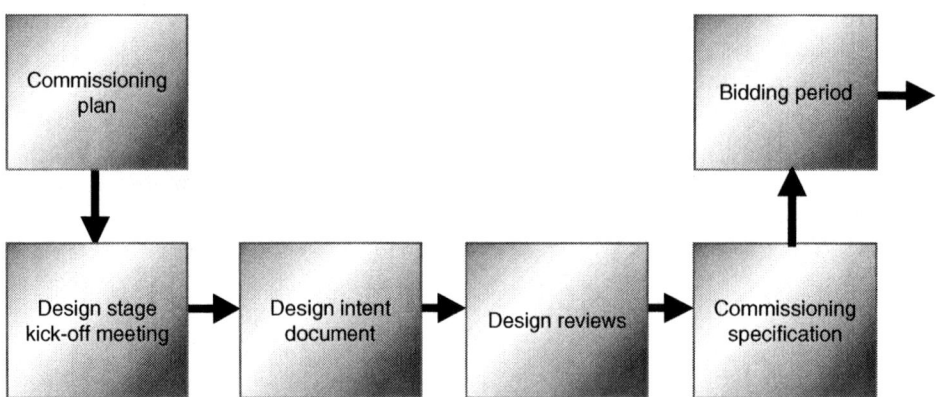

Fig. 2 Design stage commissioning activities.
Source: From Sebesta Blomberg and Associates, Inc., and Current-C Energy Systems, Inc. (see Ref. 1).

Turnover criteria additionally include the personnel performance criteria (operational and maintenance), training criteria (operational and maintenance), documentation criteria, and the resolution of any outstanding items such as study or research items. Once all of the turnover criteria are achieved, the system is turned over to the operational personnel and the project personnel's work is nearly complete; it remains only to complete any outstanding follow-up activities and conduct a project postmortem.

Also during the definition stage, a baseline project schedule should be developed with broad participation by the project team, operational personnel, outside resources, logistics resources, etc. The process used is to list the project stages and individual activities in each stage until all of the project activities are identified. The activities should then be linked to one another by identifying each activity's predecessors and successors and finally, durations should be assigned to each activity using 3-pt duration estimates (minimum, probable, maximum) to support the use of schedule risk analysis (see Ref. 2). This critical step yields an assessment of the statistical probability of completing the project by a certain time and is extremely helpful when analyzing the impact of scope or resource changes on a project.

The final steps in the definition stage are to update the cost estimate, finalize the funding strategy and methodology for the project, and determine the preferred bid structure.

Design Stage

The design stage goal involves development of designs for all equipment, processes, and systems. This stage frequently involves outside resources with expertise in detailed engineering design and often requires additional study or research in an effort to predict the operation and interaction of the full energy project on a small scale. Techniques such as pilot projects and system modeling may provide critical information as well as initial training for both project and operational and maintenance personnel and can also gain valuable input and commitment into the project. When engineers, who design the equipment and systems and know how they should operate, work with operational and maintenance personnel, who know how the equipment will really be operated, the results are sometimes surprising.

For energy projects, the interplay between initial cost and long-term expenses (utility, maintenance, and other) is also a critical part of the design stage. For example, a component with a 1-year life costing $1000 may be upgraded to a component with a 5-year life that costs $2000.

Energy system commissioning is the next step in design—it is the systematic process of assuring by verification and documentation from the design stage to a minimum of one year after construction that all building facility systems perform interactively in accordance with the design documentation and intent and with the owner's operational needs. Commissioning ensures that the energy project gives the business the promised results from the project including documentation, testing, and training. Specific commissioning activities are included as appropriate in the start-up, full-going, and turnover plans. Integrated design stage commissioning activities as practiced by companies such as Sebesta Blomberg are shown in Fig. 2.

Bid Preparation

It should be noted that bid preparation is an ongoing task that is parallel to the definition phase. It is treated separately here because of its complexity and importance.

The main types of bids in energy projects are design/build, fixed bid, not-to-exceed, performance bids, and hybrids. The appropriate type of bid depends upon such criteria as the type of energy project, the level of design detail, technological complications, the schedule, funding available, and complexity of the project.

Design/build bids are the best choice when scope is not well defined and there are many unknowns—either in scope or personnel experience. The disadvantage is that

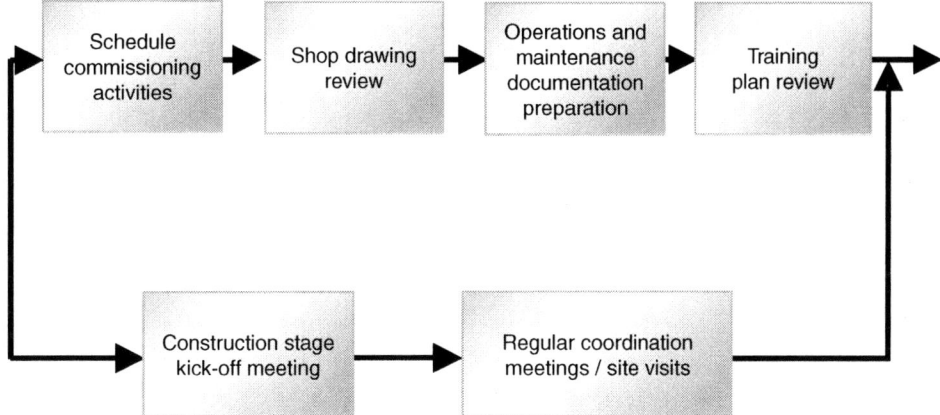

Fig. 3 Construction stage commissioning activities.
Source: From Sebesta Blomberg and Associates, Inc., and Current-C Energy Systems, Inc. (see Ref. 1).

design/build costs grow proportionally as the scope changes, with the costs passed through to the project. Consequently, the design/build bid can be the highest priced, but it provides greater flexibility to handle the issues or scope changes during the project.

Fixed bid is the best choice when the scope is well defined. There are few unknowns and project personnel are experienced enough to strongly defend and enforce the agreed upon scope of work and remove excessive contingency fees from the bid price. Scope additions or changes not requested by the operation result in those additional costs coming out of the bidder's pocket. Consequently, the fixed bid can be the lowest priced option while also providing the least flexibility.

The not-to-exceed bid is a hybrid. A project with a well-defined scope uses the fixed bid methodology and one with a less defined scope uses the design/build methodology. There is an added line item for the estimated cost of scope changes. A not-to-exceed bid is usually higher than a fixed bid, but lower than the eventual design/build bid.

Performance bid (or performance contract, "PC") is also a hybrid and combines any of the above methodologies with energy efficiency performance, which gives the supplier an incentive to ensure that the project meets the promised business objectives. In the energy project industry, an endless variety of different techniques have evolved for particular situations or needs, but they share the characteristic that the project is paid for partly based upon its adherence to prespecified performance criteria, and the contractor covers part of the risk for the project performing as promised.

Finally, the bid can also be affected by regulatory compliance and energy efficiency standards or regulations. New energy project equipment must meet the current regulatory requirements and existing equipment may have to be upgraded if it will be linked to the new equipment. ASHRAE standards for energy efficiency are used as standards in some jurisdictions, and as the U.S. Green Building Council's LEED certification and EnergyStar plaques for "high performance" or "green" buildings become more popular, those standards, processes, and procedures should also be included in the bid process.

Construction Stage

Traditionally this is when 'project management' starts, although as we have seen that it is really the fifth stage in the process. The goal of the construction stage is to build and install the equipment and processes for the project and the deliverables include schedules, resource plans, and refinements and adjustments to the communication and delivery plans.

For the detailed construction plan, the operator must decide whether to manage the construction in-house or to employ a general contractor for construction management and then choose the format for the construction schedule (number of days/week and hours/day). Once these are decided, the construction personnel needs are broken down by specific trade skills; union and nonunion workers; roles and responsibilities of core construction personnel, sub-contractors, and operational resources; and a schedule of the resources is produced.

The communication plan is updated with a plan for communicating daily and weekly construction activities and other information, including a feedback loop for rapid adjustments to construction schedule changes. Energy system commissioning activities continue, and are shown in Fig. 3.

Delivery Stage

The delivery stage can be the most chaotic stage as it includes the initial start-up, training, and turnover of the equipment and processes to the operation. The goal is to get everything started up as designed (rate, efficiency, and quality), provide training for all operational and maintenance personnel, turn over the equipment and processes, and close out the project. In a manner similar to the construction

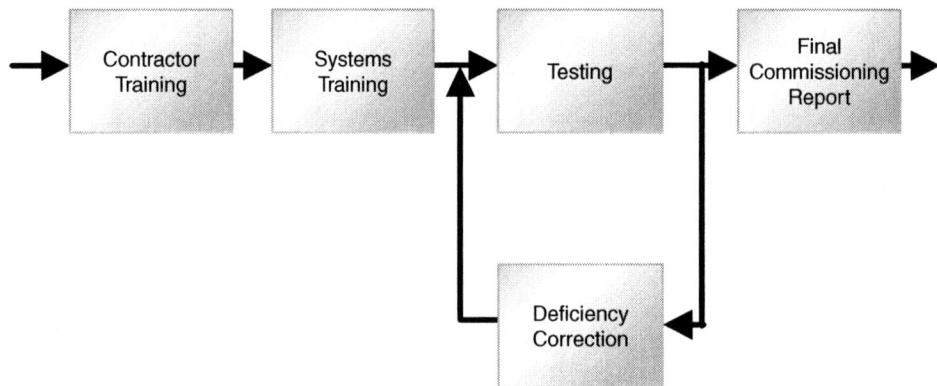

Fig. 4 Delivery stage commissioning activities.
Source: From Sebesta Blomberg and Associates, Inc., and Current-C Energy Systems, Inc. (see Ref. 1).

stage, the project delivery personnel needs are broken down by specific trade skills, roles and responsibilities, and a schedule of when each set of the personnel resources will be needed. The personnel requirements at this point are very large, especially from the ongoing operation.

The project schedule will have previously identified training activities, personnel, and schedules; start-up, full-going, and turnover activities personnel and schedules; and delivery stage commissioning activities (illustrated in Fig. 4). During the project delivery stage, those plans are followed as they have been adjusted during the previous project management stages, and the project is then completed.

Upon completion of the delivery activities, the achievement of the delivery criteria, and the turnover of the project to the operation, the project team conducts a postmortem on the project. The purpose of the postmortem is to celebrate the successes, identify the lessons learned, and to document both of these so that the next project has an even greater chance of success.

The commissioning activities extend—with the operational personnel—past the closure of the main project during the warranty period for the equipment. The postproject commissioning activities are illustrated in Fig. 5.

TOOLS FOR ENERGY PROJECT MANAGEMENT

Tools can make it easier and faster to execute project tasks, provide a standard methodology, enhance communication, and minimize confusion. The following are examples of typical tools used during energy projects.

Scope Tools

The first set of tools help manage energy project scope and provide documentation for both the equipment included in the project and for the activities that the project team will perform during the project. They include:

- Project brief.
- Project charter.
- Start-up, full-going, turnover.
- Scope change management.

The project charter, introduced in "Conceptual Stage", defines the project and provides a means to secure both understanding and commitment from all stakeholders. A sample format is shown in Fig. 6.

As the project proceeds, the start-up, full-going, turnover tool is the most important communication tool between the formal project team and the business operation. It documents the agreements between the two organizations regarding equipment performance, system performance, training completion, etc. and it is the basis for declaring portions of the project complete and for releasing project personnel. This tool must be created specifically for each project, but an example is shown in Fig. 7. For each process, the tool lists the project equipment and processes and determining performance criteria for the start-up, full-going, and turnover.

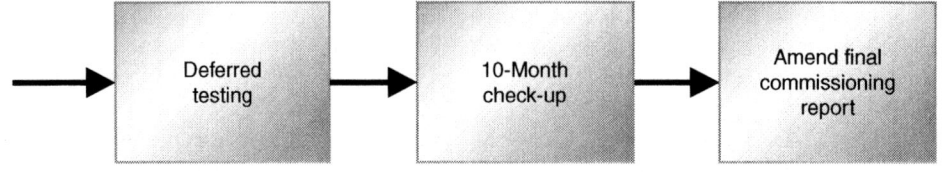

Fig. 5 Warranty period commissioning testing.
Source: From Sebesta Blomberg and Associates, Inc., and Current-C Energy Systems, Inc. (see Ref. 1).

PROJECT CHARTER	
Date:	
Project:	
Project objective:	
Business basis for project:	
Initial cost:	
Initial return:	
Initial timing:	
Major risks:	
Project sponsor(s):	
Project team:	
Project authorization:	

Fig. 6 Project charter.
Source: From Interface Consulting, LLC, 2005.

Scope management is an ongoing challenge as scope change requests come from all personnel involved in the project. Criteria should be established to sort out small, inconsequential change requests from larger ones having greater impact and a standardized scope change tool should be developed for approval and documentation. For larger change requests, a formal review should evaluate the change's impact on the project's cost, schedule, and other related systems. Its impact on the schedule can then be quantitatively evaluated using a schedule risk analysis.

START-UP, FULL-GOING, TURNOVER TOOL			
	Start-up criteria	Full-going criteria	Turnover criteria
Process #1- includes Equipment #1 and Equipment #2	Run 8 continuous hours at 80% efficiency, producing 90% quality	Run 7 continuous days, 24 hours/day at 85% efficiency, producing 95% quality	- Start-Up and Full-Going criteria are met. Operational and maintenance personnel trained and qualified on required skills. - Equipment documentation and files established and in the operation including (drawings, manuals, spare parts lists, etc.).

Fig. 7 An example of the start-up, full-going, turnover tool.
Source: From Interface Consulting, LLC 2005.

Scheduling Tools

The scheduling tools for energy projects are generally computerized and include two basic types—project scheduling software and schedule risk analysis software. Project scheduling programs project the project delivery date by linking together individual activity durations and dates from the project team to create a critical path schedule. With such a program, the impact on the schedule end date caused by interim changes becomes very clear.

After a project schedule has been created, it can be analyzed for schedule risk. Even though traditional critical path schedules are designed to show the most important activities in the project schedule, they do not take uncertainty into account. Consequently, if activities do not occur as predicted, the critical path may change, and with it the focus of the project team (see Ref. 2).

Schedule risk analysis uses 3-pt duration estimates for each activity (minimum, probable, and maximum), creating a distribution of durations for each activity. The analysis is conducted by using statistical techniques to create a new distribution of project end dates and probabilities, quantifying the probability of completing the project on time.

More importantly, for each simulation, the critical path, the critical path activities, and the activities that most effect the project end date are noted and summarized as they are calculated. From schedule risk analysis, they remain the important activities even as the project schedule changes. The project manager therefore has fewer key activities to focus on for the successful delivery of the project.

Collaboration and Alignment Tools

One of the challenges during an energy project where systems and variables are interactive is the volume of information and the number of decisions handled each and every day. No individual can be involved in every aspect of the project, and a tool such as RACI helps prevent gaps and overlaps in project team activities. RACI stands for:

- Responsible—who owns the project or the problem.
- Accountable—who must sign-off or approve the activity, and who the responsible person reports to.
- Consult—who provides input but isn't responsible for the activity.
- Inform—who needs be told but not consulted about the activity.

To use the RACI tool, identify the areas of decisions or tasks/activities involved in the energy project, identify the roles involved in the project, and complete the tool by filling in the Rs, As, Cs, and Is. Ideally, for any individual activity, only one person should have the 'R.' As the tool is

	Project sponsor	Project manager	Operation resource	Technical resource	Support resource
Project activity #1	A	R	I	C	I
Project activity #2	R	C	C	C	I
Project activity #3	C	C	R	I	C

Fig. 8 RACI chart.
Source: From Legacy business records project-library and archives Canada, 2005.

completed for each activity, task gaps and overlaps become quickly visible—allowing the project team to resolve them (Fig. 8).

SKILLS FOR ENERGY PROJECT MANAGEMENT

The energy project manager needs both technical project management skills and people project management skills to deliver a project on time, on budget, and with excellence.

Technical Project Management Skills

The technical project management skills have a long history and are well understood, and they include:

- Scope definition.
- Scope management.
- Scheduling.
- Cost management.

These skills are critical to managing the nuts and bolts of the project. Training for the technical project management skills is wide-spread and readily available from such organizations as the Association of Energy Engineers (AEE) and others. For this reason, technical skills are not addressed in this paper.

Personnel Management Skills

As project circumstances change, people skills can add a new layer of effectiveness for a project manager. For energy efficiency projects where multiple technical disciplines are included as core parts of the team, people skills are even more critical. Energy project managers must be skilled at choosing their project team, gaining commitment to the project, and resolving conflict (see Ref. 3).

The first critical decision is the composition of the project team. Each individual's qualifications, skills, attitude, and commitment should all be considered. For example, a person who has great technical skills but who treats team members poorly should be considered only after careful thought as to the amount of time the project manager is willing to spend listening to complaints and refereeing between team members, and contingency plans should be developed to minimize problems. Unresolved people issues tie up personnel and their productivity, slowing down both the quality and the completion of the project.

The energy project manager should also evaluate the team's communication and people skills. A good project team can produce trust and handle confrontation. It can almost be said that the team becomes the human representation of the project and it must be trained and prepared to handle people responsibilities as competently as technical responsibilities.

One key for energy project managers is gaining the commitment and cooperation of the operational personnel. Gaining commitment to a project is a process, not an initial announcement followed by informational updates. People go through a predictable process in becoming committed to something and the faster the project manager can recognize the process, the faster they can foster commitment.

When a new energy project is first proposed, the stakeholders, project team, and operational people all have an emotional reaction, although the reactions vary and change as they find out more about the project and its impact. If the project provides them with tangible benefits, they become committed very quickly. For many people, however, the stated benefit is a positive business benefit with no obvious personal impact. Consequently, the typical first reactions are either indifference or anger, and the energy manager should never confuse indifference with commitment.

Indifference is characterized by a "wait and see" attitude and when comments like "I've seen all of this before and I'll see it again" or "Hmm—we'll just see if they really can do what they say they can do" are made. As people progress through indifference, they often become angry and focus that anger on the energy project manager who is the visible symbol of the project. Complaints fly and the project is described as "terrible". The project manager must at this point remain positive and committed, which can be a challenge as it often seems few people are on his or her side. The project manager can even be tempted to avoid conversations with those who are angry, hoping that avoidance and time will smooth away their anger. Unfortunately, this is usually counterproductive, as motives are created to explain the project manager's avoidance and the situation spirals down.

Paradoxically, it is at this point that project managers have an opportunity to listen and learn. At the time, listening is likely to seem like the long, slow way to move the project forward but, once people feel they have been heard, they often move from saying "the entire project is awful", to "the project won't work because…". The "because" is very important, and it is important that the project manager hear and recognize this change in order to actively involve people and bring them the rest of the way to commitment. Once they agree to help be part of solving the "because", they are committed to the project. Energy projects bring change, and change always causes conflict. The project manager must be the first one to pull out a hidden conflict and set it squarely on the table for resolution. He or she must be completely intolerant of arguments vs. resolution and expect the same from the entire team.

CONCLUSION

Today's energy project manager needs to understand the plan, the tools, and the skills necessary to successfully manage the complete energy project—from costs and schedules to conflict resolution and commitment. A well-defined project management plan sets out a path that defines, creates, measures, and documents success. Using project management tools makes the process easier and leaves a trail of documentation regarding project decisions. A project manager with both technical and human skills is able to effectively manage both scope and people issues, saving time and money. Whether the project is managed in-house or a contract energy project manager is used, these three areas provide the energy project manager with what is needed to successfully deliver an energy project.

REFERENCES

1. Heneghan, T.; Sandra, M., *Commissioning Activities for Energy Project Stages*, Sebesta Blomberg and Associates, Inc., 2005.
2. Tietze, L.; *Why Schedule Risk Analysis?*, True Project, LLC, Fall/Winter, 2005.
3. Tietze, L.; *Integrating the Human Side of a Technology*, Interface Consulting LLC, 2004.

Energy Service Companies: Europe

Silvia Rezessy
Energy Efficiency Advisory, REEEP International Secretariat, Vienna International Centre, Austria, Environmental Sciences and Policy Department, Central European University, Nador, Hungary

Paolo Bertoldi
European Commission, Directorate General JRC, Ispra (VA), Italy

Abstract

Energy service companies (ESCOs) are important agents to promote energy efficiency improvements. This entry attempts to address a major gap identified in the process of conducting the first in-depth survey of ESCO businesses in Europe: the lack of common understanding and terminology related to types of energy service providers, contractual and financing terms, and the implications of different models. It provides the key terminology related to energy services, ESCOs, contractual models, and financing structure.

INTRODUCTION

Energy service companies (ESCOs) are important agents to promote energy-efficiency improvements. The initial ESCO concept started in Europe more than 100 years ago and then moved to North America. In the past decade, Europe has seen an increased interest in the provision of energy services that has been driven by electricity and gas restructuring and the push to bring sustainability to the energy sector. The actual market for energy-efficiency services in Western Europe was estimated to be 150 million euro per annum in 2000, while the market potential for Western Europe was estimated to be 5–10 billion euro per annum (these estimates are according to our definitions of ESCO, energy performance contracting (EPC), and third-party contracting (TPF) (see definitions later), and do not include energy supply) (Bertoldi et al. 2003 and references herein).[2,14] A survey of the U.S. ESCO industry estimates ESCO industry project investment reaching $1.8–$2.1 billion U.S. in 2000, with industry revenue growing at 9% per year in the period 1996–2000—down from almost 25% in the previous 5 years.[6]

This entry attempts to address a major gap identified in the process of conducting the first in-depth survey of ESCO businesses in Europe: the lack of common understanding and terminology related to types of energy service providers, contractual and financing terms, and the implications of different models. It provides the key terminology related to energy services, ESCOs, contractual models, and financing structure, pointing at the implications of different choices.

Keywords: Energy service companies; Energy performance contracting; Financing models.

This entry uses terminology agreed upon after a long process of consultation undertaken in 2004 (the entire report is available for free download at http://energyefficiency.jrc.cec.eu.int/pdf/ESCO%20report%20final%20revised%20v2.pdf) and early 2005 with major stakeholders in Europe and the United States. The entry builds on the status report "ESCOs in Europe," published in 2005 by Directorate General Joint Research Center of the European Commission. The report has covered the 25 member states of the European Union (EU), the New Accession countries Bulgaria and Romania, Switzerland, and Norway.[1]

ENERGY SERVICES, ENERGY SERVICE COMPANIES, AND PROJECT ELEMENTS

Energy services include a wide range of activities, such as energy analysis and audits, energy management, project design and implementation, maintenance and operation, the monitoring and evaluation of savings, property/facility management, energy and equipment supply, and provision of service (space heating/cooling, lighting, etc.).

Consultant engineering companies specialized in efficiency improvements, equipment manufacturers, energy suppliers or utilities, which provide energy services for a fixed fee to final energy users or as added value to the supply of equipment or energy, are referred to as *Energy Service Provider Companies (ESPCs)* (note that this is a very different use of the acronym ESPC common in North America, where it means energy service performance contracting). Energy service provider companies may have some incentives to reduce consumption, but these are not as clear as in the ESCO approach (see explanations below). Often, the full cost of energy services is recovered in the fee, so the ESPC does not assume any risk in case of underperformance. Energy service provider companies are paid a fee for their advice/service rather than being paid based on the results of their recommendations.[13] Principally, projects implemented by ESPCs are related to primary energy conversion equipment (boilers, combined heat and powers [CHPs]). In such projects, the ESPC is unlikely to guarantee a reduction in the delivered energy consumption because it may have no control of or ongoing responsibility for the efficiency of secondary conversion equipment (such as radiators, motors, and drives) and no control of the demand for final energy services (such as space heating, motive power, and light).[12]

Energy service companies also offer these same services. Energy service companies are fundamentally different from ESPCs, and ESCOs' activities can be distinguished from ESPCs' activities in the following ways:

- Energy service companies in Europe guarantee the energy savings and the provision of the same level of energy service at a lower cost by implementing an energy-efficiency project (in North America, ESCOs do not necessarily guarantee energy savings. If they do, the document that specifies this guarantee is called an energy savings performance contract). A performance guarantee can take several forms. It can revolve around the actual flow of energy savings from a project; it can stipulate that the energy savings will be sufficient to repay monthly debt service costs for an efficiency project; or it can stipulate that the same level of energy service will be provided for less money.
- The remuneration of ESCOs is tied directly to the energy savings achieved.
- Energy service companies typically finance (or assist in arranging financing) the installation of an energy project that they implement by providing a savings guarantee.
- Energy service companies may retain an ongoing operational role in measuring and verifying the savings over the financing term.

Typically, a project developed and implemented by an ESCO includes the following elements/steps:

- Site survey and preliminary evaluation
- Investment-grade energy audit
- Identification of possible energy-saving and efficiency-improving actions
- Financial presentation and client decision
- Guarantee of the results by proper contract clauses
- Project financing
- Comprehensive engineering and project design and specifications
- Procurement and installation of equipment; final design and construction
- Project management, commissioning, and acceptance
- Facility and equipment operation and maintenance for the contract period
- Purchase of fuel and electricity (to provide heat, comfort, light, etc.)
- Measurement and verifications (M&V) of the savings results.

The investment-grade audit (IGA) deserves special attention. The traditional energy audit does not sufficiently consider how implemented measures will behave over time. Because auditors must consider the conditions under which measures will function during the life of the project, an IGA builds on the conventional energy audit. Unlike the traditional energy audit, which assumes that all conditions (related to system, payback, and people) remain the same over time, an IGA attempts to predict a building's energy use more accurately by adding the dimension of a risk assessment component, which evaluates conditions in a specific building or process. Aspects of the IGA include risk management, the "people" factor, M&V, financing issues, report presentation guidelines, and master planning strategies.[9]

FINANCING OPTIONS

Three broad options for financing energy-efficiency improvements can be distinguished. The approach to financing is just one factor that shapes the structure of an EPC. Other factors relevant for the type of financing arrangements and repayment structures include the allocation of risks, the services contracted, and the length of the contract.[11]

Energy service company financing refers to financing with internal funds of the ESCO; it may involve use of its own capital or funding through debt or lease instruments. Energy service companies rarely use equity for financing, as this option limits their capability to implement projects on a sustainable basis.

Energy-user/customer financing usually involves financing with internal funds of the user/customer backed by an energy savings guarantee provided by the ESCO. For instance, a university can use its endowment fund to finance an energy project, in which the energy savings are guaranteed by an ESCO. Energy-user/customer financing may also be associated with borrowing in the case when the energy user/customer, as a direct borrower, has to provide a guarantee (collateral) to the finance institution. The provision of collateral by the consumer itself is what distinguishes customer financing from third-party financing (TPF).

Third-party financing refers mostly to debt financing. As its name suggests, project financing comes from a third party (e.g., a financial institution) and not from internal funds of the ESCO or of the customer. The finance institution may either assume the rights to the energy savings or take a security interest in the project equipment.[13] Two conceptually different TPF arrangements are associated with EPCs, and the key difference between them is which party borrows the money: the ESCO or the client.

- The first option is that the ESCO borrows the financial resources necessary for project implementation.
- The second option is that the energy-user/customer takes a loan from a financial institution, backed by an energy savings guarantee agreement with the ESCO. The purpose of the savings guarantee is to demonstrate to the bank that the project for which the customer borrows will generate a positive cash flow (i.e., that the savings achieved will certainly cover the debt repayment). Thus, the energy savings guarantee reduces the risk perception of the bank, which has implications for the interest rates at which financing is acquired. The cost of borrowing is strongly influenced by the size and credit history of the borrower.

When the ESCO is the borrower, the customer is safeguarded from financial risks related to the project's technical performance because the savings guarantee provided by the ESCO is either coming from the project value itself or is appearing on the balance sheet of the ESCO; hence, the debt resides on someone else's balance sheet (ESCO's or financial institution's). Both public and private customers can benefit from off-balance-sheet financing because the debt service is treated as an operational expense and not a capital obligation; therefore, debt ratings are not impacted. For highly leveraged companies, this is important, because the obligation not showing up on the balance sheet as debt means that company borrowing capacity is freed.[10] However, different countries apply various conditions that need to be met for financing to be viewed as an operating lease, for example. Unless those conditions are met, financing is automatically considered to be a capital lease. Therefore, parties seeking financing first need to inquire about the country-specific conditions for operational financing.

Large ESCOs with deep pockets (hence, high credit ratings) have started to prefer TPF to their own funds because their costs of equity financing and long-term financing are often much greater than what can be accessed in the financial markets. Also, if an ESCO arranges TPF, its own risk is smaller. This would allow for a lower cost of money for the same level of investment, so more money would be assigned to the project.[8] The cost associated with nonrecourse project financing by a third party (e.g., one in which project loans are secured only by the project's assets) is the highest, as it entails more risk and, hence, higher interest rates.[13]

Furthermore, as already mentioned, equity contributions from the ESCO are often deemed undesirable by ESCOs, as they tie up capital in a project. This emphasizes the fundamental concept that an ESCO is a service company and not a bank or a leasing company. The primary reason that ESCOs do not and should not provide project financing with internal funds is that it makes their balance sheets look like they are banks and not a service companies. Local practices, the inability of customers to meet financiers' creditworthiness criteria, and costs of equity financing are some of the factors that determine whether ESCOs will provide financing (debt or equity).[8] Small or undercapitalized ESCOs, which cannot borrow significant amounts of money from the financial markets, prefer their role not to be financing energy-efficiency investment.

ENERGY PERFORMANCE CONTRACTING MODELS

Energy performance contracting is a form of creative financing for capital improvement that allows funding energy-efficiency upgrades from cost reductions. Under an EPC arrangement, an external organization (ESCO) develops, implements, and finances (or arranges financing for) an energy-efficiency project or a renewable energy project and uses the stream of income from the cost savings or the renewable energy produced to repay the costs of the project, including the costs of the investment. Essentially, the ESCO will not recover all of its costs unless the project delivers all of the energy savings guaranteed. The approach is based on the transfer of technical risks from the client to the ESCO based on performance guarantees given by the ESCO. In EPC, ESCO remuneration is based on demonstrated performance; a measure of performance is the level of energy or cost savings, or the level of energy service. Therefore, EPC is a means to deliver infrastructure improvements to facilities that lack energy-engineering skills, manpower or management time, capital funding, an understanding of risk, or technology information. Cash-poor yet creditworthy customers, therefore, are good potential clients for EPC. Fig. 1 illustrates the EPC concept.

Energy performance contracting is usually distinguished from energy supply contracting (delivery contracting) that is focused on the supply of a set of energy services (e.g., heating, lighting, motive power) mainly via outsourcing the energy supply. Chauffage (see details later in this entry) includes supply contracting. In contrast, EPC typically targets savings in production and distribution.

Energy performance contracting is risk management and effective ESCOs have learned to use project financial structures to help manage the risks. Below we examine different contracting models and the conditions under which they deliver at their potential.

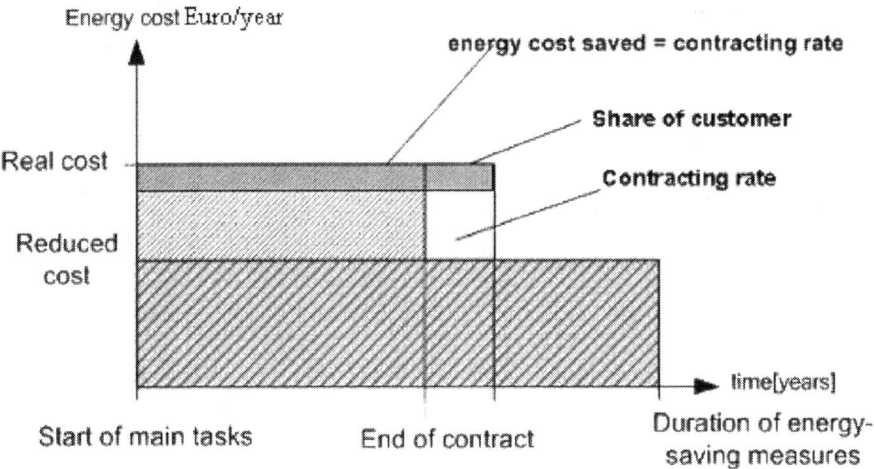

Fig. 1 Energy performance contract.
*Note: Note that on this graph, real cost refers to the initial cost, while the contracting rate depicts the cost savings, which in this case are shared between the customer and the energy service company (ESCO) (see the explanation on shared savings).
Source: From Berlin Energy Agency.

Guaranteed Savings and Shared Savings

Fig. 2 illustrates the relationships and risk allocations among the ESCO, customer, and lender in the two major performance contracting models: shared savings and guaranteed savings. Brief descriptions are also given. An important difference between guaranteed and shared savings models is that in the former case, the performance guarantee is the level of energy saved, whereas in the latter, this is the cost of energy saved.[7,11]

Under a guaranteed savings contract, the ESCO assumes the entire design, installation, and savings performance risks but does not assume credit risk of repayment by the customer. Consequently, guaranteed savings contracts are not usually applicable to ESCO financing provided internally or through TPF with ESCO borrowing. The projects are financed by the customers, who can also obtain financing from banks, from other financing agencies, or a TPF entity. The key advantage of this model is that it provides the lowest financing cost because it limits the risks of the financial institutions to their area of expertise, which is assessing and handling customers' credit risk. The customer repays the loan and assumes the investment repayment risk (the financing institution [FI], of course, always has some risk for loan nonpayment. The assessment of a customer's credit risk is done by the FI; it is one of the factors that define interest rates). If the savings are not enough to cover debt service, the ESCO has to cover the difference. If savings exceed the guaranteed level, the customer typically pays an agreed-upon percentage of the savings to the ESCO (however, changes in energy consumption (e.g., business expansion or changes of processes or production lines) are likely to bring increased energy that can deteriorate the targets. Conversely, a contraction of business (e.g., an empty wing of a hotel) or a smaller production output will result in energy savings. Therefore, crucial issues to consider involve setting the baselines and associated growth projections, setting the system boundary and conditions, and preventing leakages. A clause in the contract that allows either party to reopen and renegotiate the baseline to reflect current conditions can solve this problem).

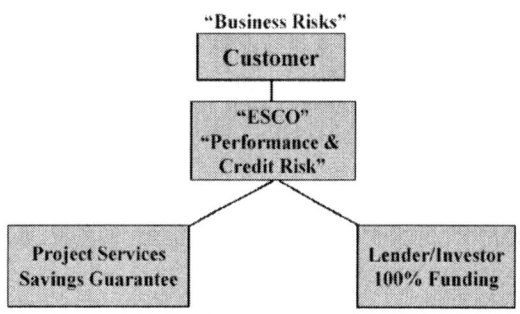

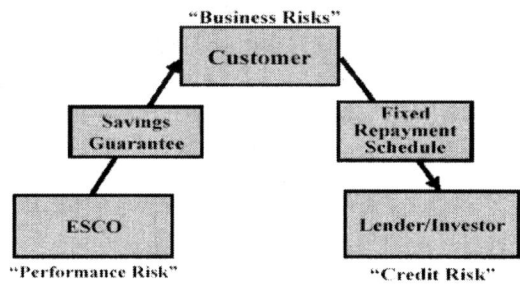

Fig. 2 Major types of performance contracting models/repayment options. Under a shared savings contract, the cost savings are split for a predetermined length of time in accordance with a prearranged percentage. There is no standard split, as this depends on the cost of the project, the length of the contract, and the risks taken by the energy service company (ESCO) and the consumer. Under a guaranteed savings contract, the ESCO guarantees a certain level of energy savings and in this way shields the client from any performance risk.
Source: From Joint Research Centre, European Commission (see Ref. 4).

Usually, the contract also contains a proviso that the guarantee is good (i.e., the value of the energy saved will be enough to meet the customer's debt obligation) provided that the price of energy does not go below a stipulated floor price (performance contracting is risk management, and dropping fuel prices—such as those experienced in North America in 1986—gave rise to this provision. We are indebted to Shirley Hansen for this clarification).[8] A variation on guaranteed savings contracts is pay-from-savings contracts, whereby the payment schedule is based on the level (percentage) of savings: the more the savings, the quicker the repayment.[13]

The guaranteed savings scheme is likely to function properly only in countries with an established banking structure, high degree of familiarity with project financing, and sufficient technical expertise, also within the banking sector, to understand energy-efficiency projects (e.g., in the United Kingdom; Austria; and, more recently, Hungary). The guaranteed savings concept is difficult to use in introducing the ESCO concept in developing markets because it requires customers to assume investment repayment risk. However, guaranteed savings foster long-term growth of ESCO and finance industries.[4] Newly-established ESCOs with no credit histories and limited resources would be unable to invest in the projects they recommend and might enter the market only if they guarantee the savings and the client secures the financing on its own.

In the United States, the guaranteed savings model evolved from the shared savings model in response to customers' desire to significantly reduce financing costs in exchange for accepting more risk due to their increased comfort with energy savings technologies. It also was initiated by smaller ESCOs and fostered by financial institutions to allow them to grow their respective industries. The primary benefit of this structure is that its reduced financing cost enables many more project investments to be made for the same debt service level. The public sector normally prefers this structure to maximize the amount of infrastructure investment made in its facilities from an EPC.

Conversely, under a shared savings model, the ESCO assumes both performance and credit risk. This is why a shared savings contract is more likely to be linked with TPF; ESCO financing; or a mixed scheme, with financing coming from the client and the ESCO whereby the ESCO repays the loan and takes over the credit risk. The ESCO therefore assumes both performance and the underlying customer credit risk; if the customer goes out of business, the revenue stream from the project will stop, putting the ESCO at risk.[13] Unfortunately, such a contractual arrangement may create leveraging and increased capital requirement problems for ESCOs because ESCOs become too indebted, and at some point financial institutions may refuse lending to an ESCO due to a high debt-to-equity ratio (experience in the United States shows that lenders tend to require a variety of credit enhancements for this type of financing, such as bonding or insurance[13]). In effect, the ESCO collateralizes the loan with anticipated savings payments from the customer based on a share of the energy cost savings. The financing in this case goes off the customer's balance sheet (under off-balance-sheet financing, also called nonappropriation financing, financiers hold title to equipment during the term of the agreement. Furthermore, to avoid the risk of energy price changes, it is possible to stipulate in the contract a single energy price. In this situation, the customer and the ESCO agree on the value of the service up front, and neither side gains from changes in energy prices. If the actual prices are lower than the stipulated floor value, the consumer has a windfall profit, which compensates the lower return of the project; conversely, if the actual prices are higher than the stipulated ceiling, the return on the project is higher than projected, but the consumer pays no more for the project. In effect, this variation sets performance in physical terms with fixed energy prices, which makes the approach resemble guaranteed savings[11]). A situation where savings exceed expectations should be taken into account in a shared savings contract. This setting may create an adversarial relationship between the ESCO and the customer[7] whereby the ESCO may attempt to lowball the savings estimate and then receive more from the excess savings (deliberate estimation of the lower value of savings is not restricted to the shared savings model; it is a standard practice for the ESCO to secure itself for the guaranteed performance with some buffer. The real questions are how big this buffer/cushion is and how the "excess" savings above the estimated ones are split between the client and the ESCO).[11]

The shared savings concept is a good introductory model in developing markets because customers assume no financial risk (the customers may have different reasons to be reluctant to assume financing, even if the cost of capital is higher for ESCOs than for customers. Among the reasons are adversity to assuming debt, borrowing limits, and budgetary restraints[11]). From the ESCO's perspective, the shared savings approach has the added value of the financing service.[11] However, this model tends to create barriers for small companies; small ESCOs that implement projects based on shared savings rapidly become too highly leveraged and unable to contract further debt for subsequent projects.[8,11] Shared savings concepts therefore may limit long-term market growth and competition between ESCOs and between FIs. For instance, small or new ESCOs with no previous experience in borrowing and with few resources are unlikely to enter the market if such agreements dominate.[3,4] It focuses the attention on projects with short payback times ("cream skimming"). Table 1 summarizes the features of the guaranteed and shared savings models.

Table 1 Guaranteed savings and shared savings: a comparison

Guaranteed savings	Shared savings
Performance related to level of energy saved	Performance related to cost of energy saved
Value of energy saved is guaranteed to meet debt service obligations down to a floor price	Value of payments to energy service companies (ESCO) is linked to energy price
Energy service companies carries performance risk energy user/customer carries credit risk	Energy service companies carries performance and credit risk as it typically carries out the financing
If the energy user/customer borrows, then debt appears on its balance sheet	Usually off the balance sheet of energy user/customer
Requires creditworthy customer	Can serve customers that do not have access to financing, but still requires a creditworthy customer
Extensive measurement and verification (M&V)	Extensive M&V
Energy service companies can do more projects without getting highly leveraged	Favors large ESCOs; small ESCOs become too leveraged to do more projects
More comprehensive project scope due to lower financing costs	Favors projects with short payback ('cream skimming') due to higher financing costs

Source: From Refs. 4,7,8 and 11.

Other Contracting Models

While there are numerous ways to structure a contract—hence, any attempt to be comprehensive in describing contracting variations is doomed—other contractual arrangements deserve some attention. Here, we describe the chauffage contract, the first-out contract, the build-own-operate-transfer (BOOT) contract, and the leasing contract.

A very frequently used type of contract in Europe is the chauffage contract, in which an ESPC or an ESCO takes over complete responsibility for the provision to the client of an agreed set of energy services (e.g., space heat, lighting, motive power). This arrangement is a type of supply-and-demand contract, and in effect it is an extreme form of energy management outsourcing. Where the energy supply market is competitive, the ESCO in a chauffage arrangement also takes over full responsibility for fuel/electricity purchasing. The fee paid by the client under a chauffage arrangement is calculated on the basis of its existing energy bill minus a percentage saving (often in the range of 5%–10%), or a fee may be charged per square meter of conditioned space. Thus, the client is guaranteed an immediate saving relative to its current bill. The ESCO takes on the responsibility of providing the improved level of energy service for a reduced bill. The more efficiently and cheaply it can do this, the greater its earnings. Chauffage contracts give the strongest incentive to ESPCs or ESCOs to provide services in an efficient way.

Chauffage contracts are typically very long (20–30 years), and the ESCO provides all the associated maintenance and operation during the contract. Chauffage contracts are very useful where the customer wants to outsource facility services and investment.[13] Such contracts may have an element of shared savings in addition to the guaranteed savings element to provide incentives for the customer. For instance, all savings up to an agreed figure would go to the ESCO to repay project costs and return on capital; above this, savings will be shared between the ESCO and the customer.

Another variation is the first-out approach, whereby the ESCO is paid 100% of the energy savings until the project costs—including the ESCO profit—are fully paid. The exact duration of the contract will actually depend on the level of savings achieved: the greater the savings, the shorter the contract.[5]

A BOOT model may involve an ESCO designing, building, financing, owning, and operating the equipment for a defined period of time and then transferring this ownership across to the client. This model resembles a special-purpose enterprise created for a particular project. Clients enter into long-term supply contracts with the BOOT operator and are charged accordingly for the service delivered. The service charge includes capital and operating-cost recovery and project profit. BOOT schemes are becoming an increasingly popular means of financing CHP projects in Europe.

Leasing can be an attractive alternative to borrowing because lease payments tend to be lower than loan payments. This method is commonly used for industrial equipment. The client (lessee) makes payments of principal and interest, and the frequency of the payments depends on the contract. The stream of income from the cost savings covers the lease payment. The ESCO can bid out and arrange an equipment lease-purchase agreement with a FI. If the ESCO is not affiliated with an equipment manufacturer or supplier, it can bid out, make suppliers' competitive analysis, and arrange the equipment.

There are two major types of leases: capital and operating. Capital leases are installment purchases of equipment. In a capital lease, the lessee owns and depreciates the equipment and may benefit from associated tax benefits. A capital asset and associated liability appears on the balance sheet. In an operating lease, the owner of the asset (lessor—the ESCO) owns the equipment and essentially rents it to the lessee for a fixed monthly fee. This is an off-balance-sheet financing source. It shifts the risk from the lessee to the lessor but tends to be more expensive for the lessor. Unlike in capital leases, the lessor claims any tax benefits associated with the depreciation of the equipment. The nonappropriation clause means that the financing is not seen as debt.

CONCLUSION

This entry has provided a concise overview of the key terminology related to energy service provision in Europe. It is the belief of the authors that the existence of standard, undisputed, and commonly used terms will facilitate the demand for, and ultimately the deployment of, more energy services.

An important step to securing the long-term credibility of energy efficiency is the ability to verify reduced consumption. To help end users and the financial community better understand EPC and gain confidence in the return on investment, it would be extremely beneficial to standardize savings M&V procedures. While the development of standard M&V has been an elusive task in Europe, as various companies consider their approaches unique and proprietary rather than developing a single standard, energy-service agreement efforts may be channeled toward agreeing upon a standard language for a set of key contract provisions, such as insurance, equipment ownership, and purchase options, which will allow standard contract forms to be built up gradually.

Due to space limitations, this entry has presented a fraction of the findings of the ongoing survey of ESCOs in Europe. Further information—including a concise review of the current status of the ESCO industry in Europe—on the project specifics of the most common types of activities and of the features of the ESCO industry in selected EU member states is available, together with an outline of the key factors of success of national ESCO industries and a list of suggested strategic actions needed to increase the deployment of ESCOs in Europe.[1]

ACKNOWLEDGMENTS

The authors would like to acknowledge the contributions toward working out often dubiously interpreted ESCO terminology for the "ESCOs in Europe" status report 2005 of Shirley Hansen, Thomas Dreessen, Pierre Langlois, and Anees Iqbal. The authors would like to thank all national and international experts who provided input to the survey and responded to our questions—and, of course, all ESCOs that provided information that was used for the analysis in the "ESCOs in Europe" report.

REFERENCES

1. Bertoldi, P.; Rezessy, S. *Energy Service Companies in Europe: Status Report 2005*. Published by DG Joint Research Centre of the European Commission and available for download at http://energyefficiency.jrc.cec.eu.int/pdf/ESCO%20report%20final%20revised%20v2.pdf, (accessed on December 2006).
2. Bertoldi, P.; Berrutto, V.; De Renzio, M.; Adnot, J.; Vine, E. *How are ESCOs behaving and how to create a real ESCO market*. In Proceedings of the European Council for Energy Efficient Economy 2003 Summer Study; European Council for an Energy-Efficient Economy: Stockholm, 2003.
3. Butson, J. In *The Potential for Energy Service Companies in the European Union*, Proceedings of the First International Conference on Improving Electricity Efficiency in Commercial Buildings, Amsterdam, September 1998.
4. CTI (Climate Technology Initiative). *Guide to Working with Energy Service Companies in Central Europe*; CTI Secretariat: Tokyo, 2003.
5. Dreessen, T. In *Advantages and Disadvantages of the Two Dominant World ESCO Models; Shared Savings and Guaranteed Savings*, Proceedings of the First Pan-European Conference on Energy Service Companies; Bertoldi, P., Ed.; Joint Research Centre, European Commission, 2003.
6. ECS (Energy Charter Secretariat). Third party financing. Achieving its Potential; Energy Charter Secretariat: Brussels, 2003.
7. Goldman, C.; Hopper, N.; Osborn, J. Review of U.S. ESCO industry market trends: an empirical analysis of project data. Energy Policy, **2005**, *33*, 387–405.
8. Hansen, S. In *Performance Contracting Models and Risk Management*, Presentation at the workshop developing the business of energy efficiency performance contracting organized by Ibmec and INEE, Rio de Janeiro, 2003.
9. Hansen, S. Personal Communication; Kiona International: U.S.A., 2004.
10. Hansen, S.; Brown, J. *Investment Grade Audit: Making Smart Energy Choices*; Fairmont Press: Lilbourn, 2003.
11. McGowan, J. State-of-the-art performance contracting. Article available online at http://www.energyusernews.com/CDA/ArticleInformation/features/BNP__Features__Item/0,2584,61014,00.html (accessed on December 2005).
12. Poole, A.D.; Stoner, T.H. Alternative financing models for energy efficiency performance contracting. *Sponsored by the USAID Brazilian Clean and Efficient Energy Program (BCEEP)*; 2003; administered by Winrock International.
13. Sorrell, S. The economics of energy service contracts. *Tyndall Centre Working Paper*, Environment and Energy Programme SPRU (Science and Technology Policy Research): Freeman Centre, 2005.
14. WEEA, World Energy Efficiency Association. *Briefing Paper on Energy Service Companies with Directory of Active Companies*; WEEA: Washington, 1999.

Index

2×6 wall construction, 899
3E Plus, 880

A

Abandoned steam line isolation, 1357
Absolute implementation, global temperature adjustment and, 275
Absorption heat pump configurations, 1542–1544
   heat sink configurations, 1542–1543
   heat source configurations, 1542–1543
   working fluids, 1543–1544
Absorption heat pump, 816–817, 1541–1547
   bottoming and topping cycles, 1542
   thermal design fundamentals, 1541–1542
Absorption refrigeration, 1544–1546
   applications of, 1545–1546
   components of, 1544
      compressor, 1544
      condenser, 1544
      evaporator, 1544
      throttle, 1544
   performance of, 1545
Absorption working fluids, 1543–1544
Acceptance based commissioning, 189–190
Acceptance phase, commissioning process and, 196–197
Access, electric power transmission systems, 360
Accidents, nuclear, 1131–1132
Accounting of energy, 1–7
Accreditation, management system standards and, 1030–1031
Accumulated depreciation, rate of return regulation and, 1255
Accumulator, mobile HVAC systems and, 1079
Acid deposition, air quality modeling and, 719
Acid gas removal processes for syngas, 908
Acid rain, 873
Actionable information, interactive access and, 619–622
Active server pages, 537
Active space heating, 1322
Activity based costing, 1042–1048
   advantages and disadvantages of, 1048
   allocation via cost drivers, 1042–1043
   background of, 1042–1043
   calculations for
      cooling and lighting, 1044
      motors and machines, 1045
   example of, 1043–1048
   implementation cost, 1048
   tracing overhead, 1042
Additional energy, 1090–1091
   natural energy, a comparison, 1088–1094
   types, 1090–1091
      chemical fuels, 1091
      non renewable, 1090–1091
      not natural, 1091
      waste products, 1091
Adiabatic mixing, air streams and, 1188

Adjustable speed drive pumps, 1553–1554
   savings from pump modifications, 1553
Adjustable speed drives, energy cost calculations and, 74–75
Adjustment charges, utility billing and, 1502
Adobe masonry walls, 1517–1518
Advanced gas-cooled reactors, 40
Advanced shading devices
   light shelves, 1642–1643
   mini-light shelves, 1643
   prismatics and refractives, 1643
Advanced thermal technologies, 1347
   analytical semiempirical modeling, 1348–1350
   biofuels, 1353
   liquid fuel production, 1354
   output comparisons, 1347–1348
   pyrolysis, 1348
Advanced Transport Reactor, 906–907
Advertising, energy demand-side management programs and, 290
Advisors, independent power producers and, 857
AEC. *See* alternate energy credit.
AEE. *See* Association of Energy Engineers.
Aerated concrete, as insulation material, 891
Aerospace uses, heat pipe applications and, 807–808
AFC. *See* alkaline fuel cells.
Affected sources, emissions trading and, 431
Aging thermal processes, 868
Agricultural products, drying of, 332–336
Agriculture residues, coal and, cofiring of, 490
Agriculture, heat pipe applications and, 810
Air balancing, ventilation system and, 61
Air barriers, 662–663
   airtight drywall approach, 667–668
   bypasses, 664
   leakage details, 664–666
   materials, 662–663
   seal penetrations, 664
Air compressors, accounting for in energy assessment, 71
Air conditioning
   accounting for in energy assessment, 70
   capacity calculations, 1061–1066
   cycles, 1188–1191
      summer hot and
         dry mode, 1190
         humid mode, 1189–1190
      winter, 1190–1191
   electricity usage of, 583
   natural gas and, 1507
   solar, 1326–1327
   solar heating and, case study of, 1317–1320
      cooling system methodology, 1317–1318
      problems with, 1320
      psychrometric analsysis, 1318
      solar roof tile system, 1318–1319
   units, review use of, 521
Air cooled equipment, 532

Air costs, air distribution and, 212
Air density impact, fan system performance and, 1224
Air discharge volume table, 220
Air distribution adjustment
   duct static pressure setpoint reduction, 275–276
   electric demand response and, 275–276
   evaluation of, 276
   fan quantity reduction, 276
   fan speed limit, 276
Air distribution compressed air control systems and, 211–212
   zone management, 211
Air emission reductions
   energy efficiency and, 9–17
   Kyoto Treaty, 9
Air emission standards, 715–717
   vehicles, 717
Air emissions reductions
   carbon sequestration, 14–15
   energy efficiency
      emissions impacts, 11–12
      hydrogen economy, 10–11
      countries focus on, 10
      effectiveness of, 10
   United States
      alternate energy credit, 14
      attempts to regulate state by state, 12–13
      renewable energy credits, 14
Air flow measurement devices, variable air volume and, importance of, 52
Air flow rate, run around heat recovery systems and, 1280–1281
Air flow restrictions, ceiling and roof insulation, 902
Air flow, 116
Air humidification, 1579, 1581–1582
Air leak detection, compressed, 219–225
Air leakage
   details, 664–666
   facility, 662–670
   measuring, web-based compressor management and, 213
Air leaks
   common sealants, 665
   driving forces of, 666
Air Master +1.0.9, 880
Air movement, underfloor air distribution (UFAD) and, 1467
Air pollution
   air quality modeling, 719–721
   ambient air quality standards, 715–717
   emissions standards, 715–717
   fossil fuels and, 715–721
   transport and dispersion, 718–721
Air production control, compressed air and, 207
Air quality modeling
   acid deposition, 719
   photo-oxidants, 719–721
   regional haze, 719

I-1

Air quality
  ANSI/ASHRAE Standard 62.1–2004 and, 51
  coal to liquid fuels and, 169
  compressed air control systems and, 211
    monitoring of, 211
  indoor, 18–23
  monitoring of, 212
  standards, ambient, 715–717
Air storage
  compressed, 226–235
  distribution, 226–235
Air streams, adiabatic mixing of, 1188
Air traffic control (ATC), Future Air Navigation System, 1, 30
Air usage, reduced, 211
Air zone leakages, 212
Aircraft fuel
  consumption, 24–30
    available ton-kilometer measurement, 24–25
    breakthrough gains in efficiency, 29
    energy costs
      advantages of conservation, 25
      impact on, 25
    future of, 29
    government's role in, 29–30
      air traffic control, 29–30
      global positioning satellites, 29
      incremental gains in efficiency, 29
    reduction factors, 26–29
      drag reduction, 26–27
      engine efficiency, 26
      engine maintenance, 28–29
      flight controls, 28
      piloting techniques, 28
      pre-flight planning, 27–28
      weight reduction, 26–27
    totals used, 24
    trends in, 24–25
  jet, 26
  kerosene based, 25–26
  production of, 25–26
  types of, 25–26
  gasoline, 26
Airflow sensors, 58
Air-Krete, as insulation material, 894
Airports
  construction, federally funded, 1020
  Federal funding of, 591
  traffic problems, 588
Air-separation technologies, gasifiers and, 909
Airside heat transfer calculations, mobile HVAC systems and, 1062–1063
Airtight drywall
  advantages and disadvantages, 667
  air barriers and, 667–668
  installation techniques, 668
Airtightness, measuring of, blower door test equipment, 668–669
Alanates, 1446
Alaska coal basins, 158
Alignment tools, energy project management, 563–564
Alkaline fuel cells (AFC), 734–736
Alliance to Save Energy, 513
Allowance allocation, 434–435
  emissions trading and, 432

Allowance assignments, Kyoto protocol and, 141
Allowance, emissions trading and, 432
Allowance/permit allocation
  auction, 432
  baseline choosing, 432
  fixed historical baseline, 432
  historical baseline with updating, 432
Alterative transport fuels vs. gasoline, comparison of, 47–48
Alternate energy credit (AEC), 14
Alternating current induction synchronous electric motor, 349
Alternating current power system, 389
  rotary converter, 389
  Westinghouse, 389
Alternating current synchronous electric motor, 349
Alternative energy sources, regulatory issues, wind power, 1614–1615
Alternative energy
  future of, 43–46
    economically competitive, 44–45
    fossil fuel dominance, 45–46
    renewable electricity generation, 44
  market penetration of, 46
  technologies
    biomass, 31–34
    hydrogen, 35–37
    pricing of, 31–48
    renewable, 31
    solar energy, 34
    types, 31–43
    wind, 34–35
Alternative fuel, 1523–1525
  cost assumptions, nuclear energy economic issues and, 1105
  technologies
    fuel cells, 37–38
    nuclear energy, 38–43
Alternative metrics, 933
Alternative power sources
  tax incentives, wind power, 1614
  wind power, government subsidies of, 1614–1615
Aluminum alloy artificial aging, 868
Aluminum plant assessments, Dept of Energy best practices case studies and, 1283, 1285–1288
Ambient air, 117
  quality standards, 715–717
Ambient energy fraction, 819–820
Ambient temperatures, 521–522
  based utility pricing, 1506
American Council for an Energy Efficient Economy, 513
American energy master planning, 551
  models, management system for energy, 551
American National Standard Institute (ANSI), 1028–1029
American National Standards Institute. See ANSI.
American Society of Heating and Refrigeration and Air Conditioning Engineers Inc. See ASHRAE.
American Society of Heating, Refrigerating and Air Conditioning Engineers (ASHRAE), 438

U factor definition, 1620
Amtrak, failure of, 1020
Anaerobic digesters, 90, 1529–1530
  waste fuels and, 1527
Analytical semiempirical model (ASEM), 1348–1350
Analyze data, Six Sigma and, 1314–1315
Ancillary services, electric demand response and, 284
Andon signals, lean manufacturing and, 473
Annual electricity generation, electric power projects and, 955–957
ANSI (American National Standards Institute), 50–62, 1028–1029
ANSI. See American National Standard Institute.
ANSI/ASHRAE Standard 62.1–2004, 50–62
  analysis of, 50–61
  building renovations, 61
  compliance to
    indoor air quality procedure, 54, 60
    ventilation rate procedure, 54, 55–60
  construction process, 61
  definitions as used in, 51
    indoor air quality, 51
  design documentation procedures, 60
  high humidity, 53–54
  indoor air quality procedures, 60
  minimal system requirements, 51–52
  mold, 53–54
  outdoor air quality, 51
  pressurization flow, 53–54
  procedures, 54
  sensor verification, 61
  system start-up guidelines, 61
    air balancing, 61
    operations and maintenance, 61
ANSI/MSE 2000:2005 management system for energy, 1032–1034
  benefits of, 1034–1034
  case study, 1037–1041
  problems addressed, 1035–1037
  process format, 1032–1033
  standard elements, 1033–1034
Anthracite coal, 156
APACHE HVAC software, 1401
Appalachian coal
  basin, 157
  production, 159
Apparent power volt amperes, 1510
Appendix, use of in energy audit report, 79
Appliances, electricity usage of, 583
Application interface, interactive access and, 621–622
Applications, enterprise energy management systems and, 623
ArchiPhysics-Solar software, 1402
Argon, as insulating gas, 1619
Arsenic, as syngas contaminant, 909
Artic Wildlife Reserve oil exploration, 595
  royalties, 596
  nationalization of, 596
Artificial aging of aluminum alloy, 868
ASD technology, electric pump motors and, 353, 355
ASEM. See analytical semiempirical modeling, 1348–1350

Index

Ash content of syngas feedstocks, 907
Ash fusion temperature, syngas feedstocks, 907
ASHRAE, 60–62, 438, 1620
    GreenGuide, 1407–1408
    Standard 62.1–2004, 50–62
    standards, 1363
        Energy Star indoor air quality requirements and, 577
Asian coal to liquid fuel plants, 167
Assets, electricity producers and, 388
Association of Energy Engineers (AEE), 131–135
    certification process, 133–135
    continuing education programs, 135
    salary information, 131–133
ATHENA green building design software, 1403
ATK. *See* available ton-kilometer.
Atomic structure, 1127
Attic radiant barriers, 1227–1232
Attic ventilation
    powered attic ventilator, 900
    vent selection, 899–900
Auction, allowance/permit allocation and, 432
Auditing
    facility energy use, 63–68
    user friendly report, 76–80
Australian coal to liquid fuel plants, 167
Autoclased concrete, as insulation material, 891
Automatic lighting control systems, 978–979
Automation, compressor air control systems and, 210
Automobile technology
    evolution of, 673–675
        communication buses, 674
        computers, 673–674
        engineering analysis and design, 673–675
        microprocessors, 673–674
        quality control, 673
        skilled assembly workers, 675
        smart sensors, 674–675
        standard connectors, 675
        system modules, 674
        wire harnesses, 675
    facility energy controls, 671–679
        building code support, 677–678
        equipment and system modules, 677
        impose standards, 677
        modular buildings, 677
        standardized commissioning of buildings, 678–679
    reasons not found in facility energy controls, 675–677
    transferable to building use, 672–673
        display systems, 672–673
        individual control systems, 672
        operational controls, 672
        options, 673
Autopilot flight controls, 28
Available ton-kilometer (ATK), consumption per, 24–25
Average unit cost, 920
Avoided cost, 919–920
Awnings, as window shading device, 1642
Axial fans, 1220

B

Backup electric rate schedule, 1509
Backup systems, photovoltaic systems and, 1152–1153
Balance supply and demand, 503
Balanced system, 700–701
Ballasts, high intensity discharge, 830–832
Banking, emissions reduction credit and, 432–433
BAS. *See* building automation systems.
Baseline construction, energy benchmarking and, 692–693
Baseloads, 1498
    rate structures of, 1498
Basins, coal, 157–158
Batch dryers, 338
Batch mail applications, 538
Bathhouses, Roman heating designs and, 1358
Batt-attic insulation, 901
    cathedral ceilings, 902
    full, 901
    installation of, 901–902
Batteries
    lead acid, 1155
    nickel cadmium, 1155
Battery technology, 403
Batts, in wall insulation, 896
BCA. *See* benefit cost analysis.
BCR. *See* benefit to cost ratio.
BEA software, 1401
Beadboard, as insulation material, 890–891, 894
BECP. *See* DOE Building Energy Codes Program.
BEES green building design software, 1403
Benefit cost analysis (BCA)
    definition of, 81
    energy efficiency and, 81–85
    net present value calculation, 84–85
    types, 81
        benefit to cost ratio (BCR), 81
        savings to investment ratio, 81
Benefit to cost ratio (BCR), 81
    calculating of, 82–84
        biasing effects, 84
        cost placement, 83
    significance of, 81–82
Benefits
    commissioning existing buildings and, 181
    energy project life cycle costing and, 971–972
Best Practices (BP) program, in industrial energy management, 873
Biasing effects, importance in BCR, 84
Bidding, energy project management planning and, 560–561
Billet heating, 868
Billing factors, utility rate structures and, 1501–1502
Binding energy, 1127
Bio solid gasification, 492–494
Biochemical conversion
    biogas, 90
    liquid fuels, 90–91
Biodiesel, 33–34
Bioethanol, 32–33
    needed resources, 33

Biofuels, 90–91, 1353
Biogas, 90
    anaerobic digestion, 90
Biological sequestration, natural carbon sinks, 129
Biomass, 14–15, 1266–1267
Biomass biochemical conversion, 90–91
Biomass combustion
    co-combustion, 88–89
    problems with, 89
Biomass energy, 31–34, 86–92
    benefits of, 91
        GHG reduction, 91
    combustion, 88–89
        circulating fluidized bed, 88
        stationary fluidized bed, 88
    content, 86–87
        higher heating value, 86–87
        lower heating value, 86–87
    definition of, 86
        chemical composition of, 86–87
    fuel types, 32–34
    modern types, 32
    organic basis, 31–34
    problems with, high bulk volume, 87
    traditional types, 32
    types available, 87
    uses, 31–32
    wood fuels, comparison of, 88
Biomass fuels
    heating values, 482–483
    solid fuel properties and, 482–483
    types, 32–34
        biodiesel, 33–34
        bioethanol, 32–33
        ethanol, 32–33
        methanol, 33
Biomass moisture content, 87
Biomass program funding, 1531
Biomass thermochemical conversion, 89–90
Bipolar direct current transmission, 359
Bituminous coal, 156
Blade materials for turbines, 1382–1383, 1385
BLAST software, 1401
Blended utility rate, 967
Block rates
    declining, 1503
    increasing, 1503
    inverted, 1503
    sliding, 1503
Blow off reduction, compressor air control systems and, 210
Blowdown heat recovery equipment replacement, 1368
    energy and cost savings, 1369, 1370
Blower door test equipment, 668–669
Blowers, 464
Blowing agents, in insulation materials, 891–892
Blown cellulose, ceiling and roof insulation, 901
Blown foam insulation, 897
Blown loose-fill insulation, 897
BMCS. *See* boiler control system management.
Boil water reactors, 39–40
Boiler control systems, 93–102
    management, 96–98
        flame safeguard control, 97–98
        multiple boilers, 98–102

Boiler shutoff, low cost energy efficient improvement case studies and, 519
Boilers, 93–102, 531, 1360, 1589
   combustion and, 95–96
   construction of, 93
   control of multiple, 98–102
   efficiency of, 94–95
   energy intensive, 463
   ratings of, 94–95
   types, 93–94
      high pressure, 93
      low pressure, 93
      medium pressure, 93
Boiling limitations, heat pipes and, 805, 806
Boiling water nuclear reactors, 1117–1118
BOOT. See build own operate transfer.
Borehole, geothermal heat pump system development and, 758
Boric acid flame retardants, in insulation materials, 891
Bottoming cycles, 1542
Box cooking, solar, 1326
Brayton cycles, basic, 1576–1577
   advantages and disadvantages, 1576–1577
Breathing zone outdoor airflow, 57
BREEM (Building Research Establishment Environmental Assessment Method), 948
BREEZE software, 1403
Brise-soleils, as window shading devices, 1642
Brown coal, 156
BSim2002 software, 1401
BSS. See building system simulation.
Btu/sf, 967
Bubble back, as insulation material, 895
Bubble, ERC and, 430–431
Bubbling fluid bed boiler, 1528
Build own operate transfer (BOOT), 571
Building automation systems (BAS), 104–110
   basic features of, 104
      controller-level hardware, 105
   client hardware and software, 106–107
   controller
      communications network, 106
      communications protocol, 106
      software programming, 105
   design issues, 107
      new facility, 107
   direct digital control, 104–119
   energy data collection and, 617–618
   future costs of, 108
   future trends, 108–110
   machine to machine communications, 108–110
   server hardware and software, 106–107
   upgrading of, 107–108
Building codes, incorporating in facility energy controls, 677–678
Building commissioning, 179–187
   monitoring and verification, 179
   new structures, 188–199
Building cooling applications, direct fired absorption chillers, 1545–1546
Building design system, integrated, 1401
Building envelope, 528
   energy efficiency and, 1094

measurements, data collection and, 258–259
Building geometry
   building system simulation process and, 119
   energy simulator software, 111
   energy use
      effect on, 111–115
      geometric ratio, 113
      index, 111
      modeling, 112–113
      factors re energy, 112
Building location, building system simulation process and, 119
Building maintenance, retrocommissioning and, 203
Building materials, building system simulation process and, 119
Building occupancy, daylighting and, 267
Building renovations, ANSI/ASHRAE Standard 62.1-2004 and, 61
Building Research Establishment Environmental Assessment Method (BREEM), 948
Building simulation, sustainable, 1396–1403
Building standards, incorporating in facility energy controls, 677–678
Building system simulation (BSS), 116–123
   applications of, 118–119
   definitions of, 116
   energy flow paths, 116–117
   evolution of, 117–118
      energy flow, 117–118
      frequency domain, 117
      numerical methods, 118
      response factor, 117
      time domain, 117
   physical issues, 116–117
      air flow, 116
      ambient air, 117
      casual heat gain, 116
      control processes, 117
      inter-surface longwave radiation, 116
      moisture transfer processes, 117
      plant interaction with building zones, 117
      shading from sun rays, 116
      shortwave radiation, 116
      solar radiation, 117
      surface convection, 116
   process
      case studies of, 120–123
      input parameters, 119–120
         building geometry, 119
         building location, 119
         building materials, 119
         casual loads, 119
         HVAC control, 120
         infiltration, 120
         internal environment control, 120
         standing devices, 120
         transparent surfaces, 120
         weather data, 119
Building types
   high rise, 1513
   low rise, 1513
   walls and windows and, 1513
Building zones, plant interaction with, 117
Buildings

electric demand response solutions for, 281–283
   rating systems for, 948–949
      Building Research Establishment Environmental Assessment Method (BREEM), 948
      Energy Star™, 948
      Green Building Challenge program, 948
   standardized commissioning of, 678–679
   ventilation problems and, 20–21
Built in storage solar water heater, 1332–1333
   finned, 1333
   plain, 1332–1333
Bulk problem, biomass energy and, 87
Burden of proof, environmental policy and, 627–628
Burner/engine redesign, waste fuel technology and, 1527
Bus circuit breaker arrangements importance of, 359
BUS^{++} software, 1401
Buses, communication, 674
Business retention utility rates, 1505
Bypasses, air barriers and, 664

C

Cables, 357, 357
   capacity determination, 358
   electric power transmission, 358
   transmission limitations, 358
CAC. See command and control policies.
Cadmium, as syngas contaminant, 909
CAES. See compressed air energy storage.
Calciners, 464
Calcium bromide cycles, 1445
Calculated data, interactive access to, 621
Calculation methodology, space heating and, 1363–1364
Calculations
   residential building heating loads, 1272–1273
   seasonal heating demand, 1272–1273
California Energy Commission, 511
California Statewide Emerging Technology Program, 514
California, electricity deregulation results in, 385
Candela, 998–999
   definition of, 998–999
CANDU. See pressurized heavy water nuclear reactors.
Cap and trade emissions trading, 431
Capacity calculations, inverse heat transfer methods and, 1065
Capacity losses, 361
Capacity, electric power transmission systems and, 360
Capital costs, 956–957
   nuclear energy and, 1101–1103
Capital expenditure, reducing energy consumption for drying, 346–347
Capital investment, energy conservation and industrial processes, 464
Capital leasing, 571–572
Capital rationing, 84–85
Capture and storage, carbon, 125–126

Index

Carbon capture and storage (CCS), 125–129
    carbon sources, 125–126
    costs of, 128–129
    geologic sequestration, 127–128
    industrial CO_2 capture, 127
    ocean direct injection, 128
    post-combustion capture, 126
    pre-combustion capture, 126–127
Carbon capture, sequestration and, 125–129, 169
Carbon dioxide, 20
    based DCV, 56
    capture, 127
        storage of, 127
    differentials, steady-state, 52
    emissions
        gasoline fueled vehicles, 788
        hybrid vehicles, 788
        plug-in hybrid vehicles, 788–789
    levels, ventilation rate procedures and, 56–57
    reduction in, 15–17
    regulating of, 11–12
    solid fuel properties and, 484
Carbon emissions, geothermal energy and, 744
Carbon filtration of wastewater, 1561
Carbon sequestration, 125–130
    biological methods, 129
    biomass, 14–15
    carbon capture and storage, 125–129
    definition of, 125–129
    disadvantages of, 15
    future prospects of, 129
    technologies, 14–15
Carbon sources, carbon capture and storage and, 125–126
Carbonyls, metal, as syngas contaminants, 909
Careers, energy engineering and, 131–136
Carnot coefficient of performance, 815
Carnot cycle efficiency, 1171
    thermodynamic temperature, 1171
Carnot efficiency, 1574–1575
Carnot engine, 1427–1428
Carnot Law, 1541
Carnot steam cycle, 1381
Cascading Style Sheets, 537
Cash flows, 932
    IRR and, NPR, 935
Casual heat gain, 116
Casual loads, building system simulation process and, 119
Catalysts, Fischer-Tropsch process and, 164–165
Cathedral ceiling insulation
    batt-attic, 902
    exposed rafting, 902–903
    R-30 batts and, 902
    raised top plate, 902
    raised-heel trusses, 902
    scissor trusses, 902
    soffit air ventilation, 902
Cathedral ceiling insulation, techniques, 902–903
CAV. *See* constant volume of supply air.
CCGT. *See* combined cycle gas turbine.
CCS. *See* carbon capture and storage.
CDD. *See* cooling/heating day adjustment.

CEE premium efficiency motor, 352–353
CEE. Consortium for Energy Efficiency.
Ceiling and roof insulation
    air flow restrictions, 902
    attic
        floor insulation, 900901
        ventilation, 899–900
    batt attic insulation, 901
    blown cellulose, 901
    cathedral ceilings. *See* Cathedral ceiling insulation.
    loose-fill insulation, 900–901
        installation techniques, 900–901
    powered attic ventilator, 900
    recessed lights, 903
    under storage floor, 901
    vent selection, 899–900
Cellulose insulation, 890, 893
    dust inhalation and, 891
    effect on environment, 891
Center for Analysis and Dissemination of Demonstrated Energy Technologies, 515
Center of gravity, aircraft pre-flight planning and, 27
Central chiller plants, cooling system adjustment and, 276
Central receiver, 1328
Centrifugal compressor surge protector, compressed air and, 209
Centrifugal compressors, 464
Centrifugal fans, 1220
Centrifugal pumps, 464, 1213
    theory, 1215
Certification, management system standards and, 1030–1031
CFB. *See* circulating fluidized bed.
CFBC. *See* circulating fluidized bed combustor.
CGI. *See* common gateway interface, 536
Chain reaction, 1128–1129
Chaos, 659
Char reaction, 487
Charting tools, 538
Chauffage, 1136
    contracts, 571
Cheap energy, U.S. transportation and, 588
Chemical fuels, energy form, 1091
Chemical plant assessments, Dept of Energy best practices case studies and, 1283–1284
Chemical process industry
    electrical element, 1305
    electricity supply, 1305
    energy conservation, examples of, 1306
    energy efficiency, 1302–1309
    EPA regulations, 1302–1303
    fuel oil supply, 1305
    material and energy balance concept, 1303–1304
    mechanical element, 1304–1305
        chillers, 1305
        compressed air, 1305
        hydraulic systems, 1305
        microturbine power generation, 1304
        pneumatic transport, 1305
        pumping, 1304
    natural gas supply, 1305

    Occupational Safety and Health Act, 1303
    process energy optimization, 1304
    waste to energy, 1305
    water element, 1305
Chemical reactions, 462
Chicago Edison, infrastructure growth, 391–392
Chilled water storage, 1417–1418
Chiller capacity, 311–312, 1418
Chiller efficiency, higher lift, 309–310
Chiller plants, design comparison, 312
Chiller energy only analyses, 309
Chillers, 531
    chemical process industry and, 1305
    plant retrofitting, 315
    pre-packaged, 314
    series, 310
China's economic growth, impact on U.S. transportation energy, 592
Chloramine, 1559
Chlorine, 1559
Chlorofluorocarbons (CFCs), effect on environment, 891–892
Cholera, 1557, 1560
CHP. *See* combined heat and power.
Churn factor, Underfloor air distribution cost effectiveness and, 1469
CIE vocabulary for spectral regions, 1617

Circuit breakers, 360, 704
    importance of adequacy, 360
Circuit resistance, 360–361
Circulating fluid bed boiler, 1528
Circulating fluidized bed (CFB), 88
Circulating fluidized bed combustor (CFBC), 489
Clausisus statement, 1171
    Carnot cycle efficiency, 1171
Client side programs, web publishing and, 537–538
    Cascading Style sheets, 537
    Dynamic HTML, 537
    extensible markup language, 537
    Java applets, 537
Climate changes, global warming and, 723
Climate control
    common types, 1660
    differences in two story building, 1660–1661
    mobile thermostat, 1661–1663
    wireless mobile thermostat, 1660–1667
Climate policy
    European Union Emissions Trading System, 141
    Kyoto protocol, 139–141
    options for, 137–138
        emission reduction, 137–138
        emissions quota allocation, 138
    regional programs, 141
    U.N. Framework Convention on Climate Change, 139
    voluntary programs, 141
    world motivation for, 137–142
    world response to, 138–139
    worldwide adaptation of, 138
Climate Technology Initiative, 448–449
Closed loop control, electric demand response and, 273

Closed-cell, high-density polyurethane, as insulation material, 891, 894
Clothes washers, 1588
CMMS. *See* computerized maintenance management systems.
Co combustion concepts, 88–89
    direct cofiring, 88–89
    indirect cofiring, 89
    parallel combustion, 89
Coagulation, 1557, 1558
Coal basins, 157–158
    Alaska, 158
    Appalachian, 157
    Great Plains, 158
    Gulf Coast region, 158
    interior, 157–158
    Rocky Mountain states, 158
Coal boiler vs. IGCC, 912
Coal composition, solid fuel properties and, 481
Coal derived syngas contaminants, 907
Coal fired electrical generating station
    exergy analysis and, 648–651
    material flows data, 650
Coal gasifier, 163
Coal mining
    liquid fuels and, 168–169
    underground, 144
Coal power plants, 954
Coal prices in United States, 152–154
Coal production in United States, 143–154
    coal prices, 152–154
    employment, 150
    history of, 143–144, 145
    innovations in, 150–151
    preparation of, 151–152
    productivity, 147–150
        regional basis, 147–149
        regional changes, 149–150
    surface mining, 144, 146
    technology trends in mining, 146–147
    tonnages produced by location, 159, 160
        Appalachia, 159
        Texas, 159
        Wyoming, 159
        western United States, 159
    transporting of, 159, 161
    types of, 144, 146
        underground mining, 144
    uses of, 159–161
        electricity generation, 159–160, 161
        industry, 160
        steel industry, 160–161
Coal supply in United States, 156–161
    basins and types found, 157–158
    coal to liquid fuels and, 168
    mining methods, 158–159
    types, 156–157, 158
        anthracite, 156
        bituminous, 156
        lignite, 156
        subbituminous, 156
Coal to liquid fuels, 163–171
    advantages of, 165
    air quality, 169
        carbon capture and sequestration, 169
    coal gasifier, 163
    coal mining, 168–169
    coal supply in United States, 168
    current and planned world production of, 167
    diesel products, 165–166
        specifications, 165
        utilization of, 166
    electricity requirements, 170
    Fischer-Tropsch process, 163
    manufacturing of, 163
    natural gas issues, 170
    plant employment, 170
    plant land requirements, 169
    plant locations, 168
        Asia, 167
        Australia, 167
        India and Pakistan, 167
        North America, 167
    plant operation of, 168
    products from, 163
    transportation needs, 168–169
        railroad limitations, 168
    waste products, 169–170
    water quality, 169
    water neutral concept, 169
Coal transportation, 159, 161
    coal to liquid fuels, 168–169
Coal
    agriculture residues and, cofiring of, 490
    anthracite, 156
    bituminous, 156
    brown, 156
    heating fuel, 1361
    lignite, 156
    manure and, cofiring and, 490–491
    natural gas vs., 161
    RDF and, cofiring and, 490
    soft, 156
    subbituminous, 156
    wind vs., 957
Coefficient of performance, 819
Cofiring, 490–491
    coal and agriculture residues, 490
    coal and manure, 490–491
    coal and RDF, 490
    fouling during, 491, 493
    NO_x emissions, 491
Cogeneration 1203
Cogeneration applications, load usage and, 1482
Cogeneration district heating and cooling systems, 318–319
    characteristics and technical aspects of, 319
    heat recycling, 318
    use in Europe, 318
Cogeneration natural gas and, 1507
Cogeneration project funding, 1531
Cogeneration Public Utility Regulatory Policies Act (PURPA) of 1978 and, 1202
Cogeneration steam turbines, 1386
Cogeneration systems, combined heat and power and, 177–178
Cogeneration thermal output temperature, combined heat and power and, 176
Cogeneration unit location, combined heat and power and, 175
Coil designs, mobile HVAC systems and, 1061–1066
Cold air retrofit case study, HVAC and, 172–174
Cold gas cleanup technologies, 908–909
Coldfusion, 537
Coliform water test, 1557
Collaboration tools, energy project management and, 563–564
Collection of data, 255–263
Collector performance, solar water heating and, 1336
Color preference index, definition of, 999
Color preference, lighting design and, 999
Color rendering
    definition of, 999
    improvement, definition of, 999
    lighting design and, 999
Color temperature
    definition of, 999
    lighting design and, 999
Combined cycle gas turbine (CCGT), 1386
Combined cycles, 1578–1579
    basics of, 1578
    evaluation, 1578–1579
    heat recovery steam generators, 1578
Combined gas law, 228–229
Combined heat and power, 863–864
    applications
        distributed generation technologies and benefits of, 299
        resale of excess power, 865
        waste heat recovery and, 864–867
    cogeneration systems, 177–178
    cogeneration thermal output temperature and, 176
    cogeneration unit location, 175
    distributed generation and, 303–308
        advantages of, 303–304
        philosophy behind, 304
        users of, 305–307
            energy service companies, 306
    efficiencies
        electric load following cycle, 1530–1531
        integrated coal gasification/combined cycle power plants, 1531
        thermal load following cycle, 1530
        waste fuels and, 1530
    examples of usage, 176–177
    exhaust gas condensation solutions, 177
    Federal support for, 866
    generator voltage, 175
    generators, induction vs. synchronous, 175–176
    industrial
        electric power systems, 175
        plant savings and, 864–867
        processes, 175–178
        thermal considerations, 176
    interconnection standards, 866, 867
    interruptible fuel rates, 177–178
    steam vs. hot water, 176
    waste heat recovery and, 1546–1547
    water augmented gas turbine power cycles and, 1584–1585
Combuster, 489
Combustion analysis, data collection and, 260–261
Combustion capture, 126–127

Index

Combustion of biomass, 88–89
Combustion processes, waste fuels and, 1525
Combustion systems, energy intensive, 463–464
Combustion
    boilers and, 95–96
    circulating fluidized bed combustor, 489
    energy conversion and, 487–489
    first law of thermodynamics and, 1424
    fixed bed combustor, 489
    fluidized bed combustor, 489
    ignition and, 487
    stoker firing, 488–489
    suspension firing, 488
Combustor
    FSTIG cycle, 1583
    injection of steam and water, 1582–1583
    STIG cycle, 1582–1582
    VAST cycle, 1583
    VASTIG cycle, 1583–1584
Comfort control, HVAC systems and, 839
Command and control costs (CAC), 430, 435
    emissions trading vs., 435
Commercial buildings, electric demand response solutions for, 270–278, 281–282
Commercial real time pricing, 1180–1182
Commercial use of energy in U.S., 583
    space vs. energy usage, 583
Commercial utility rates, 1413
Commercial waste fuel suppliers, 1530
Commissioning existing buildings, 710–712
    case study, 180–181, 185–187
    cost and benefits, 181
    definitions, 711
    energy management process, 185
    findings from, 711
    process of, 182–185, 711–712
        implementation and verification, 184
        phase development, 183–184
        team members, 183
            responsibilities of, 183
Commissioning process, 191–193
    acceptance phase, 196–197
    agent selection, 192
    agent skill set, 192
    common mistakes, 199
    phases of, 193–197
        construction, 195–196
        design, 193–195
        installation, 195–196
        predesign, 193
    postacceptance phase, 197
    problems with, 191
    reasons for, 191
    retrocommissioning, 200–206
    success factors, 197–199
    systems to include, 191
    team composition, 192–193
    testing, adjusting and balancing submittal, 195
Commissioning types, 179–180
    continuous, 180
    recommissioning, 180
    retrocommissioning, 180
Commissioning, 179–187
    acceptance-based, 189–190
    definition of, 188–189
    definitions used in, 179–181
    history of, 190
    market acceptance of, 190–191
    new buildings, 188–199
    process-based, 189–190
Commodity charge, utility billing and, 1501
Commodity rates, 1503–1504
Common gateway interface (CGI), 536
Communication architecture, IntelliGrid^{SM} and, 917
Communication buses, automobile technology evolution and, 674
Communications
    controlling of electric demand response and, 283–284
    energy management system problems and, 1034–1035
Community development, design for energy efficiency, 1409
Comparable value, 930
Competitive energy rates, 1506
Complete mix reactor treatment, 1561
Compliance costs, electricity usage and, 401
Compliance period, emissions trading and, 432
Compressed air control systems, 207–213
    air distribution, 211–212
    air quality, 211
    centralization of, 212–213
        human machine interface workstation, 212
    components, 207–209
        air production control, 207
        compressor control, 208
        motor control, 207–208
Compressed air energy storage (CAES), 214–218
    expansion turbine and control, 215–216
    lead-acid battery alternative, 214
    thermal, 214–215, 216
    uninterruptible power supply, 214–215
    vessels, 217
Compressed air leak
    control and prevention, 224–225
    detection
        discharge volume table, 220
        methods, 219–222
        repair and, 219–225
    management of, 223–224
    repairs of, 222–223
        logistical procedures, 223
        methods, 223
        overhead associated with, 223, 225
Compressed air storage
    control and maximization, 233–234
    distribution, 226–235
Compressed air systems, 236–239
    components of, 236–237
        demand side, 238
        supply side, 237
    costs of, 239
    energy balance, 227
    energy flow, 226
        generation efficiency of, 226
    optimization of, 241–245
        controlling demand, 241–242
        minimizing energy losses, 243
        reducing demand, 242
        storing, 242–243
        supply energy reduction, 243–245
    pressure set point, 238
    storage of, 228–233
        calculating
            peak air demand, 231–232
            receiver volume, 232
            usable energy, 231
        combined gas law, 228–229
        permissive start-up time, 232–233
        pneumatic capacitance, 229–230
        usable energy, 230–233
        pressure profile, 230–231
        storage delta, 230–231
    uses of, 236
Compressed air, chemical process industry and, 1305
Compression, 1235
Compressor air control systems
    control types, 208–209
        centrifugal compressor surge protector, 209
        machine protection, 209
        micropressors, 208
        plant safety, 209
        pneumatic, 208
        programmable logic controllers, 208
        protection and safety elements, 208–209
    energy savings, 209–211
        automation, 210
        blow off reduction, 210
        intercooler control, 210
        leak loss reduction, 210
        load scheduling, 210
        networked capacity control, 209–210
        precise pressure regulation, 209
        scheduling, 210–211
        start/stop sequence, 210
        surge controls, 210
        system controllers, 210
Compressor control, 208
Compressor discharge output pressure, 209
Compressor failure, storing compressed air, 242–243
Compressor permission start-up time, 232–233
Compressor surge, 1576
Compressor systems
    common sources of energy waste, 867
    industrial plant savings and, 867
Compressor train, 1579
    water injection, 1579
        quasi isothermal compression, 1579
Compressors, mobile HVAC systems and, 1076
Computerized maintenance management systems (CMMS), 710
Computers, automobile technology evolution and, 673–674
Concentric tube heat exchanger, 801
Conceptual stage, energy project management planning and, 558–559
Concrete block core insulation, 892
    grades of, 892
Concrete block walls, 1516
    reflective insulation, 891
Concrete form insulation systems, 895–896
    cost premium, 895–896
    foams in, 895
    wall shape, 895

Concrete form walls, insulated, 1517
Concrete as insulation material
    aerated, 891
    lightweight, autoclaved, 891
Concrete products, lightweight, as insulation materials, 896
Concrete wall insulation, 892
Concrete walls, 1516–1517
    precast, 1517
Concrete, Air-Krete, 894
Condensate return system, 1367–1368
    improvement of, 1375–1378
Condensation resistance, windows, 1622–1623
Condenser, mobile HVAC systems and, 1077
Condensors, 1544
    heat exchangers and, 803–804
    maintenance of, 804
    materials used in, 804
    safe operation of, 804
Conditioned floor area corrections, 2
Conduction heat transfer
    double pane window, 1618
    windows, 1617
Conduction, 1161
    heat transfer and, 823
Conductive heat flow, in window energy performance, 1616
Conductor losses, 360–361
    circuit resistance, 360–361
Conductor sizing, 701–702
Conductors, 358
Cone density chart, 994
Conformance audits, 1030
Conservation economy, 1409
    sustainable development and, 1409–1410
Conservation of energy, 1015
    efficiency vs., 21
Conservation utility rates, 1505
Conservation, aircraft fuel consumption and, 25
Consortium for Energy Efficiency, 513
Consortium for Energy Efficiency. *See* CEE.
Constant volume of supply air (CAV), 51
Constrained equilibrium modeling, exergy analysis and, 657
Construction guidelines, ANSI/ASHRAE Standard 62.1–2004 and, 61
Construction work in progress, rate of return regulation and, 1255
Construction
    commissioning process and, 195–196
    energy project management planning and, 561
Consultants, energy service companies and, 306–307
Consumer good demand, developing countries and energy demand, 500–501
Consumption
    U.S. energy's, 608
    worldwide energy resources and, 445–446, 447
Contactors, 704
Contaminant control, 839
Contamination levels
    direct methanol fuel cells (DMFC) and, 741
    molten carbonate fuel cells and, 728
    proton exchange membrane fuel cells (PEMFC) and, 737
    solid oxide fuel cells and, 730
Continuous commissioning, 180
Continuous cyclic process, 1170–1171
Continuous dryers, 338
Contract utility rates, 1505–1506
Contracted HVAC services, 709–710
Contracting models, energy service companies in Europe and, 568–572
    energy savings, 569–571
    other types, 571–572
Contractors, independent power producers and, 857
Control designs for HVAC facilities, 532–533
Control processes, 117
Control system complexity, energy efficiency and industrial plants, 870
Control
    granularity of, 273
    resolution of, 273
    Six Sigma and, 1316
Controller communications network, building automation systems and, 106
Controller communications protocol, 106
Controller software programming, building automation systems and, 105
Controller, 413–415
    enthalpy, 415
    indicating devices, 416
    input type, 414
    level hardware, building automation systems and, 105
    modes, 415
    modulating, 415–416
    output devices, 415
    relative humidity, 414–415
    system interfaces, 416
    temperature, 414
    transducer, 416
    two-position, 415
    universal, 415
Controls
    distributed control systems, 459
    industrial processes and, 459
Convection heat transfer, in windows, 1617–1618
    double pane, 1618
Convection, 1161
Conventional Brayton cycles, 1576–1577
    basic, 1576–1577
    recuperator usage, 1577–1578
Conversion efficiencies, renewable energy and, 1266
Conversion factors
    energy and, 603
    energy units and, 1164
Conversion, uranium and, 1113
Cooker kettles, steam blowthrough rate and, 1373
Cooking, 809
    solar, 1325–1327
Cool storage system design, 1418–1420
    chiller and storage capacity, 1418
    load profile, 1418
    system layout and control, 1419–1420
Cooling applications, thermal energy storage and, system design, 1418–1420
Cooling costs, activity based costing and, 1044
Cooling equipment, solar heat effect, 1618
Cooling process, energy intensive, 462–463
Cooling system adjustment
    central chiller plants, 276
    electric demand response and, 275–276
    evaluation of, 276
    increase supply air temperature, 276
Cooling system methodology, 1317–1318
Cooling towers, 246–253, 531–532, 1589–1590
    performance and rating of, 251–253
    thermo-fluid dynamic efficiency, 250
    types, 246–251
        fill type, 249–250
        mechanical draft, 247–248
        natural draft, 247–248
        packing, 249–250
        wet, 246–247
        wet-dry, 250–251
Cooling
    dehumidification and, 1186–1187
    sensible, 1186
    solar, 1326–1327
    thermal energy storage and, 1413–1418
        commercial utility rates, 1413
        storage equipment, 1414–1418
Cooling/heating day adjustment, 2
Core losses, 360
Corrective action, 1030
Cost accounting, HVAC monitoring and, 522–523
Cost allocation
    fixed cost related charges, 1500
    least-cost planning, 1500
    market driven discrete usage charge, 1500
    utility rate structures and, 1500
Cost analysis, energy efficiency and, 81–85
Cost based negotiated rates, 1505–1506
    business retention, 1505
    conservation and load management, 1505
    economic development, 1505
    special contract rates, 1505–1506
Cost benefit analysis, environmental policy and, 628–629
Cost benefits, energy efficiency and, 15–17
Cost centers, energy accounting and,
    criteria for, 3
    establishment of, 3
Cost components, electric power projects, 954–955
Cost drivers, 1042–1043
    first stage, 1042
    second stage, 1042–1043
Cost effective analysis, environmental policy and, 629–630
Cost metering, web-based compressor management and, 213
Cost minimizing pollution abatements, 433–436
    allowance allocation, 434–435
    command and control costs, 435
    distribution of costs, 434–435
    emission taxes, 435–436
Cost of aircraft fuel, industry impact, 25
Cost of capital, 1256
Cost of electricity, 959
Cost placement, importance in BCR, 83
Cost savings, energy master planning and, 552–553

Index

Cost savings, radiant barriers and, 1228
Cost
    pumped storage hydroelectricity and, 1210–1211
    steam generation, 1366–1367
Costs
    air, 212
    commissioning existing buildings and, 181
    compressed air leaks and, 223, 225
    energy project life cycle costing and, 971–972
Cotton, as insulation material, 893
Counter flow heat exchangers, 802–803
Course bubble diffusion, 1549
CPFilms LLumar® window film, case study, 1626
    demonstration building, 1627
    DOE-2 simulation values, 1627
CPP. *See* critical peak pricing.
Credits, carbon dioxide CO_2 and, 15–17
Critical peak pricing (CPP), 1175, 1177–1180
    technologies available, 1177–1178
Crop drying, solar, 1325
Cross flow heat exchanger, 801
Cross quality, 421, 422
Cryogenic air-separation technologies, gasifiers and, 909
Crystalline SI photovoltaic (PV) systems, 1148–1149
Culture, energy use and, connection between, 1016
Cumulative sum of differences (CUSUM) analysis, 6
Current flows, 357
CUSUM. *See* cumulative sum of differences.
Cutting tools, 813

D

Damage-weighted transmittance, of windows, 1623–1624
Data analysis, energy balance and, 688
Data collection, 255–263
    building envelope measurements, 258–259
    combustion analysis, 260–261
    electrical energy measurements, 255–257
        current, 255–256
        power factor, 255–256
        voltage, 255–256
    energy information system processes and, 535
    energy management system problems and, 1037
    light measurements, 257–258
        light meters, 258
    low cost data acquisition, 262–263
    software, 262–263
    thermal imaging, 259–260
    ultrasonic leak detection, 261–262
Data compilation, energy balance and, 686–686
    building or location, 685
    equipment type, 685
    meter, 686
    processes used, 686–686
Data management, energy benchmarking and, 692
Data organization, interactive access and, 620

Data presentation, interactive access and, 620
Data sources, energy benchmarking and, 691
Data structuring, energy, 618–619
Data tabulation, energy accounting and, 3–4
    fuel, 4
    needed data, 3–4
Databases, 538
Daylight harvesting, 978
Daylight illumination from windows, as energy saver, 1624
Daylight software, 1403
Daylighting design, 267
Daylighting simulation, 267–268
    scale model testing, 268
    software, 267–268
Daylighting, 264–269
    building occupancy, 267
    illumination basics, 264–267
    word glossary, 269
DCS. *See* distributed control systems.
DCV. *See* demand control ventilation.
DDC. *See* direct digital control.
Deaerator steam jet ejectors, mechanical vacuum pumps, 1370–1371
Decentralized energy
    advantages of, 1259–1260
    renewable, 1258–1264
Declining block rates, 1503
Define, measure, analyze, improve and control, 1310
Define, Six Sigma and, 131
Defining energy project management plan, 559–560
Degree of saturation, 1185
Dehumidification
    cooling and, 1186–1187
    designing a system, 839
    heating and, 1187–1188
    HVAC systems and, 839–840
Delta Score estimator, Energy Star Portfolio Manager software and, 578
Demand billing, 1510–1511
Demand buyback program, 281
Demand charge, utility billing and, 1501
Demand control ventilation (DCV), 56
Demand control, compressed air systems and, 241–242
    reducing demand, 242
Demand forecasting, utility long term planning and, 1493–1494
Demand shifting, electric demand response and, 272
Demand side energy consumption, IEKPI and, 1140
Demand side management, 499
    case studies and
        energy efficiency and, 503–505
        Northern Cyprus, 503–504
        Turkey, 503–504
    energy programs, 286–291
Demand/commodity rates, 1503–1504
Density compensation, gas and steam and, 1057–1059
Density, land use impact and, 1450
Department of Defense, fuel cell program, 511
Department of Energy role in

    revising ASHRAE/IESNA/ANSI Standard 90.1 & 90.2, 439
    setting energy standards, 439
Depreciation, 957, 1256–1257
Dept. of Commerce, 510–511
    National Institute of Standards and Technology, 510–511
    National Technical Information Service, 510
Dept. of Energy best practices case studies, energy savings and, 1283–1301
    aluminum plant assessments, 1283, 1285–1288
    chemical plant assessments, 1283–1284
    forest product plant assessments, 1284, 1288–1291
    glass plant assessments, 1284, 1291–1293
    metal casting plant assessments, 1284, 1293–1294
    mining plant assessments, 1284–1285, 1294–1296
    petroleum plant assessments, 1285, 1296–1299
    steel plant assessments, 1285, 1299–1301
Dept. of Energy, 508–510
    Emerging Technologies, 509
    Federal Energy Management Program, 509
    Inventions and Innovation group, 509
    national laboratories, 510
    Office of Electricity Delivery and Energy Reliability, 509–510
    Office of Energy Efficiency and Renewable Energy, 508–509
    Office of Fossil Energy, 509
    Office of Scientific and Technical Information, 510
DER. *See* distributed energy resources.
Deregulation of electricity, 379–386
    arguments for the consumer, 384
    political reaction, 384–384
    regional transmission operator, 380
    regulator reaction, 384–384
    retail markets, 382–383
    third party involvement, 380
    utility rate structures and, 1511–1512
    utility reaction to, 385–386
    wholesale markets, 381–382
Desiccant dehumidification, case study, 292–293
    accuracy, 294
    computer modeling, 293–294
    field testing, 292–293
    results, 294
Desiccant drying wheels, 795
Design documentation procedures, ANSI/ASHRAE Standard 62.1–2004 and, 60
Design issues, building automation systems and, 107
    new facility, 107
Design phase, commissioning process and, 193–195
DESs. *See* district energy systems.
Developer, independent power producers and, 856
Developing countries
    barriers to energy efficiency, 501
    demand for consumer goods, 500–501

economic growth, 500
energy and, financing of, 501
energy efficiency and, 498–506
implementation of energy efficiency, 501–505
  measuring economic development, 501–502
population growth, 500
status of energy in, 501
Dew point temperature, 1184–1185
DHC. See district heating and cooling.
Die temperature, 868
Diesel engines
  reciprocating, 1238, 1239, 1241, 1242
  specifications for, 1238
  submodels used, 1239
Diesel products, coal to liquid fuels and, 165–166
Digital devices
  electricity quality and, 402
  electricity usage and, 402
Dilution, indoor air ventilation and, 19–20
Dimmers, 982
  ballast to, 982
  integrated, 982
  programmed start ballasts, 982–984
Dimming ballasts, lighting control systems, 982
Dimming electronic ceramic ballasts, 830
Dimming electronic metal halide ballasts, 830
Dimming limits, electronic ballast primer and, 837
Dimming system, 982
  lighting control systems, 983, 984
Dimming, 831, 996
  lighting based demand response and, 277
  lighting design and retrofits, 996
  switching vs., lighting controls and, 989
Direct cofiring, 88–89
Direct current electric motors, 349
Direct current power system, 389
Direct current systems, 705
Direct current transmission, 359
  bipolar type, 359
  disadvantages of, 359
  inverters, 359
  rectifiers, 359
  single pole type, 359
  thyristors, 359
Direct digital control (DDC), building automation systems, 104–119
Direct fired absorption chillers, advantages and disadvantages, 1545–1546
Direct incentives, energy demand-side management programs and, 291
Direct methanol fuel cells (DMFC), 739–743
  applications, 743
  contamination levels, 741
  principles of, 740
  problems with, 741–742
  technological status, 742
Direct solar gain design, 1322
Dirty area systems, lighting design and retrofits, 996
Dirty environments, lighting design, 996
Discount rate determination, 932–933

Discount rate issues, nuclear energy economics and, 1104–1105
Discount rate, 930
Dishwashers, 1587–1588, 1603
Disinfection of treated water, 1557, 1559
  chloramine, 1559
  chlorine, 1559
  ozone, 1559
  ultraviolet light, 1559
Dispatch stack order, 1499
Dispersed flow, 1561
Dispersion of air pollution, 718–721
Display systems, automobile technology and, transferable to building use, 672–673
Disposal cost reduction
  landfill cost, 1530
  trash collection and hauling, 1530
  waste fuels and, 1530
Disposal, nuclear energy and, 1113–1114
Dissolved oxygen mechanics, 1551–1553
Distillation process, 462
Distillation, 809
  solar, 1325
Distributed control systems (DCS), 459
  process historian, 459
Distributed energy resources (DER), 404
Distributed generation, 296–308
  combined heat and power, 303–308
    advantages of, 303–304
    philosophy behind, 304
    users of, 305–307
    energy service companies, 306
  definition of, 297
  history of, 296–297
  markets
    existing, 304
    future, 304
  technologies available, 304–305
  technologies, 297–299
    benefits of, 299–301
      combined heat and power applications, 299
      environment, 301
      postpone power plant investment, 301
      security of supply, 299–300
    cost estimates, role of
      natural gas, 299
      petroleum, 299
    fuel cells, 298–299
    microturbines, 298
    network losses and usage, 300–301
    reciprocating engines, 297–298
    renewable, 299
    simple cycle gas turbines, 298
Distribution of costs, emission trading and, 434–435
Distribution Power Quality (DPQ), 1484
District cooling systems, 309–315
  chillers, plant retrofitting, 315
  control strategies, 314
  cost savings from, 314
  design parameters, 313
  full-load efficiency improvement, 313–314
  part-load efficiency improvement, 314
  plant design energy comparison, 312

pre-packaged chillers, 314
series chillers, 310
variable flow, 310–311
  chiller capacity, 311–312
  water-cooled chiller efficiency, 309
District energy systems (DESs), 316–330
  district heating and cooling, 317
  nomenclature, 316
District heating and cooling (DHC), 317
  benefits of, 317
  case studies, 324–330
  cogeneration, 318–319
  efficiencies of, 320–321
  environmental impact of, 320–321
  exergy, 321
  history of, 319
  implementation of, 320
  market establishment of, 322
  organizations involved, 317
  performance evaluation, 322–324
    exergy analysis, 322–324
  sustainable development, 321–322
    renewable energy resources, 322
  systems, 320
    energy generation, 320
    energy transmission, 320
    energy use, 320
  technical aspects, 319–320
    energy generation, 320
    energy transmission, 320
    energy use, 320
Diversity factor, energy balance and, 688
DMAIC methodology, 1312, 1313–1316
  analyze data, 1314–1315
  control, 1316
  define, 1313
  improvement by
    new process capability, 1316
    variable relationships, 1315–1316
  measure the process, 1313–1314
DMAIC. Define, measure, analyze, improve and control, 1310
DMFC. See direct methanol fuel cells.
Documentation procedures, ANSI/ASHRAE Standard 62.1-2004 and, 60
Documented processes, 1029
Documenting management systems
  guidance documents, 1030
  specification standard, 1030
DOE Building Energy Codes Program (BECP), 443
DOE-2 energy study, solar-control film, 1631
DOE-2 software, 1402
Double pane windows, conduction and convection heat transfer, 1618
Downtime
  lean manufacturing and, 472
  preventive maintenance, 472
  set ups, 473
Drag reduction
  aircraft fuel consumption and, 26–27
  programs to reduce, 27
Dry bulb temperature, 1184
Dry cooling towers, 250
Dry steam, 745
Dryer types, 340
Dryer, mobile HVAC systems and, 1077

Index

Drying
  agricultural and forestry products, 332–336
  definition of, 338–339
  energy audits,
    basic, 344–345
    detailed, 345
  energy consumption of, 339–343
    analyses, 341–343
    reducing, 345–347
      capital expenditure, 346–347
      non capital expenditure, 346
  energy efficiency of, 334–335
  energy intensive, 462
  industry operations, 338–348
    dryer types, 338–339, 340
  new methods and technology, 336
  performance of, 334–335
  reducing electrical consumption, 347
    future of, 347–348
  research on, 335–336
  scheduling, 333–334
  steps in, 332–333
  system design and selection, 335
  thermal, 332
  types, 333
Drywall, airtight barrier, 667–668
DSM. See demand side management.
Dual fuel firm rates, natural gas and, 1507
Duct static pressure setpoint reduction, air distribution adjustment and, 275–276
Ductile metal filament lamp, 392
DUCTSIZE software, 1402
Dust inhalation, cellulose insulation and, 891
Duties, 959
Dynamic data sources, 691–692
  automated meter data collection, 691–692
  metering, 691
  temperature and humidity data, 692
  utility bills, 691
Dynamic HTML, 537

E

E Source Companies LLC, 513
EAC. See energy account centers.
Economic capital, 1410
Economic development utility rates, 1505
Economic framework, environmental policy and, 628–630
Economic growth, developing countries and energy demand, 500
Economic indicators, energy efficiency and, 501–502
Economics
  energy analysis and, 652–653
  nuclear energy and, 1101–1110
  pumped storage hydroelectricity and, 1210–1211
  solar water heating and, 1336–1337
Edison, Thomas, 379, 389
EE4 CODE software, 1402
EED software, 1402
EEM. See energy efficiency measure.
EEMS. See enterprise energy management systems.

EERE. See Office of Energy Efficiency and Renewable Energy.
Efficiencies
  boilers, 94–95
  energy use and, 1093–1094
  energy use vs. conservation, 21
  IEKPI and, 1141
EIS. See energy information systems.
Elastic potential energy, 1163
Electric bulk power operations, 361–363
  control areas, 361–362
  coordination of, 361
  NERC, 361
  open access same-time information system, 362–363
  reliability councils, 361
  transmission capacity, 362
Electric consumption, industrial plants energy efficiency and, 863
Electric demand
  measuring of, 1510
  time frame of, 1510
Electric demand response, 270–278
  air distribution adjustment, 275–276
  ancillary services, 284
  benefits of, 279–280
  closed loop control, 273
  communications, 283–284
  concepts and terminology, 271–273
  controlling of, 283–384
    communications, 283–284
    metering, 283
  cooling system adjustment, 275–276
  cost savings realized, 270–271
  daily peak load management, 272
  definition of, 279
demand shifting, 272
economic issues, 284–285
  energy efficiency, 271
  granularity of control, 273
  HVAC based, 273–274
  lighting based, 276–277
  metering, 283
  rebound, 273
  reduction in service, 272–273
  resolution of control, 273
  resources and programs, 279–285
  shared burden, 273
  solutions for commercial buildings, 281–282
  solutions for industrial facilities, 282
  solutions for residential areas, 283
  zone global temperature adjustment, 274
Electric demand response programs, 280–281
  demand buyback program, 281
  interruptible, 280
  load curtailment, 280–281
Electric energy production, 744–745
Electric energy supply by source, 608
Electric kinetic energy, 1163
Electric load following cycle, 1530–1531
Electric meters, 283
Electric motors, 349–355
  actual efficiency vs. Energy Policy Act standards, 353
  alternating current induction synchronous, 349
  alternating current synchronous, 349

applications of, 350
ASD technology, 353, 355
CEE premium efficiency motor, 352–353
classification of, 349
direct current, 349
efficiency, 352
  CEE premium efficiency motor, 352–353
  EPAct motor cost to premium efficiency motor cost, 353
  equation for, 352
energy consumption, 349, 350
EPAct motor cost to premium efficiency motor cost, 353
evaluation, 353
full load nominal efficiencies, 352
lack of efficiency, 387
nominal efficiencies for, 354
power supply issues, 352
pump usage, 353, 355
  ASD technology, 353, 355
replacement considerations, 351–352
selection criteria, 350–351
service rating of, 351
sizes, 350
starting of, 355
  technologies available, 355
type classification, 351
types, 349–352
  alternating current induction synchronous, 349
  alternating current synchronous, 349
  differences in, 349–350
Electric power losses, 360–361
  capacity losses, 361
  conductor losses, 360–361
  core losses, 360
  costs of, 360
  energy losses, 361
  high operating voltage advantage of, 361
  measurements of, 361
  reactive power, 361
    losses, 361
  types, 360–361
    conductor losses, 360–361
    core, 360
Electric power projects, life cycle costing, 953–966
  annual electricity generation, 955–957
  calculation methodology, 955
  capital and noncapital costs, 956–957
  coal power plant, 954
  coal vs. wind, 957
  cost components of, 954–955, 959
  depreciation, 957
  electricity generation, 955–957
  financing charge, 957–958
  interest during construction, 958
  loan repayment, 959
  photovoltaic power system, 954
  project cost, 958
  renewable energy systems incentives, 959–961
  taxes and duties, 959
  variable costs, 958
  wind power plant, 954
  working capital interest, 958–959

Electric power transmission systems, 356–363
- access to, 360
- biological effects, 358
- bulk operations, 361–363
- cable capacity determination, 358
- cable limitations, 358
- cables, 358
  - composition of, 358
    - high pressure fluid filled, 358
    - high pressure liquid filled pipe, 358
- capacity, 360
- characteristics of, 357
  - current flows, 357
  - resistance, 357
- components, 357–359
  - conductors, 358
  - three phase transmission lines, 357–358
- coordination of, 361–363
- direct current transmission, 359
- functions of, 356–357
- ground or shield wires, 358
- insulators, 358
- interconnection of, 356
- Kirchhoff's Laws, 360
- limitations of, 360
  - stability disturbances, 360
- North American systems, 357
- overhead cable, 357
- power losses, 360–361
- purposes of, 356
- rating types, 358
  - determination of, 358
  - emergency, 358
  - normal, 358
- ratings of, 358
  - heating factor, 358
- reactance, 357
- reactive power, 356
- real power, 356
- role of, 356
- short circuit duties, 360
- substations, 358–359
- support structures, 358
- transmission components, 357–359
- types, costs, 357
- workings of, 359–361
- workings of, synchronous operation, 359–360

Electric power transmission systems, operation of, 359–361
- access, 360
- capacity, 360
- Kirchhoff's Laws, 360
- power losses, 360–361
- short circuit duties, 360
- synchronous operation, 359–360

Electric power transmission systems, synchronous operation of, 359–360
- advantages of, 350–359
- effect of power loss, 359
- importance of network coordination, 360
- regional effects, 359

Electric rate development, time of use and, 1177
Electric rate schedules, 1507–1510
- general service, 1508
- general service heating, 1508
- interruptible, 1508
- large power, 1508
- rate riders, 1509
- real time pricing, 1508
- standby, 1509
- street lighting, 1508
- TOU large power, 1508
- TOU rates, 1508
- transmission rates, 1508

Electric supply, generation of, 374–378
- fuels, 378
- resources, conventional, 374–376
- renewable, 376–378

Electric utilities, restructuring of, 400–401
Electric utility bill, calculating of, 1472–1473
Electric utility dispatch modeling, 1506–1507
Electrical consumption reduction, drying and, 347
Electrical current, 255–256
Electrical elements, chemical process industry and, 1305
Electrical energy measuring
- data collection of, 255–257
- power quality analyzers, 256
- practice activities for, 256–257

Electrical grid, wind power effects on, 1611–1612
Electrical processes, energy intensive, 463
Electrical resistance space heater, 648
Electrical service, facility guidelines for new structures and, 528
Electrical tariff filings, 380
Electricity
- alternating current power system, 389
- chemical process industry and, 1305
- Chicago Edison, infrastructure growth, 391–392
- commodity, 392
- costs. *See* electricity pricing programs.
- definition of, 387
- deregulation of, 379–386. *See also* electricity, deregulation of.
- development in the United States, 387–398
- ductile metal filament lamp, 392
- early history of, 389–392
- efficiencies and, 387
- small motors, 387
- enterprise, modernizing of achieving goals, 403–406
- Energy Policy Act of 1992 (EPAct '92), 380
- Galvin Electricity Initiative, 404–405
- growth of, 389
- history of
  - growth in 20th century, 392–394
  - market maturity, 394–398
  - nuclear power, 394–395
  - open markets, 395
  - stasis of, 395–398
- hybrid electric vehicles and, 851–852
- impact on society, 387–389
- infrastructure obsolescence, 387–388
- initial usage of, 389–392
  - Edison, 389
- IntelliGrid1 supply system, 402–403
- meters, 392
- nuclear power and, 394–395
- portable storage, 389–390
- power system, alternating current, 389
- producers of, 387
  - assets and investment, 388
- Public Utility Holding Company Act of 1935, 392
- reasons for growth, 389
- regulation of, 379–380
- Rural Electrification Administration, 392
- Societal impact, 387–389
- supply systems, 379–380
  - origins of, 379–380
    - investor owned utilities, 379–380
    - Public Utilities Holding Act of 1935, 380
    - Edison, 379
- tax impact, electric vehicles and, 851–852
- Tesla, Nicola, 389
- transformation of. *See* electricity transformation.
- turbine generators, 392
- usage. *See* electricity usage.
- utilities profitability, 392
- utilities, consolidation of, 392
- wind power as source of, 1607

Electricity, deregulation of, 379–386
- arguments for the consumer, 384
- Energy Policy Act of 1992 (EPAct '92), 380
- Federal Energy Regulatory Commission, 380
- future of, 386
- political reaction, 384–385
- Public Utilities Regulatory Policy Act of 1978, 380
- regional transmission operator, 380
- regulator reaction, 384–385
- results in California, 385
- retail markets, 382–383
- utility reaction, 385–386
- wholesale markets, 381–382
- deregulation, results in Pennsylvania, 385
- deregulation, third party involvement, 380

Electricity consumption, worldwide, 499
Electricity costs, average, by location, 1632
Electricity enterprise, future of, 399–406
- modernizing of, 399–403

Electricity generation in U.S., 581–582
- input type, 582
- residential, 583
- coal and, 159–160, 161
- coal vs. natural gas, 161
- renewable, 44

Electricity pricing, 1175
- green, 1262–1264

Electricity pricing programs, 1175–1183
- critical peak pricing, 1175, 1177–1180
- extreme day CPP, 1178
- extreme day pricing, 1175
- green, 1262–1264
- real time pricing, 1175, 1180–1182
- time of use pricing, 1175–1177
- types, 1175

Electricity requirements, coal to liquid fuel plants and, 170
Electricity supply
- global warming and, 724
- world and U.S., 608

Electricity tax impact, plug in hybrid electric vehicles and, 851–852

Electricity transformation
architecture of, 404
assistance in achieving, 403–404
battery technology, 402–403
distributed energy resources, 404
fuel cells, 403
Galvin Electricity Initiative, 404–405
IntelliGrid1 supply system, 402–403
key steps needed, 402–403
organizational initiatives to assist, 403–404
resistance to, 404
Electricity transmission
open access, 380
effects of, 380
Electricity usage
compliance costs, 401
digital devices, growth of, 402
performance concerns, 402
future costs of, 402
future demand, 402
reinvention and transformation of, 399–406
benefits of, 401
security costs, 401–402
Electrolysis of water, 1442
high-temperature, 1443
spinning reserve concept, 1442–1443
Electrolysis produced hydrogen vehicles, greenhouse gas emissions and, 790–791
Electromagnetic field energy, 1163
Electromagnetic suspension maglev system, 1023
Electronic ballast primer, 830–832
advantages, 832–833
ceramic ballasts, 830
dimming, 831
dimming limits, 837
efficiency of, 831–832
full-light output, 834
ignition methods, 833–834
lamp ignition, 831
lamp performance, 831
life expectancy, 833
lumen depreciation, 836–837
metal halide, 830
microcontrollers, 834
problems with, 835–836
starting behavior, 831
system efficacy, 834–835
thermal cut-off, 834
types, 831
voltage, 831
Electronic components, heat pipe applications and, 808–809
Electronic control systems
applications, 416–418
basics of, 407–419
characteristics, 407
components, 408–416
controllers, 413–415
sensors, 408–413
definitions of common terms, 418
fundamentals, 416
power circuit supply, 416
Electronics, photovoltaic (PV) systems and, 1154–1155

Elemental analysis, solid fuel properties and, 479
EM programs. *See* Energy management programs.
Em$. *See* emdollar.
Emdollar (em$), 420
Emergency authority, Natural Gas Policy Act of 1978 and, 1084
Emergency lighting, 998
lighting design and retrofits, 998
Emerging Technologies Program, DOE and, 509
Emergy accounting, 420–428
arguments with, 427
definitions of, 420–421
emdollar, 420
emjoule, 420
geobiosphere, 422–423
H.T. Odum, 420, 421
net emergy and time, 426
net energy, 424–426
transformity
and quality, 421
and specific emergy, 421
unit emergy values, 420–421, 423–424
yield ratio, 424–426
Emergy inputs, geobiosphere and, 423
Emergy per unit money, 420
Emergy products, 423
Emergy yield ratio (EYR) 424–426
Emissions
Kyoto protocol targeting of, 140
waste energy, 659–660
Emission reduction credit (ERC), 430
offset policy, 430
Emission reduction, 137–138
Emission reduction mandates, 630–631
Emission taxes, emission trading vs., 435–436
Emissions impact
energy efficiency and, 11–12
regulating of carbon dioxide, 11–12
Emissions permits, tradable, 630
Emissions quota, allocation of, 138
Emissions reduction credit, 432
banking of, 432–433
Emissions tax, 630
Emissions trading, 430–436
allowance/permit allocation, 432
cap determination, 432
command and control costs vs., 435
compliance period, 432
cost-minimizing pollution abatement, 433–436
effectiveness, 436
elements of, 431–433
affected sources, 431
enforcement, 433
inventory, 431–432
measurement, 431–432
penalties and enforcement, 433
spatial trading rules, 432–433
temporal trading rules, 432–433
verification of, 431–432
emission reduction credit, 430
emission taxes vs., 435–436
enforcement, 433
firm incentives, 433

firm incentives, marginal cost of abatement, 433
measurement, 431–432
National Ambient Air Quality Standards, 430
penalties, 433
thought processes of, 430–431
command and control policies, 430
National Ambient Air Quality Standards, 430
trading rules
spatial trading rules, 432–433
temporal trading rules, 432–433
true-up, 432
types of, 431
cap-and-trade, 431
offset, 431
project-based, 431
rate-based, 431
unit of trade used, 432
Emjoule, 420
EMP. *See* energy master planning.
EMPI. *See* Energy Master Planning Institute.
Empirical determinations, 1144–1145
Employment in energy industry
coal production in United States and, 150
coal to liquid fuel plants and, 170
Empower, 421
End to end system management
air distribution and, 212
End use rates, 1504
natural gas and, 1507
End use retrofit isolation, 1051
Energy
additional vs. natural, 1088–1094
alternative technologies, 31–48
average unit cost, 920
avoided cost, 919–920
carriers and sources, differences in, 1088–1089
chronology of selected events in history, 610
conservation vs. efficiency, 21
consumption by the United States, 608
conversion factors, 603
conversion to work, 604
definition of, 1160, 1162
developing countries and, 501
balance supply and demand, 503
financing of, 501
development and, 1014–1015
disruption of, 1258–1259
diversity of, 601–605
energy transfer vs. energy property, 1163–1164
environment, and development, 873–874
equivalents of, 603, 605
forms and classification, 1165
future reserves, 608–610
global and historical background, 601–615
global overview, 601–605
energy diversity, 601–605
historical background
fossil fuels and, 606–607
industrial era, 606
modern era, 607–608
post fossil fuel era, 608–610
preindustrial era, 605–606

importance in wealth generation, 444
new and alternative sources, 880–881
physics of, 1160–1173
    energy conservation, 1164
    energy degradation, 1168–1172
    forms, 1163–164
    heat, 1161
    work, 1160–1161
post fossil fuel era, 608–610
primary sources of, 604
public policy and, 1193–1200
real time costing, 920, 927–928
renewable and decentralized, 1258–1264
sources of, 605
use of, 1091–1094
    efficiency, 1093–1094
    energy conversion technologies, 1091–1092
    selection of, 1092–1093
weighted average unit cost, 920, 921–926
work and heat units, 605, 1162
Energy account centers (EAC), 3
Energy accounting, 1–7
  cost centers, 3
    criteria for, 3
  data tabulation, 3–4
    fuel, 4
    needed data, 3–4
  definition of, 1–2
  energy account centers (EAC), 3
  energy measurement, 2
  Envision software, 3
  Fraser software, 3
  Global Mvo Asset Manager software, 3
  Meter Manager software, 3
  method types, 2
    comparison methods, 2
    comparison methods, changing production adjustment, 2
    comparison methods, conditioned floor area corrections, 2
    comparison methods, heating/cooling degree day adjustment, 2
    comparison methods, multiple year monthly average, 2
    comparison methods, present to past, 2
  monitoring, targeting and reporting analysis, 4–7
  objectives of, 1–2
  performance indicators, 4
  statistical measuring, 1–2
  tools, 2–3
    manual types, 2–3
    software, 3
      Envision, 3
      Fraser, 3
      Global Mvo Asset Manager, 3
      Meter Manager, 3
      Utility Manager, 3
Energy adjustment charge, utility billing and, 1502
Energy analysis, district heating and cooling and, 322–324
Energy and atmosphere
  LEED-C&S, rating system, 943–944
  LEED-CI, rating system, 939
  LEED-NC, rating system, 950

Energy assessments, 63–68
  accuracy of, 69–75
    energy and demand balances, 69–70
    calculating energy cost, issues with, 71
  cost containment projects, 63
  levels of, 63–64
  detailed, 65–66
    major review areas, 65
  pre-assessment procedures, 64
    energy invoice review, 64
    facility layout review, 64
    pollution and waste summaries review, 64
    tariff review, 64
  procedures, 64–66
  reporting of, 66–68
    consumption data, 67–68
    executive summary, 68
    implementation, 66
    recommendations, 66–67
    physical plant description, 67
  savings from, case study, 878
  walkthrough, 64
  See also energy and demand balances, energy benchmarks.
Energy audits
  basic drying type, 344–345
  coordination in tribal communities, 1459
  detailed drying type, 345
  drying and, 343–345
  funding in tribal communities, 1460
  geographic location, 1459
  information and tribal communities, 1460
  teamwork while in tribal communities, 1461
  technical issues in tribal communities, 1459–1460
Energy balance, 681–689, 1142–1143
  adjusting, 687
  benefits from, 688–689
  data analysis, 688
  diversity factor, 688
  historical usage patterns, 686–687
  in energy management programs, 875
  input, measuring of, 4
  inputs and outputs, 4
  load factor adjustments, 687–688
  output, energy load inventory, 4
  spreadsheet construction, 681–686
    data compilation, 685–686
    equipment energy use data, 681–685
    historical energy use data, 681
  utilization factor adjustments, 688
Energy basics, in windows. *See* Windows, energy basics.
Energy benchmarks, 690–698
  analysis using aggregate, 695–698
  anomalies, 696
    spikes, 696
    sustained changes, 696–697
  seasonal, 696
  shift work, 696
  determining users, 694–695
    departmental managers, 694
    executives, 694
    facility engineering staff, 694
    tenants, 694
  displaying of analysis, 695

  external, 690
  internal, 690
  normalized, 690–692
  organizations involved in, 691
  preparing of, 691–693
    baseline construction, 692–693
    timeline, 692
    data management, 692
    data sources, 691
      dynamic data, 691
      static data, 691
    scope of, 691
  setting of, 690–691
Energy carriers, 1088
Energy Center of Wisconsin, 512
Energy code, model, 438
Energy codes
  adoption on state and local level, 440–441
    legislation, 441
    process overview, 440
    regulation, 441
    timing of, 441
  energy standards and, differences between, 438
  enforcement of, 442
    appointed third party, 442
    local level, 442
    state level, 442
  implementation of, 441–442
    timing of, 441–442
  local government, municipal code, 441
  standards for facilities, 438–443
  support of
    DOE Building Energy Codes Program, 443
    voluntary programs, 442–443
Energy codes and standards for facilities, 438–443
Energy conservation, 444–457
  chemical process industry and, 1306–1309
  Climate Technology Initiative, 448–449
  energy efficiency, 450. *See also* energy efficiency.
  environmental issues, 446–449, 450
    environmental degradation, 447
    greenhouse gases, 446–447
    solutions, 449
  environmental protection vs., 450
  examples of, chemical process industry and, 1306–1309
  financing of, 450–451
  global warming and, 724
  implementing of, 453–454
  importance of energy efficiency, 455
  lack of success, 454–455
  lean manufacturing, 467–475
    impact on energy usage, 469–474
    lean traditional vs., 468–469
  life-cycle costing, 456, 457
  measures to take, 454–456
    parameters, 456
    sectoral, 455
  practical aspects, exergy, 450
  research and development status, 450–451
  sustainable development, 451–453
    definitions of, 451–452
    environmental concerns, 452–453

Index

technical limitations, 450
work-heat-energy principle, 1164–1168
world energy resources, 445–446
See also energy efficiency.
Energy conservation, analysis
　energy consumption allocation, 460
　industrial processes and, 459–461
　Pinch analysis, 460
　See also energy efficiency.
Energy conservation, industrial processes, 458–465
　capital investment, 464
　current applications of, 465
　differentiation, 458–459
　distributed control systems, 459
　effective management of, 464–465
　　key performance metrics, 465
　　total quality energy, 465
　energy consumption analysis, 459–461, 1141–1142
　energy intensive, 461–464
　lack of awareness, 465
　need and payback, 465
　retrocommissioning process and, 202–203, 204
　See also energy efficiency.
Energy conservation, opportunities
　chemical process industry and, 1306–1309
　Climate Technology Initiative, 448–449
　Florida's fruit industry, case study, 878
　retrocommissioning process and, 202–203, 204
　See also energy efficiency.
Energy consumption
　dryers and, 339–343
　　capital expenditure, reducing by, 346–347
　　reducing of, 345–347
　electric motors, 349, 350
　industrial plants energy efficiency and, 864
　maglev and, 1019–1020
　reducing by capital expenditure, drying and, 346–347
　reducing by non capital expenditure, drying and, 346
Energy consumption allocation, 460
Energy consumption analyses, 459–461, 1141–1142. *See also* energy efficiency analysis.
Energy consumption data
　energy assessments and, 67–68
　retrocommissioning process and, 202–203, 204
Energy conversion
　bio-solid and coal gasification, 492–494
　char reaction, 487
　cofiring, 490–491
　combustion, 487–489
　futuregen, 494–496
　general principles of, 489–490
　ignition and combustion, 487
　pyrolysis, 485–486
　volatile oxidation, 486–487
Energy conversion efficiencies, 607
Energy conversion processes
　animal waste, 476–496
　biomass fuels, 476–496

　properties, 478–485
　coal, 476–496
Energy conversion technologies, 1091–1092
　char reaction, 487
　cofiring, 490–491
　combustion, 487–489
　futuregen, 494–496
　general principles of, 489–490
　ignition and combustion, 487
　pyrolysis, 485–486
　volatile oxidation, 486–487
Energy cost savings
　adjustable speed drives, 74–75
　issues with, 71–75
　　motor belts and drives, 73–74
　　motor load factors, 73
　　motors, high-efficiency, 73
　　off-peak vs. on-peak use, 71–73
　motors, high-efficiency, 73
　off-peak vs. on-peak use, 71–73
　Operations and Maintenance Best Practices Guide and, 708
　See also energy efficiency.
Energy costs
　aircraft fuel consumption and, impact on, 25
　effect on United States economy, 590
　impact on indoor air quality, 22–23
Energy crisis, Public Utility Regulatory Policies Act (PURPA) of 1978 and, 1201
Energy currencies, 1088
Energy data collection
　building automation systems, 617–618
　creating data history, 618
　enterprise energy management systems and, 616–624
　meters, 618
　sources, 617–618
　　generated utilities, 617
　　other sources, 618
　　purchased utilities, 617
Energy data structuring, 618–619
　hierarchy, 619, 620
　time intervals, 619
Energy degradation
　entropy and exergy, 1168–1172
　heat engines, 1171
　irreversibility, 1168–1172
　reversibility, 1168–1172
Energy demand
　electricity consumption, 499
　Kyoto Protocol, 499
　trends, 499–501
　worldwide, 499–501
Energy and demand balances
　accuracy of
　　air compressors, 71
　　air conditioning, 70
　　lighting, 70
　　motors, 70–71
　　process equipment, 71
　energy assessment and, 69–70
　results verification, 71
Energy demand-side management programs, 286–291
　advertising and promotion, 290
　alternative pricing, 291
　attributes of, 286

　customer education and contact, 290
　direct incentives, 291
　end-use technology activities, 288
　load-shape changes, 287
　load-shape objectives, 288
　market implementation methods, 289–290
　selection of, 286
　technology alternatives to use, 288–289
　　choosing of, 289
　trade-ally cooperation, 290
　See also energy efficiency.
Energy Design Resources, 514
Energy efficiency, 1093–1094
　advanced energy systems, 1093
　air emissions reduction, 9–17
　　effectiveness, 10
　　emissions impacts, 11–12
　　hydrogen economy, 10–11
　benefit cost analysis, 81–85
　building envelopes, 1094. *See also* energy efficient buildings.
　chemical process industry and, 1302–1309
　community development and, 1409
　cost benefits from, 15–17
　countries' focus on, 10
　definition of, 9
　demand side management, 499. *See also* energy demand-side management programs.
　developing countries and, 498–506
　　barriers to, 501
　devices, 1093
　drying and, 334–335
　effect on GHG emissions, 12
　electric demand response and, 271
　energy conservation and, 455
　energy storage, 1093–1094
　engineering and, 136
　environmental benefits from, 15–17
　exergy analysis, 1094
　implementation. *See* energy efficiency, implementation.
　indoor air quality, 18, 21–22
　industrial plants, 862–871. *See also* energy efficiency, industrial plants.
　industry classification, links between, 868–869
　integrated energy systems, 1093
　investments in, 13
　leaks and loss prevention, 1093
　low cost improvements, 517–523
　　case studies, 517–523
　　payback analyses, 517
　matching supply and demand, 1093
　measuring economic development, 501–502
　municipal wastewater treatment plant (WWTP), 15
　passive strategies, 1094
　practical aspects of, 450, 451
　technology information sources, 507–516
　　associations and organizations, 513–514
　　Center for Analysis and Dissemination of Demonstrated Energy Technologies, 515
　technology information sources. *See* energy efficiencies, technology information sources.

tribal lands, 1456–1462
    and tribal governments, 1457–1458
    worldwide energy demand, 499–501
    See also energy demand-side management programs; energy efficiency, implementation, energy management programs.

Energy efficiency, implementation of
    balance supply and demand, 503
    demand side management, 503–505
    reducing power capacity needs, 503
    technology transfer, 502–503
    See also energy management (EM) programs.

Energy efficiency, industrial plants
    challenges to, 869–870
    combined heat and power, 864–867
    complexity of control systems, 870
    compressor systems, 867
    electric consumption, 863, 864
    energy saving opportunities, 862–869
        combined heat and power, 864–867
        compressor systems, 867
        equipment modernization, 867–868
        variable speed devices, 867
        waste heat recovery, 864–867
    energy savings, 863–864
        combined heat and power, 863–864
        lighting retrofits, 863
    equipment modernization, 867–868
    industry classification, links between, 868–869
    lighting, 870
        retrofits, 863
    monitoring of, 869–870
    pollution reduction, 870
    support of, 869–870
    variable speed drives, 862, 867
    waste heat recovery, 864–867
    See also energy efficient buildings; energy management (EM) programs.

Energy efficiency, technology information sources.
    miscellaneous groups, 515
    Portland Energy Conservation Inc., 515
    state agencies, 511–513
    United States Government Agencies, 508–511
    university programs, 511–513
    utility organizations, 514–515
    See also energy management (EM) program.

Energy efficiency measure (EEM), 967
Energy efficiency programs
    sunbelt cities and, 1391–1392
Energy efficiency ratio, 819
Energy efficient buildings
    electrical service, 528
    envelope, 528
    facility guidelines, 524–534
        for integrated design, 517
        top-down approach, 525
    HVAC, 529–530
    integrated design, 517
    irrigation water use, 527
    lighting, 528–529
    maintenance, 533
    motors and drives, 529
    new structures, 527
    overall energy use, 527
    plumbing, 533
    sustain efficiency, 533–534
    test and balance, 527
    top-down approach, 525
        integrated design, 525–526
        sustainability, 526
    See also energy management (EM) programs.

Energy engineering
    Association of Energy Engineers, 131–135
    careers in, 131–136
    future of, 135–136
        energy efficient economy, 136
        energy security, 135–136
        updating transmission grid, 136

Energy flow, 117–118
Energy forms
    elastic potential, 1163
    electrical kinetic, 1163
    electromagnetic field, 1163
    gravitational potential, 1163
    latent thermal, 1163
    mechanical kinetic, 1163
    sensible thermal, 1163
    thermal, 1163

Energy generation, 320
    district heating and cooling and, 320
Energy glossary, 1173
Energy glow paths, building system simulation and, 116–117
Energy impact on living standards and culture. *See* Living standards and culture.
Energy information system processes, 539–540
Energy information systems (EIS), 535–541
    processes, 535–536
        data collection, 535
        web publishing, 535–536
    utility report cards 539–541
    web publishing, purchase or self design, 538
Energy intensity indicator, 7
    disadvantages of, 7
Energy intensive industrial processes, 461–464
    boilers, 463
    calciners, 464
    centrifugal compressors, 464
    centrifugal pumps, 464
    chemical reactions, 462
    combustion systems, 463, 464
    distillation, 462
    drying, 462
    electrical processes, 463
    fans and blowers, 464
    fired heaters, 464
    flare systems, 463
    fractionation, 462
    furnaces, 464
    incinerator systems, 463
    kilns, 464
    liquid ring vacuum pumps, 464
    mechanical, 463
    melting and fusing, 461
    process cooling, 462–463
    process heating, 461
    steam systems, 463
    vacuum systems, 463–464

Energy invoice review, energy assessments and, 64
Energy leaks, 1093
Energy load inventory, output, 4
Energy loss, 361, 1093
    minimization of, 243
Energy management
    attributes of, 543–544
    commissioning existing buildings and, 185
    determining organization's aptitude for, 546–548
        energy marketing services, 547
        engineering protocol, 547–548
        fiscal protocol, 547
        leadership strength, 547
        leadership intensity, 547
        pride intensity, 547
        replication capacity, 546–547
        visibility of, 546
    fan systems and, 1224–1226
    measurement tools for, 1056–1060
        flow improvement, 1056–1059
        temperature accuracy, 1059–1060
        temperature multiplexing, 1059
    organizational aptitude for, 543–548
        facilitating reasons, 543–544
        management pathfinding for, 545–546
        strategies, 544–545
    prevailing strategies in, 544–545
    See also energy management (EM) programs; energy management systems.

Energy management (EM) programs
    3E Plus, 880
    audit process, 874–875
    available software
        3E Plus, 880
        Industrial Energy Management Software, 880
        Process Heating Assessment and Survey Tool (PHAST 1.1.1), 880
        PSAT: Pumping Assessment Tool, 880
        Steam System Assessment Tool 1.0.0, 880
        Steam System Scoping Tool 1.0d, 880
    current trends, 875
    definition of, 873
    energy balance in, 875
    importance of ECOs, 875
    Industrial Energy Management Software, 880
    Process Heating Assessment and Survey Tool (PHAST 1.1.1), 880
    PSAT: Pumping Assessment Tool 880
    scheme for, 875
    Steam System Assessment Tool 1.0.0, 880
    Steam System Scoping Tool 1.0d, 880
    trends, 875

Energy management systems, 1027–1041
    case study, 1037–1041
    common problems, 1035
        ineffective follow-through, 1036
        lack of communication, 1035
        lack of data, 1037
        lack of focus, 1036
        lack of management commitment, 1035–1036

Index

limited resources, 1037
market reaction, 1037
procedures, 1036
recurring issues, 1036–1037
reliance on single person, 1036
shifting priorities, 1036
drivers of, 1034–1035
establish objectives, 1034
implementation of, 1035
organizational communication, 1034
relationships with suppliers, 1035
team based approach, 1034
See also energy management; energy
   management (EM) programs.
Energy manager teams, 873.
  Japan, 873
  Latin America, 874
  See also Energy projects management.
Energy master planning (EMP), 549–555
  benefits of, 549–550
  creating a plan, 552–555
    announce success of, 554–555
    cost savings proposal, 552–553
    create goals, 554
    management support, 552
    opportunity, 552
    organization's energy use, 553
    team planning, 553
    upgrade implementation, 554
    verify savings, 554
  current practices, 550–551
    failure of, 550–551
    leadership, 551
  line management accountability, 555
  organizational optimization, 550
  origins of American approach, 551
Energy Master Planning Institute (EMPI),
   552–555
Energy measurement, 2
  key functions, 2
  other types of measurement, 2
Energy monitoring, 5–6
  cumulative sum of differences analysis, 6
  performance modeling, 5–6
  performance modeling, intensity indicator, 7
  steps of, 5
Energy performance indicators, 4, 578
  energy utilization index, 4
Energy Policy Act of 1992 (EPAct '92), 380
  effects of, 380–381
  Order 888, 380
  tariff filings, 380
Energy Policy Act standards, electric motors and
   actual efficiency, 353
Energy production adjustment, 2
Energy project management, 556–565, 874
  definitions, 556–558
  finance opportunities, 874
  reasons to use, 556
  skills for
    personnel, 564–565
    technical skills, 564
  skills needed, 564–565
    communication skills, 564
  team commitment, 564
  team composition, 564
  tools, 562–564

collaboration and alignment, 563–564
    RACI concept, 563–564
    scheduling tool, 563
    scope of, 562
  See also energy project management plan.
Energy project management plan, 558–562
  bid preparation, 560–561
  conceptual stage, 558–559
  construction stage, 561
  definition stage, 559–560
  delivery stage, 561–562
  design stage, 560
  See also energy project management.
Energy projects, life cycle analysis (LCA). *See*
   energy projects, life cycle
   costing.
Energy projects, life cycle costing
  antecedents of, 968
  blended utility rate, 967
  btu/sf, 967
  cost analysis, 967–968
  definitions, 967–968, 970–971
  energy efficiency measure (EEM), 967
  framework of, 969–972
    costs and benefits, 971–972
    current situation, 969–970
      gas station example, 969
      HVAC example, 970
    definition, 970–971
  gas station example, 969
  HVAC example, 970
  internal rate of return (IRR), 967
  LED exit lights example, 970–971
  net present value (NPV), 968
  payback period, 968
  time value of money, 968
  total owning costs, 968
Energy property, energy transfer vs., 1163–1164
Energy quality, 421
Energy recovery ventilation (ERV), 52
Energy reduction, compressed air systems and,
   243–245
Energy requirement compilations, 1142
Energy requirements, industrial processes and,
   458–459
Energy resource distribution, 917
Energy resource utilization data, 1010
Energy resources worldwide, 445–446
  fossil fuels, 445
  production and consumption, 445–446, 447
  statistical analysis of, 445
Energy saving opportunities, 868
Energy saving opportunity gap (ESOG), 503
Energy savings, 569–571
  3E Plus, 880
  Air Master +1.0.9, 880
  available software, 879–880
    3E Plus, 880
    Air Master +1.0.9, 880
    Industrial Energy Management
       Software, 880
    Motor Master (MM+4.0), 879–880
    Process Heating Assessment and Survey
       Tool (PHAST 1.1.1), 880
    Steam System Assessment Tool 1.0.0,
       880
    Steam System Scoping Tool 1.0d, 880

combined heat and power, 863–864
compressed air control systems and,
   209–211
as driver in motor optimization projects, 888
guaranteed, 569
Industrial Energy Management Software,
   880
industrial plants energy efficiency and,
   863–864
lighting retrofits, 863
Motor Master (MM+4.0), 879–880
Process Heating Assessment and Survey
   Tool (PHAST 1.1.1), 880
recognized validation of, 554
shared, 569
Steam System Assessment Tool 1.0.0, 880
Steam System Scoping Tool 1.0d, 880
tradable certificates, 1433–1439
underfloor air distribution (UFAD) and,
   1467
in variable speed drives, 864
Energy savings, case studies
  Dept. of Energy best practices case studies,
     1283–1301
    aluminum plant assessments, 1283,
       1285–1288
    chemical plant assessments, 1283–1284
    forest product plant assessments, 1284,
       1288–1291
    glass plant assessments, 1284,
       1291–1293
    metal casting plant assessments, 1284,
       1293–1294
    mining plant assessments, 1284–1285,
       1294–1296
    petroleum plant assessments, 1285,
       1296–1299
    steel plant assessments, 1285,
       1299–1301
  Florida's fruit industry, case study, 876–877
    average equipment energy usage, 877
    citrus juice production, 876
    company profiles, 876
    company profiles, prior to audits, 877
    current/new trends, 877, 879
    data analysis, 878
    descriptors for, 879
    energy assessments, 878
    energy balance, 877
    energy consumption, 876
    energy conservation opportunities, 878
    energy data, 877
    energy distribution, 877
    inspection, 876
    utility company profile, 876–877
Energy security, 135–136
Energy service companies (ESCO), 566–572
  successful projects, 307–308
  use of, 306
    consultant basis, 306–307
    utility based, 307
    users of distributed generation, 306–307
Energy service companies (ESCO), in Europe,
   566–572
  contracting models, 568–572
    build own operate transfer, 571
    chauffage, 571

energy savings, 569–571
    first out approach, 571
    leasing, 571–572
    other types, 571–572
equity contributions, 568
financing options, 567–568
    internally funded, 567
    third party, 567–568
investment grade audit, 567
measurement and verification of savings, 567, 572
project elements, 566–567
Energy service companies (ESCO), United States, 111
    savings model, 570
Energy service provider companies (ESPC), 566
Energy simulator software, 111
Energy sources, 1088
    net emergy and, 426
    non renewable, unit emergy values and, 425
    See Wind power.
Energy standards
    Department of Energy's role, 439
    development and revision of, 438
        groups involved, 439
    energy codes and, differences between, 438
    management systems and, 1031–1032
Energy Star Building label, 576–577
    applying for, 577
    ASHRAE standards, 577
    indoor air quality requirements, 577
        ASHRAE standards, 577
    professional engineer review and validation, 576–577
    statement of energy performance, 576
Energy Star Partnership, 510
    sunbelt cities and, 1393
Energy Star Portfolio Manager software, 573–578
    additional tools, 577–578
        Target Finder, 577–578
    background of, 573–574
    case studies available, 578
    data input information, 576
    Delta Score Estimator, 578
    eligibility requirements, 575–576
    energy benchmarking tool, 573–574
    Energy Performance Indicators, 578
    Energy Star building label, 576–577
    overview, 574
    required building information, 576
    rulesets, 575–576
        primary spaces, 575
        secondary spaces, 575–576
    Target Finder, 577–578
    tracking water consumption, 578
    using, 574–576
    weather normalization, 576
Energy Star window criteria, 1521
ENERGY STAR, 442
Energy Start™, 948
Energy storage, photovoltaic (PV) systems and, 1155–1156
Energy supplies
    global, 1344, 1346–1347
    primary vs. secondary, 1346
    United States, 1344, 1346–1347

Energy supply and demand, matching of, 1093
Energy supply by source, 607
    electricity, 608
    world and United States, 607–608
Energy system incentives, 959–961
Energy Tax Act of 1978, 1084
    tax disincentives, 1084–1085
Energy transfer, energy property vs., 1163–1164
Energy transmission, 320
    district heating and cooling and, 320
Energy transport systems, 530–531
Energy units, conversion factors, 1164
Energy usage
    impact on life and environment, 873
    square footage vs., 583
    wastewater plants and, 1550–1551
Energy usage patterns, energy balance and, 686–687
Energy usage reduction, 599
Energy use, 320
    building geometry and
        effect on, 111–115
        modeling, 112–113
    culture and society, connection between, 1016
    district heating and cooling and, 320
    environmental impact of, 1013
    modification of, 1015
        conservation, 1015
        efficiency increase, 1015
    past global usage, 1012
    population and, 1010
    projected global usage, 1012
    United States overview, 580–586
        electricity generation, 581–582
        major sectors usage, 580–581
        types used, 581
Energy use data, energy balance, 681–685
Energy use growth patterns, living standards and, 1010–1012
Energy use index (EUI), 111
    energy service company (ESCO), 111
    use of, 111
Energy use modifications
    strategic planning for, 1015–1016
    to improve living standards, 1015
Energy use in the United States
    commercial, 583
    industrial, 584–585
Energy use in U.S. transportation, 585, 588–599
    household vehicles, 585
Energy utilization index (EUI), 4
    energy balance, 4
    normalized performance indicator, 4
    specific energy consumption, 4
Energy waste sources, compressor systems and, 867
Energy wheels, 794–795
    See heat wheels.
EnergyPlus software, 1402
EnergyPort, 918
Energy-resource utilization data, 1010
Enforcement
    of energy codes, 442
    emissions trading and, 433
Engine analysis, reciprocating, 1237–1238

Engine redesign, waste fuel technology and, 1527
Engineering, automobile technology evolution and, 673–675
Engineering review, Energy Star Building label and, 576–577
Engines, heat pipe applications and, 811–812
Enriching, uranium and, 1113
Enterprise energy management systems (EEMS), 616–624
    business applications, 623
    data collection re, 616–624
    definition of, 616–617
    principles of, 617–623
        actionable information access, 619–622
        data collection, 617–618
        data structuring, 618–619
        industry standards, 622–623
        interactive access, 619–622
        measure and verify results, 622
    results, 622
        life-cycle costing, 622
        savings, 622
Enthalpy
    formation, 1423–1424
    reaction, 1423–1424
Enthalpy calculations, two phase heat exchanger condenser and evaporator, 803
Enthalpy controller, 415
Entrained bed gasifiers, 907
    General Electric type, 907
    Shell type, 907
Entrainment, heat pipes and, 805–806
Entropy, energy degradation and, 1168–1172
Environment
    district heating and cooling and, 320–321
    green energy and, 772–774
    minimal impact of geothermal energy, 748
    waste fuels and, 1525, 1527
Environmental agencies, independent power producers and, 856
Environmental benefits, energy efficiency and, 15–17
Environmental concerns
    3-E trilemma, 1147–1148
    sustainable development and, 452–453
Environmental degradation, 447
Environmental impact
    assessment applications, exergy and, 655–660
    energy use and, 1013
    exergy and, 658–660
    geothermal energy and, 746–747
Environmental issues
    consequences of, 448
    energy conservation and, 446–449
    solutions, 449
    nuclear energy and, 1103–1104
    pumped storage hydroelectricity and, 1211–1212
Environmental policy, 625–632
    development of, 625–626
    impact of, 626–627
        air, 627
        burden of proof, 627–628
        economic framework, 628–630

Index

cost-benefit analysis, 628–629
cost-effective analysis, 629–630
land use, 627
water, 627
policy instruments, 630–631
emission reduction mandates, 630–631
emissions tax, 630
liability rules, 630
nonuse benefits, 631
pollution tax, 630
redistributive effects, 631
renewable energy resources, 631
sustainable development, 631
tradable emissions permits, 630
Environmental Protection Agency, 510. *See* EPA.
ENERGY STAR Partnership, 510
Environmental protection, energy conservation vs., 450
Environmental regulation, wind power, 1615
Environmental uses, exergy analysis and, 651–652
Envision software, 3
EPA regulations, chemical process industry and, 1302–1303
EPAact motor cost, premium efficiency motor cost ratio, 353
EPAct '92. *See* Energy Policy Act of 1992.
Equilibrium design, heat pipes and, 805
Equilibrium modeling
constrained, 657
exergy analysis and, 657
Equipment energy use data, energy balance and, 681–685
Equipment modernization
aging thermal processes, 868
aluminum alloy artificial aging, 868
assistance for, 868–869
billet heating, 868
die temperature, 868
energy saving opportunities, 868
industrial plant savings and, 867–868
metal industry, 868
solution heat treatment, 868
Equity contributions by ESCO, 568
Equivalents of energy, 603
ERC. *See* emission reduction credit.
ESCO and ESPC
contrasts between, 567
ESCO. *See* energy service companies.
ESOG. *See* energy saving opportunity gap.
ESP-r software, 1402
Ethanol, 32–33
MTBE, 33
Ethylyne glycol operating temperature, run around heat recovery systems and, 1281
EUI. *See* energy use index.
EUI. *See* energy utilization index.
European energy service companies, 566–572
European Union
adoption of Kyoto Treaty, 13
Emissions Trading System, 141
Evacuated tube thermal collector, 1334–1335
Evaporating process, mobile HVAC systems and, 1073–1074
Evaporator, 1544

heat pipe and, 804–805
mobile HVAC systems and, 1079–1080
heat exchangers and, 803–804
maintenance of, 804
materials used in, 804
safe operation of, 804
Evaporator core identification, 1064–1065
Evolution, automobile technology and, 673–675
Excess energy consumption gap (EXEC), 503
EXEC. *See* excess energy consumption gap.
Executive summary
energy assessments and, 68
user friendly energy audit report suggested format, 78
Exergy, 321, 450, 645–653, 1428–1429, 321
definition of, 645–646, 655
energy degradation and, 1168–1172
environmental impact, 655–660
assessment applications, 655–660
chaos, 659
order destruction, 659
resource degradation, 659
waste energy emissions, 659–660
green energy and, 779
reference environment, 646
related definitions to, 646
sustainable building simulation and, 1397–1398
Exergy analysis, 322–324, 647, 655–656
applications of, 647–653
coal-fired electrical generating station, 648–651
economics, 652–653
electrical resistance space heater, 648
environmental uses, 651–652
thermal storage system, 648
efficiency, 647
energy efficiency and, 1094
geothermal energy and, 748–750
reference-environment modeling, 656–658
constrained-equilibrium, 657
equilibrium, 657
natural-environment-subsystem, 656–657
process-dependent, 657–658
reference-substance, 657
steady flow process, 748
steady state, 748
Exergy balances, 646
Exfiltration, infiltration vs., 844
Exhaust, 1235
Exhaust gas condensation solutions, combined heat and power, 177
Existing building commissioning, 179–187
Expansion, 1235
Expansion turbine and control, 215–216
Expansion valve, mobile HVAC systems and, 1077–1078
Exposed rafting, 902–903
Extensible markup language (XML), 537
Exterior insulation finish system walls, 1518
External communications, wireless applications and, 1655–1656
External energy benchmarks, 690
External wireless applications, 1654–1658
Extreme day CPP, 1178
Extreme day pricing, 1175

Extrusion polystyrene (XPS), as insulation material, 891, 894
Eye anatomy and functions, 993–995
EYR. *See* emergy yield ratio.
EZDOE software, 1402

F

Facilities, building automation system design issues and, 107
Facility air leakage, 662–670
infiltration control, 662
measuring airtightness, 668–669
Facility energy codes and standards, 438–443
Facility energy controls
automobile technology and, 671–679
automobile technology transferable to display systems, 672–673
individual control systems, 672
operational controls, 672
options, 673
incorporating automobile technology in building code support, 677–678
equipment and system modules, 677
impose standards, 677
modular buildings, 677
standardized commissioning of buildings, 678–679
lack of automobile technology in, 675–677
Facility energy use auditing, 63–68
energy assessments, 63–68
Facility energy use
benchmarking of, 690–698
energy benchmarks, 690–698
tools to assist in analysis of, 680–689
energy balance, 681–689
Facility guidelines, energy efficient building and, 524–534
Facility intranet monitoring and control, HMI compressor workstation and, 212
Facility layout review, energy assessments and, 64
Facility power distribution systems, 699–706
direct current systems, 705
fault current coordination, 704
power factor, 706
transformer ratings, 705
variable frequency, 705–706
variable speed drives, 705–706
Fan quantity reduction, air distribution adjustment and, 276
Fan speed limit, air distribution adjustment and, 276
Fan speeds, 521
Fan systems, 1220–1226
energy management, 1224–1226
flow control, 1225–1226
low cost options, 1225
maintenance, 1224–1225
laws, 1224
measurements, 1220–1223
flow, 1220–1221
performance, 1222–1223
performance, 1223
air density impact, 1224
curves, 1223–1224
power requirements, 1223

types, 1220
   axial, 1220
   centrifugal, 1220
Fans, 464
Fast breeder nuclear reactors, 1120–1121
Fast neutron reactors, 40
Fast simulation modeling
   focus on, 916
      impact on reliability, 916
      integrated model validation, 916
      multiresolution, 916
   IntelliGrid^{SM} and, 915–916
Faucets, 1596–1597
Fault current circuit breakers, 704
Fault current contactors, 704
Fault current coordination, 704
Fault current energy levels, 703
Fault current energy sources, 703
Fault current interrupting devices, 703–704
Fault current support, 703
Fault current, 703–704
Faults, power system, 702–703
Federal Aid Highway Act of 1956, 1021
Federal Energy Management Program (FEMP), 509, 873
Federal Energy Management Program, Operations and Maintenance Best Practices Guide (O&M BPG), 707
Federal Energy Regulatory Commission (FERC), 380
   transmission open access, 380
Federal Government funding, airports and roads, 591
Federally funded airport construction, 1020
Feedstock for syngas
   ash content of, 907
   ash fusion temperature of, 907
   free swelling index, 908
   moisture content of, 907
   particle size of, 908
   reactivity of, 907
FEMP. *See* Federal Energy Management Program.
FERC. *See* Federal Energy Regulatory Commission.
Ferrous electromagnets maglev system, 1022–1023
Fiberglass insulation, exterior rigid, 895
Fiberglass, as insulation material, 890, 893
   effect on environment, 891
Fill type, cooling towers and, 249–250
Filtration, water treatment process and, 1557, 1558–1559
Fin efficiency, 1336
Fin tube heat exchanger, 1540
Financing charges, 957–958
Financing, energy and developing countries, 501
Financing, energy service companies in Europe and, 567–568
Fine bubble diffusion, 1549
Finned built in storage solar water heater, 1333
Finned heat exchangers, 803
Finnish Masonry Stove, 1359–1360
Fired heaters, 464
Fireplaces, 1358–1360
Firm sales rates, natural gas and, 1507

Firm transportation rates, natural gas and, 1507
First Law of Thermodynamics, 1161, 1167, 1423–1426
   combustion, 1424
   enthalpy, 1423–1424
   formulation, 1423
   heat rates, 1424–1426
   heating values, 1424–1426
   power cycle efficiency, 1424–1426
First out contracting model, ESCO and, 571
First stage cost drivers, 1042
Fischer-Tropsch process, 163–165
   catalysts, 164–165
   providers of, 166–167
   South African use of, 163
Fission, 1128
5S's, 474
   self-discipline, 474
   simplification, 474
   sorting, 474
   standardization, 474
   sweeping, 474
Fixed bed combustor, 489
Fixed cost related charges, utilities and, 1500
Fixed history baseline, allowance/permit allocation and, 432
Flame safeguard control, 97–98
Flame temperatures, solid fuel properties and, 484
Flare systems, energy intensive, 463
Flat plate thermal collector, 1333–1334
Flight controls
   aircraft fuel consumption and, 28
   autopilot, 28
   fuel control and management computer, 28
Flight management computer (FMC), 28
Flocculation, 1557, 1558
Floor insulation
   raised floor, 892
   slab-on-grade, 892
   strategies for use, 892
Florida Solar Energy Center, 512
Florida's orange and grape industry, case study of energy savings, 876–877
FLOVENT software, 1402
Flow measurement
   common problems and fixing of, 1056–1057
      orifice plate management, 1057
      straight pipe issues, 1056–1057
   fan systems and, 1220–1221
   flow technologies, comparison of, 1059
   gas and steam
      density compensation of, 1057–1059
      minimizing permanent pressure loss, 1058–1059
   improve accuracy and repeatability, 1056–1059
   temperature, 1059
Flow technologies, comparison of, 1059
Flows data, exergy analysis and, 650
Flue gas volume, solid fuel properties and, 485
Fluidized bed combuster, 489
Fluidized bed gasifiers, 907
   Advanced Transport Reactor, 906–907
Fluidized bed
   circulating, 88
   stationary, 88

Fluids
   heat exchange and, 801
   heat pipes and, 805
   real, 1430–1431
Fluorescent dimming ballast, 984
   lighting control systems, 984
FMC. *See* flight management computer.
Foam insulation
   blown, 897
   exterior and interior, 895
   strategies for use, 892
Focus and lack of, energy management system problems and, 1036
Foil faced batt insulation, 1231
Foil-faced OSB, as insulation material, 895
Follow through, energy management system problems and, 1036
Footcandle, 999
Footlambert, 999
Forced circulation solar water heater, 1333
Forest product plant assessments, Dept. of Energy best practices case studies and, 1284, 1288–1291
Forestry products, drying of, 332–336
Formation of enthalpy, 1423–1424
Formulation, 1426–1427
Fossil fuels, 445, 606–607
   air pollution, 715–721
      ambient standards, 715–717
   dominance of, 45–46
   global warming, 721–725
   post era, 608–610
Fouling during cofiring, 491, 493
FoxWeb, 537
Fractionation process, 462
Framed wall insulation, interior, 895
Franklin stove, 1359
Fraser software, 3
Free swelling index, of syngas feedstocks, 908
Frequency domain, 117
Frequency, 1652
Fresh air, indoor air quality and, 19
Frosting, heat wheel performance factors and, 797–798
FSTIG cycle, 1583
Fuel cell program, 511
Fuel cell vehicles, greenhouse gas emissions and, 792–793
Fuel cells, 37–38, 298–299, 403, 726–732
   compared to existing technologies, 38
   current use of, 37
   efficiency of, 37
   low temperature, 733–743
   molten carbonate, 727–729
   phosphoric acid fuel cells, 726–727
   solid oxide, 729–732
   technology challenges, 37–38
      durability and reliability of, 37–38
      size, 38
   workings of, 37
Fuel consumption, aircraft and, 24–30
Fuel control and management computer (FCMC), 28
Fuel cost savings, plug in hybrid electric vehicles and, 851
Fuel cycles, nuclear energy and, 1111–1114
Fuel data tabulation, energy accounting and, 4

Index

Fuel ethanol, 1267
Fuel fabrication, uranium and, 1113
Fuel independence, 873
Fuel oil
  chemical process industry and, 1305
  savings of geothermal energy, 744
Fuel operation, nuclear energy and, 1113–1114
Fuel prices, 1364
Fuel properties, 478–485
  gaseous, 485
  liquid, 485
  solid, 479–485
Fuel quantity calculations, pre flight planning and, 28
Fuel supplier, independent power producers and, 856
Fuel Use Act, 1085
Fuels, generation of electricity and, 378
Full load nominal efficiencies, electric motors and, 352
Full-light output, 834
Full-load efficiency, district cooling systems and, 313–314
Furnaces, 464, 531, 809
Fusing, 461
Fusion nuclear reactors, 1123–1124
Fusion, 1130–1131
Future Air Navigation System 1, 30
Future reserves, energy and, 608–610
Futuregen, 494–496

G

Galvin Electricity Initiative, 404–405
  configuration levels, 405
  Perfect Power System, 404–405
Gas compression heat pump, 817
Gas cooled nuclear reactors, 1119–1120
  locations of, 1119–1120
Gas exchange process, reciprocating engine analysis and, 1237–1238
Gas supply curtailment, Natural Gas Policy Act of 1978 and, 1084
Gas transportation, Natural Gas Policy Act of 1978 and, 1084
Gas turbines, 1203
  simple cycle, 298
Gas utilities, seasonal loads, 1498
Gas utility bill, calculating of, 1472
Gas
  density compensation of, 1057–1059
  ideal, 1430
Gaseous fuel properties, 485
Gasification, 89–90, 492–494, 1525, 1527
  feedstock conversion, 906
Gasifiers, 1528–1529
  air-separation technologies in, 909
    cryogenic, 909
    high-temperature membrane, 909
  biomass gasification, 912
  contaminants, 908
  economies of scale, 912
  feedstock conversion, 906–907
    entrained bed gasifiers, 907
      General Electric type, 907
      Shell type, 907
    fluidized bed gasifiers, 906–907
      Advanced Transport Reactor, 906–907
    moving bed gasifiers, 906
      Lurgi gasifier, 906
    types of, 906–907
  overall plant integration, 910–911
    coproduction, 911
    environmental signature, 911
    IGCC overall block flow, 910
    raw gas in, 908
Gasoline engines, reciprocating, 1239–1240
Gasoline fueled vehicles, CO_2 emissions and, 788
Gasoline,
  aircraft fuel type, 26
  alternative transport fuels vs., comparison of, 47–48
GBS green building design software, 1403
General Electric Co., 390
General Electric gasifiers, 907
General service electric rate schedule, 1508
General service heating electric rate schedule, 1508
Generated utilities, energy data collection source and, 617
Generation energy flow, compressed air systems and, 226
Generation of electricity
  fuels used, 378
  resources
    conventional, 374–376
    renewable, 376–378
Generator voltage, combined heat and power and, 175
Generators, induction vs. synchronous, 175–176
Geobiosphere emergy, 422–423
  computing of, 422–423
  inputs to, 423
  products, 423
Geologic sequestration, carbon capture and storage, 127–128
Geometric ratio (GR), 113
  using in energy use reduction, 113
Geometry of buildings, energy use effect of, 111–115
Geothermal electricity, 1266
Geothermal energy, 744–752, 1364
  carbon emissions saved, 744
  case study of, 750–752
  commercial viability of, 745
  direct uses, 744–745
  efficiency of, 745
  electric energy production, 744–745
  environmental impact, 746–747
    research and development, 746–747
    standards development, 747
    technology
      assessment, 747
      transfer, 747
  exergy analysis, 748–750
    steady state, steady flow process, 748
  history of, 745, 746
  installed capacity, 744–745
  nomenclature, 744
  performance evaluation of, 748–750
  pollutants in water, 746
  savings in fuel oil, 744
  sustainability of, 747–748
    minimal environment impact, 748
    network flexibility, 748
    renewal ability, 748
  technical aspects of, 745–746
    fields, 745
Geothermal fields
  dry steam, 745
  hot water, 745
  wet steam, 745
Geothermal heat pump (GHP) systems, 753–759
  definitions, 753–754
  development of, 757
    load calculation, 757
    single borehole, 758
    thermal mass, 757–758
  earth heat exchangers, 756
  HVAC vs., 755–756
  maintenance costs, 755
  organizations, 756–757
  principles of, 754–755
  types, 756
GHG emissions, energy's efficiency effect on, 12
GHG reduction due to biomass energy, 91
GHG. *See* green house gases.
GHP. *See* geothermal heat pump.
Glare, 993
  illumination basics and, 265–267
Glass coatings, 1621
Glass plant assessments, Dept. of Energy best practices case studies and, 1284, 1291–1293
Glass, thermal conductivity of, 1619
Glazing coatings, 1621
Glazing system
  coatings and tints for energy efficiency, 1622
  general description, 1620–1621
  multiple pane, 1620–1622
Global energy supplies, 1344, 1346–1347
Global Myo Asset Manager software, 3
Global peace, green energy implementation and, 781
Global positioning satellites (GPS), aircraft fuel consumption and, 29
  local-area augmentation system, 29
  wide-area augmentation system, 29
Global temperature adjustment
  absolute vs. relative implementation, 275
  decay of shed savings, 274
  electric demand response and, 274
  evaluation of, 275
  factory vs. field implementation, 275
  impediments to, 275
  implementation of, 274
  mode transitions, 274
Global unrest, green energy implementation and, 781
Global warming
  effects of, 722–723
    climate changes, 723
    sea level rise, 722–723
  fossil fuels and, 721–725
  greenhouse gas concentration trends, 723–724

lessening of, 724–725
    conservation measures, 724
    electricity supply efficiency, 724
    non-fossil energy sources, 724–725
    workings of, 721–722
Goal gasification, 492–494
Goal oriented planning, energy master planning and, 554
GPS. *See* global positioning satellites.
GR. *See* geometric ratio.
Granularity of control, electric demand response and, 273
Gravitational potential energy, 1163
Gravity treatment of wastewater, 1561
Great Plains coal basins, 158
Green Building Challenge program, 948
Green building design software, 1403
    ATHENA, 1403
    BEES, 1403
    GBS, 1403
    RETScreen, 1403
Green buildings
    concept of, 948
    rating system, 948–949. *See also* Buildings, rating systems for.
Green electricity pricing, 1262–1264
Green energy, 771–786
    advantages and disadvantages, 774–775
    applications, 779
    environmental consequences, 772–774
    exergetic aspects, 779
    factors need for success, 778
    implementation of, 779–781
        global peace, 781
        global unrest, 781
        sustainable development ratio, 779–781
    need for, 772
    resources, 777–778
    sustainable development, 776–777
        case study, 781–786
    technologies, 777–778
Greenhouse gases (GHG), 9
Greenhouse gas concentration trends, 723–724
Greenhouse gas emissions, 446–447, 787–793
    carbon dioxide
        gasoline fueled vehicles, 788
        hybrid vehicles, 788
        plug-in hybrid vehicles, 788
    electrolysis produced hydrogen vehicles, 790–791
    hydrogen vehicles
        fueled, 789
            fuel cell vehicles, 792–793
            steam methane reforming, 790
        plug-in hybrid electric vehicles, 791–792
    reduction funding, 1531–1532
    steam methane reforming, 789–790
Ground fault detection, 703
Ground source heat pumps, 532
Ground wires, 358
Grounding, 701
Guaranteed energy savings, 569
Guidance documents, 1030
Gulf Coast region coal basins, 158

H

Habitat structures
    evolution of, 1362–1363
    space heating and, 1362–1363
Hardware, client and server, building automation systems and, 106–107
Harvester ice storage equipment, 1416–1417
HCCI. *See* homogeneous charge compression ignition.
HDD. *See* heating/degree day adjustment.
Hearths, 1358–1360
Heat buildup, radiant barriers and, 1231
Heat engines, 1171
    Second Law of Thermodynamics, 1171–1172
Heat exchangers, 801–804, 808, 1540
    analysis of, 801–803
        counter flow, 802–803
        parallel flow configuration, 802
    design considerations, 803
    finned types, 803
    fluids, 801
    two phase, 803–804
    types
        concentric-tube, 801
        cross flow, 801
        shell and tube, 801
    wet cooling towers and, 248
Heat flow, conductive, in window energy performance, 1614
Heat gain
    casual, 116
    decreasing of, 1231
Heat pipe applications, 807–812
    aerospace, 807–808
    agriculture, 810
    cutting tools, 812
    electronic components, 808–809
    engines, 811–812
    exchangers, 808
    medical, 810–811
    ovens and furnaces, 809
    solar thermal applications, 809
    transportation systems, 811
Heat pipes, 804–806, 1540
    applications of, 804
    boiling limitations, 805, 806
    definition of, 804
    entrainment, 805–806
    equilibrium design, 805
    evaporator, 804–805
    sonic limitations, 805, 806
    thermal conductivity, 805
    wick type, 805
    wicking, 805–806
    working fluid, 805
Heat pumps, 814–821
    applications of, 820–821
    Carnot coefficient of performance, 815
    fundamentals of, 814–815
    ground source, 532
    heating and, 815–816
    history of, 814
    performance parameters, 818–820
        ambient energy fraction, 819–820
        coefficient of performance, 819
        energy efficiency ratio, 819
        primary energy ratio, 819
    types, 816–818
        absorption, 816–817
        gas compression, 817
        thermoelectric, 816
        vapor compression, 817–818
Heat rates, 1424–1426
Heat recovery equipment, 1540
    fin tube heat exchanger, 1540
    heat pipes, 1540
    heat wheels, 1540
    plate and frame heat exchanger, 1540
    recuperator, 1540
    run-around coils, 1540
    shell and tube heat exchanger, 1540
    steam recovery, 1540
Heat recovery steam generators, 1578
Heat recovery systems, run around, 1278
Heat recycling, cogeneration district heating and cooling systems and, 318
Heat removal factor, 1336
Heat sink configurations, 1542–1543
Heat source configurations, 1542–1543
Heat storage, 1420
Heat transfer blocking, radiant barriers and, 1227–1228
Heat transfer calculations, 1063
Heat transfer coefficients, 1279–1280
Heat transfer method, 1063–1066
Heat transfer, 822–829
    conduction, 823
        in windows, 1617
    convection, in windows, 1617–1618
    history of, 822–823
    radiation, 828–289
    in windows, 1617
    thermal conductivity, 824–827
    three moles of, 1617
    windows, 1616–1617
Heat units, energy and, 605
Heat values of waste products, 1345
Heat wheels, 794–799, 1540
    construction and materials, 795
    economic factors, 798–799
        life cycle costs, 798
        payback, 798
    environmental factors, life cycle analysis, 798–799
    operation of, 795–796
    performance factors, 796–797
        frosting, 797–798
        pressure drop and leakage, 797
        reliability, 798
Heat, 1161
    First Law of Thermodynamics, 1161
    transfer of, 1161
        conduction, 1161
        convection, 1161
        thermal radiation, 1161
    work units and, 1162
Heater core identification, inverse heat transfer methods and, 1065–1066
Heater, electrical resistance space, 648
Heating capacity calculations, 1061–1066
Heating demand, seasonal, 1273–1276

Heating factor, electric power transmission systems rating and, 358
Heating fuels, 1361–1362
   coal, 1361
   oil, 1361
   petroleum based, 1361
   wood, 1361
Heating loads, residential buildings and, 1272–1277
Heating mode, mobile thermostat climate control and, 1663
Heating oil, 1361
Heating process, mobile HVAC systems and, 1074–1075
Heating values, 1424–1426
   biomass fuels and, 482–483
   solid fuel properties and, 479, 484
Heating
   dehumidification and, 1187–1188
   heat pumps and, 815–816
   Romans
      bathhouses, 1358
      hypocaust, 1357
   sensible, 1186
Heating/cooling degree day adjustment (HDD and CDD), 2
HERS. *See* home energy rating system.
HHV. *See* higher heating value.
HID dimming ballast, 985
   lighting control systems, 985
HID. *See* high intensity discharge.
Hierarchy structure, energy data, 619, 620
High efficiency motors, cost calculations and, 73
High humidity, control of, 843–844
High intensity discharge (HID) electronic lighting, 830–838
High intensity discharge ballasts, 830–832
   electronic ballast primer, 830–832
High operating voltage, advantage, 361
High pressure boilers, 93
High pressure cylinders (HP), 1382
High pressure fluid filled (HPFF) cable, 358
High pressure gas storage, hydrogen and, 1446
High pressure liquid filled pipe (HPLF), 358
High rise buildings, 1513
High solar gain low-e coatings, for windows, 1621, 1622
High voltage alternating current (HVAC), 357
High voltage direct current (HVDC), 359
High voltage direct current lines (HVDC), 357
Higher heating value (HHV), 86–87
Higher lift, chiller efficiency and, 309–310
High-temperature electrolysis, 1443
High-temperature membrane technologies, gasifiers and, 909
Historical baseline with updating, allowance/ permit allocation and, 432
Historical energy use data, energy balance and, 681
History of energy, chronology of selected events, 610
HMI workstation. *See* human machine interface.
Home energy rating system (HERS), 442
Homogeneous charge compression ignition (HCCI), 1240, 1242
Hot water (hydronic) heating systems, 1361
Hot water optimization, case study, 1366–1379

Hot water vs. steam, combined heat and power, 176
Hot water, geothermal field and, 745
Hottel Whillier Bliss equation, 1335–1336
Housing growth, U.S. transportation energy use and, 590
HP. High pressure cylinders.
HPFF. *See* high pressure fluid filled cable.
HPLF. *See* high pressure liquid filled pipe.
Human eye anatomy, 994
Human eye functioning, 993–995
Human machine interface (HMI) compressor workstation, 212
   facility intranet monitoring and control, 212
   web-based, 212–213
Humid climates, HVAC systems and 839–846
   mold growth, 839
Humidification, 1186
Humidity
   control of, 843–844
   relative, 19
HVAC based electric demand response, 273–274
HVAC capacity, review of, 519–520
HVAC control, building system simulation process and, 120
HVAC controllers, improvement in, 1365
HVAC solution software, 1402
HVAC summer winter cutover, low cost energy efficient improvement case studies and, 520
HVAC systems
   dehumidification, 839–840
   functions of, 839
      comfort control, 839
      contaminant control, 839
      pressurization, 839
      ventilation, 839
   humid climates and, 839–846
      mold growth, 839
      proper pressurization design, 844
   mobile, 1061–1069
   poor humidity control
      case study, 840–844
      predictability of high humidity, 843–844
      vertical stacking fan-coil air conditioning, 840
HVAC
   cold air retrofit case study, 172–174
   facility guidelines for new structures and, 529–530
      air cooled equipment, 532
      boilers and furnaces, 531
      chillers, 531
      controls, 532–533
      cooling towers, 531–532
      energy transport systems, 530–531
      ground source heat pumps, 532
      hydronic circulating systems, 531
   geothermal heat pump systems vs., 755–756
HVAC. *See* high voltage alternating current, 357
HX coil ice storage equipment, 1414–1416
Hybird vehicles, carbon dioxide emissions and, 788
Hybrid electric vehicles
   current models, 849–850
   efficiency of, 844

   plug in configuration, 847–853
   plug in type, 850–851
      electricity capacity needed, 852
      electricity tax impact, 851–852
      fuel cost savings, 851
      home electrical service needed, 852
      implementation issues, 853
      sales incentives for, 852–853
      utility issues, 852
   responsiveness of, 848
   types, 848
Hybrid solar space heating, 1322
Hybrid vehicles, plug-in, carbon dioxide emissions and, 788–789
Hydraulics, water and, 1559–1560
Hydraulic systems, chemical process industry and, 1305
Hydrid cooling towers. *See* wet-dry cooling towers.
Hydrochloro-fluorocarbons (HFCs), as insulation blowing agent, 892
Hydroelectric generation, 1207–1208
Hydroelectricity, pumped storage, 1207–1212
Hydrogen economy
   advantages of, 11
      motor vehicles, 11
   availability of, 11
   disadvantages of, 11
   wind power and, 1610
Hydrogen fuel service stations, 1447
Hydrogen fuel
   challenges of, 1441
   distribution of, 1446–1147
      plants, 1447
      service stations, 1447
   production of, 1441–1445
      steam-methane reforming, 1442
      thermochemical cracking of water, 1443–1445
      water electrolysis, 1442
   storage of, 1445–1446
      alanates, 1446
      high pressure gas, 1446
      metal hydrides, 1446
Hydrogen fueled fuel cell vehicles, greenhouse gas emissions and, 792–793
Hydrogen fueled transportation systems, 1441–1447
Hydrogen fueled vehicles, 789
Hydrogen vehicles
   greenhouse gas emissions and, 790–791
   high-temperature nuclear reactor, steam methane reforming and, emissions from, 790
   steam methane reforming and, emissions from, 790
Hydrogen, 35–37
Hydrogen, barriers to commercial use, 36
   public acceptance, 36
   safety of, 36
   storage, 36
Hydrogen, energy storage for photovoltaic systems, 1155–1156
Hydrogen, future technology, 36–37
Hydrogen, production of, 36–37
Hydronic circulating systems, 531
Hydronic heating systems, 1361

advantages in, 1361
Hydronics Design Studio software, 1402
Hypertext preprocessor, 537
Hypocaust, 1357

I

IAC (Industrial Assessment Center) program, in industrial energy management, 873
IAQ. *See* indoor air quality.
IAQP. *See* indoor air quality procedure.
Ice storage equipment, 1414–1417
    harvester, 1416–1417
    HX coil, 1414–1416
    PCM, 1414
    stratified chilled water storage, 1417–1418
ICFs. *See* Concrete form insulation systems.
Ideal gas, 1430
IECC. International Energy Conservation Code.
IEKPI. *See* industrial energy key performance indicators.
IGCC vs. pulverized-coal boiler, 912
Ignition methods, electronic ballast primers and, 833–834
Ignition, combustion and, 487
Illumination basics, 264–267
    glare, 265–267
    movement of the sun, 264–265
    photometry, 264
    radiometry, 264
    solar spectrum, 265
Improvement, Six Sigma and, 1315–1316
Incentives
    emissions trading and, 433
    energy demand-side management programs and, 291
Incinerator systems, energy intensive, 463
Increasing block rates, 1503
Incremental pricing, Natural Gas Policy Act of 1978 and, 1083–1084
Independent power producers, 854–861
    history of, 854–855
    plant
        construction, 860
        operation, 860
    project development, 858
    project definition, 858
    stakeholders, 855
        advisors, 857
        contractors, 857
        environmental agencies, 856
        fuel supplier, 856
        government authorities, 856
        lenders, 857
        off taker, 856
        owner/developer, 856
        payment structure, 857–858
        stakeholders, project, 855
        stakeholders, risks, 858, 859
    types, 855
        long-term, 855
        merchant, 855
India's economic growth, impact on U.S. transportation energy, 592
Indian coal to liquid fuel plants, 167
Indicating devices, 416

Indirect cofiring, 89
Indirect solar gain design, 1322
Individual control systems, automobile technology and, transferable to building use, 672
Indoor air quality (IAQ), 18–23
    as defined by ANSI/ASHRAE Standard 62.1–2004, 51
    energy costs, impact on, 22–23
    energy efficiency and
        mutual goals of, 21–22
        relationship between, 21
    Energy Star Building label and, 577
    fresh air, 19
    investigation of, 21–22
        sophisticated testing, 21–22
        walk-through type, 21–22
    measuring of, 51
    procedure (IAQP) compliance to ANSI/ASHRAE Standard 62.1–2004, 54, 60
    sources of problems, 22
    ventilation, 18–19
        advantages, 21
        problems and, 19–20
            dilution, 19–20
            increased outside air, 20
            pollution sources, 20–21
            relative humidity, 19
            tight buildings, 20–21
Indoor environmental quality
    LEED-C&S, rating system, 944
    LEED-CI, rating system, 940
    LEED-NC, rating system, 950
Indoor negative pressures, 844
Induction generator vs. synchronous, combined heat and power, 175–176
Induction systems, 996
    lighting design and retrofits, 996
Industrial Assessment Center (IAC) program, in industrial energy management, 873
Industrial carbon dioxide capture, 127
Industrial drying operations, 338–348
    nomenclature, 338
    types, 338–339, 340
        batch, 338
        continuous, 338
    uses of, 338
Industrial electric power systems, combined heat and power and, 175
Industrial energy key performance indicators (IEKPI), 1140–1146
    cost savings tool, 1146
    demand side energy consumption, 1140
    proper usage of, 1145–1146
    types, 1141–1145
        efficiencies, 1141
        empirical determinations, 1144–1145
        energy
            balances, 1142–1143
            consumption analyses, 1141–1142
            requirement compilations, 1142
            measurements, 1143–1144
            process analyses, 1143
Industrial Energy Management Software, 880
Industrial energy management

3M company, 873
Colombia, 874
Comision Nacional deEnergia (CNE), 874
cost of, 874
Ecuador, 874
Florida, 874
fuel independence, 873
global trends, 872–881. *See also* Industrial energy management.
    acid rain, 873
    economic competitiveness, 873
    governmental energy management policies, 872
    ozone layer effects, 873
    pollution impact, 873
    U.S.A., 872
    U.S. Executive Order 12,123, 872
Japan, 873
Kuznets Curve prediction, 874
Latin America, 874
management support, 874
Peru, 874
secondary benefits of, 873
Walt Disney World, 873
Industrial energy management. *See also* Energy savings, Florida's fruit industry, case study.
Industrial facilities, electric demand response solutions for, 281–283
Industrial key performance indicators, development of, 1145
Industrial motor system optimization projects, in the U.S. *See* Motor optimization projects.
Industrial plants, energy efficiency and, 862–871
    complexity of control systems, 870
    lighting, 870
    monitoring of, 869–870
    pollution reduction, 870
    support of, 869–870
Industrial processes, differentiation, 458–459
    controls, 459
    energy requirements, 458–459
    technology requirements, 459
Industrial processes, energy conservation and, 458–465
    analysis, energy consumption analysis, 459–461
    distributed control systems, 459
    effective management, 464–465
    energy intensive, 461–464
    need and payback, 465
Industrial real time pricing, 1180–1182
Industrial thermal considerations, combined heat and power, 176
    steam vs. hot water, 176
Industrial use of energy in U.S., 584–585
    complexity of, 584
    Manufacturing Energy Consumption Survey, 584
Industrial wastewater, 1562
Industry classification, energy efficiency, links between, 868–869
Industry era, fossil fuels, 606–607
Industry standards, enterprise energy management systems and, 622–623

Index

Industry use of coal, 160
Infiltration control, 662
    air barriers, 662–663
    vapor retarders, 662
Infiltration
    building system simulation process and, 120
    exfiltration vs., 844
Inlet supply air temperature, 1280
Innovation and design process
    LEED-C&S, rating system, 945
    LEED-CI, rating system, 941
    LEED-NC, rating system, 950
Innovations, coal production in United States and, 150–151
Input of energy, balancing of, 4
Input type controllers, 414
Installation, commissioning process and, 195–196
Insulated concrete form walls, 1517
Insulated concrete forms (ICFs). *See* Concrete form insulation systems.
Insulation blowing agents
    expanding polystyrene, 891–892
    extrusion polystyrene, 892
    hydrochloro-fluorocarbons (HFCs), 892
    open-cell polyurethane, 892
    pentane, 891–892
    polystyrene, 891–892
    polyurethane, 892
Insulation materials
    aerated concrete, 891
    blown foam, 897
    blown loose-fill insulation, 897
    bubble back, 895
    cellulose insulation, 890
        effect on environment, 891
    closed-cell, high-density spray polyurethane, 891, 894
    comparison chart, 893–895
    concrete from systems, 895–896
    concrete products, lightweight, 896
    containing blowing agents, 891
    containing chlorofluorocarbons (CFCs), effect on environment, 891
    containing pentane, 891–892
    cotton, 893
    environment and, 891–892
    extrusion polystyrene (XPS), 891, 894
    fiberglass, 890, 893
    fiberglass, effect on environment, 891
    foam
        exterior and interior, 895
        products, effect on environment, 891
    foil-faced OSB, 895
    interior framed wall, 895
    loose-fill insulation, 897
    metal framing for, 898–899
    mineral wool, 890, 893
        effect on environment, 891
    molded expansion polystyrene (MEPS), 890–891, 894
    open-cell, high-density spray polyurethane, 891, 894
    paperboard sheathing, 895
    perlite, 893
    polyisocyanurate, 891
    rating system, 890
    reflective insulation, 891
    rock wool, 890, 893
        effect on environment, 891
    slag wool, 890, 893
        effect on environment, 891
    strategies for use. *See* Insulation strategies.
    structured insulation panels, 897–898
    types of, comparison chart, 893–895
    wall, 2 x 4, 896
Insulation materials. *See also* Insulation blowing materials.
Insulation panels, structured, 897–898
Insulation strategies, 892
    critical guidelines, 892
    floor insulation, 892
    foam insulation, 892
    wall insulation, 892
Insulation, ceilings and roofs. *See* Ceiling and roof insulation.
Insulation, 1363
    exterior rigid fiberglass, 895
    regulation of, 1363
    R-factor, 1363
    types, 1363
    U-factor, 1363
Insulators, 358
Integrated building design system, 1401
Integrated coal gasification/combined cycle power plants, 1531
Integrated design, energy efficient buildings and, 525–526
Integrated dimmers, lighting control systems, 982
Integrated gasification combined cycle (IGCC), coal- and biomass-based, 906–912
Intelligent on/off devices, lighting control systems, 979
IntelliGrid^{SM}, 914–918
    advantages of, 915
    automation and, 917
    communication architecture, 917
    description of, 914
    energy resource distribution, 917
    EnergyPort, 918
    fast simulation modeling, 915
    power electronics-based controllers, 917
    power flow increases, 917
    value added services, 918
IntelliGrid1 supply system, 402–403
Intensity indicator, energy, 7
Inter surface longwave radiation, 116
Interactive access
    actionable information, 619–622
    application interface, 621–622
    calculated data, 621
    data organization, 620
    data presentation, 620
    enterprise energy management systems and, 619–622
    time and usability, 620
    data, 620–621
    user-specific information, 622
Intercity mobility, maglev and, 1021–1022
Interconnection standards, 866–867
Intercooler control, compressor air control systems and, 210
Interfacing systems, controllers and, 416
Interference using wireless, 1652
Interior coal basins, 157–158
Interior framed wall insulation, 895
Intermediate pressure cylinders (IP), 1382
Intermittent energy sources, 1268
Internal audits, 1030
Internal energy benchmark, 690
Internal environment control, building system simulation process and, 120
Internal rate of return (IRR), 967
International Energy Conservation Code (IECC), 438
International Organization for Standardization (ISO), 1028–1029
International performance measurement and verification protocol (IPMVP), 919
Internet based access, web-based compressor management and, 213
Interruptible demand response programs, 280
Interruptible electric rate schedule, 1508
Interruptible fuel rates, combined heat and power, 177–178
Interruptible rates, 1504–1505
Interruptible sales rates, natural gas and, 1507
Interruptible transportation rates, 1507
Interrupting devices, fault current, 703–704
Intervention, public policy rationale for, 1194
Inventions and Innovation group, DOE and, 509
Inventory reduction, lean manufacturing and, 470
Inverse heat transfer method, 1063–1066
    capacity calculations, 1065
    evaporator core identification, 1064
    heater core identification, 1065–1066
Inverted block rates, 1503
Inverters, 359
Investment analysis techniques, 930–936
    fundamentals of, 930–932
        alternative metrics, 933
        cash flows, 932
        comparable value, 930
        discount rate, 930
        determination, 932
        NPR and IRR
            cash flows, 935
            comparison, 933
        simple payback period, 930–931
        time value of money, 930
        variable discount rates, 935
Investment grade audit, 567
Investments
    electricity producers and, 388
    energy efficiency and, 13
Investor owned utilities, 379–380
IOF vs. non-IOF facilities, study results
    capital spending vs. re-engineering, 886–887
    energy consumption data, 885
    individual projects, 887
    industries of the future, 885
    U.S. motor systems, 885
Iowa Energy Center, 512
IP. Intermediate pressure cylinders.
IPMVP options C and D, window films, 1626–1629

IPMVP. *See* international performance measurement and verification protocol.
IRR (internal rates of return), in motor optimization projects, 883–884
IRR cash flows, 935
IRR, NPR vs., 933
IRR. *See* internal rate of return.
Irradiance, 1148
Irradiation, 1148
Irreversibility, energy degradation and, 1168–1172
Irrigation water use, 527
ISM band radios, 1654
ISO. *See* International Organization for Standardization.
Isolated solar gain design, 1322

J

Java applets, 537
Java Server Pages, 537
Java Servlets, 537
Jet aircraft fuel, types, 26
Jet engine efficiency, fuel consumption and, 26
Jet engine maintenance, aircraft fuel consumption and, 28–29
Just in Time manufacturing, 468

K

Kelvin-Plank statement, 1170
   continuous cyclic process, 1170–1171
Kerosene based aircraft fuel, 25–26
Key performance indicators (KPI), distinguishing of, 1140–1141
Key performance measurements, 1143–1144
Key performance metrics (KPM), 465
Kilns, 464
Kirchhoff's Laws
   electric power transmission systems and, 360
   importance of, 360
   loop flow, 360
   parallel path flow, 360
   window emissivity calculations, 1621
Korean maglev system, 1025
KPI. *See* key performance indicators.
KPM. *See* key performance metrics.
Krypton, as insulating gas, 1619
Kuznets Curve
   illustration of, 874
   industrial energy management assessment, 874
Kyoto Protocol, 499
   allowance assignments, 141
   emission targets, 140
   flexibility mechanisms of, 141
   implementation of, 140–141
Kyoto treaty
   European Union countries adoption of, 13
   regulating of carbon dioxide CO_2, 11–12
   requirements of, 9–10
   United States Government and, 12–14

L

Lamp ignition, 831
Lamp life, 995–996
Lamp performance, 831
Lamp switching, lighting based demand response and, 277
LAN, 1654–1655
   wireless application vs., 1658
Land development policies, U.S. transportation energy use and, 588
Land use mix, 1450
Land use
   environmental policy and, 627
   impact on transportation, 1449–1451
      cumulative impact, 1450–1451
      density, 1450
      mix of, 1450
      roadway design, 1450
      site design, 1450
      transit service quality, 1450
Landfill cost, 1530
Landscape irrigation, 1590–1592
Large power electric rate schedule, 1508
Latent heat storage, 1413
Latent thermal energy, 1163
Laws, creation of, 1193–1194
Lead acid batteries, 1155
   compressed air energy storage as alternative to, 214
Leadership in Energy and Environmental Design (LEED), 200, 206
Leadership in Energy and Environmental Design for Commercial Interiors. *See* LEED-CI.
Leadership in Energy and Environmental Design for Commercial Interiors. *See* LEED. *See also* LEED-C&S, LEED-CI, LEED-NC.
Leadership in Energy and Environmental Design for Core and Shell. *See* LEED-C&S.
Leadership in Energy and Environmental Design for New Construction. *See* LEED-NC.
Leak detection, 1592–1593
Leak loss reduction, compressor air control systems and, 210
Leakage measuring, air, 213
Leakage, heat wheel performance factors and, 797
Leaks
   air, 212, 664–666
Leaky dampers, 521
Lean manufacturing
   development of, 467
   elements of, 467
   energy conservation and, 467–475
      traditional vs., 468–469
   5S's, 474
   impact on energy use, 469–474
      Andon signals, 473
      downtime, 472
      inventory reduction, 470
      overage reduction, 471
      production flexibility, 473
      single piece flow, 470
      training, 470–471
      visual factory, 473
      workmanship, 470–471
      workplace organization, 474
      Just in Time, 468
      overage reduction, 471
      quality assurance, 470–471
      Toyota, 467, 468
Leasing
   ESCO and, 571–572
   types, 571–572
      capital, 571–572
      operating, 571–572
Least cost planning, 1500
LEED
   discussion of, 945–946
   USBBG and, 945–946
LEED. Leadership in Energy and Environmental Design.
LEED-C&S
   credit categories, 941–945
   energy and atmosphere, 943–944
   indoor environmental quality, 944
   innovation and design process, 945
   materials and resources, 944
   sustainable sites, 942
   water use efficiency, 943
LEED-CI, 937–941
   benefits of, 938
   certification, 938
   energy and atmosphere, 939
   indoor environmental quality, 940
   innovation and design process, 941
   materials and resources, 940
   overview, 937, 938
   point distribution, 938
   rating formats, 938–940
   site selection, 939
   sustainable sites, 939
   technical review, 938–941
   vs. C&S, 937
   water use efficiency, 939
LEED-NC, 947–952
   green buildings, 948
   purpose of, 947–948
   rating system, 948–949
      energy and atmosphere, 950
      indoor environmental quality, 950
      innovation and design process, 950
      materials and resources, 950
      prerequisite categories, 949–950
      sustainable sites, 949
      water use efficiency, 949
Legal issues, net metering and, 1096
Legislation, adoption of energy codes, 441
Lenders, independent power producers and, 857
Level of irreversibility, 1428
LHV. *See* lower heating value.
Liability rules, environmental policy instruments and, 630
Life cycle analysis (LCA). *See* Energy projects, life cycle costing; Life cycle costing.
Life cycle analysis, heat wheels and, 798–799
Life cycle assessment, solar water heating and, 1337–1338
Life cycle cost analysis, 967–968
   calculators and models, 975–976
   uses of, 972–974
      lighting retrofit example, 974–975

Index

Life cycle costing (LCC)
　definition of, 953–954. *See also* Electric power projects, life cycle costing; Energy projects, life cycle costing.
　electric power projects
　　analysis of 3–KWP PV system, 965
　　annual electricity generation, 955–957
　　assumption and input parameter, 964
　　calculation methodology, 955
　　capital and noncapital costs, 956–957
　　coal power plant, 954
　　coal vs. wind, 957
　　cost components of, 954–955
　　cost of electricity, 959
　　depreciation, 957
　　50–MW wind project, 965
　　financing charge, 957–958
　　520–MW coal power project, 965–966
　　interest during construction, 958
　　loan repayment, 959
　　photovoltaic power system, 954
　　project cost, 958
　　renewable energy systems incentives, 959–961
　　taxes and duties, 959
　　variable cost, 958
　　wind power plant, 954
　　working capital interest, 958–959
Life cycle costing
　EEMS and, 622
　energy projects
　　antecedents of, 968
　　blended utility rate, 967
　　btu/sf, 967
　　definitions, 967–968
　　energy efficiency measure (EEM), 967
　　framework of, 969–972
　　　costs and benefits, 971–972
　　　definition, 970–971
　　　gas station example, 969
　　　HVAC example, 970
　　　situation, 969–970
　　internal rate of return (IRR), 967
　　LED exit lights example, 970–971
　　net present value (NPV), 968
　　payback period, 968
　　time value of money, 968
　　total owning costs, 968
　heat wheels and, 798
Life expectancy, electronic ballast primer and, 833
Light level uniformity, 993
Light measurements, data collection and, 257–258
Light meters, 258
Light shelves, as advanced window shading device, 1642–1643
Light switches, manual, 522
Light water nuclear reactors, 1115–1116
　history of, 1115–1116
　locations of, 1115–1116
Lighting based demand response, 276–277
　dimming, 277
　evaluation of, 277
　fixture/lamp switching, 277
　zone switching, 277
Lighting contractors, 979

Lighting control panel, 981
Lighting controls, 977–992. *See also* Lighting design and retrofits.
　adaptability, 986
　application issues, 986–991
　automatic, 978–979
　commissioning, 990–991
　common strategies, 977
　　daylight harvesting, 978
　　lumen depreciation, 978
　　on/off, 977, 979
　　scheduling, 977, 979
　　tuning, 977
　complexity, 986
　construction documents, 990
　cost-effectiveness, 985–986
　coverage patterns, 990
　design controls, 986
　dimming ballasts, 982
　dimming system, 983, 984
　ease of use and maintainability, 990
　electrical design, 989
　energy codes, 986
　equipment location, 990
　evaluating options, 985–986
　flexibility, 986
　fluorescent dimming ballast, 984
　HID dimming ballast, 985
　installation, 990
　integrated dimmers, 982
　intelligent on/off devices, 979
　interoperability, 986
　layering control systems, 989
　light levels, 990
　lighting contractors, 979
　local wall switches, 979
　low-voltage controls, 980–981, 982
　maintainability, 986
　microprocessor-based, centralized, 984–985
　occupancy sensors, 981, 982
　　coverage patterns, 990
　product selection, 989
　programmability, 986
　programmed-start ballasts, 982–984
　questions to ask vendors, 991
　reliability, 986
　security, 990
　selection table, 987–989
　step-level HID controls, 982
　　schematic, 982
　switching vs. dimming, 989
　system dimmers, 982
　typical controls, 978–985
　user education, 991
　utility rebates, 989
　voltage control, 989–990
　wallbox dimmers, 982
Lighting costs, activity based costing and, 1044
Lighting design and retrofits, 993–1000. *See also* Lighting controls.
　case study, Macy's/Brooklyn, 999–1000
　considerations, 993
　　color quality, 993
　　eye anatomy, 994
　　glare, 993
　　human eye functioning, 993–995
　　retinal rod and cone density chart, 994

　　uniformity of light levels, 993
　cost calculations, 997–998
　definitions, 998–999
　　candela, 998–999
　　color preference index, 999
　　color rendering improvement, 999
　　color temperature, 999
　　footcandle, 999
　　footlambert, 999
　　lumen, 999
　　lux, 999
　　mesopic vision, 999
　　photopic vision, 999
　　scotopic vision, 999
　dimming, 996
　dirty area systems, 996
　emergency lighting, 998
　goal, 993
　induction systems, 996
　lamp life, 995–996
　linear T5 lamped systems, 996
　motion sensors, 995
　scotopic CIE, 995
　scotopic/photopic vision sensitivity chart, 995
Lighting Research Center of Rensselaer Polytechnic Institute, 512
Lighting retrofits, 863
Lighting
　accounting for in energy assessment, 70
　energy efficiency and industrial plants, 870
　facility guidelines for new structures and, 528–529
Light-weight autoclaved concrete, as insulation material, 891
Lignite, 156
Line management accountability, energy master planning success and, 555
Linear induction motor maglev system, 1023
Linear T5 lamped systems, 996
Liquefaction, liquefied natural gas and, 1001–1002
Liquefied natural gas (LNG), 1001
　description of, 1001
　history, 1001
　liquefaction, 1001–1002
　regasification, 1004–1006
　storage of, 1002–1004
　transportation of, 1006–1008
　vaporization, 1004–1006
Liquid fuel production, 1354
Liquid fuel properties, 485
Liquid ring vacuum pumps, 464
Living standards and culture
　definitions of, 1009
　energy and development, 1014–1015
　energy and society, 1014, 1016
　energy use modifications, 1015
　energy-resource utilization data, 1010
　energy-use growth patterns, 1010–1012
　impact of energy use on, 1013–1015
　population and energy use, 1010
LNG. *See* liquefied natural gas.
Load calculation, geothermal heat pump system development and, 757
Load control rates, 1505
Load curtailment demand response programs 280–281

Load factor adjustments, energy balance and, 687–688
Load factor for electricity, dispatch stack order, 1499
Load factor
   measurement tool, 1498–1499
   specific customer profiles and effect on, 1499–1500
   utility rate structure and, 1498–1500
Load inventory, energy, 4
Load management rates, 1505
Load management utility rates, 1505
Load profile, cool storage system design and, 1418
Load scheduling, compressor air control systems and, 210
Load shape changes, energy demand-side management programs and, 287
Load shape objectives, energy demand-side management programs and, 288
Load usage
   cogeneration applications, 1482
   multiple unit cooling applications, 1483
   peak shaving, 1482
   single unit seasonal cooling appliances, 1483
   weighted average cost of power and, 1478–1482
Loads, building system simulation process and, 119
Loan repayment, 959
Local adoption of energy codes, 440–441
Local area augmentation system, global positioning satellites and, 29
Local wall switches, lighting control systems, 979
Location efficiency
   benefits and costs, 1453
   best practices, 1453
   impact on equity, 1453
   implementation of, 1451–1453
   transportation and, 1449–1453
Long term power purchase, 855
Longwave radiation, inter surface, 116
Loop flow, 360
Loose-fill
   attic insulation, 900–901
   insulation, blown, 897
Losses of power, 360–361
Louvers, as window shading device, 1642
Low cost energy efficient improvements, 517–523
   case studies, 517–523
      ambient temperature, 521–522
      boiler shut off in summer, 519
      cost accounting monitoring, 522–523
      dedicated AC units, 521
      fan speeds, 521
      leaky dampers, 521
      manual light switches, 522
      night setback on time clocks, 518
      review
         HVAC capacity, 519–520
         scheduling, 518
         summer-winter cutover, 520
      scheduling and building usage, 518–519
      thermostat
         location, 520
      cases, 520–521
      payback analyses, 517
Low pressure boilers, 93
Low pressure cylinders (LP), 1382
Low rise buildings, 1513
Low solar gain low-e coating, for windows, 1621–1622
Low temperature fuel cells, 733–734
   alkaline, 734–736
   direct methanol, 739–743
   proton exchange membrane fuel cells, 726
   summary, 742
Low voltage controls, lighting and, 980, 981–982
Low-e glass coatings, for windows, 1621
Lower heating value (LHV), 86–87
Low-voltage controls, lighting control systems, 980–981, 982
LP. Low pressure cylinders.
Lumen depreciation, 978
Lumen, 999
Lurgi gasifier, 906
Lux, 999

M

M & V. *See* measurement and verification.
M & V. *See* monitoring and verification.
M T & R. *See* monitoring, targeting and reporting energy analysis technique.
Machine to machine communications, 108–110
   XML based, 108–110
Machinery, compressor air control protection systems, 209
Macro indicators, technology transfer and, 502–503
Maglev transportation system, 591
Maglev (magnetic levitation) system, 1018–1026
   aid to intercity mobility, 1021–1022
   as transportation system, 1018
   benefits of, 1025
   case for building, 1019–1020
   dedicated right of way, 1022
   impact on society, 1018–1020
   politics and, 1020–1022
      Amtrak, 1020
      Federal Aid Highway Act of 1956, 1021
      federally funded airport construction, 1020
   reduce energy consumption, 1019–1020
   systems in use, 1024–1025
      Korea, 1025
      Nagoya, 1025
      Shanghai Airport, 1024–1025
   technical issues, 1022–2024
   types of, 1018
      electromagnetic suspension system, 1023
      ferrous electromagnets, 1022–1023
      linear induction motor, 1023
      superconductor magnets, 1024
   use in United States, 1025
Magnetic levitation. *See* maglev system.
Main meter M&V procedures, 1051
Maintenance guidelines
   facility guidelines for new structures, 533
   fan system energy management, 1224–1225
   Maintenance Best Practices Guide and, 710
   pumps and energy management, 1218
Maintenance electric rate schedule, 1509
Maintenance programs, Operations and Maintenance Best Practices Guide and, 710
Management, energy management system problems
   ignorance, 1035–1036
   reviews, 1030
   support, energy master planning and, 552
Management system for energy (MSE), 551
   current standards, 1032–1035
      ANSI/MSE 2000:2005, 1032–1034
   defining, by documents, 1030
   definition of, 1027–1028
   energy and, 1027–1041
   energy standards, evolution of, 1031–1032
   key elements, 1029–1030
      corrective action, 1030
      defined management reviews, 1030
      documented processes, 1029
      internal audits, 1030
      ongoing training, 1029–1030
      preventive action, 1030
      responsibilities, 1029
   plan do check act cycle, 1027, 1028
   standardization process, 1028–1031
      American National Standards Institute (ANSI), 1028–1029
      International Organization for Standardization (ISO), 1028–1029
Management systems, standards
   accreditation of, 1030–1031
   American National Standards Institute (ANSI), 1028–1029
   ANSI/MSE 2000:2005, 1032–1034
   certification of, 1030–1031
   current standards, 1032–1035
   defining by documents, 1030
   International Organization for Standardization (ISO), 1028–1029
   registration of, 1030–1031
   types, 1028
Manual light switches, 522
Manufacturing Energy Consumption Survey (MECS), 584
Manufacturing, lean vs. traditional, 468–469
Manure and coal
   cofiring and, 490–491
Marginal cost of abatement, 433
Market driven discrete usage charge, utilities and, 1500
Market implementation methods, energy demand-side management programs and, 289–290
Masonry walls, 1516–1518
   adobe, 1517–1518
   concrete block, 1516
   insulated concrete forms, 1517
   poured concrete, 1516–1517
   precast concrete, 1517
Material and energy balance (MEB) concept, 1303–1304
Material flows data, 650

Materials and resources
    LEED-C&S, rating system, 944
    LEED-CI, rating system, 940
    LEED-NC, rating system, 950
MCFC. *See* molten carbonate fuel cells.
Mean shift, Six Sigma Methods and, 1313
Measure the process, Six Sigma and, 1313–1314
Measured collector performance, solar water heating and, 1336
Measurement and verification (M&V), 567, 572
Measurement and verification (M&V) procedures, 1050–1055
    options available, 1051–1052
        end use retrofit isolation, 1051
        main meter, 1051
        partially measured retrofit isolation, 1051
        whole meter, 1051
    process of, 1052
        establishing baseline, 1053
        periodic status reports, 1054
        plan development, 1052–1053
        post implementation report, 1054
        preconstruction assessment, 1052
    theory of, 1051
Measurement of energy, 2
    tools for, 1056–1060
Measurement protocol, steam system optimization and, 1378–1379
Measurement tools, energy management and, 1056–1060
Measuring emissions, 431–432
MEB. *See* material and energy balance.
MEC. *See* model energy code.
Mechanical aeration, 1549
Mechanical draft cooling towers, 247–248
Mechanical kinetic energy, 1163
Mechanical processes, energy intensive, 463
Mechanical vacuum pumps, 1370–1371
MECS. *See* Manufacturing Energy Consumption Survey, 584
Medicine, heat pipe applications and, 810–811
Medium pressure boilers, 93
Melting, 461
Membrane technologies, high-temperature, gasifiers and, 909
MEPS (molded expansion polystyrene), as insulation material, 890–891, 894
Merchant power producers, 855
Mercury, as syngas contaminant, 909
Mesh network, 1653–1654
    ISM band radios, 1654
    multiple routing paths, 1653
    self configuring nodes, 1653
    spread spectrum radios, 1653–1654
Mesopic vision, 999
    definition of, 999
Metal carbonyls, as syngas contaminants, 909
Metal casting plant assessments, Dept. of Energy best practices case studies, 1284, 1293–1294
Metal framing, for insulation materials, 898–899
Metal hydrides, 1446
Metal industry
    aging thermal processes, 868
    aluminum alloy artificial aging, 868

billet heating, 868
die temperature, 868
energy saving opportunities, 868
equipment modernization and, 868
solution heat treatment, 868
Meter, energy balance data compilation and, 686
Meter level, wireless applications and, 1650, 1652
Meter Manager software, 3
Metering, 691
    applications of, 712
    importance of, 712
    net, 1096–1100
    planning, 713
Metering point, 1511
Meters
    electric, 283, 392
    energy data collection and, 618
Methane gas recovery, 1554
Methanol, 33
Methyl-tertiary butyl ester. *See* MTBE.
Micro indicators, technology transfer and, 502–503
Microcontrollers, electronic ballast primer and, 834
Micropressors, 208
Microprocessor-based, centralized lighting control systems, 984–985
Microprocessors, automobile technology evolution and, 673–674
Microturbine power generation, 1304
Microturbines, 298
Milling, uranium and, 1112–1113
Mineral wool
    effect on environment, 891
    as insulation material, 890, 893
Minimum charge, utility billing and, 1501
Mining, uranium and, 1112–1113
Mining methods, 158–159
    surface, 158–159
    underground, 158
Mining plant assessments, Dept. of Energy best practices case studies and, 1284–1285, 1294–1296
Mining technology trends, 146–147
Mobile HVAC systems, 1061–1069
    accumulator, 1079
    airside heat transfer calculations, 1062–1063
    capacity calculations, 1061–1066
    coils design, 1061–1066
    components, 1076–1081
        electrical systems, 1080–1081
    components evaluation, 1066
    compressors, 1076
    condenser, 1077
    configurations, 1075–1076
    evaporator, 1079–1080
    expansion valve, 1077–1078
    history of, 1072–1073
    innovations to, 1066–1069
    inverse heat transfer method, 1063–1066
    nomenclature, 1070–1071
    orifice tube, 1078–1079
    physics of, 1073
        evaporating process, 1073–1074
        heating process, 1074–1075
    pressure switches, 1080

psychrometric fundamentals, 1061–1066
receiver/dryer, 1077
testing procedures, 1061–1066
thermostats, 1080
TXV system vs. OT, 1081
tube-side heat transfer calculations, 1063
Mode transitions, global temperature adjustment and, 274
Model energy code (MEC), 438, 439
    adoption process, 440–441
        groups involved, 440
        timing of, 440
    development and revision of, 440
    International Energy Conservation Code (IECC), 438
    state's right in revising, 439
Model validation, fast simulation modeling and, 916
Modeling
    building geometry and energy use, 112–113
    geometric ratio, 113
Modeling
    energy monitoring performance, 5–6
    IntelliGrid^{SM}, 915
Modes, controller, 415
Modular boilers, control of, 102
Modular buildings, 677
    water source heat pump for, case study, 1564–1573
Modulating controller, 415–416
MOIST software, 1403
Moisture, biomass energy and, 87
Moisture content, of syngas feedstocks, 907
Moisture transfer processes, 117
Mold
    ANSI/ASHRAE Standard 62.1–2004 and, 53–54
    growth, 839, 844, 845, 846
    prevention by pressurization flow, 54
Molded expansion polystyrene (MEPS), as insulation material, 890–891, 894
Molten carbonate fuel cells (MCFC), 727–729
    actual use of, 729
    applications, 729
    contamination levels, 728
    operating principle of, 727–728
    problems with, 728–729
Monitoring and verification (M & V), 179
Monitoring of energy, 5–6
Monitoring, targeting and reporting (M T & R) energy analysis technique, 4–7
    definitions of, 5
    reporting, 7
    targeting, 6–7
Motion sensors, 995
    lighting design and retrofits, 995
Motor belts and drives, cost calculations and, 73–74
Motor control
    compressed air and, 207–208
    primary, 208
    ride through, 208
Motor load factors, energy cost savings and, 73
Motor Master (MM+4.0), 879–880
Motor optimization projects, 883–889
    capital spending vs. re-engineering, 886–887

drivers of, 888
empirical measures, 883–884
energy consumption data, 886
indirect impacts, 887–888
individual project results, 887
industry-wide energy savings potential, 888
internal rates of return (IRR), 883–884
IOF vs. non-IOF facilities, 885
methodology, 883
net present value (NPV), 883–884
simple payback results, 883–884
study results, 41
U.S. motor systems, 884–885
Motor system effectiveness, as driver in motor optimization projects, 888
Motor usage, accounting for in energy assessment, 70–71
Motors, high efficiency, cost calculations and, 73
Motors and drives, facility guidelines for new structures and, 529
Moving bed gasifiers, 906
    Lurgi gasifier, 906
MSE. *See* management system for energy.
MTBE (methyl-tertiary butyl ester), 33
Multiple boilers, control of, 98–102
    functional description of, 100–102
    modular, 102
Multiple pump theory, 1217
Multiple routing paths, 1653
Multiple unit cooling applications, 1483
Multiple year monthly average comparison method, 2
Multiple zone recirculating system, 57
Multiresolution modeling, 916
Municipal code, energy code adoption and, 441

N

NAAQS. *See* National Ambient Air Quality Standards.
Nagoya maglev system, 1024–1025
National Ambient Air Quality Standards (NAAQS), 430
National Energy Act of 1978, 1082–1086
    components of, 1083–1086
    Energy Tax Act of 1978, 1084
    Fuel Use Act, 1085
    National Energy Conservation Policy Act, 1085–1086
    Natural Gas Policy Act of 1978, 1083
    Public Utility Regulatory Policies Act, 1084, 1201–1205
National Energy Conservation Policy Act, 1085–1086
National Institute of Standards and Technology (NIST), 510–511
National Technical Information Service (NTIS), 510
Natural capital, 1410
Natural carbon sinks
    biological sequestration and, 129
    ocean fertilization, 129
    terrestrial, 129
Natural draft cooling towers, 247–248
Natural energy, 1089–1090
    additional energy, a comparison, 1088–1094

types, 1089–1090
    nonsolar related, 1090
    solar related, 1090
    solar, 1089–1090
Natural environment subsystem modeling, exergy analysis and, 656–657
Natural gas issues, coal to liquid fuel plants and, 170
Natural Gas Policy Act of 1978, 1083–1084
    curtailment, 1084
    emergency authority, 1084
    incremental pricing, 1083–1084
    transportation, 1084
    wellhead pricing, 1083
Natural gas
    chemical process industry and, 1305
    coal vs., 161
    distributed generation technology and cost estimating, 299
Natural gas rate schedules, 1507
    air conditioning, 1507
    congeneration, 1507
    dual fuel firm rates, 1507
    end use, 1507
    firm sales rates, 1507
    firm transportation rates, 1507
    interruptible sales rates, 1507
    interruptible transportation rates, 1507
Natural nuclear reactor, 1111
Negotiated rates, 1505–1506
    cost based, 1505–1506
NERC (North American electricity reliability council), 361
Net emergy
    energy sources, 426
    time and, 426
Net energy, 424–426
Net metering, 1096–1100
    description of, 1096
    implementation of, 1097–1098, 1100
    by state, 1097–1100
    technologies eligible for, 1098–1100
    use of, legal and policy problems, 1096
Net present value (NPV), 968
    calculation, 84–85
    capital rationing, 84–85
    in motor optimization projects, 883–884
Networked capacity control, compressor air control systems and, 209–210
New building commissioning, 188–199
New process capability, 1316
New York State Energy Research and Development Authority, 512
Nickel cadmium batteries, 1155
NIST. *See* National Institute of Standards and Technology.
Noise level, of wind power, 1612
Non capital costs, 956–957
Non combustion processes
    gasification, 1525, 1527
    waste fuels and, 1525, 1526
Non firm service rates, 1504–1505
    interruptible, 1504–1505
    load control, 1505
    load management, 1505
    standby, 1505
Non renewable energy, 1090–1091

Non renewable energy sources, unit emergy values and, 425
Non-fossil energy sources, global warming and, 724–725
Nonsolar related energy, 1090
Nonuse benefits, environmental policy instruments and, 631
Nonutility generators, electricity deregulation and, 385–386
Normalized energy benchmarks, 690–692
Normalized performance indicator (NPI), 4
Normalized time intervals, 619
North American coal to liquid fuel plants, 167
North American electricity reliability council. *See* NERC.
Northern Cyprus, 503–504
NO_x emissions, cofiring and, 491
NPI. *See* normalized performance indicator.
NPR, IRR vs., 933
NPR cash flows, 935
NPV (net present value), in motor optimization projects, 883–884
NTIS. *See* National Technical Information Service.
Nuclear energy, 38–43
    accidents, 1131–1132
    competitive cost of, 39
    development challenges, 41–43
        long term storage, 43
        politics, 41–43
        safety, 42
        spent fuel management, 42–43
        waste management, 42–43
    disposal of, 1113–1114
    economic issues
        alternative fuel cost assumptions, 1105
        modeling results, 1105–1109
    economics of, 1101–1110
        capital costs, 1101–1103
        discount rate, 1104–1105
    environmental issues, 1103–1104
        waste disposal, 1103–1104
    exposure to, 1132–1133
    fuel cycles, 1111–1114
    fuel operation, 1113–1114
    future challenges, 41
        uranium availability, 41
        waste product disposal, 41
    future technologies, 41
        helium cooled very high temperature gas reactor, 41
        supercritical water-cooled reactor, 41
    Generation IV program, 1121–1123
    generation types, 40–41
    history of, 38–39, 1125–1127
    pebble bed modular reactor, 40–41
    power plant types, 1115–1124
        boiling water, 1117–1118
        fast breeder, 1120–1121
        future of, 1121–1124
        future of, fusion, 1123–1124
        gas-cooled, 1119–1120
        light water, 1115–1116
        pressurized heavy water, 1118–1119
        pressurized water, 1116–1117
    principles of, 39
    reactor types, 39–41

advanced gas-cooled, 40
boil water, 39–40
fast neutron, 40
pressurized heavy water, 40
pressurized water, 39
technology of, 1125–1133
  atomic structure, 1127
  binding energy, 1127
  chain reaction, 1128–1129
  fission, 1128
  fusion, 1130–1131
  radioactivity, 1127–1128
  reactor design, 1129–1130
safety, 1131
waste disposal, 1103–1104
Nuclear energy, challenges of, 41–43
uranium availability, 41
long term storage, 43
politics, 41–43
safety, 42
spent fuel management, 42–43
waste management, 42–43
waste product disposal, 41
Nuclear energy power plant types, Generation IV program, 1121–1123
Nuclear power, 394–395
Nuclear power plant types, 1115–1124
boiling water, 1117–1118
fast breeder, 1120–1121
future of, 1121–1124
fusion, 1123–1124
gas-cooled, 1119–1120
light water, 1115–1116
pressurized heavy water, 1118–1119
pressurized water, 1116–1117
Nuclear power, technology of, 1125–1133
atomic structure, 1127
binding energy, 1127
chain reaction, 1128–1129
fission, 1128
fusion, 1130–1131
radioactivity, 1127–1128
reactor design, 1129–1130
Nuclear reactor types, 39–41
advanced gas-cooled, 40
boil water, 39–40
commercially in use, 40
fast neutron, 40
natural, 1111
pressurized heavy water, 40
pressurized water, 39
uranium history, 1112
Nuclear reactors, commercially in use, 40
Numerical methods, building system simulations and, 118

O

O&M BPG. *See* Operations and Maintenance Best Practices Guide.
OASIS. *See* open access same-time information system.
Occupancy sensors, 981, 982
lighting control systems, 981, 982
coverage patterns, 990
Occupational Safety and Health Act (OSHA), 1303

Ocean carbon sinks, 129
Ocean currents, renewable energy and, 1267–1268
Ocean direct injection of captured CO_2, 128
Ocean sources, renewable energy and, 1267
Ocean thermal differences, renewable energy and, 1267–1268
Odum, H.T., 420, 421
Off peak
  energy cost calculations, on-peak vs., 71–73
  heat storage, 1420
  utility load, 1497–1498
Off season rates, time of use and, 1504
Off taker, independent power producers and, 856
Office of Electricity Delivery and Energy Reliability, 509–510
Office of Energy Efficiency and Renewable Energy (EERE), 508–509, 873
Office of Fossil Energy, 509
Office of Industrial Technologies (OIT), 873
Office of Scientific and Technical Information, 510
Offset emissions trading, 431
Offset policy, 430
  bubble, 430–431
Oil companies, nationalization of, 595
Oil consumption in United States, 595
Oil exploration, Artic Wildlife Reserve and, 595
Oil in ground, uncertainty of, 593
Oil production vulnerabilities, impact of, 592–593
Oil reserve uncertainty, 592
Oil shale deposits, 594
  disadvantages of, 594
Oil tankers, impact on U.S. energy, 591–592
Oil, U.S. over reliance on, 596–597
OIT (Office of Industrial Technology), 873
On peak energy cost calculations, off-peak vs., 71–73
Open access same-time information system (OASIS), 362–363
Open-cell, high-density spray polyurethane, as insulation material, 891, 894
Open-cell, polyurethane, as insulation blowing agent, 892
Operating expenses, 1256
Operating leasing, 571–572
Operating voltage, electric power losses and, 361
Operational controls, automobile technology and, transferable to building use, 672
Operations and Maintenance Best Practices Guide (O&M BPG), 707–714
commissioning existing buildings, 710–712
  definitions, 711
  findings from, 711
  process of, 711–712
computerized maintenance management systems, 710
definitions, 708
energy savings from, 708
important elements
  contracting outside services, 709–710
  management support, 709
  measuring program quality, 709
  program implementation, 709

maintenance programs, 710
major equipment, 713
management practices, 708–710
metering, 712–713
new technologies, 713
operational efficiency, 713
predictive maintenance, 710
purpose of, 707–708
Opportunity fuels, 1523–1525, 1526, 1530
Optimal time intervals, 619
Options, automobile technology and, transferable to building use, 673
Order, 888, 380
Order destruction, exergy and, 659
Organic materials, thin film PV systems and, 1150
Organic matter, biomass energy and, 31–34
Orifice plate management, flow measurement problems and, 1057
Orifice tube, mobile HVAC systems and, 1078–1079
OSHA. *See* Occupational Safety and Health Act.
OT system, TXV vs. mobile HVAC systems and, 1081
Outdoor air quality
  ANSI/ASHRAE Standard 62.1–2004 and, 51
  outdoor airflow rates, 51
Outdoor airflow rates, 51
  constant volume of supply air, 51
  variable air volume, 51
Output controller, 415
Output devices, controllers and, 415
Output of energy
  balancing of, 4
  load inventory, 4
Outside air, value of re indoor air quality, 20
Ovens, 809
Overage reduction, lean manufacturing and, 471
Overhead, activity based costing and tracing of, 1042
Overheating, use of spectrally selective window films, 1637
Owner, independent power producers and, 856
Ozone layer, 873
Ozone, 1559

P

Packing type, cooling towers and, 249–250
PAFC. *See* phosphoric acid fuel cells.
Pakistanian coal to liquid fuel plants, 167
Panel cooking, solar, 1326
Paperboard sheathing, as insulation material, 895
Parabolic cooking, solar, 1326
Parabolic dish, 1328
Parabolic trough, 1328
Parallel combustion, 89
Parallel flow heat exchanger, 802
Parallel path flow, 360
Parallel quality, 421–422
Partially measured retrofit isolation, 1051
Particle size, syngas feedstocks, 908
Part-load efficiency, district cooling systems and, 314
Passive solar heating, windows and, 1618–1619
Passive space heating, 1322

direct solar gain design, 1322
indirect solar gain design, 1322
isolated solar gain design, 1322
Payback analyses, low cost energy efficient improvements and, 517
Payback period, 968
Payback, heat wheels and, 798
Payment structure, independent power producers and, 857–858
PCM ice storage, 1414
Peak air demand, calculating of, 231–232
Peak load management, electric demand response and, 272
Peak shaving, 1482
Peak usage, utility billing and, 1501
Peak utility load, 1497–1498
Pebble bed modular reactor, 40–41
PEMFC. See proton exchange membrane fuel cells.
Penalties, emissions trading and, 433
Pennsylvania, electricity deregulation results in, 385
Pentane (polystyrene, expanding)
insulation blowing agent, 891–892
smog and, 891
Perfect Power System, 404–405
goals of, 404–405
Performance contracting, 1134–1139
benefits of, 1138–1139
chauffage, 1136
contemporary usage, 1135
ESCOs, 1134–1135
history of, 1134–1135
types, 1136
guaranteed savings, 1136–1137
shared savings, 1136–1137
workings of, 1135–1138
deal structuring, 1136
financial modeling, 1136–1137
model comparison, 1137
value chain management, 1137–1138
Performance curves, fan systems and, 1223–1224
Performance indicators, energy and, 4
Performance measurement, fan systems and, 1222–1223
Perl, 537
Perlite, as insulation material, 893
Permanent pressure loss, flow measurement and, 1058–1059
Permit allocation, emissions trading and, 432
Permits
emissions trading and, 432
tradable emissions, 630
Personnel skills, energy project management and, 564–565
PES. Primary energy supplies.
Petroleum based products as heating fuels, 1361
Petroleum fuels, end of, 1407
Petroleum plant assessments, Dept. of Energy best practices case studies and, 1285, 1296–1299
Petroleum, distributed generation technology and cost estimating, 299
Phase equilibrium, 1431
Phone line, 1655

wireless application vs., 1658
Phosphoric acid fuel cells (PAFC), 726–727
Photometry, 264
Photo-oxidants, air quality modeling and, 719–721
Photopic vision, 999
definition of, 999
sensitivity chart, 995
Photovoltaic electricity, 1267
Photovoltaic power system, 954
Photovoltaic (PV) systems, 1147–1157
pplications of, 1156–1157
backup to, 1152–1153
costs of, 1153
crystalline SI, 1148–1149
electronic usage, 1154–1155
energy storage
hydrogen, 1155–1156
lead acid batteries, 1155
nickel cadmium batteries, 1155
full solar spectrum usage, 1150–1151
history of, 1147–1148
power electronic converters, 1154
power plant usage, 1151–1153
satellites, 1147
solar resources, 1148
stand alone, 1153
sunlight concentration, 1151
thin films, 1149–1150
Physical plant, energy assessments and, 67
Physics of energy, 1160–1173
Piloting techniques, aircraft fuel consumption and, 28
Pinch analysis, 460
Piston devices, reciprocating, 1233–1234
Plan do check act cycle (PDCA), 1027, 1028
Plant construction, independent power producers and, 860
Plant expansion, as driver in motor optimization projects, 888
Plant interaction with building zones, 117
Plant locations, coal to liquid fuels and, 168
Plant operation, independent power producers and, 860
Plant safety, compressor air control systems and, 209
Plant size, coal to liquid fuel plant, 169
Plant valuation, rate base determination and, 1254
Plate and frame heat exchanger, 1540
PLC. See programmable logic controllers.
Plenum leakage, 1468
Plug flow, 1561
Plug in configuration, hybrid electric vehicles and, 847–853
Plug in hybrid electric vehicles, greenhouse gas emissions and, 791–792
Plug in hybrid vehicles, CO_2 emissions and, 788–789
Plumbing, facility guidelines for new structures and, 533
Pneumatic capacitance, 229–230
Pneumatic compressor control, 208
Pneumatic transport, chemical process industry and, 1305
Policy instruments, environmental policy and, 630–631

Politics
nuclear energy development and, 41–43
U.S. transportation energy use and, 594–595
Pollutants, water, 746
Pollution abatements, cost minimizing, 434–435
Pollution reduction, 870
Pollution sources
carbon dioxide, 20
indoor air quality and, 20–21
volatile organic compounds, 20
Pollution summaries review, energy assessments and, 64
Pollution tax, 630
Polyisocyanurate, as insulation blowing agent, 891, 892
Polystyrene
expanding (pentane), as insulation blowing agent, 891–892
extruded, as insulation blowing agent, 892
Polyurethane
closed-cell, high-density, as insulation material, 891, 894
insulation blowing agent, 892
open-cell
high-density, as insulation material, 891, 894
insulation blowing agent, 892
Ponds, solar, 1324–1325
Population growth, developing countries and energy demand, 500
Population, energy use and, 1010
Portland Energy Conservation, Inc., 515
Positive displacement pump, 1213–1214
theory, 1216
Post combustion capture, 126
Post fossil fuel era, 608–610
Poured concrete walls, 1516–1517
Power capacity needs, reducing in developing countries, 503
energy saving opportunity gap, 503
excess energy consumption gap, 503
Power circuit supply, 416
Power cycle efficiency, 1424–1426
Power distribution systems, 699–706
basic principles, 699–701
balanced system, 700–701
single phase system, 699–700
three phase system, 699–700
unbalanced system, 700–701
design guidelines
conductor sizing, 701–702
grounding, 701
voltage levels, 702
fault current, 703–704
system protection, 702–703
design guidelines, 702
ground fault detection, 703
power system faults, 702–703
Power electronic converters, 1154
switch mode, 1154
Power electronics based controllers, 917
Power factor, 255–256, 706
apparent power volt amperes, 1510
demand billing, 1510–1511
real power watts, 1510
volt amperes reactive, 1510
Power flow increases, 917

Power generation, solar thermal technologies and, 1327–1328
Power losses
  regional effects of, 359
  electric power transmission systems and, 360–361
Power plant types, nuclear energy and, 1115–1124
Power plants, photovoltaic (PV) systems and, 1151–1153
Power quality analyzers, 256
Power quality issues, 1484–1490
  cost tracking, 1489
  customer forums, 1489
  Distribution Power Quality (DPQ), 1484
  education re, 1490
  power sags, 1484–1487
    minimizing, 1489–1490
  solutions to, 1490
  workshops, 1489
Power rating, wireless terms and, 1652
Power requirements
  fan systems and, 1223
  pump theory and, 1216–1217
Power sags, 1484–1487
  customers response to, 1487–1488
Power sources
  alternative, government subsidies of, wind power 1614–1615
  wind. *See* Wind power.
Power supply, electric motors and, 352
Power system faults, 702–703
Power usage determination, pump theory and, 1217
Powered attic ventilator, 900
Pre flight planning
  aircraft fuel consumption and, 27–28
  calculation of fuel quantity, 28
  center of gravity, 27
Precast concrete walls, 1517
Precise pressure regulation
  compressed air control systems and, 209
  compressor discharge output pressure, 209
Pre-combustion capture, 126–127
Preconstruction assessment, M&V procedures and, 1052
Predesign phase, commissioning process and, 193
Predictive maintenance, Operations and Maintenance Best Practices Guide and, 710
Preindustrial era, energy and, 605–606
Premium efficiency motor cost to EPA motor cost efficiency, 353
Preparation of coal in United States, 151–152
Present to past energy comparison method, 2
Pressure drop, heat wheel performance factors and, 797
Pressure profile, compressed air storage systems and, 230–231
Pressure sensor, 412–413
Pressure set point, compressed air systems and, 238
Pressure switches, mobile HVAC systems and, 1080
Pressure vessel development
  boilers, 1360

  space heating and, 1360–1361
Pressurization design
  exfiltration vs. infiltration, 844
  HVAC systems and, 844
  improper, case study, 844–846
  smart air syndrome, 844
  understanding of, 844
Pressurization flow
  ANSI/ASHRAE Standard 62.1-2004 and, 53–54
  energy recovery ventilation, 52
  importance in preventing mold, 54
Pressurization, 839
  indoor negative pressures, 844
Pressurized heavy water nuclear reactors, 40
  CANDU, 1118–1119
  locations of, 1118
Pressurized water nuclear reactors, 1116–1117
Pressurized water reactors, 39
Preventive action, 1030
Preventive maintenance, downtime and, 472
Preventive service monitoring, web-based compressor management and, 213
Price cap regulation, 1245–1250
  basic price restriction, 1246–1248
  case studies, 1249–1250
  service baskets, 1248
Price restrictions, price cap regulations and, 1246–1248
Pricing, energy demand-side management programs and, 291
Primary energy forms, 1088
Primary energy ratio, 819
Primary energy supplies (PES) vs. secondary, 1346
Primary motor control, 208
Priorities, energy management system problems and, 1036
Procedures, energy management system problems and, 1036
Process analyses, 1143
Process based commissioning, 189–190
Process cooling, energy intensive, 462–463
Process dependent modeling, exergy analysis and, 657–658
Process energy optimization, 1304
  process heat integration, 1304
  temperature, low and high, 1304
Process equipment, accounting for in energy assessment, 71
Process heat integration, 1304
Process Heating Assessment and Survey Tool (PHAST 1.1.1), 880
Process heating, 461
Process historian, 459
Process reliability, 888
Product cooling heat recovery, 1373–1375
Production flexibility, lean manufacturing and, 473
Production issues, as driver in motor optimization projects, 888
Production, worldwide energy resources and, 445–446, 447
Productivity, coal production in United States and, 147–149
Programmable logic controllers (PLC), 208

Programmed start ballasts
  dimmers and, 982–984
  lighting control systems, 982–984
Project based emissions trading, 431
Project costs, electric power projects and, 958
Project definition, independent power producers and, 858
Project development, independent power producers and, 858
Promotion, energy demand-side management programs and, 290
Protection features, compressor air control systems and, 208–209
Protection, power distribution systems and, 702–703
Proton exchange membrane fuel cells (PEMFC), 726
  application of, 738–739
  contamination levels, 737
  principles of, 736–737
  problems, 737–738
  technological status, 738
Proximate analysis, solid fuel properties and, 479, 480
Prudence concept, rate of return regulation and, 1255
PSAT: Pumping Assessment Tool, 880
Psychrometric analysis, 1318
Psychrometrics, fundamentals of, mobile HVAC systems and, 1061–1066
Psychrometrics, 1184–1191
  air conditioning cycles, 1188–1191
  definition of, 1184
  degree of saturation, 1185
  dew point temperature, 1184–1185
  dry bulb temperature, 1184
  humidity ratio, 1185
  processes, 1186–1188
    air streams, adiabatic mixing of, 1188
    cooling and dehumidification, 1186–1187
    heating and dehumidification, 1187–1188
    humidification, 1186
    sensible
      cooling, 1186
      heating, 1186
  saturation pressure, 1185
  wet bulb temperature, 1184
Public policy
  creation of laws and rules, 1193–1194
  description of, 1193
  energy and, 1193–1200
  factors affecting, 1195–1198
    corporate behavior, 1197
    corporate governance, 1198
    economic conditions, 1196
    industrial conditions, 1196
    input markets, 1196
    institutional conditions, 1195
    international experience, 1195
    international risk experience, 1196
    legitimacy and credibility, 1198
    market structure, 1197
    objectives and priorities, 1195
    policy incentives, 1197
    regulatory governance, 1196–1197

sector performance, 1197–1198
  implementation of, 1198–1199
    regulatory processes, 1198–1199
    rationale for intervention, 1194
Public Utilities Holding Act of 1935, 380
Public Utilities Regulatory Policy Act of 1978 (PURPA), 380
  effectiveness of, 380
Public Utility Holding Company Act of 1935, 392
Public Utility Regulatory Policies Act (PURPA) of 1978, 1201–1205
  cogeneration technology, 1202
  energy crisis, 1201
  focus of, 1202
  implementation of, 1202–1203
  origin of, 1201–1202
  Section, 210,1202
  technological innovations and growth of, 1203
    cogeneration, 1203
    gas turbines, 1203
    solar energy, 1204
    wind, 1204
  unintended consequences of, 1204–1205
Pulverized coal boiler versus IGCC, 912
Pump motors, ASD technology and, 353, 355
  issues with, 355
Pump seals, power usage determination, 1217–1218
Pumped storage hydroelectricity, 1207–1212
  cost and economics, 1210–1211
  costs of, 1207
  efficiency level, 1209
  environmental issues, 1211–1212
  facility for, 1208–1209
  workings of, 1209–1210
Pumping, chemical process industry and, 1304
Pumps, 1213–1220
  electric motors and, 353, 355
  energy management opportunities, 1218–1220
    maintenance, 1218
    retrofitting, 1218–1220
  theory of, 1214–1218
    centrifugal, 1215
    multiple, 1217
    operating characteristics, 1214–1215
    positive displacement, 1216
    power requirements, 1216–1217
    power usage determination, 1217
    pump seals, 1217–1218
  types, 1213–1214
    centrifugal, 1213
    positive displacement, 1213–1214
Purchased utilities, energy data collection source and, 617
Pure oxygen systems, 1549
PURPA. See Public Utilities Regulatory Policy Act of 1978.
PV. See photovoltaic.
Pyrolysis, 89, 485–486, 1348

Q

Quality assurance, lean manufacturing and, 470–471
Quality control, automobile technology evolution and, 673
Quality, emergy definitions of, 421–422
  cross, 421, 422
  parallel, 421–422
Quasi isothermal compression, 1579

R

R Factor insulation, 1363
R values, 2×6 wall construction, 899
RACI concept, 563–564
Radiant barriers, 1227–1232
  airtightness, 1231
  benefits of, 1227
  cost savings, 1228
    claims of, 1228
  decreasing heat gain, 1231
  definition of, 1227
  foil side, 1229–1230
  foil-faced batt insulation, 1231
  heat buildup, 1231
  heat transfer blocking, 1227–1228
  installation, 1230–1231
    airtightness, 1231
    new construction, 1231
    placement, 1231
    safety tips, 1230
  material types, 1228–1229
    costs, 1229
  payback, 1232
  reshingling, 1231
  shingle warranties, 1231
  vs. reflective insulation, 891
Radiant flux
  spectral distribution, in window energy performance, 1616
  in window energy performance, 1616
Radiant heat barriers (RHB), 903–904
  configuration of, 904
  definition of, 903
  mechanism of operation, 903–904
Radiant heat gain, solar, in windows, 1623
Radiation heat transfer, in windows, 1617
Radiation, heat transfer and, 828–829
Radioactivity, 1127–1128
Radiometry, 264
Railroads
  coal to liquid fuel transportation needs and, 168
  ease of use, 591
  U.S. transportation energy use and, 590–591
Raised floor insulation, 892
Raised heel trusses, soffit air ventilation and, 902
Raised top plate, soffit air ventilation and, 902
Rankine steam cycle, 1381
Ratchet adjustments, utility billing and, 1501–1502
Rate base, rate of return regulation and, 1254–1255. See also rate of return regulation.
  accumulated depreciation determination, 1255
  construction work in progress determination, 1255
  plant valuation determination, 1254
  prudence concept, 1255
  working capital determination, 1255
Rate design strategies, 1502–1503
Rate of return
  assessing of, 1252
  formula, 1252–1253
Rate of return regulation, 1252–1257
  advantages and disadvantages, 1253
  cost of capital, 1256
  depreciation, 1256–1257
  formula, 1252–1253
  operating expenses, 1256
  rate base, determining of, 1254–1255
    accumulated depreciation, 1255
    construction work in progress, 1255
    plant valuation, 1254
    prudence concept, 1255
    working capital, 1255
  revenue imputation, 1253–1254
  taxes, 1257
Rate riders, 1507
  electric rate schedule, 1509
Rate structures, utilities and, 1497–1512
Rate-based emissions trading, 431
Ratings
  boilers, 94–95
  electric power transmission systems and, 358
RDF, coal and, cofiring and, 490
Reactance, 357
Reaction of enthalpy, 1423–1424
Reactive power, 356
  electric power transmission systems and, 356
  losses, 361
  production of, 356
Reactive power losses, 361
Reactivity of syngas feedstocks, 907
Reactor design, 1129–1130
Reactor treatment of wastewater, 1561
  complete mix, 1561
  dispersed flow, 1561
  plug flow, 1561
Reactors, nuclear, 39–41. See also nuclear energy; nuclear power plant types; nuclear power technology.
  advanced gas-cooled, 40
  boil water, 39–40
  commercially in use, 40
  fast neutron, 40
  natural, 1111
  pressurized heavy water, 40
  pressurized water, 39
  types of, 39–41
  uranium history, 1112
Real fluids, 1430–1431
Real power watts, 1510
Real power, 356
  electric power transmission systems and, 356
Real time costing, 920, 927–928
Real time pricing (RTP), 1175, 1178, 1180–1182
  commercial basis, 1180–1182
  electric rate schedule, 1508
  example of, 1181–1182
  industrial basis, 1180–1182
  residential, 1180
  See also real time utility pricing.
Real time pricing electric rate schedule, 1508
Real time utility pricing, 1506–1507

Index

ambient temperature based, 1506
electric utility dispatch modeling, 1506–1507
Rebound, electric demand response and, 273
REC. *See* renewable energy credits.
Receiver, mobile HVAC systems and, 1077
Receiver volume, compressed air systems and calculating of, 232
Recessed lights, ceiling and roof insulation and, 903
Reciprocating engines, 1233–1243
Reciprocating engines, analysis, 1237–1238
  compression, 1235
  diesel, 1238, 1239,1241, 1242
  distributed generation technologies and, 297–298
  exhaust, 1235
  gas exchange process, 1237–1238
  gasoline, 1239–1240
  hardware of, 1235–1237
  homogeneous charge compression ignition, 1240, 1242
  intake operating cycle, 1235
  operating cycles, 1235–1237
    compression, 1235
    compression, 1235
    exhaust, 1235
    intake, 1235
  piston devices, 1233–1234
Reciprocating engines, piston devices, 1233–1234
Recommissioning, 180
Rectifiers, 359
Rectisol, in syngas treatment, 908
Recuperator, 1540
Recuperator usage, 1577–1578
Recycling, SWEATT and, 1353–1354
Recycling centers, 1530
Recycling products, sustainable development and, 1409
Redistributive effects, environmental policy instruments and, 631
Reduced air usage, 211
Reduction in aircraft fuel consumption, 26–29
  drag reduction, 26–27
  engine efficiency, 26
  engine maintenance, 28–29
  flight controls, 28
  piloting techniques, 28
  pre-flight planning, 27–28
  weight reduction, 26–27
Reference environment, exergy and, 646
Reference environment modeling, exergy analysis and, 656–658
  constrained-equilbrum, 657
  equilbrum, 657
  natural-environment-subsystem, 656–657
  process-dependent, 657–658
  reference-substance, 657
Reference substance modeling, exergy analysis and, 657
Reflective insulation, 891
  vs. radiant barriers, 891
Regasification, liquefied natural gas and, 1004–1006
Regional coal mining
  changes in, 149–150
  productivity, 147–149

Regional haze, air quality modeling and, 719
Regional transmission operator (RTO), 380
  effect on wholesale customers, 380–381
  results of, 381
Registration, management system standards and, 1030–1031
Regulating rate of return, 1252–1257
  formula, 1252–1253
Regulation
  adoption of energy codes and, 441
  price cap, 1245–1250
  revenue cap, 1245–1250
Regulatory processes, decisions and outcomes subject to, 1199
Reheat rankine steam cycles, 1381
Relative humidity, 19
Relative humidity controller, 414–415
Relative humidity sensor, 411–412
Relative implementation, global temperature adjustment and, 275
Reliability factor, fast simulation modeling and, 916
Reliance on personnel, energy management system problems and, 1036
Relief control valves, 1371
Renewable electricity generation, 44
Renewable energy credits (REC), 14
Renewable energy resources, 453
  district heating and cooling and use of, 322
  environmental policy instruments and, 631
  program funding, 1531
Renewable energy resources program funding, 1531
Renewable energy systems incentives, 959–961
Renewable energy technologies, 31, 1265–1270
  available technologies, 1266–1267
    biomass, 1266–1267
    geothermal electricity, 1266
    ocean sources, 1267
  biomass, 1266–1267
  conversion efficiencies, 1266
  decentralized, 1258–1264
  distributed generation and, 299
  fuel ethanol, 1267
  future of, 1268–1270
    adequacy, 1269–1270
    current energy use, 1268–1269
    efficiency gains, 1269
  geothermal electricity, 1266
  green electricity pricing, 1262–1264
  intermittent sources, 1268
  minimal impact of geothermal energy, 748
  ocean currents, 1267–1268
  ocean sources, 1267
  photovoltaic electricity, 1267
  promotion of, 1260–1264
    green electricity pricing, 1262–1264
    portfolio requirements, 1261–1262
    renewable trust funds, 1260–1261
    system benefits charge, 1260–1261
  resources available, 1266
  solar industrial process heat, 1267
  solar water heating, 1267
  technology development, 1267–1268
    ocean currents, 1267–1268
    ocean thermal differences, 1267–1268
    solar industrial process heat, 1267

wave energy, 1267
  technologies not cost competitive, 1267
    fuel ethanol, 1267
    photovoltaic electricity, 1267
    solar water heating, 1267
  thermal differences, ocean, 1267–1268
  trust funds, 1260–1261
  types, 1265–1266
  wastewater plant operation and, 1554
  wave energy, 1267
Renewable trust funds, 1260–1261
Rensselaer Polytechnic Institute, Lighting Research Center, 512
Repair of compressed air leaks, 219–225
  tagging of, 222–223
Repeaters, 1652
Reporting functions, energy analysis and, 7
Resale of excess power, combined heat and power applications and, 865
Research and development
  energy conservation and, 450–451
  geothermal energy environmental impact, 746–747
Reserves, energy, 608–610
Reshingling, radiant barriers and, 1231
Residential areas, electric demand response solutions for, 283
Residential building heating loads, 1272–1277
  calculations for, 1272–1273
  seasonal heating demand, calculations, 1273–1276
Residential electricity generation in the United States, 583
  air conditioning, 583
  appliances, 583
  space heating, 583
Residential energy use in U.S., 583
Residential real time pricing, 1180
Resistance, 357
Resistance temperature devices, 409–410
  solid-state, 410–411
Resolution of control, 273
Resource availability, energy management system problems and, 1037
Resource degradation, exergy and, 659
Response factor, 117
Responsibility, management systems and, 1029
Results, measuring and verifying, enterprise energy management systems and, 622
Retail power markets, 382–383
  federal government intervention, lack of, 382
  individual state regulation of, 382
  potential savings of, 382–383
Retinal rod chart, 994
Retrocommissioning process, 180, 201–206
  benefits of, 203
  definitions of, 200–201
  energy conservation benefits, 202–203, 204
  Leadership in Energy and Environmental Design, 200, 206
  maintenance activities, 203
Retrofit isolation, 1051
Retrofitting, pumps and energy management opportunities and, 1218–1220
RETScreen green building design software, 1403

Revenue cap regulation, 1245–1250
Revenue imputation, 1253–1254
Reversibility, 1426
    energy degradation and, 1168–1172
    Second Law of Thermodynamics and, 1429
RHB. *See* Radiant heat barriers.
Ride through motor control, 208
Right of way, maglev and, 1022
Right Suite Residential software, 1402
Rigorous equations, 1429–1430
Risks, independent power producers and, 858, 859
Road infrastructure, strain on, 590
Roads, Federal funding of, 591
Roadway design, land use impact and, 1450
Rock wool, as insulation material, 890, 893
    effect on environment, 891
Rocky Mountain States coal basis, 158
Romans, heating and, 1357–1358
Rotary converter, 389
    direct current power system, 389
Royalties, Artic Wildlife Reserve oil exploration and, 596
RTO. *See* regional transmission operator.
RTP. *See* real time pricing.
Rules, creation of, 1193–1194
Rulesets, Energy Star Portfolio Manager software and, 575
Run around coils, 1540
Run around heat recovery systems, 1278
    air flow rate through duct, 1280–1281
    ethylyne glycol operating temperature, 1281
    heat transfer coefficients, 1279–1280
    inlet supply air temperature, 1280
    techniques of, 1278–1279
    two phase, 1281–1282
R-value, in insulation rating systems, 890

S

Sacramento Municipal Utility District Customer Advanced Technologies Progam, 515
Safe Drinking Water Act of 1974, 1557
Safety features, compressor air control systems and, 208–209
Safety
    hydrogen, 36
    nuclear energy and, 42, 1131
Satellites, 147
Saturation, degree of, 1185
Saturation pressure, 1185
Savings
    dollars vs. energy, EEMS and, 622
    stipulated vs. real, EEMS and, 622
Savings to investment ratio (SIR), 81
Scale model testing, daylighting simulation and, 268
Scheduling
    building usage and low cost energy efficient improvement case studies, 518–519
    compressor air control systems and, 210–211
Scheduling review, low cost energy efficiency improvement, 518
Scheduling tools, energy project management and, 563

Scissor trusses, 902
Scoping tools, 562–563
Scotopic CIE, 995
    in lighting design and retrofits, 995
Scotopic vision, 999
    definition of, 999
    vision sensitivity chart, 995
Scotopic/photopic vision sensitivity chart, 995
Screening treatment of wastewater, 1561
Sea level rise, global warming and, 722–723
Seal penetrations, air barriers and, 664
Sealants for air leaks, 665
Seals, 1382–1382
Seasonal energy consumption
    calculation of, 1273–1276
    energy benchmark analysis, 696
    heating demand, 1273–1276
Seasonal utility loads, 1498
    gas utilities, 1498
Seasonally differentiated rates, time of use and, 1504
SEC. *See* specific energy consumption.
Second Law of Thermodynamics, 1167, 1426–1429
    Carnot engine, 1427–1428
    Clausius statement, 1171
    exergy, 1428–1429
    formulation, 1426–1427
    heat engines and, 1171–1172
    Kelvin-Plank statement, 1170
    level of irreversibility, 1428
    reversibility, 1426, 1429
Second stage cost drivers, 1042–1043
Secondary energy supplies (SES) vs. primary, 1346
Sectoral energy conservation measures, 455
Security, lighting controls and, 990
Security costs, electricity usage and, 401–402
Security of supply, distributed generation technologies and benefits, 299–300
Sedimentation, 1557, 1558
Selective Amine Scrubbing, in syngas treatment, 908
Selenium, as syngas contaminant, 909
Self configuring nodes, 1653
Self-discipline, lean manufacturing and, 474
Sensible cooling, 1186
Sensible heat storage, 1412
Sensible heating, 1186
Sensible thermal energy, 1163
Sensor verification, ANSI/ASHRAE Standard 62.1-2004 and, 61
Sensors
    occupancy, 981, 982
    pressure, 412–413
    relative humidity, 411–412
    resistance temperature devices, 409–410
    solid-state, 410–411
    temperature, 409
    thermocouples, 411
    transmitter/transducer, 411
Sequestration, carbon, 169
Serial communications networks, 1652–1653
Series chillers, 310
Server side programs, web publishing and, 536–537
    Active Server Pages, 537

    ColdFusion, 537
    FoxWeb, 537
    hypertext preprocessor, 537
    Java Server Pages, 537
    Java Servlets, 537
    Perl, 537
Service baskets, price cap regulation and, 1248
Service charge, utility billing and, 1501
Service ratings, electric motors and, 351
Service reductions, electric demand response and, 272–273
SES. Secondary energy supplies.
Set ups, preventive maintenance and, 473
SFB. *See* stationary fluidized bed.
Shading devices for windows, 1641–1648.
    acoustic performance, 1646
    advanced, 1641
    advanced assemblies, 1642–1643
    architectural, 1642
    assemblies, 1642
        descriptions of various, 1642
    awnings, 1642
    brise-soleils, 1642
    classification, 1642
    control strategy for, 1644
    decision-making framework, 1645, 1647–1648
    heat regulation, 1641
    louvers and blinds, 1643
    materials for, 1642
    overhangs, 1642
    performance parameters, 1646
    position of, 1643
    selection of, 1644–1645
    thermal performance, 1646
    variables, 1646
    visual performance, 1646
    window treatments, 1642
    See also advanced shading devices; windows, shading devices for.
Shading from sun rays, 116
Shanghai Airport Maglev system, 1024–1025
Shared burden, electric demand response and, 273
Shared energy savings, 569, 570, 1136–1137
Shed savings, decay of, 274
Shell and tube heat exchanger, 801, 1540
Shell gasifiers, 907
Shield wires, 358
Shift work, energy benchmark analysis and, 696
Shingle warranties, radiant barriers and, 1231
Short circuit duties, electric power transmission systems, 360
Short circuits
    circuit breakers, 360
    electric power transmission systems and, 360
Shortwave radiation, 116
Shower flow heads, 1597–1599
Silicon, 1150
Simple cycle gas turbines, 298
Simple payback period, 930–931
Simplification, lean manufacturing and, 474
Single borehole, geothermal heat pump system development and, 758
Single pass cooling equipment, 1588–1589
Single phase system, 699–700
Single piece flow, lean manufacturing and, 470

Single pole direct current transmission, 359
Single unit seasonal cooling appliances, 1483
SIR. *See* savings to investment ratio.
Site design, land use impact and, 1450
Site selection
    LEED-C&S, rating system, 939
    LEED-CI, 939
Six Sigma Methods, 1310–1316
    background of, 1311
    data based decisions, 1313
    data source, 1312
    defect prevention, 1313
    DMAIC methodology, 1312, 1313–1316
    energy consumption targets, 1312
    financial considerations, 1311–1312
    mean shift, 1313
    nomenclature, 1310–1311
    overview, 1312–1313
    variation and mean shift, 1313
Skill sets, energy project management and, 564–565
Slab-on-grade floor insulation, 892
Slag wool
    effect on environment, 891
    as insulation material, 890, 893
Sliding block rates, 1503
Slow sand water filter, 1557
Sludge drying, 1527
Sludge processing, 1554
Smart air syndrome, 844
Smart sensors, automobile technology evolution and, 674–675. *See also* sensors.
Social capital, 1410
Society, energy use and, 1016
SOFC. *See* solid oxide fuel cells.
Soffit air ventilation, 902
    raised heel trusses, 902
    raised top plate, 902
Soft coal, 156
Software, energy accounting tools and, 3
    Envision, 3
    Fraser, 3
    Global Mvo Asset Manager, 3
    Meter Manager, 3
    Utility Manager, 3
Software, energy simulator, 111
Sofware, client and server, building automation systems and, 106–107
Solar air conditioning, 1326–1327, 1457
    feasibility of, 1547
Solar cooking, 1325–1327
    box, 1326
    panel, 1326
    parabolic, 1326
Solar cooling, 1326–1327
Solar crop drying, 1325
Solar distillation, 1325
Solar energy, 34, 1204
    future of, 34
    See also solar thermal technologies.
Solar heat, effect on cooling equipment, 1618
Solar heat gain coefficient, 1623
Solar heating, 1364
    air conditioning and, case study of, 1317–1320
    cooling system methodology, 1317–1318, 1618

    features of, 1320
    heat gain coefficient, 1623
    industrial process heat, 1267
    passive, windows and, 1618–1619
    psychrometric analysis, 1318
    solar roof tile system, 1318–1319
    space heating
        active, 1322
        hybrid, 1322
        passive, 1322
    See also solar thermal applications; solar water heating.
Solar heating, passive, windows and, 1618–1619
Solar industrial process heat, 1267
Solar ponds, 1324–1325
Solar radiant heat gain, in windows, 1623
Solar radiation, 117
Solar related energy, 1090
Solar resources
    definitions, 1148
    irradiation, 1148
    photovoltaic (PV) systems and, 1148
Solar roof tile system, 1318–1319
Solar space heating
    active, 1322
    hybrid, 1322
    passive, 1322
Solar spectrum
    illumination basics and, 265
    photovoltaic (PV) systems and, 1150–1151
Solar thermal applications, 809
    cooking, 809
    distillation, 809
Solar thermal technologies, 1321–1330
    cooking, 1325–1327
    cooling, 1326
    crop drying, 1325
    distillation, 1325
    ponds, 1324–1325
    power generation, 1327–1328
        central receiver, 1328
        parabolic dish, 1328
        parabolic trough, 1328
    present use, 1329–1330
    solar space heating, 1321–1322
    water heating, 1322–1323
Solar water heating, 1267, 1322–1323, 1331–1338
    applications of, 1331
    collector performance, 1336
    current use of, 1338
    design simulation, 1336
    design transmission-absorption product, 1336
    designs, 1332–1333
        built in storage, 1332–1333
        forced circulation, 1333
        thermosyphon, 1332
    economics of, 1336–1337
    fin efficiency, 1336
    forced circulation, 1333
    heat removal factor, 1336
    Hottel Whillier Bliss equation, 1335–1336
    life cycle assessment, 1337–1338
    system design, 1335–1336
        fin efficiency, 1336
        heat removal factor, 1336

        Hottel Whillier Bliss equation, 1335–1336
        measured collector performance, 1336
        simulation, 1336
        transmission-absorption product, 1336
    thermal collectors, 1333–1335
    thermosyphon, 1332
    workings of, 1331
Solar-control film
    analysis study
        DOE-2 energy study, 1631
        kilowatt-hour usages, 1634
        motivation for, 1631
        results of, 1632–1635
            kilowatt-hour usages, 1634
            simple payback, 1633
            summer peak hour demand, 1635
        scope of, 1631–1632
        solar performance factors, 1632
        summer peak hour demand, 1635
        thermal performance factors, 1631
    appearance of, 1630
    benefits from, 1630
    construction of, 1630
    energy savings from, 1630
    installation process, 1630
    properties of, 1630
Solid fuel boilers, 1527
Solid fuels, 1340–1341
    biomass fuels, 482–483
    CO_2 emissions, 484
    coal composition, 481
    flame temperatures, 484
    flue gas volume, 485
    heating value, 479, 484
    properties, 479–485, 1341–1344
    proximate analysis, 479, 480
    ultimate/elemental analysis, 479
Solid oxide fuel cells (SOFC), 729–732
    actual use of, 731
    applications, 731–732
    contamination levels, 730
    principles of, 729–730
    problems with, 730–731
Solid waste, 1340–1341
    properties of, 1341–1344
Solid waste to energy, advanced thermal technologies (SWEATT), 1340–1354
Solid-state resistance temperature devices, 410–411
Solution heat treatment, 868
Sonic limitations, heat pipes and, 805, 806
Sorting, lean manufacturing and, 474
South African, Fischer-Tropsch process and, 163
Southern California Edison Energy Centers, 515
Southern California Gas Company Energy Resource Center, 515
Space heater, 648
Space heating, 583, 1357–1365
    calculation methodology, 1363–1364
        ASHRAE standards, 1363
    current trends, 1364–1365
        geothermal energy, 1364
        HVAC controllers, 1365
        price of fuel, 1364
        solar heating, 1364
        window technology, 1364–1365

Finnish Masonry stove, 1359–1360
Franklin stove, 1359
fuels, 1361–1362
habitat structures, 1362–1363
hearths and fireplaces, 1358–1360
history of, 1357–1358
    Romans, 1357–1358
insulation development, 1363
pressure vessel development, 1360–1361
system types, 1361
    hot water, 1361
    steam, 1361
See also solar space heating.
Spatial trading rules, emissions and, 432–433
Specialty rates, 1505–1506
Specific emergy, 421
Specific energy, 420
Specific energy consumption (SEC), 4
    disadvantages of, 4
Specification standard, 1030
Spectral distribution of radiant flux, in window energy performance, 1616
Spectral irradiance, in window energy performance, 1616
Spectrally selective window films
    aesthetic differences from conventional, 1638–1639
    applicability on different glasses, 1638
    case study, Stamford University, 1640
    conventionally applied, 1637–1639
    definition of, 1637–1638
    effectiveness guarantees, 1639
    heat block, 1638
    mitigating heat loss, 1638
    payback, 1639
    price comparison vs. conventionally applied, 1638
    special care needs, 1638
Spent nuclear fuel management, 42–43
Spikes, energy benchmark analysis and, 696
Spinning reserve concept, 1442–1443
Spread spectrum radios, 1653–1654
Square footage, energy usage vs., 583
Stability disturbances, electric power transmission systems and, 360
Stack economizer
    installation of, 1368
    savings from, 1368
Stakeholders, independent power producers and, 855–858
Stand alone photovoltaic systems, 1153
Standard connectors, automobile technology evolution and, 675
Standard of living, energy use modifications and, 1015
Standardization, lean manufacturing and, 474
Standardized commissioning of buildings, 678–679
Standards 90.1 and 90.2, 439
    review and revision process, 439–440
    timing of, 440
Standards development, geothermal energy environmental impact and, 747
Standards, impose on facility energy control suppliers, 677
Standby electric rate schedule, 1505, 1509
    backup, 1509
    maintenance, 1509
    supplementary rates, 1509
Standby rates, 1505
Standing devices, building system simulation process and, 120
Start/stop sequence, compressor air control systems and, 210
State adoption of energy codes, 440–441
State agencies, energy efficiency information sources and, 511–513
    California Energy Commission, 511
    Energy Center of Wisconsin, 512
    Florida Solar Energy Center, 512
    Iowa Energy Center, 512
    New York State Energy Research and Development Authority, 512
Static data, 691
Stationary fluidized bed (SFB), 88
Steady state, stead flow process, 748
Steady-state CO_2 differentials, 52
Steam
    density compensation of, 1057–1059
    dry, 745
    wet, 745
Steam blowthrough rate, cooker kettles and, 1373
Steam cycles, 1381–1382
    Carnot, 1381
    Rankine, 1381
    reheat rankine, 1381
    two phase working fluid, 1382
Steam generation cost, 1366–1367
Steam generators, 1589
Steam heating systems, 1361
    energy intensive, 463
Steam vs. hot water, combined heat and power, 176
Steam line, isolation of abandoned ones, 1357
Steam methane reforming, 789–790
    hydrogen vehicles
        emissions from, 790
        high-temperature nuclear reactor, 790
Steam path, 1382–1383
    blades and seals, 1382–1383
    high pressure cylinders, 1382
    intermediate pressure, 1382
    losses, 1383
    low pressure, 1382
Steam recovery, waste heat, 1540
Steam System Assessment Tool 1.0.0, 880
Steam system optimization
    abandoned line isolation, 1375
    blowdown heat recovery equipment replacement, 1368
    case study, 1366–1379
    condensate return improvement, 1375–1378
    condensate return system, 1367–1368
    measurement and verification protocol, 1378–1379
    product cooling heat recovery, 1373–1375
    relief control valves, 1371
    replacement of deaerator steam jet ejectors, 1370
    stack economizer installation, 1368
    steam blowthrough rate on cooker kettles, 1373
    steam generation cost, 1366–67
    steam trap, 1367, 1371–1373
    steam utilization, 1367
Steam System Scoping Tool 1.0d, 880
Steam trap, 1371–1373
Steam trapping, 1367
Steam turbines, 1380–1388
    blade materials, 1385
    control, 1386
    cycles, 1381–1382
    future development of, 1387–1388
    history of, 1380–1381
    mechanical design, 1383–1386
    mechanical design, stresses on, 1383–1384
    sizes, 1386
    steam path, 1382
    types, 1386
        combined cycle gas turbine, 1386
        congeneration, 1386
Steam utilization, 1367
Steam-methane reforming, 1442
Steel framed walls, 1516
Steel industry use of coal, 160–161
Steel plant assessments, Dept. of Energy best practices case studies and, 1285, 1299–1301
Step level HID controls, 982
    lighting control systems, 982
STIG cycle, 1582–1583
Stoker firing, 488–489
Storage
    compressed air systems and, 228–233
    electricity and, 389–390
    liquefied natural gas and, 1002–1004
    nuclear energy waste and, 43
Storage capacity, cool storage system design and, 1418–1419
Storage delta, compressed air systems and, 230–231
Storage of hydrogen, 36
Storage vessels, compressed air energy storage and, 217
Storing compressed air, compressor failure, 242–243
Straight pipe, flow measurement problems and, 1056–1057
Stratified chilled water storage, 1417–1418
Straw bale walls, 1518
Street lighting electric rate schedule, 1508
Structural insulated panel walls, 1518
Subbituminous coal, 156
Submetering primer, 1649–1650
Substations, 358–359
Substations
    bus/circuit breaker arrangements, 359
    equipment make-up, 359
    functions of, 358–359
Sulfur iodine (S-I) cycle, 1444–1445
Sulfur iodine process, 1443–1444
Sulfur, in syngas, 908
Sun, illumination basics and, 264–265
Sun rays, shading from, 116
Sunbelt cities
    energy efficiency programs, 1391–1392
    energy polices, 1394
    Energy Star partnership, 1393
    government support of sustainability policies, 1392–1393

sustainability and implementation of, 1393–1394
sustainability polices and, 1389–1395
Sunlight, direct beam entry, 1619
Sunlight concentration, photovoltaic (PV) systems and, 1151
Superconductor magnet maglev system, 1024
Supercritical water-cooled reactor, 41
Supplementary electric rate schedule, 1509
Supply air temperature, increasing of, 276
Surface coal mining, 144, 146
Surface convection, 116
Surface mining methods, 158–159
Surface Water Treatment rule, 1557
Surge controls, compressor air control systems and, 210
Suspension firing, 488
Sustainability, energy efficient buildings and, 526
Sustainability policies, sunbelt cities and, 1389–1395
  energy efficiency programs, 1391–1392
  Energy Star partner, 1393
  implementation studies, 1393–1394
  local government support, 1392–1393
  significance of, 1390
  sustainability as a goal, 1390–1391
  urban, 1389–1390
Sustainable building simulation, 1396–1403
  available software, 1401–1404
    APACHE HVAC, 1401
    APACHE, 1401
    ArchiPhysics-Solar, 1402
    BEA, 1401
    BLAST, 1401
    BREEZE, 1403
    BSim2002, 1401
    BUS++, 1401
    Daylight, 1403
    DOE-2, 1402
    DUCTSIZE, 1402
    EE4 CODE, 1402
    EED, 1402
    EnergyPlus, 1402
    ESP-r, 1402
    EZDOE, 1402
    FLUENT, 1402
    HVAC solution, 1402
    Hydronics Design Studio, 1402
    MOIST, 1403
    Right Suite Residential, 1402
    TRACE 700, 1402
    TRNSYS, 1402
    VisualDOE, 1402
  available software for green buildings, 1403
  components of, 1400
  energy aspects, 1397–1398
  exergy aspects, 1397–1398
  integrated building design system, 1401
  need for, 1396–1397
  reason for, 1398–1400
Sustainable development, 1406–1410
  action being taken, 1407
  ASHRAE GreenGuide, 1407–1408
  community development, 1409
  community planning and design, 1409
  conservation economy implementation, 1409–1410
  cooperation in implementation and use, 1407–1408
  definition of, 1406
  district heating and cooling and, 321–322
  educating the population, 1408
    U.S. Partnership for The Decade of Education for, 1408
    United Nations efforts, 1408
  end of petroleum fuels, 1407
  energy conservation and, 451–453
  environmental policy instruments and, 631
  future energy types, 1407
  green energy and, 776–777
    case study, 781–786
  information sources, 1409
    conservation economy, 1409
    economic capital, 1410
    natural capital, 1410
    social capital, 1410
  reasons for, 1407
  recycling products, 1409
  renewable energy resources, 453
  reporting of, 1408
  use in developing countries, 1410
  workable technologies needed, 453
Sustainable sites, LEED-NC, rating system, 949
Sustained changes, energy benchmark analysis and, 696–697
SWEATT
  coal vs., 1344
  recycling, 1353–1354
  See solid waste to energy by advanced thermal technologies.
Sweeping, lean manufacturing and, 474
Switch mode power converters, 1154
Switching, dimming vs., lighting controls and, 989
Synchronous generator vs. induction, combined heat and power, 175–176
Synchronous operation, 359–360
  electric power transmission systems and, 359–360
Syncrude, 594
Syngas compositions, 907
Syngas contaminants, 908
  arsenic, 909
  cadmium, 909
  coal derived, 908
  mercury, 909
  metal carbonyls, 909
  selenium, 909
Syngas feedstocks, 907
  ash content of, 907
  ash fusion temperature of, 907
  characteristics of, 907
  free swelling index, 908
  moisture content of, 907
  particle size of, 908
  reactivity of, 907
Syngas treatment, 908–909
  acid gas removal, 908
  Rectisol, 908
  Selective Amine Scrubbing, 908
  sulfur, 908
System benefits charge, 1260–1261
System controllers, compressor air control systems and, 210
System dimmers, lighting control systems, 982
System modules, automobile technology evolution and, 674
System start-up guidelines, ANSI/ASHRAE Standard 62.1–2004 and, 61

T

TAB. See testing, adjusting and balancing.
TAC. Task/ambient conditioning.
TACAS. See thermal compressed air energy storage.
Tar sand extraction, 594
  disadvantages of, 594
  Syncrude, 594
Target Finder, 577–578
Targeting reductions in energy use, 6–7
Tariff filings, electricity and, 380
Tariff review, energy assessments and, 64
Task/ambient conditioning (TAC), 1463–1464
  technology description, 1464
Tax disincentives, Energy Tax Act of 1978, 1084–1085
Tax incentives, Energy Tax Act of 1978 and, 1084
Taxes, 959, 1257
  emissions, 630
  fees, utility billing, 1502
  pollution, 630
Team planning, energy master planning and, 553
Technical information, user friendly energy audit report suggested format, 78
Technical skills, energy project management and, 564
Technological growth, Public Utility Regulatory Policies Act (PURPA) of 1978 and, 1203
Technology assessment, geothermal energy environmental impact and, 747
Technology requirements, industrial processes and, 459
Technology transfer
  feasibility studies, 502–503
    macro indicators, 502–503
    micro indicators, 502–503
  geothermal energy environmental impact and, 747
  implementation of energy efficiency, 502–502
Temperature based utility pricing, 1506
Temperature controller, 414
Temperature measurement
  accuracy of, 1059
  flow measurement and, 1059
Temperature multiplexing, reduce cost, 1059
Temperature sensors, 409
Temperature, process energy optimization and, 1304
Temporal trading rules, 432–433
Terminals, 357
Terrestrial carbon sinks, 129
Tesla, Nicola, 389
Test and balance, facility guidelines for new structures and, 527

Testing, adjusting and balancing (TAB), 196
Texas coal production, 159
Thermal collectors
   evacuated tube, 1334–1335
   flat plate, 1333–1334
   solar water heating and, 1333–1335
Thermal comfort, underfloor air distribution (UFAD) and, 1467
Thermal compressed air energy storage (TACAS), 214–215, 216
   adaptations of, 217–218
Thermal conductivity
   heat pipes and, 805
   heat transfer and, 824–827
Thermal cut-off, electronic ballast primer and, 834
Thermal design
   Carnot Law, 1541
   fundamentals of, 1541–1542
Thermal discomfort, underfloor air distribution and, 1468
Thermal efficiency, turbine inlet air cooling and, 1546
Thermal energy storage (TES), 1412–1421
   cooling applications, 1413–1418
      benefits of, 1413
      chilled water, 1417–1418
      commercial utility rates, 1413
      ice storage, 1414–1417
      storage equipment, 1414–1418
      storage system design, 1418–1420
   off-peak heat storage, 1420
   types, 1412–1413
      latent heat storage, 1413
      sensible heat storage, 1412
   underground, 1420
Thermal energy, 1163
Thermal imaging, data collection and, 259–260
Thermal load following cycle, 1530
Thermal mass of walls, 1513
Thermal mass, geothermal heat pump system development and, 757–758
Thermal radiation, 1161
Thermal storage system, 648
Thermal technologies
   advanced, 1347
   solid waste to energy and, 1340–1354
Thermochemical conversion, 89–90
   gasification, 89–90
   pyrolysis, 89
   technology comparison, 90
Thermochemical cracking of water, 1443–1445
   calcium bromide cycles, 1445
   future of, 1445
   sulfur iodine (S-I) cycle, 1444–1445
   sulfur iodine process, 1443–1444
   temperature reduction, 1445
Thermocouples, 411
Thermodynamic properties, calculation of, 1429–1431
   ideal gas, 1430
   phase equilibrium, 1431
   real fluids, 1430–1431
   rigorous equations, 1429–1430
Thermodynamic temperature, 1171
Thermodynamic wet bulb temperature, 1184
Thermodynamics, 1422–1432
   first law, 1423
   nomenclature, 1422
   second law, 1426–1429
Thermoelectric heat pump, 816
Thermo-fluid dynamic efficiency, cooling towers and, 250
Thermostat cases, review if needed, 520–521
Thermostat climate control, mobile, 1661–1663
Thermostat location, 520
Thermostats, mobile HVAC systems and, 1080
Thermosyphon solar water heater, 1332
Thin film photovoltaic (PV) systems, 1149–1150
   organic materials, 1150
   silicon, 1150
Third party financing (TPF), 567–568
3-E trilemma, 1147–1148
Three phase system, 699–700
Three phase transmission lines, 357–358
   terminals, 357
   types, 357–358
      high voltage alternating current, 357
      high voltage direct current lines, 357
Throttle, 1544
Throughput, 1652
Thyristors, 359
Tight buildings, ventilation problems and, 20–21
Time clocks, HVAC system at night, 518
Time domain, 117
Time intervals
   energy data structuring and, 619
   normalized, 619
   optimal, 619
Time of use (TOU) pricing, 1175–1177
   examples of, 1176–1177
   observations from, 1176
   rate development, 1177
Time of use rates, 1504
   off season, 1504
   seasonally differentiated, 1504
Time value of money, 930, 968
Timeline, energy benchmark preparation and, 692
Toilets, 1599–1601
Tools
   energy accounting and, 2–3
   software, 3
   energy project management and, 562–564
Top down approach, energy efficient buildings and, 525
   integrated design, 525–526
   sustainability, 526
Topping cycles, 1542
Total owning costs, 968
Total quality energy management, 465
TOU electric rate schedule, 1508
TOU large power electric rate schedule, 1508
TOU. *See* time of use.
Toyota, 467, 468
TPF. *See* third party financing.
TRACE 700 software, 1402
Tradable certificates for energy savings, 1433–1439
   characteristics of, 1435–1436
   implementation and results, 1436–1439
   policies, 1433–1435
Tradable emissions permits, 630
Trade-ally cooperation, energy demand-side management programs and, 290
Trading of electricity, difficulties in, 381–382
Traditional manufacturing, lean vs., 468–469
Traffic woes, U.S. transportation energy use and, 588
Training, 1029–1030
   lean manufacturing and, 470–471
Transducer sensor, 411
Transducer, 416
Transformer ownership, 1511
Transformer ratings, 705
Transformity, 420
   quality and, emergy and, 421
   specific and, emergy, 421
Transit service, land use impact and, 1450
Transmission absorption product, 1336
Transmission capacity, 362
Transmission components
   cables, 358
   lines, 357–358
   ratings, 358
   substations, 358–359
Transmission grid upgrading, energy engineering and, 136
Transmission lines, 357–358
   three phase, 357–358
Transmission rates, electric rate schedule, 1508
Transmittance of windows, 1623–1624
   damage-weighted transmittance, 1623–1624
   UV transmittance, 1623–1624
   visible transmittance, 1623–1624
Transmitter sensor, 411
Transparent surfaces, building system simulation process and, 120
Transport of air pollution, 718–721
Transportation energy use, U.S., 588–599
Transportation issues, coal to liquid fuels and, 168–169
Transportation systems, heat pipe applications and, 811
Transportation, coal and, 159, 161
Transportation, land use impact on, 1449–1451
Transportation, liquefied natural gas and, 1006–1008
Transportation, location efficiency, 1449–1453
Trash collection, 1530
Trends, interactive access to data and, 620–621
Tribal communities
   cultural awareness, 1458
   energy audits
      coordination, 1459
      funding, 1460
      geographic location, 1459
      information, 1460
      teamwork needed, 1461
      technical issues, 1459–1460
   governments, 1457–1458
   long term orientation, 1459
   sense of community, 1458–1459
Tribal lands, energy efficiency and, 1456–1462
TRNSYS software, 1402
True up, emissions trading and, 432
Tube-side heat transfer calculations, 1063
Turbine control, 1386
Turbine generators, 392

Index

Turbine inlet air cooling, 1546
   thermal efficiency, 1546
Turbines
   gas, 298
   micro type, 298
Turkey, 503–504
Two phase heat exchangers
   analysis of, 803–804
     condensers and evaporators, 803–804
       enthalpy calculations, 803
       maintenance of, 804
       materials used in, 804
       safe operation of, 804
Two phase run around heat recovery system, 1281–1282
Two phase working fluid, 1382
Two position controller, 415
TXV system, OT vs., mobile HVAC systems and, 1081

U

U Factor insulation, 1363
U.N. Framework Convention on Climate Change, 139
U.S. energy supplies, 1344, 1346–1347
U.S. energy use
   electricity generation, 598–599
     nuclear power, 598–599
     power grid concerns, 599
     solutions, 599
       energy usage reduction, 599
U.S. overview of energy use, 580–586
U.S. Partnership for The Decade of Education for Sustainable Development, 1408
U.S. transportation energy use, 588–599
   airport traffic woes, 588
   Artic Wildlife Reserve, 595
   cheap energy, effect of, 588, 589
   consumption, 595
   conversion to mass transportation, 599
   Federal funding of airports and roads, 591
   higher energy costs, 590
   housing growth, 590
   impact of China's economic growth, 592
   impact of India's economic growth, 592
   infrastructure strain, 590
   lack of coordination, 588–589
   lack of vision, 598
   land development policies, 588
   new technologies, 591
     Maglev system, 591
   oil company nationalization, 595
   oil consumed, 590
   oil in ground, 593
   oil reserve uncertainty, 592
   oil shale deposits, 594
   oil tankers, impact of, 591–592
   over reliance of oil, 596–597
   politics and, 594–595
   railroads, 590–591
   railway systems, ease of use, 591
   reacting to, 598
   rising prices and action taken, 595
   tar sand extraction, 594
   traffic woes, 588
   vulnerabilities of oil production, 592–593
UFAD. *See* underfloor air distribution.
Ultimate/elemental analysis, solid fuel properties and, 479
Ultrasonic leak detection, data collection and, 261–262
Ultraviolet light, 1559
Unbalanced system, 700–701
Underfloor air distribution (UFAD), 1463–1469
   benefits of, 1467
     air movement, 1467
     energy savings, 1467
     thermal comfort, 1467
     worker satisfaction, 1467
   cost effectiveness, 1468–1469
     churn factor, 1469
     operations and maintenance, 1469
     productivity and health, 1469
   history of, 1463
   limitations of, 1467–1468
     applicability of, 1468
     higher energy use, 1468
     minimal information, 1468
     plenum leakage, 1468
     thermal discomfort, 1468
     unfamiliar technology, 1468
   principles of, 1464–1467
     task/ambient conditioning, 1463–1464
   technology description, 1464
Underground coal mining, 144, 158
Underground thermal energy storage, 1420
Uninterruptible power supply (UPS), 214–215
   thermal compressed air energy storage, 214–215
Unit emergy values, 420–424
   common products, 426
   computing of, 422–423
   emergy per unit money, 420
   empower, 421
   non renewable energy sources, 425
   specific energy, 420
   transformity, 420
Unit of trade
   allowance, 432
   emissions and, 432
   emissions reduction credit, 432
   permit, 432
United Nations, sustainable development education and, 1408
United States coal production, 143–154
United States coal supply, 156–161
   coal to liquid fuels and, 168
United States Government Agencies, energy efficiency information sources and, 508–511
   Dept. of Energy, 508–510
   Environmental Protection Agency, 510
   Dept. of Commerce, 510–511
   Department of Defense, 511
United States
   alternate energy credit (AEC), 14
   electricity generation and, 581–582
   energy efficiency regulation state by state, 12–13
   ESCO and savings model used, 570
   renewable energy credits (REC), 14
Universal controller, 415
University programs, energy efficiency information sources and, 511–513
   Rensselaer Polytechnic Institute, 512
   Washington State University, 512–513
Uranium, 1112–1113
   availability of, 41
   conversion and enriching, 1113
   fabrication for nuclear fuel, 1113
   history of, 1112
   mining and milling of, 1112–1113
Urban sustainability policies, 1389–1390
URC. *See* utility report card.
Urinals, 1601–1603
Usable compressed air energy in storage, 230–233
   calculating, 231
User friendly energy audit report, 76–80
   assumptions used, 77–78
   customer feedback, 79
   definition of, 76
   short form type report, 79
   suggested format, 78
     appendix, 79
     assumptions and calculations, 78
     energy management plan, 78
     executive summary, 78
     report recommendations, 78–79
     technical supplement, 78
   writing of, 76–78
User-specific interactive access, 622
USGBC, LEED and, 945–946
Utility based energy service company, 307
Utilities
   electricity deregulation and
     challenges of, 385–386
     non utility generators, 385–386
     variation in rules, 385–386
   industry consolidation of, 392
   investor owned, 379–380
   long term planning, 1491–1495
     alternative resources, 1494–1495
     benefits of, 1492–1492
     demand forecasting, 1493–1494
     effective management of, 1493
     existing resources, 1493–1494
     optimal mix, 1495
     responsible parties for, 1492–1493
   profitability, 392
   role in society, 1491–1492
Utility bill
   analysis, 1471–1483
   calculating of, 1471–1472
   electric, calculating of, 1472–1473
   gas, calculating of, 1472
   weighted average cost of power, 1473–1478
     differing load profiles, 1478–1482
Utility billing factors
   adjustment charges, 1502
   basic service charge, 1501
   commodity charge, 1501
   demand charge, 1501
   energy adjustment charge, 1502
   minimum charge, 1501
   peak usage, 1501

ratchet adjustments, 1501–1502
taxes and fees, 1502
Utility bills, 691
Utility energy centers, energy efficiency information sources and, 514–515
California Statewide Emerging Technology Program, 514
Energy Design Resources, 514
Lighting Design Lab, 514–515
Sacramento Municipal Utility District, 515
Southern California Edison Energy Centers, 515
Southern California Gas Company Energy Resource Center, 515
Utility issues, hybrid electric vehicles, plug in type and, 852
Utility loads
baseloads, 1498
seasonal, 1498
types, 1497
off-peak, 1497–1498
peak, 1497–1498
Utility manager software, 3
Utility organizations, energy efficiency information sources and, 514–515
Alliance to Save Energy, 513
American Council for an Energy Efficient Economy, 513
Consortium for Energy Efficiency, 513
E Source Companies LLC, 513
Utility rate structures, 1497–1512
baseloads and, 1498
billing factors, 1501–1502
block rates, 1503
commodity function, distribution function vs., 1511–1512
competitive, 1506
cost allocation, 1500
demand/commodity, 1503–1504
deregulation, 1511–1512
electric rate, 1507–1510
end use, 1504
load factor, 1498–1500
measuring electric demand, 1510
metering point, 1511
natural gas, 1507
negotiated, 1505–1506
non firm service, 1504–1505
power factor, demand billing, 1510–1511
rate design strategies, 1502–1503
rate riders, 1507
real time pricing, 1506–1507
specialty, 1505–1506
time of use, 1504
transformer ownership, 1511
types of loads, 1497
usage profile, 1498–1500
wholesale market, 1506
Utility rates
blended, 967
commercial, 1413
Utility report cards (URC) program, 539–540
Utilization factor adjustments, energy balance and, 688
UV transmittance, of windows, 1623–1624

V

Vacuum systems, energy intensive, 463–464
Value added electricity services, IntelliGrid^{SM} and, 918
Value chain management, 1137–1138
Vapor compression heat pump, 817–818
Vapor retarders, 662
Vaporization, liquefied natural gas and, 1004–1006
Variable air volume (VAV), 51
Variable air volume control strategies, 58
airflow sensors and, 58
Variable air volume design types, 52
Variable air volume sensors, use savings, 59
Variable air volume supply, ventilation rate procedure and, 58
Variable air volume
airflow measurement devices, importance of, 52
steady-state CO_2 differentials, 52
Variable cost calculations, 958
Variable discount rates, 935
Variable flow
chiller capacity, 311–312
district cooling systems and, 310–311
Variable frequency, 705–706
Variable load applications, 867
Variable relationships, 1315–1316
Variable speed devices (VSD)
advantages of, 867
other applications of, 867
uses, 867
variable load applications, 867
Variable speed drives, 705–706, 862, 864
Variations, Six Sigma Methods and, 1313
VAST cycle, 1583
VASTIG cycle, 1583–1584
VAV. See variable air volume.
Vehicle air emission standards, 717
Vent selection, in attic insulation, 899–900
Ventilation problems, indoor air quality and dilution, 19–20
increased outside air, 20
pollution sources, 20–21
relative humidity, 19
tight buildings, 20–21
Ventilation rate procedure (VRP), ANSI/ASHRAE Standard 62.1–2004 compliance, 54, 55–60
advanced VAV control strategies, 58
breathing zone outdoor airflow, 57
calculations for, 56
Carbon dioxide levels, 56–57
demand control ventilation, 56
multiple zone recirculating system, 57
variable air volume supply, 58
Ventilation system requirements, ANSI/ASHRAE Standard 62.1–2004 and, 51–52
Ventilation, 839
advantages of, 21
indoor air quality and, 18–19
problems with, 19–20
standards of, 19
Verification of emissions, 431–432

Verification protocol, steam system optimization and, 1378–1379
Vertical stacking fan coil HVAC unit, 840
Visible transmittance, of windows, 1623–1624
Visual factory, lean manufacturing and, 473
VisualDOE software, 1402
VOC emission, printing inks and, 891
VOC. See volatile organic compounds.
Volatile organic compounds (VOC), 20
Volatile oxidation, 486–487
Volt amperes reactive, 1510
Voltage controls, lighting and, 989–990
Voltage levels, 702
Voltage, 255–256, 831
Voluntary energy efficiency programs, 442–443
ENERGY STAR, 442
government and private organizations, 442
home energy rating system, 442
VRP. See ventilation rate procedure.
VSD. See variable speed devices.

W

Walkthrough energy assessment, 64
Wall construction, 2×6, R values, 899
Wall insulation,
2×4, 896
problems and solutions, 897
2×6, 899
batts in, 896–897
concrete, 892
block cores, 892
strategies for use, 892
Wallbox dimmers, lighting control systems, 982
Walls and windows, building types, 1513
Walls, 1513–1518
exterior insulation finish systems, 1518
straw bale walls, 1518
structural insulated panels, 1518
thermal characteristics, payback of, 1521
thermal mass of, 1513
types, 1515–1518
masonry, 1516–1518
steel framed, 1516
wood framed, 1515–1516
Washington State University energy programs, 512–513
Waste disposal, nuclear energy and, 1103–1104
Waste energy emissions, exergy and, 659–660
Waste fuel suppliers, commercial, 1530
Waste fuel technologies, 1527–1530
new equipment, 1528–1530
anaerobic digesters, 1529–1530
bubbling fluid bed boiler, 1528
circulating fluid bed boiler, 1528
gasifiers, 1528–1529
retrofit applications, 1527
burner/engine, 1527
solid fuel boilers, 1527
waste heat recovery steam generators, 1527
Waste fuels project funding for, cogeneration, 1531
Waste fuels, 1523–1534
alternative types, 1523–1525
anaerobic digestion, 1527
combustion processes, 1525

Index

disposal of, 1525
economics of, 1530–1533
    combined heat and power efficiencies, 1530
    disposal cost reduction, 1530
    opportunity fuels, 1530
    reduced cost alternates, 1530
        commercial waste fuel suppliers, 1530
        recycling centers, 1530
environmental considerations, 1525, 1527
equipment and maintenance average costs, 1532
handling of, 1525
non combustion processes, 1525, 1526
opportunity types, 1523–1525, 1526
project funding for, 1531–1534
    biomass programs, 1531
    greenhouse gas emission reduction, 1531, 1532
    renewable energy resources, 1531
Waste heat recovery steam generators, 1527
Waste heat recovery, 1536–1540
    absorption heat pumps, 1541–1547
    absorption refrigeration, 1544–1546
    building cooling applications, 1545–1546
    combined heat and power applications, 864–867, 1546–1547
    design sample calculations, 1538–1540
    engineering considerations, 1537–1538
    equipment, 1540
    feasibility of, 1536–1537
    industrial plant savings and, 864–867
    quality of, 1537–1537
    solar air conditioning, 1547
    turbine inlet air cooling, 1546
Waste heat steam recovery, 1540
Waste management of nuclear energy, 42–43
Waste management planning, 1593
Waste product disposal, nuclear energy and, 41
Waste products
    coal to liquid fuel plants and, 169–170
    energy form, 1091
    heating values, 1345
Waste summaries review, energy assessments and, 64
Waste to energy, chemical process industry and, 1305
Wastewater plants
    energy efficiency measures, 1551–1554
        adjustable speed drive pumps, 1553–1554
        dissolved oxygen mechanics, 1551–1553
        methane gas recovery, 1554
        renewable sources, 1554
        sludge processing, 1554
    energy savings in, 1548–1555
    energy usage, 1550–1551
        by amount, 1551
    goals of, 1548
    operations of, 1548–1549
    types, 1549–1550
        course bubble diffusion, 1549
        fine bubble diffusion, 1549
        mechanical aeration, 1549
        pure oxygen systems, 1549

Wastewater treatment plant (WWTP), case study of, 15
Wastewater treatment plant, carbon dioxide credits for, 15–17
    reduction in, 15–17
Wastewater, 1560–1562
    contemporary issues, 1562
    disposal of, 1561–1562
    history of, 1560
        cholera, 1560
    industrial waste, 1562
    sources and collection of, 1560–1561
    standards and monitoring, 1560
    treatment of, 1561
        carbon filtration, 1561
        gravity, 1561
        reactors, 1561
        screening, 1561
Water augmented gas turbine power cycles, 1574–1585
    air humidification, 1579, 1581–1582
    analyzing cycles, 1575
    Carnot efficiency, 1574–1575
    Combined, 1578–1579
        heat and power, 1584–1585
    combustor, 1582–1583
    compressor surge, 1576
    compressor train, 1579
    conventional Brayton cycles, 1576–1577
    design parameters, 1576
    modeling, 1575
    software for simulation and analyzing, 1575
Water conservation techniques, 1604–1605
Water consumption tracking, Energy Star Portfolio Manager software and, 578
Water cooled chiller efficiency, 309
Water electrolysis, 1442
Water element, chemical process industry and, 1305
Water injection, quasi isothermal compression, 1579
Water neutrality, coal to liquid fuel water quality and, 169
Water pollutants, geothermal energy and, 746
Water quality, coal to liquid fuels and, 169
Water source heat pump for modular building, case study, 1564–1573
Water treatment process, 1557–1559
    coagulation, 1557, 1558
    disinfection, 1557, 1559
    filtration, 1557, 1558–1559
    flocculation, 1557, 1558
    sedimentation, 1557, 1558
Water treatment
    history of, 1556–1557
        slow sand filter, 1557
    importance of, 1557
Water use efficiency
    LEED-C&S, rating system, 943
    LEED-CI, rating system, 939
    LEED-NC, rating system, 950
Water using equipment, 1587–1595
    boilers and steam generators, 1589
    clothes washers, 1588, 1603–1604
    conservation techniques, 1604–1605
    cooling towers, 1589–1590
    dishwashers, 1587–1588

    dishwashers, 1603
    faucets, 1596–1597
    landscape irrigation, 1590–1592
    leak detection, 1592–1593
    showers, 1597–1599
    single pass cooling equipment, 1588–1589
    toilets, 1599–1601
    urinals, 1601–1603
    waste management planning, 1593
Water, 1556–1560
    environmental policy and, 627
    history of, cholera, 1557
    hydraulics, 1559–1560
    runoff, 1556
    sources, 1557
    standards and monitoring, 1557
        coliform test, 1557
        Safe Drinking Water Act of 1974, 1557
        Surface Water Treatment Rule, 1557
    transportation of, 1557
Wave energy, 1267
Wealth generation, energy's importance in, 444
Weather data, building system simulation process and, 119
Weather normalization, Energy Star Portfolio Manager software and, 576
Web publishing
    available tools, 538
        batch mail applications, 538
        charting tools, 538
        databases, 538
    energy information system processes and, 535–536
    purchase or self design, 538
    programming choices, 536–538
        client side programs, 537–538
        common gateway interface, 536
        server side programs, 536–537
Web-based compressor management, 212–213
    benefits of, 213
        air leakage measurement, 213
        cost metering, 213
        internet-based access, 213
        preventive service monitoring, 213
        real-time, data driven feedback, 213
        system efficiencies, 213
Weekend energy consumption, energy benchmark analysis and, 696
Weight reduction, aircraft fuel consumption and, 26–27
Weighted average cost of power, 1473–1478
    differing load profiles, 1478–1482
Weighted average unit cost, 920, 921–926
Well head pricing, 1083
Western United States coal production, 159
Westinghouse, 389
Wet bulb temperature, thermodynamic, 1184
Wet cooling towers, 246–247
    drawbacks to, 248–249
    heat exchanging, 248
Wet steam, 745
Wet-dry (hybrid) cooling towers, 250–251
Whole meter M&V procedures, 1051
Wholesale customers, regional transmission operator and, 380–381
Wholesale market energy rates, 1506
Wholesale power markets, 381–382

creation of trading rules, 381–382
liquid market structure, 381–382
maturing of, 382
results of, 382
Wick type, 805
Wicking, heat pipes and, 805–806
Wide area augmentation system, global positioning satellites and, 29
Wind farm towers, location of, 1610
Wind power plant, 954
Wind power, 1607–1615
  bird impact, 1612
  blades, importance of, 1614
  computer control systems, 1614
  control mechanisms, 1614
  cost of, 1609, 1611
  economics of, 1607–1608
  effect on electrical grid, 1611–1612
  electricity source, 1607
  energy source ranking, 1607
  environmental regulation of, 1615
  future of, 1610
  geographical use of, 1607
  government subsidies of, 1614–1615
  growth capacity, 1608
  history of, 1607
  impact on environment, 1611
  improving information availability, 1614
  installation costs, 1608
  maximizing production, 1610
  noise level of, 1612
  sensors in, 1614
  site selection, 1609–1610
  strengths as energy source, 1611
  technology of, 1612–1612
    blade diameter, 1612–1613
    tower height, 1612
  typical turbine, 1613
  variability, 1611
  visual impact of, 1612
  weaknesses as energy source, 1611–1612
Wind technology, 34–35, 1204
  development of, 35
  usage of, 35
Wind turbine design, 1612–1613
  blade diameter, 1612–1613
  tower height, 1612
Wind turbine
  example of, 1613
  site assembly of, 1613
Wind, coal vs., 957
Window economics, 1624
Window energy performance
  conductive heat flow, 1616
  irradiance in, 1616
  radiant flux, 1616
    spectral distribution, 1616
    spectral irradiance, 1616
Window energy, 1616–1625
Window film performance
  conventional applied, 1638
  spectrally selective, 1638
Window films
  CPFilms case study, 1626
    demonstration building, 1627
      DOE-2 simulation values, 1627
      energy savings from, 1627
    ideal qualities of, 1638
    IPMVP options C and D, 1626–1629
    savings before and after installation, 1628
    solar-control and insulating types, 1630–1636
    spectrally selective vs. conventional applied, 1637–1639. *See also* Spectrally selective window films.
Window glass, heat flow through, 1619
Window heat transfers, 1619–1620
Window orientation, path of sun as factor, 1618–1619
Window sash, glazing system and, 1620
Window shading options, 1619
Window size, 1616
Window technology, 1364–1365
Window thermal and solar properties, 1627
Windows, 1518–1521
Windows
  coatings and tints for energy efficiency, 1622
  condensation resistance, 1622–1623
  damage-weighted transmittance, 1623–1624
  daylight illumination from as energy saver, 1624
  direct heat entry of sunlight, 1619
  double-pane
    condensation in, 1620
    heat transfer, 1618
    insulating gas, 1619
  economics of, 1624
  energy basics, 1616–1618
    conduction heat transfer, 1617
    convection heat transfer, 1617–1618
    heat transfer, 1616–1617
    radiation heat transfer, 1617
  energy performance of, 1616, 1624
  Energy Star criteria, 1521
  energy transport, 1518–1520
  future improvements on, 1520–1521
  interior shades and, 1618
  low solar gain low-e coatings, 1621–1622
  low-e glass coatings, 1621–1622
  purpose of, 1616, 1641
  rating systems, 1520
  shading devices, 1641–1648
    acoustic performance, 1646
    control strategy for, 1644
    curtain wall systems, 1641
    decision-making framework, 1645, 1647–1648
    greenhouse effect, 1641
    performance parameters, 1646
    position of, 1643
    selection of, 1644–1645
    thermal performance, 1646
    variables, 1646
    visual performance, 1646
  solar radiant heat gain, 1623
  thermal characteristics, payback of, 1521
  UV transmittance, 1623–1624
  visible transmittance, 1623–1624
Wired external solutions
  existing LAN, 1654–1655
  phone line, 1655
Wireless applications, 1649–1658
  external communications, 1655–1656
  LAN vs., 1658
  mesh network, 1653–1654
  meter level, 1650, 1652
  phone line vs., 1658
  primer on, 1652
  serial communications networks, 1652–1653
  submetering primer, 1649–1650
  wired external solutions, 1654–1655
Wireless mobile thermostat climate control, 1660–1667
  annual savings, 1663–1666
  future of, 1667
  heating mode, 1663
  projected energy savings, 1663–1667
Wireless terms, 1652
  frequency, 1652
  interference, 1652
  mesh networks, 1652
  power rating, 1652
  repeaters, 1652
  throughput, 1652
Wiring harnesses, automobile technology evolution and, 675
Wisconsin, Energy Center of, 512
Wood as heating fuel, 1361
Wood framed walls, 1515–1516
Wood fuels, biomass energy and, 88
Work conversion, energy and, 604
Work heat energy principle, 1164–1168
  First Law of Thermodynamics, 1167
  Second Law of Thermodynamics, 1167
Work units, energy and, 605
Work, 1160–1161
  heat units and, 1162
Worker quality, automobile technology evolution and, 675
Working capital interest, 958–959
Working capital, rate of return regulation and, 1255
Working fluids, absorption, 1543–1544
Workmanship, lean manufacturing and, 470–471
Workplace organization, lean manufacturing and, 474
WWTP. *See* wastewater treatment plant.
Wyoming coal production, 159

X

XML, 108–110
XML. *See* extensible markup language.
XPS (extrusion polystyrene), as insulation material, 891, 894

Y

Yield ratio, emergy, 424–426

Z

Zone management, air distribution and, 211
  benefits of, 211–212
    air quality monitoring, 212
    end to end system management, 212
    measure zone leakage, 212
    reduced air usage, 211
    regulate air costs, 212
Zone switching, lighting based demand response and, 277

Volume 1

Accounting: Facility Energy Use / 1
Air Emissions Reductions from Energy Efficiency / 9
Air Quality: Indoor Environment and Energy Efficiency / 18
Aircraft Energy Use / 24
Alternative Energy Technologies: Price Effects / 31
ANSI/ASHRAE Standard 62.1-2004 / 50
Auditing: Facility Energy Use / 63
Auditing: Improved Accuracy / 69
Auditing: User-Friendly Reports / 76
Benefit Cost Analysis / 81
Biomass / 86
Boilers and Boiler Control Systems / 93
Building Automation Systems (BAS): Direct Digital Control / 104
Building Geometry: Energy Use Effect / 111
Building System Simulation / 116
Carbon Sequestration / 125
Career Advancement and Assessment in Energy Engineering / 131
Climate Policy: International / 137
Coal Production in the U.S. / 143
Coal Supply in the U.S. / 156
Coal-to-Liquid Fuels / 163
Cold Air Retrofit: Case Study / 172
Combined Heat and Power (CHP): Integration with Industrial Processes / 175
Commissioning: Existing Buildings / 179
Commissioning: New Buildings / 188
Commissioning: Retrocommissioning / 200
Compressed Air Control Systems / 207
Compressed Air Energy Storage (CAES) / 214
Compressed Air Leak Detection and Repair / 219
Compressed Air Storage and Distribution / 226
Compressed Air Systems / 236
Compressed Air Systems: Optimization / 241
Cooling Towers / 246
Data Collection: Preparing Energy Managers and Technicians / 255
Daylighting / 264
Demand Response: Commercial Building Strategies / 270
Demand Response: Load Response Resources and Programs / 279
Demand-Side Management Programs / 286
Desiccant Dehumidification: Case Study / 292
Distributed Generation / 296
Distributed Generation: Combined Heat and Power / 303
District Cooling Systems / 309
District Energy Systems / 316
Drying Operations: Agricultural and Forestry Products / 332
Drying Operations: Industrial / 338
Electric Motors / 349
Electric Power Transmission Systems / 356
Electric Power Transmission Systems: Asymmetric Operation / 364
Electric Supply System: Generation / 374
Electricity Deregulation for Customers / 379
Electricity Enterprise: U.S., Past and Present / 387
Electricity Enterprise: U.S., Prospects / 399
Electronic Control Systems: Basic / 407
Emergy Accounting / 420
Emissions Trading / 430
Energy Codes and Standards: Facilities / 438
Energy Conservation / 444
Energy Conservation: Industrial Processes / 458
Energy Conservation: Lean Manufacturing / 467
Energy Conversion: Principles for Coal, Animal Waste, and Biomass Fuels / 476
Energy Efficiency: Developing Countries / 498
Energy Efficiency: Information Sources for New and Emerging Technologies / 507
Energy Efficiency: Low Cost Improvements / 517
Energy Efficiency: Strategic Facility Guidelines / 524
Energy Information Systems / 535
Energy Management: Organizational Aptitude Self-Test / 543
Energy Master Planning / 549
Energy Project Management / 556
Energy Service Companies: Europe / 566

Volume 2

Energy Star® Portfolio Manager and Building Labeling Program / 573
Energy Use: U.S. Overview / 580
Energy Use: U.S. Transportation / 588
Energy: Global and Historical Background / 601
Enterprise Energy Management Systems / 616
Environmental Policy / 625
Evaporative Cooling / 633
Exergy: Analysis / 645
Exergy: Environmental Impact Assessment Applications / 655
Facility Air Leakage / 662
Facility Energy Efficiency and Controls: Automobile Technology Applications / 671
Facility Energy Use: Analysis / 680
Facility Energy Use: Benchmarking / 690
Facility Power Distribution Systems / 699
Federal Energy Management Program (FEMP): Operations and Maintenance Best Practices Guide (O&M BPG) / 707
Fossil Fuel Combustion: Air Pollution and Global Warming / 715
Fuel Cells: Intermediate and High Temperature / 726
Fuel Cells: Low Temperature / 733
Geothermal Energy Resources / 744
Geothermal Heat Pump Systems / 753
Global Climate Change / 760
Green Energy / 771
Greenhouse Gas Emissions: Gasoline, Hybrid-Electric, and Hydrogen-Fueled Vehicles / 787
Heat and Energy Wheels / 794
Heat Exchangers and Heat Pipes / 801
Heat Pipe Application / 807
Heat Pumps / 814
Heat Transfer / 822
High Intensity Discharge (HID) Electronic Lighting / 830
HVAC Systems: Humid Climates / 839
Hybrid-Electric Vehicles: Plug-In Configuration / 847
Independent Power Producers / 854
Industrial Classification and Energy Efficiency / 862
Industrial Energy Management: Global Trends / 872
Industrial Motor System Optimization Projects in the U.S. / 883
Insulation: Facilities / 890
Integrated Gasification Combined Cycle (IGCC): Coal- and Biomass-Based / 906
IntelliGrid^{SM} / 914
International Performance Measurement and Verification Protocol (IPMVP) / 919
Investment Analysis Techniques / 930
LEED-CI and LEED-CS: Leadership in Energy and Environmental Design for Commercial Interiors and Core and Shell / 937
LEED-NC: Leadership in Energy and Environmental Design for New Construction / 947
Life Cycle Costing: Electric Power Projects / 953
Life Cycle Costing: Energy Projects / 967
Lighting Controls / 977
Lighting Design and Retrofits / 993
Liquefied Natural Gas (LNG) / 1001
Living Standards and Culture: Energy Impact / 1009
Maglev (Magnetic Levitation) / 1018
Management Systems for Energy / 1027
Manufacturing Industry: Activity-Based Costing / 1042
Measurement and Verification / 1050
Measurements in Energy Management: Best Practices and Software Tools / 1056
Mobile HVAC Systems: Fundamentals, Design, and Innovations / 1061
Mobile HVAC Systems: Physics and Configuration / 1070
National Energy Act of 1978 / 1082
Natural Energy versus Additional Energy / 1088
Net Metering / 1096
Nuclear Energy: Economics / 1101
Nuclear Energy: Fuel Cycles / 1111
Nuclear Energy: Power Plants / 1115
Nuclear Energy: Technology / 1125
Performance Contracting / 1134
Performance Indicators: Industrial Energy / 1140